for implementing curricular themes from the *Agenda for Action*

Selections from the *Mathematics Teacher*

Edited by

Christian R. Hirsch
Western Michigan University
Kalamazoo, Michigan

National Council of Teachers of Mathematics

Library of Congress Cataloging in Publication Data:

Activities for implementing curricular themes from the Agenda for action.
Bibliography: p.
1. Mathematics—Study and teaching—United States.
I. Hirsch, Christian R. II. National Council of Teachers of Mathematics. III. Mathematics teacher.
QA13.A37 1986 510'.7'1073 86-5188
ISBN 0-87353-229-5

The publications of the National Council of Teachers of Mathematics present a variety of viewpoints. The views expressed or implied in this publication, unless otherwise noted, should not be interpreted as official positions of the Council.

Printed in the United States of America

Contents

Introduction

In 1980 the National Council of Teachers of Mathematics issued *An Agenda for Action*, a set of eight recommendations intended to provide a framework for the improvement of school mathematics in the 1980s. Five of the recommendations dealt directly with curriculum content or its delivery:

Recommendation 1. Problem solving must be the focus of school mathematics in the 1980s.

Recommendation 2. The concept of basic skills in mathematics must encompass more than computational facility.

Recommendation 3. Mathematics programs must take full advantage of the power of calculators and computers at all grade levels.

Recommendation 4. Stringent standards of both effectiveness and efficiency must be applied to the teaching of mathematics.

Recommendation 6. More mathematics study must be required for all students and a flexible curriculum with a greater range of options should be designed to accommodate the diverse needs of the student population.

Beginning with the September 1984 issue of the *Mathematics Teacher* and continuing with each issue through the 1985–86 school year, the "Activities" section featured instructional materials for grades 7 through 12 specifically designed to contribute to effective implementation of actions supportive of Recommendations 1 through 4. Moreover, the nature of the activities is such that many of them could be woven into the instructional component of alternative courses developed in response to Recommendation 6. This book is a compilation of those eighteen activities together with twelve additional activities selected from those appearing in the journal since 1981 and which are, themselves, reflective of curricular recommendations in the *Agenda*.

The volume is organized in five sections around the following curricular themes: problem solving (Recommendation 1), expanding basic skills (Recommendation 2), using calculators (Recommendation 3), using computers (Recommendation 3), and using manipulatives (Recommendation 4). Although many of the included activities are related to more than one recommendation, each is placed in a section for the major theme addressed or to preserve balance among the sections.

Each activity consists of a Teacher's Guide, including specific objectives, teaching suggestions, and solutions, and (in most cases) four student worksheets. The pages of this book are perforated so that they can be easily removed and reproduced for classroom use. It is recommended that once an activity has been used, teachers maintain the masters, together with noted adaptations or extensions, in a loose-leaf binder for future use.

We are now halfway through the decade of the eighties. Whereas significant progress has been reported in several areas of the curricular recommendations, continued and expanded efforts are necessary if the goals envisioned in the *Agenda* are to be fully realized. It is hoped that this collection of activities will serve effectively to supplement the ongoing efforts of mathematics educators to translate the *Agenda*'s curricular recommendations into action.

Activities for

Problem Solving

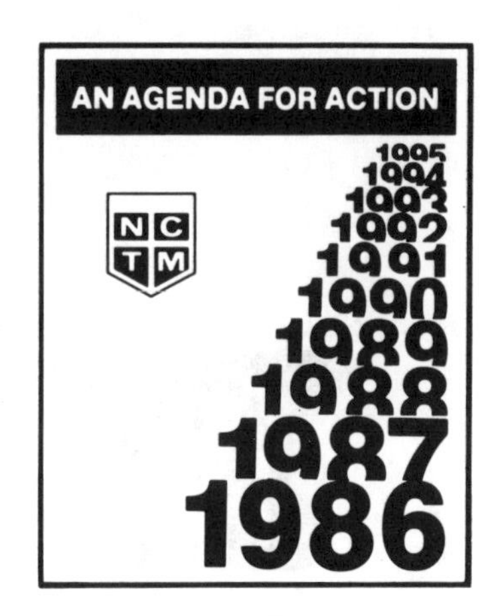

In its *Agenda for Action*, the NCTM identified the development of problem-solving ability as the primary goal of mathematics instruction during the 1980s. The first four activities in this section provide a related series of highly motivational classroom materials that involve both direct instruction on, and opportunities for practice with, several problem-solving skills and strategies. In order for these activities to be most effective, it is recommended that they be completed in the sequence given. Since problem solving is essentially a creative endeavor and thus cannot be built exclusively on routines, it is important that flexibility of thought be accepted and encouraged as students work through these activities. The fifth activity uses a mathematical game as the setting for problem solving.

The lead activity, "Teaching Problem-Solving Skills," focuses on the problem-solving skills of *make a drawing* and *work backward*. Each skill is introduced through a total-class, teacher-directed activity, and then opportunities are provided for students to solve problems, either individually or in small groups, by applying the respective skill. "Developing Problem-Solving Skills" uses a similar instructional model to introduce and reinforce the learning of two additional problem-solving skills: *make and read an organized list* and *search for a pattern*.

The activity "Extending Problem-Solving Skills" employs an instructional sequence similar to that of the first two activities to highlight the problem-solving skills of *guess and test* and *simplification* and to introduce Polya's heuristics for problem solving, with particular attention to the "looking back" phase. Follow-up problem situations provide opportunities for students to design and carry out problem-solving strategies involving one or a combination of the skills developed through these first three activities. In the activity "Teaching the Elimination Strategy," pupils are introduced to the problem-solving skill *eliminate possibilities* as well as to aspects of deductive reasoning and the use of indirect arguments. This activity also provides opportunities for the application of problem-solving strategies requiring the creative meshing of several skills; namely, *guess and test, make and read an organized list*, and *elimination*.

The final activity, "The Peg Game," guides students to use the skills of *simplification, make an organized list,* and *search for a pattern* to solve problems entailing the enumeration of the moves and jumps and the identification of the sequences of moves in interchanging five each of two different colored pegs on a game board. Questions requiring pupils to generalize from patterns together with two suggested extension activities provide further opportunities to underscore the importance of the "looking back" phase of problem solving.

TEACHING PROBLEM-SOLVING SKILLS

December 1984

By OSCAR F. SCHAAF, University of Oregon, Eugene, OR 97403

Teacher's Guide

Introduction: The recommendations for school mathematics contained in the *Agenda for Action* stress problem solving as the focus of mathematics instruction. Among the specific actions recommended in support of this position are the following (pp. 2, 3, 4):

- The mathematics curriculum should be organized around problem solving.
- Appropriate curricular materials to teach problem solving should be developed for all grade levels.
- Mathematics teachers should create classroom environments in which problem solving can flourish.

Motivational classroom materials that provide direct instruction on problem-solving skills are the most important component. This activity consists of two independent subactivities that highlight and provide opportunities for practice with two problem-solving skills: *make a drawing* (worksheets 1 and 2) and *work backward* (worksheets 3 and 4).

Grade levels: 7–10

Materials: Centimeter graph paper, a set of worksheets for each student, and a set of transparencies for demonstration of problem-solving skills and discussion of students' solutions

Objective: To develop the problem-solving skills of making a drawing and working backward

Directions: Distribute copies of the worksheets one at a time to each student. Sheets 1 and 3 are designed for a full-class, teacher-directed activity. Sheets 2 and 4 provide opportunities for pupils to practice the problem-solving skills introduced. Encourage communication and cooperation among pupils as they engage in problem-solving processes.

Sheets 1 and 2: Use a transparency of sheet 1 to introduce the skill *make a drawing*. Following completion of the two sample problems, stress the usefulness of the drawing in their solutions. Pupils should then be directed to solve the problems on sheet 2 by using the same skill. Depending on the level of your class, you may wish to help them find two or three additional examples for problem 3 before they complete the problem individually. Cutouts from centimeter graph paper or a supply of Green Stamps would help pupils better see the possibilities. Emphasize that a system for finding examples might make the search easier.

Provide centimeter graph paper for students to use in solving problem 5. A transparency of sheet 2 can be used to exhibit a sample four-block route for problem 5a. With respect to problem 6, point out that the drawing shows a wall marked in sixths on the outside and tenths on the inside. These marks correspond to the minutes needed to paint the wall.

Sheets 3 and 4: A transparency of sheet 3 should be used to introduce the skill *work backward.* Solve problems 1a, 1b, and 2a with the class. Demonstrate the working-backward procedure with problem 2a. Have pupils work problems 2b and 2c individually or with a classmate.

Solve problems 3 and 4a with the class. In the process, note that the representations for problem 4 are not unique. For example, solutions for problem 4a could be either of the following:

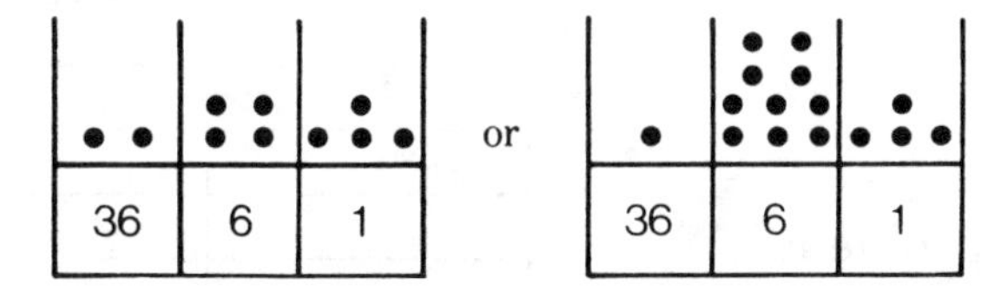

Let pupils try problems 4b and 4c on their own. Discuss their solutions and then emphasize the usefulness of working backward in solving the problems on this sheet. Next, direct students to solve the problems on sheet 4 by using this same skill.

Follow-up activities: Additional problem situations that can reinforce the problem-solving skills introduced in this activity can be found in *Problem Solving in Mathematics* (Lane County Mathematics Project 1983). This project used a similar direct approach to teaching a variety of problem-solving skills. In addition to teaching these skills directly, you can help students improve their problem-solving abilities by—

- using problem solving throughout the school year as an approach to drill and practice, laboratory activities and investigations, and the development of mathematical concepts; and
- providing pupils with appropriate nonroutine challenge problems whose solutions require the creative meshing of several skills.

A further discussion, together with examples, of this instructional approach to problem solving can be found in Brannan and Schaaf (1983).

Answers: Sheet 1: 1b. two meters, four meters, six meters; 1c. eight days. 2a. 64 feet; 2b. 64 feet, 32 feet; 2c. 368 feet

Sheet 2: 3. Nineteen different ways are possible. 4. Twenty girls. The drawing at the right should help pupils see that ten girls need to be equally spaced around each semicircle. 5b. Ten. 6. About four minutes. (If algebra is used, the solution is 3.75 minutes.)

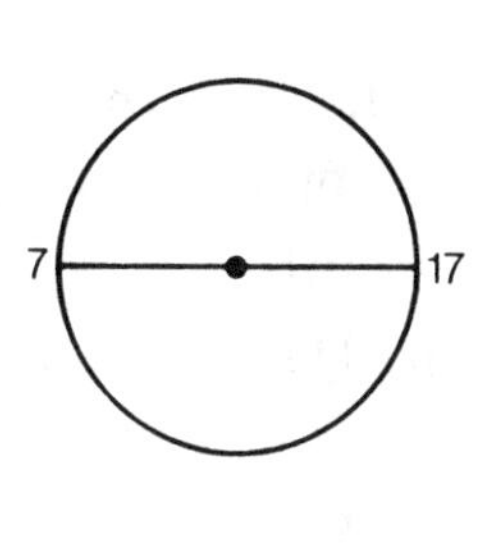

Sheet 3: 1a. fifteen; 1b. twenty-three. 2a. sixteen; 2b. twenty; 2c. twenty-seven. 3. seventy. 4. Answers may vary.

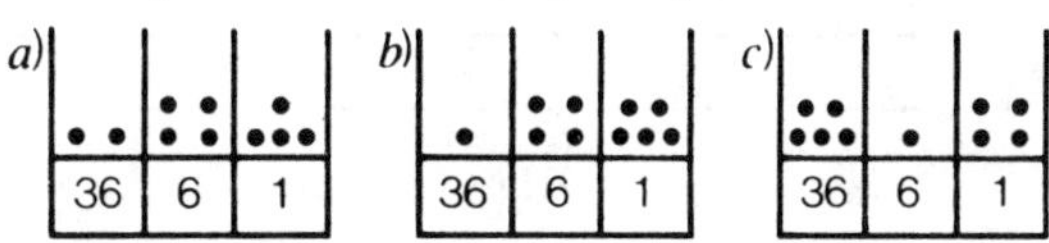

Sheet 4: 5a. 5; 5b. 11. 6. Answers may vary.

$$\text{Equation 1: } \frac{7p - 9}{18} = 3$$

$$\text{Equation 2: } 3(17n - 135) = 54$$

7. $49.00. 8. Pick one of the three-link pieces and cut each of the three links. Use them to join the remaining three pieces together.

REFERENCES

Brannan, Richard, and Oscar Schaaf. "An Instructional Approach to Problem Solving." In *The Agenda in Action,* 1983 Yearbook of the National Council of Teachers of Mathematics, edited by Gwen Shufelt, pp. 41–59. Reston, Va.: The Council, 1983.

Lane County Mathematics Project. *Problem Solving in Mathematics.* Palo Alto, Calif.: Dale Seymour Publications, 1983.

National Council of Teachers of Mathematics. *An Agenda for Action: Recommendations for School Mathematics of the 1980s.* Reston, Va.: The Council, 1980.

1. A frog is at the bottom of a ten-meter well. Each day it crawls up three meters. But at night it slips down two meters. How many days will it take the frog to get out of the well?

Problems of this sort can often be solved by using a drawing.

a. The drawing of the well at the right is divided into ten equal parts. Label each part in meters.

b. Use the drawing to help find how far the frog would move up the wall in—

two days ______

four days ______

six days ______

c. Now use the drawing to solve the problem.

____

____

Bottom

2. A ball rebounds 1/2 of the height from which it is dropped. Assume the ball is dropped 128 feet from a leaning tower and keeps bouncing. How far will the ball have traveled up and down when it strikes the ground for the fifth time?

The use of a carefully labeled drawing can also help in solving this problem. The drawing below shows the 128-foot drop of the ball from the tower.

a. Sketch in the first rebound and label the amount of rebound.

b. Sketch and indicate on the drawing the amounts of the second drop and the rebound.

c. Complete the drawing for the problem and give its solution. ______

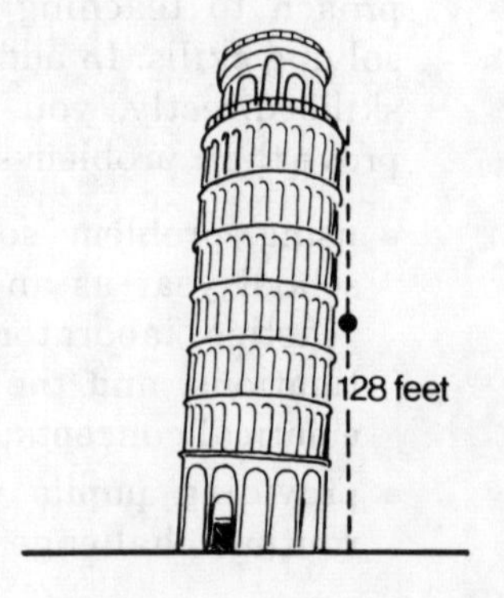

3. How many different ways can you buy four attached postal stamps? ______

   Two possible ways are shown.

   a. Make drawings to show ten additional ways.
   b. Continue this method to obtain a solution to the problem.

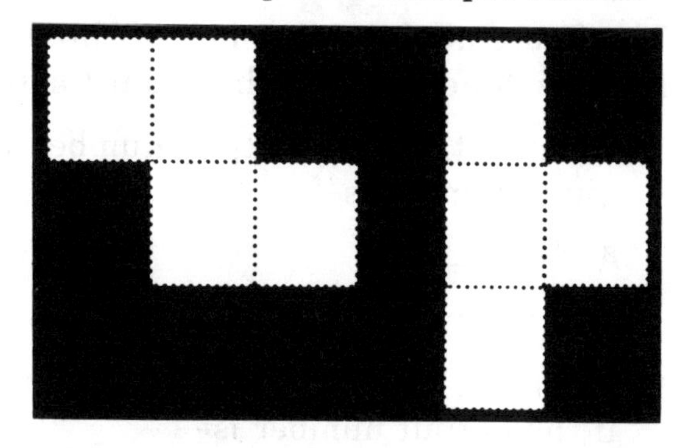

4. Some girls are standing in a circular arrangement. They are evenly spaced and numbered in order. The seventh girl is directly opposite the seventeenth girl. How many girls are in the arrangement? ______

5. a. Anne lives at point *A*. She can use six different four-block routes to walk to school. Show each possible route on a copy of this drawing.

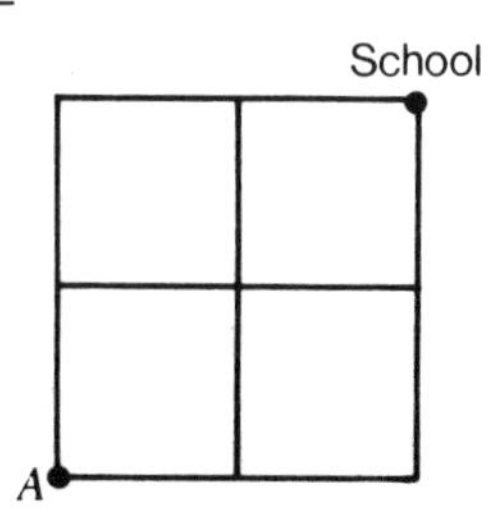

   b. Bill lives at point *B*. How many different five-block routes can he use to get to school? ______

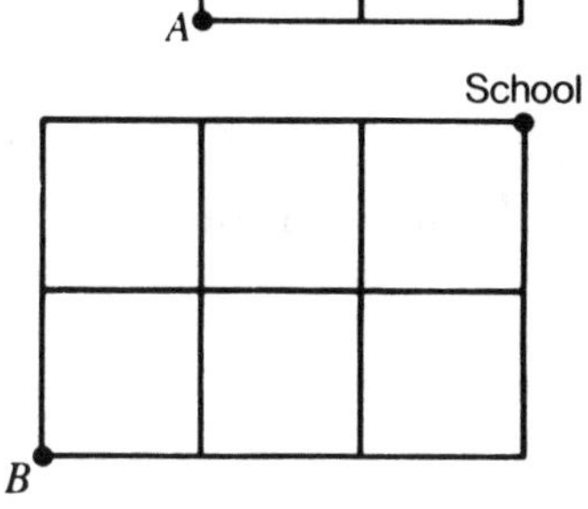

6. One painter can paint a wall in ten minutes. Another painter can paint it in six minutes. About how long will it take both painters working together to do the job? ______

Use the drawing below to help solve the problem.

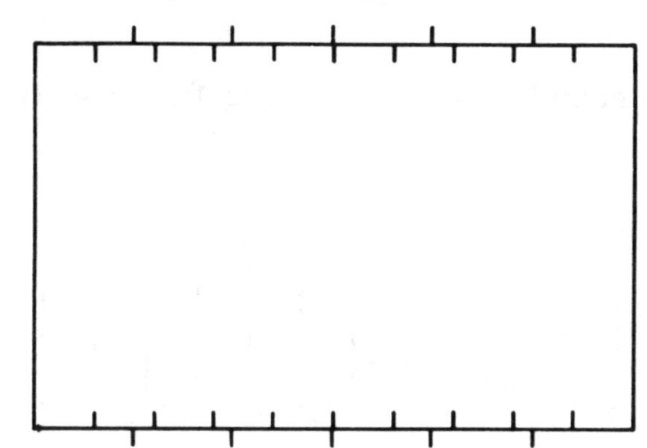

The figure at the right shows a hookup of four machines. The output from the first machine becomes the input for the next machine, and so on.

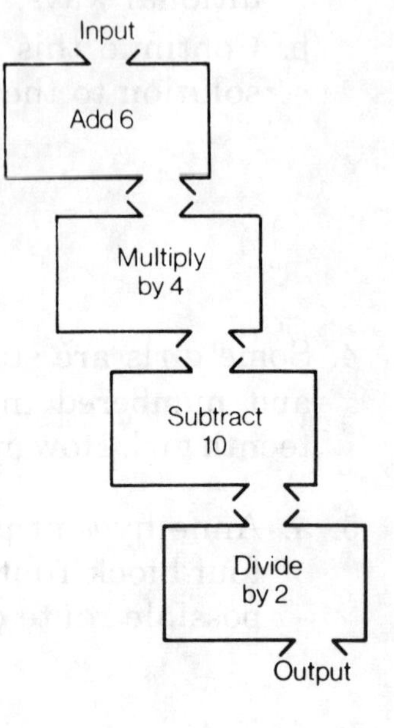

1. What is the final output number if the beginning input number is—
   a. 4? ______
   b. 8? ______

2. Work backward. Find the beginning input number if the output number is—
   a. 39 ______
   b. 47 ______
   c. 61 ______

3. Study the first counting frame shown below. It represents the number 98. What number does the second frame represent? ______

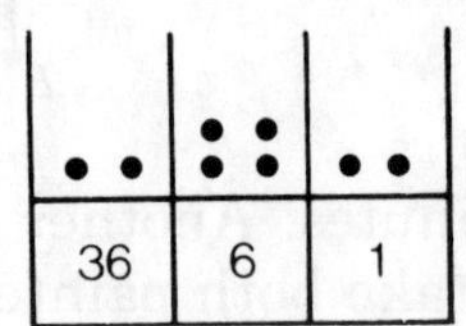

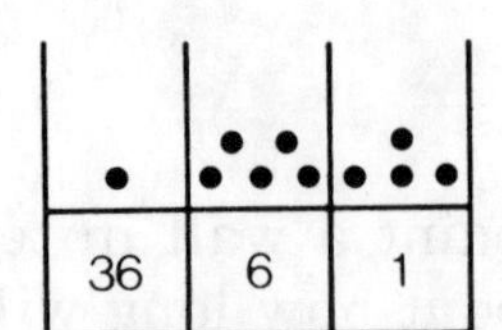

4. Work backward to complete each counting frame so that it represents the number below it.

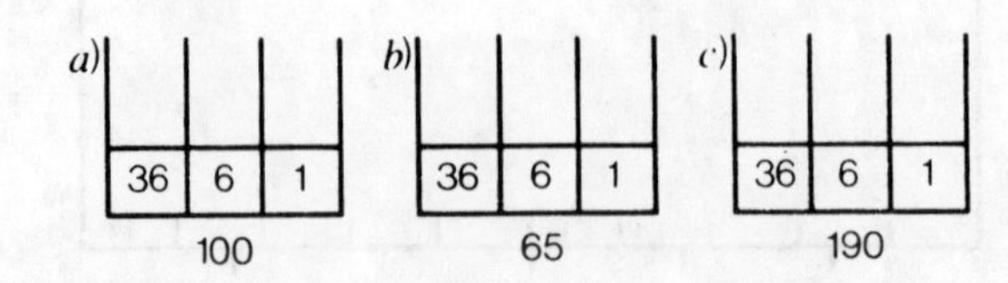

Solve each of these problems by working backward.

5. a. I'm thinking of a number. If you multiply it by 3, then subtract 5, and finally add 10, you get 20.

   What number am I thinking of? ______

   b. I'm thinking of another number. If you subtract 4 from the number, then multiply the result by 3, and then add 5, you get 26.

   What is that number? ______

6. The two equations below both have 2 as a solution:

$$\frac{7p - 2}{4} = 3 \qquad 3(17n - 30) = 12$$

Create two equations that have the same form as these but have 9 as a solution.

Equation 1 ________________ Equation 2 ________________

7. Bliksem bought a present for \$5.00, then spent 1/2 of his remaining money on jogging shoes, next bought lunch for \$2.00, and finally spent 1/2 of his remaining money on a puzzle. He had \$10.00 left.

   How much money did he start with? ______

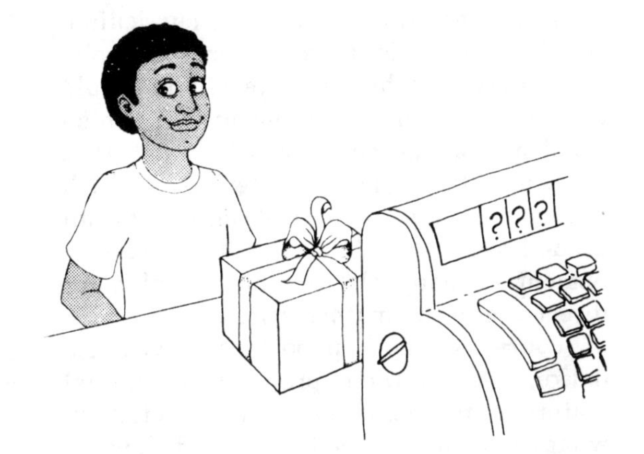

8. Below are four pieces of chain that each have three links. Explain how they can be joined into a twelve-link circular chain by cutting and rejoining only three links.

DEVELOPING PROBLEM-SOLVING SKILLS *December 1985*

By STEPHEN KRULIK and JESSE A. RUDNICK, College of Education, Temple University, Philadelphia, PA 19122

Teacher's Guide

Introduction: An Agenda for Action states that "problem solving must be the focus of school mathematics in the 1980s." It further suggests that "the mathematics curriculum should be organized around problem solving," that "mathematics teachers should create classroom environments in which problem solving can flourish," and that "appropriate curricular materials to teach problem solving should be developed for all grade levels" (NCTM 1980, 2, 4). The development of materials that are suitable for direct use in the mathematics classroom is, of course, a most important recommendation, since without proper instructional materials, teachers have little direction in what content to emphasize and how it might best be presented. Curricular materials previously offered in this series of *Agenda*-related activities focused on the problem-solving skills of *make a drawing* and *work backward* (Schaaf 1984) and *guess and test* and *simplification* (Laing 1985). The activities in this article are intended to provide suitable materials for introducing and reinforcing the learning of two additional problem-solving skills, namely, *make an organized list* and *search for a pattern.*

Grade levels: 7–10

Materials: Copies of the worksheets for each student and a set of transparencies for demonstration purposes

Objectives: (1) To develop the problem-solving skills of making and reading an organized list and searching for a pattern; (2) to provide practice in using the general heuristics of the problem-solving process

Directions: This activity consists of two subactivities: solving problems by making an organized list (worksheets 1 and 2) and solving problems by searching for a pattern (worksheets 3 and 4). Each subactivity will require approximately one instructional period and out-of-class work by the students. Allow ample class time for discussion of students' solutions to the problems. Although the second subactivity builds on the first, it is not necessary that the two subactivities be used in consecutive class periods.

Begin by distributing a copy of the first worksheet to each student. This sheet is designed to be used by the entire class in a teacher-centered activity. Using a transparency of sheet 1, present the first problem and explain how the list was constructed. Assume that the first tag chosen is a 3 and the second tag is a 6. This leaves 2, 5, and 1 to be distributed as shown in the first three vertical entries on the list. Now we have exhausted the 6s. The next choice should be a 3 first and a 2 second, with 5 and 1 distributed as shown in the next two vertical entries. Instruct the class to proceed in a similar manner to complete the list and then answer parts (b)–(d). After discussing the solution to this problem, review with the students how the use of an organized list enabled them to keep track of all possibilities as they worked. Note that the list *is* the answer.

Assign the students to complete the solution to problem 2 by using the list that has been started for them. Discuss their solutions. Then distribute the second worksheet and direct the students to complete problem 1. When discussing their solutions to this problem, note that Chen does not need to purchase *exactly* the 17 pounds of grass seed; it is more economical to buy 18 pounds. In this problem, the list is not the answer, but it leads directly to the answer. Instruct students to solve problems 2, 3, and 4 in a similar manner. Depending on the previous problem-solving experiences of the students, you may need to suggest that for problem 4, they begin by making a drawing.

To introduce the second subactivity, distribute sheet 3 to each student and use a transparency of this sheet to discuss the patterning rule for each exercise. Allow students as wide a selection of possible rules as you can, provided that their patterning rule fits all the examples given in the series.

Next distribute sheet 4. Allow time for the students to complete problem 1. In the discussion that follows, point out the value of the organized list in looking for the pattern and emphasize the usefulness of the discovered pattern in completing the solution. Assign problems 2 and 3 to be solved using the same skill.

Follow-up activities: The problem-solving skills of *make an organized list* and *search for a pattern* should be reinforced throughout the school year. Additional problems whose solutions are amenable to the use of these skills can be found in Dolan and Williamson (1983), Krulik and Rudnick (1980, 1984), and in the Lane County Mathematics Project (1983). Finally, note that although a problem can often be solved by algebra, an algebraic solution may not always be possible, and thus facility in making and reading an organized list is an important skill.

Solutions:

Sheet 1: 1.a.

First tag	3	3	3	3	3	3	6	6	6	2
Second tag	6	6	6	2	2	5	2	2	5	5
Third tag	2	5	1	5	1	1	5	1	1	1
Score	11	14	10	10	6	9	13	9	12	8

1.b. Every possible triple of numbers selected from {3, 6, 2, 5, 1} appears in the table. If, for example, an eleventh column was added consisting of the entries 5, 1, and 3, it would duplicate the entries in column six.

1.c. The completed list shows eight different scores.

1.d. The possible scores are 6, 8, 9, 10, 11, 12, 13, and 14.

2.

Clock 1	2:00	2:06	2:12	2:18	2:24
Clock 2	2:00	2:08	2:16	2:24	

The cuckoos come out together at 2:24.

Sheet 2: 1.

Number of 5-Pound Boxes	Cost at $6.58 a Box	Number of 3-Pound Boxes	Cost at $4.50 a Box	Total Pounds	Total Cost
4	$26.32	0	–0–	20	$26.32
3	19.74	1	$ 4.50	18	24.24
2	13.16	3	13.50	19	26.66
1	6.58	4	18.00	17	24.58
0	–0–	6	27.00	18	27.00

Chen should buy three 5-pound boxes and one 3-pound box; this combination is cheaper than buying exactly the 17 pounds of seed required.

2.

Number of Packages of 4	Number of Packages of 3	Number of Singleton Packages
3	1	0
3	0	3
2	2	1
2	1	4
2	0	7
1	3	2
1	2	5
1	1	8
1	0	11
0	5	0
0	4	3
0	3	6
0	2	9
0	1	12
0	0	15

The order can be filled in fifteen different ways.

3.

Number of Checks	Cost for Checks, @ $0.10	Total with $2.00 Service Charge	Cost for Checks, @ $0.05	Total with $3.00 Service Charge
1	$0.10	$2.10	$0.05	$3.05
2	0.20	2.20	0.10	3.10
3	0.30	2.30	0.15	3.15
4	0.40	2.40	0.20	3.20
5	0.50	2.50	0.25	3.25
6	0.60	2.60	0.30	3.30
7	0.70	2.70	0.35	3.35
8	0.80	2.80	0.40	3.40
9	0.90	2.90	0.45	3.45
10	1.00	3.00	0.50	3.50
11	1.10	3.10	0.55	3.55
12	1.20	3.20	0.60	3.60
13	1.30	3.30	0.65	3.65
14	1.40	3.40	0.70	3.70
15	1.50	3.50	0.75	3.75
16	1.60	3.60	0.80	3.80
17	1.70	3.70	0.85	3.85
18	1.80	3.80	0.90	3.90
19	1.90	3.90	0.95	3.95
20	2.00	4.00	1.00	4.00
21	2.10	4.10	1.05	4.05

At least twenty-one checks a month must be written if the new plan is to save you money.

(Solutions continued on next page)

REFERENCES

Dolan, Daniel T., and James Williamson. *Teaching Problem-solving Strategies.* Menlo Park, Calif.: Addison-Wesley Publishing Co., 1983.

Krulik, Stephen, and Jesse A. Rudnick. *Problem Solving: A Handbook for Teachers.* Newton, Mass.: Allyn & Bacon, 1980.

———. *A Sourcebook for Teaching Problem Solving.* Newton, Mass.: Allyn & Bacon, 1984.

Laing, Robert A. "Extending Problem-solving Skills." *Mathematics Teacher* 78 (January 1985):36–44.

Lane County Mathematics Project. *Problem Solving in Mathematics.* Palo Alto, Calif.: Dale Seymour Publications, 1983.

National Council of Teachers of Mathematics. *An Agenda for Action: Recommendations for School Mathematics of the 1980s.* Reston, Va.: The Council, 1980.

Schaaf, Oscar F. "Teaching Problem-solving Skills." *Mathematics Teacher* 77 (December 1984):694–99.

4.

Total Area	Area of Square 1	Area of Square 2	Side of Square 1	Side of Square 2	Perimeter of Square 1	Perimeter of Square 2	Sum of Perimeters
130	1	129	1	*	--	--	--
130	4	126	2	*	--	--	--
130	9	121	3	11	12	44	56
130	16	114	4	*	--	--	--
130	25	105	5	*	--	--	--
130	36	94	6	*	--	--	--
130	49	81	7	9	28	36	64
130	64	66	8	*	--	--	--
130	81	49	9	7	36	28	64
130	100	30	10	*	--	--	--
130	121	9	11	3	44	12	56

*These values will be nonintegral.

Notice that two possible answers satisfy the condition that the sum of the areas is 130 but that only one satisfies the condition that the sum of the perimeters is 64. Thus the answer is that the sides differ by 2 centimeters.

Sheet 3: 1. 5 — Patterning rule: Each term is one-half of the previous term.

2. Answers will vary. — Patterning rule: Names beginning with the letter "J"

3. 216 — Patterning rule: Each term is the cube of the number of the term.

4. Answers will vary. — Patterning rule: Names of various rock groups

5. 21 — Patterning rule: These are Fibonacci numbers. Each term is the sum of the two terms that precede it.

6. Answers will vary. — Patterning rule: Names in alphabetical order

7. ○ ○ ○ — Patterning rule: Alternating sequence of squares and circles. The number of figures of a given shape increases by one each time the shape occurs.

8. $\frac{5}{6}$ — Patterning rule: Each fraction has the denominator of the previous fraction as its numerator, and the denominator is the next integer.

9. 3750 — Patterning rule: Each term is five times the previous term.

10. — Patterning rule: The sequence of squares repeats with the upper portion of each odd-numbered term shaded and the lower portion of each even-numbered term shaded.

Sheet 4: 1.a. 4; 5

1.b.

Minutes Forward	3	7	11	15	19	23	27
Years Gained	3	8	13	18	23	28	33

By setting the timer ahead 27 minutes you will gain 33 years. The year will be 1985 + 33 = 2018.

2.

Week	Weekly Salary	Total Earnings
1	$1.00	$1.00
2	2.00	3.00
3	4.00	7.00
4	8.00	15.00
5	16.00	31.00
6	32.00	63.00
7	64.00	127.00
8	128.00	255.00
20	524 288.00	1 048 575.00

An analysis of this list suggests two patterns: the weekly salary can be expressed as 2^{n-1}, where n is the number of the week; the total earnings can be expressed as $2^n - 1$, where n is again the number of the week. From the latter pattern, it follows that the total earnings after twenty weeks of employment would be $2^{20} - 1 =$ $1 048 575.00.

3.

Stop #	1	2	3	4	5	6	7	8	9	10		16
Number Picked Up	5	0	5	0	5	0	5	0	5	0		
Number Dropped Off	0	2	0	2	0	2	0	2	0	2		
Number Aboard	5	3	8	6	11	9	14	12	17	15		24

The organized list enables us to see the pattern: 24 passengers will be aboard at the end of the sixteenth stop. Observe that the sequence of numbers of passengers actually consists of two subsequences: 5, 8, 11, 14, 17 and 3, 6, 9, 12, 15. ■

MAKE AN ORGANIZED LIST — SHEET 1

1. The five tags shown are placed in a box and mixed. Three tags are then drawn out at one time. If your score is the sum of the numbers on the three tags drawn, how many different scores are possible? What are the possible scores?

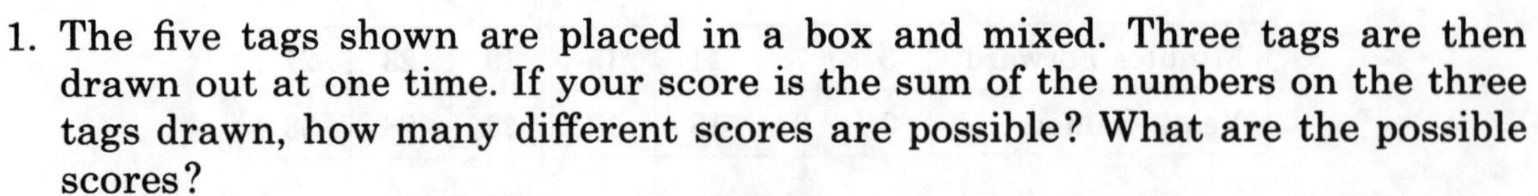

To keep track of the different scores, it is helpful to prepare an *organized list*. Let's begin by assuming that the first of the three tags drawn is the "3" and then listing the possibilities for the other two tags.

First Tag	3	3	3	3	3	3	6	6	6	2
Second Tag	6	6	6	2	2	5	2	2		
Third Tag	2	5	1	5	1		5			
Score	11	14	10	10	6					

a. Complete the chart.

b. Explain why this list accounts for all possible drawings of three tags where the order of selection is unimportant. ______________________________

c. How many *different* scores are possible? __________

d. What are the possible scores? ______________________________

2. Jim's Repair-Your-Clock Shop has two cuckoo clocks that were brought in for repairs. One clock has the cuckoo coming out every six minutes, whereas the other has the cuckoo coming out every eight minutes. Both cuckoos come out at exactly 2:00. When will they both come out together again?

Let's start to make an organized list of the times that the two cuckoos come out:

Clock 1	2:00	2:06	2:12				
Clock 2	2:00	2:08					

Finish the list and solve the problem. ______________________________

MAKE AN ORGANIZED LIST SHEET 2

1. Helen Chen wants to seed her front lawn. Grass seed can be bought in 3-pound boxes that cost $4.50, or in 5-pound boxes that cost $6.58. She needs exactly 17 pounds of seed. How many boxes of each size should she purchase to get the best buy?

The use of an organized list can also help in solving this problem.

Number of 5-Pound Boxes	Cost at $6.58 a Box	Number of 3-Pound Boxes	Cost at $4.50 a Box	Total Pounds	Total Cost
4	$26.32	0	–0–	20	$26.32
3	19.74	1	$4.50		
2					
1					
0		6			

Complete the list and solve the problem. ______________________________

Make an organized list to solve each of the following problems.

2. A customer ordered 15 blueberry muffins. If the muffins are packaged singly or in sets of 3 or 4, in how many different ways can the order be filled?

3. A bank has been charging a monthly service fee of $2.00 plus $0.10 a check for a personal checking account. To attract more customers it is advertising a new "reduced cost" plan with a monthly service charge of $3.00 plus only $0.05 a check. How many checks must you write each month for the new plan to save you money?

4. A piece of wire 64 centimeters in length is cut into two parts. The parts are then each bent to form a square. The total area of the two squares is 130 square centimeters. How much longer is a side of the larger square than a side of the smaller square? (Consider only whole-number solutions.)

For each of the following sets, give another element of the set. State in your own words what you think the patterning rule is.

1. 80, 40, 20, 10, ______
   Patterning rule: ______

2. James, Jill, Joan, John, ______
   Patterning rule: ______

3. 1, 8, 27, 64, 125, ______
   Patterning rule: ______

4. Styx, Beatles, Who, Kansas, ______
   Patterning rule: ______

5. 1, 1, 2, 3, 5, 8, 13, ______
   Patterning rule: ______

6. Alvin, Barbara, Carla, Dennis, ______
   Patterning rule: ______

7. □, ○, □□, ○○, □□□, ______
   Patterning rule: ______

8. $\frac{1}{2}, \frac{2}{3}, \frac{3}{4}, \frac{4}{5}$, ______
   Patterning rule: ______

9. 6, 30, 150, 750, ______
   Patterning rule: ______

10. ⬒, ⬓, ⬒, ⬓, ⬒, ______
    Patterning rule: ______

1. Scientists have invented a time machine. By setting the dial, you can move forward in time. Set it forward 3 minutes and you will be in the year 1988. Set it forward 7 minutes and you will be in the year 1993; set it forward 11 minutes and you will be in the year 1998; set it forward 15 minutes and you will be in the year 2003. If the machine continues in this manner, in what year will you be if you set the timer ahead 27 minutes? (This year is 1985.)

What do we know? Forward 3 minutes, gain of 3 years (1988 – 1985)
Forward 7 minutes, gain of 8 years (1993 – 1985)
Forward 11 minutes, gain of 13 years (1998 – 1985)
Forward 15 minutes, gain of 18 years (2003 – 1985)

What do we want? Forward 27 minutes, gain of how many years?

Plan: Make an organized list and search for a pattern.

Minutes Forward	3	7	11	15		27
Years Gained	3	8	13	18		?

a. Complete the patterning rule: Every time we move the timer ahead _______ minutes, we gain an additional _______ years.

b. Now complete the chart and solve the problem. ____________________

Solve the following problems by first making an organized list and then looking for a pattern.

2. Carlos was offered a part-time job that included on-the-job training. Because of the training, he was to be paid \$1.00 the first week, \$2.00 the second, \$4.00 the third, \$8.00 the fourth, and so on. How much money would Carlos have earned after twenty weeks of employment?

3. An empty streetcar picks up five passengers at the first stop, drops off two passengers at the second stop, picks up five passengers at the third stop, drops off two passengers at the fourth stop, and so on. If it continues in this manner, how many passengers will be on the streetcar after the sixteenth stop?

EXTENDING PROBLEM-SOLVING SKILLS

January 1985

By ROBERT A. LAING, Western Michigan University, Kalamazoo, MI 49008

Teacher's Guide

Introduction: Recognizing that the mathematics curriculum in grades K–12 must include more than the concepts and skills of mathematics to prepare students to be productive and contributing members of a rapidly changing technological society, the *Agenda for Action* (NCTM 1980, 3, 4) recommends that problem solving be the focus of school mathematics in the 1980s.

Implementation of this recommendation requires that curriculum developers—

- give priority to the identification and analysis of specific problem-solving strategies;
- develop and disseminate examples of "good problems" and strategies and suggest the scope of problem-solving activities for each school level.

The *Agenda* further encourages mathematics teachers to develop a classroom environment in which problem solving can flourish:

- Students should be encouraged to question, experiment, estimate, explore, and suggest explanations. Problem solving, which is essentially a creative activity, cannot be built exclusively on routines, recipes, and formulas.

The activities presented in this article are intended to reflect these recommendations. Two problem-solving skills, *guess and test* and *simplification*, are introduced and reinforced through a variety of motivating problem situations. Solutions of sample problems and directions for the teacher encourage the extension of problem-solving skills to include application of Polya's (1971) four phases of problem solving: understanding the problem, devising a plan, carrying out the plan, and looking back.

Grade levels: 7–11

Materials: Graph paper or square dot paper (optional), copies of the worksheets for each student, and a set of transparencies for class discussions

Objectives: (1) To develop problem-solving skills of guess and test and simplification; (2) to introduce the more general heuristics of problem solving suggested by Polya; (3) to provide opportunities for the design of problem-solving strategies using one or a combination of the following skills: make a drawing, work backward, guess and test, and state and solve a simpler problem.

Directions: Solutions of sample problems on the worksheets encourage students to devote sufficient time to understanding a problem before they begin searching for its solution. Teachers can reinforce this approach, when students claim a lack of understanding, through such questions as, "What steps have you taken to understand the problem?"

Begin by distributing sheets 1 and 2 to each student. Use a transparency of sheet 1 to present and discuss the sample problem together with the prescription for using guess and test. Following this discussion, assign students the problems on sheet 2, with problems 1 and 2 due at the next class period. Encourage pupils to keep an organized record of their guess-and-test results in tabular form. During the next class period, discuss these two problems using looking-back activities, such as those suggested in the following "Solutions" section. Assign problems 3 and 4 for solution and discussion in a similar fashion during the next class period.

You may wish to provide some experiences involving the problem-solving skill of looking for patterns prior to using sheets 3 and 4, which entail solving problems by simplification. Such activities as "Can You Predict the Repetend?" (Woodburn 1981), "Number Triangles: A Discovery Lesson" (Ouellette 1981), and "Pattern Gazing" (Aviv 1981) offer a good introduction to this skill.

Sheets 3 and 4 can be used in a manner similar to that used for the first two worksheets. However, do not expect students to complete more than one or two problems in each class period because of the increased level of difficulty of the problems and the amount of time needed to conduct the follow-up discussions adequately. Graph paper or dot paper can be distributed for use with problems 3 and 5.

Solutions and looking-back activities: An important component of every looking-back activity is the discussion of strategies that were used in solving the problem. This discussion is necessary to focus the students' attention on the processes used rather than on the characteristics of a particular problem. These processes will prove useful in later problem situations.

Answers: Sheet 2: 1. 40 m by 75 m

Area	Width (Guess)	Length	Amount of New Fence
3000	20	150	190
3000	30	100	160
3000	40	75	155
3000	50	60	160

Looking-back activities might include the following questions:

Could Mary do better by using such widths as 38.5 or 39? Use your calculators to explore this question.

Suppose Mary could use neighbors' fences for two adjacent sides of the pasture. How would this alter the problem? Let's solve the new problem together.

2. 31 and 33. Looking-back activities might include these questions:

Would this problem have an answer if the rooms were numbered with consecutive integers and the product were given as 1023? Why?

How might the concept of square root have been useful in solving the original problem?

Suppose we changed the term "product" to "sum." How must the rooms be numbered if the sum is odd? Why? For this new problem, can you find a quick method that will yield a good guess on the first trial? (Use the concept of average.)

3. 15 pigs. A table is quite useful in this problem for keeping track of guess-and-test attempts. Have different students show their organizational schemes on the chalkboard.

An interesting solution strategy described by Krulik and Rudnick (1980) is to picture the pigs standing on their hind legs and the chickens standing on one leg. If we count the number of legs in the air, we should get half of the total number of legs given in the problem. Fifty legs and thirty-five heads give fifteen more legs than heads, one contributed by each of the pigs.

4. 12 ft. and 28 ft. In looking back at this problem, the teacher might demonstrate how the inclusion of formulas in the table can be useful:

Shorter (S) (Guess)	Longer (L) ($2S + 4$)	Total ($S + L$)	Total Needed
10	$20 + 4 = 24$	34	40

Sheet 4: 1. Meet at the location of the innermost official. In the case of two officials, the use of some sample distances will reveal that any point between them or at either's location will yield the same minimal distance. In the case of a third official, similar reasoning will make it clear that using the location of the innermost official will result in the minimal total distance.

The general conjecture that the officials should meet anywhere between the two innermost ones in the case of an even number of officials and at the location of the innermost official in the case of an odd number of officials should be tested for the cases of four and five officials before this reasoning is applied to the case of nine officials.

2. 325 connections. Students will discover a number of patterns when simplifying this problem, making it good for sharing alternatives in class discussions. Drawing pictures of the simpler cases also reinforces this problem-solving skill.

1 classroom and office

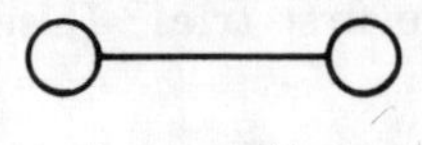

1 connection

2 classrooms and office

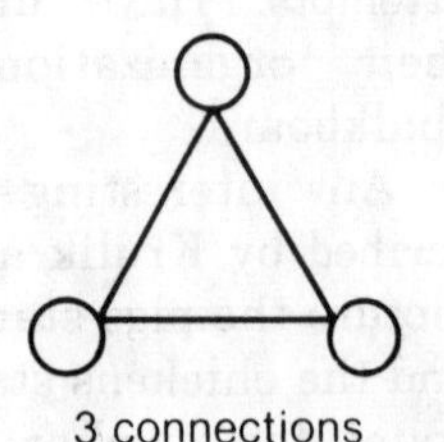

3 connections

3 classrooms and office

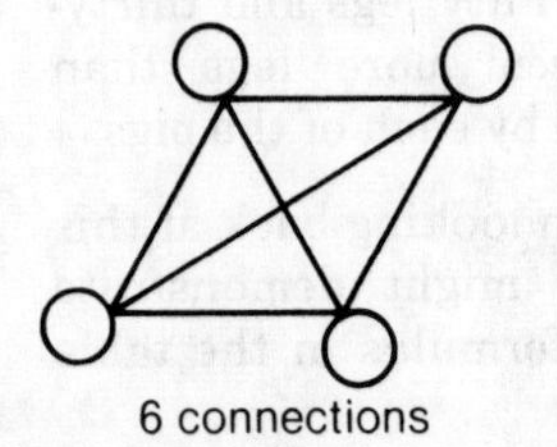

6 connections

Some pupils may recognize that the triangular numbers 1, 3, 6, 10, ... occur in the number of connections needed. The insight that results most frequently from exploring the simpler cases is that each additional classroom must be connected to each of the others, yielding the sum

$$1 + 2 + 3 + \cdots + 25 = 325$$

as the solution. You might ask students to find an easier way of finding this sum than adding all twenty-five numbers together. One discovery might be to use the average (and median, in this case) and just multiply 25 by 13. A second method would mimic a fairly standard argument found in high school textbooks for finding the sum of n consecutive counting numbers. Twice the needed sum is given by adding columns:

$$\begin{array}{l} 1 + \;\;2 + \;\;3 + \cdots + 24 + 25 \\ \underline{25 + 24 + 23 + \cdots + \;\;2 + \;\;1} \\ 26 + 26 + 26 + \cdots + 26 + 26 \end{array}$$

Therefore, the desired sum is given by $1/2 \times (25 \times 26)$.

3. 70 downward paths. In considering the simpler cases, students should discover the usefulness of the symmetry in the figures, select an appropriate notation for recording intermediate results, and note that each case is embedded in subsequent cases.

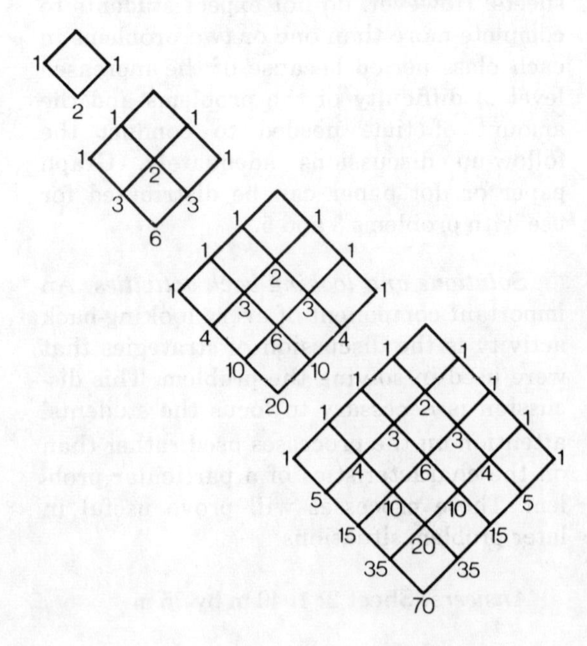

In addition to these considerations, looking-back activities should focus on number patterns that students may have discovered. These patterns can be used to extend the solution to the 5-by-5 case. The occurrence of triangular numbers, as well as of the consecutive counting numbers, should be noted. Reinforce how symmetry can be used to reduce the complexity of many problems.

4. The units digit for 7^{99} is 3. Most students soon discover that they do not have to find the entire value of each consecutive power of 7 if they are only interested in the units digit.

Power	Units Digit
7^1	7
7^2	9
7^3	3
7^4	$\underline{1}$
7^5	7
7^6	9
7^7	3
7^8	$\underline{1}$
7^9	7
7^{10}	9
7^{11}	3
7^{12}	$\underline{1}$

Although many students will discover the repeating pattern of the units digits and will see that odd powers have 7 or 3 in the units place, they may have difficulty determining which digit to use for the 99th power. The length of the repeating cycle in this example is 4 with a units digit of 1 occurring at each multiple of 4. Thus we need only to divide 99 by 4 and consider the remainder as the number of steps past the 1 that we need to go to obtain the desired digit. Follow this activity with problems that have a similar mathematical structure. The following are some examples:

Find the day of the week that is 213 days from today.

Find the time on a clock that is 100 hours from now.

5. 204 squares. In applying the simplification strategy, some students will generate the following information related to the simpler cases:

Dimension	Number of Squares
1 × 1	1
2 × 2	5
3 × 3	14
4 × 4	30
⋮	
8 × 8	?

A few students may discover that the differences between the consecutive cases are perfect squares and then may continue this pattern to the 8 × 8 case.

Another strategy that yields a more immediate solution is the discovery that the number of squares of a particular size is always a perfect square. For example, when considering the 3 × 3 case, the student discovers the following pattern:

Size	Number of Squares
1 × 1	9
2 × 2	4
3 × 3	1

$$S = 1 + 4 + 9$$
$$= 1^2 + 2^2 + 3^2$$

This pattern would suggest that the total number of squares for the checkerboard is $1^2 + 2^2 + 3^2 + \cdots + 8^2$. Circumventing the long method for obtaining this sum is difficult, since the formula for finding the sum of the squares of the first n positive integers is given by

$$S = \frac{n(n+1)(2n+1)}{6}.$$

In the looking-back activity, ask students how they organized their counting procedures so that they did not count the same square twice. One scheme would be to notice that each 2 × 2 square in the 3 × 3 case must have a bottom-left vertex. Because nine points could serve as a bottom-left vertex for a 2 × 2 square, nine such squares must exist. This problem is also one for which the answer is difficult to check; however, the validity of the process can be tested by applying it to the 5 × 5 case.

Follow-up activities: The experiences provided in this activity and that by Schaaf (1984) highlight the problem-solving skills of make a drawing, guess and test, work backward, and simplification. Solutions of the following problems will require students to develop and implement a strategy involving one or a combination of these skills. Begin to integrate problem solving into the regular curriculum by including one of these problems as part of the regular homework assignment over a period of several days.

1. A log is cut into 4 pieces in 15 seconds. At this rate, how long will it take to cut a log into 6 pieces? (Possible strategy: make a drawing; answer: 25 seconds)

2. Roosevelt wrote to 12 of his friends from music camp at a cost of $2.20 for postage. If the postage for letters is $0.23 and for postcards, $0.15, how many of each did he send? (Possible strategy: guess and test; answer: 5 letters and 7 postcards)

3. Joan decided to quit the fishing-worm business. She gave half her worms plus half a worm to Mark. Then she gave half of what was left plus half a worm to George, leaving her with a dozen worms, which she kept for herself. How many worms did Joan have when she decided to quit the business? (Possible strategy: Work backward; answer: 51 worms)

4. Fifteen people attended a party. If each person shook hands with every other person, how many handshakes were made? (Possible strategy: simplification, make a drawing, and look for patterns; answer: 105 handshakes)

5. I am a proper fraction. The sum of my numerator and denominator is 60, and their difference is 10. What is my simplest name? (Possible strategy: guess and test; answer: 5/7)

6. If 5 times a number is increased by 11 and the answer is divided by 3, the result is 32. Find the number. (Possible strategy: work backward; answer: 17)

7. To drive from Ackley to Derry, you will first pass through Booville and then through Carlton. It is five times as far from Ackley to Booville as it is from Booville to Carlton and three times as far from Booville to Carlton as it is from Carlton to Derry. If the distance from Ackley to Derry is 228 miles, how far is it from Ackley to Booville? (Possible strategy: make a drawing and guess and test; answer: 180 miles)

8. A bag of 15 silver dollars is known to contain 1 counterfeit coin, which is lighter than the other 14. What is the maximum number of comparisons that may be necessary to identify the counterfeit coin using only a two-pan balance? (Possible strategy: simplification, make a table, and look for patterns; answer: 3 weighings)

REFERENCES

Aviv, Cherie Adler. "Pattern Gazing." In *Activities from the Mathematics Teacher*, edited by Evan M. Maletsky and Christian R. Hirsch. Reston, Va.: National Council of Teachers of Mathematics, 1981.

Krulik, Stephen, and Jesse A. Rudnick. *Problem Solving—a Handbook for Teachers.* Boston: Allyn & Bacon, 1980.

National Council of Teachers of Mathematics. *An Agenda for Action: Recommendations for School Mathematics of the 1980s.* Reston, Va.: The Council, 1980.

Ouellette, Hugh. *"Number Triangles—a Discovery Lesson."* In *Activities from the Mathematics Teacher*, edited by Evan M. Maletsky and Christian R. Hirsch. Reston, Va.: National Council of Teachers of Mathematics, 1981.

Polya, George. *How to Solve It.* Princeton, N.J.: Princeton University Press, 1971.

Schaaf, Oscar F. "Activities: Teaching Problem-solving Skills." *Mathematics Teacher* 77 (December 1984):694–99.

Woodburn, Douglas. "Can You Predict the Repetend?" In *Activities from the Mathematics Teacher*, edited by Evan M. Maletsky and Christian R. Hirsch. Reston, Va.: National Council of Teachers of Mathematics, 1981.

GUESS AND TEST

SHEET 1

Sample problem: Not Himself Today!

Today, Alan Lurt (called A. Lurt by his friends) spent 20 more minutes asleep in class than he spent awake. If the class period is 52 minutes long, how long was Alan not alert?

Problems of this sort can sometimes be solved by guessing and testing.

Solution:

What do we know?	Total period is 52 minutes, and Alan spent 20 minutes more asleep than awake.
What do we want?	The number of minutes asleep
Guess the answer.	He slept 10 minutes.
Test your guess.	52 (total) − 10 (asleep) = 42 (awake). Awake more than asleep! Guess larger.

Guesses	Tests	Solution?
30 minutes asleep	52 − 30 (asleep) = 22 (awake) gives 8 minutes more asleep.	No
34 minutes asleep	52 − 34 (asleep) = 18 (awake) gives 16 minutes more asleep.	No
36 minutes asleep	52 − 36 (asleep) = 16 (awake) gives 20 minutes more asleep.	Yes

So A. Lurt was not alert for 36 minutes.

To solve problems using guess and test, try the following steps:

1. Read through the problem more than once.
2. Draw a picture of, or outline, the problem situation in some way until you feel you understand the problem.
3. Write down answers to these questions: (*a*) What is given? (*b*) What is wanted?
4. Make a reasonable guess.
5. Test your guess by substituting it into the problem.
6. If necessary, refine your guess and repeat the procedure until a guess satisfies the problem.

Problem 1: Good Horse Fence

Mary is asked by her dad to make a plan for fencing in a pasture for their new horses. They will require an area of 3000 m^2 for grazing and exercise. To reduce costs, Mary decides to use their neighbor's fence for the longest side of the rectangular area. What dimensions should she use for the pasture so that fencing costs are as low as possible?

Solution:

What do we know?	**Area (3000) = $l \times w$. Amount of new fence is $2w + l$.**
What do we want?	**Dimensions of the pasture with area 3000 m^2, so that amount of new fence is minimal**
Guess: $w = 20$	**Test! Since $l \times w = 3000$, $l = 3000 \div 20 = 150$. The amount of new fence is $2(20) + 150 = 190$.**

Can she do better? Complete the following table:

Area	Width (Guess)	Length	Amount of New Fence
3000	20	$3000 \div 20 = 150$	$2w + l = 2(20) + 150 = 190$
3000	30	$3000 \div 30 = 100$	$2w + l =$ ______ = ___
3000	40	______	= ______ = ___
3000	50	______	= ______ = ___

It appears that the best dimensions are approximately ______ by ______ .

Use the guess-and-test strategy to solve the problems that follow.

Problem 2: Julio's mathematics and English classrooms are next door to each other. The product of the room numbers is 1023. Find the room numbers.

Problem 3: Farmer Jones has pigs and chickens. They have a total of 35 heads and 100 legs. How many pigs does he have?

Problem 4: A 40-foot cable is divided into two sections. One section is 4 feet more than twice the length of the other section. How long is each section?

Sample problem: Who's Silly Now?

The famous football star Sam Bims (once called Silly Bims by the team's management) was offered a 4-year contract that included a bonus for each game in which he had at least 100 yards rushing. Sam was to receive a $1000 bonus for the first 100-yard game, $3000 for the second, $5000 for the third, $7000 for the fourth, and so on, throughout the contract. Sam had 40 such games. What was the total of his bonuses?

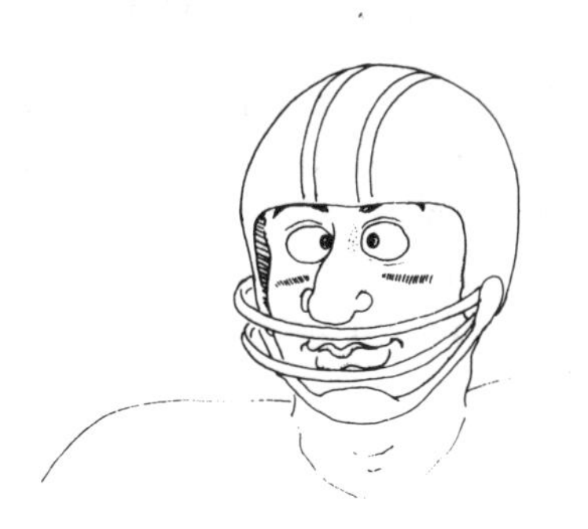

Using guess and test would not be a good choice here. Problems such as this, which are complex because of the number of conditions, are often solved more easily by solving one or more simpler problems first and looking for a pattern.

Solution:

What is known? Amount of bonus for each 100-yard game.
Sam had 40 such games.

What is wanted? Total of all the bonuses

Simplify: Let's look at some simpler problems by reducing the number of bonus games. We organize our results in a table.

Number of Bonuses	Bonuses				Total Bonus
1	1000				$1 000
2	1000	3000			$4 000
3	1000	3000	5000		$9 000
4	1000	3000	5000	7000	$16 000

Patterns: Compare the first and the last columns. Do you see a pattern? It looks like a pattern of squares. Test this conjecture: 5 bonuses would give $25 000 by the pattern. This answer checks, since $1 000 + $3 000 + $5 000 + $7 000 + $9 000 = $25 000.

Apply the pattern. 40 bonuses: $40^2 \times 1000 = 1600 \times 1000 =$ $1 600 000. Would you call him Silly Bims?

To solve problems by simplifying, try these steps:

1. Read through the problem more than once.
2. Draw a picture of, or outline, the problem situation in some way until you feel you understand the problem.
3. Write down answers to the questions: (*a*) What is given? (*b*) What is wanted?
4. Write and solve one or more simpler, but related, problems.
5. Compare the simpler situations and their solutions with the original problem.

Problem 1: Heartbreak Hill

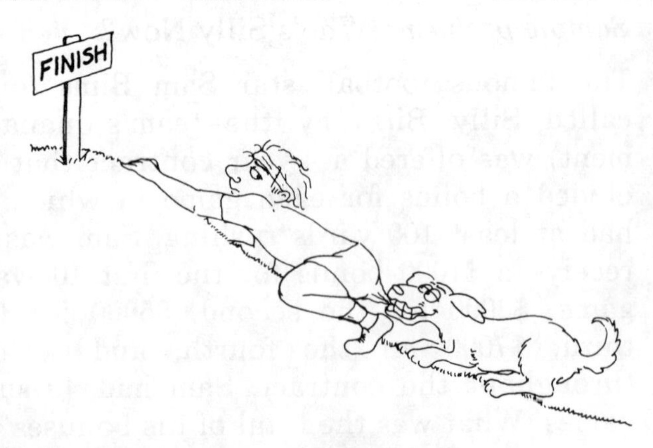

Nine officials are positioned along a straight section of a marathon route. At what location should they meet to confer so that the total distance they travel is as small as possible?

Solution:

What is known? Nine officials who will meet at one location

What is wanted? Location of meeting to minimize total distance traveled

Simplify: Answer the question for only two officials.

O_1 O_2

Next solve the problem for three officials.

O_1 O_2 O_3

Conjecture the solution and test your reasoning with one or more smaller cases.

Solve the following problems by first stating and solving a simpler problem.

Problem 2: Fataronda Middle School is planning to install an intercom system between classrooms and the main office that permits direct conversations between any pair of classrooms, as well as between a classroom and the main office. How many room-to-room and room-to-office connections will be needed if the school has 25 classrooms?

Problem 3: How many different downward paths connect P to Q in the adjacent diagram? (One such path is shown in the diagram. Remember that each section of the path must be downward.)

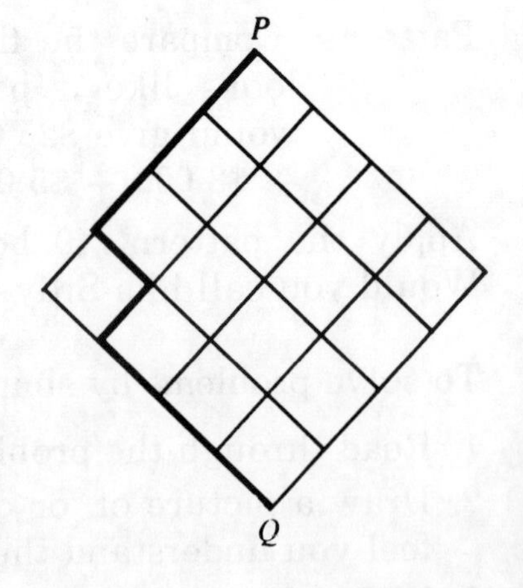

Problem 4: The units digit of 2^4 when expressed in standard form is 6. Find the units digit of 7^{99} when it is expressed in standard form.

Problem 5: How many squares are on a standard 8-by-8 checkerboard?

TEACHING THE ELIMINATION STRATEGY

January 1986

By DANIEL T. DOLAN, Office of Public Instruction, Helena, MT 59620
JAMES WILLIAMSON, Billings Public Schools, Billings, MT 59101

Teacher's Guide

Introduction: The primary curricular recommendation in *An Agenda for Action* is that "problem solving must be the focus of school mathematics in the 1980s" (NCTM 1980, 3–4). To achieve this goal, we, as classroom teachers, must create an atmosphere where problem solving flourishes, where a variety of problem-solving skills are taught in a sequential fashion that takes into account students' developmental levels, and where the skills are then applied, often in combination, in diverse problem situations.

The *Agenda* (NCTM 1980, 22) lists several recommended actions for how this atmosphere might be established. Three recommendations seem particularly relevant to the middle school and junior high school levels:

- The focus of the curriculum should be on more formal and more general problem-solving approaches and strategies themselves.
- Instruction should stress the ability to select from a range of strategies and to create new strategies by combining known techniques.
- Instruction should aid in the student's transition to more abstract reasoning.

To implement these recommendations, appropriate instructional materials must be developed to teach the strategies. Previous "Activities" have provided ideas for teaching the problem-solving skills of *make a drawing* and *work backward* (Schaaf 1984), *guess and test* and *simplification* (Laing 1985), and *make an organized list* and *search for a pattern* (Krulik and Rudnick 1985).

The activities presented here are designed to provide students with an informal introduction to deductive logic through the use of the problem-solving skill of *eliminate possibilities.* Next to *guess and check, elimination* is probably the most commonly used of all problem-solving skills. We use it daily in making a variety of decisions. A consumer's decision to purchase a particular product is often made by first gathering information and then eliminating competitive products on the basis of price, service, advice of friends, or consumer information.

Most students have their first experience with formal logic in geometry at the sophomore level. The traditional junior high curriculum has virtually no introduction to logic; this lack ignores sound educational theory. Without a foundation that offers experiences with the informal application of logic in familiar situations, it is not surprising that so many students have difficulty with the application of logic to proof in a formal geometry course.

Students in grades 7–9 are just entering what Piaget calls the formal operational stage. Because levels of maturity of students in this age group vary greatly, a complete development of logic would not be appropriate. However, it should be taught at an informal and introductory level. Teaching methods of elimination at this level can tremendously help students to make the transition to the more abstract reasoning in later courses.

The process of elimination often requires the generation of a list of possible solutions from which to eliminate possibilities systematically, the use of an indirect argument, or the use of a table to organize the information. This activity therefore will be most successful if students have previously completed the aforementioned problem-solving activities focusing on these prerequisite skills.

Grade levels: 7–11

Materials: Copies of the worksheets for each student and a set of transparencies for class discussions

Objectives: (1) To develop the problem-solving skill *eliminate possibilities;* (2) to introduce students to aspects of deductive reasoning and the use of indirect arguments; (3) to provide opportunities for the application of problem-solving strategies entailing the use of elimination and those of *guess and check* and *make an organized list or table*

Procedure: The worksheets are designed to be completed in order over a period of several days, and copies should be distributed one at a time to each student. Use a transparency of each worksheet to guide the class to a solution of the first problem and then instruct the students to solve the remaining problems on the worksheet using similar approaches. Since this may be the first time many of your students are confronted with logic problems of the type presented here, encourage students to cooperate and work together. The best results will be obtained if a follow-up discussion is held with the class after each worksheet is completed. Since each worksheet builds on ideas developed in the previous worksheet, it is essential to highlight key elements during this looking-back phase.

Worksheet 1: The most important part of the first problem is the follow-up discussion. Students will probably have little difficulty solving this problem, but each clue should be analyzed to see how a suspect is eliminated as a result of the evidence. It might be helpful to discuss whether or not clue (*a*) is needed. The fact that the killer registered at the hotel does not mean that he stayed there. Frequently in the mathematical experiences of students, all the data needed to solve a problem are given in exactly the order that leads to a solution; rarely are extra data presented that are not helpful. (Solution: Yoko Red committed the crime.)

A very important aspect of problem 2 is the concept of "first clue." Throughout pupils' elementary school experiences with arithmetic, they are instructed to read a problem and then work from the beginning. In many problems whose solution requires the elimination of possibilities, it is important that the problem be carefully analyzed to see which clue might be the first or easiest to use. This is especially true when a set of possible solutions must be generated before the elimination process can be used. In this problem, clues (*b*), (*d*), and (*e*) might be the easiest to use, but if a list had to be generated, the use of clue (*c*) as a first clue would result in a shorter list than clue (*d*). (Solution: 1375)

In problem 3, it is again important to discuss which clue is the easiest to use first and why. No single answer is correct, but it is important for students to share why they feel a clue is significant. It might even be helpful to discuss the order in which clues are used so that the entire thought process used in the solution is shared. This discussion of the strategy is the important looking-back component of problem solving stressed by Polya (1971). (Solution: 1537)

Worksheet 2: The objective is to develop the idea of using a table to organize information prior to eliminating possibilities.

The first problem provides an extension of the concept of the "best" first clue introduced in worksheet 1. The main idea is first to make a list of possible solutions. The key in doing so, as indicated in the hint, is to begin with the clue(s) that will generate the shortest list. Perhaps the best choice is to list the odd multiples of 7:

Odd multiples of 7:	~~7~~	~~21~~	~~35~~	~~49~~	~~63~~	77	91
Remainders when—							
divided by 3:	1	0	2	1	0	2	
divided by 5:	2	1	0	4	3	2	

(The solution is 77.)

With respect to problem 2, the winners are circled in the following table:

Day	Opponents
1	Spikers (Setters)
2	Spikers (Servers)
3	Setters (Servers)
4	(Spikers) Servers
5	(Setters) Servers
6	Spikers (Setters)

One approach to determining the winner (loser) of each game follows. Since the Spikers never defeated the Setters, the Setters won on days 1 and 6. Each team played each of the other teams at home and away, so we can assume that the Servers played at home on days 2 and 3. Since they never lost a home game, they won on days 2 and 3. Finally, the Servers had to lose two games. They won on days 2 and 3; therefore, they must have lost on days 4 and 5. (The following is the correct solution: Spikers won one and lost three; Setters won three and lost one; Servers won two and lost two.)

Problem 3 introduces the idea of constructing a table to organize the information during the elimination process. It is very important that discussion be focused on how the table in the hint is suggested by the information contained in the problem.

Students invariably begin completing the suggested table by using clue (*b*) to enter the American as the driver of the Audi. Either clue (*c*) or clue (*d*) can be used next. The following is one of several approaches to the solution. It is important that students be allowed to explain these alternate approaches if they discover them. (*Note:* The numbers that appear in parentheses in the table correspond to the following numbered steps that can be used to obtain the solution.)

1. Since the Audi was driven by the American, the Porsche had to be driven by the Italian or the Spaniard, but from clue (*d*), we know it wasn't the Italian.
2. The Italian is the only driver left and, thus, was the driver of the MG.
3. From clue (*c*), we know that the lady from Spain drove the red car.
4. The Porsche was red, so the MG must either be black or silver (from clue (*b*)). But from clue (*a*), we know that the Italian did not drive the black car, thus the MG must be silver. The following table shows the correct solution:

Car	Audi	MG	Porsche
Nationality Color	American black	(2) Italian (4) silver	(1) Spanish (3) red

Problem 4 is similar to problem 3. The main difference is that no hint is given for the table. The following is one possible table:

Boxer	
Class	
Medal	

However, the type of table presented in the following solution is often easier to use. Be sure to discuss how the two tables are related. Again, the numbers appearing in parentheses within the table correspond to the numbered steps used to obtain the solution.

1. We know that Todd was the heaviest winner.
2. Since Grant defeated Mitchell, they must be in the same weight class. But from clue (*a*), we know that Brown is in the 118-pound class and that Todd is in the heavyweight class, so Grant and Mitchell must be in the 145-pound class and Grant won the gold medal.
3. Clue (*b*) says that Slater was the third winner.
4. Brown weighed in at 117 pounds and thus won the silver medal in the 118-pound class.
5. O'Brien is the only fighter left and, hence, won the silver medal in the heavyweight class.

The correct solution is this:

	118-Pound	145-Pound	Heavy-weight
Gold Medal	(3) Slater	(2) Grant	(1) Todd
Silver Medal	(4) Brown	(2) Mitchell	(5) O'Brien

Worksheet 3: Solving a problem by eliminating possibilities often requires using an indirect argument, that is, assuming that something is true and showing that this assumption leads to a contradiction. The objective of the third worksheet is to introduce the concept of an indirect argument. The students are guided through portions of such an argument in problems 1 and 2 and then apply the techniques learned to solve problem 3.

Informal experiences with indirect reasoning such as those presented here provide an excellent introduction to the method of indirect proof that is so important in mathematics. We have found that students who have had this type of introduction often have far less difficulty with indirect proofs when they are formally introduced in geometry.

Students should be expected to write complete sentences as their answers to part (*e*) of the solution process in problem 1 and parts (*a*) and (*c*) of the solution of problem 2. The ability to express ideas clearly in writing is important and should be emphasized in all subject areas. Unfortunately, in mathematics classes, students are seldom required to express their thoughts in writing. Problems like these present an excellent opportunity for mathematics teachers to work with students in this critical area.

Problem 1 can be solved by examining each of the three possible solutions. Discussion of the solution should focus on the conclusion drawn in step (*e*) of each of the three cases.

Case 1: (*a*) Suppose that Adam spent \$0.60. (*b*) Since Adam spent four times as much as Carl, Carl must have spent \$0.15 (\$0.60/4). (*c*) Since Carl spent a third as much as Bruce, Bruce must have spent \$0.45 (\$0.15 × 3). (*d*) But Bruce also spent \$0.20 less than Adam, so Bruce spent \$0.40 (\$0.60 – \$0.20). (*e*) The results in parts (*c*) and (*d*) contradict each other. Thus, the assumption that Adam was the one who spent \$0.60 is incorrect.

Case 2: (*a*) Suppose Carl spent \$0.60. (*b*) Since Adam spent four times as much as Carl, Adam must have spent \$2.40 (\$0.60 × 4). (*c*) Since Carl spent a third as much as Bruce, Bruce must have spent \$1.80 (\$0.60 × 3). (*d*) But Bruce also spent \$0.20 less than Adam, so Bruce spent \$2.20 (\$2.40 – \$0.20). (*e*) The results in parts (*c*) and (*d*) contradict each other, so the assumption that Carl spent \$0.60 must be false.

Case 3: (*a*) Assume that Bruce spent \$0.60. (*b*) Since Carl spent one-third as much as Bruce, Carl must have spent \$0.20 (\$0.60/3). (*c*) Since Bruce spent \$0.20 less than Adam, Adam must have spent \$0.80 (\$0.60 + \$0.20). (*d*) But Adam also spent four times as much as Carl, so Adam spent \$0.80 (\$0.20 × 4). (*e*) No contradictions exist; therefore Bruce spent \$0.60. Had we considered the case of Bruce second, it would still have been necessary to check the third possibility (involving Carl) because the problem might have had multiple solutions. (Solution: Bruce spent $0.60.)

The following is a solution to the second problem. The most important parts of the solution are the indirect arguments in parts (*a*) and (*c*). (*a*) No. If any of the gumdrops in the bin were orange, then the bin would have contained both orange and black gumdrops, so it would have been labeled correctly. But this case would have contradicted the fact that all the bins were labeled incorrectly. (*b*) Black gumdrops. (*c*) No. If bin 2 contained orange and black gumdrops, then bin 1 would contain only orange gumdrops and would be labeled correctly. (*d*) Orange gumdrops; orange and black gumdrops. The correct solution is the following: bin 1—orange and black gumdrops; bin 2—orange gumdrops; bin 3—black gumdrops.

Problem 3 can best be solved by using a table to organize the information. Indirect arguments are used in steps 2, 3, and 6 of the solution, but the critical argument is that in step 2. The solution proceeds as follows:

	Hat	Coat
Jim	(3) Dan's	(2) Kevin's
Dan	(2) Kevin's	Peter's
Kevin	(1) Peter's	(6) Jim's
Peter	(4) Jim's	(5) Dan's

1. This fact was given in the problem.
2. Since Jim took the coat belonging to the man whose hat was taken by Dan, they must belong to the same person. The possibilities are Kevin's and Peter's, since no man has his own coat or hat. Suppose they belong to Peter. Then Dan would have Peter's hat and so would Kevin. But this situation is impossible, so they must belong to Kevin.
3. The only choices for the hat Jim has are Jim's and Dan's, but Jim can't have his own hat.
4. Jim is the only person whose hat has not been accounted for.
5. Dan's coat was taken by the person who took Jim's hat.
6. The only choices for the coat Kevin has are Jim's and Peter's, but it can't be Peter's, since no person took the coat and the hat of the same man.

(The solution is this: Jim had taken Dan's hat and Kevin's coat. Dan had taken Kevin's hat and Peter's coat.)

Worksheet 4: This worksheet provides additional experiences using indirect reasoning in conjunction with problem-solving strategies that entail the meshing of several skills, including that of eliminating possibilities.

In problem 1, students will no doubt put the names of the states in the table first. The forget-me-not can quite easily be placed under Alaska and the yellowhammer under Alabama. (Here we include a little geography in the mathematics lesson and assume some knowledge of these states.) The key clue now is "mistletoe." Since it is not the flower of Minnesota, it must be the flower of Oklahoma or Alabama. But flycatchers nest in mistletoe, and the yellowhammer has already been identified with Alabama. Therefore, the flycatcher and the mistletoe must be placed under Oklahoma. Loons and lady's slippers go together, so they must go with Minnesota. Thus the problem is solved. The table shows the correct solution:

State	Ala.	Alaska	Minn.	Okla.
Flower	camellia	forget-me-not	lady's slipper	mistle-toe
Bird	yellow-hammer	ptar-migan	loon	fly-catcher

Problem 2 is a straightforward application of an indirect argument. Assume that the first or second jogger is Fred, who always tells the truth. Both cases contain a contradiction. Therefore, the third jogger is Fred. Since he always tells the truth, the jogger in front is Joe and the one in the middle is Herman. (Solution: from left to right, the joggers are Joe, Herman, and Fred.)

In solving problem 3, students must build a table, list possible solutions, and use indirect arguments to determine correct solutions. An initial table might look like this:

Linda	Mary	Kim
not artist	not police officer	police officer
not police officer		

Clue (*b*) tells us about Linda and clue (*e*) about Mary. Therefore, Kim must be the police officer and Mary the artist. Kim beat Mary and the computer programmer at handball, so Linda is the programmer. The car salesperson offended the artist and sold the programmer a car, so Kim also sells cars. The programmer and the pilot are different people, and the problem is solved. (Solution: Linda is a computer programmer and a mechanic; Mary is an artist and a pilot; and Kim is a police officer and a car salesperson.)

Supplementary problems: The following problem can be used to reinforce the problem-solving strategies involving elimination. Its solution, however, has a slightly different twist. Students must first generate possible answers to the five questions using clues (*a*), (*b*), and (*d*).

Problem: A history exam has five true-false questions.

a) The exam has more true than false answers.

b) No three consecutive questions have the same answer.

c) Juanita knows the correct answer to the second question.

d) Questions 1 and 5 have opposite answers.

e) From this information, Juanita was able to determine all the correct answers. What are they?

Solution: Using clues (*a*), (*b*), and (*d*), the possible answers to the five questions are these:

1. T T F T F
2. T F T T F
3. F T F T T
4. F T T F T

The key clue now is (*c*). Juanita knows the correct answer to question 2. If the correct answer is true, the other questions have three combinations of answers. Clue (*e*), however, says that she was able to determine *all* the correct answers. Thus, the only possible answer for question 2 is false. The correct answers for the test are T, F, T, T, F.

Additional problems that offer opportunities to practice the skills developed here can be found in Charosh (1965), Dolan and Williamson (1983), and Hunter and Madachy (1963).

REFERENCES

Charosh, M., comp. *Mathematical Challenges.* Reston, Va.: National Council of Teachers of Mathematics, 1965.

Dolan, Daniel T., and James Williamson. *Teaching Problem-solving Strategies.* Menlo Park, Calif.: Addison-Wesley Publishing Co., 1983.

Hunter, J. A. H., and Joseph S. Madachy. *Mathematical Diversions.* Princeton, N.J.: D. Van Nostrand Co., 1963.

Krulik, Stephen, and Jesse A. Rudnick. "Activities: Developing Problem-solving Skills." *Mathematics Teacher* 78 (December 1985):685–92, 697–98.

Laing, Robert A. "Extending Problem-solving Skills." *Mathematics Teacher* 78 (January 1985):36–44.

National Council of Teachers of Mathematics. *An Agenda for Action: Recommendations for School Mathematics of the 1980s.* Reston, Va.: The Council, 1980.

Polya, George. *How to Solve It.* Princeton, N.J.: Princeton University Press, 1971.

Schaaf, Oscar F. "Teaching Problem-solving Skills." *Mathematics Teacher* 77 (December 1984):694–99.

ELIMINATE POSSIBILITIES — SHEET 1

1. Ace Detective Shamrock Bones of the City Homicide Squad is investigating a murder at the Old Grand Hotel. Five men are being held as suspects.
   - "Giant Gene" Green. He is 250 cm tall, weighs 140 kg, and loves his dear mother so much that he never spent a night away from home.
   - "Yoko Red." He is a 200-kg Sumo wrestler.
   - "Hi" Willie Brown. He is a small man—only 130 cm tall; he hates high places because of a fear of falling.
   - "Curly" Black. His nickname is a result of his totally bald head.
   - Harvey "The Hook" White. He lost both his hands in an accident.

   Use the following clues to help Detective Bones solve this crime.

   a) The killer was registered at the hotel.
   b) Before he died, the victim said the killer had served time in prison with him.
   c) Brown hair from the killer was found in the victim's hand.
   d) The killer escaped by diving from the third-floor balcony into the river running by the hotel and then swimming away.
   e) Smudges were found on the glass tabletop, indicating that the killer wore gloves.

The killer is ______.

2. Circle the number below described by the following clues:

   a) The sum of the digits is 16.
   b) The number has more than three digits.
   c) The number is a multiple of 5.
   d) It is not an even number.
   e) The number is less than 2572.

871 745 3625 2860 2582 1780 2315
1937 1485 1375 1671 1455
1075 1690 2635 2590

3. Circle the number described by these clues.

   a) No two digits are alike.
   b) The sum of the two middle digits is equal to the sum of the first and last digits.
   c) The thousands digit is the smallest.
   d) None of the digits are even.
   e) The units digit is the largest.

9315 7351 8316
1537 5713
9731 1449 3714
9371 7951
3174 1627 7624
1818 1592
1569 1963

1. Andrea collects coins. She wanted to arrange her collection in a display case in rows that all contained the same number of coins. She tried putting two coins in each row, but she had one coin left over. When she tried three or five coins in a row, two coins were left over. Finally, she tried seven coins in each row and it worked! What's the smallest number of coins Andrea could have in her collection? ______

   Hint: Use the clues to make an organized list of possibilities and then use elimination.

2. The three teams in the Western Volleyball League had a season in which each team played the other two both at home and away. The season's schedule is shown in the following table. Only one game was played on a given day.

Day	1	2	3	4	5	6
Opponents	Spikers Setters	Spikers Servers	Setters Servers	Spikers Servers	Setters Servers	Spikers Setters

   a) The Spikers never defeated the Setters.
   b) Even though the Servers lost two games, they never lost a game at home.

   What was the win-loss record of each of the three teams? ______
   ______

3. A recent international auto rally resulted in a three-way tie for first place. The three cars, an Audi, an MG, and a Porsche, were driven by three women from Italy, Spain, and the U.S. Each car was a different color. Use the following clues to determine the color of each car and the nationality of the driver.

   a) The Italian was not in the black car.
   b) The Audi, driven by the American, was not silver.
   c) The red car was driven by the woman from Spain.
   d) The Porsche was not driven by the Italian.

Hint: Set up a table like this to help organize the information.

Car	Audi	MG	Porsche
Nationality			
Color			

4. Brown, Todd, Slater, Mitchell, O'Brien, and Grant reached the final round of the Olympic boxing championships. They were the finalists in the 118-pound, 145-pound, and heavyweight classes.

   a) Brown weighed in at 117 pounds, and Todd was the heaviest winner of all.
   b) Grant defeated Mitchell, and Slater won by a knockout.

   Who were the gold and silver medal winners in each weight class? ______
   ______

1. Three boys each purchased some fruit for a snack. Adam spent four times as much as Carl. Carl spent one-third as much as Bruce, and Bruce spent $0.20 less than Adam. Which boy, if any, spent $0.60?

   Since only three possibilities exist, let's examine the possible solutions individually.

   a) Suppose Adam spent $0.60.

   b) Since Adam spent four times as much as Carl, Carl must have spent ________.

   c) Since Carl spent a third as much as Bruce, Bruce must have spent ________.

   d) But, Bruce also spent $0.20 less than Adam, so Bruce spent ________.

   e) What do the results in parts (*c*) and (*d*) suggest? ____________________

   Continue this method of *indirect reasoning* by next assuming that Carl spent $0.60; and so on.

2. Eileen was visiting her uncle's candy factory. In the storeroom, she found three bins. Bin 1 was labeled "orange gumdrops," bin 2 was labeled "black gumdrops," and bin 3 was labeled "orange and black gumdrops." Her uncle told her that she could have one gumdrop from one of the bins, but he warned her that none of the bins was labeled correctly. Eileen reached into the bin labeled "orange and black gumdrops" and pulled out a black gumdrop.

   a) Could any of the gumdrops in the bin labeled "orange and black gumdrops" be orange? ________ Why? ____________________

   b) What is the correct label for bin 3? ____________________

   c) Can the bin labeled "black gumdrops" contain orange and black gumdrops? ________ Why? ____________________

   d) What is the correct label for bin 2? ____________________

   For bin 1? ____________________

3. Four men attended the ballet with their wives. In their hurry to leave, each man mistakenly picked up another man's coat and the hat of yet another man and left. Jim took the coat belonging to the man whose hat was taken by Dan, and Dan's coat was taken by the person who took Jim's hat. Kevin took Peter's hat. Whose hat and whose coat had Jim and Dan taken? ______

   ____

1. A group of students were playing a trivia game involving states, state birds, and state flowers. Of four states—Alaska, Oklahoma, Minnesota, and Alabama—they knew the birds were the common loon, yellowhammer, willow ptarmigan, and scissortailed flycatcher, and that the flowers were mistletoe, camellia, forget-me-not, and pink-and-white lady's slipper. No one knew which bird or flower matched which state. A call to the local library resulted in the following clues.

   a) The forget-me-not is from the northernmost state.

   b) The flycatcher loves to nest in the mistletoe.

   c) The willow ptarmigan is not the bird for the camellia state.

   d) The yellowhammer is from a southeastern state.

   e) Mistletoe and Minnesota do not go together, but loons and lady's slippers do.

   Use the clues to fill in the table below.

State				
Flower				
Bird				

2. Three joggers named Fred, Herman, and Joe are jogging toward the country club. Fred always tells the truth. Herman sometimes tells the truth, whereas Joe never does.

   Determine the names of each runner and explain how you know. (*Hint:* First, determine which one is Fred.)

3. Linda, Mary, and Kim have each been employed in two of the following jobs: artist, mechanic, computer programmer, car salesperson, pilot, and police officer. No two ever had the same job. Use the following information to determine each of their jobs.

   a) The car salesperson offended the artist by criticizing her work.

   b) The artist and the police officer dated Linda's brother.

   c) The computer programmer and the pilot both got A's in mathematics in high school.

   d) The car salesperson sold the computer programmer a car.

   e) Mary got a ticket from the police officer.

   f) Kim beat both Mary and the computer programmer at handball.

THE PEG GAME

January 1982

By Charles Verhille and Rick Blake, University of New Brunswick, Fredericton, NB E3B 6E3

Teacher's Guide

Grade Level: 6–12.

Materials: Peg games may be purchased commercially or made from inexpensive materials.

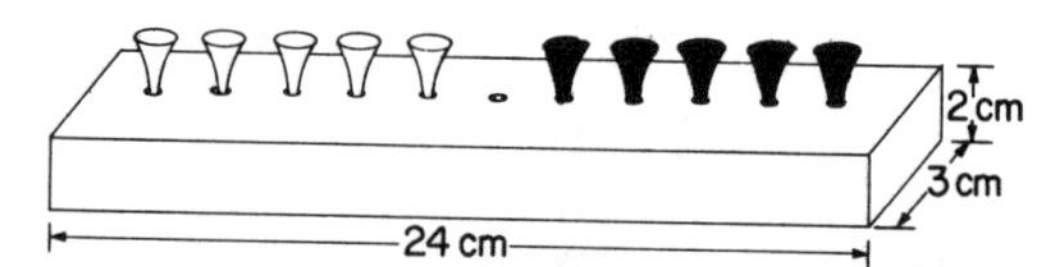

A piece of wood with the dimensions indicated can serve as the game board. Eleven equally spaced holes are drilled as illustrated. Five each of two different colored pegs are arranged, as shown, on the board. Golf tees work well for pegs.

A modified version of the game can be played using paper strips of eleven 1-inch squares for the playing board. Instead of pegs, use ten coins, five heads and five tails, to represent the colors white and black.

Objectives: To collect and organize data; to recognize, extend, and generalize patterns; and to develop a problem-solving strategy.

Procedure: The teacher should be familiar with the entire activity before attempting it with the students.

Duplicate a set of sheets for each student. The individual activities are to be done separately and in the order given. They may be presented to individual students or small groups or used in a learning center. It may be helpful to have students work in pairs so that one student can do the moving while the other counts, checks, and records.

• *Counting moves.* Distribute game boards and pegs. Students will normally complete this task at different times, allowing the teacher time to check each student's strategy before continuing. Ensure that each student demonstrates an appropriate strategy, since this is necessary for the other activities.

Some guidance may be necessary to

find patterns that lead to answers for question 5.

• *Recording the sequence of moves*. Be sure that all students understand the notation R for move a *white peg right* and L for move a *black peg left*. Note also that the sequences are set up with an R move first.

• *Counting jumps*. Here the student studies the previously developed sequences and discovers that the number of jumps needed is the square of the number of pairs of pegs at the start.

Answers:

1–3 (See answers in the table for question 7.)

4.

Number of Pairs	Number of Moves
1	3
2	8
3	15
4	24

5. *a*. 35

   b. 120

   The rule may come from patterns like the following:

   (1) Noticing how the differences for the "number of moves" are related

   3 > 5 > 2
   8 > 7 > 2
   15 > 9
   24

   (2) The next number of pairs squared minus 1: $(n + 1)^2 - 1$

   (3) The number of pairs squared plus twice itself: $n^2 + 2n$

   (4) The number of pairs times the number of pairs plus 2: $n(n + 2)$

7.

Number of Pairs	Sequence of Moves
1	*RLR*
2	*RLLRRLLR*
3	*RLLRRRLLLRRRLLR*
4	*RLLRRRLLLLRRRRLLLLRRRLLR*

8.

Number of Pairs	Sequence of Moves
1	1, 1, 1
2	1, 2, 2, 2, 1
3	1, 2, 3, 3, 3, 2, 1
4	1, 2, 3, 4, 4, 4, 3, 2, 1

9. 1, 2, 3, 4, 5, 5, 5, 4, 3, 2, 1

10. 7 pairs

11.

Number of Pairs	Number of Jumps
1	1
2	4
3	9
4	16

12. *a*. 25

    b. 100

    c. n^2

Extensions: Students might be interested in exploring the effects of changing rule C on sheet 1 of the peg game to one of these:

1. You can move a peg into an adjacent empty hole or jump over *one or two pegs* of the opposite color into an empty hole.

2. You can move a peg into an adjacent empty hole or jump over *any number of pegs* of the opposite color.

Interestingly, the number of moves needed to interchange the pegs with either of these modifications remains the same as with the original rule for a single jump.

BIBLIOGRAPHY

Duncan, Richard B. "The Golf Tee Problem." *Mathematics Teacher* 72 (January 1979):53–57.

Higginson, William. "Mathematizing 'Frogs': Heuristics, Proof, and Generalization in the Context of a Recreational Problem." *Mathematics Teacher* 74 (October 1981):505–15.

A PEG GAME

SHEET 1

Rules for the Game

The object of this peg game is to interchange the black and white pegs using the following rules:

A. You can move only one peg at a time.
B. You can move white pegs only to the right and black pegs only to the left.
C. You can move a peg into an adjacent empty hole or jump over a single peg of the opposite color into an empty hole. You are not allowed to jump over two or more pegs.

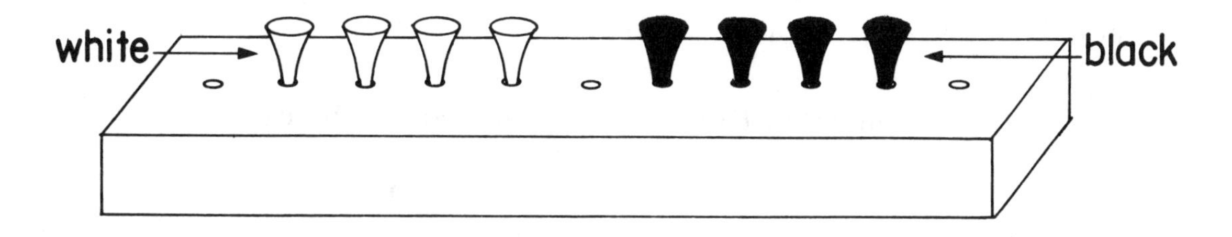

To find the number of moves needed for four pairs of pegs as shown above, first try the simpler problems that follow.

Counting Moves

1. Show how you can interchange one pair of pegs with a space between them in just three moves.

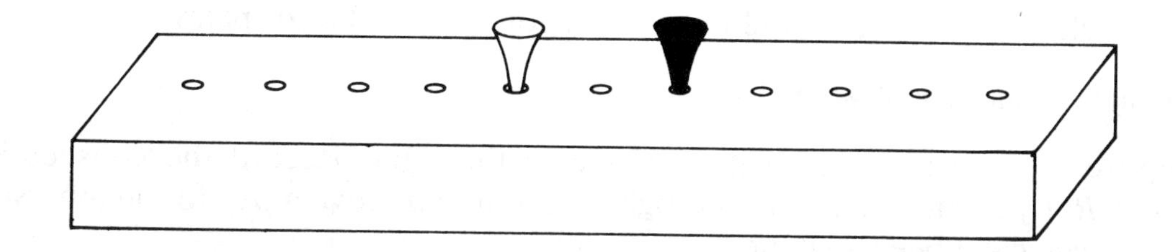

2. Now try two pairs of pegs. See if you can interchange them in eight moves.

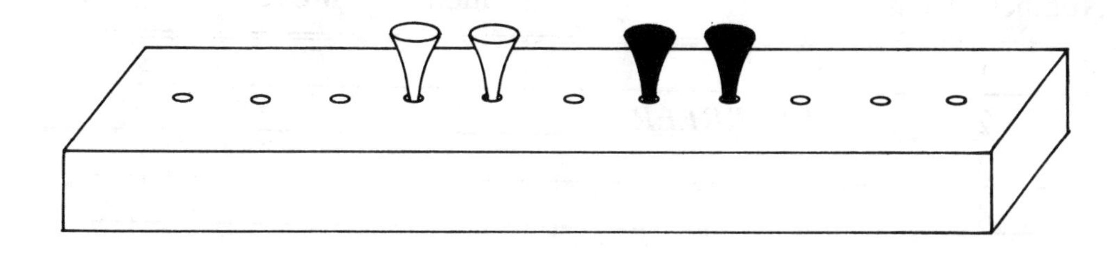

3. Next try to interchange three pairs of pegs. Look for a strategy. Count your moves carefully.

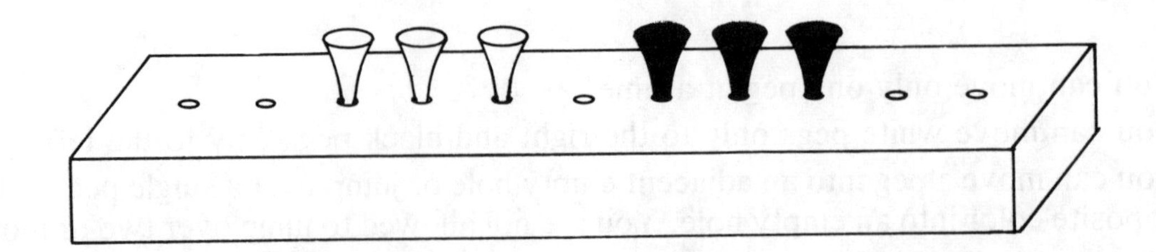

4. Complete the table below. Note that a pair consists of one peg of each color and that only one space separates the colors.

Number of Pairs	Number of Moves
1	3
2	8
3	
4	

5. Look for a pattern in your completed table that can be used to predict the number of moves for the following:

   a. five pairs of pegs?

   b. ten pairs of pegs?

   What rule can you use to find the answer for any number of pairs?

Recording the Sequence of Moves

Suppose you always begin with a move to the right. Record the consecutive moves as *R* if you move a peg to the right and *L* if you move a peg to the left. Stop when all pegs have been switched.

6. The result for two pairs of pegs is *RLLRRLLR*. Verify that this sequence of moves does interchange the pegs.

7. Complete the table.

Number of Pairs	Sequence of Moves
1	
2	*RLLRRLLR*
3	
4	

The sequence of moves can be recorded by the number of consecutive moves in a given direction, starting with a right-hand move.

Thus the sequence *RLLRRLLR* can be expressed as 1, 2, 2, 2, 1.

$$\underbrace{R}_{1}\ \underbrace{LL}_{2}\ \underbrace{RR}_{2}\ \underbrace{LL}_{2}\ \underbrace{R}_{1}$$

8. Complete the table using this numerical notation.

Number of Pairs	Sequence of Moves
1	
2	1, 2, 2, 2, 1
3	
4	
5	

9. Write the numerical sequence for five pairs of pegs.

10. How many pairs of pegs would produce this sequence of moves?

1, 2, 3, 4, 5, 6, 7, 7, 7, 6, 5, 4, 3, 2, 1

Counting Jumps

An interesting pattern emerges when you count only the jumps made in interchanging the pegs.

11. Complete the table below, recording only the number of jumps needed to switch 1, 2, 3, and 4 pairs of pegs.

Number of Pairs	Number of Jumps
1	
2	
3	
4	

12. Without using your board, how many jumps do you think will be needed for—

a. 5 pairs of pegs?__________

b. 10 pairs of pegs?__________

c. *n* pairs of pegs?__________

Activities for

Expanding Basic Skills

The eight activities in this section focus on new curricular emphases that provide mathematical methods that support the full range of problem solving and that also reflect the *Agenda*'s expanded concept of basic skills. The first three activities treat quantitative skills that are fundamental in our information-oriented society. It is recommended that these activities be completed in sequence. "Stem-and-Leaf Plots" uses interesting real-world data (e.g., best pop albums) to introduce students to effective new methods for organizing and displaying data. In addition to making stem-and-leaf plots, pupils are asked to interpret their displays and communicate, in writing, their findings. The activity "Plotting and Predicting from Pairs" evolves around the use of scatter plots to analyze possible relationships shown by such bivariate data as height and shoe size or the number of rebounds and assists by a basketball team. In the process of graphing bivariate data and analyzing their graphs, students are exposed to the concepts of positive relationship, negative relationship, and no relationship. These concepts are in turn applied to the analysis of time-related data. "Data Fitting without Formulas" introduces the *median fit* technique as a simple method to fit a straight line to a scatter plot for purposes of aiding prediction from the data. Again the emphasis is on the application of the skill to data collected from real-world situations.

The next two companion activities are devoted to estimation and mental computation, skills that are essential in a society increasingly dominated by calculator and computer output. "Estimating with 'Nice' Numbers" develops the skill of rounding to "nice numbers" (powers of 10), when the context permits, to obtain quick estimates of products and quotients and then introduces strategies for refining these estimates. Students use these skills to judge the reasonableness of given calculations as well as to estimate answers to a variety of routine problems arising in daily life. The activity "Developing Estimation Strategies" introduces the strategies of using "compatible numbers" (for estimating quotients involving whole numbers and products involving fractions and percentages), averaging, and rounding to the nearest half (to estimate totals) and provides opportunities for pupils to apply these strategies in real-life contexts.

The development of spatial skills is the focus of the next two activities. "Spatial Visualization" engages pupils in relating solids made from small cubes and their two-dimensional representations, in making isometric drawings of given solids, and in building solids from isometric drawings. The emphasis throughout is on moving back and forth visually from a solid to a drawing. In "Generating Solids" students must visualize, identify, and describe solids of revolution and then compute their surface areas and volumes.

This section ends with an activity devoted to the enhancement of the higher-order mental processes of logical reasoning and information processing. "Deductive and Analytical Thinking" provides experiences with the use of Venn diagrams to analyze numerical relationships and with the use of arrow diagrams to analyze genealogical, sociological, and propositional relationships.

STEM-AND-LEAF PLOTS

October 1985

By JAMES M. LANDWEHR, AT&T Bell Laboratories, Murray Hill, NJ 07974
ANN E. WATKINS, Los Angeles Pierce College, Woodland Hills, CA 91371

Teacher's Guide

Introduction: An Agenda for Action declares that problem solving must be the focus of school mathematics in the 1980s and that mathematics programs must include "methods of gathering, organizing, and interpreting information, drawing and testing inferences from data, and communicating results" (NCTM 1980, 3).

The American Statistical Association–National Council of Teachers of Mathematics Joint Committee on the Curriculum in Statistics and Probability is presently engaged in a three-year National Science Foundation–funded project to develop instructional materials for implementing the spirit of the recommendations cited above. This Quantitative Literacy Project is producing four booklets for teaching aspects of statistics and probability: *Exploring Data, Introduction to Probability, The Art and Techniques of Simulation,* and *Information from Samples.*

The activity sheets that follow are similar to those in *Exploring Data.* That booklet is intended to teach students effective new techniques for organizing and presenting data, interpreting these displays, and, especially, communicating the results.

One particularly effective method for displaying data is the stem-and-leaf plot, invented less than twenty years ago by John Tukey of Princeton University and AT&T Bell Laboratories. It is fast and easy to construct. If turned on its side, the stem-and-leaf plot resembles a histogram, but with an important difference—none of the original data are lost. Making a plot like this enables students to see such important features of the data as outliers, gaps, the location of the center, the range, and whether or not the distribution is skewed to smaller or larger values.

We have tried to present data that are interesting to secondary school students. For example, most students are likely to have an opinion about the best pop album of 1984. Interest in the data inspires interest in the mathematics.

The "communicating" portion of the *Agenda*'s recommendation cited earlier is one of the most neglected aspects of mathematics education. Emphasize to your students that a numerical result is of little use in the real world unless its meaning can be communicated to others. The activity sheets contain questions that ask students to summarize the information seen in a plot. They will resist writing these summaries and complain, "But this isn't an English class!" Insist that they do the writing and that they use good grammar, correct spelling, and complete sentences. Clarity in writing will encourage clarity in thought. These exercises will also give those students who have never had any success in mathematics a chance to shine. Often the students weakest in mathematics make the most perceptive comments.

Grade levels: 7–12

Materials: Calculators, a map of the United States, and a set of activity sheets for each student

Objectives: Students will (1) make stem-and-leaf plots of real-world data and (2) write accurate, perceptive analyses of the information displayed in the plots.

Directions: Distribute the activity sheets one at a time. Students can complete the activities either working independently or in small groups. Class discussion and a summary of the results should follow the completion of each sheet. The completion of the three worksheets will require two or three instructional periods, depending on the level of your class.

Sheet 1: No unique way has been found to construct stem-and-leaf plots. We have placed the smaller numbers at the top so that when the plot is turned counterclockwise ninety degrees, it resembles a histogram. Students will initially require some direction in analyzing the salient features of the displayed data. You might suggest that they consider, among other things, the range of percentages of pedestrian deaths, a percentage interval that typifies most of the cities, and any unusually high or low percentages.

Students may want to know the total number of pedestrian deaths in each city and also how vigilant the police are in citing jaywalkers and then discuss whether or not they think variables like these might be related to the percentage of pedestrian traffic deaths. This information is given in table 1 for 1983.

TABLE 1

City	Number of Pedestrian Deaths	Number of Pedestrian Citations per 100 000 Residents
New York	299	7.3
Los Angeles	131	1349
Chicago	97	0
Houston	100	21
Philadelphia	39	3
Detroit	42	389
Dallas	47	164
San Diego	33	295
Phoenix	33	100
San Antonio	24	0
Honolulu	18	224
Baltimore	19	0
Indianapolis	9	49.5
San Francisco	31	255
San Jose	20	95

Source: *Los Angeles Times*, 8 September 1984

Some questions that students can be encouraged to research are these:

1. What proportion of pedestrian deaths are children?
2. Are the pedestrian deaths usually the fault of the driver or of the pedestrian?
3. What percentage of traffic deaths are pedestrian deaths in the area of your school?

Sheet 2: After discussing solutions to the exercises on this worksheet, you may want to point out some of the problems associated with the method of balloting used by the music critics. For example, suppose two albums, *A* and *B*, are very good and only two people are voting. The first voter wants album *A* to win, so she ranks *A* first and *B* second. The second voter wants *B* to win, so he ranks *B* first and some mediocre album second. The second voter has made his choice, *B*, the winner by a rather unethical method.

Sheet 3: Some students may need assistance in completing the third stem-and-leaf plot because of the large number of stems involved. As a follow-up to this final activity, you might assign the class the project of making a stem-and-leaf plot of the populations of the fifty states. This will lead to questions of how to handle numbers of three or more digits. Possible solutions include truncating or rounding. Another useful project would be to have the class make a stem-and-leaf plot of their heights in meters.

Answers:

Sheet 1: 1. 25%; 2. 35%; 3. Answers will vary. 4. It is often helpful, but not necessary, to put the leaves in numerical order.

Stem	Leaves
1	8
2	0 5 7 9
3	2 2 4 4 5 8 9
4	0
5	7
6	0

5. a. San Francisco and New York; b. Possible explanations include the fact that both cities are densely populated, so traffic moves slowly. Consequently, a collision between two cars is unlikely to result in a death. Both cities have a large number of pedestrians, who are more likely to die than the occupant of a car if involved in an accident. San Francisco has a high alcoholism rate, which may further account for pedestrians being hit. 6. The percentage of traffic deaths that are pedestrian deaths in the fifteen largest U.S. cities ranges from 18 percent in San Antonio to 60 percent in San Francisco. Most cities have between 20 percent and 40 percent. Two cities, San Francisco and New York, have unusually high percentages. 7. The stem-and-leaf plot is better than the table for displaying the data because we can easily find the largest and smallest values; we can see whether any values are unusually large or small; we can see how spread out the values are; we can see where the majority of values are located; and we can see if any gaps occur in the data.

A weakness of the stem-and-leaf plot is that the names of the cities are missing. To find out which city corresponds to 60 percent, we have to go back to the table.

Sheet 2: 1. 17 points; 2. 40 points; 3. Answers will vary. (Actually, it was third on one list and fourth on another.) 4. Answers will vary. (This album was, in fact, first on five lists, second on two lists, and third, fourth, fifth, and sixth on one list each.)

5.

Stem	Leaves
1	2 3 4 5 7 9
2	4 4 5 6
3	
4	0 6
5	5
6	
7	
8	3
9	4

6. The top fifteen albums have between 12 and 94 points each. They cluster into three groups. Two albums, "Born in the U.S.A." and "Purple Rain," have many more points than the middle group of three albums with 40, 46, and 55 points. The nine lowest-rated albums are all clustered between 12 and 25 points. One suspects that the albums in the lowest group beat out other candidates by a very small margin. 7. It was possible to be a "top album" with only 12 points. An album could do this by being fifth on but two of the fifteen lists. "Lush Life" came in eleventh and was listed only twice. (In fact, seventy-six different albums were listed by the fifteen pop music critics. One of the critics said that such a variety were named because no other album "exhibited the unbending command or vision of the two premier works.")

Sheet 3:

1.

Stem	Leaves
5	7 8
6	4 5 6 6 6 6 7 7 8 8 9 9
7	1 3 3 5 5 5 6 6 6 6 7 7 7 7 8 8 8 8 8 9
8	0 0 1 2 2 2 3 3 3 4 4 5 5 5 8 9
9	1 5

2.

Stem	Leaves
5	
·	7 8
6	4
·	5 6 6 6 6 7 7 8 8 9 9
7	1 3 3
·	5 5 5 6 6 6 6 7 7 7 7 8 8 8 8 8 9
8	0 0 1 2 2 2 3 3 3 4 4
·	5 5 5 8 9
9	1
·	5

3\.

```
5 |
· |
· | 7
· | 8
6 |
· |
· | 4 5
· | 6 6 6 7 7
· | 8 8 9 9
7 | 1
· | 3 3
· | 5 5 5
· | 6 6 6 6 7 7 7 7
· | 8 8 8 8 8 9
8 | 0 0 1
· | 2 2 2 3 3 3
· | 4 4 5 5 5
· |
· | 8 9
9 | 1
· |
· | 5
```

4\. We prefer the last plot because the shape of the distribution is easier to see. The first and second plots have too many leaves on some lines. 5. Answers will vary. 6. The percentage of students who graduate from high school ranges from 57 percent in Louisiana to 95 percent in North Dakota. Louisiana and Washington, D. C., have unusually low graduation rates. In general, southern states have the lowest rates of graduation. The four highest rates are 95, 91, 89, and 88 percent, respectively. Three of these are from midwestern states. The rates are centered at 77 percent and are spread out rather uniformly between 64 percent and 85 percent.

BIBLIOGRAPHY

Landwehr, James M., and Ann E. Watkins, *Exploring Data*. Washington, D. C.: American Statistical Association, 1984.

MacDonald, A. D., "A Stem-Leaf Plot: An Approach to Statistics." *Mathematics Teacher* 75 (January 1982): 27–28, 25.

National Council of Teachers of Mathematics. *An Agenda for Action: Recommendations for School Mathematics of the 1980s*. Reston, Va.: The Council, 1980.

Velleman, Paul F., and David C. Hoaglin. *Applications, Basics, and Computing of Exploratory Data Analysis*. Boston: Duxbury Press, 1981.

STEM-AND-LEAF PLOTS — SHEET 1

The chart below gives the percentage of traffic deaths that were pedestrian deaths for the fifteen largest U.S. cities in 1983.

City	Percentage	City	Percentage
New York	57%	Phoenix	29%
Los Angeles	34%	San Antonio	18%
Chicago	39%	Honolulu	20%
Houston	32%	Baltimore	32%
Philadelphia	__%	Indianapolis	__%
Detroit	40%	San Francisco	60%
Dallas	27%	San Jose	38%
San Diego	34%		

Source: *Los Angeles Times*, 8 September 1984

1. Indianapolis had 36 traffic deaths. Of these, 9 were pedestrians. Fill in the percentage for Indianapolis.

2. Philadelphia had 111 traffic deaths. Of these, 39 were pedestrians. Fill in the percentage for Philadelphia.

3. a. What is the percentage for the city nearest you? ______
   b. If there were 200 traffic deaths in this city, how many were pedestrian deaths? ______

4. Below is the beginning of a stem-and-leaf plot of these data:

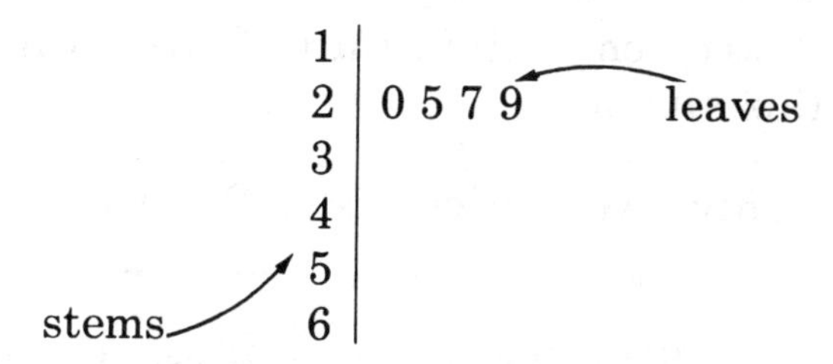

The line 2 | 0 5 7 9 represents the data for four cities: Honolulu with 20 percent, Indianapolis with 25 percent, Dallas with 27 percent, and Phoenix with 29 percent. Finish this stem-and-leaf plot for the remaining eleven cities.

5. a. Which cities have an unusually high percentage?______________
   b. What is a possible explanation for these high percentages?__________
   ____

Complete exercises 6 and 7 on a separate sheet of paper.

6. Write a one-paragraph summary of the information you can see from looking at this plot.

7. In what ways is the stem-and-leaf plot better than the chart for displaying the data? In what ways is it worse?

Fifteen of the pop music critics for the *Los Angeles Times* made a list of what they felt were the ten best albums of 1984. The albums were ranked by giving 10 points for each first place vote, 9 points for a second place vote, and so on. Below is the list of the top fifteen albums.

	Album	Artist	Points
1.	*Born in the U.S.A.*	Bruce Springsteen	94
2.	*Purple Rain*	Prince	83
3.	*How Will the Wolf Survive?*	Los Lobos	55
4.	*Reckoning*	R.E.M.	46
5.	*Private Dancer*	Tina Turner	—
6.	*Let It Be*	Replacements	26
7.	*Learning to Crawl*	Pretenders	25
8.	*Double Nickels on the Dime*	Minutemen	24
9.	*The Magazine*	Rickie Lee Jones	24
10.	*The Unforgettable Fire*	U2	19
11.	*Lush Life*	Linda Ronstadt	—
12.	*Zen Arcade*	Hüsker Dü	15
13.	*Soul Mining*	The The	14
14.	*Meat Puppets II*	Meat Puppets	13
15.	*Sparkle in the Rain*	Simple Minds	12

1. *Lush Life* was third on Paul Green's list, second on Dennis Hunt's list, and did not appear on any other list. Fill in the total points for this album.

2. *Private Dancer* was listed tenth, first, third, fourth, third, and fifth. Fill in the total points for this album.

3. *Zen Arcade* was on only two critics' lists. Describe a way it could have earned its 15 points.____________________

4. *Born in the U.S.A.* was listed by eleven critics—the most of any album. Describe a way this album could have earned its 94 points.____________________

____

Complete exercises 5 and 6 on a separate sheet of paper.

5. Make a stem-and-leaf plot of the points for the fifteen albums. The first stem should be a 1 and the last should be a 9.

6. Study your stem-and-leaf plot and then write a summary of what you learned.

7. The lists were extremely varied this year. How can you tell?____________________

____

____

The percentage of students that graduate from high school is given below for each state and the District of Columbia.

Ala.	67	Ill.	77	Mont.	83	R.I.	75
Alaska	78	Ind.	78	Nebr.	84	S.C.	66
Ariz.	68	Iowa	88	Nev.	75	S.Dak.	85
Ark.	76	Kans.	83	N.H.	77	Tenn.	65
Calif.	75	Ky.	68	N.J.	83	Tex.	69
Colo.	79	La.	57	N.M.	71	Utah	85
Conn.	78	Maine	77	N.Y.	67	Vt.	85
Del.	89	Md.	81	N.C.	69	Va.	76
D.C.	58	Mass.	78	N.Dak.	95	Wash.	76
Fla.	66	Mich.	73	Ohio	82	W.Va.	77
Ga.	66	Minn.	91	Okla.	80	Wis.	84
Hawaii	82	Miss.	64	Oreg.	73	Wyo.	82
Idaho	78	Mo.	76	Pa.	80		

Source: *USA Today*, 19 December 1984

1. On a separate sheet of paper, make a stem-and-leaf plot of the percentages using 5 as the first stem and 9 as the last stem.

2. Make a new stem-and-leaf plot using these stems. Put the percentages 50–54 on the 5 line and the percentages 55–59 on the line below. North Dakota with 95 percent and Minnesota with 91 percent are done for you.

```
5 |
· |
6 |
· |
7 |
· |
8 |
· |
9 | 1
· | 5
```

3. The stem-and-leaf plot can be spread out even further. On a separate sheet of paper, make a third plot using the stems that are begun at the right. The percentages 50–51 go on the first line, 52–53 on the second, 54–55 on the third, 56–57 on the fourth, 58–59 on the fifth, and so on.

```
5 |
· |
· |
· |
· |
6 |
· |
· |
· |
· |
```

4. Which of the three plots do you think best displays the data?__________
   Why?__________

5. a. What is the percentage of high school graduates in your state?__________
   b. Is this percentage relatively high or relatively low?__________
   c. Can you think of any reasons for this ranking?__________

6. Study your last stem-and-leaf plot. On the same sheet of paper, write a summary of conclusions you can draw from this plot.

PLOTTING AND PREDICTING FROM PAIRS

September 1984

By ALBERT P. SHULTE, Oakland Schools, Pontiac, MI 48054, and
JIM SWIFT, Nanaimo Senior Secondary School, Nanaimo, BC V9R 5K3

Teacher's Guide

Introduction: The collection of data, the selection of appropriate displays that illuminate the data, and the interpretation of those displays are all basic skills in today's information-oriented world. Recognizing the need for these skills in the information age, the *Agenda for Action* recommended curricular changes that included an increased emphasis on such activities as "locating and processing quantitative information; collecting data; organizing and presenting data; interpreting data; [and] drawing inferences and predicting from data" (p. 7). The *Agenda* also argued that not only are such processes basic skills needed by all students but they also provide necessary "mathematical methods that [help] support the full range of problem solving" (pp. 2–3). The worksheets included here involve *bivariate* data (associated pairs of numbers) and provide experiences with all the activities listed above.

The ability to work with bivariate data is an important skill, seldom treated in conventional mathematics courses. The examples in the following worksheets center around the use of a *scatter plot* to examine relationships shown by bivariate data. As such, the activities show real-life applications of coordinate graphing and provide a bridge between algebra and statistics.

Grade levels: 7–12

Materials: Graph paper, rulers, and a set of worksheets for each student. Data can be collected from printed sources or by a survey. Reference materials (e.g., the *World Almanac*) should be available.

Objectives: (1) To investigate relationships shown by bivariate data (pairs of numbers representing values of two variables); (2) to graph bivariate data and interpret the graphs; (3) to use graphs of bivariate data to aid in prediction; (4) to collect and analyze bivariate data from various sources

Procedure: The worksheets are designed to be done in sequence over a period of several days and should be distributed one at a time. The students can work independently to complete the worksheets or with the teacher's assistance when requested. Students may wish to work in pairs when gathering data.

Each worksheet should be discussed thoroughly in class before students are assigned the next worksheet. Particular attention should be given to the data collected by students. One way of highlighting the students' efforts is by using their graphs for bulletin-board displays.

Students should also be encouraged to bring in clippings from newspapers and magazines that use graphical displays of bivariate data. Opportunities should be provided for discussion of the relationships depicted.

Answers:

1.

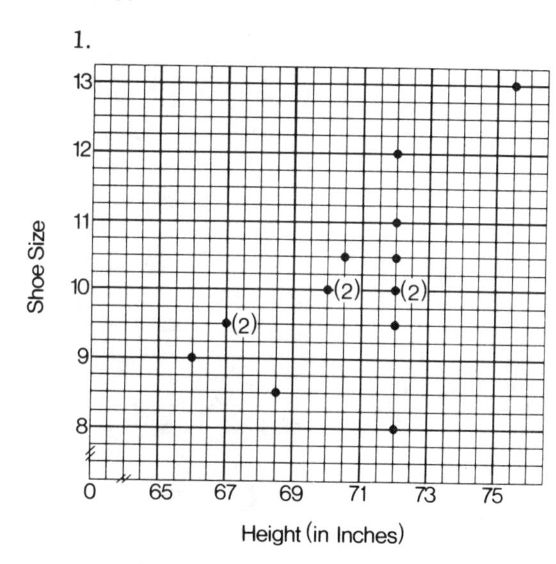

2. The shoe size also tends to increase.

3, 4. Answers will vary. (Possible variables include weight versus height for a class of students or number of points scored versus number of fouls by a school basketball team.)

5.

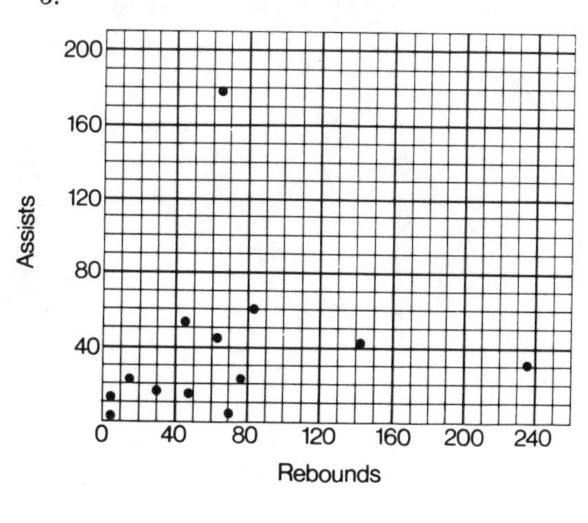

6. Players with more rebounds tend to have fewer assists. (However, if the data for Laimbeer and Thomas are removed, a slightly *positive* relationship, if anything, seems to develop.)

7, 8. Answers will vary. (Possible variables include gas mileage versus weight of an automobile.)

9.

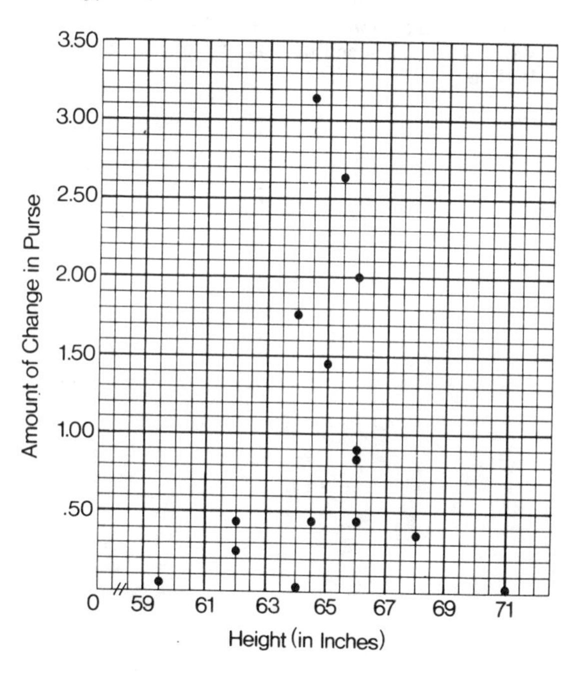

10. No particular relationship is evident. (Some students may say that people in the middle of the height range appear to have more money.)

11, 12. Answers will vary. (Possible variables include a student's height versus month of birth.)

13. 4

14. 1960–1965

15. 1967

16. 5760 ounces

17. Answers will vary. (Possible data might be school enrollments over the last twenty-five years or record-breaking times for running a mile run since 1900 graphed in increments of five years.)

BIBLIOGRAPHY

Haylock, Derek W. "A Simplified Approach to Correlation." *Mathematics Teacher* 76 (May 1983):332–36.

Holmes, Peter, ed. *Statistics in Your World.* Twenty-seven booklets and teacher's guides. Developed by the Schools Council Project on Statistical Education (Ages 11–16). Slough, Berks, England: W. Foulsham & Co., 1980.

Landwehr, James M., and Anne Watkins. "Exploring Data." Prepared for the ASA-NCTM Joint Committee on the Curriculum in Statistics and Probability, 1983.

National Council of Teachers of Mathematics. *An Agenda for Action: Recommendations for School Mathematics of the 1980s.* Reston, Va.: The Council, 1980.

Shulte, Albert P., and James R. Smart, eds. *Teaching Statistics and Probability,* 1981 Yearbook. Reston, Va.: National Council of Teachers of Mathematics, 1981. ■

PLOTTING AND PREDICTING FROM PAIRS

SHEET 1

1. Table 1 gives the height and shoe size for fifteen men. Plot the data on graph 1.

TABLE 1

Man	Height (Inches)	Shoe Size
1	70	10
2	75½	13
3	72	11
4	72	12
5	70	10
6	67	9½
7	72	9½
8	72	10
9	70½	10½
10	72	10
11	72	8
12	67	9½
13	66	9
14	68½	8½
15	72	10½

GRAPH 1

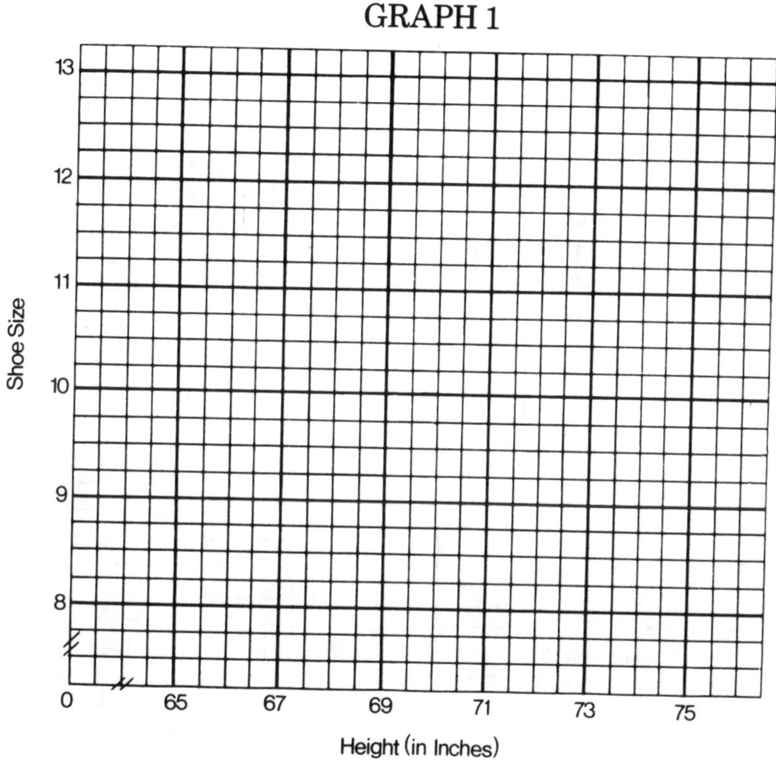

2. As the height increases, what tends to happen to the shoe size?

____

____

3. The two variables, height and shoe size, increase together. They have a *positive relationship*. What are some other pairs of variables that you think have a positive relationship?

____

____

4. Collect data for one set of pairs from question 3. Graph the data to see whether the relationship appears to be positive.

5. Table 2 gives the number of rebounds and assists made by the Detroit Pistons basketball team during their first nineteen games in 1983–84. Plot the data on graph 2.

TABLE 2

Player	Rebounds	Assists
Tripucka	81	60
Thomas	65	179
Laimbeer	237	30
Long	63	47
Tyler	78	21
Levingston	141	42
V. Johnson	45	54
Benson	70	7
Tolbert	49	15
Russell	18	21
Cureton	30	8
Thirdkill	4	11
*Austin	3	1
Totals	884	496

*No longer with team
(Data courtesy of Detroit Pistons)

GRAPH 2

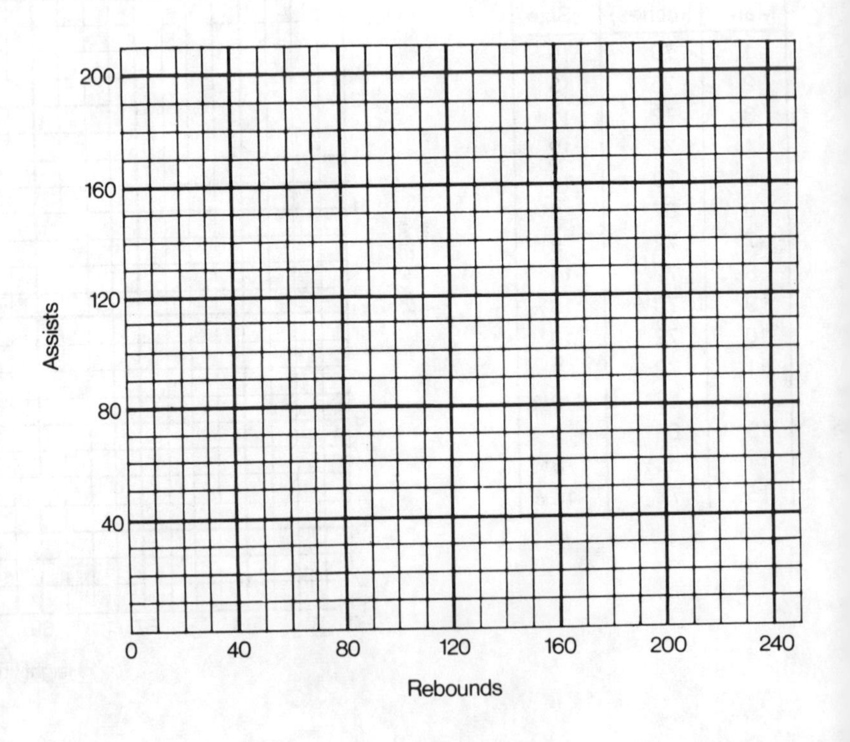

6. Do players with more rebounds tend to have more or fewer assists?

____

When one variable increases and the other decreases, they are said to have a *negative relationship*.

7. Name some pairs of variables that you think have negative relationships.

____

8. Collect data for one set of pairs from question 7. Graph the data. Does the relationship appear to be negative? ________

9. Table 3 gives the height of fifteen women and the amount of change they had in their purse on the morning of 6 December 1983. Plot the data on graph 3.

TABLE 3

Woman	Height (Inches)	Change in Purse
1	64	$0.01
2	66	0.90
3	62	0.27
4	64	1.77
5	65	1.47
6	64½	3.13
7	66	2.01
8	59½	0.05
9	71	0.01
10	68	0.38
11	66	0.46
12	65½	2.62
13	64½	0.47
14	62	0.42
15	66	0.85

GRAPH 3

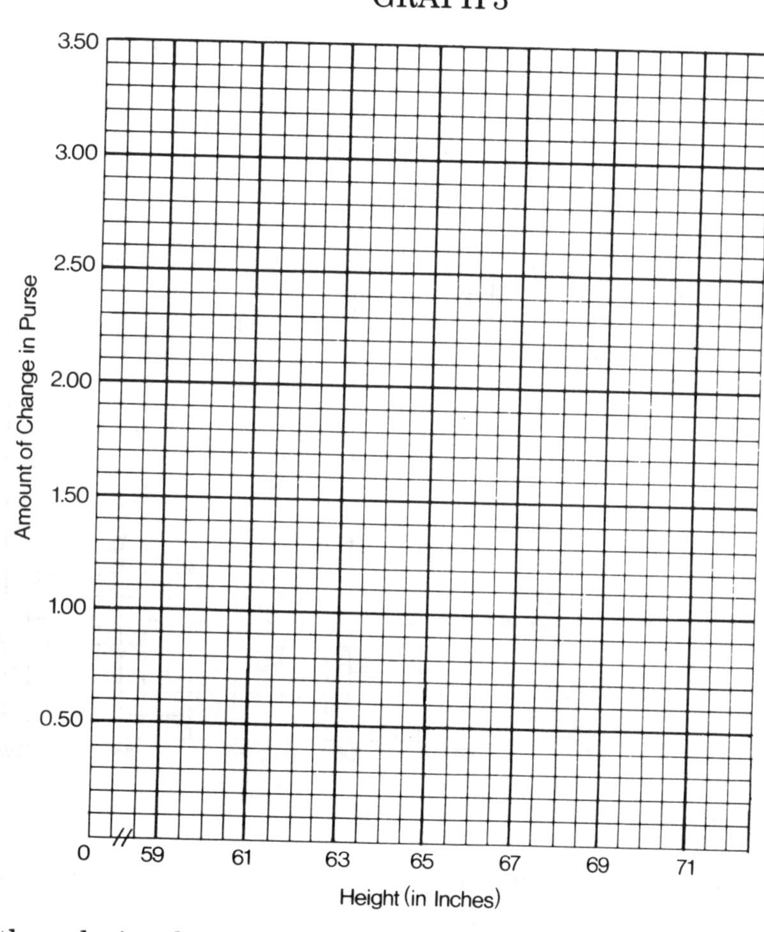

10. How would you describe the relationship between the women's height and the amount of change they had?

____

In this case, the data tend to pile up in the middle. Height and amount of change do not vary together. We can say that these variables seem to have *no relationship* or *are not related.*

11. Name some pairs of variables that you think are not related.

____

12. Collect data for one set of pairs from question 11. Graph the data. Do the pairs appear to have a relationship? ________

Some relationships change over time. Graph 4 shows the number of ounces of soft drinks consumed by the average person in the United States each year from 1945 to 1980.

GRAPH 4

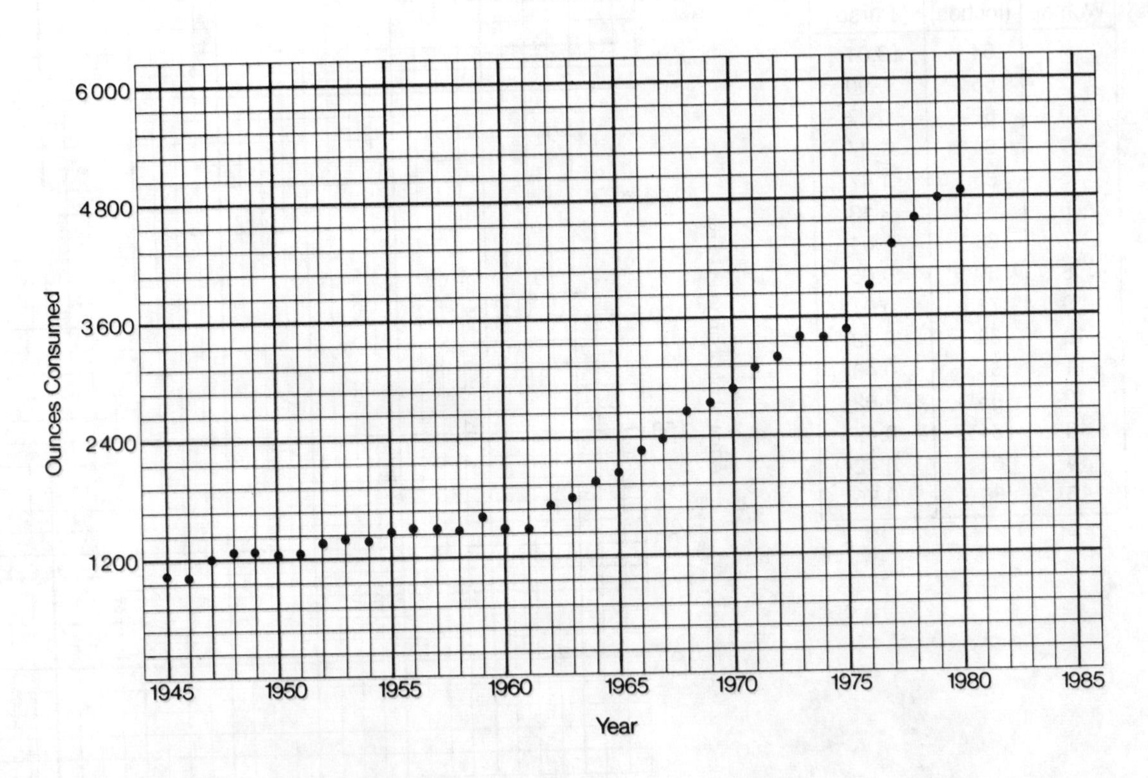

13. In 1980 people drank about ______ times as many soft drinks as they did in 1945.

14. In what five-year period did the amount consumed start to increase sharply? ______

15. In what year did people drink twice as many soft drinks as they did in 1945? ______

16. Connect the points for 1960 and 1980 with a line. Extend the line (and graph) to 1985. About how much should people be drinking then if the trend continues? ______

17. Find some time-related data. Graph the data and draw some conclusions from them. Try to answer such questions as the following:

   a) Is the relationship increasing, decreasing, or just fluctuating up and down?

   b) Can you give a possibe reason for the type of change observed?

   c) Do you expect the trend to continue in the near future?

DATA FITTING WITHOUT FORMULAS

April 1986

By ALBERT P. SHULTE, Oakland Schools, Pontiac, MI 48054
JIM SWIFT, Nanaimo District Secondary School, Nanaimo, BC V9R 5K3

Teacher's Guide

Introduction: An Agenda for Action stresses the importance of teaching students methods of collecting, organizing, and displaying data. It also urges the development of students' facility in interpreting data, drawing inferences from data, and making predictions based on the data (NCTM 1980, 7). The activity "Plotting and Predicting from Pairs" (Shulte and Swift 1984) provided a student-oriented introduction to scatterplots as a means of organizing, displaying, and interpreting bivariate data. The present activity builds on that one and develops the idea of making predictions on the basis of the information presented in a scatterplot. A simple method of fitting a straight line to a set of ordered pairs—the median-fit technique—is described. A more extensive development of this technique, which was created by James McBride of Princeton University, can be found in Landwehr and Watkins (1984).

Grade levels: 8–12

Materials: Calculators, graph paper, rulers, and a set of worksheets for each student

Objectives: (1) To investigate relationships shown by bivariate data; (2) to graph bivariate data and interpret the graphs; and (3) to fit lines to scatterplots to aid in making predictions from the data

Procedures: Reproduce a set of worksheets for each student. The five worksheets are designed to be completed in sequence, but they can be treated as related subactivities in several instructional periods as follows: worksheets 1 and 2, worksheet 3, worksheets 4 and 5. Students can complete the worksheets independently or with the assistance of the teacher as needed. Each subactivity should be discussed thoroughly in class before pupils are assigned the next worksheet(s).

Worksheets 1 and 2: The data given have been collected over several years by Stuart A. Choate, coordinator of mathematics for the Midland, Michigan, public schools. Students will need calculators to compute the totals and averages for the jar-and-candle experimental data given on worksheet 1. You may need to emphasize that the points to be graphed on worksheet 2 are of the following form: (jar capacity, average time to extinguish the candle). Note that although the individual times vary widely, the averages lie almost on a line. Fitting a line to the data plotted allows students to make predictions. To complete exercise 4, students will need to be familiar with the slope-intercept form, the point-slope form, or the two-point form of an equation for a line. Caution pupils that different scales are used on the two axes.

Worksheet 3: "Fitting" a line to a set of coordinate pairs can be done in many ways. The method introduced here is a simple one that works well if many pairs are involved. Depending on the level and background of your class, you may wish to review the method for finding the median of an ordered set of data. (If an odd number of values occurs, the median is the value appearing in the middle when the numbers are arranged in order. If an even number of values occurs, the median is the average of the two middlemost values when the numbers are arranged in order.) For some classes it may even be helpful to work through an example of finding a *median point* for a set of ordered pairs. To find the median point for the following five points, which are arranged in ascending order of the x-coordinates, (a) determine the median of the x-coordinates (5) and then (b) find the median of the y-coordinates (3):

x	2	3	5	6	8
y	1	3	5	2	6

The point (5, 3) is the *median point* of the set of five given points.

The median-fit technique requires little, if no, arithmetic and can be used to indicate if "fitting" a straight line to a set of data is appropriate. If the three median points do not lie close to a straight line, then it may not be appropriate to describe the relationship by a linear equation.

Most students discover that they can find the median point most easily with a table of coordinate pairs. But it is also easy to find the median point directly from the graph. Cover up all the points except the ones for which the particular median point is being found. Slide a transparent ruler up from the x-axis until the same number of points are above it as below it. Draw a horizontal line across the graph at this point. Repeat the procedure with the ruler parallel to the y-axis. The point where the two lines cross is the median point.

Worksheets 4 and 5: The purpose of these two worksheets is to have students consider the errors created when a scatterplot is used to predict values. After completion of these two sheets, students should be encouraged to include an estimate for the error in their prediction equations. For predicting heights, the equation can take the form $H = 2h \pm 3$. Pupils should be encouraged to compare the physical law derived from these data with the rules of thumb given at the top of worksheets 4 and 5.

BIBLIOGRAPHY

Haylock, Derek W. "A Simplified Approach to Correlation." *Mathematics Teacher* 76 (May 1983):332–36.

Holmes, Peter, ed. Statistics in Your World. Series of 27 booklets and teacher's guides. Developed by the Schools Council Project on Statistical Education (ages 11–16). Slough, Berks, England: W. Foulsham & Co., 1980.

Landwehr, James M., and Ann E. Watkins. *Exploring Data.* Washington, D.C.: American Statistical Association, 1984.

National Council of Teachers of Mathematics. *An Agenda for Action: Recommendations for School Mathematics of the 1980s.* Reston, Va.: The Council, 1980.

Shulte, Albert P., and James R. Smart, eds. *Teaching Statistics and Probability,* 1981 Yearbook of the National Council of Teachers of Mathematics. Reston, Va.: The Council, 1981.

Shulte, Albert P., and Jim Swift. "Activities: Plotting and Predicting from Pairs." *Mathematics Teacher* 77 (September 1984):442–47, 464.

Answers

Worksheet 1: 1. (a) gallon jar, 2574; half-gallon jar, 1924; quart jar, 801; pint jar, 448; half-pint jar, 225. (b) gallon jar, 73.5; half-gallon jar, 40.7; quart jar, 22.9; pint jar, 12.8; half-pint jar, 7.3.

Worksheet 2: 1.

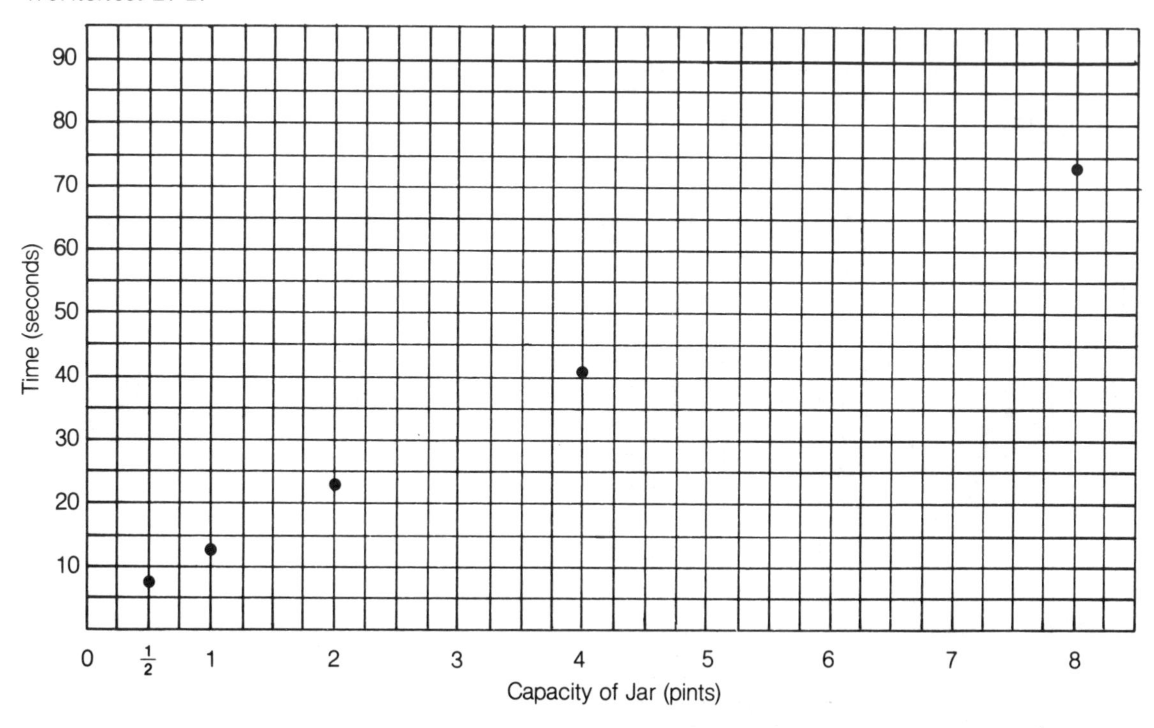

2., 3., 4. Answers will vary.

Worksheet 3: 1. (22.7, 24.2), (24.7, 26.2), (27.4, 28.9)
2., 3.

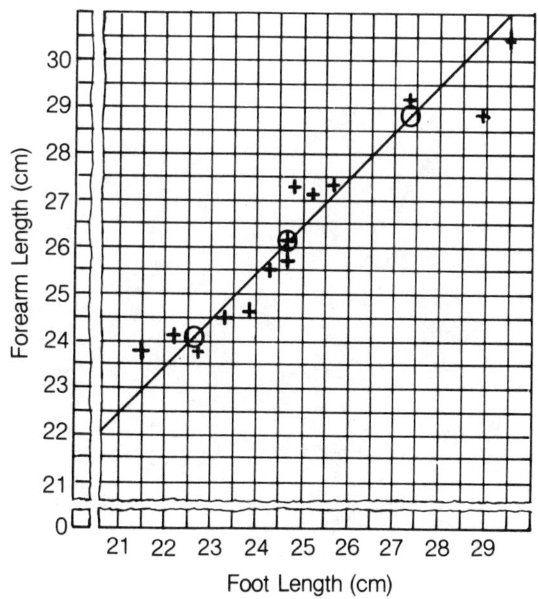

4. $a = f + 1.5$; 5. Answers will vary.

Worksheet 4: 1. (90.3, 179.6), (91.2, 182.0), (94.7, 187.4)
2.

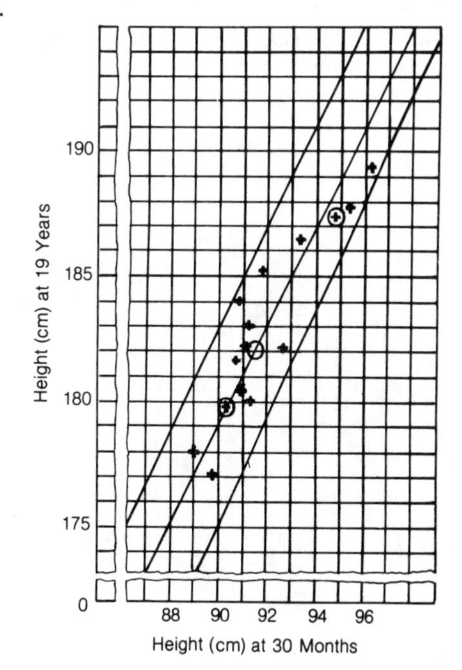

3. $H = 2h - 1$; 4. $H = 2h - 3.8$; 5. $H = 2h + 2.2$; 6. (*a*) 185 cm, (*b*) 182.2 cm, (*c*) 188.2 cm.

Worksheet 5: 1. (80.4, 160.4), (83.2, 166.5), (88.7, 176.9).

2.

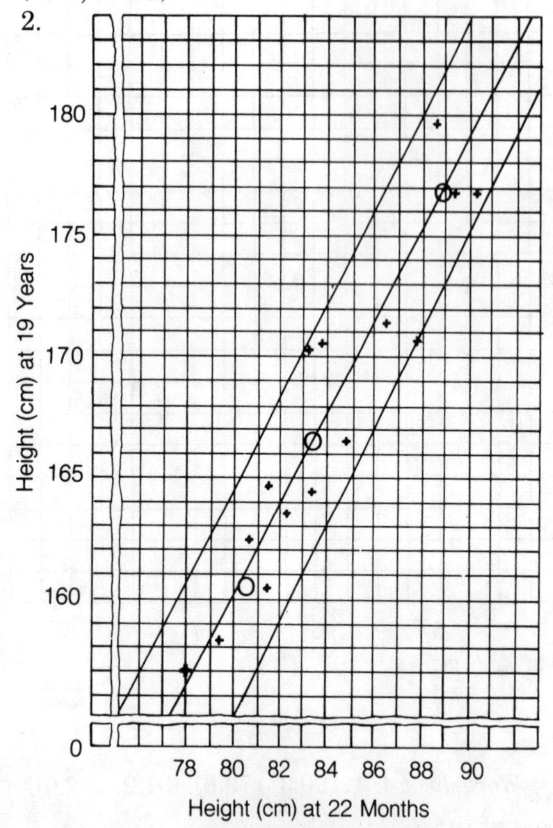

3. $H = 2h - 0.4$; 4. $H = 2h - 5.1$; 5. $H = 2h + 3.8$; 6. (*a*) 169.6 cm, (*b*) 164.9 cm, (*c*) 173.8 cm.

1. This table contains the time in seconds it took for a candle to go out after being covered with a jar of a certain size. The experiment was carried out thirty-five times with each size of jar.
   a) Find the total for each size of jar.
   b) Find the average for each size of jar.

Experiment	Time (Seconds)				
Number	Gallon Jar	½-Gallon Jar	Quart Jar	Pint Jar	½-Pint Jar
1	91	50	24	12	7
2	128	54	23	14	6
3	64	55	28	15	5
4	55	35	20	10	5
5	74	34	24	11	5
6	66	36	26	13	7
7	70	50	18	15	8
8	95	50	29	11	9
9	52	33	16	8	5
10	100	46	28	14	9
11	90	48	25	13	9
12	70	42	21	13	8
13	80	44	24	12	7
14	70	40	25	14	10
15	53	33	20	12	7
16	86	53	28	17	10
17	52	24	15	9	7
18	100	40	20	10	7
19	85	42	27	12	7
20	50	30	22	11	7
21	98	63	23	17	9
22	86	42	23	10	8
23	60	43	24	15	11
24	55	38	12	10	5
25	55	35	20	13	6
26	90	50	27	13	7
27	83	45	27	17	9
28	65	40	21	12	7
29	65	33	22	13	8
30	67	43	28	18	11
31	57	23	23	11	5
32	65	32	25	13	5
33	54	25	15	11	5
34	70	40	25	14	10
35	73	33	23	15	4
Total					
Average					

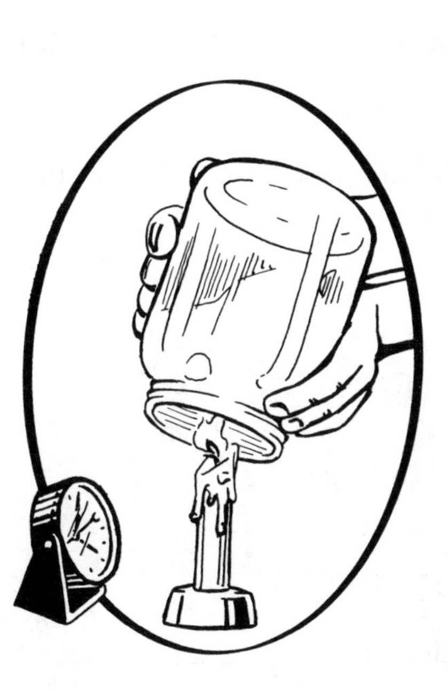

1. Plot the averages from the table on worksheet 1 on the graph below.

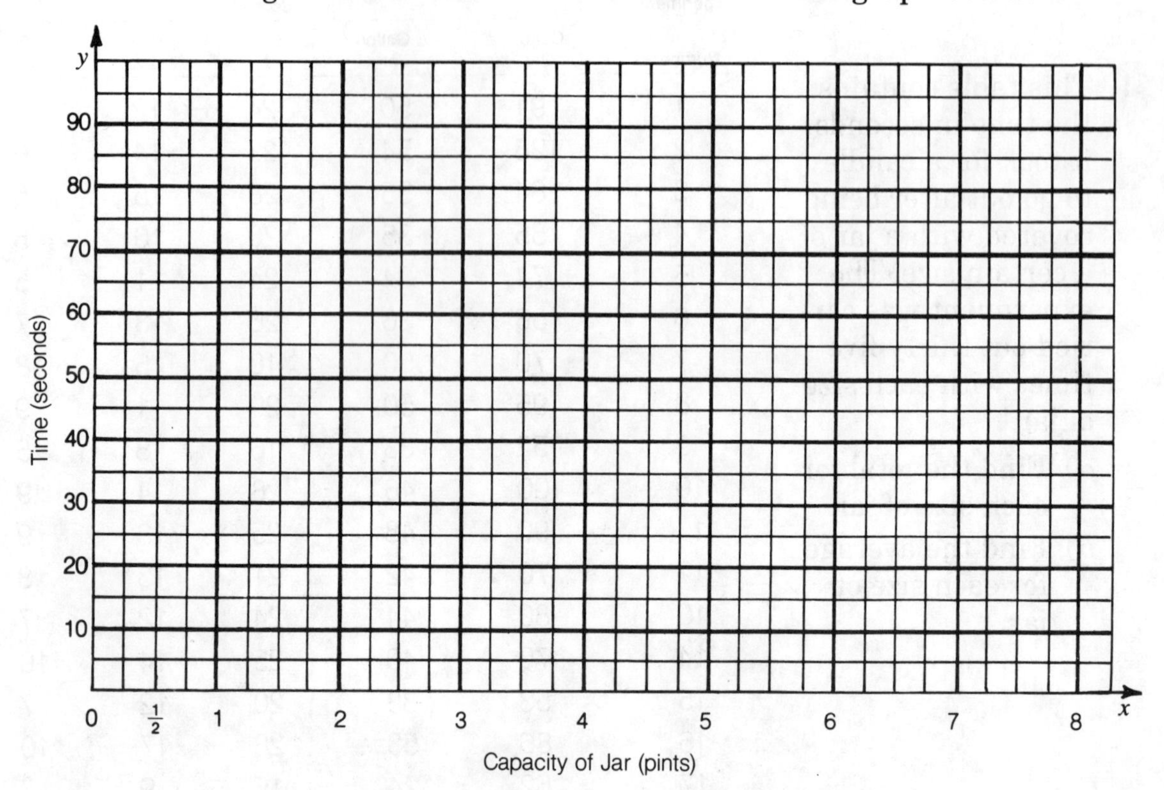

2. The points appear to lie approximately on a line. Use a ruler to draw a line that you think best fits the points.

3. Use the line you drew to predict the time it will take to extinguish a candle when covered by a jar of each of the following sizes:

   a) pint jar ________

   b) quart jar ________

   c) three-pint jar ________

   d) half-gallon jar ________

   e) three-quart jar ________

   f) gallon jar ________

4. Write an equation for the line you drew in exercise 2: ________________

Studying relationships in this manner is the way scientists discover physical laws. Your line and its equation represent your attempt to express as a physical law the relationship between the jar's size and time necessary to extinguish a candle.

Below are the lengths, in centimeters, of one foot and forearm of each of fifteen students. They have been arranged first in order of ascending foot length and then into three groups of five points.

Lower group:	foot length	21.5	22.3	22.7	23.4	23.9
	arm length	23.8	24.2	23.8	24.5	24.6
Middle group:	foot length	24.3	24.7	24.7	24.8	25.2
	arm length	25.5	25.8	26.2	27.3	27.2
Upper group:	foot length	25.6	26.5	27.4	28.9	29.5
	arm length	27.4	26.8	29.1	28.9	30.5

1. For each of the three groups, find the median of the foot lengths and the median of the arm lengths and enter them as coordinates in the spaces below. These three points are called the *median points* of each group.

   Lower group (____, ____) Middle group (____, ____) Upper group (____, ____)

2. Plot each of the individual points and the median points for each group on the following graph.

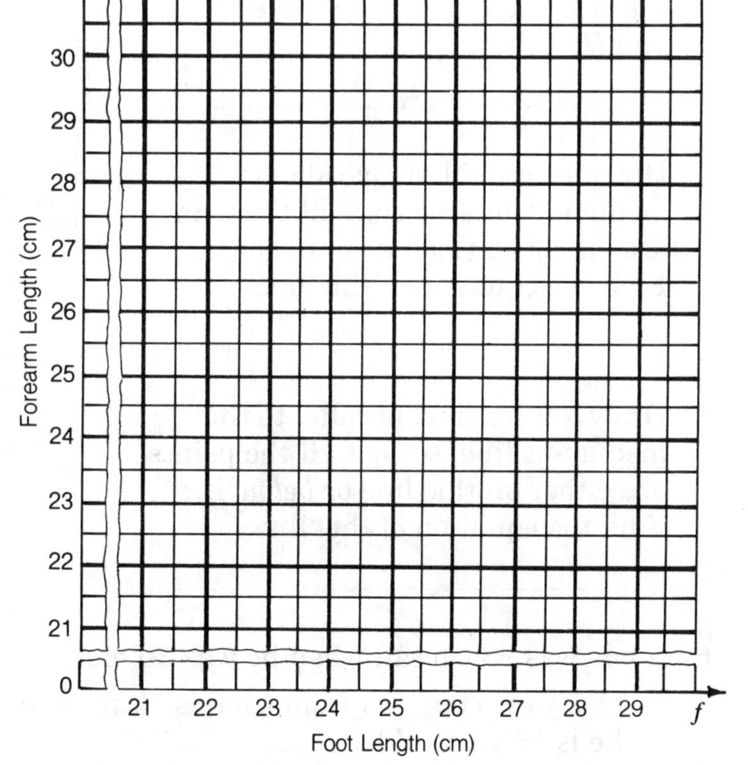

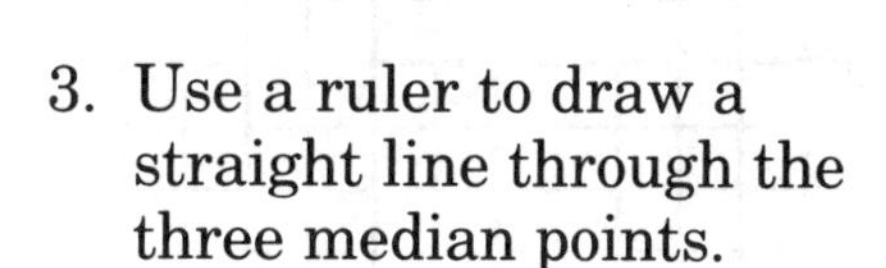

3. Use a ruler to draw a straight line through the three median points.

4. Use your line to find an equation that relates forearm length (a) to foot length (f)

   ____

5. Repeat exercises 1–4 for similar data collected from students in your class. If you cannot divide the class into three groups of equal size, make the upper and lower groups the same size.

The height of a boy at the age of 2 years and 6 months (30 months) is said to be one-half of his mature height. Below are the heights, in centimeters, of 15 boys at 30 months and at 19 years.

1. Find the median points for each of the three groups.

						Median point
Height (h) at 30 months	89.0	89.9	90.3	90.8	90.9	(____, ____)
Height (H) at 19 years	178.0	177.1	179.6	181.8	184.0	
Height (h) at 30 months	91.0	91.1	91.2	91.2	91.9	(____, ____)
Height (H) at 19 years	180.5	182.0	183.1	180.1	185.1	
Height (h) at 30 months	92.9	93.3	94.7	95.4	96.1	(____, ____)
Height (H) at 19 years	182.0	186.3	187.4	187.9	189.4	

2. Plot the points (including the median points) on the grid at the right and then draw a straight line through the median points.

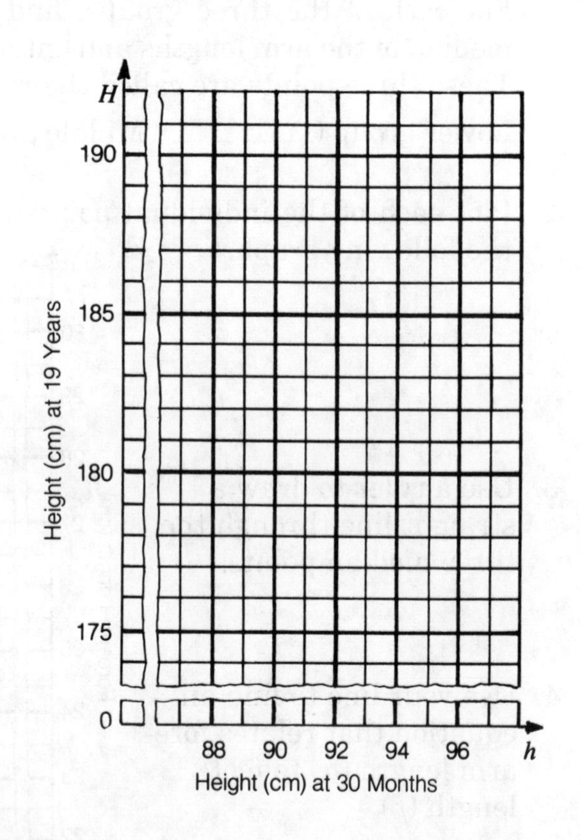

3. Use the equation of the line to express the relationship between h and H.

____

4. Draw a second line, parallel to this median-fit line, so that all the points are either on this line or *above* it. Find the equation of this line.

____

5. Draw a third line, parallel to the median-fit line, so that all the points are either on this line or *below* it. Find the equation of this line.

____

6. A boy was 93 cm tall when he was 30 months old.
   a) Use your equation from exercise 3 to predict the height of this boy when he is 19 years old. __________
   b) Use the equation from exercise 4 to find a *least* estimate of his height at 19 years. __________
   c) Use the equation from exercise 5 to find a *greatest* estimate of his height at 19 years. __________

PREDICTING FEMALE HEIGHTS SHEET 5

The height of a girl at the age of 22 months is said to be one-half of her mature height. Below are the heights, in centimeters, of 15 girls at 22 months and at 19 years.

1. Find the median points for each of the three groups.

						Median point
Height (h) at 22 months	78.0	79.4	80.4	81.3	81.3	(____, ____)
Height (H) at 19 years	157.0	158.4	161.4	164.7	160.4	
Height (h) at 22 months	82.1	83.2	83.2	83.9	84.9	(____, ____)
Height (H) at 19 years	163.7	164.4	170.2	170.5	166.5	
Height (h) at 22 months	86.2	87.9	88.7	89.4	90.1	(____, ____)
Height (H) at 19 years	171.3	170.7	179.7	176.9	176.9	

2. Plot the points (including the median points) on the grid at the right and then draw a straight line through the median points.

3. Use the equation of the line to express the relationship between h and H for this set of data.

____

4. Draw a second line, parallel to this median-fit line, so that all the points are either on this line or *above* it. Find the equation of this line.

____

5. Draw a third line, parallel to the median-fit line, so that all the points are either on this line or *below* it. Find the equation of this line.

____

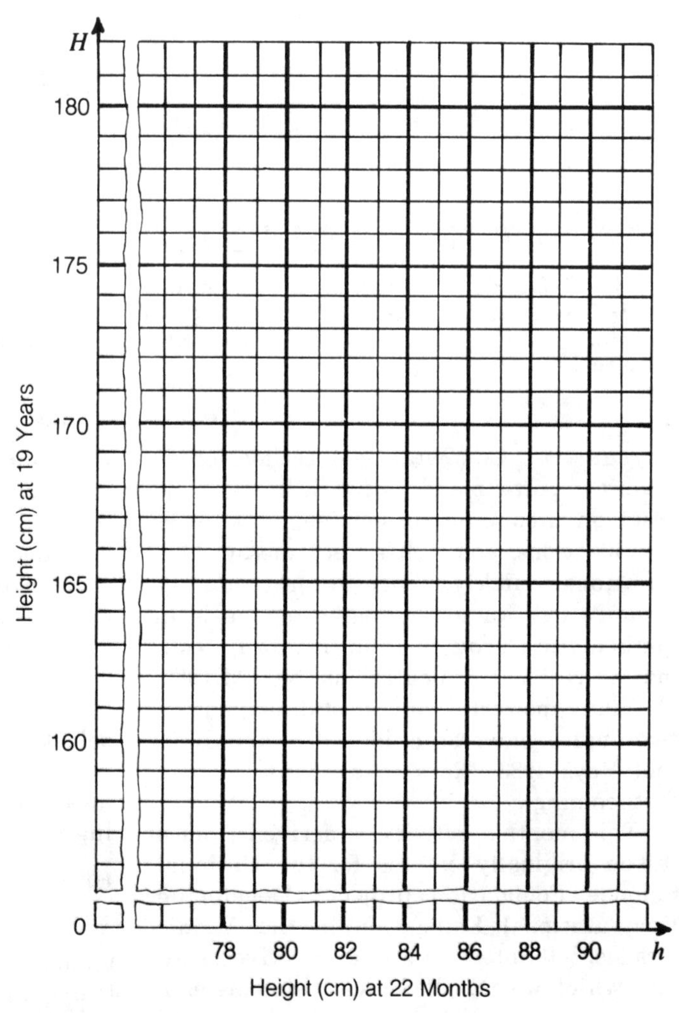

6. A girl was 85 cm tall when she was 22 months old.
   a) Use your equation from exercise 3 to predict the height of this girl when she is 19 years old. __________
   b) Use the equation from exercise 4 to find a *least* estimate of her height at 19 years. __________
   c) Use the equation from exercise 5 to find a *greatest* estimate of her height at 19 years. __________

ESTIMATING WITH "NICE" NUMBERS

November 1985

By ROBERT E. REYS, University of Missouri—Columbia, Columbia, MO 65211
BARBARA J. REYS, Oakland Junior High School, Columbia, MO 65202
PAUL R. TRAFTON and JUDY ZAWOJEWSKI, National College of Education, Evanston, IL 60201

Teacher's Guide

Introduction: Computational estimation is a very important and useful skill. Its value rests on producing reasonable answers to computations, sometimes very messy computations, quickly. Since *An Agenda for Action*'s call for an increased emphasis in mathematics programs on mentally estimating results of calculations (NCTM 1980, 3), many powerful and useful estimation techniques have been identified (cf. Reys and Reys 1983; Reys et al. 1984; Schoen forthcoming).

This month's activity is derived from a lesson originally created for the National Science Foundation project "Developing Computational Estimation in the Middle Grades." It introduces one effective strategy, which we call the "nice" numbers estimation strategy. "Nice" numbers, as used here, are those numbers that allow mental computation to be performed quickly and easily. The nice numbers considered here are powers of ten (e.g., 1, 10, 100).

This activity is designed to be initiated through a teacher-directed lesson developed through overhead transparencies. The transparencies highlight the concept of nice numbers and show how they can be used to obtain estimates quickly. Examples for students to try under your direction are included on each transparency. The student worksheets offer opportunities for additional practice, along with real-world applications of estimation; these sheets should be started in class and completed as homework. As time permits, a discussion of selected exercises on the following day will promote thinking about estimation and awareness of some different ways of obtaining reasonable estimates.

Grade levels: 7–12

Materials: Calculator (preferably one modified for overhead-projection use), transparencies of sheets 1–3, and a set of worksheets (sheets 4 and 5) for each student, to be used after the transparencies have been discussed

Objective: To develop estimating skills when multiplying or dividing by numbers near a power of ten

Procedure: Prior to using this activity, review with the class multiplication and division by powers of ten. The nice-numbers strategy encourages rounding to nice numbers (e.g., 1.03 ≐ 1; 9.93 ≐ 10) and then performing the operations indicated. After these operations are performed, the estimate is adjusted. The following example illustrates the entire process:

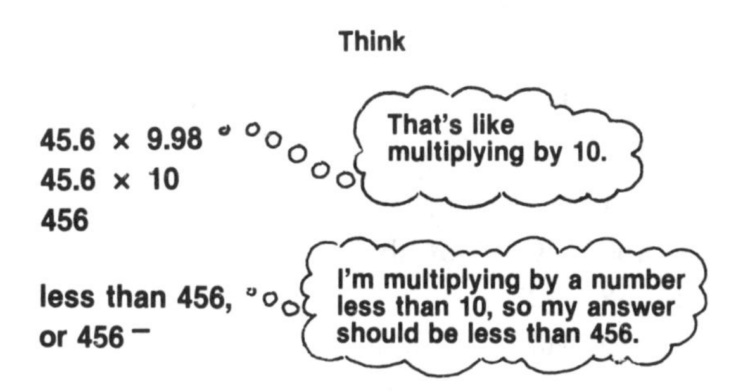

Think
45.6 × 9.98
45.6 × 10
456
That's like multiplying by 10.
less than 456, or 456 −
I'm multiplying by a number less than 10, so my answer should be less than 456.

The first transparency (sheet 1) provides some examples of nice numbers. Mention that many numbers occur near nice numbers, but the acceptable range for near-nice numbers is generally determined by the context of the problem in real-life situations. For this activity, numbers near 10 will generally be between 9 and 11, whereas numbers near 100 will be between 90 and 110. As you work through this transparency with the class, encourage students to add other nice numbers to the list. Use the bottom portion of the transparency to have students select the numbers that are near nice numbers and state which nice numbers they are near.

Use a transparency of sheet 2 to begin collecting and recording estimates for such exercises as 23 × 0.97. Ask for estimates of the product from several students. These estimates should be recorded. Encourage students to think of 0.97 as "a little less than 1." For example:

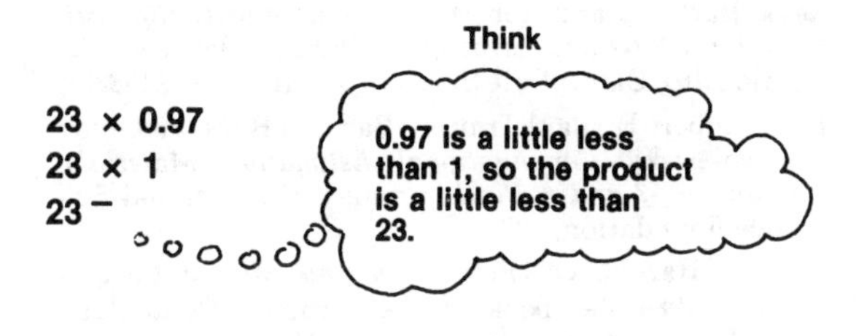

Think
23 × 0.97
23 × 1
23 −
0.97 is a little less than 1, so the product is a little less than 23.
Estimate: "A little less than 23, I'll say 22."

After some estimates have been recorded and discussed, the exact answer can be found using a calculator.

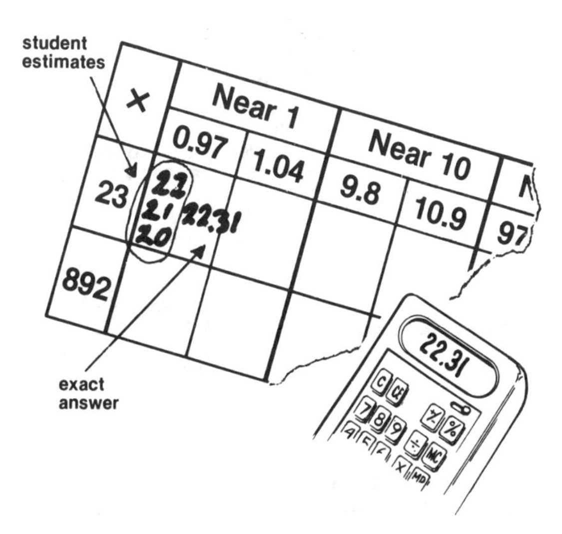

Continue in this manner, estimating 892 × 0.97. Repeat this process with several other numbers near nice numbers (e.g., 1.04, 9.8, 10.9, 97.2, 103.0). Pick and choose items from the chart as you wish. Most students will *not* need to complete the whole chart before going on.

Students should be encouraged to seek a pattern and verbalize it when observed. In general, the pattern might be described as follows:

> Given a number near n, where n is a power of ten, the product of this number and any other number is found by multiplying by n and then adding or subtracting an adjustment (compensation) determined by whether the given number was greater than or less than n.

Compensation—adding or subtracting the adjustment—is a complex process. Students need to realize that exactness is not the objective. The emphasis is on determining whether the exact answer will be more or less than the initial estimate.

Use the same process for division by near-nice numbers. Help students notice that the pattern is now reversed. When dividing by a number greater than n, the adjustment is subtracted. When dividing by a number less than n, the adjustment is added. This concept is more complex and usually requires plenty of discussion. For example:

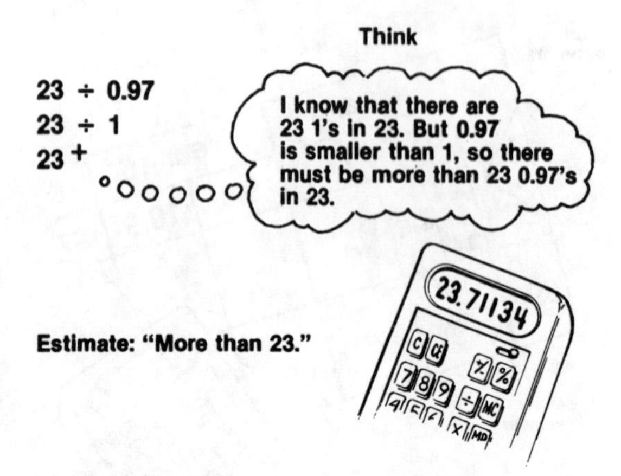

Think
23 ÷ 0.97
23 ÷ 1
23 +
I know that there are 23 1's in 23. But 0.97 is smaller than 1, so there must be more than 23 0.97's in 23.
23.71134
Estimate: "More than 23."

Students will need to see a variety of examples to help them clearly establish this idea.

The third transparency (sheet 3) highlights some work with nice numbers. Discuss the two examples given and then have the students try the six exercises. It is suggested that students share estimating strategies orally with the class. Specifically, they should be encouraged to verbalize how they move from their quick estimate to their adjusted estimate. Often the adjusted estimate can best be expressed by using the terms "more than" or "less than" or by using symbols "+" or "−."

Sheet 4 should be started immediately following the introductory lesson. It will give a quick check on how well the students have picked up the ideas presented. The remainder of this sheet and sheet 5 can be completed as a homework assignment.

Answers: The concept that a single situation involving estimation can have a number of different answers, all reasonable and falling within the acceptable range, will be foreign to many students. Some may demand to know "the answer" or the "best" answer. Encouraging several different estimates during class discussions can help students develop a greater tolerance for estimation.

Sheet 3

1. 485; adjusted estimate: 485^-, or about 480
2. 780; adjusted estimate: 780^+, or about 800
3. 1750; adjusted estimate: 1750^+, or about 1800
4. 5600; adjusted estimate: 5600^-, or about 5500
5. 0.862; adjusted estimate: 0.862^-, or about 0.840
6. 350; adjusted estimate: 350^-, or about 340

Sheet 4

1. 4600; 4600^-, or about 4500
2. 782; 782^+, or about 790
3. 840; 840^+, or about 850
4. 673; 673^+, or about 680
5. 5420; 5420^+, or about 5500
6. 29.8; 29.8^+, or about 30
7. 425 000; $425\,000^+$, or about 430 000
8. 86.7; 86.7^+, or about 87
9. 342; 342^+, or about 350
10. 168; 168^+, or about 170
11. 520; 520^-, or about 510
12. 8.46; 8.46^-, or about 8
13. 12.38; 12.38^+, or about 13
14. 436.7; 436.7^-, or about 436

Sheet 5

1. 559.55	6. 86.57	11. less
2. 8654.69	7. 53 990.74	12. $\$475^-$
3. 2.999	8. 6.999	13. 32.5^-mpg
4. 80.86	9. $1.40–$1.60	14. $\$110\,000^-$
5. 41.28	10. yes	15. more

REFERENCES

National Council of Teachers of Mathematics. *An Agenda for Action: Recommendations for School Mathematics of the 1980s.* Reston, Va.: The Council, 1980.

Reys, Barbara, and Robert Reys. *Guide to Using Estimation Skills and Strategies, GUESS, Box 1 and 2.* Palo Alto, Calif.: Dale Seymour Publications, 1983.

Reys, Robert E., Paul Trafton, Barbara Reys, and Judy Zawojewski. *Computational Estimation Materials, Grades 6, 7 and 8.* Washington, D.C.: National Science Foundation, 1984.

Schoen, Harold, ed. *Estimation and Mental Computation.* 1986 Yearbook of the National Council of Teachers of Mathematics. Reston, Va.: The Council, forthcoming.

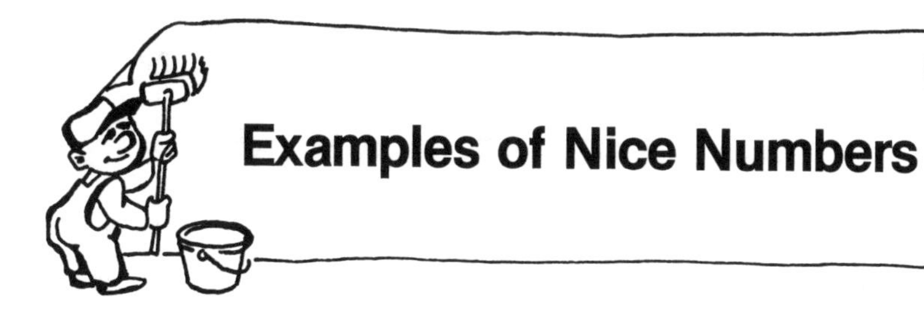

3 × <u>1</u>   40 ÷ <u>10</u>   58 × <u>100</u>

85 ÷ <u>1</u>   67 × <u>10</u>   2500 ÷ <u>100</u>

Why are the underlined numbers called “nice” numbers?

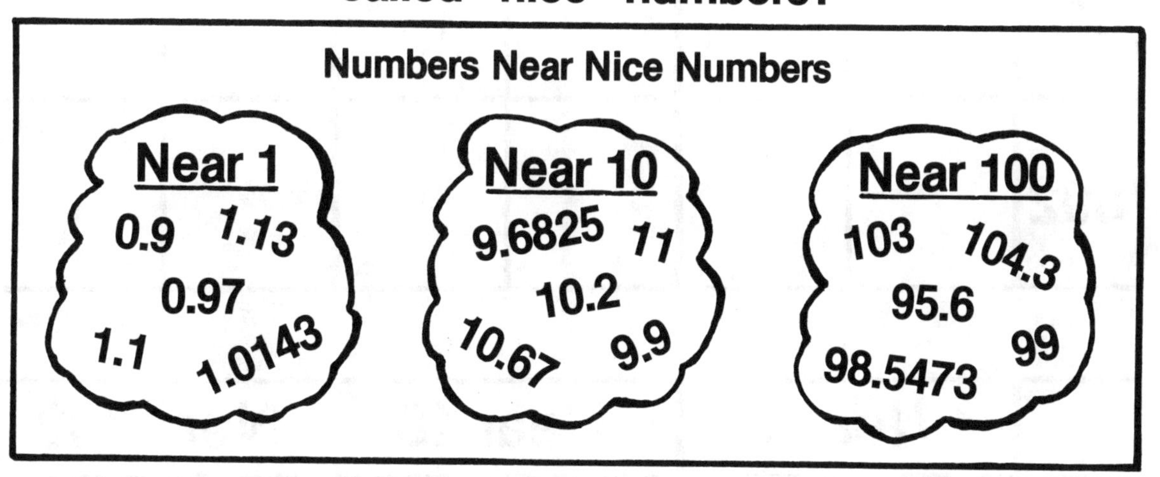

From the numbers below, select those that are near nice numbers and state which nice numbers they are near.

1.09   23.4   781   3.6   0.96

78   9.86   0.45   102.9375

97   1.127   17.2   10.4   135

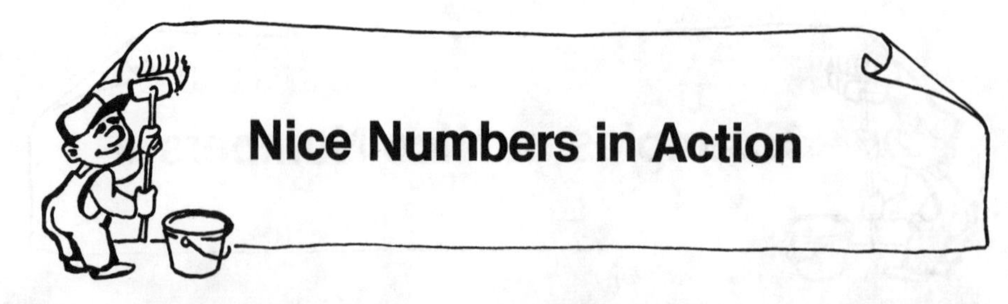

×	Near 1		Near 10		Near 100	
	0.97	1.04	9.8	10.9	97.2	103.0
23						
892						

÷	Near 1		Near 10		Near 100	
	0.97	1.04	9.8	10.9	97.2	103.0
23						
892						

To estimate 430 × 10.231

Think ... $430 \times 10.231 \doteq 430 \times 10$

4300

I'm multiplying by something bigger than 10, so it's more than 4300.

4300^+,
or about 4400

Adjusted estimate

To estimate 430 ÷ 10.231

Think ... $430 \div 10.231 \doteq 430 \div 10$

43

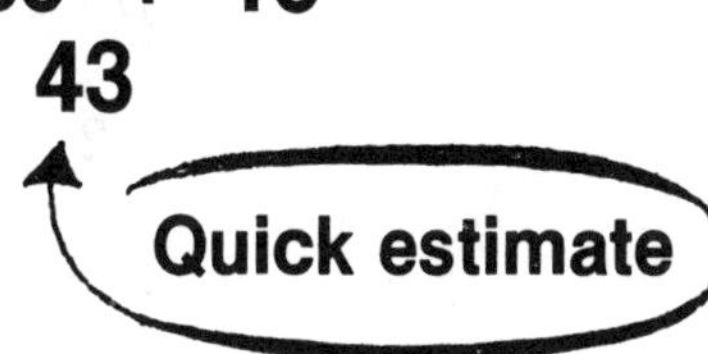

I'm dividing by a number bigger than 10, so it will be less than 43.

43^-,
or about 42

Adjusted estimate

TRY THESE

1. 485 × 0.985 ________
2. 7800 ÷ 9.61 ________
3. 175 × 10.53 ________
4. 56 × 97.8 ________
5. 0.862 ÷ 1.03 ________
6. 35 000 ÷ 104 ________

ESTIMATING WITH NICE NUMBERS SHEET 4

Example:

We need 37 Fun Day tickets. They cost \$9.75 each. Estimate the total cost of the tickets.

Make a quick estimate for each of the following and then adjust your estimate.

	Quick Estimate	Adjusted Estimate
1. 460 × 9.75	________	________
2. 782 × 1.05	________	________
3. 84 × 10.3	________	________
4. 673 × 1.08	________	________
5. 542 × 10.29	________	________
6. 29.8 × 1.1	________	________
7. 425 × 1015	________	________
8. 86.7 ÷ 0.98	________	________
9. 3420 ÷ 9.78	________	________
10. 16 800 ÷ 99	________	________
11. 52 000 ÷ 102.7	________	________
12. 8460 ÷ 1029	________	________
13. 12 380 ÷ 991	________	________
14. 4367 ÷ 10.04	________	________

ESTIMATING WITH NICE NUMBERS SHEET 5

Below are some calculator-obtained results, which in some cases have been rounded. In each case, one result is correct, the other is wrong. Use estimation to identify the correct answer and then circle it.

1. 589 × 0.95	559.55	629.945
2. 89 576 ÷ 10.35	8654.69	9271.16
3. 35.75 ÷ 11.92	426.14	2.999
4. 87 ÷ 1.076	80.86	93.61
5. 387 ÷ 9.375	36.28	41.28
6. 85 397 ÷ 986.43	86.57	83.64
7. 53.6 × 1007.29	53 990.74	52 864.50
8. 6.7 ÷ 0.9573	6.257	6.999

Use nice numbers to estimate reasonable answers for these problems.

9. Your grocery store sells hamburger for $1.39 a pound. If you buy a package of hamburger that weighs 1.08 pounds, estimate what you will pay. ____________
10. Suppose that 1-liter bottles of cola are on sale for 99¢ apiece and you are going to buy 24 bottles for the church picnic. Will $24.85 be enough to pay the bill? ____________
11. A racer averages 9.9 miles an hour when running long distances. Will a marathon race about 26 miles long take more or less than 3 hours? ____________
12. The owner of the Sound Shop has 475 tapes in inventory and makes about 95¢ on each tape sold. Estimate the profit if all the tapes are sold. ____________
13. The Brocks drove 325 miles on 10.2 gallons of gasoline. Estimate their mileage. ____________
14. If 9960 people bought tickets to see REO Speedwagon in concert, and the tickets sold for $11 each, estimate the total gate receipts for the concert. ____________
15. The typical American eats 96.3 pounds of beef each year. If the average price of beef is $2.42 a pound, would each spend more or less than $200 a year on beef? ____________

DEVELOPING ESTIMATION STRATEGIES

February 1985

By RHETA N. RUBENSTEIN, Renaissance High School, Detroit, MI 48235

Teacher's Guide

Introduction: Estimation is an important but frequently neglected objective of the mathematics curriculum (Carpenter et al. 1976). Recognizing the need for better number sense, together with mental arithmetic and estimation skills, in a society increasingly dominated by calculator and computer printouts, *An Agenda for Action* (NCTM 1980, 7–8) recommended curricular changes that included an increased emphasis on "mentally estimating results of calculations." The *Agenda* further suggested that "teachers should incorporate estimation activities into all areas of the [instructional] program on a regular and sustaining basis, in particular encouraging the use of estimating skills to pose and select alternatives and to assess what a reasonable answer may be." The following materials are intended to help students develop some computational estimation strategies.

Reys et al. (1980) characterized the processes that good estimators, from seventh graders to adults, used in computational estimation. Among these processes were "using compatible numbers," "averaging," and "rounding." These strategies are developed in this activity.

In "using compatible numbers," good estimators alter the numbers in a given situation to ones that are easier to compute mentally. This technique is frequently used in situations involving division and fractional parts, for example:

$$471 \div 6 \doteq 480 \div 6 = 80$$

$$\frac{2}{3} \text{ of } 250 \doteq \frac{2}{3} \text{ of } 240 = 2 \times 80 = 160$$

$$23\% \text{ of } 125 \doteq 25\% \text{ of } 120 = \frac{1}{4} \text{ of } 120 = 30$$

Worksheets 1 and 2 review easy arithmetic facts and ask students to identify compatible number estimates and to create their own compatible number expressions to estimate answers.

The "averaging" technique is most applicable to addition situations in which the addends are all "close" to one value, for example:

My driving record for the first four months of the year was as follows:

January	1978 miles
February	1043 miles
March	1095 miles
April	887 miles

About how many miles did I drive?

Recognizing that all four addends are about 1000 miles, then multiplying this "average" by 4 makes it easy to estimate the answer as 4000 miles. Worksheet 3 introduces averaging and provides some practice situations.

"Rounding" is a more commonly taught estimation strategy. Worksheet 3 provides practice in estimating grocery totals by rounding to the nearest half-dollar.

One of the key features of estimation is that it takes place over a short period of time. To provide ongoing opportunities to practice estimation in briefly timed periods, worksheet 4, a template for overhead transparencies, has been developed.

Grade levels: 6–9

Materials: Copies of worksheets 1–3 for each student and a transparency of worksheet 4

Objectives: To develop estimation strategies of using compatible numbers, averaging, and rounding and to provide opportunities for use of these strategies in real-life contexts

Procedures: Distribute the worksheets one at a time to each student. Work through the examples in each section with students before having them proceed individually. Success with sheet 1 depends on students' knowledge of basic facts and ability to work with rounded numbers. Success with sheet 2 requires a familiarity with fraction-percentage equivalences. Provide additional instruction on these prerequisite skills as necessary.

Make a transparency from sheet 4. Numbers have been omitted so the transparency can be reused with new numbers. For problems 1–4, present one item at a time on the overhead projector, giving students approximately fifteen seconds to read and respond. Solutions should then be discussed and strategies shared. The following are sample items, with sample estimates in parentheses:

1. 800 km, 90 km/h: ($810 \div 90 = 9$ h)
   500 km, 80 km/h: ($480 \div 80 = 6$ h)
2. \$4.95, 500 g: ($500 \div 500 = 1$¢/g)
   \$3.89, 195 g: ($400 \div 200 = 2$¢/g)
3. 1372 points, 7 minutes:
   ($1400 \div 7 = 200$ points/min.)
   1904 points, 5 minutes:
   ($2000 \div 5 = 400$ points/min.)
4. 881 students, 29%: ($3/10 \times 900 = 270$)
   591 students, 35%: ($1/3 \times 600 = 200$)

For problem 5, circle some set of items. Let students estimate the sum. Check totals on a calculator.

These materials have been developed with the expectation that they can be part of an ongoing program of integrating experiences in estimation with instruction. It is hoped that teachers and students will be continually "on the lookout" for opportunities to estimate and that the strategies developed here will be part of a growing repertoire of techniques.

Teaching estimation requires openmindedness about answers. Many good estimates are usually possible. Be ready to use judgment in evaluating students' responses. Students should be encouraged to share orally their reasons for the estimates they choose.

Answers: Sheet 1: 1. a. 50; b. 5; c. 500; d. 50; e. 90; f. 90; g. 9; h. 900; i. 700; j. 8000; k. 8; l. 5. 2. a. C, 9; b. A, 9; c. B, 5; d. C, 8; e. B, 80; f. C, 70. 3. Sample answers:

a. $720 \div 8 = 90$ b. $6300 \div 9 = 700$
c. $640 \div 80 = 8$ d. $4800 \div 6 = 800$
e. $420 \div 60 = 7$ f. $3600 \div 4 = 900$
g. $4500 \div 9 = 500$ h. $540 \div 9 = 60$

4. a. $360 \div 120 = 3$; b. $750\,000 \div 15 = 50\,000$ books; $2\,400\,000 \div 12 = 200\,000$ books; $700\,000 \div 7 = 100\,000$ books

Sheet 2:

1. See following chart.

2. a. 9; b. 18; c. 21; d. 63; e. 400; f. 2000; g. 210; h. 4000; i. 120; j. 630; k. 350; l. 800. 3. a. 30; b. 8; c. 2700; d. 600; e. 12; f. 48; g. 300; h. 60. 4. a. A, 200; b. B, 18; c. C, 16; d. A, 35; e. C, 150; f. B, 36. 5. a. $3/4 \times 80 = 60$; b. $4/5 \times 200 = 160$; c. $2/3 \times 90 = 60$; d. $9/10 \times 110 = 99$; e. $3/10 \times 50 = 15$; f. $2/5 \times 35 = 14$

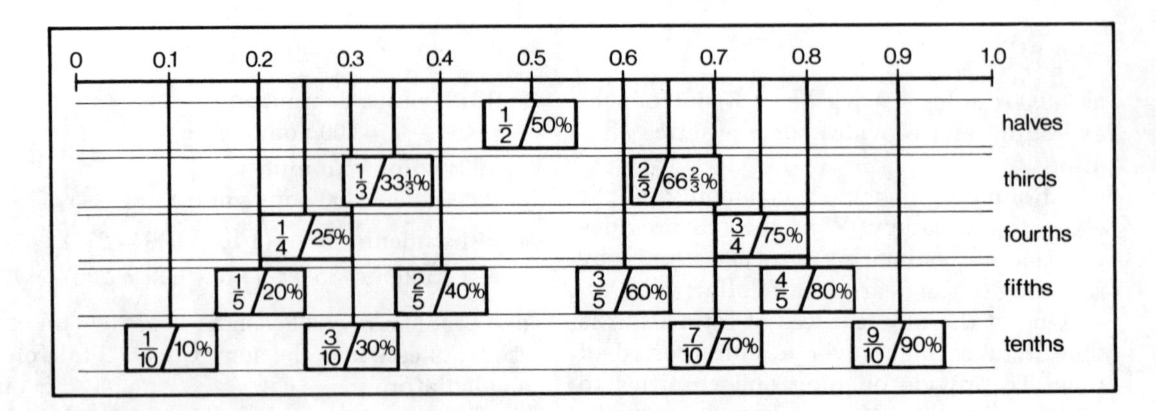

Sheet 3: 1. a. 80; b. $80 \times 10 = 800$. 2. a. $6000 \times 8 = \$48\,000$; b. $2000 \times 12 = \$24\,000$; c. $30 \times 12 = 360$ banners. 3. a. $14.50; b. $21.50–$23

REFERENCES

Carpenter, Thomas P., Terrence G. Coburn, Robert E. Reys, and James W. Wilson. "Notes from National Assessment: Estimation." *Arithmetic Teacher* 23 (April 1976):296–302.

National Council of Teachers of Mathematics. *An Agenda for Action: Recommendations for School Mathematics of the 1980s.* Reston, Va.: The Council, 1980.

Reys, Robert E., Barbara J. Bestgen, James F. Rybolt, and J. Wendell Wyatt. *Identification and Characterization of Computational Estimation Processes Used by In-School Pupils and Out-of-School Adults.* Washington, D.C.: National Institute of Education, 1980.

1. Use number facts to calculate the following:

a. $450 \div 9$ ____	e. $630 \div 7$ ____	i. $42\,000 \div 60$ ____
b. $450 \div 90$ ____	f. $6300 \div 70$ ____	j. $32\,000 \div 4$ ____
c. $4500 \div 9$ ____	g. $6300 \div 700$ ____	k. $560 \div 70$ ____
d. $4500 \div 90$ ____	h. $7200 \div 8$ ____	l. $400 \div 80$ ____

2. To estimate the quotient $5321 \div 62$, change the numbers to "compatible numbers" with which you can compute mentally.

Example: $5321 \div 62 \doteq 5400 \div 60 = 90$

Estimate each answer by choosing the expression (A, B, or C) with compatible numbers and then calculating mentally.

	A	B	C	Estimate
a. $73 \div 8$	$74 \div 8$	$73 \div 7$	$72 \div 8$	______
b. $63 \div 6.8$	$63 \div 7$	$62 \div 6$	$64 \div 7$	______
c. $46 \div 8.7$	$46 \div 8$	$45 \div 9$	$45 \div 8$	______
d. $57 \div 6.9$	$57 \div 7$	$58 \div 7$	$56 \div 7$	______
e. $6504 \div 78$	$6500 \div 80$	$6400 \div 80$	$7000 \div 80$	______
f. $291 \div 3.8$	$270 \div 3$	$300 \div 4$	$280 \div 4$	______

3. Estimate each answer by writing a compatible number expression and calculating.

Example: $2040 \div 72 \doteq 2100 \div 70 = 30$

a. $738 \div 7.9$ ______	e. $413 \div 58$ ______
b. $6205 \div 8.7$ ______	f. $3521 \div 3.8$ ______
c. $643 \div 81$ ______	g. $4483 \div 8.7$ ______
d. $4729 \div 5.9$ ______	h. $553 \div 9.1$ ______

4. Use compatible numbers to estimate each answer:

a. The Student Council sold 121 items at their rummage sale. The students collected $374. What was the approximate average cost of each item?

______

b. In 1980 the following cities had the number of libraries and books shown. Estimate the average number of books in each library for each city.

City	Libraries	Books	Estimate
Charlotte, North Carolina	15	736 343	______
Milwaukee, Wisconsin	12	2 335 485	______
Ottawa, Ontario	7	666 404	______

1. Complete the table with equivalent fractions and percentages. Note the relative location of each fraction and percentage on the number line.

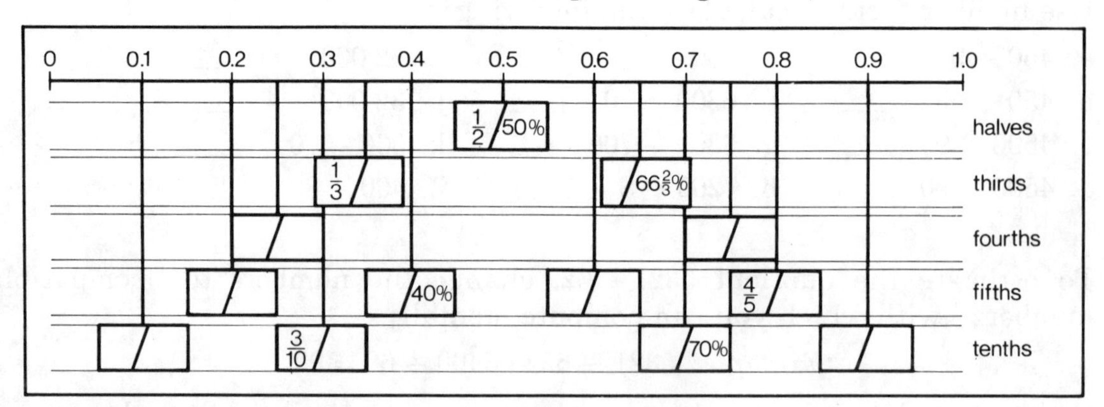

2. Use shortcuts to calculate the following problems.

Example: 4/5 of 450 = 4/~~5~~ × ~~450~~ 90 = 4 × 90 = 360

a. 1/3 of 27 ____	e. 1/6 of 2400 ____	i. 2/3 of 180 ____
b. 2/3 of 27 ____	f. 5/6 of 2400 ____	j. 9/10 of 700 ____
c. 1/4 of 84 ____	g. 3/4 of 280 ____	k. 7/8 of 400 ____
d. 3/4 of 84 ____	h. 5/6 of 4800 ____	l. 4/5 of 1000 ____

3. Rewrite the percentages as fractions, then use shortcuts to calculate.

Example: 80% of 350 = 4/5 of 350 = 4 × 70 = 280

a. 20% of 150 ________	e. 40% of 30 ________
b. 25% of 32 ________	f. 60% of 80 ________
c. 90% of 3000 ________	g. 75% of 400 ________
d. 66 2/3% of 900 ________	h. 33 1/3% of 180 ________

4. Estimate each answer by choosing the expression (A, B, or C) with compatible numbers and then calculating.

Example: 73% of 78 ≐ 75% of 80 = 3/4 × 80 = 60

	A	B	C	Estimate
a. 26% of 800	1/4 of 800	1/5 of 800	2/6 of 800	________
b. 62% of 30	2/3 of 30	3/5 of 30	7/10 of 30	________
c. 67% of 24	4/5 of 24	1/3 of 24	2/3 of 24	________
d. 5/8 of 57	5/8 of 56	5/8 of 50	5/8 of 64	________
e. 3/5 of 257	3/5 of 200	3/5 of 300	3/5 of 250	________
f. 3/4 of 47	3/4 of 50	3/4 of 48	3/4 of 40	________

5. Estimate by using compatible numbers.

Example: 77% of 41 ≐ 75% of 40 = 3/4 × 40 = 30

a. 3/4 of 79 ____	c. 2/3 of 88 ____	e. 29% of 52 ____
b. 4/5 of 203 ____	d. 88% of 110 ____	f. 42% of 36 ____

1. The eighth graders at Middleton Middle School sold booster buttons to raise funds. Here are the numbers of buttons sold by ten homerooms:

76 75 78 83 81 75 88 83 76 77

a. All the homerooms sold close to the same number of buttons. What is this number? ________

b. This number is an estimate of the average number of buttons sold by each homeroom. To estimate the total number of buttons sold, multiply the estimated average by the number of homerooms:

________ × ___10___ = ________

2. Use this averaging technique to estimate the totals in the following problems:

a. Profits for First Eight Months of 1984

$6459	6295
5784	6087
5924	5724
5924	6087

b. Monthly Family Earnings

$1872	1735	1709
2580	2261	1905
1932	2183	2218
2314	2017	2371

c. Numbers of Banners Sold

32	29	29
28	34	36
32	33	27
25	24	31

3. Grocery bill totals can be estimated by counting, to the nearest half, the approximate number of dollars spent and mentally keeping a running total.

Example	Rounded Amount	Running Total
$1.02	1	$1.
.80	1	2.
.34	.5	2.50
.53	.5	3.
2.14	2	5.
.49	.5	5.50
1.85	2	7.50 = Estimated total

Use the "count dollars" technique to estimate these totals:

a.

$0.47	.65	1.35
.45	1.59	.55
.34	2.92	1.39
.49	.95	2.59
.34		

Estimate = ________

b.

$5.89	.68	.75
1.30	1.09	1.88
1.19	.69	2.06
.72	1.00	2.20
2.52	.64	.24

Estimate = ________

1. Ms. Robertson is driving from Stayhere to Gothere, a distance of _____ km. She estimates she will average _____ km/h. About how long will the trip take?

2. Cleanzall laundry soap costs _____ for _____ grams. What is the approximate cost per gram?

3. Carolyn scored _____ points at the video game, Dream Machine. She played about _____ minutes. About how many points per minute did she average?

4. Kennedy Middle School has _____ students. The seventh grade makes up _____% of the school. About how many students are seventh graders?

5. Estimate the total grocery bill for the items circled:

$0.23	1.39	.49	1.49	.99	.78
.85	1.30	.72	1.09	1.88	2.46
.75	.95	.92	.55	.59	.35

SPATIAL VISUALIZATION

November 1984

By GLENDA LAPPAN, ELIZABETH A. PHILLIPS, and MARY JEAN WINTER, Michigan State University, East Lansing, MI 48824

Teacher's Guide

Introduction: Among the curricular recommendations in the NCTM's (1980, 3, 7) *Agenda for Action* are the following:

- Mathematics programs of the 1980s must be designed to equip students with the mathematical methods that support the full range of problem solving, including ... the use of imagery, visualization, and spatial concepts.
- There should be increased emphasis on such activities as ... using concrete representations and puzzles that aid in improving the perception of spatial relationships.

It is these two recommendations that are the focus of the four worksheets included in this activity.

Grade levels: 7–10

Materials: Ten to twelve small cubes for each student or small group of students, tape, one set of worksheets for each student, and a set of transparencies for discussion of solutions

Objectives: Students will improve their ability to visualize by building, drawing, and evaluating three-dimensional figures. These three activities are used in various combinations throughout the worksheets. A student will see how a solid and a drawing of it are related to each other and will be able to build a solid and draw a two-dimensional representation of it.

Background: Most of a student's mathematical experience with the three-dimensional world is obtained from two-dimensional pictures. Yet many students cannot "read" these two-dimensional pictures well enough to determine needed information about

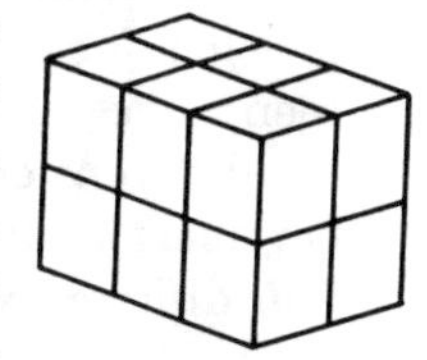

the solid objects. For example, how many cubes are needed to build the solid shown here? Students usually make two types of errors. They count the faces of every cube showing and answer "sixteen," or they count only the cubes showing and answer "ten." To answer correctly, the students must be able to visualize the hidden corner of the solid. It is important for students to explore this situation from both directions—not only reading information in two-dimensional pictures of the real world but also representing information about the real world with two-dimensional pictures. To accomplish this goal it is necessary to move back and forth from concrete experiences (solids) to abstractions (drawings).

Because of their flexibility and overall usefulness, cubes are used as the basic building units for the three-dimensional objects. As the students build solids from the cubes and look at them from different perspectives, they are developing more discriminating spatial skills. Isometric dot paper (paper with dots arranged in diagonals rather than in rows) is used to make the drawing of these views easier for the students. Using the dot paper requires that a solid be turned so that the student is looking at a corner. In the following activities, the students learn to relate a solid to a drawing of it, to make an isometric drawing of a solid, and to build a solid from an isometric drawing.

Comments: Distribute the worksheets one at a time to each student. Discuss solutions for each sheet before going on to the next one.

Sheet 1: This sheet is designed to acquaint students with representations made on isometric dot paper. The students build simple solids from cubes, copy isometric drawings of the solids, and answer questions about the numbers of cubes needed to build the solids. A discerning student will notice that cubes can sometimes be hidden by other cubes and thus not show in the isometric drawing. For example, problem 3 can be built with twelve or thirteen cubes. Notice that problems 4 and 5 are different corner views of the same solid. On sheet 3 students will be asked to draw four different views of a given solid.

Sheet 2: In this activity students look at a drawing, build a solid from the drawing, add or remove cubes, and then draw the modified solid. The emphasis is on going back and forth visually from a solid to a picture. Some students notice that solid 4 can be built in two ways. If the situation arises, tell the student to build the solid using the fewer cubes or challenge the student to draw both possibilities.

Sheet 3: In this worksheet students look at a solid from a corner to see the solid as it can be drawn on isometric dot paper. As an introduction, students look at the corners of a solid and match these to the appropriate drawings. This matching requires a great deal of eye movement back and forth from solids to drawings. Suggest to students who are having difficulty that they view the solid with one eye shut. Once these views have been matched, the students are asked to draw isometric views of a simple solid.

Sheet 4: In this activity two simple solids, called puzzle pieces, are put together to match a drawing of a solid. Students are then asked to shade the drawing to show each puzzle piece. Masking or transparent tape can be used to hold cubes together temporarily to form the puzzle pieces. To be successful, a student must put the pieces together not only in the right combination but also with the correct orientation. Finally, students are asked to create a new solid from the pieces and to draw a two-dimensional representation of this new solid. This last, open-ended problem allows the student to create, to represent, and then to evaluate the picture against the solid object.

Answers: Answers can be found on the next two pages.

Answers

Sheet 1

1. 5; 2. 8; 3. 12 or 13; 4. 10; 5. 10

Sheet 2

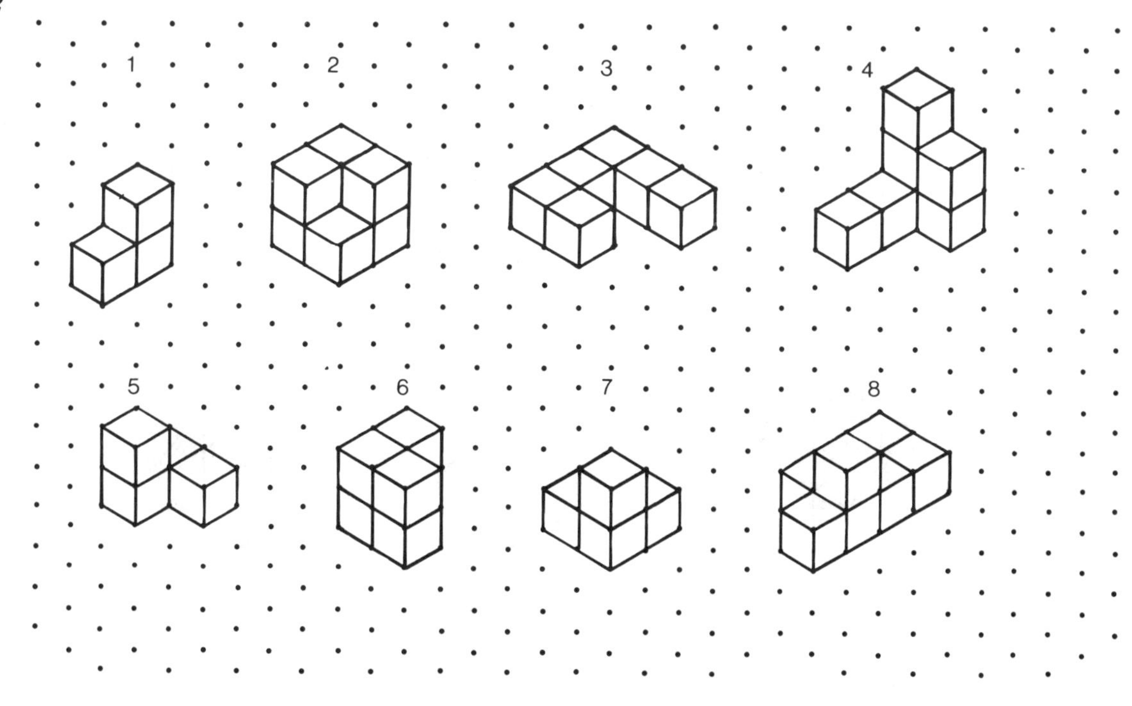

Sheet 3

Corner *C*; Corner *B*; Corner *A*; Corner *D*

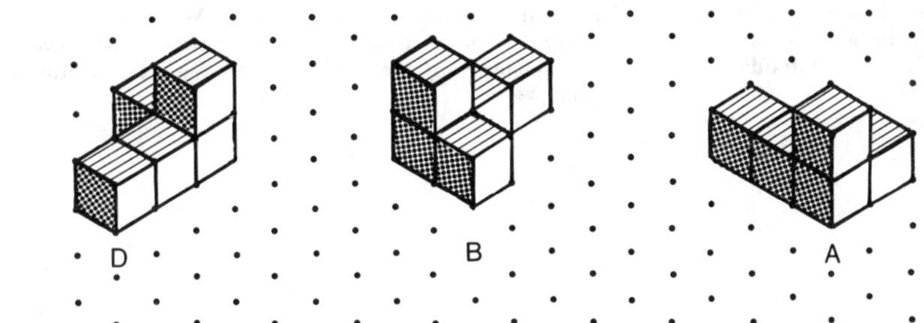

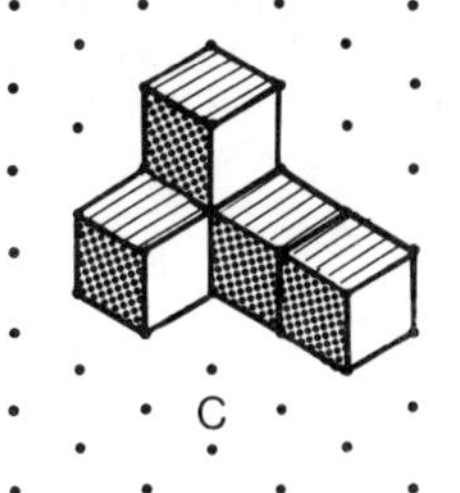

Sheet 4

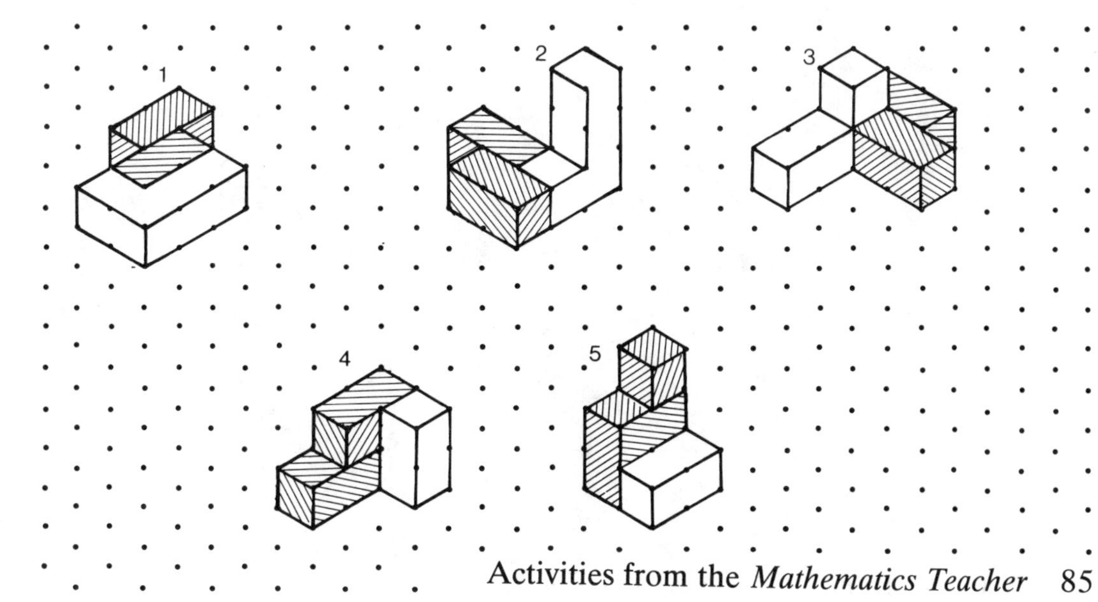

B.

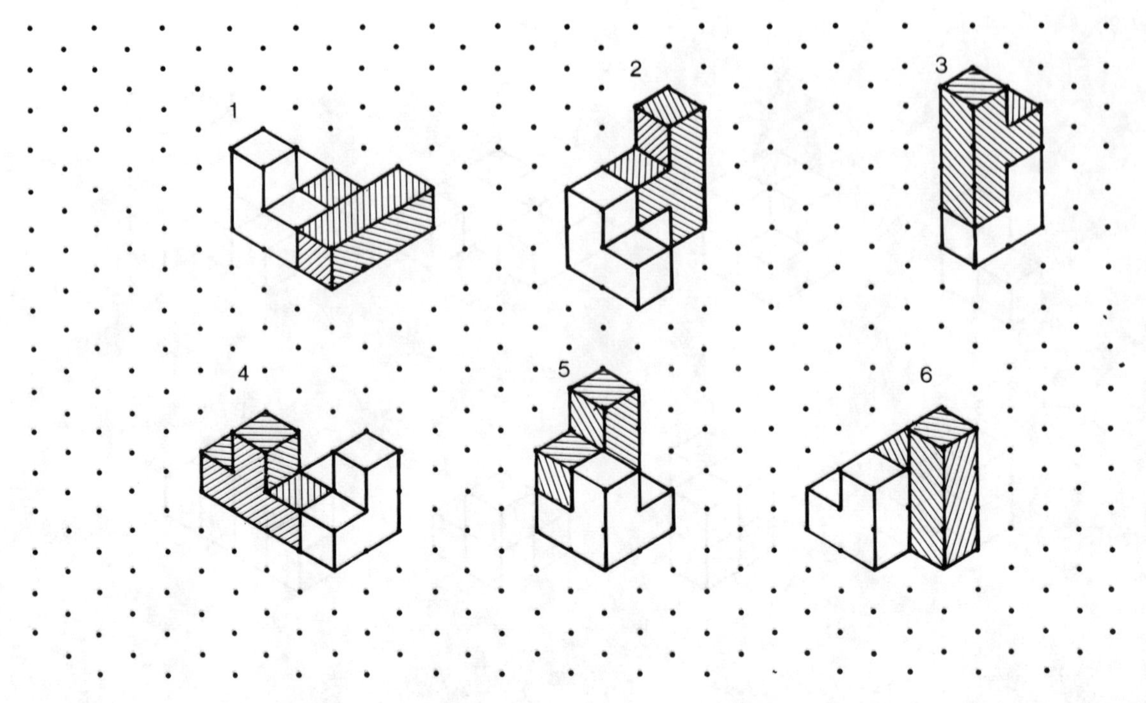

These activities are based on a unit entitled *Spatial Visualization*, developed as part of the Middle Grades Mathematics Project, a curriculum development project supported by funds from the National Science Foundation. For further information contact Glenda Lappan.

REFERENCE

National Council of Teachers of Mathematics. *An Agenda for Action: Recommendations for School Mathematics of the 1980s*. Reston, Va.: The Council, 1980.

DRAWING SOLIDS

SHEET 1

For each solid shown, do the following:

- Build the solid from cubes.
- Copy the drawing.
- Count the number of cubes used in the drawing.
- Check your count from the solid.

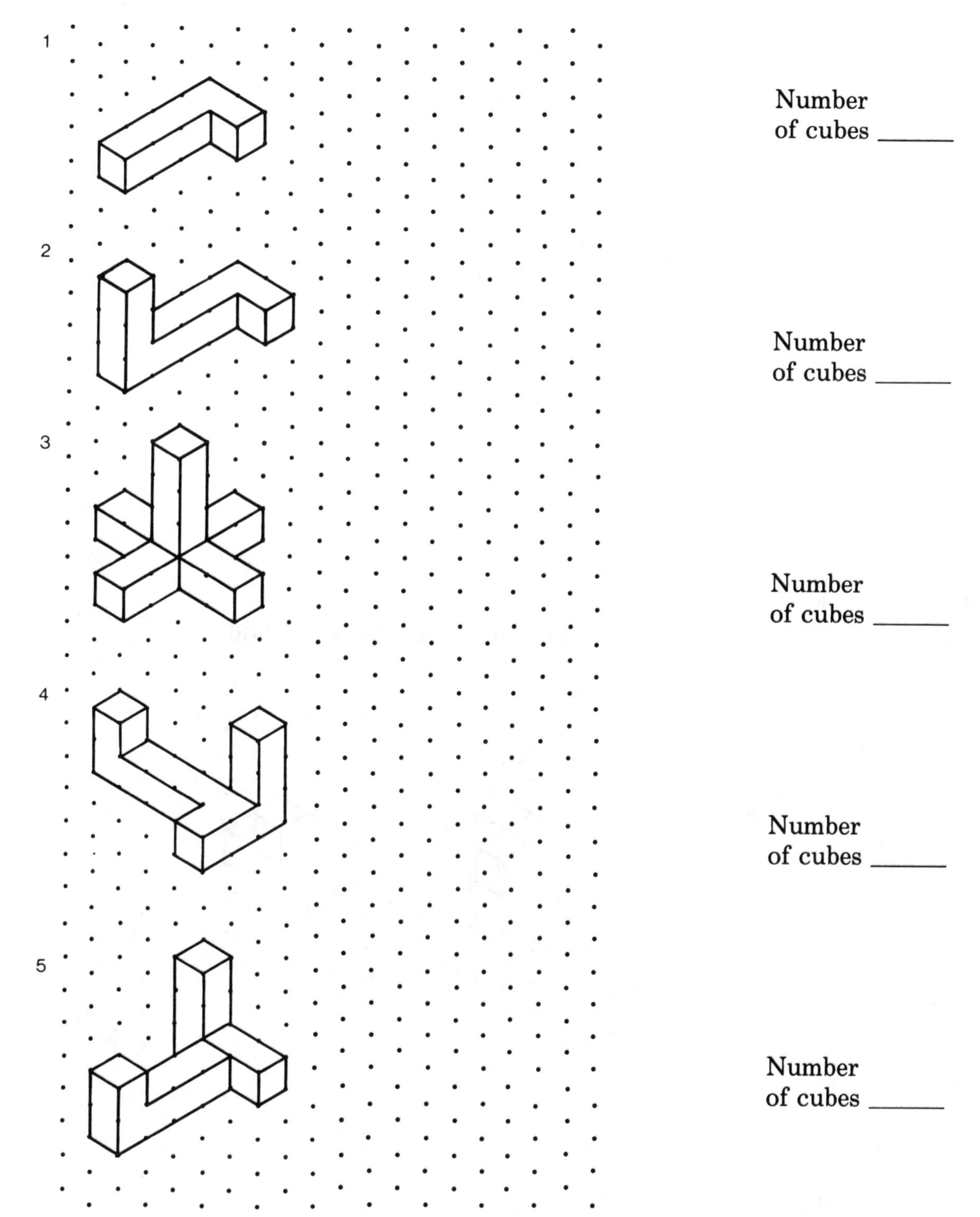

1. Number of cubes ______

2. Number of cubes ______

3. Number of cubes ______

4. Number of cubes ______

5. Number of cubes ______

For each of the four solids 1–4, do the following:

- Build the solid.
- Take away the shaded cube or cubes and then draw the remaining solid.

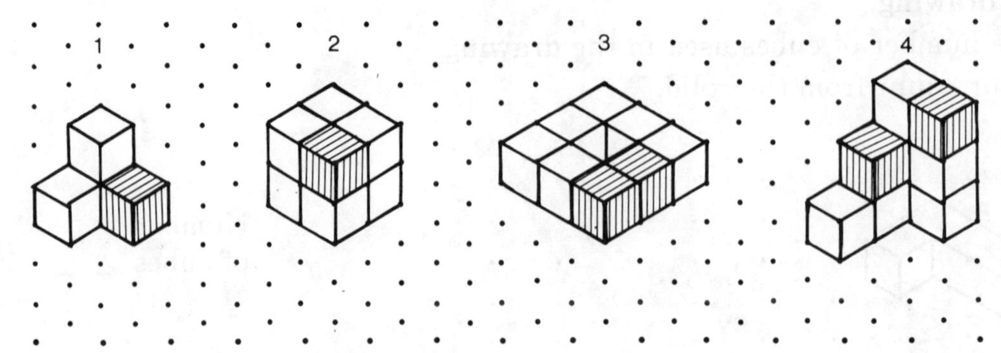

For each of the solids 5–8, do these steps:

- Build the solid.
- Add a cube to each shaded face and then draw the new solid.

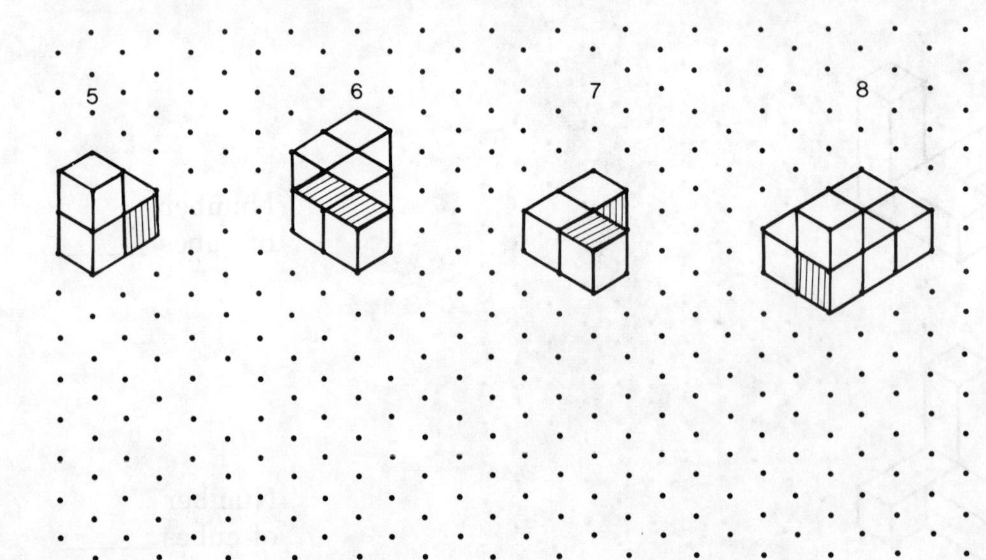

Build a solid on a piece of paper using the following plan:

- Label the corners of the paper *A*, *B*, *C*, and *D*.
- Position the paper as shown, with corners *A* and *B* at the bottom.
- Build the solid using cubes. The numbers tell you how high each stack of cubes should be.

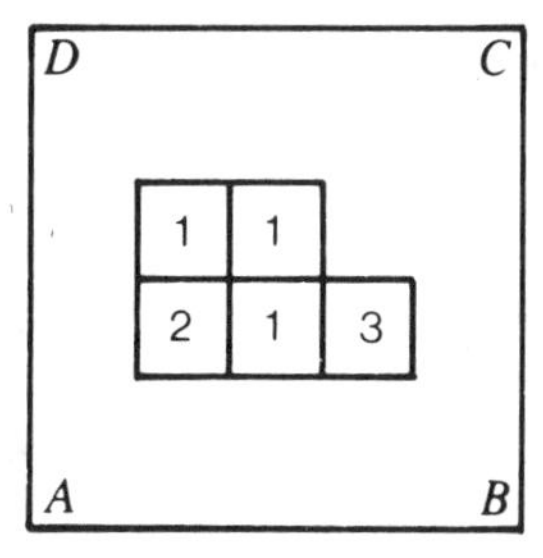

The following drawings represent the four corner views of the solid you built. Turn the paper on which your solid is built and look at the solid from each corner. Match the letter of each corner to the appropriate drawing.

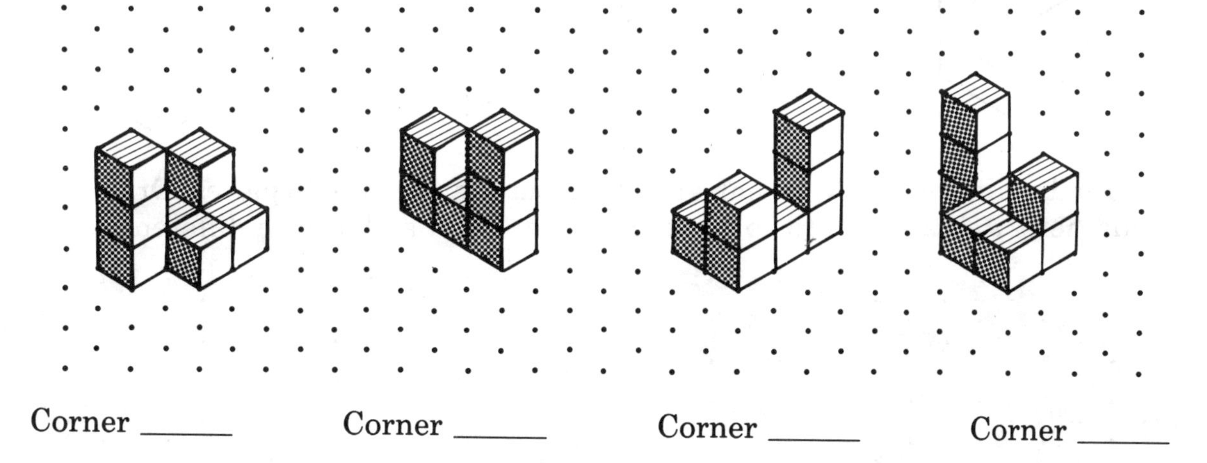

Corner ______ Corner ______ Corner ______ Corner ______

Build the solid shown below and draw views of it from two different corners. For each drawing, indicate the letter of the corner that it represents.

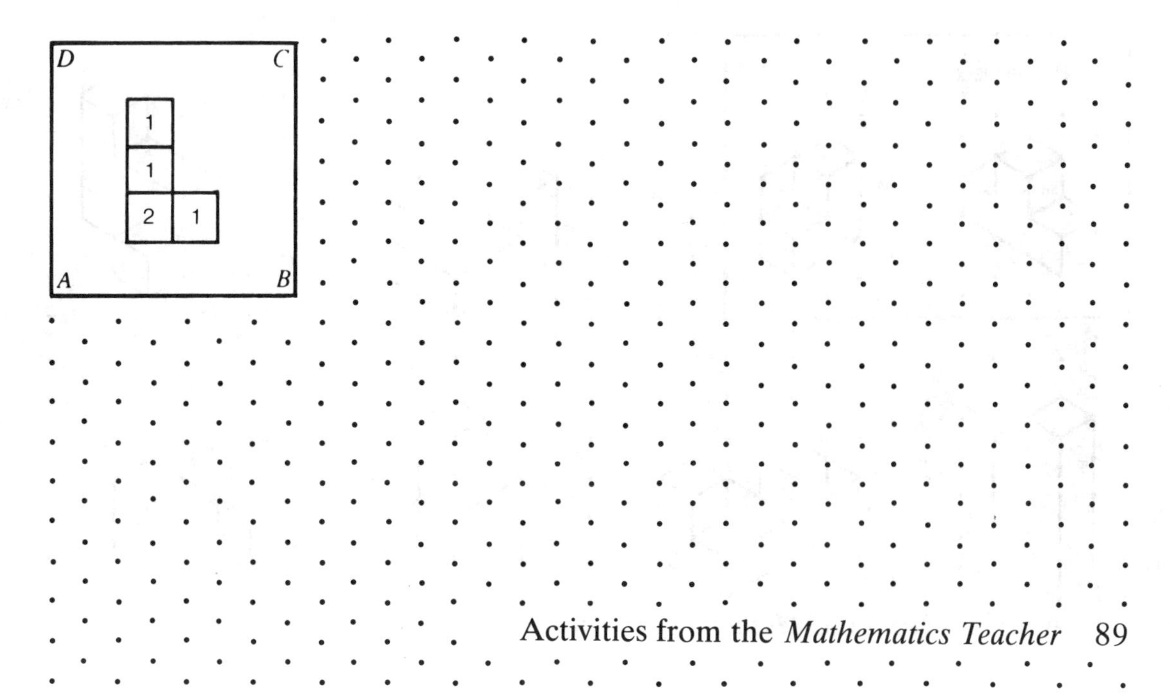

For parts A and B, construct the puzzle pieces indicated from cubes attached with tape.

- Use the two puzzle pieces to build each solid.
- Show how you built them by shading one puzzle piece in each drawing.

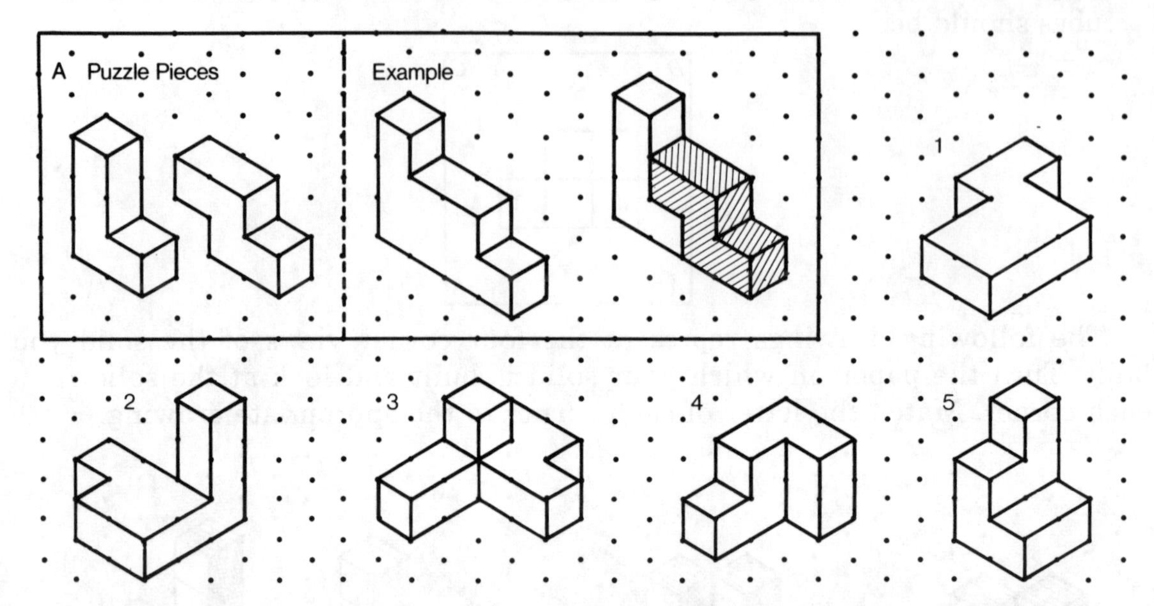

Find a different way to make a solid from the two puzzle pieces. Draw your solid here. Ask a friend to solve your puzzle. Build and draw a different solid.

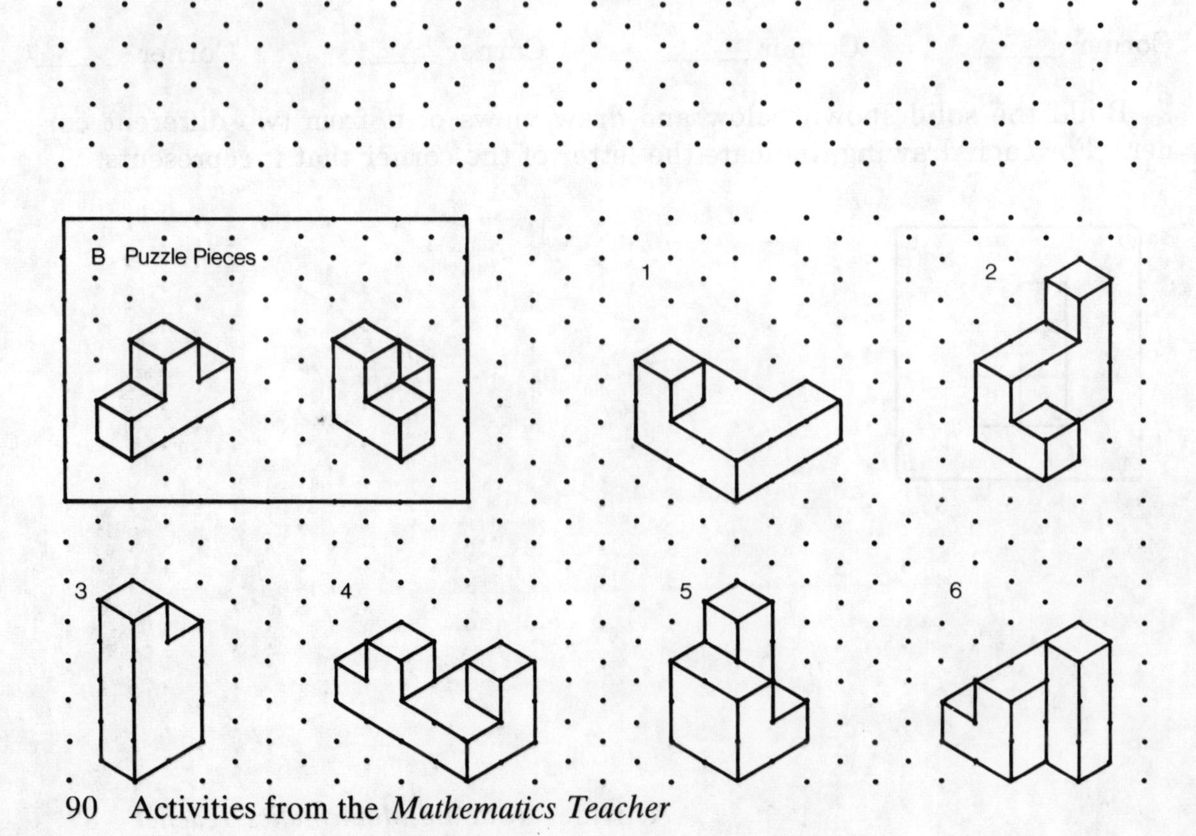

GENERATING SOLIDS

October 1983

By EVAN M. MALETSKY, Montclair State College, Upper Montclair, NJ 07043

Teacher's Guide

Grade level: 8–11

Materials: Worksheets for each student and several polygons cut from cardboard to appropriate dimensions for purposes of demonstration

Objectives: To visualize, identify, and describe the solids generated by rotating polygons about axes located in various positions; to relate these figures to cylinders and cones; and to compute their surface areas and volumes by first finding the radii and heights needed

Directions: Use cardboard models to illustrate how a rectangle rotated about a side and a right triangle rotated about a leg will generate a cylinder and a cone, respectively. Review key terms such as *radius, height,* and *slant height* as well as the related formulas for surface area and volume. Use the worksheets one at a time.

Sheet 1. Be sure each student properly identifies the radius and height for each figure generated. Discuss how changing the location of the axis of rotation changes the dimension of the solid formed. Note that when the axis cuts the rectangle and is parallel to a side, a cylinder is formed. The last illustration produces a figure that might best be described as a large cylinder with a smaller one having the same axis cut from its center.

Sheet 2. When the first right triangle is rotated about its legs, two different cones are formed. The radius (4) and height (3) of the first cone are interchanged to become the height (4) and radius (3) of the second cone. However, the slant height (5), found by the Pythagorean theorem, remains the same for both cases.

Encourage students to describe the results in the remaining two illustrations. If necessary, show the solids generated by rotating a cardboard cutout of the triangle. In the second illustration, two cones are formed extending in opposite directions from a common circular base. The third illustration might be described as a cylinder with a cone of the same radius and height cut away.

Sheet 3. More visualization skill is needed here to see the figures generated and how they are related to cylinders and cones. Encourage students to do their own thinking on these problems. However, you may want to show each generated solid to slower classes by spinning a cardboard polygon, using a wire taped in the appropriate place for the axis.

If visualization is your major emphasis, you may want students to give answers simply in terms of π, as suggested on sheet 3. For other classes, these problems can serve as valuable computational experiences using a calculator.

Answers: Decimal answers are given to two decimal places using 3.14 for π. Volumes are given in cubic units and surface areas in square units.

Sheet 1:

1. a) $r = 3, h = 4$ b) $r = 4, h = 3$
2. about the x-axis $V = 113.04$
   about the y-axis $V = 150.72$
3. a) cylinder   b) cylinder
4. about the x-axis $S = 75.36$
   about the y-axis $S = 113.04$
5. A cylinder ($r = 5, h = 3$) with another cylinder ($r = 1, h = 3$) cut from its center
6. $V = 3.14 \cdot 5^2 \cdot 3 - 3.14 \cdot 1^2 \cdot 3 = 226.08$
   $S = 2 \cdot 3.14 \cdot 5 \cdot 3 + 2 \cdot 3.14 \cdot 1 \cdot 3 + 2(3.14 \cdot 5^2 - 3.14 \cdot 1^2) = 263.76$

Sheet 2:

1. a) $r = 4, h = 3$ b) $r = 3, h = 4$
2. about the x-axis $S = 87.92$
   about the y-axis $S = 65.94$
3. A cone ($r = 3, h = 4$) joined base to base with another cone ($r = 3, h = 2$)
4. $V = \frac{1}{3} \cdot 3.14 \cdot 3^2 \cdot 4 + \frac{1}{3} \cdot 3.14 \cdot 3^2 \cdot 2 = 56.52$
5. A cylinder ($r = 4, h = 4$) with a cone ($r = 4, h = 4$) cut from it
6. $V = 3.14 \cdot 4^2 \cdot 4 - \frac{1}{3} \cdot 3.14 \cdot 4^2 \cdot 4 = 133.97$

Sheet 3:

1. A cylinder ($r = 3, h = 2$) with two cones ($r = 3, h = 1$ and $r = 3, h = 2$) joined on its bases
   $V = \pi \cdot 3^2 \cdot 2 + \frac{1}{3} \cdot \pi \cdot 3^2 \cdot 1 + \frac{1}{3} \cdot \pi \cdot 3^2 \cdot 2 = 27\pi$
2. A cylinder ($r = 3, h = 3$) with a cone ($r = 1, h = 3$) cut down the center
   $V = \pi \cdot 3^2 \cdot 3 - \frac{1}{3} \cdot \pi \cdot 1^2 \cdot 3 = 26\pi$
3. A cylinder ($r = 2, h = 4$) with a cone ($r = 2, h = 2$) cut from it
   $V = \pi \cdot 2^2 \cdot 4 - \frac{1}{3} \cdot \pi \cdot 2^2 \cdot 2 = \frac{40}{3}\pi$
4. A cone ($r = 3, h = 5$) with another cone ($r = 3, h = 2$) cut from it
   $V = \frac{1}{3} \cdot \pi \cdot 3^2 \cdot 5 - \frac{1}{3} \cdot \pi \cdot 3^2 \cdot 2 = 9\pi$

Extensions: Better classes may want to use the Pythagorean theorem with the solids on sheet 3 to find their surface areas. Other interesting solids to explore include—

1. rotating a trapezoid about the shorter of its two parallel sides;
2. rotating a rectangle about a diagonal;
3. rotating a rhombus about a diagonal. ■

ROTATING RECTANGLES

Sheet 1

When this rectangle is rotated about the line shown, it generates a cylinder. The volume and surface area of the cylinder depend on the dimensions of the rectangle.

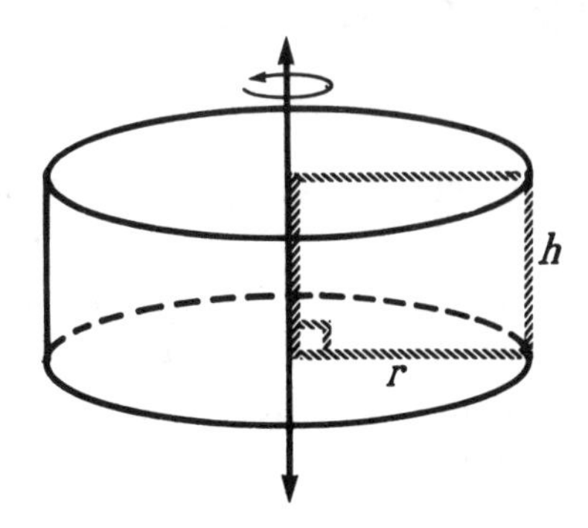

1. Find the radius and height of the cylinder formed when this rectangle is rotated about

   a) the x-axis. $r =$ ____ $h =$ ____

   b) the y-axis. $r =$ ____ $h =$ ____

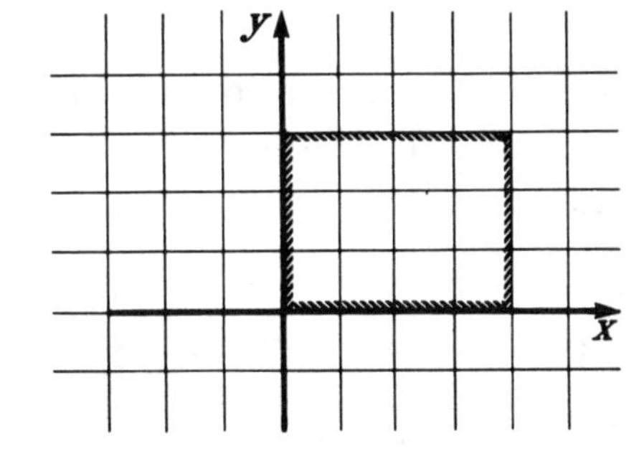

2. Which cylinder will have the greater volume? Guess first and then compute and compare.

3. Name the solid formed when this rectangle is rotated about

   a) the x-axis. ____________

   b) the y-axis. ____________

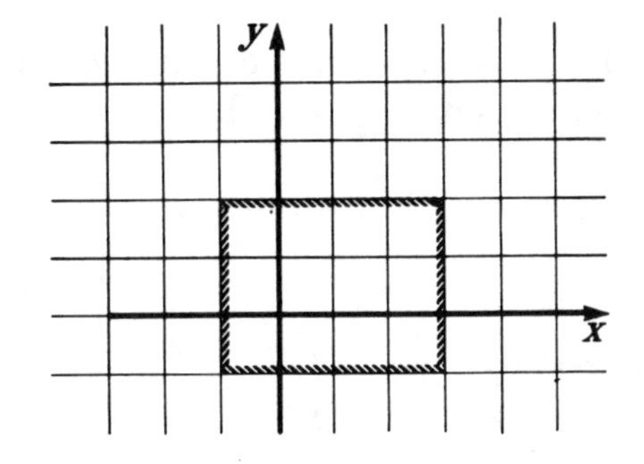

4. Which solid will have the greater surface area? Guess first and then compute and compare.

5. Describe the solid formed when this rectangle is rotated about the y-axis.

6. Find its volume and surface area.

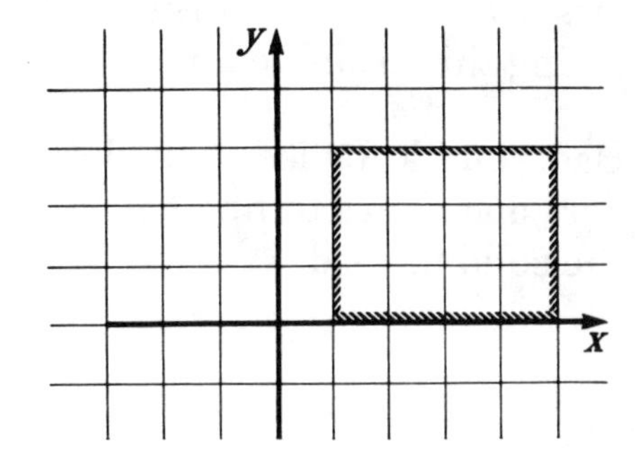

When this right triangle is rotated about the line shown, a cone is formed. The radius and height of the cone depend on the dimensions of the right triangle and the location of the axis of rotation.

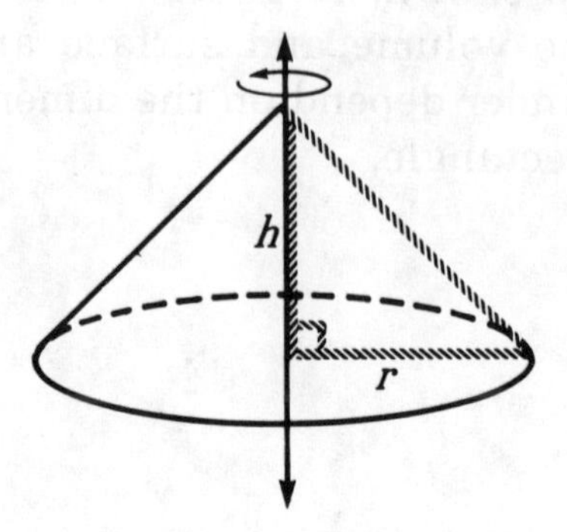

1. Find the radius and height of the cone formed when this triangle is rotated about

   a) the x-axis. $r =$ _____ $h =$ _____

   b) the y-axis. $r =$ _____ $h =$ _____

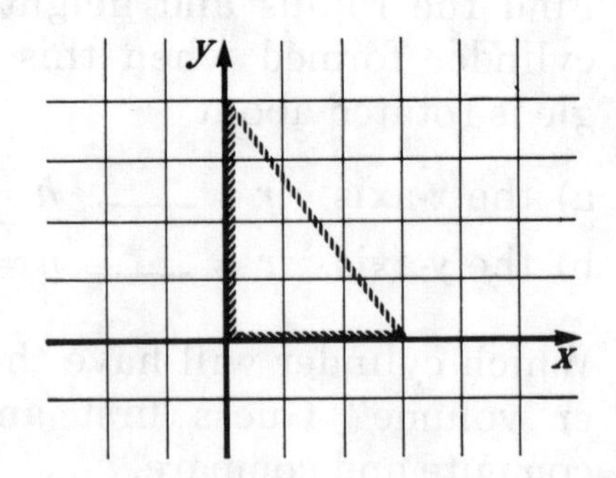

2. Which cone will have the greater surface area? Guess first and then compute and compare.

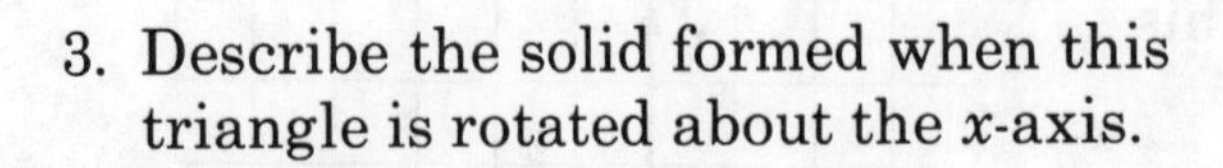

3. Describe the solid formed when this triangle is rotated about the x-axis.

4. Find the volume of the solid formed.

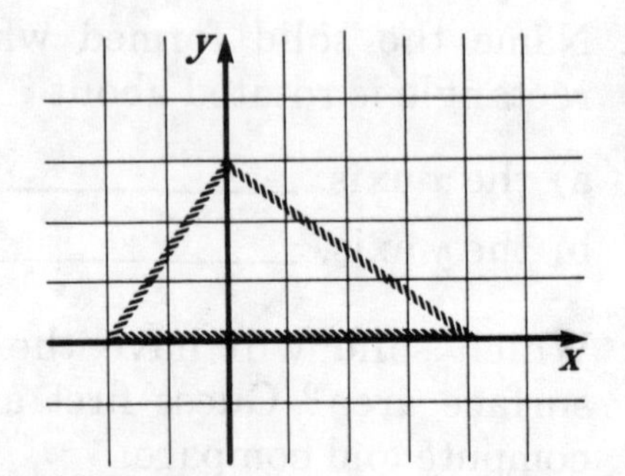

5. Describe the solid formed when rotating this triangle about the y-axis.

6. Use your knowledge of both cylinders and cones to find the volume of the solid formed.

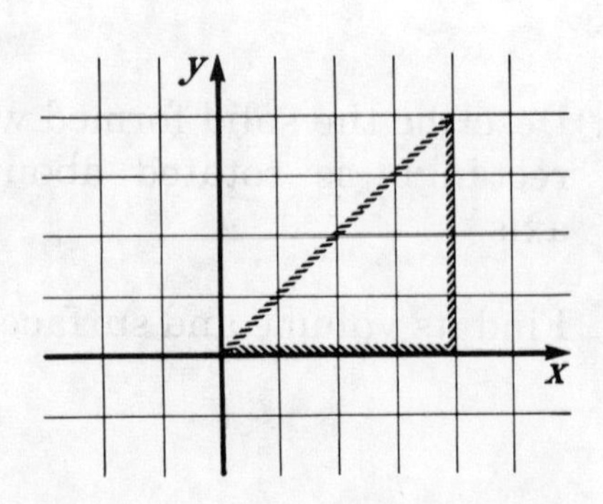

SOLIDS OF REVOLUTION

Sheet 3

Some solids of revolution are not as easy to visualize as others. These involve parts of both cylinders and cones.

Describe the shape of the solid formed by rotating the polygon about the axis shown.

Then find the necessary dimensions and use them to compute the volume. Leave your answers in terms of π.

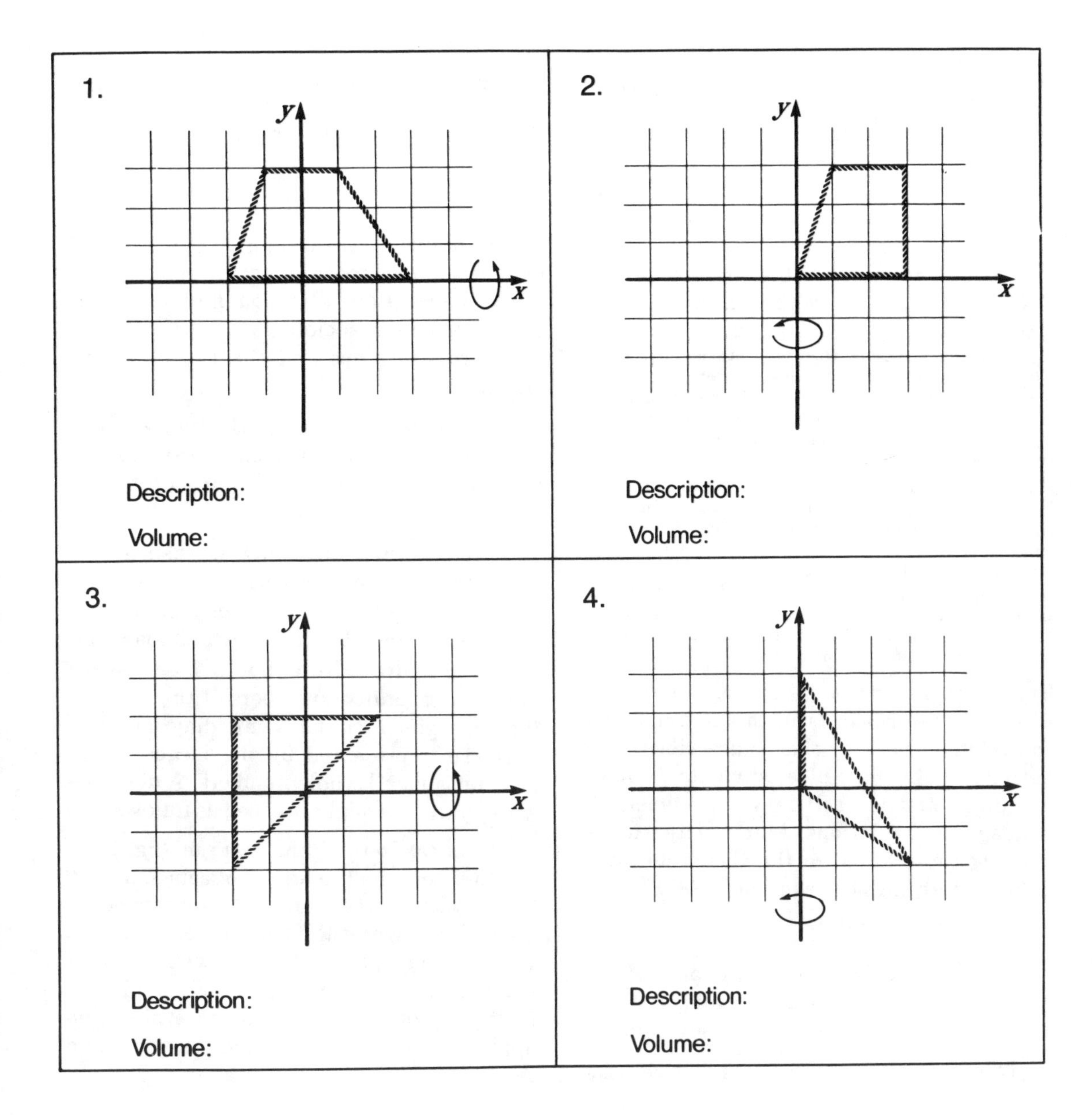

DEDUCTIVE AND ANALYTICAL THINKING

March 1985

By ROBERT L. McGINTY and JOHN G. VAN BEYNEN, Northern Michigan University, Marquette, MI 49855

Teacher's Guide

Introduction: Among the curricular recommendations in *An Agenda for Action* (NCTM 1980, 2–3, 8) are the following:

- Mathematics programs of the 1980s must be designed to equip students with the mathematical methods that support the full range of problem solving, including ... the use of mathematical symbolism to describe real-world relationships [and] the use of deductive and inductive reasoning to draw conclusions about such relationships.
- The higher-order mental processes of logical reasoning, information processing, and decision making should be considered basic to the application of mathematics. Mathematics curricula and teachers should set as objectives the development of logical processes, concepts, and language, including the identification of likenesses and differences leading to classification ... [and] the precise use of such language as *at least, at most, either-or, both-and,* and *if-then.*

These two recommendations are the focus of the four worksheets in this activity and the related pedagogy that is described. The activities also enhance students' understanding of, and facility with, numerical concepts and relations. Many of the ideas parallel those found in the Comprehensive School Mathematics Program (1978, 1979).

Grade levels: 7–9

Materials: Colored chalk, copies of the worksheets for each student, and a set of transparencies for discussion of solutions

Objectives: To enhance deductive reasoning abilities (1) through the analysis of numerical relationships and the use of Venn diagrams and (2) through the analysis of genealogical, sociological, and propositional relationships and the use of arrow diagrams

Procedures: Distribute the worksheets one at a time to each student. Discuss solutions for each sheet before going on to the next.

Sheet 1: Introduce this worksheet by first sketching intersecting sets A and B, along with the two identical lists of choices for sets A and B on the chalkboard as shown in figure 1. Note that 3, 5, and 15 are only representative members of the particular subregions in the Venn diagram. The universe, represented by the large rectangle, contains all whole numbers. A dialogue with your class might proceed as follows:

Teacher: In the Venn diagram (fig. 1) we are to determine the names of sets A and B, given the choices on the corresponding lists. If we suggest the first choice for set A, that is, "multiples of 2," then we mean that set A contains all multiples of 2 and only multiples of 2. The same interpretation applies to any choice of names for set A and set B.

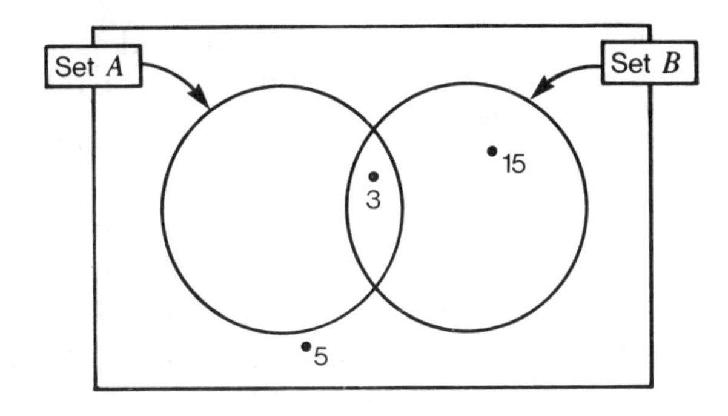

Choices for Set *A*	Choices for Set *B*
Multiples of 2	Multiples of 2
Odd Numbers	Odd Numbers
Prime Numbers	Prime Numbers
Divisors of 24	Divisors of 24
Smaller than 10	Smaller than 10
Multiples of 3	Multiples of 3
Larger than 10	Larger than 10
Multiples of 6	Multiples of 6
Multiples of 5	Multiples of 5
Divisors of 15	Divisors of 15

Fig. 1

Teacher: Could set *A* be called "multiples of 2"?

Student 1: No, because 3 is in set *A* and 3 is not a multiple of 2. [Teacher checks off "multiples of 2" on the list for set *A*.

Teacher: Could set *A* be called "odd numbers"?

Student 2: Yes, because 3 is in set *A* and 3 is odd.

Student 3: But if set *A* contains all the odd numbers, then 5 and 15 should be in set *A*, since 5 and 15 are odd. Since 5 and 15 are outside of set *A*, we cannot call it "odd numbers."

Teacher: Good thinking. We can therefore eliminate "odd numbers" as a name for set *A*.

Continue in this manner until both set *A* and set *B* have been identified. In the example just discussed, in the final analysis the students should conclude that the name of set *A* should be "divisors of 24" and the name of set *B* should be "multiples of 3."

It might be helpful if some of the students were paired to facilitate a joint effort in completing sheet 1. This procedure should provoke some interesting mathematical discussions.

Sheet 2: This activity should be preceded by a brief discussion with the entire class. Draw, preferably in colored chalk, two arrows connecting three large dots as shown in figure 2. The following is a possible dialogue.

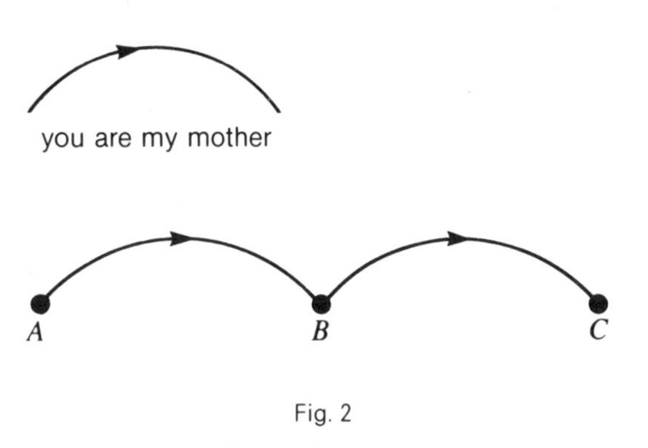

Fig. 2

Teacher: The three dots represent three different people (*A*, *B*, and *C*). The arrow tells us that *A* is related to *B* and *B* is related to *C*. In this case, *A* is saying to *B* "you are my mother" and *B* is saying to *C* "you are my mother." [Note that since both relations are the same, the color of the arrows should be the same.]

Teacher: For a fact, what could *C* say to *B*? [Draws an arrow from *C* to *B* using white chalk.]

Student 1: *C* could say to *B* "you are my daughter."

Teacher: Good. What could *A* say to *C*? [Draws an arrow from *A* to *C* using another color of chalk.]

Student 2: You are my grandmother.

Student 3: I agree, but would *C* be *A*'s maternal grandmother?

Teacher: Very good! Maybe you should explain what is meant by "maternal grandmother."

Teacher: What could *C* say to *A*? [Draws an arrow from *C* to *A*.]

Student 4: You are my granddaughter.

Student 5: Do we know for a fact that *A*

is a girl?

Student 4: No! So I guess *C* could only say to *A* "you are my grandchild."

The arrow diagram can be extended by drawing an arrow of a different color from *B* to *D*, where *B* could say to *D* "you are my brother." (See fig. 3.)

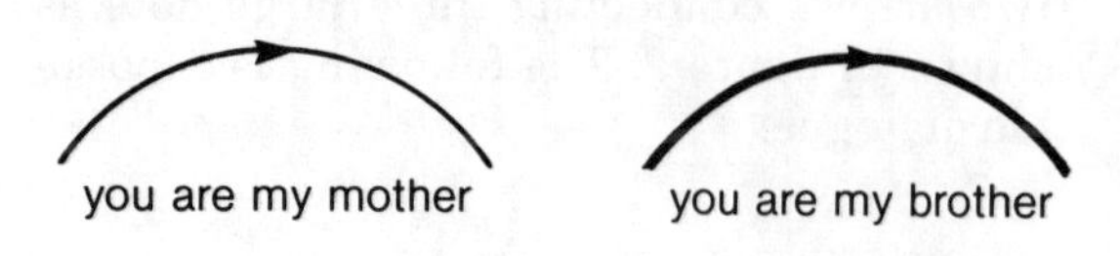

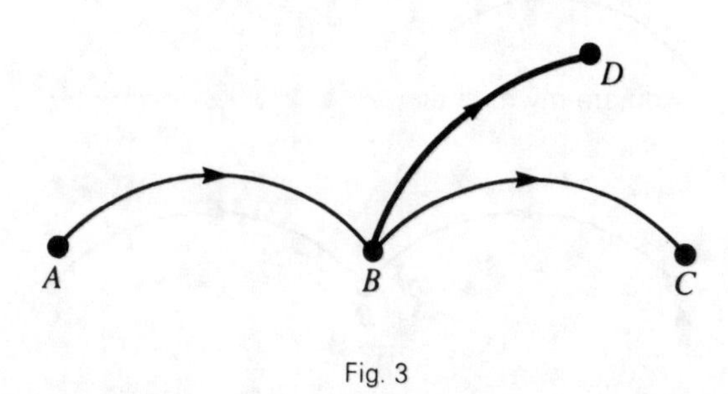

Fig. 3

Additional questions can be pursued, such as these:

1. What could *A* say to *D*? (Answer: you are my uncle)
2. What could *D* say to *B*? (Answer: you are my sister)
3. What could *C* say to *D*? (Answer: you are my son)

Other arrows of different colors can be added if you wish to extend the class activity even farther. Additional relationships can include (*a*) you are my wife, (*b*) you are my sister, (*c*) you are my cousin, and so on. When extending the arrow diagram, make sure that it remains logically, as well as visually, consistent.

Before students begin work on sheet 2, point out that we are assuming that only first marriages have occurred and that everything else is "according to Hoyle."

Sheet 3: This activity extends the idea of relations. You may wish to work through the first problem on this sheet with the entire class.

Sheet 4: Before students begin this activity, alert them to the precise nature of the wording of the statements, as well as to the mathematical distinction between the words *and* and *or*.

Answers: Sheet 1:

1. *A*—Odd Numbers, *B*—Multiples of 5;
2. *A*—Multiples of 6, *B*—Multiples of 2;
3. *A*—Multiples of 6, *B*—Prime Numbers;
4. *A*—Larger than 10, *B*—Multiples of 5;
5. *A*—Larger than 10, *B*—Odd Numbers;
6. *A*—Divisors of 24, *B*—Multiples of 3;
7. *A*—Smaller than 10, *B*—Larger than 10;
8. *A*—Prime Numbers, *B*—Multiples of 3.

Sheet 2: 1. You are my sister-in-law. 2. You are my daughter. 3. You are my aunt. 4. You are my cousin. 5. You are my cousin. 6. You are my niece or nephew. 7. You are my son-in-law. 8. You are my mother-in-law. 9. You are my father. 10. You are my uncle. 11. You are my maternal grandmother. 12. You are my brother-in-law.

Sheet 3: 1. I called your brother. 2. Your brother called me. 3. My brother called your brother. 4. Your brother called my brother. 5. You are my brother. 6. I called someone who called your brother. 7. My brother called you and me. 8. I called two of your brothers.

Sheet 4: 1. a. true; b. can't tell; c. can't tell; d. can't tell; e. can't tell; f. true. 2. a. false; b. can't tell; c. true; d. true; e. can't tell; f. can't tell. 3. a. false; b. true; c. can't tell; d. true; e. true; f. can't tell. 4. a. false; b. can't tell; c. true; d. can't tell; e. true; f. can't tell.

BIBLIOGRAPHY

Comprehensive School Mathematics Program. *CSMP Mathematics for the Intermediate Grades. Part IV: The Language of Strings and Arrows, Geometry and Measurement, Probability and Statistics.* Teacher's Guide, Experimental Version 5-36301. St. Louis, Mo.: CEMREL, 1978.

Comprehensive School Mathematics Program. *CSMP Mathematics for the Intermediate Grades. Part V: The Language of Strings and Arrows, Geometry and Measurement, Probability and Statistics.* Teacher's Guide, Experimental Version 6-37901. St. Louis, Mo.: CEMREL, 1979.

National Council of Teachers of Mathematics. *An Agenda for Action: Recommendations for School Mathematics of the 1980s.* Reston, Va.: The Council, 1980. ■

In each of the eight diagrams, identify sets A and B using the following choices:

Choices for Set A	Choices for Set B
Multiples of 2	Multiples of 2
Odd Numbers	Odd Numbers
Prime Numbers	Prime Numbers
Divisors of 24	Divisors of 24
Smaller than 10	Smaller than 10
Multiples of 3	Multiples of 3
Larger than 10	Larger than 10
Multiples of 6	Multiples of 6
Multiples of 5	Multiples of 5
Divisors of 15	Divisors of 15

1.
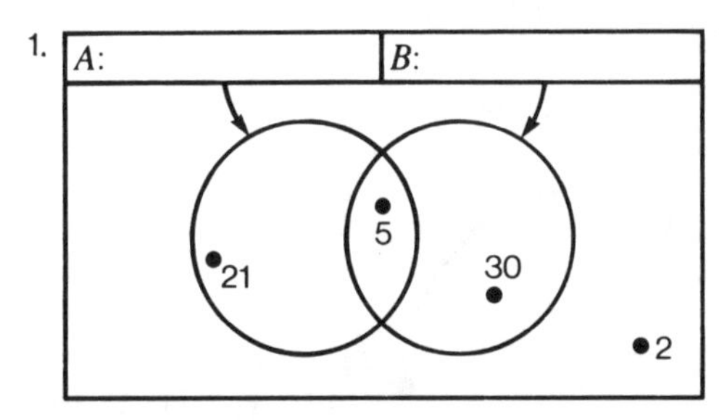

2.
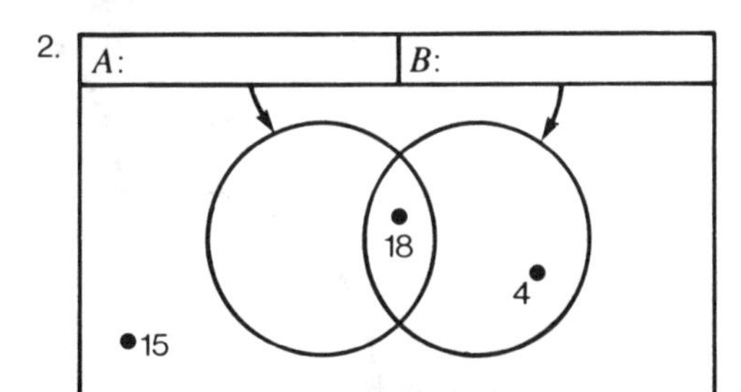

3.
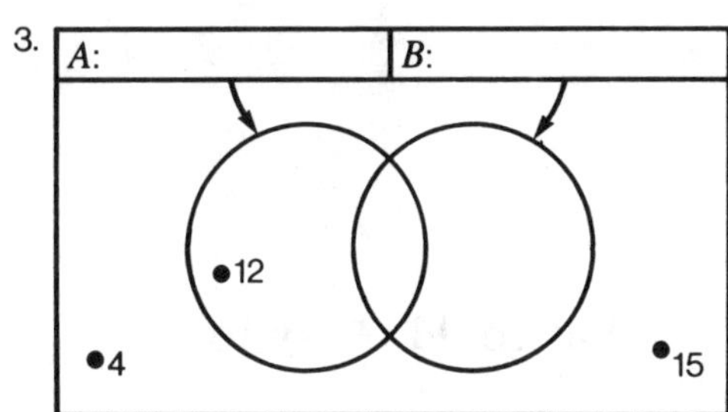

4.
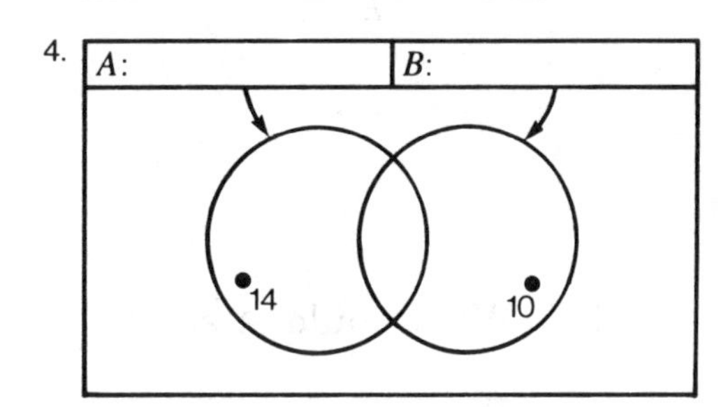

5.
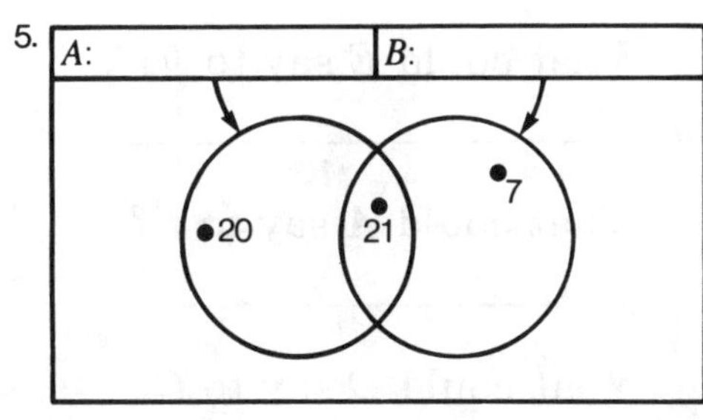

6.
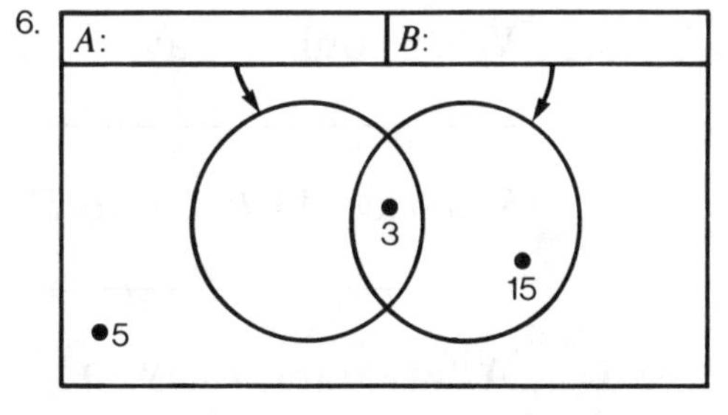

7.
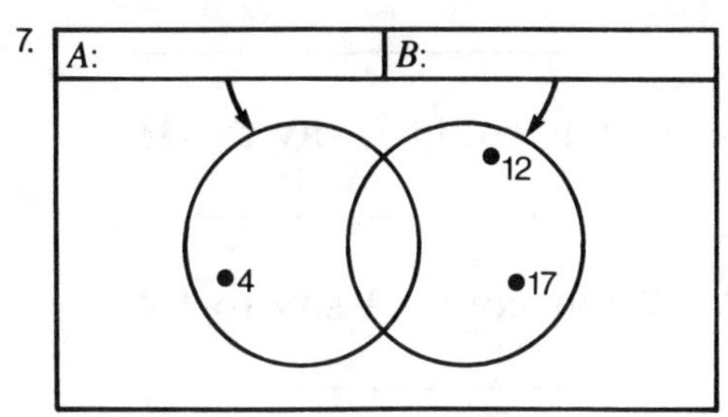

8.
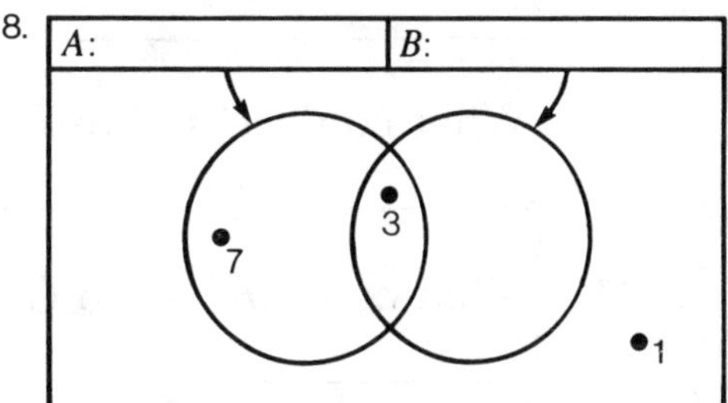

Answer the questions below, given the following relations:

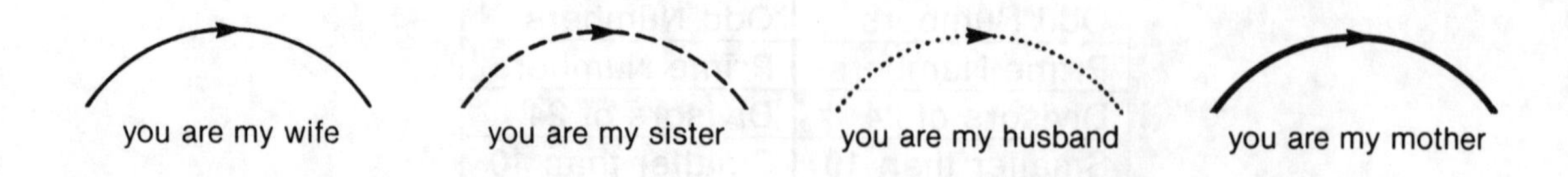

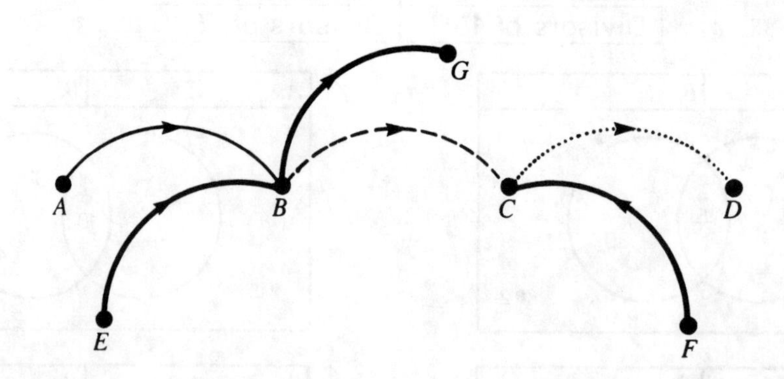

1. What could A say to C? __________
2. What could G say to C? __________
3. What could F say to B? __________
4. What could E say to F? __________
5. What could F say to E? __________
6. What could A say to F? __________
7. What could G say to A? __________
8. What could D say to G? __________
9. What could F say to D? __________
10. What could E say to D? __________
11. What could F say to G? __________
12. What could A say to D? __________

Determine what A could say to B in terms of the following relations:

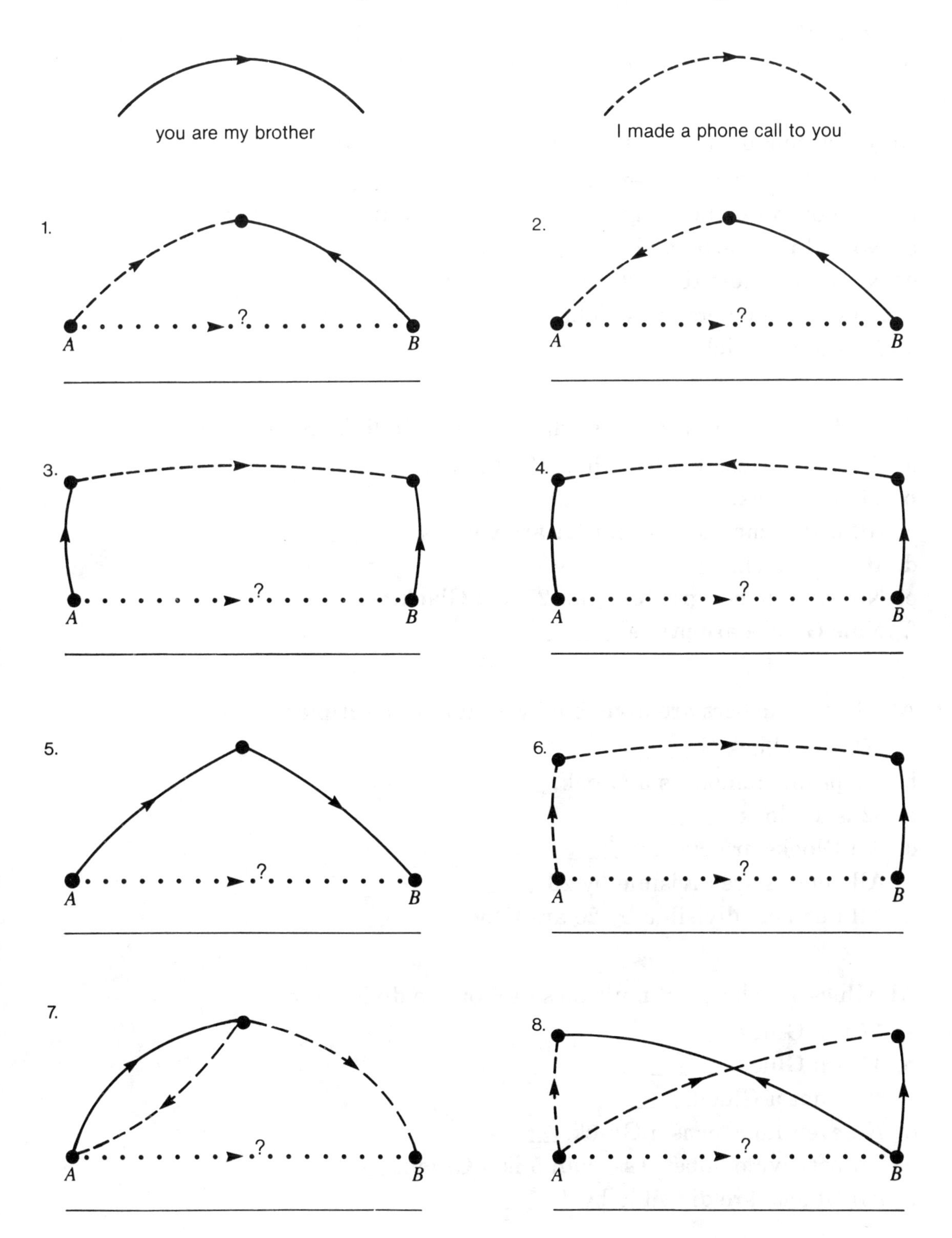

Read each of the following statements (1–4) and indicate "true," "false," or "can't tell" for each of the conclusions listed (a–f).

1. All Glick numbers that are greater than 20 are even.
   a. No odd number greater than 20 is a Glick. ______
   b. If an even number is greater than 20, then it is a Glick. ______
   c. No odd number is a Glick. ______
   d. No number less than 20 can be a Glick. ______
   e. All even numbers are Glicks. ______
   f. 33 is not a Glick. ______

2. All odd numbers that are less than 29 and all divisors of 48 are Glacks.
   a. All Glack numbers less than 15 are odd. ______
   b. 14 is a Glack. ______
   c. All odd numbers less than 21 are Glacks. ______
   d. 16 is a Glack. ______
   e. No odd number greater than 27 is a Glack. ______
   f. Nine Glacks are prime. ______

3. All Glock numbers are divisible by 2 and are multiples of 13.
   a. 169 is a Glock. ______
   b. No prime number is a Glock. ______
   c. 52 is a Glock. ______
   d. All Glocks are even. ______
   e. All Glocks are divisible by 26. ______
   f. All numbers divisible by 26 are Glocks. ______

4. All Gluck numbers are multiples of 6 or are divisible by 7.
   a. 17 is a Gluck. ______
   b. 42 is a Gluck. ______
   c. 38 is not a Gluck. ______
   d. No even number is a Gluck. ______
   e. No positive number less than 5 is a Gluck. ______
   f. All Glucks are divisible by 42. ______

Activities for

Using Calculators

Interest in the instructional uses of calculators has waned considerably over the last few years as microcomputers have become increasingly available in schools across the United States. This is unfortunate. As Recommendation 3 of *An Agenda for Action* implies, calculators and computers are equally important tools for enhancing the teaching and learning of mathematics. The activities in this section illustrate how calculators can be effectively used to assist in the discovery of number patterns and related generalizations, to support the development and reinforcement of skills and concepts, and to aid in the solution of contemporary problems that would be impractical to do with paper and pencil or a microcomputer.

The first activity, "Calculators, Quotients, and Number Patterns," provides students with opportunities to investigate and develop generalizations involving quotient patterns generated from the analysis of rectangles, parallelograms, and right triangles outlined on a hundred square.

The next two activities focus on the use of calculators in the development and refinement of estimation skills involving whole numbers, percentages, and decimals. In "Estimate and Calculate" pupils form, from a given set of digits, two numbers whose product (quotient, percentage the first is of the second) will best approximate a given target number. Their "off-scores" are analyzed and the information used to adjust their estimates so as to improve their subsequent selections. The Teacher's Guide provides a discussion of estimation strategies that should arise in discussing the worksheets. "Calculators and Estimation" begins by having students estimate a factor (divisor) so that indicated products (quotients) fall within a given range. The number of tries needed to achieve success is recorded. These experiences are then extended to situations in which a "start" number is given and then estimation and *successive* multiplication (division) are used to obtain a product (quotient) in a given range.

The activity "Percent and the Hand Calculator" capitalizes on the use of a calculator to strengthen pupils' conceptual understanding of percent and to develop their skills in estimating and mentally computing with percents. Instruction on the use of a calculator's percent key is also provided.

The last two activities in this section illustrate nicely the usefulness of calculators in the area of applications. In "Using Calculators to Fill Your Table" students calculate, organize data in tables, and then analyze their tables to solve maximization problems involving areas of three different garden layouts. "Examining Rates of Inflation and Consumption" provides an excellent setting for demonstrating the power of a calculator's automatic constant capability for multiplication. In this activity, pupils investigate the effects of inflation on consumer costs, the effects of consumption rates on our nation's coal reserves, and population growth rates.

February 1986

CALCULATORS, QUOTIENTS, AND NUMBER PATTERNS

By DAVID R. DUNCAN and BONNIE H. LITWILLER, University of Northern Iowa, Cedar Falls, IA 50614

Teacher's Guide

Introduction: Recommendation 3 of *An Agenda for Action* (NCTM 1980, 8) states that "mathematics programs must take full advantage of the power of calculators and computers at all grade levels." To achieve this goal, the use of calculators must be integrated into the core mathematics program in grades K–12. This integration should include the use of calculators "in imaginative ways for exploring, discovering, and developing mathematical concepts and not merely for checking computational values or for drill and practice" (NCTM 1980, 9). The following materials illustrate how calculators can be used effectively to aid in the discovery of number patterns.

Grade levels: 6–9

Materials: Calculators; rulers; a set of worksheets, including two copies of worksheet 4 for each student; and a transparency of the fourth worksheet for purposes of demonstration

Objective: To use a calculator in exploring and developing generalizations about quotient patterns generated from various polygons drawn on a hundred square

Directions: This activity should be introduced by working through an example with the class. Distribute calculators to those students who do not have their own. On a transparency of worksheet 4, draw two squares of different sizes, box the vertex numbers, and then circle the remaining numbers that lie on each square (fig. 1). For each square, have the class, as a group, (1) count the number of vertex numbers, (2) find the sum of the vertex numbers, (3) count the number of circled numbers, (4) find the sum of the circled numbers, (5) count the number of interior numbers, and (6) find the sum of the interior numbers. Next, for each square, have the class

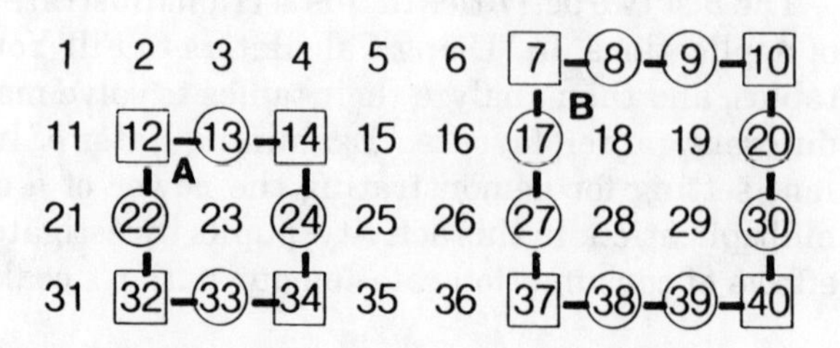

Fig. 1

find the quotient of the results of (*a*) step 2 divided by step 1, (*b*) step 4 divided by step 3, and (*c*) step 6 divided by step 5. They should observe that for square **A,** each quotient is the same, namely, 23. For square **B,** each quotient is 23.5.

Next, distribute copies of worksheets 1–3 one at a time to each student. Distribute two copies of worksheet 4 along with worksheet 1. With perhaps a clarification of the meaning of "isosceles right triangle" on worksheet 3, students should be able to complete the worksheets with minimal guidance from the teacher and successfully discover the quotient patterns.

Supplementary activities: Students who complete the worksheets before others should be given additional copies of worksheet 4 and encouraged to investigate if the patterns they discovered hold in the cases of regular pentagons and regular hexagons drawn on a hundred square. Depending on the level of the students, some might be challenged to use algebra along with a generalized hundred square to establish the patterns deductively. A generalized hundred square, as well as further activities involving a hundred square, can be found in Litwiller and Duncan (1980). For additional calculator-enhanced pattern-discovery activities, see Schmalz (1981, 1983) and Woodburn (1981).

REFERENCES

Litwiller, Bonnie H., and David R. Duncan. *Activities for the Maintenance of Computational Skills and the Discovery of Patterns.* Reston, Va.: National Council of Teachers of Mathematics, 1980.

National Council of Teachers of Mathematics. *An Agenda for Action: Recommendations for School Mathematics of the 1980s.* Reston, Va.: The Council, 1980.

Schmalz, Rosemary. "Calculator Capers." In *Activities from the Mathematics Teacher,* edited by Evan M. Maletsky and Christian R. Hirsch. Reston, Va.: National Council of Teachers of Mathematics, 1981.

———. "More Calculator Capers." *Mathematics Teacher* 76 (September 1983):414–16, 421–22, 437.

Woodburn, Douglas. "Can You Predict the Repetend?" In *Activities from the Mathematics Teacher,* edited by Evan M. Maletsky and Christian R. Hirsch. Reston, Va.: National Council of Teachers of Mathematics, 1981.

Answers:

Worksheet 1: I.

Rectangle \ Procedure	1	2	3	4	5	6
A	4	50	6	75	2	25
B	4	210	10	525	6	315
C	4	350	10	875	4	350
D	4	152	16	608	15	570

II. For rectangle **A** all the quotients are 12.5; for rectangle **B** the quotients are 52.5; for rectangle **C** the quotients are 87.5; for rectangle **D** the quotients are 38. III. For each rectangle, the three quotients are equal. IV. Yes.

Worksheet 2: I.

Parallelogram \ Procedure	1	2	3	4	5	6
A	4	52	4	52	1	13
B	4	88	8	176	4	88
C	4	264	12	792	9	594

II. For parallelograms **A, B,** and **C** the three quotients are 13, 22, and 66, respectively. III. For each parallelogram the three quotients are the same. IV. Yes.

Worksheet 3:

I.

Triangle \ Procedure	1	2	3	4	5	6
A	3	66	6	132	1	22
B	3	54	6	108	1	18
C	3	241	9	723	3	241
D	3	195	12	780	6	390

II. For isosceles right triangles **A, B, C,** and **D,** the three quotients are 22, 18, $80.\overline{3}$, and 65. III. For each triangle the three quotients are the same. IV. For each isosceles right triangle formed on a hundred square, the three quotients determined by the given algorithm are always equal. V. No. ■

I. For each rectangle drawn on the *hundred square,* carry out the indicated procedures and record your answers in the table below.

1. Count the number of vertex numbers. (A vertex number is at the vertex of the rectangle, such as the number 4.)
2. Find the sum of the vertex numbers.
3. Count the number of circled numbers. (These are all the numbers on the sides of the rectangle other than the vertices.)
4. Find the sum of the circled numbers.
5. Count the number of interior numbers, that is, the numbers inside the rectangle.
6. Find the sum of the interior numbers.

1	2	3	4	5	6	7	8	9	10
11	12	13	14	15	16	17	18	19	20
21	22	23	24	25	26	27	28	29	30
31	32	33	34	35	36	37	38	39	40
41	42	43	44	45	46	47	48	49	50
51	52	53	54	55	56	57	58	59	60
61	62	63	64	65	66	67	68	69	70
71	72	73	74	75	76	77	78	79	80
81	82	83	84	85	86	87	88	89	90
91	92	93	94	95	96	97	98	99	100

A B C D

Procedure / Rectangle	1	2	3	4	5	6
A	4	50	6	75	2	25
B						
C						
D						

II. For each rectangle find the quotient of the indicated entries and record your answers in the table at the right.

a. Column 2 divided by column 1
b. Column 4 divided by column 3
c. Column 6 divided by column 5

Procedure / Rectangle	*a*	*b*	*c*
A			
B			
C			
D			

III. What do you observe about the quotients for each rectangle in part II? ______________________

IV. Draw a differently shaped rectangle on the hundred square on worksheet 4. Repeat the procedures in parts I and II for this rectangle. Are the results consistent with your observations in part III? ______________________

PATTERNS FROM PARALLELOGRAMS — SHEET 2

I. This time parallelograms have been formed on a hundred square. For each parallelogram perform the indicated operations and record your answers in the table below.

1. Count the number of vertex numbers.
2. Find the sum of the vertex numbers.
3. Count the number of circled numbers.
4. Find the sum of the circled numbers.
5. Count the number of interior numbers.
6. Find the sum of the interior numbers.

1	2	3	4	5	6	7	8	9	10
11	12	13	14	15	16	17	18	19	20
21	22	23	24	25	26	27	28	29	30
31	32	33	34	35	36	37	38	39	40
41	42	43	44	45	46	47	48	49	50
51	52	53	54	55	56	57	58	59	60
61	62	63	64	65	66	67	68	69	70
71	72	73	74	75	76	77	78	79	80
81	82	83	84	85	86	87	88	89	90
91	92	93	94	95	96	97	98	99	100

Procedure / Parallelogram	1	2	3	4	5	6
A		52				13
B						
C						

II. For each parallelogram find the quotient of the indicated entries and record your answers in the table at the right.

a. Column 2 divided by column 1
b. Column 4 divided by column 3
c. Column 6 divided by column 5

Procedure / Parallelogram	*a*	*b*	*c*
A			
B			
C			

III. What do you observe about the quotients for each parallelogram in part II?

____

IV. Draw a differently shaped parallelogram on the hundred square on worksheet 4. Repeat the operations in parts I and II for this parallelogram. Do the results agree with your observations in part III?________________

I. For this activity isosceles right triangles have been drawn on a hundred square. For each triangle carry out the indicated procedures and record your answers in the table below.

1. Count the number of vertex numbers.
2. Find the sum of the vertex numbers.
3. Count the number of circled numbers.
4. Find the sum of the circled numbers.
5. Count the number of interior numbers.
6. Find the sum of the interior numbers.

1	2	3	4	5	6	7	8	9	10
11	12	13	14	15	16	17	18	19	20
21	22	23	24	25	26	27	28	29	30
31	32	33	34	35	36	37	38	39	40
41	42	43	44	45	46	47	48	49	50
51	52	53	54	55	56	57	58	59	60
61	62	63	64	65	66	67	68	69	70
71	72	73	74	75	76	77	78	79	80
81	82	83	84	85	86	87	88	89	90
91	92	93	94	95	96	97	98	99	100

A B C D

Procedure / Triangle	1	2	3	4	5	6
A						
B						
C						
D						

II. For each right triangle find the quotient of the indicated entries and record your answers in the table at the right.

a. Column 2 divided by column 1
b. Column 4 divided by column 3
c. Column 6 divided by column 5

Procedure / Triangle	*a*	*b*	*c*
A			
B			
C			
D			

III. What do you observe about the quotients for each triangle in part II?

IV. Draw a different isosceles right triangle on the hundred square on worksheet 4. Repeat the procedures in parts I and II for this triangle. State a generalization concerning the quotients in part II.

V. Draw a triangle that is *not* an isosceles right triangle on the hundred square on worksheet 4. Repeat the operations in parts I and II for this triangle. Do the results still agree with your generalization in part IV? ____

HUNDRED SQUARE

SHEET 4

1	**2**	**3**	**4**	**5**	**6**	**7**	**8**	**9**	**10**
11	**12**	**13**	**14**	**15**	**16**	**17**	**18**	**19**	**20**
21	**22**	**23**	**24**	**25**	**26**	**27**	**28**	**29**	**30**
31	**32**	**33**	**34**	**35**	**36**	**37**	**38**	**39**	**40**
41	**42**	**43**	**44**	**45**	**46**	**47**	**48**	**49**	**50**
51	**52**	**53**	**54**	**55**	**56**	**57**	**58**	**59**	**60**
61	**62**	**63**	**64**	**65**	**66**	**67**	**68**	**69**	**70**
71	**72**	**73**	**74**	**75**	**76**	**77**	**78**	**79**	**80**
81	**82**	**83**	**84**	**85**	**86**	**87**	**88**	**89**	**90**
91	**92**	**93**	**94**	**95**	**96**	**97**	**98**	**99**	**100**

ESTIMATE AND CALCULATE

April 1985

By EARL OCKENGA and JOAN DUEA, Price Laboratory School, Cedar Falls, IA 50613

Teacher's Guide

Introduction: An Agenda for Action (NCTM 1980) states that calculators should be available for appropriate use in *all* mathematics classrooms. The *Agenda* also recommends that calculators be used in imaginative ways for exploring, discovering, and developing mathematical concepts and skills. The activities that follow suggest ways that calculators can free students from computational restrictions and allow them to focus on the recognition of patterns and the application of numeration concepts in the development of their estimation skills.

Grade levels: 7–10

Materials: Calculators and a set of worksheets for each student

Objectives: To develop skills in estimating products, quotients, and percentages and in using a calculator to check estimates

Directions: Provide calculators for students who do not have their own. Distribute copies of the activity sheets one at a time to the students. After students have studied the example on a given worksheet and have made attempts at obtaining a better off-score, summarize for the class estimation strategies that you observed being used and perhaps identify other strategies that could have been used. The identification of the best total off-scores for the class and further discussion of associated estimation strategies should follow the completion of each sheet.

Sheet 1: Discuss the directions given at the top of the sheet. Have the students study the example. Then challenge them to see if they can arrange the digits 4 through 9 in the example to get an off-score that is less than 785 ($9 \times 7845 = 70\,605$, which has an off-score of 605).

As you observe students, you will find them using various strategies. Some students may use multiples of 1000 to estimate the product. For example, in round 1, they may think $6 \times 8000 = 48\,000$, so 6×8745 should give a product close to the target number of 50 000. Computing the product on a calculator, they would get 52 470, resulting in an off-score of 2470. To improve on this off-score, their second estimate may be 6×8574, giving a product of 51 444. A third estimate may be 6×8457, resulting in an off-score of 742.

Other students may use the strategy of working backward to find the two factors. In round 1, they would think

50 000 ÷ 5 = 10 000, so 5 × 9876 should be close to the target number.

Sheet 2: Discuss the directions. Have the students study the example, then challenge them to arrange the digits 4 through 9 in the example to get an off-score that is less than 3 (798 ÷ 4 = 199.5, which has a rounded off-score of 0).

One strategy that students might find appropriate in this setting would entail use of multiples of 100 or 1000 to estimate the quotients. For example, in round 1 a student may think 500 ÷ 5 = 100, so 498 ÷ 5 should give a quotient close to the target number. Computing the quotient on a calculator, they would get 99.6, resulting in a rounded off-score of 0.

The working-backward strategy works well for finding dividends and divisors, too. In round 1, some students may think 100 × 9 = 900, so 897 ÷ 9 should be close to the target number.

Sheet 3: Discuss the directions. Depending on the background of your students, you may also need to discuss the use of the [%] key on their calculators. For students whose calculators do not have a [%] key, you will need to discuss the use of decimal representations of percentages. Have students study the example, then challenge them to arrange the digits in the example to get an off-score that is less than 7 (51% of 784 = 399.84, which has a rounded off-score of 0).

In estimating percentages of a number, students may use easy percentages such as 10 percent, 25 percent, or 50 percent. For example, in round 1 they may think 50 percent of 400 = 200, so 49 percent of 387 should given an answer close to the target number. Computing the percentage of the number on a calculator, they would get 189.63, resulting in a rounded off-score of 10. To improve on this off-score, a second estimate may be 51 percent of 398, giving an answer of 202.98. A third estimate may be 51 percent of 392, resulting in an off-score of 0.

Additional calculator-enhanced estimation activities can be found in Miller (1981) and in Goodman (1982).

REFERENCES

Goodman, Terry. "Calculators and Estimation." *Mathematics Teacher* 75 (February 1982):137–40, 182.

Miller, William A. "Calculator Tic-Tac-Toe: A Game of Estimation." *Mathematics Teacher* 74 (December 1981):713–16, 724.

National Council of Teachers of Mathematics. *An Agenda for Action: Recommendations for School Mathematics of the 1980s.* Reston, Va.: The Council, 1980.

For each round, make three attempts, as follows:

- Place the digits 4, 5, 6, 7, 8, or 9 in the boxes to get an answer close to the target number. A digit may be used only once in each try.
- Multiply.
- Subtract to find how far your answer is from the target number. Write your "off-scores" on a separate sheet of paper.
- Record your best off-score in the column at the right.

Example. Target number: 70 000

[9] × [7][8][6][5] = 70 785

70 785 − 70 000

Off-score: 785

Round 1. Target number: 50 000

☐ × ☐☐☐☐ = ______ Best off-score: ______

Round 2. Target number: 40 000

☐ × ☐☐☐☐ = ______ Best off-score: ______

Round 3. Target number: 80 000

☐ × ☐☐☐☐ = ______ Best off-score: ______

Round 4. Target number: 60 000

☐ × ☐☐☐☐ = ______ Best off-score: ______

Round 5. Target number: 40 000

☐☐ × ☐☐☐ = ______ Best off-score: ______

Round 6. Target number: 70 000

☐☐ × ☐☐☐ = ______ Best off-score: ______

Rate yourself: Total: ______

Total off-score less than 8000 _ _ _ Super estimator
8001 to 10 000 _ _ _ _ _ _ _ _ _ _ _ _ Excellent
10 001 to 15 000 _ _ _ _ _ _ _ _ _ _ _ Good
More than 15 000 _ _ _ _ _ _ _ _ _ _ Need practice

For each round, make three attempts, as follows:

- Place the digits 4, 5, 6, 7, 8, or 9 in the boxes to get an answer close to the target number. A digit may be used only once in each attempt.
- Divide and then round your quotient to the nearest whole number.
- Subtract to find how far your answer is from the target number. Write your off-scores on a separate sheet of paper.
- Record your best off-score in the column at the right.

Example. Target number: 200

[9][8][7] ÷ [5] ≈ 197 — Off-score: 3

200 − 197

Round 1. Target number: 100

[ ][ ][ ] ÷ [ ] ≈ ______ Best off-score: ______

Round 2. Target number: 50

[ ][ ][ ] ÷ [ ] ≈ ______ Best off-score: ______

Round 3. Target number: 250

[ ][ ][ ] ÷ [ ] ≈ ______ Best off-score: ______

Round 4. Target number: 800

[ ][ ][ ][ ] ÷ [ ] ≈ ______ Best off-score: ______

Round 5. Target number: 1000

[ ][ ][ ][ ] ÷ [ ] ≈ ______ Best off-score: ______

Round 6. Target number: 200

[ ][ ][ ][ ] ÷ [ ][ ] ≈ ______ Best off-score: ______

Total: ______

Rate yourself:

Total off-score less than 40 — Super estimator
41 to 60 — Excellent
61 to 100 — Good
More than 100 — Need practice

For each round, make three attempts, as follows:

- Place the digits 1, 2, 3, 4, 5, 6, 7, 8, or 9 in the boxes to get an answer close to the target number. A digit may be used only once in each try.
- Find the percentage of the number and then round the result to the nearest whole number.
- Subtract to find how far your answer is from the target number. Write your three off-scores on a separate sheet of paper.
- Record your best off-score in the column at the right.

Example. Target number: 400

[5][1] % of [7][9][8] ≈ 407 (407−400) Off-score: 7

Round 1. Target number: 200

[ ][ ] % of [ ][ ][ ] ≈ ______ Best off-score: ______

Round 2. Target number: 100

[ ][ ] % of [ ][ ][ ] ≈ ______ Best off-score: ______

Round 3. Target number: 500

[ ][ ] % of [ ][ ][ ] ≈ ______ Best off-score: ______

Round 4. Target number: 300

[ ][ ] % of [ ][ ][ ] ≈ ______ Best off-score: ______

Round 5. Target number: 100

[ ] % of [ ][ ][ ][ ] ≈ ______ Best off-score: ______

Round 6. Target number: 50

[ ][ ][ ] % of [ ][ ] ≈ ______ Best off-score: ______

Total: ______

Rate yourself:

Total off-score less than 60 — Super estimator
61 to 80 — Excellent
81 to 100 — Good
More than 100 — Need practice

CALCULATORS AND ESTIMATION

February 1982

By Terry Goodman, Central Missouri State University, Warrensburg, MO 64093

Teacher's Guide

Grade level: 7–12.

Materials: One set of worksheets and a calculator for each student.

Objective: Students will estimate factors and divisors, using a calculator to check their estimations.

Directions: Students often have difficulty estimating when using very large or very small numbers. If the numbers are not whole numbers, this difficulty is more pronounced. These activities are designed to provide estimation practice for students. Since students are reluctant to make estimates if they are going to have to check them by hand, allow the students to use calculators for checking.

Encourage the students to work to improve their estimation skills. Problem solving can also be encouraged as students look for ways to make their estimations "better."

Sheet 1: These exercises will serve as a warm-up for the activities on succeeding sheets. Encourage students not to worry about exact answers but to try to make "good" estimates. The division key should not be used as a shortcut to estimating.

Sheet 2: Before having the students work with this sheet, do an example or two with them as a group. Starting with 15, how many successive multiplications will it take to get an answer in the interval (10 000, 10 500)? For example,

1. $15 \times 600 = 9000$
2. $9000 \times 1.1 = 9900$
3. $9900 \times 1.1 = 10\ 890$
4. $10\ 890 \times 0.95 = 10\ 345.5$

We did it in four steps. As the students work through the sheet encourage them to recognize when they need to multiply by a number greater than 1 or less than 1. Encourage them to make their estimate first, then to use the calculator. Students can make up their own problems (smaller intervals, larger numbers, and so on).

Sheet 3: Although this sheet is similar to sheet 2, you may want to work through one example with the class. Again, encourage the students to look for patterns and generalizations. They will need to focus on what happens when they divide by a number less than 1. You may want to allow the students to round off results.

In part II, students are asked to apply what they have done on sheet 2 and the first part of sheet 3. Encourage them to use both multiplication and division in the same problem. For some students, division by 2 will be more obvious than multiplication by 0.5.

Additional activities: Students can also look for other operations to use. For example,

Start	Range
5100	(50, 75)

They may decide on the following: $\sqrt{5100} = 71.414$.

As another example, consider the following:

Start	Range
3150	(100, 120)

a. $3150 \div 10 = 315$
b. $\sqrt{315} = 17.75$
c. $17.75 \times 6 = 106.49$

I. Estimate the second factor so that the product will fall within the given range. *Check* your estimate with a calculator.

Example: 12 × ___?___ Range: (5000, 5500)

12 × 400 = 4800 Too small

12 × 450 = 5400 We did it in two tries!

Try these. Indicate the number of tries in the space provided.

Start	Range	Number of Tries
1. 19 × ______	(800, 850)	______
2. 25 × ______	(2000, 2200)	______
3. 11 × ______	(550, 570)	______
4. 105 × ______	(1000, 1100)	______
5. 176 × ______	(20, 30)	______
6. 50 × ______	(4670, 4700)	______

II. Estimate the divisor so that the quotient will fall within the given range. *Check* your estimate with a calculator.

Example: 975 ÷ ___?___ Range: (10, 20)

975 ÷ 95 = 10.26 We did it in one try!!

Try these. Indicate the number of tries in the space provided.

Start	Range	Number of Tries
1. 850 ÷ ______	(90, 150)	______
2. 1050 ÷ ______	(200, 210)	______
3. 50 ÷ ______	(210, 220)	______
4. 125 ÷ ______	(4200, 4400)	______
5. 4360 ÷ ______	(150, 200)	______
6. 11 ÷ ______	(550, 600)	______

Use the start number and *successive* multiplication to obtain a product in the given range.

Example: Start: 12 Range: (5000, 5500)

1. $12 \times 400 = 4800$ Too small
2. $4800 \times 1.2 = 5760$ Too large
3. $5760 \times 0.95 = 5472$ We did it in three steps.

See if you can do the following in as few steps as possible. Record each product in the spaces provided.

	Start	Range	Number of steps
1.	11 ×	(900, 1000)	☐
2.	400 ×	(175, 195)	☐
3.	1.1 ×	(14 100, 14 200)	☐
4.	15 ×	(8000, 8100)	☐
5.	14 222 ×	(7200, 7400)	☐
6.	0.3 ×	(280, 299)	☐
7.	π ×	(100, 110)	☐
8.	12 100 ×	(36, 37)	☐

I. Use the start number and *successive* division to obtain a quotient in the given range. Try to use fewer than five steps.

Example:	Start	Range
	50	(2000, 2400)
1.	50 ÷ 0.01 = 5000	Too large
2.	5000 ÷ 2.1 = 2380.95	We did it in two steps.

Try these. Use as few steps as possible. Record each quotient in the spaces provided.

	Start	Range	Number of steps
1.	975 ÷	(10, 20)	☐
2.	143 ÷	(60, 68)	☐
3.	743 ÷	(200, 220)	☐
4.	11 ÷	(900, 1000)	☐
5.	15 ÷	(10 000, 20 000)	☐

II. Use successive multiplication and/or division to obtain a number in the given range. Record results from each step in the spaces provided.

	Start	Range	Number of steps
1.	45	(100, 110)	☐
2.	3150	(100, 120)	☐
3.	16	(3333, 3400)	☐

PERCENT AND THE HAND CALCULATOR

May 1986

By TERRENCE G. COBURN, Oakland Schools, Pontiac, MI 48054

Teacher's Guide

Introduction: In its third recommendation, *An Agenda for Action* (NCTM 1980, 9) stresses that school mathematics programs take full advantage of the power of calculators at all grade levels and then adds:

- All students should have access to calculators throughout their school mathematics program.
- The use of calculators should be integrated into the core mathematics curriculum.
- Curriculum materials that integrate and require the use of the calculator in diverse and imaginative ways should be developed and made available.

The present activity reflects these suggestions. The worksheets that follow are designed primarily to capitalize on the use of the calculator as the teacher's partner in helping students to develop an intuitive feeling for percentage. As a related outcome, students also learn how to use the calculator's percent key to compute percentages.

Grade levels: 7–10

Materials: Calculators with a percent key and a set of worksheets for each student

Objectives: To develop skills in the estimation and mental computation of percentages and in the use of the calculator's percent key to check estimates and mental computations

Prerequisites: Students should possess the following skills:

1. A familiarity with the operation of their calculators in computing with decimals. Some calculators suppress zeros at the right end of decimals, and the student should be able to interpret dollars-and-cents notation from these calculators' displays. Example:

   Find: 5 times $1.26

   Enter: 5 ⊠ 1.26 ⊟

   Display: 6.3 (which means $6.30)

2. The ability to recognize certain repeating decimals on the calculator display. For example, 0.3333333 is $\frac{1}{3}$ and 5.6666666 is $5\frac{2}{3}$.
3. The ability to round decimals to a specified place value
4. A knowledge of the fundamental ideas of percentage involving unit-region diagrams and an understanding of the following:

a. The idea of part-to-whole comparison

b. The idea that "percent" means "per one hundred"

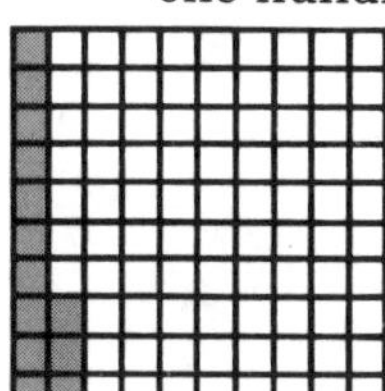

$\frac{\text{part}}{\text{whole}} = \frac{13}{100}$, or 13%.

Directions: The worksheets are design to supplement ongoing instruction involvi percentage. With each worksheet, studen are using the calculator to reinforce the skill in estimating and mentally computir percentages. Begin by providing calculatoı to those students who do not have their owı Then distribute copies of the worksheets or at a time to each student. Work through tł example on worksheet 1 with the studen before having them complete it individuall Since all calculators do not operate in tł same manner, it is important that studen have an opportunity to verify their calcu lated answer for the example. Those stu dents who keyed in the numbers and opera tions correctly but obtained an answer othe than 22.5 should be assisted in determinin the proper keying sequence for their calcu lators; the directions on the workshee should be modified accordingly. Fortunatel the keying sequence provided will work fo most calculators. After pupils have com pleted this worksheet, discuss their answer and estimation strategies before distributin worksheet 2.

It may be helpful to complete the table a the top of worksheet 2 as a teacher-directe class activity. Instruct students to use frac tions for writing percents involving repeat ing decimals. For example, $\frac{1}{12} = 8\frac{1}{3}\%$. Wher discussing solutions to exercise 2, have pupils describe their mental-computation strategies. Worksheets 3 and 4 should be treated in a manner similar to that for worksheet 1.

Variations and extensions

1. On worksheets 1, 3, and 4, you may wish to have pupils write an estimate first rather than circle a choice. If so, students should be taught to use a variety of estimation strategies while holding either the percent or base constant across several examples. For example, instruct students to do the following:

> Find 35 percent of each amount. First estimate, then use the calculator.
> *a.* 592 *b.* 412 *c.* 1187 *d.* 6.25

The student might think, "$\frac{1}{3}$ of 600 is 200, so 35 percent of 592 is about 200."

2. You might also challenge some students with "broken key" problems such as the following:

> Suppose the [÷] key on your calculator is broken. What percent of 89 is 64?

The student might explore this problem by first estimating and then entering 89 [×] 70 [%]. Next adjust—leave 89 alone and try 71 [%], 72.5 [%], and so on, until deciding that 71.9 percent is the answer to the nearest tenth of a percent.

For additional calculator-enhanced estimation activities, see Miller (1981), Goodman (1982), and Ockenga and Duea (1985).

Answers

Worksheet 1: 1. *a.* estimate, 75%; answer, 72.5%; *b.* 15%, 13.75%; *c.* 50%, 53.75%; *d.* 80%, 77.5%; *e.* 2%, 2.5%; *f.* 9%, 8.75%; *g.* 25%, 25%; *h.* 15%, 17.5%; *i.* 250%, 250%; *j.* 1%, 0.5%; 2. *a.* 25%, 25%; *b.* 15%, 13.6%; *c.* 5%, 5.2%; *d.* 150%, 158.1%

Worksheet 2: 1. $\frac{1}{3} = 33\frac{1}{3}\%$; $\frac{1}{4} = 25\%$; $\frac{1}{5} = 20\%$; $\frac{1}{6} = 16\frac{2}{3}\%$; $\frac{1}{8} = 12.5\%$; $\frac{1}{12} = 8\frac{1}{3}\%$; $\frac{1}{25} = 4\%$; $\frac{1}{50} = 2\%$. 2. *a.* 75%; *b.* 37.5%; *c.* 40%; *d.* 6%; *e.* 60%; *f.* 62.5%; *g.* $41\frac{2}{3}\%$; *h.* $83\frac{1}{3}\%$; *i.* 12%; *j.* 87.5%; *k.* 8%; *l.* 25%; *m.* 16%; *n.* 50%; *o.* $33\frac{1}{3}\%$; *p.* 100%; *q.* 80%; *r.* $66\frac{2}{3}\%$; *s.* $58\frac{1}{3}\%$; *t.* 36%; *u.* 84%; *v.* 46%; *w.* 18%; *x.* $33\frac{1}{3}\%$; *y.* 112.5%; *z.* 108%

Worksheet 3: 1. *a.* estimate, \$240; answer, \$240; answer, \$240; *b.* \$4000, \$4000; *c.* \$52, \$52; *d.* \$10, \$11.96; *e.* \$0.80, \$0.88; *f.* \$200, \$200.48. 2. *a.* \$70, \$72; *b.* \$0.80, \$0.78; *c.* \$0.03, \$0.03; *d.* \$700, \$745.20; *e.* \$20, \$23.04; *f.* \$5.00, \$4.89. 3. *a.* \$0.50, \$0.64; *b.* \$10, \$10.83; *c.* \$20 000, \$20 300; *d.* \$0.07, \$0.07; *e.* \$60 000, \$59 500.

Worksheet 4: 1. *a.* estimate, \$35; answer, \$37.50; *b.* \$200, \$200; *c.* \$7.50, \$7.50; *d.* \$75, \$75; *e.* \$750, \$750; *f.* \$0.75, \$0.75. 2. *a.* \$320, \$320; *b.* \$3200, \$3200; *c.* \$1600, \$1680; *d.* \$7000, \$6800; *e.* \$300, \$280; *f.* \$9000, \$9600. 3. *a.* \$9.00, \$10; *b.* \$6.00, \$5.88; *c.* \$10, \$11.25; *d.* \$25, \$25; *e.* \$0.20, \$0.19; *f.* \$0.04, \$0.05

REFERENCES

Goodman, Terry. "Calculators and Estimation." *Mathematics Teacher* 75 (February 1982):137–40, 182.

Miller, William A. "Calculator Tic-Tac-Toe: A Game of Estimation." *Mathematics Teacher* 74 (December 1981):713–16, 724.

National Council of Teachers of Mathematics. *An Agenda for Action: Recommendations for School Mathematics of the 1980s.* Reston, Va.: The Council, 1980.

Ockenga, Earl, and Joan Duea. "Activities: Estimate and Calculate." *Mathematics Teacher* 78 (April 1985):272–76.

1. Compare the given amount to the whole, $80. Then circle the best estimate of the percent that the given amount is of $80. Use the calculator's [%] key to check your estimate. Write the exact answer using the percent symbol, %, in the space provided.

Example:

	Estimate				Enter:	
	2%	25%	50%	75%	18 [÷] 80 [%]	22.5% (Answer)
Whole = $80						

Try these. In each case, assume that the base (or whole) is $80.

	Estimate				Estimate		
a.	90%	75%	50%	*b.*	2%	15%	30%
	______				______		
c.	90%	50%	20%	*d.*	40%	60%	80%
	______				______		
e. $2	60%	40%	2%	*f.* $7.00	9%	20%	40%
	______				______		
g. $20	10%	25%	50%	*h.* $14.00	50%	25%	15%
	______				______		
i. $200	100%	200%	250%	*j.* $0.40	1%	5%	10%
	______				______		

2. For each of the following, circle the best estimate and then use your calculator to find the percents. Round the answer on the calculator display to the nearest tenth and record your answer in the space provided.

a. 24 is what percent of 96?
25% 40% 60% ______

b. 68 is what percent of 500?
10% 15% 20% ______

c. 11 is what percent of 210?
20% 10% 5% ______

d. 6.8 is what percent of 4.3?
100% 150% 200% ______

1. Use the [%] key on your calculator to help you write each of the following fractions as a percent.

$\frac{1}{3}$ = ____%	$\frac{1}{4}$ = ____%	$\frac{1}{5}$ = ____%	$\frac{1}{6}$ = ____%
$\frac{1}{8}$ = ____%	$\frac{1}{12}$ = ____%	$\frac{1}{25}$ = ____%	$\frac{1}{50}$ = ____%

(Try to memorize these fraction-percent relationships.)

2. Without using a calculator, write the equivalent percent for each fraction below. Check your answers with your calculator when you are finished.

a. $\frac{3}{4}$ = ______ *b.* $\frac{3}{8}$ = ______ *c.* $\frac{2}{5}$ = ______

d. $\frac{3}{50}$ = ______ *e.* $\frac{3}{5}$ = ______ *f.* $\frac{5}{8}$ = ______

g. $\frac{5}{12}$ = ______ *h.* $\frac{5}{6}$ = ______ *i.* $\frac{3}{25}$ = ______

j. $\frac{7}{8}$ = ______ *k.* $\frac{4}{50}$ = ______ *l.* $\frac{2}{8}$ = ______

m. $\frac{4}{25}$ = ______ *n.* $\frac{25}{50}$ = ______ *o.* $\frac{2}{6}$ = ______

p. $\frac{4}{4}$ = ______ *q.* $\frac{4}{5}$ = ______ *r.* $\frac{4}{6}$ = ______

s. $\frac{7}{12}$ = ______ *t.* $\frac{9}{25}$ = ______ *u.* $\frac{21}{25}$ = ______

v. $\frac{23}{50}$ = ______ *w.* $\frac{9}{50}$ = ______ *x.* $\frac{4}{12}$ = ______

y. $\frac{9}{8}$ = ______ *z.* $\frac{27}{25}$ = ______

1. Circle the best estimate of 80 percent of each amount. Then find the percentage using the calculator's [%] key. Write your answer in the space provided.

Example:

	Estimate			Enter:	
What is 80 percent of $700?	$5.60	$56	$560	80 [×] 700 [%]	______

	Estimate			
a. 80% of $300	$2.40	$24	$240	______
b. 80% of $5000	$4000	$400	$40	______
c. 80% of $65	$5.20	$52	$520	______
d. 80% of $14.95	$1	$10	$100	______
e. 80% of $1.10	$80	$8	$0.80	______
f. 80% of $250.60	$2	$20	$200	______

2. Circle the best estimate of 12 percent of each amount, calculate the percentage, and record it in the space provided.

	Estimate			
a. 12% of $600	$7	$70	$700	______
b. 12% of $6.50	$0.80	$8	$80	______
c. 12% of $0.25	$3	$0.30	$0.03	______
d. 12% of $6210	$700	$70	$7	______
e. 12% of $192	$20	$200	$2000	______
f. 12% of $40.75	$0.50	$5	$50	______

3. Circle the best estimate of 3.5 percent of each amount, calculate the percentage, and record it in the space provided. Round answers to the nearest cent.

	Estimate			
a. 3.5% of $18.20	$5	$0.50	$0.05	______
b. 3.5% of $309.46	$1	$10	$100	______
c. 3.5% of $580 000	$2000	$20 000	$200 000	______
d. 3.5% of $2.06	$0.70	$0.07	$0.01	______
e. 3.5% of $1.7 million	$6000	$60 000	$600 000	______

Given the same base (the whole), find each percentage (part). Estimate first and circle your choice. Then use the calculator's [%] key. Write your answer in the space provided.

Example: What is 66 percent of $250?

Estimate: $16 $160 $1600

Enter: 250 [×] 66 [%] ________

1. Try these. The same base (or whole) is used, $250.

		Estimate			
a.	15% of $250	$3.50	$35	$350	________
b.	80% of $250	$2	$20	$200	________
c.	3% of $250	$75	$7.50	$0.75	________
d.	30% of $250	$75	$750	$0.75	________
e.	300% of $250	$7500	$750	$75	________
f.	0.3% of $250	$7.50	$0.75	$0.08	________

2. Try these. $8000 is used as the base.

		Estimate			
a.	4% of $8000	$32	$320	$3200	________
b.	40% of $8000	$32	$320	$3200	________
c.	21% of $8000	$1600	$160	$16	________
d.	85% of $8000	$700	$7000	$70 000	________
e.	3.5% of $8000	$3	$30	$300	________
f.	120% of $8000	$900	$9000	$90 000	________

3. Try these. $12.50 is used as the base. Round your answers to the nearest cent.

		Estimate			
a.	80% of $12.50	$0.90	$9	$90	________
b.	47% of $12.50	$0.60	$6	$60	________
c.	90% of $12.50	$10	$1	$0.10	________
d.	200% of $12.50	$2.50	$25	$250	________
e.	1.5% of $12.50	$2	$0.20	$0.02	________
f.	0.4% of $12.50	$0.40	$0.04	$0.01	________

USING CALCULATORS TO FILL YOUR TABLE

March 1981

By Dwayne E. Channell, Western Michigan University, Kalamazoo, MI 49008

Teacher's Guide

Grade Level: 7–10.

Materials: Copies of the three worksheets for each student, extra sheets of lined paper, and a set of calculators.

Objective: Students are to construct tables of values and use them to solve maximization problems.

Directions: Distribute copies of the three worksheets, paper, and calculators to your students.

Work with your students to be certain each understands the problem and the method being used to complete the table in problem 3. Be sure each student understands that this table lists all of the possible dimensions for gardens that can be bordered with the 120-cm sections of fence.

In problem 5, the list of possible widths for the table is not given to the students. You may want to help them complete this column before they begin work on the remainder of the table.

In problems 7 and 8, students will need to organize their own tables. Problem 8 is more difficult than the others. You may wish to work through one case (e.g., $w =$ 0.9 m where one section of fence is used for the width) to demonstrate a method for finding l and A and suggest that students use these values to begin their tables.

Additional Problems: Three different garden locations and designs are described on the activity sheets. If Tim and Kathy use a combination of the 120-cm and the 90-cm fencing sections instead of sections of just one length, could they enclose a larger garden area at any of the three locations?

A computer can be programmed to generate the tables of numbers needed to solve problems of this type. If you have access to a microcomputer or to a time-sharing terminal, you may want to try a programming approach to these problems. While at the keyboard, try this problem: Assume that an 18-m roll of fencing is available and that this fencing can be bent at 2-cm intervals. By using this single piece of fencing, what is the largest garden area that Kathy and Tim could enclose at each of the three garden locations?

REFERENCE

Watkins, Ann E. "The Isoperimetric Theorem." *Mathematics Teacher* 72 (February 1979):118–22.

Answers: Sheet 1: 1.a. 15; b. 20; 2.a. 13; b. 1.2 m × 15.6 m; c. 18.72;

3.a.

Width (m)	Length (m)	Area (m²)
1.2	15.6	18.72
2.4	13.2	31.68
3.6	10.8	38.88
4.8	8.4	40.32
6.0	6.0	36.00
7.2	3.6	25.92
8.4	1.2	10.08

b. A width of 8.4 m would use 7 sections of fence on each of two sides of a garden. This leaves one section for the third side.

Sheet 2: 4.a. 40.32 m²; b. 4.8 m × 8.4 m; c. 4; d. 7;

5.

Width (m)	Length (m)	Area (m²)
0.9	16.2	14.58
1.8	14.4	25.92
2.7	12.6	34.02
3.6	10.8	38.88
4.5	9.0	40.50
5.4	7.2	38.88
6.3	5.4	34.02
7.2	3.6	25.92
8.1	1.8	14.58

6.a. 90 cm; b. 4.5 m × 9.0 m; c. 5; d. 10

Sheet 3: 7.c. 90-cm sections; d. ten for width, ten for length; 8.a. 2.7 m × 6.3 m; b. 26.73 m²; c. no

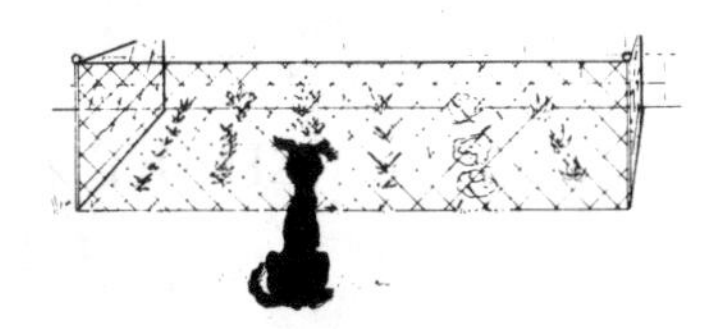

FENCING A GARDEN

Kathy and Tim are planning a small vegetable garden in their backyard. They want to enclose a rectangular garden area with a small fence so as to protect their plants from the family dog. They plan to use part of an existing wall as one of the four borders for their garden.

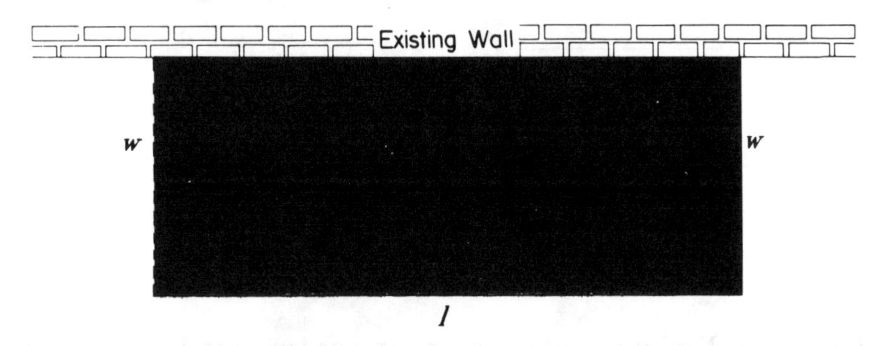

Tim and Kathy have enough money in their budget for the purchase of 18 meters (m) of fencing. Fencing sections are available in both 120-cm and 90-cm lengths. In order to grow as many vegetables as possible, they want to use the 18 m of fencing to enclose the maximum possible garden area. Which length of fencing sections should Tim and Kathy purchase?

1. a. How many 120-cm lengths of fence can they purchase? ________________
   b. How many 90-cm lengths of fence can they purchase? ________________
2. a. If they purchase the 120-cm sections and use a single section for each width of the fence, how many sections would they use for the length? (Reminder: No fencing is needed along the back of the garden.) ________________
   b. What would be the dimensions of the garden in meters? ________________
   c. How many square meters (m^2) of garden area would be enclosed by this fence? ________________
3. a. Complete the table below to find the areas enclosed by each of the different sized fences that can be made using the 120-cm sections.

Width(m)	Length(m)	Area(m^2)
1.2		
2.4		
3.6		
4.8		
6.0		
7.2		
8.4		

   b. Why is 8.4 m the longest possible width? ________________________________

____

____

4. a. What is the largest area that Kathy and Tim can enclose if they use the 120-cm sections of fence? ____________________

   b. What are the dimensions in meters of this rectangular garden?

   ____

   c. How many sections of fence would be used for each width of this garden's border? __________

   d. How many sections of fence would be used for the length of this garden's border? __________

5. Could Tim and Kathy enclose even more garden area if they use the 90-cm fencing sections? Complete the table below to find the dimensions and areas of all the possible gardens that can be bordered using the 90-cm sections of fence. The list of possible widths is started for you.

Width(m)	Length(m)	Area(m^2)
0.9		
1.8		

6. a. Which length of fencing sections should Tim and Kathy purchase if they wish to enclose a garden with the largest possible area?

   ____

   b. What are the dimensions in meters of the garden they should plant?

   ____

   c. How many sections of fence should they use for each width of their garden's border? __________

   d. How many sections should they use for the length of their garden's border?

   ____

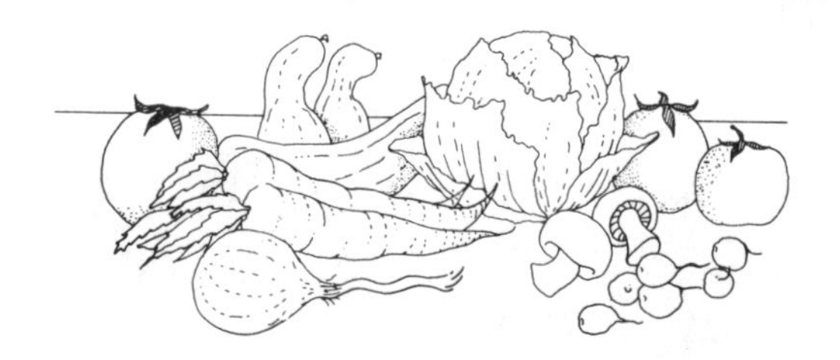

7. Kathy and Tim's parents suggested they plant their garden at a corner of the backyard so that the existing wall could serve as two of the four borders of their garden.

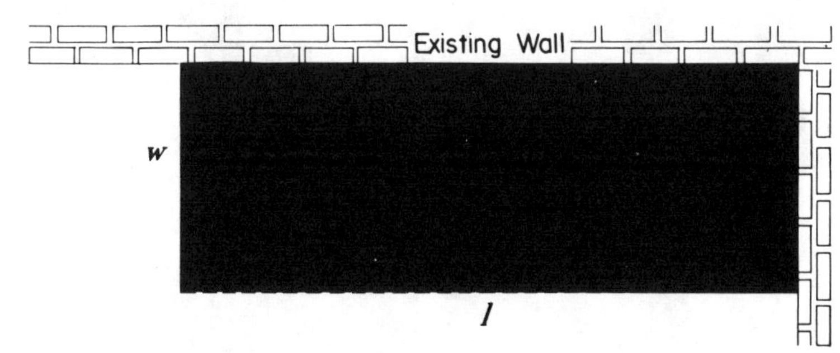

a. Construct a table of the dimensions and areas of the possible gardens bordered by the 120-cm sections of fencing.

b. Construct a table of the dimensions and areas of the possible gardens bordered by the 90-cm sections of fencing.

c. Which type of fencing should they buy to enclose the maximum garden area? ______________________________

d. How many fencing sections should they use for the width and how many for the length of this garden? ______________________________

8. Suppose Tim and Kathy used the 90-cm sections of fence and planted their garden at a corner of the house so that it was positioned as shown here.

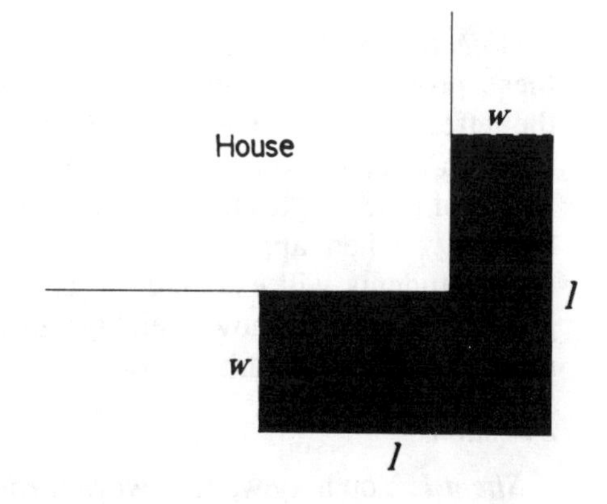

a. What are the dimensions (w and l) of the largest garden they could enclose?

____

b. What is the area of this garden? ____________________

c. Could they use the 120-cm sections to enclose a similarly shaped garden with a larger area? ____________________

EXAMINING RATES OF INFLATION AND CONSUMPTION

September 1982

By Melfried Olson, Oklahoma State University, Stillwater, OK 74078
Vincent G. Sindt, University of Wyoming, Laramie, WY 82071

Teacher's Guide

Grade level: 8–12

Materials: Student worksheets, calculators

Objectives: As students work with these problems, they will be exposed to the effects of inflation rates. Given an inflation rate, students will determine the length of time it takes for a cost to double. Similarly, when applied to consumption rates, students will examine the effect of growth rates and how quickly growth rates can deplete a finite resource.

Directions:

Sheet 1. You may want to work through the example provided before having the students complete exercise 2. It is important that the students know how to use the calculator to compute the total cost for each successive year. You may want to emphasize multiplying by 1.1 for the 10% rate, rather than multiplying by 0.1, and adding the cost increase to last year's cost to get the new cost. Also, as many calculators have an automatic constant for multiplication, you might investigate how to use it for quicker calculations. Note that the method of calculation chosen and the use of round-off or truncation can vary the amounts per year. Amounts would also change if we compounded semiannually, quarterly, monthly, or daily; so use the table as a guide to a 10% inflation rate. This is the reason for column *C* of the table on sheet 1. If we compute the 10% rate compounded at 5/6% per month, the numbers in columns *A* and *B* would be 6 and 7, respectively.

Sheet 2. Sheet 2 is similar to sheet 1 but focuses on consumption rather than inflation. The seven-year period is chosen partially for convenience but mostly because initial plans called for a 10% increase in coal production to compensate for the decline in oil supplies. Maintained at this rate per year, this means a seven-year period of time before we double consumption. Note the rules do not equate with the growth rate per year but come close enough to demonstrate clearly how a consumption rate can deplete a finite resource. The 2800 years represents the high estimate given for coal reserve supplies in the United States (Bartlett 1978).

It is informative and useful to vary the table to account for different rates of consumption, for even a 7% consumption rate will exhaust these 2800 units in under 100 years.

Sheet 3. This work is provided to help consolidate inflation and growth rates through the principle that if something doubles in cost in n years, it quadruples in cost in $2n$ years. Thus the doubling effect builds very rapidly.

Answers:

Sheet 1.

2.

R	(A)	(B)	(C)
12%	6 ($1.97)	7 ($2.21)	6
9%	8 ($1.99)	9 ($2.17)	8
8%	9 ($1.99)	10 ($2.15)	9
7%	10 ($1.97)	11 ($2.10)	10
6%	11 ($1.90)	12 ($2.01)	12

3. Note the inverse variation relationship. A general formula can be taken as $C = 70/R$ or $C = 72/R$, both of which provide a good approximation for use depending on the R values involved. Actually, the formula will vary somewhat according to the period of compounding. For a calculation of this formula, see Bartlett (1978).

Sheet 2.

2003–2009	8	56	105
2010–2016	16	112	217
2017–2023	32	224	441
2024–2030	64	448	889
2031–2037	128	896	1785
2038–2044	256	1792	3577

1. Note that in each seven-year period seven more units of coal are used than were used in total up to that period.
2. 2038–2044. Actually, in year 2041 we will have exhausted the supply.
3. This depends on the student's age now, but some students will still be alive to watch the last unit go under these rules.

Sheet 3.

1a. 5:00 P.M.

1b. Working backward from 6:00 P.M., it would be 1/64 full (1.6%)—lots of room left.

2. $23 344
3. $43 178
4. Somewhere around 43–45 years, depending on how intermediate values are rounded.

BIBLIOGRAPHY

Bartlett, A. A. "Forgotten Fundamentals of the Energy Crisis." *American Journal of Physics* 46 (September 1978):876–88.

Lange, L. H. "Some Everyday Applications of the Theory of Interest." In *Applications in School Mathematics*, 1979 Yearbook, edited by S. Sharron. Reston, Va.: National Council of Teachers of Mathematics, 1979.

Lapp, R. E. *The Logarithmic Century*. Englewood Cliffs, N.J.: Prentice-Hall, 1973.

Waits, B. K. "The Mathematics of Finance Revisited through the Hand Calculator." In *Applications in School Mathematics*, 1979 Yearbook, edited by S. Sharron. Reston, Va.: National Council of Teachers of Mathematics, 1979.

1. If one can purchase an item that costs \$1 now and inflation continues at 10% per year (compounded yearly), when will the cost double? We can make a table like the following to determine the result. Show how you would compute the remaining costs.

Year	Cost
Now	\$1.00
1	\$1.10 = 1.00 + (0.10 × 1.00) = 1.1 (1.00)
2	\$1.21 = 1.10 + (0.10 × 1.10) = 1.1 (1.10)
3	\$1.33 = 1.21 + (0.10 × 1.21) = 1.1 (1.21)
4	\$1.46 =
5	\$1.61 =
6	\$1.77 =
7	\$1.95 =
8	\$2.14 =

2. Complete the following table, starting with an item that now costs \$1. See how many years it will take for the item to cost \$2 at the various rates of inflation, compounded yearly. The 10% rate is completed for you using data from the chart above.

R Rate per Year (Compounded Yearly)	*(A)* Number of Years Just Prior to Reaching \$2	*(B)* Number of Years Just After Passing \$2	*(C)* Choice from *(A)* or *(B)* Closest to \$2
12%			
10%	7 (\$1.95)	8 (\$2.14)	7
9%			
8%			
7%			
6%			

3. Graph the results above on this coordinate system.

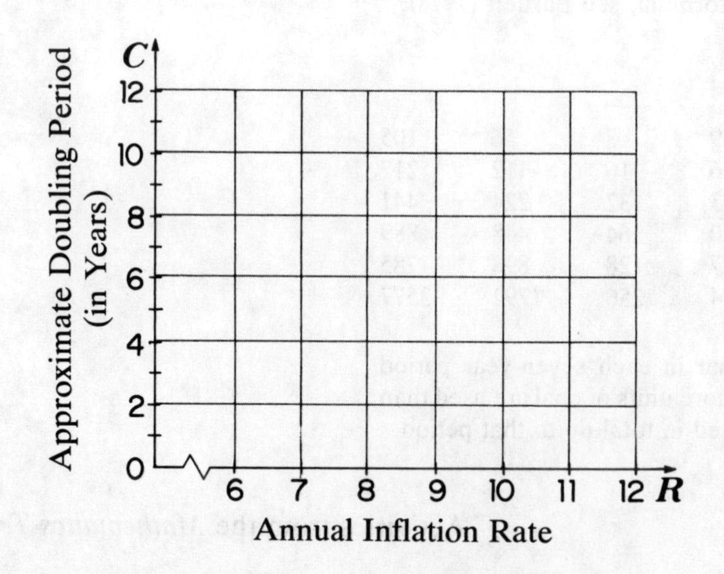

Scientists in the United States recently made two measurements concerning coal. They determined how much coal was used in the United States during 1980 and estimated the total available. Using the 1980 amount as one unit, scientists concluded that the total supply of coal reserves in the United States is 2800 units.

However, in 1981 people found more uses for coal and proposed increasing its consumption. The debate that ensued was settled by agreeing to the following rules:

Rule 1: Use one unit of coal in the first year, 1982.

Rule 2: Use the same number of units of coal per year for a seven-year period.

Rule 3: At the end of each seven-year period, double the amount of coal that can be consumed during the next seven-year period.

Use these rules to complete the following table.

Years	Number of Units of Coal Used *Each Year*	Number of Units Used during the Seven-Year Period	Total Number of Units Used Thus Far
1982–1988	1	7	7
1989–1995	2	14	21
1996–2002	4	28	49
2003–2009			
2010–2016			
2017–2023			
2024–2030			
2031–2037			
2038–2044			

1. How does the number of units used during a seven-year period compare to the total number of units used thus far? ______________________

____

2. During which seven-year period will the last unit of coal be used under these rules? ______________________

____

3. How old will you be when the last unit of coal is used? ______________________

____

1. Bacteria grow by division so that 1 bacterium becomes 2, the 2 divide to produce 4, the 4 divide to produce 8, and so on. Scientists noticed a particular strain of bacteria, Numah, for which this division time is 1 hour. They also noted that when 1 Numah is placed in a bottle of a certain size at 6:00 A.M., the bottle is full at 6:00 P.M. the same day.

a. At what time was the bottle 1/2 full?

b. How full was the bottle at noon?

2. The average price for a new car in 1981 was about $9000. At a 10% annual inflation rate, what will the average cost for a new car be in 10 years?

3. Suppose that you have an income in 1981 of $20 000. If you receive an 8% raise yearly, what will your yearly income be in 10 years?

4. Population growth in the world is approximately 1.6% per year. If that rate exists for your town and that rate continues, how many years will it take before your town doubles its population?

Activities for

Using Computers

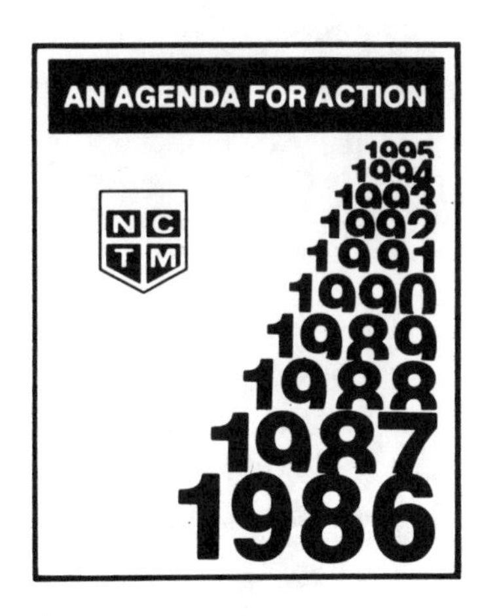

COMPUTERS

Though increasing numbers of students are gaining access to computers through courses in computer literacy and computer programming and through the use of drill-and-practice programs in mathematics courses, it is essential, if the *Agenda*'s Recommendation 3 is to be fully implemented, that computers also be integrated into the mathematics curriculum as a *tool* for developing concepts, for exploring relationships, and for problem solving. The six activities in this section reflect some of the ways in which computing technology can offer exciting new approaches to topics in the secondary mathematics curriculum.

In the first activity, "Investigating Shapes, Formulas, and Properties with Logo," students work within a computer-graphics environment, created by supplied Logo procedures, to explore relationships between areas and perimeters of squares, to derive formulas for the areas of parallelograms and triangles, and to discover the fact that the line segment connecting the midpoints of two sides of a triangle is parallel to, and one-half the length of, the third side.

The next three activities exploit the graphics capabilities of microcomputers to investigate properties of linear and absolute-value equations and the geometry of systems of linear equations. The activity "Microcomputer Unit: Graphing Straight Lines" provides a graphing program in BASIC that is to be entered, saved on disk, and then used by students to explore how the values of m and b in $y = mx + b$ affect the inclination and position of the graph of the equation and to discover relationships between the slopes of parallel lines and of perpendicular lines. "Families of Lines" provides a modification of the program in the preceding activity for pupils to use to graph pairs of linear equations and linear combinations of these pairs and in the process discover an algebraic method for solving a system of linear equations. The last worksheet requires students to complete, run, and modify a BASIC program for solving a system of equations. In "Relating Graphs to Their Equations with a Microcomputer" students use a supplied BASIC graphing program to investigate how the values of a, b, and c in the equation $y = a|x + b| + c$ affect the shape and position of its graph.

This section ends with two activities that exemplify how the use of concrete models and computers can be effectively blended to provide opportunities for students to confront, in a meaningful way, interesting and significant applied mathematical problems without a knowledge of advanced mathematics (e.g., calculus or probability theory). In the activity "Visualization, Estimation, Computation" pupils construct a model of a cone, visualize how the radius, height, area, and volume change as they alter its shape, estimate the position for maximum volume, and then analyze a BASIC program and its output to check their estimates. Students are then encouraged to modify the program to obtain a sharper estimate of the radius for maximum volume. "Problem Solving with Computers" presents two problem situations: maximizing the volume of an open-topped box to be formed by cutting equal-sized squares from the corners of a rectangular sheet and estimating the number of bags of a snack food that would need to be purchased in order to obtain a set of six different stickers, one packaged in each bag at random. Students first manipulate a physical model that represents the situation and then use computer-generated data to arrive at a solution.

INVESTIGATING SHAPES, FORMULAS, AND PROPERTIES WITH LOGO

May 1985

By DANIEL S. YATES, Mathematics and Science Center, Richmond, VA 23223

Teacher's Guide

Introduction: Recommendation 3 of *An Agenda for Action* encourages school mathematics programs to take full advantage of the power of computers at all grade levels and then adds (NCTM 1980, 8–9):

- Computers should be used in imaginative ways for exploring, discovering, and developing mathematical concepts and not merely for drill and practice.
- Curriculum materials that integrate the use of the ... computer in diverse and imaginative ways should be developed and made available.

The following activities are offered in support of these curricular recommendations. The focus of each activity is on the use of the computer as an instructional tool rather than as an object of instruction.

Grade levels: 7–10

Materials: Copies of the three worksheets for each student; Apple IIe or Apple II+ computers with 64K of memory (preferably no more than two or three students at each computer); Apple Logo disk (Krell or Terrapin Logo can be used with the changes noted near the end of this guide; the activities can be adapted for other versions of Logo to run on other machines, such as the IBM PCjr and Commodore 64); one initialized Logo file disk for each computer

Objectives: (1) To provide informal experiences and opportunities for discovery with squares, rectangles, parallelograms, and triangles; (2) to demonstrate the concepts of similarity and congruence of simple plane figures; (3) to explore relationships between areas and perimeters; (4) to develop formulas for finding the area of a parallelogram and the area of a triangle; (5) to demonstrate that the line segment joining the midpoints of two sides of a triangle is parallel to the third side and equal to one-half its length; and (6) to provide some initial experiences in using, but not programming in, the Logo language

Prerequisites: Students need not have any experience with Logo, but a familiarity with the Apple keyboard would be helpful. They should know that the area of a rectangle equals its base × height, and they should know the meaning of congruent and similar figures. Exercise 6d requires that the student find the dimensions of a square whose area is 200 square units; this problem will be a challenge for many students (unless they have a calculator!). They should also know that in a parallelogram, opposite sides are equal in length and are parallel, and they should know the meaning of "superimpose."

Directions: Before the class session, load the Logo language and then type in the procedures listed at the end of this teacher's guide. Simply type each line as it appears in the listing and press RETURN. After each END statement, hold down the CTRL key while typing C. After a short delay, the computer will state that the procedure has been defined. Then type the next procedure. When all procedures have been entered, make sure that one of the initialized Logo file disks is in the disk drive and type SAVE "GEOMETRY Ⓡ. Throughout this activity the symbol Ⓡ means to press the RETURN key. Be certain that the procedures are typed exactly as they appear, including spaces. Spaces are particularly important in Logo.

Work through the activities prior to the class meeting to test that the procedures work properly. If a procedure has a typographical error and will not execute properly and you are not familiar with the editing commands in Logo, then type ER "(name of procedure), for example, ER "PAR Ⓡ. Then retype the procedure correctly.

When all the procedures are checked for correctness, insert each of the file disks in turn and type SAVE "GEOMETRY Ⓡ. Then type CATALOG to verify that the file is on the disk; it should be listed as GEOMETRY.LOGO.

Special instructions for students: (1) Diagonal lines may appear "jagged" because of the way graphics images are produced on the video screen. Imagine that they are perfectly straight. (2) If you type an incorrect character by mistake, you can erase it by pressing the left arrow key (←). Make sure that each line is typed correctly before pressing RETURN. If you get an error message, ignore it and retype the line correctly. (3) The symbols [and] have separate keys on the Apple IIe and Franklin ACE. On the Apple II+, [is made by pressing the SHIFT and N keys simultaneously and] is made by using the SHIFT and M keys; (4) CAPS LOC must be down on the Apple IIe and ON on the Franklin ACE for Logo to function properly.

Sheet 1: Before or at the beginning of class, boot the Logo language disk. Then insert the Logo file disk and type LOAD "GEOMETRY Ⓡ. This step loads the needed procedures into the computer's "work space." Do this step for each computer.

The student begins by typing in a procedure to draw a variable-sized square. The exercises provide experiences with areas and perimeters of squares, as well as some initial exposure to the Logo language.

Sheet 2: Starting with the formula for the area of a rectangle, the student is led to conclude that the area of a parallelogram is also its base times its height. In each of many examples, the computer draws a random parallelogram and then superimposes a rectangle with the same base. (For a related treatment see Donald Schultz's article "Using Sweeps to Find Areas" in the May 1985 *MT*.) By analyzing the displays produced in multiple examples, the student should discover that the areas of the two figures are the same. To work correctly, the REC procedure must be executed after PAR is executed.

The student must correctly answer exercise 1d on sheet 2 to deduce correctly the formula for the area of a triangle in exercise 2e.

Sheet 3: These activities lead to the discovery that the line segment joining the midpoints of two sides of a triangle is parallel to and equal to one-half the length of the third side.

The computer draws a random triangle (TRI) and then connects the midpoints (TRI2). This display reveals a variety of congruent triangles, similar triangles, and parallelograms. Using many examples and the properties of a parallelogram, the student should be able to discover the intended result.

Changes for Krell or Terrapin Logo:

1. When entering the procedures, make these changes: Change SETPOS [−100−50] to SETXY (−100) (−50) in the procedures SETUP and PAR2.

   Change SETPC to PC in the procedures

REC, PAR2, and TRI2. ERNS can be omitted from the procedures REC and TRI2.

Change the second line of PAR2 to read: MAKE "Z ATAN (140 – (80 * SIN :X)) (80 * COS :X)

2. Change LOAD "GEOMETRY to READ "GEOMETRY in the teacher's guide.
3. Throughout the teacher's guide and the student activity sheets, change the left arrow key (←) to the ESC key to erase the previously typed character and change CS to DRAW to clear the screen.

Procedure listings:

```
TO SETUP
HT PU SETPOS [-100 -50] PD
END
TO PAR
SETUP
MAKE "X 10 + RANDOM 70
RT :X
REPEAT 2 [FD 80 RT 90 - :X FD 140
        RT 90 + :X]
LT :X
END
TO REC
SETPC 4
REPEAT 2 [FD 80 * COS :X RT 90 FD 140
        RT 90]
SETPC 1
ERNS
END
TO PAR2
PU RT :X FD 80
MAKE "Z ARCTAN ((140 - (80 * SIN :X)) /
        (80 * (COS :X)))
RT (90 - :X) + (90 - :Z)
SETPC 2 PD
FD (80 * (SIN (90 - :X))) / SIN (90 - :Z)
PU SETPOS [-100 -50] SETH 0 PD
SETPC 1
END
TO TRI
SETUP
MAKE "A RANDOM 61
MAKE "B 110 + RANDOM 40
MAKE "AB 100 + RANDOM 50
RT :A FD :AB RT :B
FD (:AB * SIN (90 - :A)) / COS (180 -
        (:A + :B))
RT (360 - (:A + :B)) SETX -100
END
TO TRI2
MAKE "BC (:AB * SIN (90 - :A)) / COS
        (180 - (:A + :B))
RT :A FD (.5 * :AB) RT (90 - :A)
SETPC 5
FD ((.5 * :AB) * COS (90 - :A)) +
        ((.5 * :BC * SIN (180 - (:A + :B)))
RT (90 + :A) FD (.5 * :AB)
RT :B FD (.5 * :BC)
RT (180 - :B) PU BK (.5 * :AB) LT :A PD
SETPC 1
ERNS
END
```

Answers:

Sheet 1: 2. b. a small square; d. a larger square with an angle in common with the first square. 3. a. 80; 400.

4. b and c.

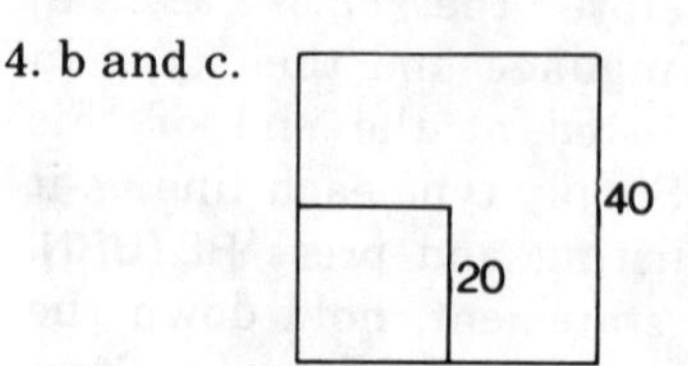

d. 160

5. SQUARE 100 Ⓡ. 6. a. SQUARE 10 Ⓡ; b. SQUARE 40 Ⓡ; c. SQUARE 10 Ⓡ; d. SQUARE 14.14 Ⓡ

Sheet 2: 1. c. equal; d. area = base × height. 2. b. two; d. height and base of triangle were equal to that of the corresponding parallelogram; area of triangle was one-half that of the corresponding parallelogram; e. area $= \frac{1}{2} \times$ base × height.

Sheet 3:

2.

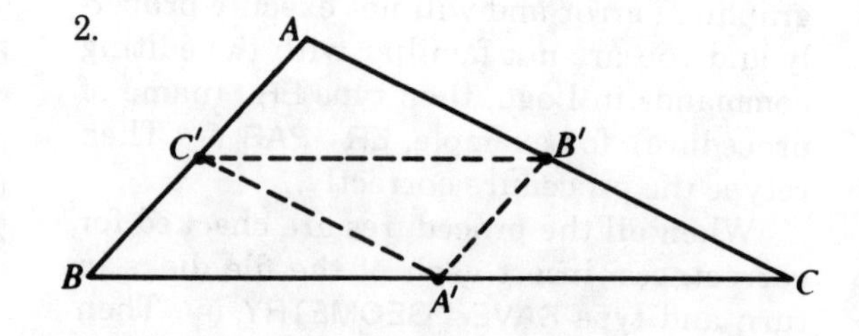

3. a. five; b. $\triangle AC'B'$ and $\triangle C'BA'$, $\triangle AC'B'$ and $\triangle B'A'C$, $\triangle AC'B'$ and $\triangle A'B'C'$, $\triangle C'BA'$ and $\triangle B'A'C$, $\triangle C'BA'$ and $\triangle A'B'C'$, $\triangle B'A'C$ and $\triangle A'B'C'$; c. $\triangle ABC$ and $\triangle AC'B'$; other possibilities include $\triangle ABC$ and $\triangle C'BA'$, $\triangle ABC$ and $\triangle B'A'C$, $\triangle ABC$ and $\triangle A'B'C'$. d. three; $BC'B'A'$, $A'C'B'C$, $B'A'C'A$. 4. a. $B'C' = \frac{1}{2}(BC)$; yes; b. $\overline{B'C'} \parallel \overline{BC}$; yes. 5. For all triangles, answers will vary from informal arguments based on the analysis of a large number of examples to a standard proof as found in a high school geometry text.

REFERENCE

National Council of Teachers of Mathematics. *An Agenda for Action: Recommendations for School Mathematics of the 1980s.* Reston, Va.: The Council, 1980.

PERIMETER AND AREA OF SQUARES — SHEET 1

In the following activities, Ⓡ means to press the RETURN key.

1. Enter the following Logo procedure. Leave spaces *exactly* as shown.

```
TO SQUARE :X Ⓡ
REPEAT 4 [FORWARD :X RIGHT 90] Ⓡ
END Ⓡ
```

2. a. Execute the procedure by typing SQUARE 20 Ⓡ.
   b. Describe the resulting display. ________________
   c. Execute the procedure again, this time typing SQUARE 50 Ⓡ.
   d. Describe the result. ________________

   These two squares are said to be *similar*. That is, they have the same shape, but different size.

3. Type CS Ⓡ to clear the screen. Now type SQUARE 20 Ⓡ. Each side of the square is 20 units long. Therefore,
   a. the perimeter (or distance around) is ______ units, and
   b. the area is ______ square units.

4. a. Without clearing the screen, draw a second square by typing SQUARE 40 Ⓡ.
   b. In the space provided at the right, draw a sketch of both squares as they appear on the screen.
   c. Label the lengths on your sketch.
   d. The perimeter of SQUARE 40 is ______ units. Thus, the perimeter of the larger square is twice the perimeter of the smaller one.

5. What would you type to draw a square whose perimeter is five times the perimeter of SQUARE 20? ________________

   Check the reasonableness of your answer by typing CS Ⓡ SQUARE 20 Ⓡ SQUARE ______ Ⓡ, where the blank contains your suggested input number.

6. What would you type to draw a square satisfying the given condition? Check your answers as in exercise 5.
   a. Perimeter one-half that of SQUARE 20 ________________
   b. Area four times that of SQUARE 20 ________________
   c. Area one-fourth that of SQUARE 20 ________________
   d. Area one-half that of SQUARE 20 ________________

To make sure that the Logo file GEOMETRY has been loaded into the computer, hold down the CTRL key while you press the T key. Now type POTS Ⓡ. You should see several procedure names listed, including TO PAR and TO REC.

1. a. Type CS Ⓡ and then PAR Ⓡ. Next type REC Ⓡ. Compare the height, base, and area of the parallelogram with the height, base, and area of the superimposed rectangle.

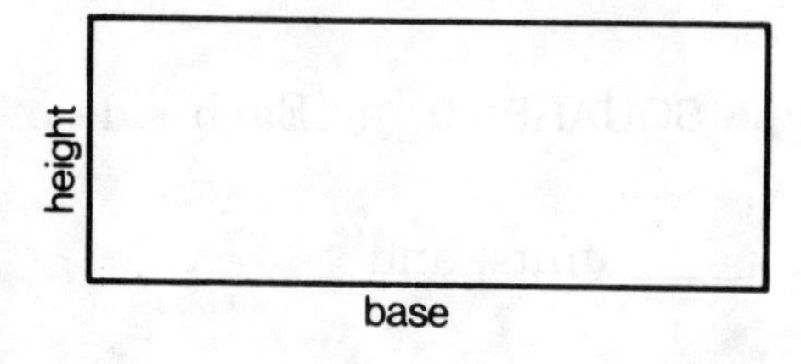

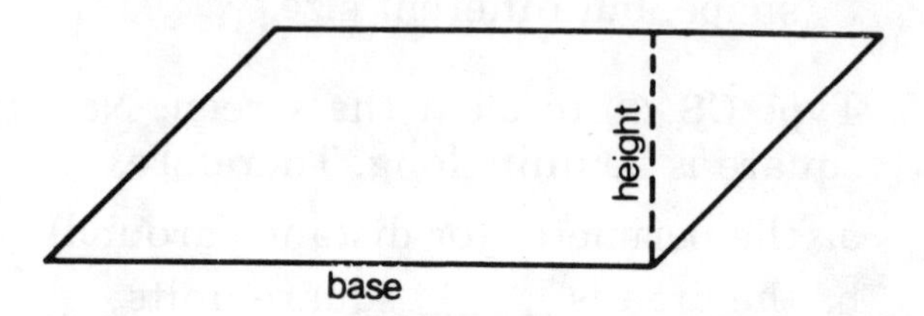

   b. Repeat step 1a several times.
   c. For each result, how did the height, base, and area of the parallelogram appear to be related to the height, base, and area of the superimposed rectangle? ____________

   Recall that the area of a rectangle is given by area = base × height.
   d. Write a formula for the area of a parallelogram in terms of its base and height. ____________

2. a. Clear the screen by typing CS Ⓡ. Now type PAR Ⓡ to draw another parallelogram.
   b. The procedure PAR2 will draw a diagonal of the parallelogram. Type PAR2 Ⓡ to see this result. The diagonal divides the parallelogram into ______ congruent triangles, that is, triangles of the same size and shape.

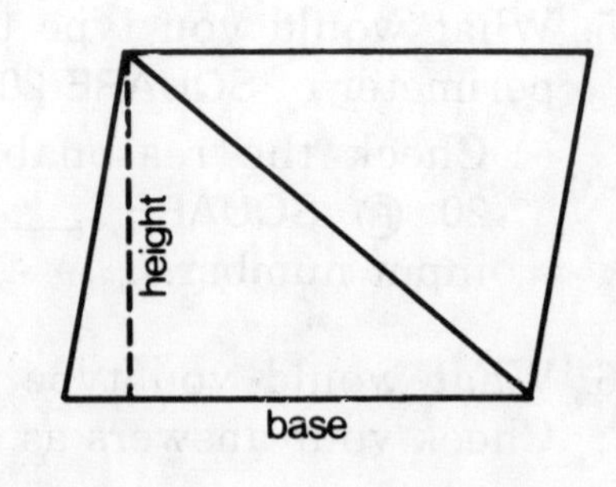

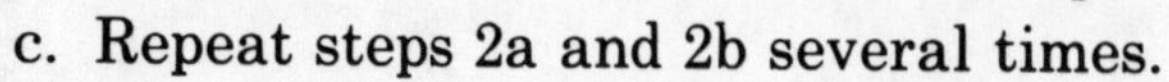

   c. Repeat steps 2a and 2b several times.
   d. For each figure, how did the height, base, and area of the lower triangle compare to the height, base, and area of the parallelogram? ____________
   e. Use your formula for the area of a parallelogram (exercise 1d) to write a formula for the area of a triangle in terms of its base and height.

   ____

Hold down the CTRL key while you press the T key. Then type POTS Ⓡ. The procedure names TO TRI and TO TRI2 should be listed. They will be used in the following activities.

1. a. Type CS Ⓡ and then TRI Ⓡ to see a random triangle. Now type TRI2 Ⓡ to see what happens when the midpoints of the sides are connected.
   b. Repeat step 1a several times. Look for properties or relationships that appear to be true for all the triangles displayed.

2. In triangle ABC at the right, A', B', and C' are the midpoints of sides $\overline{BC}$, $\overline{AC}$, and $\overline{AB}$, respectively. Complete the figure as would be done by the procedure TRI2.

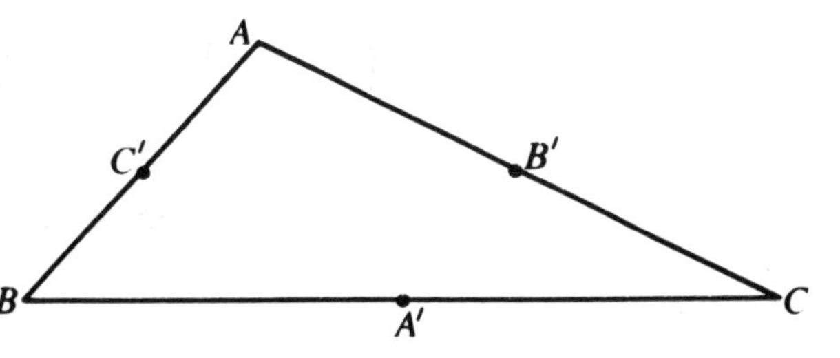

3. Refer to your sketch for exercise 2.
   a. How many triangles are in your figure? ______
   b. Name the pairs of congruent triangles. ______________________________
   ____
   c. Triangles ABC and $B'A'C$ are similar but not congruent. Name another pair of such triangles. ______________________________
   d. How many parallelograms do you see in your figure? ______________
   Name them: ______________________________
   e. Type CS Ⓡ TRI Ⓡ TRI2 Ⓡ. Verify your answers to 3a and 3d for this triangle.

4. Refer again to your sketch for exercise 2.
   a. What relationship appears to exist between the lengths of segments $\overline{BC}$ and $\overline{B'C'}$? ______________________________
   Does the same relationship appear to hold for the pairs of segments $\overline{AB}$ and $\overline{A'B'}$ and $\overline{AC}$ and $\overline{A'C'}$? ______________________________
   b. Identify another relationship between segments $\overline{BC}$ and $\overline{B'C'}$.
   ____
   Does this relationship also seem to hold for the pairs of segments $\overline{AB}$ and $\overline{A'B'}$ and $\overline{AC}$ and $\overline{A'C'}$? ______________________________
   c. Gather additional support for your answers to 4a and 4b by typing CS Ⓡ TRI Ⓡ TRI2 Ⓡ and studying the display.

5. Do you think the relationships discovered in exercise 4 are true for only certain triangles, or for all triangles? ______________________________
   Why do you think so? ______________________________

MICROCOMPUTER UNIT: GRAPHING STRAIGHT LINES

March 1983

By Ellen H. Hastings, Minnehaha Academy, Minneapolis, MN 55409
Daniel S. Yates, Mathematics and Science Center, Richmond, VA 23223

Teacher's Guide

Grade level: 8–10

Completion time: One fifty-minute period on the computer

Objectives:

1. To investigate how the value of m in the equation $y = mx + b$ affects the inclination of the graph of the equation and, specifically, to arrive at these generalizations:
   a) The larger the value of m, the steeper the slope becomes.
   b) With a positive value of m, the graph is inclined up and to the right.
   c) With a negative value of m, the graph is inclined up and to the left.
2. To investigate the role of b in the equation $y = mx + b$—
   a) b is the y-intercept of the line $y = mx + b$, and
   b) the graph of $y = b$ is a horizontal line.
3. To discover the characterizing property of parallel lines and of perpendicular lines—
   a) parallel lines have the same slope, and,
   b) for perpendicular lines, the product of their slopes is $^{-}1$.
4. To give students an opportunity to use the microcomputer as an aid to instruction.

Equipment: One or more Apple II Plus microcomputers with disk drive and monitor (preferable) or TV.

Prerequisite experience: Students should have some experience with—

1. graphing ordered pairs (x, y) on a coordinate system;
2. the linear equation; for example, they should know that $y = mx + b$ is the equation for a straight line, and they should know what it means to "plot a point;" and
3. substituting values of x to find values of y in a linear equation.

Directions: How you implement this activity will depend on the number of Apple systems you have available. If you have only one Apple, you may want to have the students work through the activity in pairs or in small groups on successive days. A larger group can participate if the screen is large enough for easy viewing. The ideal situation is to have multiple computers so that small groups can work simultaneously. It is important to give as many students as possible the experience of typing on the keyboard. The activity is sequential; so each sheet should be completed before going on to the next one.

The program that appears in table 1 in this teacher's guide should be entered into the computer and then saved on a disk (type SAVE SLOPE) ahead of time. This program, SLOPE, is used for all exercises

TABLE 1

```
10  TEXT : HOME
20  VTAB 12
30  PRINT "THIS PROGRAM GRAPHS STRAIGHT LINES"
40  PRINT : PRINT "IN THE GENERAL FORM
      Y = M*X + B."
50  FOR N = 1 TO 5000: NEXT : REM     PAUSE LOOP
60  HGR
65  HCOLOR = 3
70  PRINT : POKE 37,20: PRINT
80  PRINT : PRINT "GIVEN THE GENERAL FORM
      Y = M*X + B"
90  PRINT : PRINT "INPUT M,B     ";
100 INPUT M,B
110 PRINT : PRINT : PRINT
120 REM   DRAW AXES
130 HPLOT 140,0 TO 140,159
150 HPLOT 0,80 TO 279,80
170 FOR H = 0 TO 270 STEP 10
180 HPLOT H,77 TO H,83
190 NEXT H
210 FOR K = 0 TO 160 STEP 10
220 HPLOT 137,K to 143,K
230 NEXT K
240 HPLOT 271,65 TO 277,71
250 HPLOT 277,65 TO 271,71
260 HPLOT 147,0 to 151,6
270 HPLOT 155,0 TO 147,12
280 REM     PLOT THE STRAIGHT LINE
290 FOR X = - 14 TO 13.9 STEP .1
300 LET Y = M * X + B
310 IF Y > 8 THEN 340
320 IF Y < - 7.9 THEN 340
330 HPLOT 140 + 10 * X,80 - 10 * Y
340 NEXT X
345 REM     PRINT LINEAR EQUATION
350 PRINT : POKE 37,20: PRINT
360 PRINT "THIS IS THE GRAPH OF Y = "M" * X + "B
370 PRINT
380 PRINT "1=ANOTHER GRAPH  0=QUIT";
390 INPUT Z
400 PRINT : PRINT
410 IF Z = 1 THEN 60
420 TEXT : HOME
```

on sheets 1 and 2. On sheet 3, the students type a one-line modification to the program and then use this modified program for exercises 7 and 8. Students should be able to complete all three sheets in turn without further teacher supervision.

Before beginning the activity, students should be familiar with these computer conventions:

1. The symbol * is used to represent multiplication.
2. Ø is the numeral for zero and O is the letter.
3. The negative sign is found on the same key with the "=" sign.
4. Fractions will have to be entered as decimals.
5. The RETURN key must be pressed after each response.

To begin sheet 1, some students might need the following directions: Insert the diskette in the disk drive, close the door, and turn on the monitor and the Apple. When the red light goes out and you see the language symbol] and the flashing cursor, type RUN SLOPE and press the RETURN key. You will be asked to supply values for m and b, in this order. Type the value for m, then a comma, and then the value for b. Fractions will have to be typed as decimals (e.g., enter 1/4 as .25). Note that Ø is the number zero, whereas O is the letter. Note also that the computer uses the symbol * to represent multiplication.

Since the purpose of this activity is to learn about the slope of a line and its general orientation, the student is expected to sketch the lines freehand on the axes provided rather than trying to produce precision drawings.

You should be advised that some TV screens may slightly distort the spacing on the axes so that perpendicular lines do not appear to be exactly perpendicular. Monitors seem to give much better results. If your screen does not show perpendicularity well, and there is no other alternative, you may want to omit exercise 8 on sheet 3.

Students who are knowledgeable in programming in Applesoft may want to go through the program SLOPE and see what each line does. An extension of the activity for these students would be to modify the program to graph the special case $x = a$. A related programming challenge might be to write a program that would determine if two lines intersect and, if so, to announce the coordinates of the point of intersection.

Selected answers:

1. (d) It becomes steeper or more vertical in a counterclockwise direction.
2. (d) It becomes more vertical in a clockwise direction.
4. (d) y-intercept
6. (d) horizontal (or parallel to the x-axis)
7. (d) parallel
   (e) . . . parallel [if] they have the same slope.
8. (b) $y = -4x - 3$
   (c) . . . product of their slopes is $^{-}1$.
   (d) (The relationship always holds.)

1. In the equation $y = m * x + b$, let $b = \emptyset$ so that $y = m * x$. When the computer prompts you, enter the values for m and b and press RETURN. When the computer graphs the line, sketch the graph and record the equation. Do the same for (b) and (c).

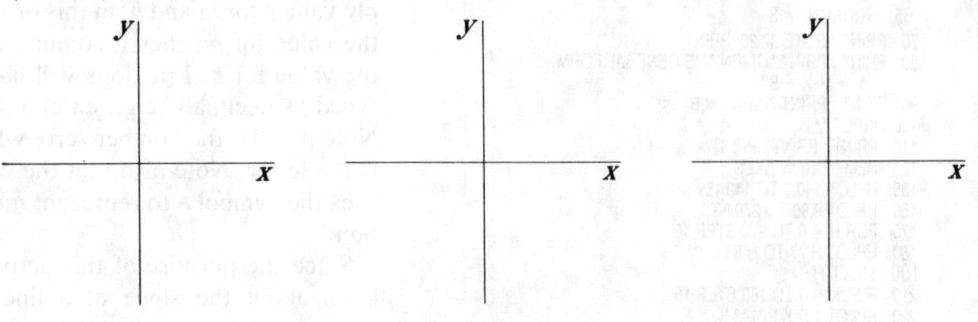

(a) $m = \frac{1}{4}$ $b = \emptyset$ **(b)** $m = 1$ $b = \emptyset$ **(c)** $m = 2$ $b = \emptyset$

equation ________ equation ________ equation ________

(d) As m increases, what happens to the graph? ________

2. Again let $b = \emptyset$ $(y = m * x)$ and substitute for m each of the following numbers in turn. Sketch the graph, as shown on the monitor, and record the equation.

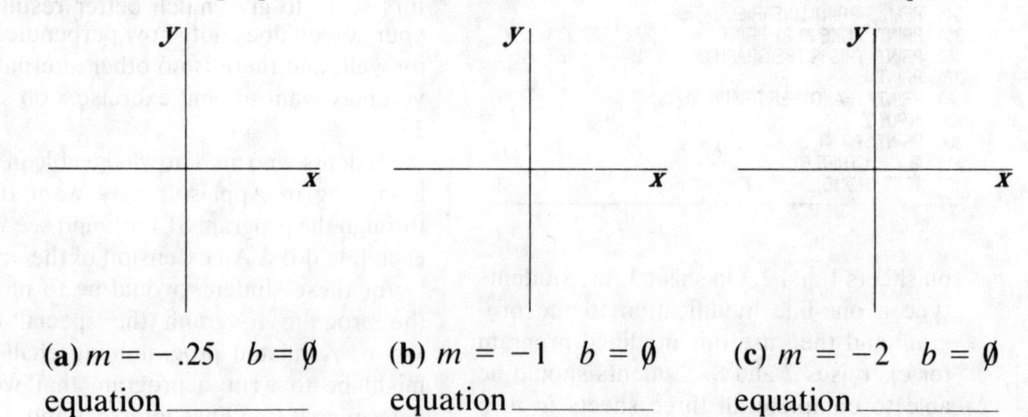

(a) $m = -.25$ $b = \emptyset$ **(b)** $m = -1$ $b = \emptyset$ **(c)** $m = -2$ $b = \emptyset$

equation ________ equation ________ equation ________

(d) As m decreases (i.e., gets larger negatively), what happens to the graph?

____

3. Without using the computer, sketch what you think each of the following equations looks like. Will the graph of the equation fall or rise from left to right?

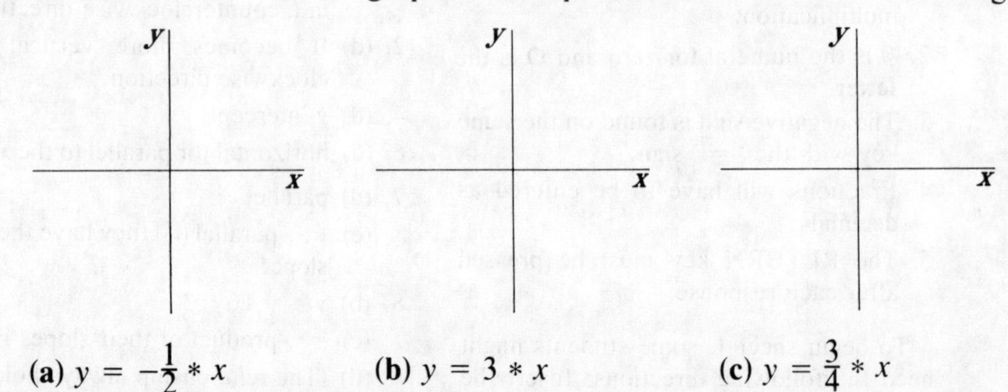

(a) $y = -\frac{1}{2} * x$ **(b)** $y = 3 * x$ **(c)** $y = \frac{3}{4} * x$

(d) Now have the Apple plot each equation in turn to see if you were right. Correct your sketches if necessary.

4. Have the computer plot each of the three lines below. For each line, sketch the graph and record the *y-intercept* (the y value where the line crosses the y-axis).

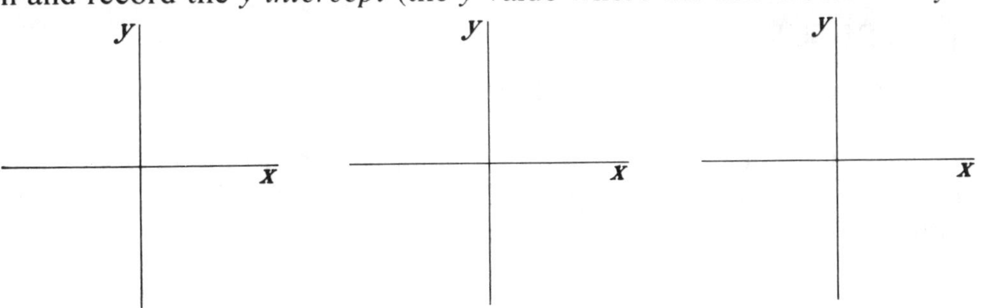

(a) $y = -x + 3$ **(b)** $y = 2x - 2$ **(c)** $y = \frac{1}{2}x + 1$

y-intercept ______ y-intercept ______ y-intercept ______

(d) In the equation $y = m * x + b$, b is the ____________________.

5. In the equation $y = mx + b$, the coefficient of x (i.e., m) is called the *slope* of the line. Study the equations below and, *without using the computer*, determine the slope and the y-intercept; then sketch the line.

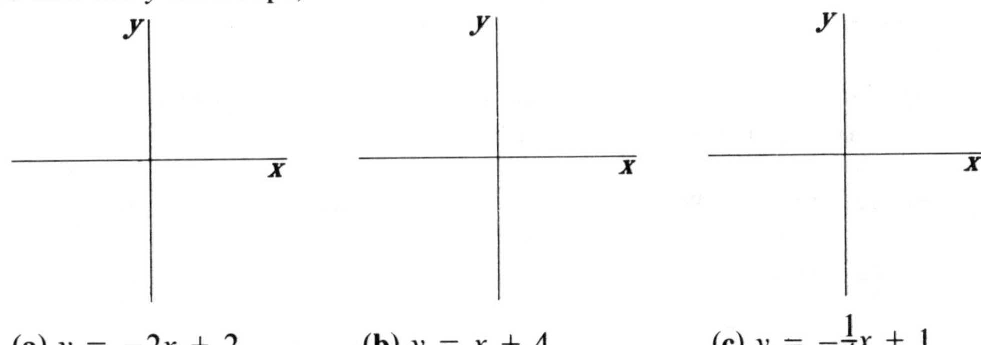

(a) $y = -2x + 2$ **(b)** $y = x + 4$ **(c)** $y = -\frac{1}{4}x + 1$

(d) Now have the computer plot each equation to see if you were right. Correct your sketches if necessary.

6. In the equation $y = mx + b$, let $m = \emptyset$ so that $y = b$. Run the program to have the computer graph each equation below. Sketch the graphs.

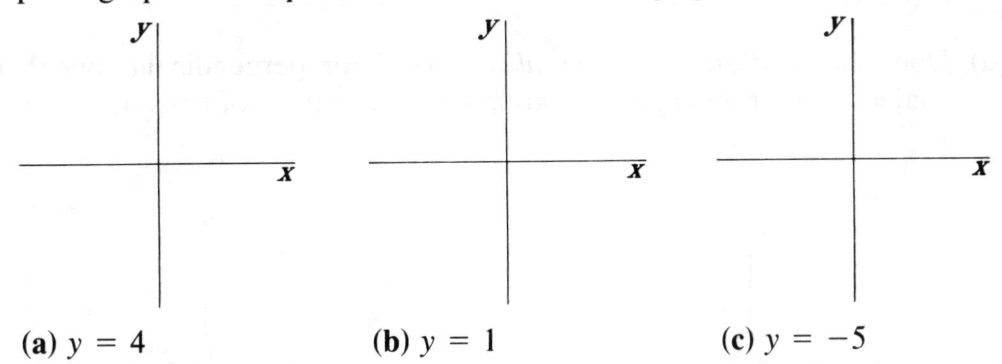

(a) $y = 4$ **(b)** $y = 1$ **(c)** $y = -5$

(d) Generalize: When the slope is $\emptyset$, how are the lines positioned?______

____

QUIT the program SLOPE and type this new program line: 39Ø IF Z = 1 THEN 8Ø. Then press RETURN.

This modified program will now plot several lines on the same axes.

7. Type RUN and then have the computer plot each pair of lines on the same axes. After each pair, QUIT the program and RUN again for the next pair. Sketch the results below.

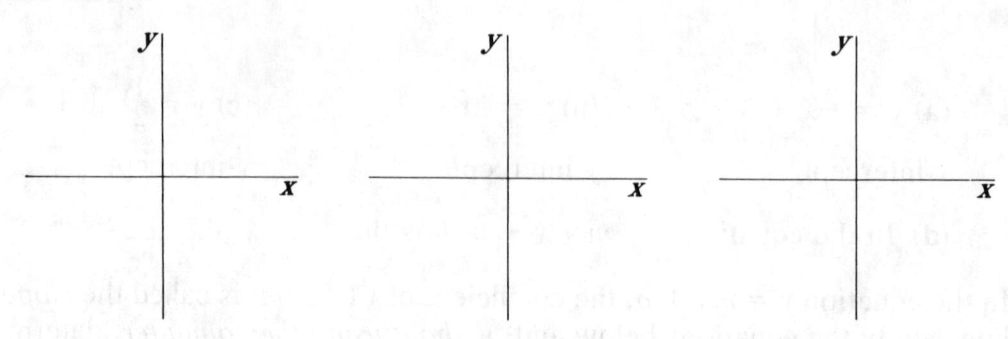

(a) $y = x + 3$
$y = x + 5$

(b) $y = -.5x + 2$
$y = -.5x + 4$

(c) $y = 1.6x - 3$
$y = 1.6x + 2$

(d) The two lines in each pair are (perpendicular) (parallel) (not related). [Cross out wrong answers.]

(e) Generalize: Two lines are ____________________ if ____________

____.

8. **(a)** RUN this same modified program to plot the line $y = \frac{1}{4}x - 1$.

(b) Then experiment with different values of m in the equation $y = mx - 3$ to find the line through (0, $^-3$) that is perpendicular to the first line.

(c) The lines $y = \frac{1}{4}x - 1$ and $y = \square x - 3$ are perpendicular. The product of their slopes is ______________________________.

(d) Does the relationship in **(c)** *always* hold for perpendicular lines? Test the conjecture by running the program again with pairs of lines of your choice.

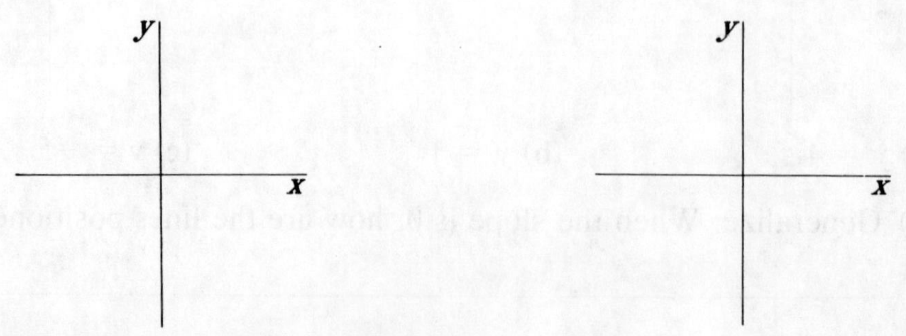

FAMILIES OF LINES

November 1983

By **Christian R. Hirsch,** *Western Michigan University, Kalamazoo, MI 49008*

Teacher's Guide

Grade Level: 8–11

Materials: A set of activity sheets for each student. Access to a microcomputer is necessary for part of the last worksheet.

Objectives: Students will (1) graph pairs of linear equations, linear combinations of these pairs, and use the results to discover an algebraic technique for solving a system of equations; and (2) complete, run, and modify a BASIC program for solving a system of linear equations.

Prerequisites: Students should have some facility in graphing linear equations of the form $ax + by = c$ as well as of the special forms for horizontal and vertical lines. The third worksheet assumes that students are familiar with the essential elements of computer programming in the BASIC language—specifically, the form and use of INPUT, LET, IF-THEN, and PRINT statements.

Directions: This activity is designed for use in a unit on systems of linear equations. The first two worksheets should be used in conjunction with or immediately following a lesson on solving systems of equations graphically. These two sheets provide exploratory graphing exercises through which students can discover the addition-with-multiplication method for solving a system of equations together with the geometric reasons behind the method. The third worksheet can be used toward the end of the unit as a means to clarify, reinforce, and extend the algebraic concepts and methods developed in the unit.

Sheets 1 and 2: Encourage pupils to draw accurate graphs. Labeling the graphs with the corresponding equations is also helpful. Depending on the level and ability of the class, on completion of sheet 1 you might show that, in general, if (x_1, y_1) is a point of intersection of the graphs of the system

$$(1)\qquad \begin{aligned} ax + by &= c \\ dx + ey &= f, \end{aligned}$$

then $ax_1 + by_1 = c$, $dx_1 + ey_1 = f$, and thus, by the addition property of equality, $(ax_1 + by_1) + (dx_1 + ey_1) = c + f$ or $(a + d)x_1 + (b + e)y_1 = c + f$. Hence, the graph of the sum of the equations in (1) also contains the point (x_1, y_1). In a similar manner, on completion of sheet 2 you might demonstrate that if (x_1, y_1) belongs to the graphs of the equations in (1), then it also belongs to the graph of the sum of $m(ax + by = c)$

and $n(dx + ey = f)$, where m and n are any two nonzero integers.

Since the primary intent of sheets 1 and 2 is geometrically to motivate an algebraic method, students with access to a microcomputer could also complete these worksheets by using computer-generated graphs of the given and derived equations. The Apple II+ BASIC program given by Hastings and Yates (1983) can be used in this manner if the following modifications and additions are made:

```
40    PRINT : PRINT "IN THE FORM
        AX + BY = C."
65    HCOLOR = 3
80    PRINT : PRINT "GIVEN THE FORM
        AX + BY = C"
90    PRINT : PRINT "INPUT A,B,C";
100   INPUT A,B,C
285   IF B = 0 THEN 343
300   LET Y = -A / B * X + C / B
342   GOTO 345
343   HPLOT 140 + 10 * C / A,
        0 TO 140 + 10 * C / A,159
360   PRINT "THIS IS THE GRAPH OF
        "A"X + "B"Y = "C
410   IF Z = 1 THEN 80
```

Sheet 3: This worksheet should be introduced by considering the solution of the general system of equations in (1). Solve the system for x by multiplying both sides of the first equation by e; then multiplying both sides of the second equation by $-b$; and finally adding the resulting two equations together and simplifying. Relate the form of the solution to lines 190 and 210 of the program given. The completion of the program will require students to solve the general system for y, translate the algebraic expression into a BASIC expression, and then supply the missing keywords and statement numbers.

Answers: Sheet 1: 1.b. (0, 4); e. They all pass through the point (0, 4). 2.b. $(-2, 4)$; e. They all pass through the point $(-2, 4)$. 3.b. $(-3, -1)$; e. They all pass through the point $(-3, -1)$. 4. The graph of E_3 passes through the point of intersection of the graphs of E_1 and E_2.

Sheet 2: 5.d. The graph is the same as that of $2x + 3y = 6$. 6.e. Each line passes through $(3, -2)$, the point of intersection of the graphs of E_1 and E_2. f. $3x = 9$ (or $x = 3$), and $3y = -6$ (or $y = -2$). 7.e. $7x = 14$ (or $x = 2$), and $-7y = -28$ (or $y = 4$). 8. Each member of the family of lines $m \cdot E_1 + n \cdot E_2$ passes through the point (a, b).

Sheet 3: 9.a. $(-4.5, 5)$; b. $(5, -2)$; 10. line 200, 260; line 220, (A*F − C*D)/P; line 240, PRINT; line 310, INPUT; line 320, 110. 12. One possible modification is as follows:

```
260   LET Q = B * F - C * E
265   IF Q = 0 THEN 275
270   PRINT "PARALLEL LINES, NO
        SOLUTION."
272   GOTO 280
275   PRINT "SAME LINE, INFINITELY
        MANY SOLUTIONS."
```

BIBLIOGRAPHY

Hastings, Ellen H., and Daniel S. Yates. "Microcomputer Unit: Graphing Straight Lines." *Mathematics Teacher* 76 (March 1983):181–86.

Moser, James M. "A Geometric Approach to the Algebra of Solutions of Pairs of Equations." *School Science and Mathematics* 67 (March 1967):217–20.

1. a) Graph the following pair of equations:

$$x + 2y = 8$$
$$x + \quad y = 4$$

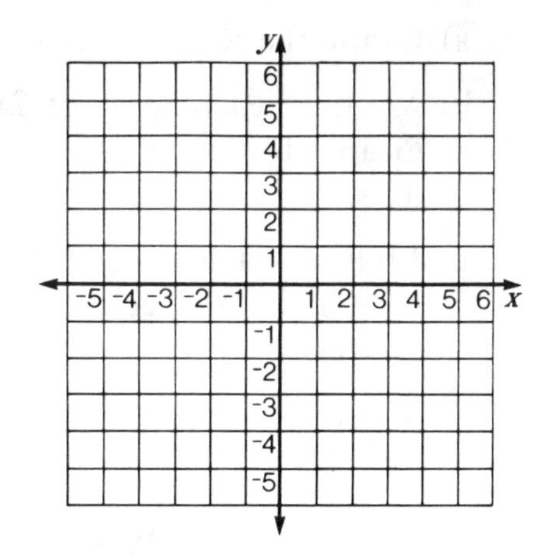

   b) What are the coordinates of the point of intersection? ____________

   c) Verify that these coordinates satisfy both equations.

   d) Add together the two equations above. Graph the new equation on the same set of axes.

   e) Do you notice anything special about these three lines? ____________

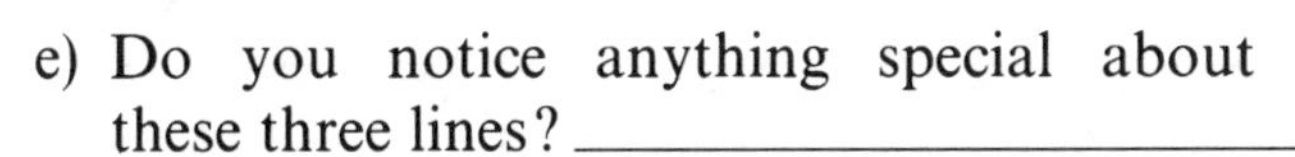

2. a) Graph the following pair of equations:

$$2x - y = -8$$
$$x + y = 2$$

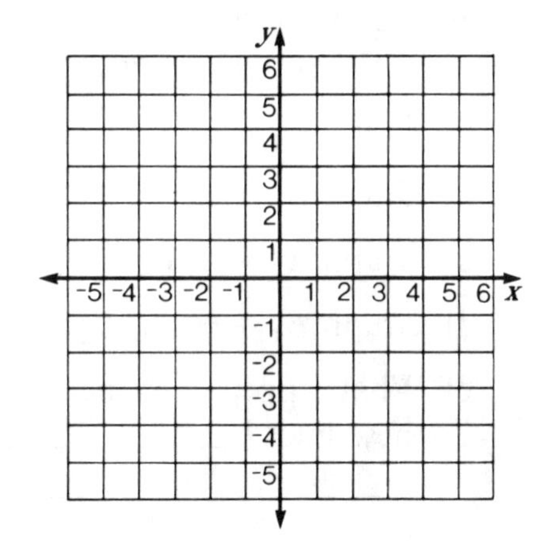

   b) What are the coordinates of the point of intersection? ____________

   c) Verify that these coordinates satisfy both equations.

   d) Add these two equations together and graph the resulting equation on the same set of axes.

   e) What appears to be true about the graphs of these three equations? ____________

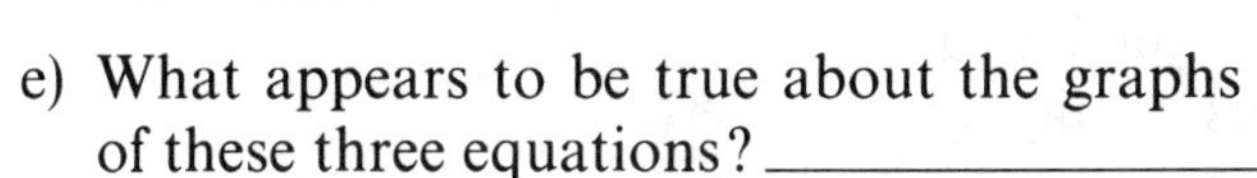

3. a) Graph the following pair of equations:

$$x - 4y = 1$$
$$-x - \quad y = 4$$

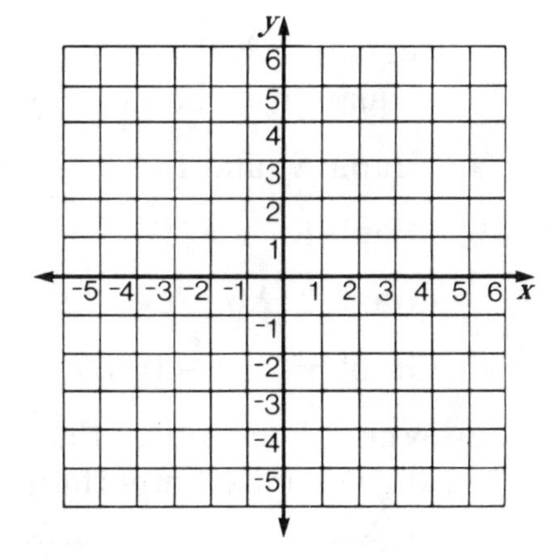

   b) What are the coordinates of the point of intersection? ____________

   c) Add these two equations together and graph the resulting equation on the same set of axes.

   d) What appears to be true about the graphs of these three equations? ____________

   ____

4. If E_1 and E_2 are equations of two intersecting lines and $E_3 = E_1 + E_2$, what do you think is true about the graphs of E_1, E_2, and E_3? ____________

____

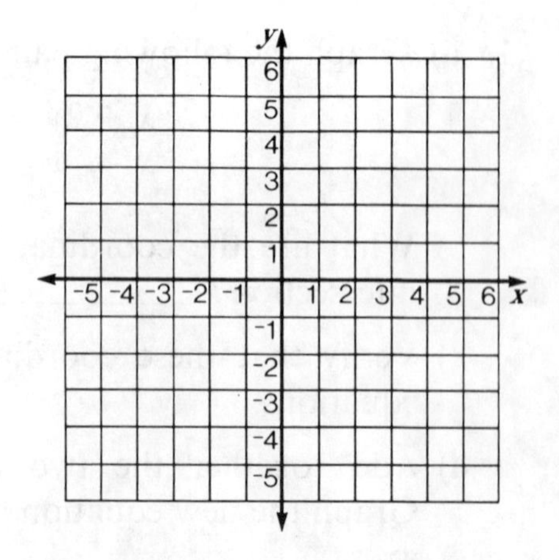

5. a) Graph the equation $2x + 3y = 6$.

   b) Multiply each term of $2x + 3y = 6$ by 5. Graph the new equation on the same set of axes.

   c) Multiply each term of $2x + 3y = 6$ by -4. Graph the new equation on this same set of axes.

   d) What do you think is true about the graph of

   $$k(2x + 3y = 6)$$

   for each nonzero integer k? ______________________________

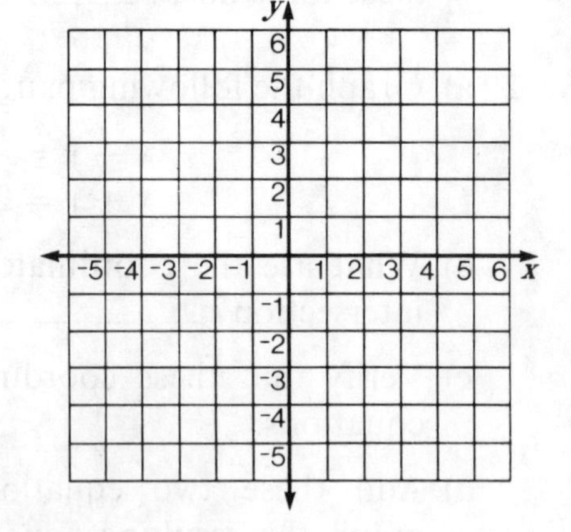

6. Let equation E_1 be $2x + y = 4$ and equation E_2 be $x - y = 5$. Use this set of axes for each part.

   a) Graph E_1 and E_2.

   b) Graph $E_1 + E_2$.

   c) Graph $E_1 + 2 \cdot E_2$.

   d) Graph $E_1 + (-2) \cdot E_2$.

   e) What appears to be true about each of these lines? ______________________________

   f) Which lines, other than the original two, are the most important of this family of lines? ______________________________

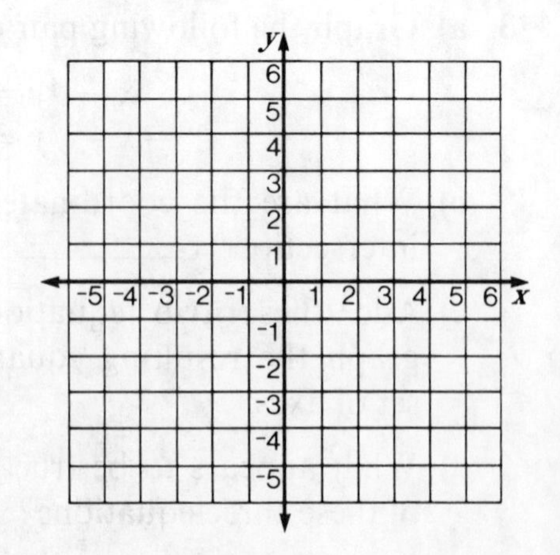

7. Let equation E_1 be $3x - y = 2$ and equation E_2 be $x + 2y = 10$.

   a) Graph E_1 and E_2.

   b) Graph $E_1 + E_2$.

   c) Graph $2 \cdot E_1 + E_2$.

   d) Graph $E_1 + (-3) \cdot E_2$.

   e) Which lines, other than the original two, are the most important of this family of lines? ______________________________

8. If E_1 and E_2 are equations of lines intersecting at a point (a, b), what do you think is true about the family of lines with equations $m \cdot E_1 + n \cdot E_2$, where m and n are any two nonzero integers? ______________________________

9. Without graphing, try to find the coordinates of the point of intersection of the pairs of lines with these equations:

a) $2x + 3y = 6$
   $2x + y = -4$

b) $x + 3y = -1$
   $2x - y = 12$

10. The BASIC program below is designed to solve a system of linear equations. Complete the program by supplying the missing keywords, algebraic expressions, and statement numbers.

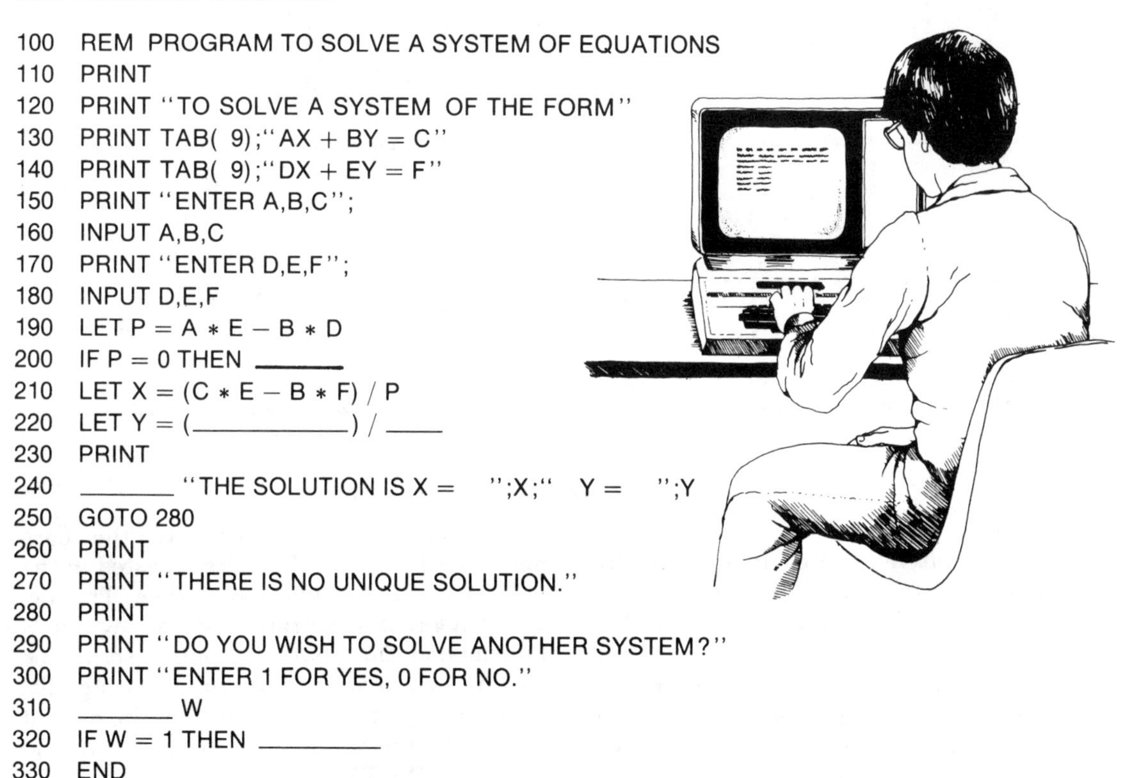

```
100  REM PROGRAM TO SOLVE A SYSTEM OF EQUATIONS
110  PRINT
120  PRINT "TO SOLVE A SYSTEM OF THE FORM"
130  PRINT TAB( 9);"AX + BY = C"
140  PRINT TAB( 9);"DX + EY = F"
150  PRINT "ENTER A,B,C";
160  INPUT A,B,C
170  PRINT "ENTER D,E,F";
180  INPUT D,E,F
190  LET P = A * E - B * D
200  IF P = 0 THEN ______
210  LET X = (C * E - B * F) / P
220  LET Y = (__________) / ____
230  PRINT
240  ______ "THE SOLUTION IS X =   ";X;"   Y =   ";Y
250  GOTO 280
260  PRINT
270  PRINT "THERE IS NO UNIQUE SOLUTION."
280  PRINT
290  PRINT "DO YOU WISH TO SOLVE ANOTHER SYSTEM?"
300  PRINT "ENTER 1 FOR YES, 0 FOR NO."
310  ______ W
320  IF W = 1 THEN ________
330  END
```

11. Copy and run the completed program for the following systems of equations:

a) $2x + 5y = 8$
   $3x - y = -5$

b) $2x + 3y = 6$
   $x - 5y = 1$

c) $4x + y = 8$
   $8x + 2y = 5$

d) $2x + 6y = 5$
   $5x - 2y = 6$

e) $3x - 6y = 6$
   $4x - 8y = 8$

12. a) Modify the program so that it checks whether the given equations are for the same line or parallel lines. For parallel lines, have the program print "**NO SOLUTION**."

For the same line, have the program print "**SAME LINE, INFINITELY MANY SOLUTIONS**."

b) Run the modified program for the systems of equations in part 11.

RELATING GRAPHS TO THEIR EQUATIONS WITH A MICROCOMPUTER

March 1986

By JOHN C. BURRILL, Whitnall High School, Greenfield, WI 53228
HENRY S. KEPNER, JR., University of Wisconsin—Milwaukee, Milwaukee, WI 53201

Teacher's Guide

Introduction: The study of absolute-value equations is often slighted in the algebra sequence because of its apparent difficulty due to students' confusion about rules. The activity presented here stresses a conceptual approach to absolute value, which is based on the visualization of the graphs of absolute-value equations. Students are encouraged to compare absolute-value equations that have similar constants by picturing their graphs. An important by-product is the experience students gain in predicting a graph on the basis of the choice of constants in the generalized equation $y = A|x + B| + C$. The relationships explored here are a prelude to all later work in the visualization of functions involving reflection, dilation, and translations. With well-chosen sets of constants, students can observe how the basic V-shaped graph closes up or flattens out, opens upward or downward, or changes the position of the vertex. The simple form of the graph of an absolute-value equation allows the student to make conjectures more easily than with more complex functions, such as the hyperbola or the tangent.

This topic provides an ideal setting for integrating the computer into the core mathematics curriculum as suggested in the NCTM's *An Agenda for Action.* The activity itself is illustrative of the *Agenda*'s recommendation that "computers should be used in imaginative ways for exploring, discovering, and developing mathematical concepts and not merely for ... drill and practice" (NCTM 1980, 9). Program 1 is designed to give students a tool with which they can produce instant graphs for absolute-value equations, thereby allowing them to make and confirm conjectures about the form and position of the graph of a given equation.

Grade levels: 9–11

Materials: One or more Apple IIe (preferably) or Apple II^+ microcomputers with disk drive and monitor; graph paper; and a set of worksheets for each student

Objectives: To allow students to investigate how the values of A, B, and C in the equation $y = A|x + B| + C$ affect the shape and position of its graphs and, in particular, to predict (a) the changes in slope of the rays when the value of A is varied, (b)

the shifts or translations of the graph in terms of its vertex when B and C are varied, (*c*) the shape and position of the graph for an equation in which reflection, dilation, and translation(s) are involved, and (*d*) an equation for a given graph

Directions: This activity can be used in any of three forms in both introductory and intermediate algebra courses. It can be used as a demonstration that is controlled by the teacher with a large screen monitor. The teacher's questions and the opportunity for students to sketch their conjectures before seeing the computer's graph are critical features of this presentation.

The activity can also be used during class when a computer laboratory is available. Students can work individually or in small groups. Periodically, the class can stop for a general discussion or summary of their observations and conjectures in the laboratory.

Finally, students can use the worksheets as several guided-discovery assignments to be completed at a computer between classes. In this format, it is important that students follow directions that call for their conjectures before verifying their work with computer-generated graphs.

The BASIC program (program 1) should be entered into the computer and saved on a disk ahead of time. Then make additional copies of the program as needed. If you enter this program on an Apple II^+, the absolute-value symbol in the REMARK and PRINT statements should be replaced by an exclamation point (!).

Before beginning this activity, students should graph the absolute-value equation $y = |x|$ using the standard approach with ordered pairs and graph paper. Class discussion should focus on the basic characteristics of the graph and include a comparison with linear equations and their corresponding graphs.

Distribute copies of the worksheets one at a time to each student. The computer program prints the values of A, B, and C in a standard form. Thus, negative constants may require some interpretation. For example, if $A = -2$, $B = -3$, and $C = -4$, the display will look like this:

$$Y = -2|X + -3| + -4$$

This form is not unlike the notation for signed numbers often given in mathematics texts for junior high school. Remind students that fractions must be entered as decimals.

The focus of this activity is not the students' mastery of graphing absolute-value equations. A more important goal is the development of students' understanding of the relationships between an equation and its graph. The ability to read an equation and form a visual image of its graph and the ability to write an equation that approximately describes a given graph are critical tools for serious students of mathematics. This skill is another form of "equivalence" in mathematics. Students must recognize several fractional and decimal forms of the same number; they should be able to recognize equivalent equations. It is also important that they have a visual image of standard equations. This activity is valuable for students' understanding of this equivalency.

Additional computer-enhanced graphing activities can be found in Hastings and Yates (1983) and in Hirsch (1983).

Answers:

Worksheet 1: 1. *a.* $y = 3|x + 0| + 0$, $y = 3|x|$; *b.* $y = -3|x + 0| + 0$, $y = -3|x|$; *c.* $y = 0.5|x + 0| + 0$, $y = 0.5|x|$. 2. *a.* $y = 2|x|$; *b.* $y = -6|x|$; *c.* $y = -\frac{1}{3}|x|$; *d.* $y = 5|x|$; *e.* $y = -2|x|$; *f.* $y = -4|x|$. 3. If A is positive, the graph will open upward and the slope of the right ray will always be A, whereas the slope of the left ray will be $-A$. If A is negative, the graph will open downward.

Worksheet 2: 1. *a.* $y = 1|x + 0| + 3$, $y = |x| + 3$; *b.* $y = 1|x + 0| + (-6)$, $y = |x| + (-6)$; *c.* $y = 1|x + 0| + 0.6$, $y = |x| + 0.6$. 2. *a.* $y = |x| + 2$; *b.* $y = |x| + (-4)$; *c.* $y = |x| + 4.5$; *d.* $y = -3|x| + 5$; *e.* $y = 2|x| + (-6)$; *f.* $y = 0.5|x| + (-2)$. 3. *a.* The value of C affects the vertical shift (translation) of the vertex. If C is positive, the vertex is C units above the origin. If C is negative, the vertex is $|C|$ units below the origin. *b.* (0, −7), (0, C). 4. The values of A and C each affect the graph in their own way. The value of A determines the slope of the two rays of the V shape. The value of C places the vertex of the graph $|C|$ units above or below the origin.

Worksheet 3: 1. *a.* $y = 1|x+4| + 0$, $y = |x+4|$; *b.* $y = 1|x+(-5)| + 0$, $y = |x+(-5)|$; *c.* $y = 1|x+(-6)| + 0$, $y = |x+(-6)|$. 2. *a.* $y = |x+(-2)|$; *b.* $y = |x+3|$; *c.* $y = |x+(-4)|$. 3. *a.* $y = |x+6|$; *b.* $y = |x+(-3)|$; *c.* $y = |x+(-7)|$. 4. The value of B affects the horizontal shift (translation) of the vertex. If B is positive, the vertex is located B units to the left of the origin; if B is negative, the vertex is $|B|$ units to the right of the origin. Thus, the vertex has coordinates $(-B, 0)$.

PROGRAM 1

```
10  REM    TRANSLATION AND DILATION OF
      ABSOLUTE VALUE GRAPHS
20  HOME
30  GOSUB 480
40  GOSUB 700
50  PRINT " A MORE GENERAL EQUATION IS
      Y=A|X+B|+C"
60  FOR J = 1 TO 3000: NEXT J
70  FOR J = 1 TO 1000: NEXT J
80  PRINT : PRINT : PRINT
90  PRINT " FOR THIS GRAPH A=1, B=0 AND
      C=0"
100 FOR J = 1 TO 3000: NEXT J
110 PRINT : PRINT : PRINT
120 PRINT " WOULD YOU LIKE TO CHANGE
      THE VALUE OF A, B OR C? (Y/N) ";
130 INPUT Y$
140 PRINT : PRINT : PRINT : PRINT
150 IF Y$ = " Y " THEN 180
160 IF Y$ = " N " THEN 470
170 GOTO 110
180 PRINT : PRINT : PRINT : PRINT : PRINT :
      PRINT : PRINT
190 INPUT " A MUST BE CHOSEN BETWEEN
      -0.1 AND -20 OR +0.1 AND +20 A= ";A
200 PRINT : PRINT : PRINT : PRINT
210 INPUT " B SHOULD BE CHOSEN
      BETWEEN -9 AND 9. B= ";B
220 PRINT : PRINT : PRINT : PRINT
230 INPUT " C MUST BE CHOSEN BETWEEN
      -7 AND +7. C= ";C
240 B1 = B * 12
250 PRINT : PRINT : PRINT : PRINT
260 PRINT " WE WILL NOW PLOT
      Y= "A" |X+ "B" |+ "C
270 FOR J = 1 TO 2000: NEXT J
280 PRINT : PRINT : PRINT : PRINT :
      HCOLOR= 3
290 FOR T = 0 TO 80 STEP .5
300 X = 120 - B1 + T:X1 = 120 - B1 - T
310 Y = 80 - 8 * C - (A * T)
320 IF (A * T) > 80 THEN 370
330 ONERR GOTO 370
340 IF Y < 0 GOTO 370
350 HPLOT X,Y: HPLOT X1,Y
360 NEXT T
370 PRINT : PRINT : PRINT : PRINT
380 PRINT " Y= "A" |X+ "B" |+ "C
390 FOR J = 1 TO 4000: NEXT J
400 PRINT : PRINT : PRINT : PRINT
410 PRINT " WOULD YOU LIKE TO CHANGE
      A, B OR C AGAIN? (Y/N) "
420 INPUT Y$
430 IF Y$ = " Y " THEN 450
440 IF Y$ = " N " THEN 470
450 GOSUB 480
460 GOSUB 700
470 END
480 HGR : HCOLOR= 3: PRINT : PRINT :
      PRINT : VTAB (21)
490 PRINT : PRINT : PRINT : PRINT
500 REM  PLOTS AXES
510 HPLOT 120,0 TO 120,160
520 HPLOT 0,80 TO 240,80
530 REM       INCREMENTS
540 FOR A = - 120 TO 120 STEP 12
550 FOR B = - 2 TO 2
560 X = 120 + A:Y = 80 + B
570 HPLOT X,Y
580 NEXT B: NEXT A
590 FOR A = 0 TO  - 80 STEP  - 8
600 FOR B = - 2 TO 2
610 X1 = 120 + B:Y1 = 80 + A
620 HPLOT X1,Y1
630 NEXT B: NEXT A
640 FOR A = 0 TO 80 STEP 8
650 FOR B = - 2 TO 2
660 X1 = 120 + B:Y1 = 80 + A
670 HPLOT X1,Y1
680 NEXT B: NEXT A
690 RETURN
700 REM       PLOTS      |X|
710 HCOLOR= 6
720 FOR T = 0 TO 120
730 X = 120 + T:X1 = 120 - T
740 Y = 80 - T
750 IF T > 80 THEN 790
760 HPLOT X,Y
770 HPLOT X1,Y
780 NEXT T
790 PRINT : PRINT
800 PRINT " THIS GRAPH REPRESENTS
      Y=|X| "
810 FOR J = 1 TO 3500: NEXT J: PRINT
820 IF Y$ = " Y " THEN 180
830 GOTO 50
```

Worksheet 4: 1. *a.* $y = 2|x + 2| + (-6)$; *b.* $y = 2|x + (-4)| + (-7)$; *c.* $y = -0.5|x + 2| + 2$; *d.* $y = -3|x + (-3)| + 5$; *e.* $y = 0.5|x + 1| + (-4)$; *f.* $y = 3|x + (-6)| + 3$. 2. *a.* vertex $(-6, -4)$, slope 2, opens upward; *b.* vertex (2, 0), slope -3, opens downward; *c.* vertex (6, 1), slope 0.5, opens upward; *d.* vertex (4, 5), slope -5, opens downward; *e.* vertex $(-1, 3)$, slope 0.25; opens upward; *f.* vertex $(-4, -2)$, slope -0.25, opens downward. 3. *a.* $y = |x + (-5)|$; *b.* $y = |x + 3|$; *c.* $y = |x + (-2)| + 2$; *d.* $y = -2|x + (-4)| + 4$; *e.* $y = -7|x + (-1)| + 7$; *f.* $y = 0.5|x + 4| + 1$.

Note: For individuals with access to the Minnesota Educational Computing Consortium's utility program *Small Characters,* we have an enhanced version of program 1 that calls *Small Characters* and places numerical values on the axes for easier use by students. We also have another version of program 1 that permits graphing equations of the form $y - D = A|B(x - (-C/B))|$ and thereby facilitates experimentation with horizontal translations through $|(-C/B)|$ units. Copies of one or both of these programs can be obtained by writing either of the authors.

REFERENCES

Hastings, Ellen H., and Daniel S. Yates. "Microcomputer Unit: Graphing Straight Lines." *Mathematics Teacher* 76 (March 1983):181–86.

Hirsch, Christian R. "Families of Lines." *Mathematics Teacher* 76 (November 1983):590–94.

National Council of Teachers of Mathematics. *An Agenda for Action: Recommendations for School Mathematics of the 1980s.* Reston, Va.: The Council, 1980.

In this activity you will investigate relationships between equations of the form $y = A|x + B| + C$ and their graphs.

1. Consider first the case where $B = 0$ and $C = 0$ so that you will be graphing $y = A|x + 0| + 0$, which has a simplified form of $y = A|x|$. For each set of constants below, enter the values of A, B, and C when prompted by the computer. After the computer generates the graph and presents the general equation, sketch the graph and write both the general and simplified forms of the equation.

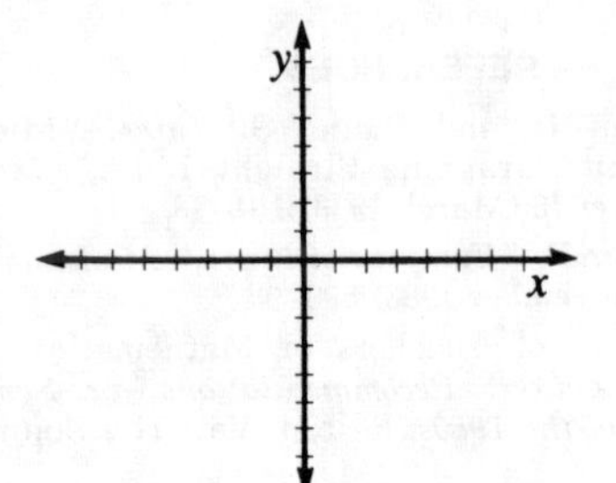

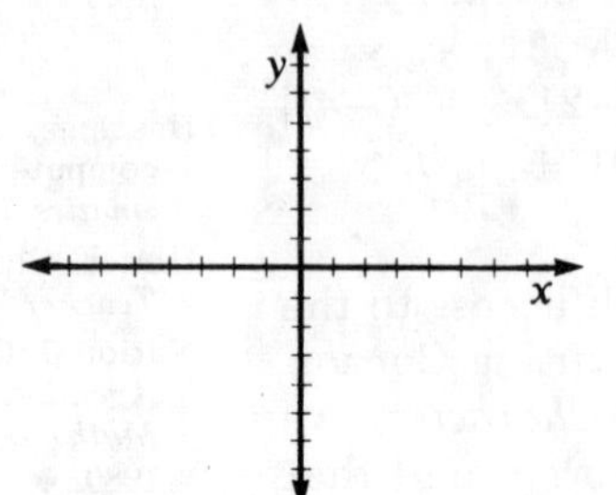

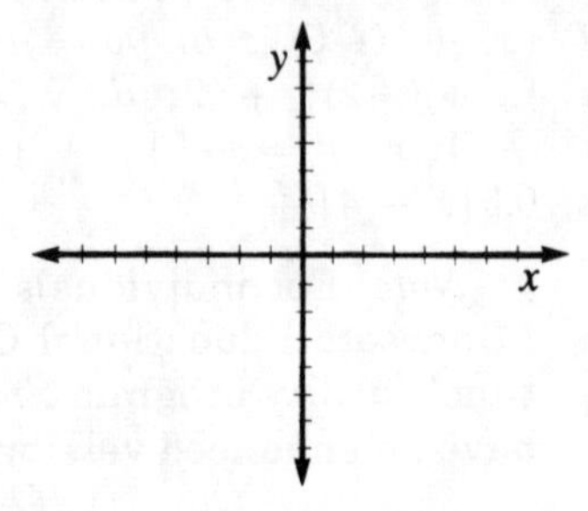

a. $A = 3, B = 0, C = 0$
general ___________
simplified ___________

b. $A = -3, B = 0, C = 0$
general ___________
simplified ___________

c. $A = 0.5, B = 0, C = 0$
general ___________
simplified ___________

2. Without using the computer, sketch the graphs of the equations with the following constants and write the simplified equations. Then enter your constants in the computer program and compare graphs. Correct your sketches if necessary.

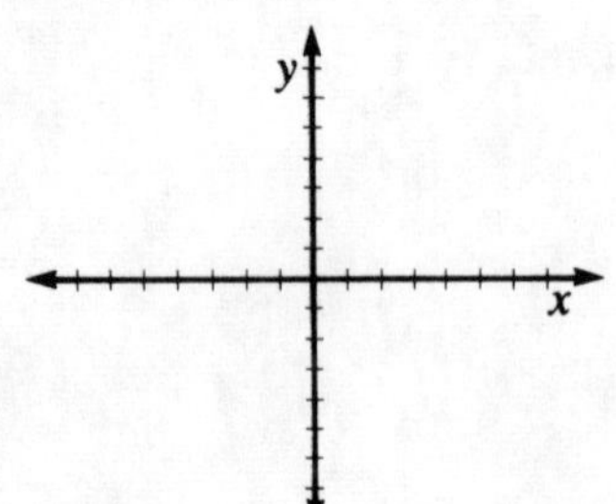

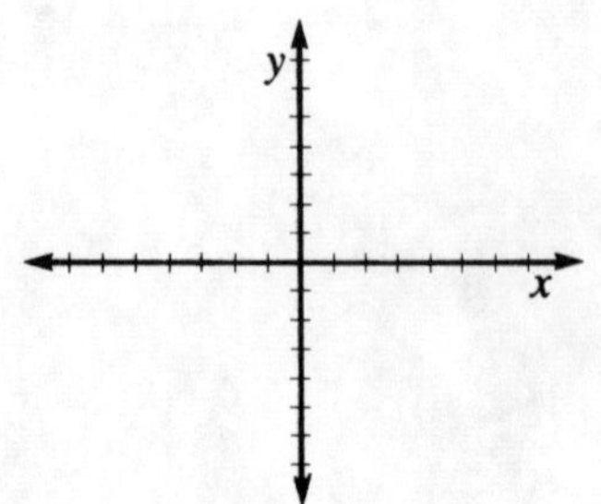

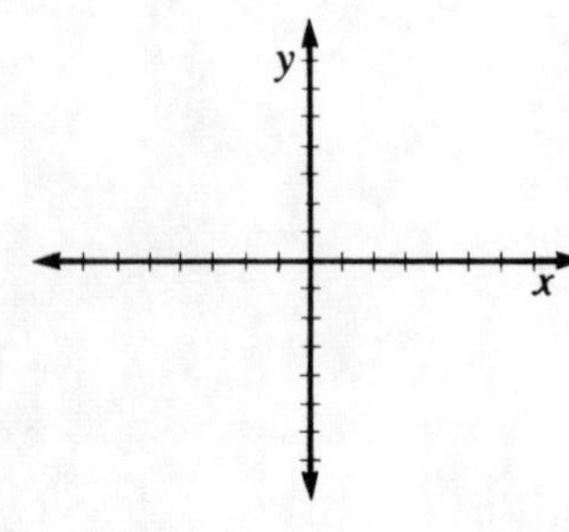

a. $A = 2, B = 0, C = 0$
equation ___________

b. $A = -6, B = 0, C = 0$
equation ___________

c. $A = -\frac{1}{3}, B = 0, C = 0$
equation ___________

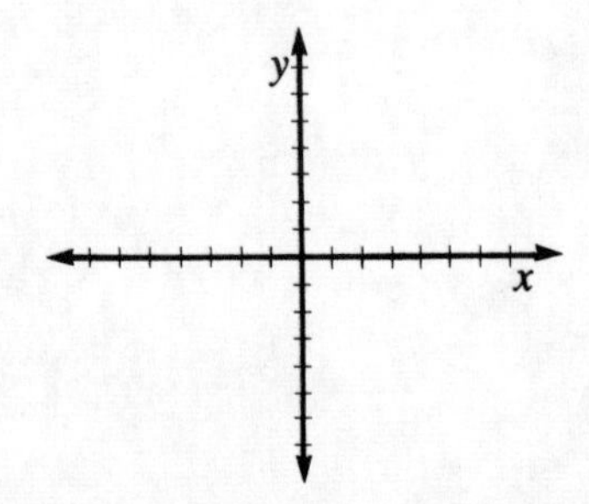

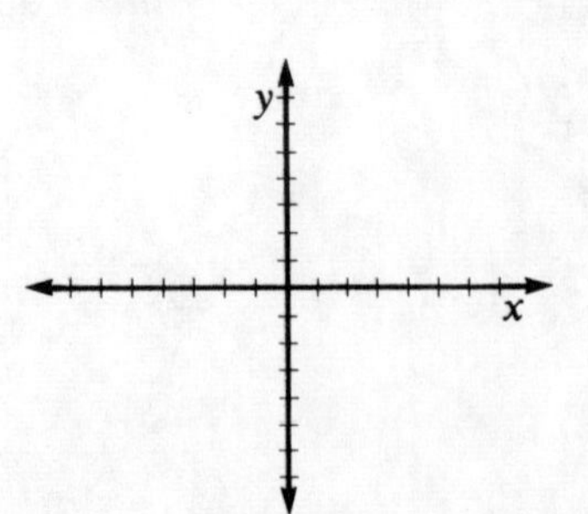

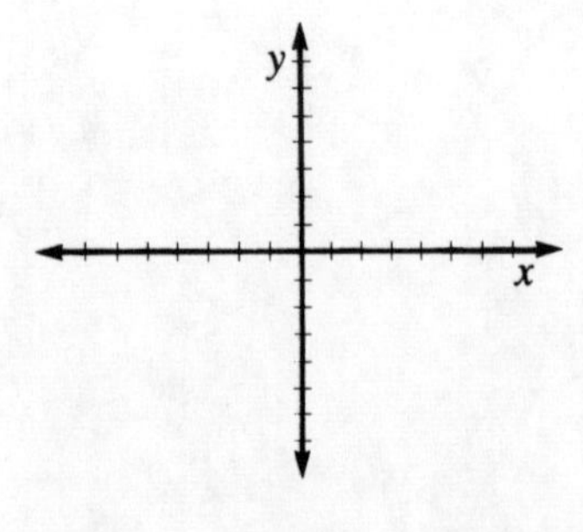

d. $A = 5, B = 0, C = 0$
equation ___________

e. $A = -2, B = 0, C = 0$
equation ___________

f. $A = -4, B = 0, C = 0$
equation ___________

3. In your own words describe how the graph of the equation $y = A|x|$ changes as the value of A changes. (Can you use your knowledge of slope to aid in your description?) ___________________________

____

RELATING GRAPHS AND THEIR EQUATIONS—*Continued* SHEET 2

1. In the general equation $y = A|x + B| + C$, let $A = 1$ and $B = 0$ so that the program will be using $y = 1|x + 0| + C$, or simply $y = |x| + C$. For each set of constants, run the program, copy the graph and general equation, and write the simplified form of the equation.

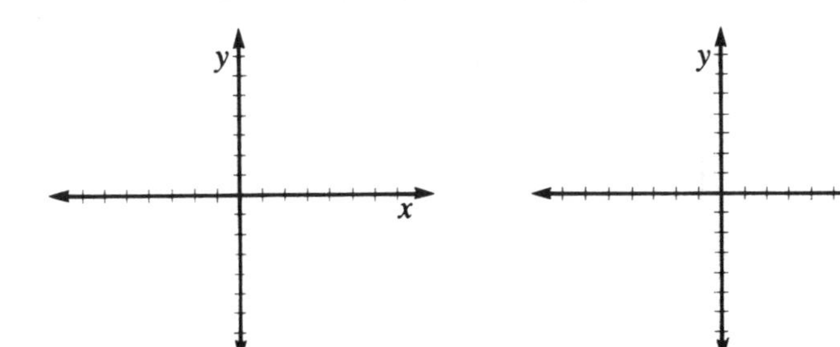
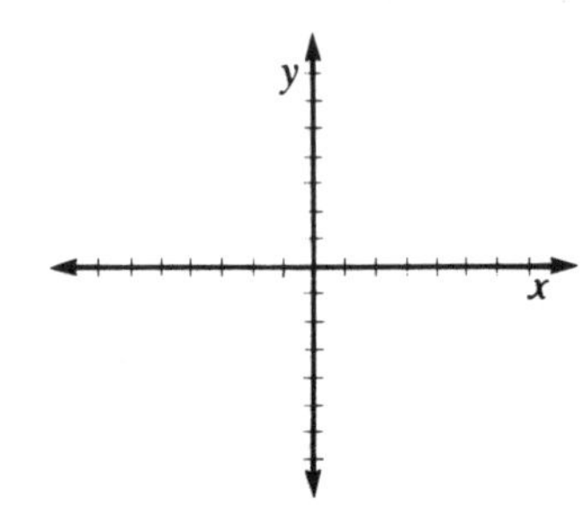

a. $A = 1, B = 0, C = 3$
general ____________
simplified ____________

b. $A = 1, B = 0, C = -6$
general ____________
simplified ____________

c. $A = 1, B = 0, C = 0.6$
general ____________
simplified ____________

2. Without using the computer, sketch the graph and write the simplified equation for each set of constants. Then enter your constants in the program and compare your graphs with the computer graphs. Correct your sketches if necessary.

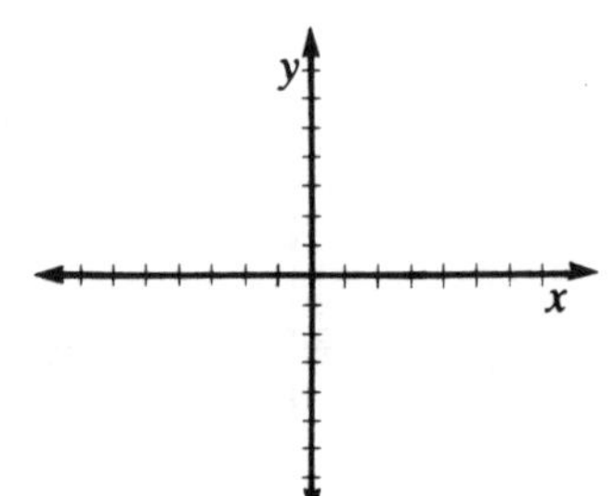
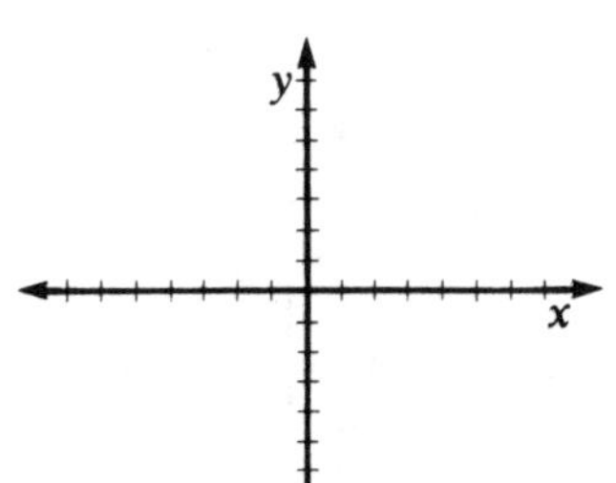
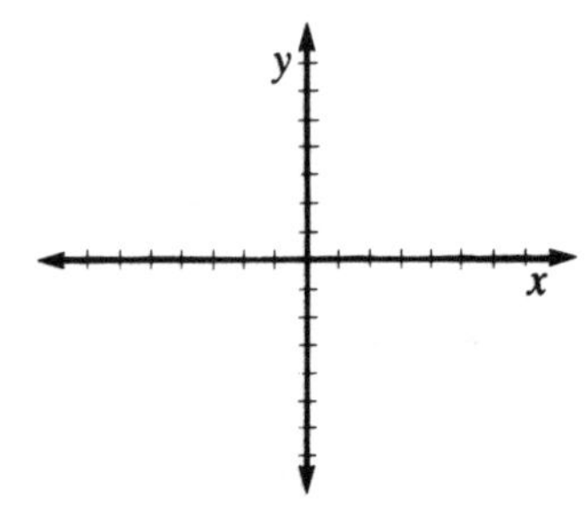

a. $A = 1, B = 0, C = 2$
equation ____________

b. $A = 1, B = 0, C = -4$
equation ____________

c. $A = 1, B = 0, C = 4.5$
equation ____________

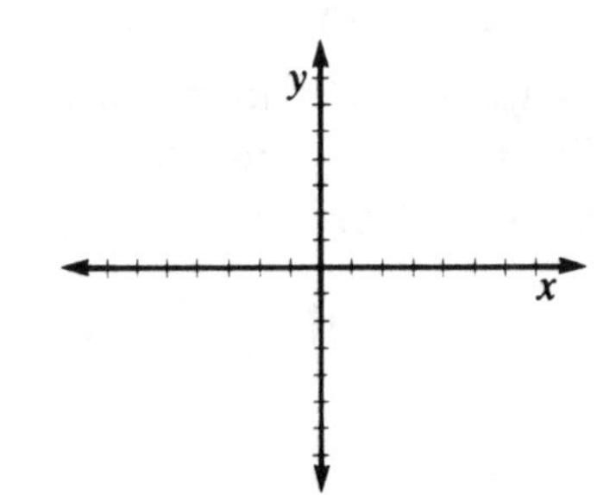
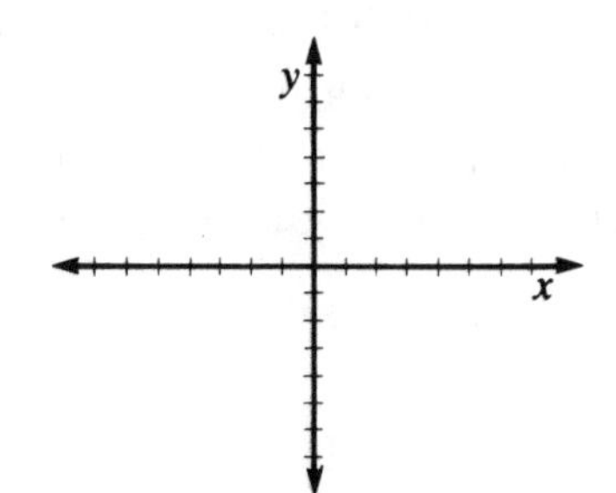
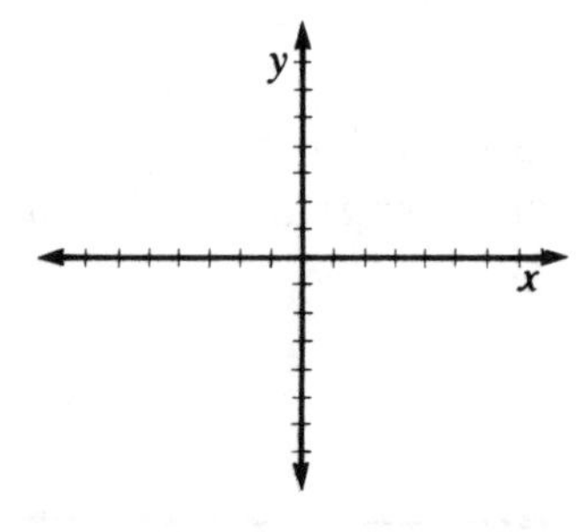

d. $A = -3, B = 0, C = 5$
equation ____________

e. $A = 2, B = 0, C = -6$
equation ____________

f. $A = 0.5, B = 0, C = -2$
equation ____________

3. *a.* Write in your own words how the graph of $y = |x| + C$ changes as the value of C is changed. ____________

b. Predict the coordinates of the vertex of the graph of $y = |x| + (-7)$ ____________ and of $y = |x| + C$ ____________.

4. Write in your own words how the graph of $y = A|x| + C$ is changed when the two values A and C are changed. Use the idea of slope and the location of the vertex. ____________

1. In the general equation $y = A|x + B| + C$, let $A = 1$ and $C = 0$ so that the program will be using $y = 1|x + B| + 0$, or simply $y = |x + B|$. For each set of constants below, after the computer generates the graph and presents the general equation, sketch the graph and write both the general and simplified forms of the equation.

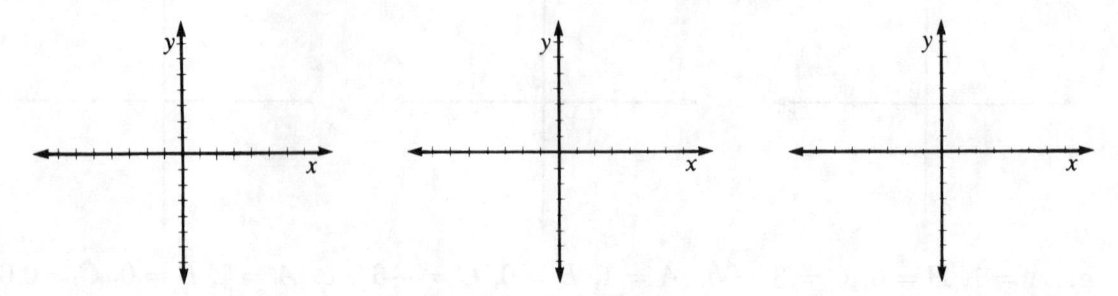

a. $A = 1, B = 4, C = 0$
general ___________
simplified ___________

b. $A = 1, B = -5, C = 0$
general ___________
simplified ___________

c. $A = 1, B = -6, C = 0$
general ___________
simplified ___________

2. Without using the computer, make a sketch of the graphs of the equations with the following constants and write the simplified equations. Then enter the constants in the program and compare the results. Correct your sketches if necessary.

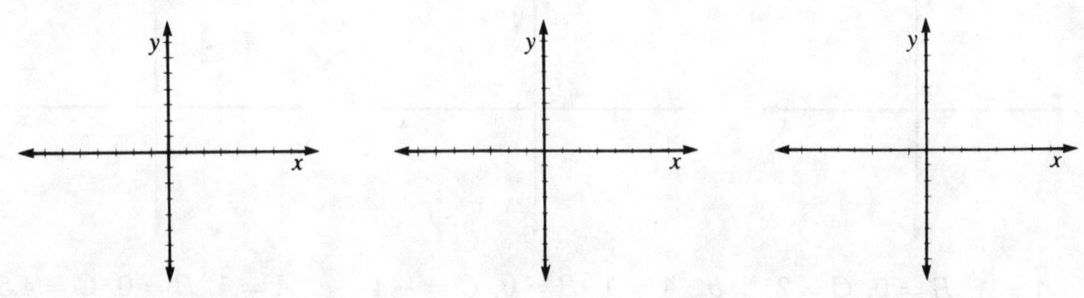

a. $A = 1, B = -2, C = 0$
equation ___________

b. $A = 1, B = 3, C = 0$
equation ___________

c. $A = 1, B = -4, C = 0$
equation ___________

3. Without using the computer, sketch the graph and write the simplified equation for each of the following sets of constants. Then enter the constants in the program and compare graphs. Correct your sketches where necessary.

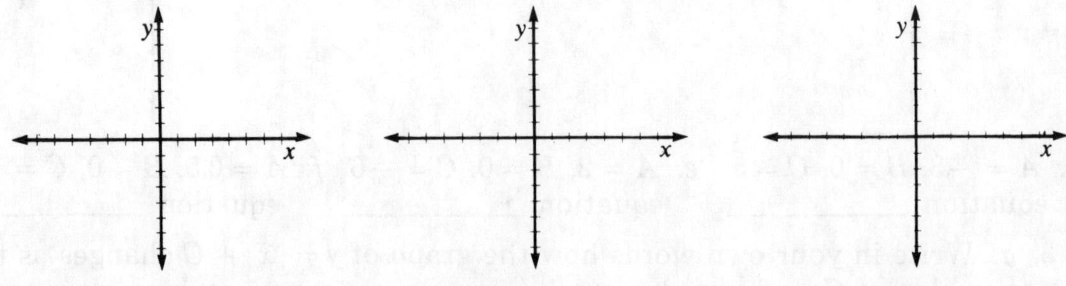

a. $A = 1, B = 6, C = 0$
equation ___________

b. $A = 1, B = -3, C = 0$
equation ___________

c. $A = 1, B = -7, C = 0$
equation ___________

4. Write in your own words how the graph of $y = |x + B|$ is changed when the value of B is changed. (*Hint:* Consider the location of the vertex.)

____

____

1. For each set of constants, write the corresponding equation below and then sketch the graph on a sheet of graph paper.

	A	B	C	Equation
a.	2	2	−6	________
b.	2	−4	−7	________
c.	−0.5	2	2	________

	A	B	C	Equation
d.	−3	−3	5	________
e.	0.5	1	−4	________
f.	3	−6	3	________

For each exercise, check your graph by entering the constants in the program or by comparing your results with those of other students.

2. For each equation, determine the coordinates of the vertex, the slope of the right ray, and whether the V opens upward or downward. Sketch the graph for each equation on a sheet of graph paper.

	Vertex	Slope	Upward/Downward
a. $y = 2\|x + 6\| + (-4)$	________	________	________
b. $y = -3\|x + (-2)\|$	________	________	________
c. $y = 0.5\|x + (-6)\| + 1$	________	________	________
d. $y = -5\|x + (-4)\| + 5$	________	________	________
e. $y = 0.25\|x + 1\| + 3$	________	________	________
f. $y = -0.25\|x + 4\| + (-2)$	________	________	________

For each exercise, check your graph by entering the constants in the program or by comparing your results with those of other students.

3. Write an equation for each graph.

a. equation__________ *b.* equation__________ *c.* equation__________

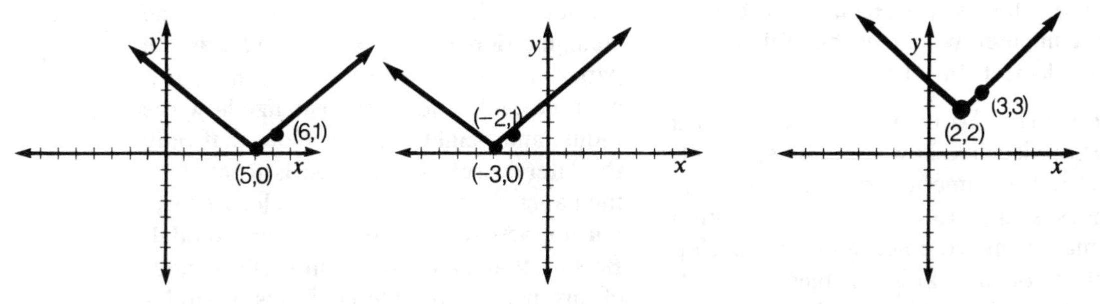

d. equation__________ *e.* equation__________ *f.* equation__________

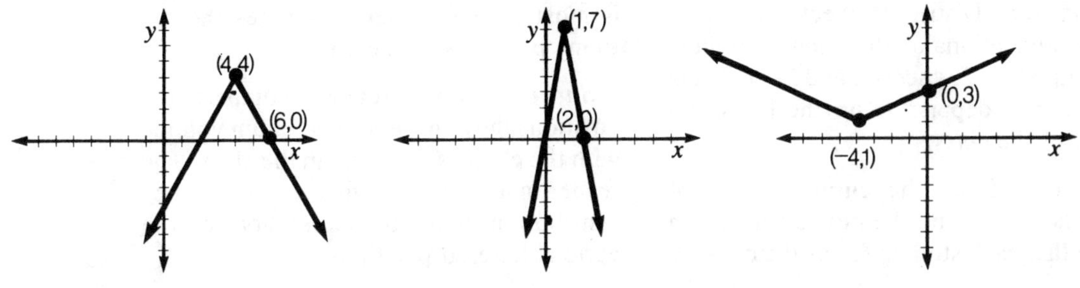

To check your work, enter the constants from your equation in the program and see if the computer-generated graph matches the given graph.

VISUALIZATION, ESTIMATION, COMPUTATION

December 1982

By Evan M. Maletsky, Montclair State College, Upper Montclair, NJ 07043

Teacher's Guide

Grade Level: 7–12

Materials: Scissors, paper clips, and copies of the worksheets. Access to a microcomputer would be helpful for the last worksheet, but it is not necessary.

Objectives: To construct a model of a cone; to visualize how the radius, height, area, and volume change as the cone changes shape; to estimate the position for maximum volume; and to develop problem-solving skills by measurement, computation, graphing, and analysis of data.

Directions: Distribute sheet 1 and have every student make the model. Follow this with sheet 2, sheets 2 and 3, or sheets 2, 3, and 4, depending on the level and ability of the class.

Sheet 1. Once the circular piece of paper is cut out and the radius cut, watch to see that each student forms the cone by placing the edge marked "top" over the other side lettered *A*, *B*, *C*, ...

Sheet 2. This activity focuses on the changing dimensions of the model as the paper cone is curled up and opened. Encourage each student to visualize how the radius and height change. Note that only the lateral area of the cone is formed by the paper. The circular base is located by, but not actually part of, the paper model. Be sure that each student moves the model through all possible positions in studying the changing volume, as it first will increase but then begin to decrease. This is most obvious when one notes the extreme positions of the cone.

Sheet 3. Have students compare the positions they chose for maximum volume with the choices of others in the class. The reader may want to make a frequency distribution of the numbers choosing the various lettered positions.

The model has been marked so that the diameters for the lettered positions are in centimeters and the radii in 0.5 centimeters. Be sure the heights are also recorded in 0.5 centimeters so that the volume will be in cubic centimeters.

Plot enough points on the graph that some trend in the changing volume can be observed. This will help give a better collective estimate of the maximum position. If you do not use sheet 4, be sure to note here that the best choice of letters for the maximum volume occurs at C, radius 6.5 cm. This is certain to surprise most students.

Sheet 4. The major goal here is to have students interpret the program in BASIC and analyze the printed output but not to write a program. Access to a microcomputer is not necessary. However, the suggestion at the end would be interesting to follow with those who have access to, and are familiar with, a microcomputer. Having the radius move from 6.0 to 7.0 cm by increments of 0.1 cm will give these results. The program is written for the Apple but can be easily modified for other computers.

RADIUS	HEIGHT	VOLUME
6	5.29	199.49
6.1	5.18	201.69
6.2	5.06	203.51
6.3	4.93	204.93
6.4	4.8	205.89
6.5	4.66	206.34
6.6	4.52	206.23
6.7	4.37	205.5
6.8	4.21	204.07
6.9	4.05	201.84
7	3.87	198.73

Answers:

Sheet 2

1. Decrease
2. Remain fixed at 8 cm
3. 8 cm
4. 8 cm
5. 0 cm
6. 8 cm

(Items 3–6: These are limiting, degenerate cases. Accept answers near these values.)

7. increases; increases
8. Yes; no. The total area approaches twice the area of the original 8-cm circle.
9. The radius first increases and then decreases.

Sheet 3

Choices will vary. The correct values for each letter, however, can be found in the printout on sheet 4.

Sheet 4

1. 30 (and 100)
2. 50
3. 60
4. 70 and 80 (using the greatest integer function)
5. 206.34 cm^3 at letter C

One possible modification of the program as suggested would be to change line 30 to

```
30 FOR R = 6 TO 7 STEP 0.1
```

The letter identification would then have to be dropped by deleting lines 40 and 110 and dropping the string variable X$ from line 90.

Modifications and Extensions: In certain classes you might want to fill the variously shaped cones with sand or rice and compare their volumes by that method.

For classes familiar with programming microcomputers, you might want to have them write and run their own programs rather than use sheet 4.

The exact maximum volume possible for a cone with a slant height of 8 cm can be found using calculus. The greatest volume is 206.37 cm^3, and it occurs at a radius of 6.53 cm.

For this activity you'll need a pair of scissors and a paper clip.

Step 1: Cut out the circle below. Then cut through to the center on the 8-cm radius line.

Step 2: Place the side of the cut line on top of the lettered surface of the circle.

Step 3: Move the edge of the cut line to one of the lettered positions and use a paper clip to hold it in place.

The circle now forms the lateral surface of a cone.

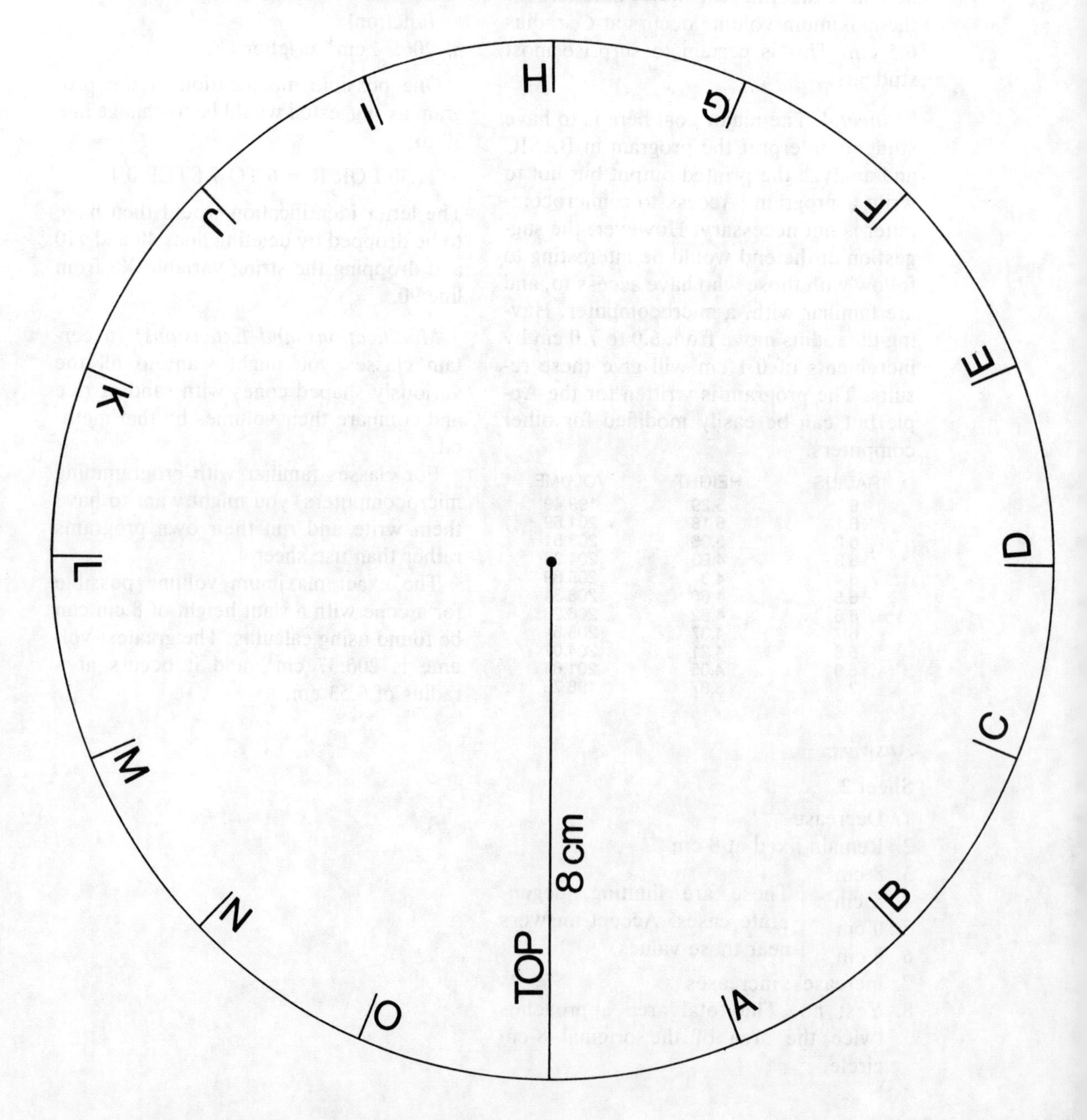

A cone has three important linear dimensions. They are the *radius*, *height*, and *slant height*.

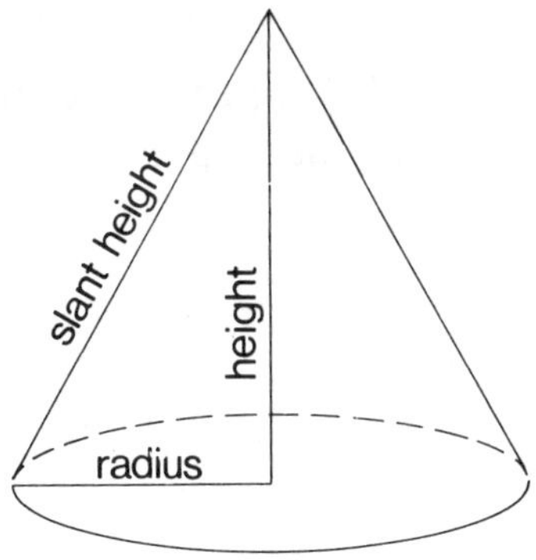

slant height
height
radius

Slowly move the cut edge from letter A to B to C, and so on. The height of the cone will increase.

1. Does the radius increase, decrease, or remain the same? ________________
2. What does the slant height do? ________________
3. What is the longest radius possible? ________________
4. What is the greatest height possible? ________________

This time open the model of the cone from a tightly overlapped position by moving the edge back toward the letters C, B, and A. The height of the cone will decrease.

5. At what height is the lateral area the greatest? ________________
6. At what height is the area of the base the least? ________________
7. As the height of the cone decreases, the lateral area increases.
   What happens to the area of the base? ________________
   What happens to the total area? ________________
8. As the cone is opened, the area of the base approaches the area of the original 8-cm circle.
   Does the lateral area approach the same value? ________________
   Does the total area approach the same value? ________________

Make the paper cone tighter again, moving the edge from letter A to B to C, and so on. The height will increase and the total area will decrease.

9. Describe in your own words what appears to happen to the volume.

____

____

As the paper model moves to narrower and wider positions, the volume of the cone formed changes.

1. Take your model and set it at the lettered position that you think forms the cone with the greatest volume. Paper clip it in place. LETTER ________

2. Measure, to the nearest 0.5 centimeters, the diameter of the cone in that position. Divide by 2 to find the radius, r. RADIUS ________

3. Carefully measure to the nearest 0.5 centimeters the height, h, of the cone in that same position. HEIGHT ________

4. Substitute your values in the formula to find the volume for that position. Use 3.14 for π. VOLUME ________

$$V = \frac{1}{3}\pi r^2 h$$

5. Plot your radius and volume on the same axes below. Then plot the values for some other radii. Use results others in the class have recorded or compute more volumes yourself. Do the different results help you better estimate the position for the maximum volume?

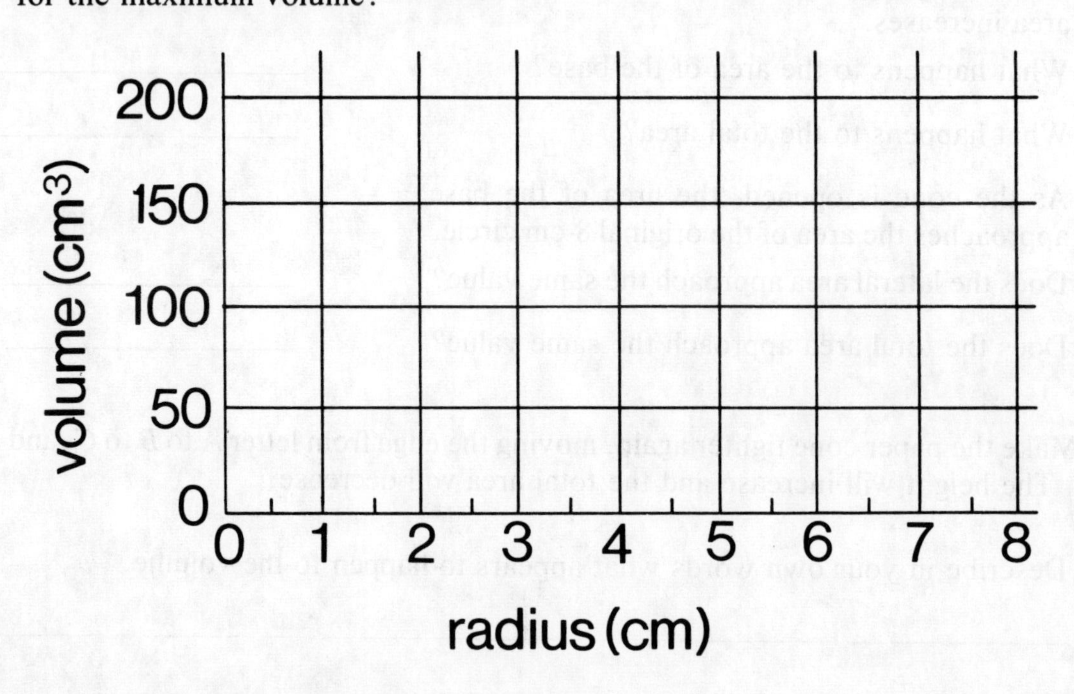

A microcomputer can be especially helpful in solving problems of this type. With just one program, it can compute many different volumes very rapidly. Here is one such program written in BASIC.

```
10  PRINT "    RADIUS","HEIGHT","VOLUME"
20  PRINT
30  FOR R = 0 TO 8 STEP .5
40  READ X$
50  LET H =  SQR (64 - R * R)
60  LET V = 3.1416 * R * R * H / 3
70  LET H =  INT (H * 100 + .5) / 100
80  LET V =  INT (V * 100 + .5) / 100
90  PRINT X$;"    ";R,H,V
100  NEXT R
110  DATA  -,O,N,M,L,K,J,I,H,G,F,E,D,C,B,A,-
120  END
```

Give the number of the line(s) that—

1. selects values of the radius R from 0 to 8 by increments of 0.5 cm, ____________
2. computes the height H using the Pythagorean theorem, ____________
3. computes the volume V using the formula $V = 1/3\ \pi R^2 H$, ____________
4. rounds the values of H and V to two decimal places. ____________

Here is the printout when the program is run.

	RADIUS	HEIGHT	VOLUME
-	0	8	0
O	.5	7.98	2.09
N	1	7.94	8.31
M	1.5	7.86	18.52
L	2	7.75	32.45
K	2.5	7.6	49.74
J	3	7.42	69.9
I	3.5	7.19	92.28
H	4	6.93	116.08
G	4.5	6.61	140.26
F	5	6.24	163.49
E	5.5	5.81	184.03
D	6	5.29	199.49
C	6.5	4.66	206.34
B	7	3.87	198.73
A	7.5	2.78	163.98
-	8	0	0

5. What is the greatest volume listed? ____________ cm^3. At what letter does it occur? ________

6. Compare these results with yours from sheet 3. Then plot all the radii and volumes listed on the graph on sheet 3.

The maximum volume shown occurs between a radius of 6 and 7 cm. Try modifying the program to print the volumes for radii from 6 to 7 cm by increments of 0.1 cm, and thereby obtain a more accurate estimate of the critical radius for maximum volume.

PROBLEM SOLVING WITH COMPUTERS

October 1984

BY DWAYNE E. CHANNELL, Western Michigan University, Kalamazoo, MI 49008

Teacher's Guide

Introduction: In its *Agenda for Action,* the NCTM (1980, 2–3) made the following recommendation:

> Mathematics programs of the 1980s must be designed to equip students with the mathematical methods that support the full range of problem solving, including . . . the use of the problem-solving capacities of computers to extend traditional problem-solving approaches and to implement new strategies of interaction and simulation.

This activity consists of two problem situations, each of which illustrates how a computer can be used as a tool to assist pupils in solving mathematical problems. Following an organizational scheme used by Channell and Hirsch (1984), each situation is presented using a four-stage Polya-type problem-solving model: problem statement, analysis, solution, and looking back. The first problem, "Maximizing Volume," uses the computer to perform straightforward numerical calculations that would be tedious without the computer. The computer program does not provide the solution to the problem but rather tabular data to be analyzed by the student. The second problem, "Collecting Stickers," introduces the use of a Monte Carlo model to simulate a physical action. Again, the computer program does not give the solution to the problem. Students must collect, tabulate, and summarize data generated by the program to arrive at an approximation of the solution. Each situation illustrates important and powerful problem-solving techniques and increases students' understanding of the role computers can play in problem solving.

Grade levels: 7–11

Materials: Paper (preferably centimeter graph paper), scissors, tape, and copies of the four worksheets. Access to a microcomputer is essential.

Objectives: To develop problem-solving skills involving the generation, summarization, and interpretation of computer-generated data

Directions: Distribute copies of sheets 1 and 2 to each student. The completion of these worksheets will require one or two class periods, depending on the level and background of your students.

Sheet 1: Have students read the problem carefully. To be certain that students understand the procedure used in constructing the open-topped boxes, have each student cut square corners from a piece of rectangular paper and form the resulting shape into a box. Don't insist that every student cut squares of the same size, since the nature of the problem will be much clearer if a variety of boxes are available for comparison. Allow students to complete the remainder of sheet 1.

Sheet 2: Have students enter and run the program on a microcomputer. The program was written for the Apple II using Applesoft BASIC, but it should run with little or no modification on other computers. Students should use the output generated by the program to complete exercises 5 through 7. In exercise 8, it may not be obvious to students that the optimal solution falls somewhere between 6 cm and 8 cm. It might be helpful to point out that the table of volumes increases as h moves from 1 cm to 7 cm and the volumes decrease as h moves from 7 cm to 17 cm. The actual maximum volume can occur on either side of 7 cm, and so the search must be made between 6 cm and 8 cm. Students with little or no programming experience will need some assistance in completing exercise 9. Point out that a 44 cm × 36 cm rectangle and a 66 cm × 24 cm rectangle have equal areas, 1584 cm^2. However, the volume of the largest open-topped boxes are not equal. More able students might be challenged to find the dimensions of a rectangular sheet with an area of 1584 cm^2 that would produce the open-topped box with maximum volume.

Distribute copies of sheets 3 and 4 to each student. Students should be able to complete them in a single class period.

Sheet 3: Have students read the problem carefully. To be certain that the simulation process is clear to all students, simulate this problem using labeled slips of paper in a box or hat before discussing the computer program. Use the following procedure:

(*a*) Place the name of each arcade game on a separate slip of paper or cardboard, making certain that all slips are of the same size.

(*b*) Place the six slips into a container so that no one can see them.

(*c*) Draw a single slip from the container and record the name using a tally mark in a table similar to the following:

Q-Bert	Burger Time	Ms. Pac-Man
Zaxxon	Donkey-Kong	Frogger

(*d*) Replace the slip that was drawn and mix the six slips well before drawing again.

(*e*) Continue drawing, recording, and replacing slips until one for each of the six arcade games has been drawn.

(*f*) The sum of the tally marks gives one approximation to the theoretical expectation.

Discuss the program. Have students enter the program, run it on a microcomputer, and organize the output in the table provided. The program was written for the Apple II using Applesoft BASIC and will most likely need slight modification for other computers. Line 100 serves the function of the RANDOMIZE command found in many BASIC dialects. The RND(X) function used in line 110 may require an argument other than 1, or it may require no argument at all in BASIC dialects other than Applesoft.

Sheet 4: Students should use the results from sheet 3 to answer exercise 1 at the top of sheet 4. The computer program should be run four more times and an average of the results computed. Exercise 5 might best be handled as a teacher-directed activity, since it involves collecting and summarizing data generated by the entire class.

Supplementary activities: In the problem about maximizing volume, students can be asked to plot the data generated by the program at the top of sheet 2 on a coordinate system. If the seventeen points are plotted and a smooth curve sketched to connect them, a graph similar to figure 1 results. A discussion of the relationships between the tabular data and the increasing-decreasing nature of the graph may prove instructive.

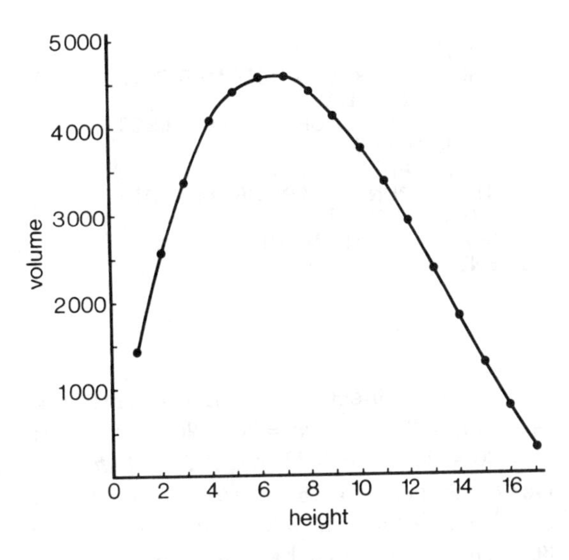

Fig. 1. Plotting points to suggest the maximum volume

For the problem about collecting stickers, students with a good programming background may enjoy writing a program that simulates this problem *and* summarizes the results for many customers. The program in table 1 summarizes results for 100 customers. In addition, it allows the user to INPUT the number of stickers in the collection. Students might enjoy running this program for the cases of 2, 3, 4, . . . , 10 stickers. The results can be plotted on a coordinate system to provide a graphic representation of the increasing expected values.

TABLE 1

```
100  REM STICKER COLLECTION SIMULATION
110  HOME : POKE (203), PEEK (78)
120  INPUT "HOW MANY STICKERS TO
       COLLECT?";N
130  PRINT : PRINT "NOW BUYING MUNCHIES
       FOR 100 CUSTOMERS."
140  PRINT : PRINT "PLEASE WAIT."
150  PRINT : PRINT "EACH * REPRESENTS A
       FINISHED CUSTOMER.": PRINT
160  LET C = 0
170  FOR I = 1 TO 100
180  FOR S = 1 TO N
190  LET F(S) = 0
200  NEXT S
210  LET T = 0
220  LET S = INT (N * RND (1)) + 1
230  LET F(S) = F(S) + 1
240  LET T = T + 1
250  FOR S = 1 TO N
260  IF F(S) = 0 THEN 220
270  NEXT S
280  LET C = C + T
290  PRINT "*";
300  NEXT I
310  PRINT : PRINT : PRINT
320  PRINT "AVERAGE CUSTOMER PURCHASED
       ";C / 100;" BAGS"
330  PRINT "OF MUNCHIES TO COLLECT ";N;"
       STICKERS."
340  PRINT : PRINT : PRINT
350  INPUT "RUN THE PROGRAM AGAIN
       (Y OR N)?";A$
360  IF A$ = "Y" THEN 110
370  END
```

Answers: Sheet 1: 1. $h = 4$ cm; $l = 44 - 2(4) = 36$ cm; $w = 36 - 2(4) = 28$ cm; $V = 36 \times 28 \times 4 = 4032$ cm^3. 2. (*a*) $h = 10$ cm; $l = 24$ cm; $w = 16$ cm; (*b*) $V = 3840$ cm^3. 3. $l = 44 - 2h$; $w = 36 - 2h$. 4. Since $36 - 2h > 0$, we must have $h < 18$, and thus the maximum integral value for h is 17 cm.

Sheet 2: 5. Yes (it is hoped). 6. (*a*) 4620 cm^3; (*b*) 7 cm 7. 7 cm 8. (*a*) 6.6 cm; (*b*) The volume corresponding to a height of 6.6 cm is 4634.8 cm^3, which is larger than the 4620 cm^3 volume resulting from a height of 7 cm. (It can be shown using calculus that the maximum volume of 4634.87 cm^3 occurs with a height of 6.58 cm.) 9. Lines 110, 120, and 130 of the program should be changed as follows:

```
110  FOR H = 1 TO 11
120  LET L = 66 - 2 * H
130  LET W = 24 - 2 * H
```

This program gives a maximum volume of 3920 cm^3 when $h = 5$ cm. Changing line 110 to read

```
110  FOR H = 4 TO 6 STEP .1
```

gives a maximum volume of 3934.7 cm^3 when $h = 5.4$ cm.

Sheet 4: 1. Answer will vary; no; the result from one customer is probably not a good estimate of the actual solution. 2. (*a*) 6 bags; (*b*) The maximum number depends on the number of bags produced by the company. For example, if 600 000 bags were made, it is theoretically possible (but highly improbable) that a very determined, but unlucky, customer would need to buy 500 001 bags to collect all six stickers! 3, 4, 5. Answers will vary. The expected (theoretical) result for the six-sticker problem is 14.7 purchases. 6. Results will vary. The expected result for the eight-sticker problem is 21.7 purchases.

BIBLIOGRAPHY

Channell, Dwayne E., and Christian Hirsch. "Computer Methods for Problem Solving in Secondary School Mathematics." In *Computers in Mathematics Education,* 1984 Yearbook of the National Council of Teachers of Mathematics, edited by Viggo P. Hansen, pp. 171–83. Reston, Va.: The Council, 1984.

Lappan, Glenda, and M. J. Winter. "Probability Simulation in the Middle School." *Mathematics Teacher* 73 (September 1980):446–49.

National Council of Teachers of Mathematics. *An Agenda for Action: Recommendations for School Mathematics of the 1980s.* Reston, Va.: The Council, 1980.

Travers, Kenneth J., and Kenneth G. Gray. "The Monte Carlo Method: A Fresh Approach to Teaching Probabilistic Concepts." *Mathematics Teacher* 74 (May 1981):327–34.

Ms. Hawes needs several open-topped boxes for storing laboratory supplies. She has given the industrial arts class several rectangular pieces of sheet metal to form the boxes. The pieces measure 44 cm × 36 cm. The class plans to make the boxes by cutting equal-sized squares from each corner of a metal sheet, bending up the sides, and welding the edges. Ms. Hawes has asked that each box have the largest possible volume. What size of squares should be cut from the corners of the metal sheets?

Analysis

To help you visualize the problem, take a sheet of paper and cut equal-sized squares from its corners. You will get a shape like the figure shown below.

Fold the side tabs up to form a box and tape the edges together. Compare your open-topped box with those made by your classmates. Note that the size of square cut from the corners of the paper determines the height (h) of the box as well as the length (l) and width (w).

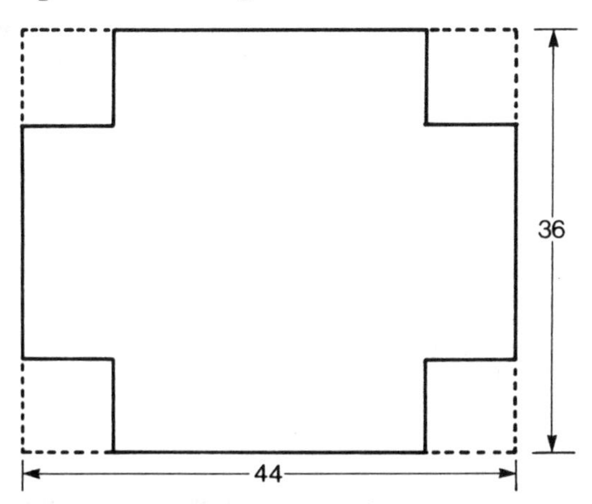

1. Assume that Ms. Hawes suggested cutting 4-cm squares from the metal sheets. What would be the dimensions of the box formed?

$h =$ ____ cm, $l = 44 - 2(4) =$ ____ cm, $w = 36 - 2(\_)$ = ____ cm

The volume (V) of this box is given by $V = lwh =$ ____ cm^3.

2. Assume that 10-cm squares are removed from the corners.

(*a*) Give the dimensions of the corresponding box:

$h =$ ____ cm, $l =$ ____ cm, $w =$ ____ cm

(*b*) Find the volume of the box. ______

3. If squares of size h cm are removed from the corners, then the height of the box formed would be h cm. Complete the expressions that represent the length and width of the box.

$l = 44 - 2(\_)$ cm and $w =$ ____ cm

4. Many different-sized squares can be cut from the metal sheets. Let h represent the size of the square cut. If h is measured to the nearest whole centimeter, the smallest possible h is 1 cm. What is the largest possible value for h? (*Hint:* The length and width of the box must both be positive values.)

Solution

You have found that the height h of the box can range from 1 cm to 17 cm depending on the size of the square cut from the corners of the rectangular metal sheets. The following BASIC program computes the dimensions and corresponding volumes of all seventeen boxes. Study the program, enter it into a computer, and RUN it.

```
100  PRINT "HEIGHT", "VOLUME"
110  FOR H = 1 TO 17
120  LET L = 44 - 2 * H
130  LET W = 36 - 2 * H
140  LET V = L * W * H
150  PRINT H,V
160  NEXT H
170  END
```

Looking Back

5. Examine the table of numbers produced by the program. Do the volumes corresponding to heights of 4 cm and 10 cm agree with your computations in exercises 2 and 3 on sheet 1? ____

6. (*a*) What is the largest volume listed in the table? ____
   (*b*) What is the corresponding height? ____

7. What size of square should be cut from the metal sheets to produce a box with this maximum volume? ____

8. To the nearest whole centimeter, a box with $h = 7$ cm gives the maximum volume. If the value of h is measured to the nearest tenth of a centimeter, the height of the box with maximum volume would still lie somewhere between 6.0 cm and 8.0 cm but not necessarily at 7.0 cm. Change line 110 of the program to read as follows:

```
110 FOR H = 6 TO 8 STEP .1
```

RUN this modified program and study the output.

(*a*) To the nearest tenth of a centimeter, what is the height of the box that gives the maximum volume? ____

(*b*) Is this an improvement over the whole-number solution of $h = 7$ cm?

____

9. Assume that the industrial arts class had sheets measuring 66 cm × 24 cm. Modify the computer program to help you determine the height of the open-topped box with maximum volume that could be made from these sheets.

Junk Foods, Inc., introduced a new cheese-flavored snack food called "Cheese Munchies." To encourage people to buy their new product, they included a colorful sticker depicting a popular arcade game in each large-sized bag of the snack food. The games depicted are Q-Bert, Burger Time, Ms. Pac-Man, Zaxxon, Donkey-Kong, and Frogger. Equal numbers of each of the six stickers were randomly distributed in the first production of Cheese Munchies. If you, as an average customer, wanted to collect all six stickers, about how many bags of Cheese Munchies would you have to buy? Your guess: ____

Analysis

One approach to solving this problem is to simulate, or imitate, buying a bag of Cheese Munchies and checking which one of the six stickers was included. Would you believe that a computer can be used to simulate the buying process?

Let's define a correspondence between the names of the arcade games depicted on the stickers and the numbers from 1 to 6 inclusive:

Q-Bert	Burger Time	Ms. Pac-Man	Zaxxon	Donkey-Kong	Frogger
↕	↕	↕	↕	↕	↕
1	2	3	4	5	6

The following program generates random whole numbers between and 1 and 6 inclusive:

```
100  POKE(203), PEEK(78)
110  PRINT INT(6 * RND(1)) + 1,
120  INPUT "ANOTHER NUMBER (Y OR N)?";A$
130  IF A$ = "Y" THEN 110
140  END
```

Each number generated by the program can represent the purchase of a bag of Cheese Munchies. The value of the number generated determines which sticker was in the bag. Numbers should be generated by the program until each of the numbers 1–6 has been printed.

Solution

Enter and RUN the program. Use the following table to keep a tally of the number of each different sticker collected by customer A. Stop when you have collected all six numbers. Find the total number of bags purchased.

Customer A

Sticker Number	1	2	3	4	5	6
Bag Tally						

Total bags purchased ____

Looking Back

1. How many bags of Cheese Munchies did customer A buy before obtaining all six stickers? ____ Do you think that this answer represents the exact solution to the problem? ____

2. (*a*) What is the minimum number of bags a really lucky person would have to buy to collect all six stickers? ____

(*b*) Is there a maximum number of bags that a very unlucky person would have to buy? (*Hint:* Assume the company produced 600 000 bags of Cheese Munchies in their first production run.) ____

3. The results for customer A do not necessarily represent the expectations of the "average" customer. To better approximate the number of bags purchased by the average customer, the problem should be simulated several times. RUN the program again for each of the following customers. Keep a tally of the results in the tables and then find the total number of bags purchased by each customer.

Customer B

1	2	3	4	5	6

Total bags ____

Customer C

1	2	3	4	5	6

Total bags ____

Customer D

1	2	3	4	5	6

Total bags ____

Customer E

1	2	3	4	5	6

Total bags ____

4. (*a*) What is the total number of bags of snack food purchased by all five customers (A, B, C, D, and E)? ____

(*b*) What is the average number of bags purchased by these five customers? ____

5. The last answer in exercise 4 represents an estimate of the solution to the sticker problem. A better estimate can be found by simulating the purchases of a larger number of customers and averaging the number of bags they purchased. With the help of your teacher, collect information from classmates who have worked on this problem. You will want to know how many customers they have simulated and how many bags were purchased by each of these customers. Calculate the average number of bags purchased by all these customers.

What is a reasonable solution to the problem stated at the top of sheet 3? ____

6. Suppose the company decided to include eight different stickers instead of six. (Centipede and Buck Rogers were added.) How many bags would the typical customer expect to buy to collect all eight stickers? Change line 110 of the program to read `110 PRINT INT(8 * RND(1)) + 1`, What is a reasonable estimate of the solution to this problem? ____

Activities for

Using Manipulatives

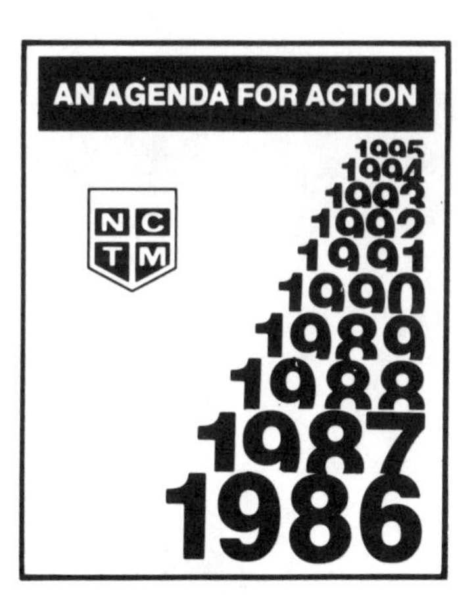

Recognizing that what is learned relative to a given mathematical topic, how long it is retained, and how readily it is applied are all functions of the manner in which learners encounter the topic, the *Agenda for Action* called for greater attention to efficient techniques for effectively engaging pupils in the learning process, specifically encouraging "the provision of situations that provide discovery and inquiry as well as basic drill [and] the use of manipulatives, where suited, to illustrate or develop a concept or skill." The activities in this final section are characterized by both of these attributes.

The first two activities use a sequence of three representation modes—concrete, pictorial, and symbolic—to develop concepts and procedures relative to topics in the algebra curriculum. In "Solving Linear Equations Physically" students use a balance-scale model together with strips and small squares to represent and solve linear equations and in the process discover methods for solving such equations using standard algebraic manipulations. The activity "Finding Factors Physically" uses an extension of the strip-square model to provide pupils opportunities to investigate factoring quadratic polynomials and thereby discover relationships among the coefficients. This discovery subsequently permits them to factor such polynomials at the symbolic level. Suggestions for modifying the model to permit the exploration of factoring cubic polynomials are also given.

Topics from geometry are the focus of the next two activities. In "Those Amazing Triangles" students practice using the basic measuring and drawing instruments of geometry in the context of discovering several recent and remarkable generalizations about triangles, including the theorems of Morley and Napoleon. The activity "Semiregular Polyhedra" begins with an analysis of models of the five regular polyhedra leading to the discovery of Euler's theorem relating the numbers of faces, vertices, and edges. Pupils are next introduced through models to a method for truncating a regular polyhedron and then use this method together with visualization and deductive counting techniques to discover patterns that permit them to predict the number of faces, vertices, and edges on a truncated polyhedron.

The section concludes with a laboratory activity devoted to the topic of geometric probability. "Out of Thin Air" uses the Monte Carlo method to simulate a parachutist jumping from an airplane and landing in a field. Experimental and theoretical probabilities that the parachutist will land in a particular region are determined and compared.

SOLVING LINEAR EQUATIONS PHYSICALLY *September 1985*

By BARBARA KINACH, Lesley College, Cambridge, MA 02138

Teacher's Guide

Introduction: The "Activities" section of the *Mathematics Teacher* first appeared in 1972 as a means of providing classroom teachers ready-to-use discovery lessons and laboratory experiences in worksheet form. The importance and efficacy of these alternative instructional methodologies were reaffirmed in the NCTM's *An Agenda for Action* (1980, 12). The *Agenda* specifically recommended that teachers employ diverse instructional strategies, materials, and resources, including the following:

- The provision of situations that provide discovery and inquiry as well as basic drill
- The use of manipulatives, where suited, to illustrate or develop a concept or skill

The following materials provide opportunities for students to investigate solving linear equations using both a physical and a pictorial model and in the process discover a method that will permit them to solve such equations at the symbolic level.

Grade levels: 7–10

Materials: Scissors, cardboard, a lighter-weight tagboard, and a set of activity sheets for each student. Transparency cutouts of the strips, squares, and balance scale on sheet 4 would be useful for purposes of demonstration and for discussion of students' solutions.

Objectives: Students will (1) represent first-degree polynomials physically (with strips and squares) and pictorially; (2) solve linear equations both physically through the use of a balance scale and pictorially; and (3) discover a method for solving linear equations using algebraic manipulation.

Directions: On the day before this lesson distribute a copy of sheet 4 to each student. As indicated on sheet 1, instruct them to glue the row of strips on a piece of cardboard, glue the rows of squares on a piece of lighter-weight tagboard, and then cut out the pieces. Provide students with an envelope in which to keep their equation-solving pieces.

On the day of the lesson, distribute the worksheets one at a time. All students should be able to complete the first two sheets during a single forty-five-minute class period if the strips and squares have been precut. Depending on the time available, sheet 3 can be completed during the next class period.

Sheet 1 emphasizes the distinction between a variable and a constant. It is important that students realize that the strips were purposely constructed so that their weights were a variable quantity in relation to the weights of the squares. In addition, this first worksheet uses the strip-square diagrams to clarify the use of grouping symbols in algebraic expressions. Students thus distinguish the difference between $2S + 3$ and $2(S + 3)$ pictorially.

Sheet 2 establishes the analogy between balancing a scale and solving a linear equation of the form $ax + b = cx + d$ where a, b, c, $d \geq 0$. In this activity pupils first attempt to maintain the scale's balance while replacing each strip with the same number of squares. This experience reinforces the conditional nature of an equation. Specifically, it demonstrates that the asserted equality between two expressions need not hold for all replacements of the variable S. The second portion of this sheet provides an algorithm for physically determining the value for S that will maintain the scale's balance. Students may solve the equations in exercise 8 by making pictorial representations of their physical manipulations or by using standard symbolic methods. At this stage, either method should be accepted. You might also have students verify their solutions for exercises 7 and 8 by actually substituting the values obtained for S into the equations and then performing the arithmetic indicated.

Finally, sheet 3 introduces a pictorial method for solving linear equations as a bridge between the physical model and the ultimate goal of algebraic manipulation. It would be instructive to demonstrate how the equation $4S + (-3) = 3S + (-4)$ in the example can also be solved pictorially by adding three white squares to both sides of the configuration in step 2 and using the fact that a white square and a gray square cancel each other. The solution of exercise 9c will require students to add three gray squares to each side of the configuration. Encourage pupils to verify their solutions for exercises 9 and 10 by substituting the values obtained for S into the equations and then performing the arithmetic indicated.

After all students have completed sheet 3, carefully establish the algebraic methods for solving linear equations in terms of the corresponding pictorial manipulations. In the example at the top of sheet 3, this correspondence can be illustrated as follows.

Step 1: Represent the equation. See (1).

Step 2: Subtract (remove) 3 gray squares. See (2).

Step 3: Subtract (remove) 3 strips. See (3).

$$
\begin{array}{llcl}
(1) & 4S + (-3) & = & 3S + (-4) \\
(2) & \underline{\quad -(-3)} & & \underline{\quad -(-3)} \\
 & 4S & = & 3S + (-1) \\
(3) & \underline{-3S \quad} & & \underline{-3S \quad} \\
 & S & = & -1
\end{array}
$$

Students should note the pictorial distinction made between a negative number and the operation of subtraction. A negative number is represented as a gray square. The operation of subtraction is indicated by crossing out with an "X" the strip or square to be removed. This distinction of notation should remind pupils that the symbol $(-)$ has different meanings.

Supplementary activities: Introduce notation for subtracting a constant from a variable. For example, to indicate the subtraction $3S - 4$, place (or draw) four *white* squares on top of the strips,

Note the distinction with the representation of $3S + (-4)$,

Challenge students to solve each of the following equations first pictorially and then using the corresponding algebraic manipulations.

a. $4S - 3 = S + 3$

b. $5(S - 1) = 2S + 7$

c. $-3 + 2S = 3(S - 2)$

Later in the year you may wish to use a modification of this strip-square model as described by Hirsch (1982) to factor quadratic polynomials physically.

Answers:

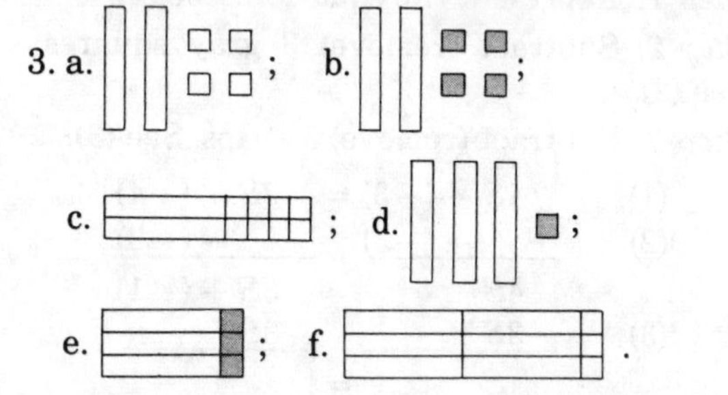

Sheet 1: 2. b. ; c. The rectangular region formed consists of two rows of identical shapes. Since the weight of one row is $S + 3$, the weight of the region is $2(S + 3)$.

3. a. ; b. ; c. ; d. ; e. ; f. .

4. a. $3S + 6$; b. $4(S + 1)$ or $4S + 4$; c. $3S + (-9)$.

Sheet 2: 5. b. No; f. No. 6. three. 7. a. $S = 1$; b. $S = 7$; c. $S = 5$; d. $S = 2$; e. $S = 3$. 8. a. $S = 4$; b. $S = 7$; c. $S = 2$.

Sheet 3: 9. a.

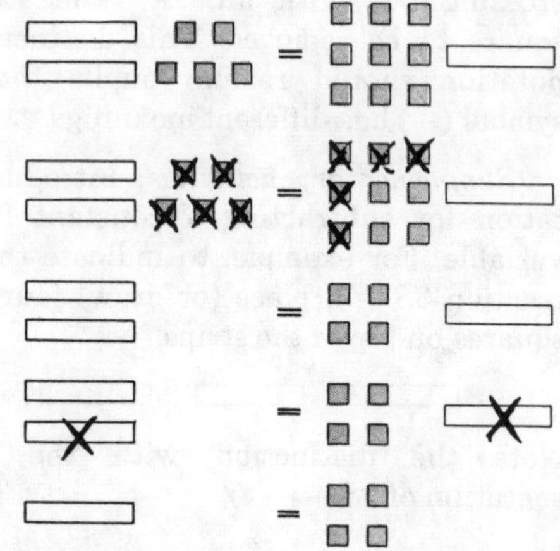

The solution is $S = -4$.

b.

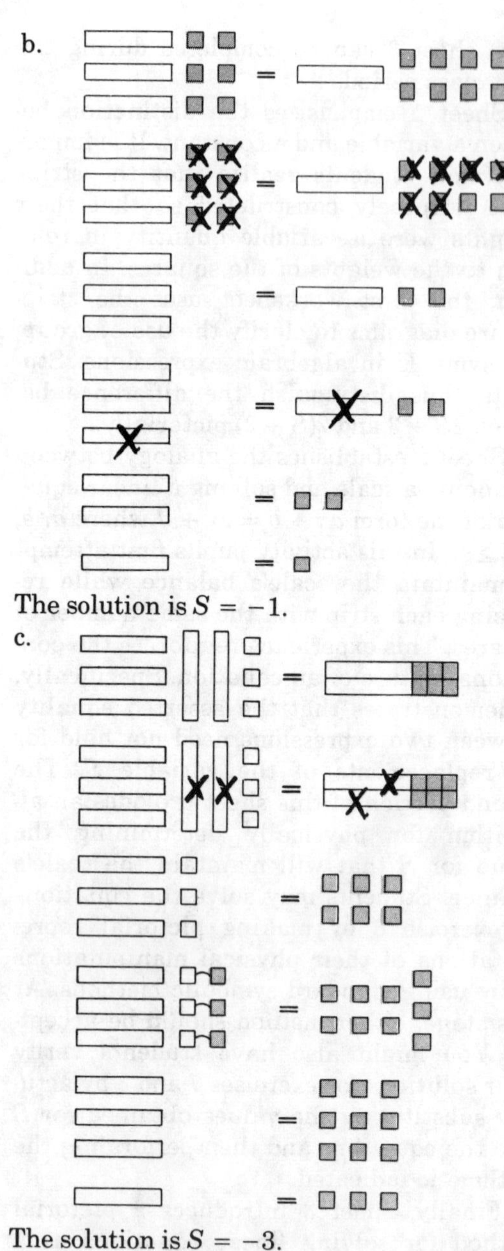

The solution is $S = -1$.

c.

The solution is $S = -3$.

10. a. $S = -4$; b. $S = 14$

REFERENCES

Hirsch, Christian R. "Finding Factors Physically." *Mathematics Teacher* 75 (May 1982):388–93, 419.

National Council of Teachers of Mathematics. *An Agenda for Action: Recommendations for School Mathematics of the 1980s*. Reston, Va.: The Council, 1980.

1. Sheet 4 of this activity consists of a series of strips, white and gray squares, and a diagram of a balance scale.
   a. Cut the sheet along the two dashed lines.
   b. Glue the series of strips onto a piece of a cardboard and then carefully cut out the strips.
   c. Glue the series of squares onto a piece of light-weight tagboard, such as a file folder, and then carefully cut out the white and the gray squares.

   For this activity, assume that the weight of—
   a. a *white square* is a *positive one* $(+1)$ unit,
   b. a *gray square* is a *negative one* (-1) unit,
   c. a *strip* is a *variable* quantity S (depending on the cardboard backing used).

2. a. The expression $2S + 6$ can be represented physically by two strips and six white squares. Place these eight pieces on your desk.
   b. Rearrange the pieces to form a rectangle.
   c. Explain why the weight of these eight pieces can also be expressed as $2(S + 3)$. ________________________

3. Represent each algebraic expression with strips and squares. Draw diagrams of your solutions in the spaces provided.

   a. $2S + 4$ b. $2S + (-4)$ c. $2(S + 4)$

   d. $3S + (-1)$ e. $3[S + (-1)]$ f. $3(2S + 1)$

4. Write an algebraic expression for each diagram.

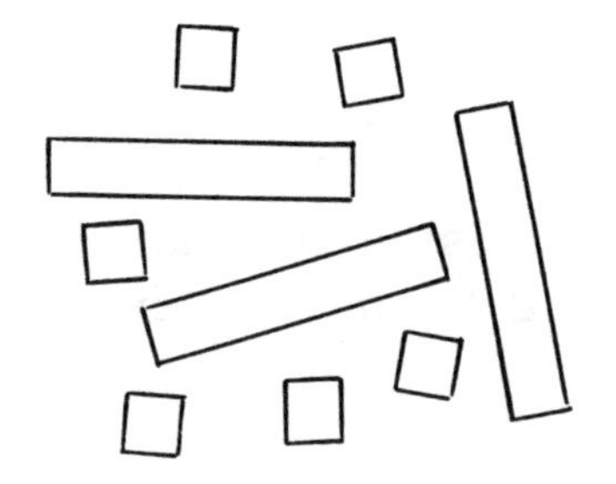
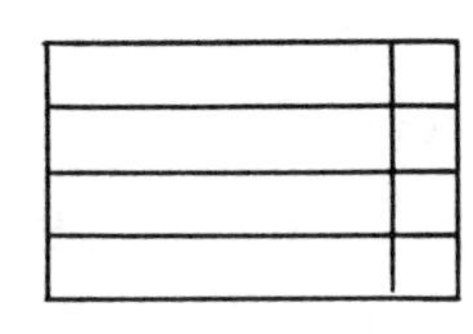
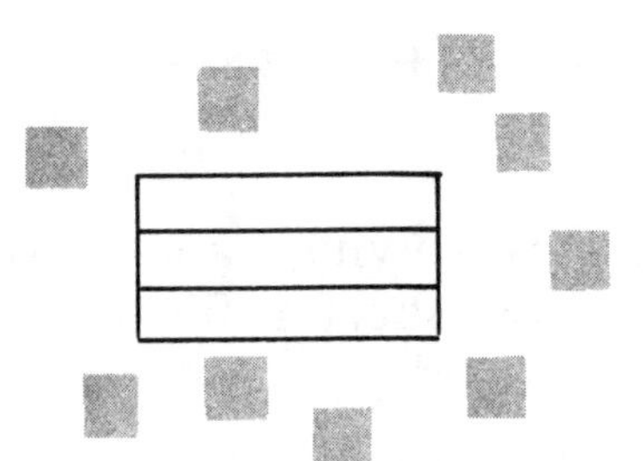

a. ____________ b. ____________ c. ____________

5. To represent the linear equation $3S + 2 = 2S + 5$ physically, place the strip-square representation of $3S + 2$ on the left pan in your diagram of a balance scale and the corresponding representation of $2S + 5$ on the right pan.
   a. Replace each strip with one white square (that is, assume $S = 1$).
   b. Would your scale remain balanced? ______
   c. Restore the original pieces to the scale.
   d. Now replace each strip with two gray squares (that is, assume $S = -2$).
   e. Simplify by using the fact that a gray square and a white square cancel each other out, since $(-1) + (+1) = 0$.
   f. Describe the contents of the left pan: ______________________
   The right pan: ______________________ Does the scale balance? ______

6. To solve the linear equation $3S + 2 = 2S + 5$ physically, we must determine the number of squares that can be used to replace each strip and keep the scale balanced. This can be accomplished by removing *equally weighted* pieces from each side of the scale until you have only one strip remaining on the scale.
   a. Represent the equation $3S + 2 = 2S + 5$ on your balance scale.
   b. Remove two squares from each side.
   c. Now remove two strips from each side.
   d. The weight of one strip equals the weight of ______ white squares. Thus, the solution of the equation is $S = 3$.

7. Use the method in exercise 6 to solve each of the following linear equations. Write your solutions in the spaces provided.
   a. $2S + 4 = S + 5$ $S =$ ______
   b. $4S = 3S + 7$ ______
   c. $5S + 3 = 4S + 8$ ______
   d. $3S = 6$ ______
   e. $4S + 1 = 2S + 7$ ______

8. Try solving each of the following equations without using the materials from sheet 4. Physically check your solutions by using the method in exercise 6.
   a. $S + 5 = 9$ $S =$ ______
   b. $2S + 6 = S + 13$ ______
   c. $5S + 2 = 3S + 6$ ______

Since we do not usually think of putting a weight of -1 on a scale, it is helpful also to look at pictorial methods for solving linear equations. For example, to solve the equation $4S + (-3) = 3S + (-4)$ pictorially, we again use the process of removing *equally valued* pieces from (or adding *equally valued* pieces to) each side of the configuration.

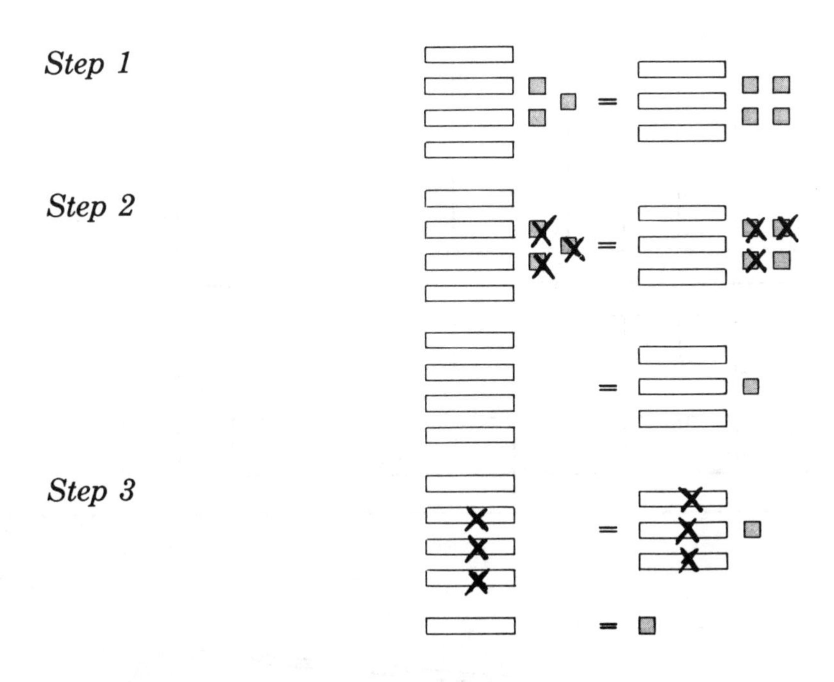

The solution is $S = -1$.

Note that to remove (or subtract) a square or strip, we cross it out with an "X." In Step 2 we could have added three white squares to each side instead of removing three gray ones.

9. Use the method described above to solve each of the following linear equations.

   a. $2S + (-5) = -9 + S$ $S =$ ______________

   b. $3S + (-6) = S + (-8)$ ______________

   c. $5S + 3 = 2[S + (-3)]$ ______________

10. Try solving each of the following equations without using pictorial representations. Check your solutions by using the method shown at the top of this sheet.

    a. $4S + (-2) = 2S + (-10)$ $S =$ ______________

    b. $5S + (-6) = 4S + 8$ ______________

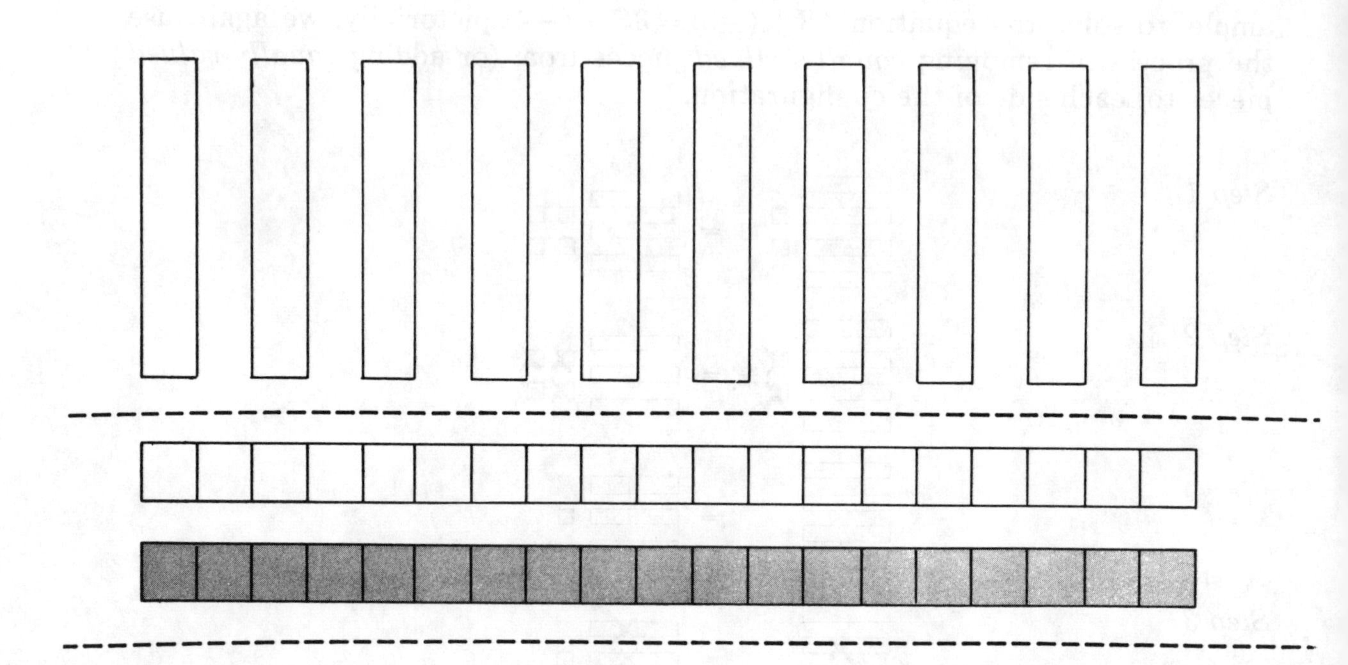

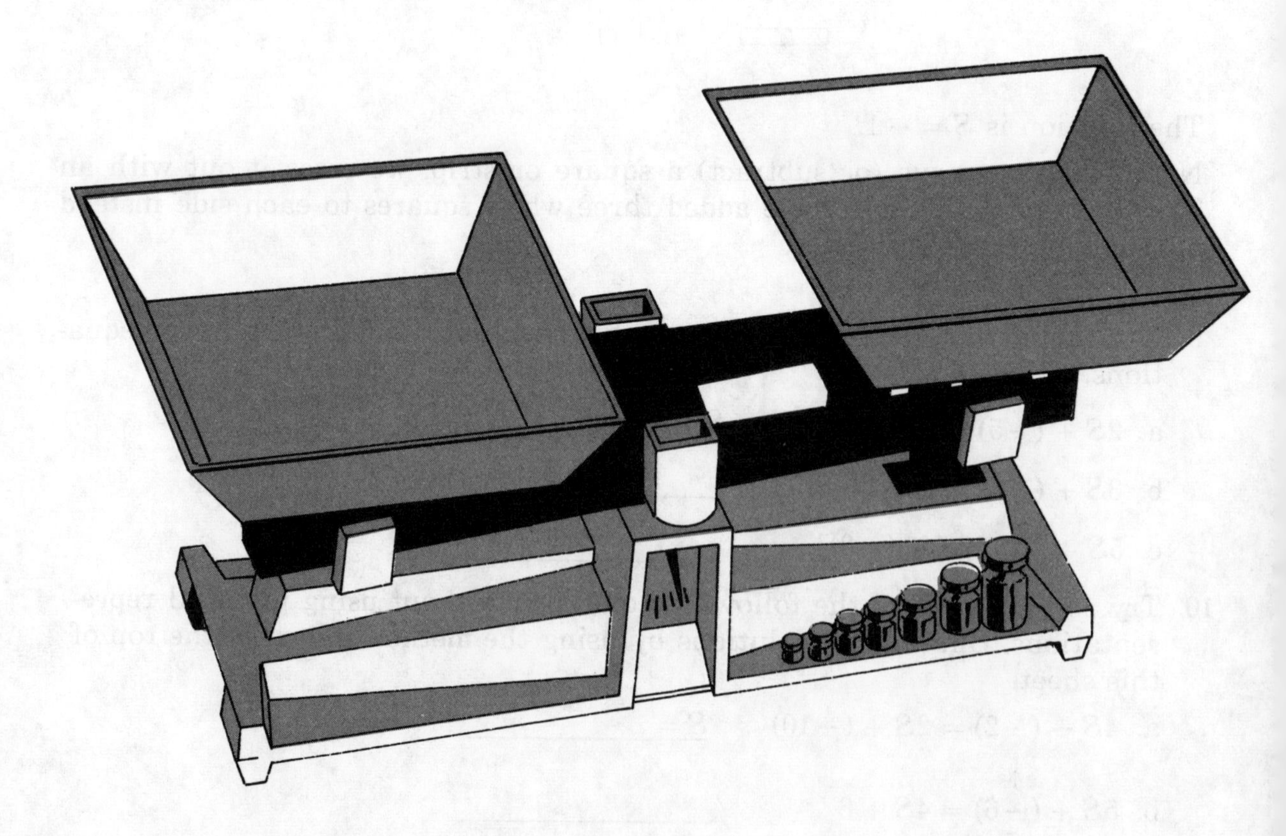

FINDING FACTORS PHYSICALLY

May 1982

By Christian R. Hirsch, Western Michigan University, Kalamazoo, MI 49008

Teacher's Guide

Grade level: 8–10.

Materials: Scissors, red pens or markers, and a set of the activity sheets for each student. Cutouts from two transparencies of sheet 4 would be helpful for discussing student solutions to the problems.

Objectives: Students will investigate factoring quadratic polynomials over the integers using a physical model and thereby discover regularities that will permit them to factor such polynomials at the symbolic level.

Background: A quadratic trinomial of the form $ax^2 + bx + c$, where $a > 0$, can be factored over the integers if and only if a rectangular shape can be formed using the corresponding pieces (with additions if necessary) from sheet 4. The dimensions of the rectangle formed are the factors of the trinomial. For example, the factors of $x^2 + 3x + 2$ can be found by arranging the pieces for x^2, $3x$ and 2 into a rectangle that measures $x + 1$ by $x + 2$. Perfect-*square* trinomials are always represented by *square* arrangements of the pieces.

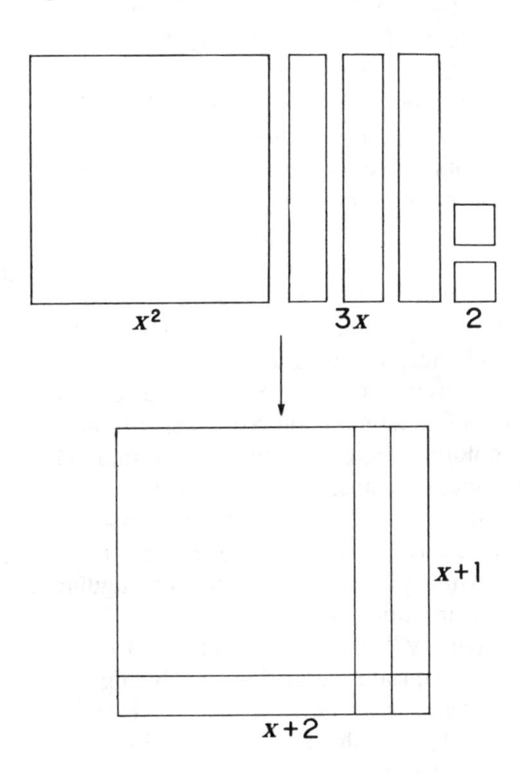

Directions: Activity sheets 1, 2, and 3 should be distributed one at a time. Students will need two copies of sheet 4—one to be distributed together with sheet 1, the other with sheet 3.

Sheets 1 and 2 introduce factoring of quadratic polynomials over the positive integers, and both can be completed during a single class session. To ensure adequate class time for the completion of these two sheets, you may wish to have pupils cut out the pieces on sheet 4 the evening prior to beginning the activity. You might suggest that students glue their copies of sheet 4 onto tagboard before cutting the pieces out. Cutouts from a transparency of sheet 4 can be used to illustrate the solution to problem 3 and thereby help ensure that all students understand how the pieces are to be manipulated to form a rectangle. You might also use the transparency cutouts to show that the x-length is not a multiple of the unit length and, in particular, the area of six of the small squares is not the same as the area of one of the rectangles. Pupils who complete sheet 2 more quickly than others might be challenged to find the factors of expressions such as $4x^2 + 20x + 25$, $6x^2 + 13x + 6$, or $9x^2 + 30x + 16$. When all students have completed sheet 2, you might have them verify their answers for problem 8 by actually multiplying the factors.

Sheet 3 focuses on factoring quadratic polynomials over the integers. The coloring of the rectangles and smaller squares can be avoided by reproducing the second copy of sheet 4 on colored paper. Similarly colored film can be used when making the second transparency of sheet 4. Use the transparency cutouts to demonstrate how the red pieces are placed on top of the white pieces to form the rectangular representation for $x^2 - 3x + 2$. Since students will find it more difficult to find the appropriate rectangles when using the red pieces, you may wish to have them complete this sheet by working in small groups. Encourage students to verify their answers for problem 12 by actually forming the corresponding rectangles (using the appropriate pieces) or by multiplying the factors. Again, some students might be challenged to find the factors of expressions such as $x^2 - 3x - 4$ or $6x^2 - 5x - 21$.

Supplementary activity: An interesting follow-up project for several students would be to investigate factoring cubic polynomials by forming rectangular solids from wooden blocks of the following sizes:

Block	Dimensions	Volume
small cube	1 by 1 by 1	1
rectangular solids	1 by 1 by x	x
	1 by x by x	x^2
larger cube	x by x by x	x^3

To ensure uniqueness of factorization, the x-length should again not be a multiple of the unit length. By using this model, students can easily see that $x^3 + 6x^2 + 12x + 8$ is the volume of a cube with edge $x + 2$, and thus $x^3 + 6x^2 + 12x + 8 = (x + 2)^3$. The article by Hendrickson (1981) illustrates nicely how Cuisenaire materials can be used for this model.

Answers:

Sheet 1: 1.b. x, c. 1; 3.c. $x + 1$ by $x + 2$; 4.a. $(x + 1)(x + 5)$, b. $(x + 2)(x + 4)$, c. $(x + 3)(x + 5)$, d. $(x + 3)(x + 4)$, e. $(x + 4)(x + 4)$, f. $x(x + 2)$, g. $(2x + 1)(x + 3)$, h. $(2x + 3)(x + 1)$, i. $(3x + 2)(x + 1)$, j. $(2x + 3)(x + 2)$.

Sheet 2: 5.a. $pq = c$, b. $p + q = b$; 6.a. $(x + 2)(x + 3)$, b. $(x + 3)(x + 3)$, c. $(x + 4)(x + 5)$, d. $(x + 1)(x + 9)$; 7.a. $pq = c$, b. $p + aq = b$; 8.a. $(2x + 3)(x + 2)$, b. $(2x + 1)(x + 4)$, c. $(2x + 3)(x + 5)$, d. $(3x + 4) \cdot (x + 2)$.

Sheet 3: 10.a. $x - 1$ by $x - 2$, b. $(x - 1)(x - 2)$; 11.a. $(x - 1)(x - 1)$, b. $(x - 2) \cdot (x - 3)$, c. $(x - 2)(x - 4)$, d. $(2x - 1)(x - 2)$, e. $(2x - 3)(x - 2)$, f. $(x - 1)(x + 2)$; 12.a. $(x - 2)(x - 6)$, b. $(x - 2)(x + 3)$, c. $(2x - 3)(x - 5)$.

BIBLIOGRAPHY

Bidwell, James K. "A Physical Model for Factoring Quadratic Polynomials." *Mathematics Teacher* 65 (March 1972):201–5.

Bruner, Jerome S. *Towards a Theory of Instruction*. New York: W. W. Norton & Co., 1968.

Flax, Rosabel. "A Squeeze Play on Quadratic Equations." *Mathematics Teacher* 75 (February 1982):132–34.

Hendrickson, A. Dean. "Discovery in Advanced Algebra with Concrete Models." *Mathematics Teacher* 74 (May 1981):353–58.

1. Sheet 4 of this activity consists of a series of large squares, rectangles, and smaller squares. The dimensions of the pieces are given below. The area of a large square is $x \cdot x = x^2$. Determine the areas of the remaining pieces.

Piece	Dimensions	Area
a. large square	x by x	x^2
b. rectangle	1 by x	______
c. smaller square	1 by 1	______

2. Carefully cut out the pieces on sheet 4.

3. a. The expression $x^2 + 3x + 2$ can be represented physically by one large square, x^2, three rectangles, $3x$, and two smaller squares, 2. Place these six pieces on your desk.

   b. Now rearrange the pieces to form a rectangle.

   c. What are the dimensions of the rectangle you formed? ______

   d. The dimensions of the rectangle formed are called the *factors* of $x^2 + 3x + 2$. We can write $x^2 + 3x + 2 = (x + 1)(x + 2)$ since both sides represent the same amount of area.

4. Form a rectangle with the pieces representing each expression given below. If a rectangle can be formed for the expression, write its dimensions in the column headed Factors.

Expression	Factors
a. $x^2 + 6x + 5$	$(x + 1)(\quad)$
b. $x^2 + 6x + 8$	______
c. $x^2 + 8x + 15$	______
d. $x^2 + 7x + 12$	______
e. $x^2 + 8x + 16$	______
f. $x^2 + 2x$	______
g. $2x^2 + 7x + 3$	______
h. $2x^2 + 5x + 3$	______
i. $3x^2 + 5x + 2$	______
j. $2x^2 + 7x + 6$	______

5. In problems 4a through 4e, you formed rectangles to represent expressions of the form $x^2 + bx + c$. The dimensions of the rectangles were of the form $(x + p)$ by $(x + q)$. That is,

$$x^2 + bx + c = (x + p)(x + q).$$

Use the results of problems 4a through 4e to answer the following questions.

a. What relationship exists between p, q and the number c?

____

b. What relationship exists between p, q and the number b?

____

6. Use your answers to 5a and 5b to find the factors of each expression below. Check your answers by actually forming the corresponding rectangles with the pieces from sheet 4.

Expression	Factors
a. $x^2 + 5x + 6$	______________
b. $x^2 + 6x + 9$	______________
c. $x^2 + 9x + 20$	______________
d. $x^2 + 10x + 9$	______________

7. In problems 4g through 4j, you formed rectangles to represent expressions of the form $ax^2 + bx + c$ where a was a prime number. The dimensions of the corresponding rectangles were of the form $(ax + p)$ by $(x + q)$; that is,

$$ax^2 + bx + c = (ax + p)(x + q).$$

Use the results of problems 4g through 4j to answer the following questions.

a. What relationship exists between p, q and the number c?

____

b. What relationship exists between a, p, q and the number b?

____

8. Use your answers to 7a and 7b to find the factors of each expression below. Check your answers by actually forming the corresponding rectangles with the pieces from sheet 4.

Expression	Factors
a. $2x^2 + 7x + 6$	______________
b. $2x^2 + 9x + 4$	______________
c. $2x^2 + 13x + 15$	______________
d. $3x^2 + 10x + 8$	______________

9. Using a red pencil or marker, color the rectangles and smaller squares on the second copy of sheet 4, and then carefully cut out these pieces.

10. The red pieces are assumed to have "negative" area. The area of a red rectangle is ^{-}x, and the area of a small red square is $^{-}1$. Thus, the areas of two congruent shapes of differing colors add to zero when placed one on top of the other. This fact can be used to find factors of an expression such as $x^2 - 3x + 2$.

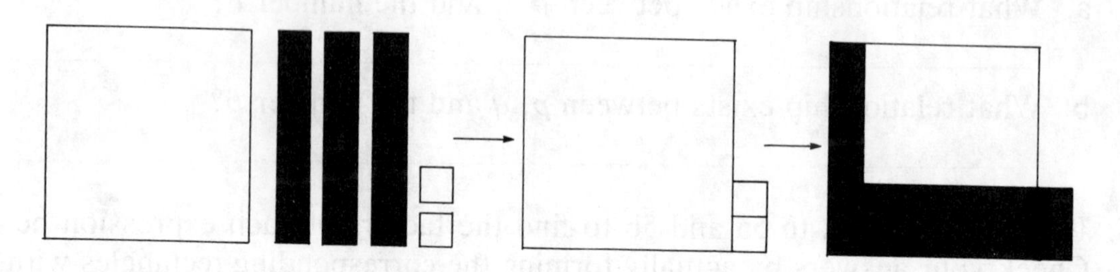

a. What are the dimensions of the white rectangle that was formed? ________

b. In factored form, $x^2 - 3x + 2 =$ (________)(________) .

11. Use your white and red pieces and the method illustrated above to form a white rectangle representing each expression below. In each case, begin by first placing the white pieces next to each other as was done above. In the column headed Factors, write the dimensions of the white rectangle that is formed.

Expression	Factors
a. $x^2 - 2x + 1$	________
b. $x^2 - 5x + 6$	________
c. $x^2 - 6x + 8$	________
d. $2x^2 - 5x + 2$	________
e. $2x^2 - 7x + 6$	________
f. $x^2 + x - 2$	________

(*Hint:* Add one white and one red rectangle to your collection of pieces representing the expression.)

12. Analyze your results for problem 11 in a manner similar to that suggested in problems 5 and 7. Now try finding the factors of each expression below without using the pieces from sheet 4.

Expression	Factors
a. $x^2 - 8x + 12$	________
b. $x^2 + x - 6$	________
c. $2x^2 - 13x + 15$	________

FACTORING KIT

SHEET 4

THOSE AMAZING TRIANGLES *September 1981*

By Christian R. Hirsch, Western Michigan University, Kalamazoo, MI 49008

Teacher's Guide

Grade Level: 7–12.

Materials: Compass, protractor, ruler, extra sheets of paper, and a set of activity sheets for each student.

Objectives: Students will practice using the basic measuring and drawing devices of geometry in the process of discovering several surprising generalizations about arbitrary triangles.

Directions: The three activity sheets are independent of one another and thus can be used individually at different times to reinforce understanding of the related topics as they are presented in your curriculum. Accurate drawings and/or constructions are a must with each of the activities. The completion of each sheet should heighten motivation for further study of geometric ideas.

Sheet 1: In this activity, pupils discover Morley's theorem: The points of intersection of the adjacent trisectors of the angles of any triangle are the vertices of an equilateral triangle (fig. 1). This surprising result was first discovered about 1899 by Frank Morley. Some students may need help interpreting the definition of adjacent trisectors given in exercise 1b. Depending on the level of your students, you may wish to suggest that for exercise 2 they draw triangles whose angle measures are integral multiples of 3, for example, a 30°-60°-90° triangle and a 30°-45°-105° triangle. A proof of Morley's theorem as given in Coxeter and Greitzer (1967) could be provided for better high school students. Other students might be encouraged to investigate the figures formed by the adjacent tri-

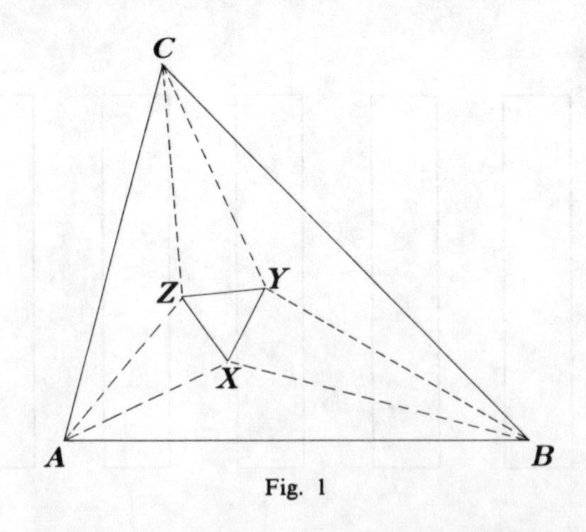

Fig. 1

sectors of the angles of *regular* polygons. For a discussion of this generalization of Morley's theorem see Hirsch et al. (1979).

Sheet 2: Students in high school geometry could be instructed to use a straightedge and compass to construct the parallel lines for this activity. Younger students might place a sheet of lined paper under the activity sheet to assist them in drawing the parallel lines. Pupils will be amazed to find that points X and X'' always coincide (fig. 2). Of course if X is chosen to be the midpoint of $\overline{AB}$ then X, X', and X'' will coincide. Tenth-grade geometry students might be challenged to prove these results. The proof of the first result depends only on noting that parallelograms $AXYZ$ and $X'BYZ$ have a common side $\overline{YZ}$ and thus $AX = X'B$. Using the definition of betweenness and subtraction, we can see that $AX' = XB$ and hence X and X' are equidistant from the midpoint M of $\overline{AB}$. If we repeat steps a–c starting with point X', then X' and X'' will be equidistant from M and thus $X = X''$. An entertaining account of this problem of closure may be found in Court (1953).

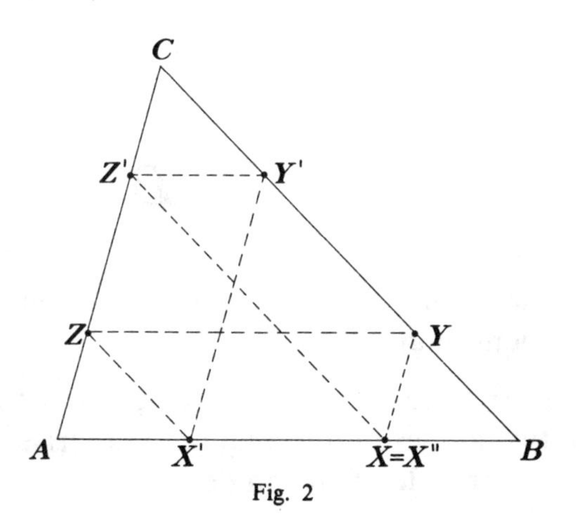

Fig. 2

Sheet 3: Depending on the background of your students, you may wish to have them construct the medians in exercise 1d. Students should compare their answers to 1g. In all cases $\triangle XYZ$ should be equilateral (fig. 3). This triangle is sometimes called the *outer Napoleon triangle* of $\triangle ABC$. Some students might be encouraged to carry out a similar investigation in which the equilateral triangles are constructed inward on the sides.

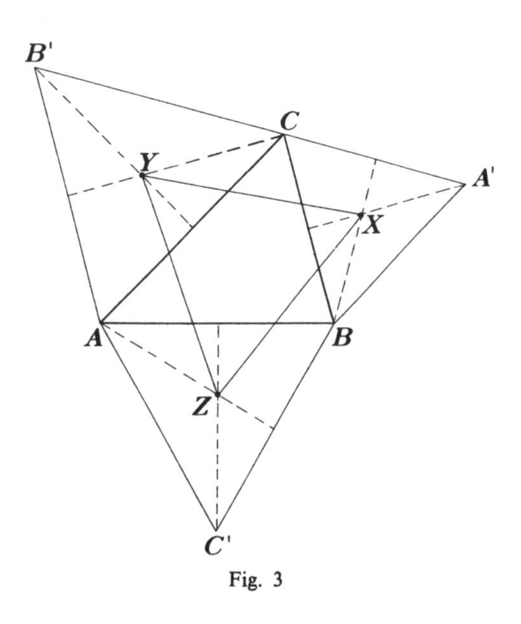

Fig. 3

In exercise 2 pupils should find that $\overline{AA'}$, $\overline{BB'}$, and $\overline{CC'}$ are concurrent and the same length. You might ask them to find the measures of the six angles around the point of concurrency to uncover yet another interesting result. If no angle of $\triangle ABC$ has a measure greater than 120°, then the point of concurrency of $\overline{AA'}$, $\overline{BB'}$, and $\overline{CC'}$ provides the solution to the following real-world problem. Suppose points A, B, and C locate consumption centers that are to be supplied materials from a single supply depot D. Where should the depot be located so that the sum of the distances from D to the three centers is as small as possible? A justification of this solution appropriate for high school students may be found in Hirsch et al. (1979).

REFERENCES

Court, Nathan A. "Geometrical Magic." *Scripta Mathematica* 19 (June–September 1953): 198–200.

Coxeter, H. S. M., and S. L. Greitzer. *Geometry Revisited.* New York: Random House, 1967.

Hirsch, Christian R., Mary Ann Roberts, Dwight O. Coblentz, Andrew Samide, and Harold L. Schoen. *Geometry.* Glenview, Ill.: Scott, Foresman & Co., 1979.

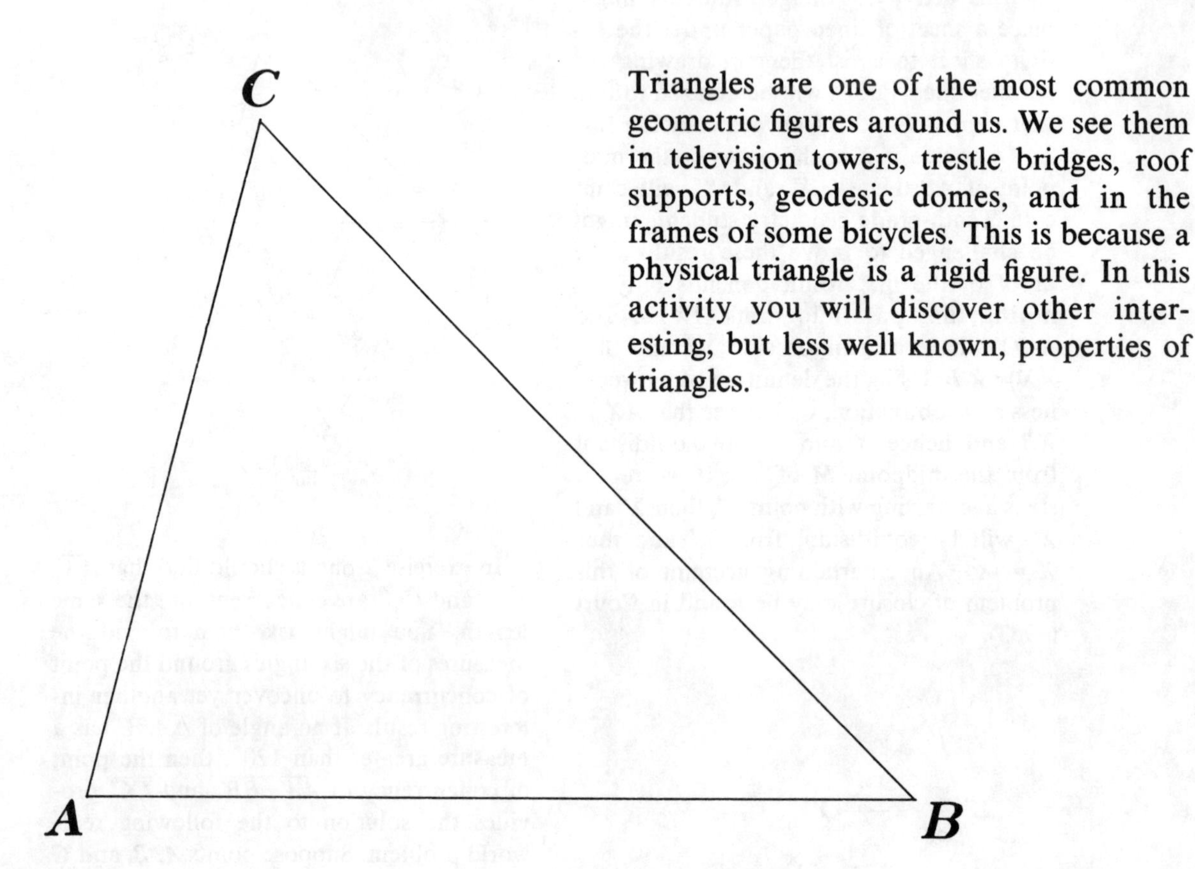

Triangles are one of the most common geometric figures around us. We see them in television towers, trestle bridges, roof supports, geodesic domes, and in the frames of some bicycles. This is because a physical triangle is a rigid figure. In this activity you will discover other interesting, but less well known, properties of triangles.

1. a. Use a protractor to trisect each angle of the triangle above. (To trisect an angle means to divide it into three congruent angles.)
   b. Label the points where the adjacent trisectors of different angles intersect X, Y, and Z. (The adjacent trisectors of two angles of a triangle are the trisectors, one at each vertex, "closest" to the common side of the angles.)
   c. Draw $\triangle XYZ$.
   d. Use a ruler or a compass to compare the lengths of the sides of $\triangle XYZ$.
   e. What kind of triangle is $\triangle XYZ$? ______________________
2. On a separate sheet of paper, draw at least two more large, differently shaped triangles and repeat steps a–e in each case.
3. It's amazing! For each triangle, the three points of intersection of the adjacent trisectors of its angles form ______________________.

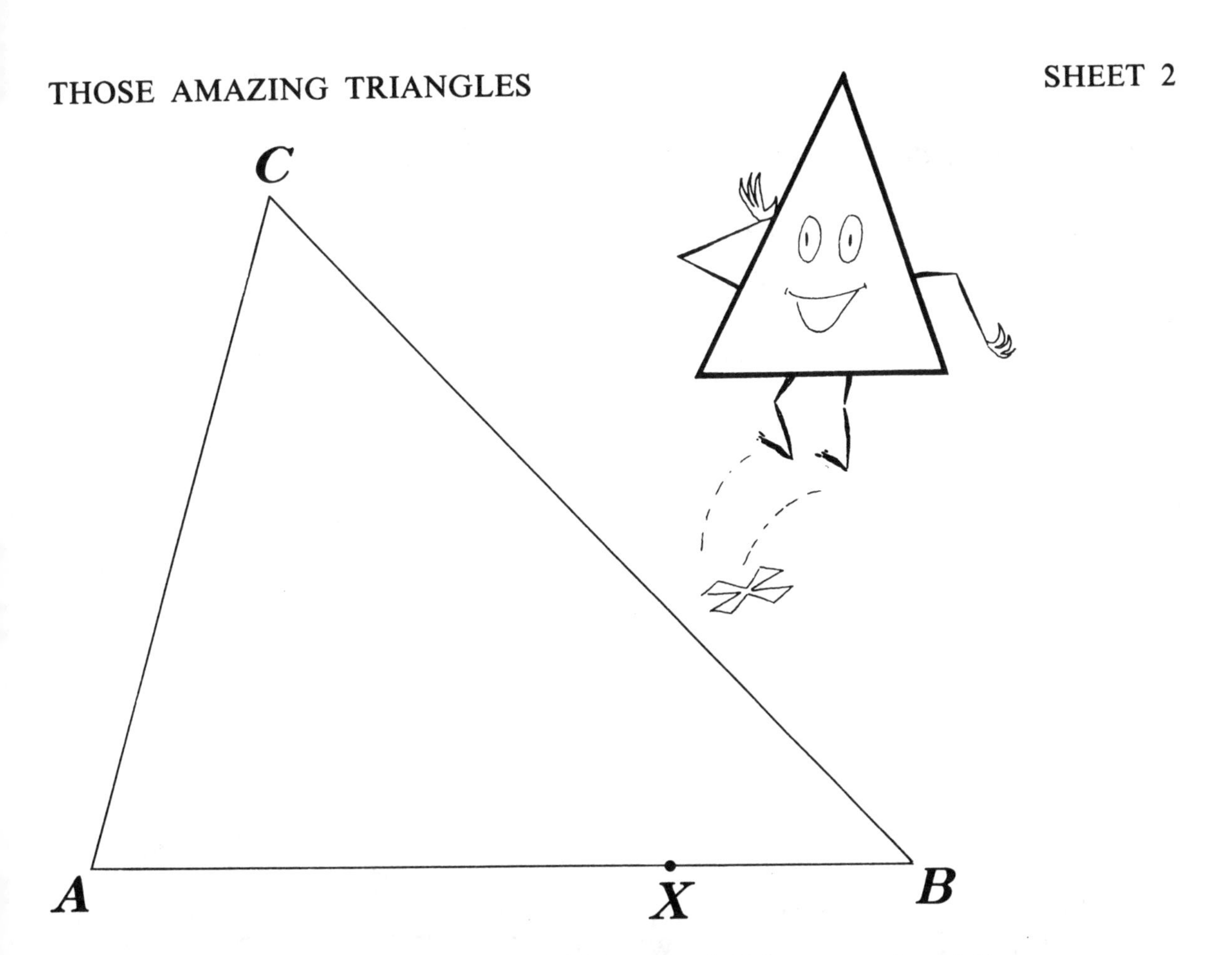

1. On the triangle above, a point X has been marked between A and B.
   a. Use a ruler to draw a line through X parallel to side $\overline{AC}$. Label the intersection with $\overline{BC}$ point Y.
   b. Through Y, draw a line parallel to side $\overline{AB}$. Label the intersection with $\overline{AC}$ point Z.
   c. Through Z, draw a line parallel to side $\overline{BC}$ and label the intersection with $\overline{AB}$ point X'.
   d. Repeat steps a–c, starting this time with point X'. Label the intersection points on the sides of the triangle Y', Z', and X'', respectively.

   What appears to be true about points X and X''? ____________________

2. Trace $\triangle ABC$ on a sheet of paper. Choose any point between A and B and label it X. Follow steps a–d. What appears to be true about points X and X'' this time? ______

   ____

3. Check to see if this amazing result holds for any triangle you choose to draw.

4. Try to find a point X on side $\overline{AB}$ of $\triangle ABC$ above so that when steps a through c are completed points X and X' coincide.

1. a. On a sheet of paper draw a large triangle. Label its vertices *A*, *B*, and *C*.
   b. Use a compass to construct an equilateral $\triangle A'BC$ outward on side $\overline{BC}$ as in the diagram below.
   c. Similarly construct an equilateral $\triangle AB'C$ on side $\overline{AC}$ and an equilateral $\triangle ABC'$ on side $\overline{AB}$.
   d. Draw two medians in $\triangle A'BC$, in $\triangle AB'C$, and in $\triangle ABC'$. (A median of a triangle is a segment from any vertex to the midpoint of the opposite side.)
   e. Label the points of intersection of the three pairs of medians *X*, *Y*, and *Z*.
   f. Use a different color pen or pencil to draw $\triangle XYZ$.
   g. What kind of triangle is $\triangle XYZ$? ______________________

That's amazing! It is believed that this result was first discovered by Napoleon. Here are a few other surprising results he missed.

2. a. Repeat steps 1a, 1b, and 1c.
   b. Draw $\overline{AA'}$, $\overline{BB'}$, and $\overline{CC'}$. What appears to be true about these segments? __________
   c. Use a ruler or a compass to compare the lengths of $\overline{AA'}$, $\overline{BB'}$, and $\overline{CC'}$. What appears to be true? __________

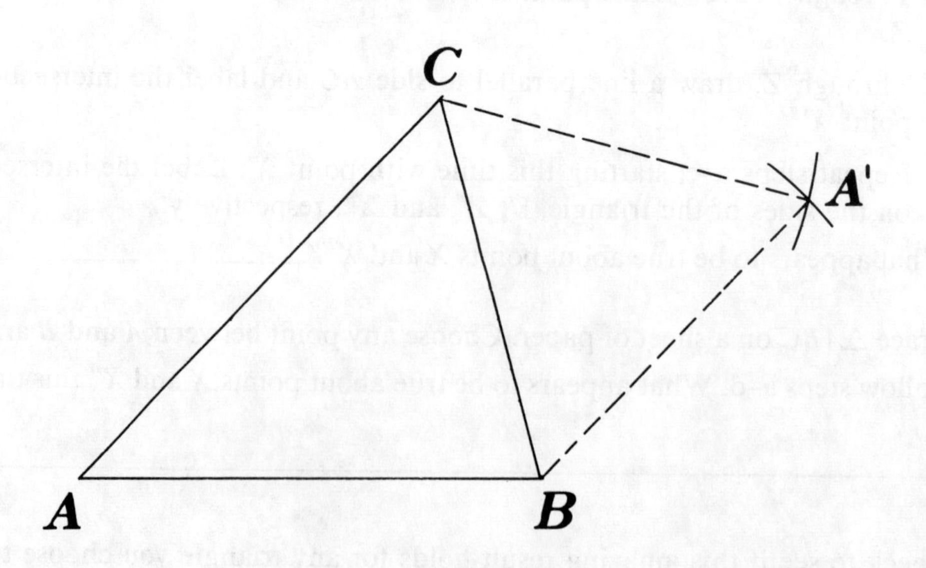

SEMIREGULAR POLYHEDRA

October 1982

By Rick N. Blake and Charles Verhille, University of New Brunswick, Fredericton, NB E3B 6E3

Teacher's Guide

Grade level: 7–12

Objectives:

1. To provide students with the opportunity to count deductively, to collect data, and to look for patterns using polyhedra
2. To check and modify, if necessary, a given pattern after obtaining further data
3. To establish patterns and test their validity

Materials: One set of worksheets for each student; models of the five regular polyhedra for each student or for small groups of students. Models of the truncated regular polyhedra should be available for classroom demonstrations and for some students who have difficulty visualizing the faces of truncated polyhedra. Templates for those models can be found in Wenninger (1975).

Prerequisites: Students should have had some experience with three-dimensional objects. These include the following:

1. Using models of the regular polyhedra to—
   a) count faces, vertices, and edges
   b) discover Euler's formula, $F + V = E + 2$
2. Being aware that the faces of regular polyhedra are congruent
3. Finding cross sections of three-dimensional figures

Procedure: Prior to distributing the worksheets, use a model of a hexahedron to identify with the class the faces, vertices, edges, and shapes of faces. Discuss truncating the hexahedron and then use a truncated model with the class to identify the vertices, edges, and the shapes of the faces.

Distribute all three worksheets to each student. Models of the regular polyhedra should be available to individual students or to small groups of students. Students should be encouraged to use the models to help them complete the worksheets.

Copies of the completed worksheets (except questions 10, 12, 14, and 15) could be posted about the room to aid students in checking question 1 as well as data they collect when completing the rest of the sheets. It is important that the information for problem 1 be correct before proceeding.

Allow students sufficient time to find the pattern in problem 10. If students cannot find the patterns in problems 10, 12, and 14, the teacher could work through the reasoning in problem 10 and have students do problems 12 and 14.

Extension:

1. Star polyhedra can be formed by placing pyramids on the faces of the regular polyhedra. Thinking strategies similar to the ones used in this activity can be used to find their faces, vertices, and edges. See Wenninger (1975) for templates and models.

2. Some patterns involving faces, vertices, and edges can be found by truncating regular prisms and pyramids. These can be generalized to prisms and pyramids with *n*-gonal bases.

Solutions:

1. See table. $F + V = E + 2$ (Euler's formula).
3. *a*) 6, square; *b*) becomes an octagon; *c*) 8, truncating the vertices, triangle; *d*) $6 + 8 = 14$.
4. *a*) 8; *b*) zero; *c*) 3; *d*) $8(3) = 24$.
5. *a*) 12; *b*) 12; *c*) 3; *d*) $12 + 8(3) = 36$.
6. *a*) 4, triangle; *b*) becomes a hexagon; *c*) 4, truncating the vertices, triangle; *d*) $4 + 4 = 8$.
7. *a*) 4; *b*) zero; *c*) 3; *d*) $4(3) = 12$.
8. *a*) 6; *b*) 6; *c*) 3; *d*) $6 + 4(3) = 18$.
9. *a*) decagons and triangles; *b*) $12 + 20 = 32$.
10. Number of faces plus number of vertices on a regular polyhedron equals the number of faces on the truncated polyhedron.
11. $20(3) = 60$.
12. (Number of vertices on a regular polyhedron) × (number of vertices per truncated vertex)
13. $30 + 20(3) = 90$.
14. (Number of edges of a regular polyhedron) + (number of vertices of the regular polyhedron) × (number of vertices per truncated vertex)
15. *a*) $F + V = E + 2$; *b*) same.
16.

Polyhedron	*F*	*V*	*E*	*F*	*V*	*E*
Tetrahedron	4	4	6	8	12	18
Hexahedron	6	8	12	14	24	36
Dodecahedron	12	20	30	32	60	90
Octahedron	8	6	12	14	24	36
Icosahedron	20	12	30	32	60	90

17. Yes

BIBLIOGRAPHY

Butler, Ruth, and Robert W. Clark. "Faces of a Cube." *Mathematics Teacher* 72 (March 1979): 199–202.

Davis, Edward J., and Don Thompson. "Sectioning a Regular Tetrahedron." *Mathematics Teacher* 73 (February 1980):121–25.

O'Daffer, Phares, and S. Clements. *Geometry: An Investigative Approach*. Menlo Park, Calif.: Addison-Wesley, 1976.

Wenninger, Magnus. *Polyhedron Models for the Classroom*. 2d ed. Reston, Va.: National Council of Teachers of Mathematics, 1975.

1. *a*. Use your models of regular polyhedra to complete the first half of the table at the bottom of sheet 3.

   b. Write a formula relating the numbers F, V, and E. ______________
   (Check your results before proceeding.)

2. If you cut off the vertices of a polyhedron, you obtain a *truncated polyhedron*. Let the cuts be made approximately one-third the distance from each vertex.

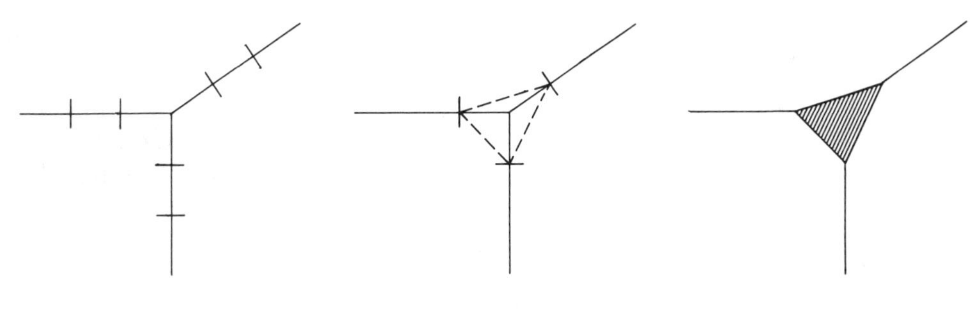

STEP 1 STEP 2 STEP 3

3. *a*. How many faces are on a hexahedron? ___

   What is the shape of each face? ________

   b. What happens to the shape of the face of the hexahedron when it is truncated?

   ____

   c. When you truncate a hexahedron, how many additional faces do you get? __________

   Where do these faces come from?

   ____

   What is their shape? ______________

   d. What is the total number of faces on a truncated hexahedron? ____________

   e. Record your information in the table on sheet 3.

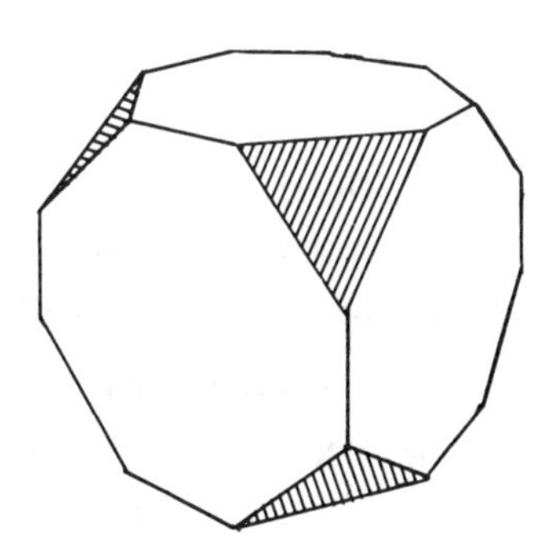

TRUNCATED HEXAHEDRON

4. *a*. How many vertices are on a hexahedron? ______________

   b. How many of these are on a truncated hexahedron? ______________

   c. How many vertices are formed at each additional face? ______________

   d. What is the total number of vertices on a truncated hexahedron? ________

   e. Record your information in the table on sheet 3.

5. *a*. How many edges are on a hexahedron? ______________

   b. How many of these are on a truncated hexahedron? ______________

   c. How many edges are formed with each additional face? ______________

   d. What is the total number of edges on a truncated hexahedron? ________

   e. Enter your data in the table on sheet 3.

6. *a*. How many faces are on a tetrahedron? _____
   What is the shape of each face? ____________
   b. What happens to the shape of the face of a tetrahedron when it is truncated?
   ____
   c. When you truncate a tetrahedron, how many additional faces do you get? ____________
   Where do these faces come from? ____________
   What is their shape? ____________
   d. What is the total number of faces on a truncated tetrahedron? ____________
   e. Record your information in the table on sheet 3.

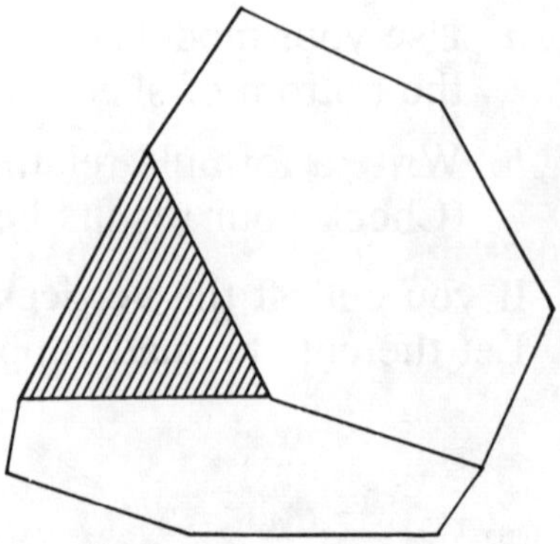

TRUNCATED TETRAHEDRON

7. *a*. How many vertices are on a tetrahedron? ____________
   b. How many of these are on a truncated tetrahedron? ____________
   c. How many vertices are formed with each additional face? ____________
   d. What is the total number of vertices on a truncated tetrahedron? ________
   e. Record your information in the table on sheet 3.
8. *a*. How many edges are on a tetrahedron? ____________
   b. How many of these are on a truncated tetrahedron? ____________
   c. How many edges are formed with each additional face? ____________
   d. What is the total number of edges on a truncated tetrahedron? ____________
   e. Record your information in the table on sheet 3.
9. Use the reasoning in problems 3 and 6 to determine the following:
   a. What is the shape of each face of a truncated dodecahedron? ____________
   b. How many faces does it have? ____________
   c. Enter this information in the table on sheet 3.

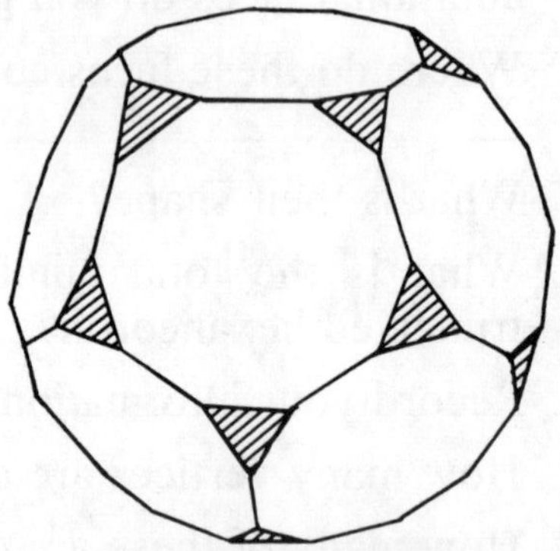

TRUNCATED DODECAHEDRON

10. What is a rule for finding the number of faces when you truncate regular polyhedra?
    ____
    ____
11. Use the reasoning in problems 4 and 7 to determine the number of vertices of a truncated dodecahedron and enter this information in the table on sheet 3.
12. What is a rule for finding the number of vertices of truncated regular polyhedra?
    ____

13. Use the reasoning in problems 5 and 8 to determine the number of edges of a truncated dodecahedron and enter this information in the table on sheet 3.
14. What is a rule for finding the number of edges of truncated regular polyhedra? ____
15. *a*. What is the relationship among the number of faces, vertices, and edges for a truncated polyhedron? ____

    b. How does it compare to the relationship you found in problem 1? ____
16. Determine and record the number of faces, vertices, and edges for a truncated octahedron and icosahedron.
17. Does the relationship from problem 15 still hold? ____
18. Test your rules from problems 10, 12, and 14 for both the truncated octahedron and icosahedron.

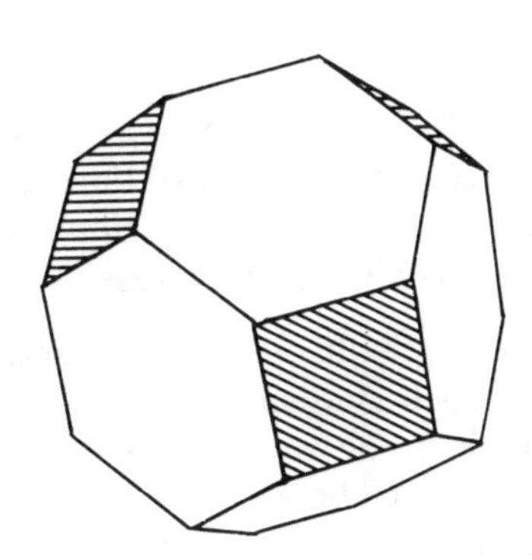

TRUNCATED OCTAHEDRON

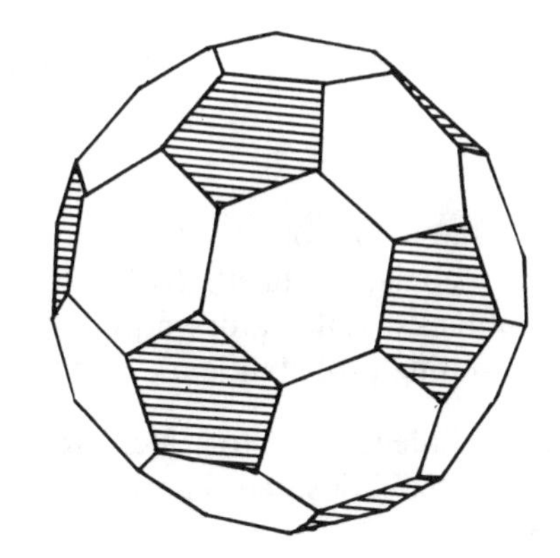

TRUNCATED ICOSAHEDRON

Polyhedron	Regular				Truncated			
	Faces F	Vertices V	Edges E	Shapes of Faces	Faces F	Vertices V	Edges E	Shapes of Faces
Tetrahedron								
Hexahedron								
Dodecahedron								
Octahedron								
Icosahedron								

OUT OF THIN AIR

April 1982

By Vicki Schell, Pensacola Junior College, Pensacola, FL 32504

Teacher's Guide

Grade level: 7–9.

Materials: One copy of sheet 1 and sheet 2 and two copies of sheet 3 for each student; thumbtacks. Calculators may be helpful in computing probabilities.

Objectives: Students are to calculate experimental probabilities and theoretical probabilities.

Background: The term *geometric probability* is used to mean that one can use geometric concepts (in particular, length, area, and volume) in order to make appropriate probability computations.

A model or a simulation is a *representation* of something else. It rarely reflects all the details of the concept it represents; the function of such a model is to give us an idea of how the concept works.

In our simulations, we are making an assumption that the points are chosen at random, that is, that they are all equally likely to be chosen.

The law of large numbers says, in effect, that we can make the experimental results agree as closely as we like with the theoretical probability of our simulation by having a sufficiently large number of trials. The experimental probability is calculated by dividing the number of times a thumbtack lands in a particular region by the number of times it has been tossed.

Directions: Distribute worksheets and thumbtacks to each student.

Before beginning the worksheets, have a group discussion to be sure that everyone understands the problem, the simulation, and the procedures and can calculate the experimental and theoretical probabilities. Agree that the position of the thumbtack is determined by its center point.

Allow sufficient time for students to carry out the trials and to organize their data and calculate the probabilities. (It may be appropriate before going on to sheet 2 to discuss as a class why the results of the theoretical probabilities may differ from those of the experimental probabilities on sheet 1.) Sheet 2 encourages the discovery that merely because we have two events, we do not necessarily have equally likely outcomes, and that other factors may have to be considered.

Additional problems: Suggest minor changes in our original simulation. Rearrange the grid into new divisions of fields, varying the number of fields or varying the area of each field. Resultant probabilities can be calculated.

For example, divide the field in half, like this:

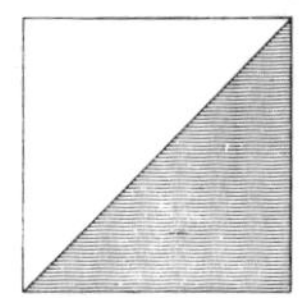

or like this:

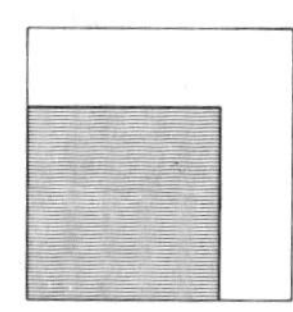

Are the experimental results essentially the same for each of these patterns?

Solutions

Sheet 1

1. Yes
2. Answers will vary.
3. Answers will vary.
4. 1/4
5. 1/2
6. 3/4

Sheet 2

7. 63 square units; 45 square units; 108 square units
8. Answers will vary.
9. Answers will vary.
10. $63/108 = 0.58\overline{3}$; $45/108 = 0.41\overline{6}$.
11. Answers will vary.
12. Experimental chance
13. Answers will vary.
14. Answers will vary.
15. The law of large numbers tells us that we could make our results agree more closely if we had a sufficiently large number of trials.

BIBLIOGRAPHY

Carnahan, Walter H. "Pi and Probability." *Mathematics Teacher* 46 (February 1953):65–66.

Dahlke, Richard, and Robert Fakler. "Geometrical Probability." In *Teaching Statistics and Probability*, 1981 Yearbook of the National Council of Teachers of Mathematics, edited by Albert P. Shulte, pp. 143–53. Reston, Va.: The Council, 1981.

Hirsch, Christian R. "Probability and Pi." *Mathematics Teacher* 70 (September 1977):519–22.

Smith, J. Philip. "Probability, Geometry, and Witches." *Mathematics Student Journal* 19 (May 1972):1–3.

PROBLEM: A PARACHUTIST JUMPS FROM AN AIRPLANE AND LANDS IN A FIELD. (See diagram.)

What are the chances that she will land in a particular numbered plot?

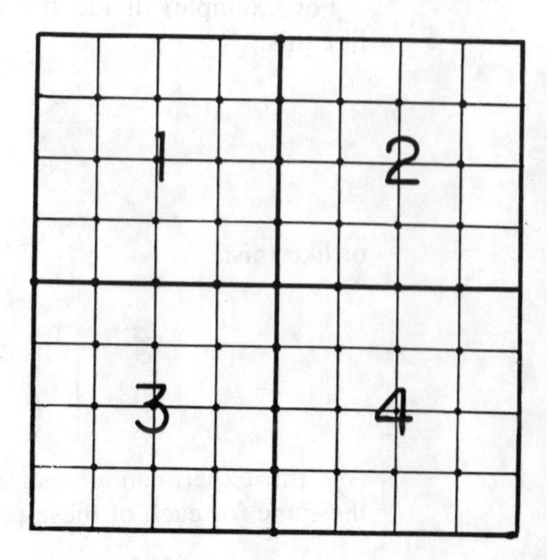

Directions:

Divide your grid into squares with sides 4 units as shown.

Model the situation by randomly tossing a thumbtack onto the grid from several feet away. Record your results in the chart below. (If the thumbtack bounces off the field, do not record that toss.)

Plot	Tally	Total
Plot 1		
Plot 2		
Plot 3		
Plot 4		

1. Does she have an equal chance to land in all parts of the field?__________
2. How many times did she land in plot 1?__________
3. What is her experimental probability of landing in plot 1?__________
4. What is the theoretical probability that she will land in plot 1?__________
5. What is the expected (theoretical) probability of landing in plots 1 or 4?__________
6. What is the expected (theoretical) probability of *not* landing in plot 2?__________

PROBLEM: ANOTHER PARACHUTIST JUMPS FROM AN AIRPLANE AND LANDS IN A DIFFERENT FIELD.

Where will he land?

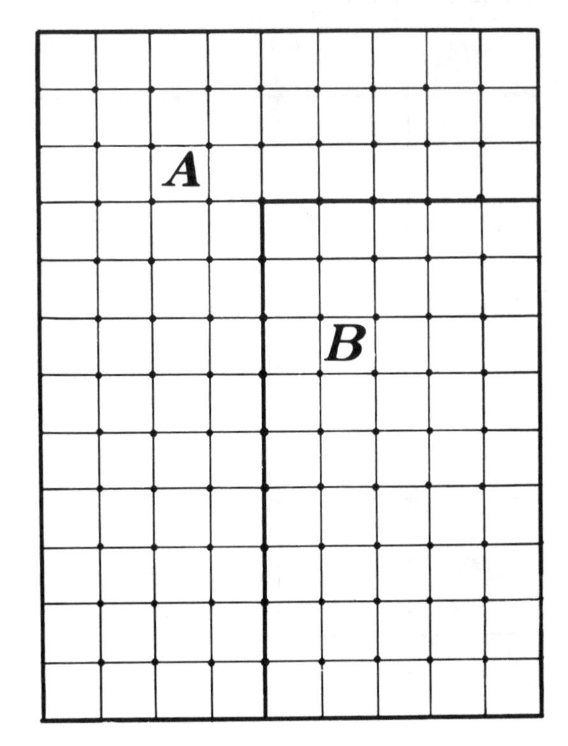

Directions:

Divide your grid into two areas as shown on the diagram.

Model the situation by tossing a thumbtack onto the grid fifty times from several feet away. Record your results in the chart below. (If the thumbtack bounces off the field, do not record that toss.)

Plot	Tally	Total
Plot *A*		
Plot *B*		

7. What is the area of plot *A*? ________ of plot *B*? ________

   What is the total area of the two plots? ________

8. How many times did he land in plot *A*? ________

9. What is the experimental probability that he will land in plot *A*? ________

10. What is the theoretical probability the parachutist will land in plot *A*? ________ in plot *B*? ________

11. Does the experimental probability differ from the theoretical probability? ________

12. How can you explain this difference? ________________________________

13. Tabulate the experimental probability using the total results of the entire class. ________

14. Does this change your experimental probability? ________

15. How can you explain this difference? ________________________________

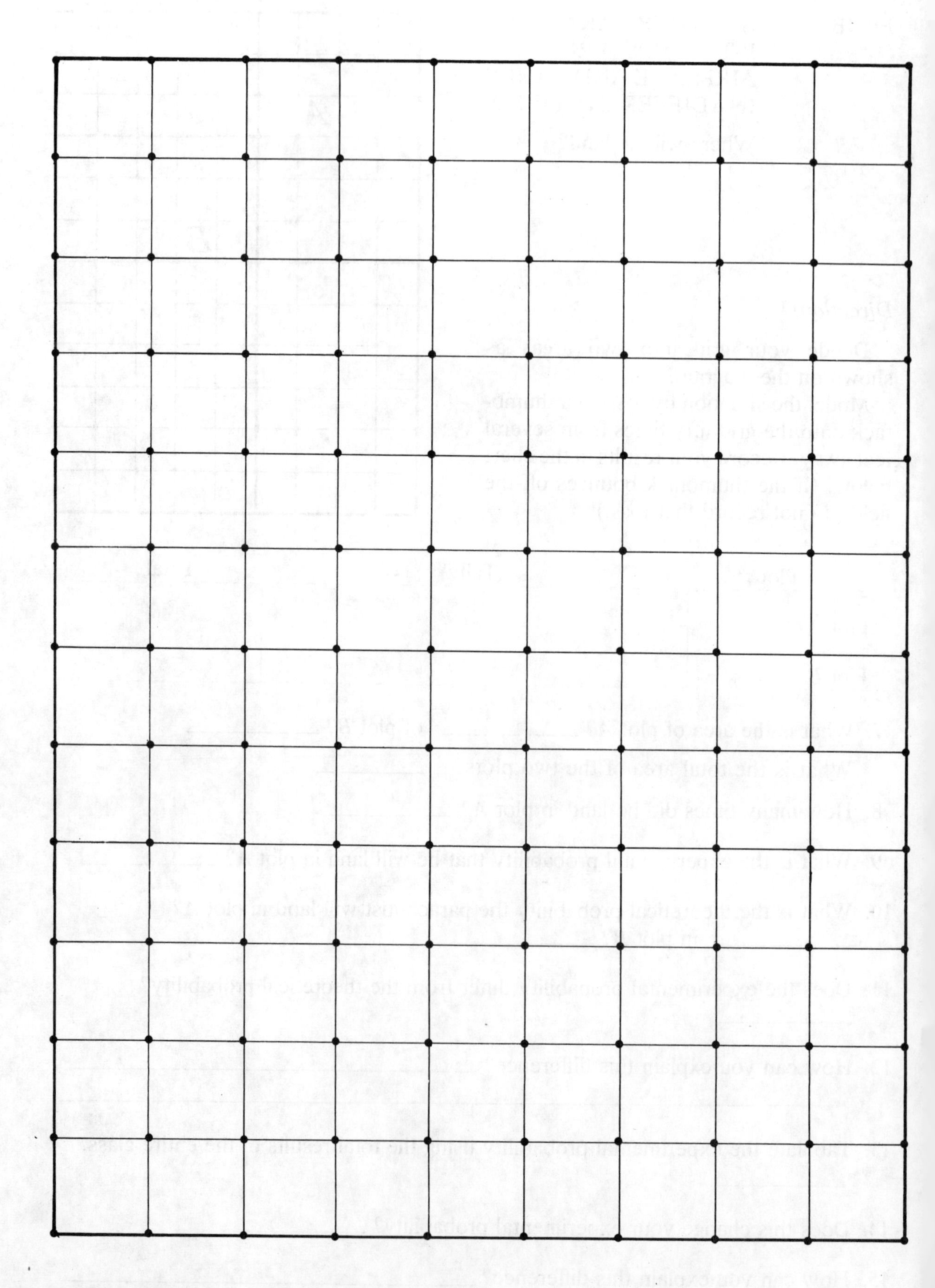

vessels continue along the superficial inguinal nodes. The glans penis and penile urethral lymphatics traverse the penis beneath Buck's fascia and usually terminate in the deep inguinal lymph nodes of the femoral triangle or occasionally in the external iliac nodes.

The male urethra consists of anterior and posterior portions and its length varies from 17.5 to 20 cm (12). While the anterior urethra is subdivided into bulbous and penile (or pendulous) segments, the posterior urethra is subdivided into prostatic and membranous segments.

The anterior urethra extends from the external meatus to the inferior edge of the urogenital diaphragm, coursing through the CS (12). The bulbous portion, the widest segment (1.5–2 cm), is surrounded by the bulb of the CS, does not have a stiff fascia, and extends up to the penoscrotal junction, delineated by the suspensory ligament of the penis (13,14). The proximal part of the bulbous urethra, termed the "sump" of the bulbous urethra, assumes a conical shape at the bulbomembranous junction (12). The penile urethra lies within the free part of the penis, extending from the penile ligament to the external urethral meatus. This part terminates in the glans penis, and there is an ampullar dilatation that is 1 to 1.5 cm long and is called the fossa navicularis (12,13). Many small glands (of Littré) are located at penile urethral lumen; they are more numerous in the bulbar urethral area and near the fossa navicularis (13).

The posterior urethra extends from the bladder neck to the lower edge of the urogenital diaphragm. The prostatic urethra is approximately 3 to 4 cm long, begins at the bladder neck, and passes through the prostate gland slightly anterior to midline, assuming an arcuate course as far as the prostate apex. It has an elevation in the midportion, called the verumontanum, which is 1 cm long. In the center of the verumontanum lies the prostatic utricle, a small saccular depression that is a remnant of the müllerian duct. The orifices of the paired ejaculatory ducts are located just distal and lateral to the utricle. The prostatic glands empty directly into the prostatic urethra via multiple small openings that surround the verumontanum. The prostatic urethra tapers distally and forms the membranous urethra, which is approximately 1 to 2 cm long and passes through the urogenital diaphragm and is surrounded by skeletal muscle forming the external urinary sphincter (12–14).

The Cowper glands are located within the urogenital diaphragm on each side of the membranous urethra. The 2-cm-long ducts of the Cowper glands drain into the bulbous urethral sump on either side of the midline (12).

SONOGRAPHIC TECHNIQUE FOR EVALUATION OF THE PENIS

Sonographic examination of the penis is performed with the patient in either the supine or lithotomy (frog leg) position, with the penis lying on the anterior abdominal wall or supported with towels between the thighs. Since the structures being imaged lie close to the surface of the transducer, high-frequency (7.5–12 MHz) linear array transducers are used to obtain high-resolution images of the penis (1–3). A sufficient amount of sonographic acoustic gel should be used on the surface of the penis in order to obtain good-quality images, and excessive compression by the transducer should be avoided, especially in trauma patients. Examination is performed in transverse and longitudinal planes, starting at the level of the glans and moving toward the base of the penis (3,15). A transperineal approach may be used, if required, to assess the base of the penis. The two CC are homogeneous in echotexture and are identified as two hypoechoic circular structures. The tunica albuginea is visualized as a linear hyperechoic structure covering the CC (normal images) (Fig. 1) (3,15). The cavernosal artery is visualized on the medial portion of the CC. The normal cavernosal artery diameter ranges from 0.3 to 1.0 mm (mean 0.3–0.5 mm) in the flaccid state (14,16). The CS is often compressed and is difficult to visualize optimally from the ventral aspect (3). Color Doppler examination of the penis should be performed in both transverse and longitudinal planes. Peak systolic velocities (PSVs) of the cavernosal arteries should also be recorded. The cavernosal artery velocity in normal healthy volunteers is 10 to 15 cm/sec in the flaccid state (15,17).

ULTRASOUND EVALUATION OF THE URETHRA (SONOURETHROGRAPHY)

The anterior urethra can be examined with a high-resolution linear array transducer (as used in a penile examination) after distension of the lumen. Distension can be achieved using either an antegrade or a retrograde approach. The antegrade technique requires the patient to void and if a full stream is achieved, a clamp is placed to distal penis to preserve distension (15). A retrograde traditional technique is the introduction of normal saline solution into the urethra via either a Foley catheter or a syringe placed directly in the urethral meatus (13). Alternatively, intraluminal lubricant and anesthetic jelly, i.e., lidocaine urologic jelly, can be used rather than saline solution to distend the urethra in sonourethrography (13,15,18). It is injected directly into the meatus until the urethra appears fully distended on sonography (13). After distension, a flexible penile clamp is applied to the glans to maintain urethral distension (13,15). Sonourethrography using anesthetic jelly is especially helpful in patients with distal urethral stenosis and stenosis of the meatus, because it is not possible to introduce the Foley catheter into the urethra (13). The retrograde technique is generally preferred. The penile urethra is examined from the dorsal surface of the penis; if needed, scanning can also be performed from the ventral surface. The distal bulbar urethra can be examined through the scrotum, and the proximal bulbous urethra can be imaged transperineally (15). Longitudinal scans are more useful than transverse scans because of their wide field of view (13).

The posterior urethra can be evaluated using a transrectal approach similar to the study of the prostate and the anal canal (13,19). After the transducer is inserted into the rectum and directed anteriorly toward the prostate, the patient is asked to urinate. The main disadvantage of this approach is the patient's inability to initiate urination. This problem can be solved by changing the patient position to upright or by using a special stool with a hole at the center through which the probe can be inserted into the rectum (13,20).

Real-time examination is performed with anesthetic jelly before distension of the urethra, in order to detect any stones or foreign bodies that might otherwise dislodge during injection and to evaluate any fluid-filled structures for change in size during injection. The normal urethra is visualized on sonography as a tubular anechoic structure with a thin, smooth echogenic wall; the lumen distends to at least 4 mm in diameter (15).

PHYSIOLOGY OF PENILE ERECTION

Normal erectile function requires psychological health, normal endocrine balance, an intact nervous system, adequate penile blood supply, and normal cavernosal sinusoidal tissue (15). In the flaccid penis, a balance exists between in- and outflow of blood from the erectile bodies (11). The smooth-muscle tissue of the cavernous sinusoids and arterioles control tumescence and detumescence of the penis (21). The sympathetic and parasympathetic nervous systems coordinate to regulate the tonicity of the cavernous smooth muscle (8). Penile erection can be elicited by central psychogenic and reflexogenic mechanisms, which interact during normal sexual activity (10). Psychogenic erections are initiated centrally in response to auditory, visual, olfactory, or imaginary stimuli. Reflexogenic erections result from stimulation of sensory receptors on the penis, which, through spinal interactions, cause somatic and parasympathetic efferent actions (10,11). Cholinergic transmission supports erection, and a noncholinergic nonadrenergic neurotransmitter also seems to have a central role in erection (8). Nitric oxide has been demonstrated to be the nonadrenergic, noncholinergic neurotransmitter responsible for this mediated mechanism of erection (22,23). On arousal, parasympathetic activity triggers a series of events starting with the release of nitric oxide and ending with increased levels of the intracellular mediator, cyclic guanosine monophosphate (cGMP). Increases in cGMP cause penile vascular and trabecular smooth-muscle relaxation (10,11,24). Erection occurs when the smooth muscles of the sinusoids and the arterioles relax in response to nitric oxide and cGMP, resulting in a large increase in arterial flow and rapid filling of the sinusoidal spaces (8). The rapid filling of the cavernosal spaces compresses the venules and emissary veins, thus leaving the corpora against the fibrous tunica albuginea, resulting in decreased venous outflow (8,11,25). The combination of increased inflow and decreased outflow rapidly raises intracavernosal pressure, resulting in progressive penile rigidity and full erection (11). Adrenergic impulses control detumescence. Norepinephrine released from postganglionic sympathetic nerve endings, causes contraction of the sinusoidal smooth muscle, thereby opening the subtunical venular plexus. Detumescence is the result of the prompt drainage of intracavernous blood (8).

ERECTILE DYSFUNCTION

Etiology and Pathophysiology of ED

ED is defined as the persistent or recurrent inability to achieve or maintain penile erection sufficient for satisfactory sexual performance. This disorder is very common worldwide and it is estimated to affect up to 30 million men in the United States (26). ED increases with age, with an incidence of 39% in men in their 40s and 67% in men in their 70s (26,27). ED can be divided into two major categories, psychogenic and organic. Previously, psychogenic ED was considered to be the most common type, but current evidence suggests that the organic cause is more common and is present in as many as 80% patients with ED (26). Psychogenic ED is common in younger patients with a mean age of 35 years (28).

Organic causes of ED are subdivided into vasculogenic (arteriogenic, venous, cavernosal, and mixed), neurogenic, anatomic, pharmacological, and endocrinologic etiologies (Table 1) (29). A psychological component usually coexists regardless of the primary etiology (30).

Vasculogenic etiologies represent the largest group, with either arterial insufficiency or venous incompetence (31–33). Since each corpus cavernosum receives arterial blood via its cavernosal artery, a branch of the internal pudendal artery arising from the internal iliac artery, ED

TABLE 1 ■ Organic Causes of Erectile Dysfunction

Organic Causes of Erectile Dysfunction
Vasculogenic
Arterial
Cavernosal
Mixed
Neurogenic
Spinal cord injury
Multiple sclerosis
Herniated disk
Anatomic
Peyronie's disease
Endocrinologic
Hypogonadism
Hyperprolactinemia
Hypothyroidism
Hyperthyroidism
Diabetes mellitus
Others
Cigarette smoking
Chronic alcohol abuse
Medication

will occur from stenosis or occlusion of any of these three arteries, thereby preventing the necessary increase in arterial flow required to generate or maintain an erection (31–34). Progressive arteriosclerosis is the most common cause of arterial ED. Risk factors for the development of atherosclerosis in the cavernosal arteries are the same as for peripheral or other vascular disease in general and include hypertension, hypercholesterolemia, tobacco use, and diabetes mellitus (8). In the presence of adequate arterial inflow, ED due to venous insufficiency is termed venous or veno-occlusive ED (8). Venous drainage from the CC is through small emissary veins that drain into the crural and cavernosal veins at the base of the penis as well as through the dorsal vein. ED will result from venous incompetence if the engorged cavernosal tissue does not expand to compress the emissary veins against the tunica albuginea (31,34). Occasionally, the presence or development of excessively large venous channels through the CC can cause venous leakage. Degeneration of the tunica albuginea causes inadequate compression of the subtunical and emissary veins and may occur with aging or Peyronie's disease (8). Congenital or acquired venous shunts between the corpus cavernosum and the spongiosum also can result in ED (8). Vasculogenic ED occurs in approximately 50% to 70% of patients with organic ED. While arterial insufficiency alone accounts for approximately 30% of these cases, isolated venous leaks with normal arterial function are found in about 15% of the cases (31,32,34).

Occasionally, organic ED can occur due to anatomic or morphological abnormalities of the CC or the tunica albuginea. Focal or diffuse scarring of cavernosal tissue prevents dilatation and expansion of the sinusoidal spaces. Peyronie's disease, i.e., fibrotic or calcified plaques along the tunica albuginea, causes painful curvature of the penis during erection, leading to rapid detumescence (31,35).

Endocrine abnormalities causing ED include hypogonadism, hyperprolactinemia, hypothyroidism, and hyperthyroidism. Diabetes mellitus is the most common endocrine disease causing ED, but it is caused by vasculogenic or neurologic ED rather than by hormonal deficiency (11,27). The most common neurological causes of ED are spinal cord injury, multiple sclerosis, and herniated disc (11). Cigarette smoking and chronic alcohol abuse also cause ED (27). Cigarette smoking may induce vasoconstriction and penile venous leakage because of its contractile effect on the cavernous smooth muscle (36). Chronic alcohol abuse may result in liver dysfunction, decreased testosterone and increased estrogen levels, and alcoholic polyneuropathy that may also affect penile nerves (37).

Many medications have been associated with various types of sexual dysfunction (38). About 25% of cases of ED are related to medication side effects (26). Generally, drugs that interfere with central neuroendocrine or local neurovascular control of penile smooth muscle have the potential to cause ED (11,38,39). Medications used to treat hypertension (diuretics such as thiazides and spironolactone, sympatholytics such as methyl dopa, clonidine, reserpine, and α blockers, and β blockers, especially nonselective agents), depression (tricyclic antidepressants, monoamino oxidase inhibitors, and selective serotonin reuptake inhibitors), and other psychiatric disorders (antipsychotic agents and anxiolytic agents such as benzodiazepines) are most commonly associated with ED (11,38,39).

Diagnostic Tests for the Evaluation of ED

Nocturnal Penile Tumescence Test

After excluding neurological and endocrinologic abnormalities, sleep monitoring of penile tumescence and rigidity [nocturnal penile tumescence (NPT) test] can be performed to exclude psychogenic causes of ED. This test is performed in the sleep laboratory during the rapid-eye-movement stage of sleep (2,31,40). A normally healthy male develops NPT during this phase of sleep. In the absence of drugs, alcohol or a sleep disorder, patients who fail to achieve erections during sleep most likely have an organic cause of impotence (31,40). Various tests have been designed to evaluate NPT to diagnose ED, but all the tests have limitations and there is no universal agreement on which test to perform for NPT; hence, these tests are not currently favored (2).

Penile–Brachial Pressure Index

Is also used to screen ED patients. The penile–brachial pressure index (PBI) represents the ratio of penile systolic blood pressure divided by the brachial arterial systolic blood pressure. A PBI of 0.7 or less has been used to indicate arteriogenic impotence (2). However, this test is nonspecific and inconsistent (41,42).

Duplex Doppler Sonography

Duplex Doppler sonography is highly reliable and is recommended as a first-line test to evaluate penile arterial and veno-occlusive function (43). The diameters of the cavernous arteries are measured before and after intracavernous injection of vasodilator agent. After injecting a vasodilator, spectral waveforms of the cavernosal arteries and their PSVs are measured at five-minute intervals for 25 minutes (2,8).

Doppler Evaluation Technique of ED

After a detailed clinical evaluation, patients should be informed about the exam procedure and its risks (Table 2).

TABLE 2 ■ Erectile Dysfunction—Sonographic Technique

- Image the penis in long and transverse views
- Identify the cavernosal arteries
- Measure the anteroposterior dimension of cavernosal arteries
- Record the baseline velocities in cavernosal arteries on both sides
- Inject prostaglandin E in one corpus cavernosum laterally with 30-gauge needle
- Observe for tumescence and rigidity
- Record the angle corrected velocities in both cavernosal arteries at 5, 10, 15, 20, 25, and 30 min
- Record the velocities in deep dorsal vein

Doppler ultrasound assessment should be performed in a private setting in a quiet room. To obtain better image quality and spectral data, a 7 to 14 MHz linear array transducer is used for penile Doppler examination. Gray-scale examination of the penis is performed to exclude Peyronie's disease. On the ventral side, the corpus cavernosum and cavernosal arteries within the corpus cavernosum are identified on the transverse or longitudinal images. The diameter of the cavernosal artery is measured on both sides, and the spectral Doppler waveform is recorded. A vasoactive agent is used to stimulate penile erection. Injection of the vasoactive substance results in the physiologic response of erection in normal males and helps to separate patients with neurogenic or psychogenic dysfunction from those with vascular disturbances.

Several vasoactive agents, including papaverine, phentolamine, and prostaglandin E1 (PGE1) alone or in combination can be used through the intracavernous route (Table 3) (2,44,45).

PGE1 is preferred due to a lower risk of priapism. Prostaglandin results in physiological erection in 87% of the subjects studied.

PGE1 is injected into the distal two-thirds of the shaft of penis in one corpus cavernosum, using a small, 27- to 30-gauge needle (2). PGE1 is the most accurate diagnostic drug with the lowest prolonged erection rate of 0.1% (45). The total quantity of PGE1 injected is 10 to 20 μg (2). Some authors prefer to use PGE1 at a dose of 10 μg followed by a further 10 μg dose 15 minutes later if there is a suboptimal clinical response. A total dose of 20 μg PGE1 produces minimal side effects, but "stepwise" use is likely to reduce the risk of priapism (46). After intracavernosal injection of prostaglandin E, 96% of the prostaglandin is locally metabolized within 60 minutes. Priapism is seen in 1% and penile fibrotic lesions are seen in 2% of patients. Papaverine use has fallen out of favor because of its high incidence of postinjection fibrosis.

Intracavernous agents can cause numerous complications such as pain, ecchymosis, penile hematoma, and prolonged erection in up to 7% to 11% of cases (14,16,47). When an oral vasoactive agent, sildenafil citrate, plus visual sexual stimulation are used as an alternative to intracavernosal injection of vasoactive agents for penile Doppler evaluation, similar results, such as those with PGE1 or papaverine injection, are obtained (48).

TABLE 3 ■ Commonly Used Drugs with Doses for Penile Doppler Evaluation

Drug	Dose
Papaverine	30–60 mg
Phentolamine	0.25–1.25 mg[a]
Prostoglandin E1 (Alprostadil)	10–20 μg
Trimix (10 mg papaverine + 0.4 mg phentolamine + 10 μg prostaglandin E1/mL)	0.5–1 mL
Sildenafil citrate (oral)	50 mg[b]

[a]This drug may be used with papaverine or prostaglandin E1.
[b]This is used with visual sexual stimulation.

Cavernosal arterial anatomy varies among individuals; hemodynamic parameters differ at various sites of measurement (49). The PSVs of the cavernosal arteries should be measured at a constant location, preferably at the junction of the proximal one-third and distal two-thirds junction of the cavernosal artery.

After injection of vasoactive agents, the cavernosal artery diameter is measured and starting at five minutes postinjection, spectral Doppler waveform in both cavernosal arteries should be recorded every five minutes until 25 minutes elapse. Angle corrected PSV (30° and 60°) of the cavernosal artery should be recorded near the proximal third of the penile shaft because the velocities are greatest at this level (2). The dorsal penile arteries and deep dorsal vein are assessed and PSVs are then recorded.

Stages of Penile Erection

In the flaccid state, there is limited blood supply to the penis. Spectral Doppler waveform of the cavernosal artery demonstrates the high-resistance waveform (Fig. 2). PSVs are in the 5 to 15 cm/sec range. In the filling phase, there is increased blood flow to the penis via the cavernous arteries with characteristic variations in spectral waveforms (15,17,50). The cavernosal arteries dilate and the spectral Doppler waveform is characterized by increasing systolic and diastolic velocities. The helicine arteries also dilate and are seen as tortuous vessels branching from the cavernous arteries that split into several arterioles of smaller size (Fig. 3). In the tumescent phase, there is progressive decrease in both the diastolic velocity and the PSV (Fig. 4). As a result of venous occlusion, the diastolic flow decreases to zero and then undergoes flow reversal. With maximal rigidity, decreased systolic velocity can be observed (rigid phase) (Fig. 5). When venous occlusion occurs, the helicine arteries become progressively less visible and disappear with maximum rigidity (1,51,52). Three-dimensional (3D) power Doppler provides better image quality than two-dimensional (2D) image in demonstrating the helicine arteries and their branches (Fig. 6).

During detumescence, diastolic flow appears again in the cavernosal arteries and flows are appreciable in the CS, in the circumflex veins, and in the dorsal veins (1,15,17,50).

Evaluation of Arteriogenic Erectile Dysfunction

Diagnosis of arterial ED is made based primarily on measurements of diameters of cavernosal arteries and their velocities. The thickness of the cavernosal artery wall can be a factor in the assessment of arteriogenic erectile dysfunction (AED). Normal cavernosal arteries have strong, thin walls with strong pulsation, whereas arteries with diffuse atherosclerosis have thick walls with diminished pulsation (8). Various parameters, such as peak systolic flow velocity, degree of arterial dilatation, and acceleration time, have been suggested, but peak systolic flow

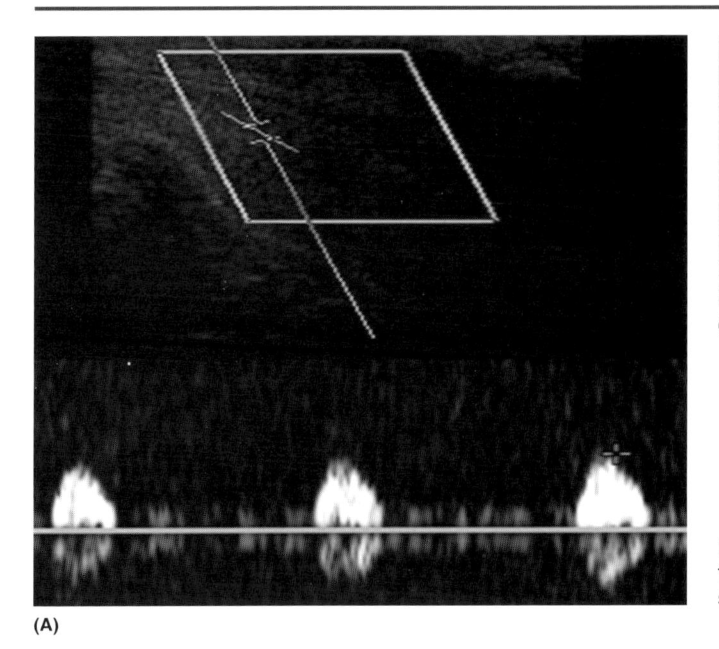

(A)

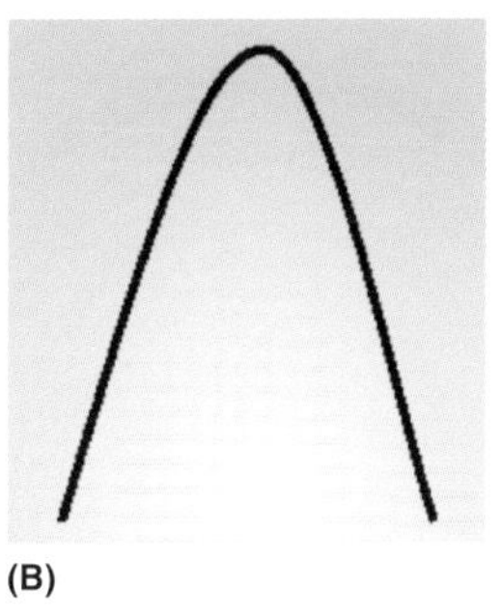

(B)

FIGURE 2 A AND B ■ Flaccid phase of penile erection. (**A**) Spectral waveform demonstrates high-resistance waveform. Velocities in the 5 to 15 cm/sec range. (**B**) Diagrammatic representation of spectral Doppler waveform.

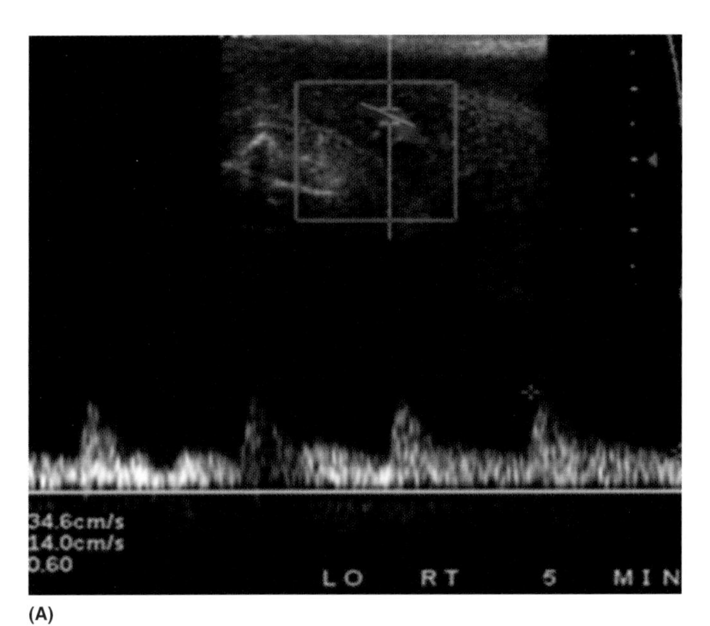

(A)

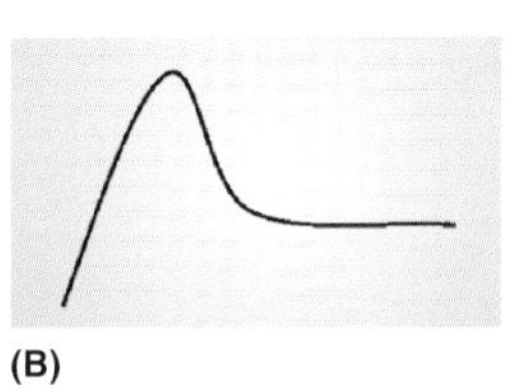

(B)

FIGURE 3 A AND B ■ Filling phase of penile erection. (**A**) Spectral waveform demonstrates increased systolic velocity and increased diastolic flow. (**B**) Diagrammatic representation of spectral Doppler waveform.

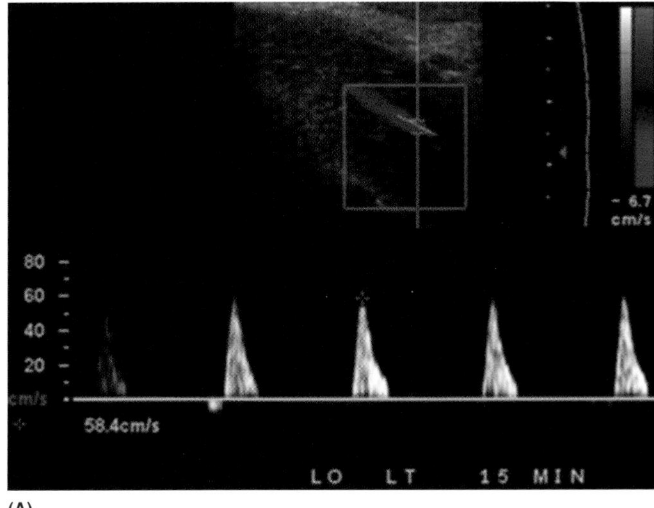

(A)

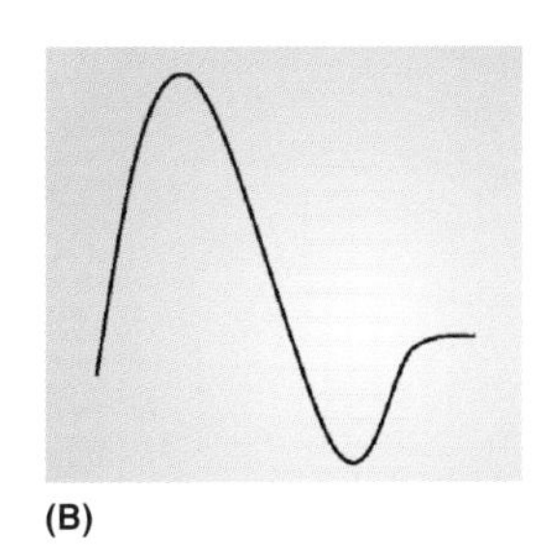

(B)

FIGURE 4 A AND B ■ Tumescent phase of penile erection. (**A**) The systolic velocities stabilize and diastolic flow decreases or becomes reversed. (**B**) Diagrammatic representation of spectral Doppler waveform.

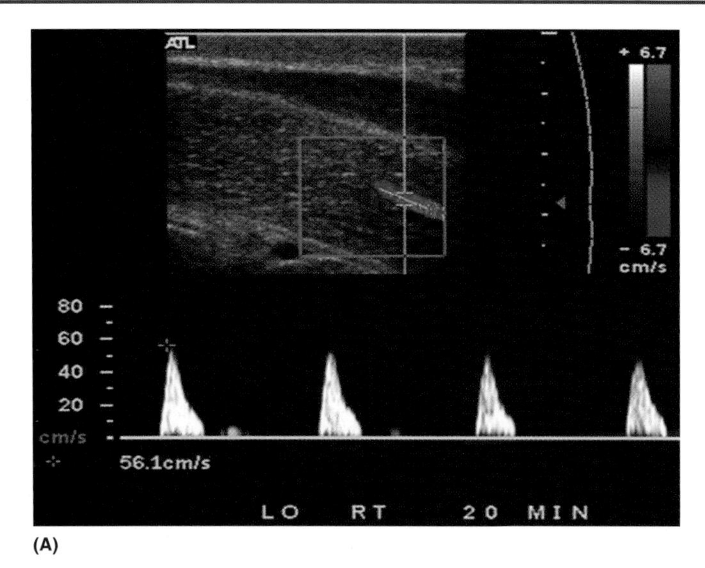

(A)

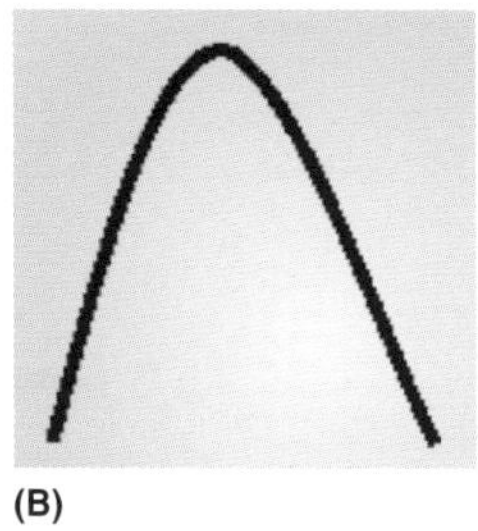

(B)

FIGURE 5 A AND B ■ Rigid phase of penile erection. (**A**) Spectral waveform demonstrates very high-resistance waveform with no diastolic flow and minimal systolic flow. The intracavernosal pressure reaches about 200 mm of mercury. (**B**) Diagrammatic representation of spectral Doppler waveform.

velocity is generally accepted to be a more accurate indicator of arterial disease (14–17,25). Because arterial diameter and flow rate change during the various phases of erection, the parameters are measured five minutes after injection and measurement of PSV is repeated at five-minute intervals for 25 minutes (2,8).

The average PSV after cavernosal injection of vasoactive agents has been found to be 30 to 40 cm/sec in normal volunteers (17). PSVs of cavernosal arteries differ and depend on the vasoactive agent used. Velocities greater than 25 cm/sec probably are adequate if papaverine is injected; velocities in the 35 to 40 cm/sec range are probably normal if prostaglandins are injected (2). A PSV less than 25 cm/sec in the cavernosal artery and dampened waveform are standard diagnostic criteria for arterial insufficiency (Fig. 7). The angiographic correlation has shown that a velocity threshold of 25 cm/sec has 92% accuracy in the diagnosis of arterial integrity. When penile angiography is compared with duplex Doppler examination of the same patients, PSV < 25 m/sec is consistently associated with severe arterial disease (2,47,53).

If a patient's PSV falls between 25 to 30 cm/sec, diagnosis of AED is suspicious and secondary criteria of arterial disease should be taken into consideration. The secondary diagnostic criteria of arterial insufficiency include failure of cavernosal artery dilatation of less than 75%, the presence of focal stenosis, occlusion or retrograde arterial flow (54), and asymmetry of cavernous peak systolic flow velocities greater than 10 cm/sec (2,55). A resistance index (RI) greater than 0.9 is associated with normal results in 90% of patients (16). Table 4 shows the cavernosal artery PSVs after prostaglandin injection.

The cutoff point for the acceleration time required to discriminate between atherosclerotic and nonatherosclerotic ED have been defined as equal to or higher than 100 msec for atherosclerotic ED (56,57). It has been reported that spectral Doppler analysis of the cavernosal artery in the flaccid penis provides a noninvasive method to assess arterial disease. A cutoff PSV value of 10 cm/sec in the flaccid state in the cavernosal artery has been reported to have the best accuracy for predicting arterial insufficiency, with a 96% sensitivity, 92% specificity, and a 93% overall accuracy (58). However, the normal velocity criteria have not been definitely established, and no data regarding venous abnormalities can be obtained without vasoactive agent injection (59).

If properly performed, duplex Doppler sonography is equal to or may even be superior to pharmacologic arteriography for the diagnosis of arteriogenic impotence (60).

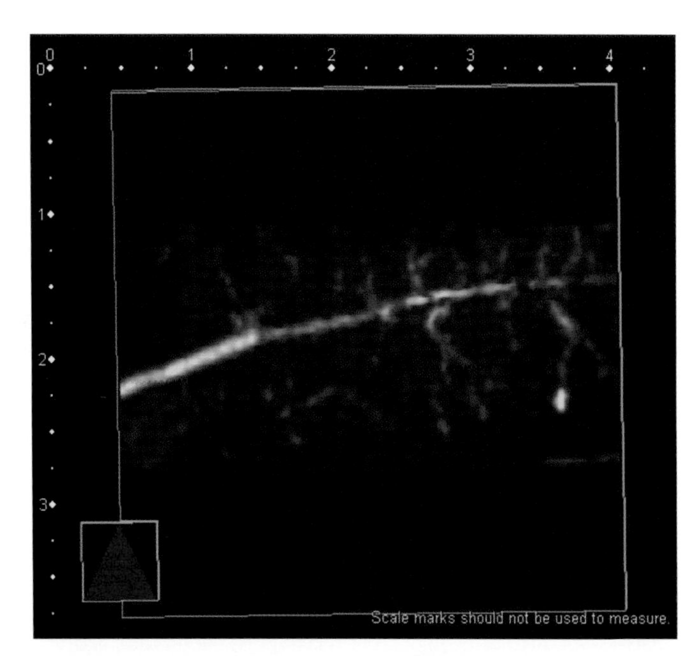

FIGURE 6 ■ Three-dimensional power Doppler sonogram with maximum intensity projection algorithm shows cavernosal artery and helicine branches of cavernous arteries.

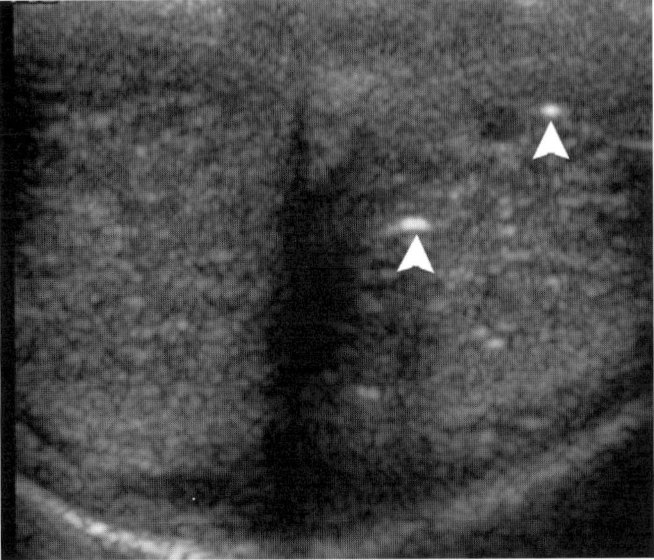

(A)

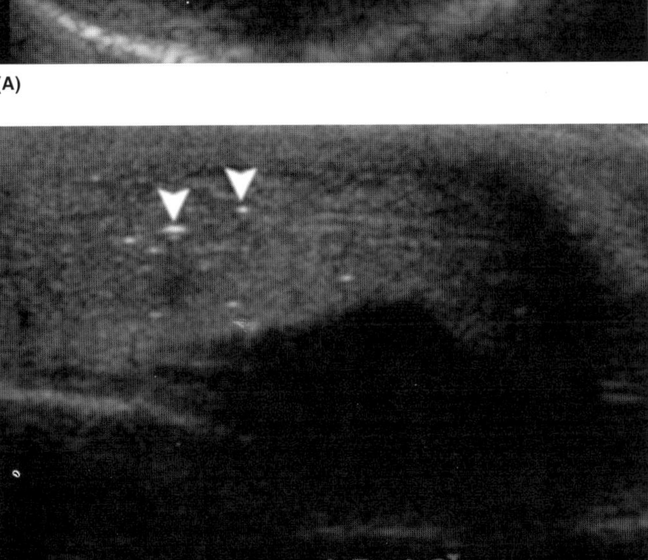

(B)

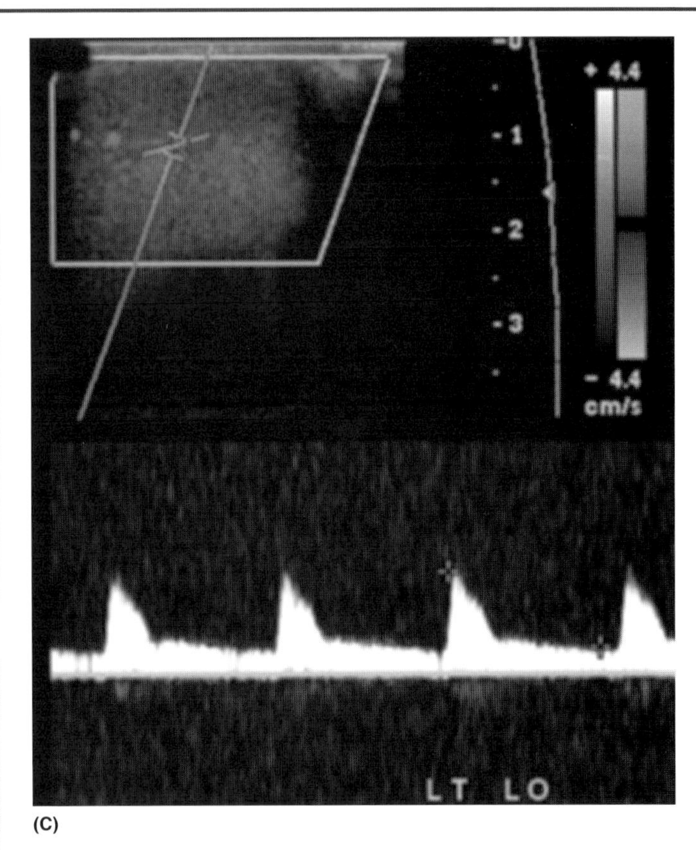

(C)

FIGURE 7 A–C ■ Arterial insufficiency in a diabetic patient. Transverse (**A**) and longitudinal (**B**) sonograms demonstrate multiple calcifications (*arrowheads*) in the corpora cavernosa. (**C**) Spectral waveform demonstrates maximum peak systolic velocity at 30 minutes post prostaglandin injection to be about 20 cm/sec and diastolic velocity of >5 cm. Patient had tumescence but no erection or rigidity.

Evaluation of Venous Erectile Dysfunction

A venous leakage is suspected when there is adequate arterial inflow and erection is obtained but the duration is short and there is persistent antegrade diastolic flow throughout the examination (Table 5) (2). Using Doppler sonography, diagnosis of venous erectile dysfunction (VED) can be made only if a patient had normal arterial function, i.e., normal PSV. An arterial diastolic velocity greater than 5 cm/sec (angle-corrected velocity) throughout all phases of erection constitutes persistent arterial diastolic flow and suggests venous leak (2,47) (Fig. 8). If a diagnosis of venous leak is made by sonography, further evaluation can be performed by cavernosography and cavernosometry. Persistent blood flow in the dorsal vein also represents venous insufficiency (55). Demonstration of early blood flow in the dorsal vein has a sensitivity of 80% and specificity of 100% on cavernosography for diagnosing VED. RI measurement is also a reliable, noninvasive method for diagnosing venous incompetence. RI is measured at 20 minutes after injection of the vasoactive agent. An RI value of less than 0.75 is associated with venous leakage in 95% of patients, whereas an RI value greater than 0.9 is associated with normal results in 90% of patients (2).

TABLE 4 ■ Cavernosal Artery Peak Systolic Velocities after Prostaglandin Injection

Normal (range 25–40 cm/sec)
25–30 cm/sec—borderline
Peak systolic velocities < 25 cm/sec suggest arterial disease

TABLE 5 ■ Deep Dorsal Vein Velocities

Normal <3 cm/sec
Moderately increased 10–20 cm/sec
Markedly increased >20 cm

Source: From Ref. 50.

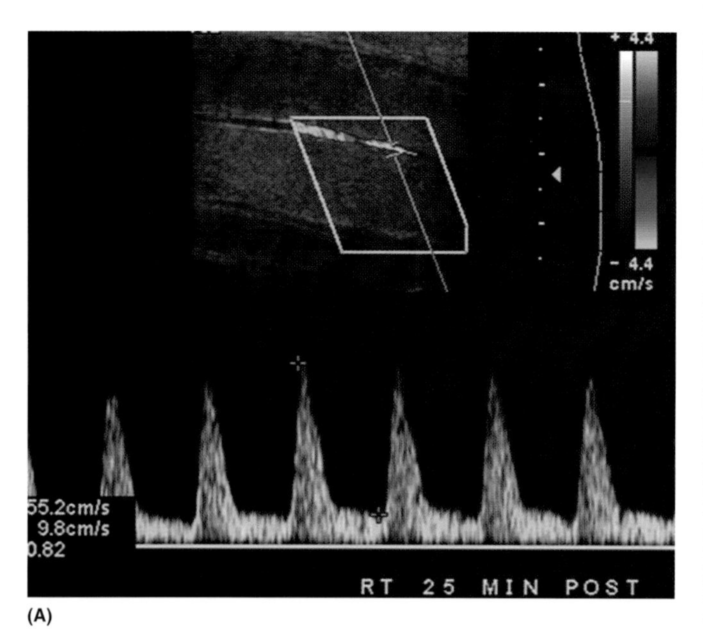

(A)

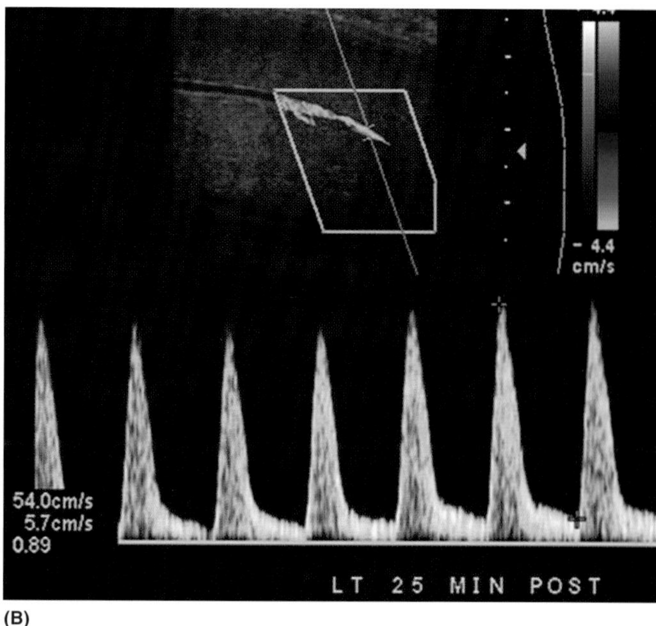

(B)

FIGURE 8 A AND B ■ Venous erectile dysfunction. Both cavernosal arteries (**A** and **B**) demonstrate normal arterial velocities but persistent diastolic velocity of >5 cm/sec post prostaglandin injection at 25 minutes. This is suggestive of venous leak. This is the most common abnormality detected in patients with erectile dysfunction.

Pulsatility index reflects the presence of retrograde diastolic flow below the baseline; some authors prefer the pulsatility index to assess venous incompetence and a pulsatility index value less than four is used to predict venous incompetence (61).

Occasionally, a suboptimal response to PGE1 injection may be seen and presents as continued forward flow in the diastole due to inadequate veno-occlusion, especially in young patients (62). In this instance, a 2 mg intracavernosal phentolamine, a selective α-adrenergic receptor antagonist, can be injected (62,63). Phentolamine blocks the increased sympathetic neuronal activity, thereby producing smooth-muscle relaxation, which often occurs in the anxious patient during penile ultrasound examination (62,63). After phentolamine injection, many patients, especially in the younger age group, exhibit statistically significant increases in grade of erection and PSV and a decrease in end diastolic velocity. Therefore, the use of intracavernosal phentolamine is recommended for ultrasound assessment of venous incompetence, especially in young patients (62,63). Alternatively, a rubber band can also be placed at the base of the penis to occlude the dorsal vein and PSVs in the dorsal vein will then show a decrease and the patient will have a longer period of erection; this is an indirect sign of VED (50).

Evaluation of Mixed (Indeterminate) ED

The diagnosis of mixed arterial ED and VED cannot be made using duplex Doppler sonography because venous competence cannot be assessed in a patient with arterial insufficiency. In patients with impaired arterial flow, therefore, the cavernosal sinusoids may never fill to the point of occluding small emissary veins draining the CC, and the Doppler findings of venous incompetence, persistent cavernosal arterial diastolic flow, and flow in the dorsal vein may be seen even if the veins are intrinsically normal (15,25,31). When arterial inflow is normal but the erectile response is poor and there is antegrade diastolic flow throughout the examination, it is determined to be an indeterminate result (2).

Cavernosometry and Cavernosography

When venous leakage is suspected after the combined injection and stimulation test, pharmacologic cavernosometry may be performed to demonstrate the degree of veno-occlusive dysfunction. As this is an invasive test, cavernosometry should be performed only in patients in whom surgery is planned. If cavernosometry has abnormal results, cavernosography is performed to display the site of venous leakage. This procedure has been abandoned at many institutions.

PRIAPISM

Priapism is a relatively uncommon condition that is defined as persistent engorgement or erection of the penis (or clitoris) for more than three hours in circumstances unrelated to sexual desire or stimulation (3). The term "priapism" is derived from the word, Priapus, a minor god of fertility and luck and deity of gardens and fields in Greek mythology (64). This condition typically involves the paired CC, although rare exceptions with involvement of the CS and sparing of the cavernosal spaces have been reported (65).

Priapism is mainly classified as low-flow (ischemic) and high-flow (arterial or nonischemic) priapism. Low-flow priapism is a medical emergency that may cause irreversible ischemic tissue changes due to a severe decrease in venous drainage from the CC. High-flow priapism is a less common form of priapism and involves unregulated inflow of blood that is usually secondary to some form of arterial trauma (3). This type of priapism is not considered an emergency, the patient does not have pain, and spontaneous resolution occurs in more than half of the reported cases (3).

Incidence and Etiology of Priapism

The overall incidence rate of priapism is 1.5 per 100,000 person-years and 2.9 per 100,000 person-years in men 40-years-old or older (66). The incidence of priapism is much higher in special-risk populations such as men with cocaine drug use, advanced pelvic or hematological malignancy, and those on antipsychotic medications (3,67,68). Causes of priapism are idiopathic in one-third of cases, due to alcohol abuse or medication in 21% of cases, due to perineal trauma in 12% of cases, and due to sickle-cell anemia in 11% of cases (69). Priapism associated with sickle-cell disease is classically described as ischemic, although rare exceptions of high-flow priapism associated with sickle-cell disease have been reported (3,70). Drug-induced priapism has been reported with the use of a variety of medications, most commonly the antihypertensive drugs guanethidine, prazosin, and hydralazine and psychotropic medications (3,71). Antipsychotics are associated with a small but definite risk of priapism, and the most commonly causative agents are trazodone, thioridazine, and chlorpromazine (3,72). Trazodone and cocaine may have synergistic effects in promoting priapism, and their combination may increase the risk of priapism (73). Priapism has also been reported in association with the recreational drug, Ecstasy (74). Neurologic etiologic factors of priapism are degenerative lumbar canal stenosis, cauda equina syndrome, and herniated disk. Malignant priapism is a rare cause, with 20% to 53% of cases of penile metastasis from other primary tumors initially presenting with malignant priapism (3,75).

Nonischemic priapism can result from perineal trauma, including bicycling and other straddle injuries (3,76). Injury to the arterial system and formation of an arteriolacunar fistula is most often implicated as the causative factor in nonischemic high-flow priapism. The venous outflow system is typically preserved in these conditions, and the blood in the corpora remains well oxygenated (3). The condition may also be iatrogenic following deep dorsal vein arterialization for vasculogenic impotence (77).

Table 6 shows causes of priapism.

TABLE 6 ■ Causes of Priapism

- ***Idiopathic***
  - Alcohol abuse
  - Ecstasy
  - Medication
    - Antihypertensive drugs
      - Guanethidine
      - Prazosin
      - Hydralazine
    - Antipsychotics
      - Trazodone
      - Thioridazine
      - Chlorpromazine
  - Perineal trauma
    - Bicycling or other straddle injuries
  - Sickle cell anemia
- ***Neurologic***
  - Degenerative lumbar canal stenosis
  - Cauda equina syndrome
  - Herniated disk
- ***Malignant priapism***
  - Primary or metastatic tumors of penis
- ***Vasoactive agents***
  - Papaverine
  - Prostaglandin E1
  - Sildenafil
- ***Iatrogenic***
  - Following deep dorsal vein arterialization

Low-Flow Priapism

Ischemic or veno-occlusive priapism is the most common form of priapism and is a medical emergency. It is characterized by a painful, rigid erection, absent cavernosal blood flow, and acidotic corpora (3). Low-flow priapism is characterized by inadequate venous outflow leading to hypoxia, acidosis, and tissue ischemia (78,79). Blood gas analysis of the aspirate reveals low oxygen, high carbon dioxide content, and low pH (3). This disease is a urologic emergency because prolonged cavernosal ischemia causes corporeal fibrosis and permanent ED; it is usually seen after intrapenile injections of vasoactive drugs or in men with sickle-cell disease and disseminated malignancy (78,79). Low-flow priapism can also result from dorsal vein or sinusoidal thrombosis; in low-flow priapism, the CC contain venous blood (1).

Priapism lasting more than 24 hours can cause severe cellular damage and widespread necrosis. Necrosis occurs in cases lasting more than 48 hours and eventually results in irreversible ED (80). The failure of detumescence seen in low-flow priapism may be secondary to failed α-adrenergic neurotransmission, endothelin deficit, or deactivation of intracellular cofactors of smooth-muscle contraction caused by hypoxia or hypercarbia (3).

In Doppler sonography, cavernosal arterial flow is usually absent, but high-resistance, low-velocity spectra may also be observed in cavernosal arteries of patients with low-flow priapism (1,3).

Treatment of Low-Flow Priapism

The initial diagnostic penile aspiration is used as a therapeutic measure (3). This should be combined with intracavernosal instillation of a sympathomimetic agent, i.e., phenylephrine injection after aspiration, to induce detumescence. If patients do not respond to this conservative treatment, surgical shunts are used for diversion of blood away from the corpus cavernosum (3).

High-Flow Priapism

High-flow priapism is usually secondary to penile or perineal trauma and is often characterized by a fistula formation between the cavernosal artery and the lacunae in the corpus cavernosum, known as an arterial-lacunar fistula (2,4). These patients present with painless partial tumescence and are able to increase the rigidity of the penis with sexual stimulation so that sexual intercourse may be possible. This is not a urologic emergency, and corporeal aspiration of oxygenated blood is confirmatory for high-flow priapism (2,3). The venous outflow is maintained, thereby preventing complete erection, stasis, and hypoxia (81). High-flow priapism may present days or even weeks after the original injury (3).

Color duplex Doppler sonography has replaced arteriography as the imaging modality of choice for the diagnosis of priapism (3,79) because it is sensitive, has a noninvasive nature, and is widely available. Currently, arteriography is used only for embolization of cavernosal arteries (82).

In patients with recent arterial laceration, gray-scale ultrasound reveals an irregular hypoechoic region secondary to tissue injury or distended lacunar spaces in the corpus cavernosum. This irregular area appears with well-circumscribed margins mimicking a pseudoaneurysm analogous to a capsule formation if the injury has been of longstanding duration (14). The arterial-lacunar fistula seen in high-flow priapism essentially bypasses the helicine arteries and appears as a characteristic color blush extending into the cavernosal tissue and as turbulent high-velocity flows on color duplex sonography (Fig. 9). A second ultrasound manifestation of high-flow priapism is increased cavernosal artery flow without sexual stimulation. The

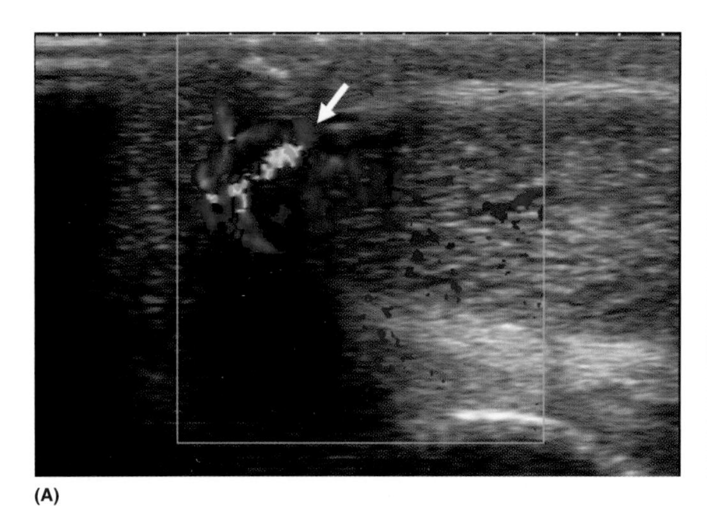

(A)

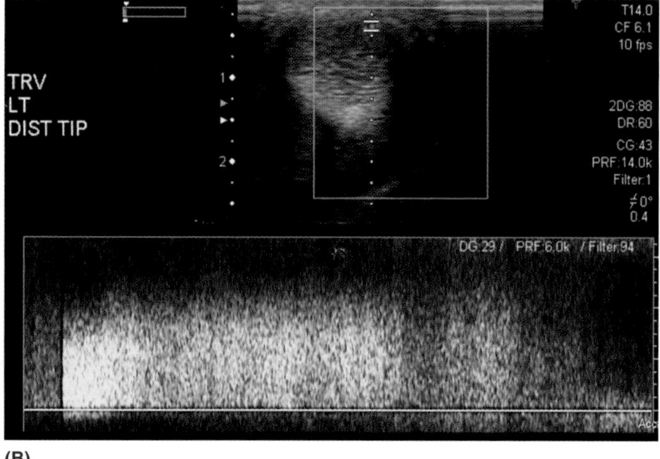

(B)

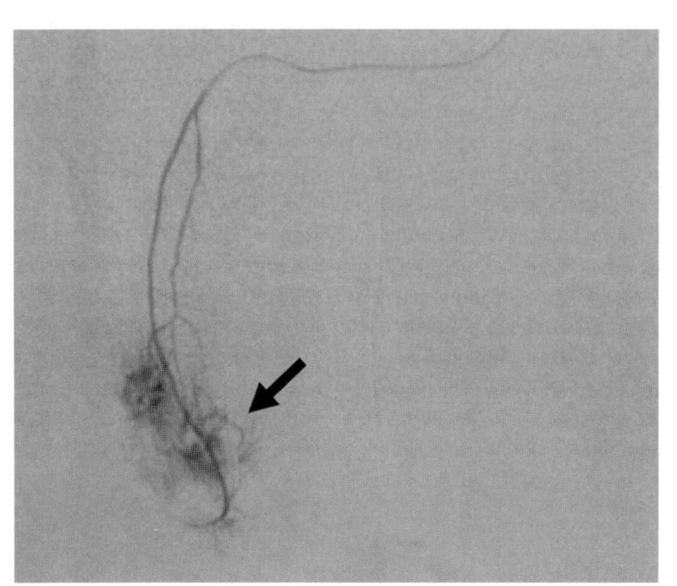

(C)

FIGURE 9 A–C ■ High-flow priapism. The patient injured his penile shaft while riding a bull 15 days prior to admission. The patient presented with painless persistent erection. (**A**) Longitudinal color-flow sonogram of the penis shows an area of color-flow aliasing (*arrow*) within the left corpus cavernosum near the tip of the penis. This is consistent with the presence of an arterial-lacunar fistula. (**B**) Corresponding color-flow Doppler with spectral tracing reveals marked turbulence of blood flow, thereby confirming the arterial-lacunar fistula. (**C**) The patient was treated with arterial embolization. Preembolization arteriography reveals early blush (*arrow*) corresponding to the site of the fistula. After embolization, the patient's priapism subsided and normal erectile function returned. *Source*: From Ref. 4.

TABLE 7 ■ Differences Between High- and Low-Flow Priapism

Low-flow priapism	High-flow priapism
Ischemic and medical emergency	Nonischemic and not an emergency
Blood gas analysis shows low oxygenated blood	Blood gas analysis shows normal oxygenated blood
In Doppler sonography, absent or low velocity spectra is observed in the cavernosal arteries	In Doppler sonography, normal or high velocity spectra is observed in the cavernosal arteries
None	Arterial-lacunar fistula and turbulent high-flow velocity at the fistula site
Treatment is blood aspiration with intracavernosal sympathomimetic agent or surgical shunts	Treatment is embolization or followup without any intervention

draining veins are usually prominent and may exhibit arterialized waveforms. In longstanding cases of high-flow priapism, penile edema may be a prominent feature (62).

Treatment of High-Flow Priapism

It is embolization or followup without any intervention (3,79). If embolization is preferred for the treatment, serial penile duplex Doppler studies should be performed as followup to assure complete resolution of the arterial lacunar fistula and ultimate restoration of normal cavernosal blood flow (83). Color Doppler ultrasound can also be used to localize the fistula and subsequently apply external compression in order to achieve permanent fistula occlusion and resolution of priapism in certain cases (84,85). Table 7 shows the differences between high- and low-flow priapism.

PEYRONIE'S DISEASE

A French surgeon, François de la Peyronie, first described Peyronie's disease in 1743. Peyronie's disease is a localized connective tissue disorder of the penis that affects the tunica albuginea and erectile tissue and is characterized by the formation of fibrous tissue plaques within the tunica albuginea, generally causing a penile deformity (86–88). The erect penis generally angulates toward the inelastic scar tissue (86). In the acute phase of the disease that continues approximately 12 to 18 months, clinical manifestations include pain on erection and palpable indurations (plaque) on the penile shaft (86,89). In the chronic phase, penile deformity stabilizes and pain decreases (86).

Incidence, Risk Factors, and Etiology of Peyronie's Disease

Peyronie's disease occurs usually in men in their 50s; its prevalence varies from 0.4% to 23%, depending on the population studied and the definition of Peyronie's disease (86,90,91). A recent survey of 8000 German men revealed that 3.2% of them had palpable plaque in the penis (92).

Common risk factors for developing Peyronie's disease include hypercholesterolemia (the most common), diabetes mellitus, and hypertension. Patients with at least one risk factor had a significantly higher risk of severe penile deformity (>60°) compared to those with no risk factor (86).

Risk factors for systemic vascular disease, such as diabetes mellitus, serum lipid abnormalities, and hypertension, also have significant effects on erectile function (85,93,94). Vascular and neural pathways of erection may be impaired by diabetes mellitus, whereas the vascular system is primarily affected by hypercholesterolemia and hypertension (93,94). Erectile capacity is also severely decreased in the presence of these risk factors in patients with Peyronie's disease.

Peyronie's disease is caused due to multifactorial etiology (89,95,96). Trauma to the erect or semierect penis is the primary initiating factor of plaque formation (95,97). Repetitive microtrauma or excessive bending of the erect penis may cause bleeding into the subtunical spaces or tunical delamination at the point where the septum integrates into the inner circular layer of the tunica albuginea (91,95,97). Peyronie's disease is caused by localized aberration of the wound-healing process. The initial result of microvascular injury is fibrin deposition, which may be a precursor to Peyronie's plaque formation (98). Peyronie's plaques consist of dense collagenous connective tissue with reduced and fragmented elastic fibers (91). Radiographically visible dystrophic calcifications are present in approximately 30% of these patients (99).

The significant correlation between Peyronie's disease and aging and marital status has also been reported (100). Some authors have speculated that Peyronie's disease "may result from the inevitable accumulated sexual activity that increases the risk of repetitive penile trauma or from the increased vulnerability of the tunica albuginea" with aging (92).

Minimal repetitive trauma to the erect penis during normal sexual intercourse may induce Peyronie's disease, and those patients with partial ED may be even more prone to Peyronie's disease because of the increased mechanical stress associated with the buckling that occurs while attempting sexual intercourse with a partial erection (95,97,101).

Cytokines and growth factors, especially transforming growth factor-β, have a major role in the development of penile fibrosis (86,91,102). Peyronie's disease is also associated with Dupuytren's contracture, plantar fascial contracture, tympanosclerosis, diabetes mellitus, urethral instrumentation, autoimmune disorders, and an inherited form of these disorders (HLA-B7 antigens) (96,100,103). In addition to these factors, some medications such as propranolol and methotrexate may result in Peyronie's disease (104,105). Approximately 0.01% of patients with Peyronie's disease have a positive family history of the disease (86).

Clinical Presentation and Imaging of Peyronie's Disease

Patients usually present with a nodule or plaque in the penis, painful erection, and/or penile deformity during erection in the early phases of the disease. The presence of harder plaque, permanent penile deformity during erection, and ED are seen in the chronic phase of the disease (91). Penile pain, deformity, or curvature results in ED. Impotence and loss of erection occur in approximately 30% of patients with Peyronie's disease.

Physical examination reveals a well-defined penile plaque or an area of palpable induration (91). The plaque is generally located in the midline of the dorsal surface and near the base of the penis with dorsal penile deformity. Lateral and ventral plaques are less common but cause more coital difficulty due to greater deviation from the natural coital angle. Penile pain may be present with erection or during sexual intercourse. Spontaneous improvement in pain usually occurs within six months due to reduced inflammation (91).

Most patients with Peyronie's disease have rigid erections but many develop difficulty attaining and/or maintaining erections. The probable mechanism for ED in these patients is veno-occlusive dysfunction (106).

The natural evolution of the disease process results in deterioration in the penile deformity and ED in 30.2%, no change in 66.7%, and complete improvement in 3.2% of patients (86). Some patients are not aware of the penile deformity and are diagnosed during ED evaluation (86).

Ultrasound is a useful diagnostic technique for evaluation of Peyronie's disease. Sonography may demonstrate the presence and extent of fibrotic plaques and calcification (15). Noncalcified plaques appear as focal hyperechoic areas of thickening of the tunica albuginea with echogenicity equal to or higher than that of tunica albuginea (Fig. 10) (15,88). Calcified plaques are hyperechoic in appearance and may have acoustic shadow if the calcium content is high, and such shadows are detected in about one-third of Peyronie's disease patients (14–16,88,107). If fibrosis or scarring occurs within the CC, the involved sinusoids cannot distend during the development of an erection and these lesions appear as scattered echogenic foci with irregular margins within the homogeneous corpus cavernosum. These lesions are generally visible when the penis is flaccid, but they are more prominent with erection when the surrounding sinusoids are distended with blood. Usually, multiple lesions are present and both CC are involved (14,15). Rarely, Peyronie's plaque may involve the entire tunica albuginea of the penis (107). Sometimes, the spongy tissue of the CS is involved (14).

Sonography is the perfect tool for followup of patients either under conservative treatment or after surgery because exact measurement of the plaque size is possible (88).

Power Doppler sonography can reveal hyperperfusion around the plaques as a sign of inflammation in the active state of the disease. Inactive plaques do not show signs of hyperperfusion (88).

PENILE TUMORS AND MASS LESIONS

Penile carcinoma is rare in the United States and Europe but is a significant health problem in developing countries and constitutes up to 10% of malignant diseases in men in African and South American countries (108). Poor genital hygiene and phimosis are the main etiologic factors of penile carcinoma, and circumcision is a well-established prophylactic factor (108,109). Low socioeconomic status, viral infections, radiation, and use of tobacco in any form are other contributory risk factors for developing penile carcinoma. Squamous-cell carcinoma is the most common primary tumor of the penis (108).

Penile ultrasonography is a noninvasive and easy method for assessing penile cancer. Ultrasonography provides information regarding invasion of the tunica albuginea, urethra, or CC by penile carcinoma (110). Sonography displays penile involvement more accurately

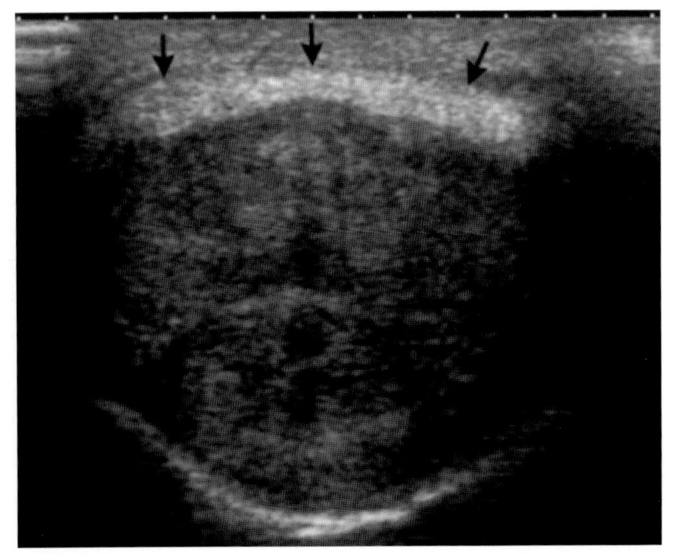

(A)

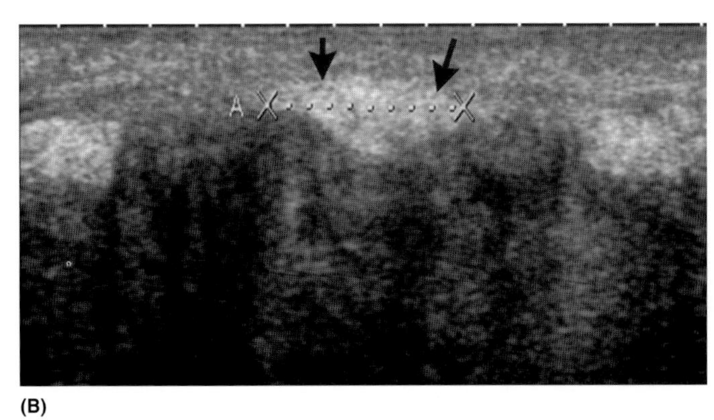

(B)

FIGURE 10 A AND B ■ Peyronie's disease. Transverse (**A**) and longitudinal (**B**) gray-scale sonogram of penis demonstrates hyperechoic plaques (*arrows*) of Peyronie's disease involving the tunica albuginea of penis.

TABLE 8 ■ Jackson Classification of Penile Cancer

Stage	Description
Stage I (A)	Tumor confined to glans, prepuce, or both
Stage II (B)	Tumor extending onto penile shaft
Stage III (C)	Tumor with operable inguinal metastases
Stage IV (D)	Tumor involving adjacent structures; tumor associated with inoperable inguinal metastasis or distant metastasis

Source: From Ref. 111.

than clinical examination, which usually overestimates the extent of the carcinoma (110). The main drawback of ultrasound is its inability to demonstrate lesions of the base of the penis (108,110).

Sonography is also used to demonstrate nonpalpable groin lymph nodes and to detect femoral vessel involvement (108). The heterogeneous appearance of the lymph nodes on sonography is considered suspicious for metastatic disease.

For staging, two main staging systems, the Jackson (Table 8) and TNM (tumor–node–metastases) classifications are used (108,111). The TNM system describes the depth of invasion and extent of nodal involvement.

Metastases to the Penis

These are uncommon but can result from advanced prostate cancer (112). Gastrointestinal system tumors and rarely lymphoma, either primary or metastatic, melanoma, and lung cancer can metastasize to the penis (112,113). Penile metastases usually invade the CC, but they can infiltrate the CS, the glans, and the urethra. Patients with penile metastases usually have widespread disease from aggressive tumors (112,113). The most common presenting symptom of penile metastasis is priapism, and patients presenting with priapism should be evaluated for primary or metastatic penile tumors (112,113).

Penile Lymphoma

This is also a very rare tumor of the penis (114). Penile lymphoma may appear as a mass, as plaques or ulcers in the skin of the penis, or as diffuse penile swelling. Focal lesions are usually localized in the shaft of the penis involving the CC, but the glans penis and prepuce may also be involved (112,114). Other rare benign solid tumors of the penis are angioma and fibroma (15).

Sonographically, penile tumors, primary or metastatic, usually appear as hypoechogenic or heterogeneous solid masses and are sometimes poorly defined, either disrupting or replacing the normally homogeneous CC (110,112,115). Differential diagnosis should be between neoplasms and inflammatory lesions (abscess or granuloma) or a hematoma (114). Color Doppler examination may help in the differentiation and usually reveals hypervascularity in the tumor masses (112). Differentiation of various types of tumors or differentiation between malignant and benign tumors cannot be done by sonography (115). In addition, there is no significant association between echogenicity and tumor morphology or grade (110). Extensive lesions can disrupt or obliterate the septum penis and can break through the tunica albuginea (112). Sonourethrography can be used if clinical symptoms suggest urethral involvement (15).

PENILE TRAUMA

Etiology of Penile Injury

Penile fracture is an uncommon injury caused by exertion of the axial force on the erect penis, thereby resulting in a tear of the tunica albuginea with subcutaneous hemorrhage (116–120). This injury usually occurs during vigorous sexual intercourse when the rigid penis slips out of the vagina and is misdirected against the partner's pubic bone or perineum, resulting in buckling trauma (116). Another common cause of penile fracture is self-inflicted abnormal downward bending of the erect penis to achieve detumescence due to patient misinformation about penile tissues (117). A blunt trauma to the flaccid penis usually does not lead to penile fracture and usually causes extratunical or cavernosal hematomas (118). Other causes of penile fracture include rolling over in bed with an erect penis and direct blunt injury to the penis (118,119).

When external force is applied to erect penile tissue, it causes a sudden rise in the intracorporeal pressure, resulting in further distension of the already thinned tunica albuginea of the erect penis and thereby causing it to tear (119,120). The thickness of the tunica albuginea decreases from 2 to 0.5 to 0.25 mm during erection. As the thinned tunica albuginea is prone to injury, a sudden, direct strain on the dorsum can cause a penile fracture (119,120). Previous penile injury or periurethral infection can increase the possibility of fracture due to decreased elasticity (121).

Penetrating penile trauma is rare and may result from farm accidents, animal or human bites, and missile and zipper injuries (119). Penile ischemia or gangrene can occur secondary to penile prosthesis, priapism, corporeal injections of pharmacotherapeutic agents, diabetes, and chronic dialysis (119,122). Gradual necrosis has been observed due to low cardiac output and renal failure (122). Penetrating penile injuries or gangrene of the glans may lead to cavernous spongiosum fistula, which may cause ED in later years (119).

Penile Fracture (Corpus Cavernosal and Tunica Albuginea Rupture)

Penile fractures are rare urologic emergencies. Most patients report hearing a cracking or popping sound with a sharp pain followed by rapid detumescence, swelling, discoloration, and deformity of the penis.

Generally, a tear occurs in only one of the CC and its surrounding tunica albuginea; however, corpus spongiosal and urethral injuries may also occur (120).

Fractures usually occur in the proximal shaft or midshaft of the penis (120,123). Corpus cavernosal ruptures are generally transverse in orientation and are located in the ventral portion of the corpus cavernosum adjacent to the urethra (119,120,123). Sonography can demonstrate an irregular hypoechoic or hyperechoic defect at the cavernosal rupture site (Fig. 11).

The integrity of the tunica albuginea is the most important factor in determining the necessity for surgical intervention (4,118,120). Sonography can detect the exact site of a tear as an interruption of the thin echogenic line of the tunica albuginea and can show evidence of associated hematoma (4,118,123,124). Surgical repair is generally recommended for patients with a suspected tear of the tunica albuginea or with urethral injury. Tunica albuginea rupture is associated with urethral rupture in 10% to 20% of cases (118). Table 9 shows the American Association for the Surgery of Trauma (AAST) injury scale of the penis (125).

Intracavernosal Hematomas without Fracture

Injury to the subtunical venous plexus or to the smooth-muscle trabeculae in the absence of complete tunical disruption can lead to cavernosal hematomas (4,126). Intracavernosal hematomas are usually bilateral and result from injury to the cavernosal tissue when the base of the penile shaft is crushed against the pelvic bones (120). The sonographic appearance of a penile hematoma varies with the age of the lesion. Hematomas appear as hyperechoic or complex masses in the acute phase and then become cystic, often with septation (15). Cavernosal damage can cause fibrosis, which appears as an ill-defined echogenic scar replacing erectile tissue (1). If sonography demonstrates an intact tunica albuginea, patients can be

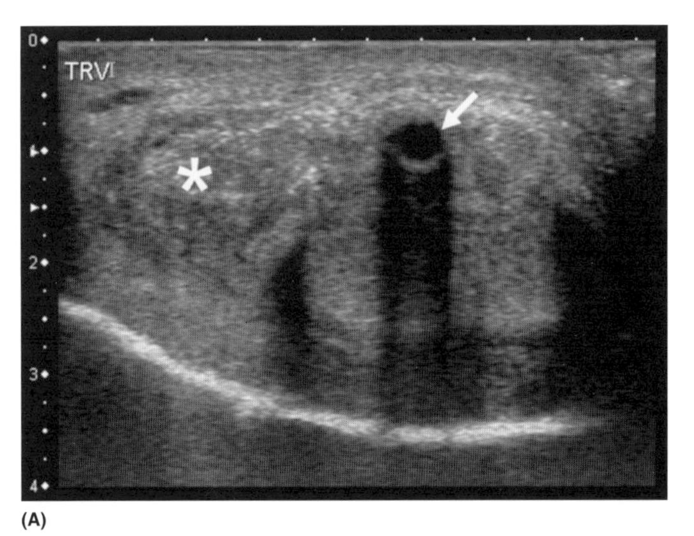

(A)

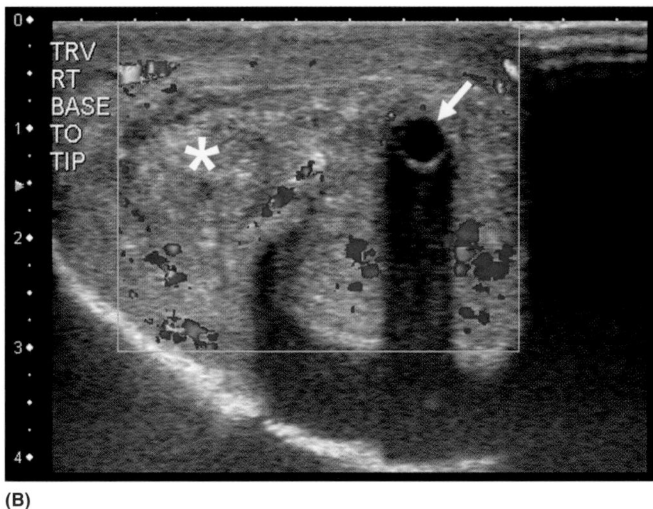

(B)

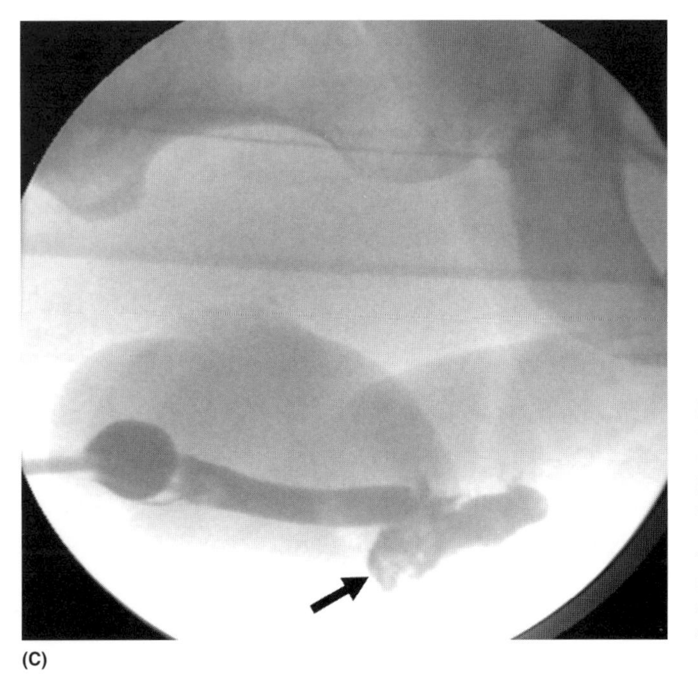

(C)

FIGURE 11 A–C ■ Fracture of the penis. Following sexual activity, the patient presented with loss of penile rigidity, pain, and hematuria. (**A**) Transverse gray-scale sonogram of the penis demonstrates subcutaneous hematoma of variable echotexture in continuity with the right corpus cavernosa, suggesting a break in the continuity of the tunica albuginea (*asterisk*). This appearance suggests penile fracture. The Foley catheter is within the urethra (*arrow*). (**B**) A corresponding color-flow image demonstrates an absence of color flow within the area of the penile fracture. (**C**) Retrograde urethrogram reveals associated rupture of the bulbar urethra (*arrow*). The patient subsequently underwent surgical exploration and repair. *Source*: From Ref. 4.

TABLE 9 ■ AAST Organ Injury Severity Scale for the Penis

Grade[a]	Description
I	Cutaneous laceration/contusion
II	Buck's fascia laceration without tissue loss
III	Cutaneous avulsion
	Laceration through glans/meatus
	Cavernous or urethral defect less than 2 cm
IV	Partial penectomy
	Cavernous or urethral defect 2 cm or greater
V	Total penectomy

[a]Advance 1 grade for bilateral injuries up to grade III.
Abbreviation: AAST, American Association for the Surgery of Trauma.
Source: From Ref. 125.

TABLE 10 ■ Anatomical Classification of Urethral Injury Following Blunt Trauma

Type I	Posterior urethra intact but stretched Type I
Type II	Partial or complete pure posterior urethral injury with tear of membranous urethra above the urogenital diaphragm (Type II) (130)
Type III	Partial or complete combined anterior and posterior urethral injury with disruption of the urogenital diaphragm (Type III) (130)
Type IV	Bladder neck injury with extension into the urethra
Type IVa	Injury of the base of the bladder with periurethral extravasation simulating a true type IV urethral injury
Type V	Partial or complete pure anterior urethral injury

Source: From Refs. 130 and 131.

treated conservatively (4). Intracavernosal hematoma in a long-distance mountain biker and spontaneous idiopathic hemorrhage of the corpus cavernosum have been described (121,127).

Avulsion of the Dorsal Penile Vessels and Thrombosis

Rupture of the dorsal penile vessels may mimic penile fracture (128), but deformation and immediate detumescence do not occur because of the intact tunica albuginea (118). The hematoma may be superficial or remain under Buck's fascia depending upon the site of involvement of the penile veins. If an arterial lesion is present, a posttraumatic arteriovenous fistula with increased venous pressures and dilatations of the injured vessel may result (118). Thrombosis of the superficial and deep dorsal penile veins is also a rare urologic emergency, and the clinical presentation and sonographic appearance can mimic penile fracture. Sonography demonstrates a noncompressible dorsal vein and, if ruptured, associated hematoma (4,129).

URETHRAL AND SPONGIOSAL TRAUMA

Urethral trauma can occur due to external, contusive, or penetrative injury (Table 10) (13). External trauma can be a frequent event and can occur in the penile urethra as a result of road or work accidents, sporting activities, or sex (13). Posterior urethral injury is usually caused by a crushing force to the pelvis and is associated with pelvic fractures (12). Posterior urethral injury has been found in 4% to 14% of patients with pelvic fracture (130). An anterior urethral injury usually occurs due to a straddle pelvic injury and is most often isolated (12). The urethra can be compressed by subcutaneous or intraspongiosal hematomas and may present complete or incomplete mucosal interruption (13). Retrograde urethrography is the most useful exam to detect urethral injury. Sonography can be used to demonstrate perilesional fluid collections (13).

Approximately 20% of penile fractures are associated with lesions of the CS and the urethra. The inability to urinate and in cases of late referral, extravasation of the urine can be observed although the absence of such findings does not exclude urethral damage (124).

Evaluation of the urethra with sonography can help to identify interruption of the urethral wall, but urethrography may still be needed (120). Sonourethrography may be useful to demonstrate the continuity of the anterior urethra. Real-time examination of the urethra after instillation of ultrasound jelly (within the urethra) may increase the possibility of detecting extravasations through a ruptured urethral wall (Fig. 12) (14).

The presence of air in the cavernosal bodies in the absence of external penetrating trauma may be an indirect sign of urethral injury (118). Sonography may be able to demonstrate edema or hematoma of the CS following penile trauma (13).

OTHER URETHRAL PATHOLOGIES

Urethral Tumors

Benign tumors of the urethra are very rare (12). Common benign urethral masses are polyps, papillomas, and condylomata acuminata; they may manifest as filling defects in which case biopsy is required to establish the correct diagnosis (12,15). Malignant tumors of the urethra usually occur after 50 years of age and are rare (less than 1% of all urological tumors). The most common histology type (80% of cases) is squamous-cell carcinoma. Other malignant tumors of the urethra are transitional-cell carcinoma that constitutes 15% of cases and arises in the posterior urethra and adenocarcinoma that represents 5% of cases and arises in the bulbous urethra (12,15). Sonourethrography shows single or multiple endoluminal filling defects that are associated with stenosis, especially if the lesions are papillary (13). On the other hand, infiltrating tumors may trigger extensive and tight stenosis, which often prevents opacification

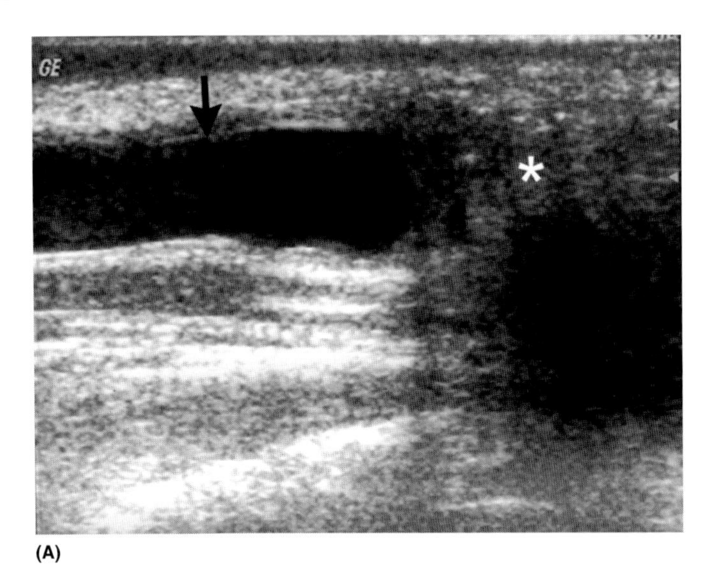

(A)

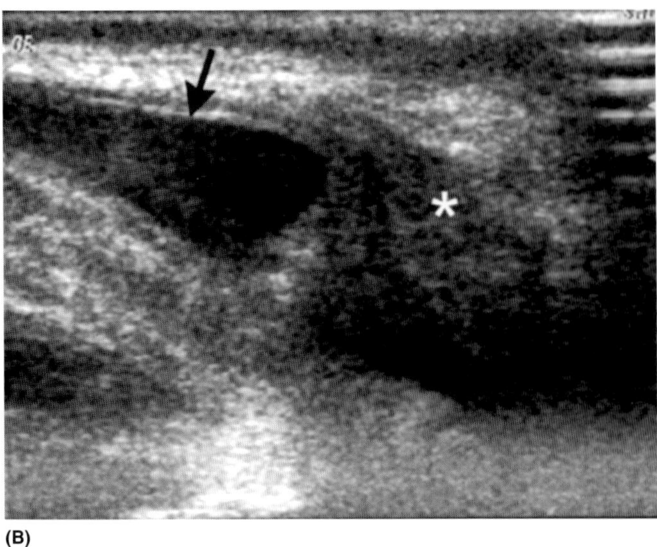

(B)

FIGURE 12 A AND B ■ Complete urethral rupture. Longitudinal (**A**) and longitudinal oblique (**B**) sonourethrogram reveal part of the distal penile urethra (*arrow*) with complete disruption and nonvisualization of proximal penile urethra with associated edema and hematoma (*asterisk*).

of the urethral lumen superior to the lesion. In these cases, sonourethrography is particularly useful as it can detect spread into the surrounding periurethral tissue (13).

Male urethral carcinoma can spread either by direct extension to adjacent structures or by metastasis to regional lymph nodes. Staging of urethral tumors is summarized in Table 11.

The lymphatics of the anterior urethra drain into the superficial and deep inguinal nodes and sometimes into the external iliac nodes. Tumors of the posterior urethra most commonly spread to the pelvic lymph nodes. Hematogenous dissemination is uncommon until advanced local disease is present or in primary transitional-cell carcinoma of the prostatic urethra. Treatment of urethral carcinoma is surgical excision (12).

Periurethral Cysts

These may develop in the ducts of Cowper glands in the bulbous urethra or in the ducts of Littré more distally (15).

TABLE 11 ■ Staging of Urethral Tumors

Stage	Explanation
Stage I	Tumor is confined to subepithelial connective tissue
Stage II	Tumor invades the corpus spongiosum, prostate, or periurethral muscle
Stage III	Tumor invades the corpus cavernosum and bladder neck or beyond the prostatic capsule
Stage IV	Tumor invades other adjacent organs

Source: From Ref. 12.

Urethral Diverticulas

These are congenital or acquired outpouchings from the urethral wall and appear as simple or complex fluid collections related to the urethra. Diverticulas may change in size with distension and emptying of the urethra (14–16). They arise mostly from the midurethra, initially extend posteriorly, and can develop stone or malignancy.

Urethral Stricture

This refers to a fibrous scarring of the urethra due to collagen and fibroblast proliferation (12,132). Associated scar in the surrounding CS is called spongiofibrosis. The scarring process can extend through the tissue of the CS and into adjacent structures. Contraction of this scar reduces the width of the urethral lumen (12). The most common causes of anterior urethral strictures are infection (gonococcus and rarely tuberculosis), trauma (instrumentation, straddle injury, and iatrogenic), and congenital anomalies (12). Posterior urethral strictures are usually caused by trauma or surgery (transurethral resection of the prostate or open prostatectomy) (12,13). Reiter's syndrome and Wegener's granulomatosis also are rare causes of urethral strictures (15). Retrograde urethrography is the primary imaging technique for evaluation of anterior or posterior urethral stricture (12,132). However, sonourethrography is more accurate than retrograde urethrography for estimating the length and diameter of urethral stricture (13,132,133). This technique is more accurate in the bulbar urethra because the hand-held transducer is positioned in the midsagittal plane directly perpendicular to the diseased urethral segment (133). Sonourethrography also allows evaluation of the involvement of the periurethral spongy tissue (12,13,133,134). Strictures appear as constrictions with hyperechoic submucosa that may be due to involvement of the adjacent spongy tissue (133,134). Periurethral fibrosis appears as thickened, irregular, nondistensible tissue encroaching into the otherwise anechoic urethral lumen (12,134). Sonourethrography allows spongiofibrosis to be determined objectively by simply measuring luminal diameter. If a nondistensible portion of the urethral lumen measures less than 3 mm in diameter during maximal retrograde distension, the spongiofibrosis is considered to be severe (134). Posterior shadowing represents severe fibrosis produced by extremely dense collagen deposition; it is usually associated with complete or near-complete strictures due to trauma. Ultrasound reveals focal obliteration of the proximal part with posterior shadowing (134). 3D ultrasound allows better demonstration of urethral stricture (Fig. 13). The treatment of spongiofibrosis is surgical resection (12). To demonstrate and assess posterior urethral stricture, a transrectal approach can be used (13). As ultrasound does not use ionizing radiation, repeated sonograms can be performed without risk of radiation exposure (15).

Urethral Stones and Foreign Bodies

Urethral stones are rare and most originate from the upper urinary tract or the bladder; they are usually small and are referred to as migrant stones (12,13). Rarely, primary stone formation occurs in the urethra when stricture is present or it may be associated with a urethral diverticulum (12). They are usually radio opaque and are visible radiographically in the vesicourethral area. Sonography is a useful technique as urethral stones are easily demonstrated by high-resolution sonography with a typical hyperechoic feature (13).

Urethral foreign bodies may be introduced into the urethra through the meatus. The presence of a foreign body in the urethra can cause infection, obstruction, or damage to the urethral wall. Solid objects may appear as echogenic structures along the course of the urethra. Continuity of the urethral wall can be evaluated by sonourethrography after removal of the foreign body (15).

Posterior Urethral Valve

Posterior urethral valves are rare (1 per 5000–8000 male) congenital urinary tract malformations, but they are the most common urethral anomaly and the most frequent cause of urinary tract obstruction-related, end-stage renal failure in boys (135). Transabdominal sonography may reveal a thick-walled bladder, a dilated posterior urethra, some dilatation of the renal pelvis, and dilatation or tortuosity as well as thickening of the ureters as indirect signs of urethral anomalies (135–137). Urethral sonography can improve imaging of the posterior urethra. Using transperineal sonography during voiding, posterior urethral valves can appear as hyperechoic linear structures in the dilated posterior urethra. Prominent localized hypertrophy of the bladder neck musculature is also visible in most of these patients (138). Dilatation of the urinary bladder and proximal urethra are associated with a characteristic "keyhole" deformity at the bladder base and thickening of the bladder wall. Direct visualization of the posterior urethral valves is sometimes possible without voiding. Negative sonographic findings are highly suggestive of the absence of urethral anomalies (135). Currently, posterior urethral valves are usually diagnosed during the routine antenatal scans.

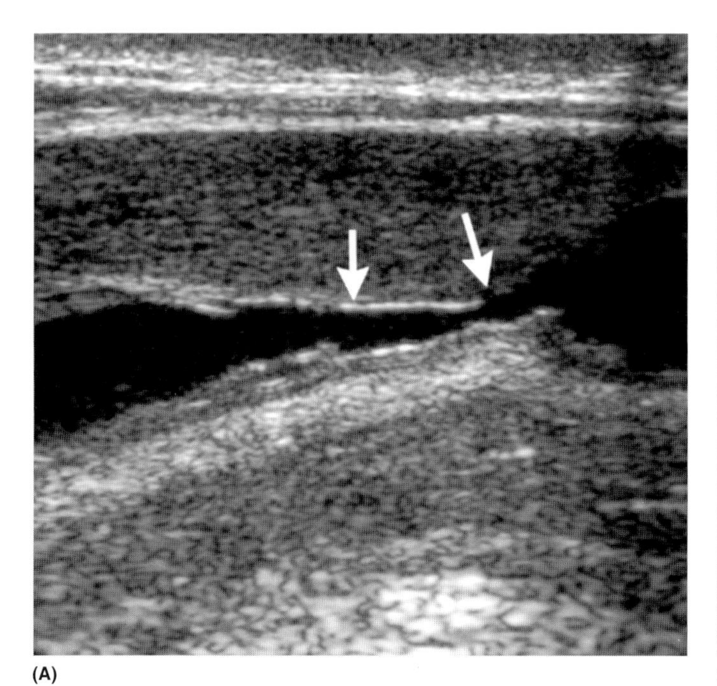

(A)

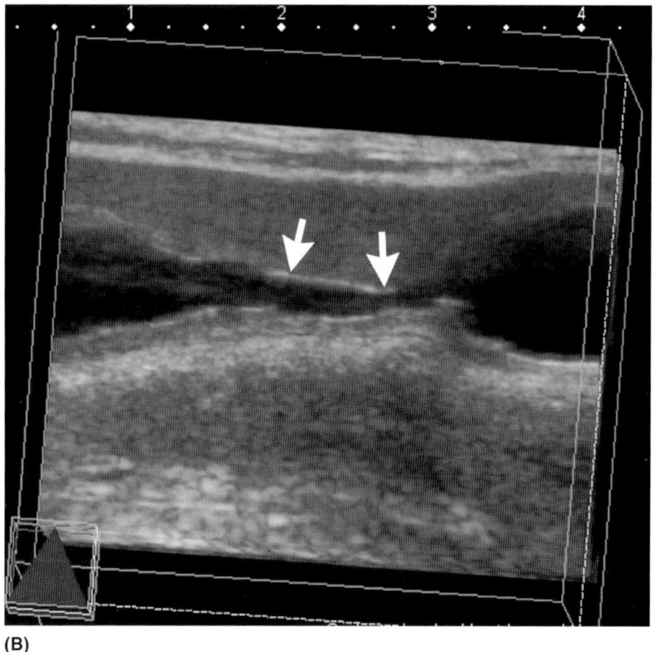

(B)

FIGURE 13 A AND B ■ Urethral Stricture. Longitudinal (**A**) gray-scale sonourethrogram demonstrates a 2-cm-long urethral stricture (*arrows*) with mucosal thickening and irregularity. (**B**) Corresponding longitudinal three-dimensional sonourethrogram.

REFERENCES

1. Bertolotto M, Neumaier CE. Penile sonography. Eur Radiol 1999; 9(suppl 3):S407–S412.
2. Dogra V, Bhatt S. Erectile dysfunction and priapism. In: Dogra V, Rubens D, eds. Ultrasound Secrets. 1st ed. Philadelphia: Hanley & Belfus, 2004:420–424.
3. Sadeghi-Nejad H, Dogra V, Seftel AD, Mohamed MA. Priapism. Radiol Clin North Am 2004; 42:427–443.
4. Bhatt S, Kocakoc E, Rubens DJ, Seftel AD, Dogra VS. Sonographic evaluation of penile trauma. J Ultrasound Med 2005; 24: 993–1000.
5. Breza J, Aboseif SR, Orvis BR, Lue TF, Tanagho EA. Detailed anatomy of penile neurovascular structures: surgical significance. J Urol 1989; 141:437–443.
6. Aboseif S, Shinohara K, Breza J, Benard F, Narayan P. Role of penile vascular injury in erectile dysfunction after radical prostatectomy. Br J Urol 1994; 73:75–82.
7. Kim ED, Blackburn D, McVary KT. Post-radical prostatectomy penile blood flow: assessment with color flow Doppler ultrasound. J Urol 1994; 152:2276–2279.
8. Rao DS, Donatucci CF. Vasculogenic impotence. Arterial and venous surgery. Urol Clin North Am 2001; 28:309–319.
9. Fitzpatrick TJ. The penile intercommunicating venous valvular system. J Urol 1982; 127:1099–1100.
10. Andersson KE, Wagner G. Physiology of penile erection. Physiol Rev 1995; 75:191–236.
11. Miller TA. Diagnostic evaluation of erectile dysfunction. Am Fam Physician 2001; 61:95–104.
12. Kawashima A, Sandler CM, Wasserman NF, LeRoy AJ, King BF Jr, Goldman SM. Imaging of urethral disease: a pictorial review. Radiographics 2004; 24(suppl 1):S195–S216.
13. Pavlice P, Barozzi L, Menchi I. Imaging of male urethra. Eur Radiol 2003; 13:1583–1596.
14. Doubilet PM, Benson CB, Silverman SG, Gluck CD. The penis. Semin Ultrasound CT MR 1991; 12:157–175.
15. Benson CB, Doubilet PM, Vickers MA Jr. Sonography of the penis. Ultrasound Q 1991; 9:89–109.
16. Benson CB, Doubilet PM, Richie JP. Sonography of the male genital tract. AJR Am J Roentgenol 1989; 153:705–713.
17. Schwartz AN, Wang KY, Mark LA, et al. Evaluation of normal erectile function with color flow Doppler sonography. AJR Am J Roentgenol 1989; 153:1155–1160.
18. Desser TS, Nino-Murcia M, Olcott EW, Terris MK. Advantages of performing sonourethrography with lidocaine hydrochloride jelly in a prepackaged delivery system. AJR Am J Roentgenol 1999; 173:39–40.
19. Rifkin MD. Sonourethrography: technique for evaluation of prostatic urethra. Radiology 1984; 153:791–792.
20. Chang SC, Yeh CH. The applications of dynamic transrectal ultrasound in the lower urinary tract of men and women. Ultrasound Q 1992; 10:185–223.
21. Fournier GR Jr, Juenemann KP, Lue TF, Tanagho EA. Mechanisms of venous occlusion during canine penile erection: an anatomic demonstration. J Urol 1987; 137:163–167.
22. Azadzoi KM, Kim N, Brown ML, Goldstein I, Cohen RA, Saenz de Tejada I. Endothelium-derived nitric oxide and cyclooxygenase products modulate corpus cavernosum smooth muscle tone. J Urol 1992; 147:220–225.
23. Rajfer J, Aronson WJ, Bush PA, Dorey FJ, Ignarro LJ. Nitric oxide as a mediator of relaxation of the corpus cavernosum in response to nonadrenergic, noncholinergic neurotransmission. N Engl J Med 1992; 326:90–94.
24. Burnett AL. Role of nitric oxide in the physiology of erection. Biol Reprod 1995; 52:485–489.
25. Kim SH. Doppler US evaluation of erectile dysfunction. Abdom Imaging 2002; 27:578–587.
26. NIH Consensus Conference. Impotence. NIH Consensus Development Panel on Impotence. JAMA 1993; 270:83–90.
27. Feldman HA, Goldstein I, Hatzichristou DG, Krane RJ, McKinlay JB. Impotence and its medical and psychosocial correlates: results of the Massachusetts Male Aging Study. J Urol 1994; 151:54–61.
28. Mellinger BC, Weiss J. Sexual dysfunction in the elderly male. Am Urol Assoc Update Series 1992; 11:146–152.
29. Lizza EF, Rosen RC. Definition and classification of erectile dysfunction: report of the Nomenclature Committee of the International Society of Impotence Research. Int J Impot Res 1999; 11:141–143.
30. Rosen RC, Leiblum SR, Spector IP. Psychologically based treatment for male erectile disorder: a cognitive-interpersonal model. J Sex Marital Ther 1994; 20:67–85.
31. Benson CB, Vickers MA Jr, Aruny J. Evaluation of impotence. Semin Ultrasound CT MR 1991; 12:176–190.
32. Lue TF, Mueller SC, Jow YR, Hwang TI. Functional evaluation of penile arteries with duplex ultrasound in vasodilator-induced erection. Urol Clin North Am 1989; 16:799–807.
33. Shabsigh R, Fishman IJ, Quesada ET, Seale-Hawkins CK, Dunn JK. Evaluation of vasculogenic erectile impotence using penile duplex ultrasonography. J Urol 1989; 142:1469–1474.
34. Krysiewicz S, Mellinger BC. The role of imaging in the diagnostic evaluation of impotence. AJR Am J Roentgenol 1989; 153: 1133–1139.
35. Lue TF, Tanagho EA. Physiology of erection and pharmacological management of impotence. J Urol 1987; 137:829–836.
36. Juenemann KP, Lue TF, Luo JA, Benowitz NL, Abozeid M, Tanagho EA. The effect of cigarette smoking on penile erection. J Urol 1987; 138:438–441.
37. Miller NS, Gold MS. The human sexual response and alcohol and drugs. J Subst Abuse Treat 1988; 5:171–177.
38. Brock GB, Lue TF. Drug-induced male sexual dysfunction. An update. Drug Saf 1993; 8:414–426.
39. Finger WW, Lund M, Slagle MA. Medications that may contribute to sexual disorders. A guide to assessment and treatment in family practice. J Fam Pract 1997; 44:33–43.
40. Melman A. The evaluation of erectile dysfunction. Urol Radiol 1988; 10:119–128.
41. Abber JC, Lue TF, Orvis BR, McClure RD, Williams RD. Diagnostic tests for impotence: a comparison of papaverine injection with the penile-brachial index and nocturnal penile tumescence monitoring. J Urol 1986; 135:923–925.
42. Mueller SC, Lue TF. Evaluation of vasculogenic impotence. Urol Clin North Am 1988; 15:65–76.
43. Broderick GA, Arger P. Duplex Doppler ultrasonography: noninvasive assessment of penile anatomy and function. Semin Roentgenol 1993; 28:43–56.
44. Moemen MN, Hamed HA, Kamel II, Shamloul RM, Ghanem HM. Clinical and sonographic assessment of the side effects of intracavernous injection of vasoactive substances. Int J Impot Res 2004; 16:143–145.
45. Porst H, van Ahlen H, Block T, et al. Intracavernous self-injection of prostaglandin E1 in the therapy of erectile dysfunction. Vasa Suppl 1989; 28:50–56.
46. Kim SC, Seo KK, Park BD, Lee SW. Risk factors for an early increase in dose of vasoactive agents for intracavernous pharmacotherapy. Urol Int 2000; 65:204–207.
47. Quam JP, King BF, James EM, et al. Duplex and color Doppler sonographic evaluation of vasculogenic impotence. AJR Am J Roentgenol 1989; 153:1141–1147.
48. Basar MM, Batislam E, Altinok D, Yilmaz E, Basar H. Sildenafil citrate for penile hemodynamic determination: an alternative to intracavernosal agents in Doppler ultrasound evaluation of erectile dysfunction. Urology 2001; 57:623–626.
49. Chiou RK, Alberts GL, Pomeroy BD, et al. Study of cavernosal arterial anatomy using color and power Doppler sonography: impact on hemodynamic parameter measurement. J Urol 1999; 162:358–360.
50. Fitzgerald SW, Erickson SJ, Foley WD, Lipchik EO, Lawson TL. Color Doppler sonography in the evaluation of erectile dysfunction. Radiographics 1992; 12:3–17.
51. Montorsi F, Sarteschi M, Maga T, et al. Functional anatomy of cavernous helicine arterioles in potent subjects. J Urol 1998; 159: 808–810.
52. Sarteschi LM, Montorsi F, Fabris FM, Guazzoni G, Lencioni R, Rigatti P. Cavernous arterial and arteriolar circulation in patients with erectile dysfunction: a power Doppler study. J Urol 1998; 159:428–432.

53. Lue TF, Hricak H, Marich KW, Tanagho EA. Vasculogenic impotence evaluated by high-resolution ultrasonography and pulsed Doppler spectrum analysis. Radiology 1985; 155:777–781.
54. Hattery RR, King BF Jr, Lewis RW, James EM, McKusick MA. Vasculogenic impotence. Duplex and color Doppler imaging. Radiol Clin North Am 1991; 29:629–645.
55. Benson CB, Vickers MA. Sexual impotence caused by vascular disease: diagnosis with duplex sonography. AJR Am J Roentgenol 1989; 153:1149–1153.
56. Speel TG, van Langen H, Wijkstra H, Meuleman EJ. Penile duplex pharmaco-ultrasonography revisited: revalidation of the parameters of the cavernous arterial response. J Urol 2003; 169: 216–220.
57. Kim SC. Recent advancement in diagnosis of vasculogenic impotence. Asian J Androl 1999; 1:37–43.
58. Roy C, Saussine C, Tuchmann C, Castel E, Lang H, Jacqmin D. Duplex Doppler sonography of the flaccid penis: potential role in the evaluation of impotence. J Clin Ultrasound 2000; 28: 290–294.
59. Mancini M, Bartolini M, Maggi M, Innocenti P, Villari N, Forti G. Duplex ultrasound evaluation of cavernosal peak systolic velocity and waveform acceleration in the penile flaccid state: clinical significance in the assessment of the arterial supply in patients with erectile dysfunction. Int J Androl 2000; 23:199–204.
60. Mueller SC, von Wallenberg-Pachaly H, Voges GE, Schild HH. Comparison of selective internal iliac pharmaco-angiography, penile brachial index and duplex sonography with pulsed Doppler analysis for the evaluation of vasculogenic (arteriogenic) impotence. J Urol 1990; 143:928–932.
61. Fitzgerald SW, Erickson SJ, Foley WD, Lawson TL, Lipchik EO. Color Doppler ultrasound in the evaluation of erectile dysfunction: prediction of venous incompetence. Radiology 1990; 177(P):129.
62. Wilkins CJ, Sriprasad S, Sidhu PS. Colour Doppler ultrasound of the penis. Clin Radiol 2003; 58:514–523.
63. Gontero P, Sriprasad S, Wilkins CJ, Donaldson N, Muir GH, Sidhu PS. Phentolamine re-dosing during penile dynamic colour Doppler ultrasound: a practical method to abolish a false diagnosis of venous leakage in patients with erectile dysfunction. Br J Radiol 2004; 77:922–926.
64. Papadopoulos I, Kelami A. Priapus and priapism. From mythology to medicine. Urology 1988; 32:385–386.
65. Taylor WN. Priapism of the corpus spongiosum and glans penis. J Urol 1980; 123:961–962.
66. Eland IA, van der Lei J, Stricker BH, Sturkenboom MJ. Incidence of priapism in the general population. Urology 2001; 57:970–972.
67. Compton MT, Miller AH. Priapism associated with conventional and atypical antipsychotic medications: a review. J Clin Psychiatry 2001; 62:362–366.
68. Altman AL, Seftel AD, Brown SL, Hampel N. Cocaine associated priapism. J Urol 1999; 161:1817–1818.
69. Pohl J, Pott B, Kleinhans G. Priapism: a three-phase concept of management according to aetiology and prognosis. Br J Urol 1986; 58:113–118.
70. Ramos CE, Park JS, Ritchey ML, Benson GS. High flow priapism associated with sickle cell disease. J Urol 1995; 153:1619–1621.
71. Rubin SO. Priapism as a probable sequel to medication. Scand J Urol Nephrol 1968; 2:81–85.
72. Ankem MK, Ferlise VJ, Han KR, Gazi MA, Koppisch AR, Weiss RE. Risperidone-induced priapism. Scand J Urol Nephrol 2002; 36:91–92.
73. Myrick H, Markowitz JS, Henderson S. Priapism following trazodone overdose with cocaine use. Ann Clin Psychiatry 1998; 10:81–83.
74. Dublin N, Razack AH. Priapism: ecstasy related? Urology 2000; 56:1057.
75. Chan PT, Begin LR, Arnold D, Jacobson SA, Corcos J, Brock GB. Priapism secondary to penile metastasis: a report of two cases and a review of the literature. J Surg Oncol 1998; 68:51–59.
76. Golash A, Gray R, Ruttley MS, Jenkins BJ. Traumatic priapism: an unusual cycling injury. Br J Sports Med 2000; 34:310–311.
77. Wolf JS Jr, Lue TF. High-flow priapism and glans hypervascularization following deep dorsal vein arterialization for vasculogenic impotence. Urol Int 1992; 49:227–229.
78. Pautler SE, Brock GB. Priapism. From Priapus to the present time. Urol Clin North Am 2001; 28:391–403.
79. Bertolotto M, Quaia E, Mucelli FP, Ciampalini S, Forgacs B, Gattuccio I. Color Doppler imaging of posttraumatic priapism before and after selective embolization. Radiographics 2003; 23:495–503.
80. Spycher MA, Hauri D. The ultrastructure of the erectile tissue in priapism. J Urol 1986; 135:142–147.
81. Broderick GA, Gordon D, Hypolite J, Levin RM. Anoxia and corporal smooth muscle dysfunction: a model for ischemic priapism. J Urol 1994; 151:259–262.
82. Montague DK, Jarow J, Broderick GA, et al; Members of the Erectile Dysfunction Guideline Update Panel; American Urological Association. American Urological Association guideline on the management of priapism. J Urol 2003; 170:1318–1324.
83. Hakim LS, Kulaksizoglu H, Mulligan R, Greenfield A, Goldstein I. Evolving concepts in the diagnosis and treatment of arterial high flow priapism. J Urol 1996; 155:541–548.
84. Mabjeesh NJ, Shemesh D, Abramowitz HB. Posttraumatic high flow priapism: successful management using duplex guided compression. J Urol 1999; 161:215–216.
85. Shapiro RH, Berger RE. Post-traumatic priapism treated with selective cavernosal artery ligation. Urology 1997; 49:638–643.
86. Kadioglu A, Tefekli A, Erol B, Oktar T, Tunc M, Tellaloglu S. A retrospective review of 307 men with Peyronie's disease. J Urol 2002; 168:1075–1079.
87. Brock G, Hsu GL, Nunes L, von Heyden B, Lue TF. The anatomy of the tunica albuginea in the normal penis and Peyronie's disease. J Urol 1997; 157:276–281.
88. Fornara P, Gerbershagen HP. Ultrasound in patients affected with Peyronie's disease. World J Urol 2004; 22:365–367.
89. Hellstrom WJ, Bivalacqua TJ. Peyronie's disease: etiology, medical, and surgical therapy. J Androl 2000; 21:347–354.
90. Lindsay MB, Schain DM, Grambsch P, Benson RC, Beard CM, Kurland LT. The incidence of Peyronie's disease in Rochester, Minnesota, 1950 through 1984. J Urol 1991; 146:1007–1009.
91. Gholami SS, Gonzalez-Cadavid NF, Lin CS, Rajfer J, Lue TF. Peyronie's disease: a review. J Urol 2003; 169:1234–1241.
92. Schwarzer U, Sommer F, Klotz T, Braun M, Reifenrath B, Engelmann U. The prevalence of Peyronie's disease: results of a large survey. BJU Int 2001; 88:727–730.
93. Chew KK, Earle CM, Stuckey BG, Jamrozik K, Keogh EJ. Erectile dysfunction in general medicine practice: prevalence and clinical correlates. Int J Impot Res 2000; 12:41–45.
94. Romeo JH, Seftel AD, Madhun ZT, Aron DC. Sexual function in men with diabetes type 2: association with glycemic control. J Urol 2000; 163:788–791.
95. Jarow JP, Lowe FC. Penile trauma: an etiologic factor in Peyronie's disease and erectile dysfunction. J Urol 1997; 158:1388–1390.
96. Chilton CP, Castle WM, Westwood CA, Pryor JP. Factors associated in the aetiology of Peyronie's disease. Br J Urol 1982; 54:748–750.
97. Devine CJ Jr, Somers KD, Jordan SG, Schlossberg SM. Proposal: trauma as the cause of the Peyronie's lesion. J Urol 1997; 157: 285–290.
98. Somers KD, Dawson DM. Fibrin deposition in Peyronie's disease plaque. J Urol 1997; 157:311–315.
99. Gelbard MK. Dystrophic penile calcification in Peyronie's disease. J Urol 1988; 139:738–740.
100. Mulhall JP, Creech SD, Boorjian SA, et al. Subjective and objective analysis of the prevalence of Peyronie's disease in a population of men presenting for prostate cancer screening. J Urol 2004; 171: 2350–2353.
101. Devine CJ Jr, Somers KD, Ladaga LE. Peyronie's disease: pathophysiology. Prog Clin Biol Res 1991; 370:355–358.
102. El-Sakka AI, Hassoba HM, Pillarisetty RJ, Dahiya R, Lue TF. Peyronie's disease is associated with an increase in transforming growth factor-beta protein expression. J Urol 1997; 158:1391–1394.
103. Nyberg LM Jr, Bias WB, Hochberg MC, Walsh PC. Identification of an inherited form of Peyronie's disease with autosomal dominant inheritance and association with Dupuytren's contracture and histocompatibility B7 cross-reacting antigens. J Urol 1982; 128:48–51.
104. Osborne DR. Propranolol and Peyronie's disease. Lancet 1977; 1(8021):1111.

105. Phelan MJ, Riley PL, Lynch MP. Methotrexate associated Peyronie's disease in the treatment of rheumatoid arthritis. Br J Rheumatol 1992; 31:425–426.
106. Lopez JA, Jarow JP. Penile vascular evaluation of men with Peyronie's disease. J Urol 1993; 149:53–55.
107. Pourbagher MA, Turunc T, Pourbagher A, Guvel S, Koc Z. Peyronie disease involving the entire tunica albuginea of the penis. J Ultrasound Med 2005; 24:387–389.
108. Misra S, Chaturvedi A, Misra NC. Penile carcinoma: a challenge for the developing world. Lancet Oncol 2004; 5:240–247.
109. Moses S, Bailey RC, Ronald AR. Male circumcision: assessment of health benefits and risks. Sex Transm Infect 1998; 74:368–373.
110. Agrawal A, Pai D, Ananthakrishnan N, Smile SR, Ratnakar C. Clinical and sonographic findings in carcinoma of the penis. J Clin Ultrasound 2000; 28:399–406.
111. Jackson SM. The treatment of carcinoma of the penis. Br J Surg 1966; 53:33–35.
112. Nakayama F, Sheth S, Caskey CI, Hamper UM. Penile metastasis from prostate cancer: diagnosis with sonography. J Ultrasound Med 1997; 16:751–753.
113. Perez-Mesa C, Oxenhandler R. Metastatic tumors of the penis. J Surg Oncol 1989; 42:11–15.
114. Bunesch Villalba L, Bargallo Castello X, Vilana Puig R, Burrel Samaranch M, Bru Saumell C. Lymphoma of the penis: sonographic findings. J Ultrasound Med 2001; 20:929–931.
115. Andipa E, Liberopoulos K, Asvestis C. Magnetic resonance imaging and ultrasound evaluation of penile and testicular masses. World J Urol 2004; 22:382–391.
116. Rosenstein D, McAninch JW. Urologic emergencies. Med Clin North Am 2004; 88:495–518.
117. Zargooshi J. Penile fracture in Kermanshah, Iran: report of 172 cases. J Urol 2000; 164:364–366.
118. Bertolotto M, Mucelli RP. Nonpenetrating penile traumas: sonographic and Doppler features. AJR Am J Roentgenol 2004; 183:1085–1089.
119. Mydlo JH, Harris CF, Brown JG. Blunt, penetrating and ischemic injuries to the penis. J Urol 2002; 168:1433–1435.
120. Choi MH, Kim B, Ryu JA, Lee SW, Lee KS. MR imaging of acute penile fracture. Radiographics 2000; 20:1397–1405.
121. Brotzman GL. Penile fracture. J Am Board Fam Pract 1991; 4:351–353.
122. Lowe FC, Brendler CB. Penile gangrene: a complication of secondary hyperparathyroidism from chronic renal failure. J Urol 1984; 132:1189–1191.
123. Kervancioglu S, Ozkur A, Bayram MM. Color Doppler sonographic findings in penile fracture. J Clin Ultrasound 2005; 33:38–42.
124. Forman HP, Rosenberg HK, Snyder HM III. Fractured penis: sonographic aid to diagnosis. AJR Am J Roentgenol 1989; 153:1009–1010.
125. Moore EE, Malangoni MA, Cogbill TH, et al. Organ injury scaling VII: cervical vascular, peripheral vascular, adrenal, penis, testis, and scrotum. J Trauma 1996; 41:523–524.
126. Matteson JR, Nagler HM. Intracavernous penile hematoma. J Urol 2000; 164:1647–1648.
127. Zandrino F, Musante F, Mariani N, Derchi LE. Partial unilateral intracavernosal hematoma in a long-distance mountain biker: a case report. Acta Radiol 2004; 45:580–583.
128. Nicely ER, Costabile RA, Moul JW. Rupture of the deep dorsal vein of the penis during sexual intercourse. J Urol 1992; 147:150–152.
129. Schmidt BA, Schwarz T, Schellong SM. Spontaneous thrombosis of the deep dorsal penile vein in a patient with thrombophilia. J Urol 2000; 164:1649.
130. Colapinto V, McCallum RW. Injury to the male posterior urethra in fractured pelvis: a new classification. J Urol 1977; 118:575–580.
131. Goldman SM, Sandler CM, Corriere JN Jr, McGuire EJ. Blunt urethral trauma: a unified, anatomical mechanical classification. J Urol 1997; 157:85–89.
132. Gallentine ML, Morey AF. Imaging of the male urethra for stricture disease. Urol Clin North Am 2002; 29:361–372.
133. Morey AF, McAninch JW. Role of preoperative sonourethrography in bulbar urethral reconstruction. J Urol 1997; 158:1376–1379.
134. Morey AF, McAninch JW. Sonographic staging of anterior urethral strictures. J Urol 2000; 163:1070–1075.
135. Schoellnast H, Lindbichler F, Riccabona M. Sonographic diagnosis of urethral anomalies in infants: value of perineal sonography. J Ultrasound Med 2004; 23:769–776.
136. Avni EF, Ayadi K, Rypens F, Hall M, Schulman CC. Can careful ultrasound examination of the urinary tract exclude vesicoureteric reflux in the neonate? Br J Radiol 1997; 70:977–982.
137. Williams CR, Perez LM, Joseph DB. Accuracy of renal-bladder ultrasonography as a screening method to suggest posterior urethral valves. J Urol 2001; 165:2245–2247.
138. Good CD, Vinnicombe SJ, Minty IL, King AD, Mather SJ, Dicks-Mireaux C. Posterior urethral valves in male infants and newborns: detection with US of the urethra before and during voiding. Radiology 1996; 198:387–391.

Female Pelvis

Anna S. Lev-Toaff and Deborah Levine

37

TECHNIQUES AND NORMAL ANATOMY

The primary indications for pelvic sonography are the evaluation of pelvic masses, pelvic pain, and abnormal bleeding. Views of the pelvis are obtained transabdominally through a full urinary bladder and transvaginally with an empty bladder. These two methods complement each other and allow for complete evaluation of the pelvic organs (Table 1). Longitudinal and transverse views of the uterus, cervix, cul-de-sac, and adnexa are obtained. Measurements of the uterus, ovaries, and any pathologic findings should be documented in three dimensions. The anteroposterior (AP) diameter of the endometrium should be measured on a longitudinal view of the uterus. This measurement includes both layers of the endometrium (Fig. 1). Views of the kidneys are obtained to exclude hydronephrosis. Color Doppler is helpful in the evaluation of pelvic masses, especially in distinguishing between cysts and vessels, in documenting the solid nature of a lesion, and in the evaluation of ovarian torsion.

In selected cases transperineal imaging may be useful to evaluate the vagina, cervix, uterus, urethra, and bladder. This technique is especially helpful in patients with vaginal atresia; it can also be used in children.

Sonohysterography

The instillation of sterile saline into the uterine cavity through a transcervical catheter has greatly expanded the diagnostic capability of transvaginal sonography. By distending the uterine cavity, the two layers of endometrium are separated and permit detection of diffuse and/or focal abnormalities of the endometrium, such as endometrial polyps and hyperplasia. Sonohysterography (SHG) is useful to help distinguish between endometrial lesions and myometrial lesions such as uterine myomas and adenomyosis (1). The precise location of myomas with respect to the uterine cavity is important clinical information to help determine therapeutic approach. In addition, demonstration of a normal endometrium and normal uterine cavity serves to obviate endometrial sampling and other invasive procedures in many cases. In addition, SHG can be used to evaluate fallopian tube patency by demonstrating the appearance of saline in the pelvis or, more directly, by demonstrating the echogenic bubbles of saline (or contrast material) traversing the tubes and spilling into the pelvis. Color Doppler or air has also been used effectively to evaluate tubal patency.

Three-Dimensional Ultrasound

With three-dimensional ultrasound (3DUS), any desired plane through a pelvic organ can be obtained, regardless of the orientation of the sound beam during acquisition. For example, the coronal "face-on" plane through the uterus is routinely obtained with 3DUS but only rarely seen with conventional two-dimensional (2D) transvaginal sonography. This view of the uterus is essential for assessing the external uterine contour, particularly the fundus, for the diagnosis of uterine anomalies (2). The evaluation of tubal versus ovarian lesions and uterine versus adnexal lesions is often facilitated by the views obtainable with 3DUS. SHG can be combined with 3DUS to yield highly detailed images of the uterine cavity

TABLE 1 ■ Transvaginal and Transabdominal Scanning

Transabdominal advantages
■ View of entire pelvis
■ Evaluate large masses
■ Evaluate masses out of range of transvaginal probe
■ Can be used in patients with intact hymen
Transabdominal disadvantages
■ Full bladder requires time for patient to fill and may cause pain during examination
■ Some patients cannot adequately fill bladder
■ Difficult to evaluate retroverted uterus
Transvaginal advantages
■ Proximity to pelvic organs with higher frequency transducer allows for higher resolution imaging and better tissue characterization
■ Empty bladder scanning typically is less painful than with a distended bladder
■ Good for obese patients and patients with abdominal wall scars, which limit ability to scan transabdominally
Transvaginal disadvantages
■ Limited field of view; masses out of the range of the transducer are missed
■ Cannot be used in patients with intact hymen and in some postmenopausal women
■ Some patients will not be comfortable with examination

and myometrium (3). This view of the uterus is also readily comprehensible by referring physicians and patients because it is close to the familiar concept of uterine shape. The coronal view of the uterus facilitates assessment of the endometrium for focal alterations in echotexture and of the uterine cavity for symmetry and distortion.

With 3D or volume ultrasound, a volume (rather than a slice) of ultrasound data is acquired and stored. The stored data can be reformatted and analyzed in numerous ways; navigation through the saved volume can demonstrate innumerable arbitrary planes. In the multiplanar display, three perpendicular planes are displayed simultaneously. Correlation between these three planes is used to confirm a given desired plane, such as the midsagittal or mid-coronal plane. In addition, all or part of the saved volume can be processed into a rendered image that can be displayed alone or in correlation with the multiplanar display. Although these expanded applications are relatively new in the field of sonography, there seems to be little doubt that these tools will become a routine part of pelvic sonography. Among the advantages of 3DUS in gynecology compared with conventional two-dimensional ultrasound (2DUS) are (*i*) the ability to digitally store ultrasound volume data which can be retrieved and studied at a later time either on-site or remotely, thus facilitating consultation and networking; (*ii*) the ability to interact with the volume data and obtain any plane (including those not obtainable with 2DUS) through the uterus, adnexa, and pelvic floor, even after the patient is no longer present; (*iii*) simultaneous correlation of three orthogonal planes in the multiplanar display, along with rendered images, as desired; (*iv*) accurate volume calculations, even of irregular structures such as the endometrium.

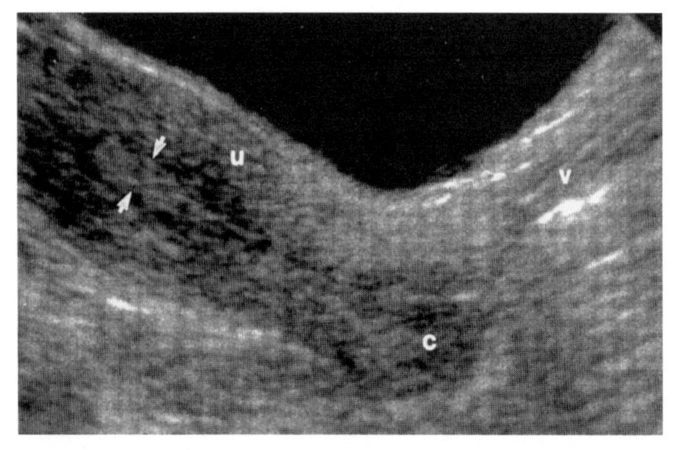

FIGURE 1 ■ Normal postmenarchal uterus. The uterine body is larger than the cervix. The endometrium (*arrows*) is the region of relatively bright central linear echoes. v, vagina; u, uterine body; c, cervix.

Uterus

Normal Uterus

Normal uterine size varies with age. Because of maternal hormone stimulation, the neonatal uterus is relatively large, with the body being larger than the cervix. The size rapidly decreases so that in early childhood the uterus has a tubular shape with the uterine body being smaller than the cervix (Fig. 2) (1,4,5). As the child approaches menarche, the uterine body again increases in size. In the postmenarchal period, the body is typically twice the size of the cervix. The dimensions of the normal uterus in women of childbearing age are 8 cm long × 4 cm AP × 4 cm wide. The multiparous uterus is larger than the nulliparous uterus by up to 1 cm in each dimension.

The myometrium should be homogeneous with smooth margins. The arcuate vessels are in the periphery of the uterus and can be seen as a segmented hypoechoic band in the outer third of the myometrium (Fig. 2).

The body of the uterus is separated from the cervix by the isthmus at the level of the internal os. The uterus assumes a variety of different positions described in relation to the angle of the long axis of the uterine body to the long axis of the cervix (flexion) and the long axis of the uterus to the long axis of the vagina (version). The most common position is anteverted. When the uterus is retroverted or retroflexed, it is difficult to evaluate transabdominally and should be scanned with the vaginal probe (Fig. 3).

The cervix is homogeneous in echotexture with a hypoechoic central canal. Nabothian cysts are commonly seen in the cervix of parous women. These cysts measure less than 2 cm and are usually anechoic, but occasionally contain debris. They are probably caused by prior inflammation.

Uterine Anomalies

Failure of complete fusion of the mullerian ducts during embryonic development can result in a variety of

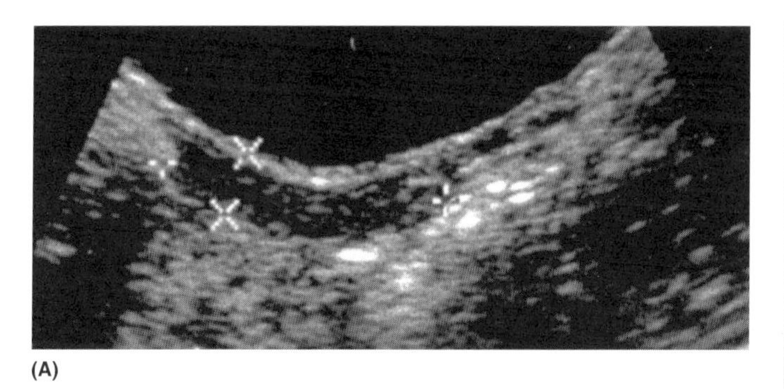

(A)

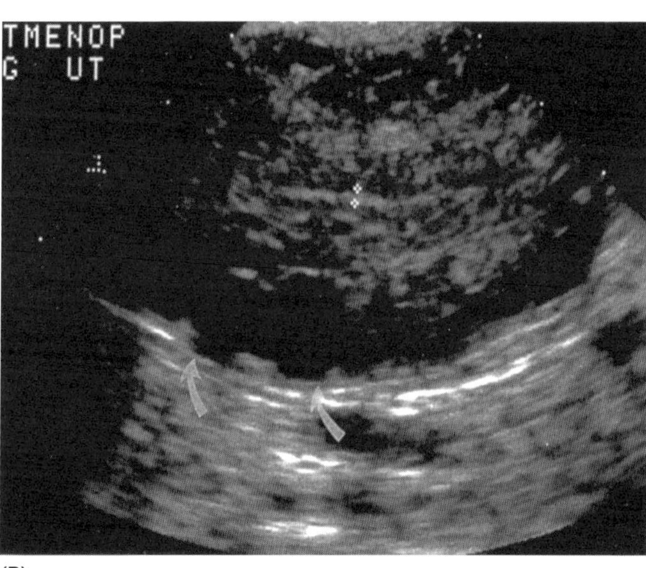

(B)

FIGURE 2 A AND B ■ (**A**) Transabdominal view of the uterus in a four-year-old girl. The cervix is larger than the body of the uterus. (**B**) Transvaginal view of the uterus in a postmenopausal woman. The endometrium is a thin linear hyperechoic band (*calipers*). This patient also has prominent arcuate vessels (*curved arrows*).

structural anomalies of the uterus, cervix, and vagina, both symmetric and asymmetric. Symmetric anomalies include uterus didelphys, bicornuate uterus, septate or subseptate uterus, and the arcuate uterus. All of the above lesions can also be asymmetric if one of the two mullerian ducts is underdeveloped. For example, one of the two horns of a bicornuate uterus may be small with or without an endometrial cavity. Likewise, one side of a septated vagina may be atretic. If one mullerian duct fails entirely to develop, a unicornuate uterus will result. There is an association between uterine anomalies and urinary tract anomalies, more so in the asymmetric uterine anomalies.

The frequency of uterine anomalies in the general population is difficult to discern, as many of these anomalies are asymptomatic. Estimates of frequency in the literature range from 1% to 10% with even higher rates in infertility populations (6).

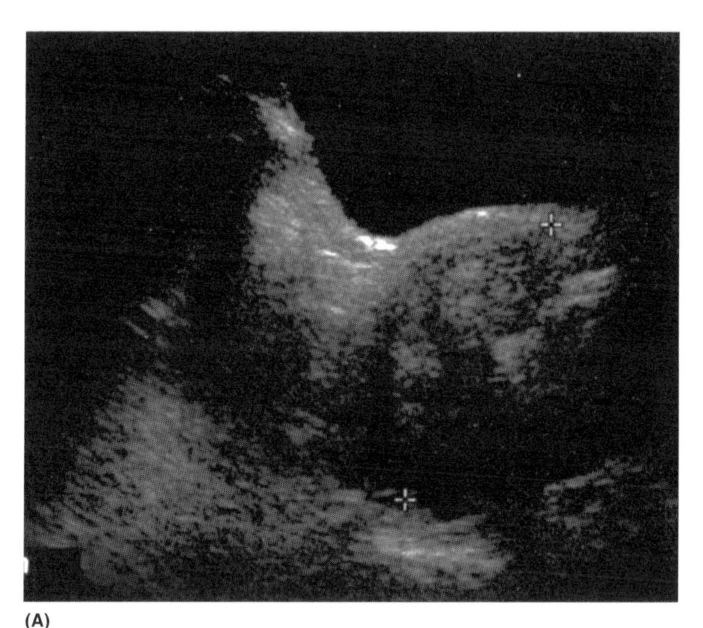

(A)

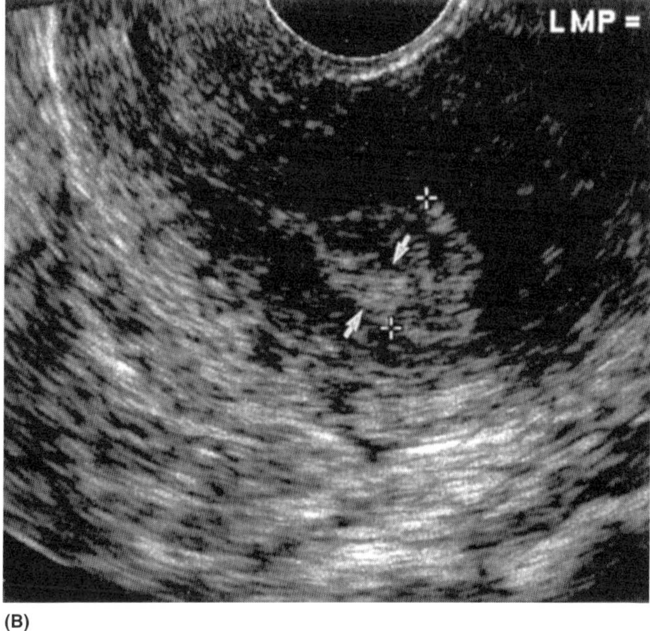

(B)

FIGURE 3 A AND B ■ Retroflexed uterus in a woman with intermenstrual bleeding. (**A**) Transabdominal examination shows a retroflexed uterus, but it is difficult to evaluate the fundus and the endometrium. (**B**) Transvaginal examination shows a thickened endometrium that measures 18-mm (*calipers*) with a focal area of increased echogenicity (*arrows*), which was a polyp. Transvaginal examination is necessary to completely evaluate the uterus in patients with retroverted or retroflexed uterus and to evaluate the endometrium in women with abnormal bleeding.

TABLE 2 ■ Uterine Anomalies

	Horns	Ultrasound diagnosis
Unicornuate	1 horn	Difficult to diagnose without 3DUS Small uterus Displaced from midline
Bicornuate One cervix (bicornis unicollis) Two cervices (bicornis bicollis)	2 horns	Fundal sagittal groove over 1 cm in depth
Didelphys	Two horns; hemi-uteri are widely separated	Two distinct cervices; may have duplication of vagina
Arcuate		External contour normal Fundal myometrium convex toward the lumen; distance between intercornual line and lowest point of fundus ≤1 cm
Septate		External fundal contour normal or minimally indented. Cavity divided by internal septum of variable length and thickness
DES exposure		Hypoplasia of uterine body with small or T-shaped uterine cavity

Abbreviations: DES, diethylstilbestrol; 3DUS, three-dimensional ultrasound.

Uterine anomalies have been classified by the American Society of Reproductive Medicine (7). Table 2 is adapted from this classification. Pelvic sonography is useful in screening for uterine anomalies. Transabdominal scanning is valuable in identifying the degree of separation of the uterine horns and to a limited extent in defining the external uterine contour. Transabdominal scanning is also important to detect an associated renal anomaly (Fig. 4). The finding of renal agenesis is a useful clue to the presence of an anomaly, more likely asymmetric. Endovaginal scanning during the secretory phase (when the endometrium is most prominent) is helpful in defining the presence of one or two separate endometrial stripes and in determining where the cavities become unified. The presence of a single or duplicated cervical canal can also be determined. Three-dimensional volume ultrasound has greatly enhanced the capability of sonography to define in detail the morphology of uterine anomalies (8,9). The use of SHG in conjunction with three-dimensional transvaginal sonography provides highly detailed imaging of the anomalous uterus and any coexistent abnormalities (3). The mid-coronal plane of the uterus is of particular value to assess the external and internal uterine contour in order to distinguish between milder anomalies such as an arcuate uterus, the various types of septate uteri, and the less common bicornuate uterus. Magnetic resonance imaging (MRI) has been shown to be effective in demonstrating the morphology of uterine anomalies and may be helpful when sonographic expertise is unavailable or in complicated cases.

Uterine anomalies are usually asymptomatic. Clinical symptoms vary with the type of anomaly (9,10). The risk of first trimester spontaneous miscarriage in women with septate uteri is significantly increased. Overall, uterine anomalies have been associated with poor reproductive outcomes including increased risk of first and second trimester miscarriage, preterm delivery, placental abruption, and intrauterine growth restriction. Uterine anomalies may also present with a variety of gynecologic problems including abnormal bleeding, pelvic pain, obstructed menstruation and secondary endometriosis, and pelvic mass. Sonographically, the early gestation in an anomalous uterus may mimic an abnormal gestation such as a cornual pregnancy; the nonpregnant horn may also mimic a uterine or adnexal mass.

Endometrium

The endometrium is visualized as a linear band, usually hyperechoic to myometrium, in the center of the uterus. The total thickness of the endometrium represents the anterior and posterior opposed layers. Endovaginal scanning is required to optimally visualize the endometrium. When endometrial fluid is present, this should not be included in the endometrial thickness measurement. The hypoechoic layer around the endometrium represents an inner compact layer of myometrium and should not be included in endometrial thickness measurements (11).

Normal endometrial thickness and appearance vary with the phase of the menstrual cycle (Fig. 5) (Table 3) (12). During the menstrual phase, the endometrium is very thin and minimally hyperechoic and may be interrupted. A small amount of hypoechoic fluid may be seen within the cavity. Normal menses are not associated with sizable fluid collections in the uterine cavity or retained echogenic blood clots. However, these findings may be seen with obstructed menstrual flow (e.g., adhesions) or excessive bleeding (submucous myomas or adenomyosis). After

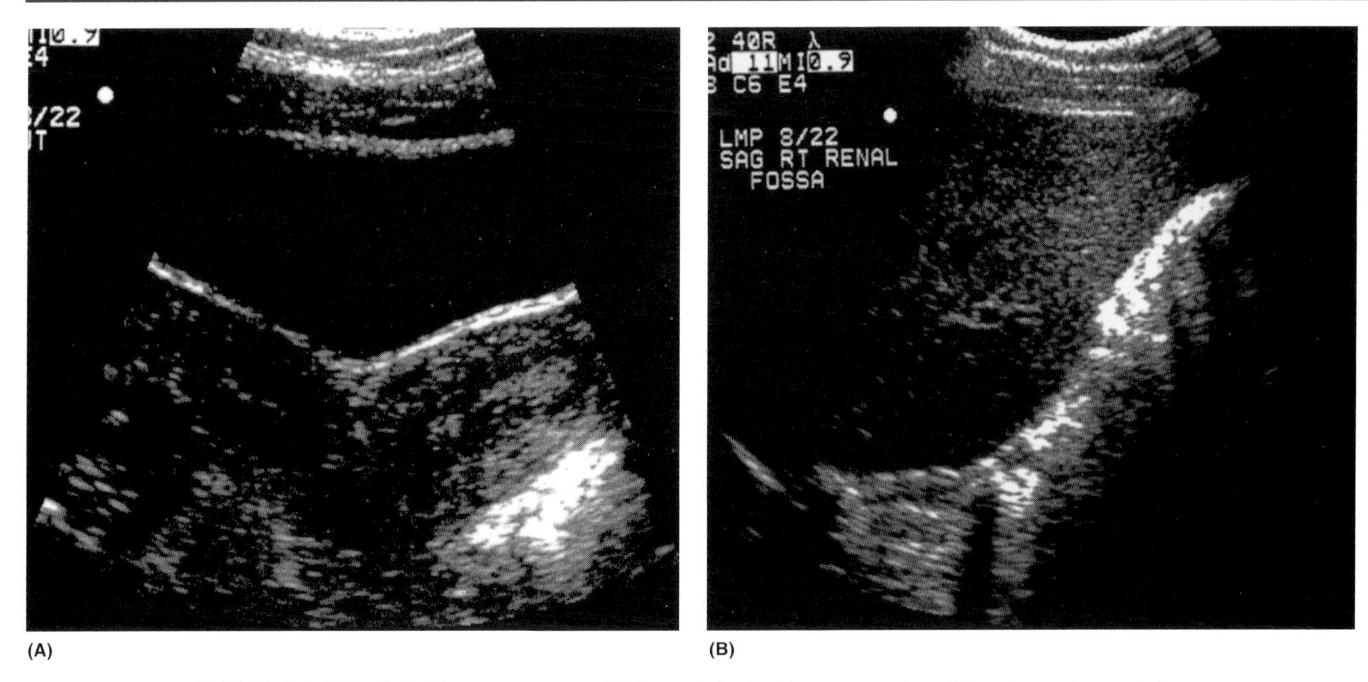

(A) (B)

FIGURE 4 A AND B ■ Bicornuate uterus. (**A**) Transabdominal transverse view of the uterus demonstrates two horns that are widely separated. Only one cervix was seen on vaginal scanning. (**B**) View of the right renal fossa demonstrates an absent right kidney.

the menses, the endometrium gradually thickens. In the periovulatory phase, the endometrium has a striated appearance; the inner endometrium is relatively hypoechoic and is surrounded by a more hyperechoic peripheral endometrium. A central thin echogenic line does not represent a layer of endometrium but rather the interface between the two opposing layers of endometrium. This "central white line" is useful sonographically; when it is interrupted, an intracavitary lesion, such as a polyp, myoma, or adhesion is suggested. After ovulation, there is progressive thickening of the secretory phase endometrium. In the late secretory phase, the endometrium is at its greatest thickness, with homogeneously increased echogenicity and increased through transmission (5).

Fallopian Tube

The normal fallopian tube is difficult to distinguish from surrounding vessels and ligaments. It usually is not visualized unless it is abnormal or surrounded by fluid.

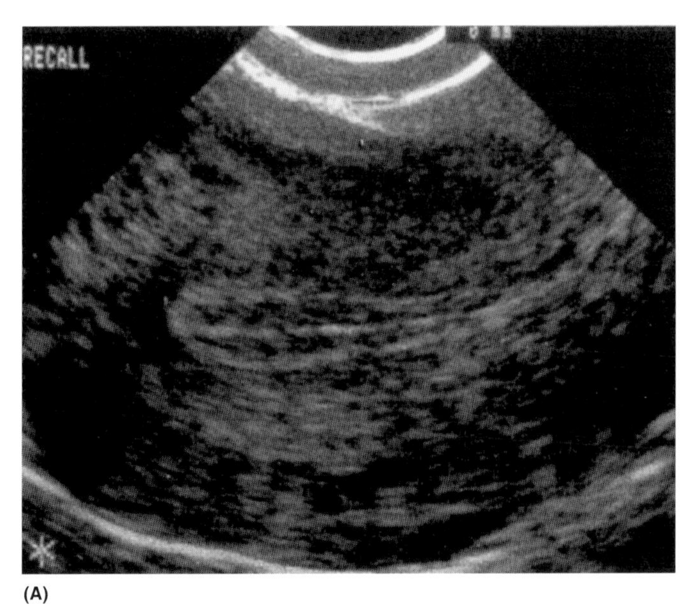

(A)

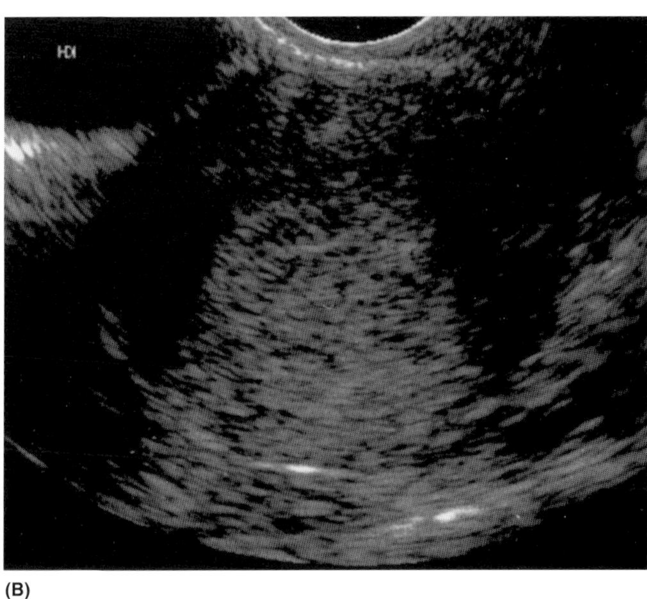

(B)

FIGURE 5 A AND B ■ Normal endometrium. (**A**) "Triple line" endometrium in midcycle. (**B**) Secretory phase endometrium that is thick and echogenic with posterior acoustic enhancement.

TABLE 3 ■ Normal Endometriumv

Day of cycle	Phase	Thickness (mm)	Appearance
1–4	Menstrual phase	1–4	Thin interrupted echocomplex, slightly hyperechoic Small amounts of fluid may be seen
5–14	Proliferative phase	4–8	Mildly echogenic surrounded by thin hypoechoic band of inner myometrium
12–15	Periovulatory		Striated with inner hypoechoic endometrium surrounded by peripheral echogenic endometrium; central echogenic line is interface between opposing endometrial surfaces
15–28	Secretory phase	8–16	Thick Echogenic with through transmission

Ovaries

Ovarian Size

Ovaries in girls younger than two years of age are on average 1 mL in volume; in neonates they can be slightly larger. There is a considerable range of normal ovarian volume between birth and two years of age because of the normal presence of cyst and follicles. A majority of ovaries imaged in the first three months of life were found to contain follicles and the finding of ovarian follicles throughout childhood is not abnormal or uncommon; most commonly these follicles are less than 5 mm in diameter (13).

The ovaries increase in size in prepubertal girls with follicles up to 1 cm in size; a multicystic appearance predominates in premenarchal girls. After menarche, the ovaries are ovoid in shape; the mean volume is at least 6 mL, equivalent to measurements of 3 cm × 2 cm × 2 cm (Fig. 6) (14). Normal ovarian volume in the menstruating females is 5 to 15 mL, with an approximate mean of 10 mL; however, measurements as high as 22 mL have been reported in normal ovaries (14). In Table 4, the size and appearance of ovaries with respect to patient age are listed.

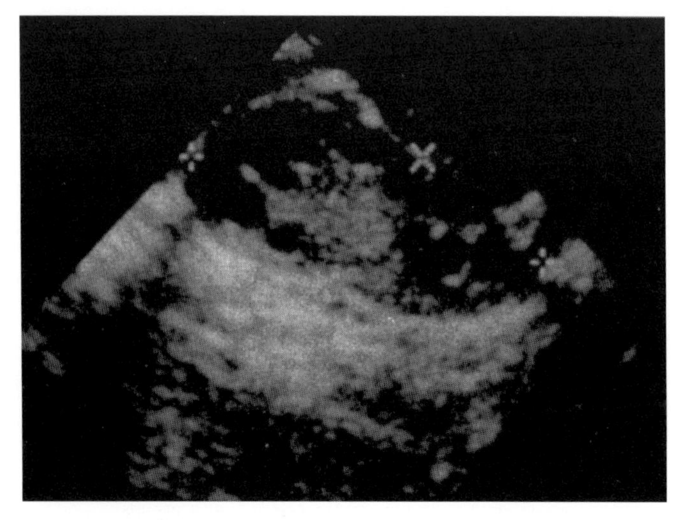

FIGURE 6 ■ Normal ovary with multiple small follicles.

Functional Ovarian Cysts

In the proliferative (follicular) phase of the menstrual cycle, multiple small follicles are visualized, usually 1 cm in diameter or less. A dominant follicle emerges in the second week of the follicular phase and reaches a maximum size of 2 to 2.5 cm before ovulation. After ovulation, the corpus luteum develops. The corpus luteum may appear solid, cystic, or hemorrhagic, but there is usually evidence of enhanced through transmission because of the fluid content. It is distinguishable from the mature follicle by the presence of a thicker wall, slightly irregularity in contour, internal echoes, and a surrounding ring of neovascularity, accompanied by the luteal (or secretory) echogenic appearance of the endometrium. When there is more than minimal bleeding into the corpus luteum, a hemorrhagic cyst becomes apparent. The appearance of the hemorrhagic cyst varies with the stage of hemorrhage and the amount of fluid within the cyst (15). In Table 5, the appearance of ovarian follicles with respect to phase of the menstrual cycle is described, and in Table 6, the various appearances of hemorrhagic cysts are listed.

Occasionally a corpus luteum is formed (i.e., dominant follicle is luteinized) despite the fact that the follicle did not rupture and release an egg. This is known as luteinized unruptured follicle and is one type of anovulation found in infertile women (16). In such cases the corpus luteum appears as a simple cyst. Likewise a dominant follicle, in the absence of the luteinizing hormone (LH) surge (stimulus for ovulation), may persist as an anechoic follicular cyst.

TABLE 4 ■ Ovary Appearance with Respect to Age

Age	Size	Appearance
Neonate	May be larger than 1 cm	Follicles <1 cm
<2 yr	<1.0 cm^3	Follicles <5 mm
Prepubertal	<2.5 cm^3	Follicles <1 cm
Postpuberty	9.8 ± 5.8 cm^3	Follicles present
Postmenopause	Decrease in size	Cysts seen in up to 15%

TABLE 5 ■ Normal Ovary

Days of cycle	Phase	Description	Follicle size	Appearance
1–14	Follicular	Multiple small follicles	<1 cm	Thin wall anechoic
10–14	Late follicular	Dominant follicle	<2.5 cm	Thin wall anechoic
15–28	Luteal	Corpus luteum	2–3 cm, up to 10 cm for hemorrhagic cyst	Thicker wall Irregular contour internal echoes, vascular rim ±Hemorrhage (Table 6)

Usually the corpus luteum persists in a functional state for 14 days and then degenerates and ceases to secrete estrogen and progesterone. This allows the start of a new menstrual cycle. Remnants of the corpus luteum may be detectable sonographically in the next cycle as a small complex cystic structure within the ovary, known as atretic corpus luteum. A corpus luteum cyst (hemorrhagic or simple) that persists beyond its normal functional 14 days may lead to delay in the onset of the next menses by several days or several weeks; this is known as persistent corpus luteum cyst.

Ovarian cysts are common in all age groups, but especially in women of menstrual age. Typical benign cysts are completely anechoic with enhanced through transmission. They have a thin wall, no septations, and no solid elements. Because dominant follicles and corpus luteum cysts frequently are up to 3 cm in size, benign-appearing lesions in this size range require no follow-up. Hemorrhagic cysts have a more complex appearance (Table 6) with internal septations and retractile clot (Fig. 7). However, if a cyst is small (<3 cm) and has the classic appearance of a hemorrhagic cyst in a woman of reproductive age, it can be treated as a benign lesion and does not require follow-up.

If a lesion has a questionable appearance or is larger than 3 cm, serial scans are helpful since hemorrhagic cysts undergo rapid change in their internal characteristics. A follow-up sonogram shortly (within the first 10 days) after the onset of the next menses typically demonstrates resolution of the cyst. The timing of the follow-up study is intended to avoid confusion by a newly formed dominant follicle or corpus luteum. The management of benign-appearing premenopausal ovarian cysts is shown in Table 7.

TABLE 6 ■ Various Appearances of Hemorrhagic Cysts

- Simple cyst (thin wall, anechoic)
- Septated cyst
- Irregular wall
- Echogenic mass that simulates a solid mass
- Strands of internal densities
- Retractile clot

Free Fluid

A minimal amount of fluid is present in the cul-de-sac of asymptomatic women throughout the menstrual cycle. The largest quantity of normal free fluid occurs in midcycle. This midcycle increase in pelvic fluid is not a good indicator of ovulation because it is present before, during and after ovulation secondary to transudation from the ovary.

Complex fluid (with debris or septations) is abnormal and results from hemorrhage, infection, or neoplasm.

SONOGRAPHIC PITFALLS

Nonovarian Solid Pelvic Masses

A variety of nonovarian tumors and other abnormalities can masquerade as solid ovarian masses (Table 8). The most common of these is the pedunculated fibroid (Fig. 8). This diagnosis is made sonographically when the ovaries are seen separate from the concerning solid mass. A vascular stalk can often be shown extending from the

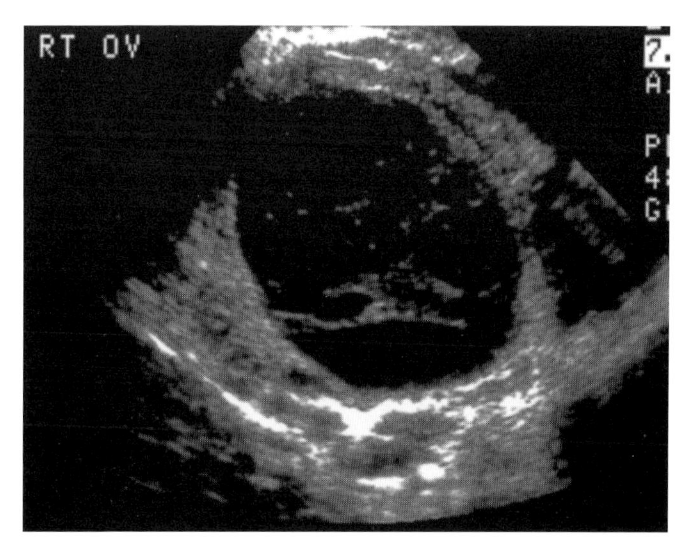

FIGURE 7 ■ Hemorrhagic cyst. Transvaginal view of the right ovary demonstrates a cyst with multiple internal echoes and strands of internal echoes. This is the classic appearance of a hemorrhagic cyst.

TABLE 7 ■ Management of Benign-Appearing[a] Premenopausal Ovarian Cysts

Size	Follow-up
<3 cm	No follow-up
3–6 cm	Follow-up ultrasound shortly after the start of the next menses
>6 cm	Likely neoplastic
	If anechoic, likely benign

[a]Benign appearance of anechoic cyst: thin wall, no septations, no solid elements. Benign appearance of hemorrhagic cyst: cyst with strands of internal densities, retractile clot, may have irregular walls.

pedunculated fibroid to the uterus. Three-dimensional sonography or MRI is helpful in difficult cases. Nongynecologic conditions that may mimic solid ovarian masses include pelvic kidney (Fig. 9A), diverticulitis (Fig. 9B), rectosigmoid carcinoma, vascular masses, pelvic lymph nodes, an inflamed appendix, and pelvic hematomas.

Nonovarian Cystic Pelvic Masses

Multiple lesions within the pelvis can masquerade as an ovarian cyst (Table 9). The etiology of extraovarian cysts is suggested by visualization of a separate ipsilateral ovary (Fig. 10) and in some cases by connection with the organ of origin, such as in the case of a bladder diverticulum (17) or a Tarlov cyst (18).

Bowel loops frequently mimic ovarian cysts. Therefore, watch for peristalsis when a questionable lesion is visualized. Clusters of Nabothian cysts in the cervix have also been confused with ovarian cysts.

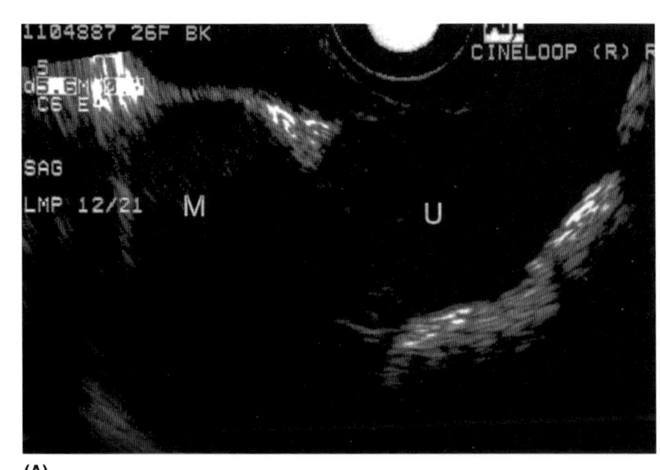

(A)

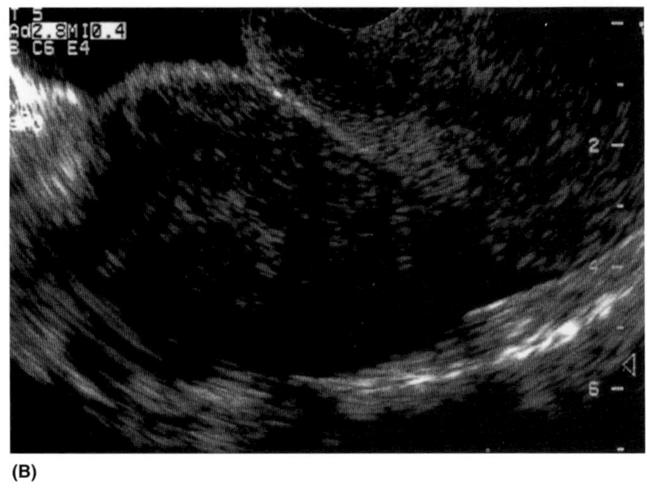

(B)

FIGURE 8 A AND B ■ Pedunculated fibroid. (**A**) Transabdominal view of the pelvis demonstrates a mass (M) adjacent to the uterus (U). (**B**) Transvaginal examination demonstrates a tissue plane between the uterus and the mass.

TABLE 8 ■ Mimics of Solid Ovarian Masses[a]

Diagnosis	Appearance	Comments
Pedunculated fibroid	±Stalk connecting to uterus	MRI for best specificity
Fallopian tube masses		
■ TOA	Ill-defined mass when acute Well-defined echogenic mass when chronic	Other signs of pelvic inflammatory disease Pain
		Vaginal discharge
■ Cancer	Complex or solid mass	
Appendicitis	Right side	Elevated white count
	Adjacent to cecum	Fever
		Tender
Pelvic kidney	Collecting system	Empty ipsilateral renal fossa
	Reniform shape	.
Rectosigmoid carcinoma	Hypoechoic, heterogeneous mass	Arises from bowel
Vascular mass/malformation	Doppler flow	
Hematoma	Heterogeneous mass	History of trauma or surgery
Lymph node	Homogeneous solid mass	

[a]In all of these, the ipsilateral ovary should be seen as a separate structure.
Abbreviations: TOA, tubo-ovarian abscess; MRI, magnetic resonance imaging.

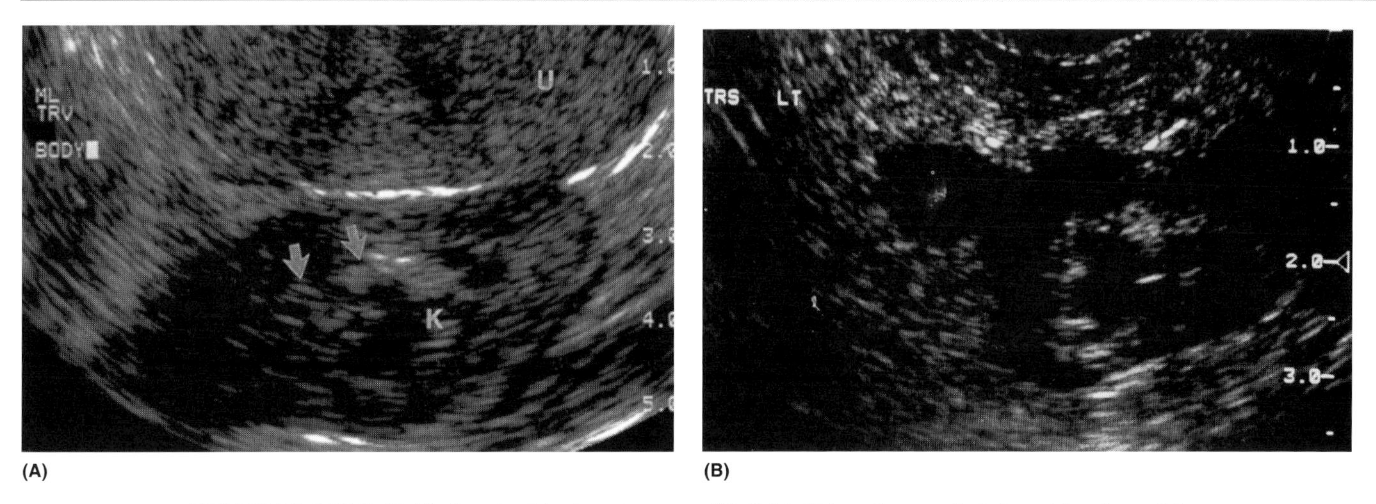

(A) (B)

FIGURE 9 A AND B ■ Mimics of solid adnexal masses. (**A**) Pelvic kidney. Transvaginal view of the uterus (U) with a mass (K) seen behind the uterus with a tissue plane between the two. Echogenic fat (*arrows*) can be seen in the center of this pelvic kidney. (**B**) Diverticulitis. Transvaginal view in the left pelvis shows a hypoechoic mass arising from sigmoid colon in this patient with diverticulitis.

TABLE 9 ■ Ovarian Cyst Mimics

Diagnosis	Appearance	Comments
Adnexal cyst		
Paraovarian/paratubal	Separate from ipsilateral ovary	Comprise 10% of adnexal masses at pathologic examination
Hydrosalpinx	Elongated tubular structure	
Tubo-ovarian abscess	Complex, hypoechoic mass	Echogenic fluid in cul-de-sac
	Septations	
	Irregular margins	
Ectopic pregnancy	Living extrauterine embryo	Positive pregnancy test result
	Extrauterine sac or ringlike structure	Decidual reaction in uterus
Endometrioma	Cyst with low-level echoes	Fluid in the cul-de-sac
Extra-adnexal cyst		
Lymphocele	±Septations	Prior surgery
Mesenteric cyst	Adjacent to bowel	Arise from bowel mesentery
Peritoneal inclusion cyst	Multiloculated	History of prior surgery or PID
	Septations with blood flow simulating neoplasm	Adhesions
	±Thick wall	
Bladder diverticulum	See connection with bladder	
Hydroureter	Follows course of ureter	
Tarlov cyst	Adjacent to spine	
Miscellaneous		
Varices	Color Doppler fills structure	
Adnexal hypoechoic solid mass	±Internal vascularity Internal echoes seen endovaginally	
Bowel	Peristalsis	
Abscess	Adjacent to bowel	Appendicitis
		Diverticulitis

Abbreviation: PID, pelvic inflammatory disease.

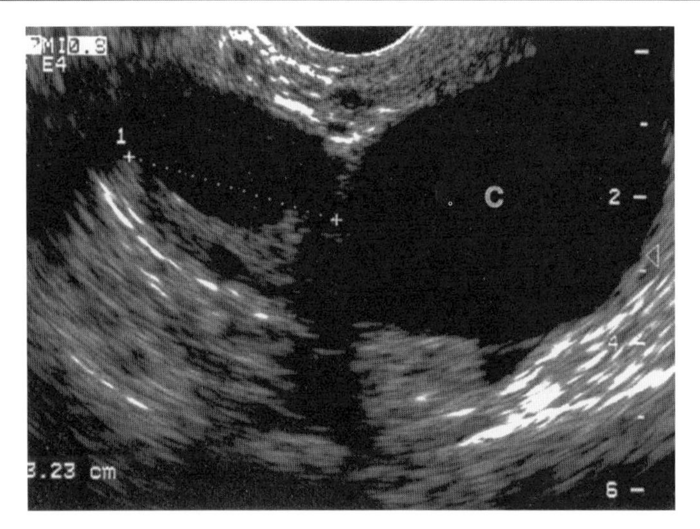

FIGURE 10 ■ A 6 cm paraovarian cyst (C) is seen medial to the right ovary (*calipers*).

Nonvisualization of a Palpable Pelvic Mass

Dermoid cysts have a variety of appearances because of their complex nature. Frequently, the dermoid cyst mimics bowel contents and is seen only as an echogenic area with shadowing. In a patient with a palpable pelvic mass in whom no abnormality is visualized on sonography, consider an echogenic dermoid (Fig. 11) and carefully scan in the region of the palpable mass. A combination of transabdominal and transvaginal imaging is ideal for evaluation of dermoid cysts.

Don't Stop after One Lesion Is Found

Many benign ovarian tumors occur bilaterally (dermoids, serous cystadenomas, and metastases). In addition, women with one gynecologic malignancy are at increased risk for a second malignancy (Fig. 12). Some ovarian tumors, such as endometrioid tumors and estrogen-producing thecoma and granulosa cell tumors, are associated with endometrial hyperplasia and endometrial cancer (Fig. 13). There also are rare syndromes in which gynecologic malignancies are grouped such as the Lynch cancer family syndrome, in which there is an association between ovarian cancer, colon cancer, and endometrial cancers (20).

SONOGRAPHIC ABNORMALITIES OF THE PELVIS

Abnormal Uterus

Uterine Enlargement

Causes of uterine enlargement are listed in Table 10. These include fibroids, adenomyosis, pregnancy and pregnancy-related conditions, uterine sarcoma, endometrial carcinoma, and cervical or lower uterine segment obstruction (congenital or acquired) resulting in a fluid-filled uterus.

Fibroids ■ Fibroids occur in approximately 25% of women of reproductive age (Fig. 14). They consist of round well-circumscribed nodules of smooth muscle and typically cause an enlarged uterus with multiple masses that are echo-attenuating. They are sensitive to estrogen stimulation and may increase in size during pregnancy (14). Cystic areas are second ary to degeneration. Clumps of calcification cause echogenic foci with shadowing. Peripheral rim-like calcification is also common.

Fibroids may be located within the myometrial wall, mural; bulge into the endometrial cavity, submucosal; be located almost entirely within the endometrial cavity, intracavitary; or deform the serosal surface of the uterus,

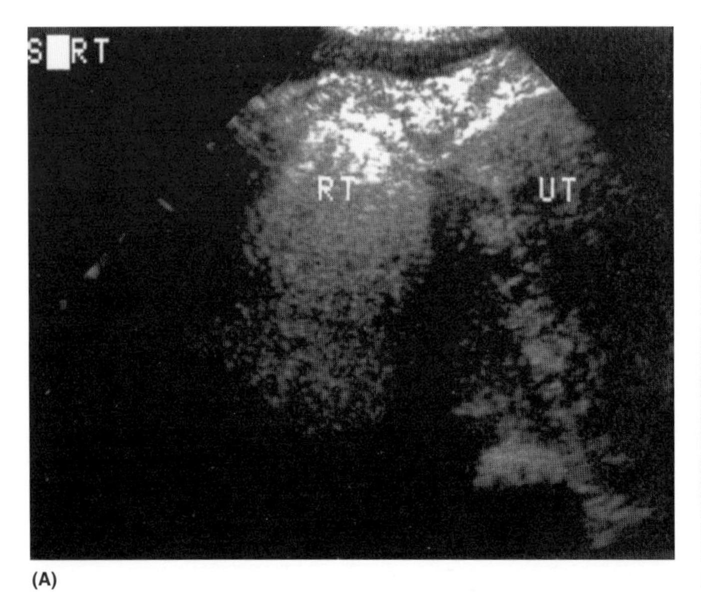

(A)

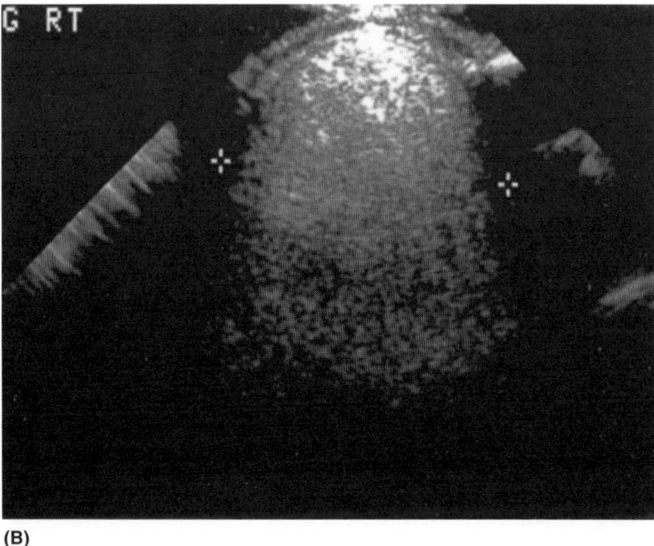

(B)

FIGURE 11 A AND B ■ Dermoid. (**A**) Transabdominal view of the uterus (UT) demonstrates a questionable right adnexal mass (RT). (**B**) Endovaginal scan demonstrates extremely echogenic nature of this mass, which was not recognized on two prior sonograms.

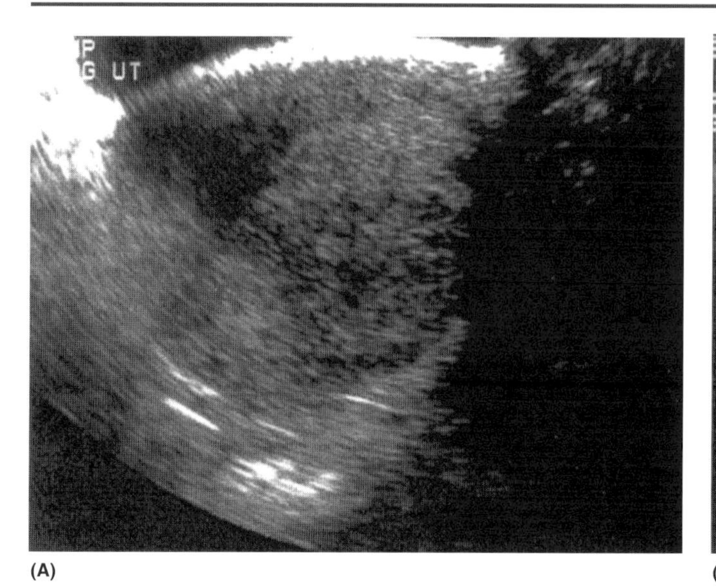

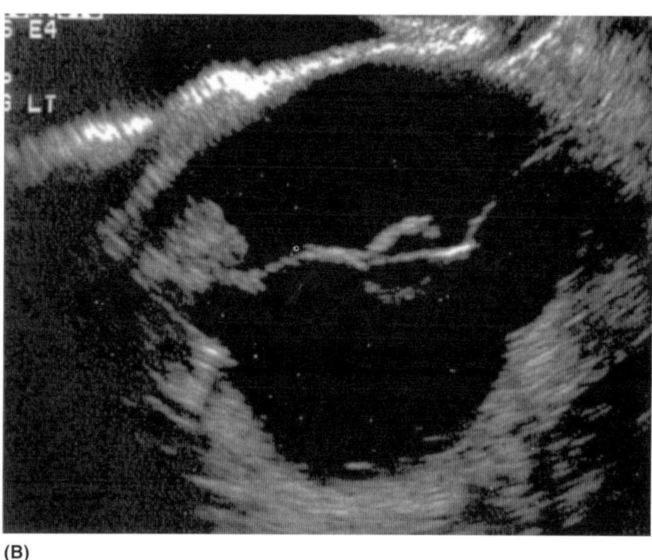

(A) (B)

FIGURE 12 A AND B ■ Concurrent lesions: a 90-year-old woman with endometrial cancer and ovarian cancer. (**A**) Transabdominal view of the uterus demonstrates ill-definition of the endometrium with invasion of the endometrium into the myometrium. (**B**) A 6-cm left adnexal cyst with multiple septations and solid nodules from ovarian cancer.

subserosal (Fig. 8). Intracavitary and subserosal fibroids may be pedunculated. Fibroids, especially submucosal, frequently cause abnormal uterine bleeding, infertility, miscarriage (early and late), premature delivery, and fetal malpresentation (21). Submucosal fibroids are the cause of postmenopausal bleeding in 10% of cases (22). SHG is very helpful in defining the intracavitary extent of a submucosal myoma. Fibroids that have at least 50% of their volume in the cavity are potentially amenable to hysteroscopic resection.

Most fibroids arise from the uterine body. Cervical and broad ligament fibroids are rare. Findings in patients with fibroids are summarized in Table 11.

Small fibroids may not be detected on transabdominal sonography; only a heterogeneous echotexture of the myometrium may be noted. At times, only a contour distortion along the interface between the uterus and bladder is seen. In these cases, transvaginal sonography will usually provide a more specific diagnosis because even subcentimeter fibroids can be delineated. In some

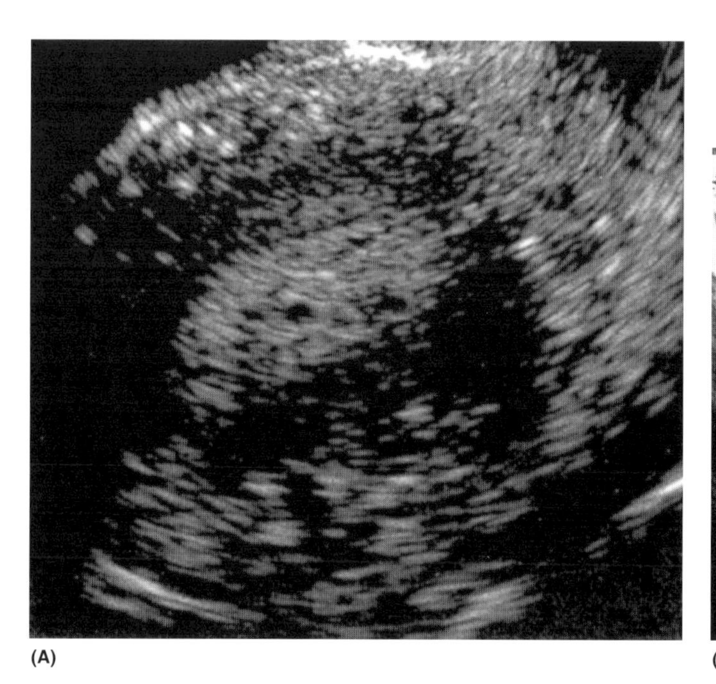

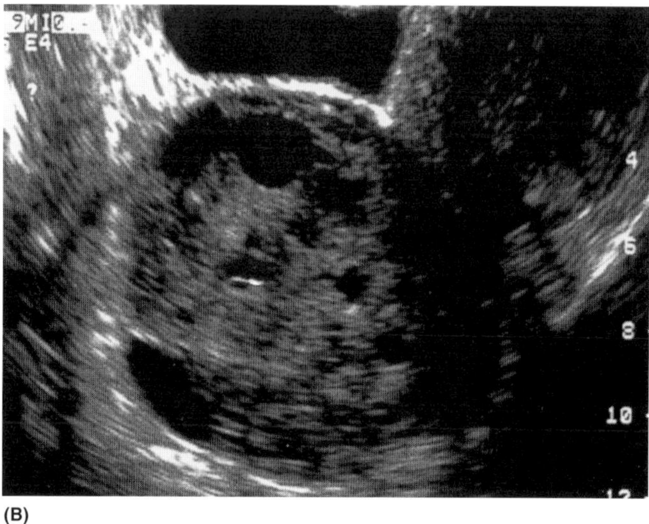

(A) (B)

FIGURE 13 A AND B ■ Concurrent lesions: granulosa cell tumor with endometrial hyperplasia. (**A**) Thickened endometrium (15 mm) with a small cyst. The histologic type was endometrial hyperplasia, probably secondary to the estrogenic effect of the granulosa cell tumor (**B**). *Source*: From Ref. 19.

TABLE 10 ■ Diffuse Uterine Enlargement[a]

Diagnosis	Comments
Normal parous uterus	Multiparous women can have uterine size 1–2 cm larger than "normal" in each dimension
Fibroids	See Table 11
Adenomyosis	Uterus diffusely enlarged with areas of heterogeneous uterine echotexture, ± small cysts, and areas of striped shadowing
	Asymmetric thickening of anterior/posterior uterine wall
	Focal or diffuse indistinctness of the endometrial–myometrial junction ±nodular contour to endometrium
	Focal adenomyosis may mimic a heterogeneous fibroid
Endometrial carcinoma	Early: endometrial thickening or mass with focal or diffuse of loss of endometrial–myometrial interface
	Late: enlargement of uterus
Sarcoma	Rapid increase in size of uterus
	Difficult to distinguish from fibroids, unless serial examinations are available
Congenital uterine obstruction	Hematometra or hematometrocolpos due to imperforate hymen, vaginal atresia, transverse vaginal septum, cervical agenesis or stenosis, rudimentary uterine horn with functioning endometrium
Acquired uterine obstruction	Cervical stenosis—postsurgical or postradiotherapy, cervical or endometrial cancer
Pregnancy	
■ Normal pregnancy	Size varies with gestational age
■ Missed abortion	Retained failed pregnancy
■ Gestational trophoblastic disease	Endometrial cavity enlarged with abnormal tissue, classically with multiple cystic spaces
Recent postpartum	Size varies with time since delivery

[a]Assumes a normal size of the uterus to be 8 × 4 × 4 cm.

cases, transvaginal sonography will show that there is no discrete mass and the diagnosis of adenomyosis should be considered.

Uterine Sarcoma ■ Sarcomas comprise less than 5% of all uterine malignancies. Fewer than 1 in 200 surgically removed fibroids are found to be sarcomatous. Sonographically, uterine sarcomas resemble heterogeneous fibroids or endometrial carcinoma (22). When a rapid change in the size of fibroids is noticed, concern should be raised about the possibility of uterine sarcoma (Fig. 15).

Adenomyosis ■ Adenomyosis denotes the presence of ectopic endometrial tissue within the myometrium, surrounded by hypertrophic myometrium. Usually adenomyosis results in uterine enlargement. Adenomyosis is frequently asymptomatic and found incidentally in hysterectomy specimens. However, this condition may present clinically with uterine tenderness, dyspareunia, abnormal uterine bleeding (meno- or metrorrhagia), painful menses, low back pain, or crampy pelvic pain. Adenomyosis may be a cause of infertility.

On sonography, the uterus is usually enlarged with a globular shape (Fig. 16). Typically the thickness of the anterior and posterior uterine corpus is asymmetric, more commonly the posterior wall being thicker than the anterior (23). Additional sonographic features of adenomyosis are best evaluated on transvaginal examination. The endometrial–myometrial junction is often indistinct and the endometrial margin may appear nodular. The myometrium contains ill-defined foci of heterogeneity with areas of increased and decreased echogenicity. Tiny cystic spaces (1–3 mm) may be found within the heterogeneous myometrium (24). The foci of heterogeneous myometrium are usually contiguous with the endometrium and more extensive in the posterior uterine wall. In diffuse disease, the entire myometrium may be involved. Occasionally, adenomyosis presents as focal lesion, known as adenomyoma. Adenomyoma is less well circumscribed and has a more heterogeneous echotexture than most fibroids. Polypoid adenomyomas are occasionally found in the uterine cavity, mimicking intracavitary fibroids with cystic change (25).

The preoperative distinction between fibroids and adenomyosis is important in women who are being treated for infertility or abnormal bleeding since myomas can be readily enucleated from the surrounding myometrium; however, adenomyosis typically requires a

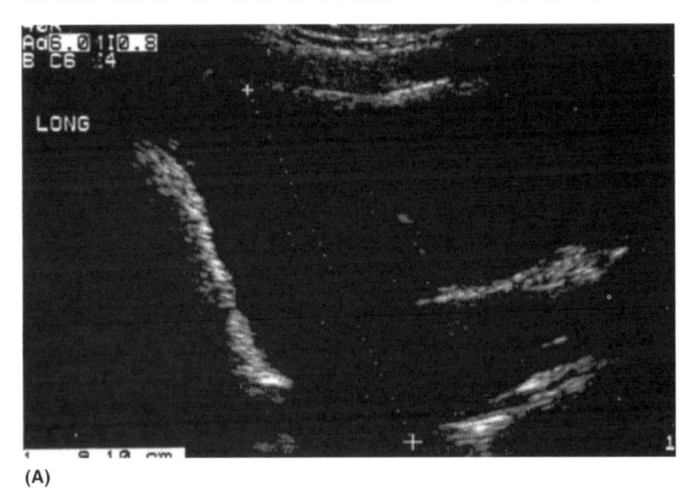

(A)

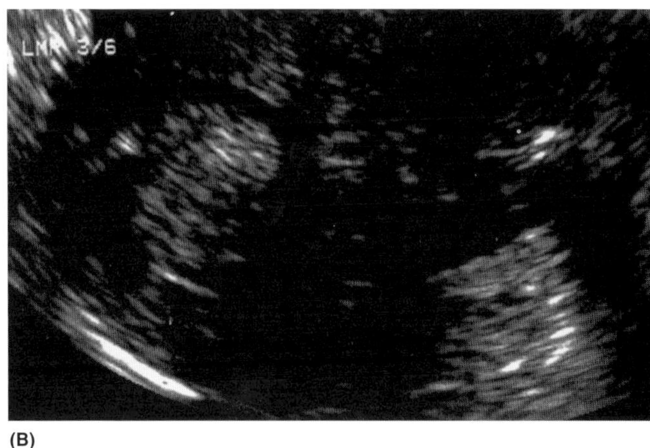
(B)

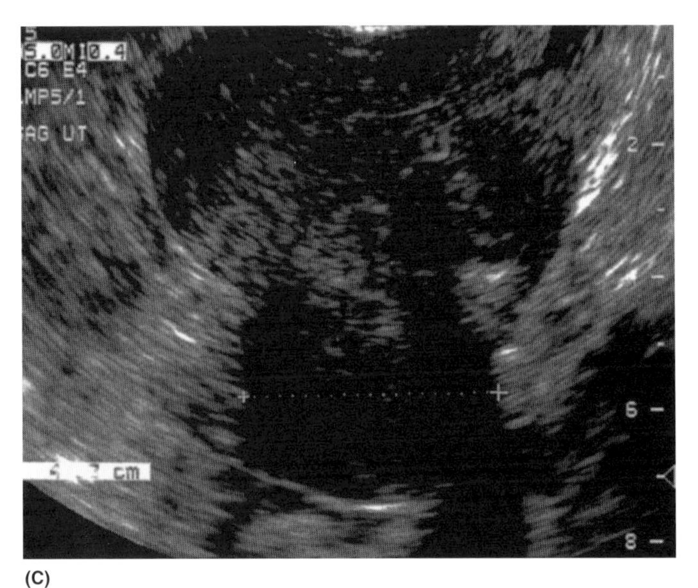
(C)

FIGURE 14 A–C ■ Fibroids. (**A**) Transabdominal view of a fibroid uterus. The uterus is enlarged with a heterogeneous echotexture and a lumpy contour caused by fibroids. (**B**) Submucosal fibroids surrounded by fluid during a sonohysterogram. (**C**) Subserosal fibroid with broad attachment to the myometrium and an exophytic component.

hysterectomy or more extensive resection of uterine tissue. In addition, uterine artery embolization appears to be more effective for fibroids than for adenomyosis. MRI is helpful when this distinction is critical in clinical decision making (23). Not infrequently, fibroids and adenomyosis coexist; sonographic diagnosis is more difficult in the presence of fibroids, and MRI can be helpful in these cases.

Obstruction ■ Obstruction to the outflow of the uterus may be congenital or acquired. In the neonate, developmental anomalies may cause obstruction to the outflow of normal uterine and vaginal secretions which are secondary to maternal hormonal stimulation. The most common anomaly is imperforate hymen which presents as a bulging perineal mass due to hydrocolpos (fluid in the vagina) or hydrometrocolpos (fluid in the vagina and uterus) (4). In adolescence, shortly after the onset of menarche, patients may present with severe cyclical low abdominal pain because of obstructed menstruation. This obstruction may be due to imperforate hymen, transverse vaginal septum or vaginal atresia, or cervical agenesis or stenosis. Obstructed menstrual flow results in hematocolpos and/or hematometra (4,26). If left untreated, this can lead to endometriosis and infertility.

Acquired cervical or lower uterine obstruction may present in adults due to a variety of benign and malignant causes. These include cervical scarring after intervention such as dilatation and curettage, cervical cone biopsy, cervical laser surgery, and radiation therapy. Cervical and endometrial cancer may also present as hemato- or pyometra (when complicated by infection) (Fig. 17).

Endometrial Cancer ■ Enlargement of the uterus is a late finding in endometrial cancer. This disease is discussed in more detail in Chapter 29.

Bright Reflectors in the Uterus

Causes of bright echoes in the uterus and endometrium are listed in Table 12.

Uterine Calcifications ■ The most common cause of dense echoes in the uterus is calcifications resulting from

TABLE 11 ■ Fibroids

Echotexture
- Hypoechoic
  - Shadowing secondary to whorls of fibrous tissue and edge artifacts
- Echogenic
- Heterogeneous (secondary to degeneration or uterine artery embolization)
- Isoechoic
- Cystic areas
  - Secondary to degeneration
- Calcifications
- Rim calcification
- Clumps of coarse calcification

Location
- Submucosal
  - Associated with menometrorrhagia
  - Distort endometrial–myometrial margins
- Intramural
  - Most common
- Subserosal
  - Distort uterine contour
- Pedunculated
  - ±Stalk
  - May present as adnexal mass
- Cervical
- Broad ligament
  - Simulate ovarian mass

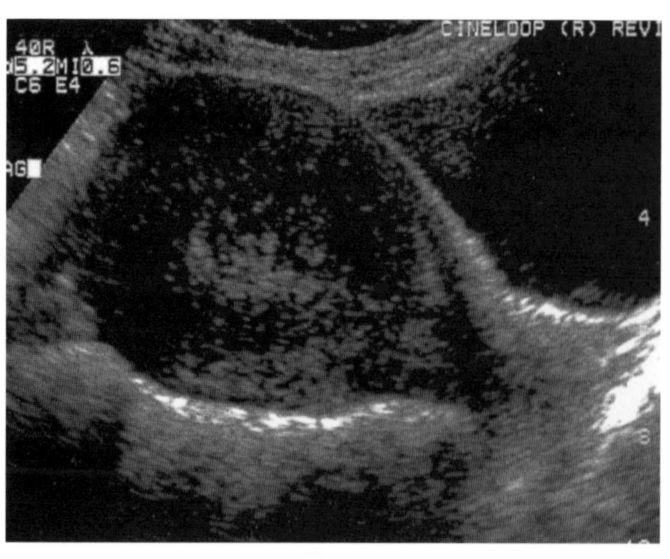

FIGURE 16 ■ Enlarged uterus in a 53-year-old woman with abnormal bleeding. The uterus is enlarged slightly and heterogeneous in echotexture but has no focal masses. Histologic examination revealed adenomyosis.

fibroids. These appear as clumps of calcification (Fig. 18A) or as rim calcification around a mass.

Arcuate artery calcifications are seen around the periphery of the uterus, usually in older women with medical problems such as diabetes, chronic renal failure, or hypertension (28). These appear as a curvilinear array of punctuate calcifications in the outer myometrium.

Tiny punctate echogenic foci are commonly seen on transvaginal sonography at the endometrial–myometrial interface and along the endocervix (Fig. 18B). Most commonly these are due to microcalcifications or other debris and are usually of no clinical significance (29).

Intrauterine Contraception Devices ■ Another cause of bright reflectors within the uterus is intrauterine

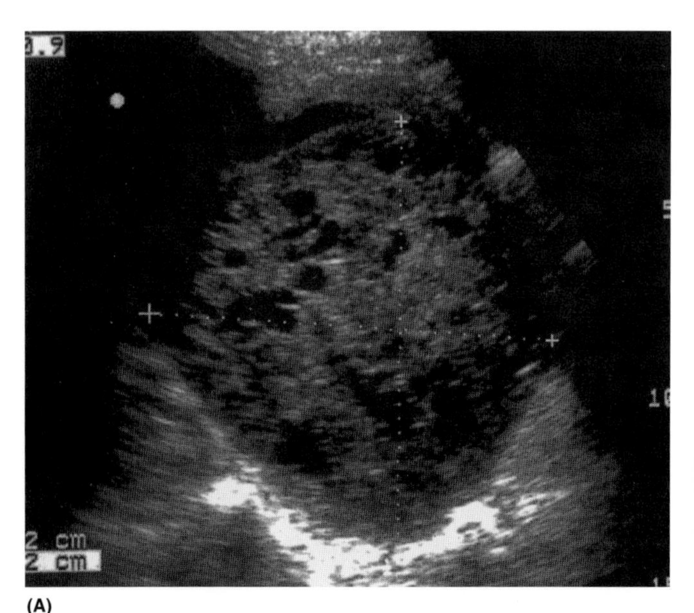

(A)

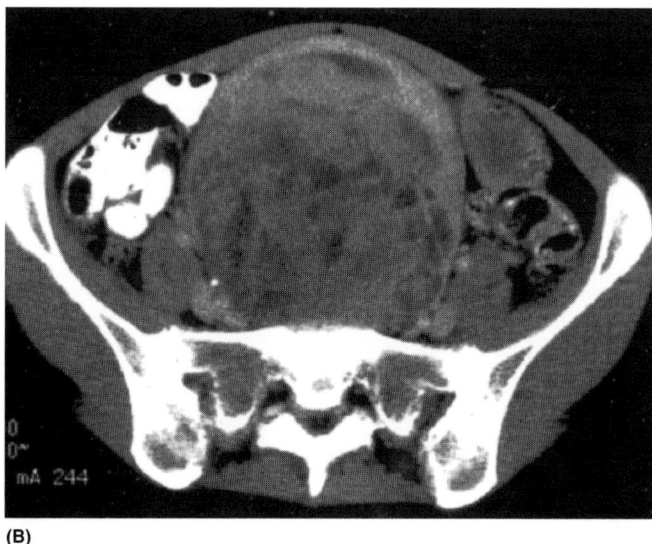

(B)

FIGURE 15 A AND B ■ Uterine sarcoma. (**A**) Transabdominal view of the uterus in a woman with a recent myomectomy demonstrates an enlarged uterus with a bizarre appearance to the myometrium with multiple cystic spaces. (**B**) CT has a similar appearance.

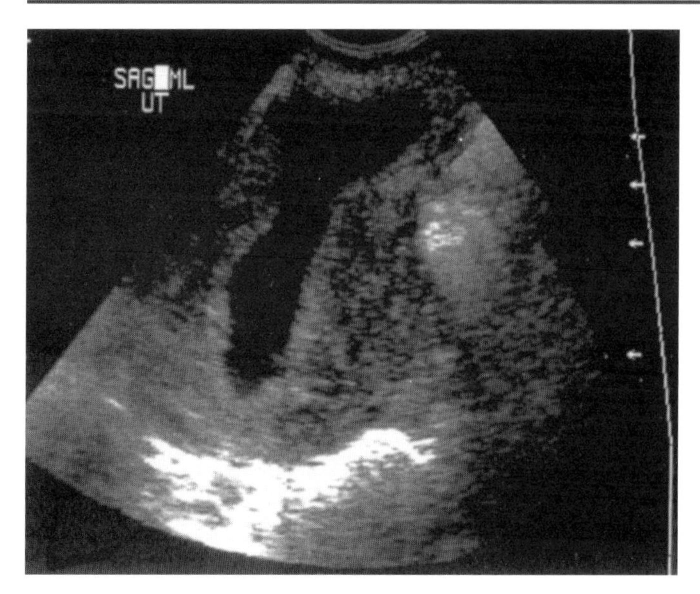

FIGURE 17 ■ Hematometra. Sagittal view of the uterus in a 63-year-old asymptomatic woman placed on cyclic hormonal replacement therapy demonstrates a large endometrial fluid collection with a thin surrounding endometrium. She subsequently underwent surgical dilation for cervical stenosis. *Source*: From Ref. 27 p. 228.

contraception devices (IUDs) (Fig. 19). Ultrasound is helpful in locating an IUD when the string cannot be felt. The IUD should be located centrally within the endometrium. A straight shaft IUD gives a bright linear reflector with entrance–exit reflections and ring-down artifacts (30). The appearance of other IUDs will depend on their shape (Fig. 19B).

If the IUD is not visualized sonographically, radiographs should be obtained to exclude an extra-uterine location of an IUD.

Abnormal Endometrium

Thick Endometrium

Causes of a thickened endometrium include the normal secretory phase endometrium, endometrial hyperplasia, endometrial polyps, fibroids, endometritis, early pregnancy, and complications of pregnancy. These are listed in Table 13.

Secretory Endometrium ■ In women of reproductive age, the upper limit of normal endometrial thickness is 14 to 16 mm at the end of the secretory phase. If a thickened endometrium is identified at sonography and it is unclear if the woman is in the secretory phase, a follow-up study performed early after the following menses will be useful to detect a persistent abnormality.

Fibroids ■ Submucosal fibroids may also give the appearance of a thick endometrium due to the endometrial–myometrial interface. SHG is helpful to clarify the exact location of the fibroid and to exclude a coexistent endometrial lesion.

Adenomyosis ■ Adenomyosis can mimic endometrial thickening. Because the endometrial–myometrial interface may be indistinct or nodular, the endometrial echocomplex may appear thickened on transvaginal sonography. The instillation of fluid during SHG will often demonstrate that there is no true endometrial lesion and point to a diagnosis of adenomyosis.

TABLE 12 ■ Uterine Calcifications and Other Echogenic Structures

Diagnosis	Appearance	Location	Comments
Fibroids	Clumps of coarse or rim-like calcification	Submucosal Intramural Subserosal Pedunculated	When pedunculated may simulate an adnexal mass
Arcuate artery calcifications	Segmental calcifications	Peripheral	Usually seen in older women with diabetes, hypertension, renal failure
Post-D&C, prior infection; idiopathic	Punctuate calcifications	Endometrial–myometrial interface	
IUD	Bright linear echoes with entrance –exit reflections (straight shaft) Segmental reflectors (Lippes loop)	Centrally located in endometrium cavity	Appearance depends on type of IUD
Retained products conception		Endometrial cavity	Calcified placental fragments lithopedion
Endometritis and/or myometritis	Ill-defined bright echoes with "dirty" shadowing ±fluid	Endometrial cavity or myometrium	Small amount of fluid and gas may be normal in early postpartum period
Postcesarean section or postsurgical findings	Punctate or ill-defined echoes with "dirty" shadowing ±fluid	Surgical bed	Due to suture material or small amount of normal postop gas and fluid (clinical correlation to distinguish from infection)

Abbreviations: Post-D&C, after dilation and curettage; IUD, intrauterine contraception device.

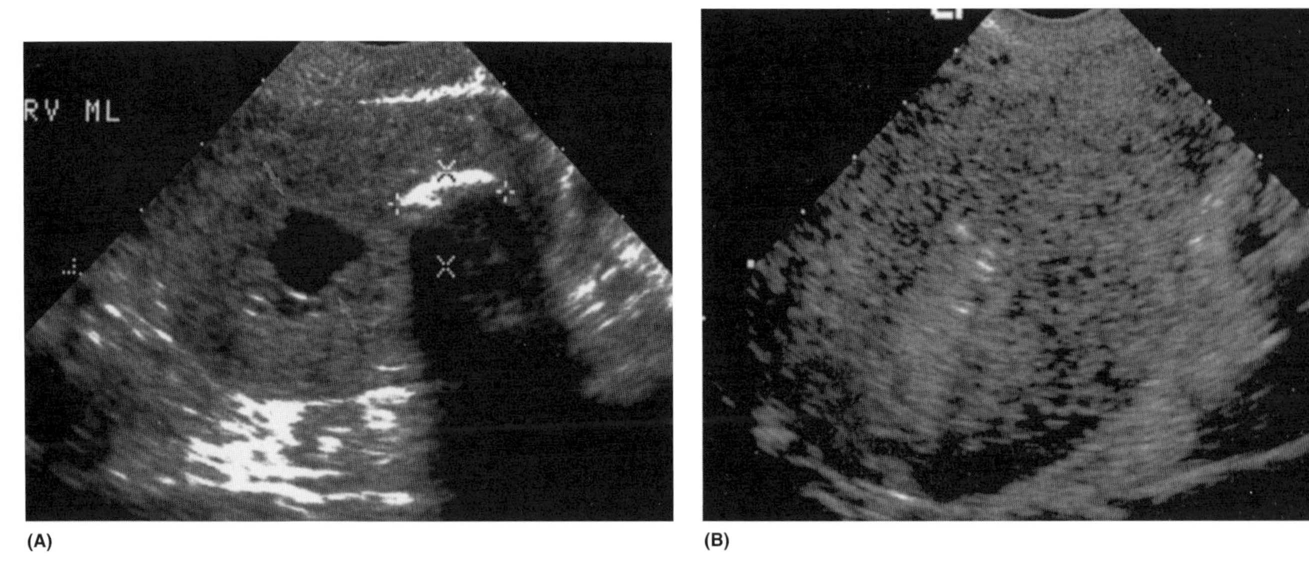

(A) (B)

FIGURE 18 A AND B ■ Uterine calcifications. (**A**) Transvaginal transverse view of the uterus in a postmenopausal woman with abnormal bleeding demonstrates a well-defined echogenic focus with shadowing secondary to a calcified fibroid. Adjacent to this area is a fluid collection in a region of thickened endometrium (*arrows*). This was endometrial hyperplasia. (**B**) Punctuate calcifications at the endometrial–myometrial interface in a patient with two prior dilatation and curettage procedures.

Endometritis ■ Endometritis occurs in association with pelvic inflammatory disease (PID) and in postpartum or postsurgical patients. The endometrium appears prominent or irregular with a small amount of endometrial fluid (31). Gas bubbles can be present; however, these are also a normal postpartum finding (32).

Synechiae ■ Synechiae are difficult to visualize in the nongravid uterus. They are found in women with a history of spontaneous abortion or curettage of the gravid and/or infected uterus. Vaginal sonography may demonstrate bright echoes within the endometrial cavity in this condition. At SHG, adhesions appear as thin or thick echogenic bands crossing the fluid-filled uterine cavity (1–3). The adhesions may be broad based and may obliterate all or parts of the uterine cavity (1–3,33).

Endometrial Hyperplasia, Polyps, and Cancer ■ As women approach menopause, the incidence of endometrial

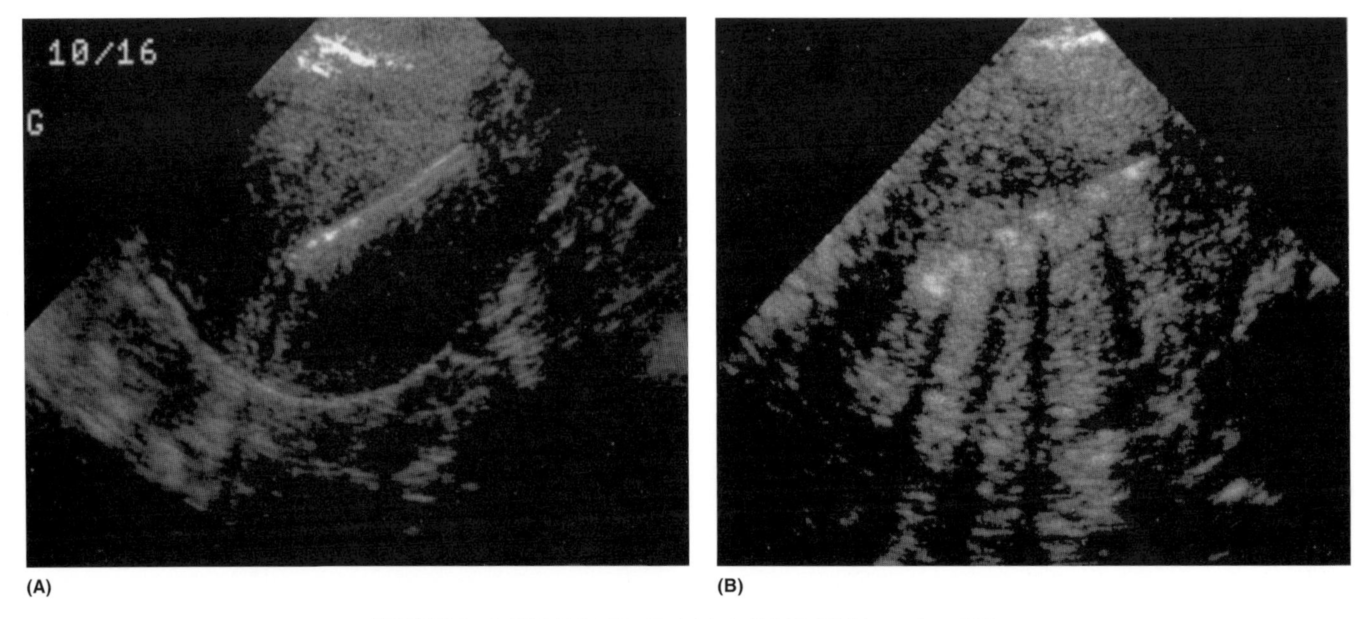

(A) (B)

FIGURE 19 A AND B ■ (**A**) Straight shaft IUD. (**B**) Lippes loop IUD.

TABLE 13 ■ Causes of a Thick Endometrium

Diagnosis	Appearance	Additional history
Normal	Homogeneously echogenic with through transmission	Secretory phase (Table 3)
Pregnancy		
Normal early pregnancy		
Missed abortion		
Ectopic pregnancy		
Hydatidiform mole		
Retained products of conception		
Hyperplasia	Focal or diffuse thickening	± Bleeding
Polyps	Echogenic, cystic areas	± Bleeding
	Sessile or pedunculated	
	± Central vessel	
Cancer	Larger, more irregular than polyp, invasion of the myometrium	± Bleeding
	Heterogeneous	
Endometritis	Endometrial fluid ± Gas	Recently postpartum or postsurgical or with PID

Abbreviation: PID, pelvic inflammatory disease.

hyperplasia, polyps, and cancer increases, causing endometrial thickening (Fig. 20). Endometrial hyperplasia is caused by estrogen unopposed by progesterone; the endometrium is thickened either diffusely or focally (Figs. 13A and 18A). Endometrial polyps usually are asymptomatic but may cause uterine bleeding. They may present on transvaginal sonography as diffuse or focal endometrial thickening (Figs. 3B and 20B) (34). In both of these conditions (endometrial hyperplasia and polyps), the interface between the endometrium and the myometrium is preserved. The presence of a polyp is suggested when transvaginal sonography demonstrates a focus of endometrial thickening which is usually homogeneous in echotexture and iso- or slightly hyperechoic relative to the endometrium. This appearance can also be seen in focal endometrial hyperplasia, with or without atypia, and even with early endometrial cancer. The presence of one or more tiny cystic spaces is another clue to the presence of a polyp. These cystic spaces may represent cystic areas within a polyp or may represent blood vessels. On color or power Doppler, a feeding vessel entering the base of a polyp is generally considered typical, but polyps are also found without feeding vessels. SHG is helpful to confirm the presence of a polyp and to distinguish between focal and diffuse endometrial abnormalities (33). Once the fluid separates the two opposing endometrial layers, it is easy to determine whether a focal polypoid lesion projects into the fluid and to assess the rest of the endometrial lining. The location of focal abnormalities should be carefully described to ensure removal of the lesion at hysteroscopy. Polyps may be single or multiple, pedunculated or sessile (broad-based). Occasionally, polyps may be heterogeneous in echotexture due to presence of infarction or hemorrhage. Rarely, a malignant focus is found within an endometrial polyp.

Endometrial cancer also is a cause of endometrial thickening. The diagnosis is suggested when there is marked or irregular endometrial thickening and loss of the endometrial–myometrial interface (Figs. 12A and 20C). These conditions are discussed in detail in Chapter 29.

Sonohysterography

When endometrial thickening is present sonographically and the etiology is unclear or if the endometrium cannot be clearly identified in a symptomatic patient (or if the result of an endometrial biopsy is negative despite an abnormal sonogram), SHG often is helpful in identifying the cause of the sonographic abnormality (33). For this procedure, a catheter is placed into the endometrial cavity under sterile conditions, and 10 to 30 mL of saline is injected into the endometrial cavity. In this manner, polyps and fibroids are outlined and better characterized (Fig. 20) (Table 14).

Endometrial Fluid

Fluid within the endometrial cavity is seen in both normal and pathologic conditions (Table 15) (31). In women in the menstrual phase of their cycle, a tiny amount of fluid is a normal finding. Fluid within the endometrium also is seen in normal early pregnancy and abnormal pregnancy (missed abortion, ectopic pregnancy, and molar pregnancy). Other causes of endometrial fluid include infection, degenerating fibroids, and obstruction (31). In older patients, fluid can be secondary to malignancy (uterine, cervical, tubal, or ovarian); however, cervical stenosis of a benign etiology (especially in women who

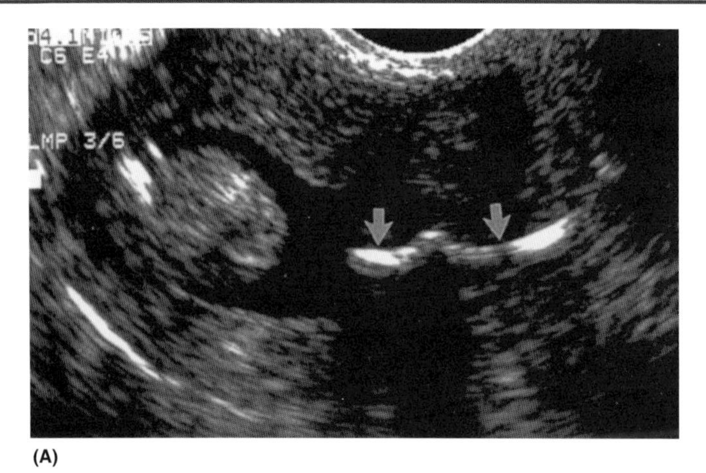

(A)

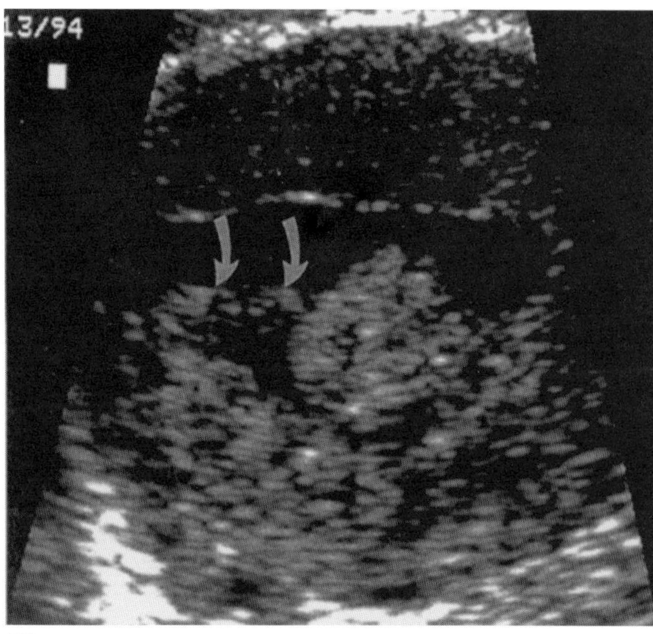

(B)

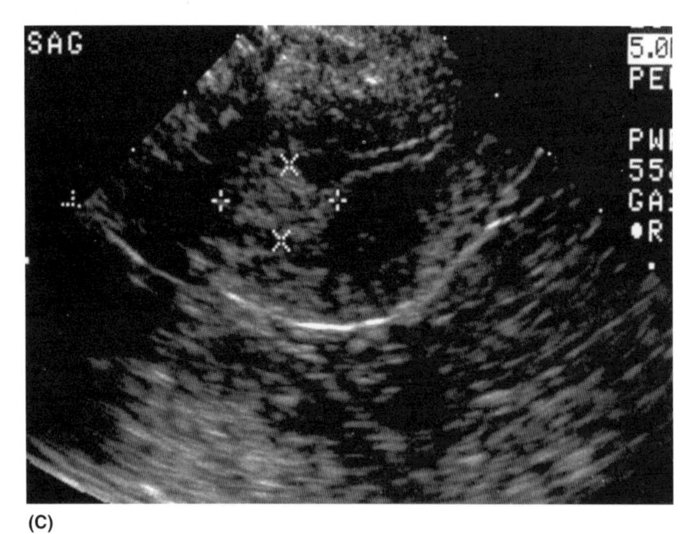

(C)

FIGURE 20 A–C ■ Utility of endometrial fluid in outlining endometrial abnormalities is demonstrated. (**A**) Sonohysterography catheter (*arrows*) is seen entering the endometrial cavity. Fluid outlines a well-defined intracavitary mass. (**B**) Fluid outlines an echogenic mass on a stalk (*curved arrows*) with an endometrial polyp. (**C**) Small amount of fluid in the endometrium outlines an ill-defined mass (*calipers*) with distortion of the endometrial–myometrial interface in this patient with endometrial cancer.

TABLE 14 ■ Endoluminal Masses Surrounded by Fluid[a]

		Endometrial–myometrial interface	
Diagnosis	**Endometrial appearance**	**Intact**	**Disrupted**
Polyps	Smooth margins	Yes	
	Echogenic mass		
	± Feeding vessel		
	Pedunculated or sessile		
Fibroids	Shadowing (even without calcification)	Mural fibroids	Submucosal fibroids
	Sessile, pedunculated		
	Isoechoic or hypoechoic to myometrium (rarely, echogenic)		
Synechiae	Thick or thin echogenic bands traversing cavity	Yes	
	± Distortion of cavity		
Hyperplasia	Focal or diffuse	Yes	
	Smooth or irregular endometrial thickening		
	Can be focal		
Cancer	Irregularly thickened endometrium	Early	All others
	Heterogeneous		

[a]Fluid either occurring spontaneously or introduced with sonohysterography.

TABLE 15 ■ Endometrial Fluid

Etiology	Examples
Normal secretions	Menstrual phase
Obstruction	
■ Congenital	Duplication anomaly with obstruction
	Vaginal or cervical atresia
	Imperforate hymen
■ Malignancy	Cervical cancer
	Endometrial cancer
■ Cervical stenosis	Previous childbirth, instrument ation, or radiation therapy
Infection	PID (pyometra)
Pregnancy-related conditions	Normal pregnancy
	Incomplete abortion
	Ectopic pregnancy
Polyps	
Degenerating fibroids	

Abbreviation: PID, pelvic inflammatory disease.

previously had children or instrumentation) is more common (Fig. 17). The presence of fever in a woman with a fluid collection suggests pyometra.

Abnormal Cervix

The most common mass within the cervix is the Nabothian cyst. Solid masses include fibroids (Fig. 21A), endocervical polyps, and malignancies (Fig. 21B). Cervical fibroids are hypoechoic and typically well defined. They may originate from the cervix and project exophytically as a deep pelvic mass. However, fibroids may also be found on a stalk in the cervical canal, originating from the cervix or having prolapsed from the uterine cavity. Cervical cancer is more likely to present with enlargement of the cervix by a mass with ill-defined margins. The cervix is rarely the site of an ectopic pregnancy (Table 16).

Abnormal Vagina

Vaginal masses are rare. The most common visualized with sonography are Gartner cysts; these are remnants of the embryonic Wolffian ducts that also give rise to paraovarian and paratubal cysts. These cysts usually are located within or near the vaginal wall and typically are palpable on physical examination. Vaginal adenocarcinoma and rhabdomyosarcoma appear as solid masses, occasionally with areas of necrosis (4,35).

Abnormal Fallopian Tube

The major causes of fallopian tube abnormalities are infection and ectopic pregnancy (Table 17). Less frequent causes of fallopian tube masses are chronic tubo-ovarian abscess (TOA) and fallopian tube cancer, both of which appear as a solid mass within the tube or adjacent to the ovary (Fig. 22).

Pelvic Inflammatory Disease

Acute PID ■ The pelvic findings in early acute PID are subtle and nonspecific (36). These may include enlargement of the uterus with indistinctness of its margins, enlargement of the ovaries with increased numbers of small cysts, increased echogenicity of the pelvic fat, and fluid in the endometrium and cul-de-sac (37). The findings in the acutely inflamed fallopian tube are more specific. Since the normal fallopian tubes are rarely identifiable

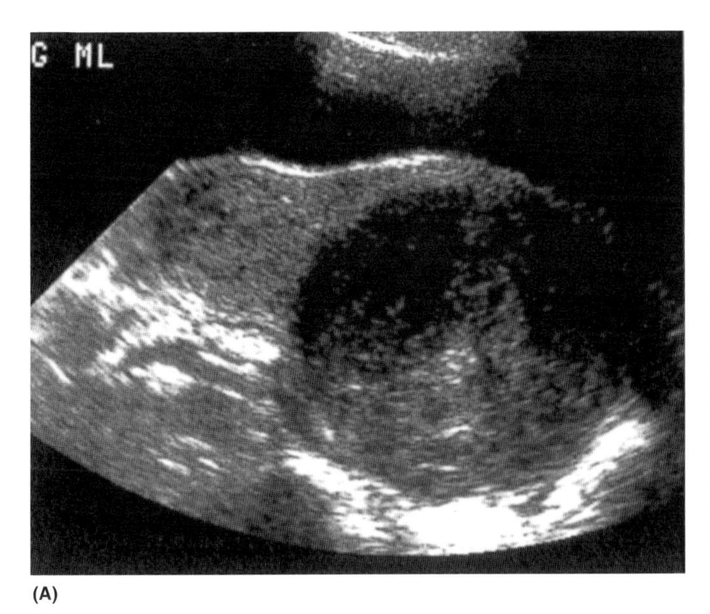

(A)

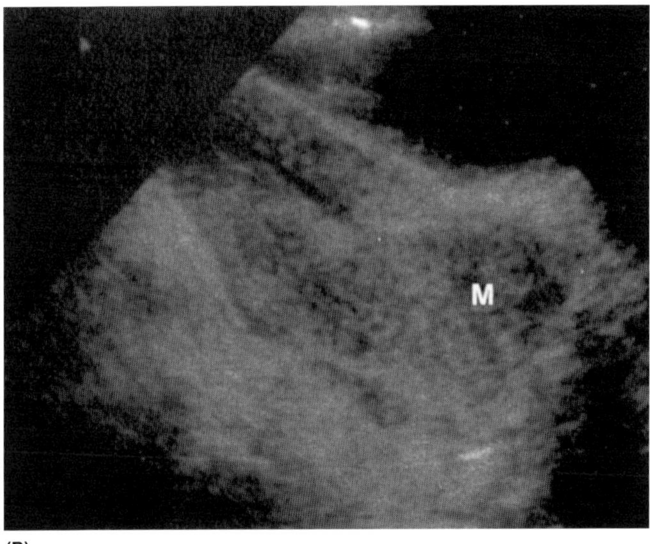

(B)

FIGURE 21 A AND B ■ Cervical masses. (**A**) Sagittal view of the cervix demonstrates a large cervical fibroid which deviates the lower uterine segment anteriorly. (**B**) Transvaginal view of the cervix demonstrates an ill-defined relatively isoechoic mass (M) in this patient with cervical cancer.

TABLE 16 ■ Cervical Mass

Diagnosis	Findings
Nabothian cyst	Usually <2 cm
	Adjacent to endocervical canal
	Incidence increases with age
Cervical fibroid	Typically hypoechoic ±stalk
Cervical cancer	Enlarged cervix, mass with ill-defined margins
Ectopic pregnancy	Bleeding and positive pregnancy test, gestational sac in cervix, empty uterine cavity
Polyps	Echogenic, located in cervical canal
	±Feeding vessel

in the absence of pelvic fluid, the presence of inflamed and edematous tubes is readily recognized. The wall of the fallopian tube is thickened (≥5 mm) as are the inner endosalpingeal folds (38). With occlusion of the distal end, the tube becomes distended with simple or complex fluid and appears oval or pear shaped; incomplete septae are formed as the tube folds back onto itself. The inflamed tube is commonly found posterior to the uterus in the cul-de-sac; fluid-debris levels may be seen within the tube. Color and power Doppler can demonstrate the hyperemia of the inflamed tubes, which show low resistance flow (36).

PID may also involve the ovary. Initially, the tube becomes adherent to the ovary. When the tube and ovary remain distinctly identifiable, this is known as a tubo-ovarian complex. If infection progresses and results in more tissue damage, a TOA may form; this can be unilateral or bilateral. A TOA appears as a heterogeneous pelvic mass. When TOA is bilateral, a large mass can be seen to fill the cul-de-sac and extend from one side of the uterus to the other (Fig. 22C). In these cases, the margins between the uterus and the adnexae are lost and the pelvis is filled with an amorphous complex inflammatory mass.

Chronic PID ■ Recurrent or incompletely treated acute PID may result in chronic PID. The most common manifestation of chronic PID is a hydrosalpinx, a fluid-filled distally obstructed fallopian tube. Uncomplicated hydrosalpinx presents as an elongated serpiginous fluid-filled tubular structure (Fig. 22A). Compared with an adnexal cyst, hydrosalpinx is more tubular, usually anechoic and often demonstrates incomplete septae. The wall of the uncomplicated hydrosalpinx is usually thin (less than 5 mm). Tiny echogenic nodules measuring 2 to 3 mm arising from the internal tubal wall can be seen projecting into the tubal fluid; these are the remnants of endosalpingeal mucosal folds (38). Pelvic adhesions caused by the same pelvic infection may surround pockets of fluid and result in peritoneal inclusion cysts. Much less commonly, hydrosalpinx is secondary to extratubal diseases which obstruct the distal end of the fallopian tube, such as endometriosis, appendicitis, and inflammatory bowel disease.

When an acute infection is superimposed on chronic PID, the tube wall becomes thickened, nodular, and hyperemic (Fig. 22B) (38). Internal debris in a hydrosalpinx can indicate the presence of pus (pyosalpinx). Alternatively, internal debris can be due to blood (hematosalpinx) related to obstruction of uterine outflow (in a uterine anomaly or uterine adhesions) or ectopic pregnancy.

TABLE 17 ■ Fallopian Tube Masses

Diagnosis	Appearance	Comments
Paratubal cyst	Anechoic cyst	Look for ovary separate from cyst
Hydrosalpinx	Fluid-filled tube	
	Polypoid projections/ folds	
Pyosalpinx	Hyperemic	Other signs/symptoms of infection
	Thick walls	
	Internal debris	
TOA	Heterogeneous mass	Other signs/symptoms of infection
	Hyperemic	
	Ill-defined margins; ovary contained within "mass"	
Chronic TOA	Solid mass	Rare
Cancer	Solid mass	Rare
Ectopic pregnancy	Echogenic or ringlike mass	+Pregnancy test result
	±Embryo, ±yolk sac	
	Blood-filled tube	

Abbreviation: TOA, tubo-ovarian abscess.

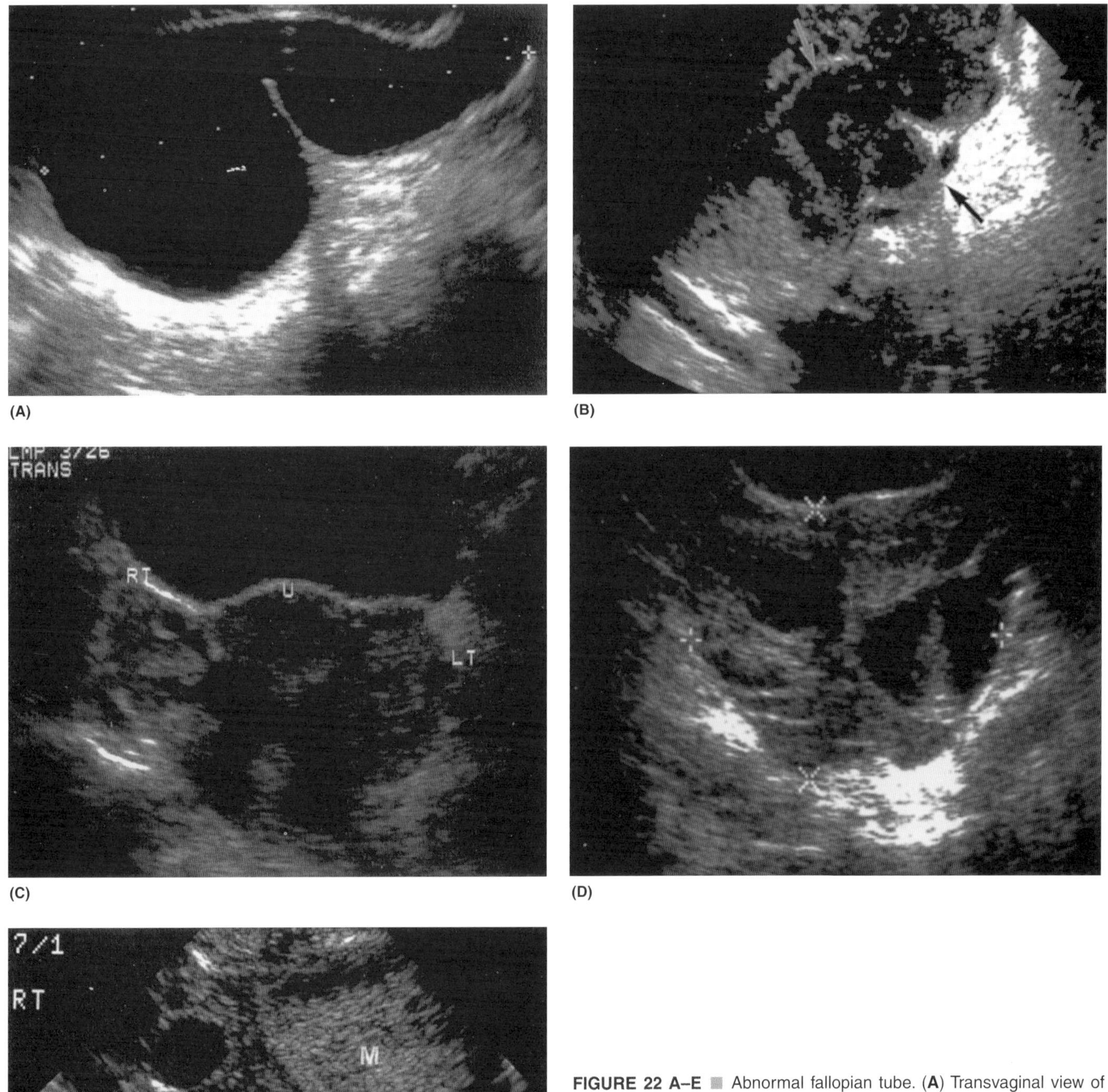

FIGURE 22 A–E ■ Abnormal fallopian tube. (**A**) Transvaginal view of the right adnexa reveals an elongated tubular structure with a thin wall and no internal debris. This is the typical appearance of a hydrosalpinx. (**B**) Thick-walled tubular structure (*arrows*) in a patient with pelvic inflammatory disease. (**C**) Transabdominal view of the uterus and adnexa (RT, LT) reveals bilateral enlarged adnexa. (**D**) Endovaginal examination reveals a heterogeneous adnexal mass mixed cystic and solid in a patient with bilateral tubo-ovarian abscess. (**E**) Well-circumscribed echogenic 2.5 cm mass adjacent to the right ovary in a patient with a chronic tubo-ovarian abscess. U, uterus; M, mass; O, right ovary.

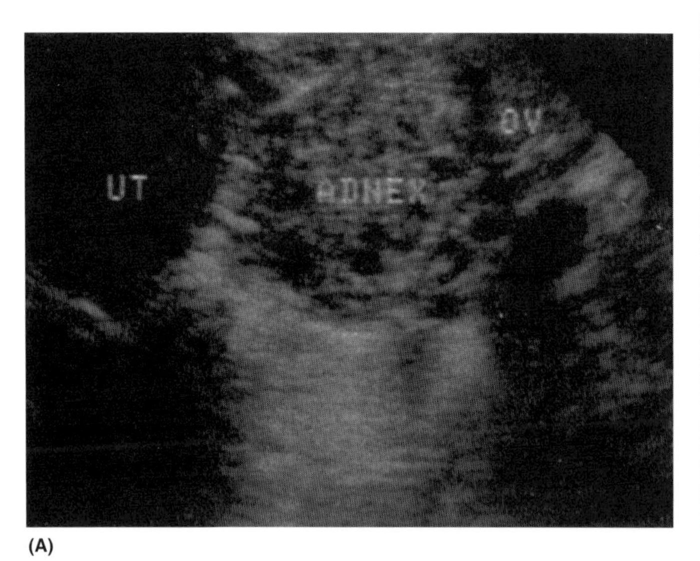

(A)

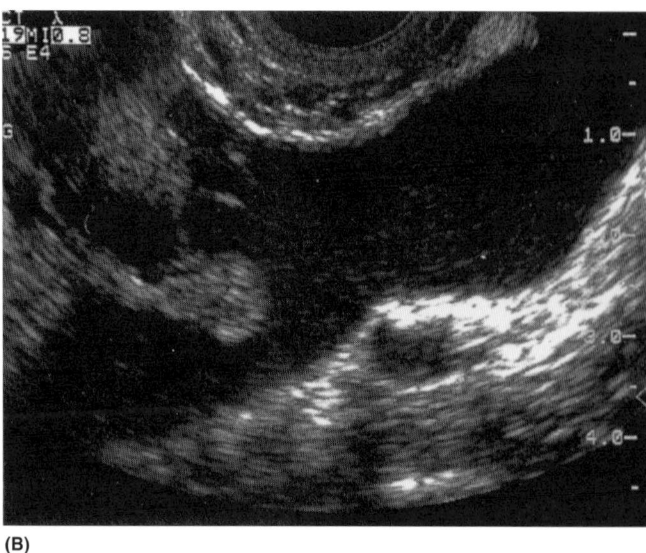

(B)

FIGURE 23 A AND B ■ Ectopic pregnancy. (**A**) A complex left adnexal mass (ADNEX) is seen between the uterus (UT) and left ovary (OV). (**B**) A different patient with ectopic pregnancy demonstrates free fluid in the cul-de-sac with multiple internal echoes caused by hemoperitoneum.

Ectopic Pregnancy

Ectopic pregnancy most commonly presents as a complex adnexal mass separate from the ovary in patients with a positive pregnancy test and, typically, unilateral pelvic pain and/or vaginal bleeding (Fig. 23). The complex adnexal mass representing an ectopic pregnancy must be distinguished from a hemorrhagic corpus luteum cyst arising from the ovary (39). The most definitive sonographic finding in ectopic pregnancy is an extrauterine gestational sac containing a yolk sac and an embryo with cardiac activity; however, this is present in only 8% to 26% of cases (40). Finding an extrauterine gestational sac with a yolk sac, with or without an embryo, is also highly definitive (41). The finding of an extraovarian echogenic ring is another presentation; on color Doppler this is associated with a characteristic "ring of fire" created by high velocity, low impedance trophoblastic flow around the extrauterine gestational sac. However, luteal flow associated with the corpus luteum of pregnancy can be confused with the trophoblastic flow in an ectopic gestation; the corpus luteum is within the ovary. Therefore, it is very important to determine that the echogenic ring of an ectopic pregnancy is indeed separate from the ovary. The findings on pelvic ultrasound in early ectopic pregnancy may be normal. Nonetheless, ectopic pregnancy should be suspected in a woman with a positive pregnancy test and no signs of an intrauterine gestation, even in the absence of a sonographic adnexal mass. Careful examination to detect any fluid in the peritoneal cavity is required, as this is a common finding in ectopic pregnancy.

Fallopian Tube Carcinoma

Fallopian tube carcinoma is rare; a typical presentation is hydrorrhea, a watery vaginal discharge. It is suggested when a solid or complex adnexal mass is visualized separate from the ovary in a woman without ectopic pregnancy (42). The tube may become distended with secretions, causing pain.

Abnormal Ovary

The differential diagnosis of an ovarian abnormality depends on a variety of factors, including patient's age (Table 18), time since last menstrual period, hormonal status, symptoms (pain), pregnancy test result, prior surgery, and findings on physical examination. Functional cysts and benign neoplasms comprise most of the adnexal lesions in women of reproductive age. Malignant lesions are more common in older women.

The differential diagnosis also is influenced by the sonographic characteristics of the lesion as well as any associated findings (Table 19).

The classic sonographic description of adnexal masses is based on the cystic nature of most adnexal lesions. They are categorized as completely cystic, complex (mixed cystic and solid), or solid. Findings that increase the likelihood of malignancy are listed in Table 20.

Doppler Analysis of Adnexal Masses

Color Doppler is helpful in determining if a questionable lesion is a cyst or a vessel. It is also helpful in establishing the solid nature of a hypoechoic solid mass. When pulsed Doppler analysis of tumor vascularity is to be performed, color Doppler aids in the placement of the Doppler gate.

A large amount of literature has been dedicated to the use of Doppler sonography for distinguishing between benign and malignant adnexal masses.

Because tumor vessels lack a muscular layer, they frequently have low resistance flow. This resistance can be quantified either with the resistive index [(peak systole – end diastole)/peak systole] or the pulsatility index [(peak systole – end diastole)/mean]. A resistive index of 0.4 or less or a pulsatility index of 1.0 or less is associated with

TABLE 18 ■ Ovarian Tumors by Age

Age	Tumor type	Comments
Premenarchal	Granulosa cell tumor	Precocious puberty
	Germ cell tumor	
	Immature teratoma	Solid mass with elements of fat and calcification
Reproductive age	Functional cyst	Usually ≤3 cm, unilocular
	Dermoid	See Table 24
	Cystadenoma/adenocarcinoma	Anechoic or complex cyst
	Granulosa cell tumor	Endometrial hyperplasia and endometrial carcinoma
Postmenopausal	Metastases	Mixed cystic and solid
		Ascites
		Omental thickening
	Fibroma/thecoma	Most common solid ovarian mass in this age group
	Cystadenocarcinoma	
	Endometrioid tumor	Endometrial hyperplasia and endometrial carcinoma
	Granulosa cell tumor	Endometrial hyperplasia and endometrial carcinoma

malignant disease (43). However, many other physiologic and benign neoplastic lesions can also have a low resistive index (Table 21). Furthermore, there is considerable overlap in the pulsatility indices of benign and malignant lesions so that Doppler sonography has severe limitations in the differentiation of benign from malignant adnexal disease on the basis of low-impedance flow (pulsatility index <1.0) (44).

The corpus luteum typically has low resistance flow; therefore, if possible, patients should be scanned in the first 10 days of the cycle to avoid confusion with luteal flow. However, even when physiologic masses are excluded from analysis, the sensitivity and specificity of Doppler resistive index are not sufficient to replace the morphologic impression of a lesion being benign or malignant (45,46). Gray scale morphologic features are more sensitive in this discrimination (45,46). Other Doppler findings associated with malignancy include lack of a diastolic notch and presence of blood flow within solid portions of the tumor (43).

TABLE 19 ■ Prevalence of Ovarian Tumors

Origin	Percentage of tumors	Types	Appearance	Associations
Epithelial	70–75	Serous	Anechoic or complex cyst	
		Mucinous	Complex cyst	Pseudomyxoma peritonei
		Endometrioid	Solid	Endometrial hyperplasia and endometrial carcinoma
		Brenner	Solid	Mucinous cystadenoma
		Undifferentiated	Solid	
Germ cell	15–20	Teratomas (95% of germ cell tumors,immature teratoma)	Complex cyst	AFP
		Dysgerminoma	Solid	HCG (5%)
		Endodermal sinus	Solid	AFP and HCG
		Choriocarcinoma	Solid	AFP and HCG
Sex-cord stromal	10	Granulosa cell	Solid with cystic areas	Estrogen: precocious puberty or endometria hyperplasia/cancer
		Thecoma	Solid	
		Fibroma	Solid	Meigs syndrome
Metastasis	5		Solid or solid with cystic areas	

Abbreviations: AFP, α-fetoprotein; HCG, human chorionic gonadotropin.

TABLE 20 ■ Adnexal Mass Findings that Increase the Likelihood of Malignancy

Larger mass size
Solid elements
Septations, especially thick septations with vascular nodularity
Thick wall, especially with vascular nodularity
Blood flow
■ In solid elements
■ Low resistive index
■ Absent diastolic notch
Abnormal serum CA-125
Older age of patient
Ascites
Metastases

Morphologic Features of Adnexal Masses

Important morphologic features of an adnexal mass include mass size, presence of solid component, complexity, papillations, loculations, echogenicity, and presence of vascular flow in solid elements (47,48). Mass size has been associated with malignancy. Masses greater than 10 cm present a high risk of malignancy; masses between 5 and 10 cm are of intermediate risk; and masses less than 5 cm are of low risk of malignancy. Complexity refers to the appearance of the mass wall and any septations. Wall thickness greater than 3 mm and septations of more than 2 to 3 mm are suggestive of malignancy. Papillary projections, especially with vascular flow, are strongly associated with malignancy. The presence of fluid in the cul-de-sac and persistence or increasing size of a lesion also favor malignancy. Given the overlap in Doppler indices between benign and malignant lesions, current preoperative assessment of adnexal masses emphasizes careful morphologic assessment combined with evaluation of vascularity using color and/or power Doppler sonography (49).

Anechoic Cysts

Anechoic cysts have a thin wall, are completely devoid of internal echoes, and demonstrate enhanced through transmission. Regardless of their size, they are unlikely to be malignant (Fig. 24) (50–52). During the reproductive years, the most common anechoic cyst is the functional cyst. These usually are small and measure less than 2 cm in diameter; however, they may enlarge up to 10 cm. They typically regress spontaneously over one or two menstrual cycles. Oral contraceptives are sometimes used to suppress ovarian function in order to promote cyst involution. Anechoic cysts larger than 6 cm are more likely to be neoplastic than functional. If there are no wall irregularities or septations, the cyst most likely represents a benign cystadenoma.

TABLE 21 ■ Low Resistive Index (<0.4)

■ Functioning ovary
■ Luteal phase
■ Benign lesions
■ Luteal cyst
■ Dermoid
■ Endometrioma
■ Adenofibroma
■ Pedunculated fibroid
■ Inflammation
■ Tubo-ovarian abscess
■ Ectopic pregnancy
■ Malignant tumor

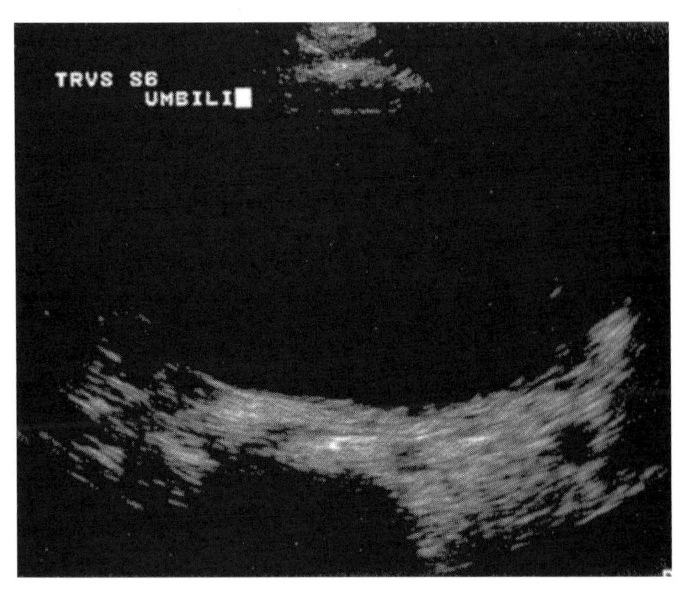

FIGURE 24 ■ A 10-cm anechoic cyst in a 59-year-old woman. There is a thin wall and internal debris or septations. This was a benign cystadenoma.

Other causes of simple adnexal cysts are paraovarian (Fig. 10) and paratubal cysts, peritoneal inclusion cysts (Fig. 25), and luteal cysts (Fig. 26) (Table 22). When an adnexal cyst is identified, it is important to determine whether it is part of the ovary (look for rim of ovarian tissue) or separate from it. In the case of paraovarian and paratubal cysts, the ovary is seen separately or adjacent to the cyst. This is also the case for pelvic cysts of nongynecologic etiology (53).

Paraovarian Cysts ■ Paraovarian cysts account for 10% of adnexal masses (Fig. 10). They arise from embryonic remnants in the broad ligament. Their size does not change during the menstrual cycle. The ovary is visible separate from the cyst, although the cyst may be immediately adjacent to the ovary.

Theca Lutein Cysts ■ Theca lutein cysts are large bilateral ovarian cysts that can appear as multiloculated masses (Fig. 26). They are associated with high levels of human chorionic gonadotropin, usually secondary to ovarian stimulation by gonadotropic drugs used to achieve multiple ovulations. Other causes are gestational trophoblastic disease, choriocarcinoma, or multiple gestations.

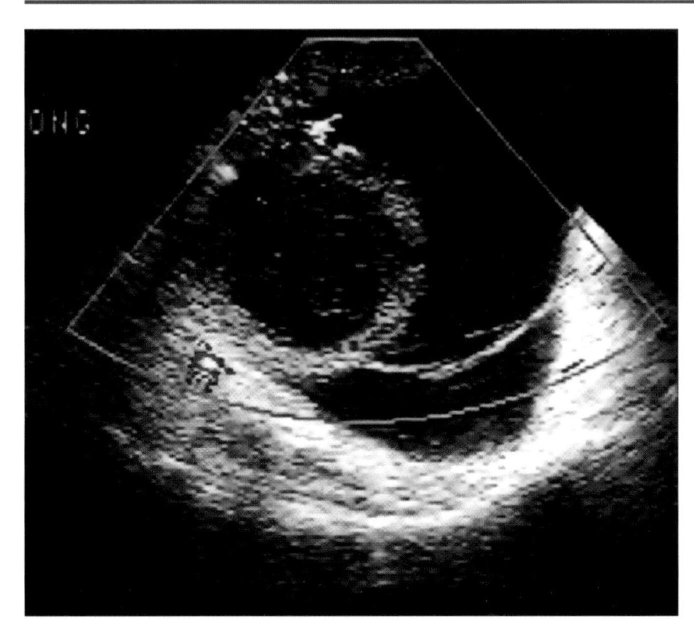

FIGURE 25 ■ Peritoneal inclusion cyst: a 6-cm right adnexal mass in a patient with multiple prior pelvic surgeries. There were areas of septation with blood flow. This was a peritoneal inclusion.

TABLE 22 ■ Simple Adnexal Cyst

Diagnosis	Comments
Functional cyst	Usually <2 cm in diameter, may be large
	Spontaneous regression
Paraovarian/paratubal cyst	Separate from ovary, does not regress
Hydrosalpinx	Elongated cyst
Cystadenoma	Stable or increase in size overtime
Peritoneal inclusion cyst	History of prior surgery or PID
Theca lutein cysts	Occur with elevated HCG levels (infertility drugs or gestational trophoblastic disease)
	Bilateral and multiple

Abbreviations: PID, pelvic inflammatory disease; HCG, human chorionic gonadotropin.

Complex Cysts

Complex cysts are composed of cystic and solid elements. In women of reproductive age, the hemorrhagic cyst is the most common type of complex ovarian cyst. The internal architecture of a complex cyst is important in establishing an appropriate differential diagnosis. In many cases, it allows for a specific diagnosis (Table 23). This is usually true for hemorrhagic cysts, endometriomata, and dermoid cysts. If a complex ovarian cyst does not demonstrate the typical appearance of one of these three common lesions, concern must be raised about other etiologies including benign and malignant neoplasms (Fig. 23).

Endometriosis ■ Endometriosis is a common condition in which endometrial tissue is present outside of the uterus. In most cases, sonography cannot demonstrate the limited form of this disease when it consists of tiny implants. However, deep endometriotic nodules in the rectovaginal septum may be imaged with the transvaginal probe, appearing as solid nodules of varying size which are tender on palpation with the probe (54,55). Likewise, vesicouterine implants can be detected as solid tender nodules involving the posterior bladder wall when transvaginal sonography is performed with a partially filled bladder (54,55).

Ovarian endometriomas are present in 13% to 38% of women with endometriosis (55). The typical appearance of an endometrioma or "chocolate cyst" is a thick-walled cyst filled with homogeneously low-level echoes (ground-glass appearance), unilateral or bilateral. Echogenic foci may be detected within the cyst wall. Occasionally, the internal wall of the cyst is irregular in contour and covered with a sludge-like material (Fig. 27). These complex cysts are frequently adherent to the uterus or pelvic sidewall.

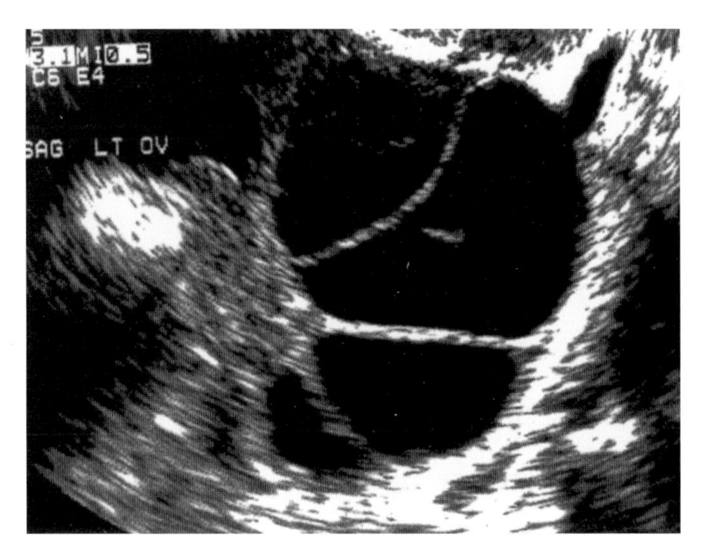

FIGURE 26 ■ Hyperstimulated ovary in a woman being treated for infertility.

Dermoid Cysts ■ Dermoid cysts are the most common ovarian neoplasm and the most common benign ovarian neoplasm diagnosed in the reproductive years; only 1% contain malignant elements (usually squamous cell carcinoma), typically in older women. They are most often asymptomatic, and bilateral tumors are present in 10% to 15% of cases. The typical clinical presentation is that of an adnexal mass that does not resolve on follow-up. Because dermoid cysts are derived from the three germ-cell layers, they are pleomorphic and show a spectrum of sonographic appearances. These include a completely cystic mass, a cystic mass with an echogenic mural nodule, a fat-fluid level, echogenic foci with shadowing (teeth or bone), or a complex mass with internal septations and bright linear echoes (Figs. 11 and 28) (56,57). Rare forms of dermoids include the specialized tumors of struma ovarii (with thyroid tissue) and carcinoid tumors.

Immature teratomas are rare, about 1% of all teratomas, and occur in young women 10 to 20 years of age, usually unilateral. α-Fetoprotein is elevated in 50% of cases.

TABLE 23 ■ Complex Cyst

Diagnosis	Appearance	Comments
Hemorrhagic cyst	See Table 6	Rapid change and resolution
Endometrioma	Diffuse homogeneous low level echoes, thick wall	Persistent, immobile
	±Peripheral punctuate calcification	
Tubo-ovarian abscess	Septations	Cervical motion tenderness
	Irregular margins	Pain, fever
	Free fluid	
Dermoid	Echogenic mural nodule	Bilateral in 10–15%
	Fluid–fluid level	
	Echogenic foci with shadowing (teeth)	
	Echogenic mass can be mistaken for bowel	
Hydrosalpinx	Elongated serpiginous tubular structure with partial septations	Persistent
Serous or mucinous cystadenoma/ cystadenocarcinoma	Anechoic (serous cysts) or with echoes (mucinous cyst)	Solid elements suggest malignancy
	Septations	
Ectopic pregnancy	Echogenic or ring-like mass	Positive pregnancy test
	±Embryo, ±yolk sac	Adjacent to ovary
	Ring of vascular flow	
Peritoneal inclusion cyst	Irregular or polygonal shape	Persistent
	±Thick walls with blood flow	History of prior surgery or PID
		Ovary engulfed by cysts or at margin of complex cysts

Abbreviation: PID, pelvic inflammatory disease.

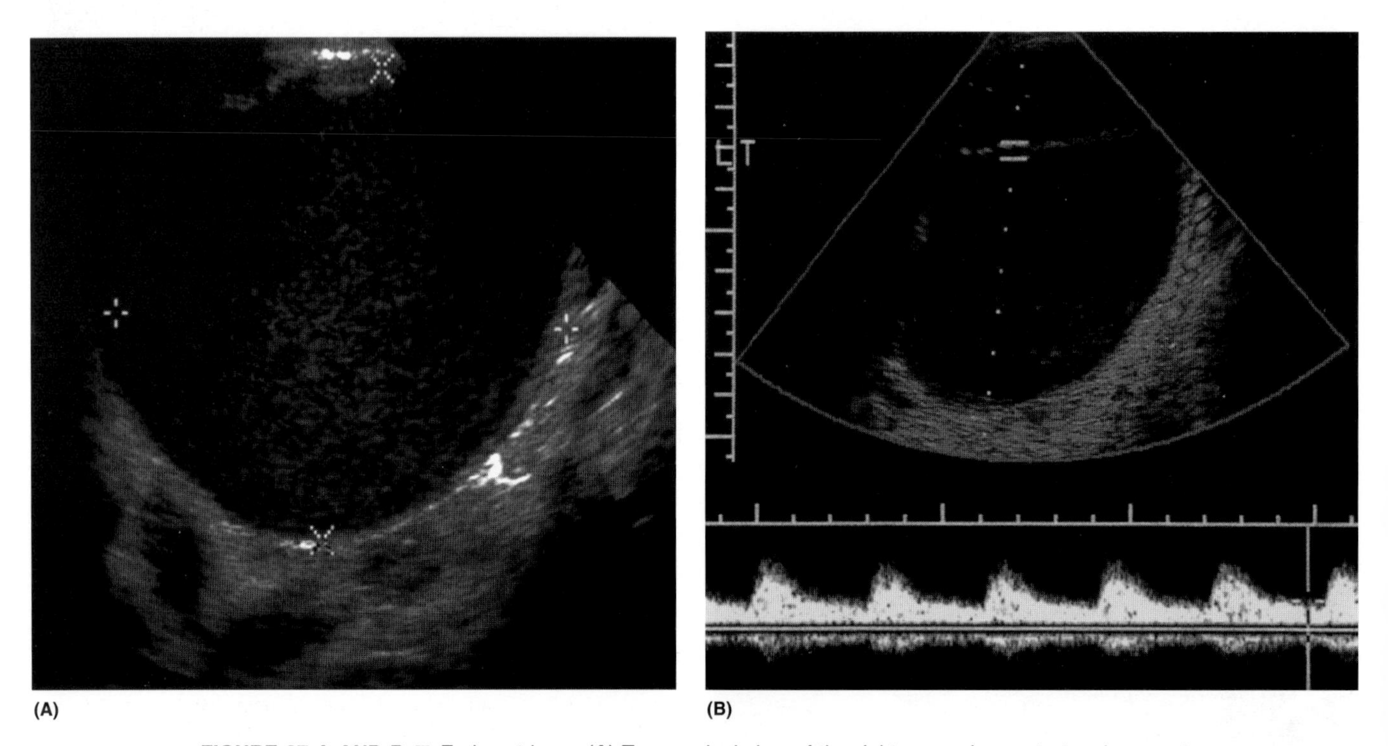

(A) (B)

FIGURE 27 A AND B ■ Endometrioma. (**A**) Transvaginal view of the right ovary demonstrates 4-cm cyst with diffuse low-level internal echoes. Wall thickening is seen in a few areas. This is the classic appearance for an endometrioma. (**B**) At times, a thin septation can be seen within an endometrioma. Low-resistance flow can be seen.

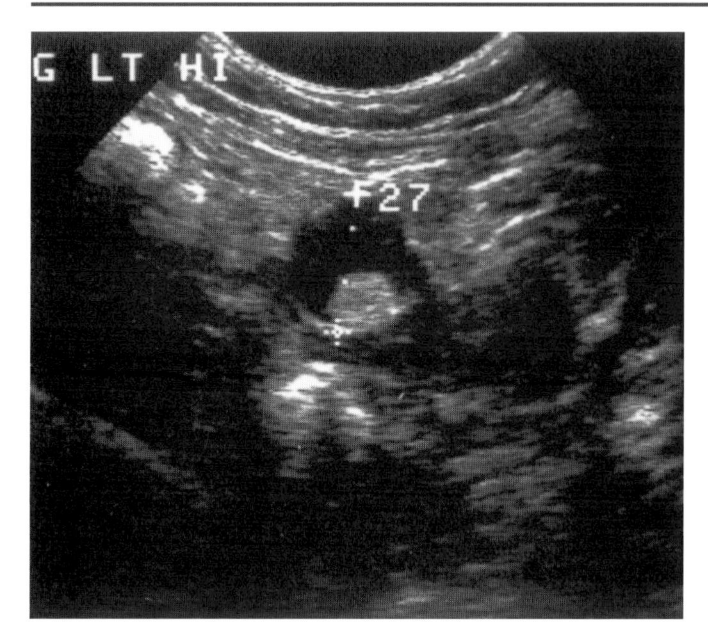

FIGURE 28 ■ Dermoid: a 2.7-cm left adnexal mass in a 30-year-old woman demonstrates a cyst with a 7-mm echogenic mural nodule. *Source*: From Ref. 27.

Epithelial Tumors ■ The cystadenoma and cystadenocarcinoma are the most common types of epithelial tumors. Serous tumors tend to be anechoic with septations. Mucinous tumors tend to have internal debris. The number and thickness of septations as well as the presence of mural nodularity increase the likelihood of malignancy (Fig. 29). Less common varieties of epithelial tumors are endometrioid, clear cell, Brenner, and undifferentiated carcinoma.

Peritoneal Inclusion Cysts ■ Peritoneal inclusion cysts may be complex with numerous loculations. The cysts occur in women who have had previous pelvic surgery or PID; in this setting the diagnosis should be considered and serial follow-up study used to demonstrate stability. Chronic pelvic pain is a common clinical feature. Conservative management can also be supported by MRI which shows a mass that fills the pelvic spaces but does not have mass effect on adjacent organs. Peritoneal inclusion cysts may have thick walls and septations with blood flow, as such they may mimic malignancy (Fig. 25). The ovary can be trapped in the middle of the multilocular mass or at the edge of it (58).

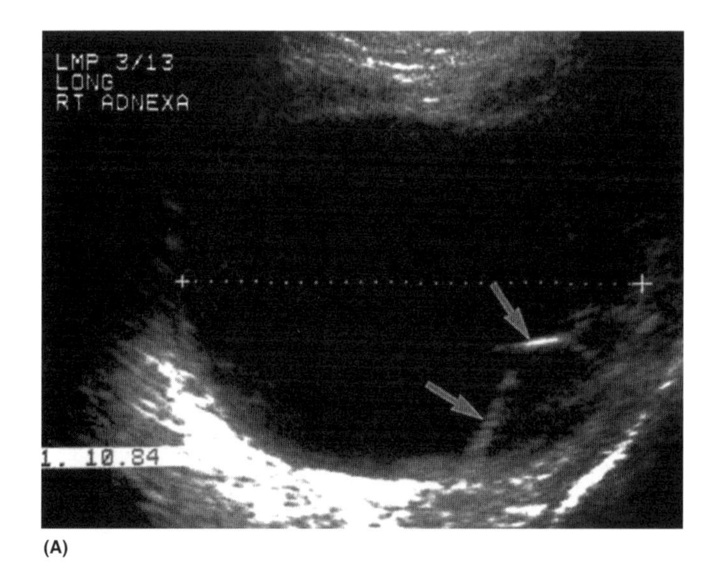

(A)

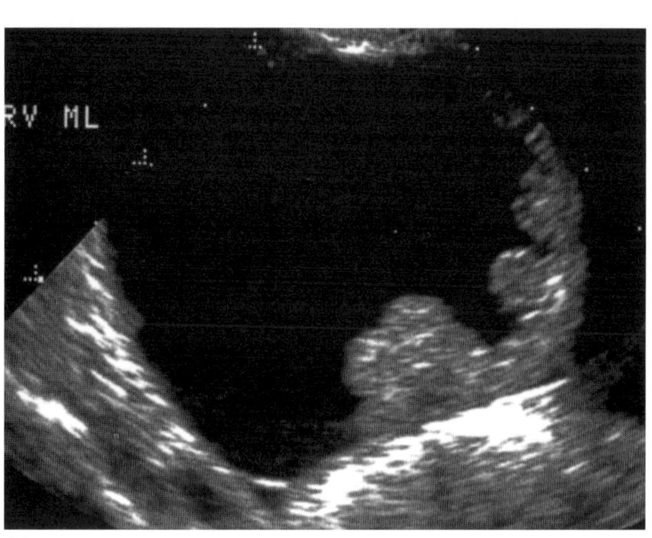

(B)

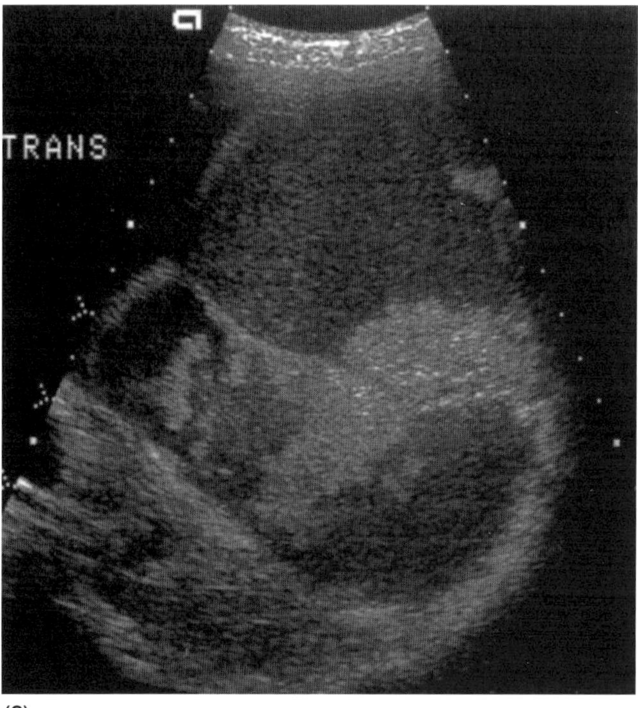

(C)

FIGURE 29 A–C ■ Serous and mucinous tumors. (**A**) Transvaginal view of the right adnexa demonstrates a 10.8-cm mass with a few septations (*arrows*). This was a serous cystadenoma. (**B**) A 4-cm adnexal mass that is predominantly cystic but with multiple mural nodules. Histologic diagnosis was serous cystadenocarcinoma. (**C**) A 10-cm right adnexal mass with diffuse internal echoes and irregular solid elements. Histologic diagnosis was mucinous cystadenocarcinoma.

TABLE 24 ■ Large Ovary Without Mass

Diagnosis	Unilateral/bilateral	Comments
Isoechoic hemorrhagic cyst	Unilateral	
Polycystic ovarian disease	Bilateral	Ovarian enlargement (>10 cc) with 12 or more peripheral follicles in each ovary, measuring 2–9 mm in diameter; increased central echogenic stroma (>25% cross- ectional area)
		History of anovulation and signs of androgenic excess
Torsion (some cases)	Unilateral	Patient presents with pain on the affected side
		Absent blood flow on affected side (not an absolute finding)
		Associated with mass/cyst or excessive mobility of adnexal supporting ligaments
		Small peripheral follicles
Pelvic inflammatory disease or adjacent inflammation, such as diverticulitis	Usually bilateral (in PID)	Fever
		Vaginal discharge
		Cervical motion tenderness
Solid ovarian neoplasm	Unilateral	Primary
	Bilateral	Metastic disease
		Lymphoma

Abbreviation: PID, pelvic inflammatory disease.

Enlarged Ovaries without Focal Mass

Causes of bilateral enlarged ovaries include polycystic ovary disease, PID, and, rarely, neoplasms (either primary or metastatic), especially lymphoma (Table 24). Unilateral causes of an enlarged ovary without a focal mass include ovarian torsion, an isoechoic hemorrhagic cyst, PID, and, rarely, malignancy.

Polycystic Ovaries ■ Polycystic ovaries are found in association with several endocrine disorders (59). Most commonly, polycystic ovaries are part of the polycystic ovary syndrome which comprises anovulation, a spectrum of menstrual abnormalities (ranging from amenorrhea, oligomenorrhea, metrorrhagia, menometrorrhagia to normal menses), and androgenic hyperactivity (hirsutism, acne, central obesity). Polycystic ovaries are enlarged (>10 cc), with 10 or more peripheral follicles measuring 2 to 9 mm in diameter (60). The central stroma in these ovaries is prominent (>25% of the cross-sectional area) and echogenic (61). The criterion for diagnosis is controversial; the American approach relies more on anovulation and biochemical proof of androgenic excess, while the Europeans rely more on the presence of the sonographic finding of polycystic ovaries (Fig. 30) (60).

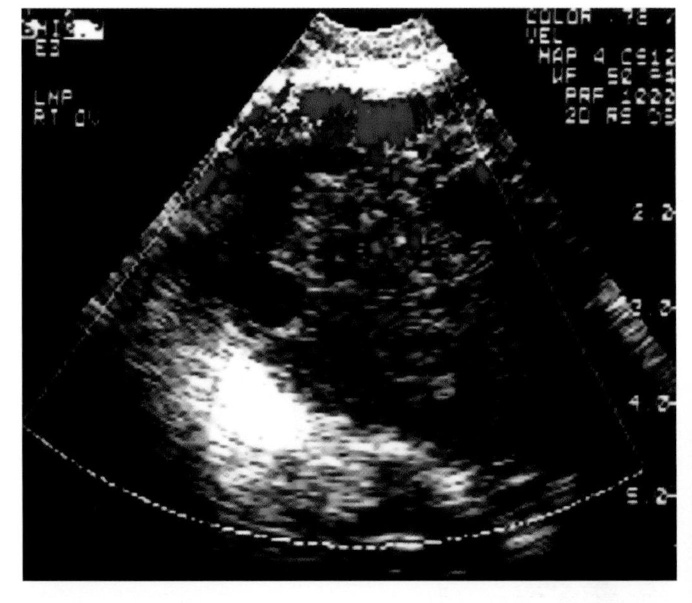

FIGURE 30 ■ Polycystic ovary. A 24-year-old woman presented with right adnexal pain. Vaginal examination demonstrated bilateral enlarged ovaries with multiple small peripheral cysts. Doppler demonstrated symmetric and equal flow in both ovaries. The patient had the stigmata of polycystic ovary syndrome with obesity, hirsutism, and oligomenorrhea.

Torsion ■ Most commonly ovarian torsion occurs in prepubertal and adolescent girls; however, a significant number of cases occur during pregnancy. It is often secondary to an ovarian mass, either a functional cyst or a tumor (62). The diagnosis of torsion is suggested by the clinical presentation of sudden onset of severe unilateral pelvic pain. The torsed ovary is enlarged and edematous; it may have a cystic, solid, or complex echotexture, and free fluid is often present in the pelvis (62,63).

Doppler findings in ovarian torsion are variable and may cause a delay in diagnosis. In the classic case of complete torsion, arterial blood flow is absent on the affected side. It is helpful if blood flow can be visualized on the unaffected side, to ensure that Doppler controls are set appropriately. In many cases normal arterial and venous Doppler signals have been found; in 60% of patients in Pena's series, ovarian torsion was surgically proven despite normal Doppler examination (64). This is probably the result of intermittent or partial torsion in an organ with a dual blood supply from the ovarian and uterine arteries. Therefore, Doppler findings may be used to support a diagnosis of torsion but a normal Doppler evaluation does not exclude the diagnosis. Rapid diagnosis and treatment is important in order to salvage the ovary prior to necrosis.

Isoechoic Hemorrhagic Cyst ■ When a hemorrhagic cyst is isoechoic compared with the remainder of the ovary, it is difficult to distinguish from a large ovary without a focal mass. Follow-up demonstrates resolution of the hemorrhagic cyst with decreased size of the affected ovary.

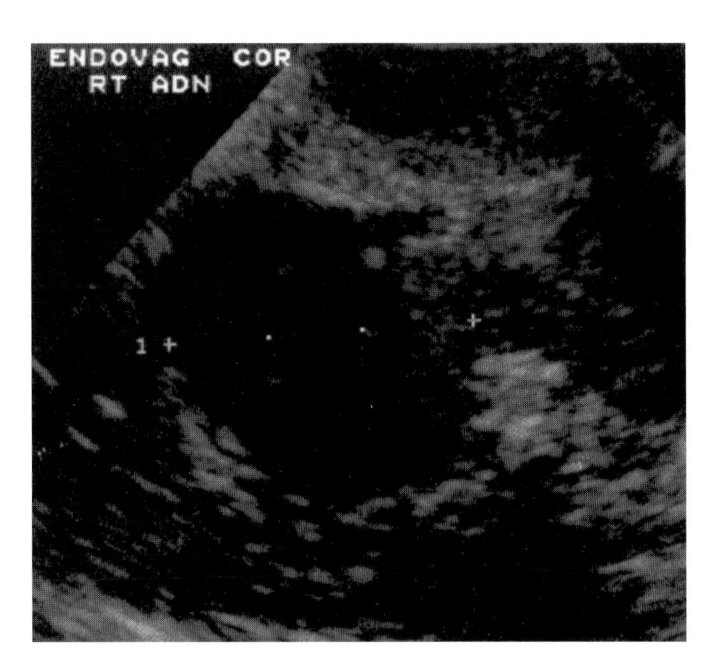

FIGURE 31 ■ A 4-cm solid hypoechoic mass in a 52-year-old woman. This was a fibrothecoma.

TABLE 25 ■ Solid Ovarian Mass

Neoplasm
Fibroma
Thecoma
Adenocarcinoma
Endometrioid tumor
Nonteratomous germ cell tumors
Sex-cord stromal tumors (thecoma and fibroma)
Granulosa cell tumor
Dysgerminoma
Brenner tumor
Metastatic disease
TOA
Ovarian torsion

Abbreviation: TOA, tubo-ovarian abscess.

Solid Ovarian Masses

The most worrisome cause of unilateral enlarged ovary is a solid ovarian neoplasm (Fig. 31). Whereas the serous and mucinous tumors tend to present as cystic and solid masses, there are less common ovarian neoplasms that present as solid ovarian masses. These are listed in Table 25.

REFERENCES

1. Lev-Toaff AS. Sonohysterography: evaluation of endometrial and myometrial abnormalities. Semin Roentgenol 1996; 31:288–298.
2. Bega G, Lev-Toaff AS, Becker E, et al. Three-dimensional ultrasonography in gynecology: technical aspects and clinical applications. J Ultrasound Med 2003; 22:1249–1269.
3. Lev-Toaff AS, Pinheiro LW, Bega G, et al. Three-dimensional multiplanar sonohysterography: comparison with conventional two-dimensional sonohysterography and x-ray hysterosalpingography. J Ultrasound Med 2001; 20:295–306.
4. Siegel MJ, Surratt JT. Pediatric gynecologic imaging. Obstet Gynecol Clin North Am 1992; 19:103–127.
5. Spivak MR, Cohen HL. Ultrasonography of the adolescent female pelvis. Ultrasound Q 2002; 18:275–288.
6. Byrne J, Nussbaum-Blask A, Taylor WS, et al. Prevalence of Mullerian Duct Anomalies detected at ultrasound. Am J Med Genet 2000; 94:9–12.
7. The American Fertility Society. The American Fertility Society Classifications of adnexal adhesions, distal tubal occlusions, tubal occlusions secondary to tubal ligation, tubal pregnancies, Mullerian anomalies and intrauterine adhesions. Fertil Steril 1988; 49:944–955.
8. Salim R, Woelfer B, Backos M, et al. Reproducibility of three-dimensional ultrasound diagnosis of congenital uterine anomalies. Ultrasound Obstet Gynecol 2003; 21:578–582.
9. Kupesic S. Clinical implications of sonographic detection of uterine anomalies for reproductive outcome. Ultrasound Obstet Gynecol 2001; 18:387–400.
10. Patton PE, Novy MJ. Reproductive potential of the anomalous uterus. Semin Reprod Endocrinol 1988; 6:217–233.
11. Fleischer A, Kalemeris G, Machin J, et al. Sonographic depiction of normal and abnormal endometrium with histopathologic correlation. J Ultrasound Med 1986; 5:445–452.
12. Forrest TS, Elyaderani MK, Muilenburg MI, et al. Cyclic endometrial changes: US assessment with histologic correlation. Radiology 1988; 167:233–237.

13. Cohen HL, Eisenberg P, Mandel F, et al. Ovarian cysts are common in premenarchal girls: a sonographic study of 101 children 2–12 years old. AJR Am J Roentgenol 1992; 159:89–91.
14. Cohen HL, Tice HM, Mandel FS. Ovarian volume measured by US: bigger than we think. Radiology 1990; 177:189–192.
15. Jain K. Sonographic spectrum of hemorrhagic ovarian cysts. J Ultrasound Med 2002; 21:879–886.
16. Zaidi J, Jurkovic D, Campbell S, et al. Luteinized unruptured follicle: morphology, endocrine function and blood flow changes during the menstrual cycle. Hum Reprod 1995; 10:44–49.
17. Levine D, Filly R. Using color Doppler jets to differentiate a pelvic cyst from a bladder diverticulum. J Ultrasound Med 1994; 13:575–577.
18. McClure MJ, Atri M, Haider MA, et al. Perineural cysts presenting as complex adnexal cystic masses on transvaginal sonography. AJR Am J Roentgenol 2001; 177:1313–1318.
19. Levine D. Sonography of the postmenopausal pelvis. In: Anderson J, ed. Gynaecological Imaging. London, Churchill Livingstone. 1999; 483–499.
20. Lynch HT, Lynch JF. What the physician needs to know about Lynch syndrome: an update. Oncology 2005; 19:455–463.
21. Gronlund L, Hertz J, Helm P, et al. Transvaginal sonohystero graphy and hysteroscopy in the evaluation of female infertility, habitual abortion, or metrorrhagia: a comparative study. Acta Obstet Gynecol Scand 1999; 78:415–418.
22. O'Connoll LP, Fries MH, Aeringue E, et al. Triage of abnormal postmenopausal bleeding: a comparison of endometrial biopsy and transvaginal sonohysterography versus fractional curettage with hysteroscopy. Am J Obstet Gynecol 1998; 178:956–961.
23. Reinhold C, McCarthy S, Bret PM, et al. Diffuse Adenomyosis: comparison of endovaginal ultrasound and MR imaging with histopathologic correlation. Radiology 1996; 199:151–158.
24. Devlieger R, D'Hooghe T, Timmerman D. Uterine adenomyosis in the infertility clinic. Human Reprod Update 2003; 9:139–147.
25. Lee EJ, Han JH, Rye HS. Polypoid Adenomyomas: sono-hysterographic and color Doppler findings with histopathologic correlation. J Ultrasound Med 2004; 23:1421–1429.
26. Cohen H, Haller J. Pediatric and adolescent genital abnormalities. Clin Diagn Ultrasound 1989; 24:187–216.
27. Levine D. The postmenopausal pelvis. In: Nyberg DA, ed. Transvaginal Ultrasound. St. Louis, MO: Mosby Year Book, 1992.
28. Occhipinti K, Kutcher R, Rosenblatt R. Sonographic appearance and significance of arcuate artery calcification. J Ultrasound Med 1991; 10:97–100.
29. Duffield C, Gerscovich EO, Gillen MA, et al. Endometrial and endocervical micro echogenic foci: sonographic appearance with clinical and histologic correlation. J Ultrasound Med 2005; 24:583–590.
30. Callen P, Filly R, Manyer T. Intrauterine contraceptive devices: evaluation by sonography. AJR Am J Roentgenol 1980; 135:797–800.
31. Nalaboff KM, Pellerito JS, Ben-Levi E. Imaging the endometrium: disease and normal variants. Radiographics 2001; 21:1409–1424.
32. Wachsberg R, Kurtz A. Gas within the endometrial cavity at postpartum US: a normal finding after spontaneous vaginal delivery. Radiology 1992; 183:431–433.
33. Berridge DL, Winter TC. Saline infusion sonohysterography. J Ultrasound Med 2004; 23:97–112.
34. Langer RD, Pierce JJ, O'Hanlan KA, et al. Transvaginal ultrasonography compared with endometrial biopsy for the detection of endometrial disease. N Engl J Med 1997; 337:1792–1798.
35. Siegel M. Pediatric gynecologic sonography. Radiology 1991; 179:593–600.
36. Horrow MM. Ultrasound of pelvic inflammatory disease. Ultrasound Q 2004; 20:171–179.
37. Cacciatore B, Leminen A, Ingman-Friberg S, et al. Transvaginal sonographic findings in ambulatory patients with suspected pelvic inflammatory disease. Obstet Gynecol 1992; 80:912–916.
38. Timor-Tritsch IE, Lerner JP, Monteagudo A, et al. Transvaginal sonographic markers of tubal inflammatory disease. Ultrasound Obstet Gynecol 1998; 12:56–66.
39. Atri M. Ectopic pregnancy versus corpus luteum cyst revisited. J Ultrasound Med 2003; 22:1181–1184.
40. Nyberg DA, Mack LA, Jeffrey RB, et al. Endovaginal sonographic evaluation of ectopic pregnancy: a prospective study. AJR Am J Roentgenol 1987; 149:1181–1186.
41. Russell SA, Filly RA, Damato N. Sonographic diagnosis of ectopic pregnancy with endovaginal probes: what really has changed? J Ultrasound Med 1993; 12:145–151.
42. Ajjimakorn S, Bhamarapravati Y. Transvaginal ultrasound and the diagnosis of fallopian tube carcinoma. J Clin Ultrasound 1991; 19:116–119.
43. Fleischer A, Rodgers W, Rao B, et al. Assessment of ovarian tumor vascularity with transvaginal color Doppler sonography. J Ultrasound Med 1991; 10:563–568.
44. Salem S, White LM, Lai J. Doppler sonography of adnexal masses: the predictive value of the pulsatility in benign and malignant disease. AJR Am J Roentgenol 1994; 163:1147–1150.
45. Levine D, Feldstein VA, Babcook CJ, et al. Sonography of ovarian masses: poor sensitivity of resistive index for identifying malignant lesions. AJR Am J Roentgenol 1994; 162:1355–1359.
46. Brown DL, Frates MC, Laing FC, et al. Ovarian masses: can benign and malignant lesions be differentiated with color and pulsed Doppler US. Radiology 1994; 190:333–336.
47. Brown DL, Doubilet PM, Miller FH, et al. Benign and malignant ovarian masses: selection of the most discriminating gray-scale and Doppler sonographic features. Radiology 1998; 208:103–110.
48. Timmerman D, Bourne TH, Tailor A, et al. A comparison of methods for preoperative discrimination between malignant and benign adnexal masses: the development of a new logistic regression model. Am J Obstet Gynecol 1999; 181:57–65.
49. Fleischer AC. Sonographic assessment of the morphology and vascularity of ovarian masses. Ultrasound Q 2002; 18:81–88.
50. Levine D, Gosink BB, Wolf SI, et al. Simple adnexal cysts: the natural history in postmenopausal women. Radiology 1992; 184:653–659.
51. Nardo LG, Kroon ND, Reginald PW. Persistent unilocular ovarian cysts in a general population of postmenopausal women: is there a place for expectant management? Obstet Gynecol 2003; 102:589–593.
52. Knudsen UB, Tabor A, Mosgaard B, et al. Management of ovarian cysts. Acta Obstet Gynecol Scand 2004; 83:1012–1021.
53. Kim JS, Woo SK, Suh SJ, et al. Sonographic diagnosis of paraovarian cysts: value of detecting a separate ipsilateral ovary. AJR Am J Roentgenol 1995; 164:1441–1444.
54. Okaro E. The role of ultrasound in the management of woman with acute and chronic pelvic pain. Best Pract Res Clin Obstet Gynecol 2004; 18:105–123.
55. Moore J, Copley S, Morris J, et al. A systematic review of the accuracy of ultrasound in the diagnosis of endometriosis. Ultrasound Obstet Gynecol 2002; 20:630–634.
56. Kim HC, Kim SH, Lee HJ, et al. Fluid–fluid levels in ovarian teratomas. Abdom Imaging 2002; 27:100–105.
57. Jermy K, Luise C, Bourne T. The characterization of common ovarian cysts in premenopausal women. Ultrasound Obstet Gynecol 2001; 17:140–144.
58. Guerriero S, Ajossa S, Mais V, et al. Role of transvaginal sonography in the diagnosis of peritoneal inclusion cysts. J Ultrasound Med 2004; 23:1193–1200.
59. Franks S. Polycystic ovary syndrome. Arch Dis Child 1997; 77:89–90.
60. The Rotterdam ESHREI/ASRM-Sponsored PCOS Consensus Workshop Group. Revised 2003 consensus on diagnostic criteria and long-term health risks related to polycystic ovary syndrome. Fertil Steril 2004; 81:19–25.
61. Adams J, Polson DW, Franks S. Prevalence of polycystic ovaries in women with anovulation and idiopathic hirsutism. Br Med J Clin Res Ed 1986; 293:355–359.
62. Stark JE, Siegel MJ. Ovarian torsion in prepubertal and pubertal girls, sonographic findings. AJR Am J. Roentgenol 1994; 163:1479–1482.
63. Albayram F, Hamper UM. Ovarian and adnexal torsion: spectrum of sonographic findings with pathologic correlation. J Ultrasound Med 2001; 20:1083–1089.
64. Pena JE, Ufberg D, Cooney N, et al. Usefulness of Doppler sonography in the diagnosis of ovarian torsion. Fertil Steril 2000; 73:1047–1050.

Postmenopausal Pelvis

Anna S. Lev-Toaff and Deborah Levine

38

Approximately one-third of a woman's expected life span occurs after the age of 50 years, the mean age of menopause. With the growth of the population and an increase in life expectancy, the number of postmenopausal women is increasing. An estimated 45 million women in the United States will surpass the age of 55 by the year 2020 (1).

The three major indications for sonography in this population are (*i*) to evaluate the endometrium in patients with abnormal bleeding, (*ii*) to evaluate the ovaries and uterus in women with a palpable pelvic mass, and (*iii*) to screen for endometrial or ovarian cancer in high-risk women.

Knowledge of the expected changes in the uterus and adnexa secondary to aging and of common disease processes of the postmenopausal pelvis is important in caring for postmenopausal women.

ANATOMY AND PATHOLOGY

Sonography of the postmenopausal pelvis should be performed as in premenopausal women. Transabdominal and transvaginal sonography are complementary techniques. Views of the uterus, adnexae, and cul-de-sac are obtained through a full bladder with a transabdominal probe; transvaginal imaging is performed on an empty bladder. Once an enlarged uterus or large pelvic mass has been excluded by means of initial transabdominal sonography, transvaginal sonography is the optimal method for detailed imaging of the uterus and ovaries. The uterus, ovaries, and any abnormal masses are measured in three planes. The endometrial thickness is measured in the sagittal plane. This anteroposterior measurement includes both endometrial layers. The hypoechoic layer surrounding the endometrium represents the inner layer of myometrium and should not be included in this measurement. Endometrial fluid, when present, also should be excluded from this measurement (2).

The transvaginal examination is especially important in women who have postmenopausal bleeding because it allows for accurate measurement of the endometrium and improved visualization of endometrial (as well as adnexal) pathologic processes. The transvaginal probe is well accepted by most women after menopause. If a woman is hesitant about its use, time should be taken to explain how the probe improves the quality of the examination. Extra gel and topical anesthetic gel can be used as a lubricants in women with atrophic vaginitis.

Uterus

The uterus decreases in size after menopause, with the most rapid decrease in the first five years, followed by a more gradual decline. The body of the uterus diminishes in size relative to the cervix (Fig. 1). Late in life, an infantile configuration can be present in which the cervix is larger than the body of the uterus (Fig. 2). This decline in size is affected by parity (3) and by the presence of fibroids.

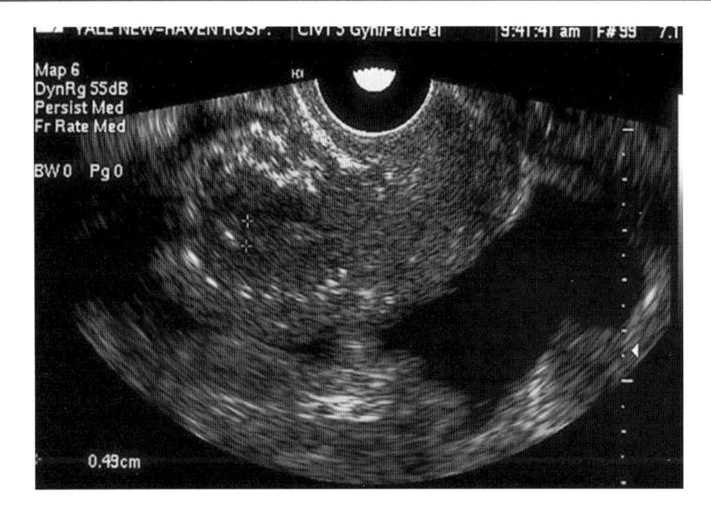

FIGURE 1 ■ Transvaginal longitudinal view of the uterus in a 62-year-old diabetic woman. The peripheral echogenic foci are arcuate artery calcifications.

Endometrium

Endometrial carcinoma is the most common gynecologic malignant disease: 40,100 new cases and 6800 deaths were reported in the United States in 2003 (4). Diagnosed mainly in postmenopausal women, 75% to 85% of cases occur after the age of 50 years (5). Unopposed estrogen therapy, tamoxifen therapy, and ovarian tumors that secrete estrogen increase the risk of endometrial cancer. Risk factors include diabetes, obesity, hypertension, and a high-fat diet; hereditary nonpolyposis colorectal cancer syndrome is associated with a 20% to 30% increased risk (4,6). Most women with endometrial cancer present with vaginal bleeding. It is important to understand that there is an enormous difference in endometrial cancer risk between postmenopausal women with bleeding (high risk) and those that are asymptomatic (low risk). This fact influences our approach to endometrial thickness as an indicator for invasive procedures (see below) (7).

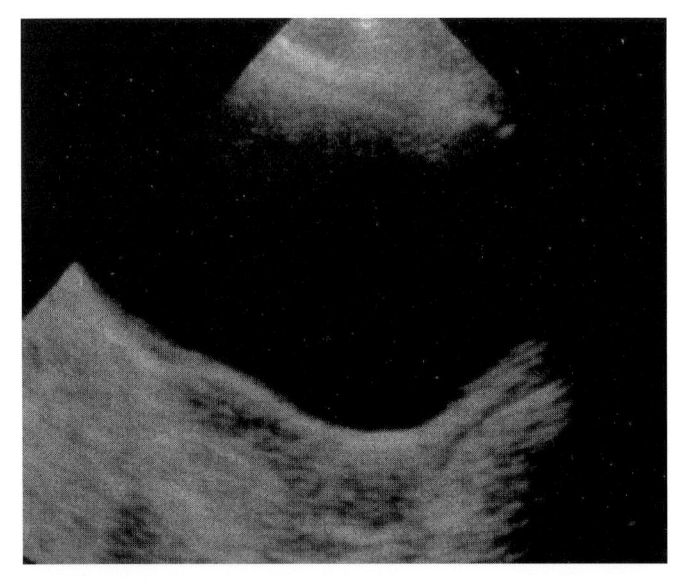

FIGURE 2 ■ Transabdominal longitudinal view of the uterus in a 90-year-old woman. The uterus has an infantile configuration, with the uterine body smaller than that of the cervix.

Transvaginal sonography is important in evaluating for endometrial carcinoma because the endometrium can be accurately measured and well visualized. The normal postmenopausal endometrium is atrophic; the thickness of the endometrium in normal postmenopausal women diminishes with increasing time since the onset of menopause (8). The normal postmenopausal endometrium is seen as a thin, echogenic line of homogeneous texture. In women without vaginal bleeding (without additional risk factors for endometrial cancer—such as obesity, diabetes, and hypertension), a threshold of 11 mm has been suggested. This threshold has been shown to discriminate between women with a risk of endometrial cancer of 6.7% that warrants further diagnostic intervention (thickness >11 mm) and women with a very low risk of 0.002% (thickness ≤11 mm) in whom biopsy is not necessary (7). In the presence of other factors that increase the risk of endometrial cancer, such as obesity, unopposed estrogen or tamoxifen, and age above 70, a threshold of 8 mm is suggested in asymptomatic women (7).

A threshold for biopsy of 5 mm is widely used for postmenopausal women with bleeding. In the presence of postmenopausal bleeding, the risk of endometrial cancer is 7.3% when the endometrium is thick (>5 mm) and less than 0.07% if the endometrium is thin (≤5 mm) (7). Estrogenic hormones, when given alone, cause increased endometrial thickness as well as an increased incidence of endometrial hyperplasia, polyps, and cancer. Therefore, the patient's symptoms (bleeding or not bleeding) and history of hormone use are important in the evaluation of the postmenopausal endometrium.

Abnormal Bleeding and Endometrial Cancer

Uterine bleeding is an early indicator of endometrial cancer; bleeding occurs in 80% to 90% of women with endometrial cancer. However, usually, postmenopausal bleeding is due to benign causes, most commonly secondary to endometrial atrophy. It is generally accepted that there is only a 10% prevalence of endometrial cancer among bleeding postmenopausal women (9).

When sonography demonstrates an endometrial thickness of 4 mm or less, endometrial atrophy is the most likely histologic diagnosis and endometrial sampling is widely considered not necessary (10–12). If abnormal bleeding continues, follow-up transvaginal sonography or endometrial biopsy is suggested. Some authors have suggested a threshold as low as 3 mm because endometrial carcinoma has also been reported at an endometrial thickness of 3 mm (13). If the endometrium measures above 4 mm, or is not clearly visualized, sonohysterography is recommended. At sonohysterography, single layer endometrial thickness values are used because the endometrial layers are separated by fluid. If the single layer thickness is below 2.5 mm, endometrial atrophy is the probable diagnosis. If the endometrium is diffusely thicker than 2.5 mm, a diffuse endometrial lesion is suggested and endometrial biopsy should follow. When sonohysterography reveals a focal endometrial thickening, hysteroscopically guided resection is required in order to ensure removal of the lesion (10–12). This approach helps to avoid unnecessary invasive procedures while reducing the chance of missing the cause of bleeding due to sampling error. Blind endometrial biopsy has been shown to miss 15% of endometrial cancers; dilation and curettage has an 11% false negative rate for endometrial carcinoma (14). Besides endometrial lesions, sonohysterography is also helpful in revealing other causes of postmenopausal bleeding such as uterine fibroids. Women with submucosal fibroids, who are taking sequential hormones have a higher risk of abnormal withdrawal bleeding than women without submucosal fibroids (15). A decision tree for the evaluation of endometrial thickness in postmenopausal women with and without postmenopausal bleeding is given in Figure 6.

Hormones and the Endometrium

The use of hormone replacement therapy has diminished in recent years primarily due to concern regarding possible increased incidence of breast cancer related to its use (16). Currently, estrogenic hormones are prescribed for shorter periods of time and primarily for alleviation of vasomotor symptoms and vaginal dryness (17,18). Even for these indications, current guidelines recommend use of the lowest effective dose for the shortest duration of time (17,18). Prevention of heart disease is no longer considered an indication for hormone replacement therapy (18). Regarding prevention of osteoporosis, estrogen has been relegated to a secondary role when other medications are not suitable (19). Although there are fewer women receiving hormone replacement therapy, their potential deleterious effects on the endometrium and the sonographic changes they induce must be recognized.

The type of hormone replacement therapy prescribed depends, in part, on whether or not a woman has her uterus. In the presence of the uterus, unopposed estrogen was shown to induce endometrial hyperplasia in 62% of women, 34% of which were complex or atypical (potentially premalignant) by 36 months of treatment (20,21). Addition of progesterone diminished this risk to levels similar to that in women not on hormone replacement therapy (20,21). Therefore, in women with a uterus, progesterone therapy must be added to estrogen treatment. Progesterone may be given in a continuous daily fashion along with estrogen. Alternatively, progesterone may be given in a sequential fashion, typically for 12 days every month or two to counteract the trophic effect of estrogen on the endometrium. Estrogen therapy may be given orally, transdermally, or vaginally (estrogen ring or cream) (22). Regardless of the mode of administration, estrogen therapy alone has a stimulating effect on the endometrium, which results in a thickened appearance on transvaginal sonography. Progesterone may be given either transdermally along with estrogen or sequentially by oral means. Progesterone may also be delivered locally by means of an intrauterine device containing the hormone. Vaginal rings containing both estrogen and progesterone are also available.

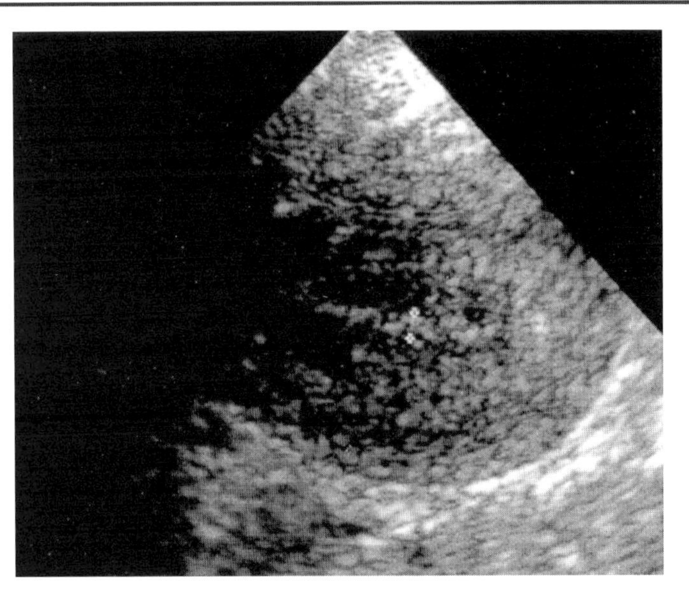

FIGURE 3 ■ Transvaginal sagittal view of the uterus demonstrates a normal, thin postmenopausal endometrium (*calipers*). Both endometrial layers are atrophic and appear as a single thin echogenic line.

In postmenopausal women who do not take exogenous estrogens, the endometrium is typically thin (less than 4 mm) and atrophic (Fig. 3). The thickness of the postmenopausal endometrium diminishes with increasing time since the onset of menopause (8). It should be emphasized that in postmenopausal women without bleeding, who do not have risk factors for endometrial carcinoma (such as obesity, diabetes, and age >70), an endometrial thickness of up to 11 mm carries a very low risk of endometrial cancer (7). Women who take continuous daily estrogen and progesterone for three to four months also have endometrial thinning and atrophy (23). In contrast, 55% of women who use unopposed estrogen have a thickened, slightly heterogeneous endometrium (24). This thickening should be regarded with concern because of estrogen's trophic effect on the endometrium; these women should undergo further diagnostic evaluation.

Sequential Hormones ■ Over half the postmenopausal women who use sequential hormones have a thick (greater than 8 mm) endometrium at some point in their cycle (24). In women on sequential regimens, who receive progestogens only during part of the month or bi- or trimonthly, the endometrium measured soon after withdrawal bleeding (that occurs after cessation of progesterone) is not significantly different from the thickness in women on combined continuous regimens (25). Therefore, it is important to measure endometrial thickness shortly after the withdrawal bleeding. If a woman presents with endometrial thickening at some other time in her treatment cycle, endometrial measurement should be repeated after the withdrawal bleeding (Fig. 4). If the endometrium remains thickened at this time, further evaluation is warranted.

The effect of hormones on the sonographic appearance of the endometrium is summarized in Table 1.

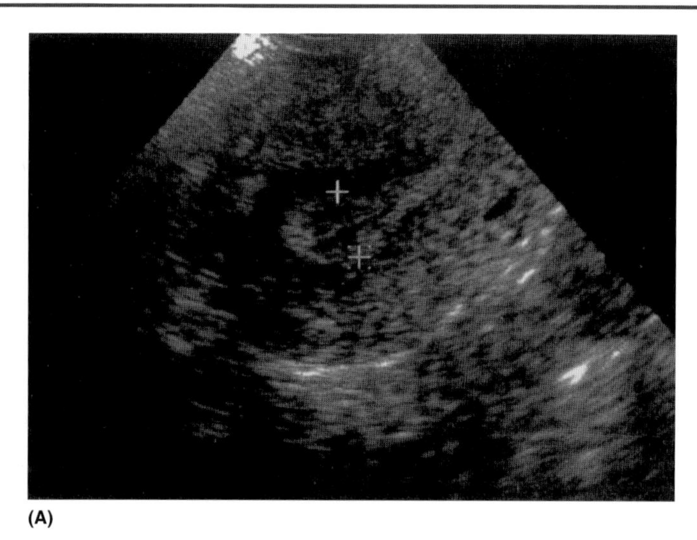

(A)

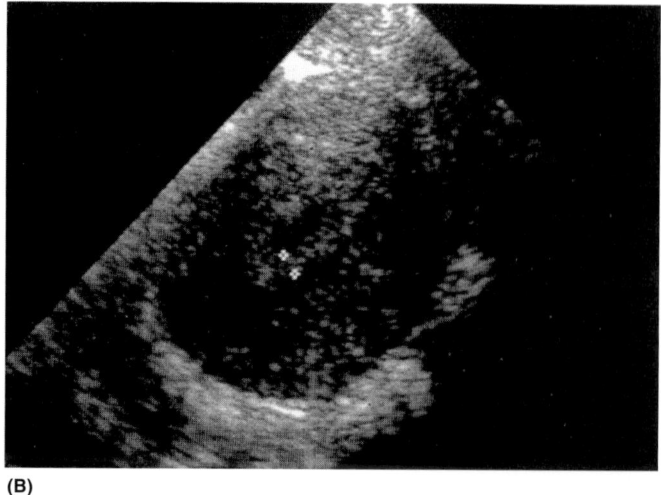

(B)

FIGURE 4 A AND B ■ Changing endometrial thickness in women using sequential hormone therapy. (**A**) Transvaginal sagittal view of the uterus in the mid-cycle demonstrates an endometrial thickness of 11 mm. (**B**) The patient was rescanned at the beginning of the cycle, and the endometrial thickness was 5 mm. *Source*: From Ref. 24.

Tamoxifen ■ Tamoxifen, a selective estrogen receptor modulator, is extensively used for the treatment of breast cancer and for its prevention in women at high risk (26). In breast tissue, tamoxifen acts as an estrogen antagonist; however, it has an estrogenic effect on the postmenopausal endometrium and myometrium. Long-term treatment with tamoxifen may induce a number of uterine abnormalities including endometrial hyperplasia, polyps, and endometrial carcinoma (27). In the myometrium, tamoxifen may stimulate subendometrial cystic changes and adenomyosis (28). The findings on transvaginal sonography in women on tamoxifen may be difficult to interpret because an apparently thickened endometrium with cystic changes may be due to endometrial hyperplasia, endometrial polyps, or endometrial cancer (Figs. 5–7). In addition, these same findings on transvaginal sonography may reflect subendometrial changes due to reactivation of adenomyosis in the inner layer of the myometrium. This entity is usually associated with overlying endometrial atrophy. In order to clarify the findings, sonohysterography may be used. The instilled fluid will clearly outline any polyps or focal endometrial thickening. However, even with sonohysterography, it is sometimes difficult to identify the endometrial layer because of distortion of the endometrial–myometrial interface caused by subendometrial changes; in these cases, endometrial sampling should be performed to exclude an endometrial lesion (29).

TABLE 1 ■ Effect of Hormones on the Postmenopausal Endometrium

Hormone regimen	Appearance[a]
No hormones	Thin, atrophic, usually <4 mm
Unopposed estrogen	Thick May be heterogeneous
Daily estrogen and progesterone	Thin, atrophic, usually <4 mm
Sequential estrogen and progesterone	Thickness varies with phase of cycle
Tamoxifen	Thick, cystic changes in endometrium or subendometrium

[a]These thresholds are for women who are not bleeding; for symptomatic women, see Figure 6.

Small Amounts of Endometrial Fluid

A small amount of fluid located between two thin homogeneous endometrial layers is occasionally found in normal asymptomatic postmenopausal women (Fig. 8) (24,30). This fluid accumulation may or may not be due to cervical stenosis. The significance of endometrial fluid accumulation depends upon the appearance of the surrounding endometrium. The fluid collection is likely benign as long as the endometrium itself is thin (3 mm) and homogeneous and without focal excrescences; in this case, endometrial sampling is not required (30). If the peripheral endometrium is thicker than 3 mm or not homogeneous, sampling is required. Women on hormone replacement therapy are more likely to have endometrial fluid (31).

Uterine Artery Doppler

Investigators have suggested that Doppler pulsatility index (PI) or resistive index (RI) is helpful in distinguishing benign from malignant causes of endometrial thickening (32). Others have reported a substantial overlap in Doppler indices between benign and malignant endometrial lesions (33). Therefore, uterine artery Doppler does not improve the sensitivity or specificity of endometrial

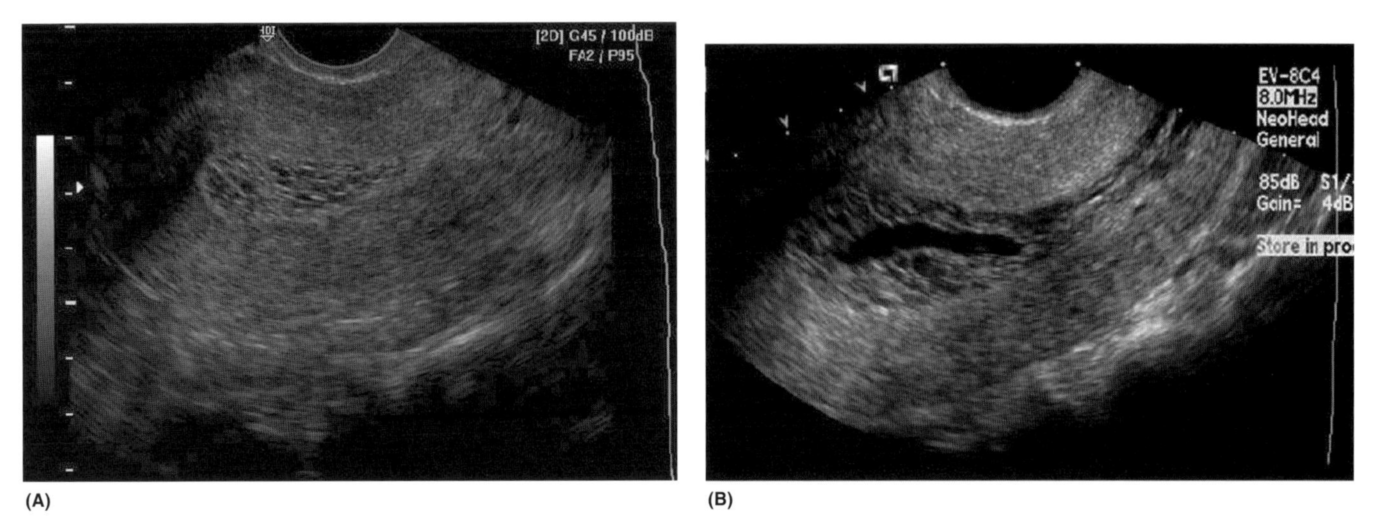

FIGURE 5 A AND B ■ Transvaginal sonography of the uterus in a 52-year-old woman who had been taking tamoxifen for three years and presented with vaginal spotting. (**A**) Sagittal sonogram shows a multicystic appearance to the thickened endometrial echocomplex. (**B**) Sonohysterography demonstrates that the findings are due to a thickened endometrium with a diffuse multicystic pattern. Although the entire endometrium is thickened, the posterior endometrium is thicker than the anterior. Biopsy revealed cystic hyperplasia without atypia.

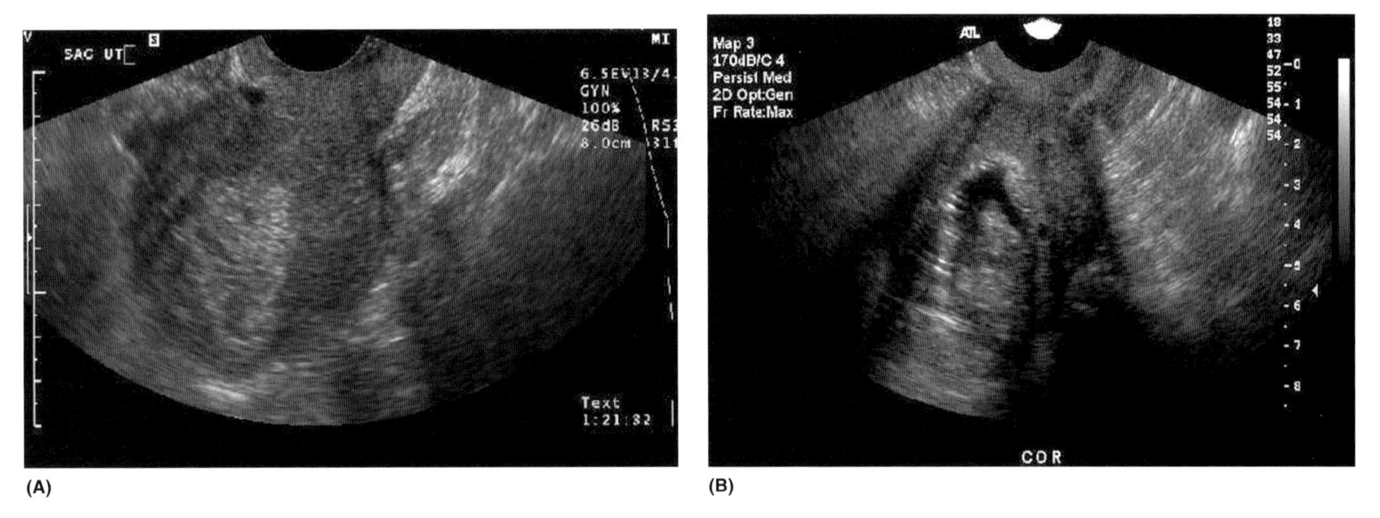

FIGURE 6 A AND B ■ Transvaginal sonography of the uterus in a 67-year-old woman who had been taking tamoxifen for four years. (**A**) Sagittal sonogram shows a markedly thickened echogenic endometrium. (**B**) Sonohysterography demonstrates that the findings are due to a large echogenic mass in the uterine cavity and slight cystic endometrial thickening. The histologic findings were benign endometrial polyp and cystic endometrial atrophy.

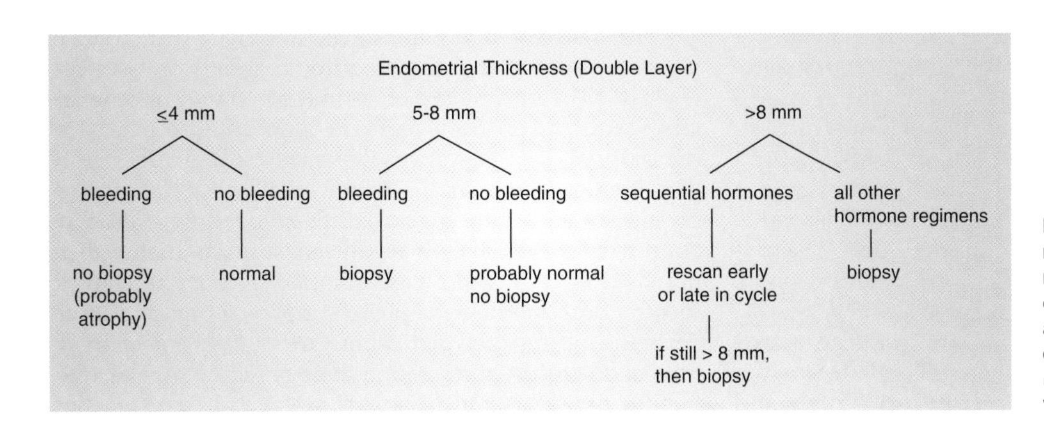

FIGURE 7 ■ Flow chart of recommendations for followup of postmenopausal women with respect to endometrial thickness, symptoms, and hormone use. "Bleeding" in this chart signifies abnormal bleeding, not the withdrawal bleeding expected with sequential hormone use.

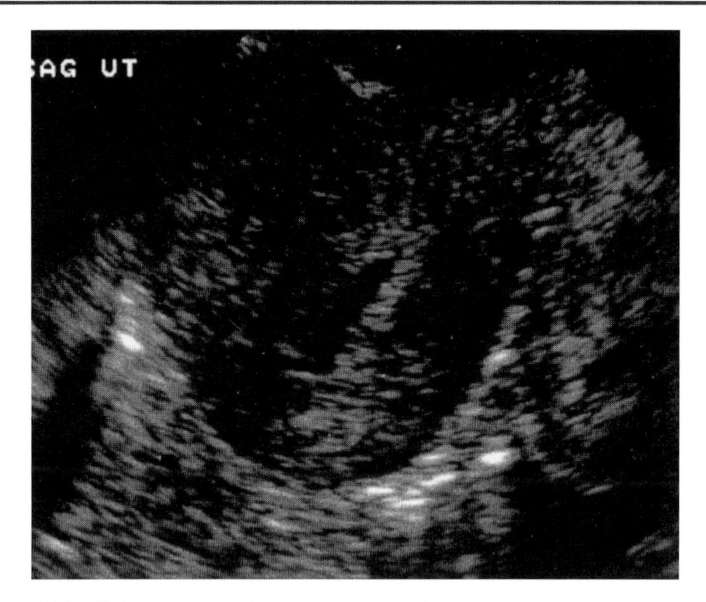

FIGURE 8 ■ Transvaginal sagittal view of the uterus in a 70-year-old woman demonstrates a small amount of endometrial fluid. The surrounding endometrium is thin and benign in appearance.

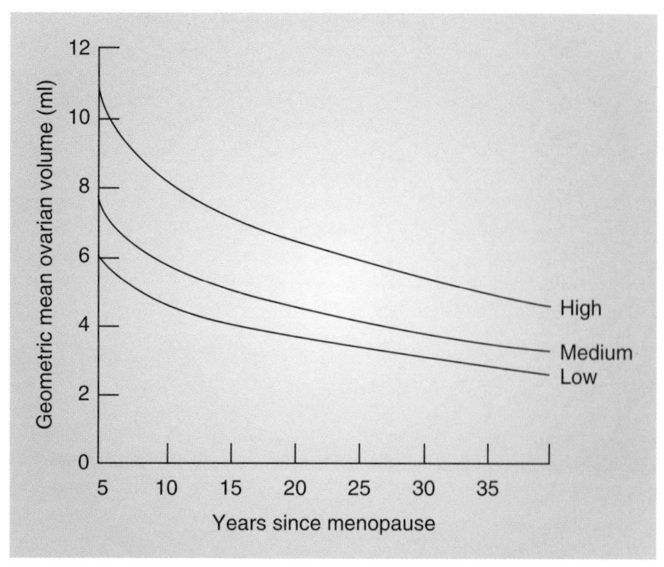

FIGURE 9 ■ Graph of ovarian volume versus years since menopause. *Source*: From Ref. 39.

cancer detection beyond endometrial thickness and morphology (33).

Ovaries

Ovarian cancer accounts for only about 3% of all cancers in women; it ranks second among gynecologic cancers, with endometrial cancer being first (34). Ovarian cancer causes more deaths than any other cancer of the female reproductive system; the risk for ovarian cancer increases with age, with the peak risk in the late 70s (34). Of malignant ovarian tumors, 80% occur in postmenopausal women (35). Continuous ovulation associated with nulliparity increases the risk of ovarian cancer; protective factors are those that suspend ovulation including pregnancy, lactation, and oral contraceptives. Hereditary syndromes account for 10% of ovarian cancer. The breast–ovarian cancer syndrome, caused by mutations in *BRCA1* and *BRCA2* genes, is associated with an 11% to 40% risk of ovarian cancer. The hereditary nonpolyposis colorectal cancer syndrome (Lynch II) carries a 12% risk of ovarian cancer (36). Routine screening for average-risk women is not currently recommended (34). There is insufficient evidence to support the introduction of ovarian cancer screening in the asymptomatic average-risk postmenopausal population; screening is associated with increased rates of surgery and patient anxiety (37). However, women who are at high risk may be monitored by a combination of transvaginal sonography and the serum tumor marker CA 125 (34). Ultrasound alone is of limited value as an independent modality for the detection of early-stage epithelial ovarian cancer in asymptomatic women who are at increased risk for this disease (38).

Transvaginal sonography is important in the assessment of postmenopausal ovaries in that (*i*) it enables the examiner to determine whether a mass is adnexal and (*ii*) it can characterize the mass as completely cystic (and therefore likely benign), as complex, or as completely solid. Like the uterus, the ovaries decrease in size after menopause (Fig. 9) (40). The mean ovarian volume declines gradually during the postmenopausal years from 8.6 ± 2.3 mL in the first postmenopausal year to 2.2 ± 1.4 in the 15th year; this data was gathered from postmenopausal women who were not on hormone replacement therapy or tamoxifen (40). This variation in size leads to difficulty in establishing a normal size for the postmenopausal ovary in any given woman. The upper limit of normal volume for postmenopausal ovaries is often given as 10 mL. However, since ovarian volume decreases gradually with each year since the onset of menopause, it may be more logical to use a nomogram relating volume to years since menopause, as published by Tepper et al. (40). The value of this nomogram in the recognition of abnormal postmenopausal ovaries has been demonstrated by Zalel et al., showing that the ovarian volume was greater than 2 standard deviations above the age-adjusted mean ovarian volume in 100% of women diagnosed with malignant ovarian tumors and in 86% of women with benign ovarian tumors (41).

Because of their small size and lack of follicles, postmenopausal ovaries are often difficult to visualize (Fig. 10). The percentage of normal postmenopausal ovaries identified on transvaginal sonography depends on the skill of the examiner. Reported results are highly variable. Rodriguez et al. identified 82% of postmenopausal ovaries with transvaginal ultrasound; all the abnormal ovaries were visualized; the mean surgical diameter of nonvisualized ovaries was 7.3 mm. Although efforts to identify both ovaries in a postmenopausal patient should be made, it appears that failure to identify an ovary is unlikely to result in missed pathology (42). However, certain maneuvers can increase detection; these include using

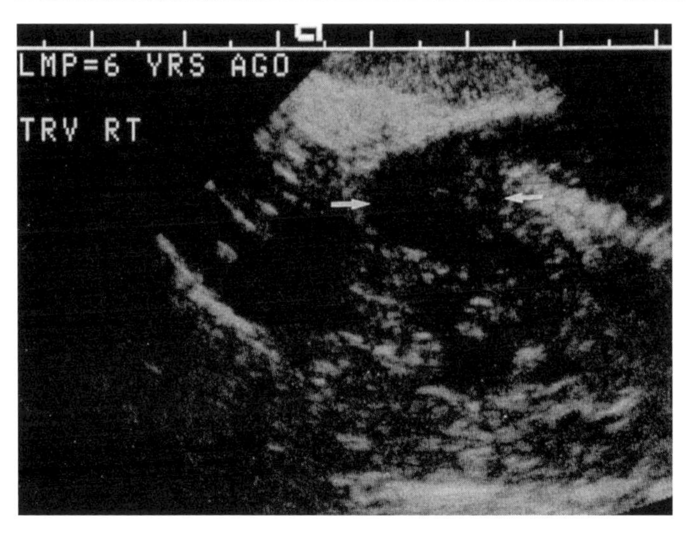

FIGURE 10 ■ Normal postmenopausal ovary (*arrows*). The normal postmenopausal ovary is homogeneous and hypoechoic.

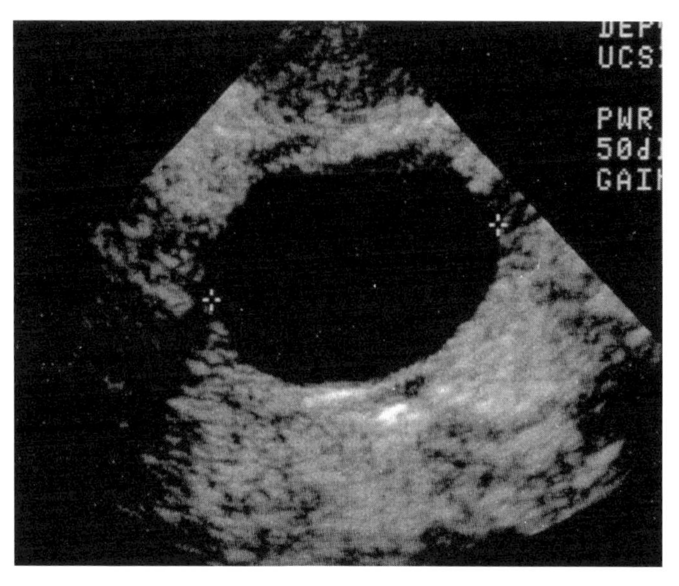

FIGURE 12 ■ Transvaginal sonogram of a 61-year-old woman demonstrates a 2.5-cm anechoic thin-walled cyst. On follow-up examination 4 months later, the cyst was no longer visualized.

the free hand to displace bowel loops by compression of the lower abdomen and holding the probe still to identify the nonmobile ovary in the peristalsing bowel. Using transabdominal sonography can also increase yield by detecting ovaries that are relatively high and lateral in the pelvis.

Postmenopausal Adnexal Cysts

Although, in premenopausal women, ovarian cysts result from cyclic hormonal stimulation, this stimulus is no longer present after menopause. In the past, the finding of an adnexal cyst in a postmenopausal woman was considered an indication for surgery. This practice was modified with the demonstration that adnexal cysts are common in postmenopausal women (43). Small (less than 5 cm), unilocular, anechoic adnexal cysts are seen in up to 17% of healthy postmenopausal women and are likely benign (43,44). These cysts can be either ovarian or paraovarian (Fig. 11).

The cysts tend to be small (less than 2 cm) and can change in size (Fig. 12) (44). A recent report including 156,106 women over 50 years of age, who were screened for ovarian lesions identified 2763 women (18%) with unilocular ovarian cysts below 10 cm in diameter (45). Of these cysts, 70% resolved, 17% developed a septum, 6% developed a solid component, and 7% persisted as a unilocular lesion. No woman with an isolated unilocular cystic ovarian lesion developed ovarian cancer (45). These and other studies lend support to the practice of sonographic followup of unilocular ovarian cysts less than 5 cm in size in postmenopausal women as long as there is no increase in size, change in morphology, or abnormal levels of CA 125 (45,46).

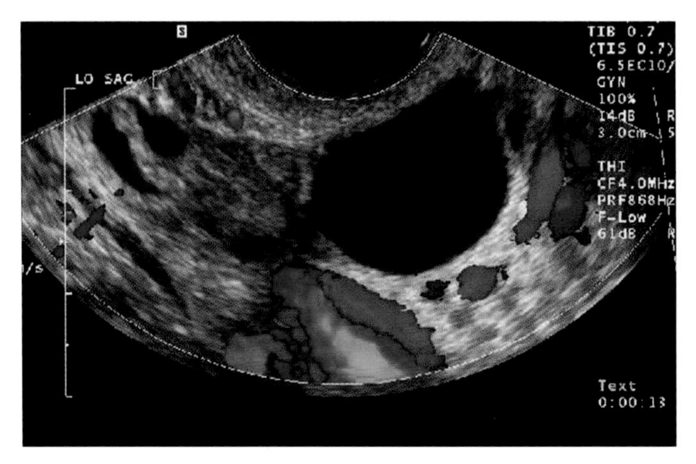

FIGURE 11 ■ A 2.5-cm postmenopausal paraovarian cyst in a 52-year-old woman shows no vascular flow.

A 5 cm upper limit for conservative management of postmenopausal unilocular cysts is prudent because large size increases the likelihood of malignancy; appropriate followup of women with postmenopausal adnexal cysts is important. A decision tree for the management of postmenopausal adnexal cysts is shown in Figure 13. As in premenopausal women, not all cystic adnexal structures are ovarian in origin. Other lesions that may be found include paraovarian cysts (which will persist), peritoneal inclusion cysts (in women with a history of surgery or pelvic inflammatory disease), and hydrosalpinges.

Doppler Evaluation

Use of color Doppler, while scanning the pelvis, allows for distinction between vessels and cysts (or hydrosalpinx) (Fig. 14) and allows for the diagnosis of homogeneous hypoechoic lesions such as fibromas that can masquerade as adnexal cysts as well. When pulsed Doppler analysis of tumor vascularity is to be performed, color Doppler aids in the placement of the Doppler gate.

Numerous articles have described Doppler indices both in screening for ovarian cancer and in distinguishing

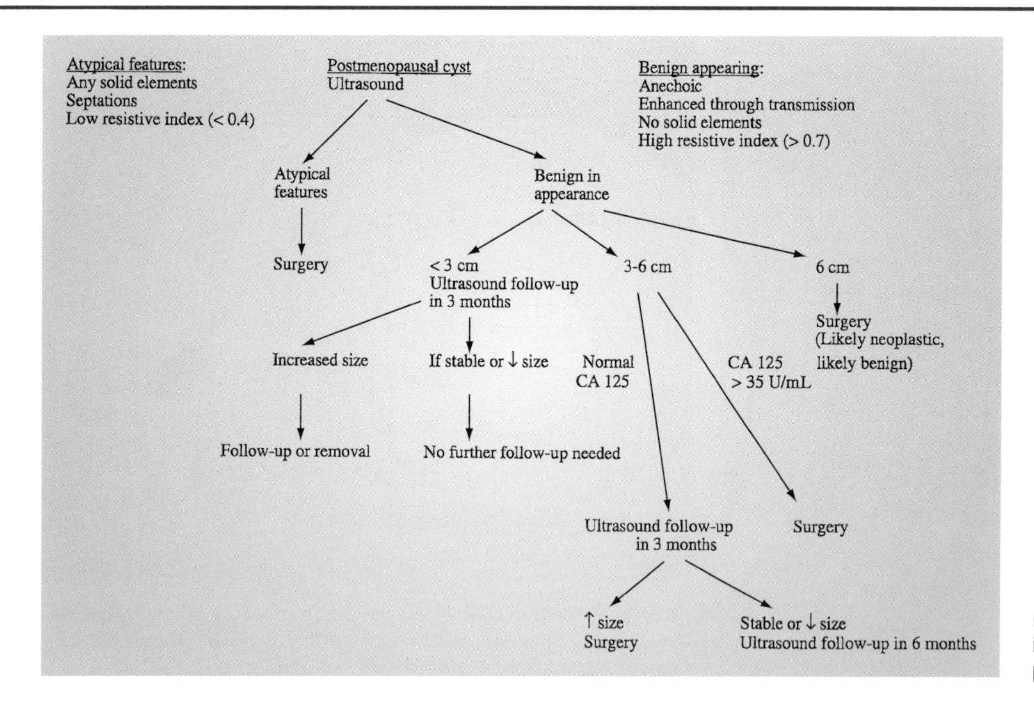

FIGURE 13 ■ Decision tree regarding the management of postmenopausal adnexal cysts.

between benign and malignant adnexal masses. Because tumor vessels lack a muscular layer, they frequently have low resistance flow. This resistance can be quantified either with the RI (peak systole–end diastole/peak systole) or with the PI (peak systole–end diastole/mean). An RI of 0.4 or less or a PI of 1.0 or less is associated with malignant disease (47). Some workers have shown that Doppler PI and RI together with CA 125 increased the sensitivity of prediction of malignancy in postmenopausal ovarian masses (48). However, there is considerable overlap in the PIs of benign and malignant lesions so that Doppler sonography alone has severe limitations in the differentiation of benign from malignant adnexal disease on the basis of low-impedance flow (PI < 1.0) (49). In general, Doppler RI and PI are not useful in the distinction between benign and malignant masses because no reliable discriminatory value with both high sensitivity and high specificity has been found due to the large overlap in values obtained for benign and malignant lesions (50). A recent meta-analysis of the literature on ultrasound characterization of ovarian masses shows that using a combination of ultrasound techniques is superior to morphologic assessment, color Doppler flow imaging, or Doppler indices alone. This was true for both pre- and postmenopausal women (51).

CA 125

CA 125, a serum marker for ovarian cancer, is elevated to above 35 U/mL in up to 85% of women with epithelial ovarian cancer as compared to 1% of healthy controls (52).

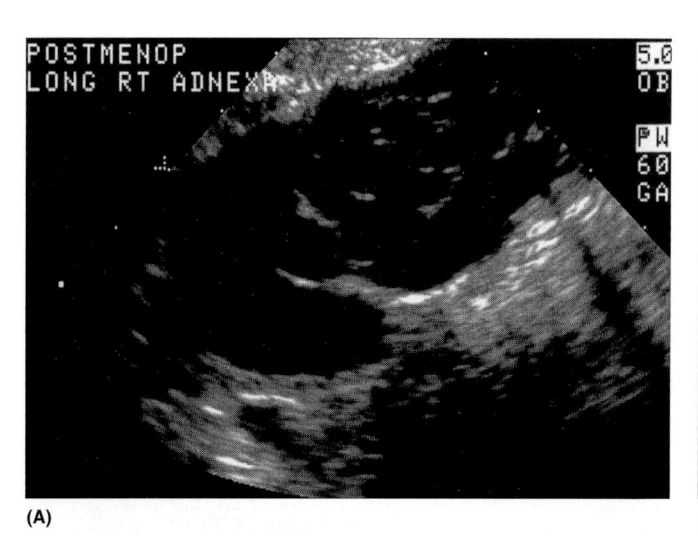

(A)

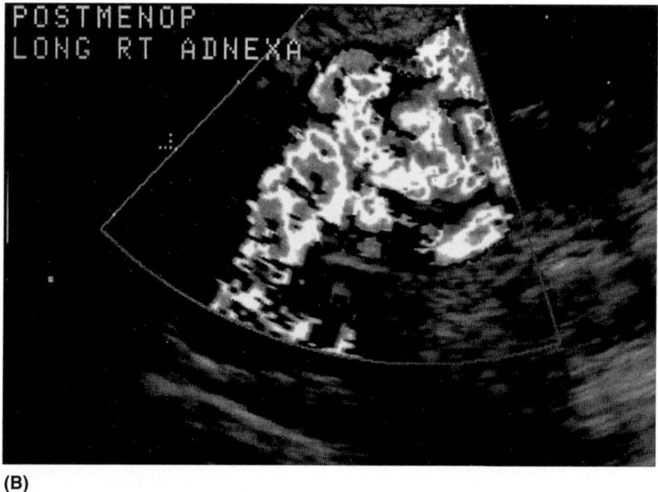

(B)

FIGURE 14 A AND B ■ (**A**) Transvaginal view of the right adnexal region demonstrates a hypoechoic region with multiple septations. (**B**) Color Doppler demonstrates flow throughout this region consistent with dilated pelvic veins.

In some women, CA 125 levels are elevated up to 60 months before the clinical detection of ovarian cancer (53). This marker has not proved effective in screening for ovarian cancer, however, because only 50% of patients with stage 1 malignant ovarian tumors have preoperative CA 125 levels higher than 35 U/mL. Furthermore, this test suffers from poor specificity with a high false-positive rate, usually attributable to nonmalignant gynecologic disease. As a result of its low specificity, this marker is not helpful in premenopausal women. In postmenopausal women, however, the specificity is 98.5% and the resulting positive predictive value of a CA 125 level higher than 35 U/mL is 14%, which meets the minimum requirement of 10% for a successful screening test (54,55). When combined with an adnexal mass on ultrasound examination, an elevated CA 125 level increases the suspicion of ovarian cancer (55).

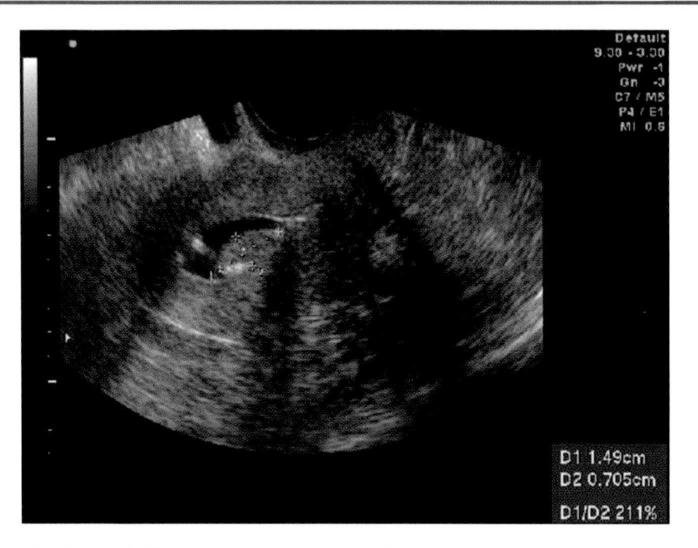

FIGURE 15 ■ SHG demonstrates a 1.2-cm echogenic endoluminal mass with a single cystic focus; the remainder of the endometrium is normal. Initial office endometrial biopsy in this patient was negative. After sonohysterography, hysteroscopic resection was performed; this mass proved to be a benign endometrial polyp.

SONOGRAPHIC PITFALLS

Focal Endometrial Disease

Focal endometrial disease is less of a problem for sonographers than for the gynecologist attempting to determine the correct histologic diagnosis in a woman with abnormal uterine bleeding. Sonography frequently identifies focal endometrial lesions missed by endometrial biopsy and D&C (56–58). These focal lesions usually are secondary to endometrial polyps and focal hyperplasia, but they may be due to endometrial cancer (56). In women with postmenopausal bleeding and endometria thicker than 5 mm, it has been shown that if there are focal lesions in the uterine cavity, hysteroscopy with endometrial resection is superior to D&C for obtaining a representative endometrial sample (58). When a patient has a thickened endometrium or a focal endometrial lesion and the histologic diagnosis is "scant tissue" or atrophic endometrium, a focal lesion was probably missed during biopsy (Fig. 15) (58). In these patients, sonohysterography identifies the lesion for biopsy localization, or the biopsy can be performed under hysteroscopic guidance. Therefore, when a polyp or focal hyperplasia is suspected, the referring physician must be informed of this possibility.

Endometrial Atrophy

An important concept in evaluating the postmenopausal endometrium is that histologic endometrial disease need not be present for postmenopausal women to experience bleeding. As the endometrium atrophies, the vessels become friable and easily damaged, leading to bleeding. Scans in such patients show a thin endometrium, and biopsy reveals atrophic endometrium. As previously mentioned, reports suggest that visualization of a thin endometrium (4 mm or less) can be used to decrease the need for endometrial biopsy.

Nabothian Cysts

Nabothian cysts commonly are seen in the cervix of postmenopausal women (Fig. 16). These cysts are thought to be secondary to obstruction of endocervical mucus glands, possibly due to infection. These cysts are usually anechoic; occasionally they contain debris. Because the ovaries can lie adjacent to the cervix, care must be taken not to mistake a cluster of nabothian cysts for adnexal lesions. This mistake is more likely to be made in the reproductive years when nabothian cysts can mimic normal follicles in their size and appearance. Nabothian cysts are located in the cervix and not in the endocervical canal. If a cystic structure is identified centrally in the canal, this should suggest the presence of an endocervical polyp.

Identification of the Ovaries

A major challenge in scanning the postmenopausal pelvis is proper identification of the ovaries. Lack of follicles and

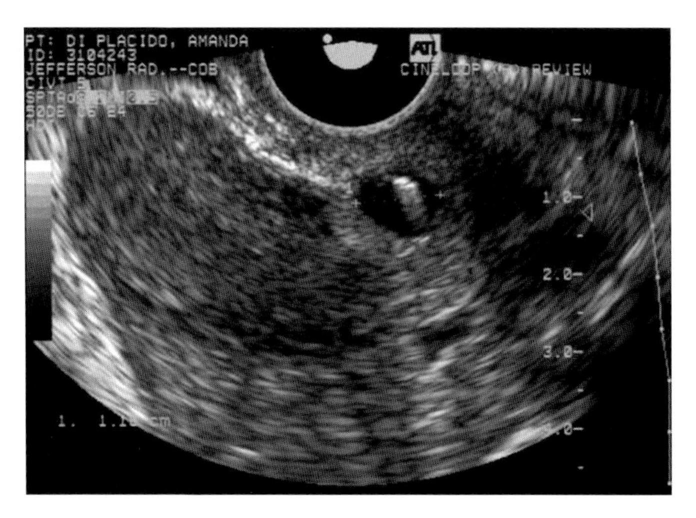

FIGURE 16 ■ Transvaginal sagittal view of the uterus in a 59-year-old woman demonstrates a single, large nabothian cyst with echogenic debris. Most nabothian cysts are anechoic and smaller than this one.

small size often render the postmenopausal ovary difficult to visualize. Use of both transabdominal and transvaginal scanning aids in visualizing the ovaries. The transabdominal approach allows for visualization of ovaries high in the pelvis and is especially helpful in patients who have undergone hysterectomy. In these women, the position of the adnexa varies. The adnexa may be retracted into a higher position than usual, or they may lie in the cul-de-sac. Vaginal ultrasound allows for better definition of ovarian masses. Loops of bowel within the pelvis can complicate visualization of the ovaries. It is often helpful for the examiner to apply a small amount of pressure on the lower abdomen with the free hand in order to displace bowel loops out of the true pelvis and bring the ovaries (or any abnormality) closer to the vaginal probe. Structures resembling ovaries should be observed for a period of time to ensure the absence of peristalsis.

Hypoechoic Solid Mass

Both fibromas and thecomas are hypoechoic solid adnexal masses that occur in postmenopausal women. When homogeneous and hypoechoic, these lesions can masquerade as adnexal cysts. With transabdominal scanning, it can be difficult to discern internal echoes. Endovaginal scanning demonstrates the solid nature of these masses and confirms the lack of enhanced through transmission (Fig. 17). Color Doppler is helpful in these patients by demonstrating flow within a solid mass, which would not be expected in an adnexal cyst. An exophytic pedunculated fibroid or a fibroid located between the layers of the broad ligament (intraligamentous) can mimic a solid adnexal mass. Identification of a separate ovary is critical to prove the myometrial origin of such a fibroid. When this is not possible, magnetic resonance imaging (MRI) is helpful. Other pelvic lesions that can mimic adnexal solid masses include neurogenic tumors, lymphadenopathy, Tarlov cysts, pelvic kidney, round ligament fibroids, and occasional masses of bowel or mesenteric origin.

SONOGRAPHIC ABNORMALITIES OF THE PELVIS

Enlarged Uterus

In postmenopausal women, an enlarged uterus is usually secondary to fibroids. Other causes include adenomyosis, malignant tumors, and benign or malignant obstruction with hydrocolpos or hydrometrocolpos. Table 2 lists the causes of postmenopausal uterine enlargement. Extension of other pelvic tumors to the cervix, such as rectal carcinoma, can also result in uterine obstruction.

Fibroids

Uterine enlargement in postmenopausal women is most likely secondary to fibroids. In postmenopausal women, fibroids are likely to demonstrate degenerative changes. These changes can result in calcifications or heterogeneity of the fibroid echotexture. Sometimes a degenerated fibroid will appear echogenic and, if centrally located, could mimic an endometrial lesion (Fig. 18). Fibroids usually regress in size after menopause, although the use of estrogenic hormones may slow or even reverse this process. Selective estrogen receptor modulators, such as raloxifene and tamoxifen, and selective progesterone receptor modulators (RU486) may promote the shrinkage of fibroids (59). Calcified myomas do not usually continue to shrink in size during menopause and may continue to cause bulk or pressure symptoms. In addition, intracavitary or submucosal myomas are not an uncommon cause of postmenopausal bleeding; these are best evaluated with sonohysterography.

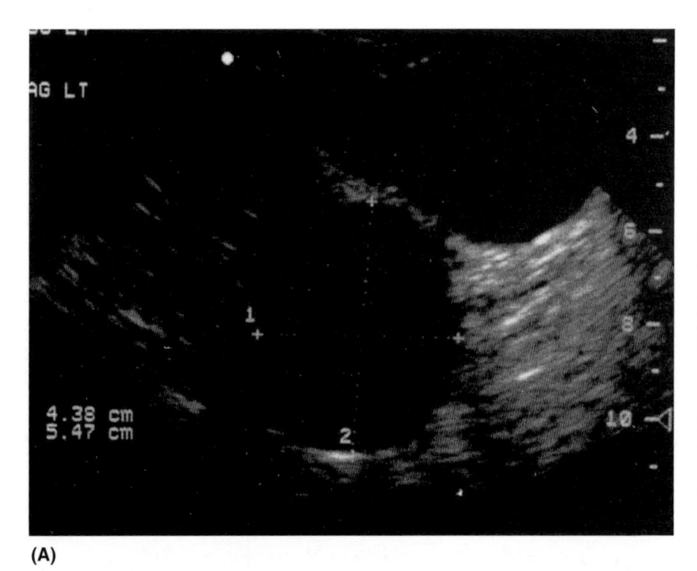

(A)

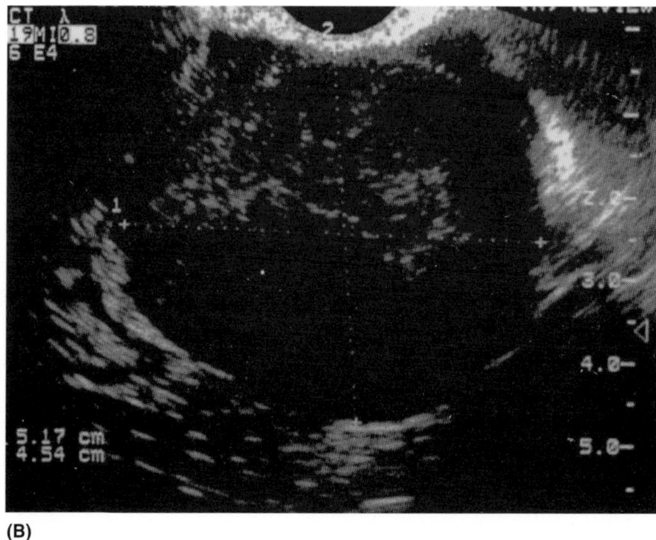

(B)

FIGURE 17 A AND B ■ **(A)** Transabdominal scan in an 80-year-old woman demonstrates a 5.4-cm hypoechoic mass. From the transabdominal scan, it is not clear whether this mass is cystic or solid. **(B)** Transvaginal view demonstrates heterogeneous echoes within the mass that confirm its solid nature. This was an ovarian fibroma.

TABLE 2 ■ Postmenopausal Uterine Enlargement

Diagnosis	Appearance	Additional history
Fibroids	See Table 11 in Chap. 36	Bleeding, pain
Sarcoma	Similar to fibroids	Rapid growth
Adenomyosis	Diffuse enlargement without focal mass	Pain, bleeding
	±Small myometrial cysts	
	±Focal masses	
Endometrial cancer	Thickened endometrium	Endometrial bleeding
	Poorly defined endometrial/myometrial junction	
	±Mass invades myometrium	
Obstructed uterus	Fluid-filled uterus	Usually secondary to benign cervical stenosis, but possibly resulting from cervical or endometrial cancer

Sarcoma

Uterine sarcomas comprise less than 5% of malignant uterine tumors. The sonographic appearance is similar to that of a fibroid or endometrial carcinoma (60). Rapid enlargement of the uterus with the sonographic appearance of fibroids should increase the suspicion of sarcoma in any postmenopausal woman. As mentioned previously, however, this may also be due to the use of estrogens. Sonographic appearance including Doppler has not been helpful in differentiating between leiomyoma and leiomyosarcoma; therefore, a high index of suspicion in the presence of rapid interval change is most important (60).

Adenomyosis

As the time since the onset of menopause increases, adenomyosis becomes a less common cause of uterine enlargement. In the transitional years, however, and in early menopause, adenomyosis is a common, frequently overlooked cause of uterine enlargement, frequently accompanied by uterine tenderness on examination, dyspareunia, and/or bleeding. It may also be found as an incidental finding in an asymptomatic postmenopausal woman with a bulky enlarged uterus (Fig. 19). Adenomyosis should be suspected when the uterus is diffusely enlarged without a focal mass, especially when small myometrial cysts are present. Adenomyosis is associated with a distinctive pattern of striated shadowing arising from the myometrium. Although the endometrium may appear normal, adenomyosis may also result in a poorly defined endometrium, which may be measured as abnormally thick because the endometrial/myometrial junction is ill defined. The endometrial margins may appear nodular, suggesting endometrial disease; sonohysterography is helpful for further evaluation in these cases (61).

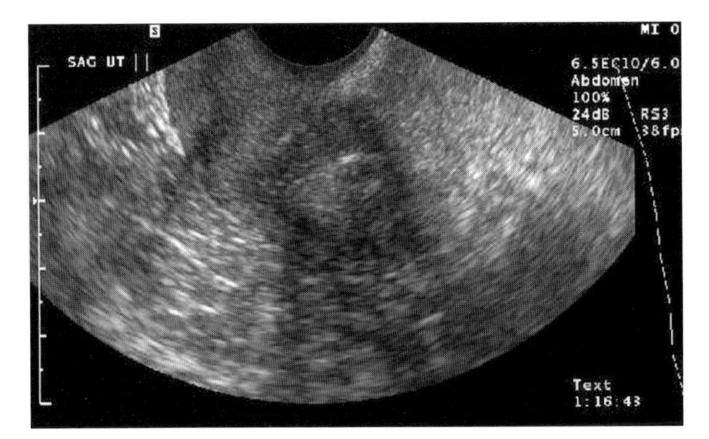

FIGURE 18 ■ Echogenic central myoma mimics endometrial lesion. Transvaginal sagittal view of the uterus in a 71-year-old woman suggests the possibility of an endometrial abnormality. Sonohysterography demonstrated a normal endometrium with a posterior mural–submucosal echogenic fibroid.

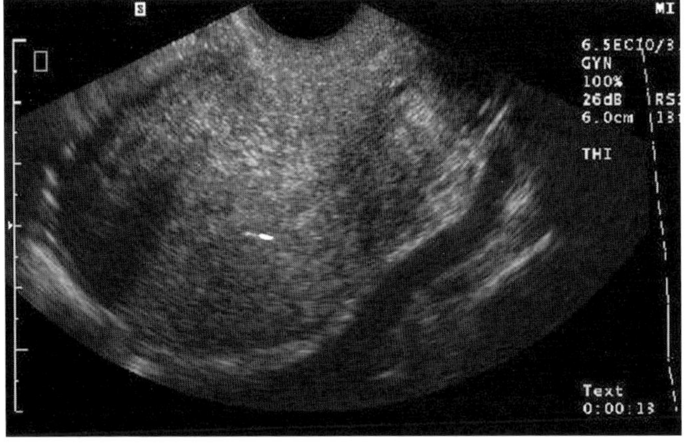

FIGURE 19 ■ Transvaginal sagittal view of the uterus in a 54-year-old asymptomatic postmenopausal woman shows a bulky uterus with a grossly heterogeneous myometrial echotexture. The endometrial echocomplex cannot be clearly discerned. MRI showed diffuse adenomyosis with an atrophic endometrium.

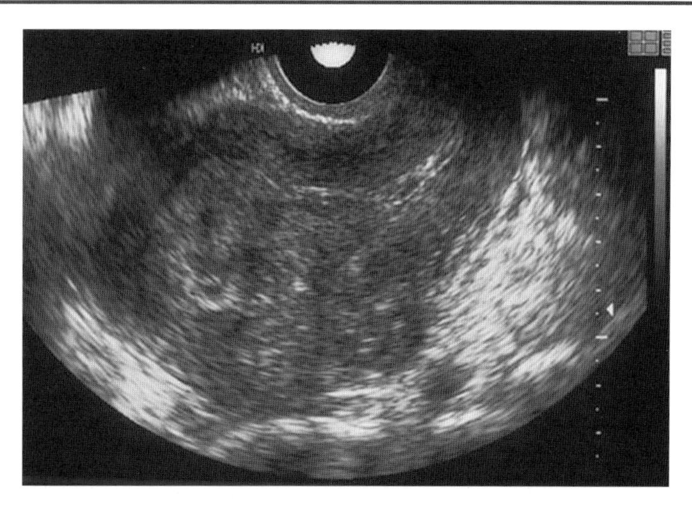

FIGURE 20 ■ Endometrial carcinoma. Transvaginal sagittal sonogram demonstrates a grossly thickened heterogeneous endometrium with loss of the endometrial–myometrial interface posteriorly consistent with deep myometrial invasion by endometrial carcinoma.

Endometrial Cancer

Uterine enlargement is a late finding in endometrial cancer (Fig. 20); patients typically note bleeding at an earlier stage of the disease. Sonography demonstrates loss of the endometrial–myometrial margin with invasion of the endometrial tumor into the myometrium.

Uterine Calcifications

Uterine calcifications are seen in women with degenerated fibroids and in women with arcuate artery calcification. These arterial calcifications are segmental, peripheral calcifications occurring in the course of the arcuate arteries (Fig. 1) (62). Because of shadowing, these calcifications can obscure visualization of the endometrium. In these patients, transvaginal scanning is especially important because, with the higher-frequency vaginal probe, it often is possible to view between calcifications to diagnose any endometrial lesion. Punctate endometrial and subendometrial calcifications are also commonly seen; these are usually of no clinical significance (63).

Thickened Endometrium

In postmenopausal women, a thick endometrium always should be regarded with suspicion because of the risk of endometrial carcinoma. Although endometrial thickening is a normal finding in women who use exogenous estrogens, further evaluation is needed to exclude more serious causes, specifically endometrial hyperplasia, polyps, and cancer. A list of causes of postmenopausal endometrial thickening is given in Table 3.

In addition to endometrial thickness, the morphologic appearance of the endometrium is important. A homogeneous endometrium with a sonographically depictable central echo between symmetric endometrial layers is associated with a benign endometrium histologically, whereas heterogeneity is associated with pathologic processes (polyps, hyperplasia, and cancer).

Endometrial Hyperplasia

Endometrial hyperplasia may be either diffuse or focal. There are various types of endometrial hyperplasia (64). In cystic hyperplasia, cysts are sonographically visible

TABLE 3 ■ Postmenopausal Thick Endometrium

		Endometrial–myometrial interface	
Diagnosis	**Endometrial appearance**	**Intact**	**Disrupted**
Polyps	Smooth margins Echogenic mass ±Vessels On SHG: pedunculated or sessile	X	–
Fibroids	Shadowing (even without calcification) Sessile, pedunculated Variable echotexture—commonly hypoechoic	Mural/subserosal fibroids	Submucosal fibroids
Hormone use	Dependent on hormone regimen (Table 1)	X	–
Hyperplasia	Diffuse or focal thickening; may also be found within a polyp	X	–
Cancer	Irregularly thickened heterogenous endometrium or large polypoid mass; occasionally within a smooth polyp	Early	All others

Abbreviation: SHG, sonohysterography.

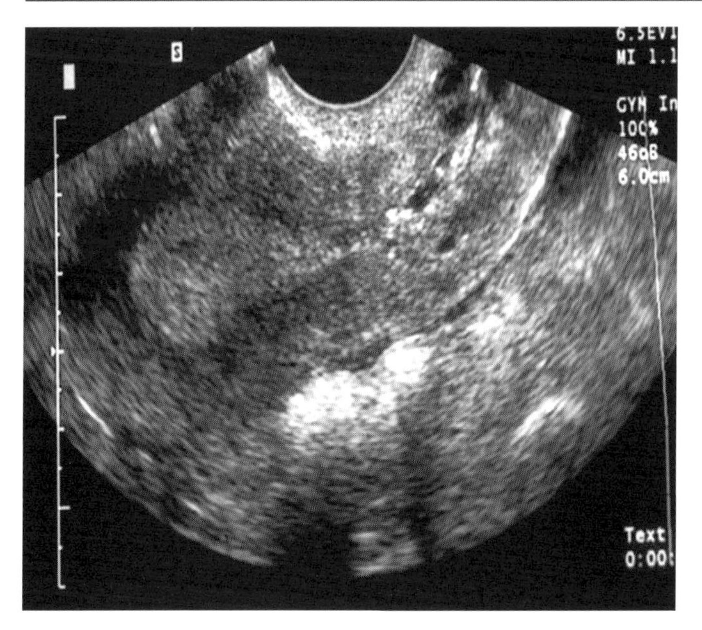

FIGURE 21 ■ Endometrial hyperplasia. Transvaginal sagittal view of the uterus demonstrates a slightly heterogeneous endometrium that is 18-mm thick. Histologic examination demonstrated adenomatous endometrial hyperplasia with atypia. Note multiple nabothian cysts.

within the endometrium (Fig. 5). This hyperplasia has a low rate of progression to endometrial cancer (65). Adenomatous endometrial hyperplasia, especially if atypical, is often a precursor to endometrial carcinoma (Fig. 21). Simple hyperplasia is less frequently atypical, but when it is atypical, it may be a precursor to endometrial cancer as well. Focal endometrial hyperplasia is difficult to distinguish sonographically from an endometrial polyp. Endometrial hyperplasia is a histologic diagnosis that can only be included in the differential diagnosis when endometrial thickening of a diffuse or focal nature is imaged.

Endometrial Polyps

Endometrial polyps appear on transvaginal sonography as either diffuse or focal thickening of the endometrium. Occasionally, the endometrium can be seen to be splayed around an intraluminal mass. They are easier to visualize when surrounded by endometrial fluid during sonohysterography. Usually, polyps are seen to be more echogenic than the surrounding endometrium. A stalk may be visualized or inferred, with blood flow extending through the pedicle to the polyp (Fig. 22). Macrocysts are also present in some endometrial polyps. Although most endometrial polyps are benign, occasionally endometrial hyperplasia or carcinoma is found within the polyp.

Endometrial Carcinoma

The diagnosis of endometrial carcinoma is suggested when the endometrium is thickened with a heterogeneous echotexture and ill-defined margins between the endometrium and the myometrium (66). An irregularly marginated polypoid mass may be seen projecting within the uterine cavity, often surrounded by fluid or blood (Fig. 23) (67). These findings are best visualized with transvaginal scanning; therefore, transvaginal studies are crucial in the evaluation of postmenopausal bleeding. Color or power Doppler can be useful to distinguish between vascularized tissue and blood clots.

Demonstration of myometrial invasion is clear evidence of endometrial cancer. Patients with advanced endometrial cancer have enlargement of the uterus, a lobular uterine contour, and mixed echogenicity of the

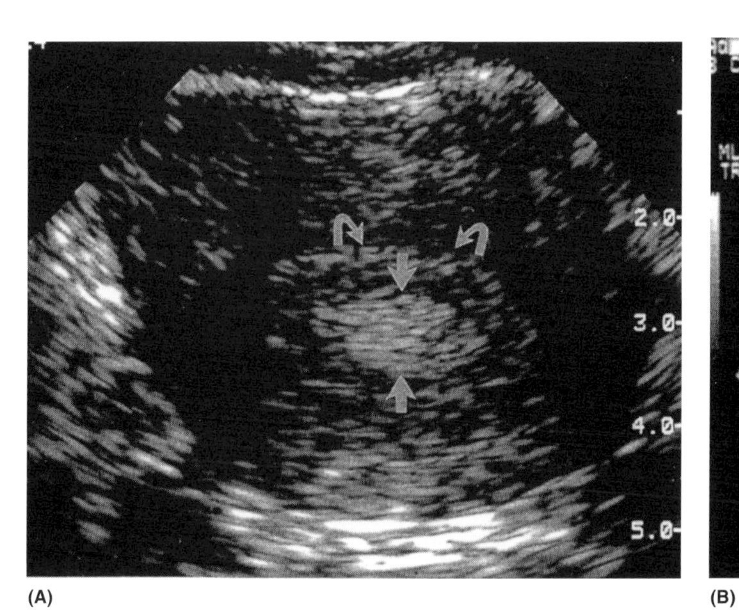

(A)

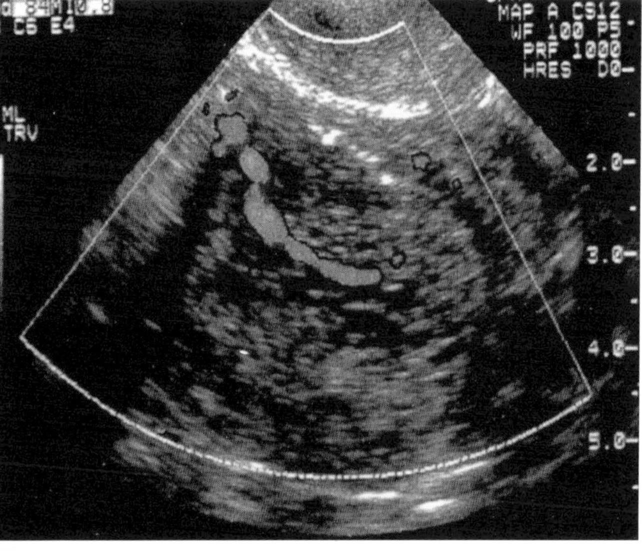

(B)

FIGURE 22 A AND B ■ Endometrial polyp. (**A**) Transvaginal transverse view of the uterus demonstrates a 1-cm echogenic mass (*straight arrows*) that is well defined and is slightly more echogenic than the surrounding endometrium (*curved arrows*). (**B**) A heterogeneous cystic-appearing endometrium with a single vessel entering centrally is consistent with a vascular pedicle of an endometrial polyp.

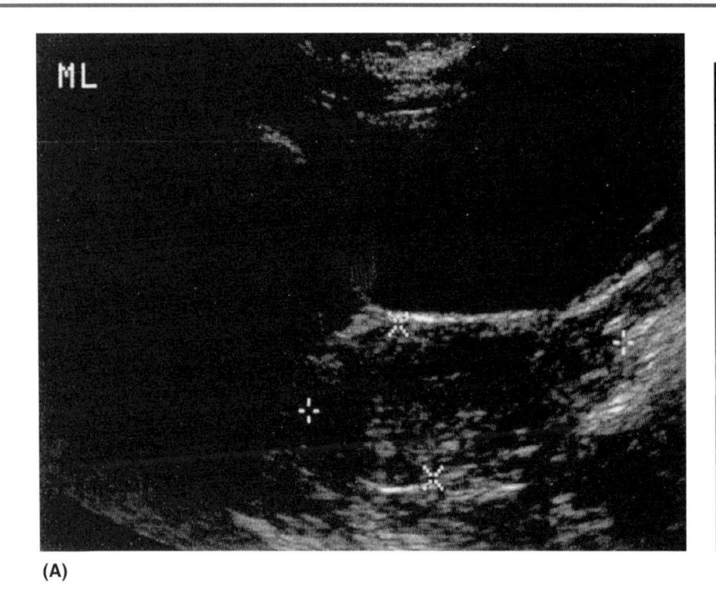

(A)

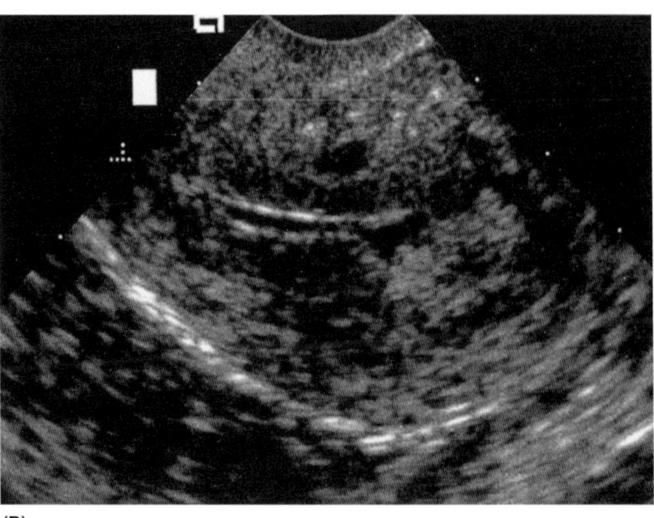

(B)

FIGURE 23 A AND B ■ Endometrial cancer. (**A**) Transabdominal view of the uterus demonstrates a normal-sized uterus in a 62-year-old woman with postmenopausal bleeding. (**B**) Vaginal examination in the same patient demonstrates an echogenic mass with invasion into the myometrium and a small amount of fluid within the endometrial cavity. Transvaginal examination is critical in the complete evaluation of postmenopausal bleeding. This was a grade 2–3 endometrial cancer.

myometrium (Fig. 20). The level of invasion (superficial vs. deep) can be assessed by transvaginal ultrasound (68). Contrast-enhanced MRI is generally regarded as the most efficacious single imaging test for staging once the diagnosis of endometrial carcinoma has been made (69,70).

Endometrial Fluid

A small amount of endometrial fluid with an otherwise benign-appearing endometrium is occasionally seen in normal postmenopausal women. These small fluid collections may be seen in women who are on a sequential hormone therapy regimen during withdrawal bleeding; in other women, they may be secondary to a minor degree of cervical stenosis. Larger amounts of fluid with a normal endometrium are more clearly indicative of cervical stenosis, due to benign or malignant causes (Fig. 24) (30,31). Endometrial fluid is also associated with malignant endometrial and cervical disease, however, so the surrounding myometrium and the cervix should be carefully evaluated. Endometrial irregularities or masses suggest focal hyperplasia, polyps, or cancer. Table 4 describes the appearance of endometrial fluid resulting from various causes.

Cervical Masses

As previously mentioned, nabothian cysts are the most common cervical masses in postmenopausal women. These cysts usually are anechoic, but they may have internal debris (Fig. 16). Solid cervical masses are either due to fibroids or due to malignant disease. Cervical cancer typically is hypoechoic with ill-defined margins, whereas fibroids present as a discrete mass with well-defined margins. Cervical cancer frequently causes obstruction and hydro- or hematometra as well as vaginal bleeding. A list of cervical masses is given in Table 5.

Ovaries

Because ovarian size decreases after menopause, the distinction between a small solid ovarian mass and a normal postmenopausal ovary is sometimes difficult. When this question arises, the examiner should visualize the contralateral ovary. An ovary twice the size of a contralateral ovary should be regarded with suspicion.

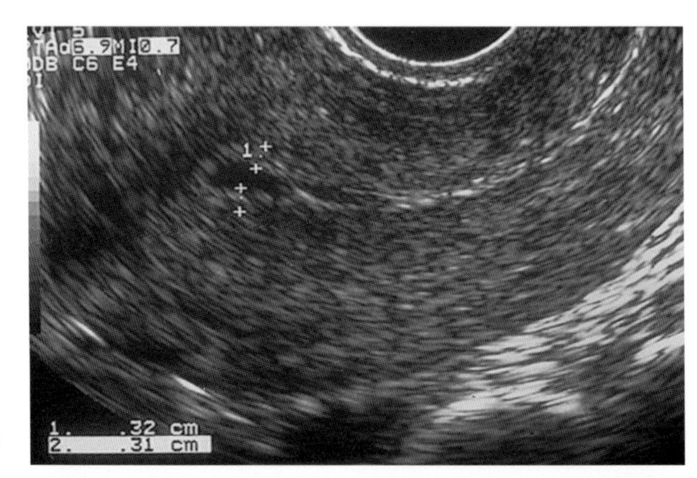

FIGURE 24 ■ Endometrial fluid. Transvaginal sagittal view of the uterus in a 65-year-old woman who used sequential hormones and who had not recently experienced withdrawal bleeding. A large amount of endometrial fluid is seen, and the surrounding endometrium is thin and well defined. The collection of fluid was secondary to cervical stenosis. *Source*: From Ref. 71.

TABLE 4 ■ Postmenopausal Endometrial Fluid

Diagnosis	Appearance	History
Normal	<2 mL Thin endometrium	
Recent dilation and curettage	±Irregular endometrium	
Infection	±Irregular endometrium	Other signs or symptoms of infection Recent instrumentation
Cervical stenosis	Thin endometrium surrounding fluid	Previous childbirth, cervical surgery (cone biopsy, cerclage) Previous radiation therapy
Cervical cancer	Irregular mass in cervix	
Endometrial hyperplasia	Focal or diffuse thickening ±Cystic spaces	Abnormal bleeding
Endometrial polyp	±Vascular stalk	Abnormal bleeding
Endometrial cancer	Thick endometrium Loss of endometrial–myometrial interface	Abnormal bleeding
Uterine adhesions—lower segment	Distorted cavity	Previous myomectomy Endometrial ablation

Solid adnexal masses that occur in postmenopausal women include fibromas, thecomas, and metastatic tumors (Table 6).

The incidence of ovarian cancer increases with age. Eighty percent of affected women are older than 50 years of age (36). The sonographic appearance of an adnexal mass remains most important in distinguishing between benign and malignant lesions, however. An anechoic cyst with a thin wall and enhanced fluid transmission is likely to be benign in a postmenopausal woman, regardless of the size of the cyst (45,46). Mural nodularity and septations increase the likelihood of malignancy (Fig. 25). Dermoids, although more common in women of menstrual age, also occur in postmenopausal women. The likelihood of malignant degeneration of a dermoid increases with the patient's age (Fig. 26). Bilateral solid adnexal masses are probably fibromas, ovarian cancer, or metastatic disease (Fig. 27). Postmenopausal adnexal masses are listed in Table 6.

Fallopian Tube

Hydrosalpinx and pyosalpinx occur infrequently in postmenopausal women (Fig. 28). Hydrosalpinx is visualized as a dilated, serpiginous elongated cystic structure between the uterus and the ovary. Color Doppler is helpful in distinguishing hydrosalpinx from adjacent vessels, especially pelvic varices. Hydrosalpinx in the postmenopausal patient is often asymptomatic and is commonly related to

TABLE 5 ■ Postmenopausal Cervical Mass

Diagnosis	Findings
Cervical fibroid	Typically hypoechoic Well-defined margins ±Calcifications ±Areas of cystic or echogenic degeneration
Cervical cancer	±Ill-defined margins
Nabothian cyst	Usually <2 cm Adjacent to endocervical canal Incidence increases with age
Polyp	Centrally located within the canal ±Stalk May be arising from the uterus

TABLE 6 ■ Postmenopausal Adnexal Mass

Sonographic appearance	Types of lesions	Comments
Anechoic cyst	Ovarian cyst	Likely benign
	Paraovarian cyst	If large, more likely neoplastic
Complex cyst	Cystadenoma Cystadenocarcinoma Dermoid	Nodularity and septation increase likelihood of malignancy
	Fallopian tube cancer	Complex mass
Solid mass	Fibroma, thecoma Undifferentiated carcinoma Metastatic disease Inflammatory (secondary to adjacent appendicitis or diverticulitis)	

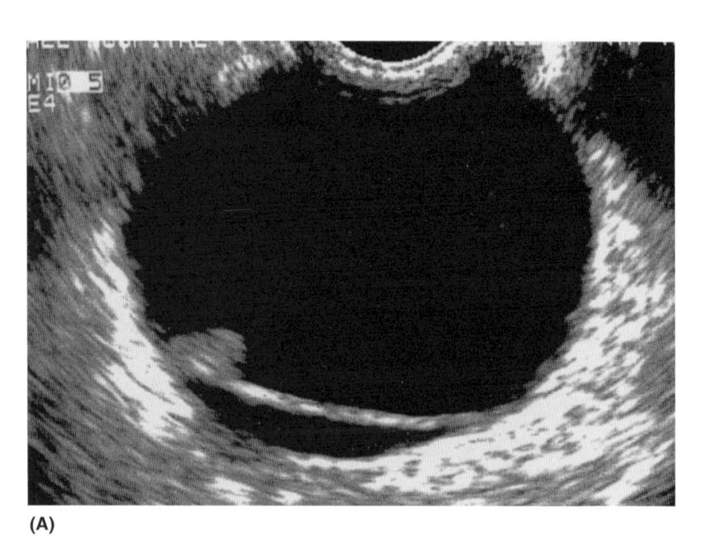
(A)

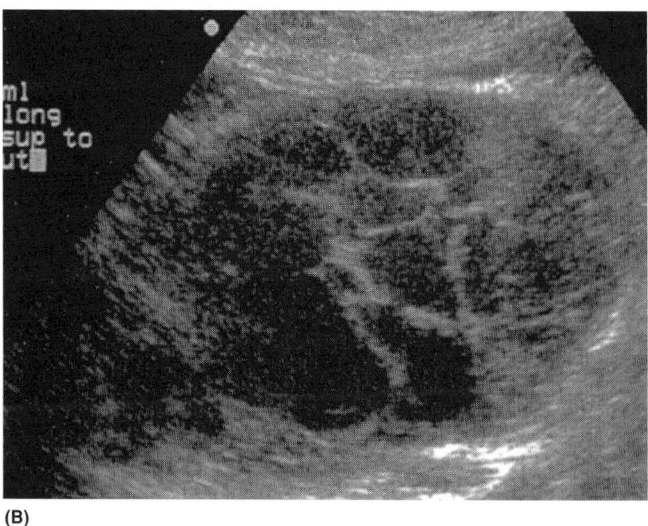

(B)

FIGURE 25 A AND B ■ (**A**) A 6-cm cyst with thin septation in a single solid mural nodule in a 52-year-old woman. This was a papillary serous cystadenoma of low malignant potential. (**B**) An 8-cm multiseptated cystic mass in a 65-year-old woman. This was a serous cystadenocarcinoma.

prior pelvic inflammatory disease or operative procedures such as hysterectomy or tubal ligation.

Postmenopausal tubo-ovarian abscess accounts for 2% of all such abscesses. It is most common in patients with recent instrumentation (dilation and curettage) and malignant disease (both gynecologic and from the adjacent gastrointestinal tract) (72). From 15% to 18% of postmenopausal women with tubo-ovarian abscess also have diabetes (73,74). Unlike younger women, there is a relatively high incidence of concomitant genital tract malignancies (74).

Vaginal Cuff

In patients studied after hysterectomy, attention should be directed to imaging the vaginal cuff, especially when the hysterectomy was performed for malignant disease, as this is a common site for recurrence. This structure is usually small (less than 2.1 cm) (75) and homogeneous (Fig. 29). If the region of the vaginal cuff appears prominent, this may be due to the presence of the cervix if only a supracervical hysterectomy was performed. Alternatively, an enlarged or irregular appearance to the vaginal cuff should suggest the possibility of locally recurrent malignant disease or peritoneal metastatic deposits to this region. Postradiation fibrosis or adjacent inflammatory processes (such as

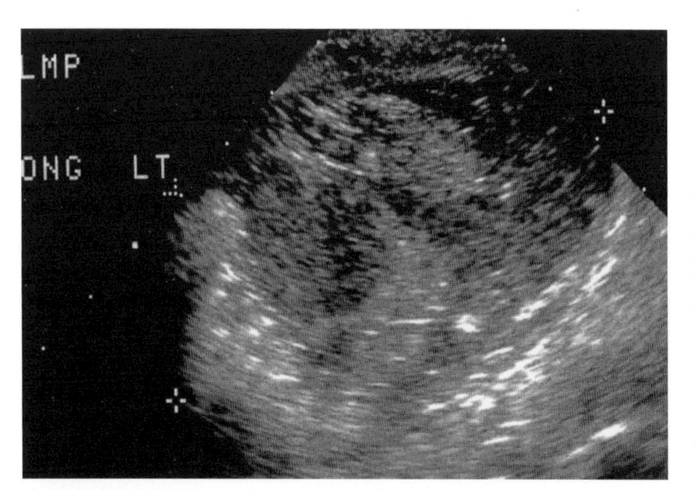

FIGURE 26 ■ A 7-cm heterogeneous left adnexal mass in a 65-year-old woman. This mass has linear bright echoes and echogenic mural nodules suggestive of a dermoid. Histologically, the diagnosis was confirmed. No evidence of malignant degeneration was seen.

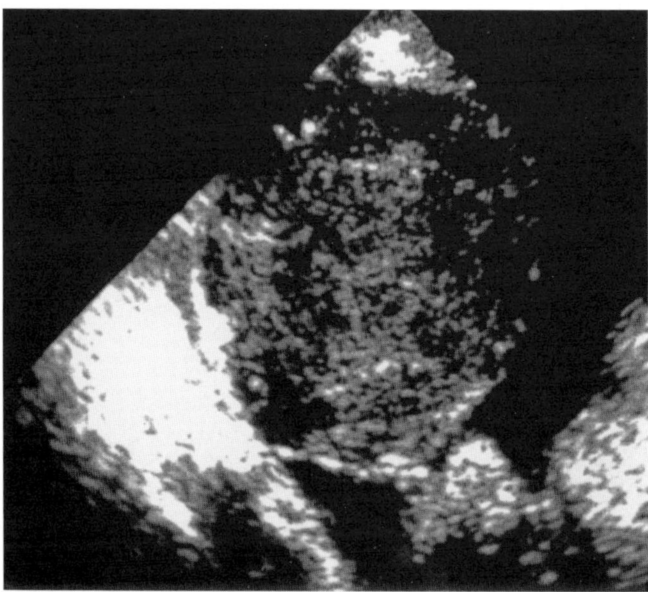

FIGURE 27 ■ A 5-cm solid right adnexal mass in a 78-year-old woman with pancreatic cancer and visible ascites. This patient had metastatic disease.

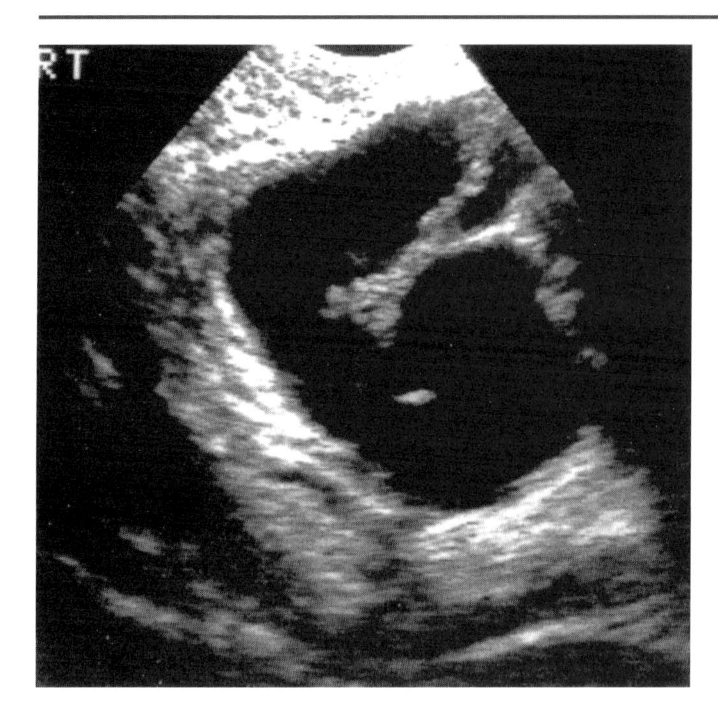

FIGURE 28 ■ Transvaginal sagittal view of a hydrosalpinx in a 65-year-old postmenopausal woman.

diverticulitis) also can cause nodular areas to form within the vaginal cuff (75).

Free Fluid

Small amounts of free fluid are occasionally seen in the cul-de-sac of asymptomatic postmenopausal women; however, a large amount of fluid is associated with various disease processes including malignant tumors. Fluid with septations indicates malignant disease, hemorrhage, infection, or encysted fluid trapped by adhesions from previous surgery or inflammation (peritoneal inclusion cysts).

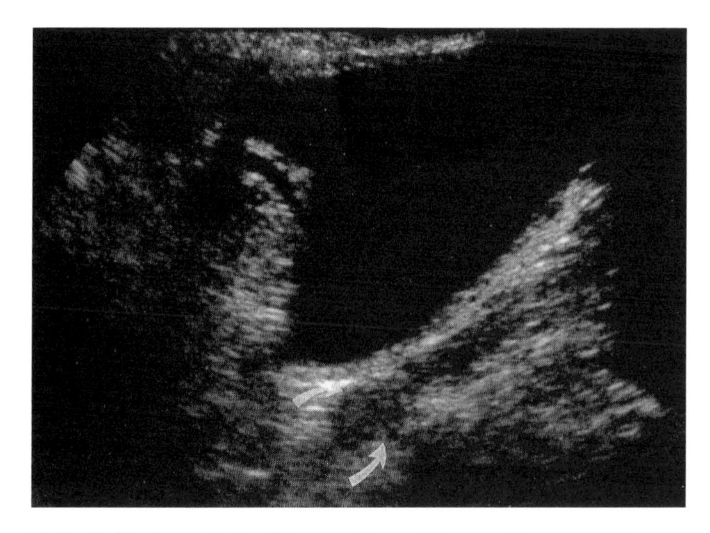

FIGURE 29 ■ Vaginal cuff (*arrows*) in an 80-year-old woman 15 years after hysterectomy.

REFERENCES

1. Lobo RA. Benefits and risks of estrogen replacement therapy. Am J Obstet Gynecol 1995; 173:982–989.
2. Fleischer A, Gordon A, Entman S, et al. Transvaginal Sonography (TVS) of the endometrium: current and potential clinical applications. Crit Rev Diagn Imaging 1990; 30:85–110.
3. Platt JF, Bree RL, Davidson D. Ultrasound of the normal nongravid uterus: correlation with gross and histopathology. J Clin Ultrasound 1990; 18:15–19.
4. American Cancer Society. Cancer Facts and Figures: 2003. Atlanta (GA): ACS, 2003.
5. Platz C, Benda J. Female genital tract cancer. Cancer 1995; 75:270–294.
6. Watson P, Vasen HF, Mecklin JP, et al. The risk of endometrial cancer in hereditary nonpolyposis colorectal cancer. Am J Med 1994; 96:516–520.
7. Smith-Bindman R, Weiss E, Feldstein V. How thick is too thick? When endometrial thickness should prompt biopsy in postmenopausal women without vaginal bleeding. Ultrasound Obstet Gynecol 2004; 24:558–565.
8. Bruchim I, Biron-Shental T, Altaras MM, et al. Combination of endometrial thickness and time since menopause in predicting endometrial cancer in women with postmenopausal bleeding. J Clin Ultrasound 2004; 32:219–224.
9. Granberg S, Wikland M, Karisson B, et al. Endometrial thickness as measured by endovaginal ultrasonography for identifying endometrial abnormality. Am J Obstet Gynecol 1991; 164:47–52.
10. Bree RL, Bowerman RA, Bohm-Velez M, et al. US evaluation of the uterus in patients with post menopausal bleeding: a positive effect on diagnostic decision making. Radiology 2000; 216:260–264.
11. Goldstein SR, Zeltser I, Horan CK, et al. Ultrasonography-based triage for perimenopausal patients with abnormal uterine bleeding. Am J Obstet Gynecol 1997; 177:102–108.
12. O'Connell LP, Fries MH, Aeringue E, et al. Triage of abnormal postmenopausal bleeding: a comparison of endometrial biopsy and transvaginal sonohysterography versus fractional curettage with hysteroscopy. Am J Obstet Gynecol 1998; 178:956–961.
13. Buyuk E, Durmusoglu F, Erenus M, et al. Endometrial disease diagnosed by transvaginal ultrasound and dilation and curettage. Acta Obstet Gynecol Scand 1999; 78:419–422.
14. Dijkhuizen FP, Mol BW, Brolmann HA, et al. The accuracy of endometrial sampling in the diagnosis of patients with endometrial cancer and hyperplasia. Cancer 2000; 89:1765–1772.
15. Akkad AA, Habiba MA, Ismail N, et al. Abnormal uterine bleeding on hormone replacement: the importance of intrauterine structural abnormalities. Obstet Gynecol 1995; 86:330–334.
16. Rossouw JE, Anderson GL, Prentice RL, et al. Risks and benefits of estrogen plus progestin in healthy postmenopausal women. Principal results from the Women's Health Initiative randomized controlled trial. JAMA 2002; 288:321–333.
17. Barrett-Connor E, Hendrix S, Ettinger B, et al. Best clinical practices: a comprehensive approach. In: Wenger NK, ed. International Position Paper on Women's Health and Menopause: A Comprehensive Approach, National Institutes of Health Publication No.: 02-3284, July 2002.
18. Anderson GL, Limacher M, Assaf AR, et al. Effects of conjugated equine estrogen in postmenopausal women with hysterectomy. JAMA 2004; 291:1701–1712.
19. The Practice Committee of the American Society for Reproductive Medicine. Estrogen and progestogen therapy in postmenopausal women. Menopausal Med 2004; 12:7–16.
20. The writing group for the PEPI trial. Effects of hormone replacement therapy on endometrial histology in postmenopausal women: the postmenopausal estrogen-progestin interventions trial. JAMA 1996; 275:370–375.
21. Pickar JH, Yeh I, Wheeler JE, et al. Endometrial effects of lower doses of conjugated equine estrogens and medroxyprogesterone acetate. Fertil Steril 2001; 76:25–31.

22. Baker VL. Alternatives to oral estrogen replacement. Transdermal patches, percutaneous gels, vaginal creams and rings, implants, other methods of delivery. Obstet Gynecol Clin North Am 1994; 21:271–297.
23. Magos A, Brincat M, Studd J, et al. Amenorrhea and endometrial atrophy with continuous oral estrogen and progestogen therapy in postmenopausal women. Obstet Gynecol 1985; 65:496–499.
24. Levine D, Gosink B, Johnson L. Change in endometrial thickness in postmenopausal women on hormone replacement therapy. Radiology 1995; 197:603–608.
25. Omodei U, Ferrazzia E, Ruggeri C, et al. Endometrial thickness and histological abnormalities in women on hormonal replacement therapy: a transvaginal ultrasound/hysteroscopic study. Ultrasound Obstet Gynecol 2000; 15:317–320.
26. Osborne CK. Tamoxifen in the treatment of breast cancer. N Engl J Med 1998; 339:1609–1618.
27. Van Leeuwen FE, Benraadt TH, Coebergh JW, et al. Risk of endometrial cancer after tamoxifen treatment of breast cancer. Lancet 1994; 343:448–452.
28. Cohen I, Beth Y, Tepper R, et al. Adenomyosis in postmenopausal breast cancer treated with tamoxifen: a new entity? Gynecol Oncol 1995; 58:86–91.
29. Elhelw B, Ghorab MN, Farrag SH. Saline sonohysterography for monitoring asymptomatic postmenopausal breast cancer patients taking tamoxifen. Int J Gynecol Obstet 1999; 67:81–86.
30. Goldstein SR. Postmenopausal endometrial fluid collections revisited: look at the doughnut rather than the hole. Obstet Gynecol 1994; 83:738–740.
31. Vuento MH, Pirhonen JP, Makinen JI, et al. Endometrial fluid accumulation in asymptomatic postmenopausal women. Ultrasound Obstet Gynecol 1996; 8:37–41.
32. Bourne TH, Campbell S, Steer CV, et al. Detection of endometrial cancer by transvaginal ultrasonography with color flow imaging and blood flow analysis: a preliminary report. Gynecol Oncol 1991; 40:253–259.
33. Develioglu OH, Bilgin T, Yalcin OT, et al. Transvaginal ultrasonography and uterine artery Doppler in diagnosing endometrial pathologies and carcinomas in postmenopausal bleeding. Arch Gynecol Obstet 2003; 268:175–180.
34. American Cancer Society. Cancer Facts and Figures 2005. Atlanta: American Cancer Society, 2005:15–16.
35. Weiss N, Homonchuk T, Young J. Incidence of the histologic types of ovarian cancer: the U.S. Third National Cancer Survey, 1969–1971. Gynecol Oncol 1977; 5:161.
36. Bandera CA. Advances in the understanding of risk factors for ovarian cancer. J Reprod Med 2005; 50:399–406.
37. Fung MF, Bryson P, Johnston M, et al. Screening postmenopausal women for ovarian cancer: a systematic review. J Obstet Gynaecol Can 2004; 26:717–728.
38. Fishman DA, Cohen L, Blank SV, et al. The role of ultrasound evaluation in the detection of early-stage epithelial ovarian cancer. Am J Obstet Gynecol 2005; 192:1214–1221.
39. Goswamy RK, Campbell S, Royston JP, et al. Ovarian size in postmenopausal women. Br J Obstet Gynaecol 1988; 95:795.
40. Tepper R, Zalel Y, Markov S, et al. Ovarian volume in postmenopausal women—suggestions to an ovarian size nomogram for menopausal age. Acta Obstet Gynecol Scand 1995; 74:208–211.
41. Zalel Y, Tepper R, Altaras M, et al. Transvaginal sonographic measurements of postmenopausal ovarian volume as a possible detection of ovarian neoplasia. Acta Obstet Gynecol Scand 1996; 75:668–671.
42. Rodriguez MH, Platt LD, Medearis AL, et al. The use of transvaginal sonography for evaluation of postmenopausal ovarian size and morphology. Am J Obstet Gynecol 1988; 159:810–814.
43. Wolf SI, Gosink BB, Feldesman MR, et al. Prevalence of simple adnexal cysts in postmenopausal women. Radiology 1991; 180: 65–71.
44. Levine D, Gosink BB, Wolf SI, et al. Simple adnexal cysts: the natural history in postmenopausal women. Radiology 1992; 184: 653–659.
45. Modesitt SC, Pavlik EJ, Ueland FR, et al. Risk of malignancy in unilocular ovarian cystic tumors less than 10 centimeter in diameter. Obstet Gynecol 2003; 102:594–599.
46. Nardo LG, Kroon ND, Reginald PW. Persistent unilocular ovarian cysts in a general population of postmenopausal women: is there a place for expectant management? Obstet Gynecol 2003; 102:589–593.
47. Fleischer A, Rodgers W, Rao B, et al. Assessment of ovarian tumor vascularity with transvaginal color Doppler sonography. J Ultrasound Med 1991; 10:563–568.
48. Botsis D, Kassanos D, Kalogirou D, et al. Transvaginal color Doppler and CA 125 as tools in the differential diagnosis of postmenopausal masses. Maturitas 1997; 26:203–209.
49. Salem S, White LM, Lai J. Doppler sonography of adnexal masses: the predictive value of the pulsatility in benign and malignant disease. AJR Am J Roentgenol 1994; 163:1147–1150.
50. Stein SM, Laifer-Narin S, Johnson MB, et al. Differentiation of benign and malignant adnexal masses: relative value of grey-scale, color Doppler and spectral Doppler sonography. AJR Am J Roentgenol 1995; 164:381–386.
51. Kinkel K, Hricak H, Lu Y, et al. US characterization of ovarian masses: a meta-analysis. Radiology 2000; 217:803–811.
52. Jacobs I, Bast RC. The CA-125 tumour-associated antigen: a review of the literature. Hum Reprod 1989; 4:1–12.
53. Zurawski VR, Knapp RC, Einhorn N, et al. An initial analysis of preoperative serum Ca 125 levels in patients with early stage ovarian carcinoma. Gynecol Oncol 1988; 30:7–14.
54. Einhorn N, Sjovall K, Knapp RC, et al. Prospective evaluation of serum CA-125 levels for early detection of ovarian cancer. Obstet Gynecol 1992; 80:14–18.
55. Garner EIO. Advances in early detection of ovarian carcinoma. J Reprod Med 2005; 50:447–453.
56. Dubinsky T, Parvey R, Gormaz G, et al. Transvaginal hysterosonography: comparison with biopsy in the evaluation of postmenopausal bleeding. J Ultrasound Med 1995; 14:887–893.
57. Van den Bosch T, Vandendael A, Van Schoubroeck D, et al. Combining vaginal ultrasonography and office endometrial sampling in the diagnosis of endometrial disease in postmenopausal women. Obstet Gynecol 1995; 85:349–352.
58. Epstein E, Ramirez A, Skoog L, et al. Dilatation and curettage fails to detect most focal lesions in the endometrial cavity in women with postmenopausal bleeding. Acta Obstet Gynecol 2001; 80: 1131–1136.
59. Ohara N. Selective estrogen receptor modulator and selective progesterone receptor modulator: therapeutic efficacy in the treatment of uterine leiomyoma. Clin Exp Obstet Gynecol 2005; 32:9–11.
60. Aviram R, Ochshorn Y, Markovitch O, et al. Uterine sarcomas versus leiomyomas: grey scale and Doppler sonographic findings. J Clin Ultrasound 2005; 33:10–13.
61. Reinhold C, McCarthy S, Bret PM, et al. Diffuse adenomyosis: comparison of endovaginal ultrasound and MR imaging with histopathologic correlation. Radiology 1996; 199:151–158.
62. Occhipinti K, Kutcher R, Rosenblatt R. Sonographic appearance and significance of arcuate artery calcification. J Ultrasound Med 1991; 10:97–100.
63. Duffield C, Gerscovich EO, Gillen MA, et al. Endometrial and endocervical micro echogenic foci: sonographic appearance with clinical and histologic correlation. J Ultrasound Med 2005; 24:583–590.
64. Montgomery BE, Daumgs GS, Dunton CJ. Endometrial hyperplasia: a review. Obstet Gynecol Surv 2004; 59:368–378.
65. Woodruff J, Pickar J. Incidence of endometrial hyperplasia in postmenopausal women taking conjugated estrogens (Premarin) with medroxyprogesterone acetate or conjugated estrogen alone. Am J Obstet Gynecol 1994; 170:1213–1222.

66. Sheth S, Hamper U, Kurman R. Thickened endometrium in the postmenopausal woman: sonographic-pathologic correlation. Radiology 1993; 187:135–139.
67. Hulka C, Hall D, McCarthy K, et al. Endometrial polyps, hyperplasia, and carcinoma in postmenopausal women: differentiation with endovaginal sonography. Radiology 1994; 191:755–758.
68. Yazbeck C, Poncelet C, Crequat J, et al. Preoperative endovaginal ultrasound in the assessment of myometrial invasion of endometrial adenocarcinoma. Gynecol Obstet Fertil 2003; 31:1024–1029.
69. Kinkel K, Kaji Y, Yu KK, et al. Radiologic staging in patients with endometrial cancer: a meta-analysis. Radiology 1999; 212:711–718.
70. Akaeda T, Isaka K, Takayama M, et al. Myometrial invasion and cervical invasion by endometrial carcinoma: evaluation by CO2-volumetric interpolated breathhold examination (VIBE). J Magn Reson Imaging 2005; 21:166–171.
71. Levine D. The postmenopausal pelvis. In: Nyberg D, ed. Transvaginal Ultrasound. St Louis: Mosby-Year Book, 1992:228.
72. Kremer S, Kutcher R, Rosenblatt R, et al. Postmenopausal tubo-ovarian abscess: sonographic considerations and clinical significance. J Ultrasound Med 1992; 11:613–616.
73. Hoffman M, Molpus K, Roberts W, et al. Tubo-ovarian abscess in postmenopausal women. J Reprod Med 1990; 35:525–528.
74. Jackson SL, Soper DE. Pelvic inflammatory disease in postmenopausal women. Infect Dis Obstet Gynecol 1999; 7:248–252.
75. Schoenfeld A, Levavi H, Hirsch M, et al. Transvaginal sonography in post-menopausal women. J Clin Ultrasound 1990; 18:350–358.

First Trimester Ultrasound

Clifford S. Levi, Edward A. Lyons, and Sidney M. Dashefsky

39

INTRODUCTION

During the first trimester, the conceptus progresses from a single-celled zygote to a fetus. Major changes occur during the intervening period, during which time the corresponding sonographic appearances depend on the stage of development and size of the conceptus. Most examinations are requested because the patient has presented with vaginal bleeding or pelvic pain, or a palpable mass has been identified on physical examination. The goals of a sonographic examination should be clinically relevant and appropriate to the developmental stage (1).

Early in the first trimester, the major clinical concerns of the referring physician are the following:

1. What is the site of implantation? (Is the pregnancy intrauterine or ectopic?)
2. Is the embryo-fetus alive?
3. What is the likelihood of subsequent demise of a live embryo-fetus?

The primary goals of a first trimester ultrasound examination should be to address these concerns.

Endovaginal sonography (EVS) has been widely available since the mid-1980s and has radically altered sonographic practice in the first trimester. The improved resolution of EVS over transvesical sonography (TVS) has resulted in earlier and more accurate diagnosis of ectopic pregnancy (2) and embryonic demise (3,4). EVS has also reduced the need for serial sonography in first trimester diagnosis. The diagnosis of early pregnancy failure based on a single ultrasound examination is, however, a controversial issue (5,6). Often, more than one examination is necessary to diagnose early pregnancy failure with certainty.

Current practice is based on the use of reliable sonographic indicators of ectopic pregnancy and embryonic demise. The accuracy of some of the sonographic signs used as indicators of the presence of a live embryo or of embryonic demise is contingent on the use of a high-resolution scanner (7) and on the expertise of the sonographer or sonologist. Published values in the literature based on data acquired using high frequency transducers (3,4,8) cannot simply be applied to lower-resolution 5.0 MHz transducers without prior verification (9). The endovaginal sonographic signs published in this article assume the use of modern equipment with a transducer frequency of at least 7 to 8 MHz.

Endovaginal color flow Doppler (EVCFD) became available in the early 1990s. Some authors suggest that EVCFD provides improved diagnostic accuracy over EVS in the identification of early intrauterine and ectopic pregnancies and may allow more definitive diagnoses to be made at the initial sonographic examination (10–13). EVCFD may also play a role in the diagnosis of early pregnancy failure (14).

In our practice, sonographic diagnosis is usually based on comparison of a sonographic finding with an internal sonographic control or with biochemical data. For example, nonvisualization of the secondary yolk sac at a specific mean gestational sac diameter (MSD) is suggestive of early pregnancy failure.

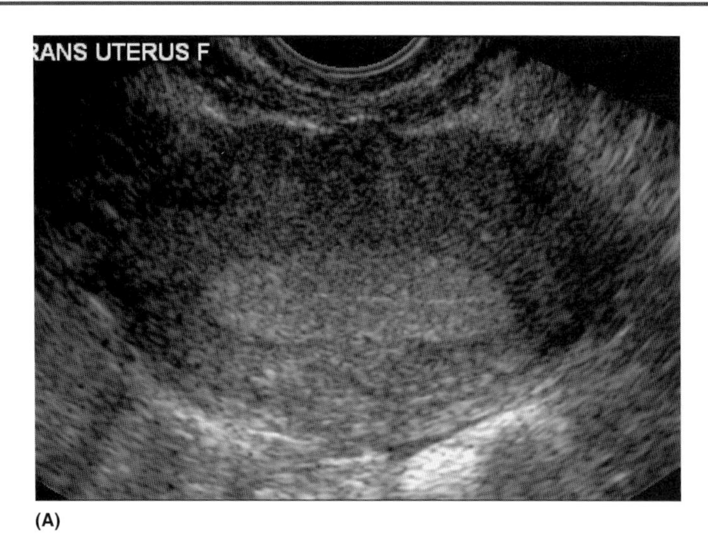

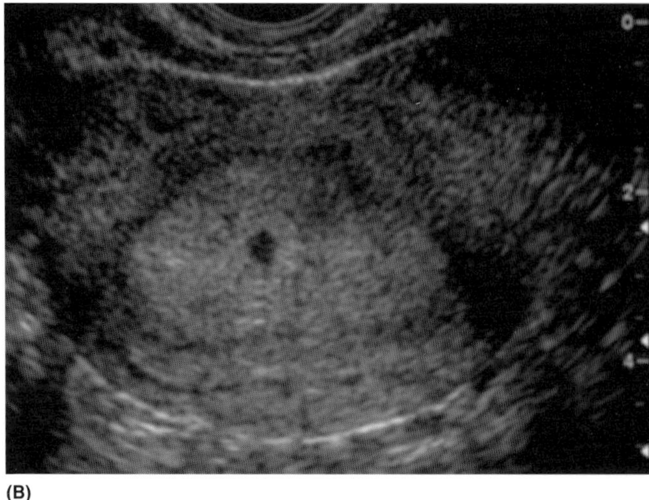

(A) (B)

FIGURE 1 A AND B ■ Unreliable menstrual history in normal pregnancy. The initial exam (**A**) was done at six weeks' menstrual age but no sac is seen on this axial scan of the uterus. One week later the patient felt she was seven weeks by last menstrual period but the scan (**B**) shows only a four-week sac.

In general, although the patient's menstrual history is helpful in guiding management, it is not used as a major determinant of early pregnancy failure because of variability in the time of ovulation and because the menstrual history may be unreliable (Fig. 1). In some patients, however, the menstrual age can provide information instrumental in decision making in first trimester diagnosis. This group includes patients in whom the gestational age is known with certainty because of ovulation induction or in vitro fertilization. In other patients, a minimum gestational age can be established by extrapolation based on a history of a previous positive serum β-human chorionic gonadotropin (β-hCG) concentration.

Although decisions regarding embryonic demise and ectopic pregnancy are usually made based on internal sonographic controls rather than on menstrual history, the patient's menstrual history provides useful information and should not be disregarded in first trimester ultrasound diagnosis. A significant discrepancy between the patient's menstrual history and the gestational age as predicted by sonographic landmarks or measurements may be the first indicator of early pregnancy failure or first trimester growth retardation (15).

Other important goals of first trimester diagnosis include the following:

- Assessment of gestational age
- Determination of the number of embryos and assessment of chorionicity and amnionicity of multiembryonic pregnancy
- Detection of embryonic-fetal anomalies
- Assessment of uterine or adnexal masses

EMBRYOLOGY PERTINENT TO EARLY SONOGRAPHIC FINDINGS

To understand the normal and abnormal sonographic findings in the first trimester, it is necessary to review basic embryology. Only a brief discussion of embryology is provided in this section, emphasizing embryologic facts pertinent to corresponding sonographic findings. For those who wish a more detailed discussion, most of the information provided in this section is referenced from *The Developing Human: Clinically Oriented Embryology* by K.L. Moore and T.V.N. Persaud (16–21).

All dates presented in this chapter are in menstrual or gestational age in keeping with the obstetric and radiologic literature rather than embryologic age as used by embryologists.

NORMAL SONOGRAPHIC APPEARANCE

A brief outline of the normal sonographic landmarks in the early first trimester is provided in Table 1.

Gestational Sac

Implantation of the blastocyst is complete by day 23 of menstrual age (17). At that time, however, the entire conceptus measures 0.1 mm and is beyond the resolution of current ultrasound equipment. The earliest sonographic sign of intrauterine pregnancy was described by Yeh et al. (23), who identified a focal echogenic zone of decidual thickening at the site of implantation at 3.5 to 4 weeks' menstrual age. This sign may be difficult to appreciate, and the diagnostic value in terms of predicting the presence of an intrauterine pregnancy has not been published.

EVCFD may provide the first reliable evidence of the presence of an intrauterine pregnancy (12,13). The EVCFD diagnosis of intrauterine pregnancy is based on the demonstration of peritrophoblastic flow characterized by a high-velocity, low-impedance signal (10,11) in the spiral arteries, possibly related to vascular shunts in the peritrophoblastic myometrium (24). In early pregnancy, the syncytiotrophoblast invades and plugs the maternal spiral arteries and erodes the decidua to create lacunae which form the intervillous space. Maternal blood flow from the

TABLE 1 ■ Normal Endovaginal Sonographic Findings: 3.5 to 6.5 Weeks' Menstrual Age

Approximate menstrual age (wk)	Sonographic signs	Sonographic features	Comments
3.5–4	Decidual thickening	Focal thickening of the echogenic decidua at the site of implantation	Sign difficult to appreciate; predictive value never published
After 4.5	Peritrophoblastic flow	High-velocity, low-impedance signal at the implantation site	Peak velocity of 8–30 cm/sec prior to visualization of the gestational sac. Spectral and color Doppler may produce high acoustic intensities and the risk to the embryo should be considered
4.5–5.5	Intradecidual sign	Gestational sac within the decidua abutting the endometrial canal	Should always be seen when maternal serum β-hCG is ≥1700–2000 mIU/mL (First International Reference Preparation or Third International Standard, see text)
After 5.5	DDS	Echogenic ring formed by decidua capsularis/chorion laeve surrounded by echogenic decidua vera	DDS is diagnostic of an intrauterine pregnancy. Vague or absent sign is nondiagnostic (of either an intrauterine or ectopic pregnancy)
After 4.5	Yolk sac sign	Visualization of the yolk sac within the gestational sac	Yolk sac often seen when MSD is 5–6 mm and almost always seen when MSD is 8 mm
5.5	Double-bleb sign	Visualization of the amnion as a 2 mm bleb adjacent to the yolk sac	Transient finding; after this stage, visualization of the amnion in the absence of a visible embryo is abnormal
After 5.5 weeks	Visualization of the embryo	Visualization of the embryo adjacent to the yolk sac	Embryo almost always seen when the MSD is 16 mm
After 5.5 weeks	Cardiac activity	Cardiac activity within embryo immediately adjacent to the yolk sac	Nonvisualization of cardiac activity may be completely normal in embryos with 4–5 mm crown-rump length

Abbreviations: β-hCG, β-human chorionic gonadotropin; MSD, mean gestational sac diameter; DDS, double-decidual sign.
Source: From Ref. 22.

spiral arteries into the intervillous space is minimal at eight to nine weeks of gestation (25–27). Maternal blood flow from the spiral arteries results in intervillous circulation beginning in the periphery of the placenta at about nine weeks and spreading throughout the placenta after 12 weeks (25).

Emerson et al. found that the peak velocity of peritrophoblastic flow in a normal intrauterine pregnancy ranged from 8 to 30 cm per second before EVS visualization of the gestational sac, from 10 to 30 cm per second with a gestational sac of 1 to 5 mm MSD, and from 10 to 60 cm per second with gestational sacs of 6 to 10 mm MSD (13). In the series of Emerson et al., the sensitivity of diagnosis of intrauterine pregnancy was improved from 90% with EVS alone to 99% using EVCFD. The specificity of diagnosis of intrauterine pregnancy with EVCFD was 99% to 100% (13). In a separate study, EVCFD correctly identified 47 of 47 intrauterine pregnancies, compared with EVS, which identified 38 of 47 (81%) (28). The findings suggest that one can demonstrate peritrophoblastic flow with EVCFD before EVS demonstration of the gestational sac.

Color Doppler with a small region of interest or pulsed Doppler may generate high acoustic energies. In using Doppler in early pregnancy, it is important to adhere to the ALARA (as low as reasonably achievable) principle and to observe the thermal and mechanical indices (TI and MI). Exposure to Doppler should be brief, with the lowest possible acoustic output to achieve diagnostic images (29).

The first reliable gray-scale sonographic evidence of intrauterine pregnancy is visualization of the gestational sac (chorionic cavity surrounded by the trophoblast) within the thickened decidua (23). This sign, first described by Yeh et al. (23), is referred to as the intradecidual sign. Using EVS, it is usually possible to identify the gestational sac within the decidua by approximately 4.5 to 5 weeks' menstrual age, when the MSD should be approximately 2.5 mm (Fig. 4). Early data suggested that the earliest that the gestational sac could be visualized was at four weeks three days, and that it should always be visible by five weeks two days (31). A recent study with high-resolution EVS (7.5–10 MHz), demonstrated a gestational sac in all 67 patients scanned between 28 and 42 days menstrual age (32).

To distinguish an intrauterine gestational sac from a decidual cyst, the examiner must ensure that the sac abuts the endometrial canal. The gestational sac is round at this stage in pregnancy and becomes elliptical in shape as pregnancy progresses.

The double-decidual sign (33) is based on visualization of the gestational sac as an echogenic ring formed by the decidua capsularis and chorion laeve eccentrically located within the decidua vera (Figs. 5–7). The decidua basalis-villous chorion complex (future placenta) may also be visualized as an area of eccentric echogenic thickening. The double-decidual sign was originally described by Nyberg et al. (34) and can often be identified by about 5.5 to 6 weeks' menstrual age at approximately the same time that the yolk

Dates	Events
Weeks 3–4	
Day 14	Ovulation: postovulatory changes: (*i*) ovary: formation of the corpus hemorrhagicum/luteum; (*ii*) uterus: decidual reaction; functional layer of the endometrium becoming thick, soft, and edematous in response to progesterone
Day 14–15	Fertilization
Day 18	Morula stage
Day 20	Blastocyst stage; beginning of implantation (Fig. 2)
Day 22–23	Formation of the amniotic cavity and bilaminar embryonic disc
Day 23	Implantation complete; formation of primary yolk sac
Days 21–28	Formation of the extraembryonic coelom
Day 27–28	Formation of secondary yolk sac (Fig. 3); secondary rather than primary yolk sac identifiable by ultrasound (for the remainder of this chapter, "yolk sac" used to refer to the secondary yolk sac)
Days 21–28	Proliferation of syncytiotrophoblast (Fig. 2); lacunar network formed as maternal arterial and venous channels communicate with lacunae in the syncytiotrophoblast (Fig. 2); formation of primary chorionic villi by day 28; formation of the chorion from the syncytiotrophoblast, chorionic villi, and extraembryonic somatic mesoderm; formation of the chorionic cavity from the extraembryonic coelom
Week 5	
Day 29–30	Gastrulation: formation of the three primary germ layers: ectoderm, endoderm, and mesoderm; formation of primitive streak (mesenchyme) and notochord; notochord completed by end of fifth week
Days 31–42	Neurulation: formation of the neural plate and neural tube, which gives rise to the central nervous system; neural tube formation beginning by approximately day 35; closing of the rostral end of the neural tube (rostral neuropore) at day 40; closing of the caudal neuropore on day 42 (21); failure of closure of the neural tube resulting in neural tube defects
Days 34–44	Formation of somites
Day 35	During the fifth week, formation of primitive blood vessels; formation of the heart as paired tubes that fuse and pump blood by the end of the fifth week; at this stage the primitive cardiovascular system consists of the heart and a vascular network in the embryo, yolk sac, connecting stalk, and chorion; tertiary chorionic villi (with central vascular networks) are present; fetal blood in the tertiary villi is separated from maternal blood circulating in the intervillous spaces by capillary endothelium, mesenchyme, cytotrophoblast, and syncytiotrophoblast
Weeks 6–10: embryonic period (19,20)	
Weeks 6–10	Essentially all internal and external structures present in the adult form during this period *Cardiovascular system* ■ 6 wk: unidirectional blood flow ■ 8 wk: attainment by the heart of its definitive form ■ 10 wk: completion of development of the peripheral vascular system *Gastrointestinal system* ■ 6 wk: formation of the primitive gut ■ 8–12 wk: herniation of the midgut into the umbilical cord ■ 8 wk: separation of the rectum from the urogenital sinus ■ 10 wk: perforation of the anal membrane *Genitourinary system* ■ 8 wk: ascension of the metanephros, or primitive kidneys, from the pelvis ■ 11 wk: kidneys in adult position, external genitalia still in a sexless state; the external genitalia not in mature fetal form until the end of the 14th week *Musculoskeletal system* ■ 5.5–6 wk: formation of limb buds ■ 7.5–8 wk: digital rays distinct; upper limbs bent at elbow ■ 8 wk: center of ossification present in clavicle ■ 9 wk: center of ossification present in mandible and palate ■ 9 wk: vertebral center and neural arches beginning to ossify ■ 10–11 wk: frontal bone beginning to ossify ■ 11 wk: onset of ossification in all long bones beginning with femora and humeri

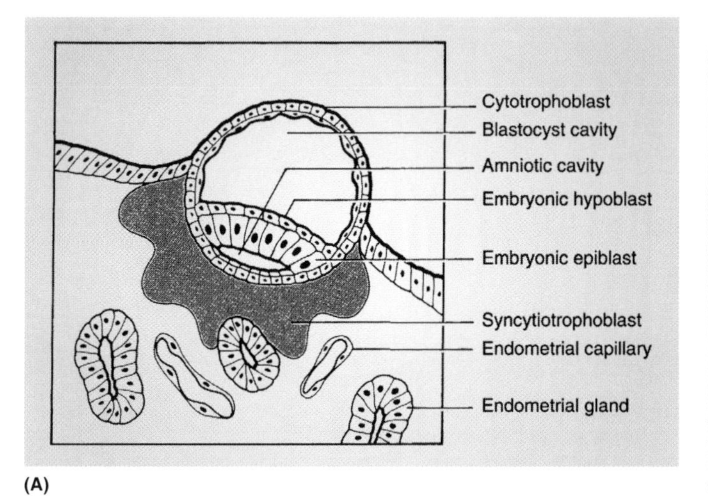

(A)

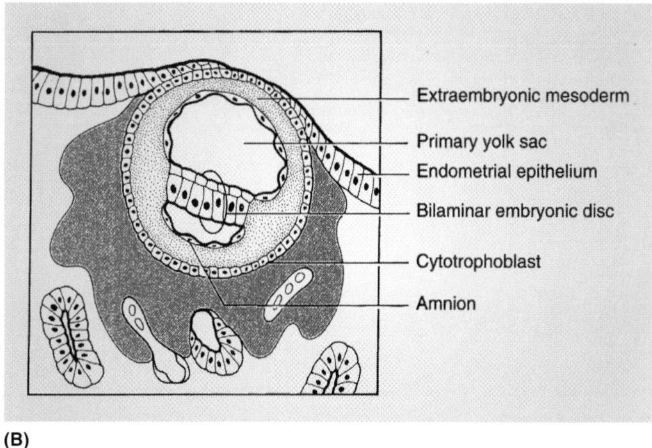

(B)

FIGURE 2 A AND B ■ Implantation of the blastocyst into the endometrium. The entire conceptus is approximately 0.1 mm at this stage. (**A**) Partially implanted blastocyst at approximately 22 days' menstrual age. (**B**) Almost completely implanted blastocyst at approximately 23 days' menstrual age. *Source*: Modified from Ref. 30.

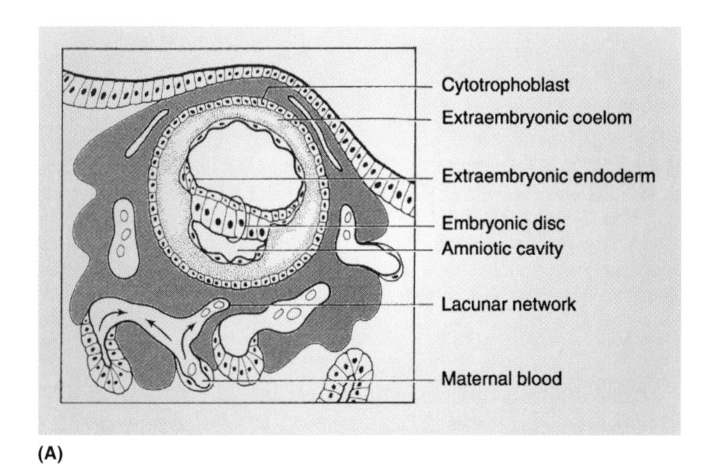

(A)

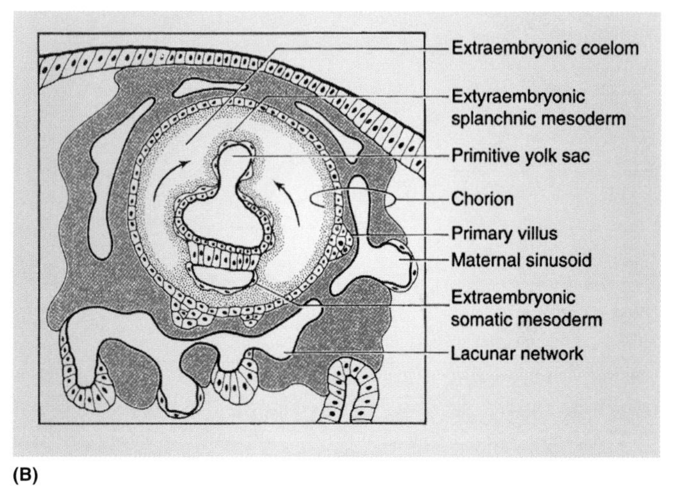

(B)

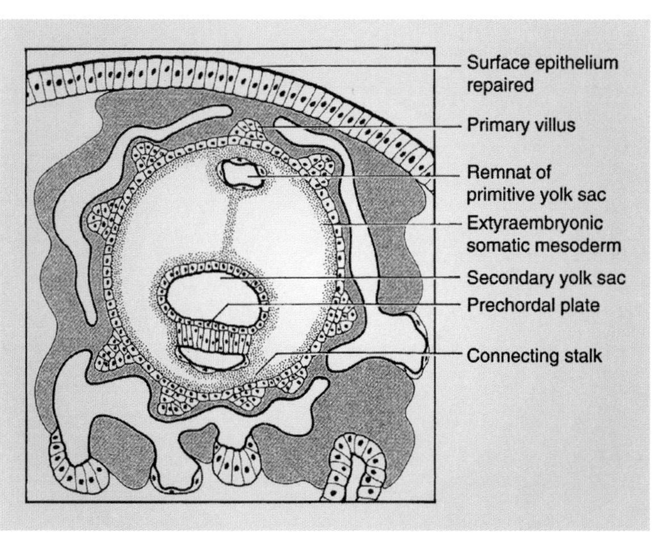

(C)

FIGURE 3 A–C ■ Formation of the secondary yolk sac. (**A**) Day 26 menstrual age. Formation of cavities within the extraembryonic mesoderm that coalesce to form the extraembryonic coelom. (**B** and **C**) As the extraembryonic coelom enlarges, the secondary yolk sac forms, and the primary yolk sac is pinched off and extruded. *Source*: From Ref. 30.

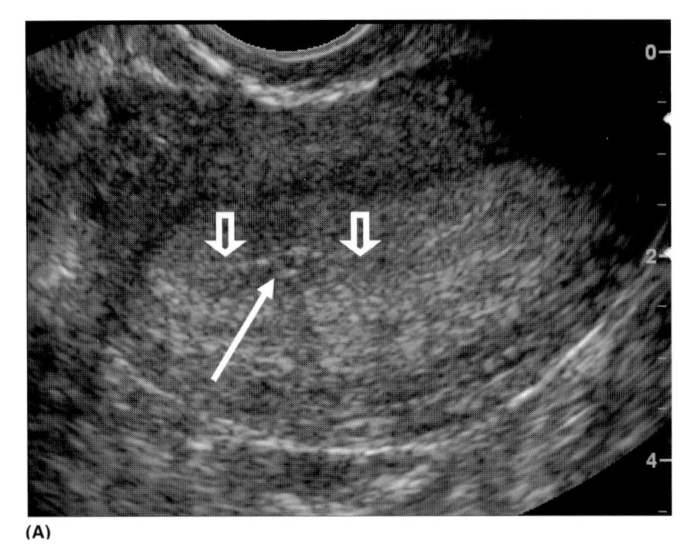

(A)

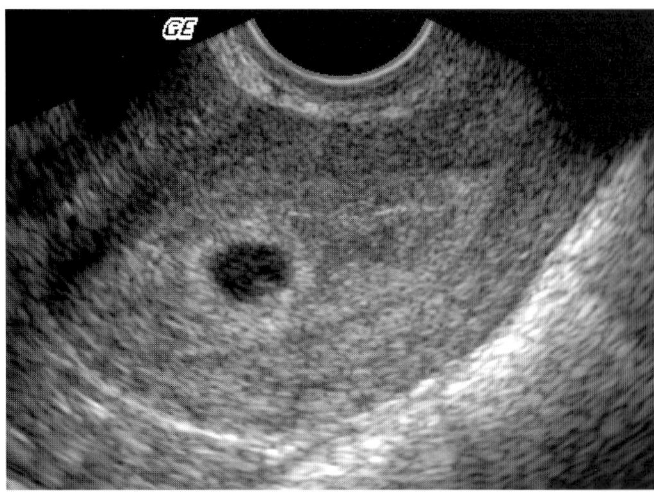
(B)

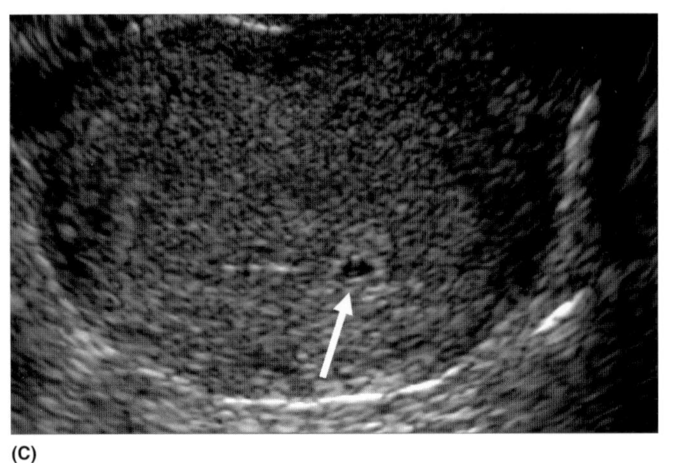
(C)

FIGURE 4 A–C ■ Intradecidual sign. Endovaginal sonogram in the sagittal plane in a patient at (**A**) 4.5 weeks' and (**B**) 5.5 weeks' menstrual age. The 3 mm gestational sac (*arrow*) is demonstrated within the decidua abutting and slightly displacing the echogenic endometrial canal (*open arrows*). An endovaginal scan in the coronal plane of another patient at 4.5 weeks with similar findings to **A**.

sac becomes visible with EVS (33). A well-defined double-decidual sign is accurate in predicting the presence of an intrauterine pregnancy. A vague or absent double-decidual ring may be seen in some patients with ectopic pregnancies and should be considered nondiagnostic (23).

The gestational sac can often be identified at low serum β-hCG levels. The *threshold level* (lowest β-hCG level at which one can identify a normal intrauterine gestational sac) has much less clinical significance than the *discriminatory level* (the β-hCG level above which it is

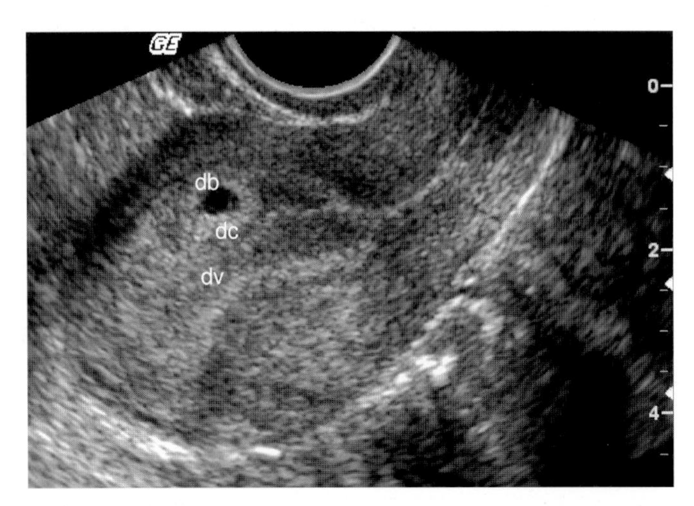

FIGURE 5 ■ Double-decidual sign (5 weeks' menstrual age). The decidua vera can be visualized as separate from the echogenic decidua capsularis and chorion laeve of the gestational sac. dv, decidua vera; dc, decidua capsularis; db, decidua basalis.

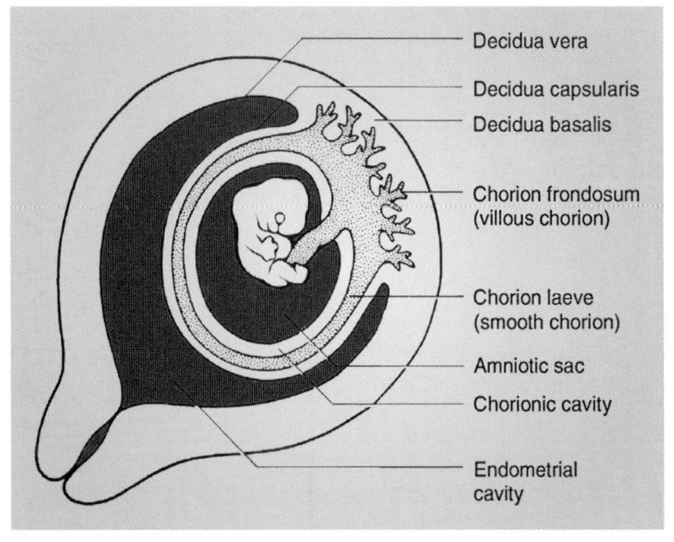

FIGURE 6 ■ Diagram of the anatomic basis for the double-decidual sign. *Source*: From Ref. 22.

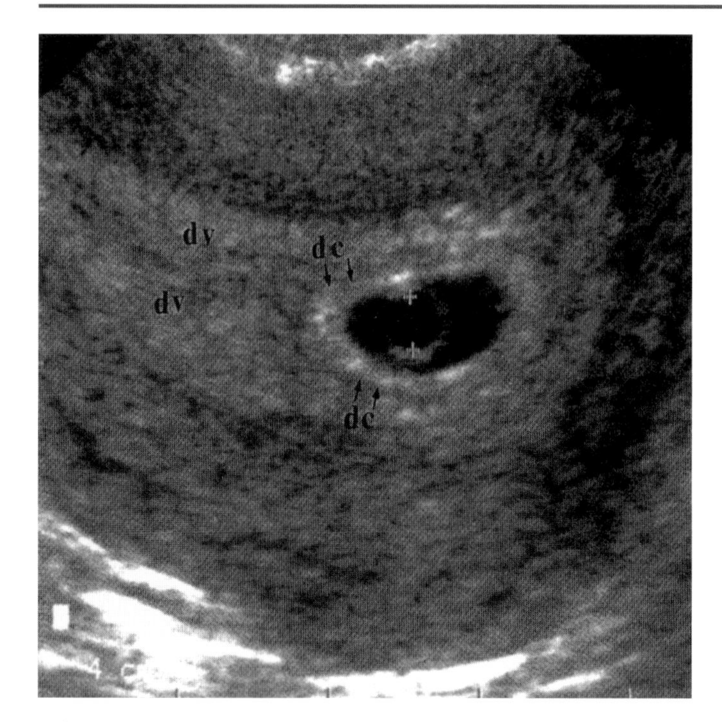

FIGURE 7 ■ Gestational sac and yolk sac (five weeks' menstrual age). A normal yolk sac (*electronic calipers*) measuring 3.1 mm in internal diameter is visualized. The embryo is not identified. The decidua vera and decidua capsularis (double-decidual sign) are identified. dv, decidua vera; dc, decidua capsularis. *Source*: From Ref. 22.

abnormal to be unable to identify a gestational sac). Two early studies identified β-hCG–discriminatory levels for EVS. Bree et al. (35) identified a discriminatory level of 1000 mIU/mL (First International Reference Preparation). Nyberg et al. (36) identified a discriminatory level of 1000 mIU/mL (Second International Standard), which converts to 1700 to 2000 mIU/mL First International Reference Preparation. More recently, Sengoku et al. (37) and Keith et al. (38) identified discriminatory levels of 2000 mIU/mL (First International Reference Preparation) and 1161 mIU/mL (Third International Standard), respectively. The First International Reference Preparation and Third International Standard are equivalent. The actual discriminatory level depends on the resolution of the ultrasound scanner, the patient's body habitus and position of the uterus, and the type of hormonal assay. (The above discriminatory levels were obtained using endovaginal ultrasound and cannot be applied in transabdominal ultrasound examinations.)

In the series by Keith et al. (38), the discriminatory level for sonographic identification of a gestational sac increased to 1556 mIU/mL for twins and 3372 mIU/mL (Third International Standard) for triplets. When applying discriminatory levels in clinical practice, it is important to remember that the numbers generally used are for singleton pregnancies, and that by using discriminatory levels in isolation to effect clinical decision making, ignores multifetal gestations. Nyberg and Filly emphasize that experienced sonologists rarely rely on a single piece of information and simultaneously consider multiple variables to create a diagnostic impression (31).

Yolk Sac

The yolk sac plays an important role in early embryonic life. It is involved in transfer of nutrients to the embryo, in hematopoiesis, and in formation of the primitive gut (39). The yolk sac remains connected to the midgut by the vitelline duct, which can often be demonstrated sonographically on a two-dimensional and three-dimensional scan (Fig. 8).

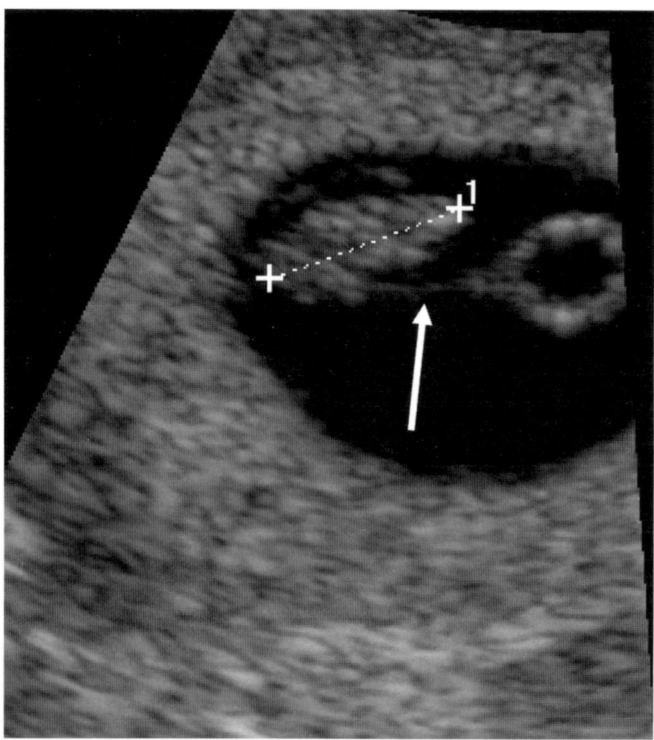

(A)

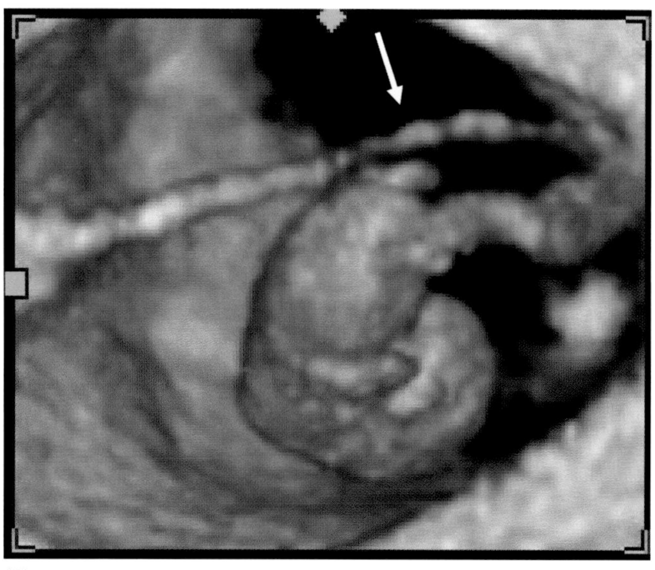

(B)

FIGURE 8 A AND B ■ Vitelline duct. (**A**) Normal gestation at seven weeks' menstrual age. The yolk sac is seen connected to the vitelline duct (*arrow*). (**B**) 3D view of a normal gestation at 8.5 weeks' menstrual age. The vitelline duct (*arrow*) is seen connecting with the umbilical cord.

The yolk sac, which is normally round or oval and has a uniformly thick, echogenic wall (8), can often be demonstrated by EVS when the MSD is 5 to 6 mm, and it is often seen before visualization of the embryo or amnion (Fig. 7). Although the double-decidual sign can be equivocal, the presence of a yolk sac within the gestational sac (the yolk sac sign) is diagnostic of an intrauterine pregnancy (40).

Using EVS (3), the yolk sac should always be visualized when the MSD is at least 8 mm. The yolk sac grows at a rate of 0.1 mm per millimeter of growth of the MSD before 15 mm MSD, after which it grows at a rate of 0.03 mm per millimeter of growth of the MSD (8). Bree et al. demonstrated the yolk sac in eight of eight patients with a serum β-hCG of at least 7200 mIU/mL (First International Reference Preparation) (35).

Embryo and Amnion

At approximately 5.5 weeks' menstrual age, the amnion may normally be visualized before the embryo as a 2 mm bleb adjacent to the yolk sac. This transient finding was originally described by Yeh and Rabinowitz and is referred to as the double-bleb sign (41). Visualization of the amnion without an embryo after the double-bleb stage is abnormal (1,31).

Shortly after the double-bleb stage, the amnion becomes difficult to visualize and then reappears when the crown-rump length (CRL) is about 8 to 12 mm as a thin, filamentous, rounded membrane surrounding the embryo (33). The amnion is, in turn, completely surrounded by the thick, echogenic chorion (42,43). The thinnest portion of the chorion is composed of the decidua capsularis and chorion laeve. The yolk sac is situated between the amnion and the chorion (Fig. 9). Chorionic fluid is often more echogenic than the essentially anechoic amniotic fluid (Fig. 10). Sonographic differentiation between the amnion and the chorion is usually not difficult in the first trimester and allows for reliable determination of amnionicity and chorionicity in multifetal pregnancies (Figs. 11 and 12) (42,43).

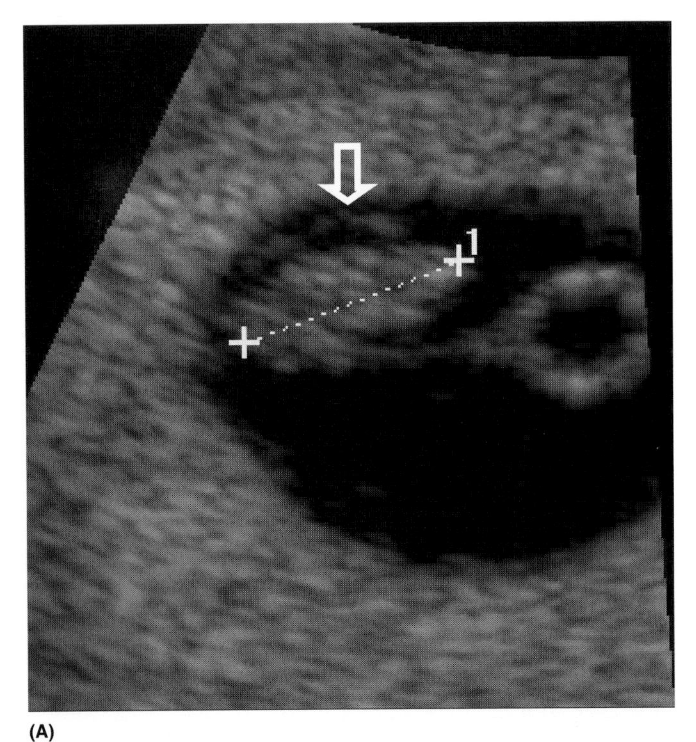

(A)

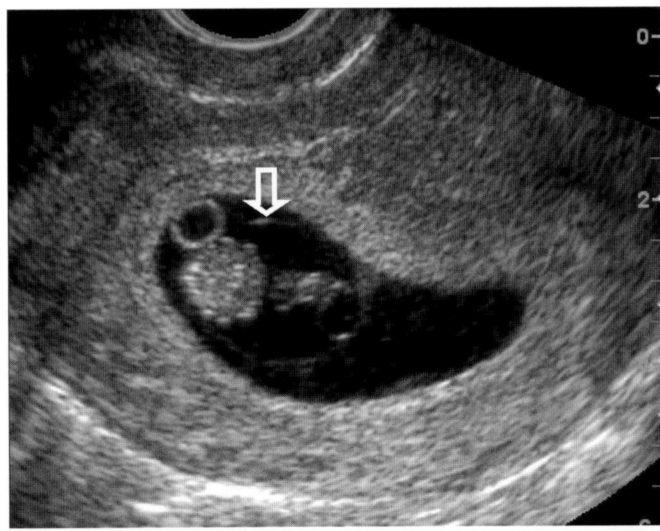

(B)

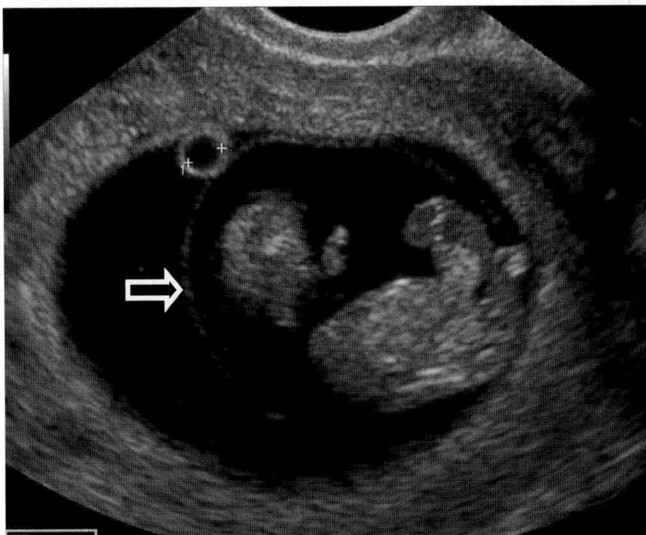

(C)

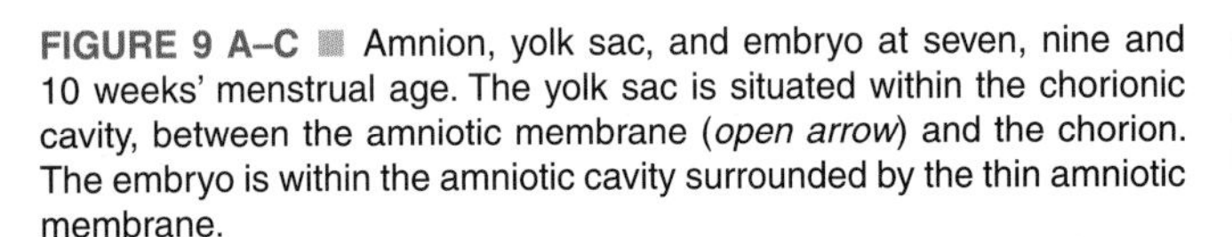

FIGURE 9 A–C ■ Amnion, yolk sac, and embryo at seven, nine and 10 weeks' menstrual age. The yolk sac is situated within the chorionic cavity, between the amniotic membrane (*open arrow*) and the chorion. The embryo is within the amniotic cavity surrounded by the thin amniotic membrane.

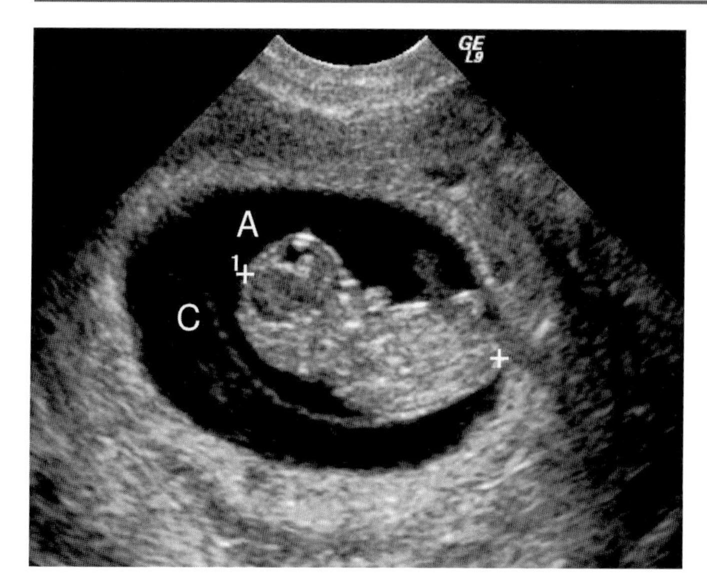

FIGURE 10 ■ Relative echogenicity of amniotic and chorionic fluids. Normal gestation at 9.5 weeks' menstrual age. A live embryo is seen (*electronic calipers*) surrounded by the amniotic membrane. The chorionic fluid is more echogenic than the amniotic fluid. a, amniotic fluid; c, chorionic fluid.

Using EVS, embryos as small as 1 to 2 mm in CRL can be identified routinely (4). In normal pregnancies, the embryo can be identified in gestational sacs as small as 10 mm (31) and should always be identified when the MSD is 16 to 18 mm or larger with optimal scanning parameters using high-resolution EVS. Cardiac activity may be identified immediately adjacent to the yolk sac (Fig. 13) and is indicative of a live embryo (44). The absence of cardiac activity, however, does not necessarily indicate embryonic demise. Using EVS, absent cardiac activity may be completely normal in embryos up to 4 to

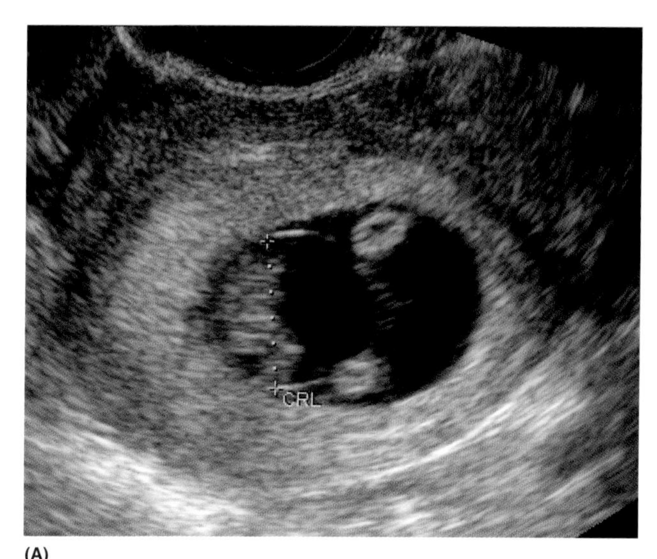

(A)

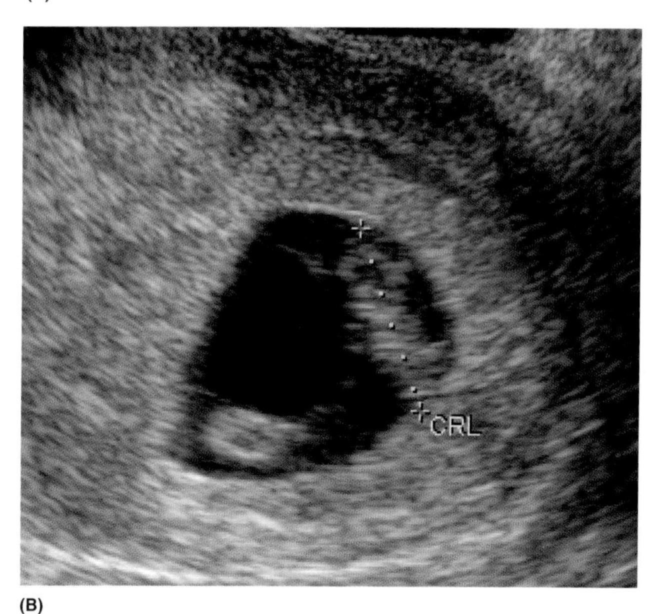

(B)

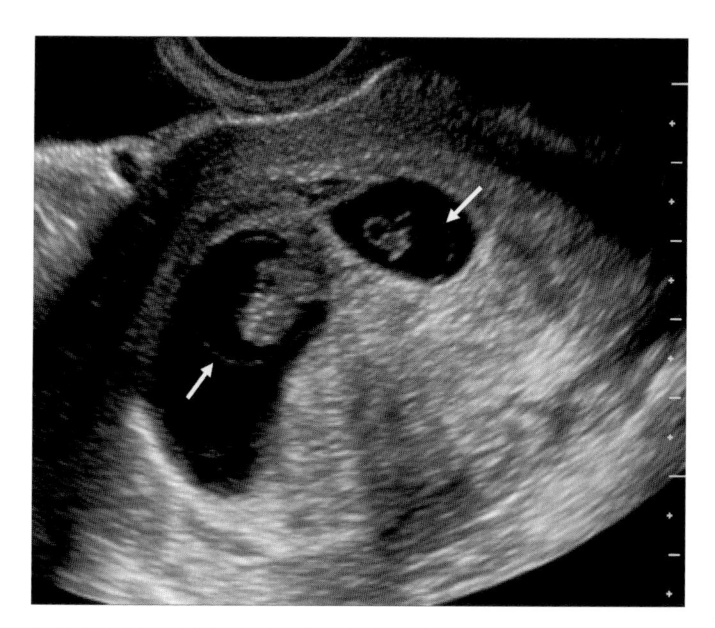

FIGURE 11 ■ Dichorionic diamniotic twin pregnancy (eight weeks' menstrual age). Endovaginal sonography shows two gestational sacs and an amniotic membrane surrounding each of the fetuses (*arrows*).

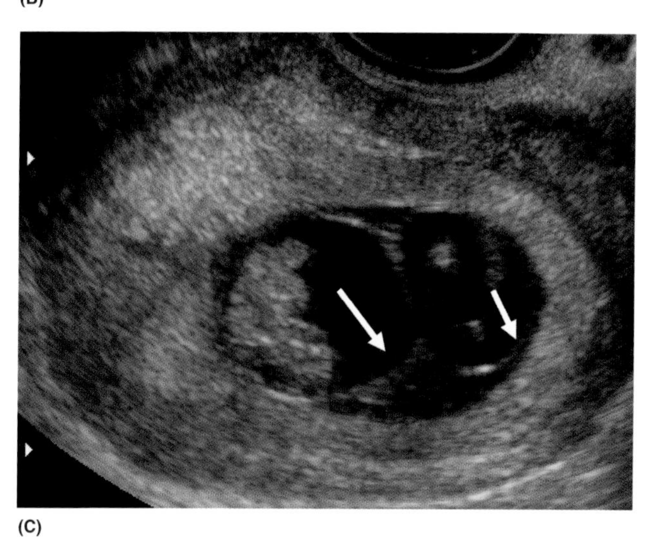

(C)

FIGURE 12 A–C ■ Monochorionic–diamniotic twins (menstrual age: eight weeks three days). (**A** and **B**) A single chorionic sac is present containing two live embryos (*electronic calipers*) and two round echogenic yolk sacs (A) outside of the amnion. (**C**) An amniotic membrane (*arrows*) is identified surrounding each embryo. CRL, crown-rump length.

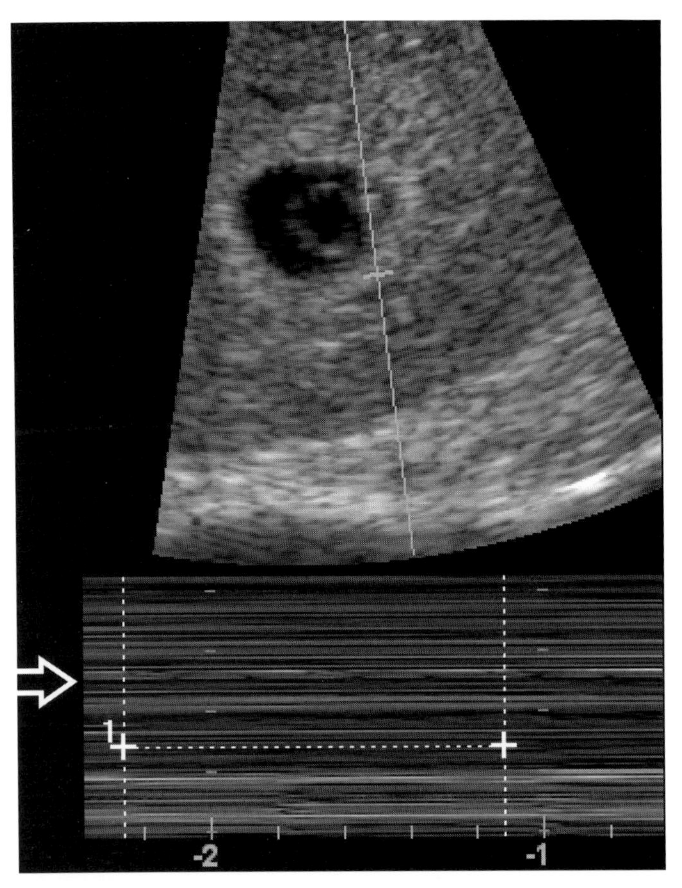

FIGURE 13 ■ Embryonic cardiac activity (six weeks' menstrual age). An endovaginal sonogram in the sagittal plane shows an embryo between the yolk sac and the gestational sac wall. An M-mode tracing through the embryo demonstrates cardiac activity (*arrow*). M-mode is used to document cardiac activity (for the patient's record), but the presence or absence of cardiac activity is determined at real time.

TABLE 2 ■ Factors Associated with an Increased Incidence of Ectopic Pregnancy

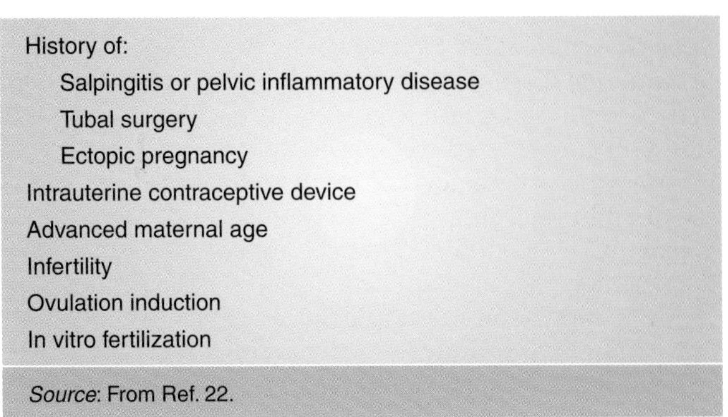

History of:
Salpingitis or pelvic inflammatory disease
Tubal surgery
Ectopic pregnancy
Intrauterine contraceptive device
Advanced maternal age
Infertility
Ovulation induction
In vitro fertilization

Source: From Ref. 22.

5 mm in CRL (4,45). Using TVS, absent cardiac activity may be normal in embryos up to 9 mm in CRL (45).

In two separate studies (46,47), cardiac activity was first demonstrated at 40 days' menstrual age and was noted in all normal embryos by 46 days' menstrual age.

ECTOPIC PREGNANCY

Ectopic pregnancy accounts for 1.4% of all pregnancies and for approximately 15% of maternal deaths. An increase in the prevalence of risk factors has led to a parallel increase in the incidence of ectopic pregnancy. The case–fatality rate, however, has declined from 3.5 in 1000 in 1970 to less than 1 in 1000 (48,49), probably because of improved diagnostic accuracy in the early stages of ectopic pregnancy and resultant earlier intervention.

All patients of reproductive age are at risk for ectopic pregnancy. The prevalence of ectopic pregnancy varies according to the patient population and their inherent risk factors (approximately 10–40%).

Table 2 describes factors associated with an increased incidence of ectopic pregnancy. The common element in these risk factors is prevention or delay of transit of the zygote through the fallopian tube. Chlamydia-induced salpingitis has been implicated as a major cause of the increasing incidence of ectopic pregnancy (50).

A strong association exists between infertility and ectopic pregnancy, probably because of the shared tubal abnormalities in both conditions. The increased incidence of multiple pregnancy resulting from ovulation induction and in vitro fertilization further increases the risk of both ectopic and heterotopic (intrauterine and ectopic) gestation. The hydrostatic forces generated during embryo transfer may also contribute to the increased risk. The frequency of heterotopic pregnancy is approximately 1 in 7000 in the general population (51), and it is greater in high-risk groups (52).

Sonographic Diagnosis

Regardless of clinical presentation, the primary goal of early first trimester sonographic diagnosis should be to identify the location of the gestational sac. The most important contribution of EVS in the evaluation of suspected ectopic pregnancy is its ability to identify either a normal or an abnormal intrauterine gestational sac earlier and more reliably than TVS.

Because of the low incidence of heterotopic pregnancy, sonographic demonstration of an intrauterine pregnancy reduces the likelihood of ectopic pregnancy to an almost insignificant level. Heterotopic pregnancy, however, should be suspected in the appropriate clinical setting, especially in high-risk groups. Evaluation of the adnexa should be routine in all patients, including those with documented intrauterine pregnancies.

The demonstration of a live embryo in the adnexa is diagnostic of ectopic pregnancy. In early intrauterine pregnancy, incomplete abortion, or ectopic pregnancy, one cannot always identify the gestational sac. In the absence of specific sonographic findings, the probability of ectopic pregnancy can be predicted by identification of nonspecific sonographic features and by correlation with the discriminatory level of the serum β-hCG (35,36). The relative risk of ectopic pregnancy and the clinical status of the patient determine the need for surgical intervention or repeat sonography and conservative management.

Specific Sonographic Findings

Intrauterine Pregnancy ■ As described in the section entitled Normal Sonographic Appearance, the intradecidual sign (23) can be used to demonstrate the presence of an intrauterine pregnancy before visualization of the yolk sac or embryo. Using EVS, the double-decidual sign (34) is usually demonstrated at approximately the same time that the yolk sac is visualized.

In patients with ectopic pregnancies, the decidua may slough, resulting in a fluid collection within the endometrial canal that is referred to as a *decidual cast* or *pseudogestational sac* (Fig. 14). EVS improves differentiation of the pseudogestational sac from the choriodecidual reaction of the intradecidual and double-decidual signs (53). A pseudogestational sac is a fluid collection within the endometrial canal surrounded by a single decidual layer, as opposed to a sac within the decidua abutting the endometrial canal (intradecidual sign) or the two concentric rings of the double-decidual sign (Figs. 4–7).

Small cysts within the decidua may appear as sac-like structures in patients with ectopic pregnancy (54). These decidual cysts may be distinguished from gestational sacs in that the cysts do not abut the endometrial canal and do not have an echogenic trophoblastic ring (Fig. 15). In our experience, *endometrial cysts* occur most commonly in ectopic pregnancy but are not specific for its diagnosis. Decidual cysts may also occur in patients with normal early intrauterine pregnancies or in non-pregnant patients. In patients with a normal intrauterine pregnancy, the decidual cyst may predate sonographic identification of the gestational sac. The important point is that the decidual cyst should not be mistaken for an intrauterine gestational sac, an interpretation that would decrease the index of suspicion of an ectopic pregnancy (54).

Live Embryo in the Adnexa ■ The sonographic demonstration of a live embryo in the adnexa is specific for the diagnosis of ectopic pregnancy (Fig. 16) (55). A live extrauterine fetus has been detected with EVS in 17% to 28% of patients with ectopic pregnancies (2,53,56,57) compared with approximately 10% with TVS (58).

Nonspecific Findings

Adnexal Mass ■ An adnexal mass can be found in conditions other than ectopic pregnancy and is therefore not diagnostic. The presence of an adnexal mass in patients with a positive β-hCG who have no sonographic evidence of an intrauterine pregnancy, however, has a positive predictive value of 70% to 75% for ectopic pregnancy (Table 3) (55).

Tubal Ring ■ Fleischer et al. report finding a tubal ring in 49% of patients with ectopic pregnancy and in 68% of those with unruptured tubal pregnancies using EVS (53). A tubal ring is an echogenic adnexal ring separate from the ovary created by the trophoblast of the ectopic pregnancy surrounding the gestational sac (Fig. 17). In the series of Nyberg et al. (55), the positive predictive value of a tubal ring for ectopic pregnancy was 100%.

Free Fluid ■ The presence of free fluid is a nonspecific finding that suggests the presence of an ectopic pregnancy in the appropriate clinical setting (Table 4). The amount of fluid and the echogenicity of the fluid are

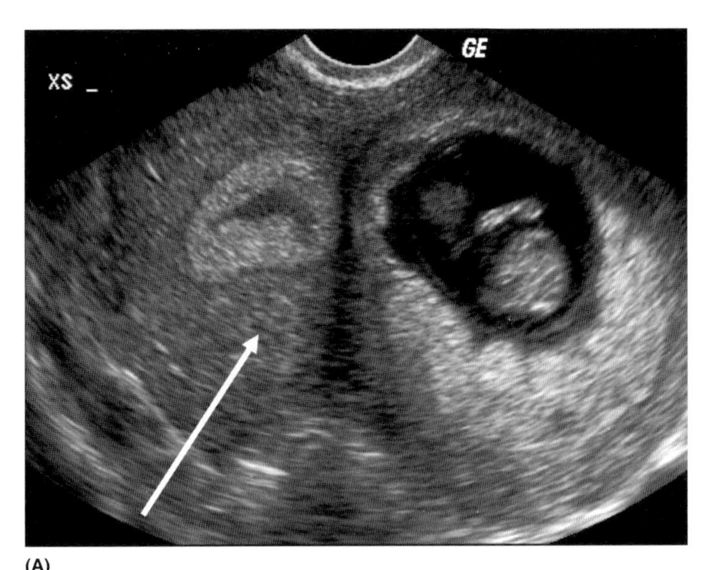

(A)

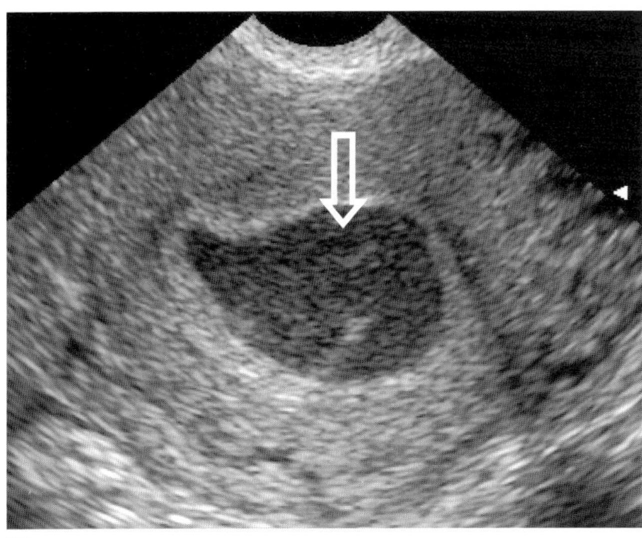

(B)

FIGURE 14 A AND B ■ Decidual cast. (**A**) Coronal sonogram of a didelphic uterus with a live 10-week embryo in the uterus on the patients left and a decidual cast in the uterus on the right (*arrow*). (**B**) A sagittal sonogram of the uterus of another patient. The "pseudogestational sac" is composed of fluid and debris (*open arrowhead*) within the endometrium, surrounded by a single layer of decidua.

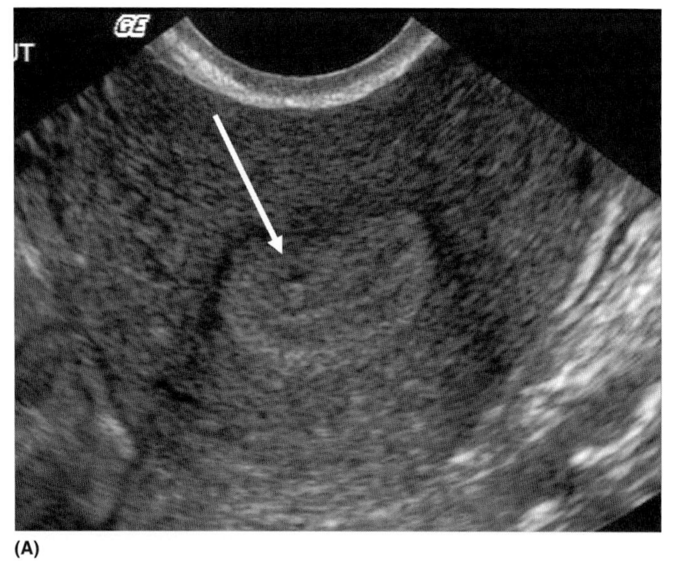

(A)

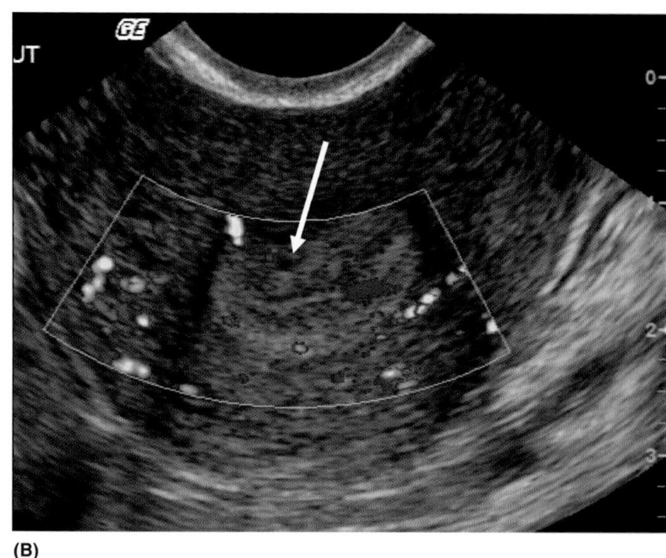

(B)

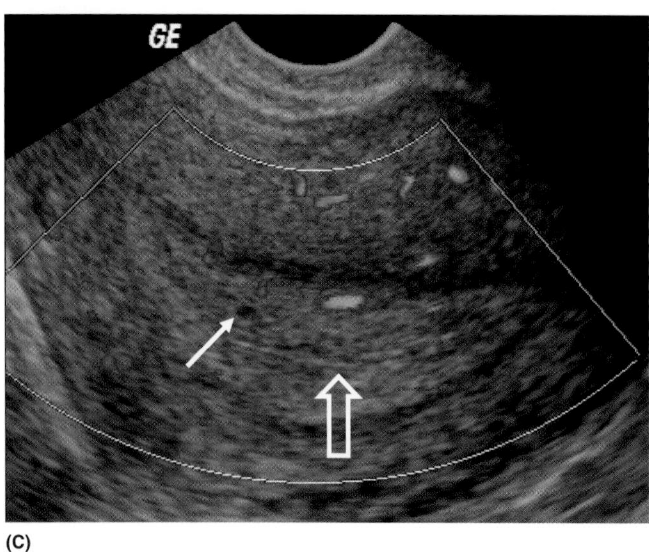

(C)

FIGURE 15 A–C ■ Decidual cyst in two patients with ectopic pregnancy. (**A**) A 3 mm cyst (*arrow*) is identified within the decidua. This cyst is not an intradecidual gestational sac in that it is peripherally located within the decidua and does not abut the endometrial canal. (**B**) When color Doppler is added, the cyst is shown to be avascular. (**C**) Another example of a decidual cyst (*arrow*) lying deep to the endometrial canal (*open arrow*).

important clues in predicting the presence of an ectopic pregnancy. A large amount of fluid and increased echogenicity of the fluid are both more indicative of an ectopic pregnancy. The presence of a large amount of free fluid or of echogenic free fluid increases the positive predictive value from 63% to 86% (55). Although the presence of a large amount of free fluid suggests tubal rupture, free fluid is not a specific finding because the tube may be intact in the presence of a large hemoperitoneum (59). In patients with suspected ectopic pregnancy, the combination of an adnexal mass and echogenic free fluid is associated with a 97% positive predictive value for ectopic pregnancy (Figs. 17 and 18) (55).

Normal Sonogram ■ Patients with ectopic pregnancy may have a completely normal pelvic ultrasound examination. In the series of Nyberg et al., 34% of patients with ectopic pregnancy had no evidence of either an adnexal mass or free fluid (55).

Endovaginal Color Flow Doppler ■ EVCFD diagnosis of ectopic pregnancy is based on the identification of adnexal peritrophoblastic flow defined as high-velocity, low-resistance flow separate from the ovary (Figs. 19 and 20) (12). Some studies suggest that EVCFD increases the diagnostic sensitivity for diagnosis of ectopic pregnancy compared with EVS alone (12,13). The addition of color Doppler flow imaging to EVS allows increased sensitivity in the detection of ectopic pregnancy in those select cases where a tubal ring or live ectopic is not identified. Furthermore, these studies suggest that EVCFD increases the percentage of initial examinations that are diagnostic of either intrauterine or ectopic pregnancy compared with EVS alone (13). In our experience, the presence of low-resistance flow within a mass separate from the ovary further suggests the diagnosis of ectopic pregnancy. More recently an absent or decreasing Doppler signal in an adnexal mass has been used as a criterion toward conservative management of ectopic pregnancy.

Interstitial (Cornual) Pregnancy

Approximately 95% of ectopic pregnancies occur in the ampullary or isthmic portions of the fallopian tube (60). The next most common site is an interstitial ectopic

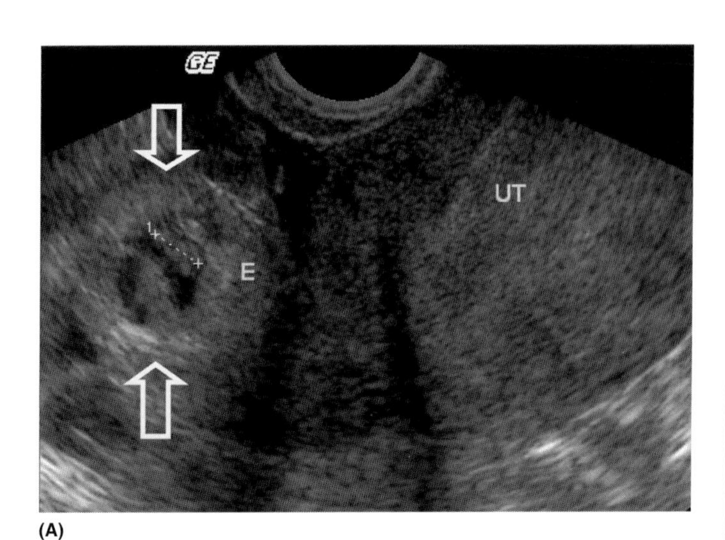

(A)

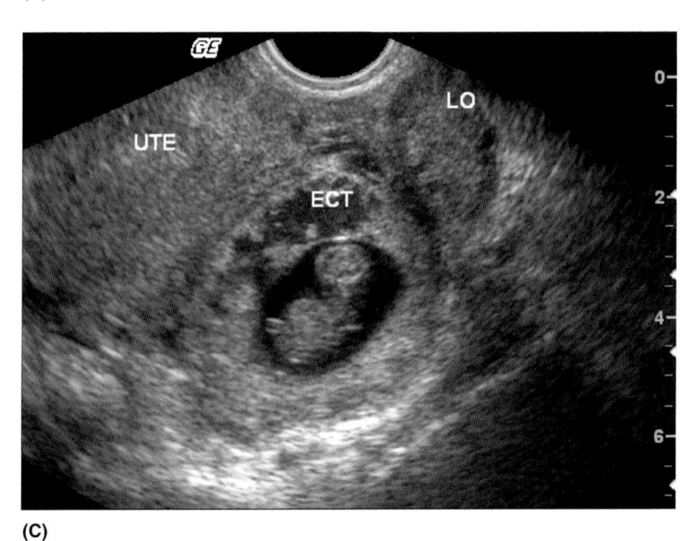

(C)

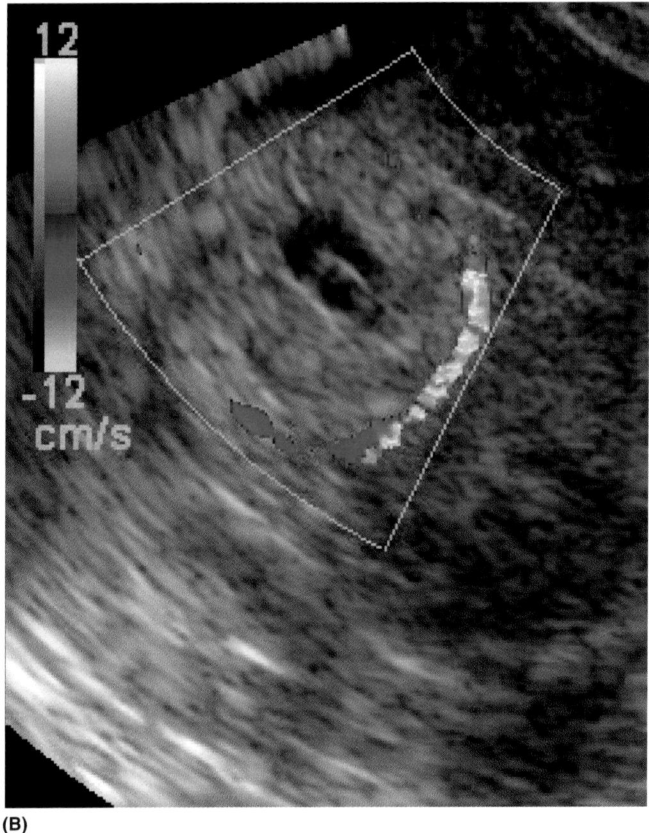

(B)

FIGURE 16 A–C ■ Live ectopic pregnancy. (**A**) A 7 mm six-week embryo (*calipers*) is visualized within a well-defined trophoblastic ring (*open arrows*). The sac is situated in the right tube, adjacent to the empty uterus. Embryonic cardiac activity was identified at sonography. ut, uterus; E, embryo (*within electronic calipers*). (**B**) Trophoblastic flow is seen on this color scan of the sac. (**C**) Another patient with a live nine-week tubal ectopic situated between the empty uterus and the normal left ovary (lo). ect, tubal ectopic; ute, empty uterus; lo, left ovary.

pregnancy occurring in the intramural portion of the tube where it traverses the wall of the uterus to enter the endometrial canal. Ovarian, cervical, and abdominal sites of ectopic pregnancy are extremely rare.

Because of its intramural location, an interstitial ectopic pregnancy ruptures later than other tubal gestations and often causes massive intraperitoneal hemorrhage from dilated uterine vessels. The mortality of interstitial pregnancy is therefore higher than that of other tubal ectopic pregnancies. Although not specific, the sonographic diagnosis of an interstitial pregnancy is suggested by an eccentric location of the gestational sac surrounded by an incomplete myometrial mantle. If the gestational sac encroaches to within 5 mm of the uterine serosa, an

TABLE 3 ■ Ectopic Pregnancy: Extrauterine Findings

Sonographic findings	Sensitivity (%)	Specificity (%)	Positive predictive value (%)	Negative predictive value (%)
Adnexal mass	21	93	70	58
Free fluid	63	69	63	69
Moderate-to-large amount of free fluid	29	96	86	62
Echogenic fluid, no mass	15	98	85	58
Adnexal mass, no fluid	22	94	75	57
Echogenic fluid plus adnexal mass	42	99	97	67

Source: From Ref. 55, pp. 824–826.

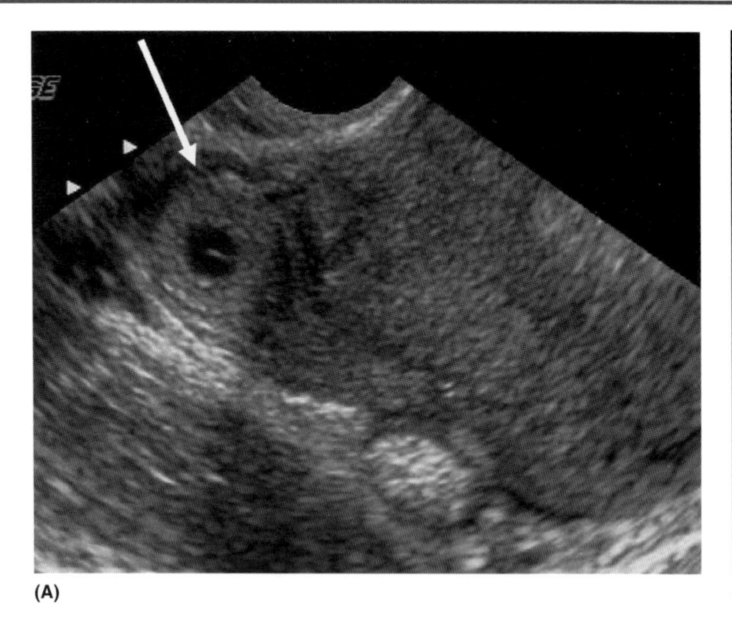

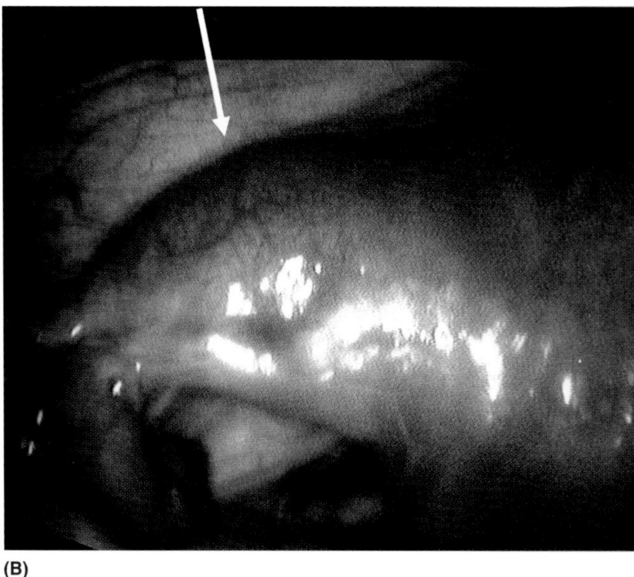

(A) (B)

FIGURE 17 A AND B ■ Ectopic pregnancy: tubal ring. (**A**) An extrauterine trophoblastic ring (*arrow*) is visualized adjacent to the uterus. (**B**) A laparoscopic picture of the distended proximal end of the fallopian tube (*arrow*). The ectopic sac was removed following a salpingostomy.

interstitial ectopic pregnancy should be suspected (53). The absence of a well-defined double-decidual sac sign and direct extension of the endometrial canal up to the midportion of the gestational sac or cornual mass improves differentiation of interstitial ectopic from eccentric intrauterine implantation. This finding is referred to as the *interstitial line sign* (Fig. 20) (61). The interstitial line sign may be helpful in the diagnosis of a cornual ectopic pregnancy even in the absence of a demonstrable gestational sac in the cornua (Fig. 20).

Management

Conservative management of ectopic pregnancy has become more common with earlier detection through endovaginal ultrasound and sensitive hCG assays. Conservative management includes laparoscopic salpingectomy or salpingostomy, methotrexate administration, and expectant management for spontaneous resolution (62).

Early sonographic diagnosis of ectopic pregnancy likely identifies some cases that would have escaped diagnosis because of spontaneous resolution without intervention. Atri et al. reported that 24% of ectopic pregnancies sonographically diagnosed over a 19-month period resolved spontaneously (63). Pregnancies that are more likely to resolve demonstrate findings such as a longer time interval from the last menstrual period (64), small adnexal masses of less than 3.5 cm and preferably less than 2 cm, low serum hCG levels of less than 1000 mIU/mL, rapidly

TABLE 4 ■ Diagnostic Table for Patients with Suspected Ectopic Pregnancy and Positive Pregnancy Test

Sonographic findings (uterine and adnexa)	β-Human chorionic gonadotropin (1st IRP) (mIU/mL)	Diagnosis
Intrauterine pregnancy with normal adnexa		Intrauterine pregnancy
Intrauterine pregnancy with adnexal mass or free fluid		Possible heterotopic pregnancy; clinical correlation required
No intrauterine pregnancy with no adnexal abnormality	>1700–2000	Probable ectopic pregnancy; possible incomplete abortion, normal or abnormal intrauterine pregnancy. Management depends on clinical findings
No intrauterine pregnancy with no adnexal abnormality	<1700–2000	Likelihood of ectopic pregnancy depends on clinical findings
No intrauterine pregnancy with echogenic free fluid, adnexal mass		Probable ectopic pregnancy
No intrauterine pregnancy with live ectopic or trophoblastic ring		Ectopic pregnancy

Abbreviation: Ist IRP, First International Reference Preparation.
Source: From Ref. 22.

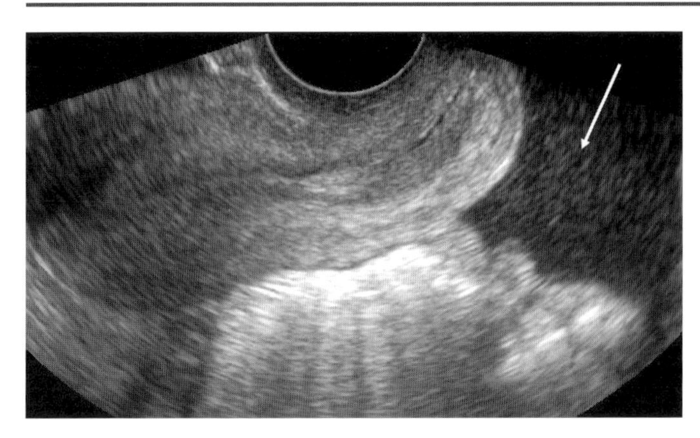

FIGURE 18 ■ Ectopic pregnancy: free fluid. Echogenic free fluid (*arrow*) is present in the posterior cul de sac in a patient with a ruptured ectopic pregnancy.

decreasing hCG levels, lack of a gestational sac, and no cardiac activity detected. The more advanced and vascular the adnexal mass caused by the ectopic pregnancy, the less likely it is to resolve spontaneously (63,64). While resolving, the adnexal mass of an ectopic pregnancy may increase in size and become more vascular on ultrasound (63). Ultrasound is not reliable for identifying cases that are undergoing spontaneous resolution. The diagnosis is based on continuing clinical stability of the patient and decreasing serial hCG levels (64).

Although spontaneous resolution of ectopic pregnancy may occur, expectant management of patients with confirmed diagnoses remains controversial. The benefit of expectant management over other conservative treatments, such as medical treatment with methotrexate or minimally invasive laparoscopic surgery, has not been established (65,66).

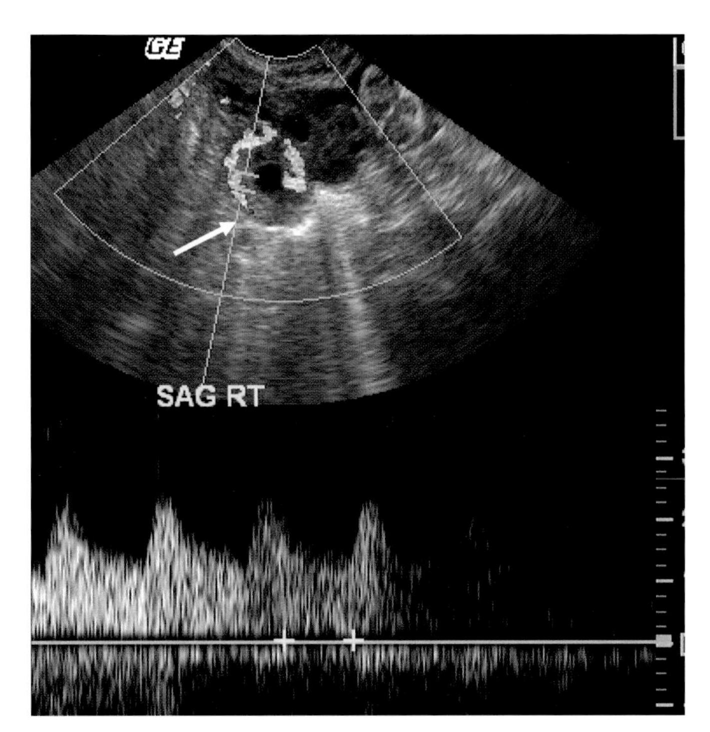

FIGURE 19 ■ Ectopic pregnancy: gray-scale and Doppler findings. Sagittal sonogram through the right adnexa shows a tubal ring (*arrow*) adjacent to the ovary. Color and spectral Doppler interrogation of the right adnexa demonstrates high-velocity, low-resistance flow.

Ectopic pregnancy has been effectively treated on an outpatient basis with methotrexate therapy. The treatment results in decreased patient morbidity and increased preservation of reproductive capability (62). Methotrexate causes tubal abortion and resorption of the ectopic. Most protocols limit candidates to women with no free fluid, an adnexal mass of less than 2 to 3 cm, lack of embryonic cardiac activity, and hCG levels of less than 5000 to 10,000 mIU/mL. Failure of treatment is most closely linked to high hCG levels and the presence of embryonic cardiac activity (67). At follow-up of patients treated with methotrexate, ultrasound may demonstrate that the adnexal mass or affected fallopian tube has grown in size and become more vascular. Treatment success is best monitored with clinical factors and declining hCG levels (67,68). Unruptured live ectopic pregnancies (tubal, cornual, and cervical) have also been successfully treated with local injection of potassium chloride or methotrexate under ultrasound guidance (69).

EARLY PREGNANCY FAILURE

One of the most important roles of ultrasound in early pregnancy is to determine if the embryo-fetus is alive. Early pregnancy failure is common, even in patients with no clinical symptomatology. The presence of spotting or bleeding and cramping with a closed cervical os (threatened abortion) occurs in approximately 25% of patients and increases the risk of early pregnancy failure. About 50% of patients presenting with threatened abortion have a normal outcome, and 50% subsequently abort (33,46). In patients presenting with threatened abortion who subsequently abort, the embryo is usually already dead at the time of presentation. Spontaneous expulsion may be delayed. Appropriate patient management depends on the status of the embryo.

Hormonal assays provide useful ancillary information regarding the status of the pregnancy. Low maternal serum progesterone levels are associated with a poor prognosis in early pregnancy (70). Although the data appears to be preliminary, the combination of low serum progesterone, an empty gestational sac, and maternal age may improve diagnostic accuracy from current methods (70). Serum β-hCG levels may be equivocal or normal in the presence of embryonic demise or anembryonic gestation if there is persistent function of chorionic tissue (33) and although helpful, cannot reliably discriminate between live pregnancies and early pregnancy failure (31,70). Currently, the most accurate indicator of a live embryo or of embryonic demise is sonography.

The sonographic diagnosis of early pregnancy failure depends on the stage of development.

- *Loss within the first 2 weeks after conception* (*3 to 4 weeks' menstrual age*): This type of pregnancy failure is called "subclinical or preclinical loss" or loss before the patient has missed a menstrual period. Often, the patient has

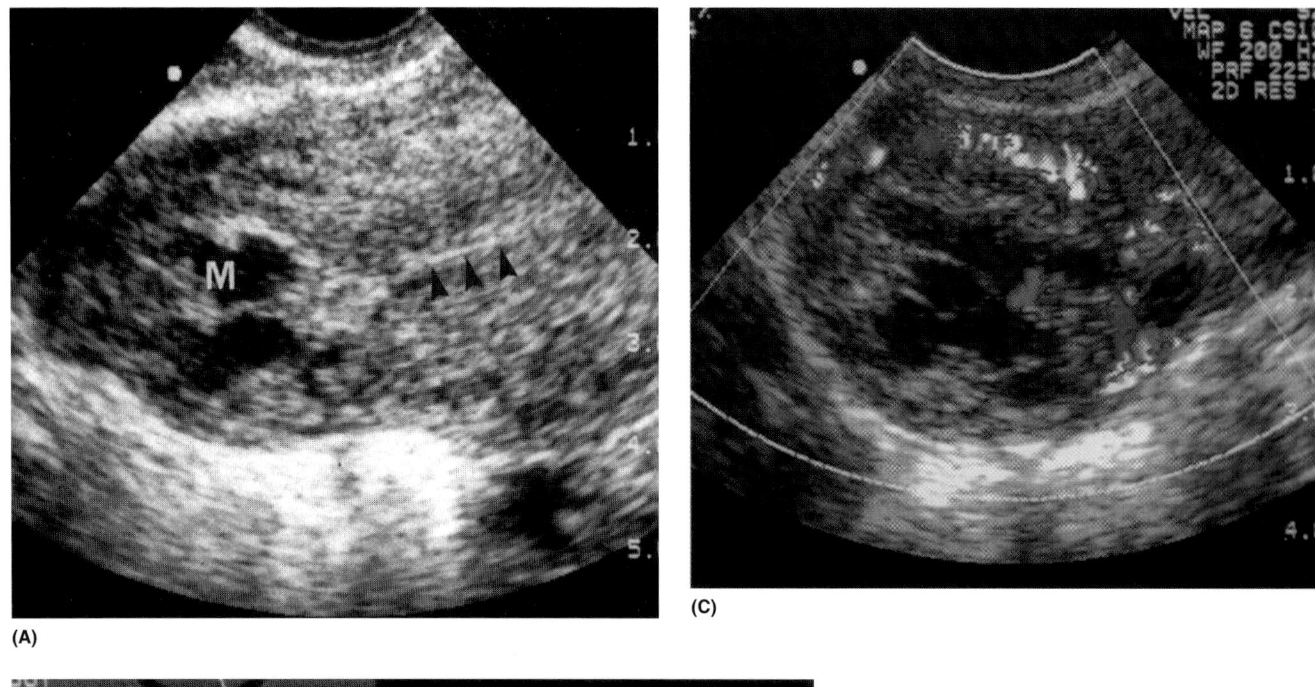

(A)

(C)

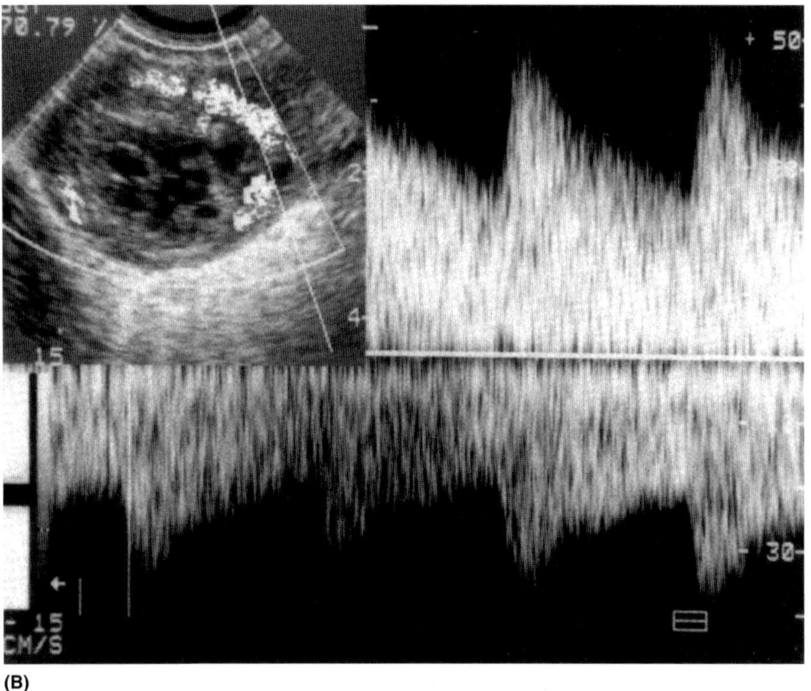

(B)

FIGURE 20 A–C ■ Interstitial ectopic pregnancy. (**A**) Coronal sonogram through the right cornu of the uterus shows the endometrium (*arrowheads*) extending to an inhomogeneous mass. (**B**) Endovaginal color flow Doppler and (**C**) pulsed Doppler demonstrate peritrophoblastic flow. *Abbreviation*: M, inhomogeneous mass. *Source*: From Ref. 22.

no sonographic evidence of pregnancy at this stage. If pregnancy failure occurs prior to implantation, there will be no clinical evidence of pregnancy and the serum β-hCG will not become positive.

- *Loss at 5 and 6 weeks' menstrual age:* The sonographic diagnosis of pregnancy failure is usually based on gestational sac findings.
- *Loss at 7 to 12 weeks' menstrual age:* The sonographic diagnosis of embryonic demise is usually based on absent embryonic cardiac activity or abnormal gestational sac findings in the absence of a demonstrable embryo.

Investigators have long suspected that preclinical loss rates are high in humans, and in the past two decades, data concerning these losses have become available. Cohort studies indicate that many women who show positive β-hCG assays never have any clinical evidence of pregnancy. A study by Wilcox et al. (71) showed a 31% rate of pregnancy loss after implantation in normal healthy volunteers. In their series, 22% of all pregnancies aborted early, resulting in subclinical loss. Cytogenetic abnormalities have also been documented in 20% of ostensibly normal embryos from in vitro fertilization (72). The foregoing are consistent with the pivotal studies of Hertig et al. (73,74), who showed high frequencies of morphologic abnormalities in preimplantation embryos. Loss rates are influenced by fetal factors including chromosomal abnormalities and abnormal development of chromosomally normal fetuses, and maternal factors including maternal age, smoking, alcohol consumption, immunological factors, uterine anomalies and acquired abnormalities, and other variables. Approximately 50% to 60% of early abortuses are chromosomally abnormal, with the spontaneous pregnancy loss rate dropping off later in the first trimester (75).

Sonographic studies also show a higher incidence of early spontaneous pregnancy loss decreasing as pregnancy progresses. In a study of 232 first-trimester patients with an endovaginal scan at the first visit, Goldstein (76) determined the incidence of subsequent pregnancy loss by following all of them to delivery or spontaneous abortion. All patients had a positive urinary pregnancy test and none had vaginal bleeding. This study seeks to establish the "baseline noise" of early pregnancy failure by taking a group of normal, healthy women with good nutrition, medical and prenatal care and following them from very early gestation to outcome. This group had an overall pregnancy loss rate of 11.5% in the embryonic period (i.e., <70 days from last menstrual period) and 1.7% in the fetal period. The loss rate diminished as the pregnancy progressed and more structures were identified. The loss rate was 8.5% when a yolk sac was seen, 7.2% with an embryo of a CRL of <5 mm, 3.3% with a CRL of 6 to 10 mm, and 0.5% with a CRL of >10 mm. The loss rate leveled off at 2% from 14 to 20 weeks, the fetal period.

Ultrasound Diagnosis

Once the gestational sac becomes demonstrable by ultrasound, the diagnosis of early pregnancy failure can be made reliably using sonographic criteria. Whether the diagnosis of early pregnancy failure should be made based on a single ultrasound examination, or on serial sonographic assessment is a controversial issue. The policy issued by the Faculty of Clinical Radiology of the Royal College of Radiologists and the Council of the Royal College of Gynaecologists states that when the death of an embryo is suspected, two transvaginal scans separated by a minimum of seven days should be performed (5). Their guidelines state that an embryo of CRL greater than 10 mm with no evidence of heart pulsations on two separate occasions of at least seven days apart is suggestive of a missed abortion. When the MSD is less than 15 mm or the CRL is less than 10 mm, the examination should be repeated two weeks later to assess growth of the gestational sac and embryo and any evidence of cardiac activity (5). Other authors suggest that while serial examination improves diagnostic certainty in difficult cases, in many patients this may create unnecessary anxiety (31,70). Many authors who advocate the use of sonographic criteria to diagnose early pregnancy failure on a single examination suggest that the examiner use more conservative sonographic criteria than the published data to avoid incorrect diagnosis (allowing a few millimeters of leeway prior to making the diagnosis of embryonic demise) (1,31). While the decision on whether to make the diagnosis based on single or serial examination is controversial, two important points should be considered: It has been shown repeatedly that the discriminatory levels quoted in the literature cannot be reproduced with 5 MHz endovaginal probes (6,9,70) and that it is necessary to use a high-frequency endovaginal probe with meticulous technique to make the diagnosis of embryonic demise on a single examination. If there is any doubt, follow-up examination should be performed.

Embryonic Cardiac Activity

The presence of cardiac activity is the single most important diagnostic finding, indicating that the embryo-fetus is alive. In addition to demonstrating that the embryo-fetus is alive at the time of the examination, the presence of cardiac activity changes the prognosis in patients presenting with threatened abortion from a 50% rate of pregnancy failure to much more favorable odds. The predictive value with respect to the ultimate viability of the fetus depends on the menstrual age at the time of the examination, presenting symptomatology, and other sonographic predictors of abnormal outcome (see below). After 7 weeks' menstrual age, the presence of cardiac activity is associated with a pregnancy loss rate of 2% to 2.3% (77,78) and after 16 weeks the rate is only 1% (79). Most pregnancy failure occurs early, either as subclinical loss or in embryos less than seven weeks' menstrual age. In a series of 96 patients with an embryonic CRL of less than 5 mm (six weeks' and one-day menstrual age), EVS demonstration of cardiac activity was associated with a pregnancy loss rate of 24% (4).

As noted in the section entitled Embryo and Amnion, the absence of sonographically demonstrable cardiac activity is not necessarily abnormal. Using TVS, normal embryos less than 9 mm in CRL may have absent cardiac activity (45). Using EVS, normal embryos less than 4 or 5 mm in CRL (depending on the series) may have absent cardiac activity (4,45). Allowing a few millimeters of leeway, the absence of cardiac activity in larger embryos is diagnostic of embryonic demise, assuming that the scan has been performed on modern equipment with a high-resolution endovaginal probe of at least 7 MHz, using a high frame rate, the frame-averaging mode has been turned off, and the focal zone is set appropriately (Fig. 21). In our lab, in patients with embryos of borderline CRL, we perform follow-up examination in one week's time.

Gestational Sac Characteristics

In many patients, the embryo cannot be visualized at the time of initial sonographic examination, and the diagnosis of embryonic demise cannot be made on the basis of embryonic cardiac activity. In those patients, the diagnosis of pregnancy failure may possibly be made based on gestational sac characteristics (80).

The most reliable indicator of abnormal outcome based on gestational sac characteristics is abnormal gestational

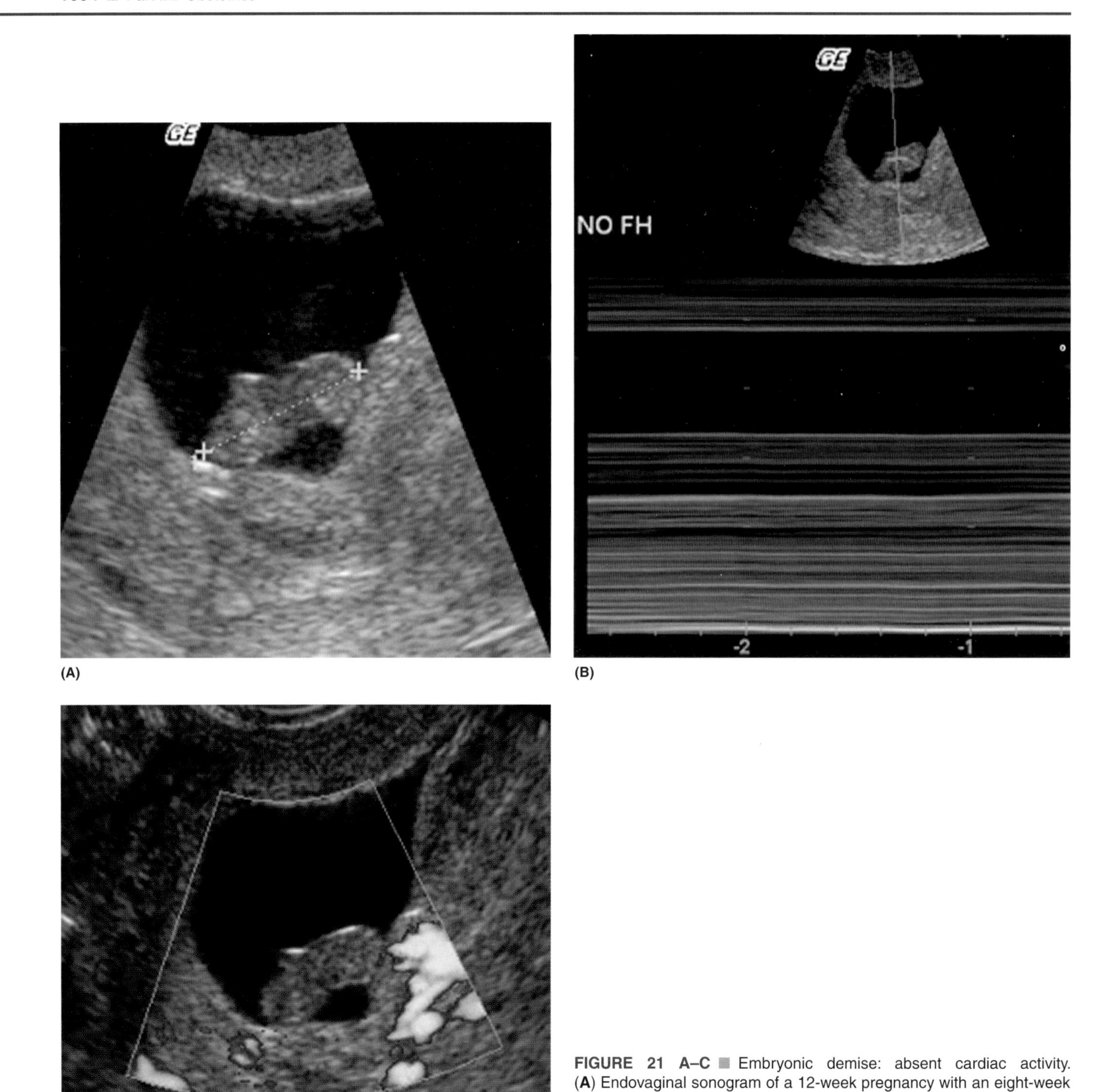

FIGURE 21 A–C ■ Embryonic demise: absent cardiac activity. (**A**) Endovaginal sonogram of a 12-week pregnancy with an eight-week embryo having a crown-rump length of 1.62 cm (*calipers*). (**B**) Cardiac activity is absent on the M-mode tracing (*arrow*). M-mode is used to document cardiac activity (for the patient's record), but the presence or absence of cardiac activity is determined at real time. (**C**) Absent cardiac activity on power Doppler. FH, fetal heart (a commonly used acronym for embryonic cardiac activity).

sac size (3,80). In 1985, Bernard and Cooperberg observed that a sac larger than 2 cm in diameter without an embryo had a poor outcome (81). In 1986, using TVS, Nyberg et al. defined an abnormally large gestational sac as a gestational sac of at least 25 mm MSD lacking an embryo, or a gestational sac of at least 20 mm MSD lacking a yolk sac (80).

These criteria were re-evaluated for EVS (3,6,9,31). Using EVS, if the MSD is 8 mm or larger without a demonstrable yolk sac, or 16 mm or larger without a demonstrable embryo, the gestational sac is considered to be abnormally large, suggesting the possibility of early pregnancy failure. Most authors allow 2 to 3 mm of leeway in

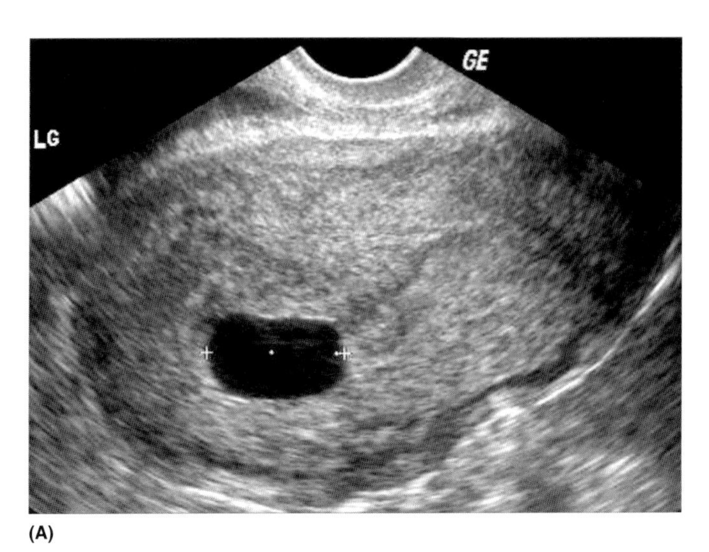

(A)

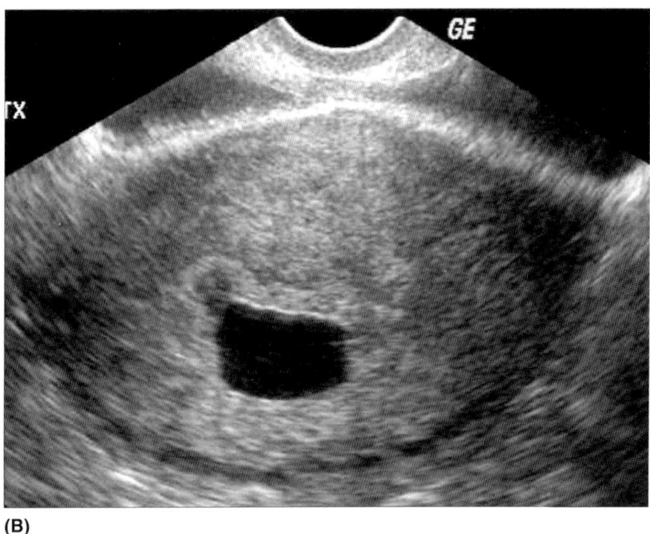

(B)

FIGURE 22 A AND B ■ Early pregnancy failure: abnormal gestational sac size. Patient with 12-week gestation with brown spotting. (**A**) Sagittal and (**B**) coronal endovaginal sonogram shows a gestational sac (*electronic calipers*) with a mean diameter of 2.2 cm. Neither a yolk sac nor an embryo is identified.

MSD measurements (31) as a margin of error, and many do not use the absent yolk sac as a sign of pregnancy failure (Fig. 22). Furthermore, these parameters only apply to high-resolution EVS and cannot be used for examinations performed with a 5 MHz endovaginal probe (Fig. 23) (6,7,9,68). Because of variation in resolution and technique, the validity of these data should be verified in each lab before clinical use. For borderline cases, follow-up examination should be performed in one week's time.

We consider other gestational sac criteria to be less reliable in the diagnosis of embryonic demise and use them as ancillary findings. These criteria include a distorted gestational sac shape (Fig. 24), a thin decidual reaction (less than 2 mm), weak decidual echogenicity, an absent double-decidual sign, or an abnormally low position of the gestational sac within the endometrial cavity (80).

Normal gestational sac growth is at least 1.1 mm per day. Nyberg et al. found that patients with early pregnancy failure had MSD growth rates of less than 0.7 mm per day (82). In practice, we use this information as a guideline in planning for follow-up ultrasound examinations in patients in whom a diagnosis cannot be made on the initial examination and as an indicator of inadequate growth on follow-up assessment.

Amnion and Yolk Sac Criteria

Visualization of the amnion in the absence of a sonographically demonstrable embryo is abnormal and is diagnostic of an anembryonic gestation or embryonic demise with resorption of the embryo (Fig. 25) (33). In small embryos prior to visualization of cardiac activity, an amniotic cavity that is disproportionately large relative to the CRL and size of the chorionic cavity is suggestive of embryonic demise (83).

Other findings useful in the diagnosis of embryonic demise include a collapsing, irregularly marginated or collapsing amnion (Fig. 26) and yolk sac calcification (84). In general, however, other signs of embryonic demise are present when these findings are positive.

Sonographic Predictors of Abnormal Outcome

Sonographic findings may be used to predict abnormal outcome in the presence of a live embryo or before visualization of the embryo. These findings can be used to identify a subgroup of embryos that are at high risk of embryonic demise or subsequent diagnosis of fetal anomaly and that require close follow-up.

Embryonic Bradycardia

Although embryonic cardiac activity indicates that the embryo is alive at the time of the examination, an abnormally slow heart rate may predict impending demise (Fig. 27). In a study by Doubilet and Benson, a heart rate of less than 80 bpm in embryos with a CRL less than 5 mm was universally associated with subsequent embryonic demise (85), a heart rate of 80 to 90 bpm was associated with a 64% risk of demise, a heart rate of 90 to 99 bpm was associated with a 32% risk, and a heart rate of 100 bpm or more was associated with an 11% risk. In embryos under 5 mm in CRL, 1 00 bpm represented a plateau above which increases in heart rate were not associated with decreased risk of subsequent embryonic demise, a finding indicating that heart rates of 100 bpm or higher are normal in these embryos.

In embryos 5 to 9 mm in CRL, a heart rate lower than 100 bpm was always associated with abnormal outcome, with the normal heart rate plateau identified at 120 bpm. In embryos 10 to 15 mm in CRL, a heart rate of less than 110 bpm appears to be associated with a poor prognosis (85).

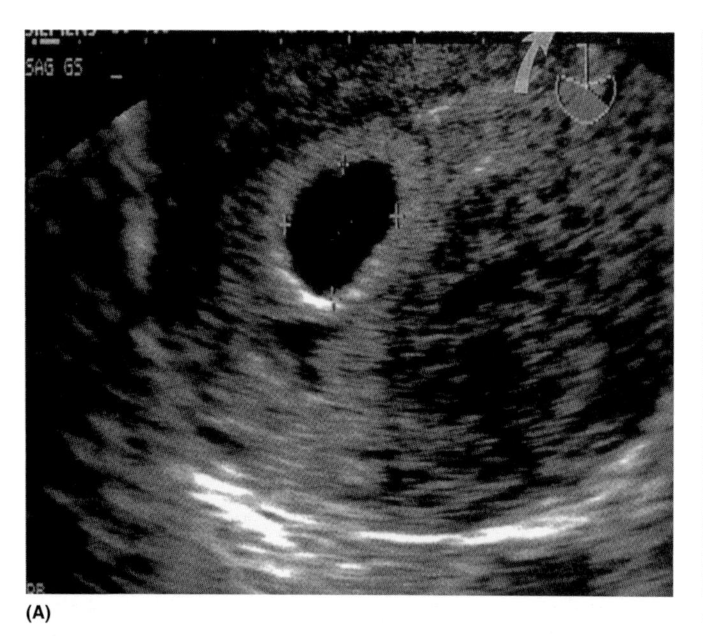

(A)

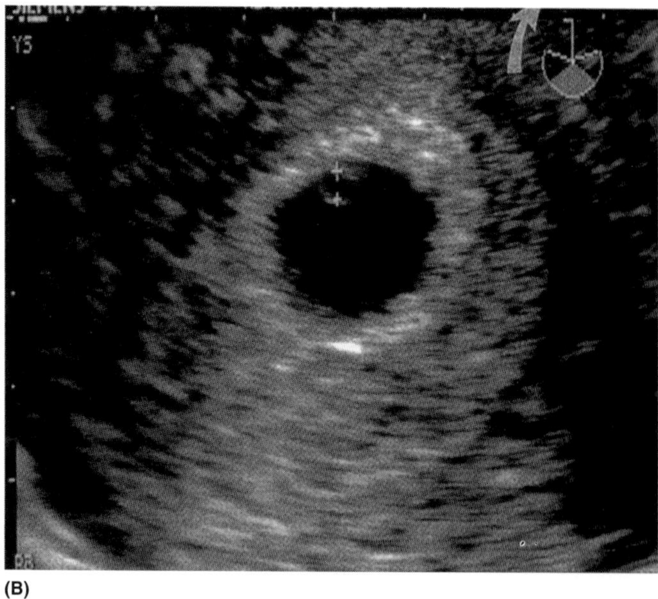

(B)

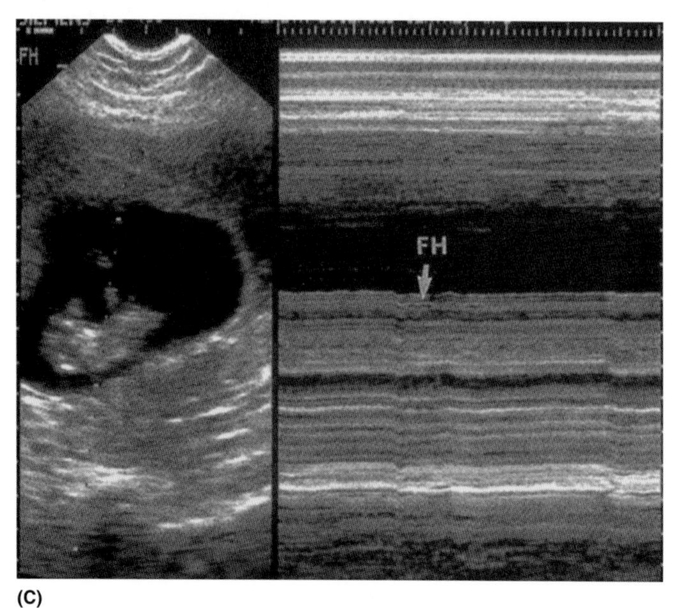

(C)

FIGURE 23 A–C ■ Borderline abnormal sac size using a 5 MHz transducer: live fetus on follow-up. (**A**) Endovaginal sonogram in the sagittal plane in a patient who was unsure of the date of conception. The mean gestational sac diameter (calculated from this scan and from a measurement in the coronal plane) was 18 mm. The frequency of the probe was 5.0 MHz (*curved arrow*). (**B**) The yolk sac was demonstrated. No embryo was detected. Because of the borderline measurement and the 5 MHz probe, a follow-up examination was suggested. (**C**) The patient failed to return until the menstrual age of the conceptus was 11 weeks, when a live embryo was demonstrated. FH, fetal heart. (See text for discussion.)

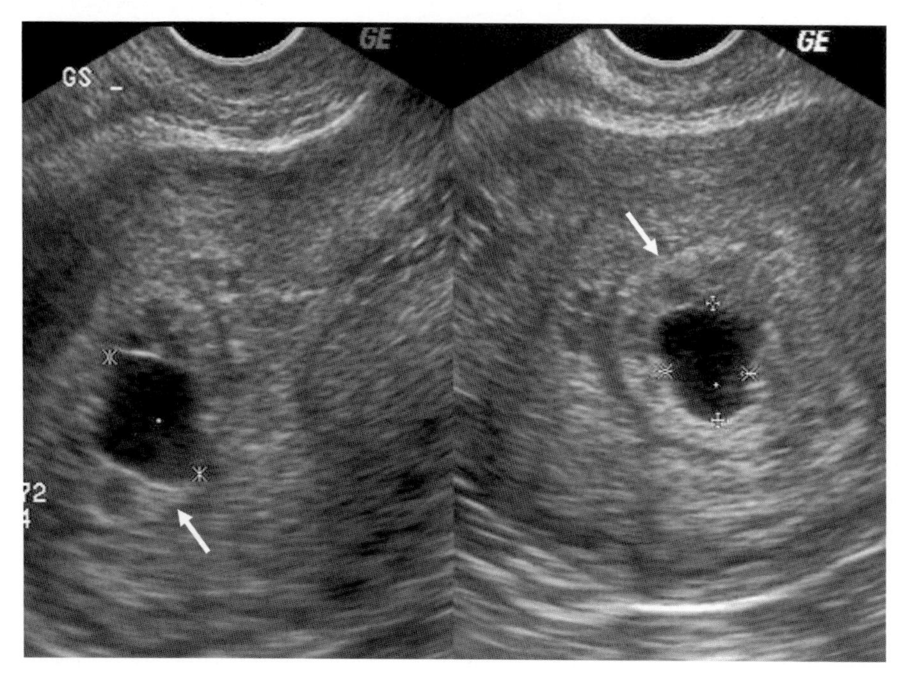

FIGURE 24 A AND B ■ Early pregnancy failure: distorted gestational sac, thin decidual reaction, weak decidual echogenicity. Sagittal (**A**) and coronal (**B**) sonograms in a patient with recurrent abortion, now at 11 weeks' gestation. The sac is irregular in shape and a decidual reaction that varies in thickness and is inhomogeneous (*arrows*).

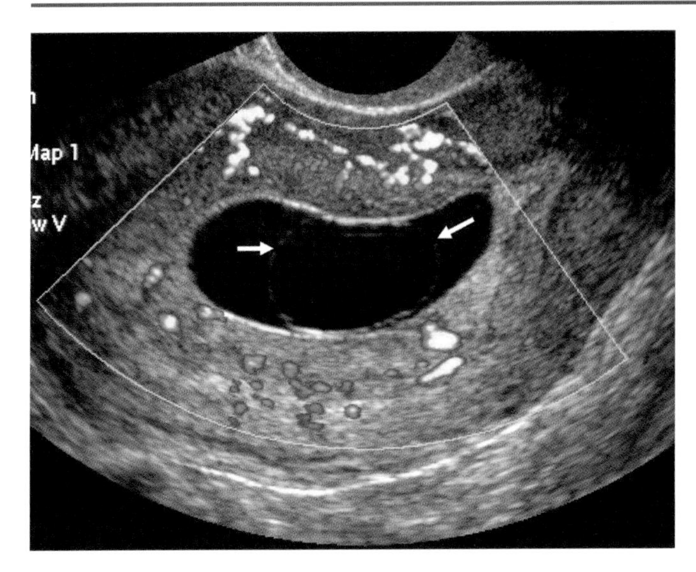

FIGURE 25 ■ Early pregnancy failure: endovaginal sonogram in the sagittal plane in a patient presenting with bleeding. The amnion (*arrows*) is visualized in the absence of a detectable embryo.

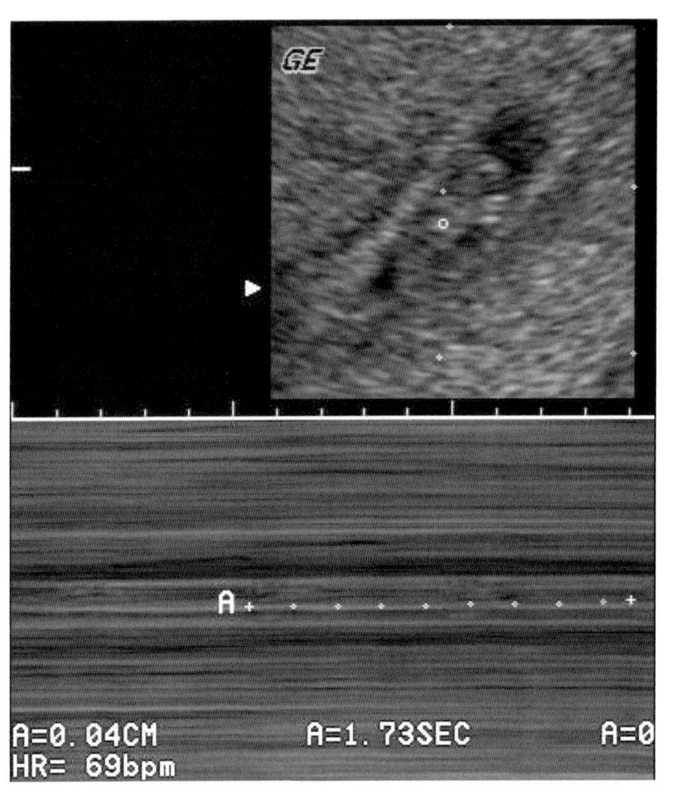

FIGURE 27 ■ Predictors of abnormal outcome: embryonic bradycardia. M-mode tracing in a patient with an embryonic crown-rump length of 10 mm. The heart rate was 69 beats per minute. The patient aborted spontaneously five days later.

MSD in Relation to the CRL

The MSD should be 5 mm greater than the CRL. Bromley et al. (86) found that, in 16 patients with pregnancies between 5.5 and 9 weeks' menstrual age in which the MSD was <5 mm more than the CRL (i.e., MSD – CRL <5 mm), 15 had spontaneous first trimester abortions (Fig. 28).

Yolk Sac Size and Shape

Although the subject is still controversial (87), in our experience yolk sac size and shape may be useful as relative predictors of abnormal outcome (8). Perhaps the most important consideration is that yolk sac abnormalities may predict abnormal outcome in pregnancies that appear completely normal by all other ultrasound criteria.

We found that, between 5 and 10 weeks' menstrual age, when the yolk sac diameter is outside the 5% and

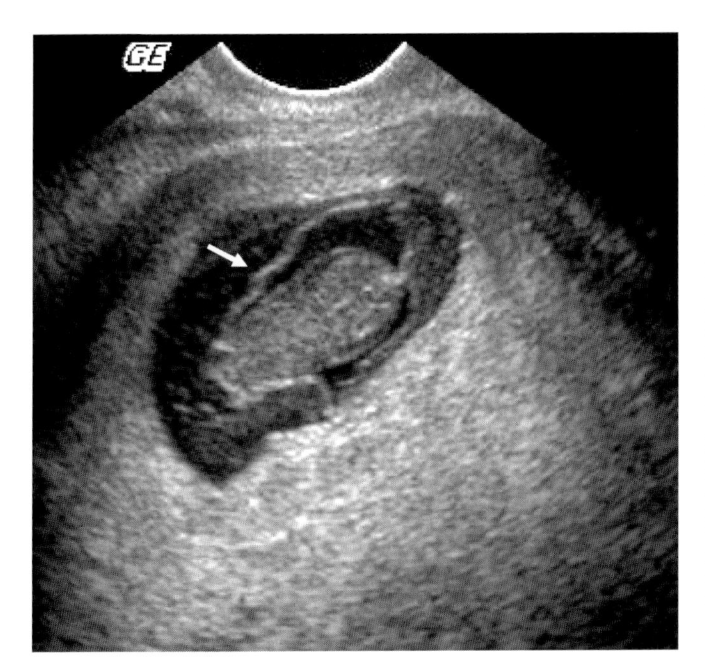

FIGURE 26 ■ Collapsing amnion. Early pregnancy failure at 12 weeks: endovaginal sonogram in the sagittal plane in a patient presenting with bleeding and no embryonic cardiac activity. The amnion (*arrow*) is collapsing around the embryo and is irregular in outline or floppy.

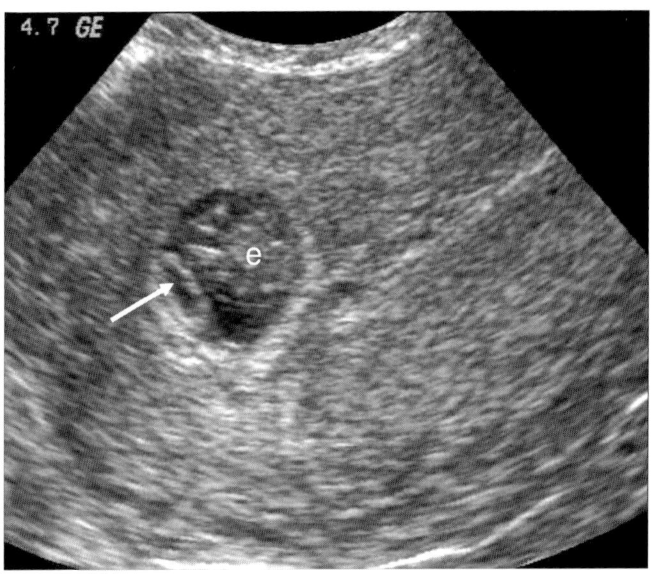

FIGURE 28 ■ Predictors of abnormal outcome: mean sac diameter minus crown-rump length is less than 5 mm. Sagittal sonogram shows a 12 mm embryo within a relatively small gestational sac and a compressed yolk sac (*arrow*). The embryo was dead on follow-up examination. e, embryo.

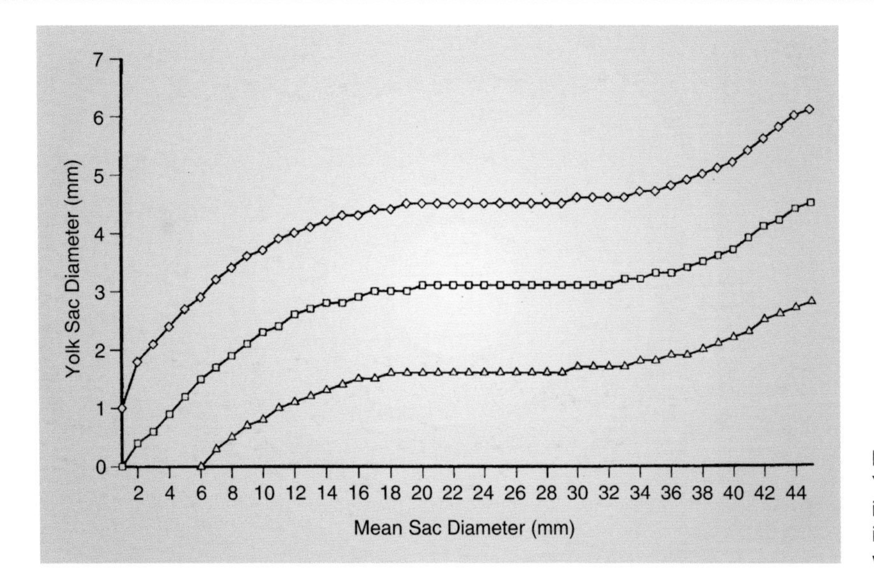

FIGURE 29 ■ Normal first trimester obstetric data Yolk sac diameter versus mean sac diameter. Squares indicate the mean; triangles represent a 5% confidence interval, diamonds represent a 95% confidence interval. *Source*: From Ref. 22.

95% confidence intervals compared with MSD (Fig. 29), the risk of abnormal outcome increases from approximately 24%, to between 44.4% and 60% (8). In our experience, yolk sac diameters of greater than 5.6 mm between 5 and 10 weeks' menstrual age (Fig. 30) are always associated with abnormal outcome (either embryonic demise or fetal anomaly) (8).

A persistently abnormal yolk sac shape is also a predictor of abnormal outcome (8). Although our data are based on a small number of patients with abnormal-appearing yolk sacs, when the yolk sac shape is crenated or irregular, outcome appears to depend on the appearance of the yolk sac on follow-up examination in one week. In our series, when the yolk sac shape reverted to normal, outcome was always normal. When the yolk sac shape remained abnormal, the embryos were at increased risk of embryonic demise or fetal anomaly (Fig. 31).

We use yolk sac data to determine which pregnancies are at increased risk of abnormal outcome. In the absence of other indicators of abnormal outcome, patients with abnormalities of yolk sac size or shape are followed closely in the first trimester. If the embryo survives the first trimester, sonographic follow-up is performed at 18 weeks' menstrual age to exclude a sonographically demonstrable anomaly. Genetic counseling is also offered.

Relation of β-HCG to MSD

Nyberg et al. found that 65% of abnormal pregnancies had a disproportionately low serum β-hCG for gestational sac size (88). Thus, if the β-hCG value is low for a given MSD, the pregnancy may be considered at risk for loss.

Subchorionic Hemorrhage

Subchorionic accumulation of blood is common in the first trimester and may be associated with vaginal bleeding. In a group of patients presenting with vaginal bleeding between 10 and 20 weeks' menstrual age, identification of a subchorionic hemorrhage was associated with a 50% fetal loss rate (89). When the volume of the hematoma was less than 40% the volume of the gestational sac, the outcome tended to be favorable. Some authors have suggested that small subchorionic hematomas in pregnancies of less than 12 weeks' gestational age are of no clinical significance and do not affect outcome (90).

The size of the subchorionic hemorrhage appears to be predictive of outcome. In a series of 516 patients with first trimester vaginal bleeding, a live embryo, and a subchorionic hematoma, Bennett et al. demonstrated an increase in the spontaneous abortion rate from 9.3% to 18.8% in patients with large subchorionic hematomas (defined as a hematoma separating more than two-thirds of the chorionic sac circumference from the decidua) (91). Small subchorionic hematomas (defined as separation of less than one-third of the chorionic sac circumference) and moderate subchorionic hematomas (defined as separation of one-third to one half of the chorionic sac circumference) were associated with spontaneous abortion rates of 7.7% and 9.2%, respectively.

Subchorionic hemorrhage may be due to abruption of the edge of the chorion frondosum-decidua basalis complex or to marginal sinus rupture (92). Although the hemorrhage usually abuts or elevates the edge of the chorion frondosum-decidua basalis complex, the bulk of the hemorrhage is usually situated between the decidua capsularis and chorion laeve and the decidua vera. Acute hemorrhage may be hyperechoic or isoechoic relative to chorionic fluid, and it becomes isoechoic with the chorionic fluid in one to two weeks (Fig. 32A–C).

EVCFD Predictors of Pregnancy Failure

In normal pregnancies, EVCFD demonstrates increased vascularity in the decidua immediately surrounding the

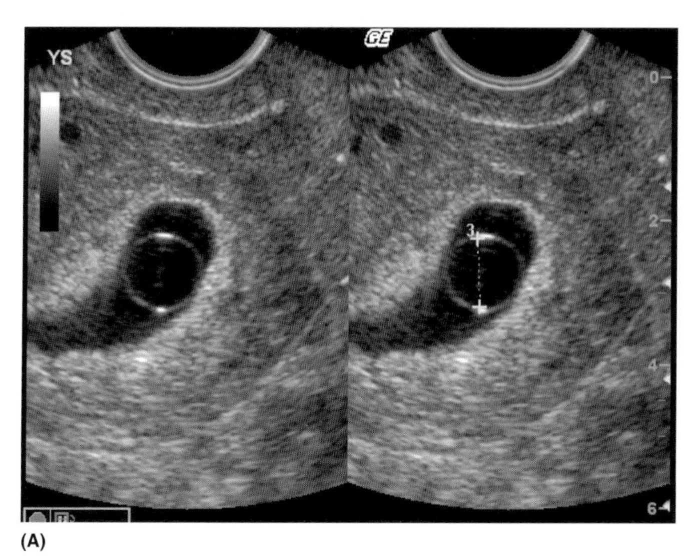

(A)

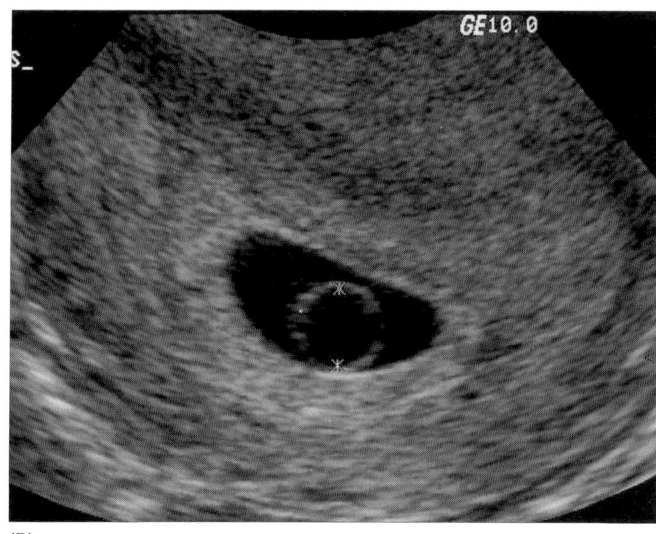

(B)

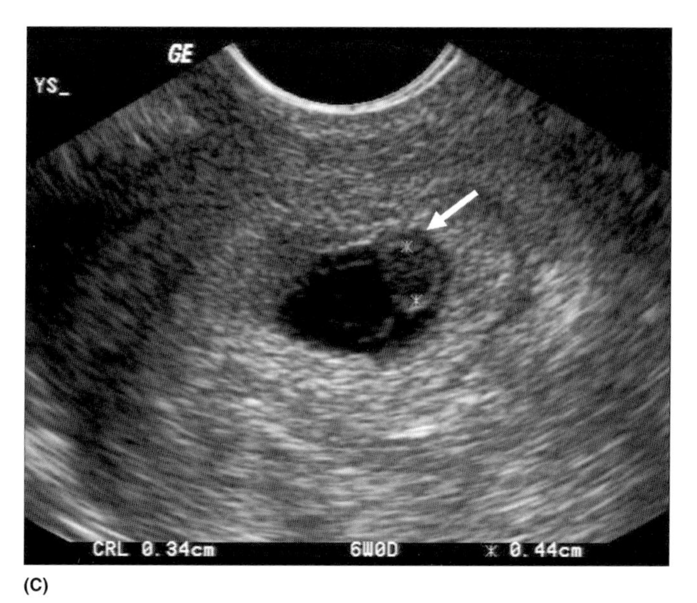

(C)

FIGURE 30 A–C ■ Large yolk sac. (**A**) Endovaginal sonography (sagittal plane) at 6.5 weeks shows an abnormally large yolk sac (*with and without calipers*) with an internal diameter of 9.4 mm. The patient aborted spontaneously four days later. (**B**) Another patient with a large yolk sac of 5.9 mm (*electronic calipers*) that within a week (**C**) got smaller and more echogenic (*arrow*) and then aborted.

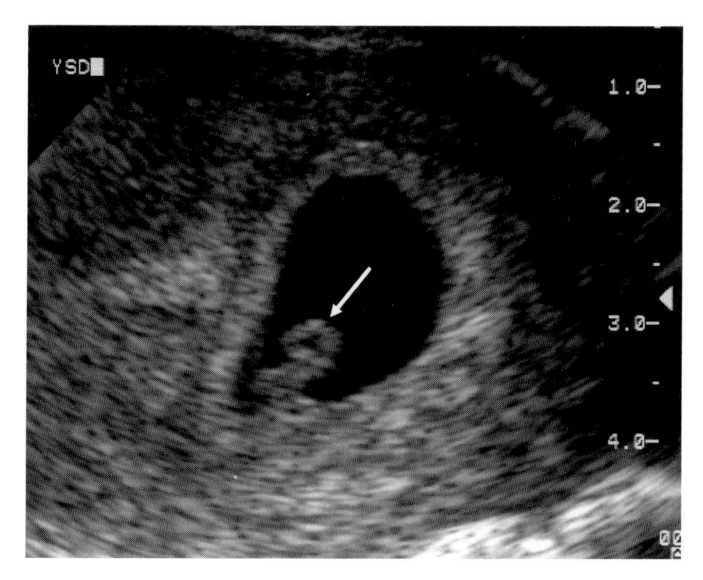

FIGURE 31 ■ Irregular yolk sac. Endovaginal sonography shows an irregular yolk sac (*arrow*) adjacent to a small embryo.

echogenic rim of the trophoblast (26). As noted previously, low-resistance arterial flow is present normally in the decidual spiral arteries (11,14). Extensive vascular flow in the trophoblast, however, indicates maternal flow in the intervillous space and may be associated with an increased risk of early pregnancy failure and pregnancy complications (14,26). The embryo and early placenta normally develop in a low oxygen environment. Prior to nine weeks, embryonic nutrition appears to depend on maternal uterine glandular secretions delivered to the intervillous space, and maternal plasma proteins and other molecules that diffuse into the intervillous space (25). Excessive maternal blood flow to the intervillous space may result in mechanical damage and increased oxidative stress that may cause cellular dysfunction and damage (93).

The decidual spiral arteries are invaded by extravillous trophoblastic cells. Inadequate trophoblastic invasion of the spiral arteries may be seen in early pregnancy failure and may be associated with increased resistance to

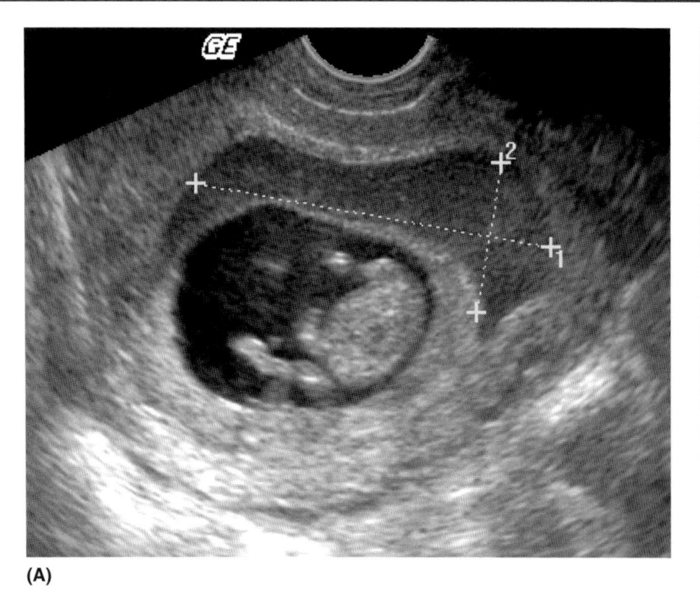

(A)

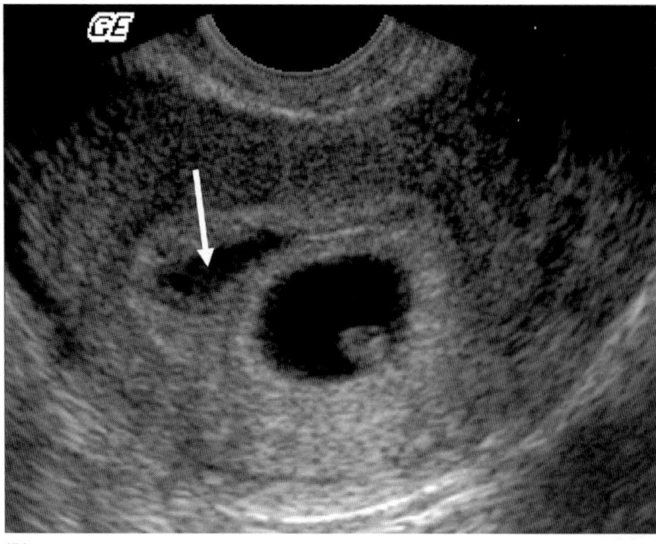

(B)

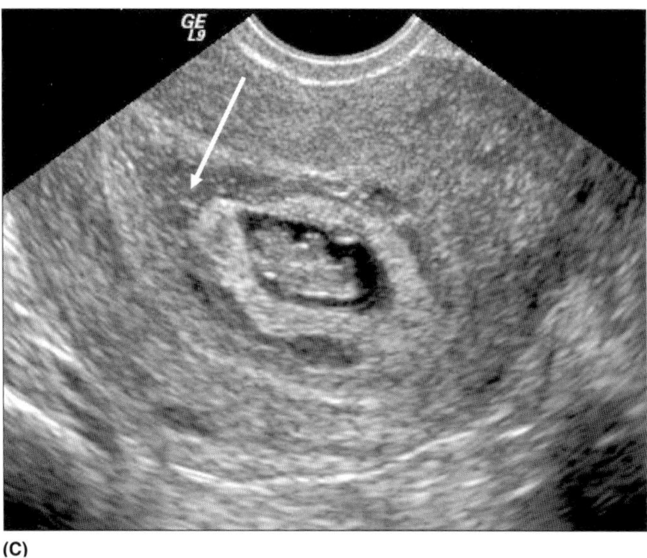

(C)

FIGURE 32 A–C ■ Subchorionic hemorrhage. (**A**) Endovaginal sonogram demonstrates a subchorionic hemorrhage (*electronic calipers*) within the endometrial canal adjacent to a gestational sac that contains a live embryo with a crown-rump length of 38mm (10.5 weeks' menstrual age). (**B**) Coronal scan of a seven-week sac with a small subchorionic hemorrhage (*arrow*). (**C**) A large subchorionic hemorrhage at eight weeks' menstrual age. The blood (*arrow*) is echogenic and surrounds the sac and embryo. The patient aborted one week later.

flow in the spiral arteries (14). One study suggests that an abnormal resistive index (higher than 0.55) in the decidual spiral arteries and active arterial blood flow in the intervillous space may be associated with an increased incidence of early pregnancy failure (14). The authors speculate that abnormal high-pressure blood flow in the spiral arteries may increase the pressure to the immature villi, with resulting detachment of the early villi and subsequent miscarriage.

Fibroids

Fibroids are associated with almost twice the spontaneous loss rate in early singleton pregnancies with documented cardiac activity. Benson et al. (94) noted a loss rate of 14% in women with fibroids compared with 7.6% in a control group. The groups were matched both for maternal and gestational age. Multiple fibroids had a higher loss rate than did single ones (23.6% vs. 8%) but there was no association with size or location.

FETAL ANOMALIES

With EVS, one can image the embryo in its earliest stages of development. Although the embryo can be visualized even before the onset of cardiac pulsation (4), the actual diagnosis of anomalies is limited by the resolution of current ultrasound equipment and by the morphologic and physiologic appearance of a specific anomaly at the time of sonographic examination. In general, the diagnosis of gross anomalies such as cystic hygromas (Fig. 33) and acrania (Fig. 34) can be made in the first trimester with either TVS or EVS (95). Case reports of first trimester diagnoses of polydactyly, ectopia cordis (96), and other anomalies have been published, but most diagnoses of anomalies are made in the second or third trimester. The advent of the 11- to 14-week scan has resulted in the sonographic evaluation of more first trimester fetuses than ever before. Evaluation of fetal anatomy is now commonplace in the late first trimester and

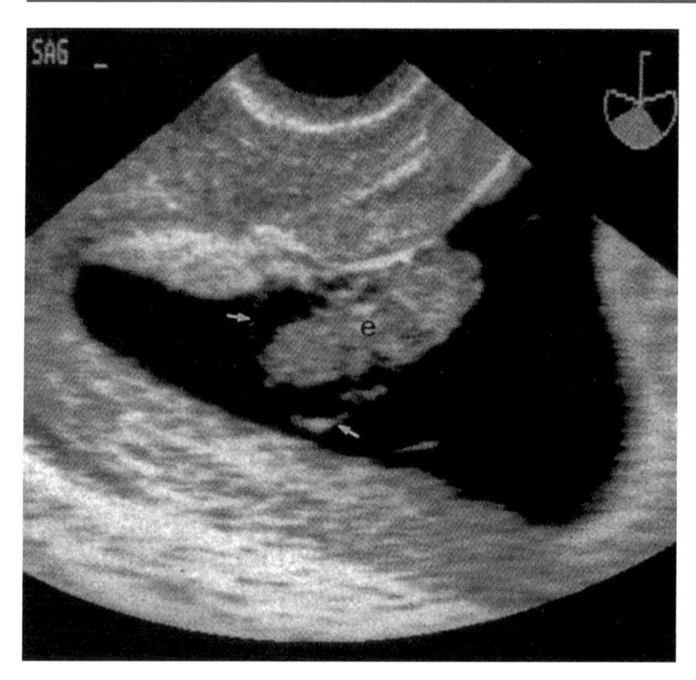

FIGURE 33 ■ Cystic hygromas (9.5 weeks' menstrual age). Endovaginal sonography shows bilateral cystic hygromas (*arrows*) in an embryo (e) with Turner syndrome. *Source*: From Ref. 22.

has resulted in case reports and series documenting a wide spectrum of fetal anomalies.

Many severe anomalies may have a normal sonographic appearance early in the first trimester. The most dramatic example is anencephaly (97), which may only become obvious after ossification of the calvarium occurs at 10 weeks' menstrual age. Anencephaly results from failure of closure of the rostral neuropore at 42 days' (six weeks') menstrual age (21). At approximately eight weeks' menstrual age, a fetal head may be identified in an anencephalic embryo because abnormal neural tissue is present superior to the orbits and skull base and the absence of the calvarium may not be recognized (98). The abnormal neural tissue superior to the orbits and skull base (referred to as exencephaly) wears down and progresses to anencephaly by the second trimester (99), with the associated classical appearance of absence of the calvarium above the orbital line. The diagnosis of exencephaly/anencephaly can be made in the first trimester after 11 weeks by visualization of the abnormal, protruding cerebral tissue in the absence of a surrounding calvarium, referred to as the "Mickey Mouse" sign (Fig. 33) (98,99).

Similarly, even severe renal anomalies are usually beyond the resolution of current equipment in the first trimester. Although recent studies suggest that it is possible to identify the fetal kidneys and bladder in almost all fetuses at 11 to 12 weeks (99), amniotic fluid volume does not depend on fetal renal function until the second trimester, making the diagnosis of oligo/anhydramnios impossible in the first trimester.

Conversely, in the first trimester, normal embryologic anatomy may mimic the sonographic appearances of fetal anomalies because of developmental stage. The fetal rhombencephalon (Fig. 35) appears as a cystic structure in the posterior fossa beginning at seven weeks' menstrual age and should not be mistaken for an intracranial cyst or hydrocephalus (100). The telencephalic and mesencephalic vesicles can also be seen at this stage (101).

Physiologic midgut hernia is often demonstrated as a small (6–9 mm) echogenic mass protruding into the umbilical cord at approximately eight weeks' menstrual

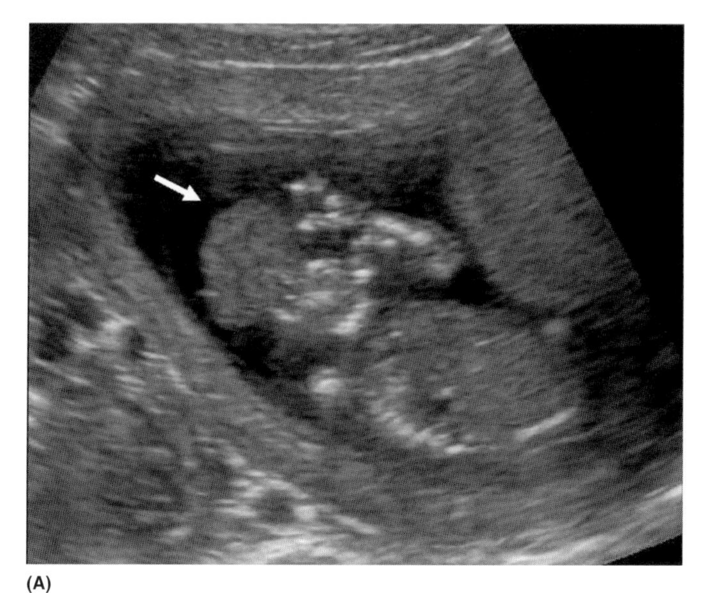

(A)

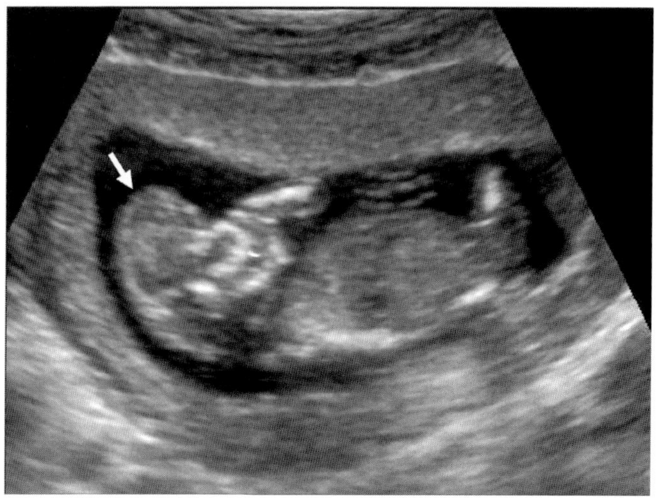

(B)

FIGURE 34 A AND B ■ Acrania. (**A**) Coronal and (**B**) sagittal sonogram of a 13-week fetus with no bony calvarium and a featureless brain (*arrow*) bulging from the calcified skull base.

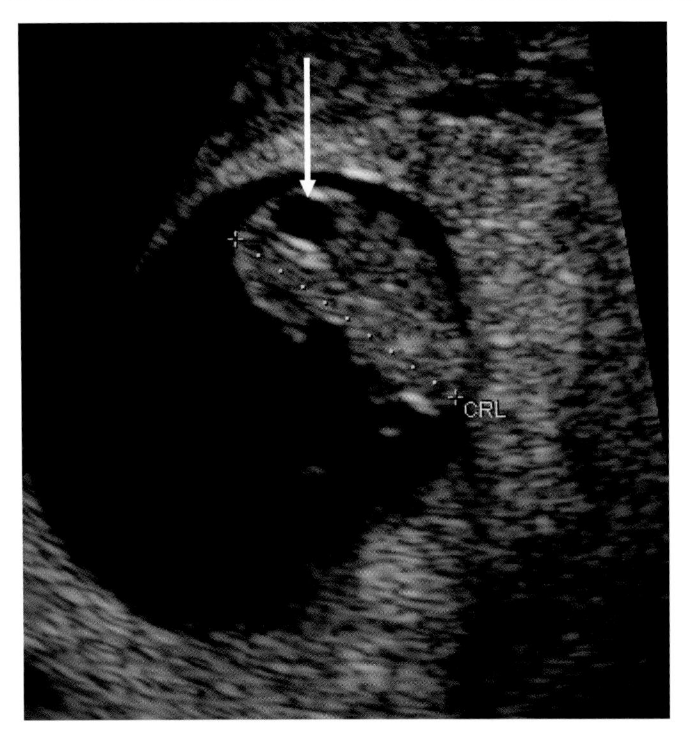

FIGURE 35 ■ Normal embryonic intracranial anatomy at 8.5 weeks' menstrual age. The embryonic rhombencephalon (*arrow*) is well visualized in this fetus with a crown-rump length (CRL) of 17 mm.

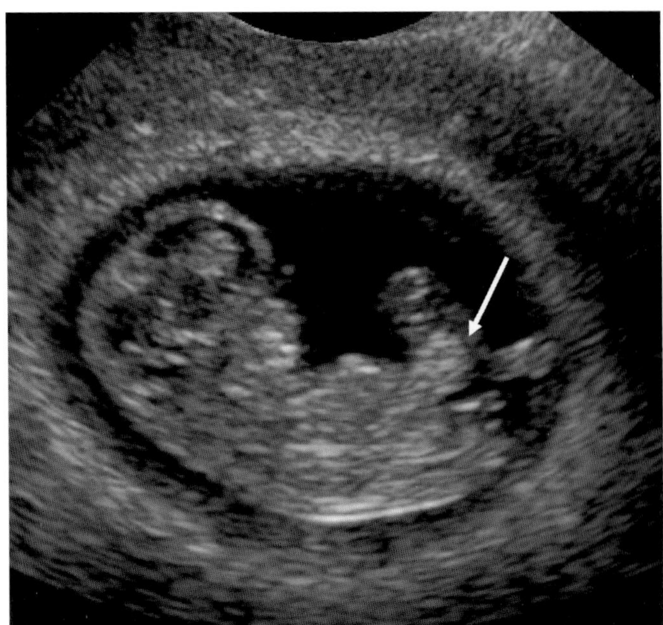

FIGURE 37 ■ Physiological herniation of the bowel. This 10-week embryo has echogenic bowel (*arrow*) protruding into the base of the umbilical cord.

age and is still present in 20% of normal fetuses at 12 weeks (Fig. 36) (102). The presence of herniated liver or heart is clearly abnormal, but because of overlap with the appearance of physiologic midgut hernia, the diagnosis of omphalocele may not be possible prior to 12 weeks if only a small amount of bowel is herniated into the umbilical cord. Small cysts measuring 2 to 7.5 mm are present in the umbilical cord in 0.4% to 2.1% of first trimester pregnancies (Fig. 37) (103,104). Ghezzi et al. identified two different subgroups of cord cysts: single cysts which were not associated with abnormal outcome, and multiple cysts which were associated with missed abortion and fetal anomalies (104).

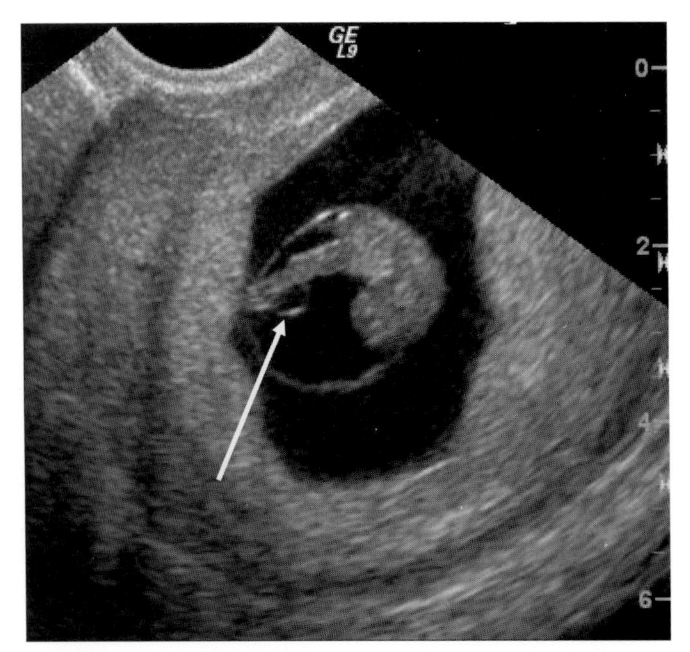

FIGURE 36 ■ Umbilical cord cyst. A 2 mm cyst (*arrow*) is present in the umbilical cord of this normal embryo at 8.5 weeks' menstrual age.

FIRST TRIMESTER MASSES

Ovarian Masses

The commonest mass seen in the first trimester of pregnancy is the corpus luteum cyst (105). This cyst secretes progesterone to support the pregnancy until the placenta can take over its hormonal function.

The corpus luteum forms in the secretory phase of the menstrual cycle and increases in size if pregnancy occurs. The corpus luteum of pregnancy is usually less than 5 cm in diameter and most commonly appears as a thin-walled, unilocular cyst (Fig. 38) (105). The cysts typically have prominent vascularity around the periphery which has been referred to as a "ring of fire."

The appearance of the corpus luteum cyst, however, may vary considerably. Corpus luteum cysts may be much larger, occasionally more than 10 cm in diameter. Internal septations and echogenic debris may be present secondary to internal hemorrhage (Fig. 39). The cyst wall and septations may be thick.

Clearly, a hemorrhagic corpus luteum cyst may be impossible to differentiate from a pathologic cyst on the basis of a single ultrasound examination. Corpus luteum

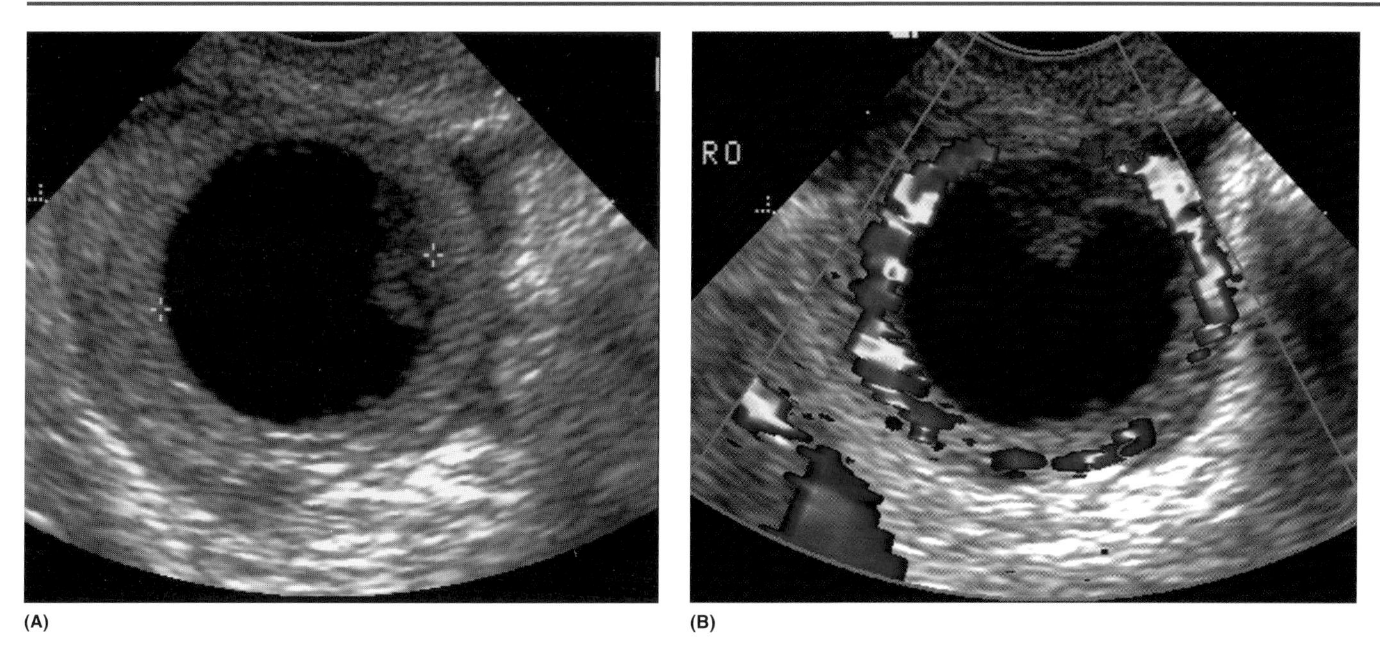

(A) (B)

FIGURE 38 A AND B ■ Corpus luteum cyst of pregnancy. (**A**) Endovaginal sonogram in the coronal plane through the right adnexa of a patient in early pregnancy shows a 2 cm corpus luteum cyst (*calipers*). (**B**) Color Doppler of the same cyst shows the typical ring of fire of vascular flow peripheral to the cyst.

cysts usually regress or have decreased in size on follow-up sonographic examination at 16 to 18 weeks' menstrual age. Cystic masses that persist should be followed. Surgical intervention is often indicated for large cysts that do not regress by the middle of the second trimester. Not all corpus luteum cysts regress, however, and differentiation from a pathologic cyst may be impossible.

Other cystic masses may appear initially in the first trimester of pregnancy because of displacement by the enlarged uterus. Although malignant ovarian neoplasm associated with pregnancy is rare (106), torsion, rupture, or dystocia is not. When elective surgical intervention is indicated, it is usually performed in the second trimester, when the likelihood of inducing premature labor is considered to be lowest. Dermoid cysts may present a characteristic appearance of a cystic mass with focal calcification and a fluid–fluid level. Other cystic masses may be more difficult to differentiate from corpus luteum cysts, although

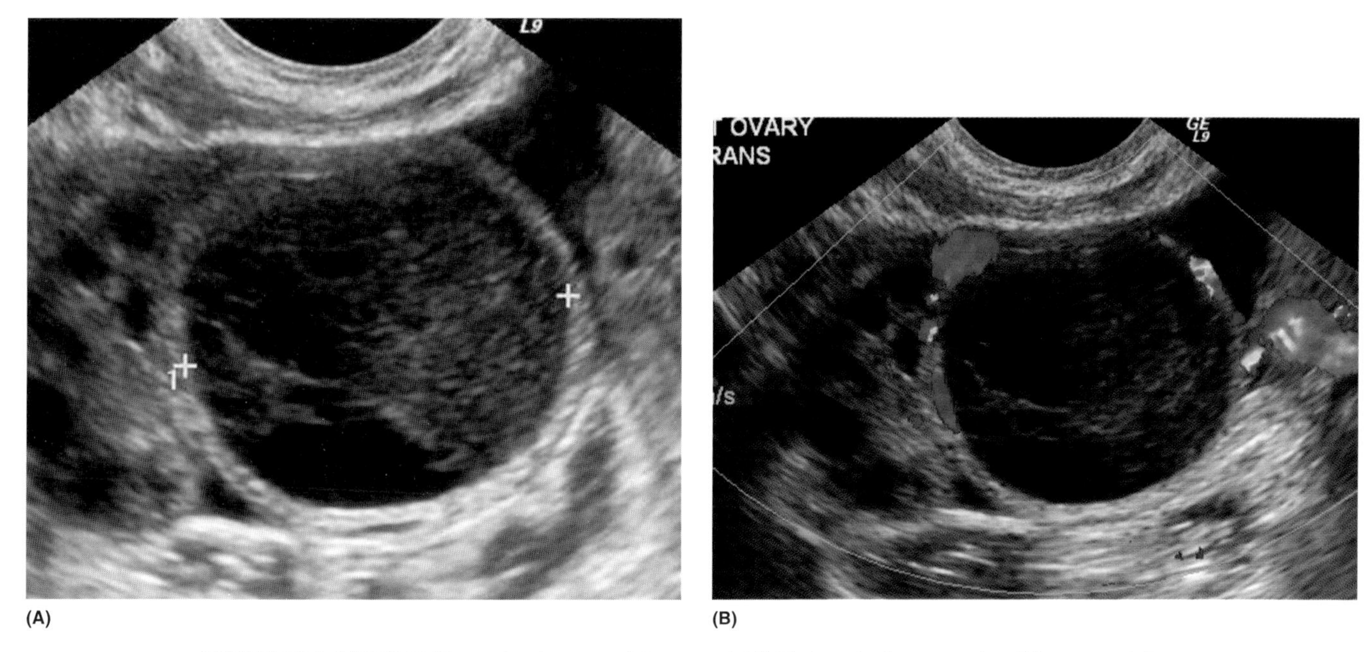

(A) (B)

FIGURE 39 A AND B ■ Hemorrhagic corpus luteum cyst. (**A**) Endovaginal sonography of the corpus luteum cyst (*electronic calipers*) shows internal echogenic material consistent with hemorrhage. (**B**) A Color Doppler scan also shows the hemorrhage as well as the vascular flow peripheral to the cyst.

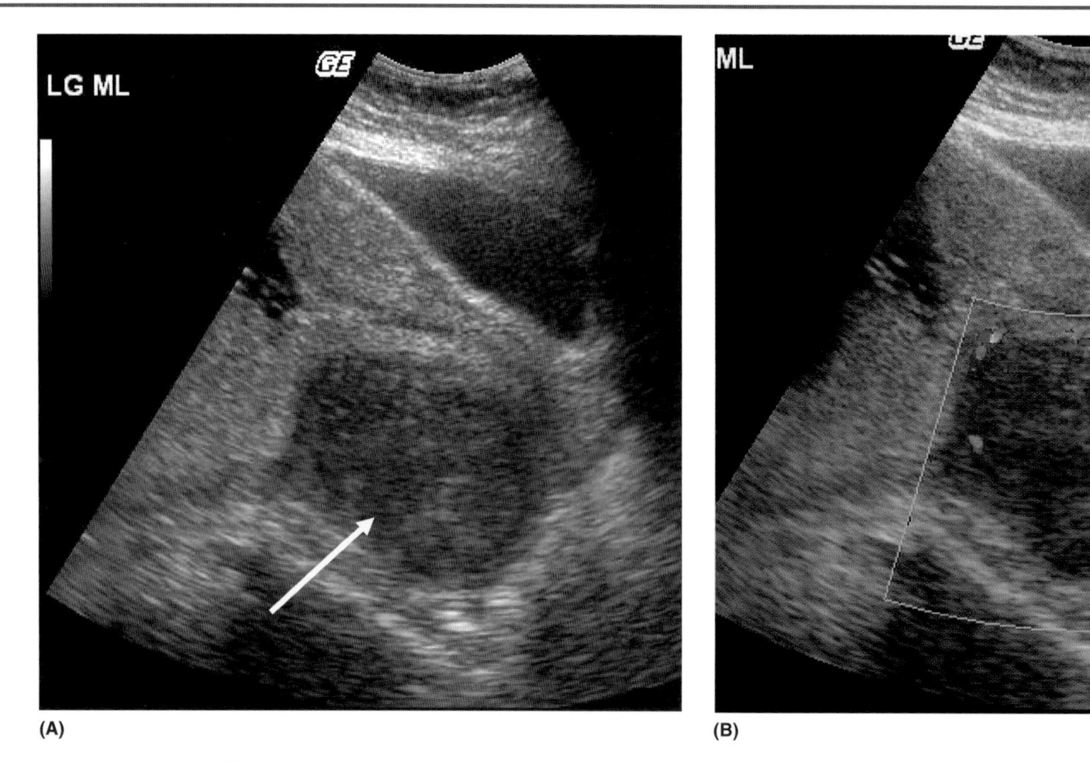

(A) (B)

FIGURE 40 A AND B ■ Uterine fibroid in pregnancy. (**A**) Sagittal sonogram of the lower uterine segment in a 25-week gestation shows a 6 cm hypoechoic fibroid (*arrow*) arising from the posterior aspect of the cervix. (**B**) Color Doppler of the fibroid shows vascular flow only in the periphery of the fibroid.

endometriomas (endometrial cysts) may have characteristic low-level echoes throughout. All cysts should be followed carefully to assess change in size (105).

Uterine Masses

Uterine fibroids are common pelvic masses often identified during pregnancy. Most fibroids do not change in size during pregnancy. Some fibroids, however, may enlarge rapidly because of stimulation by estrogen. Infarction and necrosis may result from rapid growth. These patients often experience pain.

Sonographically, uterine fibroids appear as solid uterine masses that attenuate sound to a variable degree (Figs. 40 and 41); they may or may not be calcified and they may have focal cystic areas related to necrosis.

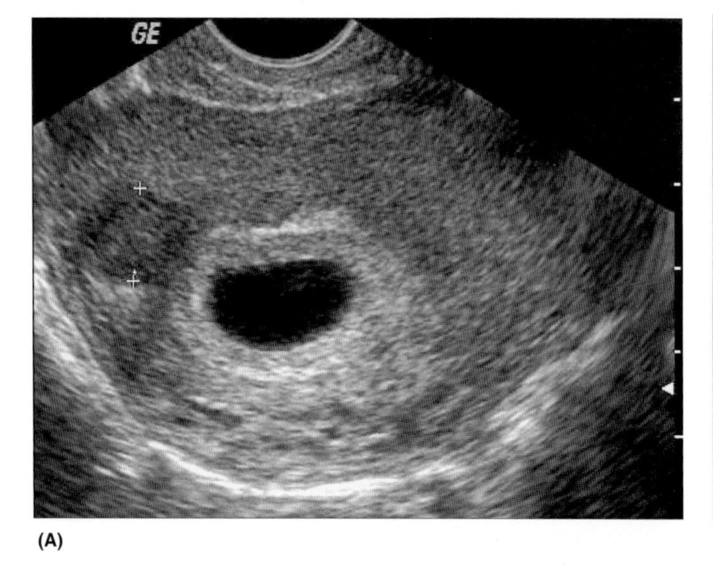

(A)

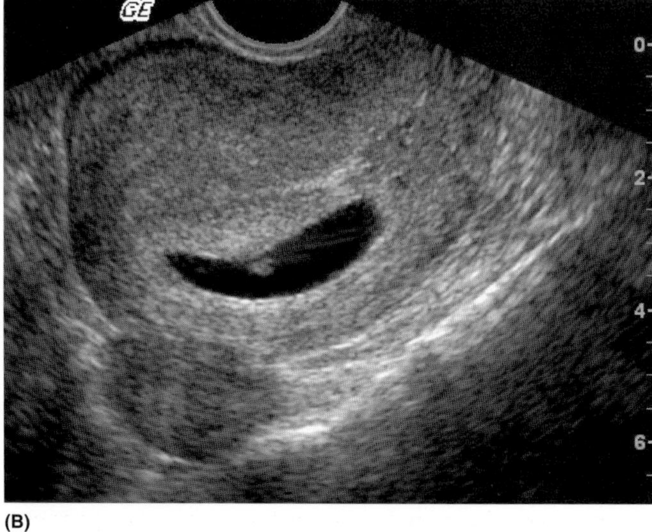

(B)

FIGURE 41 A–C ■ Uterine fibroids in pregnancy. Endovaginal sonogram in the sagittal plane of three different pregnancies shows (**A**) a small fibroid (*calipers*) abutting the sac, (**B**) a hypoechoic subserosal fibroid not near the sac. (*Continued*)

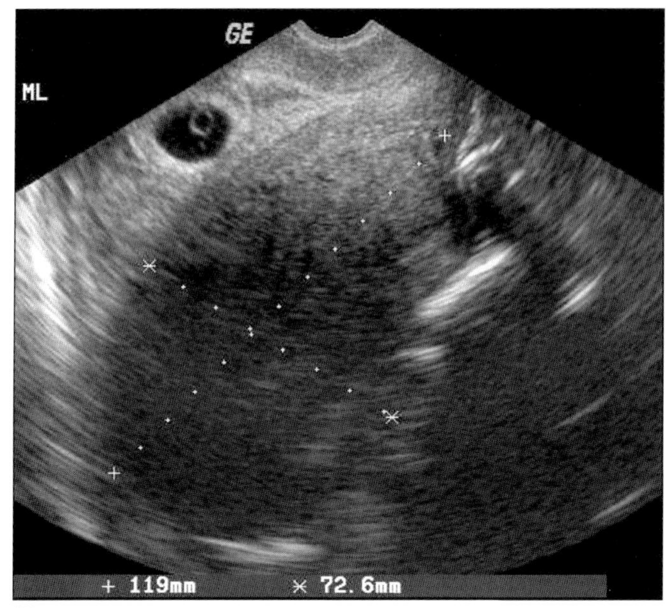

(C)

FIGURE 41 A–C ■ (*Continued*) (**C**) A large 11.9 × 7.2 cm fibroid (*electronic calipers*) distorting the posterior body of the uterus.

Fibroids may be differentiated from focal myometrial contractions by the transient nature of myometrial contractions. A repeat examination 20 to 30 minutes after the initial examination reveals the absence of a focal myometrial contraction, whereas fibroids are still present. Fibroids also may distort the uterine contour (serosal surface), whereas focal myometrial contractions usually do not.

REFERENCES

1. Levi CS, Lyons EA, Dashefsky SM. The first trimester. In: Rumack CM, Wilson SR, Charboneau JW, eds. Diagnostic Ultrasound. St Louis: Mosby–Year Book, 1991:692.
2. Dashefsky SM, Lyons EA, Levi CS, et al. Suspected ectopic pregnancy: endovaginal and transvesical ultrasound. Radiology 1988; 169:181.
3. Levi CS, Lyons EA, Lindsay DJ. Early diagnosis of nonviable pregnancy with endovaginal ultrasound. Radiology 1988; 167:383.
4. Levi CS, Lyons EA, Zheng XH, et al. Endovaginal ultrasound: demonstration of cardiac activity in embryos of less than 5.0 mm in crown rump length. Radiology 1990; 176:71.
5. Hately W, Case J, Campbell S. Establishing the death of an embryo by ultrasound: report of a public inquiry with recommendations. Ultrasound Obstet Gynecol 1995; 5:393.
6. Rowling SE, Coleman BG, Langer JE, et al. First trimester US parameters of failed pregnancy. Radiology 1997; 203:211.
7. Levi CS. Prediction of early pregnancy failure on the basis of mean gestational sac size and the absence of a sonographically demonstrable yolk sac. (Letter). Radiology 1995; 195:873.
8. Lindsay DJ, Lovett IS, Lyons EA, et al. Yolk sac diameter and shape at endovaginal US: predictors of pregnancy outcome in the first trimester. Radiology 1992; 183:115.
9. Rowling SE, Langer JE, Coleman BG, et al. Sonography during early pregnancy: dependence of threshold and discriminatory values on transvaginal transducer frequency. Am J Roentgenol 1999; 172:983.
10. Dillon EH, Feycock AL, Taylor KJW. Pseudogestational sacs: Doppler US differentiation from normal or abnormal intrauterine pregnancies. Radiology 1990; 176:359.
11. Taylor KJW, Ramos IM, Feycock AL, et al. Ectopic pregnancy: duplex Doppler evaluation. Radiology 1989; 173:93.
12. Pellerito JS, Taylor KJW, Quedens-Case C, et al. Ectopic pregnancy: evaluation with endovaginal colour flow imaging. Radiology 1992; 183:407.
13. Emerson DS, Cartier MS, Altieri LA, et al. Diagnostic efficacy of endovaginal colour flow imaging in an ectopic pregnancy screening program. Radiology 1992; 183:413.
14. Jaffe R, Dorgan A, Abramowicz JS. Color Doppler imaging of the uteroplacental circulation in the first trimester: value in predicting pregnancy failure or complication. Am J Roentgenol 1995; 164:1255.
15. Benacerraf BR. Intrauterine growth retardation in the first trimester associated with triploidy. J Ultrasound Med 1988; 7:153.
16. Moore KL, Persaud TVN. The beginning of human development: the first week. In: The Developing Human: Clinically Oriented Embryology. 4th ed. Philadelphia: WB Saunders, 1998:17–46.
17. Moore KL, Persaud TVN. Formation of the bilaminar embryonic disc and chorionic sac: the second week. In: The Developing Human: Clinically Oriented Embryology. 4th ed. Philadelphia: WB Saunders, 1998:47–62.
18. Moore KL, Persaud TVN. Formation of germ layers and early tissue and organ differentiation: the third week. In: The Developing Human: Clinically Oriented Embryology. 4th ed. Philadelphia: WB Saunders, 1998:63–82.
19. Moore KL, Persaud TVN. Organogenic period: the fourth to eighth weeks. In: The Developing Human: Clinically Oriented Embryology. 4th ed. Philadelphia: WB Saunders, 1998:65.
20. Moore KL, Persaud TVN: The fetal period: ninth week to birth. In: The Developing Human: Clinically Oriented Embryology. 4th ed. Philadelphia: WB Saunders, 1998:107–128.
21. Moore KL, Persaud TVN. The nervous system. In: The Developing Human: Clinically Oriented Embryology. 4th ed. Philadelphia: WB Saunders, 1998:451–490.
22. Levi C, Dashefsky D, Lyons E, et al. First trimester ultrasound: a practical approach. In: McGahan J, Porto M, eds. Diagnostic Obstetrical Ultrasound. Philadelphia: JB Lippincott, 1994.
23. Yeh H-C, Goodman JD, Carr L, et al. Intradecidual sign: a US criterion of early intrauterine pregnancy. Radiology 1986; 161:463.
24. Schaaps JP, Tsatsaris V, Goffin F, et al. Shunting the intervillous space: new concepts in human uteroplacental vascularization. Am J Obstet Gynecol 2005; 192:323–332.
25. Jauniaux E, Gulbis B, Burton GJ. The human first trimester gestational sac limits rather than facilitates oxygen transfer to the foetus—a review. Placenta 2003; (suppl A):S86–S93.
26. Jauniaux E, Greenwold N, Hempstock J, Burton GJ. Comparison of ultrasonographic and Doppler mapping of the intervillous circulation in normal and abnormal early pregnancies. Fertil Steril 2003; 79:100–106.
27. Carbillon L, Ziol M, Challier JC, et al. Doppler and immunohistochemical evaluation of decidual spiral arteries in early pregnancy. Gynecol Obstet Invest 2005; 59:24–28.
28. Cartier MS, Altieri LA, Emerson DS, et al. Diagnostic efficacy of endovaginal color flow Doppler in an ectopic pregnancy screening program [abstr]. Radiology 1990; 177(P):117.
29. Abramowicz JS, Kossoff G, Marsal K, Ter Haar G. Safety statement, 2000 (reconfirmed 2002). International society for ultrasound in obstetrics and gynecology (ISUOG). Ultrasound Obstet Gynecol 2002; 19:105.
30. Moore KL, ed. The Developing Human: Clinically Oriented Embryology. 4th ed. Philadelphia: WB Saunders, 1988:13.
31. Nyberg DA, Filly RA. Opinion: predicting pregnancy failure in empty gestational sacs. Ultrasound Obstet Gynecol 2003; 21:9–12.
32. Oh JS, Wright G, Coulam CB. Gestational sac diameter in very early pregnancy as a presictor of fetal outcome. Ultrasound Obstet Gynecol 2002; 20:267–269.
33. Filly RA. Ultrasound evaluation during the first trimester. In: Callen PW, ed. Ultrasonography in Obstetrics and Gynecology. 3rd ed. Philadelphia: WB Saunders, 1994:63.
34. Nyberg DA, Laing FC, Filly RA, et al. Ultrasonographic differentiation of the gestational sac of early intrauterine pregnancy from the pseudogestational sac of ectopic pregnancy. Radiology 1983; 146:755.

35. Bree RL, Edwards ML, Bohm-Velez, et al. Transvaginal sonography in the evaluation of normal early pregnancy: correlation with HCG level. Am J Roentgenol 1989; 153:7.
36. Nyberg DA, Laing FC, Filly RA, et al. Ultrasonographic differentiation of the gestational sac with quantitative HCG levels. J Ultrasound Med 1987; 6:145.
37. Sengoku K, Tamate K, Ishikawa M, et al. Transvaginal ultrasonographic findings and hCG levels in early intrauterine pregnancies. Nippon Sanka Fujinka Gakkai Zasshi 1991; 43:535–540.
38. Keith SC, London SN, Weitzman GA, et al. Serial transvaginal ultrasound scans and beta-human chorionic gonadotropin levels in early singleton and multiple pregnancies. Fertil Steril 1993; 59:1007–1010.
39. Moore KL, Persaud TVN. The placenta and fetal membranes. In: The Developing Human: Clinically Oriented Embryology. 4th ed. Philadelphia: WB Saunders, 1998:129–166.
40. Nyberg DA, Mack LA, Harvey D, et al. Value of the yolk sac in evaluating early pregnancies. J Ultrasound Med 1988; 7:129.
41. Yeh H-C, Rabinowitz JG. Amniotic sac development: ultrasound features of early pregnancy—the double bleb sign. Radiology 1988; 166:97.
42. Levi CS, Lyons EA, Lindsay DJ, et al. The sonographic evaluation of multiple gestation pregnancy. In: Fleischer AC, Romero R, Manning FA, et al., eds. The Principles and Practice of Ultrasonography in Obstetrics and Gynecology. 4th ed. Norwalk, CT: Appleton & Lange, 1991:359.
43. Mahony BS, Filly RA, Callen PW. Amnionicity and chorionicity in twin pregnancies: prediction using ultrasound. Radiology 1985; 155:205.
44. Cadkin AV, McAlpin J. Detection of fetal cardiac activity between 41 and 43 days of gestation. J Ultrasound Med 1984; 3:499.
45. Pennell RG, Needleman L, Pajak T, et al. Prospective comparison of vaginal and abdominal sonography in normal early pregnancy. J Ultrasound Med 1991; 10.
46. Rempen A. Diagnosis of viability in early pregnancy with vaginal sonography. J Ultrasound Med 1990; 9:711.
47. Howe RS, Isaacson KJ, Albert JL, et al. Embryonic heart rate in early human pregnancy. J Ultrasound Med 1991; 10:367.
48. Atrash HK, Friede A, Hogue CJR. Ectopic pregnancy mortality in the United States, 1970–1983. Obstet Gynecol 1987; 70:817.
49. Lawson HW, Atrash HK, Saftlas AF, et al. Ectopic pregnancy surveillance, United States, 1970–1985. MMWR CDC Surveill Summ 1988; 37:9.
50. Coupet E. Ectopic pregnancy: the surgical epidemic. J Natl Med Assoc 1989; 81:567.
51. Hann LE, Bachman DM, McArdle CR. Coexistent intrauterine and ectopic pregnancy: a reevaluation. Radiology 1984; 152:151.
52. Rein MS, Di Salvo DN, Friedman AJ. Heterotopic pregnancy associated with in vitro fertilization and embryo transfer: a possible role for routine vaginal ultrasound. Fertil Steril 1989; 51:1057.
53. Fleischer AC, Pennell RG, McKee MS, et al. Ectopic pregnancy: features at transvaginal sonography. Radiology 1990; 174:375.
54. Ackerman TE, Levi CS, Lyons EA, et al. Decidual cyst: endovaginal sonographic sign of ectopic pregnancy. Radiology 1993; 189:727.
55. Nyberg DA, Hughes MP, Mack LA, et al. Extrauterine findings of ectopic pregnancy at transvaginal US: importance of echogenic fluid. Radiology 1991; 178:823.
56. Timor-Tritsch IE, Yeh MN, Peisner DB, et al. The use of transvaginal ultrasonography in the diagnosis of ectopic pregnancy. Obstet Gynecol 1989; 161:157.
57. Thorsen MK, Lawson TL, Aiman EJ, et al. Diagnosis of ectopic pregnancy: endovaginal vs. transabdominal sonography. Am J Roentgenol 1990; 155:307.
58. Mahony BS, Filly, RA, Nyberg DA, et al. Sonographic evaluation of ectopic pregnancy. J Ultrasound Med 1985; 4:221.
59. Frates MC, Brown DL, Doubilet PM, et al. Tubal rupture in patients with ectopic pregnancy: diagnosis with transvaginal US. Radiology 1994; 191:769.
60. Cartwright PS. Ectopic pregnancy. In: Jones HW III, Wentz AC, Burnett LC, eds. Novak's Textbook of Gynecology. 11th ed. Baltimore: Williams & Wilkins, 1988:479.
61. Ackerman TE, Levi CS, Dashefsky SM, et al. Interstitial line: sonographic finding in interstitial (cornual) ectopic pregnancy. Radiology 1993; 189:83.
62. Centers for Disease Control. Current trends ectopic pregnancy: United States, 1990–1992. MMWR Morb Mortal Wkly Rep 1995; 44:46–48.
63. Atri M, Bret PM, Tulandi T. Spontaneous resolution of ectopic pregnancy: initial appearance and evolution at transvaginal ultrasound. Radiology 1993; 186:83–86.
64. Atri M, Chow C-M, Kintzen G, et al. Expectant treatment of ectopic pregnancies. Am J Roentgenol 2001; 176:123–127.
65. Zacur HA. Expectant management of ectopic pregnancy. Radiology 1993; 186:11–12.
66. Frates MC, Laing FC. Sonographic evaluation of ectopic pregnancy: an update. Am J Roentgenol 1995; 165:251–259.
67. Lipscomb GH, McCord ML, Stovall TG, et al. Predictors of success of methotrexate treatment in women with tubal ectopic pregnancies. N Engl J Med 1999; 341:1974–1978.
68. Atri M, Bret PM, Tulandi T, Senterman MK. Ectopic pregnancy: evolution after treatment with transvaginal methotrexate. Radiology 1992; 185:749–753.
69. Plancher S, Conway C, Zalud I. Transvaginal color Doppler ultrasound in the conservative treatment and surveillance of three ectopic pregnancies. Croat Med J 1998; 39(2):216–219.
70. Elson J, Salim A, Tailor A, et al. Prediction of early pregnancy viability in the absence of an ultrasonically detectable embryo. Ultrasound Obstet Gynecol 2003; 21:57–61.
71. Wilcox AJ, Weinberg CR, O'Connor JF, et al. Incidence of early loss of pregnancy. N Engl J Med 1988; 319:189.
72. Bateman BG, Felder R, Kolp LA, et al. Subclinical pregnancy loss in clomiphene citrate-treated women. Fertil Steril 1992; 57:25.
73. Hertig AT, Rock J. A series of potentially abortive ova recovered from fertile women prior to the first missed menstrual period. Am J Obstet Gynecol 1949; 58:968.
74. Hertig AT, Rock J, Adams BC, et al. Thirty-four fertilized human ova, good, bad and indifferent, recovered from 210 women of known fertility: a study of biologic wastage in early human pregnancy. Pediatrics 1959; 23:202.
75. Cunningham FG, MacDonald PC, Gant NF, Leveno KJ, Gilstrap LC III. Abortion. In: Cunningham FG, MacDonald PC, Gant NF, Leveno KJ, Gilstrap LC III, eds. Williams Obstetrics. 19th ed. Norwalk, CT: Appleton & Lange, 1993:661–690.
76. Goldstein SR. Embryonic death in early pregnancy: a new look at the first trimester. Obstet Gynecol 1994; 84:294–297.
77. Cashner KA, Christopher CR, Dysert GA. Spontaneous fetal loss after demonstration of a live fetus in the first trimester. Obstet Gynecol 1987; 70:827.
78. Wilson RD, Kendrick V, Wittmann BK, et al. Spontaneous abortion and pregnancy outcome after normal first-trimester ultrasound examination. Obstet Gynecol 1986; 67:352.
79. Simpson JL. Incidence and timing of pregnancy losses: relevance to evaluating safety of early prenatal diagnosis. Am J Med Genet 1990; 35:165.
80. Nyberg DA, Laing FC, Filly RA. Threatened abortion: sonographic distinction of normal and abnormal gestation sacs. Radiology 1986; 158:397.
81. Bernard KG, Cooperberg PL. Sonographic differentiation between blighted ovum and early viable pregnancy. Am J Roentgenol 1985; 144:597.
82. Nyberg DA, Mack LA, Laing FC, et al. Distinguishing normal from abnormal gestational sac growth in early pregnancy. J Ultrasound Med 1987; 6:23–26.
83. Horrow MM. Enlarged amniotic cavity: a new sonographic sign of early embryonic death. Am J Roentgenol 158:359–362.
84. Harris RD, Vincent LM, Askin FB. Yolk sac calcification: a sonographic finding associated with intrauterine embryonic demise in the first trimester. Radiology 1988; 166:109.
85. Doubilet PM, Benson CB. Embryonic heart rate in the early first trimester: what rate is normal? J Ultrasound Med 1995; 14:431.
86. Bromley B, Harlow BL, Laboda LA, et al. Small sac size in the first trimester: a predictor of poor fetal outcome. Radiology 1991; 178:375.

87. Kurtz AB, Needleman L, Pennell RG, et al. Can detection of the yolk sac in the first trimester be used to predict the outcome of pregnancy? A prospective sonographic study. Am J Roentgenol 1992; 158:843.
88. Nyberg DA, Filly RA, Filho DRD, et al. Abnormal pregnancy: early diagnosis by US and serum chorionic gonadotropin levels. Radiology 1986; 158:393.
89. Sauerbrei EE, Pham DH. Placental abruption and subchorionic hemorrhage in the first half of pregnancy: US appearance and clinical outcome. Radiology 1986; 160:109.
90. Stabile I, Campbell S, Grudzinskas JG. Threatened miscarriage and intrauterine hematomas: sonographic and biochemical studies. J Ultrasound Med 1989; 8:289.
91. Bennett GL, Bromley B, Lieberman E, Benacerraf BR. Subchorionic hemorrhage in first-trimester pregnancies: prediction of pregnancy outcome with sonography. Radiology 1996; 200: 803–806.
92. Nyberg DA, Cyr DR, Mack LA, et al. Sonographic spectrum of placental abruption. Am J Roentgenol 1987; 148:161.
93. Jauniaux E, Burton GJ. Pathophysiology of histological changes in early pregnancy loss. Placenta 2005; 26:114–123.
94. Benson CB, Chow JS, Chang-Lee W, Hill JA III, Doubilet PM. Outcome of pregnancies in women with uterine leiomyomas identified by sonography in the first trimester. J Clin Ultrasound 2001; 29:261–264.
95. Cullen MT, Green J, Whetham J, et al. Transvaginal ultrasonographic detection of congenital anomalies in the first trimester. Am J Obstet Gynecol 1990; 163:466.
96. Bennett TL, Burlbaw J, Drake CK, et al. Diagnosis of ectopia cordis at 12 weeks' gestation using transabdominal ultrasonography with color flow Doppler. J Ultrasound Med 1991; 10:695.
97. Goldstein RB, Filly RA. Prenatal diagnosis of anencephaly: spectrum of sonographic appearances and distinction from the amniotic band syndrome. Am J Roentgenol 1988; 151:547.
98. Chatzipapas IK, Whitlow BJ, Economides DL. The "Mickey Mouse" sign and the diagnosis of anencephaly in early pregnancy. Ultrasound Obstet Gynecol 1999; 13:196–199.
99. Peregrine E, Pandya P. Structural anomalies in the first trimester. In: Rumack CM, Wilson SR, Charboneau JW, Johnson JM, eds. Diagnostic Ultrasound. 3rd ed. St. Louis, MO: Elsevier Mosby, 2005:1127–1157.
100. Cyr DR, Mack LA, Nyberg DA, et al. Fetal rhombencephalon: normal US findings. Radiology 1988; 166:691.
101. Timor-Tritsch IE, Monteagudo A, Warren WB. Transvaginal ultrasonographic definition of the central nervous system in the first and early second trimesters. Am J Obstet Gynecol 1991; 164:497.
102. Schmidt W, Yarkoni S, Crelin ES, et al. Sonographic visualization of physiologic anterior abdominal wall hernia in the first trimester. Obstet Gynecol 1987; 69:911.
103. Skibo LK, Lyons EA, Levi CS. First-trimester umbilical cord cysts. Radiology 1992; 182:719.
104. Ghezzi F, Raio L, Di Naro E, et al. Single and multiple umbilical cord cysts in early gestation: two different entities. Ultrasound Obstet Gynecol 2003; 21:215–219.
105. Fleischer AC, Boehm FH, James AE Jr. Sonographic evaluation of pelvic masses and maternal disorders occurring during pregnancy. In: Sanders RC, James AE Jr, eds. The Principles and Practice of Ultrasonography in Obstetrics and Gynecology. 3rd ed. Norwalk, CT: Appleton-Century-Crofts, 1985:435.
106. Pennes DR, Bowerman RA, Silver TM. Echogenic adnexal masses associated with first-trimester pregnancy: sonographic appearance and clinical significance. J Clin Ultrasound 1985; 13:391.

Fetal Measurements

Lyndon M. Hill

40

INTRODUCTION

The most important role of ultrasound in obstetrics is to accurately assess the duration of pregnancy. For purposes of clarity, menstrual age (time from the last menses) will be used throughout this chapter. Gestational age is frequently used interchangeably with menstrual age but is more accurately defined as "time in utero." In patients with regular menstrual cycles and ovulation on day 14, menstrual age is two weeks greater than true embryonic age.

Reliance on the last menstrual period (LMP) to determine the length of pregnancy is fraught with error. Although a menstrual cycle may have a mean length of 29.1 days, cycles of 25 to 31 days are recorded by only 77% of menstruating women. Variability in cycle lengths is greatest for women under 25 years of age and over 39 years. Whereas the optimal time for conception is between day 10 and day 15, women can become pregnant on almost any day of the cycle. Hence, conception cannot be predicted with certainty even in women with regular menstrual cycles (1,2). In one study, the ultrasonic estimation of menstrual age differed from LMP age by more than 14 days in 15% of all second-trimester pregnancies examined (3). A change of this degree in the patient's estimated date of confinement can have a significant effect on subsequent obstetric management.

During the first half of pregnancy, the pattern of fetal development is remarkably constant across geographic, ethnic, and socioeconomic lines. Fetal growth during this part of pregnancy appears to reflect an inherent genetic drive (4–6). For the average Caucasian fetus, a weight of 1.1 kg at 28 weeks, 2.2 kg at 34 weeks, and 3.3 kg at 40 weeks represents the "normal" rate of growth (7). However, fetal growth rate and length of gestation are not closely related. Individual fetal growth curves (Fig. 1) (8) may,

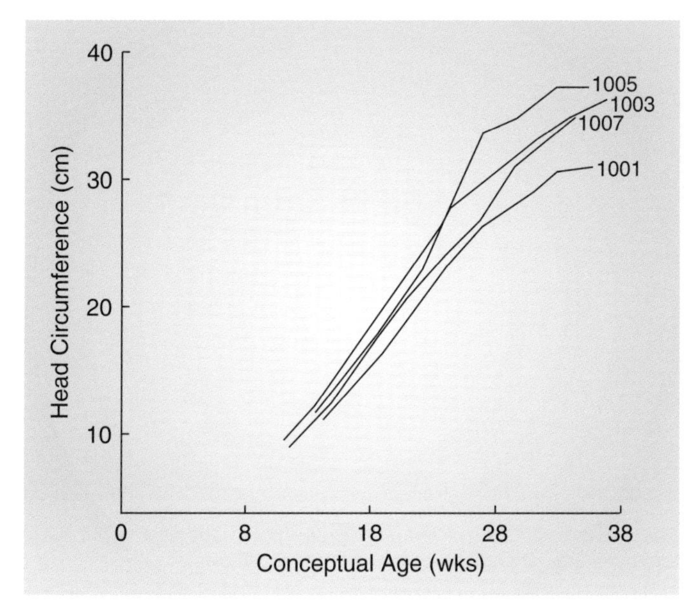

FIGURE 1 ■ Longitudinal studies of the change in head circumference with gestational age in four fetuses. The head circumference was measured at three-week intervals. *Source*: From Ref. 8.

(*Continued*)

therefore, be a more accurate way to assess growth and indirectly determine menstrual age in the third trimester (9). Finally, the variability of any fetal measurement increases with advancing gestational age. Hence, the range of a sonographic assessment of menstrual age should be reported.

FIRST TRIMESTER

Gestational Sac

The first sonographic evidence of intrauterine pregnancy is a thickened and echogenic endometrial stripe. The implantation of the blastocyst into the uterine wall results in its eccentric position relative to the endometrial lining (Fig. 2). In the early part of the first trimester, the gestational sac is composed predominantly of a chorionic cavity. At four weeks' menstrual age, the chorionic cavity measures 1.0 mm in average diameter. The gestational sac can be visualized with transvaginal sonography at four weeks and four days (average sac diameter of approximately 3.0 mm) (10). The gestational sac increases in mean diameter by approximately 1.0 mm per day (range, 0.71–1.75 mm per day) (11).

A mean gestational sac diameter is obtained from three dimensions (length, width, and depth), measured from inner edge to inner edge (Fig. 3). The shape of a gestational sac may vary from round to ovoid depending on maternal bladder volume (Fig. 4) uterine contractility (Fig. 5) or the presence of myomas immediately adjacent to the sac.

Three-dimensional (3D) volumetric measurements of the gestational sac are superior to a calculated two-dimensional (2D) volume (12). However, the measurements obtained using electronic calipers on a 2D image are sufficiently accurate for clinical use.

Up to eight weeks' menstrual age, an excellent correlation exists among human chorionic gonadotropin (hCG) levels, menstrual age, and gestational sac size (Figs. 6 and 7, Table 1[a]) (13–15). Because concentrations of hCG rise exponentially with time, a range of hCG levels is observed for any given gestational sac size (Fig. 8).

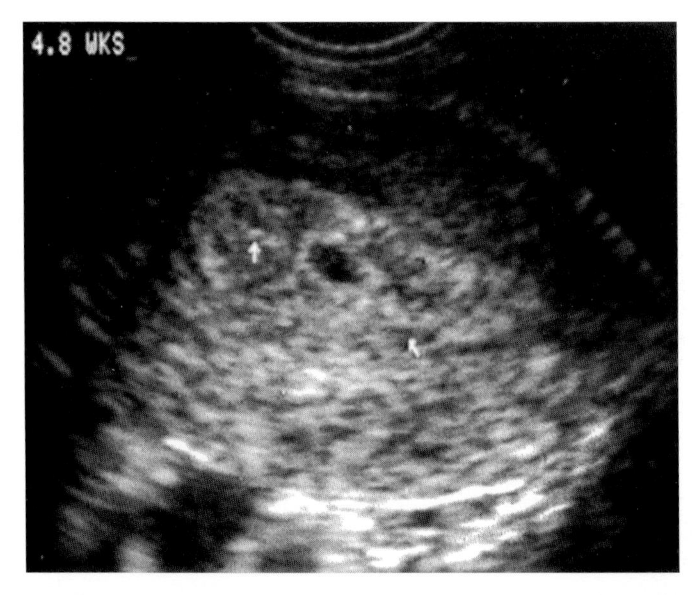

FIGURE 2 ■ A 4.8-week menstrual age embryo. The gestational sac is embedded superior to the endometrial stripe (*arrows*).

[a] Tables 1 to 23 appear at the end of the chapter.

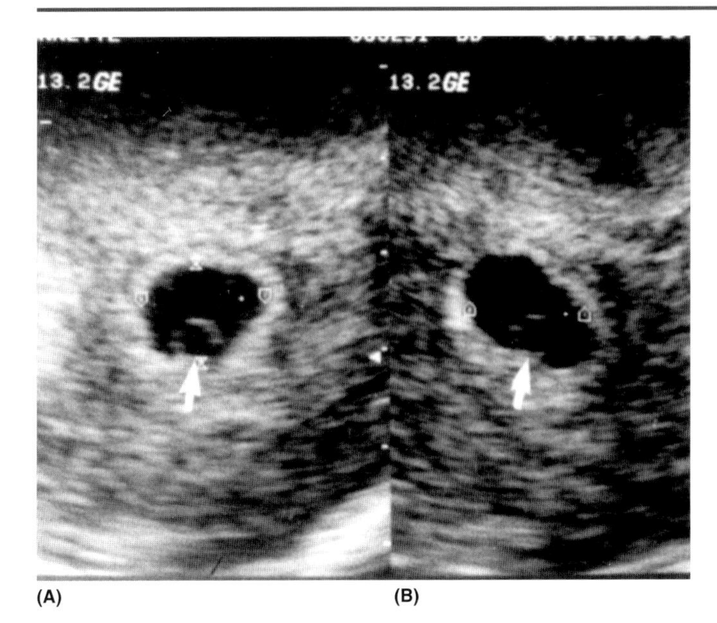

FIGURE 3 A AND B ■ A 5.2-week gestational sac: (**A**) longitudinal; (**B**) transverse. The yolk sac (*arrows*) is visualized within the gestational sac.

The relationships among menstrual age, hCG levels, and transvaginal ultrasound findings are provided in Table 2. Values for hCG are reported in terms of the World Health Organization third international standard, which is similar to the internal reference preparation (IRP). The second international standard is no longer available; its values were approximately 50% of IRP values (17).

The overall accuracy of a gestational sac measurement in estimating embryonic age is approximately one week (18). Hellman et al. (19) determined menstrual age from 5 to 12 weeks by mean sac diameter (Fig. 9). As menstrual age advances beyond seven weeks, the 95% confidence interval of gestational sac measurements is greater than for crown–rump length (CRL) (Fig. 10).

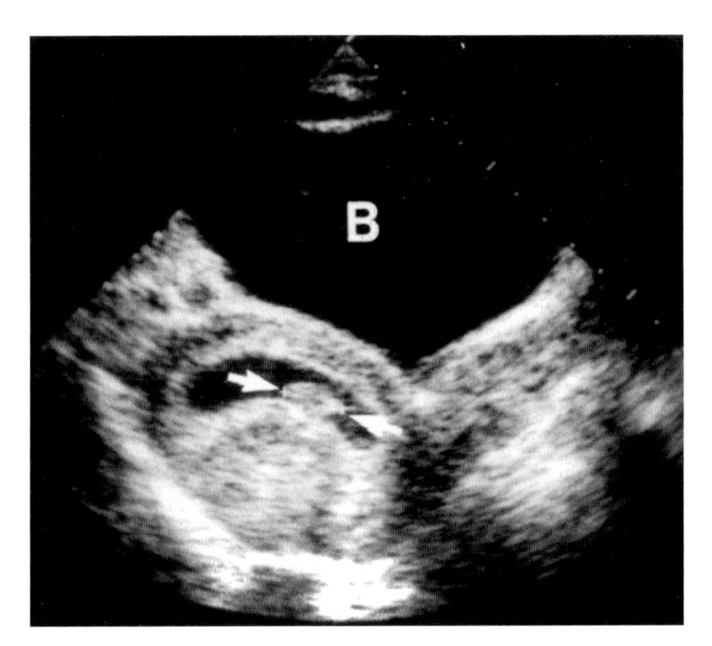

FIGURE 4 ■ An 8.0-week gestational sac compressed by a full bladder (B). Arrows outline the crown–rump length.

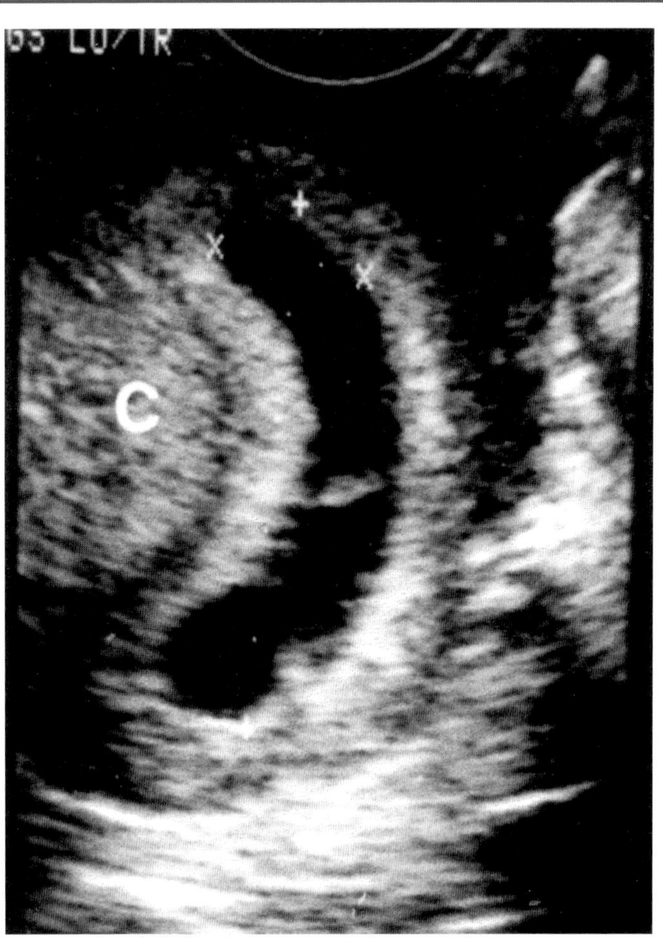

FIGURE 5 ■ Transvaginal scan of an ovoid seven-week gestational sac. The effect of the uterine contraction (C) on the contour of the gestational sac can be appreciated.

Hence, whenever the CRL can be measured, it should be used for menstrual age assessment, rather than the mean gestational sac diameter.

Pitfall

One must always assess the nature of the fluid in the uterus. Fluid may accumulate within the endometrial cavity in patients with an ectopic pregnancy. Whereas a gestational sac implants eccentrically into the endometrium (Fig. 2), endometrial fluid associated with an ectopic gestation is located centrally.

Comparison of Gestational Sac with Crown–Rump Length

A comparison of the mean sac diameter with the CRL provides helpful prognostic information relative to pregnancy outcome. In approximately 2% of pregnancies scanned between 5.3 and 9.4 menstrual weeks, the gestational sac is smaller than expected for the gestational age by the CRL (Figs. 11 and 12). When the difference between the mean sac diameter and the CRL was less than 5 mm, between 5 and 7.9 mm, and more than 8 mm, Dickey et al. (22) reported a pregnancy loss rate of 80%, 26.5%, and 10.6%, respectively. At our facility, a mean sac diameter minus CRL of 5 mm and below has been associated with

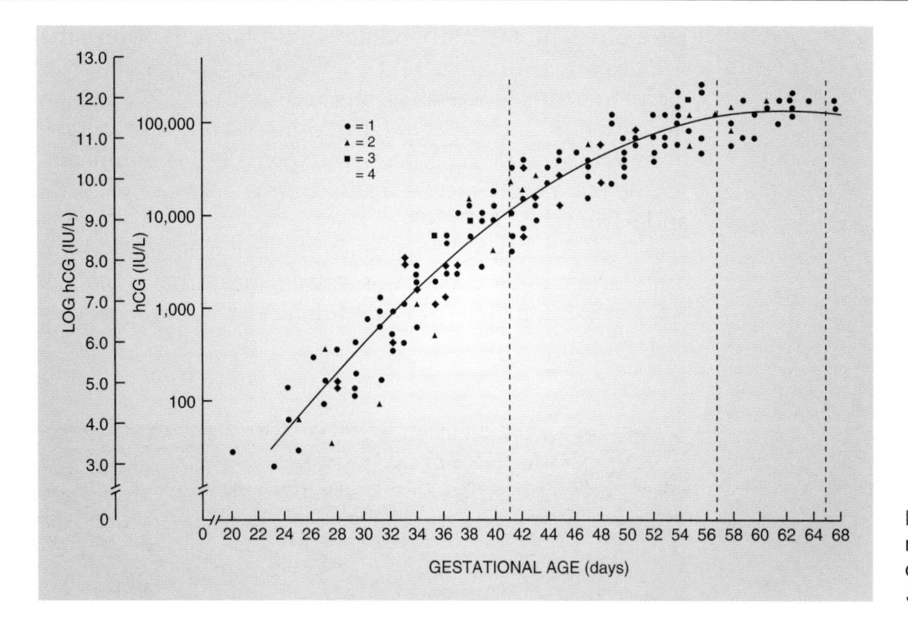

FIGURE 6 ■ hCG increase in normal early pregnancy. Symbols correspond to the indicated number of data points. hCG, human chorionic gonadotropin. *Source*: From Ref. 13.

a miscarriage rate of 50% (unpublished data). This small-sac syndrome occurs more often with karyotypically normal, in contrast to karyotypically abnormal, fetuses. Meegdes et al. (23) reported that chorionic villous vascularization is deficient in cases of embryonic death or blighted ovum. Hence, suboptimal maternal–embryo exchange may cause embryonic demise up to eight weeks' gestation. These results are consistent with the findings of Stern and Coulam (24) who reported that the CRL was smaller than expected in 86% of pregnancies lost after the documentation of cardiac activity.

Crown–Rump Length

The sonographic measurement of the CRL was introduced by Robinson in 1973 (25). The advent of transvaginal sonography has permitted the visualization and measurement of embryonic poles of above 2mm. The embryonic heart begins to pulsate at 5.5 to 6.5 menstrual weeks (CRL: 1.5–3.0mm) (10). Hence, an embryo may be visualized before cardiac activity has begun (Fig. 13) (26).

The 2.0- to 3.0-mm embryonic disc is relatively straight. Because neither the crown nor the rump can be

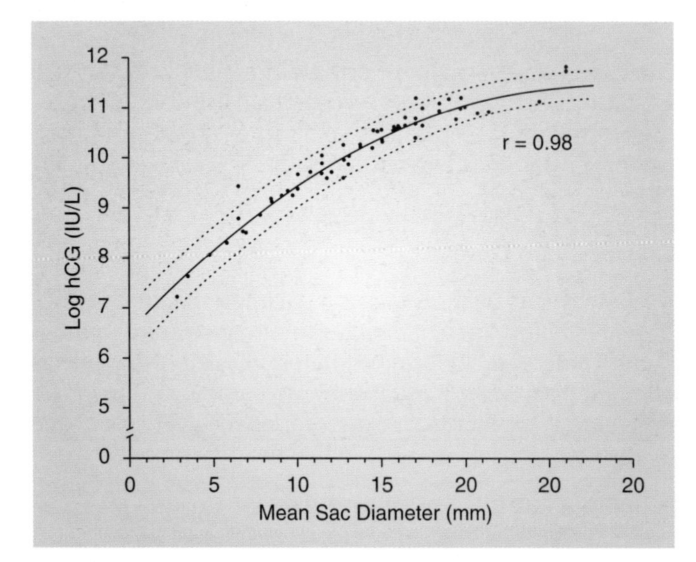

FIGURE 7 ■ The increase in hCG seen in normal early pregnancy with respect to mean gestational sac diameter. hCG, human chorionic gonadotropin. *Source*: From Ref. 14.

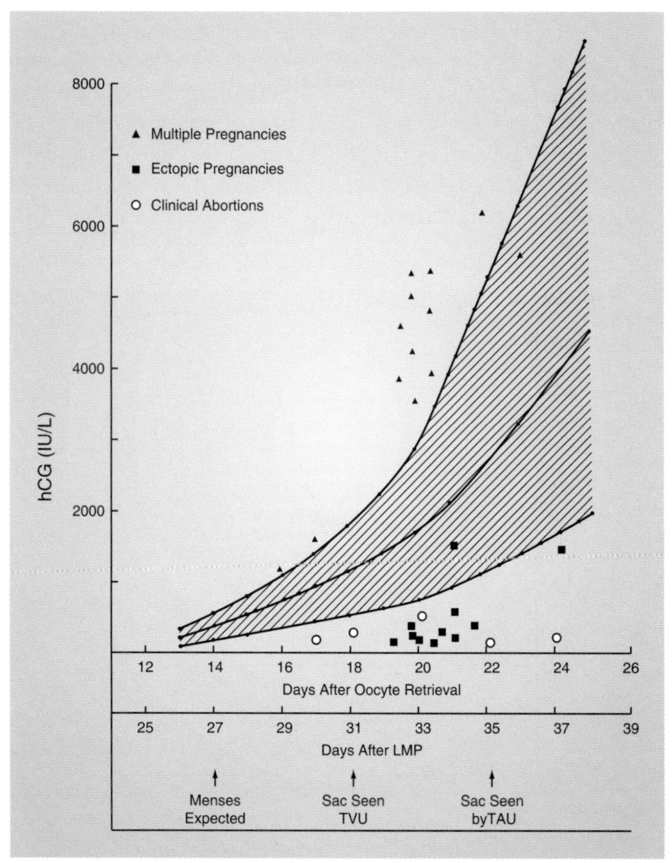

FIGURE 8 ■ Range and median hCG with gestational age. hCG, human chorionic gonadotropin; LMP, last menstrual period; TVU, transvaginal ultrasonography; TAU, transabdominal ultrasonography. *Source*: From Ref. 16.

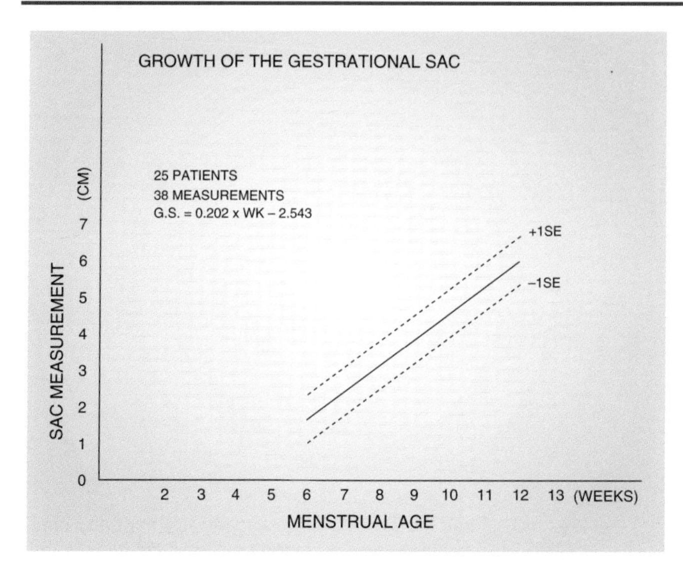

FIGURE 9 ■ Relation between the diameter of the gestational sac and the duration of pregnancy. *Source*: From Ref. 19.

identified at this gestational age, a measurement is actually along the embryo's greatest length. The embryo attains a C-shape by six weeks' gestation. The sonographic measurement (4.0 mm) obtained is therefore from the neck to the rump (Fig. 14) (27). Once a gestational age of eight menstrual weeks is reached, a true CRL can be measured (Fig. 15a and b) (28). Although reflex activity is observed by 7.5 weeks' menstrual age (29), spontaneous motor activity characteristically begins between weeks 8 and 9 from the LMP (30).

The CRL (Table 3) has generally been considered the most accurate method of assessing menstrual age (31). Kalish et al. (32) in a study population of 104 pregnancies

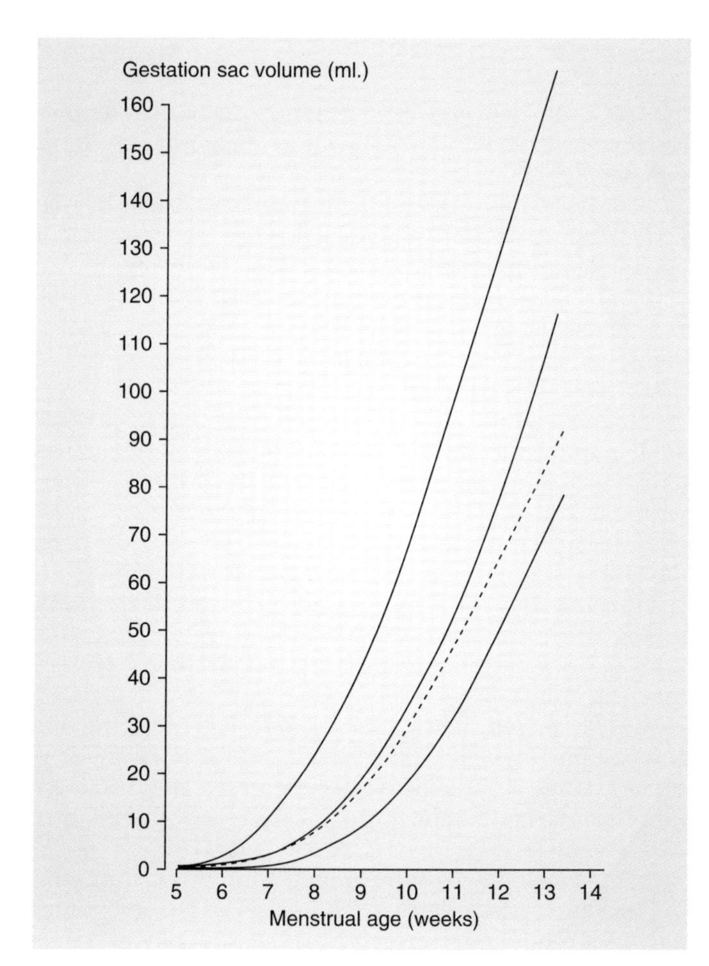

FIGURE 10 ■ The smoothed mean and 2 SD values (*continuous lines*) of 319 gestation sac volume estimations from 5 to 14 weeks of amenorrhea. The broken line represents the mean sac volume after subtraction of the estimated fetal volume. *Source*: From Ref. 20.

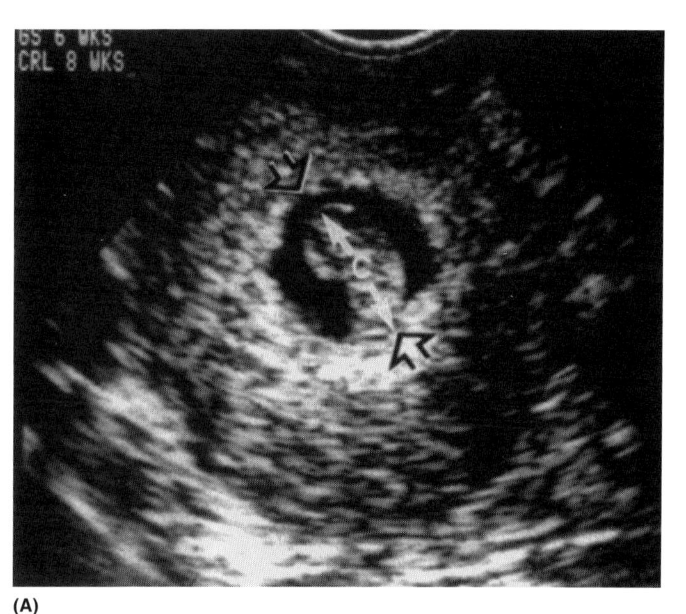

(A)

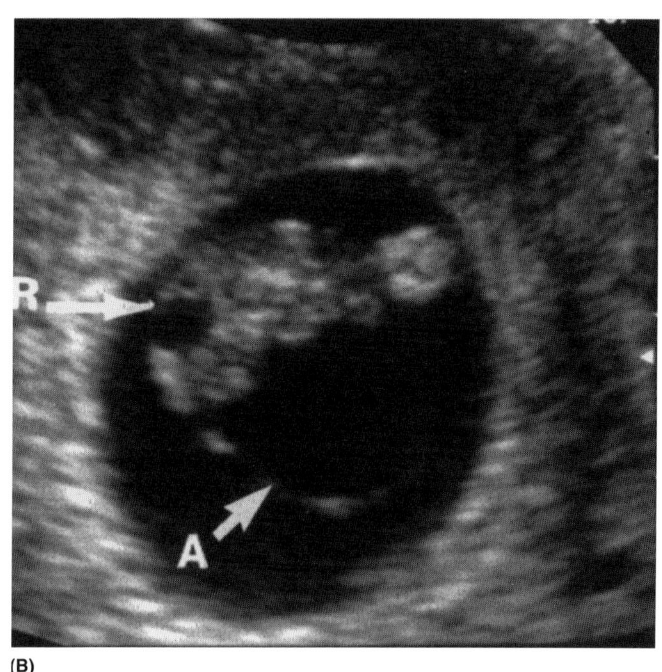

(B)

FIGURE 11 A AND B ■ Small gestational sac syndrome. (**A**) The gestational sac (*open arrows*) measures 1.7 cm (six weeks) and the crown–rump length of 1.6 cm is consistent with an eight-week intrauterine gestation. (**B**) The normal relationship of the gestational sac with the crown–rump length at eight weeks' gestation. The normal rhombencephalon and the amnion are visible. c, crown–rump length; R, rhombencephalon; A, amnion.

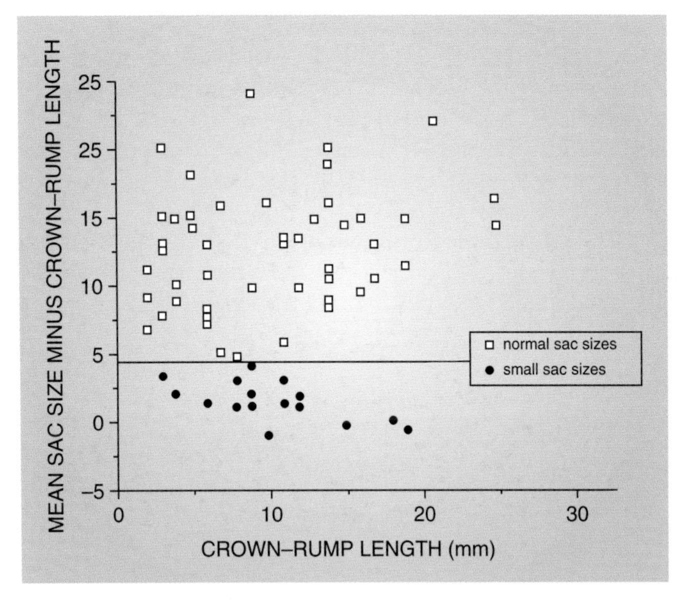

FIGURE 12 ■ Graph of distribution of mean sac size minus crown–rump length in patients with small gestational sacs and those with normal-sized sacs. *Source*: From Ref. 21.

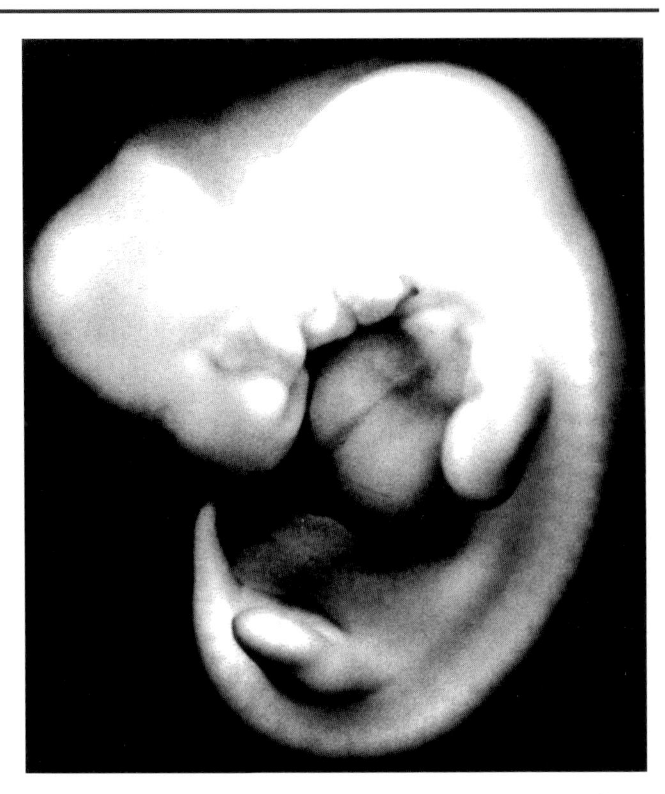

FIGURE 14 ■ An embryo at 6.4 weeks' menstrual age (0.8 cm). *Source*: From Ref. 27.

conceived by in vitro fertilization reported that a first- and second-trimester ultrasound examination can determine fetal age to within five days and below and seven days an below, respectively in over 95% of cases. Hence, menstrual age assessment is only marginally better in the first, in contrast to the second, trimester.

When evaluating a formula or table that assesses menstrual age from the CRL, it is important to evaluate the study population. Crown–rump dating based on even a sure LMP is inherently less reliable than a population derived from in vitro fertilization (IVF) pregnancies (32). The CRL is more reliable than the LMP in predicting the date of delivery. Sonographic pregnancy dating has also been shown to reduce the number of post term pregnancies (33).

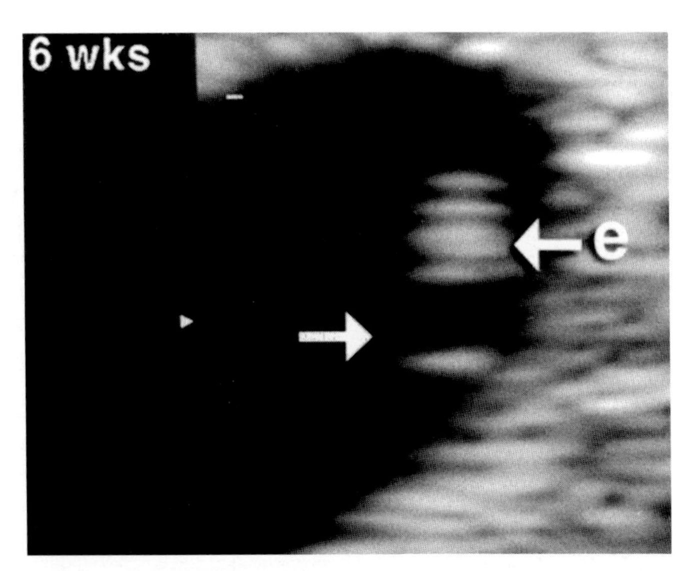

FIGURE 13 ■ A 3 mm embryo (e) (six weeks menstrual age) is superior to the yolk sac (*arrow*). The amnion is closely applied to the embryo.

The variability in predicting menstrual age by the CRL is approximately 8% (2 SD) when expressed as a percentage of the predicted value (31). For example, the 95th percent confidence interval increased from 0.7 weeks at a menstrual age of 9 weeks to 1.1 weeks at 14 weeks' menstrual age. The CRL of male and female fetuses in the first trimester is the same (34).

A CRL significantly smaller than the dates is associated with an increased risk of chromosomal anomalies (35) or structural malformations (36). In addition, intrauterine factors (i.e., abnormal placentation) may affect embryo/fetal growth. A singleton CRL below 2 standard deviations from the mean based upon accurate dating criteria has been associated with an increased risk of both miscarriage and small-for-gestational-age infants (37). With dichorionic–diamniotic twins, a discrepancy of more than three days in CRLs between the twins is associated with an increased risk (likelihood ratio: 5.9) of having discordant birth weights (38). Hence, one should not routinely change a patients' dates based on a first trimester ultrasound examination without carefully evaluating her dating criteria.

The development of the human embryo occurs along a continuum, particularly in the first seven weeks of pregnancy. The sequential appearance of embryonic structures visualized with transvaginal sonography may, therefore, be used to approximately date first trimester pregnancies. For example, the initial visualization of a gestational sac,

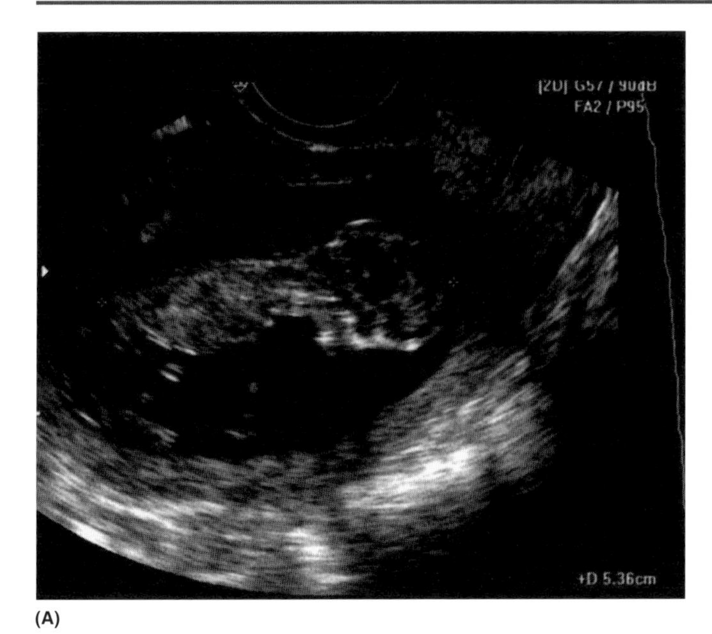

(A)

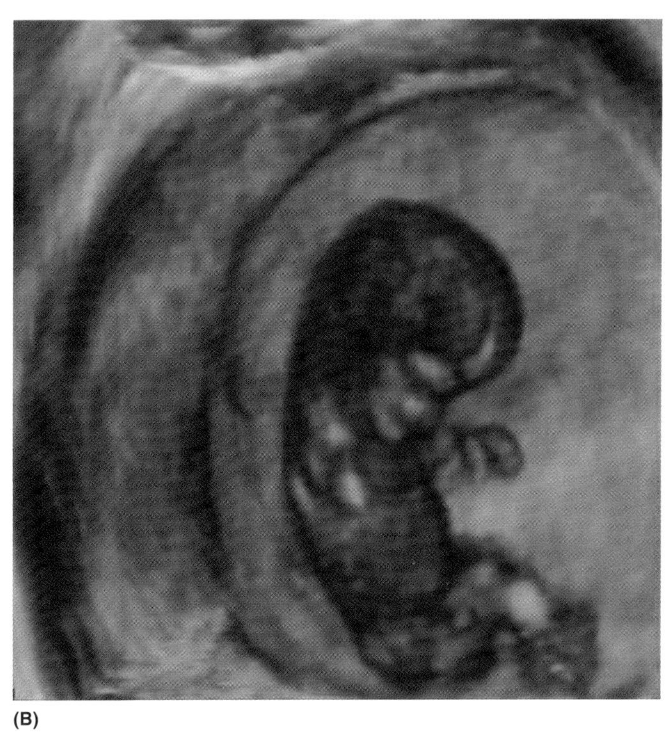

(B)

FIGURE 15 A AND B ■ Crown–rump length. (A) A crown–rump length of 5.36 cm at 12 weeks' menstrual age. (B) 3D image of a crown–rump length at 10 weeks' menstrual age.

the yolk sac, and an embryonic pole with cardiac activity occurs at 4.5, 5, and 6 weeks from the LMP, respectively. Other structural changes evident on transvaginal sonography include the detection and resolution of mid-gut herniation (Fig. 16) and the presence of a midline falx dividing the lateral ventricles at nine weeks' gestation (Table 4) (39). A 6.5-MHz transvaginal probe was used to develop the time line for the detection of the foregoing embryonic structures. Use of a lower-frequency probe would have different results. Consequently, menstrual age assessment based on these particular landmarks is not widely used.

Pitfall

During measurement of the CRL, the embryo normally rests with a kyphotic curvature to the spine. This measurement leads to mild shortening. During periods of motor activity, however, the fetus straightens, thereby permitting an accurate CRL measurement (Table 5).

Umbilical Cord Length

Fetal movement and the amount of space in the amniotic cavity determine the length of the umbilical cord. In the first trimester, the length of the umbilical cord is approximately equivalent to the CRL (Figs. 17 and 18) (40).

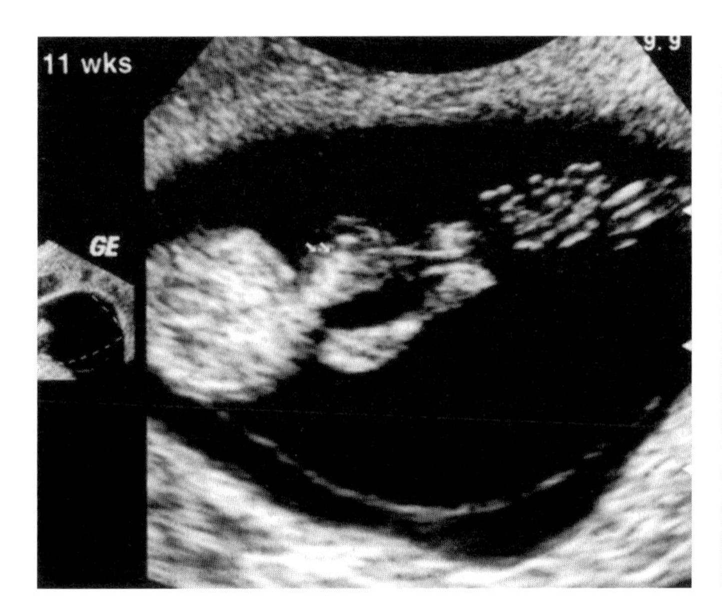

FIGURE 16 ■ Mid-gut herniation of bowel (*arrows*) at 11 weeks' menstrual age.

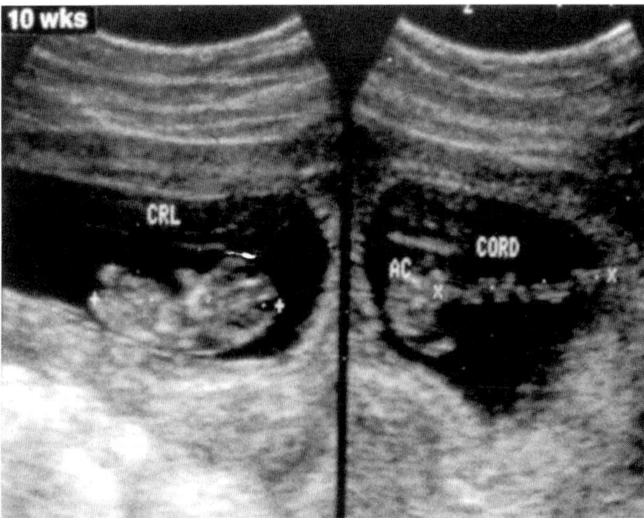

FIGURE 17 ■ First trimester cord length is equivalent to the crown–rump length (CRL).

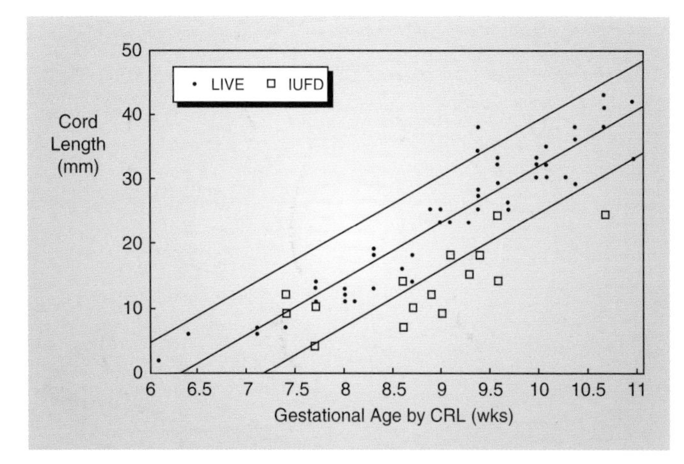

FIGURE 18 ■ Cord length (mm) ± 2 SD versus menstrual age (weeks). Squares indicate IUFD. N, live fetus; CRL, crown–rump length; IUFD, intrauterine fetal demise. *Source*: From Ref. 40.

SECOND AND THIRD TRIMESTERS

Biparietal Diameter

The utility of a sonographically derived biparietal diameter (BPD) in assessing menstrual age has been recognized for over 30 years (Fig. 19, Table 6) (41,42). A BPD between 15 and 24 weeks' menstrual age is comparable to the CRL in determining gestational age (43). Although a BPD can be obtained before 13 weeks' menstrual age (Fig. 20), the CRL is generally used in the first trimester for pregnancy dating. The reliability of the BPD in determining menstrual age decreases as menstrual age advances. This change is due, in part, to inherent differences between fetal sizes. Furthermore, the mean BPD growth rate declines steadily from 3.4 mm per week at 17 weeks to 1.2 mm per week at 39 weeks. Hence, a small increase in the BPD is associated with a larger increase in estimated menstrual age later in pregnancy (Fig. 21) (44).

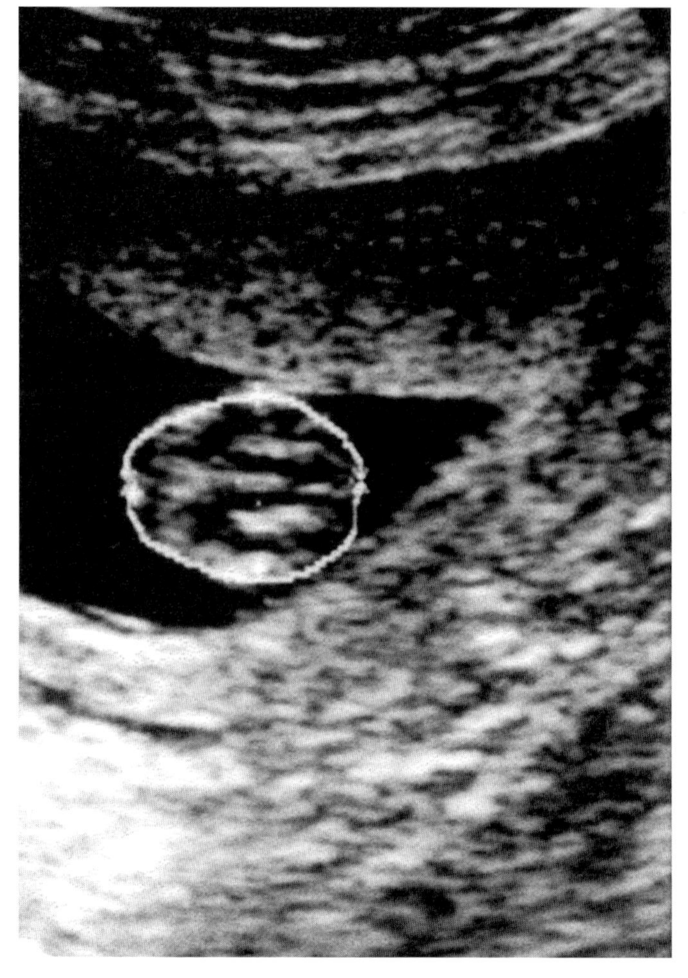

FIGURE 20 ■ Biparietal diameter and head circumference at 11.4 weeks' gestation.

The BPD is highly correlated with the head circumference (HC). However, the random error in menstrual age assessment is larger for the BPD (4.26 days) than for the

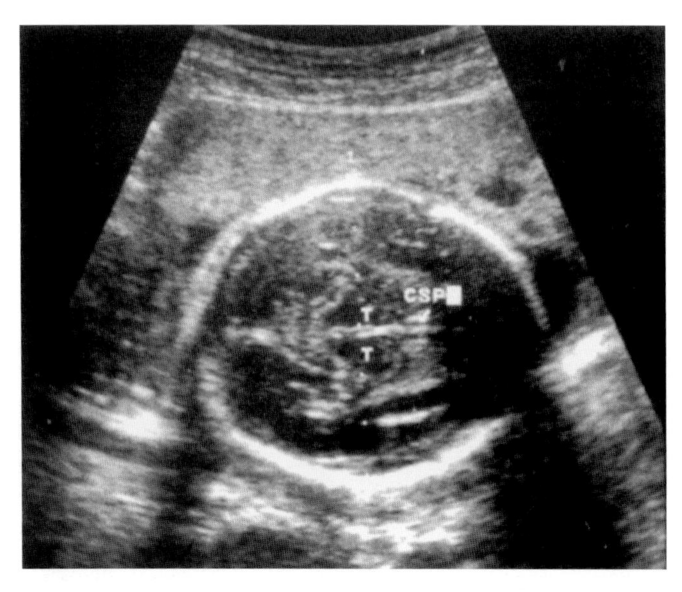

FIGURE 19 ■ Biparietal diameter (*calipers*) at 25 menstrual weeks. T, thalami; CSP, cavum septum pellucidum.

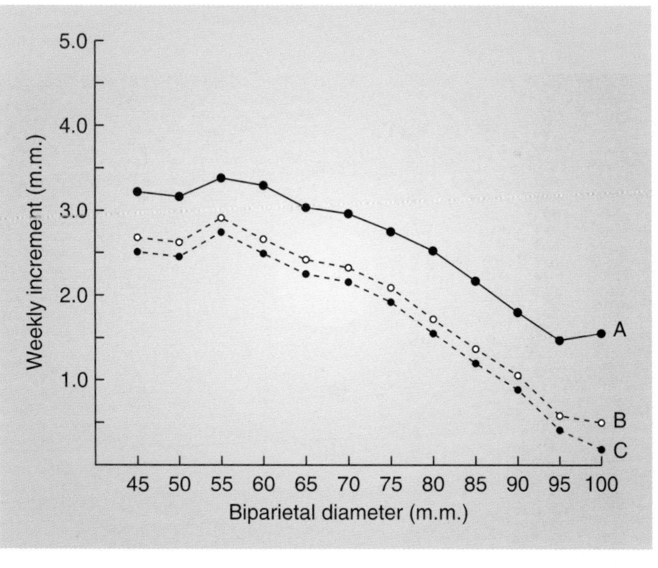

FIGURE 21 ■ Mean growth rate of the biparietal diameter (with lower tolerance limits) in relation to the size of the diameter: (**A**) mean; (**B**) 10% limit; (**C**) 5% limit. *Source*: From Ref. 44.

HC (3.77 days) (45). The BPD does not add any predictive information to menstrual age assessment by the HC. Chervenak et al. (45), therefore, suggested that the HC should replace the BPD as a standard dating parameter.

Pitfall

Several pitfalls can occur when measuring the BPD (Table 7). To obtain an accurate measurement, minimal pressure with the real-time transducer is frequently sufficient to change an obliquely oriented fetal head to an occiput transverse position. A measurement from leading edge to leading edge is obtained. For BPD measurements, gain controls should be adjusted so the width of the parietal echo is no greater than 5 mm. The measurement error inherent in caliper placement is 1 to 2 mm (46).

The anatomic landmarks that should be visualized in order to obtain a correct BPD include symmetrically positioned thalami and third ventricle and visualization of the septum pellucidum at one-third the fronto-occipital distance from the frontal bone. The imaging of midline echo alone is not sufficient to obtain an accurate BPD. Johnson et al. (47) produced variations of the BPD of up to 19 mm by scanning at various angles through the fetal head even with the midline still imaged.

An abnormal fetal position, such as breech (48), or a significant reduction in amniotic fluid (49) can affect the shape of the calvarium and can thereby reduce the reliability of the BPD in assessing menstrual age. The cephalic index (CI) (Fig. 22) is used to evaluate the shape of the newborn head. It is derived from the following formula:

$$CI = \frac{\text{Short axis (outer-to outer BPD)} \times 100}{\text{Long axis (frontooccipital diameter)}}$$

A normal CI ranges between 74% and 83%. When the CI is less than 74%, the head is dolichocephalic (Fig. 23); values higher than 83% indicate brachycephaly (Fig. 24). The HC rather than the BPD should be used for menstrual age assessment whenever dolichocephaly or brachycephaly occurs (50). The above formula requires the measurement of an outer-to-outer BPD. Jeanty et al. (51) derived the CI using the standard outer-to-inner BPD as the short axis. With this formula, the mean CI is 80.64% ± 4.97. Hence, dolichocephaly and brachycephaly are defined by a CI lesser than 75.67% and greater than 85.61%, respectively.

With oligohydramnios, the fetal head molds to the contour of the uterus. This phenomenon may reduce both the BPD and the occipitofrontal diameters (52). Consequently, menstrual age assessment by either the BPD or the HC has a greater variation in the presence of oligohydramnios than when the amniotic fluid volume is normal.

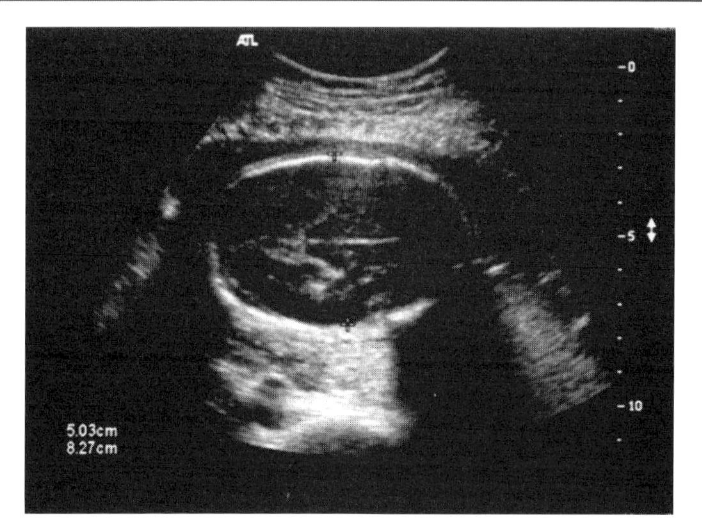

FIGURE 23 ■ Dolichocephaly (cephalic index: 60%). The cephalic index (60%) is derived from the BPD and outer-to-outer occipital frontal diameter (see text). BPD, biparietal diameter.

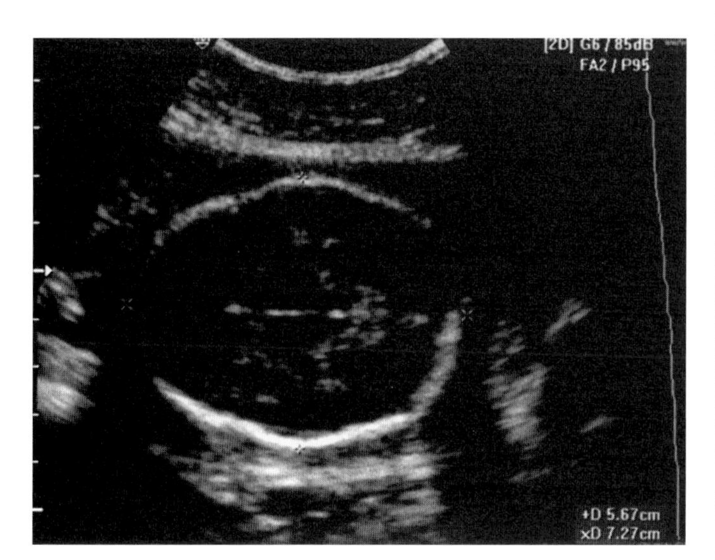

FIGURE 22 ■ Measurement of the cephalic index: outer-to-outer biparietal diameter divided by the outer-to-outer occipital frontal diameter.

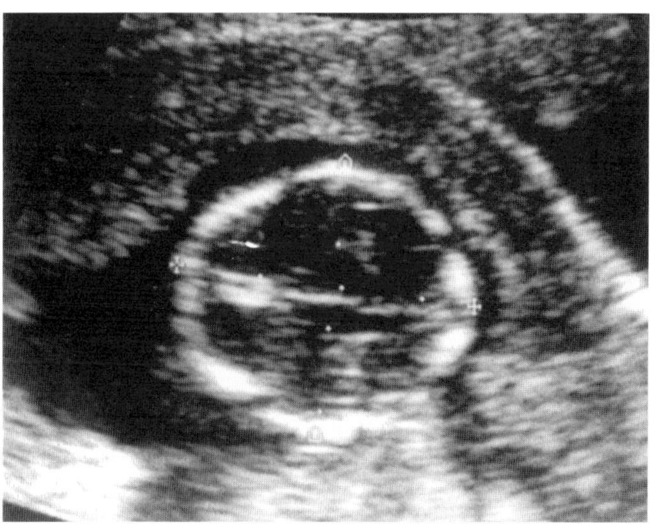

FIGURE 24 ■ Brachycephaly (cephalic index: 92%) at 15 weeks' gestation. *Left to right*, occipital-frontal diameter; *top to bottom*, outer-to-outer biparietal diameter.

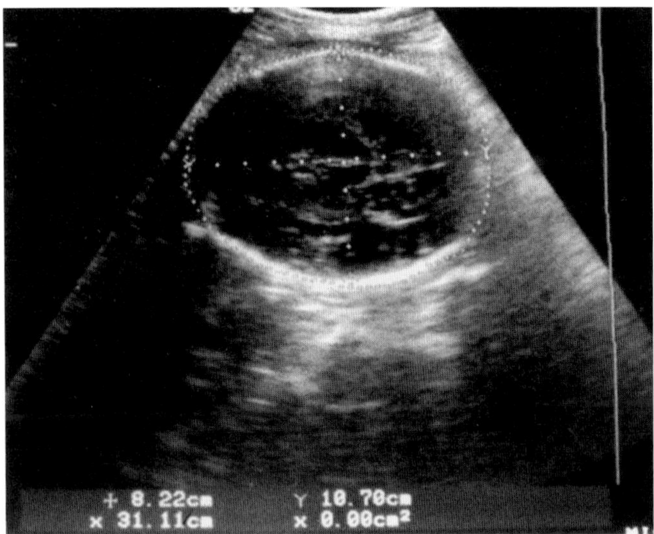

FIGURE 25 ■ Head circumference at 35 menstrual weeks.

FIGURE 26 ■ Abdominal circumference at 24 menstrual weeks. STOM, stomach; UV, umbilical vein; SP, spine.

Head Circumference

The landmarks for the HC are the same as for the BPD. The circumference measurement should be taken around the calvarium, not around the scalp. The HC can be measured directly by tracing around the calvarium or by a formula that utilizes the two long axes of the HC image.

The ellipse mode on ultrasound units calculates the HC from the two long axes. The continuous trace method is consistently 1% larger than a derived HC measurement from the two long axes (53). The measurement technique employed should be consistent with the HC table adopted by a facility for menstrual age assessment.

Because of the variations in head shape, measurement of the HC has become an integral part of assessing menstrual age in the second and third trimesters (Fig. 25, Table 8) (54). The values obtained for the HC at a given menstrual age by different investigators are similar (55). The percentile value for the HC (56) by menstrual age has been derived from the original data of Hadlock et al. (54) (Table 9) (57).

The variability inherent in estimating fetal age from the HC is generally less than with the BPD (45). However, the variability is not a constant throughout pregnancy, but increases with menstrual age (53,54).

The HC is part of several formulas to estimate fetal weight and is also used to monitor growth between examinations. Finally, the HC can be compared with other measurements, such as abdominal circumference (AC) and femur length (FL), to assess the possible effects of growth disturbances (e.g., macrosomia and growth restriction) on the fetus.

Abdominal Circumference

The sonographic measurement of the AC at the level of the stomach and umbilical vein was initially described by Campbell and Wilkin (58) to estimate fetal weight. Hadlock et al. (59) (Table 10), as well as other investigators (60,61), have used the AC to estimate gestational age. Variability is greater in predicting menstrual age from the AC than with the HC (45). When the variability associated with determining gestational age by the AC is expressed as a percentage, it remains constant at about 13% and is independent of gestational age (55).

The normal fetal abdomen is round to slightly ovoid (62). As with the HC, specific anatomic landmarks must be visible to obtain an accurate AC measurement (Fig. 26).

1. A 90° transverse view of the spine at 3 or 9 o'clock
2. The stomach bubble in the left fetal abdomen
3. A single rib should be visible on either side of the abdomen. The presence of more than one rib indicates that the plane of section is oblique
4. Visualization of the junction of the umbilical segment of the left portal vein and the right portal vein

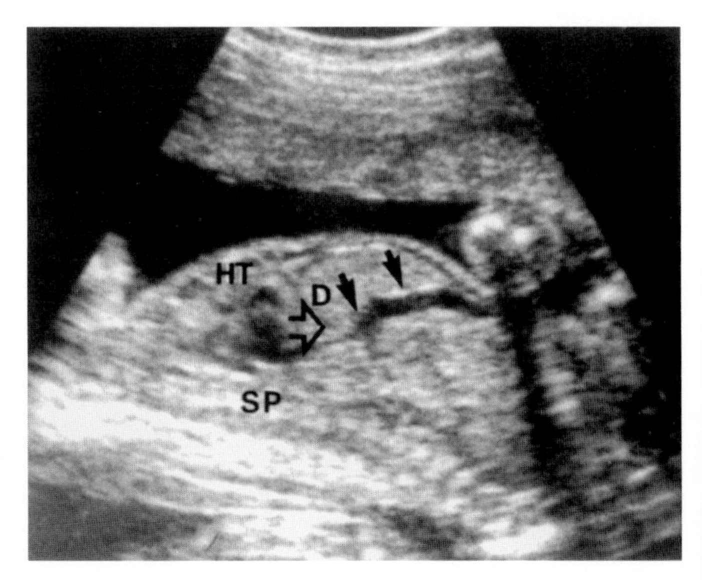

FIGURE 27 ■ Course of the umbilical vein (*arrows*) to the ductus venous. HT, heart; SP, spine; D, ductus venous.

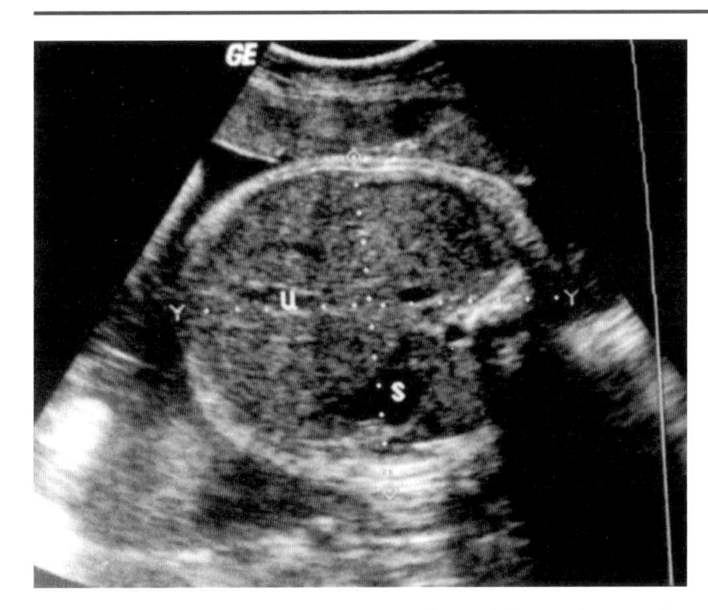

FIGURE 28 ■ Outer-to-outer diameters of the abdomen that may be used to obtain a circumference. S, stomach; U, umbilical vein.

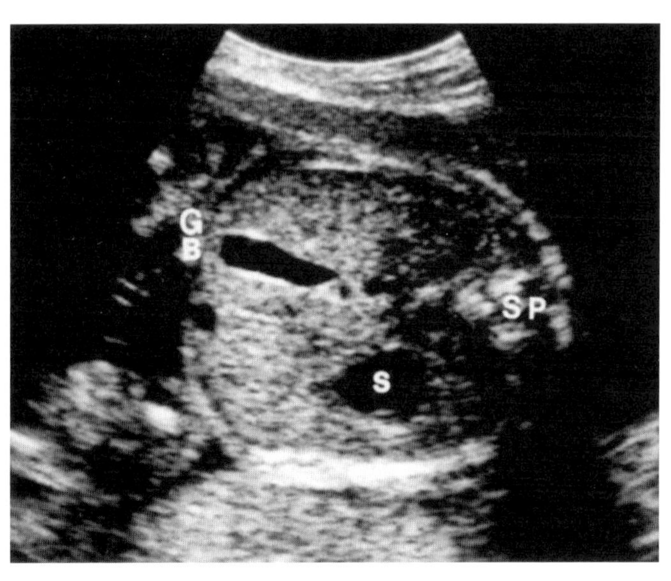

FIGURE 30 ■ Fetal abdominal circumference at the level of the gallbladder. GB, gallbladder; S, stomach; SP, spine.

Since the umbilical vein runs obliquely to the fetal long axis, its mere visualization on a transaxial view does not define a specific anatomic position (Fig. 27).

The AC can either be measured directly or be calculated from the two axes (Fig. 28). As previously mentioned, the ellipse function on ultrasound units calculates the AC from the two diameters. When the AC is traced, the values obtained are 3.5% larger than from the derived technique (63). The percentile values for the AC by menstrual age (56) have been derived from the original data of Hadlock et al. (59) (Table 11). Most AC charts are based on a tracing of the AC.

Pitfall

Several pitfalls may occur when measuring the AC (Table 12). If a long segment of the umbilical vein is visualized, the plane of section is too oblique (Fig. 29). An AC measurement at the level of the umbilical cord insertion is too low. The fetal gallbladder (Fig. 30) is in the right upper quadrant and should not be mistaken for the more medially positioned umbili cal vein. Because the umbilical vein cannot be visualized with the spine in the 12 o'clock position (Fig. 31), an AC measurement should not be taken with the fetus in this position. If the shape of the AC is elliptical (Fig. 32) rather than circular, the error in

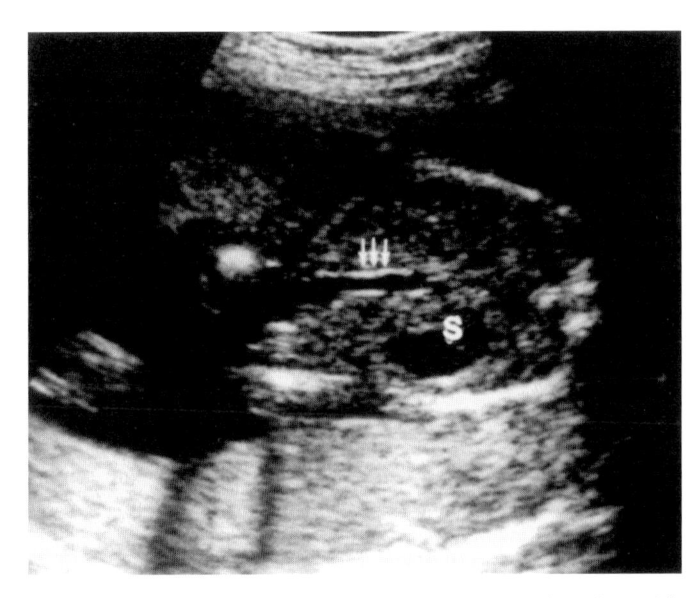

FIGURE 29 ■ An oblique section of the fetal abdomen along the umbilical vein (*arrows*). S, stomach.

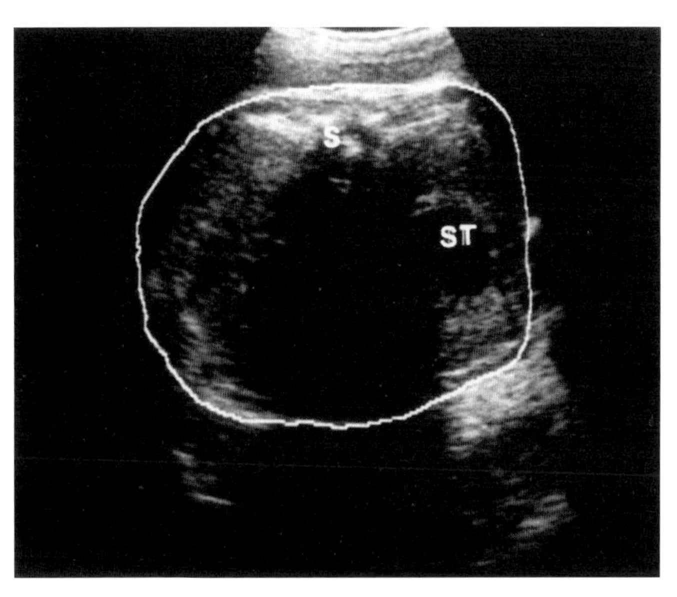

FIGURE 31 ■ An incorrect abdominal circumference. Although the stomach (ST) is visualized, the spine (S) is at 12 o'clock, and the umbilical vein cannot be visualized.

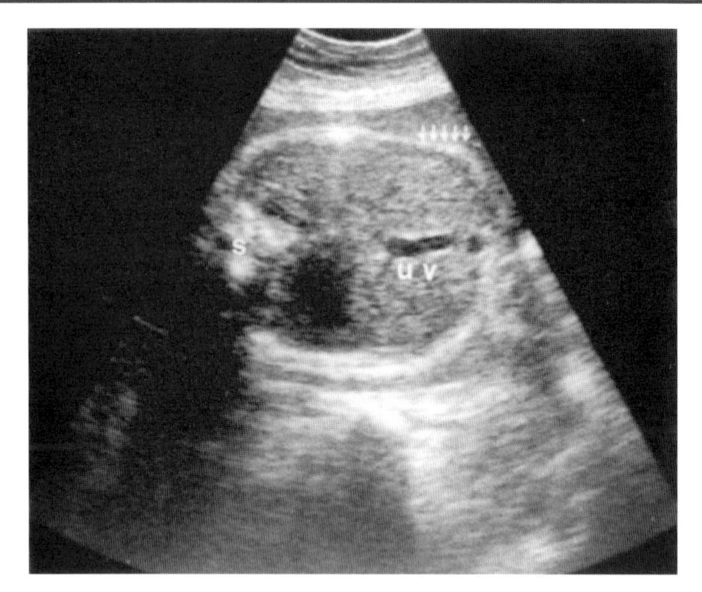

FIGURE 32 ■ An elliptic abdominal circumference may be produced when too much pressure (*arrows*) is placed on the fetus with the transducer. s, spine; uv, umbilical vein.

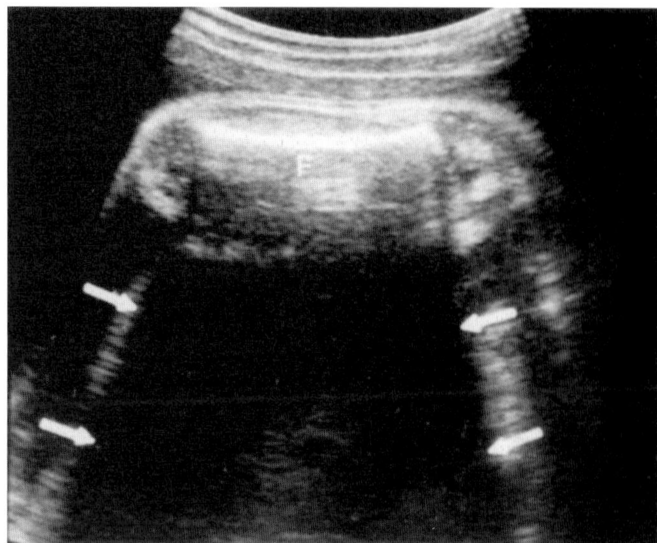

FIGURE 34 ■ Femur length (F) at 32 weeks' gestation. Arrows demarcate acoustic shadow.

weight estimation or gestational age assessment can be substantial (64).

Femur Length

The FL was originally measured to detect short-limbed dysplasias (65). Numerous articles have been published on the relationship between FL and menstrual age (Table 13) (66). The percentile values from the 3rd to the 97th percentile for menstrual age are provided in Table 14.

To measure the femur appropriately, it should be visualized in a plane parallel to that of the transducer; only the ossified femoral shaft is measured. The normal diaphysis of the femur has a curved medial border and a straight lateral border (Fig. 33) (67). The appropriate borders can usually be recognized by the underlying acoustic shadow (Fig. 34). The reported interobserver variation in the measurement of the femur is 4.4 mm (66).

Some studies have found the FL superior to the BPD in assessing menstrual age (4,57,68); in other studies, the two parameters are comparable. As with other fetal parameters, the standard deviation of the menstrual age estimate by FL increases with advancing gestation.

Pitfall

When measuring the FL, the cartilaginous ends of the femur (Fig. 35) should not be included in the measurements (69,70).

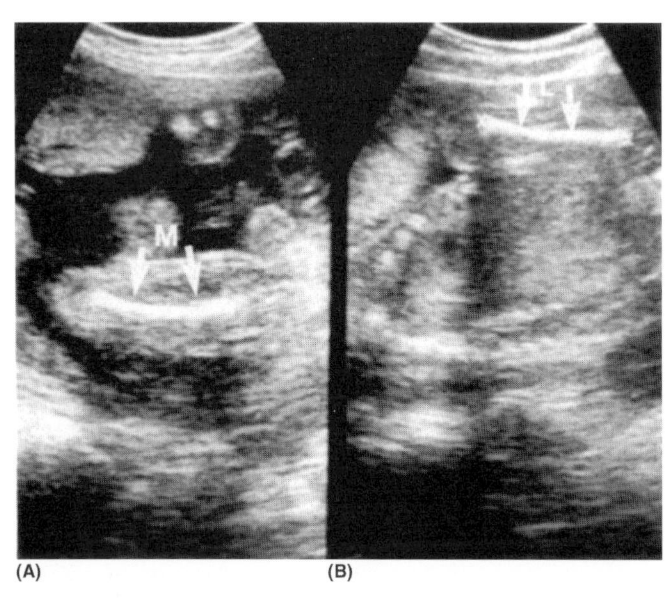

FIGURE 33 A AND B ■ The markers outline the appropriate length of the femur. (**A**) Curved medial (M) border (*arrows*). (**B**) Straight lateral (L) border (*arrows*).

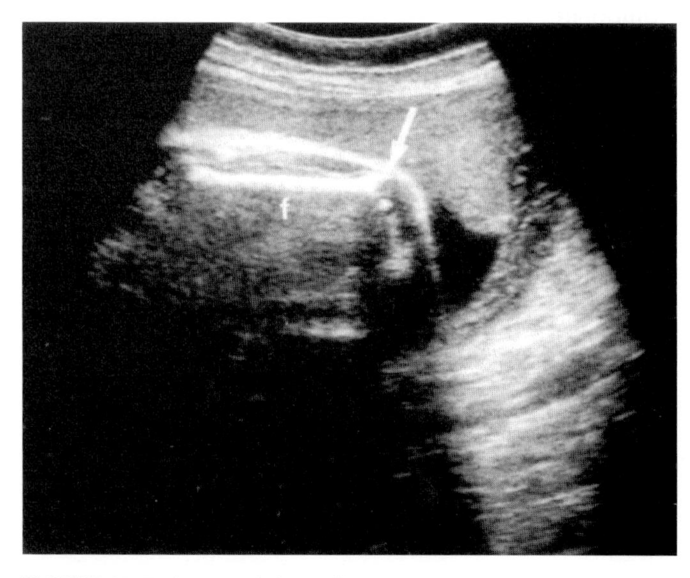

FIGURE 35 ■ A 37-week femur illustrating the distal cartilage (*arrow*) that should not be included in the (f) femoral measurement.

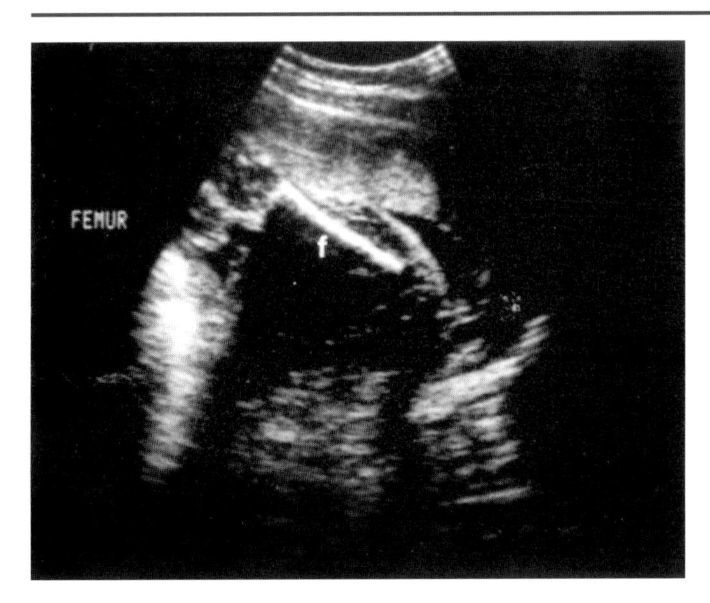

FIGURE 36 ■ The femur (f) length should not be measured unless it is perpendicular to the transducer.

Potential sources of error in the measurement of the FL include the use of different ultrasound units, the type of transducer (sector, linear, or curvilinear), and the angle of inclination with which the ultrasound beam strikes the bone. For example, Lessoway et al. (69) found a variation of 2.4 to 3.4 mm in femoral measurements among different ultrasound units in their department. To control for the effect of angling, the femur should be measured perpendicular to the transducer (70). A femur measured in the axial plane (Fig. 36) is significantly shorter than one measured in the horizontal plane (Fig. 33), with up to a 2.6-week difference in gestational age prediction (71–73). The femur closest to the transducer should be measured to minimize the variation associated with measuring at different depths (Table 15).

Composite Assessment of Menstrual Age

In the normal fetus, any of the standard fetal parameters (BPD, HC, AC, and FL) may be larger or smaller than the mean value anticipated for gestational age (e.g., a 75th percentile HC and a 25th percentile AC). A composite assessment of menstrual age, using all four parameters, gives a lower systematic random error than any single parameter (57,74,75). For example, the overall variability is decreased by 33% (1.3–1.0 weeks) when a composite assessment of menstrual age is used rather than only the BPD (57). Of the four standard parameters for assessing menstrual age, the combination of HC and FL results in the best coefficient of determination with the least variability. The addition of the BPD and AC does not significantly improve menstrual age assessment by the HC and FL (4).

When the menstrual age predictions determined by these four measurements do not agree, the clinician must critically evaluate each image. The simple averaging of values may lead to a falsely low or a falsely high estimation of menstrual age because of a single erroneous measurement. Specific menstrual age–independent ratios, i.e., CI (50), FL/BPD (76), and FL/AC (77) are helpful in determining the reason for the discrepancy in menstrual age assessment from different parameters. For example, the FL/BPD ratio between 23 and 40 weeks menstrual age is 79 ± 8% (90% confidence interval). If the FL/BPD ratio is above normal, then the FL may be overestimated or the BPD may be underestimated. As a general rule, the parameters used in the composite assessment of menstrual age during the third trimester should not vary by more than three weeks (78).

The use of multiple parameters to assess menstrual age in the second and third trimesters has a variability of approximately 7% (±2 SD). Hence, the sonographic menstrual age range at 20 and 30 weeks' gestation would be about 1.4 weeks and about 2.1 weeks, respectively (79).

Occasionally, maternal habitus or fetal lie complicates biometry. Transvaginal or transumbilical sonography may be helpful in these cases (Fig. 37) (80–82).

Pitfall

The sonologist or sonographer must determine that a specific measurement is not affected by a pathologic process or growth disturbance before incorporating it into the composite assessment of menstrual age. For example, in the presence of fetal macrosomia, the HC and the AC overestimate age (83).

Gender and population differences may affect menstrual age. At birth, the male infant is approximately 58 g heavier than the female (84). Parker et al. (85) found a slower rate of HC and AC growth in female fetuses compared with males after 28 weeks' gestation. However, the difference between the genders at 18 weeks' gestation is within the range of measurement error and, therefore, is of little practical value for accurate pregnancy dating (86).

Hadlock et al. (5) have demonstrated that the use of multiple parameters to assess gestational age in a middle-class white population is also applicable to fetuses with different socioeconomic and racial backgrounds.

Other Long Bones

Any of the long bones can be used to assess fetal age (Fig. 38) (87,88). The humerus and FL are of comparable lengths until 16 weeks' menstrual age; the femur then becomes progressively longer. The ulna is differentiated from the radius by its greater extension into the elbow. Distally, the radius is linked to the thumb and the ulna to the fifth finger. The ulna and tibia have similar lengths until 22 weeks' gestation (89).

The percentile values for a given menstrual age have been established for the humerus, ulna, and tibia (Tables 16–18). However, the 5th and 95th percentile range for a specific bone length (humerus, ulna, and tibia) is 7 to 11 days greater than for FL (Tables 16–18). The range for radius length is 18 days in the second trimester and 60 days in the third trimester (Table 19). As a result, the FL is the most reliable long bone to use in the evaluation of menstrual age.

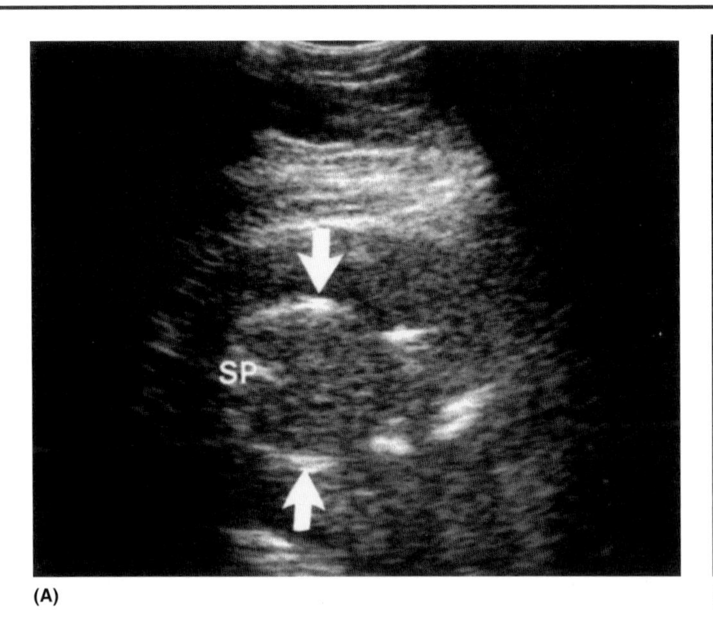

(A)

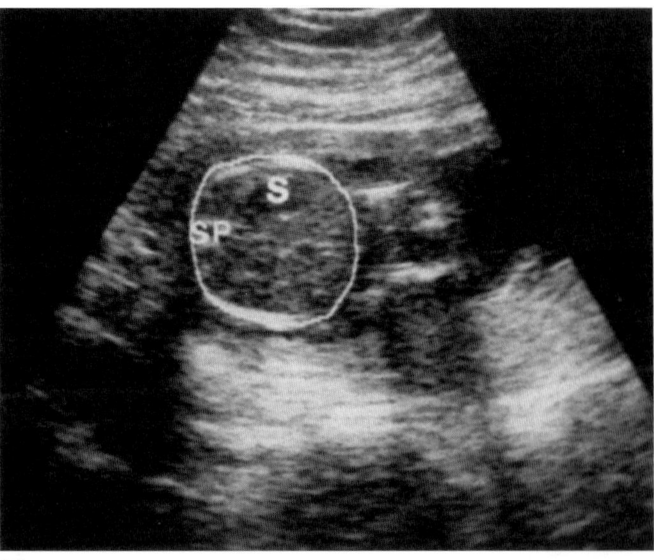

(B)

FIGURE 37 A AND B ■ A 19-week abdominal circumference in an obese patient. (**A**) Transabdominal (*arrows*). (**B**) Transumbilical. S, stomach, SP, spine.

Foot Length

Fetal foot length has been correlated with menstrual age (Fig. 39A, Table 20) (90).

The 95% confidence interval for foot length is 1.6, 2.4 and 3.2 weeks at 20, 30, and 40 weeks' menstrual age, respectively (90). The centile values for fetal foot length at specific menstrual ages are outlined in Table 21 (91).

The fetal foot length is approximately equivalent to the FL throughout the three trimesters. The fetal foot is not affected by most skeletal dysplasias. As a result, an FL/foot ratio of below 0.84 has been used to diagnose severe skeletal dysplasias (92).

Transverse Cerebellar Diameter

Between 14 and 20 weeks' menstrual age, the transverse cerebellar diameter (TCD) (Fig. 39B) in millimeters is roughly equivalent to the menstrual age in weeks (Table 22) (83,93). The percentile values for specific TCDs are provided in Table 19. Chavez et al. (93) have reported that the TCD predicts menstrual age within ±5 days in 98.7% of second and third-trimester pregnancies.

Alternative Parameters to Assess Menstrual Age

Every measurable fetal parameter correlates with menstrual age. However, in terms of menstrual age assessment, the standard deviation or, more precisely, the confidence intervals of a specific measurement are as important as the correlation coefficient. Adequate sample sizes are required in order to obtain menstrual age percentiles for the specific fetal parameter being measured. Retrospectively obtained charts of a parameter versus menstrual age should be validated prospectively. Finally, appropriate statistical methods must be used. For example, the reported menstrual age prediction from clavicular length (Fig. 39B) (94) and scapular length (Fig. 39C) (95) had a fixed standard deviation throughout the second and third trimesters. All fetal measurements have an increasing variation as menstrual age advances (96). It, therefore, seems that the model selected by the authors of these two studies did not allow the standard deviation to change with advancing gestation. As a result, the ±2 SD values would be too far apart in early pregnancy and too close together in the third trimester. Because of the wide confidence intervals, as well as the other methodological problems outlined above, the alternative parameters listed in Table 23 have not gained wide acceptance for menstrual age assessment.

THIRD TRIMESTER

In the third trimester, sonography should not be used with the expectation of accurately determining menstrual age. However, a serial assessment of fetal size over three weeks or more can ensure adequate interval growth. When the menstrual age estimate at the second examination agrees with the first, the confidence interval of the sonographic evaluation of menstrual age is reduced (75). For example, if a patient is 34 weeks from her LMP, but the composite assessment of menstrual age by sonography is 30 weeks, a second examination should be performed in three to four weeks. If the second ultrasound examination agrees with the first, then fetal growth has been normal. This finding suggests that the patient's dating by LMP was incorrect.

Because of the large biologic variability in third trimester fetal biometry, sonologists have attempted to assess fetal maturation by other sonographic criteria. Qualitative maturity indices that have been evaluated include

1. placenta grade (101,102),
2. colonic diameter (103,104),
3. humeral cap (103),
4. distal femoral epiphysis (105), and
5. proximal tibial epiphysis (105).

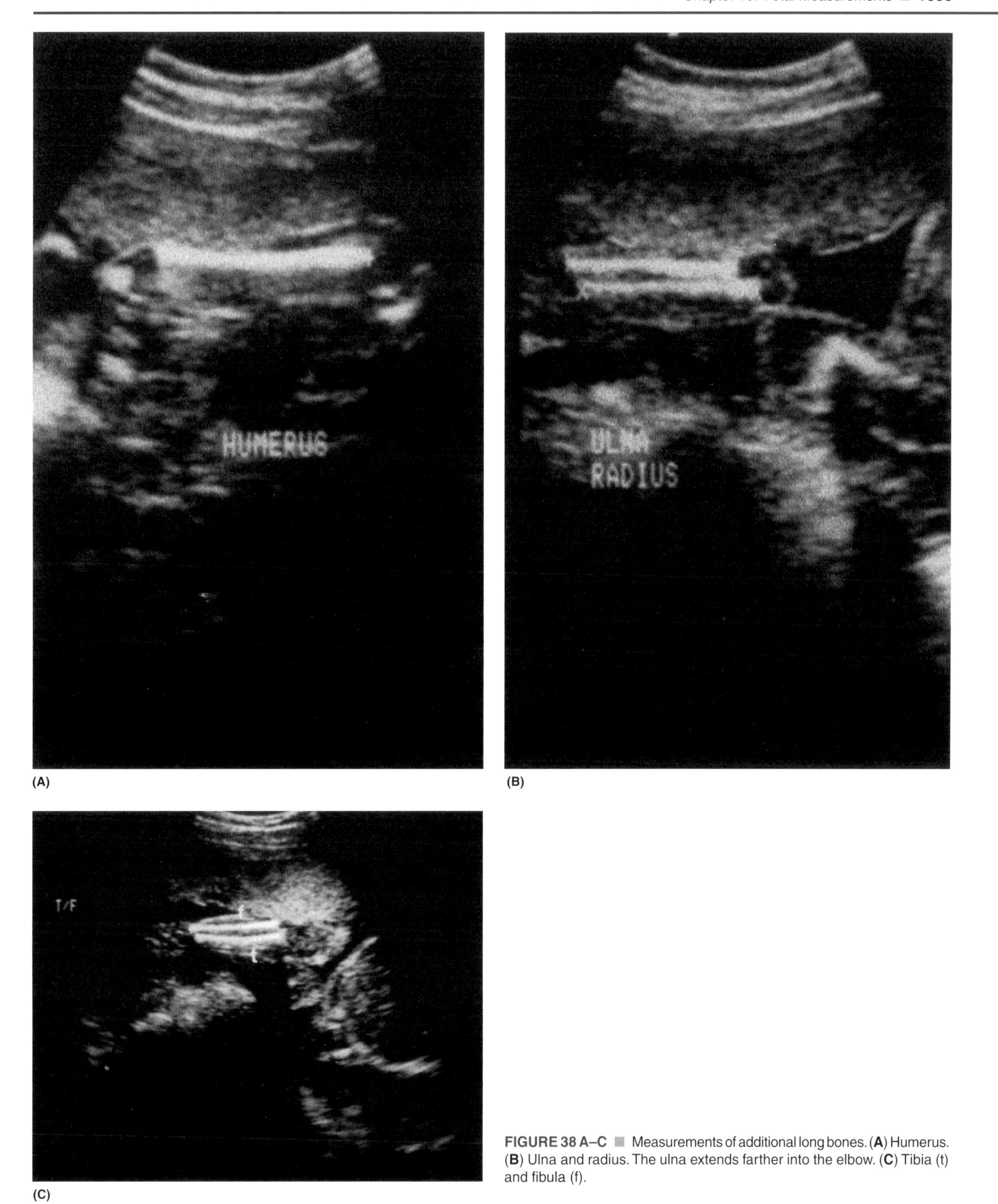

(A)

(B)

(C)

FIGURE 38 A–C ■ Measurements of additional long bones. (**A**) Humerus. (**B**) Ulna and radius. The ulna extends farther into the elbow. (**C**) Tibia (t) and fibula (f).

Unfortunately, these sonographic parameters are as variable in the third trimester as fetal biometry. As a result, they have not been widely used in the evaluation of menstrual age.

CONCLUSION

An estimation of menstrual age, fetal size, or fetal growth rate depends on an accurate measurewment of specific fetal dimensions. The earlier in pregnancy the embryo/fetus is measured, the narrower the confidence limits of the subsequent prediction of fetal age.

The assignment of menstrual age requires more than accurately performing biometry. An evaluation of the fetal environment, a careful fetal anatomic survey, and an assessment of maternal history are also required. Antepartum factors contribute an important independent parameter that may significantly affect the eventual assignment of menstrual age. All sonographic predictors of menstrual age are imprecise in the third trimester. In specific cases, a second ultrasound examination in three to four weeks may be required to assess the rate of fetal growth and to refine the sonographic prediction of menstrual age.

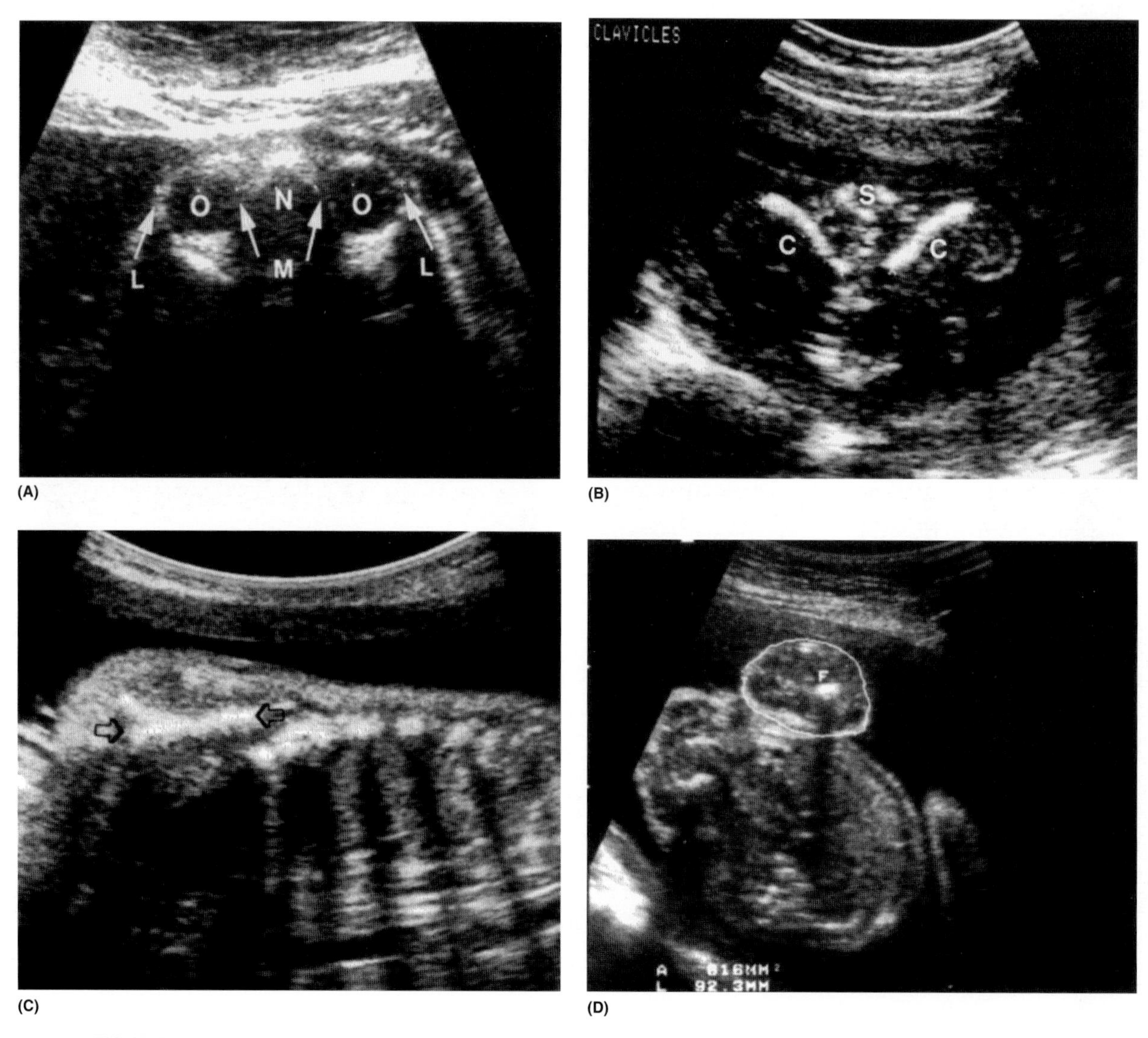

FIGURE 39 A–H ■ Alternative parameters to assess gestational age are outlined by the markers. (**A**) Axial scan through the level of the normal orbits (O) at 29 menstrual weeks demonstrates the nasal bridge (N), the medial wall of each orbit (M), and lateral wall of each orbit (L). The ocular diameter is the distance from the medial (M) to the lateral (L) wall of the orbit and is approximately equal to the interocular distance, which is measured from one medial wall to the other. The binocular distance is the sum of the two ocular diameters and the interocular distance or the measurement from one lateral wall to the other lateral wall of the orbit. (**B**) Clavicular length (C) (19 weeks). The clavicle is measured vertically to include the greatest clavicular length. (**C**) Scapular length (35 weeks) is measured in sagittal plane to include the longest measurements. (**D**) Thigh circumference at the junction of the upper and middle third of the femur at 26 weeks. (*Continued*)

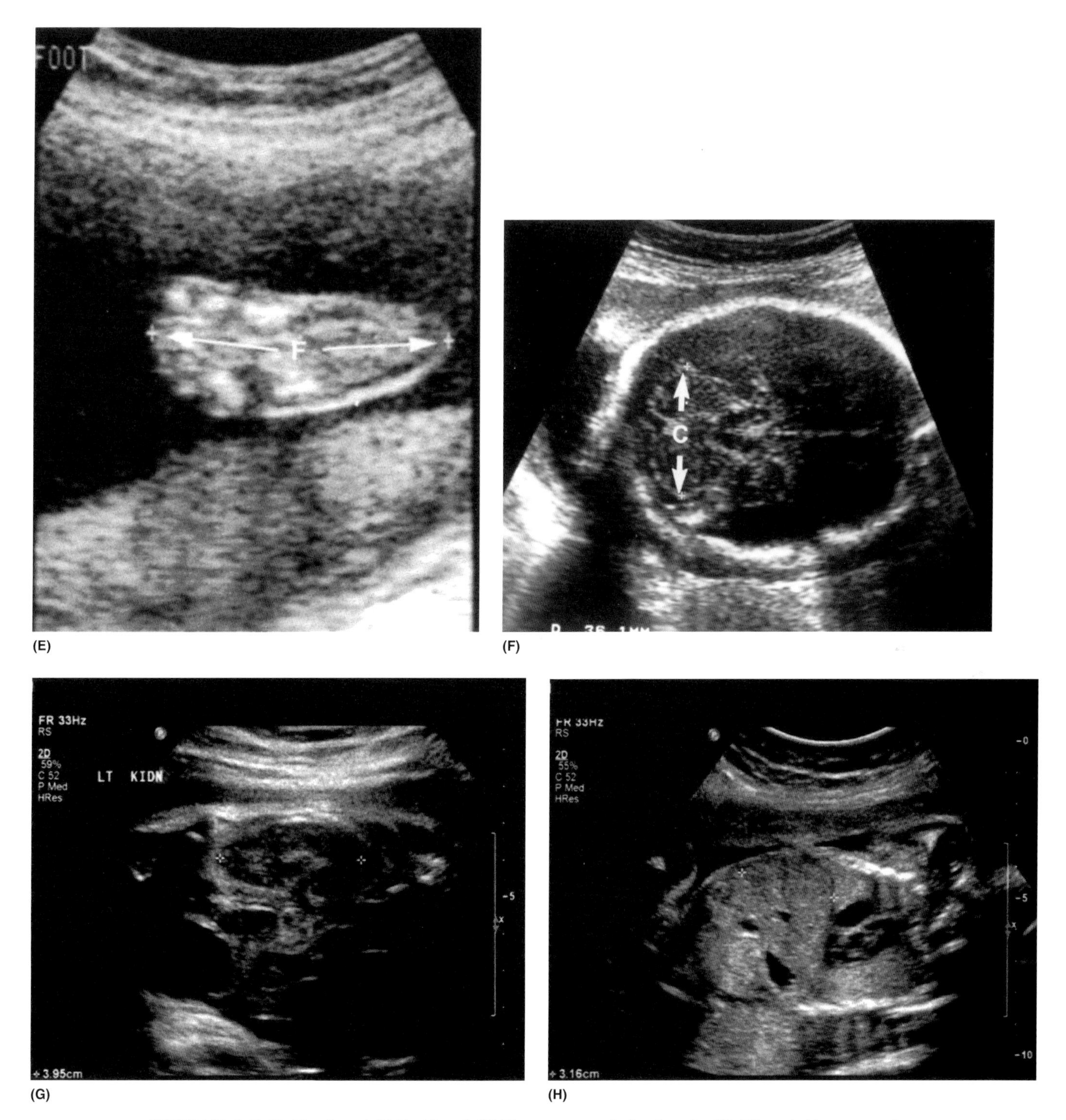

(E) (F) (G) (H)

FIGURE 39 A–H ■ (*Continued*) (**E**) Foot length (F). Transverse cerebellar diameter (C) (30 weeks) is measured in the axial plane. (**G**) Renal length at 33 weeks. (**H**) Liver length at 24 weeks' menstrual age. Since the right lobe is larger than the left, tables have been developed utilizing the right lobe. A parasagittal view is obtained with the tip of the liver and the hemidiaphragm imaged. This image also provides a coronal image of the fetal stomach.

TABLE 1 ■ Expected hCG Levels for Increasing Mean Sac Size Measurements

Mean sac size (mm)	Expected mean (U/L)	Serum hCG levels[a] 95% prediction interval (IU/L)
3	710	1050–2800
4	2320	1440–3760
5	3100	1940–4980
6	4090	2580–6530
7	5340	3400–8450
8	6880	4420–10810
9	8770	5680–13660
10	11040	7220–17050
11	13730	9050–21040
12	16870	11230–25640
13	20480	13750–30880
14	24560	16650–36750
15	29110	19910–43220
16	34100	23530–50210
17	39460	27470–57640
18	45120	31700–65380
19	50970	36130–73280
20	56900	40700–81150
21	62760	45300–88790
22	69390	49810–95990
23	73640	54120–102540
24	78350	58100–108230
25	82370	61650–112870
26	85560	64600–116310
27	87820	66900–118420
28	89050	68460–119130
29	89230	69220–118420
30	88340	69150–16310

Note: The 95% prediction intervals (rounded to the nearest 10 IU/L) were calculated from the best fitting quadratic regression lines obtained from the lower and upper confidence limits, respectively, for the individual predicted values for each original observation in the study.

[a]Expected mean serum hCG values were calculated from the following regression equation and have been rounded off to the nearest 10 IU/L.

Abbreviation: hCG, human chorionic gonadotropin.

Source: From Ref. 14.

TABLE 2 ■ Relationship of Gestational Age, hCG Levels, and Transvaginal Ultrasound Findings

		β-hCG (mIU/mL)	
Ultrasound findings	Days from last menstrual period	First international reference preparation	Second international standard
Sac	34.8 ± 2.2	1398 ± 155	914 ± 106
Fetal pole	40.3 ± 3.4[a]	5113 ± 298[a]	3783 ± 683
Fetal heart motion	46.9 ± 6.0[a]	17208 ± 3772[a]	13178 ± 2898[a]

$P < 0.5$ when compared with sac.

Abbreviation: hCG, human chorionic gonadotropin.

Source: From Ref. 10.

TABLE 3 ■ Predicted MA from CRL Measurements[a]

CRL (cm)	MA (wk)	CRL (cm)	MA (wk)	CRL (cm)	MA (wk)	CRL (cm)	MA (wk)	CRL (cm)	MA (wk)	CRL (cm)	MA (wk)
0.2	5.7	2.2	8.9	4.2	11.1	6.2	12.6	8.2	14.2	10.2	16.1
0.3	5.9	2.3	9.0	4.3	11.2	6.3	12.7	8.3	14.2	10.3	16.2
0.4	6.1	2.4	9.1	4.4	11.2	6.4	12.8	8.4	14.3	10.4	16.3
0.5	6.2	2.5	9.2	4.5	11.3	6.5	12.8	8.5	14.4	10.5	16.4
0.6	6.4	2.6	9.4	4.6	11.4	6.6	12.9	8.6	14.5	10.6	16.5
0.7	6.6	2.7	9.5	4.7	11.5	6.7	13.0	8.7	14.6	10.7	16.6
0.8	6.7	2.8	9.6	4.8	11.6	6.8	13.1	8.8	14.7	10.8	16.7
0.9	6.9	2.9	9.7	4.9	11.7	6.9	13.1	8.9	14.8	10.9	16.8
1.0	7.1	3.0	9.9	5.0	11.7	7.0	13.2	9.0	14.9	11.0	16.9
1.1	7.2	3.1	10.0	5.1	11.8	7.1	13.3	9.1	15.0	11.1	17.0
1.2	7.4	3.2	10.1	5.2	11.9	7.2	13.4	9.2	15.1	11.2	17.1
1.3	7.5	3.3	10.2	5.3	12.0	7.3	13.4	9.3	15.2	11.3	17.2
1.4	7.7	3.4	10.3	5.4	12.0	7.4	13.5	9.4	15.3	11.4	17.3
1.5	7.9	3.5	10.4	5.5	12.1	7.5	13.6	9.5	15.3	11.5	17.4
1.6	8.0	3.6	10.5	5.6	12.2	7.6	13.7	9.6	15.4	11.6	17.5
1.7	8.1	3.7	10.6	5.7	12.3	7.7	13.8	9.7	15.5	11.7	17.6
1.8	8.3	3.8	10.7	5.8	12.3	7.8	3.8	9.8	15.6	11.8	17.7
1.9	8.4	3.9	10.8	5.9	12.4	7.9	13.9	9.9	15.7	11.9	17.8
2.0	8.6	4.0	10.9	6.0	12.5	8.0	14.0	10.0	15.9	12.0	17.9
2.1	8.7	4.1	11.0	6.1	12.6	8.1	14.1	10.1	16.0	12.1	18.0

[a]The 95% confidence interval is ±8% of the predicted age.

Abbreviations: MA, menstrual age; CRL, crown–rump length.

Source: From Ref. 31.

TABLE 4 ■ Percentage of Embryonic Structures Present or Absent[a]

Weeks of gestation	4	5	6	7	8	9	10	11	12
Gestational sac only	100	100	100	100	100	100	100	100	100
Yolk sac	0	91	100	100	100	100	100	100	100
Fetal pole w/heart motion	0	0	86	100	100	100	100	100	100
Single ventricle	0	0	6	82	70	25	0	0	0
Falx	0	0	0	0	30	75	100	100	100
Mid-gut herniation	0	0	0	0	100	100	100	50	0
Total cases	6	11	15	17	10	13	15	11	6

[a]Summary of detection of six embryonic structures in the first trimester of pregnancy. The solid line differentiates weeks of gestation in which most embryos displayed changes in ultrasonographic appearance. The white areas indicate that a structure was present, whereas the shaded areas indicate that a structure was absent.

Source: From Ref. 39.

TABLE 5 ■ Pitfalls in Determination of Crown–Rump Length

Pitfall	Comments
Kyphotic spine	Wait until fetus straightens spine

TABLE 6 ■ Predicted Menstrual Ages[a] for BPD Values from 2.0 to 10.0 cm

BPD (cm)	Menstrual age (wk)	BPD (cm)	Menstrual age (wk)
2.0	12.2	6.1	25.0
2.1	12.5	6.2	25.3
2.2	12.8	6.3	25.7
2.3	13.1	6.4	26.1
2.4	13.3	6.5	26.4
2.5	13.6	6.6	26.8
2.6	13.9	6.7	27.2
2.7	14.2	6.8	27.6
2.8	14.5	6.9	28.0
2.9	14.7	7.0	28.3
3.0	15.0	7.1	28.7
3.1	15.3	7.2	29.1
3.2	15.6	7.3	29.5
3.3	15.9	7.4	29.9
3.4	16.2	7.5	30.4
3.5	16.5	7.6	30.8
3.6	16.8	7.7	31.2
3.7	17.1	7.8	31.6
3.8	17.4	7.9	32.0
3.9	17.7	8.0	32.5
4.0	18.0	8.1	32.9
4.1	18.3	8.2	33.3
4.2	18.6	8.3	33.8
4.3	18.9	8.4	34.2
4.4	19.2	8.5	34.7
4.5	19.5	8.6	35.1
4.6	19.9	8.7	35.6
4.7	20.2	8.8	36.1
4.8	20.5	8.9	36.5
4.9	20.8	9.0	37.0
5.0	21.2	9.1	37.5
5.1	21.5	9.2	38.0
5.2	21.8	9.3	38.5
5.3	22.2	9.4	38.9
5.4	22.5	9.5	39.4
5.5	22.8	9.6	39.9
5.6	23.2	9.7	40.5
5.7	23.5	9.8	41.0
5.8	23.9	9.9	41.5
5.9	24.2	10.0	42.0
6.0	24.6		

[a]Menstrual age = 6.8954 + 2.6345 (BPD) + 0.008771 (BPD)^3 [r^2 = 98.7%].

Abbreviation: BPD, biparietal diameter.

Source: From Ref. 41.

TABLE 7 ■ Pitfalls in Determination of BPD

Pitfall	Comments
Misplaced calipers	Measure BPD from leading edge to leading edge
Angled BPD	Follow strict criteria for BPD including Midline echo Thalami symmetric Cavum septi pellucidi one-third of fronto-occipital diameter
Normal head shape (brachycephalic or dolichocephalic)	Check cephalic index Obtain head circumference

Abbreviation: BPD, biparietal diameter.

TABLE 8 ■ Predicted-Menstrual Age[a] for Head Circumference

Head circumference (cm)	Menstrual age (wk)	Head circumference (cm)	Menstrual age (wk)
8.0	13.4	22.5	24.4
8.5	13.7	23.0	24.9
9.0	14.0	23.5	25.4
9.5	14.3	24.0	25.9
10.0	14.6	24.5	26.4
10.5	15.0	25.0	26.9
11.0	15.3	25.5	27.5
11.5	15.6	26.0	28.0
12.0	15.9	26.5	28.1
12.5	16.3	27.0	29.2
13.0	16.6	27.5	29.8
13.5	17.0	28.0	30.3
14.0	17.3	28.5	31.0
14.5	17.7	29.0	31.6
15.0	18.1	29.5	32.2
15.5	18.4	30.0	32.8
16.0	18.8	30.5	33.5
16.5	19.2	31.0	34.2
17.0	19.6	31.5	34.9
17.5	20.0	32.0	35.5
18.0	20.4	32.5	36.3
18.5	20.8	33.0	37.0
19.0	21.2	33.5	37.7
19.5	21.6	34.0	38.5
20.0	22.1	34.5	39.2
20.5	22.5	35.0	40.0
21.0	23.0	35.5	40.8
21.5	23.4	36.0	41.6
22.0	23.9		

[a]Menstrual age = 8.8 + 0.55 (head circumference) + 2.8 × 10^{-4} (head circumference)^3; r^2 = 97.9% 1 SD − 1.18 wk.

Source: From Ref. 54.

TABLE 9 ■ Percentile Values for Fetal Head Circumference from the 3rd Through the 97th Percentile for Menstrual Age in Weeks

Menstrual week	Head circumference (cm)				
	3rd	10th	50th	90th	97th
14	8.8	9.1	9.7	10.3	10.6
15	10.0	10.4	11.0	11.6	12.0
16	11.3	11.7	12.4	13.1	13.5
17	12.6	13.0	13.8	14.6	15.0
18	13.7	14.2	15.1	16.0	16.5
19	14.9	15.5	16.4	17.4	17.9
20	16.1	16.7	17.7	18.7	19.3
21	17.2	17.8	18.9	20.0	20.6
22	18.3	18.9	20.1	21.3	21.9
23	19.4	20.1	21.3	22.5	23.2
24	20.4	21.1	22.4	23.7	24.3
25	21.4	22.2	23.5	24.9	25.6
26	22.4	23.2	24.6	26.0	26.8
27	23.3	24.1	25.6	27.1	27.9
28	24.2	25.1	26.6	28.1	29.0
29	25.0	25.9	27.5	29.1	30.0
30	25.8	26.8	28.4	30.0	31.0
31	26.7	27.6	29.3	31.0	31.9
32	27.4	28.4	30.1	31.8	32.8
33	28.0	29.0	30.8	32.6	33.6
34	28.7	29.7	31.5	33.3	34.3
35	29.3	30.4	32.2	34.1	35.1
36	29.9	30.9	32.8	34.7	35.8
37	30.3	31.4	33.3	35.2	36.3
38	30.8	31.9	33.8	35.8	36.8
39	31.1	32.2	34.2	36.2	37.3
40	31.5	32.6	34.6	36.6	37.7

Source: Adapted from Refs. 56, 57.

TABLE 10 ■ Predicted Menstrual Age[a] for AC Values

AC (cm)	Menstrual age (wk)	AC (cm)	Menstrual age (wk)
10.0	15.6	23.5	27.7
10.5	16.1	24.0	28.2
11.0	16.5	24.5	28.7
11.5	16.9	25.0	29.2
12.0	17.3	25.5	29.7
12.5	17.8	26.0	30.1
13.0	18.2	26.5	30.6
13.5	18.6	27.0	31.1
14.0	19.1	27.5	31.6
14.5	19.5	28.0	32.1
15.0	20.0	28.5	32.6
15.5	20.4	29.0	33.1
16.0	20.8	29.5	33.6
16.5	21.3	30.0	34.1
17.0	21.7	30.5	34.6
17.5	22.2	31.0	35.1
18.0	22.6	31.5	35.6
18.5	23.1	32.0	36.1
19.0	23.6	32.5	36.6
19.5	24.0	33.0	37.1
20.0	24.5	33.5	37.6
20.5	24.9	34.0	38.1
21.0	25.4	34.5	38.7
21.5	25.9	35.0	39.2
22.0	26.3	35.5	39.7
22.5	26.8	36.0	40.2
23.0	27.3	36.5	40.8

[a]Menstrual age = 7.6070 + 0.7645 (AC) + 0.00393 $(AC)^2$; $r^2 = 97.8\%$; 1 SD = 1.2 wk.

Abbreviation: AC, abdominal circumference.

Source: From Ref. 59.

TABLE 11 ■ Percentile Values for Fetal Abdominal Circumference

Menstrual week	Abdominal circumference (cm)				
	3rd	10th	50th	90th	97th
14	6.4	6.7	7.3	7.9	8.3
15	7.5	7.9	8.6	9.3	9.7
16	8.6	9.1	9.9	10.7	11.2
17	9.7	10.3	11.2	12.1	12.7
18	10.9	11.5	12.5	13.5	14.1
19	11.9	12.6	13.7	14.8	15.5
20	13.1	13.8	15.0	16.3	17.0
21	14.1	14.9	16.2	17.6	18.3
22	15.1	16.0	17.4	18.8	19.7
23	16.1	17.0	18.5	20.0	20.9
24	17.1	18.1	19.7	21.3	22.3
25	18.1	19.1	20.8	22.5	23.5
26	19.1	20.1	21.9	23.7	24.8
27	20.0	21.1	23.0	24.9	26.0
28	20.9	22.0	24.0	26.0	27.1
29	21.8	23.0	25.1	27.2	28.4
30	22.7	23.9	26.1	28.3	29.5
31	23.6	24.9	27.1	29.4	30.6
32	24.5	25.8	28.1	30.4	31.8
33	25.3	26.7	29.1	31.5	32.9
34	26.1	27.5	30.0	32.5	33.9
35	26.9	28.3	30.9	33.5	34.9
36	27.7	29.2	31.8	34.4	35.9
37	28.5	30.0	32.7	35.4	37.0
38	29.2	30.8	33.6	36.4	38.0
39	29.9	31.6	34.4	37.3	38.9
40	30.7	32.4	35.3	38.2	39.9

Source: Adapted from Refs. 56, 57.

TABLE 12 ■ Pitfalls in Determination of AC

Pitfall	Comments
Elliptic shape	Too oblique a scan; follow strict criteria for AC including
	Spine at 3 or 9 o'clock
	Transaxial spine
	Stomach in left midabdomen
	Small umbilical vein segment
Too long a segment of umbilical vein	Too oblique a scan plane
Measurement at umbilical cord insertion	Too low for measurement
	Follow foregoing criteria
Spine up	Cannot obtain true AC; wait until spine is at 3 or 9 o'clock
In utero environment	Increase in AC (macrosomia)
	Decrease in AC (intrauterine growth retardation)
	Early ultrasound estimates are helpful to predict true menstrual age

Abbreviation: AC, abdominal circumference.

TABLE 13 ■ Predicted Menstrual Age in Weeks for Specific Femur Lengths (1.0–7.9 cm)

Femur length (cm)	Menstrual age (wk)	Femur length (cm)	Menstrual age (wk)
1.0	12.8	3.1	19.2
1.1	13.1	3.2	19.6
1.2	13.4	3.3	19.9
1.3	13.6	3.4	20.3
1.4	13.9	3.5	20.7
1.5	14.2	3.6	21.0
1.6	14.5	3.7	21.4
1.7	14.8	3.8	21.8
1.8	15.1	3.9	22.1
1.9	15.4	4.0	22.5
2.0	15.7	4.1	22.9
2.1	16.0	4.2	23.3
2.2	16.3	4.3	23.7
2.3	16.6	4.4	24.1
2.4	16.9	4.5	24.5
2.5	17.2	4.6	24.9
2.6	17.6	4.7	25.3
2.7	17.9	4.8	25.7
2.8	18.2	4.9	26.1
2.9	18.6	5.0	26.5
3.0	18.9	5.1	27.0
5.2	27.4	6.6	33.8
5.3	27.8	6.7	34.2
5.4	28.2	6.8	34.7
5.5	28.7	6.9	35.2
5.6	29.1	7.0	35.7

(*Continued*)

TABLE 13 ■ Predicted Menstrual Age in Weeks for Specific Femur Lengths (1.0–7.9 cm) (*Continued*)

Femur length (cm)	Menstrual age (wk)	Femur length (cm)	Menstrual age (wk)
5.7	29.6	7.1	36.2
5.8	30.0	7.2	36.7
5.9	30.5	7.3	37.2
6.0	30.9	7.4	37.7
6.1	31.4	7.5	38.3
6.2	31.9	7.6	38.8
6.3	32.3	7.7	39.3
6.4	32.8	7.8	39.8
6.5	33.3	7.9	40.4

Source: Adapted from Refs. 54, 56.

TABLE 14 ■ Percentile Values of Fetal Femur Length from the 3rd Through the 97% Percentile for Given Menstrual Ages (in Weeks)

Menstrual week	Femur length (cm)				
	3rd	10th	50th	90th	97th
14	1.2	1.3	1.4	2.5	1.6
15	1.5	1.6	1.7	1.9	1.9
16	1.7	1.8	2.0	2.2	2.3
17	2.1	2.2	2.4	2.6	2.7
18	2.3	2.5	2.7	2.9	3.1
19	2.6	2.7	3.0	3.3	3.4
20	2.8	3.0	3.3	3.6	3.8
21	3.0	3.2	3.5	3.8	4.0
22	3.3	3.5	3.8	4.1	4.3
23	3.5	3.7	4.1	4.5	4.7
24	3.8	4.0	4.4	4.8	5.0
25	4.0	4.2	4.6	5.0	5.2
26	4.2	4.5	4.9	5.3	5.6
27	4.4	4.6	5.1	5.6	5.8
28	4.6	4.9	5.4	5.9	6.2
29	4.8	5.1	5.6	6.1	6.4
30	5.0	5.3	5.8	6.3	6.6
31	5.2	5.5	6.0	6.5	6.8
32	5.3	5.6	6.2	6.8	7.1
33	5.5	5.8	6.4	7.0	7.3
34	5.7	6.0	6.6	7.2	7.5
35	5.9	6.2	6.8	7.4	7.8
36	6.0	6.4	7.0	7.6	8.0
37	6.2	6.6	7.2	7.9	8.2
38	6.4	6.7	7.4	8.1	8.4
39	6.5	6.8	7.5	8.2	8.6
40	6.6	7.0	7.7	8.4	8.8

Source: From Refs. 56, 57.

TABLE 15 ■ Pitfalls in Determination of Femur Length

Pitfall	Comments
Femur perpendicular or oblique to transducer	Obtain femur length parallel to transducer
Including cartilaginous ends of femur	Measure ossified femoral shaft only
Measuring femur at greatest distance from the transducer	Measure femur that is closest to the transducer (when possible)
Use of sector or linear array transducers	Use single type of transducer to reduce errors between examinations

Source: From Refs. 56, 57.

TABLE 16 ■ The Fifth Through 95% Percentile for Given Menstrual Ages (in Weeks) for Humeral Length (in mm)

Bone length (mm)	Humerus percentile		
	5th	50th	95th
10	9 + 6	12 + 4	15 + 2
11	10 + 1	12 + 6	15 + 4
12	10 + 3	13 + 1	15 + 6
13	10 + 6	13 + 4	16 + 1
14	11 + 1	13 + 6	16 + 4
15	11 + 3	14 + 1	16 + 6
16	11 + 6	14 + 4	17 + 2
17	12 + 1	14 + 6	17 + 4
18	12 + 4	15 + 1	18
19	12 + 6	15 + 4	18 + 2
20	13 + 1	15 + 6	18 + 5
21	13 + 4	16 + 2	19 + 1
22	13 + 6	16 + 5	19 + 3
23	14 + 2	17 + 1	19 + 6
24	14 + 5	17 + 3	20 + 1
25	15 + 1	17 + 6	20 + 4
26	15 + 4	18 + 1	21
27	15 + 6	18 + 4	21 + 3
28	16 + 2	19	21 + 6
29	16 + 5	19 + 3	22 + 1
30	17 + 1	19 + 6	22 + 4
31	17 + 4	20 + 2	23
32	18	20 + 5	23 + 4
33	18 + 3	21 + 1	23 + 6
34	18 + 6	21 + 4	24 + 2
35	19 + 2	22	24 + 6
36	19 + 5	22 + 4	25 + 1
37	20 + 1	22 + 6	25 + 5
38	20 + 4	23 + 3	26 +1
39	21 + 1	23 + 6	26 + 4
40	21 + 4	24 + 2	27 + 1
41	22	24 + 6	27 + 4
42	22 + 4	25 + 2	28
43	23	25 + 5	28 + 4
44	23 + 4	26 + 1	29

(*Continued*)

TABLE 16 ■ The Fifth Through 95% Percentile for Given Menstrual Ages (in Weeks) for Humeral Length (in mm) (*Continued*)

Bone length (mm)	Humerus percentile		
	5th	50th	95th
45	24	26 + 5	29 + 4
46	24 + 4	27 + 1	30
47	25	27 + 5	30 + 4
48	25 + 4	28 + 1	31
49	26	28 + 6	31 + 4
50	26 + 4	29 + 2	32
51	27 + 1	29 + 6	32 + 4
52	27 + 4	30 + 2	33 + 1
53	28 + 1	30 + 6	33 + 4
54	28 + 5	31 + 3	34 + 1
55	29 + 1	32	34 + 5
56	29 + 6	32 + 4	35 + 2
57	30 + 2	33 + 1	35 + 6
58	30 + 6	33 + 4	36 + 3
59	31 + 3	34 + 1	36 + 6
60	32	34 + 6	37 + 4
61	32 + 4	35 + 2	38 + 1
62	33 + 1	35 + 6	38 + 5
63	33 + 6	36 + 4	39 + 2
64	34 + 3	37 + 1	39 + 6
65	35	37 + 5	40 + 4
66	35 + 4	38 + 2	41 + 1
67	36 + 1	38 + 6	41 + 5
68	36 + 6	39 + 4	42 + 2
69	37 + 3	40 + 1	42 + 6

Source: From Ref. 88.

TABLE 17 ■ The Fifth Through 95% Percentile for Given Menstrual Ages (in Weeks + Days) for Ulna Length (in mm)

Bone length (mm)	Ulna percentile		
	5th	50th	95th
10	10 + 1	13 + 1	16 + 1
11	10 + 4	13 + 4	16 + 4
12	10 + 6	13 + 6	16 + 6
13	11 + 1	14 + 1	17 + 2
14	11 + 4	14 + 4	17 + 5
15	11 + 6	15	18
16	12 + 2	15 + 3	18 + 3
17	12 + 5	15 + 5	18 + 6
18	13 + 1	16 + 1	19 + 1
19	13 + 4	16 + 4	19 + 4
20	13 + 6	16 + 6	20
21	14 + 2	17 + 2	20 + 3
22	14 + 5	17 + 5	20 + 6
23	15 + 1	18 + 1	21 + 1
24	15 + 4	18 + 4	21 + 4
25	16	19	22 + 1

(*Continued*)

TABLE 17 ■ The Fifth Through 95% Percentile for Given Menstrual Ages (in Weeks + Days) for Ulna Length (in mm) (*Continued*)

Bone length (mm)	Ulna percentile		
	5th	50th	95th
26	16 + 3	19 + 3	22 + 4
27	16 + 6	19 + 6	22 + 6
28	17 + 2	20 + 2	23 + 3
29	17 + 5	20 + 6	23 + 6
30	18 + 1	21 + 1	24 + 2
31	18 + 4	21 + 5	24 + 6
32	19 + 1	22 + 1	25 + 1
33	19 + 4	22 + 5	25 + 5
34	20 + 1	23 + 1	26 + 1
35	20 + 4	23 + 4	26 + 5
36	21 + 1	24 + 1	27 + 1
37	21 + 4	24 + 4	27 + 5
38	22 + 1	25 + 1	28 + 1
39	22 + 4	25 + 4	28 + 5
40	23 + 1	26 + 1	29 + 1
41	23 + 4	26 + 5	29 + 5
42	24 + 1	27 + 1	30 + 2
43	24 + 5	27 + 5	30 + 6
44	25 + 1	28 + 2	31 + 2
45	25 + 6	28 + 6	31 + 6
46	26 + 2	29 + 3	32 + 3
47	26 + 6	29 + 6	33
48	27 + 3	30 + 4	33 + 4
49	28	31 + 1	34 + 1
50	28 + 4	31 + 4	34 + 5
51	29 + 1	32 + 1	35 + 2
52	29 + 5	32 + 6	35 + 6
53	30 + 2	33 + 3	36 + 3
54	30 + 6	34	37
55	31 + 4	34 + 4	37 + 5
56	32 + 1	35 + 1	38 + 2
57	32 + 6	35 + 6	38 + 6
58	33 + 3	36 + 3	39 + 4
59	34	37 + 1	40 + 1
60	34 + 4	37 + 5	40 + 6
61	35 + 2	38 + 2	41 + 3
62	35 + 6	39	42
63	36 + 4	39 + 4	42 + 5
64	37 + 1	40 + 2	43 + 2

Source: From Ref. 88.

TABLE 18 ■ The Fifth Through 95% Percentile for Given Menstrual Ages (in Weeks + Days) for Tibial Length (in mm)

Bone length (mm)	Tibia percentile		
	5th	50th	95th
10	10 + 4	13 + 3	16 + 2
11	10 + 6	13 + 5	16 + 4
12	11 + 1	14 + 1	17
13	11 + 4	14 + 3	17 + 2
14	11 + 6	14 + 6	17 + 5
15	12 + 1	15 + 1	18

(*Continued*)

TABLE 18 ■ The Fifth Through 95% Percentile for Given Menstrual Ages (in Weeks + Days) for Tibial Length (in mm) (*Continued*)

Bone length (mm)	Tibia percentile		
	5th	50th	95th
16	12 + 4	15 + 4	18 + 3
17	13	15 + 6	18 + 6
18	13 + 2	16 + 1	19 + 1
19	13 + 5	16 + 4	19 + 4
20	14 + 1	17	19 + 6
21	14 + 4	17 + 3	20 + 2
22	14 + 6	17 + 6	20 + 5
23	15 + 1	18 + 1	21 + 1
24	15 + 4	18 + 4	21 + 3
25	16	18 + 6	21 + 6
26	16 + 3	19 + 2	22 + 1
27	16 + 6	19 + 5	22 + 4
28	17 + 1	20 + 1	23
29	17 + 4	20 + 4	23 + 4
30	18 + 1	21	23 + 6
31	18 + 4	21 + 3	24 + 2
32	18 + 6	21 + 6	24 + 5
33	19 + 2	22 + 1	25 + 1
34	19 + 5	22 + 4	25 + 4
35	20 + 1	23 + 1	26
36	20 + 4	23 + 4	26 + 3
37	21	23 + 6	26 + 6
38	21 + 4	24 + 3	27 + 2
39	21 + 6	24 + 6	27 + 5
40	22 + 3	25 + 2	28 + 1
41	22 + 6	25 + 5	28 + 4
42	23 + 2	26 + 1	29 + 1
43	23 + 5	26 + 4	29 + 4
44	24 + 1	27 + 1	30
45	24 + 4	27 + 4	30 + 4
46	25 + 1	28	30 + 6
47	25 + 4	28 + 4	31 + 3
48	26 + 1	29	31 + 6
49	26 + 4	29 + 3	32 + 2
50	27	29 + 6	32 + 6
51	27 + 4	30 + 3	33 + 2
52	28	30 + 6	33 + 6
53	28 + 4	31 + 3	34 + 2
54	29	31 + 6	34 + 6
55	29 + 4	32 + 3	35 + 2
56	30	32 + 6	35 + 6
57	30 + 4	33 + 3	36 + 2
58	31	33 + 6	36 + 6
59	31 + 4	34 + 3	37 + 2
60	32	34 + 6	37 + 6
61	32 + 4	35 + 3	38 + 2
62	33	35 + 6	38 + 6
63	33 + 4	36 + 4	39 + 3
64	34 + 1	37	39 + 6
65	34 + 4	37 + 4	40 + 3
66	35 + 1	38	41
67	35 + 5	38 + 4	41 + 4
68	36 + 1	39 + 1	42
69	36 + 6	39 + 5	42 + 4

Source: From Ref. 88.

TABLE 19 ■ Menstrual Age for 5th, 50th, and 95th Percentile Prediction Intervals for Radius Length in a Caucasian Population (Values for African Americans in Parentheses)

Radius (cm)	5th Percentile	50th Percentile	95th Percentile
0.6	11.7 (12.3)	12.9 (12.9)	14.3 (13.6)
0.7	11.9 (12.6)	13.2 (13.2)	14.6 (13.9)
0.8	12.2 (12.9)	13.5 (13.5)	15.0 (14.2)
0.9	12.5 (13.1)	13.8 (13.8)	15.4 (14.5)
1.0	12.7 (13.4)	14.2 (14.1)	15.7 (14.8)
1.1	13.0 (13.7)	14.5 (14.4)	16.1 (15.1)
1.2	13.3 (14.0)	14.8 (14.7)	16.4 (15.5)
1.3	13.6 (14.3)	15.1 (15.1)	16.8 (15.8)
1.4	13.9 (14.7)	15.5 (15.4)	17.2 (16.2)
1.5	14.3 (15.0)	15.8 (15.7)	17.6 (16.5)
1.6	14.6 (15.3)	16.2 (16.1)	18.0 (16.9)
1.7	14.9 (15.7)	16.6 (16.4)	18.4 (17.2)
1.8	15.3 (16.0)	17.0 (16.8)	18.8 (17.6)
1.9	15.6 (16.4)	17.3 (17.2)	19.3 (18.0)
2.0	16.0 (16.7)	17.7 (17.5)	19.7 (18.4)
2.1	16.3 (17.1)	18.1 (17.9)	20.1 (18.8)
2.2	16.7 (17.5)	18.6 (18.3)	20.6 (19.3)
2.3	17.1 (17.8)	19.0 (18.7)	21.0 (19.7)
2.4	17.5 (18.2)	19.4 (19.2)	21.5 (20.1)
2.5	17.9 (18.6)	19.9 (19.6)	22.0 (20.6)
2.6	18.3 (19.1)	20.3 (20.0)	22.5 (21.0)
2.7	18.7 (19.5)	20.8 (20.5)	23.1 (21.5)
2.8	19.1 (19.6)	21.2 (20.9)	23.5 (22.0)
2.9	19.6 (20.3)	21.7 (21.4)	24.1 (22.4)
3.0	20.0 (20.8)	22.2 (21.9)	24.7 (22.9)
3.1	20.5 (21.3)	22.7 (22.3)	25.2 (23.4)
3.2	20.9 (21.7)	23.2 (22.8)	25.8 (24.0)
3.3	21.4 (22.2)	23.8 (23.3)	26.4 (24.5)
3.4	21.9 (22.7)	24.3 (23.8)	27.0 (25.0)
3.5	22.4 (23.2)	24.9 (24.4)	27.6 (25.6)
3.6	22.9 (23.7)	25.5 (24.9)	28.3 (26.2)
3.7	23.5 (24.2)	26.0 (25.5)	28.9 (26.7)
3.8	24.0 (24.8)	26.6 (26.0)	29.6 (27.3)
3.9	24.5 (25.3)	27.2 (26.6)	30.2 (27.9)
4.0	25.1 (25.9)	27.9 (27.2)	30.9 (28.6)
4.1	25.7 (26.5)	28.5 (27.8)	31.6 (29.2)
4.2	26.3 (27.0)	29.1 (28.4)	32.4 (29.8)
4.3	26.9 (27.6)	29.8 (29.0)	33.1 (30.5)
4.4	27.5 (28.2)	30.5 (29.7)	33.9 (31.2)
4.5	28.1 (28.9)	31.2 (30.3)	34.6 (31.9)
4.6	28.7 (29.5)	31.9 (31.0)	35.4 (32.6)
4.7	29.4 (30.2)	32.6 (31.7)	36.2 (33.3)
4.8	30.1 (30.8)	33.4 (32.4)	37.1 (34.0)
4.9	30.7 (31.5)	34.1 (33.1)	37.9 (34.8)
5.0	31.4 (32.2)	34.9 (33.8)	38.8 (35.5)
5.1	32.2 (32.9)	35.7 (34.6)	39.7 (36.3)
5.2	32.9 (33.6)	36.5 (35.3)	40.5 (37.1)
5.3	33.6 (34.4)	37.4 (36.1)	41.5 (38.0)
5.4	34.4 (35.1)	38.2 (36.9)	42.5 (38.8)
5.5	35.2 (35.9)	39.1 (37.7)	43.4 (39.7)
5.6	36.0 (36.7)	40.0 (38.6)	44.4 (40.5)
5.7	36.8 (37.5)	40.9 (39.4)	45.4 (41.4)
5.8	– (38.3)	– (40.3)	– (42.2)

Source: From Ref. 87.

TABLE 20 ■ Comparison of Mean Postpartum and Ultrasonographic Foot Length with Streeter's Pathologic Data (1920)

Gestation (wk)	Streeter's data (mm)	US foot length (mm)	Postpartum foot length (mm)
11	7	8	–
12	9	9	–
13	11	10	–
14	14	16	–
15	17	16	–
16	20	21	–
17	23	24	–
18	27	27	–
19	31	28	–
20	33	33	33
21	35	35	–
22	40	38	–
23	42	42	–
24	45	44	–
25	48	47	48
26	50	51	–
27	53	54	52
28	55	58	–
29	57	57	57
30	59	61	60
31	61	62	60
32	63	63	66
33	65	67	68
34	68	68	71
35	71	71	72
36	74	74	74
37	77	75	78
38	79	78	78
39	81	78	80
40	83	82	81
41	–	–	82
42	–	–	82
43	–	–	84

Source: From Ref. 90.

TABLE 21 ■ Fetal Foot Length Percentiles by Menstrual Age[a]

Menstrual age (wks)	Fetal foot length percentiles (smoothed)						
	N	CV (%)	5th	10th	50th	90th	95th
15	18	12.7	1.4	1.5	1.8	2.2	2.3
16	146	10.4	1.6	1.7	2.1	2.5	2.6
17	375	9.7	1.9	2.0	2.4	2.8	2.9
18	613	9.8	2.2	2.3	2.7	3.1	3.2
19	1160	8.9	2.5	2.6	3.0	3.3	3.4
20	929	9.3	2.8	2.9	3.2	3.6	3.7
21	552	8.5	3.1	3.2	3.5	3.9	4.0
22	360	8.9	3.4	3.5	3.9	4.2	4.3
23	222	8.1	3.7	3.8	4.2	4.6	4.7
24	177	7.0	4.0	4.1	4.5	4.9	5.0

(*Continued*)

TABLE 21 ■ Fetal Foot Length Percentiles by Menstrual Age[a] (*Continued*)

Menstrual age (wks)	Fetal foot length percentiles (smoothed)						
	N	CV (%)	5th	10th	50th	90th	95th
25	125	7.1	4.3	4.4	4.8	5.1	5.2
26	123	7.0	4.6	4.7	5.1	5.4	5.5
27	108	6.3	4.8	4.9	5.3	5.7	5.8
28	74	5.4	5.1	5.2	5.6	5.9	6.0
29	66	6.2	5.3	5.4	5.8	6.2	6.3
30	65	5.2	5.6	5.7	6.1	6.4	6.5
31	62	5.7	5.8	5.9	6.3	6.7	6.8
32	65	5.3	6.0	6.1	6.5	6.9	7.0
33	39	4.4	6.3	6.4	6.8	7.1	7.2
34	37	6.8	6.5	6.6	7.0	7.4	7.5
35	24	6.2	6.8	6.9	7.3	7.6	7.7
36	15	5.5	7.0	7.1	7.5	7.9	8.0
37	17	5.3	7.3	7.4	7.7	8.1	8.2

[a]Values for percentiles are in centimeters. *N*, number of fetuses; CV, coefficient of variation.

Source: From Ref. 91.

TABLE 22 ■ GA Percentile by TCD

TCD (cm)	Percentiles of GA (wk)				
	5th	10th	50th	90th	95th
1.2	13.3	13.6	14.5	15.6	15.9
1.3	13.9	14.2	15.1	16.2	16.5
1.4	14.6	14.9	15.7	16.8	17.1
1.5	15.2	15.5	16.4	17.4	17.7
1.6	15.9	16.1	17.0	18.1	18.4
1.7	16.5	16.8	17.7	18.7	19.0
1.8	17.2	17.4	18.3	19.4	19.7
1.9	17.8	18.1	19.0	20.1	20.4
2.0	18.4	18.7	19.7	20.8	21.1
2.1	19.1	19.4	20.3	21.5	21.8
2.2	19.7	20.0	21.0	22.2	22.5
2.3	20.3	20.7	21.7	22.9	23.2
2.4	21.0	21.3	22.3	23.6	23.9
2.5	21.6	21.9	23.0	24.3	24.6
2.6	22.2	22.5	23.7	25.0	25.4
2.7	22.8	23.2	24.3	25.7	26.1
2.8	23.4	23.8	25.0	26.4	26.8
2.9	24.0	24.4	25.6	27.1	27.5
3.0	24.6	25.0	26.3	27.7	28.2
3.1	25.2	25.6	26.9	28.4	28.9
3.2	25.8	26.2	27.5	29.1	29.5
3.3	26.3	26.8	28.1	29.7	30.2
3.4	26.9	27.3	28.7	30.4	30.8
3.5	27.4	27.9	29.3	31.0	31.5
3.6	28.0	28.4	29.9	31.6	32.1
3.7	28.5	29.0	30.4	32.2	32.7
3.8	29.0	29.5	31.0	32.8	33.3
3.9	29.5	30.0	31.5	33.3	33.8
4.0	30.0	30.5	32.0	33.8	34.4
4.1	30.5	31.0	32.5	34.3	34.9
4.2	31.0	31.5	33.0	34.8	35.3

(*Continued*)

TABLE 22 GA Percentile by TCD (*Continued*)

TCD (cm)	Percentiles of GA (wk) 5th	10th	50th	90th	95th
4.3	31.4	31.9	33.4	35.3	35.8
4.4	31.9	32.4	33.9	35.7	36.2
4.5	32.3	32.8	34.3	36.1	36.6
4.6	32.7	33.2	34.7	36.4	36.9
4.7	33.1	33.6	35.0	36.8	37.2
4.8	33.5	34.0	35.4	37.0	37.5
4.9	33.9	34.3	35.7	37.3	37.8
5.0	34.3	34.7	36.0	37.5	38.0
5.1	34.6	35.0	36.2	37.7	38.1
5.2	34.9	35.3	36.5	37.8	38.2

Abbreviations: GA, gestational age; TCD, transverse cerebellar diameter.
Source: From Ref. 93.

TABLE 23 Additional Fetal Parameters Associated with Menstrual Age

Parameter	Figure	Ref.
Orbital diameter	39A	97
Clavicular length	39B	94
Scapular length	39C	95
Thigh circumference	39D	98
Foot	39E	91
Transverse cerebellar diameter	39F	93
Renal length	39G	99
Liver length	39H	100

REFERENCES

1. Chiazze L, Brayer FT, Macisco JJ, et al. The length and variability of the human menstrual cycles. JAMA 1968; 203(6):377–80.
2. Treloar AE, Boynton RE, Behn BC, et al. Variation of the human menstrual cycle throughout reproduction life. Int J Fertil 1967; 12(1 Pt 2):77–126.
3. Persson P-H, Kullander S. Long-term experience of general ultrasound screening in pregnancy. Am J Obstet Gynecol 1983; 146(8):942–947.
4. Hill LM, Guzick D, Hixson J, et al. Composite assessment of gestational age: a comparison of institutionally derived and published regression equations. Am J Obstet Gynecol 1992; 166(2):551–555.
5. Hadlock FP, Harrist RB, Shah YP, et al. Sonographic fetal growth standards: are current data applicable to a racially mixed population? J Ultrasound Med 1990; 9:157–160.
6. Liggins GC. The drive to fetal growth. In: Beard RW, Nathanielz PW, eds. Fetal Physiology and Medicine. Philadelphia: WB Saunders, 1976:254.
7. Dunn PM, Butler NR. Intrauterine growth: a discussion of some of the problems besetting its measurement. In: Brass W, ed. Symposium of the Society of the Study of Human Biology. Vol. 10: Biological Aspects of Demography. London: Taylor & Francis, 1971:147.
8. Deter RL, Harrist RB, Hadlock FP, et al. The use of ultrasound in the assessment of normal fetal growth: a review. J Clin Ultrasound 1981; 9(9):481–493.
9. Deter RL, Harrist RB. Growth standards for anatomic measurements and growth rates derived from longitudinal studies of normal fetal growth. J Clin Ultrasound 1992; 20(6):381–388.
10. Fossum GT, Davajan V, Kletzky DA. Early detection of pregnancy with transvaginal ultrasound. Fertil Steril 1988; 49(5):788–791.
11. Nyberg DA, Mack LA, Laing FC, et al. Distinguishing normal from abnormal gestational sac growth in early pregnancy. J Ultrasound Med 1987; 6(1):23–27.
12. Muller T, Sütterlin M, Pöhls U, et al. Transvaginal volumetry of first trimester gestational sac: a comparison of conventional with three-dimensional ultrasound. J Perinat Med 2000; 28(3):214–220.
13. Daya S. Human chorionic gonadotropin increased in normal early pregnancy. Am J Obstet Gynecol 1987; 156(2):286–290.
14. Daya S, Woods S, Ward S, et al. Transvaginal ultrasound scanning in early pregnancy and correlation with human chorionic gonadotropin levels. J Clin Ultrasound 1991; 19(3):139–142.
15. Nyberg DA, Filly RA, Filho DLD, et al. Abnormal pregnancy: early diagnosis by ultrasound and serum chorionic gonadotropin levels. Radiology 1986; 158(2):393–396.
16. Hay DL et al. Monitoring early pregnancy with transvaginal ultrasound and choriogonadotrophic levels. Aust NZ J Obstet Gynecol 1989; 29(2):165.
17. Kadar N, Bohrer M, Kemmann E, et al. The discriminatory human chorionic gonadotropin zone for endovaginal sonography: a prospective, randomized study. Fertil Steril 1994; 61(6):1016–1020.
18. Jouppila PC. Length and depth of the uterus and the diameter of the gestational sac in normal gravidas during early pregnancy. Acta Obstet Gynecol Scand 1971; 50(suppl 15):29–31.
19. Hellman LM, Kobayashi M, Fillisti L, et al. Growth and development of the human fetus prior to the twentieth week of gestation. Am J Obstet Gynecol 1969; 103(6):789–800.
20. Robinson HP. Gestation sac volumes as determined by sonar in the first trimester of pregnancy. Br J Obstet Gynaecol 1975; 82:100.
21. Bromley B et al. Small sac size in the first trimester: a predictor of poor fetal outcome. Radiology 1991; 178:375.
22. Dickey RP, Olar TX, Taylor SN, et al. Relationship of small gestational sac crown-rump length differences to abortion and abortus karyotypes. Obstet Gynecol 1992; 79(4):554–557.
23. Meegdes BH, Ingenhoes R, Peeters LL, et al. Early pregnancy wastage: relationship between chorionic vascularization and embryonic development. Fertil Steril 1988; 49(2):216–220.
24. Stern JJ, Coulam CB. Mechanism of recurrent spontaneous abortion. I. Ultrasonographic findings. Am J Obstet Gynecol 1992; 166(6 Pt 1):1844–1850.
25. Robinson HP. Sonar measurement of fetal crown-rump length as means of assessing maturity in first-trimester pregnancy. Br Med J 1973; 4(5883):28–31.
26. Levi CS, Lyons EA, Zheng XH, et al. Endovaginal ultrasound: demonstration of cardiac activity in embryos of less than 5.0 mm in crown-rump length. Radiology 1990; 176(1):71–74.
27. Moore KL, Persaud TVN, Shiota K, eds. Color Atlas of Clinical Embryology. Philadelphia: WB Saunders, 1994:40.
28. Goldstein SR. Embryonic ultrasonographic measurements: crown-rump length revisited. Am J Obstet Gynecol 1991; 165(3): 497–501.
29. Reinhold E. Ultrasonics in early pregnancy: diagnostic scanning and fetal motor activity VII. Fetal movements and fetal behavior. Contrib Gynecol Obstet 1976; 1:102–148.
30. Timor-Tritsch IE, Farine D, Rosen MG. A close look at early embryonic development with the high-frequency transvaginal transducer. Am J Obstet Gynecol 1988; 159(3):676–681.
31. Hadlock FP, Shah YP, Kanon DJ, et al. Fetal crown-rump length: re-evaluation of relation to menstrual age (5–18 weeks) with high-resolution real-time US. Radiology 1992; 182(2):501–505.

32. Kalish RB, Thaler HT, Chasen ST, et al. First-and second-trimester ultrasound assessment of gestational age. Am J Obstet Gynecol 2004(3);191:975–978.
33. Taipale P, Hiilesmaa V. Predicting delivery date by ultrasound and last menstrual period in early gestation. Obstet Gynecol 2001; 97(2):189–194.
34. Seibing A, McKay K. Ultrasound in first trimester shows no difference in fetal size between the sexes. Br Med J 1985; 290(6470):750.
35. Drugan A, Johnson MP, Isada NB, et al. The smaller than expected first-trimester fetus is at increased risk for chromosome anomalies. Am J Obstet Gynecol 1992; 167(6):1525–1528.
36. Tchobrutsky C, Breart GL, Rambaud DC, et al. Correlation between fetal defects and early growth delay observed by ultrasound (Letter) Lancet 1985; 1(8430):706–707.
37. Reljic M. The significance of crown-rump length measurement for predicting adverse pregnancy outcome of threatened abortion. Ultrasound Obstet Gynecol 2001; 17(6):510–512.
38. Kalish RB, Chasen ST, Gupta M, et al. First trimester prediction of growth discordance in twin gestations. Am J Obstet Gynecol 2003; 189(3):706–709.
39. Warren WB, Timor-Tritsch I, Peisner DB, et al. Dating the early pregnancy by sequential appearance of embryonic structures. Am J Obstet Gynecol 1989; 161(3):747–753.
40. Hill LM, DiNofrio DM, Guzick D. Sonographic determination of first trimester umbilical cord length. J Clin Ultrasound 1994; 22(7):435–438.
41. Hadlock FP, Deter LR, Harrist RD, et al. Fetal biparietal diameter: a critical re-evaluation of the relation to menstrual age by means of real-time ultrasound. J Ultrasound Med 1982; 1(3):97–104.
42. Campbell S. The prediction of fetal maturity by ultrasonic measurement of the biparietal diameter. J Obstet Gynaecol Br Commonwlth 1969; 76(7):603–609.
43. Campbell S, Warsof SL, Little D, et al. Routine ultrasound screening for the prediction of gestational age. Obstet Gynecol 1985; 65(5): 613–620.
44. Campbell S, Newman GB. Growth of the fetal biparietal diameter during normal pregnancy. J Obstet Gynaecol Br Commonwlth 1971; 78(6):513–519.
45. Chervenak FA, Skupski DW, Romero R, et al. How accurate is fetal biometry in the assessment of fetal age? Am J Obstet Gynecol 1998; 178(4):678–687.
46. Davison JM, Lind T, Farr V, et al. The limitations of ultrasonic cephalometry. J Obstet Gynaecol Br Commwlth 1973; 80(9):769–775.
47. Johnson ML, Dunne MG, Mach LA, et al. Evaluation of fetal intracranial anatomy by static and real-time ultrasound. J Clin Ultrasound 1980; 8(4):311–318.
48. Kasby CB, Poll V. The breech head and its ultrasound significance. Br J Obstet Gynaecol 1982; 89(2):106–110.
49. Wolfson RN, Zador I, Halvorsen P, et al. Biparietal diameter in premature rupture of membranes: errors in estimating gestational age. J Clin Ultrasound 1983; 11(7):371–374.
50. Hadlock FP, Deter RL, Carpenter RJ, et al. Estimating fetal age: effect of head shape on BPD. AJR Am J Roentgenol 1981; 137(1):83–85.
51. Jeanty P, Cousaert E, Hobbins JC, et al. A longitudinal study of fetal biometry. Am J Perinatol 1984; 1(2):118–128.
52. Hill LM, Breckle R, Gehrking WC. The variable effects of oligohydramnios on the biparietal diameter and the cephalic index. J Ultrasound Med 1984; 3(2):93 95.
53. Chitty LS, Altman DG, Henderson A, et al. Charts of fetal size. II. Head measurements. Br J Obstet Gynaecol 1994; 101(1):35–43.
54. Hadlock FP, Deter RL, Harrist RB, et al. Fetal head circumference: relation to menstrual age. AJR Am J Roentgenol 1982; 138(4):649–653.
55. Deter RL, Harrist RB, Hadlock FP, et al. Fetal head and abdominal circumferences. II. A critical re-evaluation of the relationship to menstrual age. J Clin Ultrasound 1982; 10(8):365–372.
56. Callen PW. Ultrasonography In Obstetrics and Gynecology. 3rd ed. WB Saunders Co, 1994:133.
57. Hadlock FP, Deter RL, Harrist RB, et al. Estimating fetal age: computer-assisted analysis of multiple fetal growth parameters. Radiology 1984; 152(2):497–501.
58. Campbell S, Wilkin D. Ultrasonic measurement of fetal abdominal circumference in the estimation of fetal weight. Br J Obstet Gynaecol 1975; 82(9):689–697.
59. Hadlock FP, Deter RL, Harrist RB, et al. Fetal abdominal circumference as a predictor of menstrual age. AJR Am J Roentgenol 1982; 139(2):367–370.
60. Hoffbauer H, Arabin PB, Baumann ML. Control of fetal development with multiple ultrasonic body measures. Contrib Gynecol Obstet 1979; 6:147–156.
61. Tamura RK, Sabbagha RE. Percentile ranks of sonar fetal abdominal circumference measurement. Am J Obstet Gynecol 1980:138(5): 475–479.
62. Shields JR, Medearis AL, Bear MB. Fetal head and abdominal circumferences: effect of profile shape on the accuracy of ellipse equations. J Clin Ultrasound 1987; 15(4):241–244.
63. Chitty LS, Altman DG, Henderson A, et al. Chart of fetal size: 3. Abdominal measurements. Br J Obstet Gynaecol 1994; 101(2): 125–131.
64. Rossavik IK, Deter RL. The effect of abdominal profile shape changes on the estimation of fetal weight. J Clin Ultrasound 1984; 12(1):57–59.
65. Mahoney MJ, Hobbins JC. Prenatal diagnosis of chondroectodermal dysplasia (Ellis-Van Creveld Syndrome) with fetoscopy and ultrasound. N Engl J Med 1977; 297(5):258–260.
66. Hadlock FP, Harrist RB, Deter RL, et al. Fetal femur length as a predictor of menstrual age: sonographically measured. AJR Am J Roentgenol 1982; 138(5):875–878.
67. Abrams SL, Filly RA. Curvature of the fetal femur: a normal sonographic finding. Radiology 1985; 156(2):490.
68. Yagel S, Adoni A, Oman S, et al. A statistical examination of the accuracy of combining femoral length and biparietal diameter as an index of fetal gestational age. Br J Obstet Gynaecol 1986; 93(2):109–115.
69. Lessoway VA, Schulzer M, Wittmann BL. Sonographic measurement of the fetal femur: factors affecting accuracy. J Clin Ultrasound 1990; 18(6):471–476.
70. Goldstein RB, Filly RA, Simpson G. Pitfalls in femur length measurements. J Ultrasound Med 1987; 6(4):203–207.
71. Wells PNT. Biomedical Ultrasonics. New York: Academic Press, 1977:150.
72. Abramowicz J, Jaffe R. Comparison between lateral and axial ultrasonic measurements of the fetal femur. Am J Obstet Gynecol 1988; 159(4):921–922.
73. Pretorius DH, Nelson TR, Manco–Johnson M. Fetal age estimation by ultrasound: the impact of measurement errors. Radiology 1984; 152(3):763–766.
74. Ott WJ. Accurate gestational dating. Obstet Gynecol 1985; 66(3):311–315.
75. Rose BI, Lamb EJ. Multiple simultaneous predictors of gestational age: an application of Bayes' theorem. Am J Perinatol 1988; 5(1):44–50.
76. Hohler CW, Quetel TA. Comparison of ultrasound femur length and biparietal diameter in late pregnancy. Am J Obstet Gynecol 1981; 141(7):759–762.
77. Hadlock FP, Deter RL, Harrist RD, et al. A date-independent predictor of intrauterine growth retardation: femur length/ abdominal circumference ratio. AJR Am J Roentgenol 1987; 141(5):979–984.
78. Jones TB, Wolfe HM, Zador IE. Biparietal diameter and femur length discrepancies: are maternal characteristics important? Ultrasound Obstet Gynecol 1991; 1(6):405–409.
79. Hadlock FP. Sonographic estimation of fetal age and weight. Radiol Clin North Am 1990; 28(1):39–50.
80. Kossoff G, Griffiths KA, Dixon CE. Is the quality of transvaginal images superior to transabdominal ones under matched conditions? Ultrasound Obstet Gynecol 1991; 1(1):29–35.
81. Wolfe HM, Zador IE, Bottoms SF, et al. Trends in sonographic fetal organ visualization. Ultrasound Obstet Gynecol 1993; 3(2): 97–99.
82. Rosenberg JC, Guzman ER, Vintzileos AM, et al. Transumbilical placement of the vaginal probe in obese pregnant women. Obstet Gynecol 1995; 85(1):132–134.
83. Hill LM, Guzick D, Fries J, et al. The transverse cerebellar diameter in estimating gestational age in the large for gestational age fetus. Obstet Gynecol 1990; 75(6):981–985.
84. Gardosi J, Mongelli M, Wilcox M, et al. An adjustable fetal weight standard. Ultrasound Obstet Gynecol 1995; 6(3):168–174.

85. Parker AJ, Davies P, Mayho AM, et al. The ultrasound estimation of sex-related variations of intrauterine growth. Am J Obstet Gynecol 1984; 149(6):665–669.
86. Moore WMO, Ward BS, Jones VP, et al. Sex difference in fetal head growth. Br J Obstet Gynaecol 1988; 95(3):238–242.
87. Hill LM, Guzick D, Thomas ML, et al. Fetal radius length: a critical evaluation of race as a factor in gestational age assessment. Am J Obstet Gynecol 1989; 161(1):193–199.
88. Jeanty P, Rodesch F, Delbeke D, et al. Estimation of gestational age from measurements of fetal long bones. J Ultrasound Med 1984; 3(2):75–79.
89. Bagnall KM, Harris PF, Jones PRM. A radiographic study of the longitudinal growth of primary ossification centers in limb long bones of the human fetus. Anat Rec 1982; 203(2):293–299.
90. Mercer BM, Sklar S, Shariatmadar A, et al. Fetal foot length as a predictor of gestational age. Am J Obstet Gynecol 1987; 156(2):350–355.
91. Meirowitz NB, Ananth CV, Smulian JC, et al. Foot length in fetuses with abnormal growth. J Ultrasound Med 2000; 19(3):201–205.
92. Brons JTJ, van der Harten JJ, VanGeijn HP, et al. Ratios between growth parameters for the prenatal ultrasonographic diagnosis of skeletal dysplasias. Eur J Obstet Gynecol Reprod Biol 1990; 34(1–2):37–46.
93. Chavez MR, Ananth CV, Smulian JC, et al. Fetal transcerebellar diameter measurement with particular emphasis in the third trimester: a reliable predictor of gestational age. Am J Obstet Gynecol 2004; 191(3):979–984.
94. Yarkoni S, Schmidt W, Jeanty P, et al. Clavicular measurement: a new biometric parameter for fetal evaluation. J Ultrasound Med 1985; 4(9):467–470.
95. Sherer DM, Plessinger MA, Allen TA. Fetal scapular length in the ultrasonographic assessment of gestational age. J Ultrasound Med 1994; 13(7):523–528.
96. Altman DG, Chitty LS. Charts of fetal size: I. Methodology. Br J Obstet Gynaecol 1994; 101(1):29–34.
97. Mayden KL, Tortora M, Berkowitz RL, et al. Orbital diameters: a new parameter for prenatal diagnosis and dating. Am J Obstet Gynecol 1982; 144(3):289–297.
98. Deter RL, Warda A, Rossavik IV, et al. Fetal thigh circumference: a critical evaluation of its relationship to menstrual age. J Clin Ultrasound 1986; 14(2):105–110.
99. Cohen HC, Cooper J, Eisenberg P, et al. Normal length of fetal kidneys: sonographic study in 397 obstetric patients. AJR Am J Roentgenol 1991; 157(3):545–548.
100. Vintzileos AM, Neckles S, Campbell WA, et al. Fetal liver ultrasound measurements during normal pregnancy. Obstet Gynecol 1985; 66(4):477–480.
101. Grannum PA, Berkowitz RL, Hobbins JC. The ultrasonic changes in the maturing placenta and their relation to fetal pulmonic maturity. Am J Obstet Gynecol 1979; 133(8): 915–922.
102. Destro F, Calcagnile F, Ceccarello P. Placental grade and pulmonary maturity in premature fetuses. J Clin Ultrasound 1985; 13(9):637–639.
103. Goldstein I, Reece EA, O'Connor TZ, et al. Estimating gestational age in the term pregnancy with a model based on multiple indices of fetal maturity. Am J Obstet Gynecol 1989; 161(5): 1235–1238.
104. Goldstein I, Lockwood C, Hobbins JC. Ultrasound assessment of fetal intestinal development in the evaluation of gestational age. Obstet Gyencol 1987; 70(5):682–686.
105. Goldstein I, Lockwood C, Belanger K, et al. Ultrasonographic assessment of gestational age with the distal femoral and proximal tibial ossification centers in the third trimester. Am J Obstet Gynecol 1988; 158(1):127–130.

Placenta and Cervix

Beverly A. Spirt and Lawrence P. Gordon

41

INTRODUCTION

A thorough evaluation of the placenta should be an integral part of every obstetric ultrasound examination (Table 1). Knowledge of the normal anatomy of the placenta as well as of placental pathology is necessary to correctly interpret the sonographic images.

DEVELOPMENT

The early gestational sac is first visible at transvaginal sonography at about four weeks' menstrual age. Its hyperechoic rim contains developing villi composed of fetal vessels surrounded by the lacunar space, which is the precursor of the intervillous space. At about five weeks' menstrual age, those villi situated opposite the implantation site begin to atrophy, forming a smooth surface (*chorion laeve*). The remaining villi, the *chorion frondosum*, become the placenta, which may be identified at sonography from about eight weeks (Fig. 1). At this stage, the amniotic sac is smaller than the chorionic cavity, and the amniotic membrane is visible at sonography. The amnion and chorion fuse at approximately 12 weeks.

Failure of villous regression results in abnormalities of placental shape. Placenta membranacea is a rare anomaly in which almost all the chorion is diffusely covered by villi. A variant of this condition occurs when aberrant villous atrophy results in a ring-shaped (annular) placenta (1,2). Both of these entities are associated with recurrent antepartum bleeding. A more common result of failure of villous regression is the succenturiate (accessory) lobe, which is present in 3% to 7% of patients (Figs. 2 and 3) (1). Recognition of succenturiate lobes is important because they may result in complications such as placenta previa, vasa praevia, and retained placenta after delivery.

SIZE

The size of the placenta is proportional to the size of the fetus. A small placenta is usually associated with a small-for-dates baby; this is not indicative of a cause-and-effect relationship (1,3).

Placentas that are too large or too small may be seen in various fetal and maternal conditions (Fig. 4) (Tables 2 and 3). In practice, the size of the placenta is usually assessed visually. The following pitfalls may occur when assessing the thickness of the placenta: (*i*) contractions may simulate placental thickening (see later); (*ii*) severe polyhydramnios may cause

TABLE 1 ■ Sonographic Evaluation of the Placenta

■ Location	■ Texture
■ Shape	■ Retroplacental structures
■ Size	

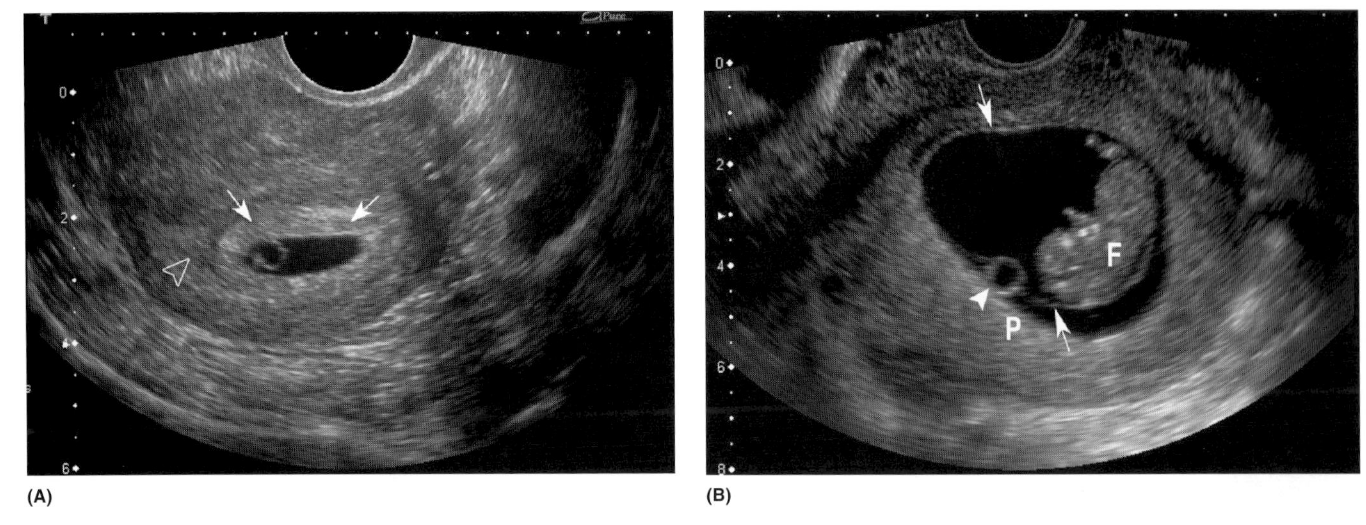

(A) (B)

FIGURE 1 A AND B ■ (**A**) Transvaginal scan at six weeks' menstrual age shows gestational sac (*arrows*) surrounded by echogenic villi. The endometrial cavity is filled with decidual reaction (*arrowhead*). (**B**) Transvaginal scan at 10 weeks shows an early placenta (P) with chorion laeve opposite. *Arrowhead* indicate amnion; *arrowhead*, yolk sac; F, fetus.

the placenta to appear artifactually small; and (*iii*) oligohydramnios may make the placenta seem large.

According to Fox, the placenta attains its definite form and full thickness by the end of the fourth month and continues to grow circumferentially to term (1). However, several ultrasound studies have shown that placental thickness increases with age until term. A measurement of placental thickness obtained by placing the transducer perpendicular to the plane of the placenta, at the midpoint of the placenta, has been used to predict perinatal outcome (4–9). Hoddick et al. considered 4 cm the upper limit of normal for any gestational age (4). Other authors have used cutoff values of 3 cm at 18 to 21 weeks (7,8), 3.5 cm at 20 to 22 weeks (9), and 5.1 cm at 32 to 34 weeks (9) to predict perinatal outcome. In the case of Hb Bart disease, an abnormally thick placenta may be the first sonographic sign of this condition (7).

TEXTURE

Normal Features

Echoes from the branching villi, which are bathed in maternal blood in the intervillous space (Fig. 5), produce

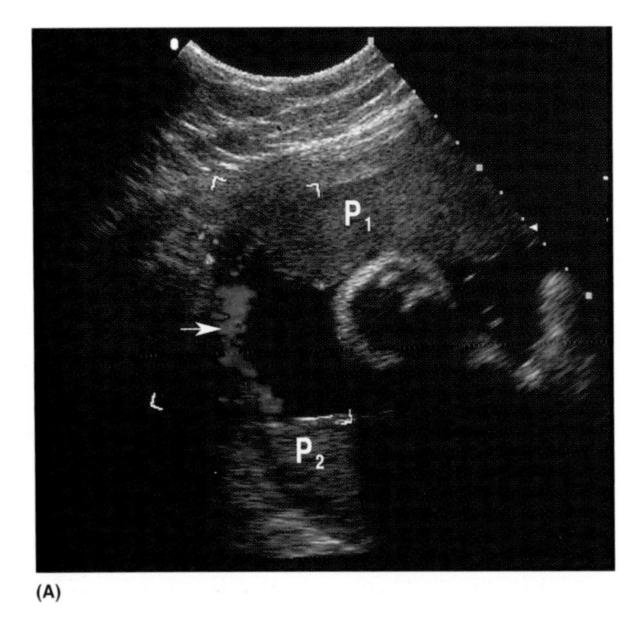

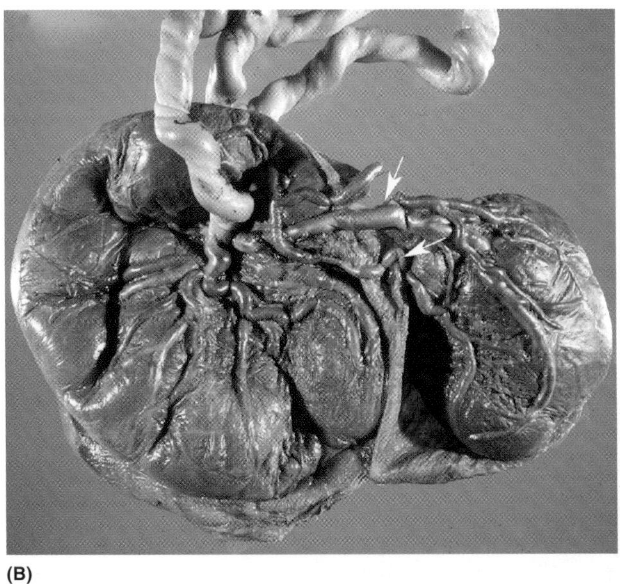

(A) (B)

FIGURE 2 A AND B ■ Succenturiate lobe. (**A**) At 33 weeks, separate placental tissue is seen anteriorly (P_1) and posteriorly (P_2) with a vessel (*arrow*) between the two lobes. (**B**) Maternal surface of the placenta after delivery. Smaller succenturiate lobe is connected to the main placenta by a bridge of vessels (*arrows*) and membranes.

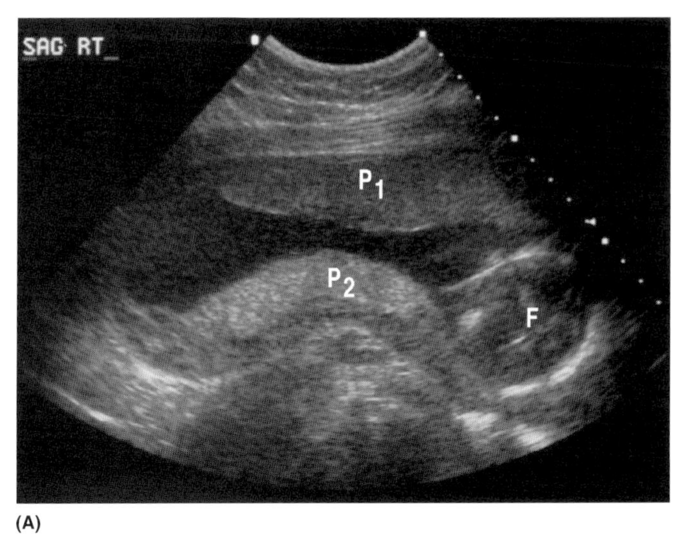

(A)

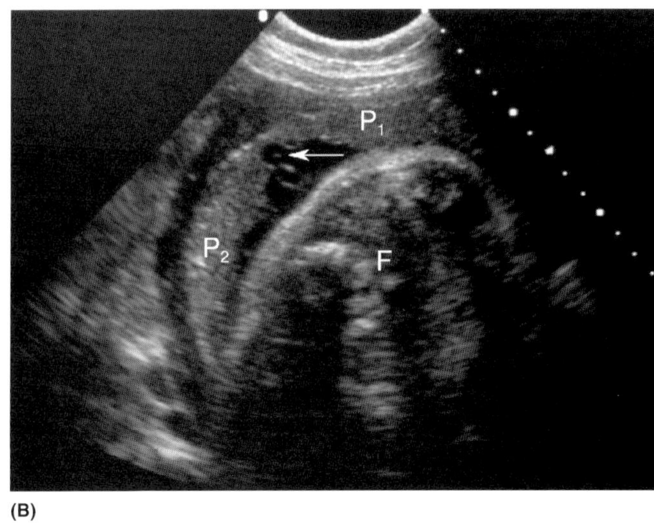

(B)

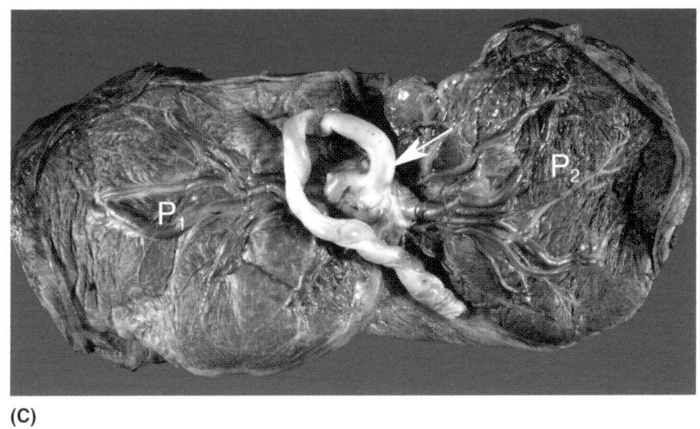

(C)

FIGURE 3 A–C ■ Bilobate placenta. (**A**) Sagittal scan at 17 weeks shows anterior (P_1) and posterior (P_2) placental lobes that are similar in size. (**B**) Transverse scan at 35 weeks. Note interplacental cord insertion (*arrow*). (**C**) Maternal surface of the placenta after delivery. The umbilical cord is typically inserted between the two lobes (*arrow*). F, fetus.

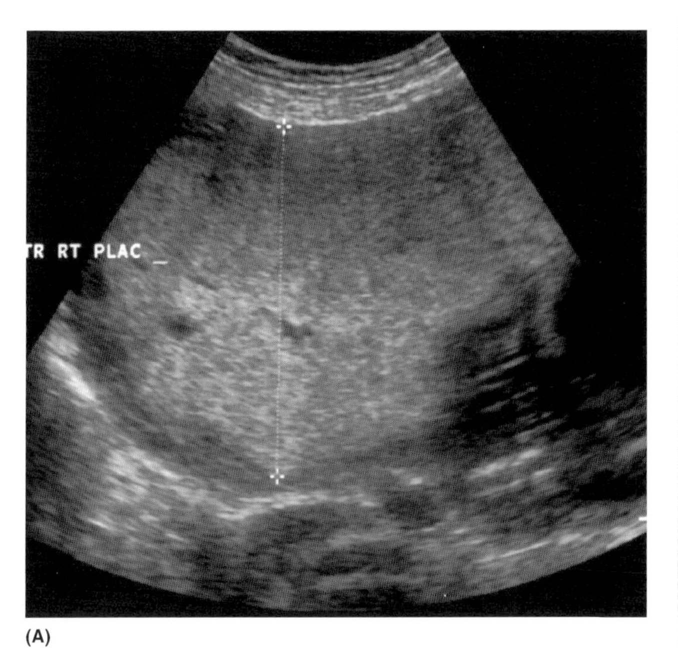

(A)

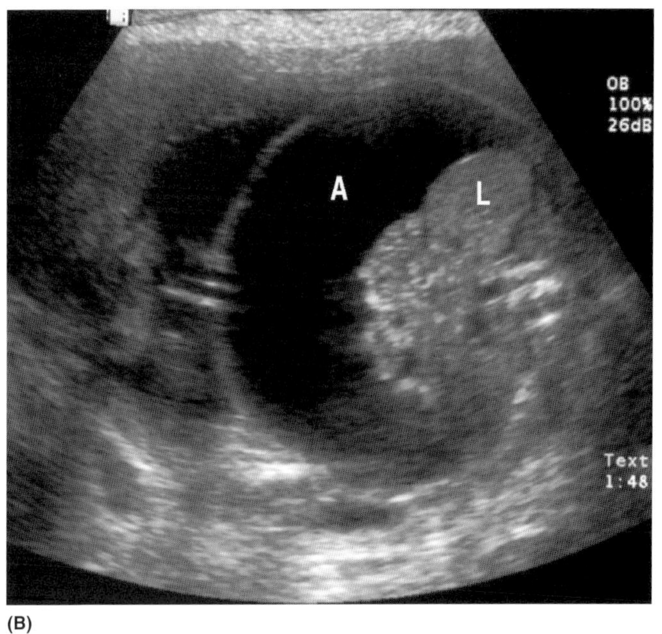

(B)

FIGURE 4 A AND B ■ Enlarged placenta; fetal hydrops due to cytomegalovirus. (**A**) Thick placenta is seen at 21 weeks. (**B**) Fetal ascites is present. A, fetal ascites; L, fetal liver.

TABLE 2 ■ Conditions Associated with Small Placentas

- Toxemia
- Small-for-dates fetus
- Maternal Hypertension
- Chromosomal abnormality
- Severe maternal diabetes mellitus
- Chronic infection

TABLE 3 ■ Conditions Associated with Large Placentas

- Blood group incompatibilities
- Diabetes mellitus
- Maternal anemia
- Fetal hydrops
- Fetal neoplasm
- Triploidy
- Homozygous α-thalassemia (Hb Bart's Disease)
- High altitude
- Maternal congestive heart failure
- Beckwith-Wiedemann syndrome

the diffuse granular echotexture of the placenta. This texture remains constant throughout gestation. The most notable exception is calcium deposition. Placental calcium deposition is a normally occurring, nonpathologic process that is believed to result from supersaturation of the placental environment with calcium and phosphate (10). It occurs microscopically during the first two trimesters and becomes macroscopic at about 29 weeks. Calcification may be found along the basal plate, in the intraplacental septa, and in collections of fibrin in the intervillous and subchorionic spaces (Fig. 6). Chemical, radiographic, and sonographic studies have shown an exponential increase in placental calcification with increasing gestational age (11–14); more than 50% of placentas contain some degree of calcification after 33 weeks (13,14). Placental calcification is more common in women of lower parity (13,14), and it is not increased in postmature placentas (1,11–13). Placental calcification is of no known clinical significance (1,2); it is not useful to grade placentas according to calcium content.

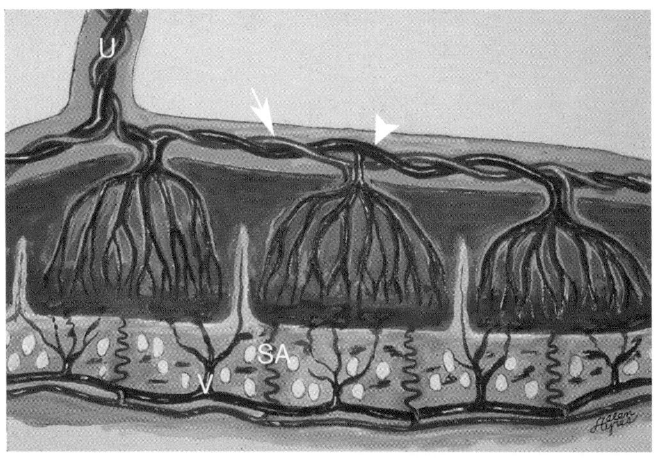

(A)

(B)

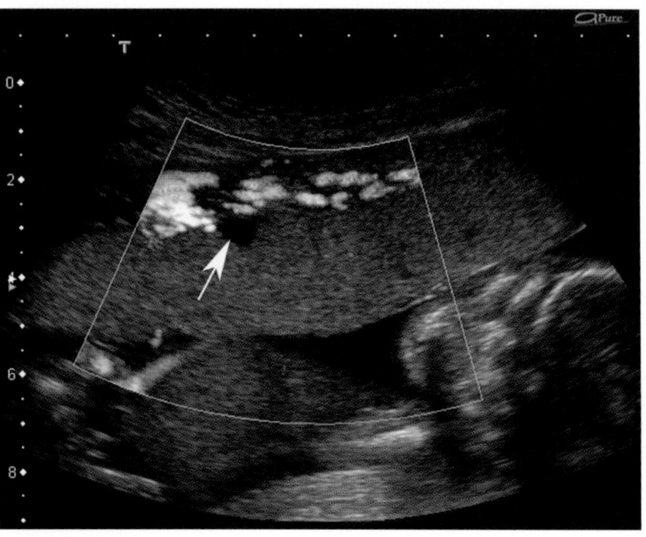

(C)

FIGURE 5 A–C ■ Anatomy of the placenta. (**A**) Drawing of the placental circulation. U, umbilical cord; *arrow*, umbilical artery branching in chorion; *arrowhead*, umbilical vein in chorion; SA, spiral arteriole; V, maternal draining veins. *Source*: Spirt BA, Kagan EH. Sonography of the placenta. Semin Ultrasound 1980; 1:293–310. (**B**) Anterior placenta at 19 weeks. The placenta (P) maintains a typical granular echotexture. Retroplacental draining veins (*arrowheads*) are visible. Vessels are seen along the fetal surface, particularly adjacent to the umbilical cord insertion (*arrow*). F, fetus. (**C**) Color flow Doppler shows umbilical cord vessels and retroplacental vessels. Note anechoic space (*arrow*) that does not show color flow (placental lake, see later).

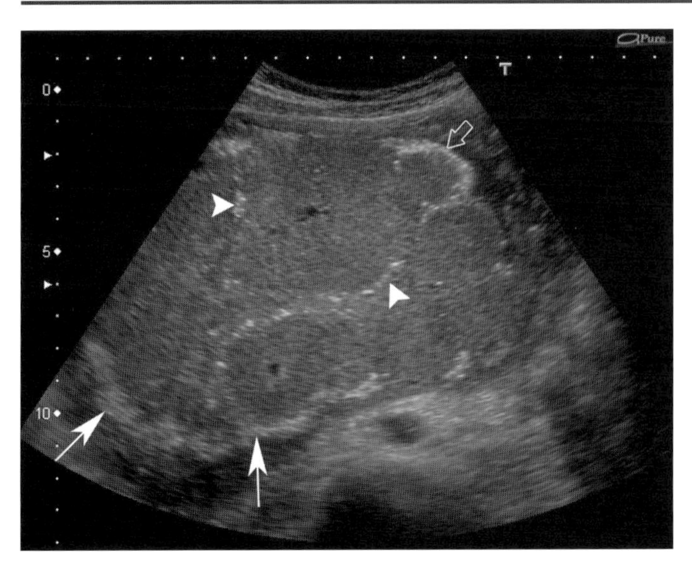

FIGURE 6 ■ Placental calcification. At 38 weeks, calcification is seen along the basal plate (*arrows*), chorionic plate (*open arrow*), and septa (*arrowheads*).

Common Macroscopic Lesions

The interactions between the maternal and fetal circulations that occur in the placenta are reflected in its sonographic appearance. Subchorionic fibrin deposition, intervillous thrombosis, and perivillous fibrin deposition are common macroscopic lesions resulting from these interactions (Table 4). All these lesions contain varying amounts of fibrin, which is hypoechoic, and blood (Fig. 7). These lesions are usually of no clinical significance. The placenta has a large reserve capacity; it can lose up to 30% of its villi with no obvious effect on the fetus (1).

Subchorionic Fibrin Deposition

Except for the subchorionic space, most areas of the placenta are densely packed with villi. Blood tends to pool and eddy in the subchorionic space. At delivery, plaques of laminated subchorionic fibrin are found in about 20% of placentas from uneventful pregnancies (1). These plaques correlate with subchorionic anechoic–hypoechoic lesions at sonography (Figs. 8 and 9) (15). In our experience, they may be seen as early as 7.5 weeks (Fig. 10). They usually decrease in size during the course of the pregnancy, as fibrin is laid down. Slow flow or turbulence is often visible at real-time sonography; in many such lesions, color Doppler imaging fails to demonstrate flow.

Perivillous Fibrin Deposition

Perivillous fibrin deposition is caused by pooling and stasis of blood in the intervillous space. It appears as an anechoic–hypoechoic intraplacental lesion at sonography (Fig. 11). A hyperechoic rim is often seen around these lesions at sonography, caused by compression of the villi bordering the lesion (16). This lesion is found in 22% of placentas from uncomplicated term pregnancies and usually has no clinical significance (1,2). Rarely, massive perivillous fibrin deposition may occur, involving up to 70% to 80% of the placenta; this results in fetal demise.

Intervillous Thrombi

Up to 48% of term placentas from uncomplicated pregnancies contain intervillous thrombi (1); these lesions result from fetal hemorrhage into the intervillous space (17). Intervillous thrombi appear sonographically as intraplacental anechoic–hypoechoic lesions (Fig. 12) (18). These are usually few in number and small (1–2 cm), and are not considered clinically significant. Because they are a site of fetal bleeding into the maternal circulation, however, such lesions possibly could lead to isoimmunization.

Placental "Lakes"

Anechoic–hypoechoic intraplacental "lakes" are not uncommon and may contain flow (Fig. 13). At delivery, these correlate with blood-filled spaces that presumably represent a stage in the evolution of either perivillous fibrin deposition or intervillous thrombi.

TABLE 4 ■ Common Anechoic and Hypoechoic Lesions in the Placenta

Lesion	Etiology	Microscopic description	Clinical significance
Subchorionic fibrin deposition	Pooling of blood in the subchorionic space	Laminated subchorionic fibrin; fresh blood may be present	None
Intervillous thrombosis	Fetal hemorrhage into intervillous space	Laminated fibrin, fetal and maternal erythrocytes	None (theoretically could lead to isoimmunization)
Perivillous fibrin deposition	Pooling of blood in the intervillous space	Fibrotic villi surrounded by nonlaminated fibrin; fresh blood may be present	None
"Maternal lake"	Probable precursor of perivillous fibrin deposition, or of intervillous thrombosis	Empty space	None
Septal cyst	Secretion of fluid by entrapped nests of trophoblasts (2)	Small (5–10 mm) cyst in septum	None

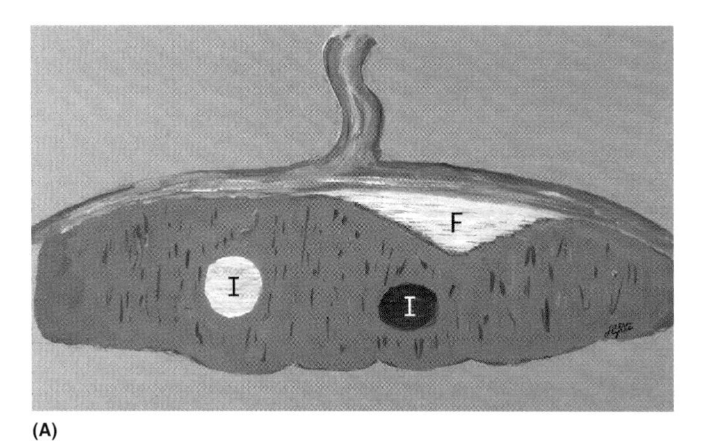

(A)

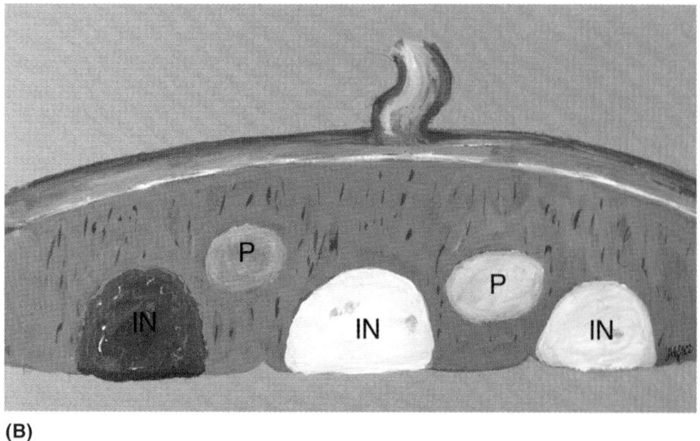

(B)

FIGURE 7 A AND B ■ Common macroscopic lesions in the placenta. (**A**) Subchorionic fibrin deposits (F) contain laminated, yellow–white fibrin. Intervillous thrombi (I) also contain laminated fibrin. The lesions appear round to oval; they range from red to laminated white, depending on the age. (**B**) Perivillous fibrin (P) deposits contain unlaminated fibrin. Infarcts (IN) are dark red to white nonlaminated lesions adjacent to the basal plate. *Source*: Spirt BA, Gordon LP. Sonography of the placenta. In: Fleischer AC, Manning FA, Jeanty P et al., eds. Sonography in Obstetrics and Gynecology: Principles and Practice, sixth edition. New York: McGraw-Hill, 2001: 195–224.

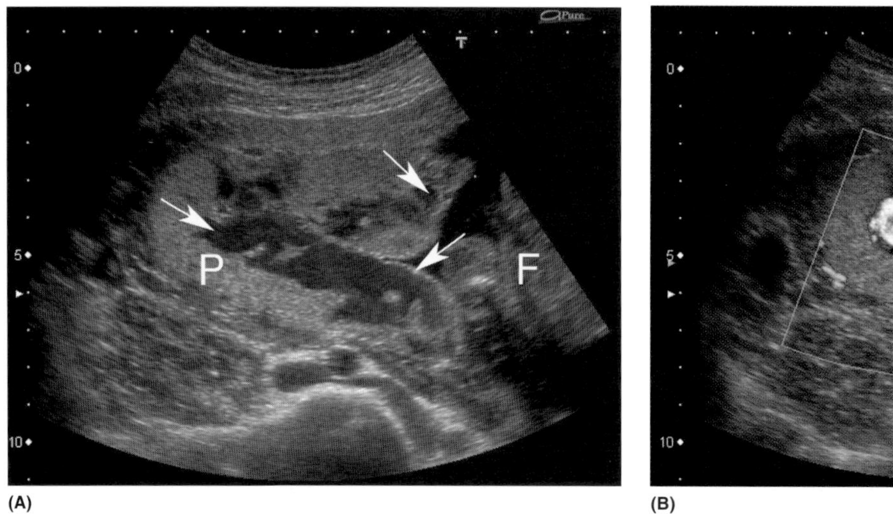

(A)

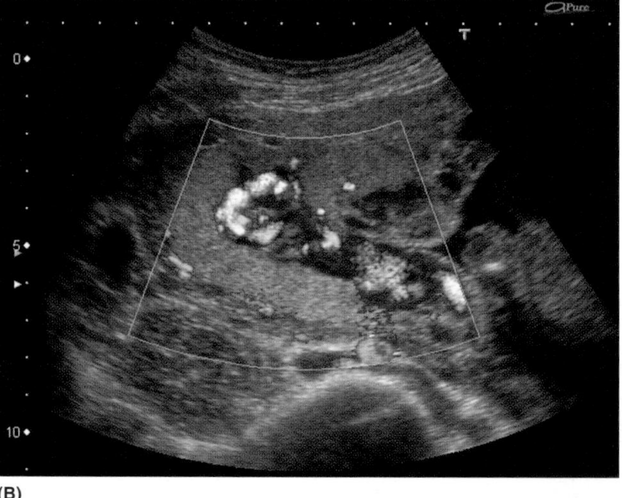

(B)

FIGURE 8 A AND B ■ Subchorionic fibrin deposition. (**A**) Prominent subchorionic hypoechoic areas are present at 21 weeks (*arrows*); the echoes within represent slow flow, seen at real-time sonography. (**B**) Flow is demonstrated with color flow Doppler. P, placenta; F, fetus.

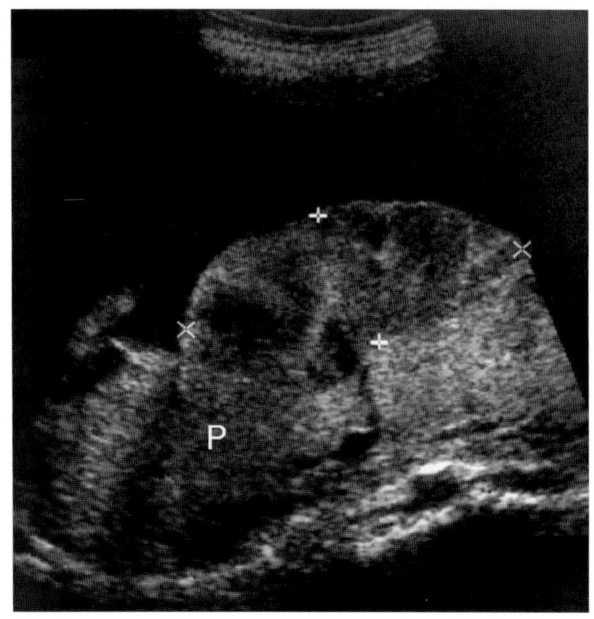

(A)

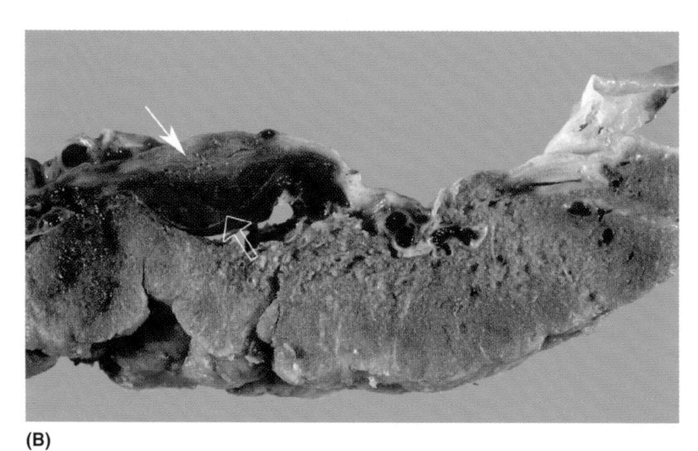

(B)

FIGURE 9 A AND B ■ Subchorionic fibrin deposition. (**A**) Sagittal scan of posterior placenta (P) at 26 weeks shows prominent hypoechoic subchorionic lesion (*cursors*) containing areas of slow flow. (**B**) Corresponding slice of term placenta shows that the lesion is composed of laminated fibrin (*arrow*) and blood (*open arrow*). *Source*: Spirt BA, Gordon LP. Imaging of the placenta. In: Taveras JM, Ferrucci JT, eds. Radiology: Diagnosis, Imaging, Intervention. Philadelphia: J. B. Lippincott, 1992:1–18.

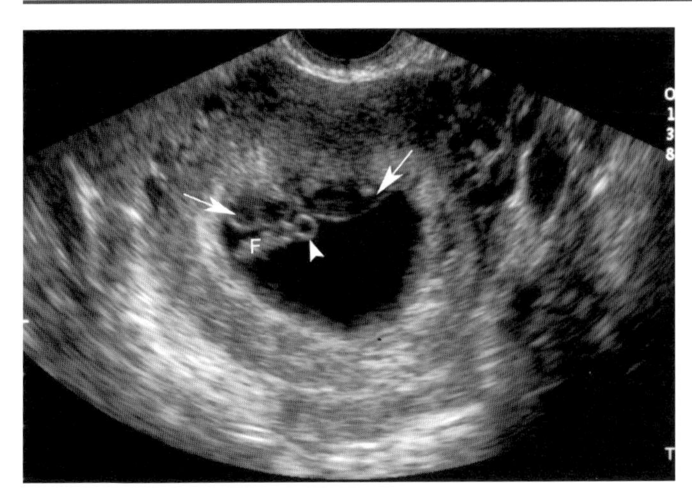

FIGURE 10 ■ Subchorionic fibrin deposition. Transvaginal scan at 7.5 weeks shows prominent hypoechoic subchorionic blood-pooling (*arrows*). *Arrowhead*, yolk sac. F, fetus.

Septal cysts are another anechoic intraplacental lesion that may be seen at antenatal sonography (19). These cysts measure 5 to 10 mm in diameter and are found in up to 19% of term placentas from uncomplicated pregnancies (1). They occur at the apex of placental septa, which are decidua-containing folds of the basal plate formed during the third month of gestation. The septa divide the maternal surface into 15 to 20 lobules; neither the septa, the lobules, nor the cysts have any known physiologic significance.

Septal cysts, perivillous fibrin deposition, and intervillous thrombi are not sonographically distinguishable from each other.

Infarcts

Infarcts result from disruption of the maternal blood supply to the placenta, due to either an underlying maternal vascular disorder or a retroplacental hematoma. Small, macroscopically visible infarcts are found in 25% of term placentas from uncomplicated pregnancies and have no clinical significance (1). However, extensive infarction involving more than 10% of the villi has been associated with intrauterine growth restriction, fetal hypoxia, and fetal demise (1). In these situations, the underlying maternal vascular disorder is the root of the problem. Infarcts cannot be identified at sonography unless they are complicated by hemorrhage (19), probably because infarcts contain necrotic villi, whereas the anechoic–hypoechoic lesions described previously contain blood, fibrin, or fluid.

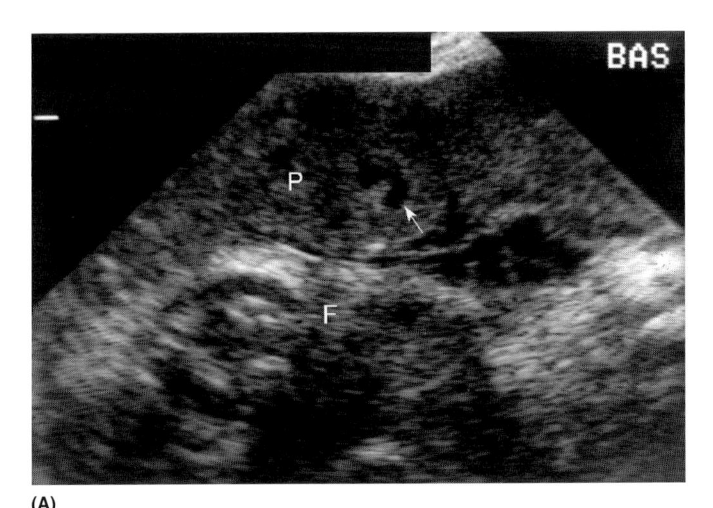

(A)

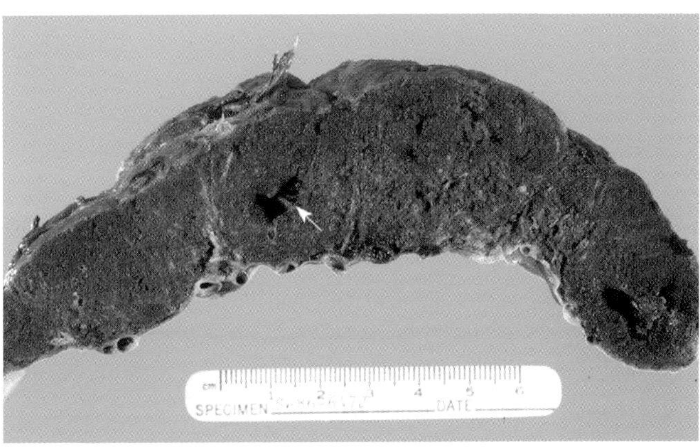

(B)

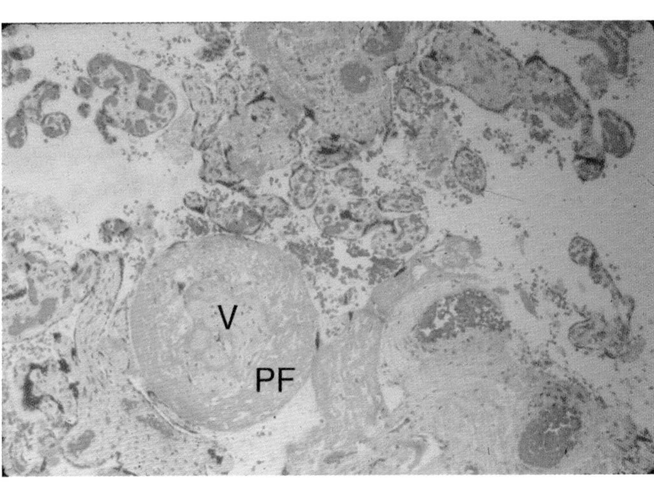

(C)

FIGURE 11 A–C ■ Perivillous fibrin deposition. (**A**) Sector scan at 35 weeks shows an anechoic intraplacental lesion with an associated echogenic area (*arrow*). P, placenta; F, fetus. (**B**) Cross-section of the placenta shows the corresponding lesion (*arrow*). The empty space contained blood that drained on sectioning. (**C**) At microscopy, the solid component contained perivillous fibrin (PF) surrounding villi (V). P, placenta; F, fetus. *Source*: Spirt BA, Gordon LP. Sonographic evaluation of the placenta. In: Rumack C, Charbonneau W, Wilson S, eds. Diagnostic Ultrasound, 2nd ed. St. Louis: Mosby, 1998:1337–1358.

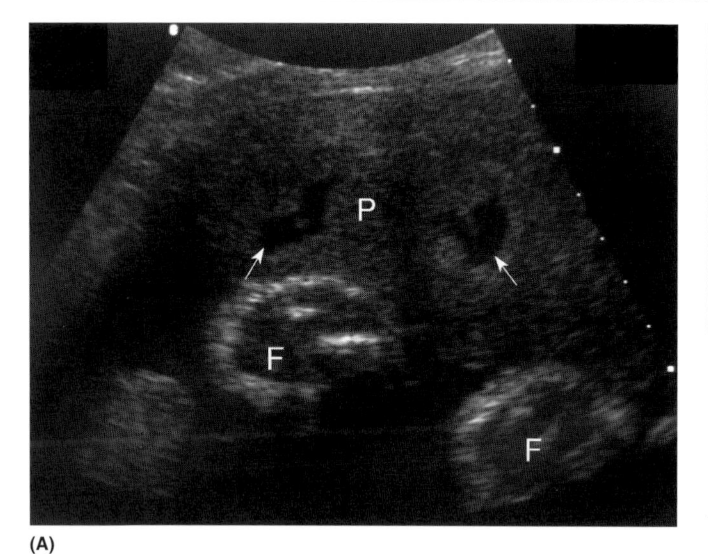

(A)

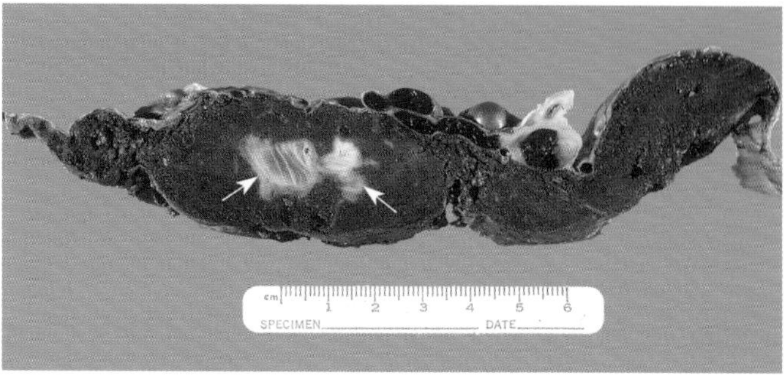

(B)

FIGURE 12 A AND B ■ Intervillous thrombi. (**A**) At 33 weeks, two irregular shaped hypoechoic lesions (*arrows*) are seen in this anterior placenta (P). Several similar lesions were present in this placenta. F, fetus. (**B**) Laminated fibrin is present in two adjacent lesions (*arrows*) at term; this is characteristic of intervillous thrombosis.

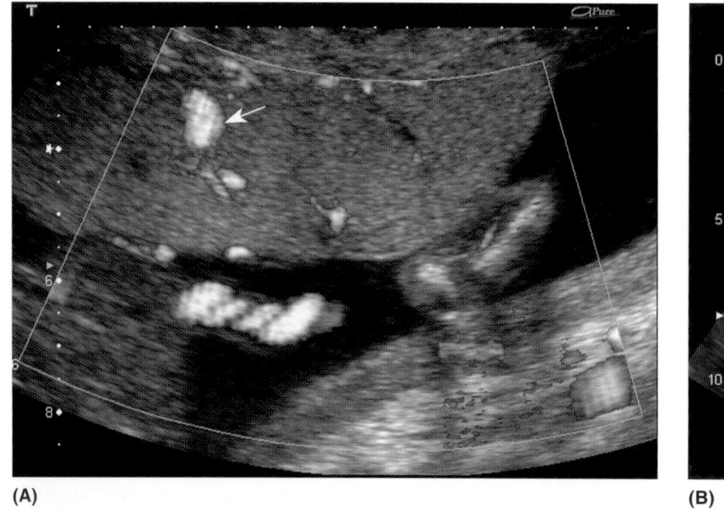

(A)

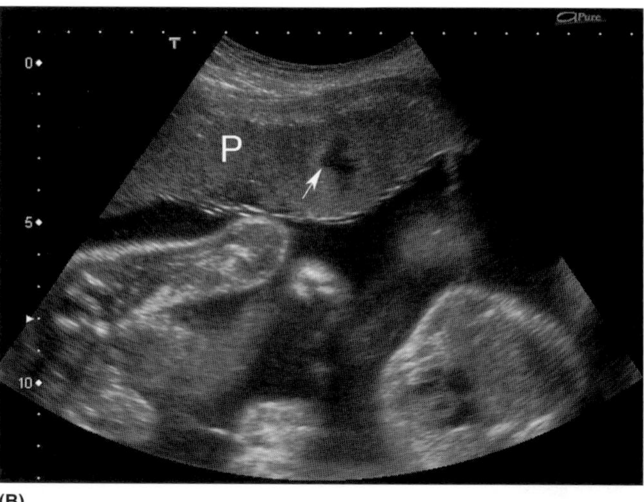

(B)

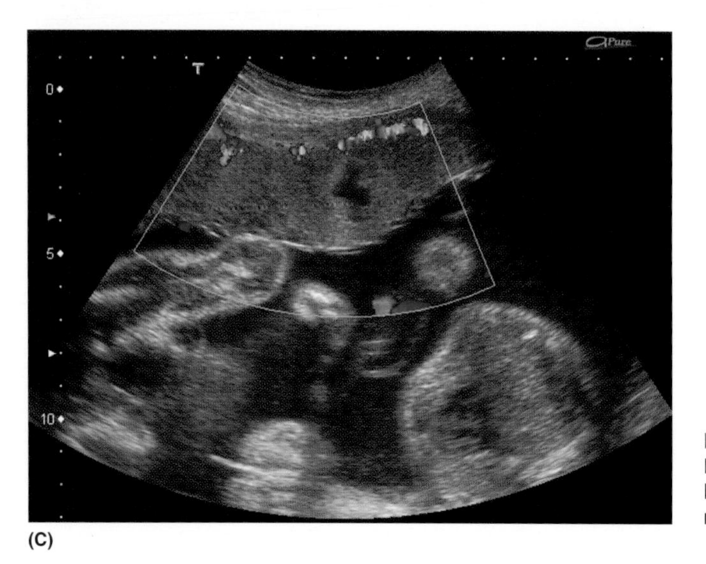

(C)

FIGURE 13 A–C ■ Intraplacental "lake." (**A**) At 19 weeks, color flow Doppler demonstrates blood flow within a "lake" (*arrow*). (**B** and **C**) Hypoechoic lesion (*arrow*) in anterior placenta (P) at 25 weeks shows minimal flow with color Doppler.

TABLE 5 ■ Abnormal Intraplacental Lesions

Lesion	Incidence	Sonographic appearance	Clinical significance
Hydatidiform mole	0.07% in the United States and Europe, 0.2% in Japan (6)	Multiple anechoic lesions of varying sizes	Premalignant (choriocarcinoma)
Partial mole	?	Enlarged placenta containing multiple anechoic lesions of varying sizes	Persistent trophoblastic disease may develop; usually associated with triploidy
Chorioangioma	1% (small lesions) (20)	Solid well-circumscribed intraplacental mass that may have visible vessels	Small: none; large: fetal hydrops
Teratoma	Rare	Complex solid mass located between amnion and chorion, either on fetal surface of the placenta or adjacent to it	None
Metastases from maternal neoplasms	Very rare	Multiple solid intraplacental nodules of varying sizes are seen macroscopically	Transplacental spread to fetus (two documented cases of transplacental spread of melanoma) (20)

Abnormalities

In Table 5 abnormal intraplacental lesions that may be seen at sonography are listed. These include trophoblastic disease (complete hydatidiform mole and partial mole) and primary and secondary nontrophoblastic neoplasms.

Trophoblastic Disease

An enlarged uterus containing material with multiple anechoic vesicles of varying sizes, in the absence of a fetus, is seen with complete hydatidiform mole (Fig. 14). The vesicles represent dilated, hydropic villi that enlarge with advancing gestational age; no normal placental tissue is found. Moles are believed to result from the abnormal fertilization of an empty ovum by a single sperm with a duplicated haploid genome (46,XX karyotype) or, less commonly, dispermy (46,XY) (21,22). A coexistent fetus may occur along with a mole in the case of a twin pregnancy with one empty ovum (Fig. 15).

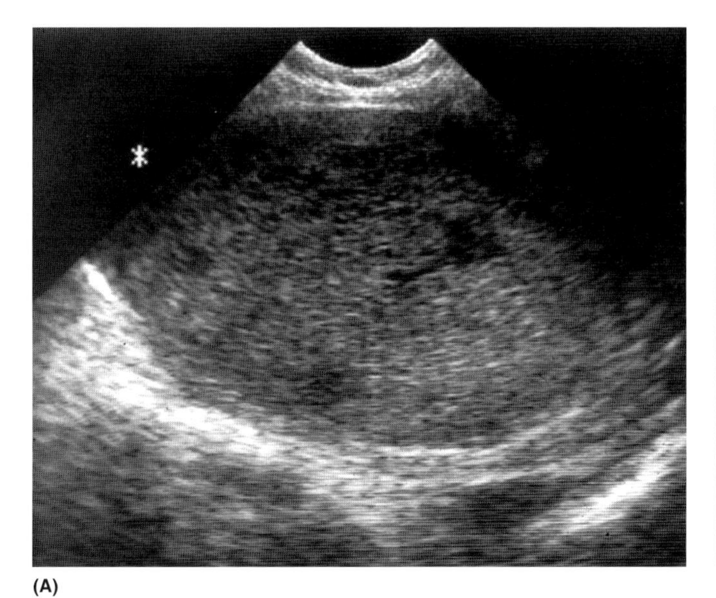

(A)

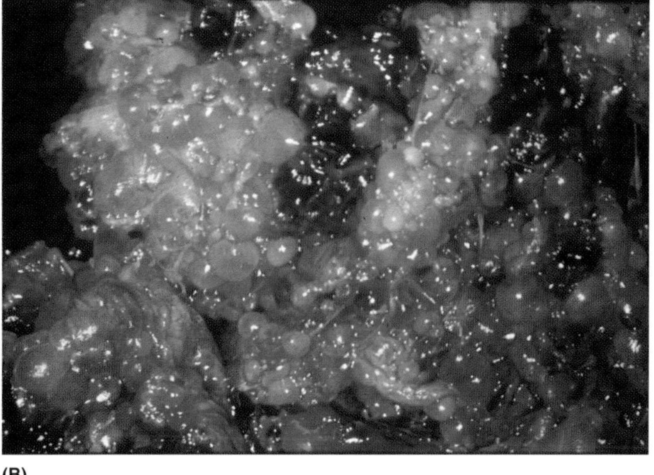

(B)

FIGURE 14 A AND B ■ (A) Hydatidiform mole. Sagittal scan shows an enlarged uterus filled with solid material containing multiple small anechoic lesions. This is the typical appearance of a hydatidiform mole. *Source*: From Spirt BA, Gordon LP. Practical aspects of placental evaluation. Semin Roentgenol 1991; 26:32–91. (B) Gross specimen of a hydatidiform mole shows multiple grape-like cysts. *Source*: From Ref. 15.

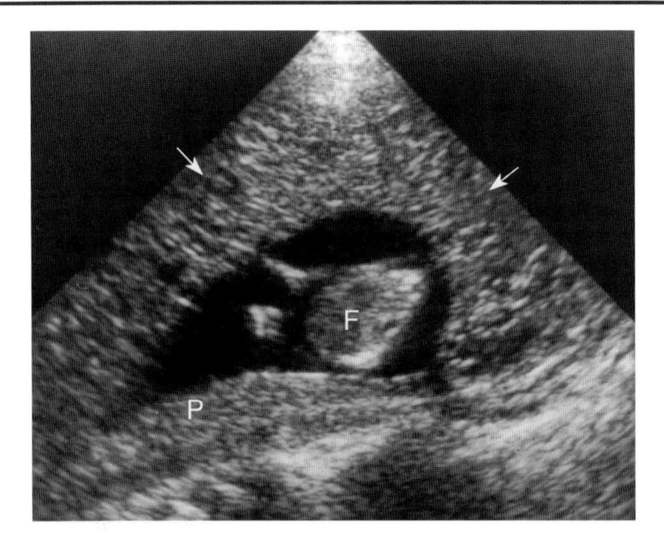

FIGURE 15 ■ Hydatidiform mole with coexistent fetus (F). Sector scan at 13 weeks shows a large anterior mass (*arrows*) containing multiple anechoic vesicles of varying sizes, along with a fetus and a separate posterior placenta (P). At delivery, a hydatidiform mole was found separate from the placenta and fetus. *Source*: Spirt BA, Gordon LP. Imaging of the placenta. In: Taveras JM, Ferrucci JT, eds. Radiology: Diagnosis, Imaging, Intervention. Philadelphia: J. B. Lippincott, 1992:1–18.

An enlarged placenta containing multiple anechoic lesions usually represents a partial mole (Fig. 16). In this entity, normal villi are interspersed with hydropic villi; the fetus is usually abnormal. Most partial moles are triploid (69 chromosomes) (23). Triploid pregnancies that do not spontaneously abort in the first trimester frequently cause symptoms of preeclampsia at about 18 weeks.

Rarely, an enlarged placenta with large edematous villi may be a manifestation of placental mesenchymal dysplasia, in which case the fetus is usually normal. However, there is a high rate of IUGR and of fetal demise with this entity (24). Beckwith–Weidemann syndrome occurs in approximately 20% of cases. Placental mesenchymal dysplasia is a recently described condition; its cause is not known.

Persistent trophoblastic disease (choriocarcinoma and invasive mole) occurs in about 10% of patients with hydatidiform mole. Triploid as well as nontriploid partial moles have also developed into persistent trophoblastic disease requiring chemotherapy (25–28). Thus, monitoring serum β-human chorionic gonadotropin levels in

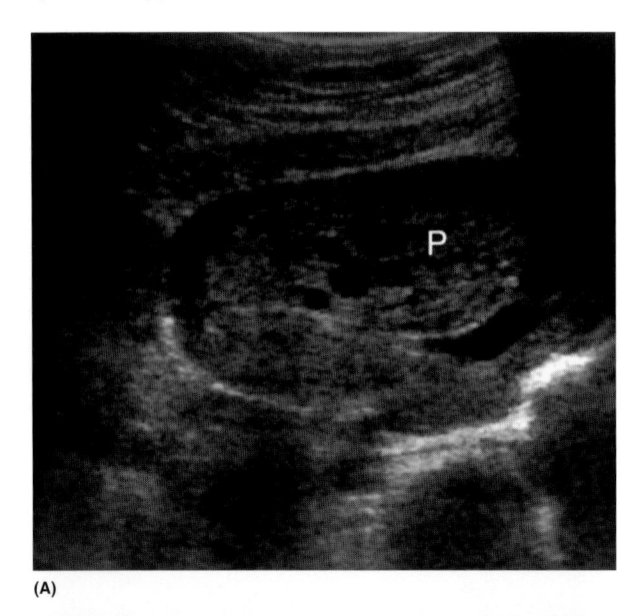

(A)

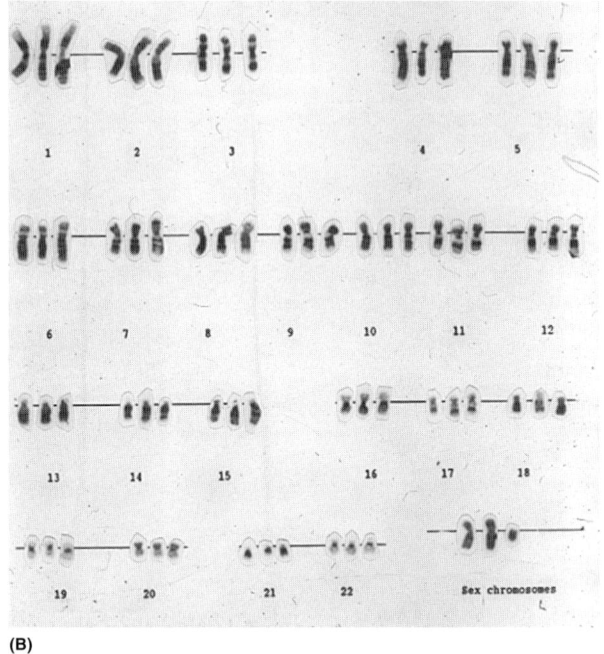

(B)

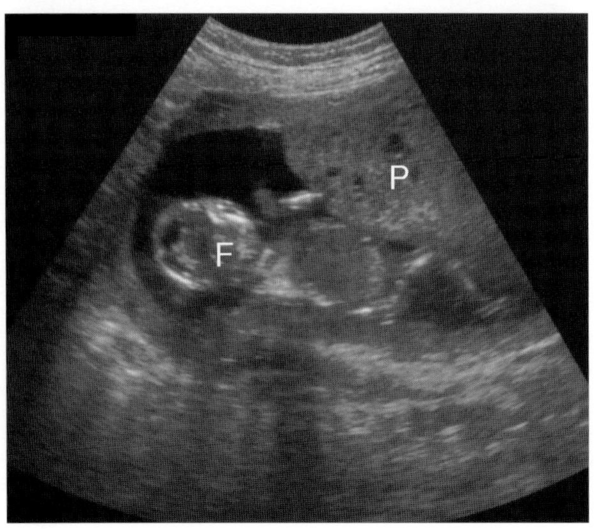

(C)

FIGURE 16 A–C ■ Partial mole (triploidy). (**A**) Sagittal scan at 15 weeks shows a large placenta (P) with multiple anechoic lesions. Fetal demise was documented. Gross specimen showed multiple vesicles interspersed with normal placental tissue. *Source*: Spirt BA, Gordon LP. The placenta and cervix. In: McGahan JP, Porto M, eds. Diagnostic Obstetrical Ultrasound. Philadelphia, J. B. Lippincott, 1994:83–103. (**B**) Karyotype demonstrates triploidy. *Source*: Spirt BA, Gordon LP. The placenta and cervix. In: McGahan JP, Porto M, eds. Diagnostic Obstetrical Ultrasound. Philadelphia, J. B. Lippincott, 1994:83–103. (**C**) Triploidy with multiple fetal anomalies. A large placenta with multiple anechoic lesions was seen at 14 weeks, along with holoprosencephaly, omphalocele, and cystic hygroma. F, fetus.

all patients with any form of hydatidiform mole is important.

Primary Nontrophoblastic Tumors

The only known primary nontrophoblastic tumors of the placenta are chorioangioma and teratoma. Small chorioangiomas are found in approximately 1% of systematically sectioned placentas and have no clinical consequence (1). Large chorioangiomas are uncommon; they often cause significant vascular shunting, leading to fetal cardiac failure and hydrops. Polyhydramnios has been associated with large chorioangiomas.

Chorioangiomas are usually subchorionic in location. They may be single or multiple and often have a distinct fibrous capsule or a pseudocapsule of compressed villi (1). At sonography, a well-marginated mass is visualized, usually subchorionic (Figs. 17 and 18). Flow may be demonstrated with Doppler imaging. The fetus should be monitored for the development of cardiac complications and hydrops.

Placental teratomas are rare. They lie between the amnion and chorion, either on the fetal surface of the placenta or, less commonly, in the membranes adjacent to the margin of the placenta. They have no clinical consequence (1).

RETROPLACENTAL AREA

Examination of the retroplacental area is an important part of obstetric sonography. The retroplacental myometrium and decidua appear hypoechoic compared with the placenta. Tubular structures representing the draining veins are found along the length of the maternal surface (Fig. 19). The spiral arterioles, which supply maternal blood to the intervillous space, may be seen with color flow Doppler imaging.

TABLE 6 ■ Causes of Retroplacental Masses

- Contractions
- Myomas
- Retroperitoneal hematomas
- Abruptio placentae

Contractions

Transient changes in the appearance of the retroplacental myometrium and decidua are seen with contractions, which occur throughout pregnancy and are imperceptible to the mother. These are most commonly seen in the latter part of the first trimester and the early part of the second trimester (Figs. 20 and 21). Contractions are a source of considerable confusion because they often mimic retroplacental myomas and hematomas (Table 6). Contractions are responsible for many false diagnoses of early placenta previa (see later). Rescanning the patient 20 minutes to an hour later should suffice to distinguish between a contraction and a true retroplacental abnormality. Alternatively, color flow Doppler may show increased vascularity in the retroplacental myometrium during a contraction (Fig. 21C).

Retroplacental Mass Lesions

Retroplacental Myomas

Retroplacental myomas (Fig. 22) tend to be well circumscribed and hypoechoic. Sometimes flow is demonstrated with Doppler examination. The diagnosis is easily confirmed if multiple myomas are present. Large myomas may have a complex echotexture as a result of degeneration and/or hemorrhage. They may increase or decrease in size during the course of the pregnancy.

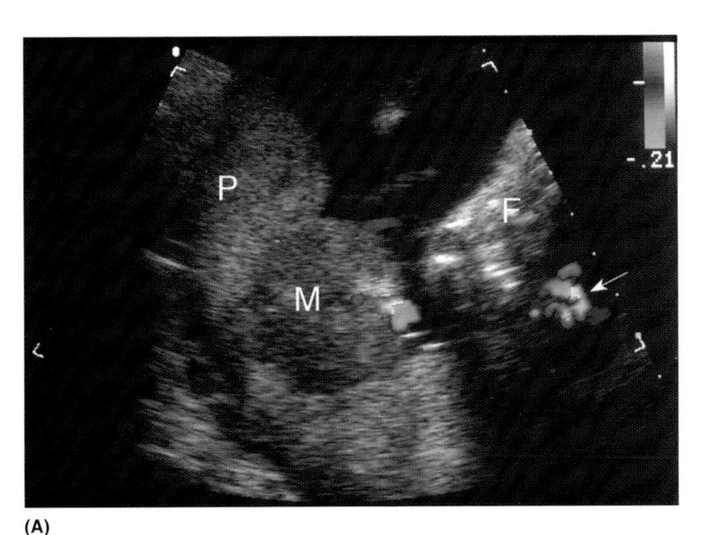

(A)

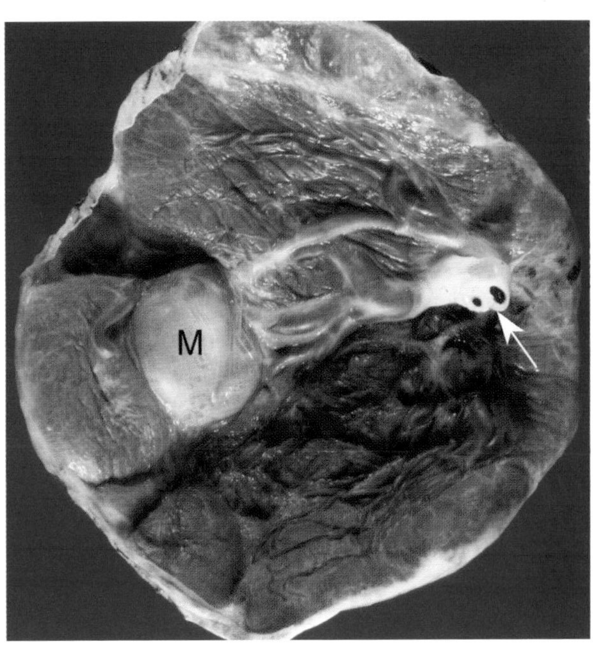

(B)

FIGURE 17 A AND B ■ Chorioangioma. (**A**) At 24 weeks, a 4 cm chorioangioma (M) is seen at a distance from the umbilical cord insertion (*arrow*). This was confirmed at delivery (**B**). P, placenta; F, fetus.

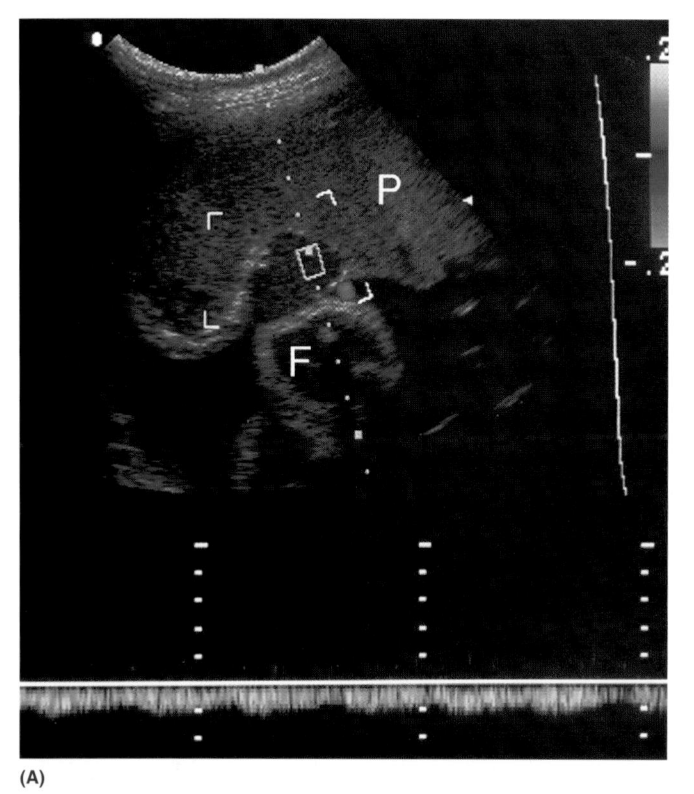

(A)

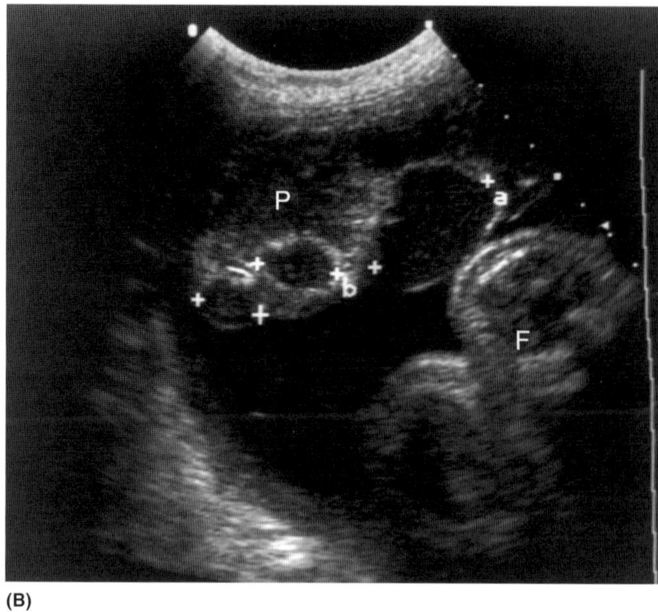

(B)

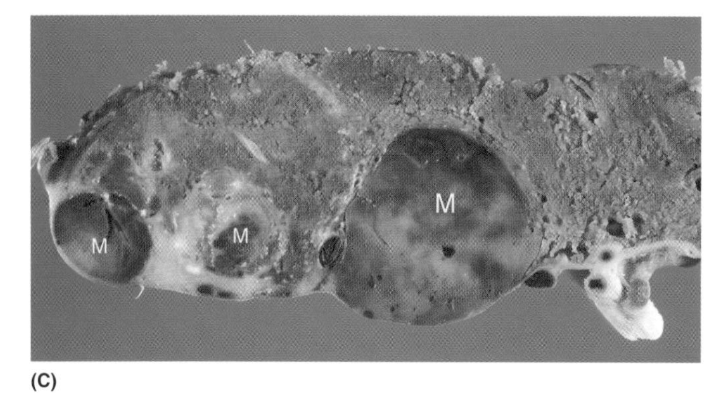

(C)

FIGURE 18 A–C ■ Multiple chorioangiomas. (**A**) Flow is demonstrated in 36 mm chorioangioma (*Doppler cursor*) at 36 weeks. P, placenta; F, fetus. (**B**) This placenta (P) contained three chorioangiomas (*cursors*) measuring 16, 20, and 36 mm, respectively, seen here at 40 weeks. F, fetus. (**C**) Gross specimen shows three chorioangiomas (M) beneath the fetal surface of the placenta. The fetus did not develop hydrops, and vaginal delivery at term was uneventful.

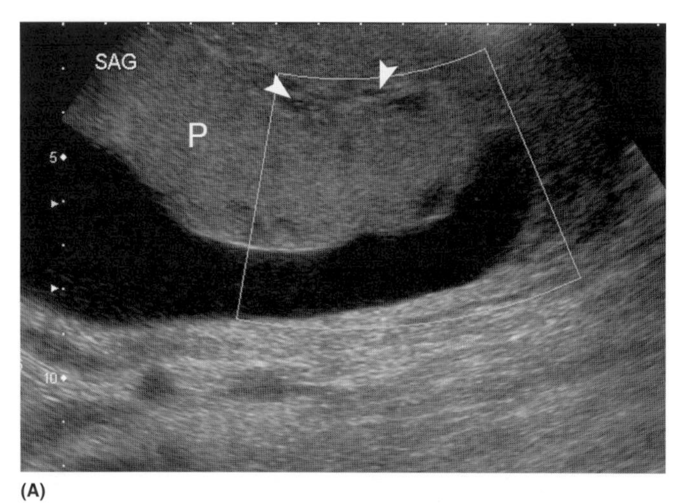

(A)

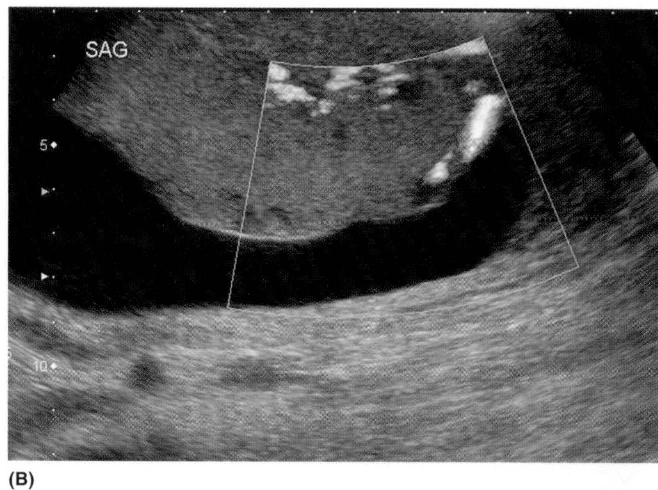

(B)

FIGURE 19 A AND B ■ Retroplacental anatomy. (**A**) Anterior placenta (P) at 23 weeks with retroplacental draining veins (*arrowheads*), demonstrated with color flow Doppler (**B**).

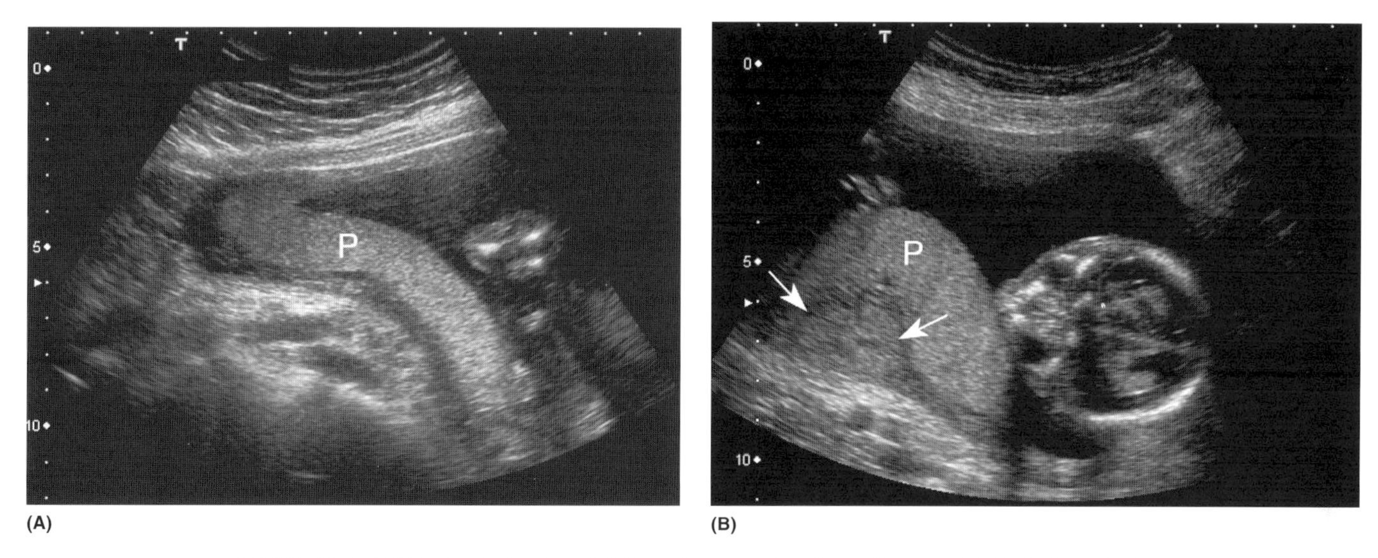

FIGURE 20 A AND B ■ Contraction. (**A**) Posterior placenta (P) at 18 weeks. (**B**) Forty-five minutes later, the placenta (P) is distorted and a hypoechoic retroplacental "pseudomass" is seen (*arrows*).

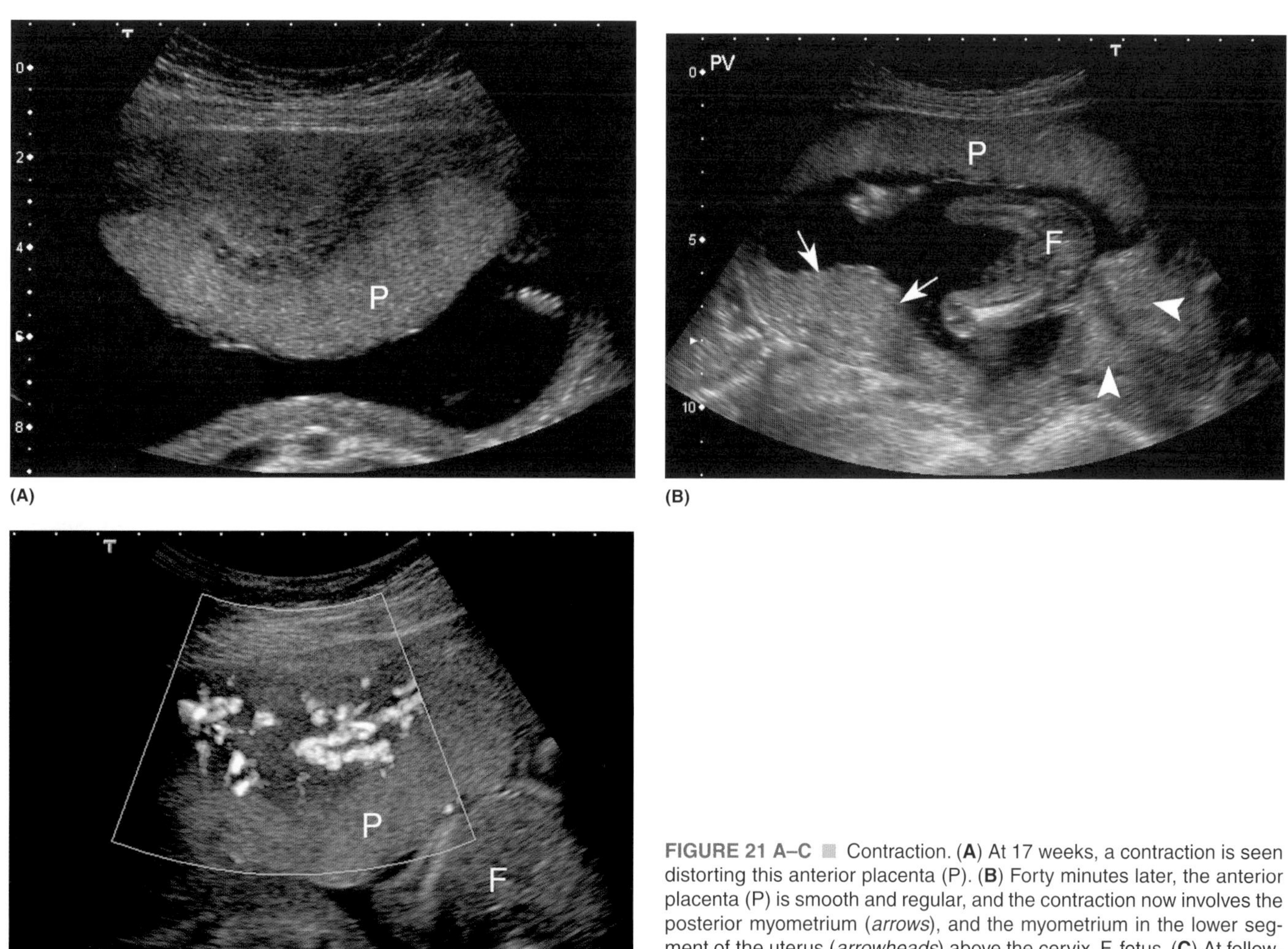

FIGURE 21 A–C ■ Contraction. (**A**) At 17 weeks, a contraction is seen distorting this anterior placenta (P). (**B**) Forty minutes later, the anterior placenta (P) is smooth and regular, and the contraction now involves the posterior myometrium (*arrows*), and the myometrium in the lower segment of the uterus (*arrowheads*) above the cervix. F, fetus. (**C**) At follow-up examination at 22 weeks, color flow Doppler shows increased vascularity in the retroplacental myometrium during a contraction. This finding, when present, may be used to distinguish a contraction from a retroplacental myoma or hematoma. P, placenta; F, fetus.

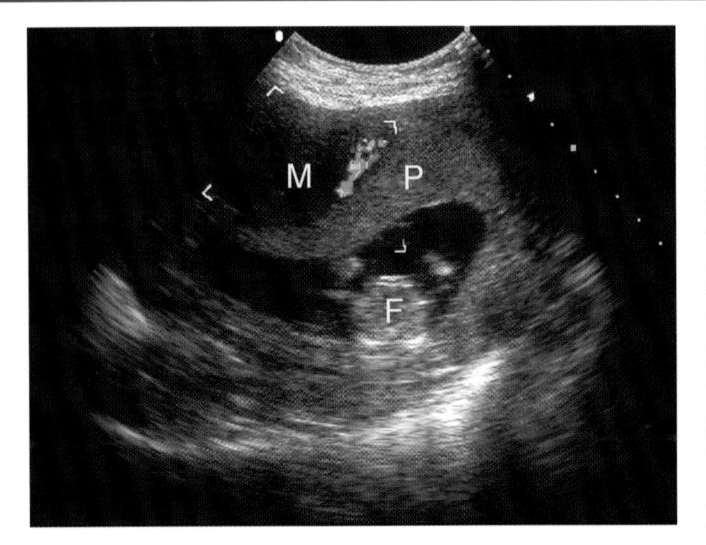

FIGURE 22 ■ Myoma. A retroplacental myoma (M) is seen at 13 weeks. Normal retroplacental vessels are seen between the myoma and the placenta (P). F, fetus.

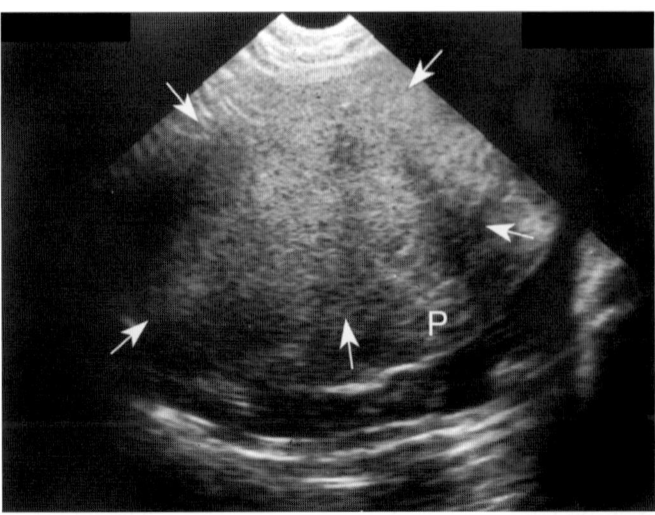

FIGURE 24 ■ Abruptio placentae. Transverse scan at 35 weeks shows a large, hyperechoic retroplacental hematoma (*arrows*). The patient was hypotensive, with acute vaginal bleeding. At cesarean section, a 75% abruption was found. P, placenta.

Retroplacental and Submembranous Hematomas

Distinguishing small retroplacental hematomas from retroplacental myomas is occasionally difficult in the absence of appropriate clinical history, particularly if the patient has no vaginal bleeding (29). Rather than forming a retroplacental hematoma, blood may dissect along the basal plate and collect beneath the chorionic membrane adjacent to or at a distance from the placenta (Fig. 23). While some authors feel that a small subchorionic hematoma carries no increased risk of miscarriage (30,31), others believe that subchorionic hemorrhage, along with antepartum vaginal bleeding as part of the same process, is associated with an adverse outcome (32). Serial follow-up examinations are important to ensure that subchorionic hematomas decrease in size.

Abruptio Placentae

Abruptio placentae is an acute event in which premature separation of the placenta causes severe vaginal bleeding, pain, and hypovolemic shock. These patients usually go right to delivery, and antepartum sonography may not be performed. In patients who are sufficiently stable to have an ultrasound examination, a large echogenic retroplacental collection may be seen (Fig. 24).

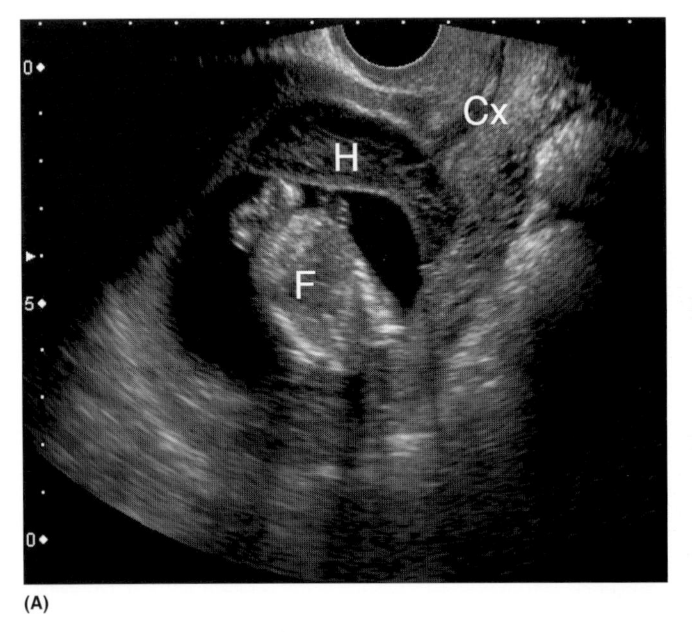

(A)

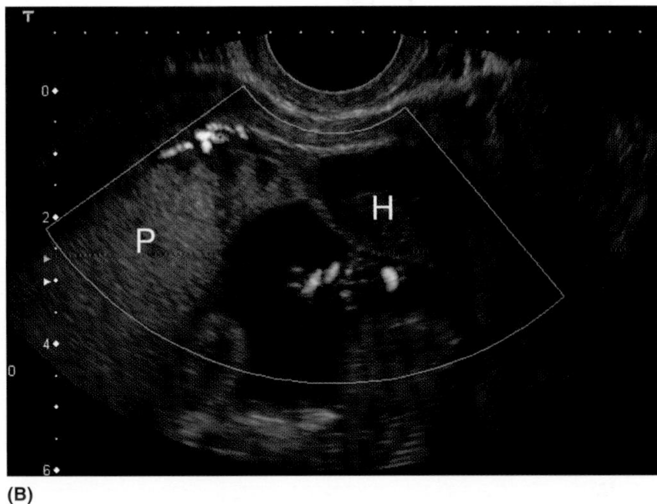

(B)

FIGURE 23 A AND B ■ Subchorionic hematoma. (**A**) Transvaginal scan at 14 weeks shows a subchorionic hematoma (H) overlying the cervix (Cx) in this patient with vaginal bleeding. F, fetus. (**B**) The hematoma (H) is directly adjacent to the placenta (P). At follow-up transvaginal sonography two weeks later, the hematoma had decreased in size.

TABLE 7 ■ Classification of Placenta Creta

- Placenta accreta: adheres to myometrium
- Placenta increta: invades myometrium
- Placenta percreta: extends through uterine wall

PLACENTA CRETA

Placenta creta refers to a focal or diffuse defect of the decidua, which causes the placenta to abnormally adhere to or invade the myometrium, depending on the degree of involvement (Table 7). The normal retroplacental structures may be absent in the case of placenta creta (33,34). At sonography, an increased number of intraplacental "lakes" may be visualized (Fig. 25) (35–38). These lakes may be seen as early as 15 weeks, and are probably the most reliable sign of placenta creta (38). If the decidual defect is minimal, and if no invasion into the myometrium has taken place, the ultrasound examination may appear normal.

The incidence of placenta creta is higher in patients who have a history of cesarean section, multiparity, manual removal of a placenta, and uterine scars of other origin. Placenta creta is associated with placenta previa in over 30% of cases. Thus, placenta creta should always be considered in patients with placenta previa. Rupture of the uterus occurs in approximately 14% of patients with placenta creta, most commonly before delivery (1). A hysterectomy is usually necessary in cases of placenta creta because of the extensive bleeding that occurs, and failure of the placenta to separate from the uterus.

Abnormal implantation of the gestational sac in the lower uterine segment, and specifically in the area of cesarian section scar, has been documented sonographically in patients with placenta creta (39). However, patients with this finding may not necessarily have placenta creta (Fig. 26).

PLACENTA PREVIA

Placenta previa refers to a placenta that covers part or all of the internal os of the cervix (Fig. 27). It occurs in less than 1% of deliveries and necessitates a cesarean section. The incidence of placenta previa is higher in older mothers and in women who smoke (40).

Placenta previa is often misdiagnosed at first or second trimester sonography because of overfilling of the maternal urinary bladder (Fig. 28) or because of contractions (Fig. 29). If placenta previa is suspected, the diagnosis should be confirmed by rescanning the patient after she has voided and/or after at least 20 to 30 minutes have elapsed (41).

CERVIX

The appearance of the cervix should be documented in every antenatal ultrasound examination. It is important in the identification of placenta previa and in the evaluation and management of cervical incompetence. Cervical incompetence, resulting in recurrent spontaneous abortions, is usually due to previous trauma such as termination of pregnancy or surgery. Other risk factors include exposure to diethylstilbestrol, and congenital anomalies. A short cervix at transvaginal sonography is highly predictive of preterm birth, the leading cause of perinatal morbidity and mortality.

The normal cervix has a near vertical orientation, using the transabdominal approach (Fig. 30A and B) (42). The endocervical canal may appear as an echogenic line, or a hypoechoic line representing a mucous plug. A surrounding

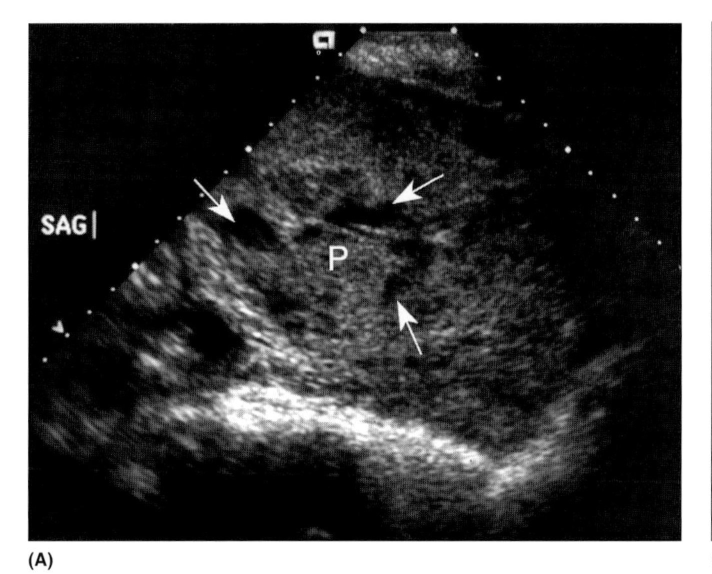

(A)

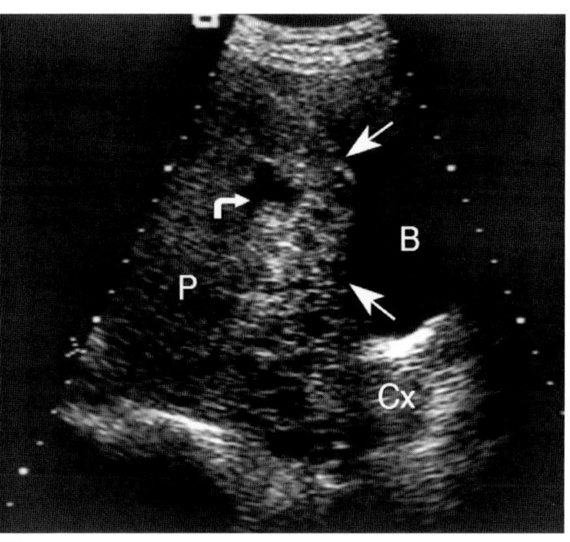

(B)

FIGURE 25 A AND B ■ Placenta percreta. (**A**) Multiple lakes (*arrows*) are present in this large placenta (P) at 27 weeks. The patient had three prior cesarian sections. (**B**) Sagittal scan at 35 weeks shows absence of the retroplacental myometrium; the placenta had invaded through the myometrium to the bladder wall (*straight arrows*). A prominent "lake" is visible in the placenta (*curved arrow*). B, maternal bladder; Cx, cervix.

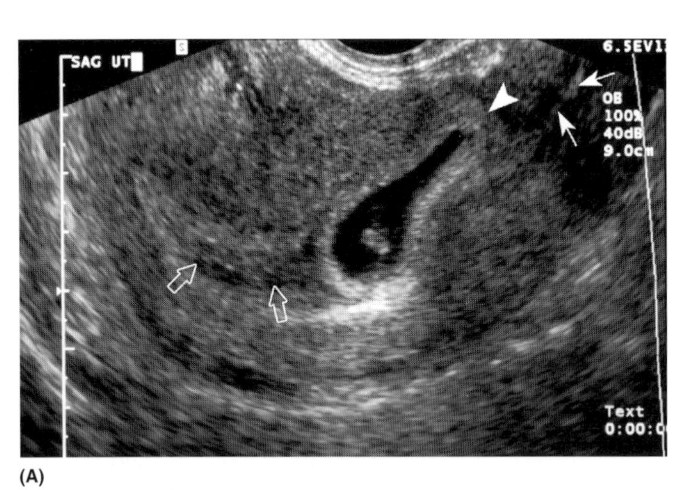

(A)

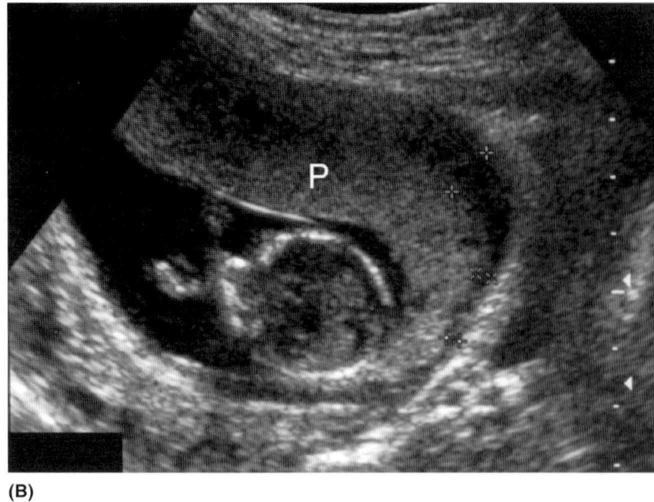

(B)

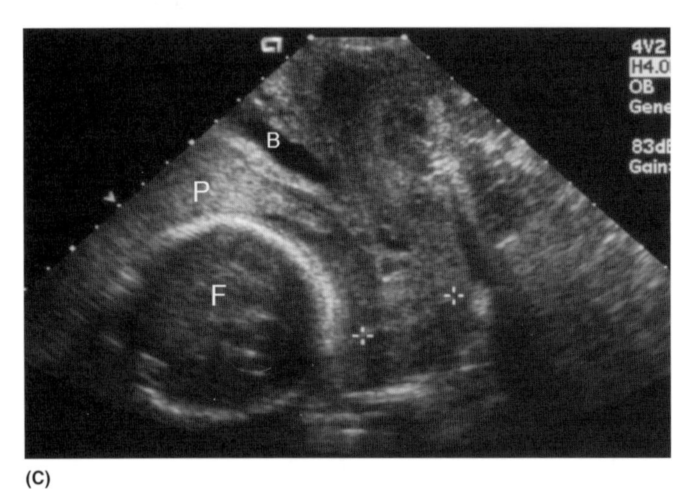

(C)

FIGURE 26 A–C ■ Implantation in cesarian section scar. Patient was status post seven cesarian sections. (**A**) Sagittal transvaginal scan at 6.5 weeks shows low implantation of gestational sac in cesarian section scar (*arrowhead*). *Arrows*, endocervical canal; *open arrows*, decidual reaction in endometrial cavity. (**B**) At 14 weeks, the anterior placenta (P) is of normal thickness, the retroplacental stripe is visible (*cursors*), and no lakes are seen. (**C**) Translabial scan at 29 weeks shows anterior placenta, with no evidence for placenta previa, and intact maternal bladder (B). Abruptio placenta was found at delivery. There was no placenta creta. *Cursors*, endocervical canal; F, fetus.

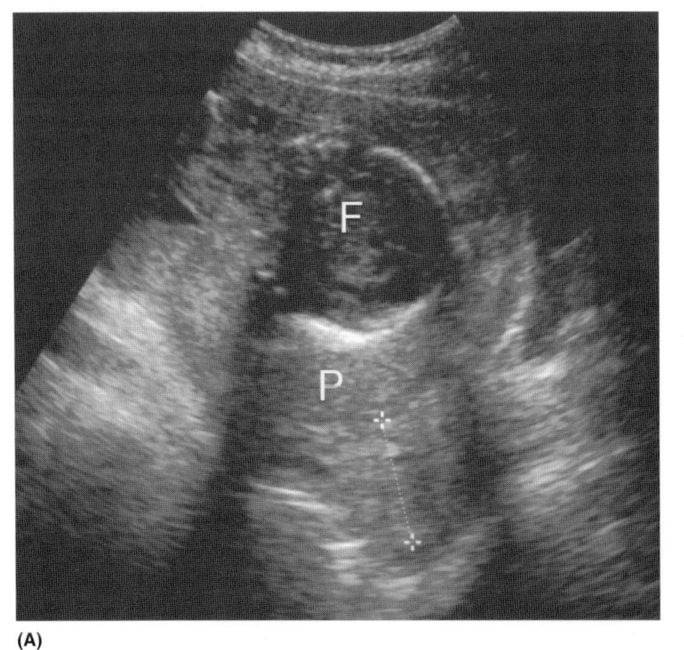

(A)

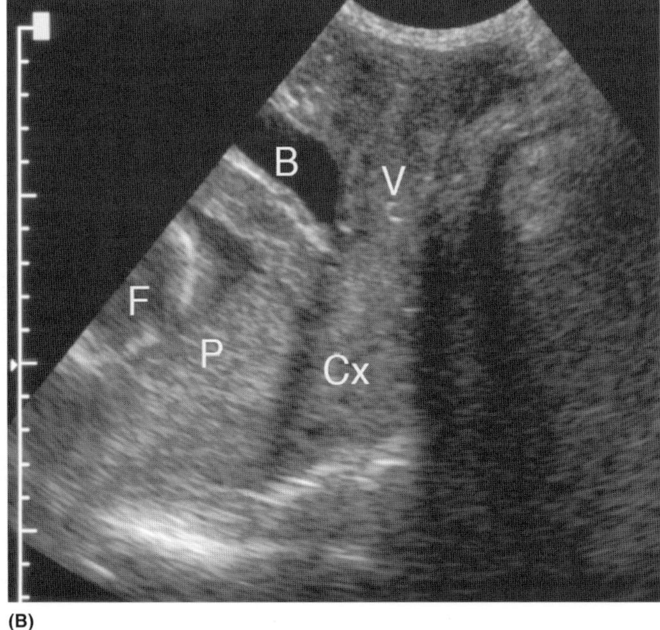

(B)

FIGURE 27 A–C ■ Placenta previa. (**A**) Midline transabdominal sagittal scan shows the posterior placenta (P) completely covering the cervix (*cursors*). (**B**) Translabial scan confirms placenta previa. B, bladder; P, placenta; V, vagina; F, fetus; Cx, cervix. (*Continued*)

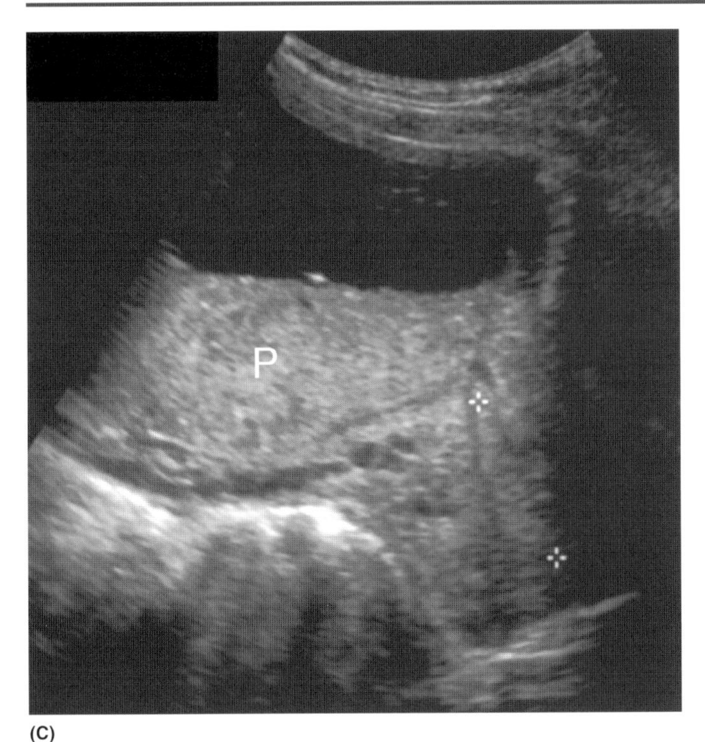

(C)

FIGURE 27 A–C ■ (*Continued*) (**C**) Asymmetric previa. Sagittal scan at 30 weeks shows placental margin covering cervix. *Cursors*, endocervical canal. P, placenta.

hypoechoic zone likely represents endocervical glands. The endocervical canal is often curved (Fig. 30C and D). The presence of a curved cervix usually indicates that the length is normal; a short cervix is always straight (43).

Ideally, the cervix should be evaluated when the patient's bladder is empty because an overfull bladder may artifactually elongate the cervix. Contractions also make it difficult to visualize the cervix accurately. After 20 weeks, the cervix is often obscured by fetal parts with the transabdominal approach. Maternal obesity is another factor that may preclude transabdominal examination of the cervix. Transvaginal and translabial (transperineal) sonographic techniques are useful to image the cervix when the transabdominal approach is unsuccessful (44–46).

Translabial scanning is performed when the patient's bladder is empty. The transducer, over which a glove or a sterile cover has been placed, is positioned between the labia majora in a sagittal orientation. The relationship of the cervix to the vagina and the lower uterine segment and presenting parts is easily assessed (Figs. 31 and 32). When fetal parts obscure the internal cervical os at transabdominal sonography, the translabial and the transvaginal approaches are both useful to exclude placenta previa. When performing transvaginal ultrasound to evaluate the cervix, it is important to use minimal pressure so as not to artificially elongate the cervix.

Using the transabdominal approach, a cervical length of 3.0 cm is considered the lower limit of normal (47,48). In our practice, if the cervical length at transabdominal sonography is less than 3.0 cm, or if the patient is at risk for cervical incompetence, translabial or transvaginal sonography is then performed.

Using the transvaginal approach, the mean length of the cervix is 35 mm at 24 weeks and 33.7 mm at 28 weeks (49). The lower limit of normal at transvaginal sonography

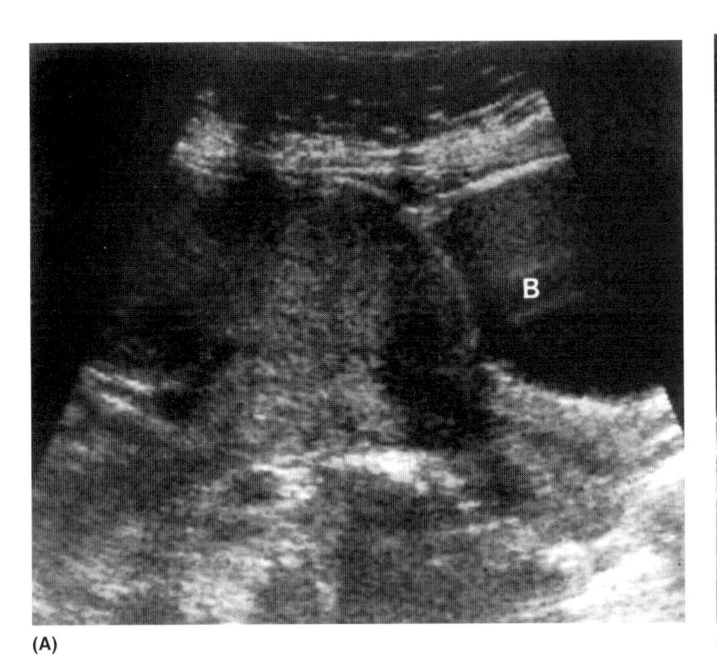

(A)

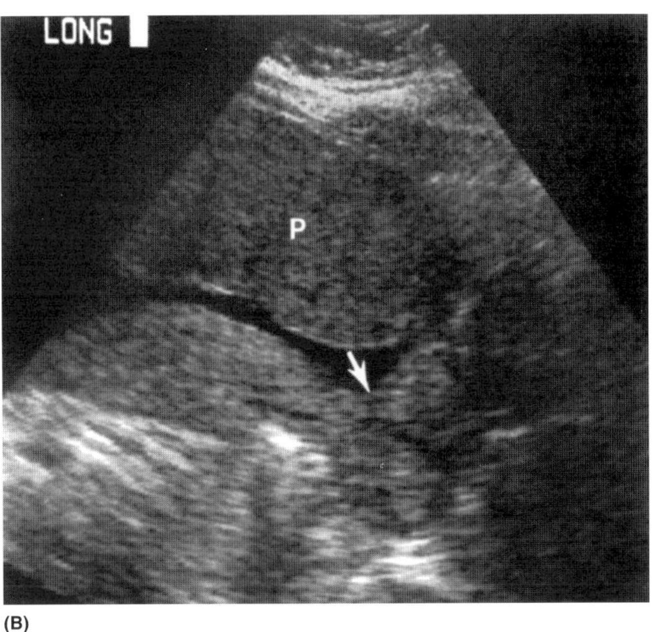

(B)

FIGURE 28 A AND B ■ Bladder effect mimicking placenta previa. (**A**) Sagittal midline scan at 22.5 weeks shows an apparent placenta previa. B, maternal bladder. (**B**) Repeat scan after voiding shows the anterior placenta (P) to be well away from the internal cervical os (*arrow*). *Source*: Spirt BA, Gordon LP. Imaging of the placenta. In: Taveras JM, Ferrucci JT, eds. Radiology: Diagnosis, Imaging, Intervention. Philadelphia: J. B. Lippincott, 1992:1–18.

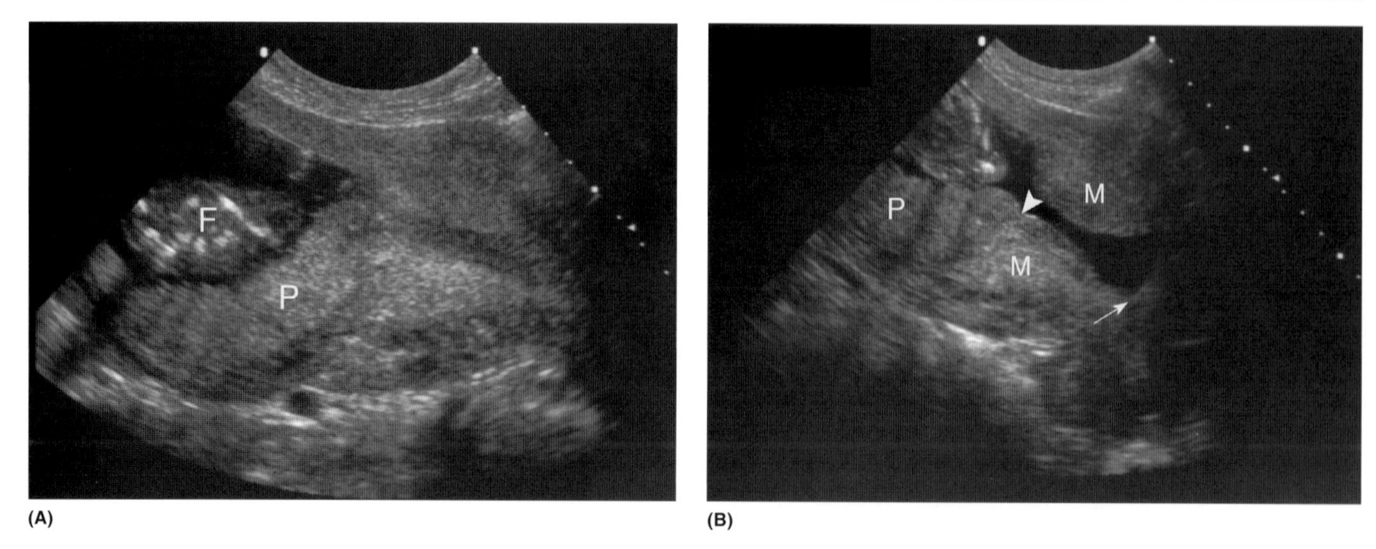

FIGURE 29 A AND B ■ Contraction mimicking placenta previa. (**A**) Midline transabdominal sagittal scan at 17 weeks shows an apparent placenta previa. P, placenta; F, fetus. (**B**) Repeat scan seven minutes later shows that the posterior placenta (P) is well away from the internal os of the cervix (*arrow*). Note the contraction of the anterior and posteroinferior myometrium (M). *Arrowhead*, edge of placenta.

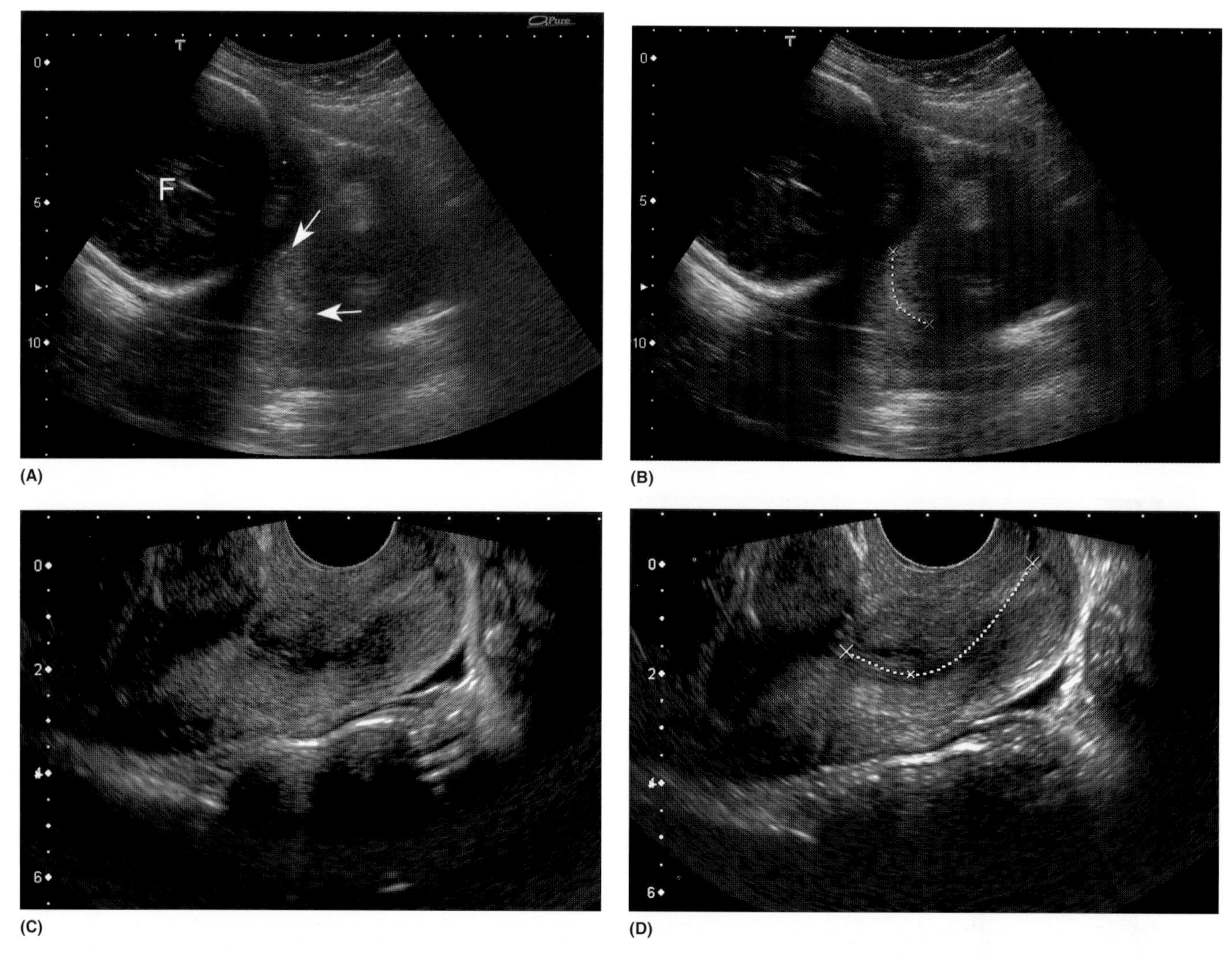

FIGURE 30 A–D ■ Normal cervix. (**A** and **B**) Sagittal transabdominal scan at 30 weeks shows normal, almost vertically oriented cervix. Cervical length is 32 mm. *Arrows*, endocervical canal; F, fetus. (**C** and **D**) Sagittal transvaginal scan at 10 weeks shows normal curved cervix, measuring 40 mm.

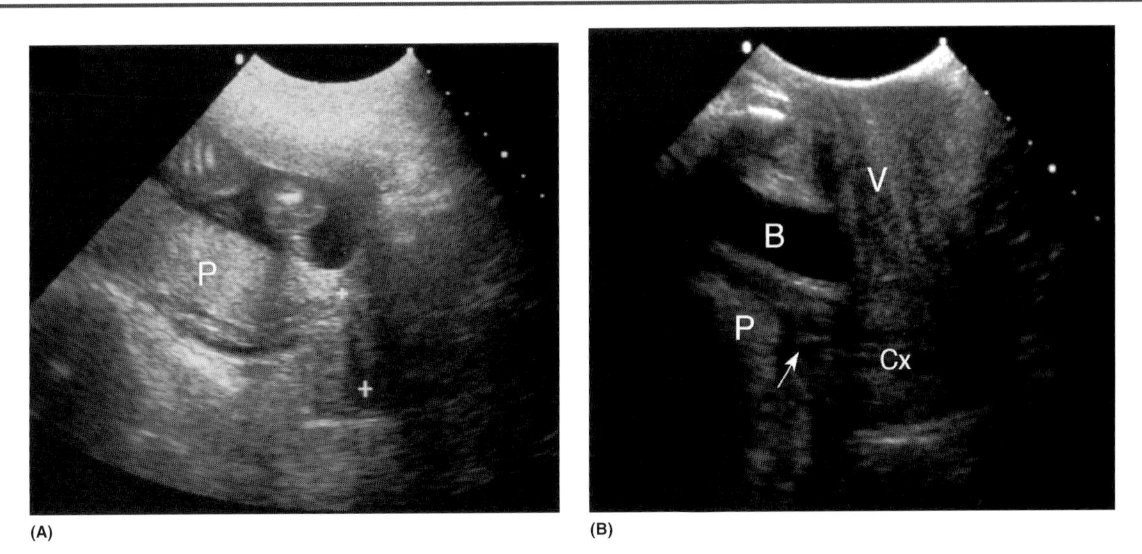

FIGURE 31 A AND B ■ Translabial scan of placenta previa. (**A**) Sagittal transabdominal scan at 20 weeks shows edge of posterior placenta (P) covering the cervix (*cursors*). (**B**) At 35 weeks, sagittal translabial scan confirms placenta previa (P) overlying the internal os (*arrow*). B, maternal bladder; V, vagina; Cx, cervix.

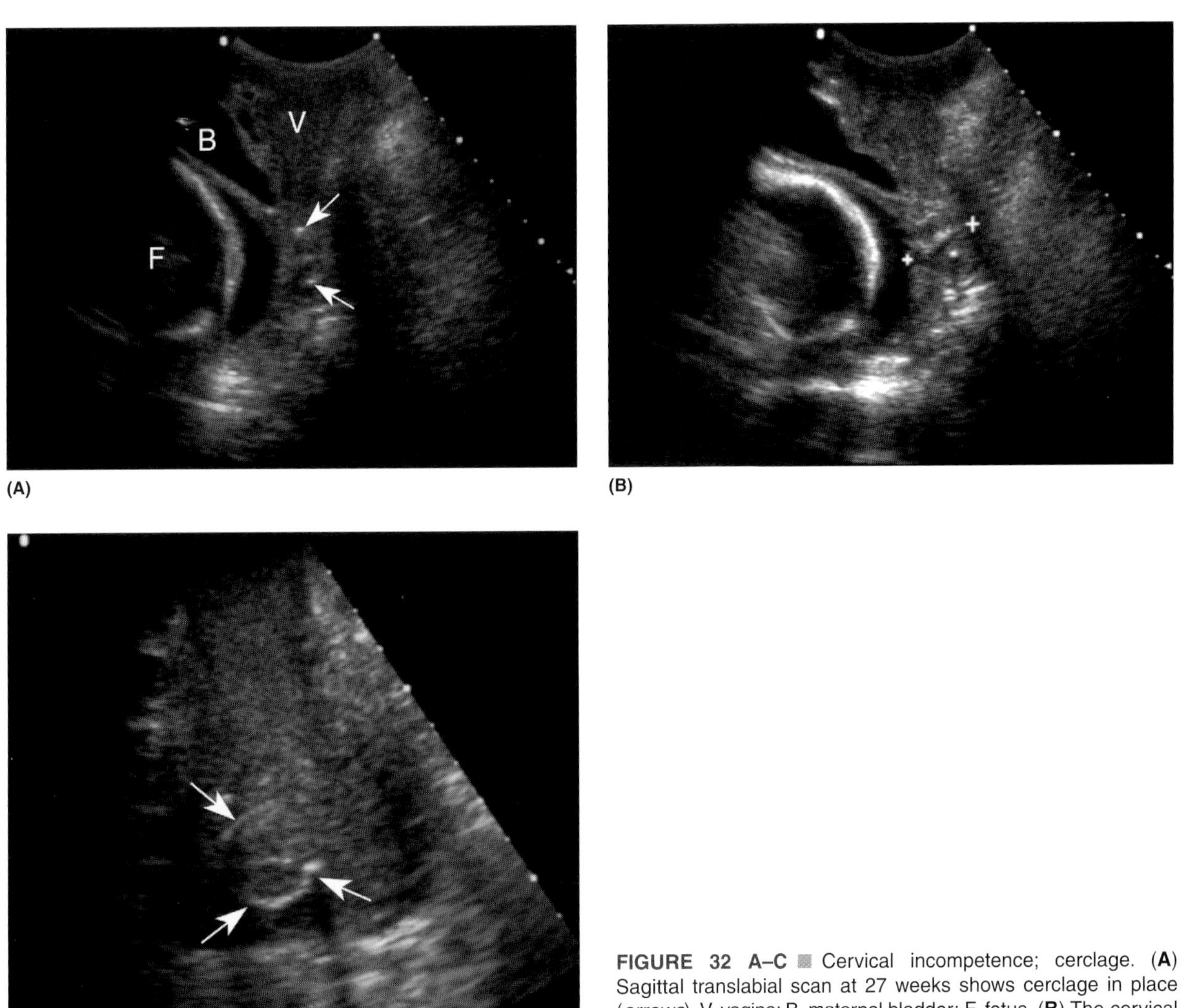

FIGURE 32 A–C ■ Cervical incompetence; cerclage. (**A**) Sagittal translabial scan at 27 weeks shows cerclage in place (*arrows*). V, vagina; B, maternal bladder; F, fetus. (**B**) The cervical length was 28 mm. (**C**) Cerclage (*arrows*) is demonstrated in transverse plane.

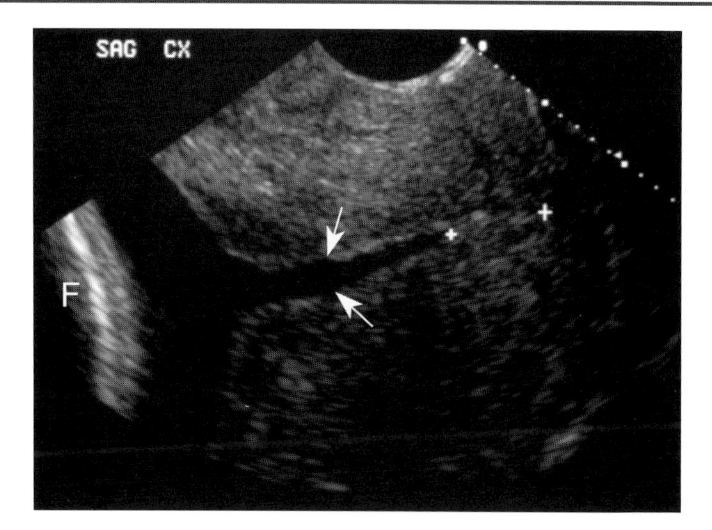

FIGURE 33 ■ Cervical incompetence. Transvaginal scan at 27 weeks shows short cervix (*cursors*); the cervical length was 10 mm. Note fluid in endocervical canal (*arrows*). F, fetus.

is 2.5 cm (43,49). A cervical length of less than 25 mm at transvaginal sonography is associated with an increased risk of preterm birth. A cervical length of 15 mm or less from 14 to 24 weeks is associated with a 48% risk of spontaneous delivery by 32 weeks (48). The sonographic appearance of cervical incompetence may include dilatation of the endocervical canal, or bulging and protrusion of the amniotic membranes (Figs. 33 and 34).

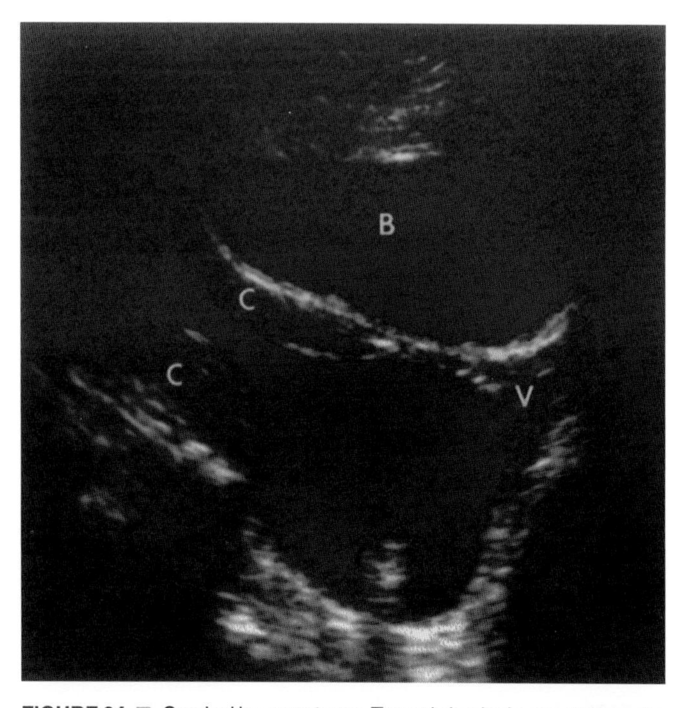

FIGURE 34 ■ Cervical incompetence. Transabdominal scan at 16 weeks shows the amniotic sac protruding through the open cervix (C) into the vagina (V). B, maternal bladder. The patient miscarried several hours after the examination. *Source*: Spirt BA, Gordon LP. The placenta and cervix. In: McGahan JP, Porto M, eds. Diagnostic Obstetrical Ultrasound. Philadelphia, J. B. Lippincott, 1994:83–103.

Whereas cerclage is often used in high-risk patients with recurrent spontaneous abortions, the efficacy of cerclage in the prevention of preterm labor in patients with short cervix found at transvaginal sonography has not been established (43,50).

REFERENCES

1. Fox H. Pathology of the Placenta. Philadelphia: WB Saunders, 1978.
2. Kraus FT, Redline RW, Gersell DJ, et al. Placental Pathology. Washington, DC: American Registry of Pathology, Armed Forces Institute of Pathology, 2004.
3. Gruenwald P. The supply line of the fetus: definitions relating to fetal growth. In: Gruenwald P, ed. The Placenta and its Maternal Supply Line. Lancaster, England: Medical and Technical Publishing, 1975.
4. Hoddick WK, Mahoney BS, Callen PW, et al. Placental thickness. J Ultrasound Med 1984; 4:479–482.
5. Ghosh A, Tang MHY, Lam YH, et al. Ultrasound measurement of placental thickness to detect pregnancies affected by homozygous alpha-thalassemia-1. Lancet 1994; 334:988–989.
6. Ko TM, Tseung LH, Hsu PM, et al. Ultrasonographic scanning of placental thickness and the prenatal diagnosis of homozygous alpha-thalassemia 1 in the second semester. Prenat Diagn 1995; 15:7–10.
7. Tongsong T, Wanapirak C, Sirichotiyakul S. Placental thickness at mid-pregnancy as a predictor of Hb Bart's Disease. Prenat Diagn 1999; 19:1027–1030.
8. Tongsong T, Wanapirak C, Sirichotiyakul S, et al. Sonographic markers of Hemoglobin Bart Disease at midpregnancy. J Ultrasound Med 2004; 23:49–55.
9. Elchalal U, Ezra Y, Levi Y, et al. Sonographically thick placenta: a marker for increased perinatal risk—a prospective cross-sectional study. Placenta 2000; 21:268–272.
10. Poggi SH, Bostrom KI, Demer LL, et al. Placental calcification: a metastatic process? Placenta 2001; 22:591–596.
11. Jeacock MK. Calcium content of the human placenta. Am J Obstet Gynecol 1963; 87:34–40.
12. Wentworth P. Macroscopic placental calcification and its clinical significance. J Obstet Gynaecol Br Commonwlth 1965; 72:215–222.
13. Tindall VR, Scott JS. Placental calcification: a study of 3,025 singleton and multiple pregnancies. J Obstet Gynaecol Br Commonwlth 1965; 72:356–373.
14. Spirt BA, Cohen WN, Weinstein HM. The incidence of placental calcification in normal pregnancies. Radiology 1982; 142: 707–711.
15. Spirt BA, Kagan EH, Rozanski RM. Sonolucent areas in the placenta: sonographic and pathologic correlation. AJR Am J Roentgenol 1978; 131:961–965.
16. Spirt BA, Gordon LP. Sonography of the placenta. In: Fleischer AC, Romero R, Manning FA, et al., eds. The Principles and Practice of Ultrasonography in Obstetrics and Gynecology. 4th ed. Norwalk, CT: Appleton & Lange, 1991:133–157.
17. Kaplan C, Blanc WA, Elias J. Identification of erythrocytes in intervillous thrombi: a study using immunoperoxidase identification to hemoglobins. Hum Pathol 1982; 13:554–557.
18. Spirt BA, Gordon LP, Kagan EH. Intervillous thrombosis: sonographic and pathologic correlation. Radiology 1983; 147:197–200.
19. Harris RD, Simpson WA, Pet LR, et al. Placental hypoechoic/anechoic areas and infarction: sonographic–pathologic correlation. Radiology 1990; 176:75–80.
20. Fox H. Pathology of the Placenta. 2nd ed. London: WB Saunders, 1997.
21. Elston CW. Gestational trophoblastic disease. In: Fox H, ed. Haines and Taylor Obstetrical and Gynaecological Pathology. 3rd ed. Edinburgh: Churchhill Livingstone, 1987:1045–1078.
22. Mazur MT, Kurman RJ. Gestational trophoblastic disease. In: Kurman RJ, ed. Blaustein's Pathology of the Female Genital Tract. 3rd ed. New York: Springer, 1987:835–875.
23. Szulman AE, Surti U. The syndromes of hydatidiform mole. I. Cytogenetic and morphologic correlations. Am J Obstet Gynecol 1978; 131:665–671.

24. Pham T, Steele J, Stayboldt C, et al. Placental mesenchymal dysplasia is associated with high rates of intrauterine growth restriction and fetal demise: a report of 11 new cases and a review of the literature. Am J Clin Pathol 2006; 126(1):67–78.
25. Szulman AE, Wong LC, Hsu C. Residual trophoblastic disease in association with partial hydatidiform mole. Obstet Gynecol 1981; 57:392–394.
26. Szulman AE, Surti U. The clinicopathologic profile of the partial hydatidiform mole. Obstet Gynecol 1982; 59:597–602.
27. Heifetz SA, Czaja J. In situ choriocarcinoma arising in partial hydatidiform mole: implications for the risk of persistent trophoblastic disease. Pediatr Pathol 1992; 12:601.
28. Gardner HA, Lage JM. Choriocarcinoma following a partial hydatidiform mole: a case report. Hum Pathol 1992; 23:468.
29. Spirt BA, Kagan EH, Aubry RH. Clinically silent retroplacental hematoma: sonographic and pathologic correlation. J Clin Ultrasound 1981; 9:203–205.
30. Stabile I, Campbell S, Grudzinskas JG. Threatened miscarriage and intrauterine hematomas: sonographic and biochemical studies. J Ultrasound Med 1989; 8:289–292.
31. Pedersen JF, Mantoni M. Prevalence and significance of subchorionic hemorrhage in threatened abortion: a sonographic study. AJR Am J Roentgenol 1990; 154:535–537.
32. Ball RH, Ade CM, Schoenborn JA, et al. The clinical significance of ultrasonographically detected subchorionic hemorrhages. Am J Obstet Gynecol 1996; 174(3):996–1002.
33. Pasto ME, Kurtz AB, Rifkin MS, et al. Ultrasonographic findings in placenta increta. J Ultrasound Med 1983; 2:155–159.
34. deMedonca LK. Sonographic diagnosis of placenta accreta: presentation of six cases. J Ultrasound Med 1988; 7:211–215.
35. Hoffman-Tretin JC, Koenigsberg M, Rabin A, et al. Placenta accreta: additional sonographic observations. J Ultrasound Med 1992; 11:20–34.
36. Finberg HJ, Williams JW. Placenta accreta: prospective sonographic diagnosis in patients with placenta previa and prior caesarean section. J Ultrasound Med 1992; 11:333–343.
37. Lerner JP, Deane S, Timor-Tritsch IE. Characterization of placenta accreta using transvaginal sonography and color Doppler imaging. Ultrasound Obstet Gynecol 1995; 5:198–201.
38. Comstock CH, Love JJ, Bronsteen RA, et al. Sonographic detection of placenta accreta in the second and third trimesters of pregnancy. Am J Obstet Gynecol 2004; 190:1135–1140.
39. Comstock CH, Lee W, Vettraino IM, et al. The early sonographic appearance of placenta accreta. J Ultrasound Med 2003; 22: 19–23.
40. Naeye RL. Functionally important disorders of the placenta, umbilical cord, and fetal membranes. Hum Pathol 1987; 18:680–691.
41. Artis AA, Bowie JD, Rosenberg ER, et al. The fallacy of placental migration: effect of sonographic techniques. AJR Am J Roentgenol 1985; 144:79–81.
42. Bowie JD, Andreotti RF, Rosenberg, ER. Sonographic appearance of the uterine cervix in pregnancy: the vertical cervix. Am J Roentgenol 1983; 140:737–740.
43. Berghella V, Bega G, Tolosa JE, et al. Ultrasound assessment of the cervix. Clin Obstet Gynecol 2003; 46:947–962.
44. Hertzberg BS, Bowie JD, Carroll BA, et al. Diagnosis of placenta previa during the third trimester: role of transperineal sonography. Am J Roentgenol 1992; 159:83–87.
45. Mahony BS, Nyberg DA, Luthy DA, et al. Translabial ultrasound of the third-trimester uterine cervix: correlation with digital examination. J Ultrasound Med 1990; 9:717–723.
46. Hertzberg BS, Bowie JD, Weber TM, et al. Sonography of the cervix during the third trimester of pregnancy: value of the transperineal approach. AJR Am J Roentgenol 1991; 157:73–76.
47. Riley L, Frigoletto FD, Benacerraf BR. The implications of sonographically identified cervical changes in patients not necessarily at risk for preterm birth. J Ultrasound Med 1992; 11:75–79.
48. Hassan SS, Romero R, Berry SM, et al. Patients with an ultrasonographic cervical length <= 15 mm have a nearly 50% risk of early spontaneous preterm delivery. Am J Obstet Gynecol 2000; 182: 1458–1467.
49. Iams JD, Goldenberg RL, Meis PJ, et al. The length of the cervix and the risk of spontaneous premature delivery. N Engl J Med 1996; 334:567–572.
50. Berghella V, Odibo AO, Tolosa JE. Cerclage for prevention of preterm birth in women with a short cervix found on transvaginal ultrasound examination: a randomized trial. Am J Obstet Gynecol 2004; 191:1311–1317.

Umbilical Cord and Chorioamniotic Membranes

42

Beverly A. Spirt, Lawrence P. Gordon,
and Harris J. Finberg

INTRODUCTION

The umbilical cord is of critical importance as the "lifeline" of the fetus. This chapter reviews the development and normal sonographic appearance of the umbilical cord and chorioamniotic membranes, and the abnormalities that may occur.

UMBILICAL CORD

Embryology

About two weeks after fertilization (Fig. 1A), an embryonic disc curves around the small, developing amniotic sac, which faces the portion of the chorion that will differentiate into the placenta. The primary yolk sac is present on the opposite surface of the embryo. These structures are surrounded by extraembryonic mesoderm, which fills the remainder of the gestational sac. By the third week this mesoderm has formed a cavity, except in one area, where a bridge of tissue (the connecting stalk) attaches the embryo, amnion, and yolk sac to the outer rind of chorion. This stalk becomes the umbilical cord (Fig. 1B).

As the connecting stalk develops, a series of coordinated events takes place (Fig. 1C):

1. The amniotic cavity undergoes rapid enlargement, grows around the edges of the embryo, eventually fills the extraembryonic cavity, and fuses to the surface of the connecting stalk, enveloping the entire embryo.
2. The embryo rotates so the primary yolk sac ends up facing the connecting stalk and developing placenta.
3. The embryonic disc begins to flex and fold around the primary yolk sac, which is largely incorporated into the fetus to become the primitive gut. This is accomplished by the growth and fusion of cephalic, caudal, and paired lateral body folds, creating a cylindric fetus.
4. An outpouching of the primary yolk sac grows down into the connecting stalk to form the secondary yolk sac, which bulges into extraembryonic coelom outside the amnion. This is recognized as the structure called the yolk sac on sonograms performed between 5 and 11 weeks of gestation. It remains connected to the fetus by a thin tubular structure within the developing umbilical cord (the omphalomesenteric duct). After 10 weeks of gestation, as the yolk sac shrinks, this duct usually involutes. (The fetal end persists in about 2% of people, remaining as a Meckel diverticulum on the antimesenteric surface of the distal ileum.)
5. An additional, smaller outpouching develops from the caudal end of the primitive gut where the bladder is differentiating. This diverticulum (the allantois) also extends into the developing umbilical cord and its blood vessels become the umbilical arteries and vein. Persistance of the allantois (also referred to as the urachus) may cause midline abnormalities between bladder and umbilicus as well as cystic remnants in the cord.

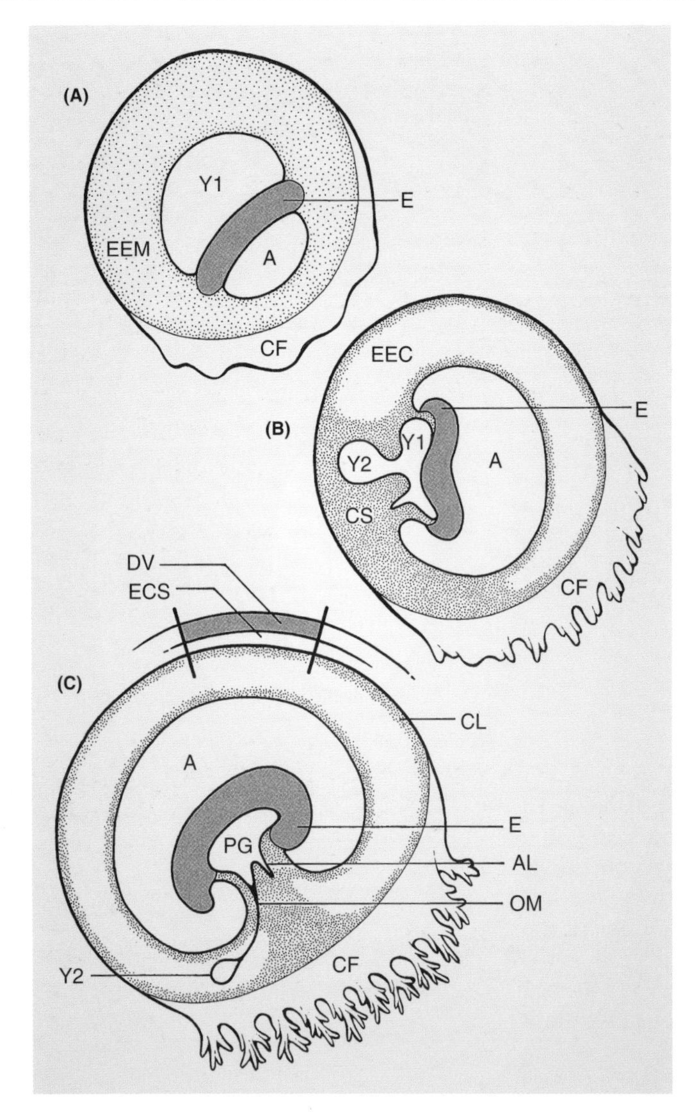

FIGURE 1 A–C ■ Development of the umbilical cord and membranes. (**A**) 13 days; (**B**) 21 days; (**C**) 28 days. A, amniotic cavity; AL, allantois; CF, chorion frondosum; CL, chorion laeve; CS, connecting stalk; DV, decidua vera (along endometrium); E, embryo; ECS, extrachorionic space (endometrial cavity); EEC, extraembryonic coelom; EEM, extraembryonic mesoderm; OM, omphalomesenteric duct; PG, primary gut; Y1, primary yolk sac; Y2, secondary yolk sac. *Source*: Finberg H. Umbilical cord and amniotic membranes. In: McGahan J, Porto M (eds.) Diagnostic Obstetrical Ultrasound. Philadelphia, JB Lippincott, 1994.

Normal Anatomy

The umbilical cord normally has two arteries of similar caliber and a single vein (Fig. 2). The diameter of the vein is usually twice that of the artery. The umbilical arteries arise from the hypogastric arteries in the fetus, and carry deoxygenated blood from the fetus to the placenta (Fig. 3). The umbilical vein proceeds in a cephalic midline direction from the umbilicus into the left branch of the portal vein and then into the ductus venosus, bringing oxygenated blood back from the placenta.

The blood vessels of the cord are surrounded by Wharton's jelly, a watery gel that provides turgor and resistance to compressibility. The cord is covered by amnionic epithelium, which is firmly adherent to the connective tissue. Unlike the amnion, which is loosely applied against the chorion of the gestational sac, the amnion of the cord cannot be elevated or stripped away.

The umbilical cord vessels normally spiral around each other, usually in a counterclockwise direction (1,2). The helical pattern is usually well established by 11 weeks menstrual age (3), although occasionally not appearing well coiled until after 20 weeks. Both the spiraling of the vessels and the overall length of the umbilical cord are directly related to fetal activity (1,4).

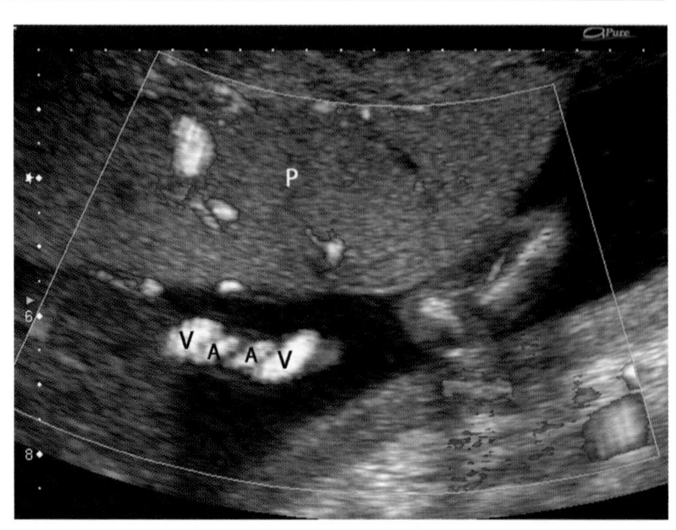

FIGURE 2 ■ Image of normal umbilical cord at 19 weeks shows two arteries (A) of similar caliber, and vein (V). P, placenta.

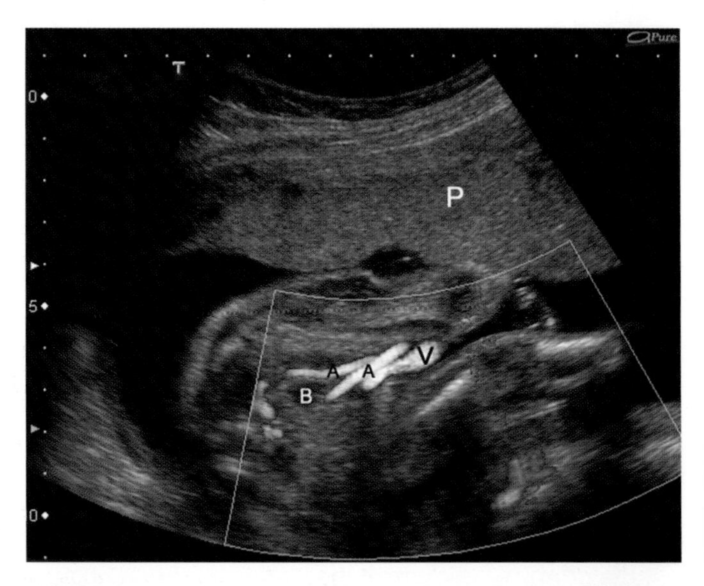

FIGURE 3 ■ The two umbilical arteries (A) enter the fetal abdomen at the umbilicus, course downward and on either side of the bladder (B), and join with the hypogastric arteries. V, umbilical vein; P, placenta.

Restriction of fetal activity, whether because of inadequate amniotic space, intrinsic fetal abnormality, or physical tethering of the fetus, may result in a cord that is shorter with few or absent helical turns (5). The normal cord averages about 60 cm in length, and has a diameter of up to 2 cm. A cord that is less than 32 cm in length is considered short (4).

The umbilical cord may be visualized at sonography by eight weeks menstrual age. In the first trimester, the length of the umbilical cord is about equal to the crown–rump length of the fetus. At eight weeks, the fetal intestines (midgut) normally herniate into the umbilical cord (Fig. 4). This should not be mistaken for an abdominal wall defect or a mass. At 12 weeks, the intestines rapidly return to the abdominal cavity.

In the second and third trimesters, the three cord vessels are clearly visualized at sonography (Fig. 5). The number of cord vessels may be difficult to determine when the cord has an unusually large number of twists, or in the case of a large maternal body habitus. The normal complement of vessels is best evaluated on a cross-section image of the cord (Fig. 6) and is most accurately determined at the fetal insertion.

Differential Diagnosis of Umbilical Cord Abnormalities

Umbilical cord abnormalities may be grouped into five categories: abnormalities of the cord vessels, developmental abnormalities, mechanical abnormalities, abnormalities of cord insertion, and cord masses (Table 1).

Abnormalities of the Cord Vessels

Single Umbilical Artery

Single umbilical artery (SUA), the most common intrinsic malformation of the umbilical cord, occurs in up to 1% of pregnancies, with higher frequencies in autopsy series and in first- and second-trimester delivery series (3,4,6,7) (Fig. 7). This condition may be due either to

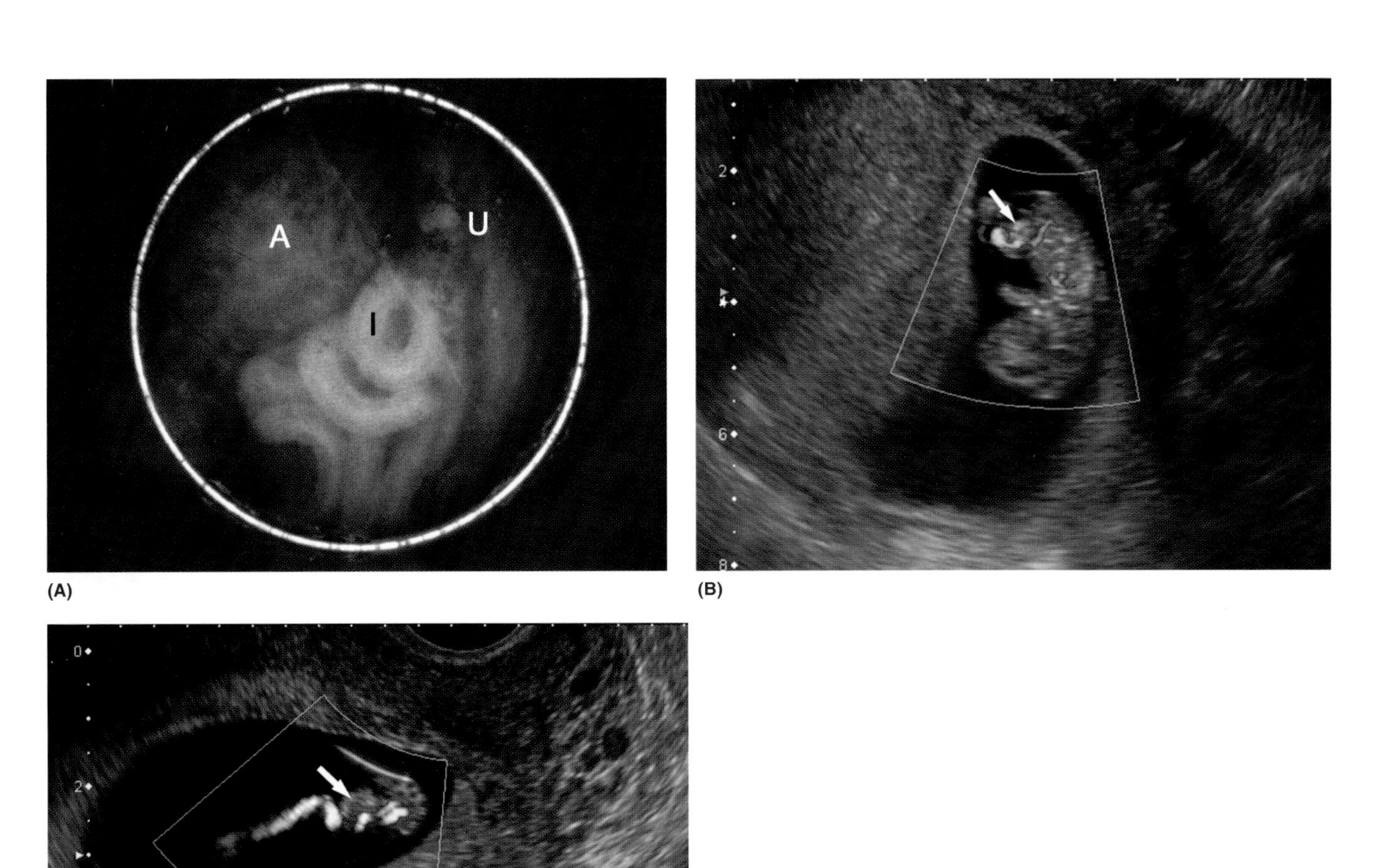

FIGURE 4 A–C ■ (**A**) Fetoscopy image at 9–10 weeks shows intestines (I) herniated into umbilical cord (U). A, abdomen. *Source*: Spirt BA, Gordon LP, Oliphant M. Prenatal Ultrasound: A Color Atlas with Anatomic and Pathologic Correlation. Churchill Livingstone, New York, 1987. (**B**) Sagittal and (**C**) transverse images of fetuses at 10 weeks show intestines (*arrow*) herniated into base of umbilical cord.

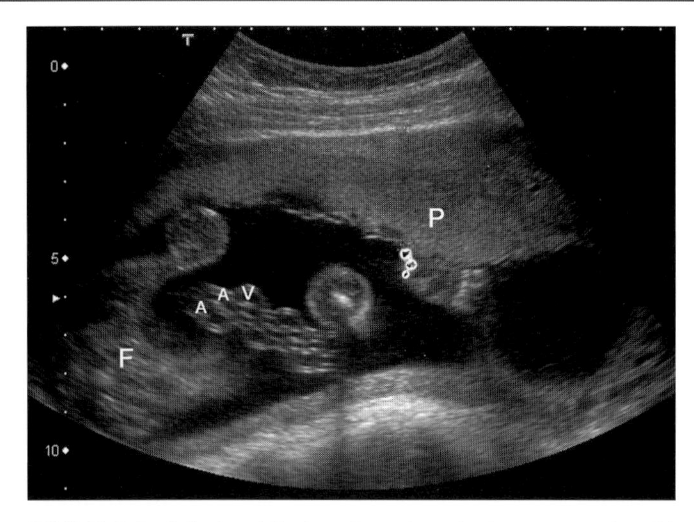

FIGURE 5 ■ Coiled umbilical cord at 20 weeks has two arteries (A) and vein (V). P, placenta.

primary aplasia or to atrophy of the second umbilical artery (5,8). In a detailed review of SUA, Heifetz found a strong association with other fetal malformations, with additional anomalies occurring in up to 20% of cases (9). Other authors have reported an incidence of other anomalies ranging from 9% in a particularly large series (7) to up to 50% (4). These abnormalities may affect any fetal organ system, and are frequently multiple (3,5). Chromosomal abnormalities, particularly Trisomies 13 and 18, and 45,X occur in approximately 10% of fetuses with SUA (Fig. 8) (6,9,10). In addition, SUA has been associated with IUGR (11).

Identification of an SUA should prompt a detailed fetal anatomic survey. In the case of an isolated SUA, if normal four-chamber and outflow tract views of the heart are obtained, fetal echocardiography is not felt to be indicated (12). Similarly, if a thorough prenatal examination was performed, a postnatal renal ultrasound examination is not considered to be necessary in these cases (13).

The umbilical arteries often fuse proximal to the placenta, and rarely at a distance from the placental insertion. Therefore, it is important to evaluate the cord vessels at or close to the fetal insertion, or at least 3 to 5 cm from the placental insertion. SUA can be confirmed within the fetal abdomen by identifying the absence of the intra-abdominal portion of one of the umbilical arteries and of the hypogastric artery on that side. The common iliac artery on that side is smaller than the opposite side, and shows a significant difference in pulsatility index (14). The left umbilical artery is more often absent in SUA than the right (15–17). The occurrence of additional abnormalities appears to be independent of the side of the single vessel (15,17,18).

There is usually a compensatory increase in the diameter of the single umbilical artery to greater than 50% of the diameter of the umbilical vein, as the blood flow to the placenta is transported entirely by that artery (19). In most cases of SUA, the single umbilical artery has normal Doppler flow velocity waveforms (14).

Hypoplastic Umbilical Artery ■ With discordant umbilical arteries, the hypoplastic artery will usually have a higher resistive index (14). Hypoplasia of the umbilical artery (Fig. 9) is of uncertain clinical significance. A higher incidence of anomalies (20) and an association with marginal or velamentous cord insertion has been reported (21), but the numbers are small. Hypoplastic umbilical artery may be part of the spectrum of SUA. As with SUA, a detailed examination of the fetal anatomy should be performed.

Persistent Right Umbilical Vein ■ A detailed fetal anatomic survey is also indicated in the case of persistent right umbilical vein, a rare anomaly that is due to involution of the left umbilical vein. This vessel may enter and enlarge the right portal vein or may enter the right atrium directly. This anatomic variant has no definite clinical significance itself, but about half of the described cases have had other anomalies or anomaly complexes that may affect any organ system (22).

Aneurysm of the Umbilical Artery ■ Umbilical artery aneurysm is rare (Fig. 10) and may lead to fetal demise due to umbilical vascular compression caused by dissection or by extensive intraaneurysmal thrombosis (23,24). Association with SUA and with Trisomy 18 has been reported (24,25).

Umbilical Vein Varix ■ Umbilical vein varix refers to the sonographic finding of a focal dilatation of the intra-abdominal, extrahepatic portion of the umbilical vein (Fig. 11), with a transverse diameter at least 50% larger than that of the intrahepatic umbilical vein (26). The etiology of this finding is unclear. As this portion of the umbilical vein has a relatively weak support structure, any change in venous pressure could cause dilatation. Umbilical vein varix has been associated with aneuploidy,

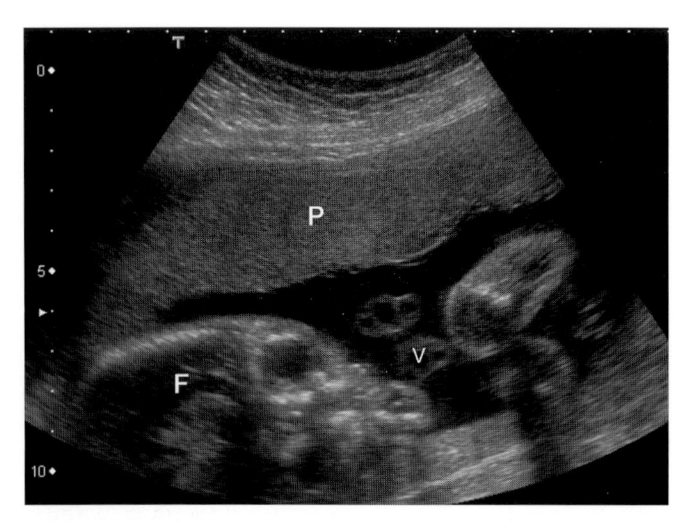

FIGURE 6 ■ Transverse images of umbilical cord clearly demonstrates the normal umbilical vein (V) between the two arteries. P, placenta; F, fetus.

TABLE 1 ■ Umbilical Cord Abnormalities and Associated Risks

Abnormality	Associated risks
Cord vessels	
Single umbilical artery	Increased risk of other anomalies
Hypoplastic umbilical artery	Possible increased risk of other anomalies
Persistent right umbilical vein	No definite risk itself Significant risk of other anomalies
Umbilical vein varix	Usually normal outcome
Umbilical artery aneurysm	Risk of fetal demise
Developmental abnormalities	
Limb-body wall complex	Lethal exteriorization of thoracoabdominal organs
Omphalocele	Associated fetal anomalies, aneuploidy
Abnormal cord length:	
Short cord	Cord compression accidents Short cord syndrome, limb body-wall complex Fetal distress
Hypocoiled or hypercoiled umbilical cord	Increased risk of fetal anomalies, IUGR
Long cord	Nuchal cord, knots, torsion
Mechanical abnormalities	
Nuchal cord	Vascular compromise uncommon
True cord knot	Vascular compromise uncommon
Stricture or torsion	Rare but lethal
Abnormalities of cord insertion	
Velamentous cord insertion	Vasa previa if vessels cross internal cervical os Severe hemorrhage
Marginal cord insertion	No clinical significance
Cord masses	
Mucoid degeneration of Wharton's jelly	No definite risk
Diffuse cord edema	Sequela of hydrops Idiopathic
Diffuse cord swelling	Urine extravasation from patent urachus
Omphalomesenteric duct cyst	Vascular compromise rare
Allantoic (urachal) duct cyst	?Risk of other anomalies or aneuploidy
Hemangioma (angiomyxoma)	Vascular compromise possible
Teratoma	Vascular compromise possible
Thrombosis, hematoma	Severe risk of vascular compromise Embolic fetal or placental events
Gastroschisis	Surgically corrected
Umbilical hernia	Surgically corrected
Omphalocele	Risk of aneuploidy, other anomalies

with multiple anomalies, with isolated cardiac anomalies, and with poor perinatal outcome (26–29). Thus, a detailed survey of the fetal anatomy should be performed in this situation, and a fetal echocardiogram should be considered. In the absence of other anomalies, the prognosis for a fetus with umbilical vein varix is generally good (26,28,30); however, the development of hydrops has been reported, and fetal death has occurred in otherwise apparently normal fetuses, likely on the basis of cardiac failure and/or thrombosis of the varix (26). Note that umbilical vein varix is not described in the current pathology literature. This may reflect the fact that it is a dynamic process which, in the absence of thrombosis, may not be grossly recognizable.

Developmental Abnormalities

Congenital absence of the umbilical cord is rare. It is thought to result from a defective embryonic folding process in which severe maldevelopment of all the folds leads to failure of formation of the umbilical cord, with the abdominal contents of the fetus within a sac

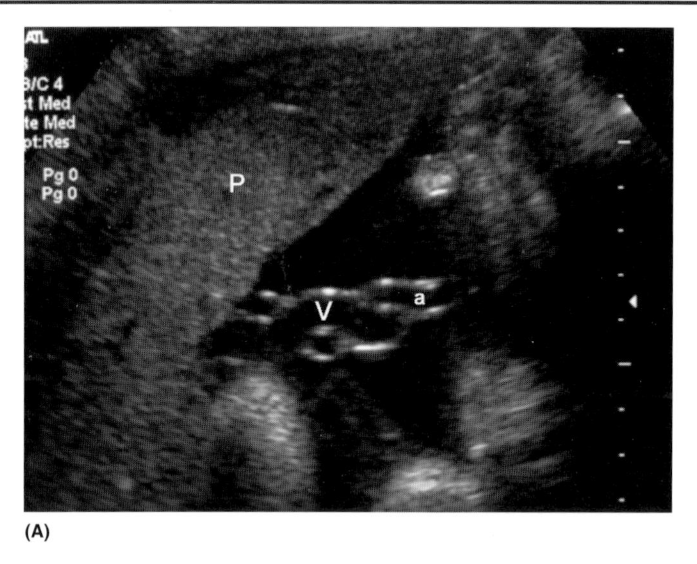

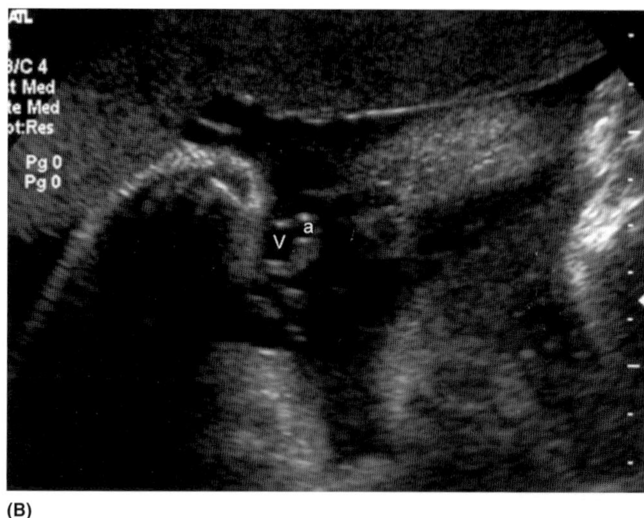

(A) (B)

FIGURE 7 A AND B ■ (A) Two-vessel umbilical cord at 32 weeks, confirmed with cross-section image (B). No other anomalies were visualized at sonography, and the infant was normal. a, artery, V, vein.

that is directly attached to the placenta or attached by a short umbilical cord (4,31,32). This abnormality is also referred to as the body stalk anomaly or as limb-body wall complex (Fig. 12). It is uniformly lethal. This anomaly should be distinguished from gastroschisis or omphalocele (Fig. 13). The short umbilical cord syndrome, which includes a large omphalocele, exstrophy of the urinary bladder, imperforate anus, and spinal deformity most likely is within the spectrum of body stalk anomaly.

Abnormalities Related to Cord Length

Cord length is thought to be related to fetal mobility, especially during the early stages of pregnancy (4). Conditions that restrict or reduce fetal movements such as oligohydramnios or amniotic bands are associated with short cords. The presence of a short cord, regardless of cause, may predispose to inadequate fetal descent and fetal heart rate abnormalities related to cord compression (33). Measuring cord length by ultrasound is not practical; however a qualitative estimate can be made. A long cord is thought to predispose to knotting, torsion, and prolapse (see below).

Abnormal Coiling of the Cord ■ The umbilical cord normally has one coil per 5 cm (34,35). Abnormal coiling of the umbilical cord, including absence of coils (Fig. 14) and an increased number of coils (hypercoiling) has been associated with increased perinatal morbidity and mortality (34–37). In live-born singleton pregnancies, 5% of umbilical cords have no spiraling (1). The frequency of noncoiled umbilical vessels in cases of intrauterine death is 18% (5). Absent twists are also more common in SUA cords and in twins.

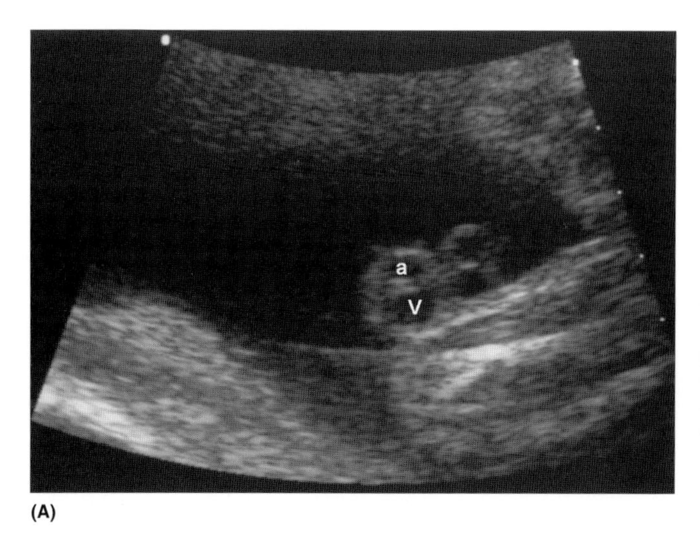

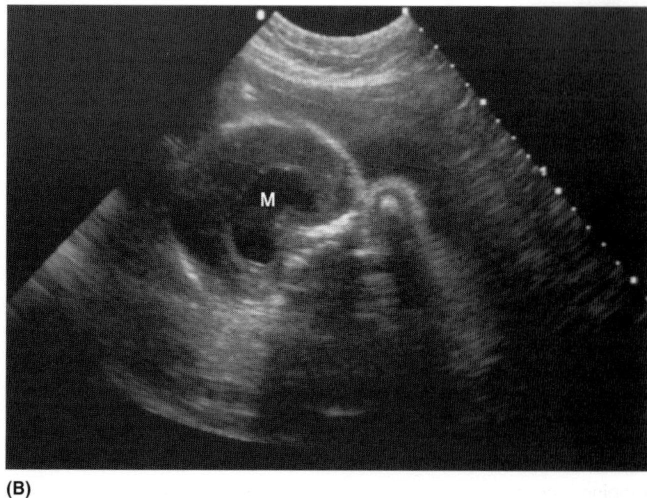

(A) (B)

FIGURE 8 A AND B ■ (A) Two vessel umbilical cord at 18 weeks. V, vein; a, artery. (B) Fetal brain shows monoventricle (M) typical of holoprosencephaly. Karyotype showed Trisomy 18.

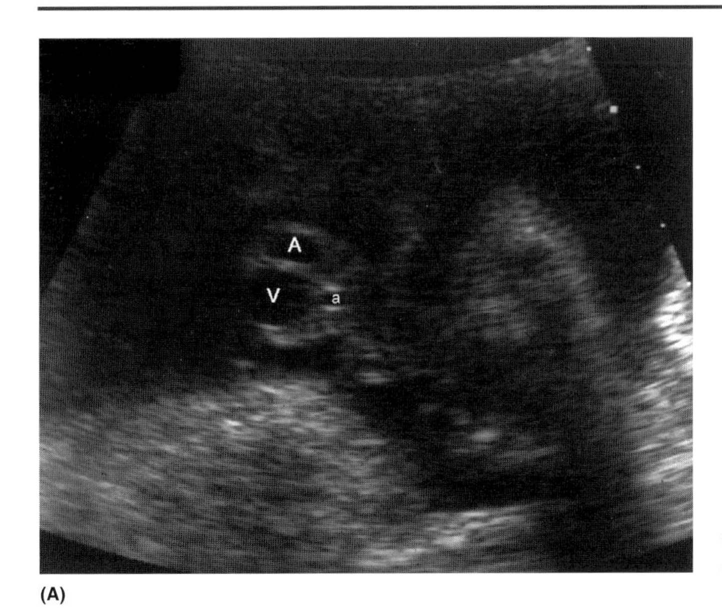

(A)

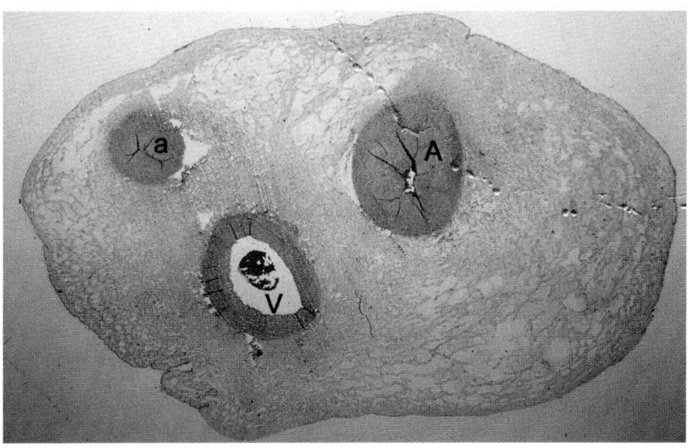

(B)

FIGURE 9 A AND B ■ Discordant umbilical arteries at 37 weeks (**A**) with pathologic correlation (**B**). The infant was normal. V, umbilical vein; a, hypoplastic artery; A, larger than usual umbilical artery.

An umbilical coiling index, obtained by dividing the total number of coils by the length of the cord in centimeters, has been devised for the delivered cord (34). The normal index among 100 consecutive live neonates was 0.21 (± 0.07 SD) coils per centimeter. This concept has been applied sonographically as well (38). Hypocoiled (below 10th percentile) and hypercoiled (above 90th percentile) umbilical cords have been associated with adverse outcomes including fetal demise and IUGR (35). Atypical uncoordinated, aperiodic coiling and springshaped "supercoiling" documented sonographically have been associated with fetal anomalies, including a case of Trisomy 18 (39).

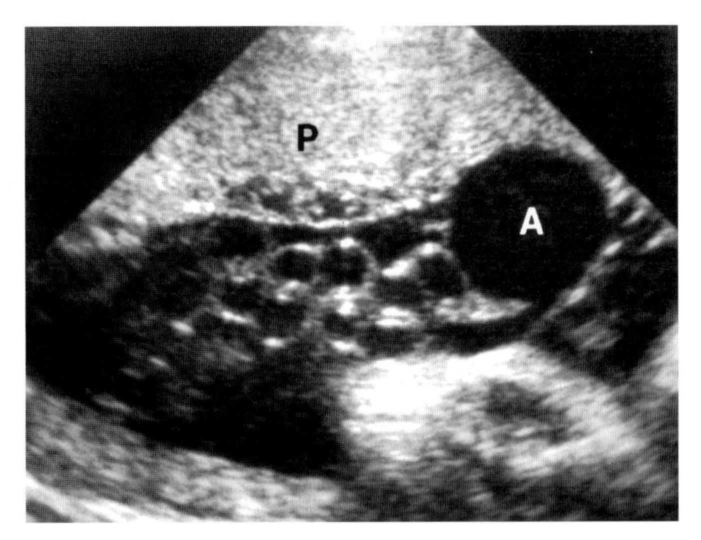

FIGURE 10 ■ Umbilical artery aneurysm (A) adjacent to the placental insertion of the cord in a 38-week pregnancy. The fetus had an endocardial cushion defect. Subsequent cordocentesis (avoiding the aneurysm) proved Trisomy 18. Fetal size and amniotic fluid volume were in the normal range, suggesting no impairment of fetal perfusion. (Doppler imaging was not available at the time of this pregnancy.) P, placenta. *Source*: Finberg H. Umbilical cord and amniotic membranes. In: McGahan J, Porto M (eds.) Diagnostic Obstetrical Ultrasound. Philadelphia, JB Lippincott, 1994.

Mechanical Abnormalities

Increased cord length may predispose the fetus to cord entanglements, including nuchal cord, to the development of true cord knots, to cord prolapse, and to cord compression events (15).

Nuchal Cord

Nuchal cord, wherein the umbilical cord completely encircles the fetal neck one or more times, is a common finding at antenatal sonography and at delivery. The incidence at delivery ranges from 23% to 35% (40–43), with the incidence of multiple loops approximately 2.5% to 8%. Color flow Doppler has greatly facilitated the diagnosis of nuchal cord in utero (Fig. 15). At antenatal sonography, the overall incidence is approximately 23% (41,42,44). Several studies have shown that a nuchal cord, whether diagnosed in utero or at term, is not associated with adverse perinatal outcome (40,41,43–46). There is no increased risk of stillbirth with a nuchal cord (42). No difference in operative delivery rates, five-minute Apgar scores, or postnatal complications were reported with single nuchal cords despite an increased incidence in the nuchal cord group of fetal bradycardia and variable decelerations. The presence of multiple nuchal cord loops has been associated with a significantly greater likelihood of meconium in the amniotic fluid, abnormal heart rate patterns in advanced labor, requirement for operative delivery, and mild acidosis at birth (44). However, no increase in adverse neonatal outcome was found. Nuchal cords detected in utero are usually transient. However, a nuchal cord that has persisted beyond four weeks in late gestation may increase the risk of fetal compromise (41,43).

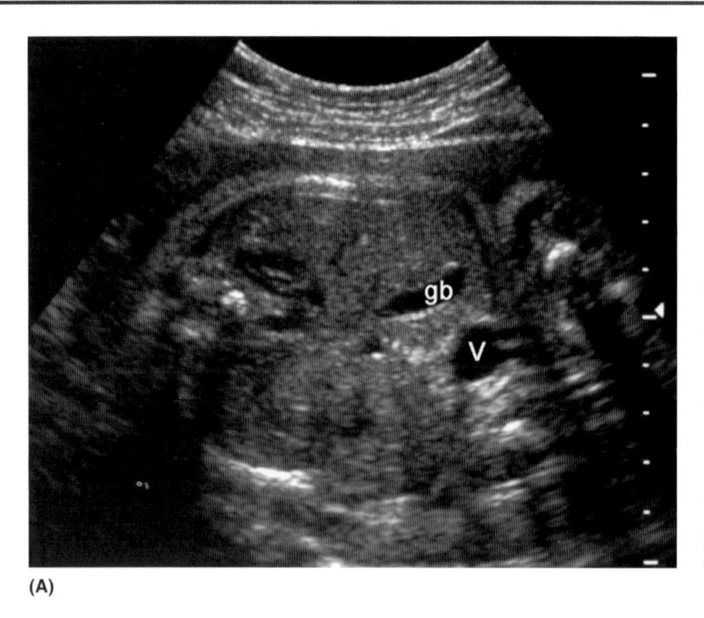

(A)

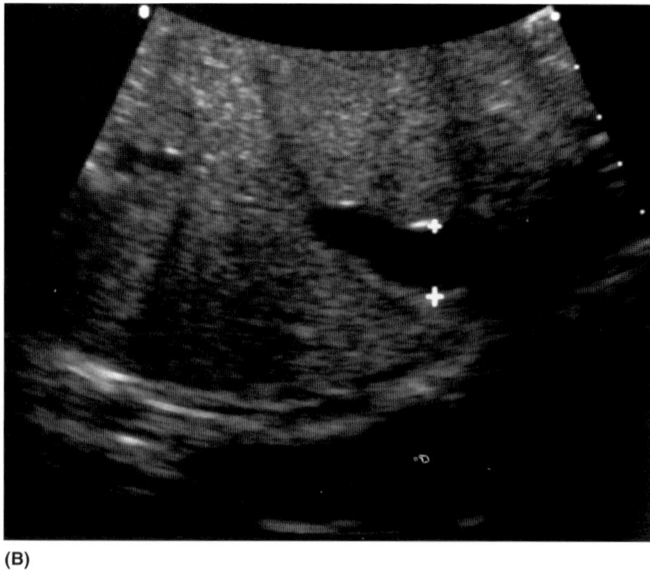

(B)

FIGURE 11 A AND B ■ Umbilical vein varix. (**A**) Intra-abdominal umbilical vein varix (V) at 28 weeks, with diameter of 12 mm. gb, gallbladder. (**B**) Intra-abdominal umbilical vein varix (*cursors*) at 38 weeks (diameter, 13 mm). In both cases, the infants were normal.

Knots

True umbilical cord knots are found in up to 0.5% of all deliveries (4). They are more common in the case of unusually long cords, polyhydramnios, or monoamniotic twins. They occur in equal numbers in abortions and in term pregnancies, implying that they occur relatively early in pregnancy (7). A knot can be recognized by the characteristic looping of cord segments in a cloverleaf pattern (Fig. 16). True knots have a perinatal mortality rate of 8% to 11%, due to obstruction of the fetal circulation from either an acutely tightened knot or a longstanding knot (4).

Torsion and Stricture

Umbilical cord strictures are rare. These are usually short, well-defined, and at the fetal end. Fibrosis and absence of Wharton's jelly are found at delivery. Recurrence of umbilical cord stricture in the same family, leading to fetal demise, has been reported (47). Strictures are often complicated by torsion (4). Umbilical cord torsion has been described at sonography as a decreased distance between cord helixes (48,49). Cord torsion usually occurs at the fetal end. This may be related to stricture, or to the fact that this area has less

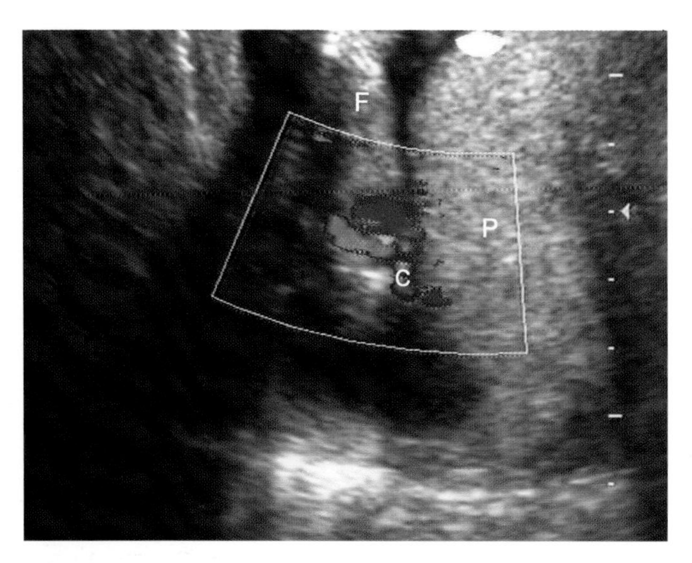

FIGURE 12 ■ Body stalk anomaly at 14 weeks. Severe scoliosis was demonstrated, along with omphalocele, clubfoot, and an extremely short umbilical cord (C) tethering the fetus (F) to the placenta (P).

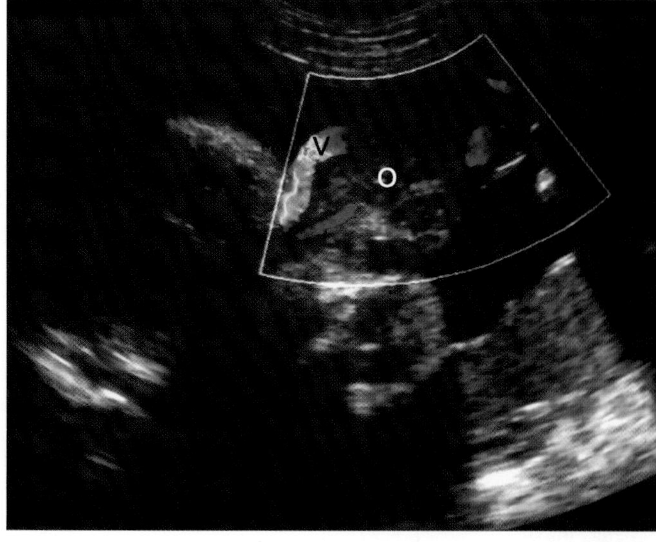

FIGURE 13 ■ Omphalocele (O) containing liver is demonstrated. Note umbilical vessels (V) deviating around omphalocele.

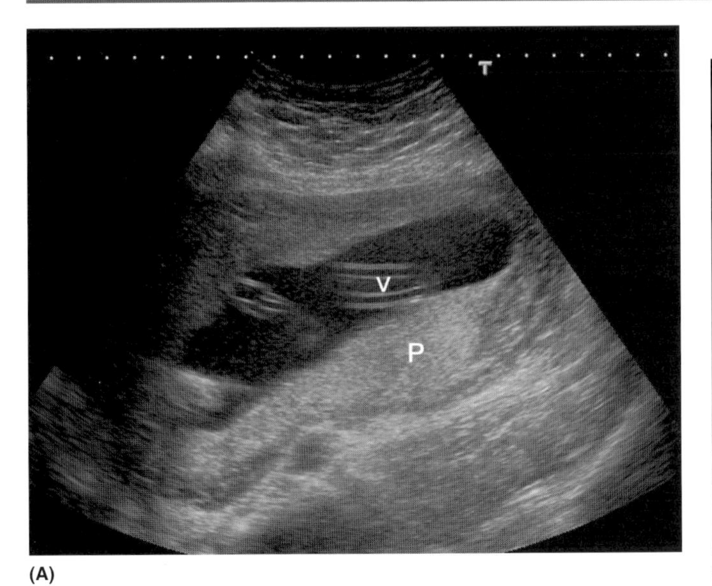

(A)

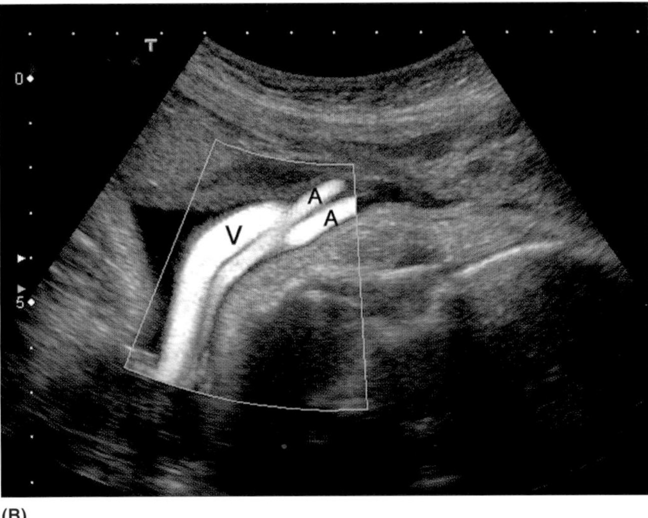

(B)

FIGURE 14 A AND B ■ Uncoiled (**A**) and hypocoiled (**B**) umbilical cords. A, artery; V, vein; P, placenta.

turgor due to decreased Wharton's jelly. It is thought that cord compression due to torsion can result in fetal arrhythmias, causing cardiac failure and nonimmune hydrops (49).

Vasa Previa

The presence of fetal vessels from the placenta crossing the internal os of the cervix is referred to as vasa previa (Fig. 17). This condition may be due either to a succenturiate placental lobe or to a velamentous cord insertion. Compression of unsupported cord vessels during labor may lead to fetal bradycardia, asphyxia, and death (50,51). Rupture of vasa previa leads to fetal exsanguination. Thus the diagnosis of vasa previa is an indication for cesarian section.

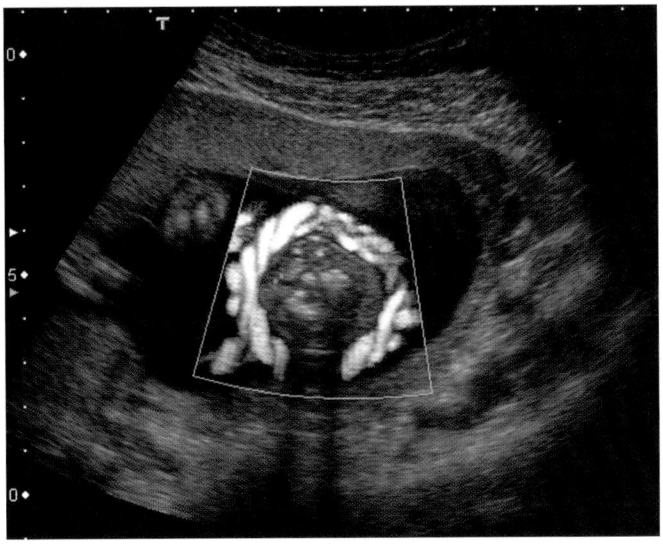

The diagnosis of vasa previa should be considered in any case of marginal or velamentous cord insertion in the lower uterine segment, and whenever a succenturiate lobe is present that could connect to the main portion of the placenta by vessels passing near the cervix. Vasa previa is visible at gray-scale sonography. With the advent of color flow Doppler, the ability to diagnose vasa previa has improved significantly; the combination of transvaginal ultrasound and color flow Doppler is most effective in making this diagnosis.

FIGURE 15 ■ Nuchal cord at 20 weeks.

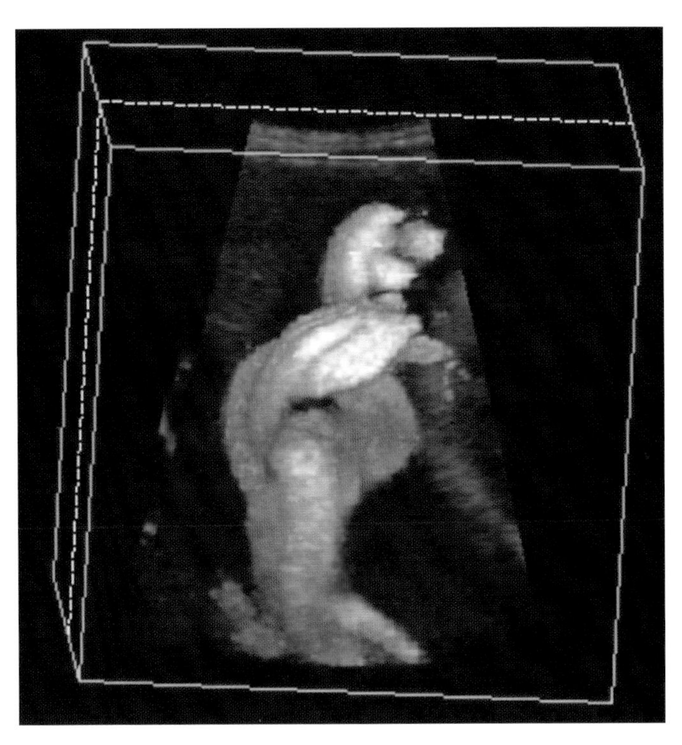

FIGURE 16 ■ Umbilical cord knot. *Source*: Courtesy of Toshiba America Medical Systems.

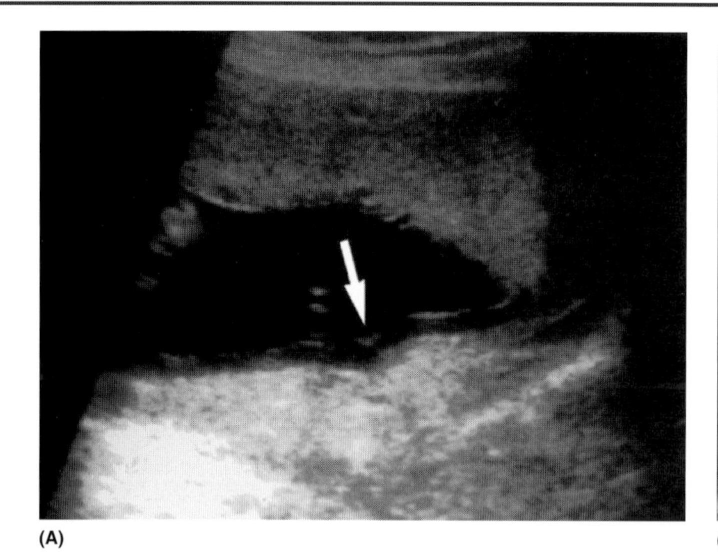

(A)

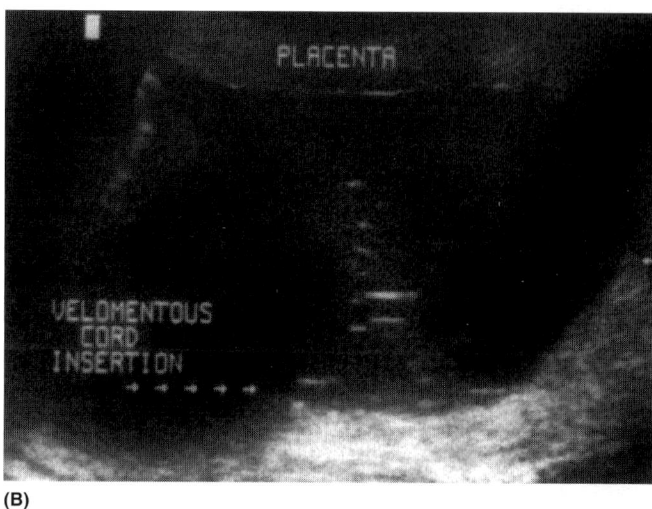

(B)

FIGURE 17 A AND B ■ Vasa previa due to velamentous umbilical cord insertion discovered incidentally on a routine 20-week sonogram. (**A**) Prominent chorionic artery, present at the caudal margin of the anterior placenta, extends across the cervix in close proximity to the internal os (*arrow*). (**B**) The aberrant chorionic vessels continue onto the posterior uterine wall to the point of origin of the umbilical cord. Findings were confirmed at elective cesarean section. *Source*: Finberg HJ. The role of sonography in obstetrical intensive care setting. In: Foley M (ed.) Obstetric Intensive Care: A Simplified Practical Approach. Philadelphia, WB Saunders, 1996.

Abnormalities of Cord Insertion

Battledore Placenta

The umbilical cord commonly inserts on the fetal surface of the placenta in an eccentric location (Fig. 18) (4). A markedly eccentric, marginal insertion of the umbilical cord is called "battledore placenta" (Fig. 19). This occurs in approximately 6% of placentas and is of no clinical significance (4,52).

Velamentous Cord Insertion

Velamentous insertion of the umbilical cord, in which the cord vessels attach to the membranes at a distance from the placenta (Fig. 20), occurs in approximately 2% of cases (4). The incidence of velamentous insertion is increased in multiple pregnancies. Velamentous insertion occurs with bilobate placentas; the umbilical cord is usually inserted on the membranes between the two lobes.

With velamentous insertion, the cord vessels are not protected by Wharton's jelly and are at risk of rupture during labor and delivery. Serious hemorrhage from velamentous vessels during labor and delivery occurs in 2% of cases of velamentous insertion (4). Bleeding from velamentous vessels may also occur in the antepartum period.

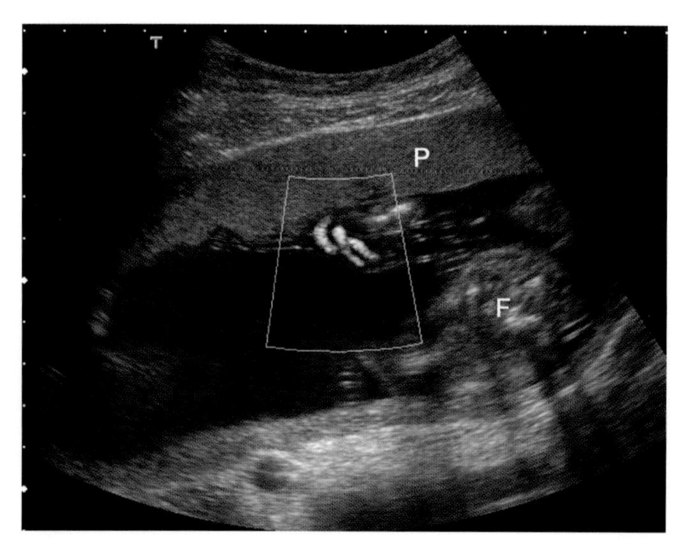

FIGURE 18 ■ Eccentric placental cord insertion at 21 weeks. P, placenta; F, fetus.

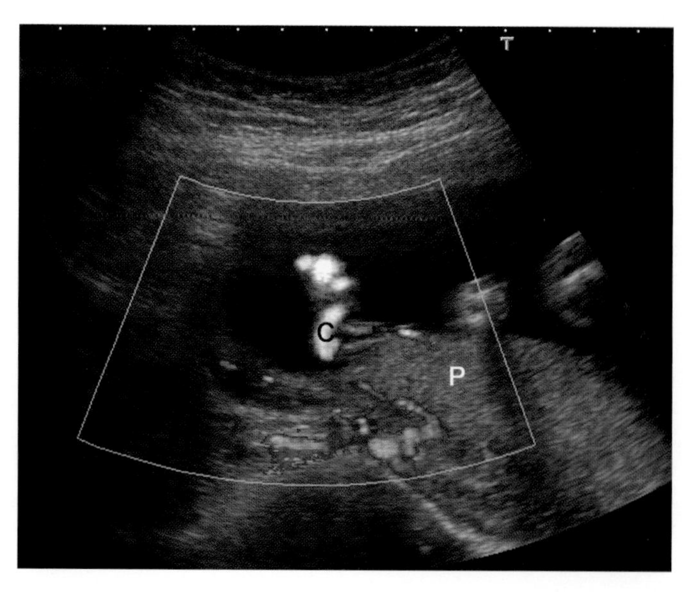

FIGURE 19 ■ Battledore placenta. Marginal insertion (C) at 22 weeks. P, placenta.

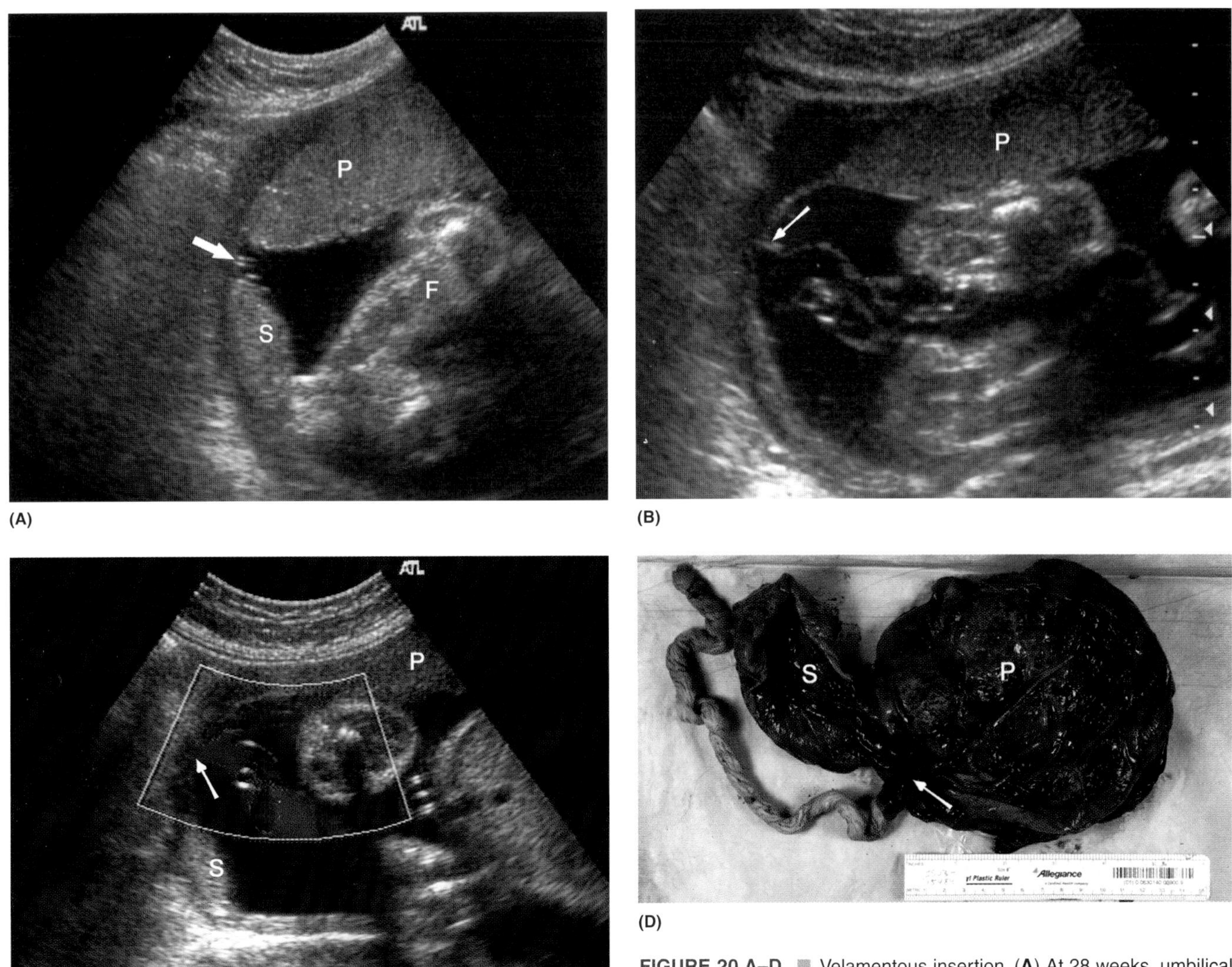

FIGURE 20 A–D ■ Velamentous insertion. (**A**) At 28 weeks, umbilical cord (*arrow*) is seen between placenta (P) and succenturiate lobe (S). F, fetus. (**B,C**) Followup scans confirm membranous insertion of umbilical cord (*arrow*). (**D**) Gross specimen shows umbilical cord insertion (*arrow*) into membranes between the smaller succenturiate lobe (S) and the main placenta (P).

Cord Masses

Umbilical cord masses usually occur at the fetal insertion site. They may be divided into the following categories: vascular lesions, cystic lesions, and solid masses.

Vascular Lesions

Hematoma ■ Iatrogenic hematomas of the umbilical cord may result from amniocentesis or cordocentesis (Fig. 21). Spontaneous hematomas of the umbilical cord, consisting of extravasation of blood into Wharton's jelly, may range from 4.0 to 40.0 cm in length. They are rare and of uncertain clinical significance (4).

Thrombosis ■ Thrombosis of umbilical vessels, an uncommon entity, may be associated with cord compression, torsion, stricture, knots, abnormal coiling, or hematoma. Most (71%) involve the umbilical vein; 18% involve both arteries and vein; and isolated umbilical artery thrombosis occurs in 11% (7). Fetal umbilical vessel thrombosis is associated with a high incidence of fetal morbidity and mortality, particularly if both umbilical arteries are occluded (7).

Cysts ■ Pseudocysts of the umbilical cord result from mucoid degeneration of Wharton's jelly (Fig. 22). These contain clear mucoid material, and lack an epithelial lining.

True umbilical cord cysts may develop from either omphalomesenteric or allantoic remnants, distinguishable only by histologic examination, or rarely from amniotic inclusions. Allantoic or omphalomesenteric duct cysts tend to be close to the fetal end of the cord. Most are small, 1 to 2 cm in diameter (Fig. 23), and of no clinical significance (4).

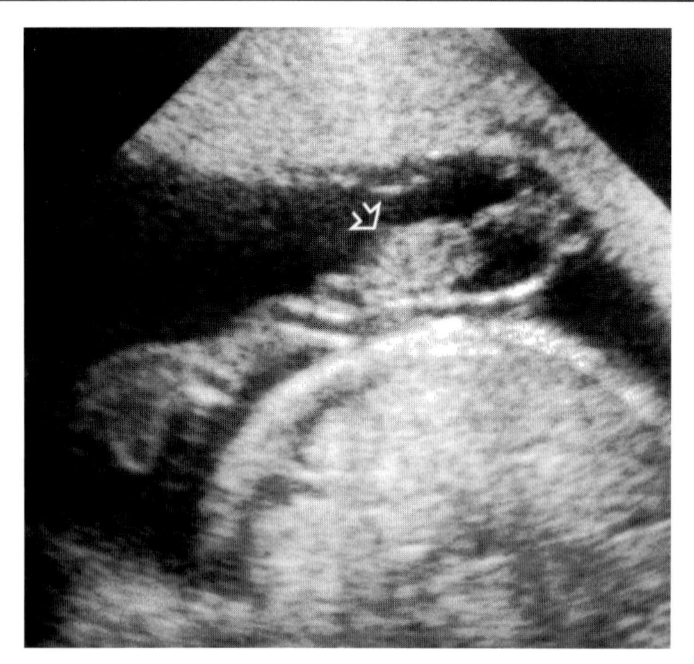

FIGURE 21 ■ Thrombus of the umbilical vein: echogenic thrombus (*arrow*) occurring as an immediate complication of a cordocentesis. A fetal transfusion was performed for erythroblastosis fetalis resulting from Rh incompatibility at about 32 weeks. The thrombus developed during the transfusion, with constriction of the umbilical vein between the clotted segment and the fetus. The fetus developed profound bradycardia and could not be resuscitated despite emergency cesarean section. *Source*: Finberg H. Umbilical cord and amniotic membranes. In: McGahan J, Porto M (eds.) Diagnostic Obstetrical Ultrasound. Philadelphia, JB Lippincott, 1994.

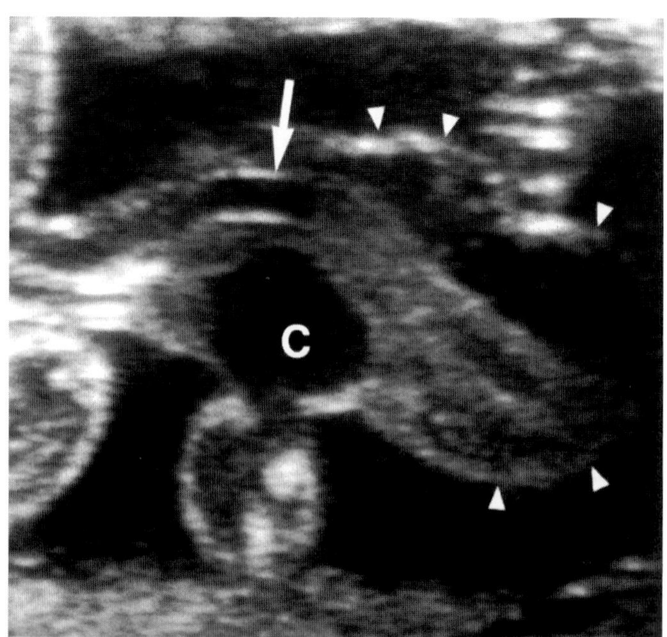

FIGURE 23 ■ A small omphalomesenteric cyst (C) near the fetal origin of the umbilical cord at 22 weeks is associated with marked watery swelling of the cord and wide splaying of the vessels. The umbilical vein (*arrow*) is deviated by the cyst; *arrowheads* show the edge of the cord. *Source*: Finberg H. Umbilical cord and amniotic membranes. In: McGahan J, Porto M (eds.) Diagnostic Obstetrical Ultrasound. Philadelphia, JB Lippincott, 1994.

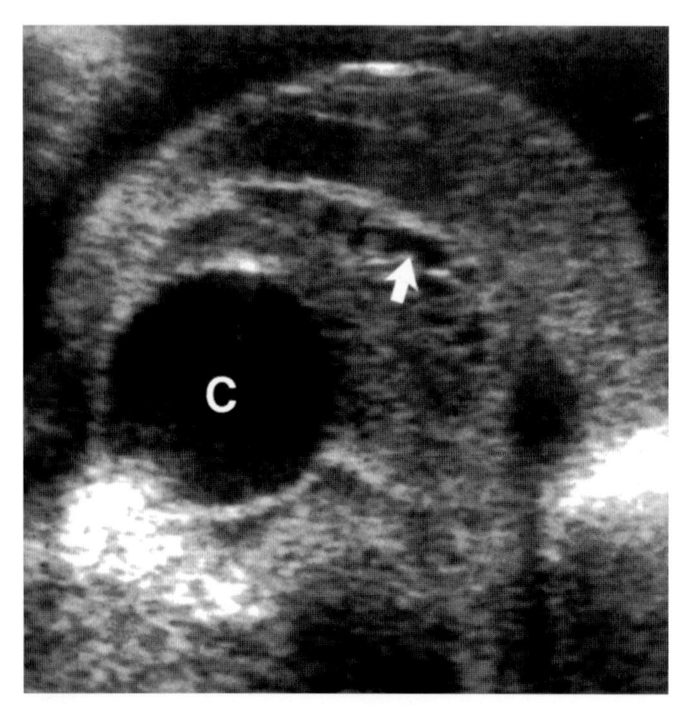

FIGURE 22 ■ A focal area of increased Wharton's jelly is seen adjacent to a 2 cm simple cyst (C) at the fetal end of the umbilical cord. Small beaded locules of fluid, cystic degeneration (*arrow*), are seen within the cord adjacent to the vein. Pregnancy outcome was normal. *Source*: Finberg H. Umbilical cord and amniotic membranes. In: McGahan J, Porto M (eds.) Diagnostic Obstetrical Ultrasound. Philadelphia, JB Lippincott, 1994.

Theoretically, larger cysts may cause vascular compromise. Cord cysts are more prevalent in the first trimester and usually resolve by the second trimester (53,54). Persistent cysts and pseudocysts have been associated with other anomalies, and with aneuploidy. (55,56).

Diffuse Cord Swelling ■ Diffuse cord edema rarely occurs as an isolated finding (Fig. 24). It is usually a consequence of fetal hydrops of any origin. Diffuse cord swelling may occur with a patent urachus, with extravasation of urine into the cord.

Solid Tumors ■ Hemangioma (angiomyxoma) (57) and teratoma are the two known primary tumors of the umbilical cord. Both appear as well-circumscribed echogenic masses (Fig. 25). As with placental chorioangioma, umbilical cord hemangioma may be associated with fetal hydrops from high-output congestive heart failure. High levels of α-fetoprotein have been reported in maternal serum and amniotic fluid (58,59).

Omphalocele and Gastroschisis ■ Omphalocele and gastroschisis result from abnormal closure of the anterior abdominal wall. With omphalocele, abdominal contents protrude into the base of the umbilical cord. Omphalocele carries a high risk of other anomalies. With gastroschisis, bowel and other intraabdominal contents protrude into the amniotic fluid through a right paramedian abdominal

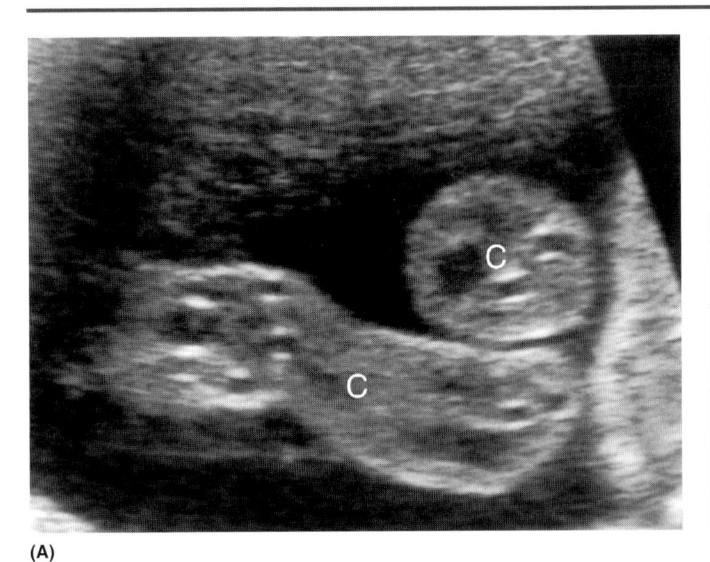

(A)

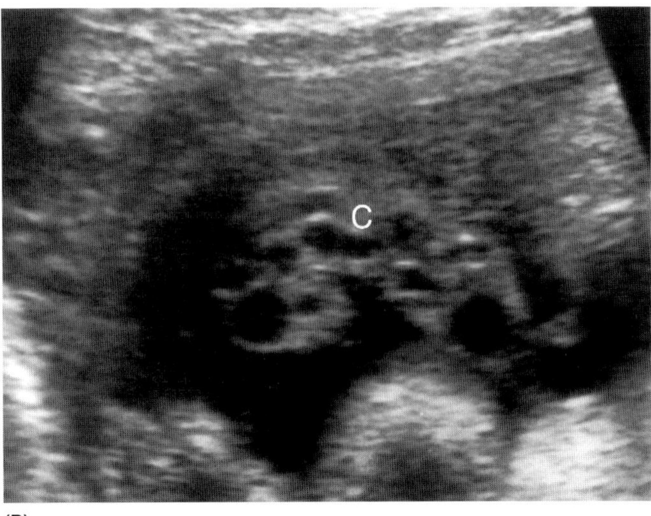

(B)

FIGURE 24 A AND B ■ (**A**) Idiopathic diffuse swelling of the umbilical cord (C), measuring up to 2.5 cm in diameter at 25 weeks, was not associated with fetal hydrops or other anomaly. (**B**) Swelling gradually regressed, and the cord (C) appeared normal by sonography at 34 weeks and by inspection at delivery of a normal infant two weeks later. *Source*: Finberg H. Umbilical cord and amniotic membranes. In: McGahan J, Porto M (eds.) Diagnostic Obstetrical Ultrasound. Philadelphia, JB Lippincott, 1994.

wall defect. The umbilical cord inserts normally just to the left of the defect.

A small omphalocele with herniated intestine can simulate a solid mass of the cord (Fig. 26). Focal thickening of Wharton's jelly may occur with omphaloceles, particularly when they are large (Fig. 27).

FIGURE 25 ■ Hemangioma of the umbilical cord. The echogenic lesion (H), at the fetal end of the cord, measured barely 2 cm in diameter when first imaged at 18 weeks. It enlarged to 4 cm by the time of this study 1 month later and then stabilized. An omphalomesenteric cyst (not shown) distal to the hemangioma grew from barely detectable at 18 weeks to 10 cm in diameter at vaginal delivery of a normal infant. Intensive monitoring throughout the pregnancy showed no evidence of fetal compromise. *Source*: Finberg H. Umbilical cord and amniotic membranes. In: McGahan J, Porto M (eds.) Diagnostic Obstetrical Ultrasound. Philadelphia, JB Lippincott, 1994.

CHORIOAMNIOTIC MEMBRANES

At about five weeks' menstrual age, the chorionic villi opposite the implantation pole of the gestational sac begin to regress, obliterating the intervillous space in that location. This leaves a smooth surface, the chorion leave. The remaining villi continue to proliferate, forming the placenta. The chorion surrounds the extraembryonic coelom within which the embryo, with its primary yolk sac and amniotic sac, develops. The amniotic cavity forms from cytotrophoblast adjacent to the dorsal aspect of the bilaminar embryonic disk, and eventually surrounds the fetus and enlarges to the size of the chorionic sac. The amnion is visualized at sonography as a thin membrane surrounding the fetus (Fig. 28). At 14 to 16 weeks, the amnion becomes adherent to the chorion, obliterating the intervening space (extraembryonic coelom).

The chorioamniotic membranes may be visualized at sonography in several situations (Table 2).

Chorioamniotic Separation

Persistent separation of the chorion and amnion after 16 weeks is rare (Fig. 29). It may be due to primary nonfusion of the membranes, or it may occur in varying degrees secondary to amniocentesis (60–62) or hysterotomy for fetal surgery (63).

Diagnosis depends on sonographic visualization of the thin amnionic membrane, which is usually seen only when it lies perpendicular to the ultrasound beam. The membrane may be visualized to the base of the umbilical

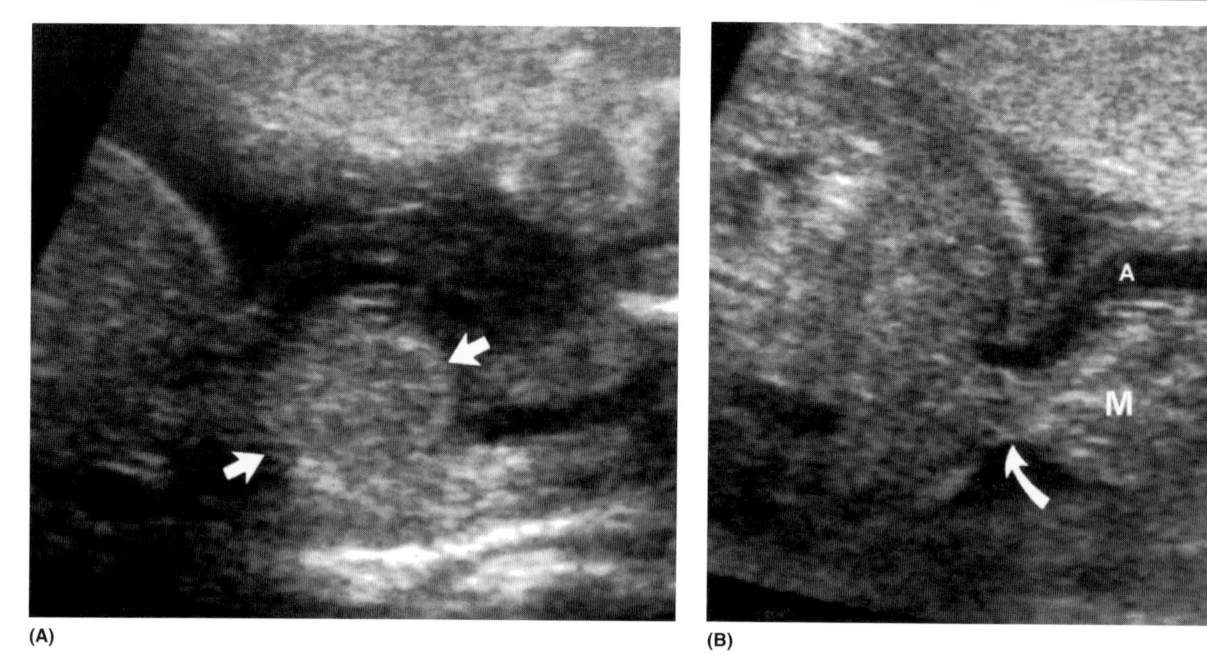

FIGURE 26 A AND B ■ Pitfall: small omphalocele simulating a solid cord mass. (**A**) An echogenic, 2 cm lesion (*arrows*) is seen within the umbilical cord at the fetal end. This could represent a teratoma or hemangioma. (**B**) A view axial to the cord origin shows a stalklike zone of similar echogenicity (*arrow*), demonstrating continuity between the cord mass (M) and the fetal intestine. An umbilical artery (A) deviating around the omphalocele is also shown. *Source*: Finberg H. Umbilical cord and amniotic membranes. In: McGahan J, Porto M (eds.) Diagnostic Obstetrical Ultrasound. Philadelphia, JB Lippincott, 1994.

cord, at which it becomes adherent. Nonfusion of the chorion and amnion has been associated with aneuploidy, notably Trisomy 21, Trisomies 13 and 18, and Turner syndrome, and with other abnormalities (61,64). Chorioamniotic separation after instrumentation has been associated with increased morbidity and mortality due to amniotic band formation (62,63).

Membranes in Multiple Gestations

The subject of multiple gestations is discussed in detail in Chapter 49. In the context of membranes that may be found in the amniotic space, identification of a dividing membrane is important in excluding the presence of monoamniotic twins, a form of twin pregnancy with a mortality rate approaching 50%. Recognition of (*i*) opposite-gender

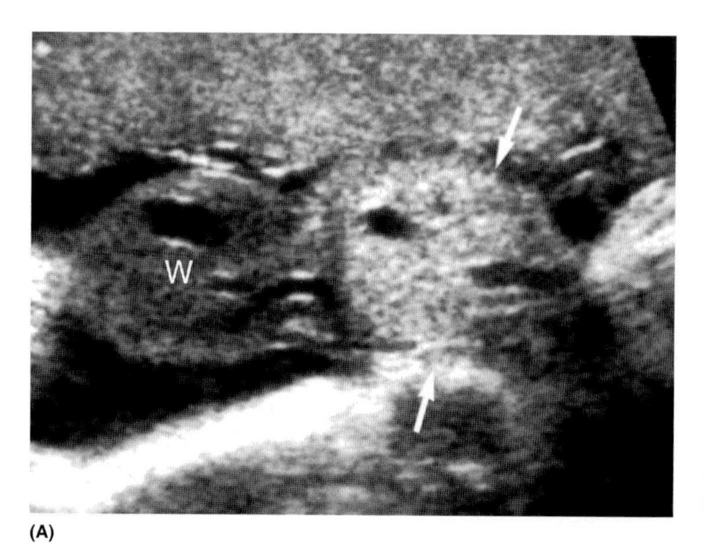

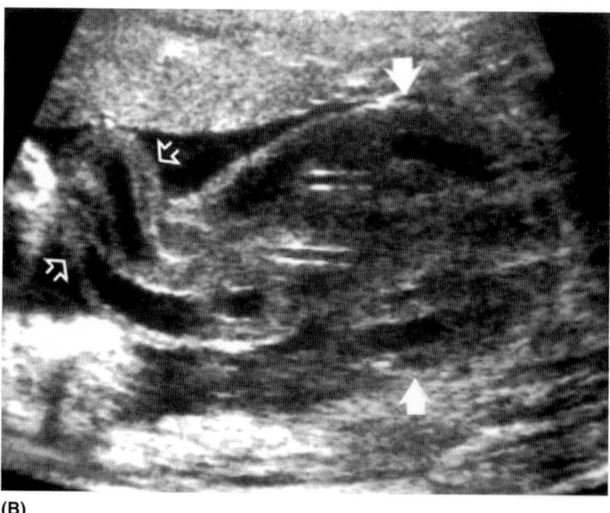

FIGURE 27 A AND B ■ (**A**) Focal thickening of Wharton's jelly (W) is seen, enlarging the first several centimeters of the umbilical cord to a 4 cm diameter at its origin along the anteroinferior surface of a larger fetal omphalocele (*arrows*). (**B**) The three cord vessels are concentrically splayed in this segment (*closed arrows*), with the cord resuming a normal size and configuration more distally (*open arrows*). *Source*: Finberg H. Umbilical cord and amniotic membranes. In: McGahan J, Porto M (eds.) Diagnostic Obstetrical Ultrasound. Philadelphia, JB Lippincott, 1994.

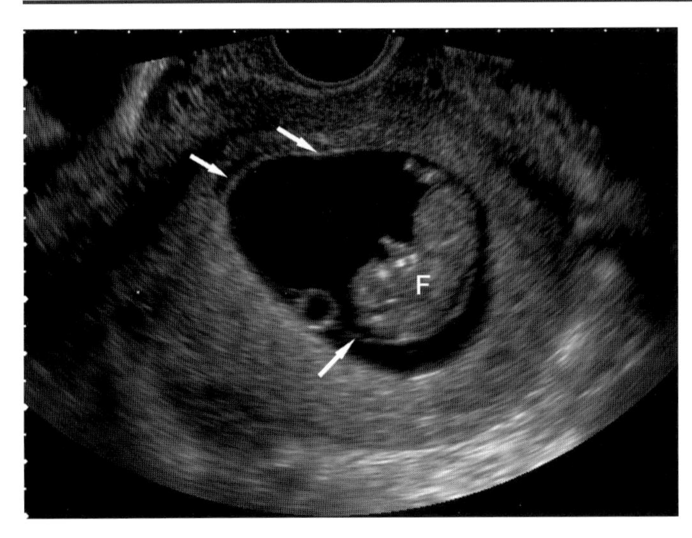

FIGURE 28 ■ Amniotic membrane (*arrows*) surrounds a 10-week fetus (F).

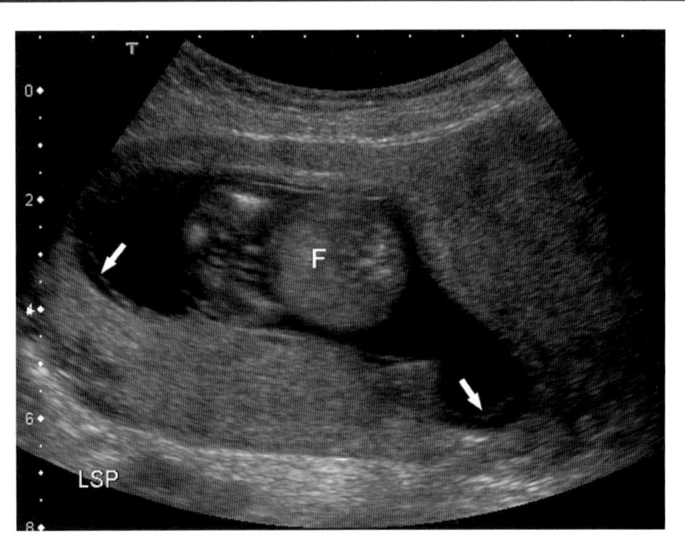

FIGURE 29 ■ At 15 weeks, the amnion (*arrows*) and chorion are not yet fused. F, fetus.

twins, (*ii*) two separate placentas, or (*iii*) a chorionic peak of placental tissue extending into the intertwin membrane at its junction with a single, fused placental zone confirms a dichorionic diamniotic twin pregnancy (65). A relatively thick intertwin membrane, measuring at least 2 mm in thickness, is seen with diamniotic dichorionic twins (66). Monochorionic diamniotic twins have a thin, hard-to-image membrane less than 2 mm thick.

TABLE 2 ■ Chorioamniotic Membranes

Diagnosis	Findings
Chorioamniotic separation	
Before 16 weeks	Unfused amnion is a normal finding
After 16 weeks	Unfused membrane parallels contour of gestational cavity May occur following amniocentesis
Intertwin dividing membrane	
Dichorionic, diamniotic	Two separate placentas or fused placentas
	Membrane > 2 mm thick
	Chorionic peak
Monochorionic, diamniotic	One placenta Thinner, wispy membrane < 2 mm Delicate T intersection of membrane and placenta
Extra-amniotic	No fetal deformity or tethering pregnancy
Early amnion rupture sequence (amniotic band syndrome)	Asymmetric fetal deformities: constrictions, amputations, abdominal or cranial defects Amniotic bands may be visible Fetus may be tethered or restricted
Uterine synechiae (amniotic sheets)	Broad-based membrane traversing the gestational cavity
Subchorionic Hematoma	Lenticular shaped collection which decreases in size on followup examinations. May be remote from placenta.

Placenta Extrachorialis

An extrachorial placenta refers to a placenta in which the fetal membranes do not extend to the edge of the placenta (Fig. 30). In this situation, the chorionic plate is smaller than the basal plate. The fetal membranes attach to the chorionic plate in a ring configuration that is either flat (circummarginate) or folded (circumvallate) (Fig. 31). This may involve part or all of the circumference of the

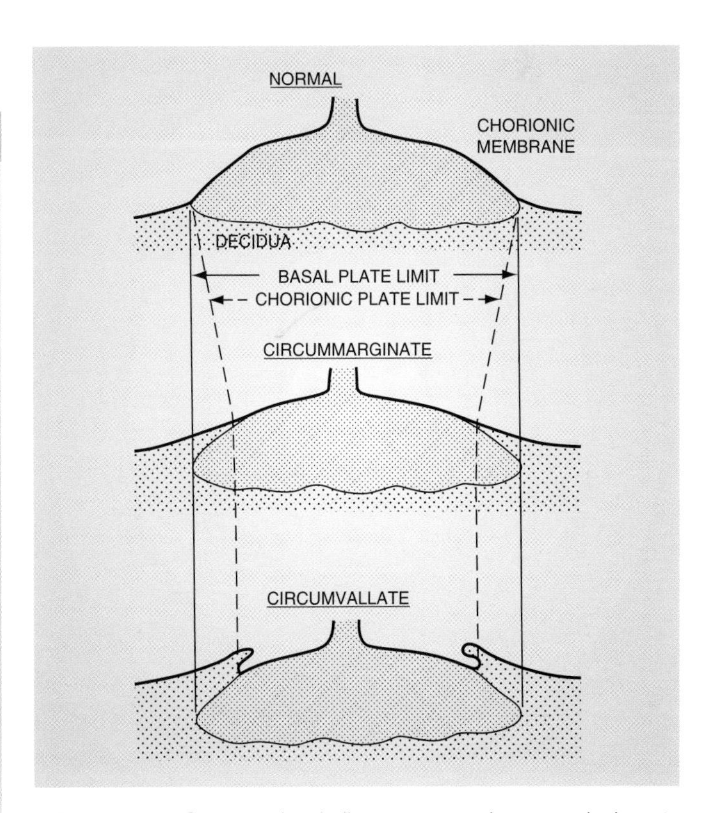

FIGURE 30 ■ Cross-sectional diagram comparing normal placenta with extrachorial placentas. With circummarginate placenta, the transition from membranous to villous chorion occurs at a distance from the edge of the placenta. The circumvallate placenta has a fold in the chorioamniotic membrane. *Source*: Spirt BA, Kagan EH. Sonography of the placenta. Semin Ultrasound 1980; 1:293–310.

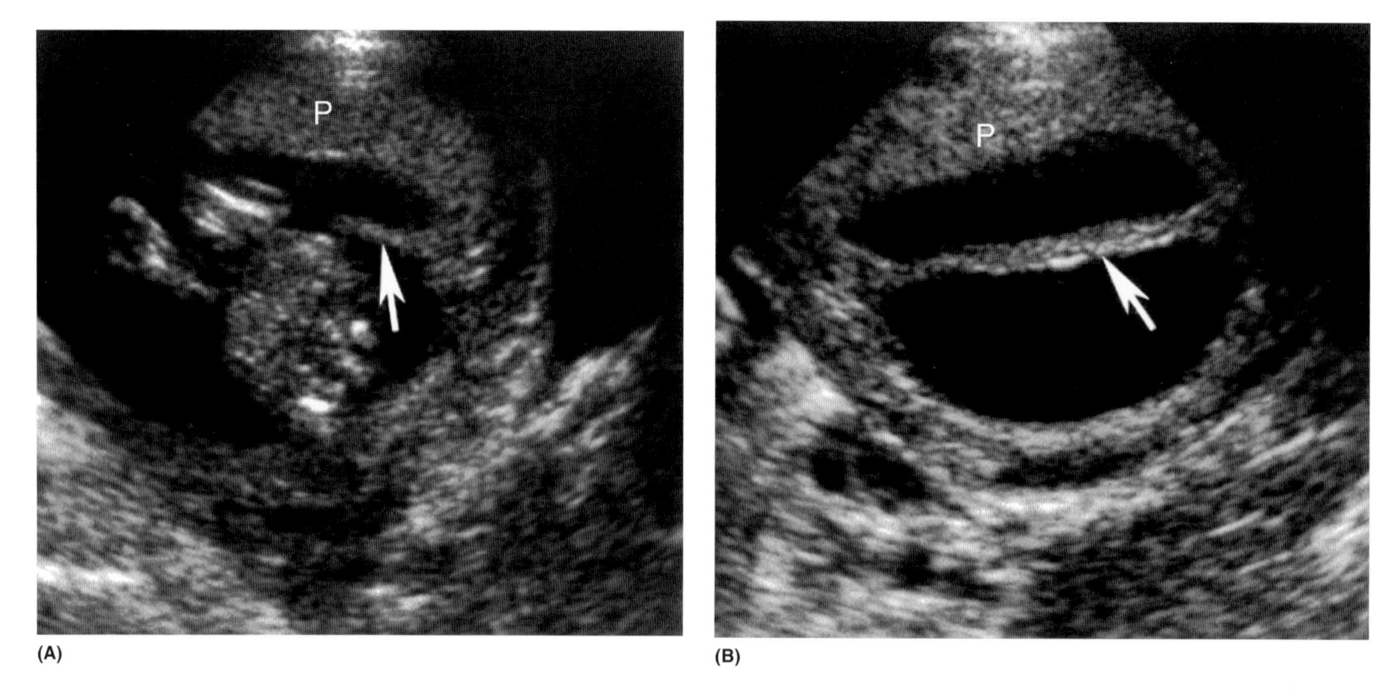

FIGURE 31 A AND B ■ Circumvallate placenta. (**A**) Transverse and (**B**) sagittal scans at 14 weeks show thick membranous fold (*arrows*) involving a portion of the placenta. A partial circumvallate placenta was confirmed at delivery. *Source*: Spirt BA, Gordon LP. Sonographic evaluation of the placenta. In: Rumack CM, Wilson SR, Charboneau JW (eds.) Diagnostic Ultrasound, 2 ed. St. Louis: Mosby Yearbook, 1998, pp. 1337–1358.

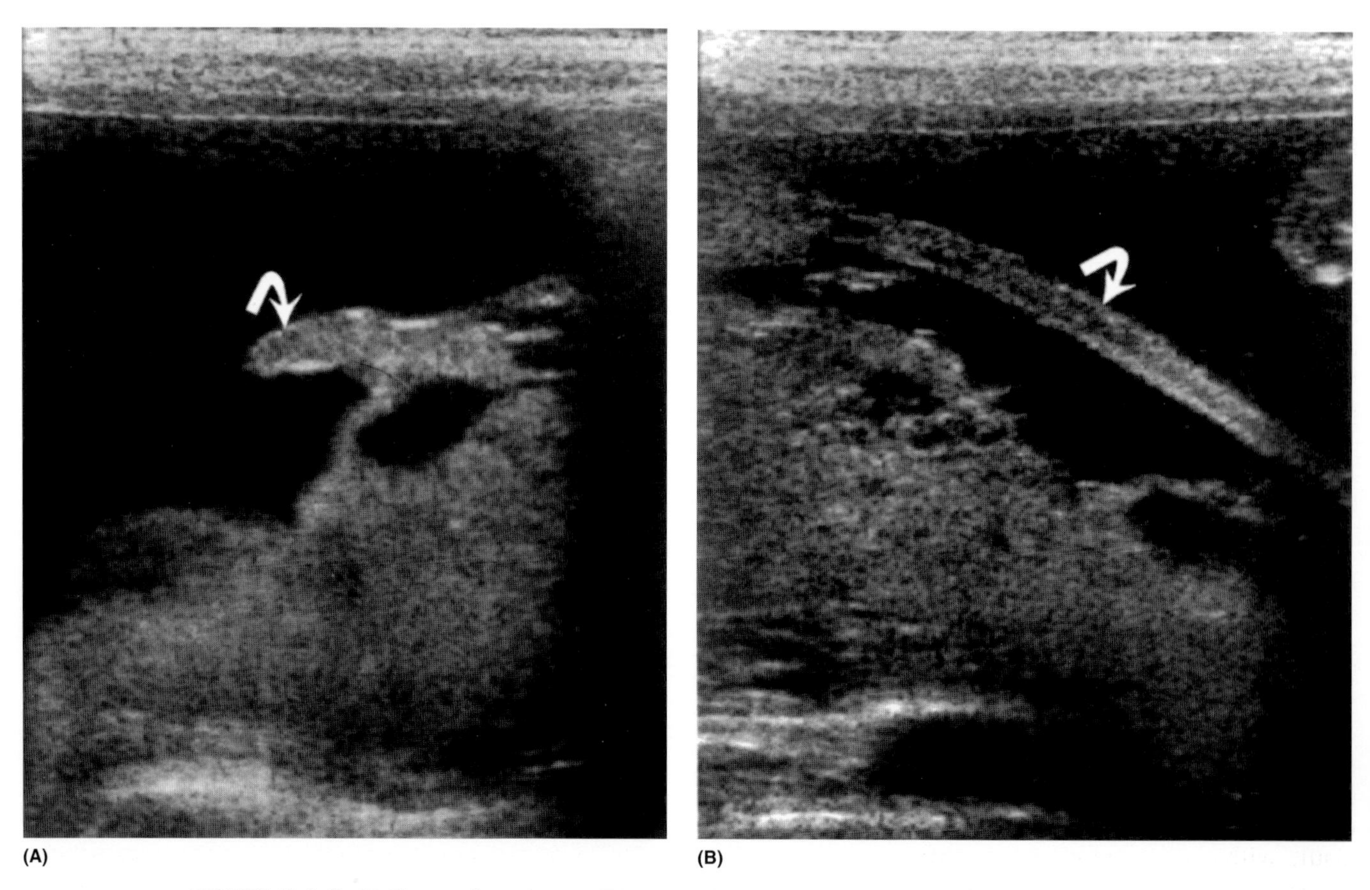

FIGURE 32 A–D ■ Circumvallate placenta. Transverse (**A**) and oblique (**B**) scans at 26.5 weeks show thick membrane fold (*arrows*) involving a portion of the placenta. (*Continued*)

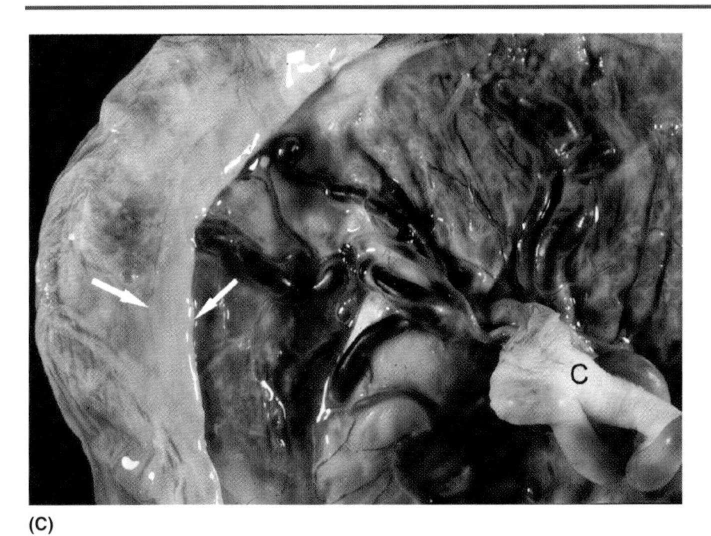

(C)

(D)

FIGURE 32 A–D ■ (*Continued*) (**C,D**) A partial circumvallate placenta was confirmed at delivery. *Arrows*, membrane fold; C, umbilical cord. *Source*: Spirt BA, Gordon LP: Imaging of the placenta. In: Taveras JM, Ferrucci JT (eds.) Radiology: Diagnosis/Imaging/Intervention. Philadelphia, JB Lippincott Company 1992, pp. 1–18.

placenta. At sonography, a thick membrane fold may be seen at the edge of the placenta (Fig. 32). The circummarginate placenta and the partial circumvallate placenta have no clinical significance. The total circumvallate placenta has been associated with threatened abortion, premature labor, and marginal hemorrhage (4,67–69).

Extra-Amniotic Pregnancy

This is a rare condition wherein the amnion ruptures early in the pregnancy, and the fetus survives in the intact chorionic cavity (extraembryoic coelom) (Fig. 33). The amniotic band syndrome has not been reported in this condition; this is attributed to the occurrence of extra-amniotic pregnancy beyond the stage when fetal deformities of amniotic band syndrome are believed to take place (70).

Amniotic Band Syndrome or Early Amnion Rupture Sequence

Amniotic band syndrome refers to asymmetric anomalies affecting the limbs, craniofacial area, and internal organs, caused by bands that may deform and/or constrict entrapped fetal structures. Resultant anomalies include ring constriction, lymphedema, limb amputations, facial clefts, open defects of the chest or abdomen wall, and cranial deformities. The pathogenesis is uncertain (71,72); the bands are thought to result from disruption of the amnion during fetal development. Differences in severity are thought to be related to the timing of the insult, with early amniotic rupture (from 6 to 8.5 weeks' menstrual age) resulting in the most severe defects. The defects may range from a single, subtle constriction ring of a finger detected postnatally to devastating, lethal body clefts and amputations. Craniofacial abnormalities occur in up to one-third of cases (Figs. 34 and 35) (73). Fetoscopic release of amniotic bands has been performed in cases of limb constriction diagnosed at antenatal ultrasound (74,75).

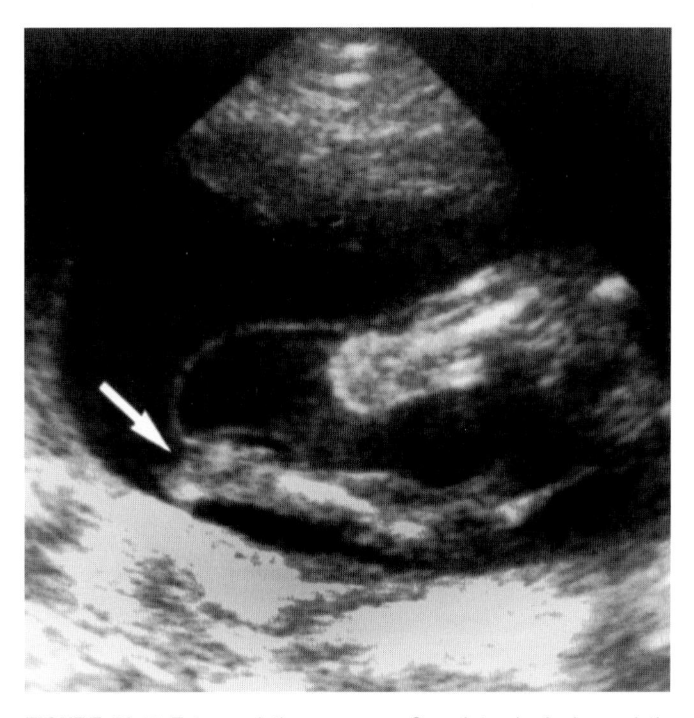

FIGURE 33 ■ Extraamniotic pregnancy. Complete chorionic-amniotic separation occurred in this 19-week, clinically uncomplicated pregnancy. Amniotic disruption is present, with one fetal leg seen outside the amnion (*arrow*). On subsequent scans, the fetus was fully extra-amniotic. A normal infant was delivered near term. *Source*: Finberg H. Umbilical cord and amniotic membranes. In: McGahan J, Porto M (eds.) Diagnostic Obstetrical Ultrasound. Philadelphia, JB Lippincott, 1994.

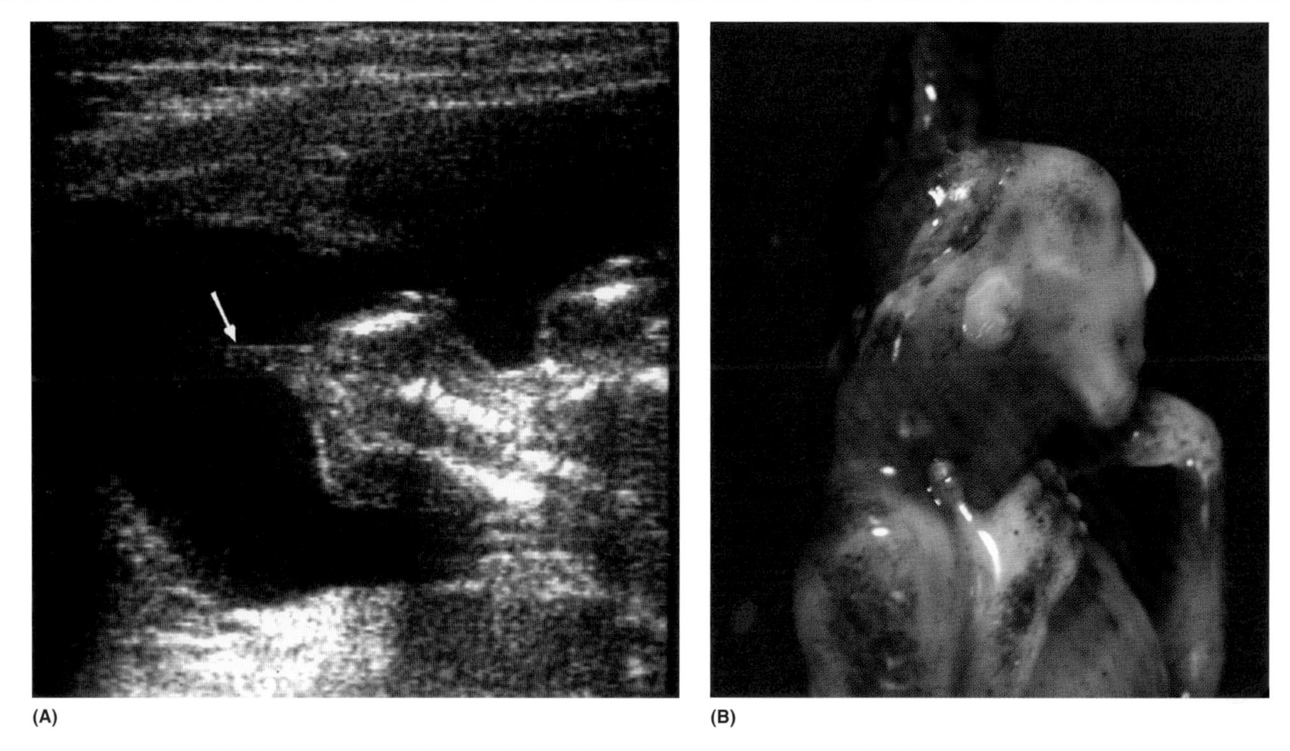

(A) (B)

FIGURE 34 A AND B ■ (**A**) Amniotic band (*arrow*) causes cranial deformity (**B**) similar to anencephaly. *Source*: (**A**) Courtesy of Dr. Robert Silverman, Perinatal Center, Syracuse, New York.

Uterine Synechia (Amniotic Sheets)

Amniotic sheets (uterine synechiae covered by a layer of amnion and chorion) occur in up to 0.5% of pregnancies (76,77). They usually develop as a consequence of dilatation and curettage. At sonography, a broad-based membrane is visualized; the base probably represents the synechia, with the remainder of the sheet composed of membranes (78). The fetus is freely mobile around the membrane (Fig. 36). The placenta implants on the amniotic sheet in about 26% of cases; this does not affect pregnancy outcome (79). Amniotic sheets have no clinical significance (77).

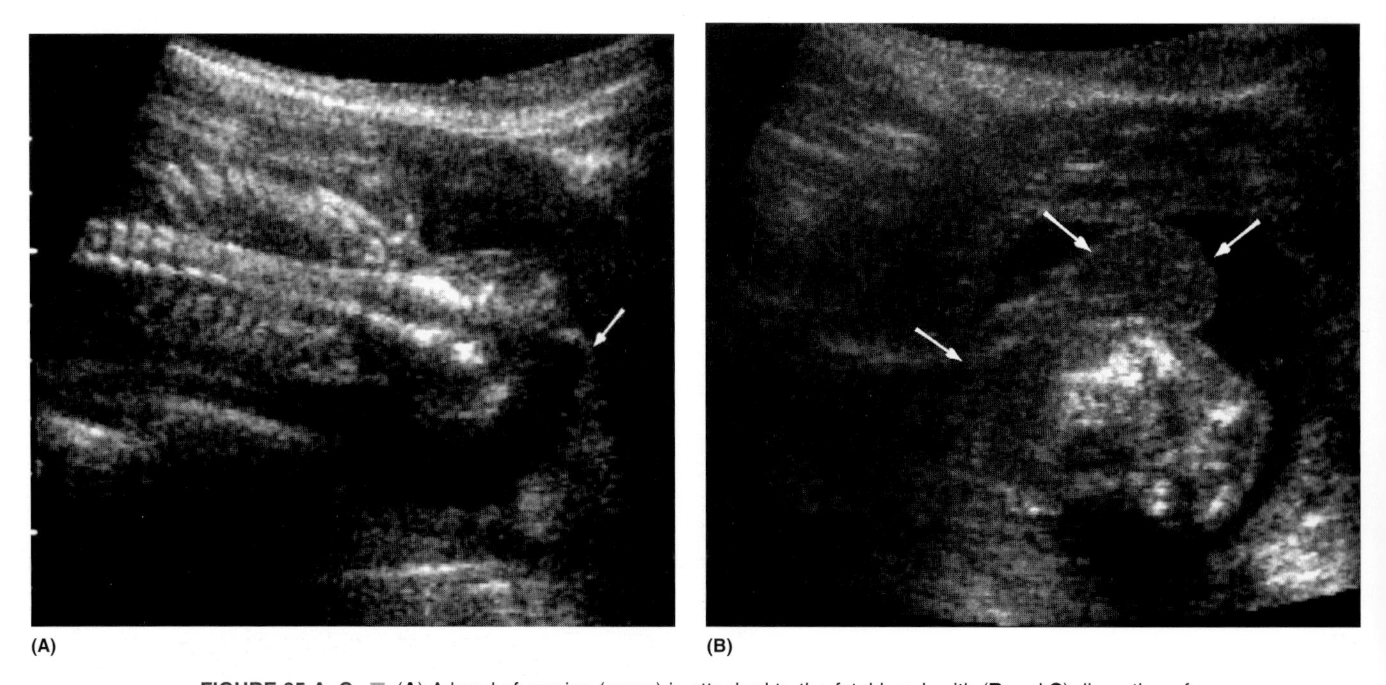

(A) (B)

FIGURE 35 A–C ■ (**A**) A band of amnion (*arrow*) is attached to the fetal head, with (**B** and **C**) disruption of formation of the cranium (acrania) and disorganization of the cranial contents (*arrows in B*). *Source*: (**A** and **B**) Courtesy of Dr. Robert Silverman, Perinatal Center, Syracuse, New York. (*Continued*)

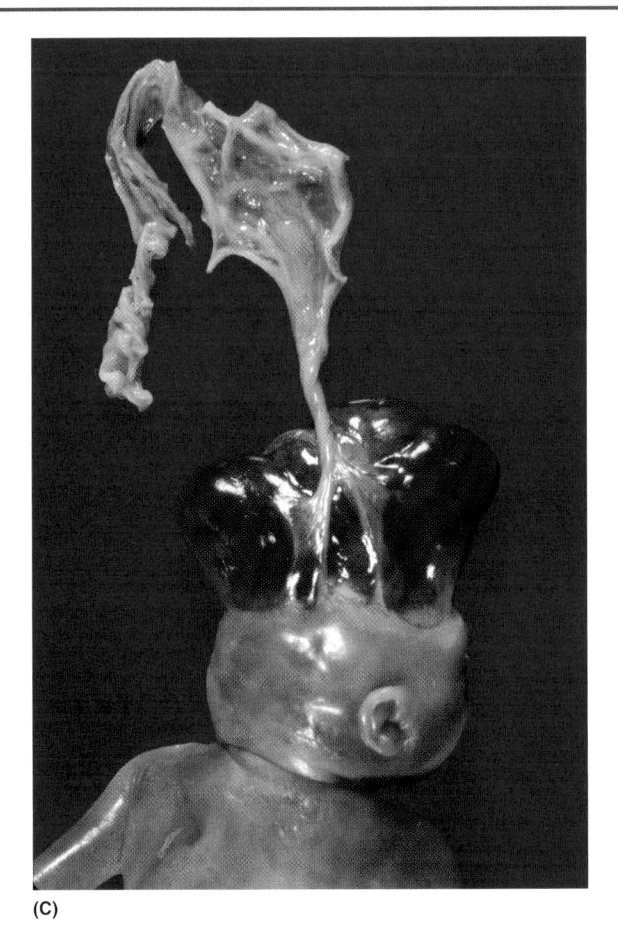

(C)

FIGURE 35 A–C ■ (*Continued*)

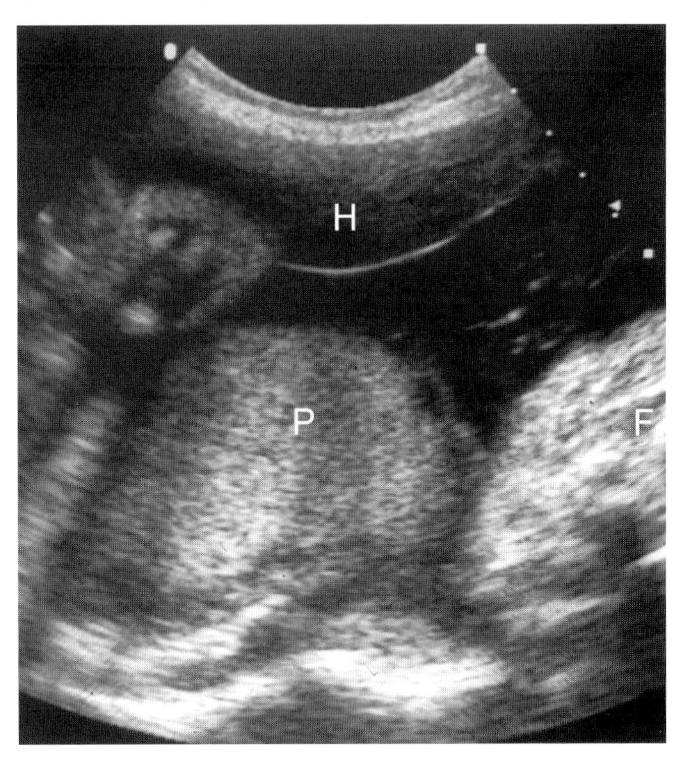

FIGURE 37 ■ Subchorionic hematoma in a patient with antepartum bleeding. At 32 weeks, a subchorionic collection (H) is identified. This decreased in size on follow-up examinations. A subchorionic hematoma was confirmed at delivery. P, placenta; F, fetus.

Subchorionic Hematoma

Bleeding that occurs in an intrauterine pregnancy may collect in the amniotic cavity, in the potential space between amnion and chorion (subamniotic), or in the endometrial cavity extrinsic to the chorionic membrane (subchorionic). Subchorionic hematoma appears at sonography as a collection of blood and/or fibrin that elevates the membranes at the placental margin, or at a distance from the placenta (Fig. 37).

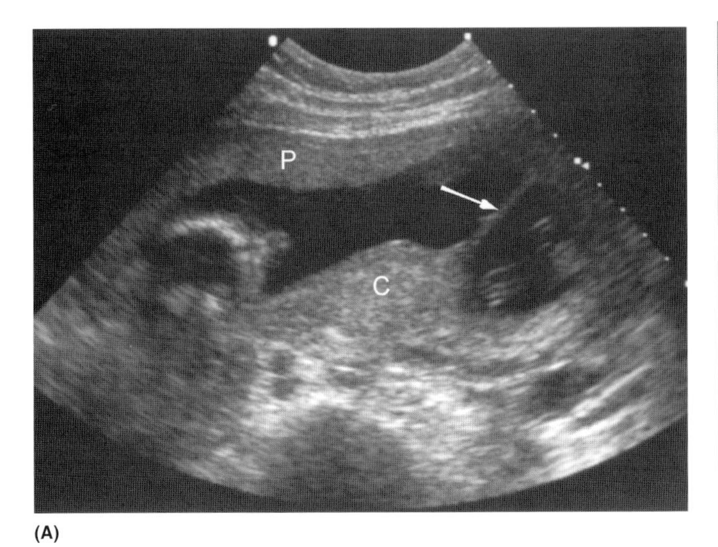

(A)

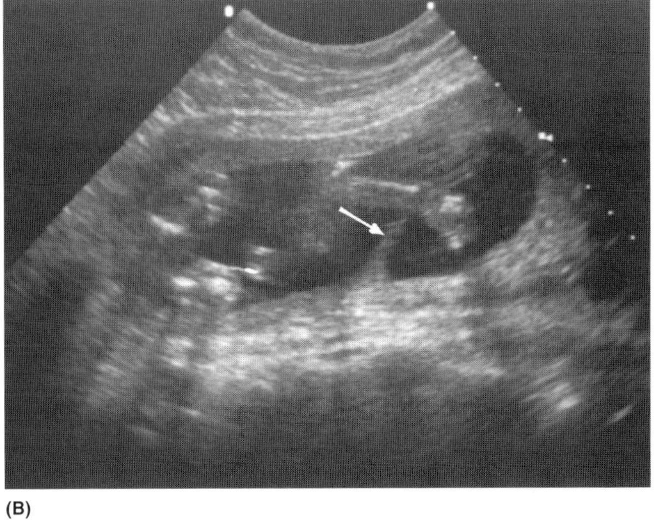

(B)

FIGURE 36 A AND B ■ Synechia (amniotic sheet) at 18 weeks. (**A**) A relatively thick broad-based membrane (*arrow*) extends across the uterus. P, placenta; C, myometrial contraction. (**B**) A fetal limb is seen traversing the membrane (*arrow*).

REFERENCES

1. Lacro RV, Jones KL, Benirschke K. The umbilical cord twist: origin, direction, and relevance. Am J Obstet Gynecol 1987; 157: 833–838.
2. Fletcher S. Chirality in the umbilical cord. Br J Obstet Gynaecol 1993; 100:971–972.
3. Heifetz SA. The umbilical cord: obstetrically important lesions. Clinical Obstetrics Gynecology 1996; 39(3):571–587.
4. Fox H. Pathology of the umbilical cord. In Pathology of the Placenta, 2nd ed. London: WB Saunders Co Ltd, 1997, pp. 418–452.
5. Benirschke K, Kaufman P. Umbilical cord and major fetal vessels. In Pathology of the Human Placenta, 2 ed. New York: Springer, 1990, p. 180.
6. Prucka S, Clemens M, Craven C, et al. Single umbilical artery: what does it mean for the fetus? a case-control analysis of pathologically ascertained cases. Genet Med 2004; 6:54–57.
7. Froehlich LA, Fujikura T. Follow-up of infants with single umbilical artery. Pediatrics 1973; 52:22–29.
8. Kraus FT, Redline RW, Gersell DJ, et al. Placental Pathology. Washington, DC. American Registry of Pathology, Armed Forces Institute of Pathology, 2004.
9. Heifetz SA. Single umbilical artery: a statistical analysis of 237 autopsy cases and review of the literature. Perspect Pediatr Pathol 1984; 8:345–378.
10. Granese R, Coco C, Jeanty P. The value of single umbilical artery in the prediction of fetal aneuploidy: findings in 12,672 pregnant women. Ultrasound Quarterly 2007:177–121.
11. Catanzarite VA, Hendricks SK, Maida C, et al. Prenatal diagnosis of the two-vessel cord: implications for patient counselling and obstetric management. Ultrasound Obstet Gynecol 1995; 5:98–105.
12. Gossett DR, Lantz ME, Chisholm CA. Antenatal diagnosis of single umbilical artery: is fetal echocardiography warranted? Obstet Gynecol 2002; 100:903–908.
13. Thummala MR, Raju TN, Langenberg P. Isolated single umbilical artery anomaly and the risk for congenital malformations: a meta-analysis. J Ped Surg 1998; 33:580–585.
14. Sepulveda W, Bower S, Flack N, et al. Discordant iliac and femoral artery flow velocity waveforms in fetuses with single umbilical artery. Am J Obstet Gynecol 1994; 171:521–525.
15. Abuhamad AZ, Shaffer W, Mari G, et al. Single umbilical artery: does it matter which artery is missing? Am J Obstet Gynecol 1995; 173:728–732.
16. Chow JS, Benson CB, Doubilet PM. Frequency and nature of structural anomalies in fetuses with single umbilical arteries. J Ultrasound in Med 1998; 17:765–768.
17. Geipel A, Germer U, Welp T, et al. Prenatal diagnosis of single umbilical artery: determination of the absent side, associated anomalies, Doppler findings and perinatal outcome. Ultrasound Obstet Gynecol 2000; 15:114–117.
18. Blazer S, Sujov P, Escholl Z, et al. Single umbilical artery—right or left? Does it matter? Prenatal Diagnosis 1997; 17:5–8.
19. Sepulveda W, Peek MJ, Hassan J, et al. Umbilical vein to artery ratio in fetuses with single umbilical artery. Ultrasound Obstet Gynecol 1996; 8:23 26.
20. Petrikowsky B, Schneider E. Prenatal diagnosis and clinical significance of hypoplastic umbilical artery. Prenatal Diagnosis 1999; 16:939–940.
21. Raio L, Chezzi F, Di Naro E, et al. The clinical significance of antenatal detection of discordant umbilical arteries. Obstet Gynecol 1998; 91:86–91.
22. Jeanty P. Persistent right umbilical vein: an ominous prenatal finding? Radiology 1990; 177:735–738.
23. Fortune DW, Oster AG. Umbilical cord aneurysm. Am J Obstet Gynecol 1978; 131:339–340.
24. Siddiqi TA, Bendon R, Schultz DM, et al. Umbilical artery aneurysm: prenatal diagnosis and management. Obstet Gynecol 1992; 80:530–533.
25. Sepulveda W, Corral E, Kottmann C, et al. Umbilical artery aneurysm: prenatal identification in three fetuses with trisomy 18. Ultrasound Obstet Gynecol 2003; 21:213–214.
26. Sepulveda W, Mackenna A, Sanchez J, et al. Fetal prognosis in varix of the intrafetal umbilical vein. J Ultrasound Med 1998; 17: 171–175.
27. Allen SL, Bagnall C, Roberts AB, et al. Thrombosing umbilical vein varix. J Ultrasound Med 1998; 17:189–192.
28. Mahony BS, McGahan JP, Nyberg DA, et al. Varix of the fetal intra-abdominal umbilical vein: comparison with normal. J Ultrasound Med 1992; 11:73–76.
29. Rahemtullah A, Lieberman E, Benson C, et al. Outcome of pregnancy after prenatal diagnosis of umbilical vein varix. J Ultrasound Med 2001; 20:135–139.
30. Estroff JA, Benacerraf BR. Fetal umbilical vein varix: sonographic appearance and postnatal outcome. J Ultrasound Med 1992; 11:69–73.
31. Giacoia GP. Body stalk anomaly: congenital absence of the umbilical cord. Obstet Gynecol 1992; 80:527–529.
32. Lockwood CJ, Scioscia AL, Hobbins JC. Congenital absence of the umbilical cord resulting from maldevelopment of embryonic body folding. Am J Obstet Gynecol 1986; 155:1049–1051.
33. Rayburn WF, Beynen A, Brinkman DL. Umbilical cord length and intrapartum complications. Obstet Gynecol 1981; 57:450–452.
34. Strong TH, Jarles PL, Vega JS, et al. The umbilical coiling index. Am J Obstet Gynecol 1994; 170:29–32.
35. Machin GA, Ackerman J, Gilbert-Barness E. Abnormal umbilical cord coiling is associated with adverse perinatal outcomes. Pediatric Develop Pathology 2000; 3:462–471.
36. Strong TH, Elliott JP, Radin TG. Non-coiled umbilical blood vessels: a new marker for the fetus at risk. Obstet Gynecol 1993; 81:409–411.
37. Strong TH, Finberg HJ, Mattox JH. Antepartum diagnosis of non-coiled umbilical cords. Am J Obstet Gynecol 1994; 170: 1729–1731.
38. Predanic M, Perni SC, Chasen ST, et al. Assessment of umbilical cord coiling during the routine fetal sonographic anatomic survey in the second trimester. J Ultrasound Med 20005; 24:185–191.
39. Cromi A, Ghezzi F, Duerig P, et al. Sonographic atypical vascular coiling of the umbilical cord. Prenat Diagn 2005; 25:1–6.
40. Miser WF. Outcome of infants born with nuchal cords. J Fam Pract 1992; 34:441–445.
41. Clapp JF, Stepanchak W, Hashimoto K, et al. The natural history of antenatal nuchal cords. Am J Obstet Gynecol 2003; 89:488–493.
42. Carey JC, Rayburn WF. Nuchal cord encirclements and risk of stillbirth. Internat J of Gynecol Obstet 2000; 69:173–174.
43. Schaffer l, Burkhardt T, Zimmermann R, et al, Nuchal cords in term and postterm deliveries—do we need to know? Obstet Gynecol 2005; 105:23–28.
44. Larson JD, Rayburn WF, Crosby S, et al. Multiple nuchal cord entanglements and intrapartum complications. Am J Obstet Gynecol 1995; 173:1228–1231.
45. Gonzalex-Quintero VH, Tolaymat L, Muller AC, et al. Outcomes of pregnancies with sonographically detected nuchal cords remote from delivery. J Ultrasound Med 2004; 23:43–47.
46. Mastrobattista JM, Hollier LM, Yeomans ER, et al. Effects of nuchal cord on birthweight and immediate neonatal outcomes. Am J Perinatology 2005; 22:83–85.
47. French AE, Gregg VH, Newberry Y, et al. Umbilical cord stricture: a cause of recurrent fetal death. Obstet Gynecol 2005; 105: 1235–1239.
48. Collins JC, Muller RJ, Collins CL. Prenatal observation of umbilical cord abnormalities: a triple knot and torsion of the umbilical cord. Am J Obstet Gynecol 1993; 169:102–104.
49. Collins JH. Prenatal observation of umbilical cord torsion with subsequent premature labor and delivery of a 31-week infant with mild nonimmune hydrops. Am J Obstet Gynecol 1993; 172:1048–1049.
50. Oyelese KO, Turner M, Lees C, et al. Vasa previa: an avoidable obstetric tragedy. Obstet Gynecol Survey 1999; 54:138–145.
51. Lee W, Lee VL, Kirk JS, et al. Vasa previa: prenatal diagnosis, natural evolution, and clinical outcome. Obstet Gynecol 2000; 95: 972–976.
52. Liu CC, Pretorius DH, Scioscia AL, et al. Sonographic prenatal diagnosis of marginal placental cord insertion: clinical importance. J Ultrasound Med 2002; 21:627–632.

53. Skibo LK, Lyons EA, Levi CS. First-trimester umbilical cord cysts. Radiology 1992; 182:719–722.
54. Sepulveda W, Leible S, Ulloa A, et al. Clinical significance of first trimester umbilical cord cysts. J Ultrasound Med 1999; 18:95–99.
55. Smith GN, Walker MM, Johnson S, et al. The sonographic finding of persistent umbilical cord cystic masses is associated with lethal aneuploidy and/or congenital anomalies. Prenat Diagn 1996; 16:1141–1147.
56. Sepulveda W, Gutierrez J, Sanchez J, et al. Pseudocyst of the umbilical cord: prenatal sonographic appearance and clinical significance. Obstet Gynecol 1999; 93:377–381.
57. Ghidini A, Romero R, Eisen RN, et al. Umbilical cord hemangioma. Prenatal identification and review of the literature. J Ultrasound Med 1990; 9:297–300.
58. Pollack MS, Bound LM. Hemangioma of the umbilical cord: sonographic appearance. J Ultrasound Med 1989; 8:163–166.
59. Resta RG, Luthy DA, Mahoney BS. Umbilical cord hemangioma associated with extremely high alpha-feto-protein levels. Obstet Gynecol 1988; 72:488–491.
60. Levine D, Callen PW, Pender SG, et al. Chorioamniotic separation after second-trimester genetic amniocentesis: importance and frequency. Radiology 1998; 209:175–181.
61. Bromley B, Shipp TD, Benacerraf BR. Amnion-chorion separation after 17 weeks gestation. Obstet Gynecol 1999; 94:1024–1026.
62. Barak S, Leibovitz Z, Degani S, et al. Extensive hemorrhagic chorion-amnion separation after second-trimester genetic amniocentesis. J Ultrasound Med 2003; 22:1283–1288.
63. Graf JL, Bealer JF, Gibbs DL, et al. Chorioamniotic membrane separation: a potentially lethal finding. Fetal Diagnosis & Therapy 1997; 12:81–84.
64. Ulm B, Ulm MR, Bernaschek G. Unfused amnion and chorion after 14 weeks of gestation: associated fetal structural and chromosomal abnormalities. Ultrasound Obstet & Gynecol 1999; 13:392–395.
65. Finberg HJ. The "twin peak" sign: reliable evidence of dichorionic twinning. J Ultrasound Med 1992; 11:571–577.
66. Stagiannis KD, Sepulveda W, Southwell D, et al. Ultrasonographic measurement of the dividing membrane in twin pregnancy during the second and third trimesters: a reproducibility study. Am J Obstet Gynecol 1995; 173:1546–1550.
67. Scott JS. Placenta extrachorialis (placenta marginata and placenta circumvallate): a factor in antepartum hemorrhage. J Obstet Gynaecol Br Commonw 1960; 67:904–918.
68. Naftolin F, Khudr G, Benirschke K, et al. The syndrome of chronic abruptio placentae, hydrorrhea, and circumvallate placentae. Am J Obstet Gynecol 1973; 116:347–350.
69. Fox H, Sen DK. Placenta extrachorialis: a clinicopathological study. J Obstet Gynaecol Br Commonw 1972; 79:32–35.
70. Jeanty P, Laucirica R, Luna SK. Extra-amniotic pregnancy: a trip to the extraembryonic coelom. J Ultrasound Med 1990; 9:733–736.
71. Schwarzler P, Moscoso G, Senat M-V, et al. The cobweb syndrome: first trimester sonographic diagnosis of multiple amniotic bands confirmed by fetoscopy and pathological examination. Human Reproduction 1998; 13:2966–2969.
72. Moerman P, Fryns J-P, Vandenberghe K, et al. Constrictive amniotic bands, amniotic adhesions, and limb-body wall complex: discrete disruption sequences with pathogenetic overlap. Am J Med Genet 1992; 42:470–479.
73. Morovic CG, Berwart F, Varas J. Craniofacial anomalies of the amniotic band syndrome in serial clinical cases. Plast Reconstr Surg 2004; 113:1556–1562.
74. Keswani SG, Johnson MP, Adzick NS, et al. In utero limb salvage: fetoscopic release of amniotic bands for threatened limb amputation. J Ped Surg 2003; 38:848–851.
75. Ronderos-Dumit D, Briceno F, Navarro H, Sanchez, N. Endoscopic release of limb constriction rings in utero. Fetal Diagn Ther 2006; 21:255–258.
76. Lazebnik N, Hill LM, Many A, et al. The effect of amniotic sheet orientation on subsequent maternal and fetal complications. Ultrasound Obstet Gynecol 1996; 8:267–271.
77. Ball RH, Buchmeier SE, Longnecker M. Clinical significance of sonographically detected uterine synechaie in pregnant patients. J Ultrasound Med 1997; 16:465–469.
78. Finberg HJ. Uterine synechiae in pregnancy: expanded criteria for recognition and clinical significance in 28 cases. J Ultrasound Med 1991; 10:547–555.
79. Korbin CD, Benson CB, Doubilet PM. Placental implantation on the amniotic sheet: effect on pregnancy outcome. Radiology 1998; 206:773–775.

Fetal Head and Brain

John P. McGahan

43

EARLY EMBRYOLOGY

Understanding of normal development of the fetal central nervous system (CNS) is important to help distinguish the normal sonographic appearance of the fetal brain from CNS abnormalities. By approximately 4.5 to 5 menstrual weeks, the primitive neural plate has developed. The neural plate then divides into the neural crest and the neural tube. The neural tube differentiates into the primitive brain and the spinal cord.

Rhombencephalon

At about seven to eight menstrual weeks, the primary vesicles of the brain are formed and can be recognized with endovaginal sonography. The three primary vesicles include the forebrain (prosencephalon), the midbrain (mesencephalon), and the hindbrain (rhombencephalon). It is important to note these cystic brain structures are normal and are not cystic brain abnormalities. The fetal rhombencephalon is recognized as a cystic structure in the posterior fossa (Fig. 1) (1).

Choroid Plexus Location

Later, the prosencephalon differentiates into the telencephalon and the diencephalon. The telencephalon forms the cerebrum and lateral ventricles, whereas the diencephalon forms the midlines structures. Before 13 weeks, choroids plexus fills the entire lateral ventricles including the frontal horns (Fig. 2). This is different in later development when the choroid does not fill the frontal horns of the lateral ventricles.

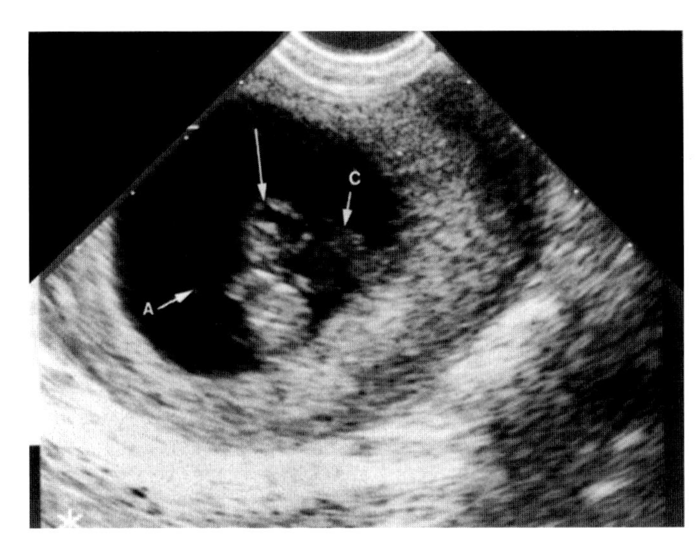

FIGURE 1 ■ Normal rhombencephalon. Sagittal scan of the fetal head at nine menstrual weeks demonstrates a cystic structure in the posterior fossa representing normal rhombencephalon (*arrow*). A, amnion; C, umbilical cord.

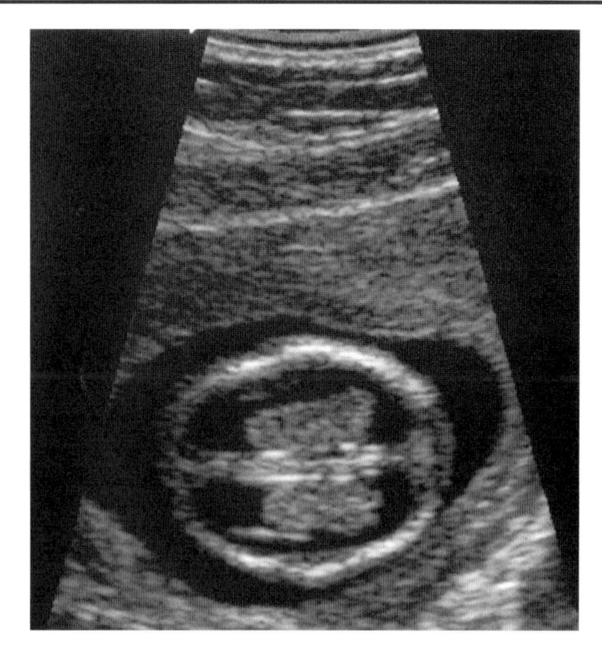

FIGURE 2 ■ Cerebral ventricles. Scan at 13 to 14 weeks demonstrates prominence of the choroids plexus, which occupies a good portion of the cerebral hemisphere.

Cavum Septum Pellucidum Development

The midline corpus callosum and the cavum septum pellucidum do not completely develop until approximately 18 weeks of gestation. Therefore, failure of recognition of the cavum septum pellucidum early in pregnancy does not indicate agenesis of the corpus callosum. Rescanning at 18 to 20 weeks should be performed.

Inferior Vermian Closure—False-Positive Dandy–Walker Variant

Finally, the rhombencephalon differentiates into the metencephalon and the myelencephalon. The metacephalon gives rise to the pons, cerebellum, and upper portion of the fourth ventricle. The myelencephalon gives rise to the medulla and lower portion of the fourth ventricle. The inferior fourth ventricle is in direct communication with the cisterna magna early in pregnancy. This inferior vermis does not close until approximately the 18th week of pregnancy (2,3). Therefore, too early scanning will demonstrate an opening between the fourth ventricle and the posterior fossa. This is usually not a Dandy–Walker variant and rescanning between 18 and 20 weeks should be performed.

NORMAL ANATOMY—SECOND TRIMESTER

Evaluation of the fetal CNS, including the cerebral ventricles, posterior fossa (including cerebellar hemispheres and cisterna magna), and the spine has been recommended as a guideline for performance of antenatal obstetric ultrasound (4). Other investigators believe that a comprehensive approach to evaluation of the fetal head in the second and third trimesters of pregnancy includes visualization of (*i*) the ventricular atrium, (*ii*) the cisterna magna, and (*iii*) the cavum septi pellucidi. Visualization of these three anatomic regions on standard imaging planes excludes the diagnosis of most intracranial abnormalities (5).

Ventricular Atrium

The "downside" ventricular atrium can be visualized in 99% of normal fetuses (5). The ventricular atrium is best visualized on a standardized transaxial plane taken through the fetal cranium. The ventricular atrium, at this level, remains stable in size throughout the second and third trimesters of pregnancy. The normal ventricular atrium has a mean diameter between 6 and 7 mm. A measurement of 10 mm is 4 standard deviations above the mean and is considered the cutoff between normal and abnormal size of the ventricular atrium (Fig. 3) (5,6). Alagappan et al. studied a larger population of 500 normal fetuses and found the mean value for the downside ventricular atrium to be 6.6 mm. The standard deviation in this study population was larger, at 1.4 mm. Therefore, using 2.5 standard deviations, these investigators established 10 mm as the upper limit of normal in this population (7).

Data have demonstrated the feasibility of visualizing the proximal or "upside" cerebral ventricle throughout pregnancy using an angled technique (8). Normal measurements of the body of the upside cerebral ventricle remain stable throughout pregnancy (Fig. 4) (9).

Cisterna Magna

An image of the posterior fossa in the transaxial plane is obtained by angling caudally from the plane used to visualize the ventricular atrium. This approach visualizes the brain stem and the vermis of the cerebellum outlined by the cisterna magna (10,11). The anterior-to-posterior measurement of the cisterna magna obtained in the midline from the posterior margin of the cerebellar vermis to the inner wall of the occipital bone is approximately 5 mm. The

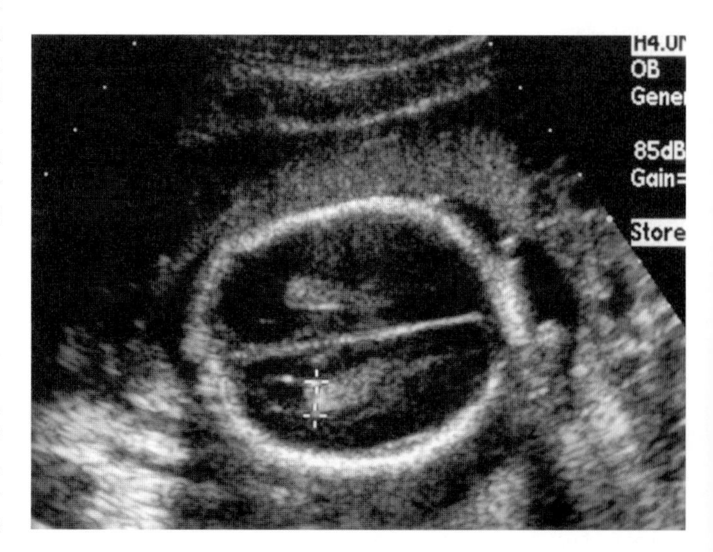

FIGURE 3 ■ Cerebral ventricles. Calipers placed in the regions of the trigone of the down side lateral ventricle. Note choroids plexus fills the entire ventricle.

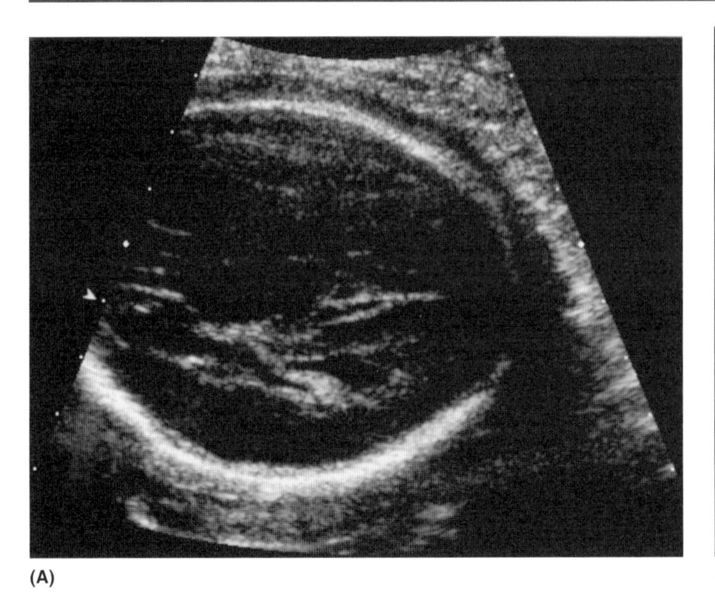

(A)

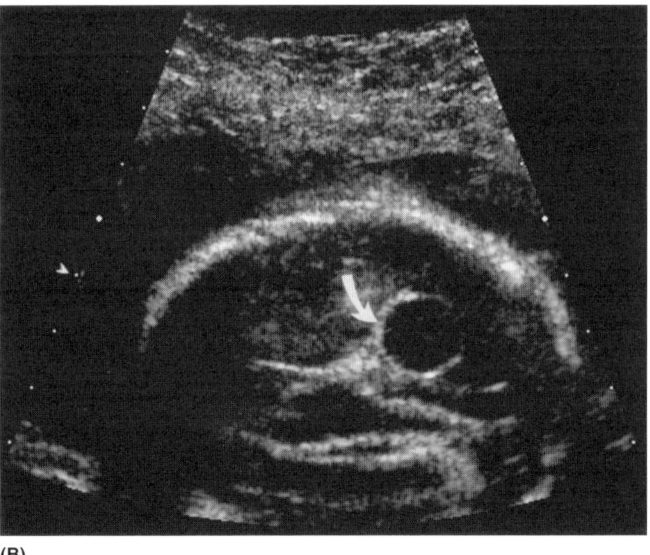

(B)

FIGURE 4 A AND B ■ Angled technique. (**A**) Routine axial scan of the fetal head. (**B**) Using the angled technique, a cyst (*arrow*) is identified in the upside cerebrum that was not seen on a routine axial scan.

largest cisterna magna measurement should be no more than 10 mm in depth and no less than 2 mm (10). If the cerebellar hemispheres are compressed and the cisterna magna is not seen, this is associated with an open neural tube defect (12). The fourth ventricle should not communicate with the cisterna magna, after 18 weeks, in a standard plane, but should be separated by the intact cerebellar vermis. Moreover, the diameter of the two cerebellar hemisphere increases by approximately 1 mm per week of pregnancy between 15 and 21 menstrual weeks (Fig. 5).

In the fetus, arachnoid septations are normally visualized within the cisterna magna and should not be confused with vascular structures (11). If a ventricular atrial view and a posterior fossa view are satisfactorily obtained and are normal, the risk of a CNS anomaly is approximately 0.005% (13).

FIGURE 5 ■ Posterior fossa. Axial scan of the fetal head demonstrating cavum septum pellucidum noted anteriorly. Posteriorly, the cerebellar hemisphere is noted. Calipers 1, note the width of the cisterna magna. Calipers 2, note cisterna magna. Nuchal thickness may be measured on this image. v, vermis; c, cavum septum pellucidum; H, cerebellar hemisphere.

Cavum Septi Pellucidi

Identification of a third anatomic structure, the cavum septi pellucidi, may be helpful to detect subtle midline abnormalities such as agenesis of the corpus callosum or lobar holoprosencephaly (5). The cavum septum pellucidi is typically identified on a transaxial plane with the anteriorly placed cavum septi pellucidi as a cerebrospinal fluid (CSF)–containing structure positioned between the frontal horns of the lateral ventricles (Fig. 5). When scanning through this region, the shape of the cranium can also be assessed (5).

ARTIFACTS AND PITFALLS

Pseudohydrocephalus

Within the early second trimester of pregnancy, the brain parenchyma on the side farthest from the transducer is hypoechoic and may be confused with hydrocephalus (14). Additionally, adjacent to the dependent portion of the bony calvarium is an echogenic interface thought to represent the fetal subarachnoid space that may be misinterpreted as the lateral ventricular wall, causing further confusion (15). Misdiagnosis of hydrocephalus may be overcome by noting the points outlined in Table 1 and depicted in Figures 6 to 8 (16).

TABLE 1 ■ Hydrocephalus vs. Pseudohydrocephalus

Misdiagnosis of hydrocephalus may be avoided by:

- Noting a specular reflection anteriorly that represents the normal sylvian cistern with the accompanying middle cerebral artery
- Scanning more cephalad in the fetus, so the true walls of the frontal horn of the lateral ventricle can be seen
- Noting that both the medial and the lateral walls of the ventricular atrium can be recognized with meticulous scanning
- Noting that the ventricular choroid plexus remains relatively parallel to the long axis of the lateral ventricle (absence of dangling choroid)

Unilateral vs. Bilateral Cerebral Abnormalities

Unilateral hydrocephalus is incorrectly diagnosed when the upside fetal cerebral ventricle is not visualized because of the artifact created by the bony calvarium in the near field (14). This misdiagnosis may be overcome by angling the transducer through the anterolateral fontanelle or the thinner squamosal portion of the temporal bone to visualize the obscured fetal cerebral ventricle (Fig. 4) (8,9). In addition, such abnormalities as a choroid plexus cyst may be unilateral, rather than bilateral, and the use of the angled technique is important in these cases (Fig. 4).

Reverberation Artifact

Reverberation artifact from the anterior cranial vault may be identified within the fetal cranium. This source of confusion may be remedied by noting the lack of distortion of the intracranial anatomy or by using such technique as the angled approach to visualize the proximal intracranial structures (Fig. 4).

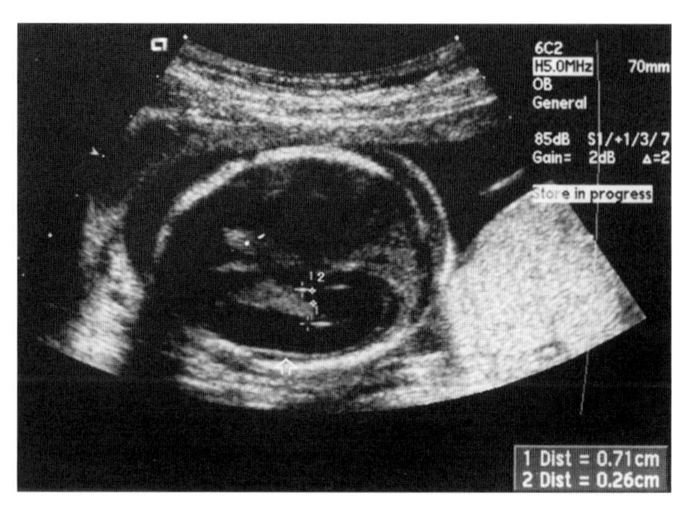

FIGURE 7 ■ Pseudohydrocephalus. Axial scan of the fetal head at 20 weeks demonstrates echogenic interface of the subarachnoid space (*open arrow*) that may be misinterpreted at a lateral wall of the lateral ventricle. However, note that the down side lateral ventricle is marked by caliper 1. Also note mild separation of the choroid from the medial wall of the ventricle, as marked by second caliper.

Pseudochoroid Plexus Cyst

An oval hypoechoic structure can be visualized in the inferior and lateral aspects of the atrium of the lateral

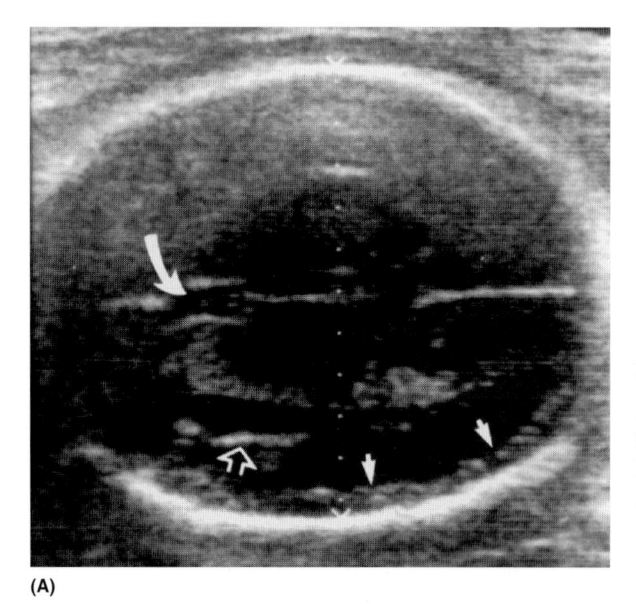

(A)

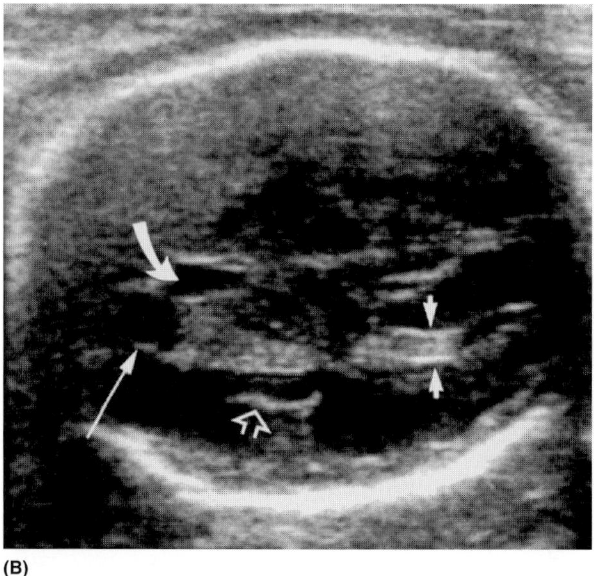

(B)

FIGURE 6 A AND B ■ Pseudohydrocephalus. (**A**) Fetal head at 26 weeks' menstrual age demonstrates a hypoechoic area adjacent to the dependent portion of the calvarium that may be misinterpreted as hydrocephalus. The echogenic interface of the subarachnoid space (*closed arrows*) may be misinterpreted as the lateral wall of the lateral ventricle. (*Curved arrow*, cavum septi pellucidi; *open arrow*, sylvian cistern). (**B**) To avoid this pitfall, look for (*i*) the normal appearance to the cavum septi pellucidi (*curved arrow*); (*ii*) the frontal horns of the lateral ventricle (*long arrow*); (*iii*) the sylvian cistern (*open arrow*); most important, (*iv*) the two walls of the trigone of the lateral ventricle (*small arrows*); and (*v*) the failure of the choroid plexus to dangle to a more dependent portion within the cerebral hemisphere.

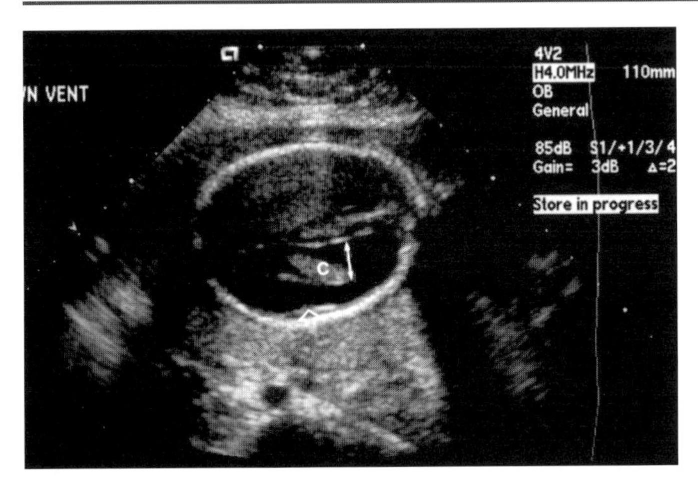

FIGURE 8 ■ Pseudohydrocephalus/mild ventriculomegaly. The echogenic interface of the subarachnoid space (*open arrow*) may be misinterpreted as the lateral wall of the lateral of ventricle. Note in this fetus, there is very mild ventriculomegaly (*arrow*) with choroid plexus "dangling" in the down side lateral ventricle. C, choroid plexus.

ventricle. This image should not be mistaken for a true choroid plexus cyst. This artifact may be recognized based on its oval shape, small size, constant location, as well as by noting that the structure is without the strong acoustic interface (wall) characteristic of a true cyst (17).

Ventricular Measurement Errors

Heiserman et al. (18) demonstrated several possible errors in measurement of the downside lateral ventricular atrium. These errors include measurement of the wrong portion of the ventricle and measurement of the atrium obliquely rather than perpendicular to the long axis of the ventricle. All these errors increase the perceived size of the ventricle.

Normal Ventricular Size: Intracranial Malformations

Certain intracranial malformations such as the Arnold–Chiari type II malformation or the Dandy–Walker malformation may have only minimal associated ventriculomegaly (12,19). Ventriculomegaly may increase with increasing menstrual age. Ventricular size may be normal in a small percentage of these intracranial malformations, and therefore thorough evaluation of the entire cranial contents is necessary.

False-Positive Enlarged Cisterna Magna

When evaluating the posterior fossa, the examiner should scan in the correct anatomic plane. Incorrect angulation in a semicoronal plane may artificially increase the anteroposterior measurement of the posterior fossa (20). This error also causes erroneous increased nuchal thickness (Fig. 9).

FETAL MAGNETIC RESONANCE IMAGING

Another imaging exam that needs to be briefly mentioned is the use of magnetic resonance imaging (MRI) for evaluation of the fetal CNS. Sonography is the main modality used for evaluation of the pregnant uterus. On occasion, there may be limited evaluation of the fetus with sonography. In certain situations using fast imaging techniques, MRI may be useful to better define abnormalities suspected by sonography or in certain situations MRI may provide diagnosis that could not be made by other means (Fig. 10).

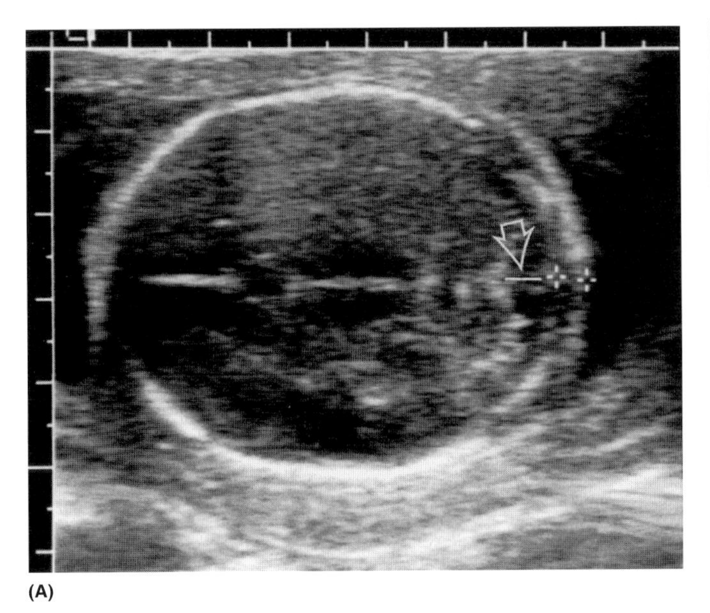

(A)

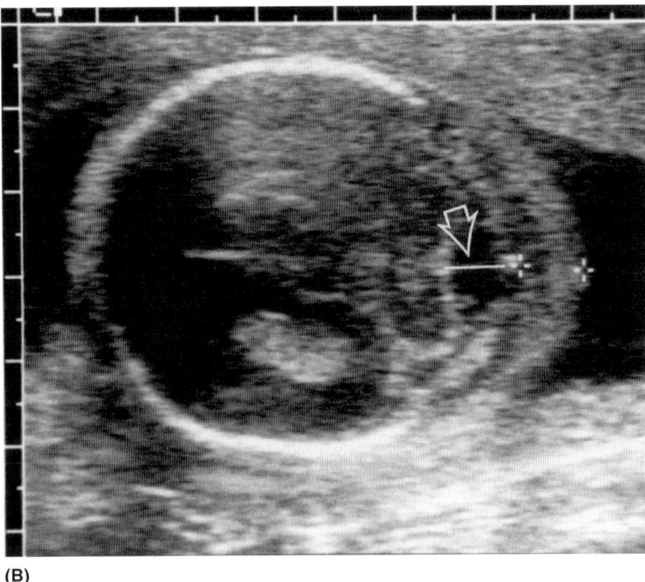

(B)

FIGURE 9 A AND B ■ False enlargement of the cisterna magna. (**A**) Routine scan, including the cavum septi pellucidi, thalami, and cerebral hemisphere, shows a normal anteroposterior measurement of the cisterna magna (*line*). Nuchal thickness is shown (*calipers*). (**B**) Scanning in a semicoronal plane artificially increases the anteroposterior dimension in the cisterna magna (*line*), possibly giving a false impression of a mega cisterna magna. Nuchal thickness is also falsely increased (*calipers*).

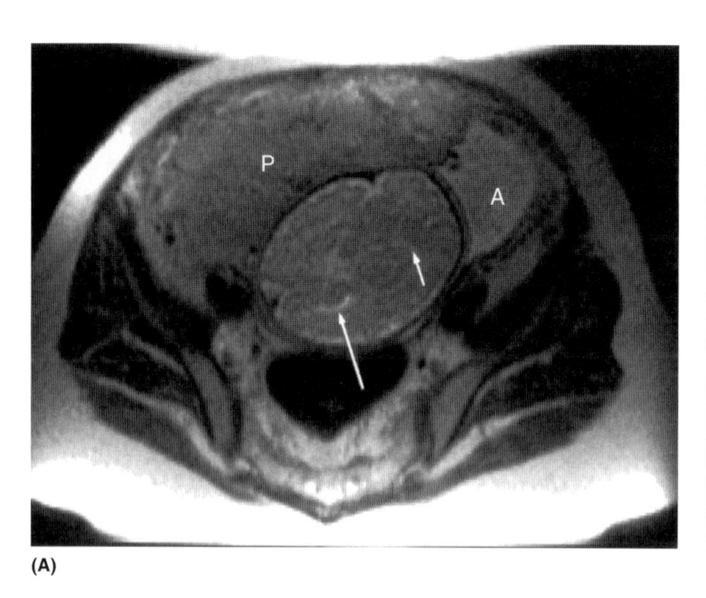

(A)

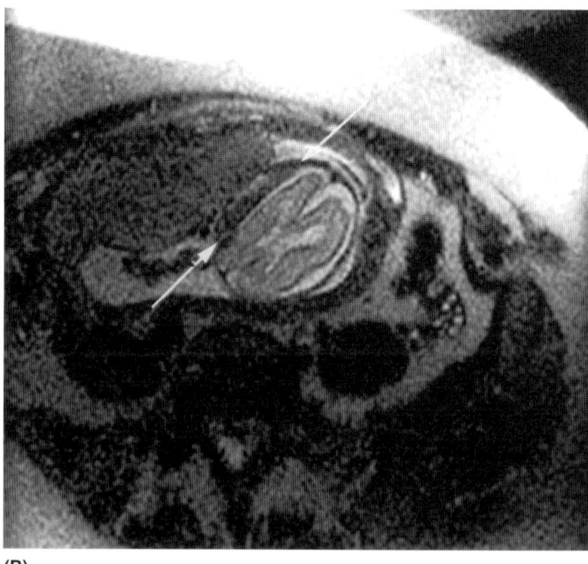

(B)

FIGURE 10 A AND B ■ MRI fetal brain. (**A**) A single shot fast spin echo technique that is T-2 weighted shows a normal fetal brain. Note bright signal from CSF in normal size occipital horn (*long arrow*) and normal size frontal horn (*short arrow*). (**B**) Using a similar technique in another fetus demonstrates extra-axial collection (*arrows*) corresponding to subdural hematoma. P, placenta; A, amniotic fluid.

Examples of where MRI has this potential include screening for tuberous sclerosis or for lissencephaly to name but two (21,22). A more detailed discussion of MRI in the brain is beyond the scope of this text, dedicated to ultrasound.

ABNORMALITIES

Mild Ventriculomegaly

Most studies accept 10 mm as the upper limit of normal for the ventricular atrial width in the second trimester of pregnancy. However, rather than a single cutoff value, there is overlap of normal and abnormal fetus with ventricular width either below or above 10 mm. Hertzberg's data defined a population with choroid plexus separation greater than 3 mm but with ventricular atrium of 10 mm or less, which illustrates this point (23). The outcome was normal in 80% and abnormal in 20%.

Mild ventriculomegaly would be considered by some to be an atrial width of 10 to 15 mm, whereas others consider 10 to 12 mm as mild ventriculomegaly (Fig. 8) (24). Also, ventriculomegaly size has shown to be slightly larger in male as compared to female fetus (25,26). In a review of isolated mild ventriculomegaly, Pilu et al. demonstrated an abnormal outcome in 23% (27). This included perinatal death (4%), chromosomal abnormalities (4%), malformations not detected on sonography (9%), and neonatal cognitive or motor delay (11%), while 77% had a normal outcome. Also, the risk of abnormal neurological outcome was greater in females (23%) compared to males (4%) and when the atrial width was greater than 12 mm (14%) compared to a width between 10 and 12 mm (4%) (27).

Moderate Ventriculomegaly

The term ventriculomegaly indicates enlargement of the cerebral ventricles and is not used synonymously with the term hydrocephalus, which implies ventriculomegaly based on obstruction of flow of CSF. Causes of ventriculomegaly can be classified into three major categories. These include hydrocephalus, brain atrophy, and abnormal development of the brain (Fig. 11).

Hydrocephalus usually implies blockage of the flow of CSF and rarely results from overproduction of CSF by tumor. CSF is produced within the choroid plexus and flows within the ventricles and over the subarachnoid

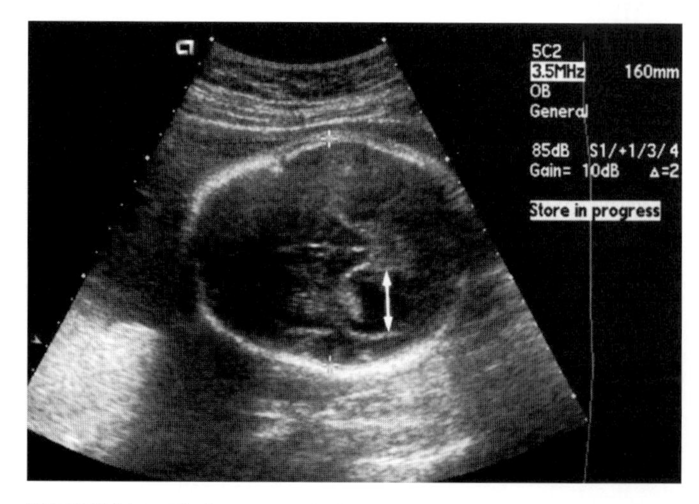

FIGURE 11 ■ Moderate ventriculomegaly. Axial scan through the fetal head at 33 weeks demonstrates moderate dilatation of the down side lateral ventricle (*arrow*).

space to be absorbed in the arachnoid granulations. Obstructive hydrocephalus is most commonly noncommunicating (blockage within ventricles) rather than communicating (extraventricular blockage) within the fetus. Some causes of hydrocephalus may be associated with various syndromes (Table 2) (28). Ultrasound is useful in identifying not only ventriculomegaly but also the site of blockage. Ultrasound may be useful to identify if ventriculomegaly involves both the lateral ventricles, the third ventricle, and the fourth ventricle, and it may be useful to localize the site of obstruction (Fig. 12).

Another cause of ventriculomegaly is underdevelopment of the brain in or about the trigone region of the lateral ventricle, called colpocephaly. This condition is seen commonly with agenesis of the corpus callosum and does not respond to ventricular shunting. Moreover, in lissencephaly (smooth brain), an absence of gyri and sulci and, occasionally, ventriculomegaly may be noted.

Brain destruction is a cause of ventriculomegaly and may include in utero damage resulting from CNS infection. CNS infections may be associated with brain atrophy, with resultant microcephaly and ventriculomegaly.

TABLE 2 ■ Causes and Associations of Hydrocephalus

Causes	Associations
Neural tube defects	Spina bifida (Arnold–Chiari type II malformation)
	Cephaloceles
Other central nervous system malformations	Dandy–Walker malformation
	Agenesis of the corpus callosum
	Lissencephaly
Aqueductal stenosis	Sporadic occurrence
	Postinflammatory status
	X-linked recessive disorders
	Autosomal disorders
Masses	Neoplasms
	Arachnoid cysts
Obstruction of cerebrospinal fluid flow in subarachnoid space	Postinflammatory status
	Posthemorrhage status
	Thanatophoric dysplasia
	Achondroplasia
	Osteogenesis imperfecta
Vascular anomalies	Vein of Galen aneurysm
Multiple anomaly syndromes	Meckel–Gruber syndrome
	Walker–Warburg syndrome
	Apert syndrome
	Smith–Lemli–Opitz syndrome
	Nasal-facial-digital syndrome
	Albers Schonberg disease
	Robert syndrome
	Fragile X syndrome
	Trisomy 18
	Trisomy 13

Source: From Ref. 28.

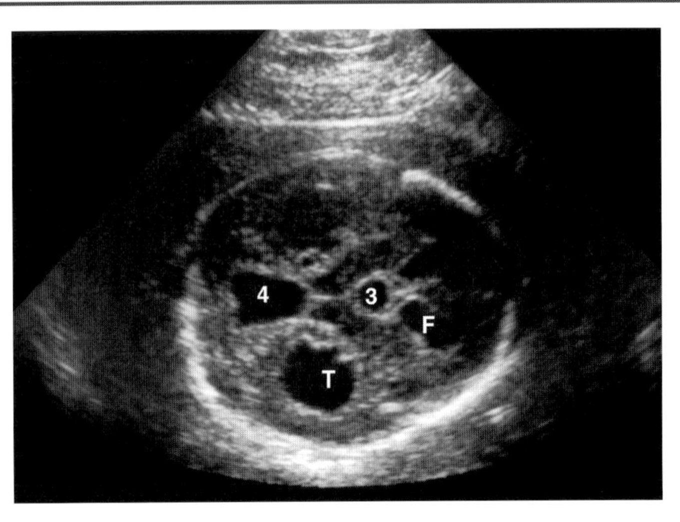

FIGURE 12 ■ Hydrocephalus involving all ventricles. Axial scan of the head demonstrates dilatation of the lateral, third (3), fourth (4) ventricle, frontal horns (F), and trigone (T).

These infections are discussed in more detail at the end of this chapter in the section entitled Intracranial Calcifications. Often a single etiologic entity produces ventriculomegaly by more than one mechanism. For instance, a fetus with intracranial infection may develop ventriculomegaly secondary to cerebral atrophy as well as meningitis with resultant blockage in the aqueduct of Sylvius and resultant hydrocephalus.

When ventriculomegaly is detected, a careful anatomic survey of the fetus is necessary to determine the presence of any associated anomalies. Furthermore, hydrocephalus is associated with a significant risk of chromosomal abnormalities. Distinguishing features and more common causes of hydrocephalus are listed in Table 3.

Open Neural Tube Defect

Ventriculomegaly related to open neural tube defect and the Arnold–Chiari type II malformation may manifest as so-called "lemon" and "banana" signs (12). The "lemon" head is caused by overlapping of the frontal bones in a lemon shape rather than an oval shape. This appearance can occur in fetuses with cephaloceles (29), but it may also be seen in normal fetuses. Therefore, a careful search for other findings is warranted. For instance, the "banana" sign is caused by downward displacement of the cerebellum, with a resultant banana shape, and it also leads to effacement of the cisterna magna (Fig. 13). An associated spinal defect occurs in this abnormality.

Dandy–Walker Malformation

Dandy–Walker malformation consists of large posterior fossa cysts of variable size, vermian agenesis, hypoplasia of the cerebellar hemispheres, and variable amounts of hydrocephalus (Figs. 14 and 15). Dandy–Walker malformation is discussed in more detail later in this chapter in the section entitled Midline Cystic Abnormalities of the Brain (19,30).

TABLE 3 ■ Causes and Distinguishing Features of Hydrocephalus

Malformation	Features
Arnold–Chiari type II	Deformed cranium ("lemon" sign), usually disappears malformation in third trimester
	Obliteration of cisterna magna ("banana" sign)
	Associated open spinal defect
	Minimal hydrocephalus
	Large massa intermedia
	Pointed frontal horns
Dandy–Walker malformation	Midline posterior fossa cyst
	Cyst communicates with fourth ventricle
	Vermian agenesis
	Hypoplastic cerebellar hemispheres
	Elevated tentorium
	Mild-to-moderate hydrocephalus
Agenesis of the corpus callosum	Third ventricle elevated
	Separation of frontal horns
	Associated colpocephaly
	Mild hydrocephalus
Aqueductal stenosis	Moderate-to-massive hydrocephalus
	Normal fourth ventricle
	Normal posterior fossa
	Normal cranial configuration
Cephalocele	Open defect, usually occipital skull base
	Obliteration of the cisterna magna
	Occasional "lemon" sign
	Mild-to-moderate hydrocephalus
	Polyhydramnios

Agenesis of the Corpus Callosum

The corpus callosum is a midline structure above the third ventricle that connects the fibers between the cerebral hemispheres. Agenesis of the corpus callosum may be complete or partial, and it may or may not be accompanied by hydrocephalus. Only cases of complete agenesis of the corpus callosum have been diagnosed in utero. In many cases, one sees isolated enlargement of the occipital horns (colpocephaly) without hydrocephalus and a high-riding third ventricle. There is a "tear drop" shape to the ventricle, with the ventricle parallel to the midline structures (Fig. 16). These cases may be more difficult to diagnose. Scanning anteriorly through the ventricles shows an absence of the cavum septum pellucidum in agenesis of the corpus callosum (Fig. 16) (31). In situations associated with hydrocephalus, the diagnosis can be made with certainty by noting the lateral displacement of both lateral ventricles by an enlarged third ventricle. Other CNS or non-CNS anomalies are often associated with agenesis of the corpus callosum. Agenesis of the corpus callosum may also be associated with an interhemispheric cyst (Fig. 16).

Lissencephaly

Lissencephaly is a term used to describe a brain with absent or poor sulci formation. Lissencephaly is also called smooth brain and may be associated with dilated trigone, occipital horns, and temporal horns of the lateral ventricles. This ventriculomegaly is probably secondary to incomplete development of the calcarine sulci and the hippocampus (32). Lissencephaly is probably not recognized prenatally until the dilated lateral ventricles are identified (33).

Aqueductal Stenosis

Aqueductal stenosis is one of the more common causes of in utero hydrocephalus. The origin of stenosis of the aqueduct is often unknown, although specific causes of aqueductal stenosis do exist. For instance, in utero infection such as viral infections with cytomegalovirus may cause adhesive arachnoiditis and may obstruct the flow of CSF at different sites including the aqueduct of Sylvius. A defective recessive gene on the X chromosome has also been associated with aqueductal stenosis. Male infants with this defect have hydrocephalus and often severe mental retardation (34). Moderate-to-massive dilatation of the lateral ventricles is easily identified. The fetus may have mild dilatation of the third ventricle with a normal fourth ventricle. Usually, the diagnosis of aqueductal stenosis is made by exclusion, by noting no other specific cause of the hydrocephalus. The ventriculomegaly in aqueductal stenosis often progresses throughout pregnancy, and other CNS anomalies or chromosomal abnormalities occur infrequently.

Massive Intracranial Fluid Collections

The three causes of massive intracranial fluid collections with intact cranial vault are hydrocephalus, holoprosencephaly, and hydranencephaly (Table 4) (35). Distinguishing features are listed in Table 5. CSF collections may be divided into generalized and focal collection. In generalized, massive CSF collections, considerations include hydrocephalus, holoprosencephaly, and hydranencephaly. If no cortical mantle is present, but a falx is noted, then this anomaly is most probably hydranencephaly. In a fetus with a cortical mantle, absent falx, and midline intracranial and facial abnormalities, holoprosencephaly is the diagnosis. Otherwise, the abnormality is probably hydrocephalus (Table 5).

Massive Hydrocephalus

The causes of hydrocephalus are listed previously. Massive hydrocephalus is usually secondary to an obstructive phenomenon such as aqueductal stenosis (Fig. 17) rather than to cerebral malformations such as the Arnold–Chiari type II malformation. The ultrasound features are listed in Table 5.

Alobar Holoprosencephaly

Holoprosencephaly may exhibit a sonographic spectrum of abnormalities classified as alobar, semilobar, or lobar

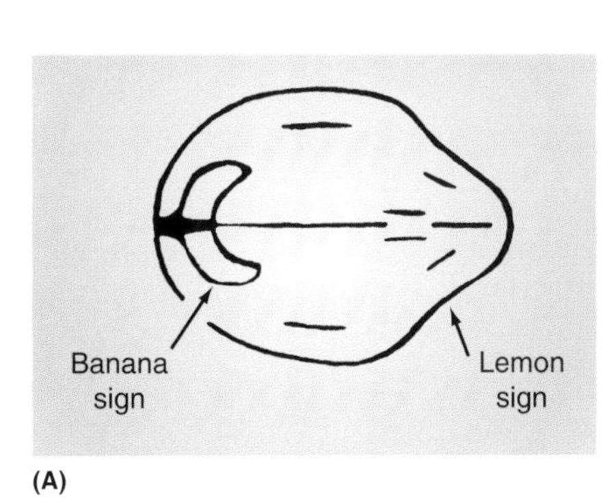

(A)

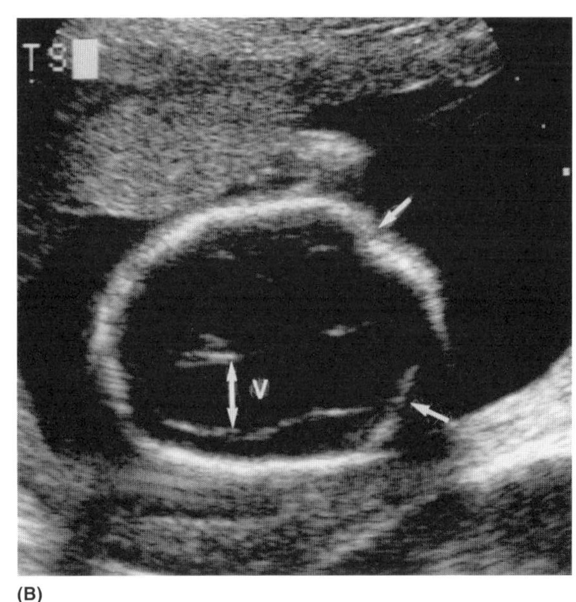

(B)

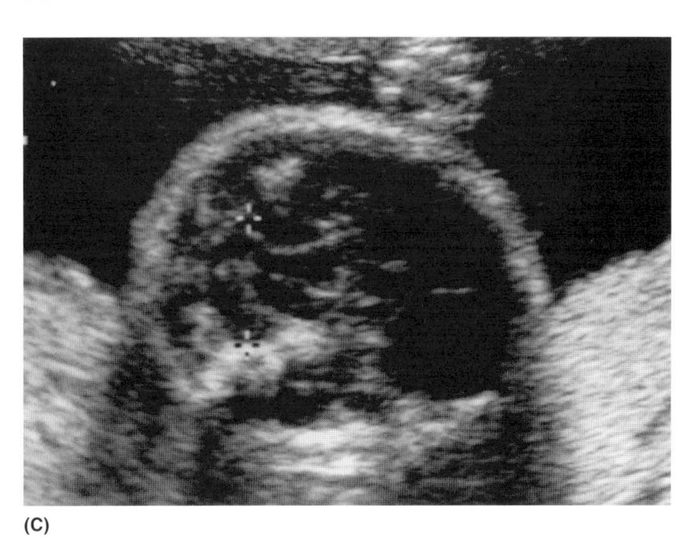

(C)

FIGURE 13 A–C ■ "Lemon–banana" sign. (**A**) Diagram from Nicolaides et al. demonstrates the overlapping coronal sutures of the fetal head termed the "lemon" sign. Moreover, the cerebellar hemispheres within the posterior fossa are compressed, termed the "banana" sign. These features are important for ultrasound screening for open neural tube defects. (**B**) "Lemon head": axial ultrasound scan of the fetal head demonstrates overlapping of the coronal sutures, so-called "lemon head" (*large arrows*), occurring with an open neural tube defect. Mild-to-moderate enlargement of the trigone of the lateral ventricle (12 mm) often occurs with open neural tube defects. (**C**) Scans of the posterior fossa fail to demonstrate the normal appearance of the cisterna magna but rather the associated compression of the cerebellar hemispheres or so-called "banana" sign (*calipers*). The cisterna magna is obliterated. *Source*: (**A**) From Ref. 12.

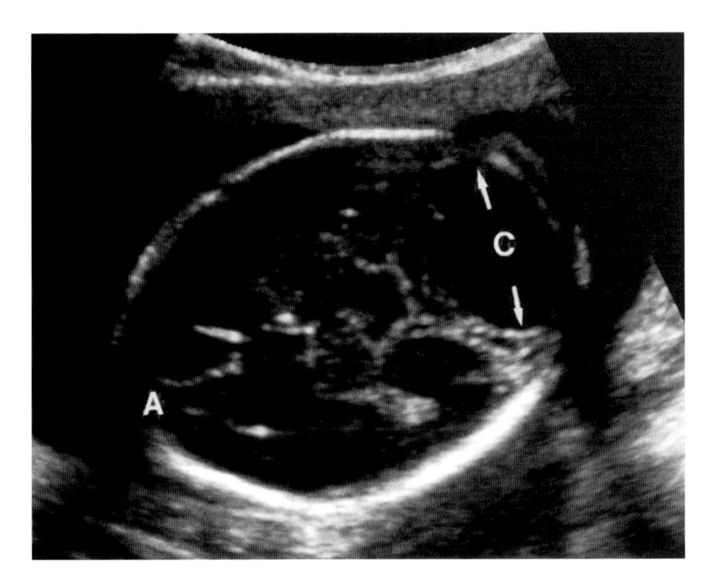

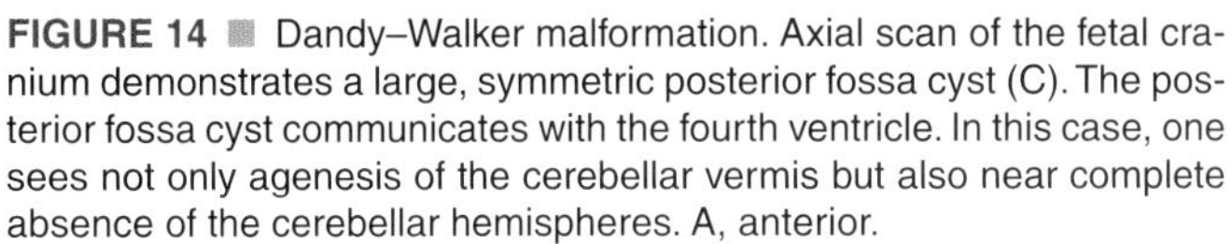

FIGURE 14 ■ Dandy–Walker malformation. Axial scan of the fetal cranium demonstrates a large, symmetric posterior fossa cyst (C). The posterior fossa cyst communicates with the fourth ventricle. In this case, one sees not only agenesis of the cerebellar vermis but also near complete absence of the cerebellar hemispheres. A, anterior.

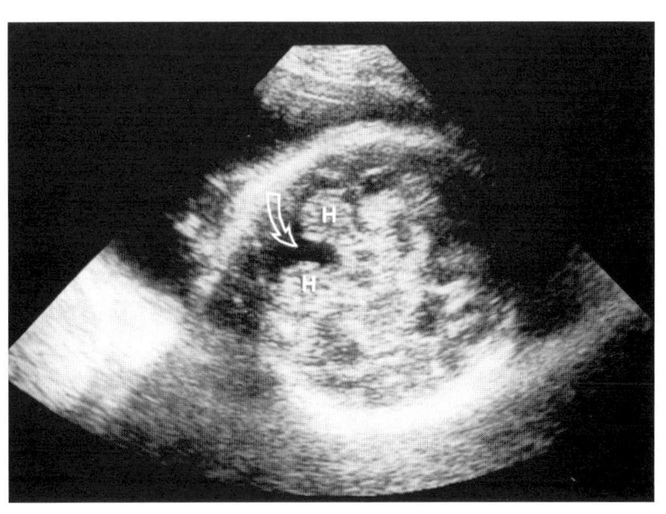

FIGURE 15 ■ Dandy–Walker variant. In utero axial ultrasound scan shows a cystic posterior fossa defect that communicates with the fourth ventricle (*open arrow*). Scanning after 18 weeks is necessary, as the inferior vermis may have delayed closure until this time. H, cerebellar hemispheres.

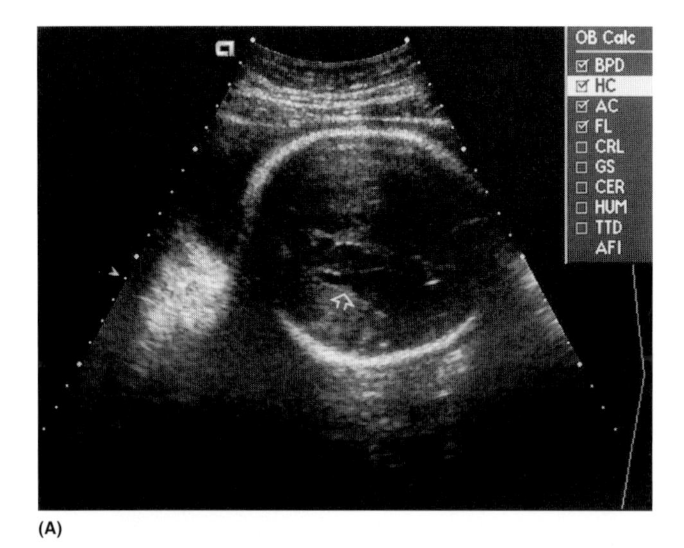

(A)

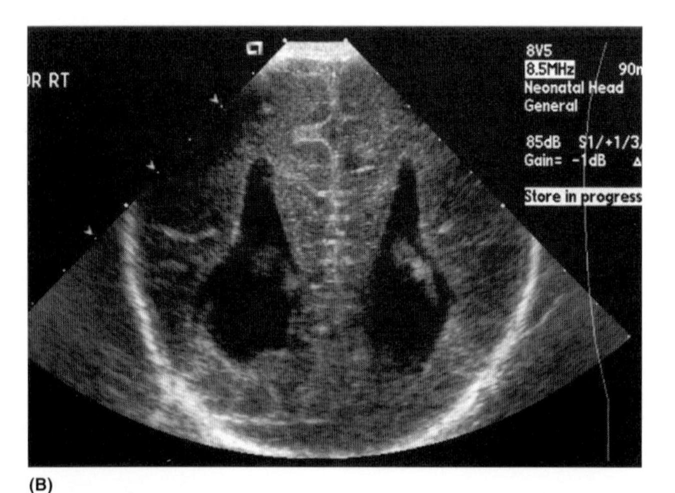

(B)

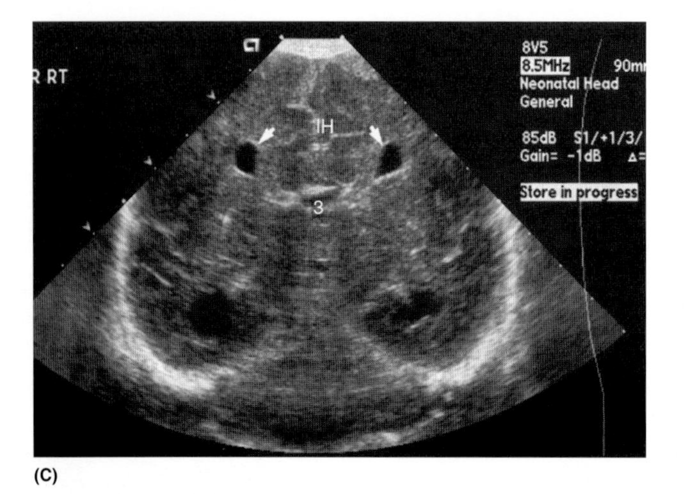

(C)

FIGURE 16 A–C ■ Agenesis of the corpus callosum with and without interhemispheric cyst. (**A**) Axial scan through the ventricles shows characteristic "teardrop" shape of the lateral ventricle (*open arrows*). (**B**) Coronal scan of the newborn head demonstrates similar "teardrop" appearance to the lateral ventricles. Also, note that the ventricles are parallel to the interhemispheric cistern rather than converging toward midline in the region of the frontal horns. (**C**) Coronal scan of the newborn head shows that the frontal horns of the lateral ventricles (*arrows*) are displaced laterally and point superiorly. The interhemispheric cistern extends to the third ventricle without intervening corpus callosum. IH, interhemispheric cistern; 3, third ventricle.

TABLE 4 ■ Causes of Massive Intracranial Fluid Collections

Common	Uncommon
Hydrocephalus	Holoprosencephaly Hydranencephaly

(Figs. 18–21) (Table 6). The lobar form of holoprosencephaly has only fused frontal horns, and the rest of the brain tissue appears grossly normal. The most common form of holoprosencephaly is alobar. This condition results in a monoventricular cavity with fused thalami (Figs. 18 and 21). Alobar holoprosencephaly is secondary to failure of the development of the cleavage of the prosencephalon (35). Separation of the prosencephalon is induced by the prenotochordal mesoderm, which is needed for normal development of the midline facial structures. Thus, alobar holoprosencephaly is often associated with concomitant significant facial abnormalities including cyclopia, ethmoidocephaly, cebocephaly, and cleft lip. Alobar holoprosencephaly is often associated with trisomy 13 (36).

Alobar holoprosencephaly may have three separate configurations, namely, "pancake," "cup," or "ball" forms (Fig. 21) (35). The "pancake" type, which is the rarest, occurs when the residual brain is minimal and compressed over the skull base. A "ball" variation occurs when the cerebral cortex covers the monoventricular

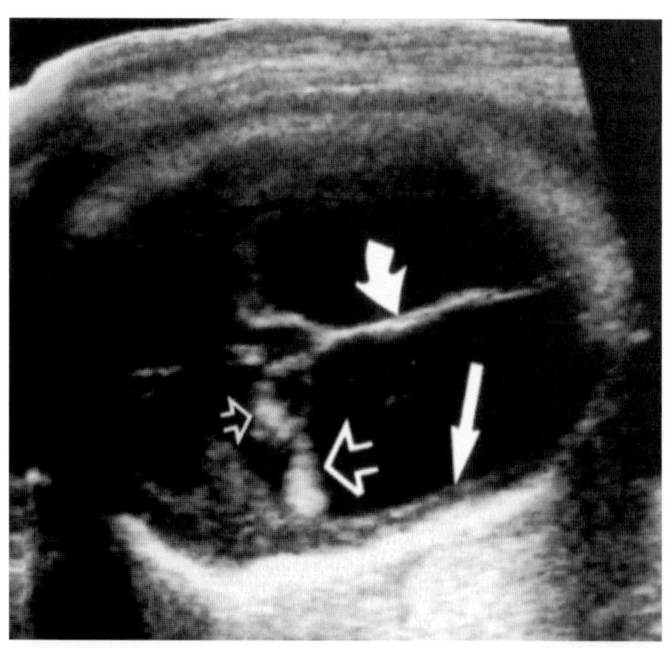

FIGURE 17 ■ Massive hydrocephalus. Transaxial scan through the fetal head demonstrates massive cerebrospinal fluid collection with intact dura and midline structures (*curved arrow*) and the presence of compressed cerebral tissue (*long arrow*). The prominent dangling choroid plexus (*large open arrow*) in the downside ventricle and the choroid plexus from the upside lateral ventricle dangle into the contralateral ventricle (*small open arrow*).

TABLE 5 ■ Distinguishing Features of Hydrocephalus, Hydranencephaly, and Alobar Holoprosencephaly

Abnormality	Cerebral tissue	Falx	Fused thalami	Midline cavity	Dangling choroid	Midface abnormalities
Hydrocephalus	Yes	Yes	No	No	Yes	No
Holoprosencephaly	Yes	No	Yes	Yes	Minimal	Yes
Hydranencephaly	No	Yes	No	No	Minimal	No

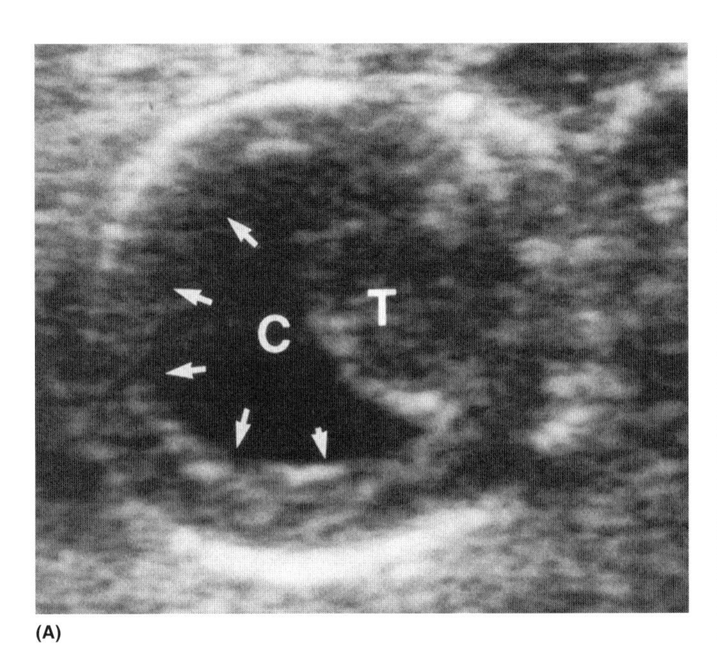

(A)

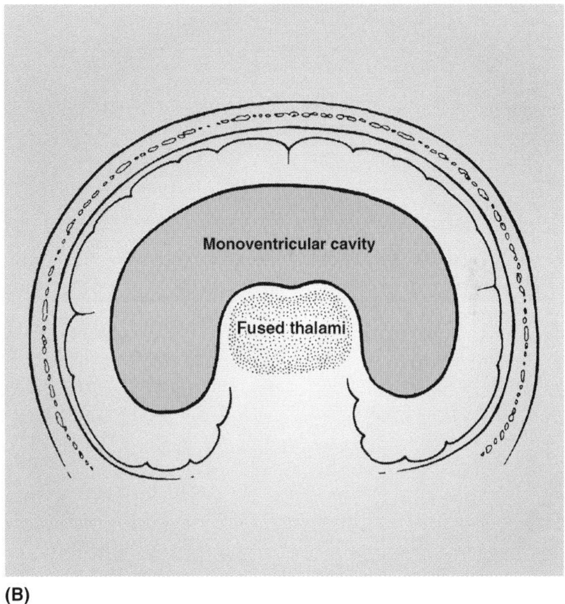

(B)

FIGURE 18 A AND B ■ Alobar holoprosencephaly: in utero postnatal. (**A**) In utero coronal ultrasound scan of the fetal head demonstrates a monoventricular cavity surrounded by cerebral tissue (*arrows*) with fused thalami diagnostic of alobar holoprosencephaly. (**B**) Artist's drawing of coronal ultrasound scan demonstrates the characteristic absence of midline structures, the large monoventricular cavity, and the fused thalami. C, monoventricular cavity; T, thalami.

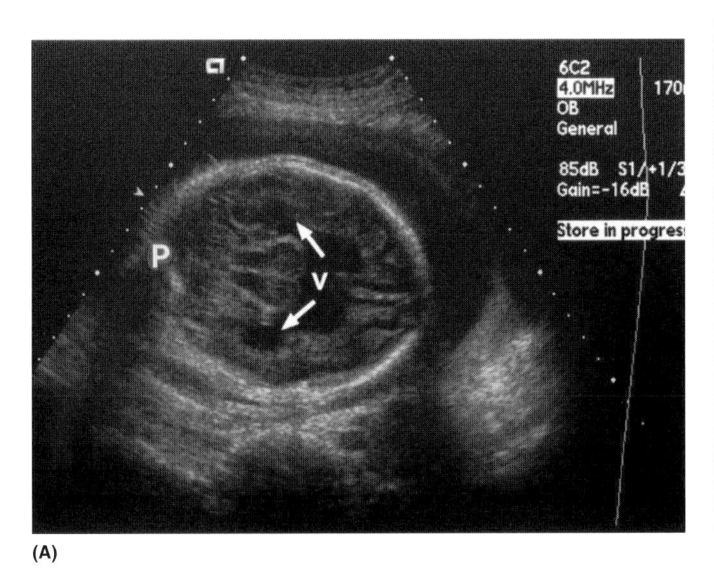

(A)

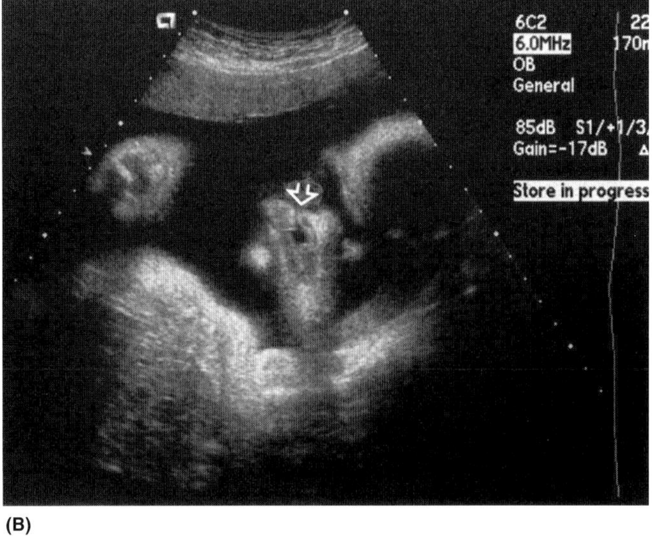

(B)

FIGURE 19 A AND B ■ Semilobar holoprosencephaly. (**A**) A transverse imaging of fetal head shows partial separation of the monoventricular cavity (V *with arrows*). P, posterior. (**B**) This was associated with cleft lip and palate as demonstrated on this coronal image of the fetal face (*open arrow*).

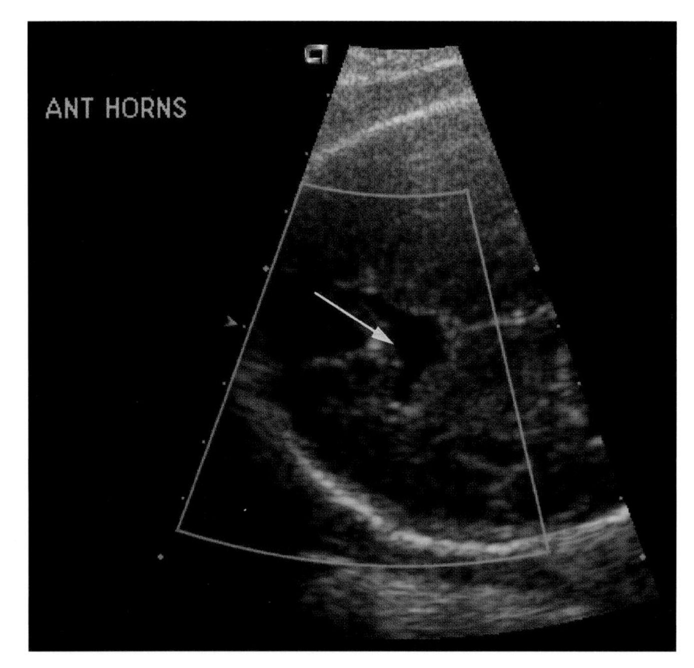

FIGURE 20 ■ Septo-optic-dysplasia. Coronal ultrasound of the region of the frontal horns demonstrating absence of the septum pellucidum, fused frontal horns with inferior pointing frontal horns (*arrow*).

cavity. The "cup" form is intermediate between the two, with the residual brain having a cup-like configuration on the sagittal view. Other investigators simply divide cases of alobar holoprosencephaly into those with or without a "dorsal sac." Thus, alobar holoprosencephaly without a dorsal sac is similar to the ball form, whereas the cup and pancake forms have a dorsal sac (35). Distinguishing ultrasound features are listed in Table 6.

Hydraencephaly

The term "hydraencephaly" is derived from the combination of the words hydrocephalus and anencephaly, although it differs from both these entities. In contrast to hydrocephalus, fetuses with hydraencephaly have a complete lack of cerebral tissue, and in contrast to anencephaly, there is covering by bone, skin, dura, and leptomeninges (Fig. 22) (37). Hydraencephaly is thought to result from bilateral in utero internal carotid artery infarction. Ultrasound features are listed in Table 5 (Fig. 22).

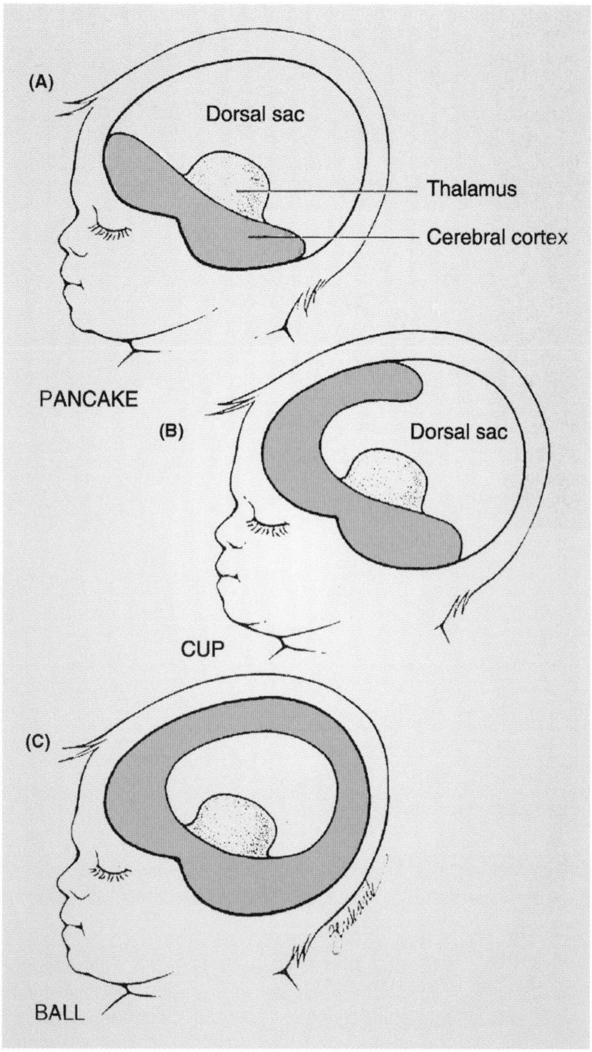

FIGURE 21 A–C ■ Alobar holoprosencephaly. Drawing of the sagittal view of the (**A**) "pancake," (**B**) "cup," and (**C**) "ball" configurations of alobar holoprosencephaly. This concept is based on the amount of cerebral tissue that covers the monoventricular cavity, residual tissue that resembles a pancake, a cup, or a ball. In both pancake and cup forms, a dorsal sac protrudes from the monoventricular cavity. *Source*: From Ref. 35.

TABLE 6 ■ Classifications and Features of Holoprosencephaly

Classification	Features
Alobar (most common in utero)	Fused thalami
	Facial abnormalities
	Monoventricular cavity
	Absent falx cerebri
Semilobar (rarest form)	Fused thalami
	Facial abnormalities
	Monoventricular cavity with increased cerebral tissue, especially in the occipital lobes
Lobar (usually identified in infants)	Squared frontal horns
	Absent cavum septi pellucidi ± ventriculomegaly
	Otherwise fairly normal intracranial anatomy

Midline Cystic Abnormalities of the Brain

Some focal cystic abnormalities of the brain may be associated with other intracranial abnormalities such as hydrocephalus, whereas others are isolated findings. Some cystic intracranial abnormalities are typically midline. These include Dandy–Walker malformation and vein of Galen aneurysm. Other cystic intracranial abnormalities are usually lateral but can occur within the midline, such as arachnoid cyst or brain tumors. In addition, complex intracranial abnormalities, such as agenesis of

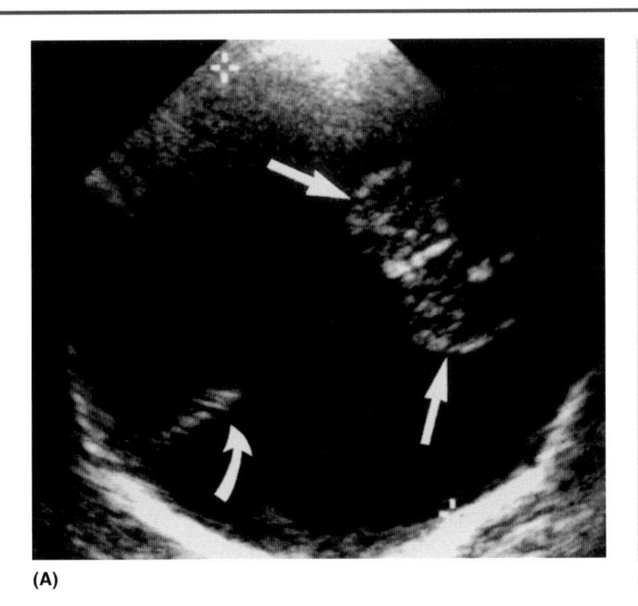

(A)

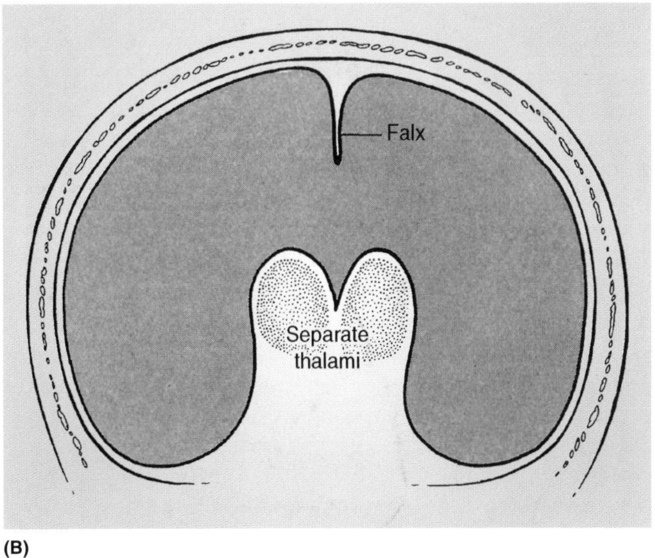

(B)

FIGURE 22 A AND B ■ Hydranencephaly in utero. (**A**) Coronal in utero ultrasound scan shows a midline falx (*curved arrow*) and no cerebral tissue with nonfused thalami (*straight arrows*), diagnostic features of hydranencephaly. (**B**) Corresponding diagram demonstrates the absence of cerebral tissue but the presence of the falx and the nonfused thalami.

the corpus callosum or alobar holoprosencephaly, may have midline cystic components (Table 7).

Dandy–Walker Malformation

One of the more common midline cystic abnormalities is the Dandy–Walker malformation, which is characterized by a midline cyst located within the posterior fossa that communicates with the fourth ventricle (Fig. 14). This cyst is posterior to the ventricle and is associated with near or complete absence of the cerebellar vermis and with an enlarged posterior fossa. Associated hydrocephalus is seen in 80%, concurrent CNS malformations occur in more than 50%, and systemic or chromosomal abnormalities are common (19,20).

In an occasional variant of Dandy–Walker malformation, only partial absence of the cerebellar vermis occurs

TABLE 7 ■ Midline Intracranial Cysts: Distinguishing Features

Cyst type	Location	Features
Dandy–Walker	Posterior fossa	Communication with fourth ventricle
		Concomitant hydrocephalus
Vein of Galen aneurysm	Supratentorial, posterior to corpus callosum	Doppler signal within cyst
		Arteriovenous shunting
Arachnoid cyst	Any location	Smooth wall
		Asymmetric location
		Mass effect
		± Hydrocephalus (dependent on location)
Cystic neoplasm	Any location	Solid mass
		Cystic components
		Irregular outline
		Mass effect
Agenesis of the corpus callosum with interhemispheric cyst	Interhemispheric between frontal horns of the lateral ventricles	Communication with cephalad-displaced third ventricle
Alobar holoprosencephaly	Posterior, supratentorial	Typical features of alobar holoprosencephaly (see prior tables)

in association with a small posterior fossa cyst (Fig. 15). Dandy–Walker malformation, also called complete vermian agenesis, is defined as complete or near-complete agenenesis of the cerebellum vermis with a midline cyst that communicates with the fourth ventricle and causes enlargement of the posterior fossa. Many investigators believe that a spectrum of posterior fossa malformations exists between Dandy–Walker malformation and Dandy–Walker variant, which is also called inferior vermian agenesis. In Dandy–Walker variant, one sees only partial vermian agenesis and no enlargement of the posterior fossa. Therefore, a communication between the fourth ventricle and the cisterna magna is identified only on the most caudal portions of the posterior fossa. Although the defect is less severe, it is often associated with other abnormalities. For instance, Estroff et al. found that 47% of these fetuses had concurrent non-CNS anomalies, and 29% had an abnormal karyotype (30). Of these fetuses, 52% were developmentally normal in the first year of life. Similarly, Chang et al. found a nearly equal incidence of morphologic abnormalities with inferior vermian agenesis (76%) as compared with complete vermian agenesis (75%), but they noted a higher incidence of chromosomal abnormalities with inferior vermian agenesis (53%) compared with complete vermian agenesis (32%) (38).

Vein of Galen Aneurysm

An uncommon but pathognomonic midline intracranial cyst is the vein of Galen aneurysm. The vein of Galen is a major draining vein that lies posterior and slightly superior to the thalami within the subarachnoid space. In this condition, abnormal rapid flow occurs between the artery and the vein, with resultant aneurysmal dilatation of the vein of Galen. Thus, a moderate-sized, midline cyst with Doppler flow is diagnostic of an arteriovenous malformation (Fig. 23) (39).

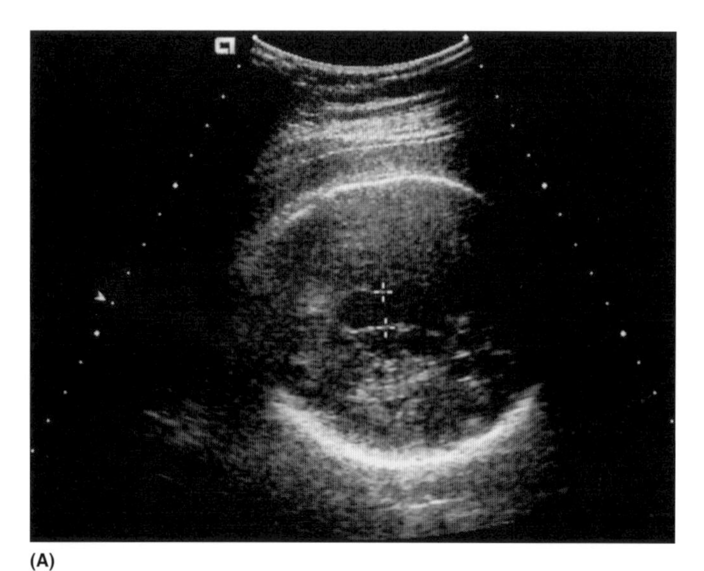

(A)

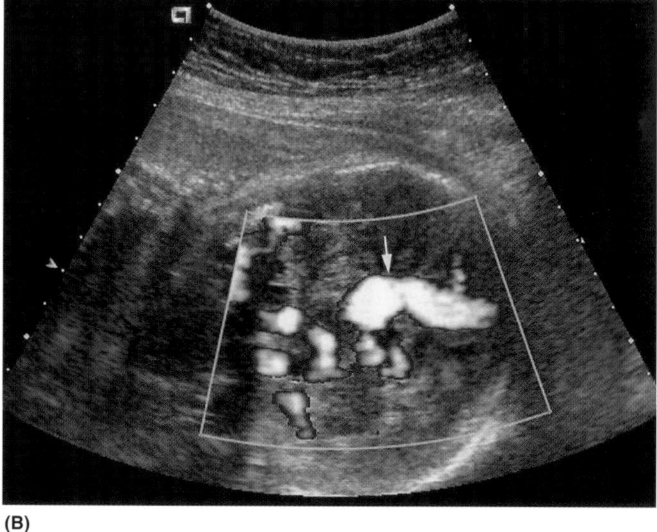

(B)

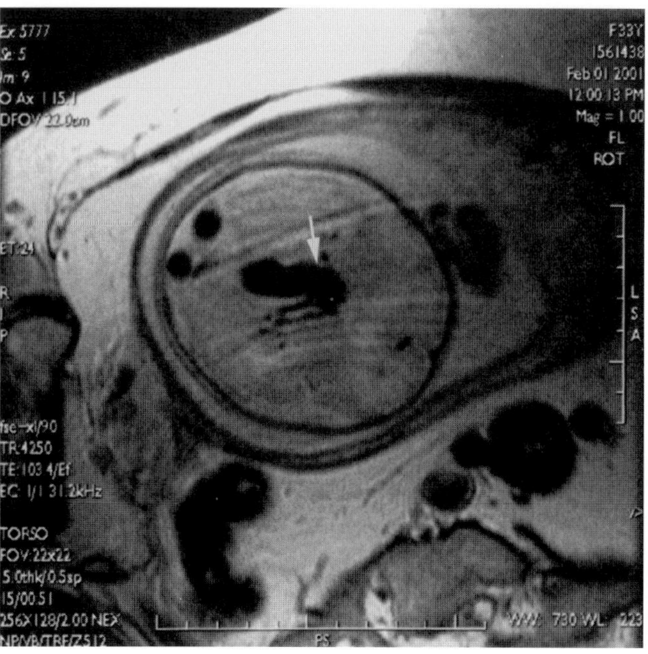

(C)

FIGURE 23 A–C ■ Dilated vein of Galen. (**A**) Transverse axial sonogram shows a central tubular anechoic structure (*calipers*). (**B**) Semicoronal power Doppler image demonstrates increase size of the vein of the arteriovenous malformation (*arrow*). (**C**) Corresponding in utero MRI showing dilated vein as central area of decreased signal intensity (*arrow*).

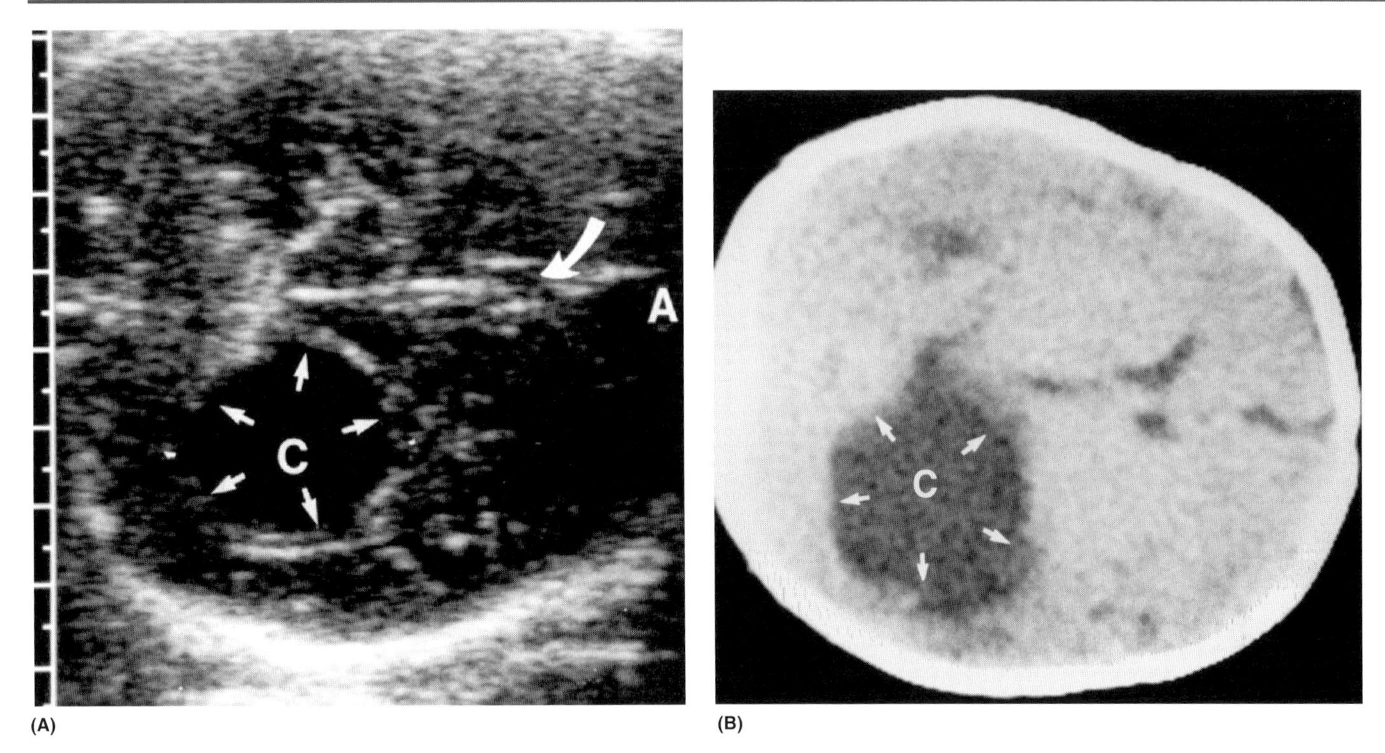

(A) (B)

FIGURE 24 A AND B ■ Arachnoid cyst. (**A**) In utero axial ultrasound scan demonstrates a cystic area in the temporal occipital area. *Curved arrow* represents cavum septi pellucidi. (**B**) Corresponding CT scan demonstrates a cystic area in the temporal occipital region corresponding to an arachnoid cyst. A, anterior; C, arachnoid cyst.

Other Midline Cystic Abnormalities

Other midline cystic abnormalities may be associated with more complex malformations, such as agenesis of the corpus callosum with an interhemispheric cyst. In agenesis of the corpus callosum, the cavum septum pellucidum is absent, and the third ventricle is elevated between the laterally displaced frontal horns. A midline interhemispheric cyst may communicate with the third ventricle (31).

A posterior midline fluid collection may be related to holoprosencephaly, as previously described (Fig. 21). This dorsal sac originates from a monoventricular cavity and extends posteriorly (35).

Other cystic abnormalities are typically located laterally, but they may be in the midline. Such abnormalities include arachnoid cyst (Fig. 24) and, rarely, cystic neoplasm (Fig. 25) (40).

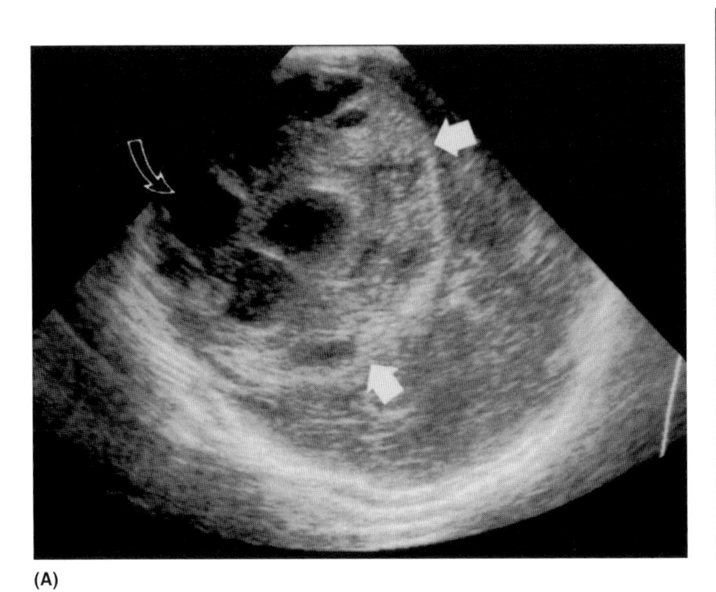

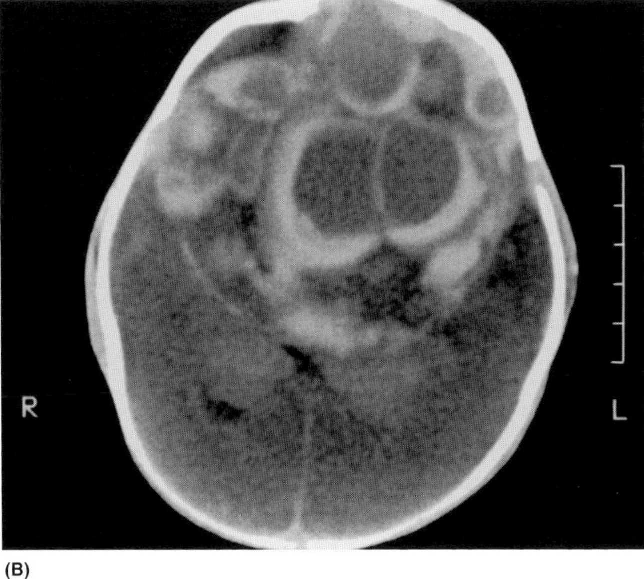

(A) (B)

FIGURE 25 A AND B ■ Cystic teratoma of the brain. (**A**) Axial scan demonstrates a large complex mass (*arrows*) with cysts in the frontal region (*curved arrow*). (**B**) CT scan shows an enhancing mass with cystic components and solid components.

TABLE 8 ■ Causes of Lateral Intracranial Cysts

Common	Uncommon
Choroid plexus cysts	Arachnoid cyst
	Porencephaly
	Schizencephaly
	Unilateral hydrocephalus
	Cystic neoplasm
	Intracranial hemorrhage

Lateral or Asymmetric Cyst Abnormalities

Lateral cystic abnormalities are listed in Tables 8 and 9. Most common is the choroid plexus cyst. Other lateralizing cystic brain abnormalities are listed in Table 8, with distinguishing features listed in Table 9.

Choroid Plexus Cysts

Choroid plexus cysts are frequently identified prenatally. Most of these cysts are benign and regress spontaneously (Fig. 4). Reports have associated these cysts with trisomy 18 (41,42). Therefore, the examiner should be familiar with the features of trisomy 18 (43). These features include certain abnormalities that may be recognized sonographically, such as rocker-bottom feet or overlapping fingers with clenched hands (Fig. 26). Nyberg et al. have (44) presented data with major sonographic findings identified prenatally in trisomy 18, including choroid plexus cyst, cystic hygroma, intrauterine growth retardation, nuchal

TABLE 9 ■ Lateral Intracranial Cysts (Distinguishing Features)

Abnormality	Features
Choroid plexus cyst	Cyst present within the choroid plexus of the lateral ventricle
	Bilateral or unilateral
Arachnoid cyst	Smooth wall
	Asymmetric
	Mass effect
	± Hydrocephalus (dependent on location)
Porencephaly	Usually unilateral
	Communicates with ventricle
	Cavity size increases with distance from ventricle
	Decreased size of ipsilateral bony calvarium
	Cavity lined by white matter (postnatal MRI)
Schizencephaly	Unilateral or bilateral
	May or may not communicate with ventricle
	Cavity size increases with distance from ventricle
	Defect lined by gray matter (postnatal MRI)
Unilateral hydrocephalus	Unusual: check opposite ventricle after changing fetal position or perform "angled" scanning (see text)
Cystic neoplasm	Mass effect
	Solid or irregular with cystic components
Intracranial hemorrhage	Initial echogenic clot in ventricle
	Later echogenic clot plus ventriculomegaly on side of hemorrhage

Abbreviation: MRI, magnetic resonance imaging.

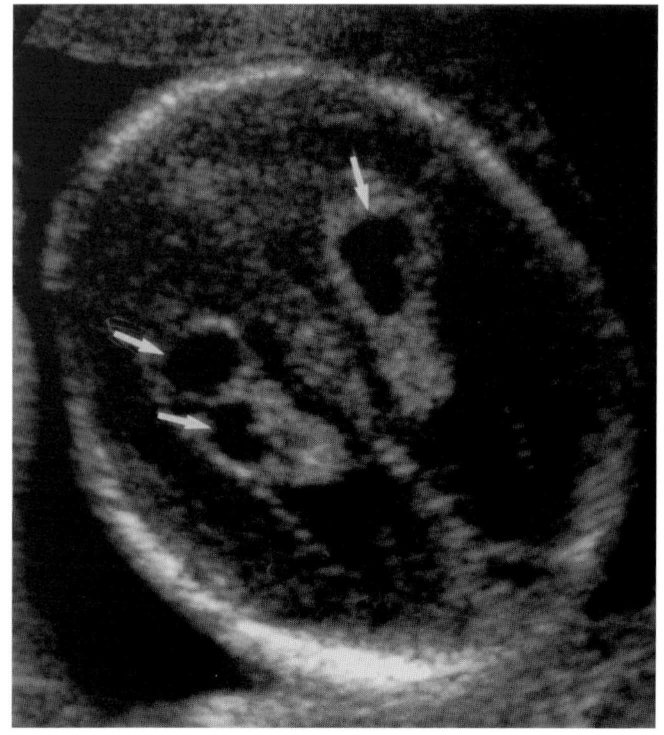

(A)

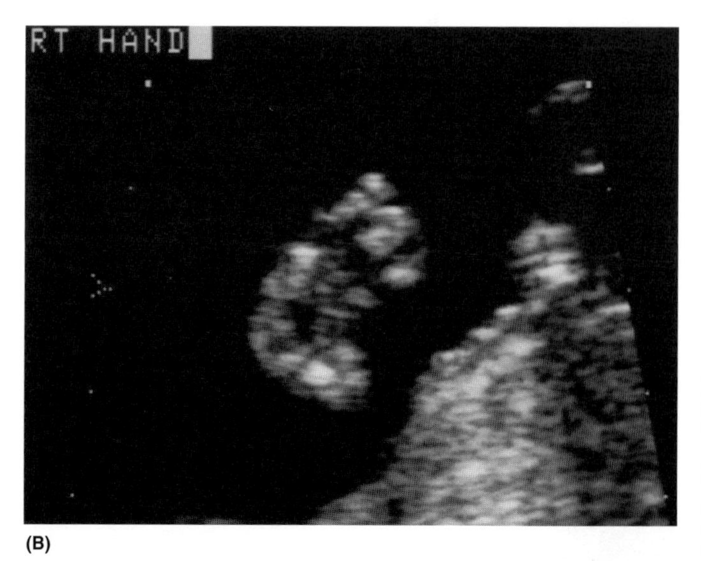

(B)

FIGURE 26 A AND B ■ Choroid plexus cyst—trisomy 18. (**A**) Prenatal ultrasound demonstrating multiple bilateral choroid plexus cysts (*arrows*). (**B**) Ultrasound image of the fetal hand demonstrating clenched hand. The hand was never observed to open during imaging. There were other associated fetal abnormalities.

thickening, meningomyocele, and cardiac defect. Extremity abnormalities most commonly identified include clubfeet, rocker-bottom feet, and clenched hands. Identification of the fetal hands opening and closing is an important feature in helping to distinguish a normal fetus and one with trisomy 18. The adequacy of ultrasound in detecting features of chromosomal abnormalities such as trisomy 18 when choroid plexus cysts are detected is debated. Some investigators have believed that amniocentesis should be offered, based on certain criteria such as size (larger than 5mm or larger than 10mm), bilaterality of the cysts, and lack of regression of these cysts. At least a detailed ultrasound assessment should be offered to all patients with fetuses in which a choroid plexus cyst is identified (44–48). Most investigators agree that even if one additional ultrasound abnormality is detected, amniocentesis should be offered. Appropriate genetic counseling must be given to the patient based on the most recent published data, prior personal experience, and familiarity of possible abnormalities associated with trisomy 18 (44–50).

Porencephaly

Two abnormalities, porencephaly and schizencephaly, are often considered together because of their similar appearance. Some investigators believe that these two entities have separate origins, however. Porencephaly (Fig. 27) results from localized brain destruction during early gestation. This destructive process results in a smooth-walled, fluid-filled cavity that communicates directly with the cerebral ventricle and extends to the cranial vault. Porencephaly is usually unilateral, with the defect increasing in size the further the distance from the ventricle. The destructive process involves the entire thickness of the cerebral cortex (35,51). Therefore, the lining of the cavity contains white matter that may be imaged by postnatal MRI (Fig. 28) (51).

Schizencephaly

Schizencephaly may resemble porencephaly on ultrasound examination. These defects may be either unilateral or bilateral. Schizencephaly is considered to be a migrational abnormality rather than a destructive process (53). The walls of the schizencephalic defect are usually separate, but they may be fused (52,53). Because schizencephaly is a migrational abnormality, the cavity is completely lined by gray matter, a feature that distinguishes the condition from porencephaly when postnatal MRI is performed (Fig. 28) (51).

Arachnoid Cysts

Arachnoid cysts may be midline or lateral. These cysts are thought to arise by either separation of the arachnoid into two layers that secrete CSF and form a cyst or formation of a cyst between the arachnoid and the pia mater. The cells lining the cyst wall produce CSF, which is trapped and thus produces a well-demarcated cystic cavity. Sonographically, these cysts are well circumscribed, often

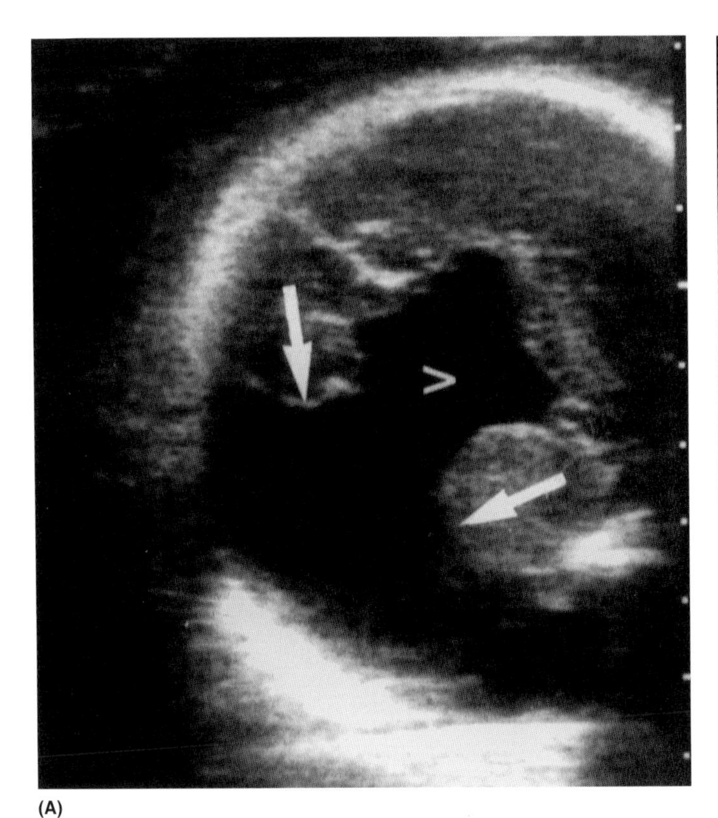

(A)

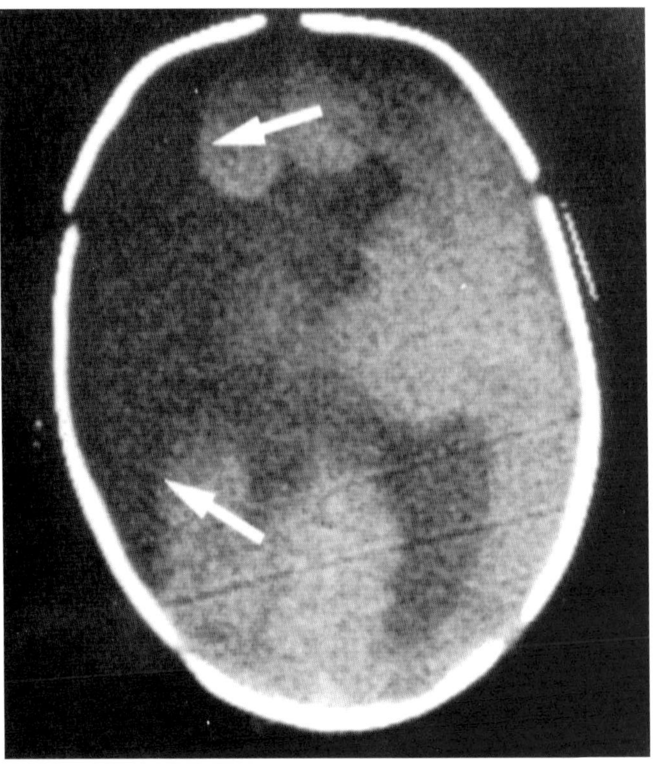

(B)

FIGURE 27 A AND B ■ Porencephaly. (**A**) In utero coronal ultrasound scan demonstrates the absence of the septum pellucidum with dilated ventricles (V) and a large, wedge-shaped cystic area originating from one ventricle (*arrows*) that extends to the bony calvarium (**B**). Corresponding axial CT scans of the newborn demonstrate findings similar to those described in **A** (*arrow*).

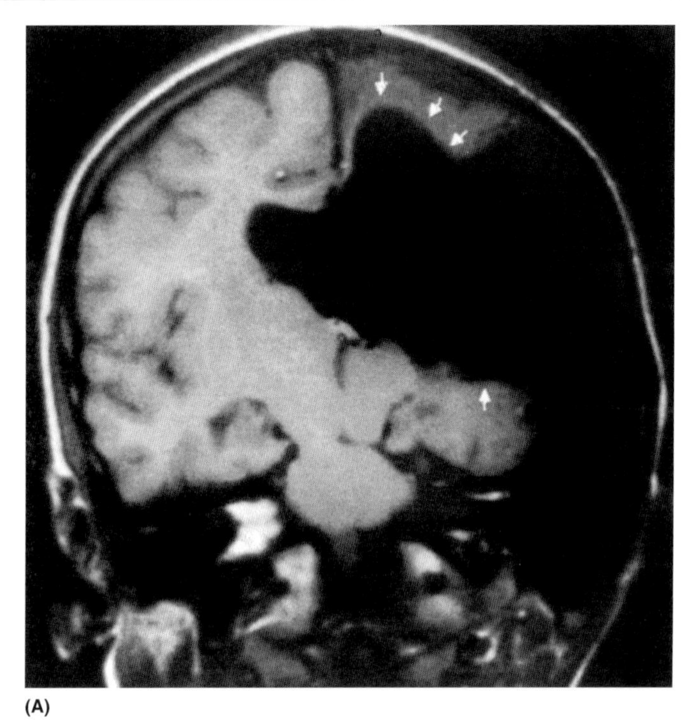

(A)

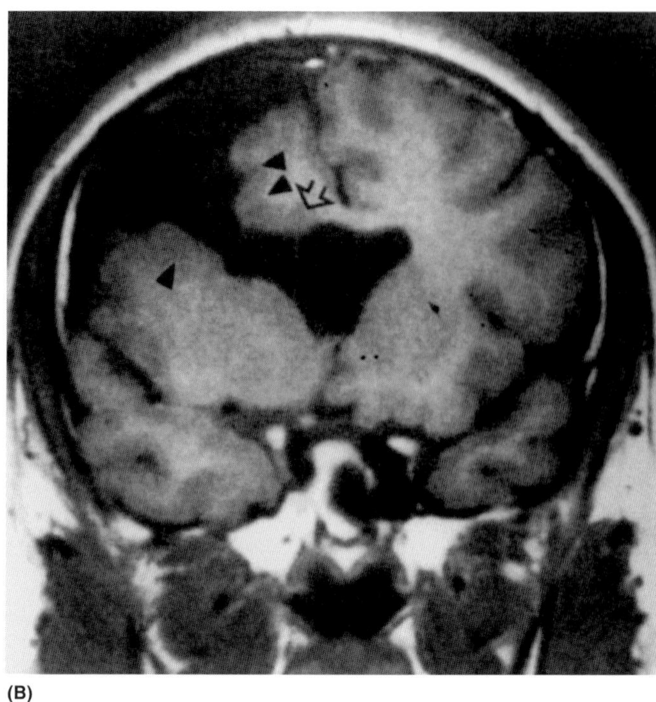

(B)

FIGURE 28 A AND B ■ MRI of porencephaly versus schizencephaly. (**A**) Coronal spin-echo 600/20 image shows a large cavity extending from the left lateral ventricle through the entire cerebral hemisphere. The cavity is covered by white matter (*arrows*). Thus, MRI defines this to represent porencephaly. (**B**) Unilateral schizencephaly visualized with a coronal spin-echo 600/20 MRI scan shows a defect similar to that seen in porencephaly, except gray matter rather than white matter covers the entire cleft (*arrow* and *arrowheads*). Thus, MRI defines this to represent schizencephaly, which is a form of polymicrogyria, a developmental abnormality, rather than porencephaly, which is thought to represent an in utero cerebral infarction. *Source*: From Ref. 52.

circular, and they may occur anywhere within the subarachnoid space. They may produce hydrocephalus because of their mass effect (Fig. 24) (35).

Intracranial Neoplasms

Intracranial neoplasms are rare in the newborn or fetus. Intracranial tumors in the perinatal period are most commonly teratomas (54). Gliomas are the second most common intracranial tumors in newborns (55). Intracranial neoplasms with cystic components are usually teratomas (Fig. 25) (54). These tumors may appear anywhere within the brain, usually have an irregular appearance, and are mostly echogenic, with occasional cystic portions.

Unilateral Hydrocephalus

Unilateral hydrocephalus is a rare abnormality caused by focal obstruction of the CSF pathway at the level of the foramen of Monro. Examination of the fetal cerebral ventricle using the angled technique is helpful to distinguish between unilateral and bilateral hydrocephalus.

In Utero Intracranial Hemorrhage

Fetal intracranial hemorrhage has been reported in the early third trimester of pregnancy. Intracranial hemorrhage is thought to occur in the germinal matrix of newborns. The origin of germinal matrix hemorrhage is explained by the extreme capillary permeability of the vascular germinal matrix that predisposes it to hemorrhage. Elevation in blood pressure caused by fetal hypoxia may lead to hemorrhage in the germinal matrix. The mechanism for in utero intraventricular hemorrhage is less clear and is often secondary to maternal factors such as pancreatitis, hepatitis, preeclampsia, or bleeding disorders. Often, the cause of in utero intraventricular hemorrhage is unknown (56). Unlike neonatal hemorrhage, fetal intracranial hemorrhage is most often intraventricular. Initial sonographic examination shows an echogenic clot filling the ventricles that later may result in hydrocephalus (56). In utero hemorrhage usually occurs in the beginning of the third trimester of pregnancy. Later, one may see hydrocephalus with mixed echogenic debris layering within the ventricles. Occasionally, mixed echogenicity within the enlarged choroid is identified, resulting from hemorrhage into the choroid plexus (Fig. 29). Rarely in utero hemorrhages may occur in other regions (Figs. 10 and 30) (57).

Cranial Defects (Partial or Complete)

Causes and distinguishing features of cranial defects or deformed cranium are listed in Table 10. These deformities may overlap. Cranial defects may be partial or complete, and partial defects are usually secondary to cephaloceles, amniotic band syndrome (ABS), or limb–body wall complex (LBWC). Complete defects are secondary to exencephaly with acrania or to anencephaly. Moreover,

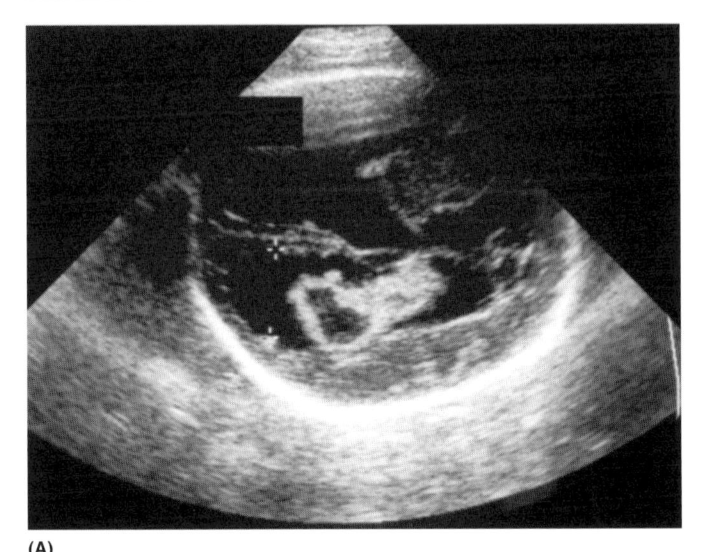

(A)

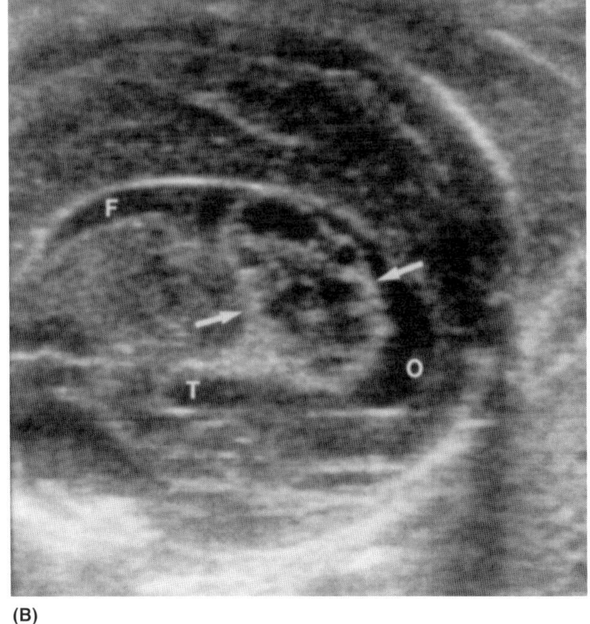

(B)

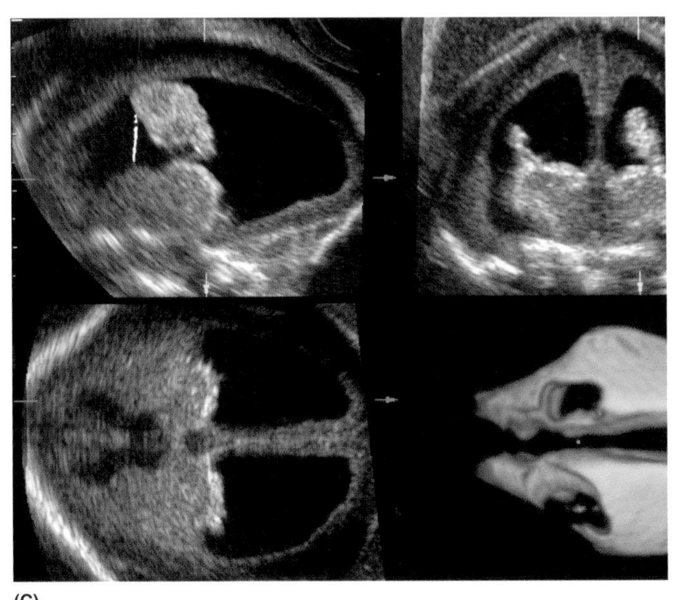

(C)

FIGURE 29 A–C ■ Sonographic spectrum of intracranial hemorrhage. (**A**) Ultrasound scan of the fetal head shows a well-identified clot in the irregular choroid plexus in the dilated lateral ventricle (*caliper*). (**B**) Sagittal ultrasound scan of a dilated ventricle in another fetus with probable 10-day-old hemorrhage shows a mixed echogenic clot (*arrows*) adhered to the trigone region of the lateral ventricle. (**C**) 3D image in another patient with intraventricular hemorrhage as echogenic clot and resultant ventriculomegaly. F, frontal horn; O, occipital horn; T, temporal horn. *Source*: Courtesy of General Electric Medical.

severe bone demineralization occurring with such abnormalities as osteogenesis imperfecta may resemble a complete cranial defect.

The differential diagnosis of deformed cranium is comprehensive and includes everything from physiologic head compression to open neural tube defects or fetal demise. Distinguishing features of the more common abnormalities are cited in Table 10. For instance, the "lemon" sign was originally described with open neural tube defects. This sign may not be present in open neural tube defects identified in the third trimester of pregnancy, however, whereas a mild lemon-head appearance may be identified in normal fetuses (Table 11).

Anencephaly

Anencephaly, the single most common open neural tube defect, is characterized by the absence of the cerebral hemispheres and the lack of bony calvarium above the orbits (Fig. 31). The fetus has a normal midbrain and posterior fossa but lacks normal development of the cerebral hemisphere. Approximately one-third of cases have a variable amount of angiomatous stroma that may mimic rudimentary brain (58). Anencephaly may be accompanied by polyhydramnios, which is thought to be secondary to severe brain dysfunction resulting in ineffective fetal swallowing.

Exencephaly (Acrania)

Acrania is a developmental abnormality characterized by partial or complete absence of the cranial vault. Exencephaly is acrania with protrusion of brain tissue into the amniotic cavity. Exencephaly is different from but similar to anencephaly (59). As in anencephaly, the fetal cranium is absent; however, unlike in anencephaly, brain tissue is always present (Fig. 32). In exencephaly, the brain tissue appears heterogenous and disorganized. The hypothesis of a continuum of findings between exencephaly and anencephaly is supported by the observation of residual brain tissue (angiomatous stroma) in about one-third of cases of anencephaly (58).

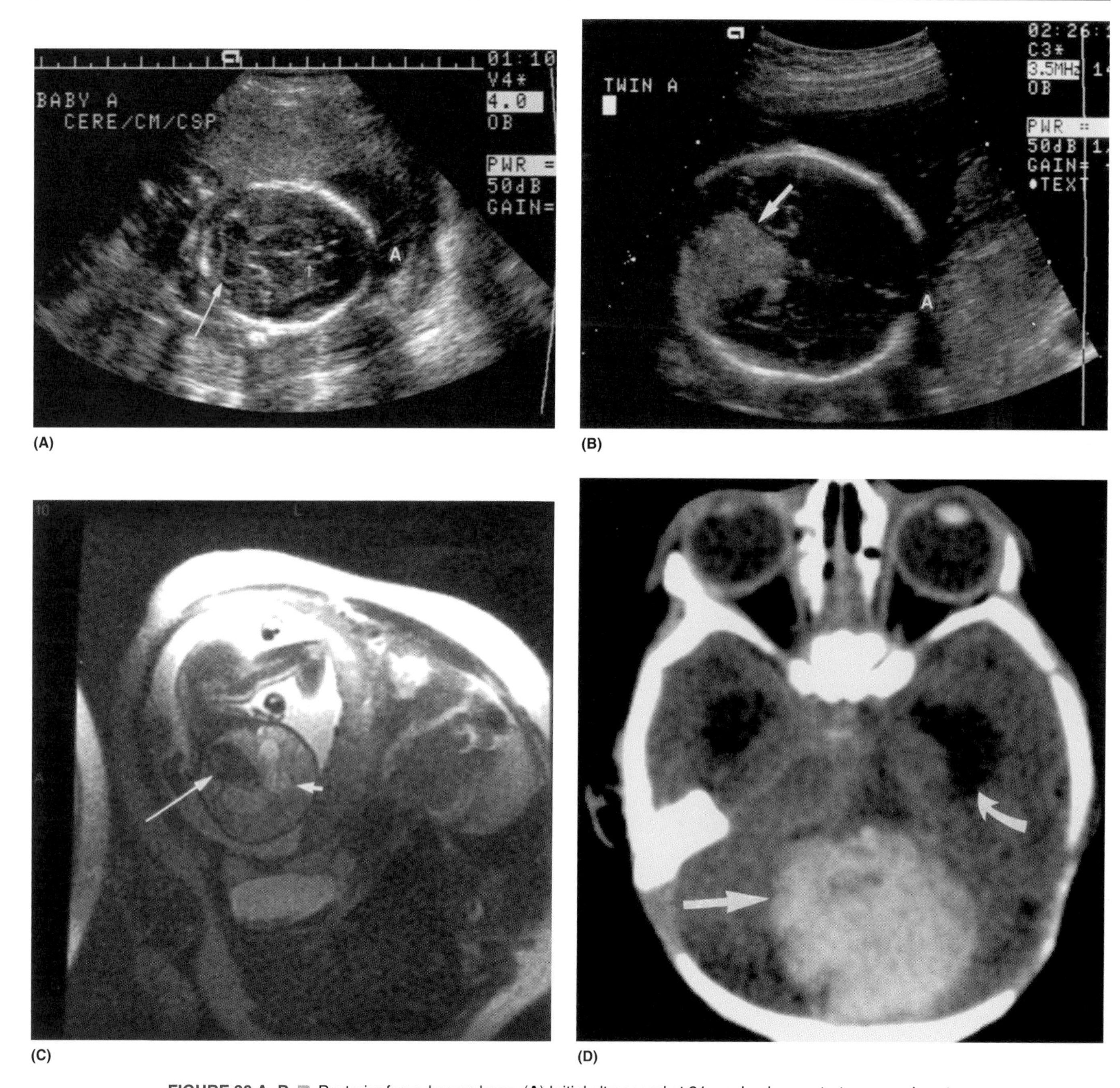

FIGURE 30 A–D ■ Posterior fossa hemorrhage. (**A**) Initial ultrasound at 21 weeks demonstrates normal posterior fossa (*arrow*). (**B**) Follow-up sonogram at 28 weeks demonstrates large echogenic posterior fossa mass (*arrow*). A, anterior. (**C**) In-utero single-shot fast-spin echo MRI shows region of decreased signal intensity (*arrow*) corresponding to cerebellar hemorrhage. Arrow points to corresponding ventriculomegaly. (**D**) Postnatal CT shows area of hemorrhage as region of increased density (*arrow*). Also note mild ventriculomegaly (*curved arrow*). *Source*: From Ref. 57.

Cephaloceles

The sonographic appearance of cephaloceles is variable. Most cephaloceles are midline occipital extracranial masses (29,60). If the cephaloceles are large, the residual brain size is small. Cephaloceles may be associated with the lemon-head appearance of the bony calvarium seen in open neural tube defects of the Arnold–Chiari type II malformation (Fig. 33). Cephaloceles may also occur in other intracranial locations. These lesions are discussed in more detail in the chapter on the Fetal Neck and Spine.

ABS and LBWC

ABS (61) or LBWC (62) should be considered in a fetus with an extremely bizarre and asymmetric cranial defect (Fig. 34). Both ABS and LBWC may be associated with certain abnormalities thought to result from disruption of the amnion in utero. In LBWC, the more severe defect, the fetal body is usually adhered to the placenta, with absent or shortened umbilical cord. Many investigators consider LBWC and ABS to be separate entities, whereas others consider these to be a spectrum of the same syndrome.

TABLE 10 ■ Distinguishing Features of Cranial Defects of a Deformed Cranium

Anencephaly	Absent bony calvarium above orbits
	Orbits well visualized
	Absence of supratentorial brain
	Residual brain (angiomatous stroma)
	± Spina bifida
	Polyhydramnios
Exencephaly secondary to acranium	Calvarium absent
	Disorganized supratentorial brain tissue
ABS	Asymmetric cephalocele
	Fixed fetal parts
	Other deformities, including limb amputation
LBWC	Asymmetric encephaloceles
	Similar to ABS except fetal body to placenta
	Usually more severe than ABS, associated bizarre defects of fetal body, scoliosis, omphaloceles
Cephalocele	Midline defect
	Extracranial cyst or brain tissue
	± Ventriculomegaly
	"Lemon" sign possible
Open neural tube defects	"Lemon" sign
	"Banana" sign
	Mild hydrocephalus
	Spinal defect
Fetal demise	Overlapping sutures (Spalding's sign)
	Poor visualization of intracranial structures
	Associated findings of fetal demise
Microcephaly	Calvarium present
	Decreased brain tissue
	Head circumference 2–3 SD below expected for menstrual age
Craniosynostosis	Complete or partial
	Deformed skull
	± Microcephaly (see above)
	Abnormal cephalic index

Abbreviations: ABS, amniotic band syndrome; LBWC, limb–body wall complex; SD, standard deviation.

TABLE 11 ■ Differential Diagnosis of "Lemon" Sign of Fetal Head

- Common: normal fetus
- Less common: open neural tube defect (Arnold–Chiari type II)
- Uncommon: cephalocele

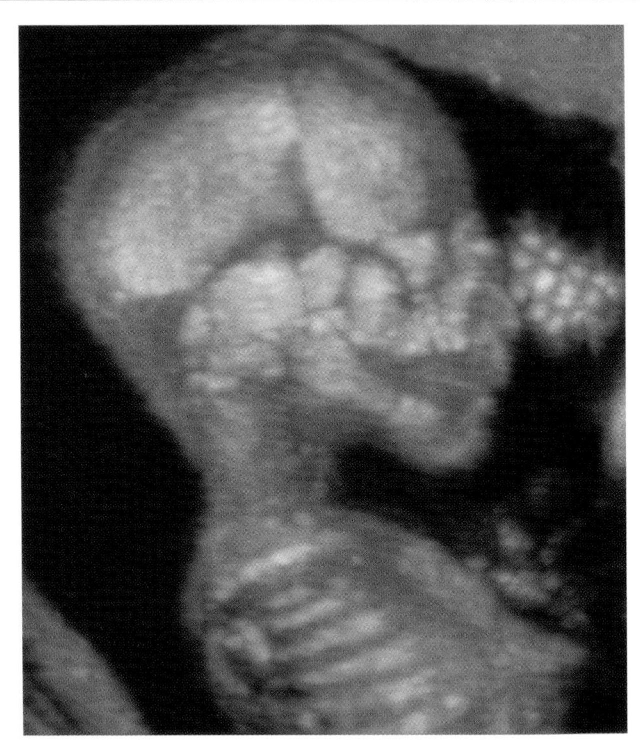

(A)

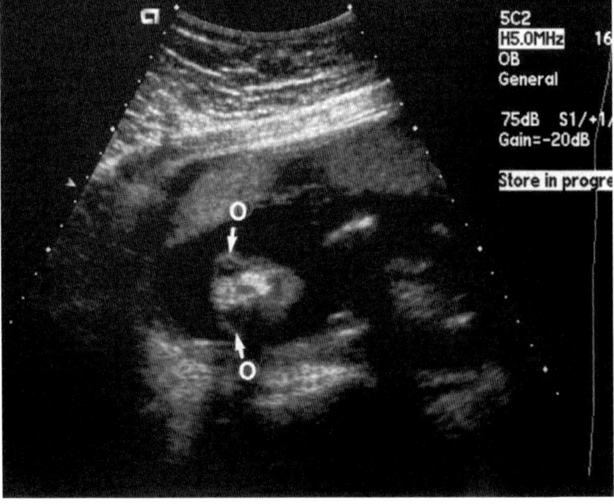

(B)

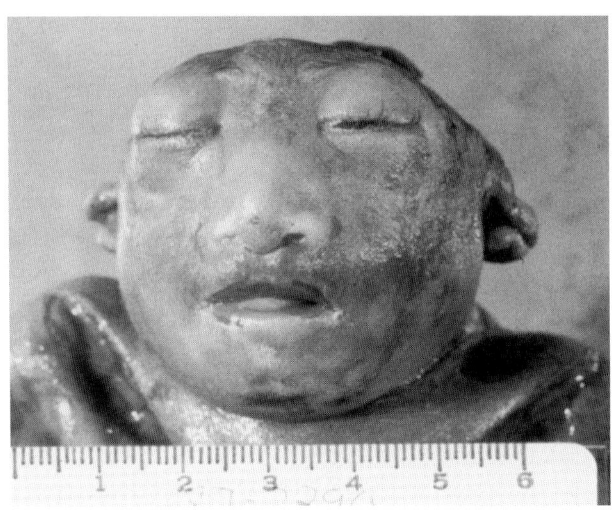

(C)

FIGURE 31 A–C ■ Normal ossification versus anencephaly. (**A**) 4D image showing normal ossification of fetal cranium. (**B**) Coronal scan of another fetus shows absence of the brain and calvarium cephalad to the orbits (O). (**C**) Anatomical specimen demonstrating absent cranial vault, in this second fetus; essentially no normal cerebral tissue above the orbits. *Source*: (**A**) Courtesy of General Electric Medical.

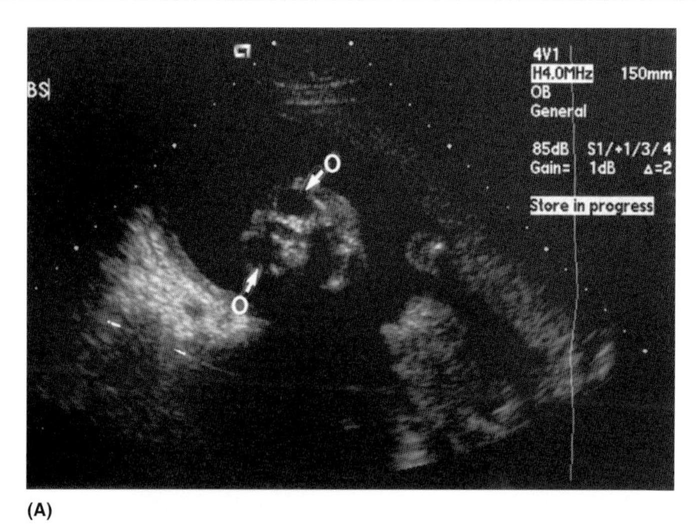

(A)

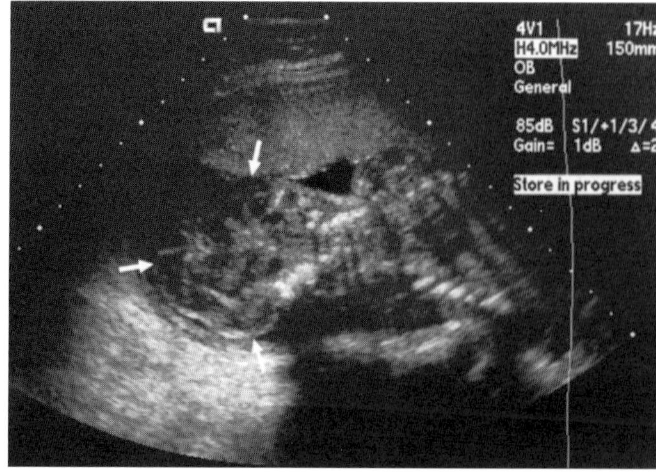

(B)

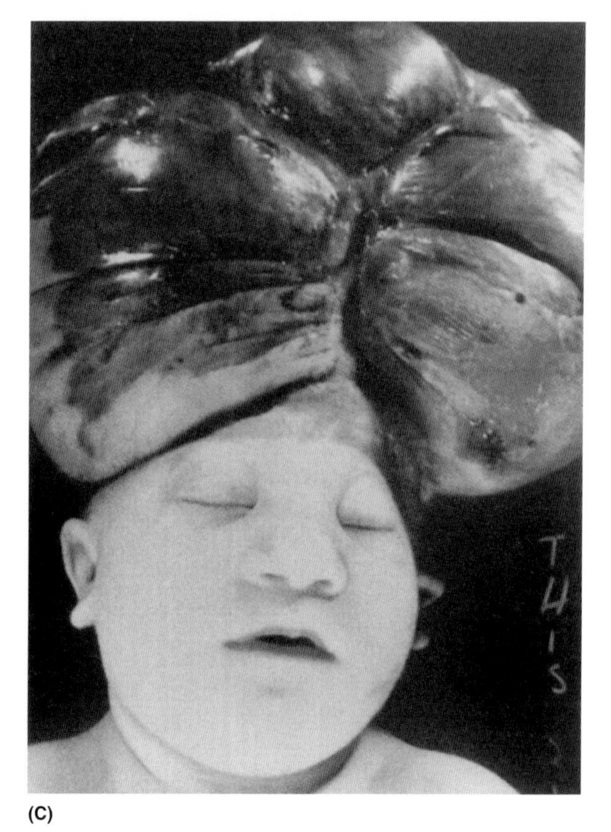

(C)

FIGURE 32 A–C ■ Acrania. (**A**) Coronal scan through the fetal face demonstrates apparent absence of the fetal brain and calvarium cephalad to the orbits. (**B**) Coronal scan through the fetal head demonstrates protruding brain (*arrows*) and lack of bony calvarium. (**C**) Anatomical specimen of a different 38-week stillborn shows a large multilobed cerebral tissue covered by a highly vascular layer of skin. *Source*: From Ref. 60.

These may be similar entities with a common origin, that is, disruption of the amnion, but with different features. The LBWC is also called the body-stalk anomaly, usually with a defect of the amnion extending from the placenta to the fetal body wall incorporating the umbilical cord. This severe anomaly is associated with bizarre intracranial defects, abdominal wall defects, and limb amputations.

In ABS, the amnion is disrupted, with the fetal body part adhering to the chorion. The resultant amputational defects may be minor to severe, depending on the timing of the amnion disruption. In the ABS, the fetus is fixed in the amniotic defect at a site separate from the placenta. The amnion may be thought to protect the fetus from the chorion. Thus, with disruption of the fetal amnion, the fetal part adheres to the chorion. During continual growth, the fetal part is constricted, resulting in asymmetric amputational defects.

Bony Demineralization

Diffuse demineralization may occur in osteogenesis imperfecta, hypophosphatasia, or achondrogenesis type I. This process is presented in more detail in the chapter on Fetal Skeleton. In cases of severe demineralization of the bony calvarium, supervisualization of intracranial structures may be noted. This increased visualization of the intracranial structures may be confused with such abnormalities as exencephaly resulting from acrania. Unlike in exencephaly, however, the fetus has an intact

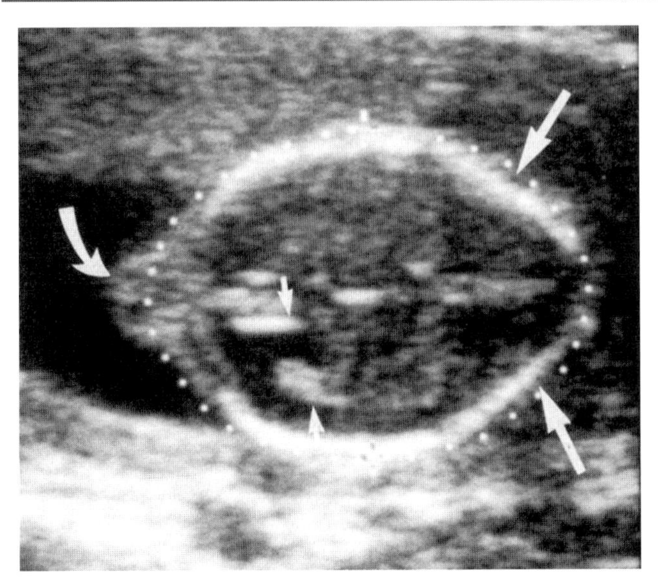

FIGURE 33 ■ Cephalocele: "banana" sign. Ultrasound scan of the fetal head demonstrates a small encephalocele (*curved arrow*), the lateral ventricle dilated to 12 mm (*small arrows*), and mild overlapping of the cranial sutures (*large straight arrows*), or so-called lemon sign.

but poorly mineralized cranial vault. Careful scanning reveals concomitant limb abnormalities. For instance, osteogenesis imperfecta is associated with bowed limbs combined with supervisualization of the intracranial structures (Fig. 35) (63).

Small or Deformed Cranium

Microcephaly

Severe microcephaly may be difficult to distinguish from anencephaly. Microcephaly, characterized by a reduction of brain mass and head size with an intact bony calvarium, is defined as a fetal head circumference either 2 (64) or 3 (65) standard deviations below the mean of that expected for menstrual age. Most clinicians define microcephaly as being less than 3 standard deviations below the mean of expected head circumference. The head circumference, rather than the biparietal diameter, is used as the measurement for microcephaly, because the biparietal diameter is affected by normal head compression (Tables 12–14) (65). Chervenak et al. also found that the ratio of head circumference (perimeter) to abdominal circumference at 4 standard deviations was associated with no false-positive diagnoses, whereas the ratio of femur length to head circumference at 3 standard deviations was a sensitive threshold without a false-positive diagnosis of microcephaly.

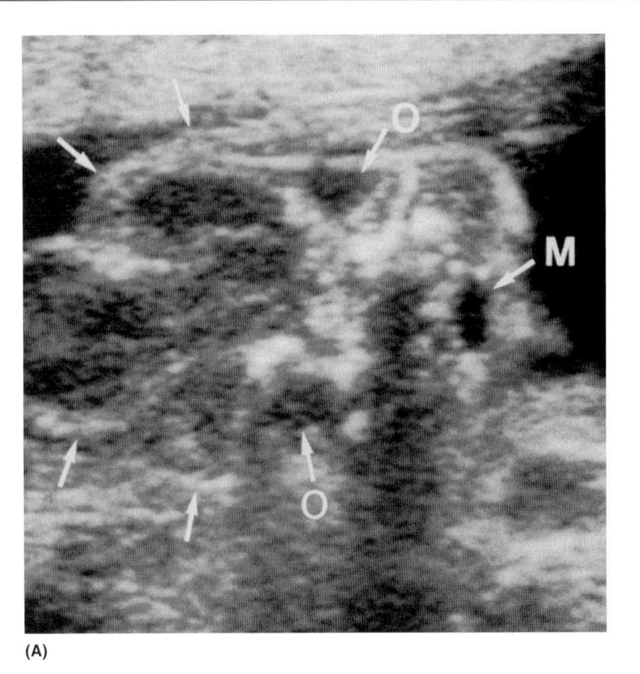

(A)

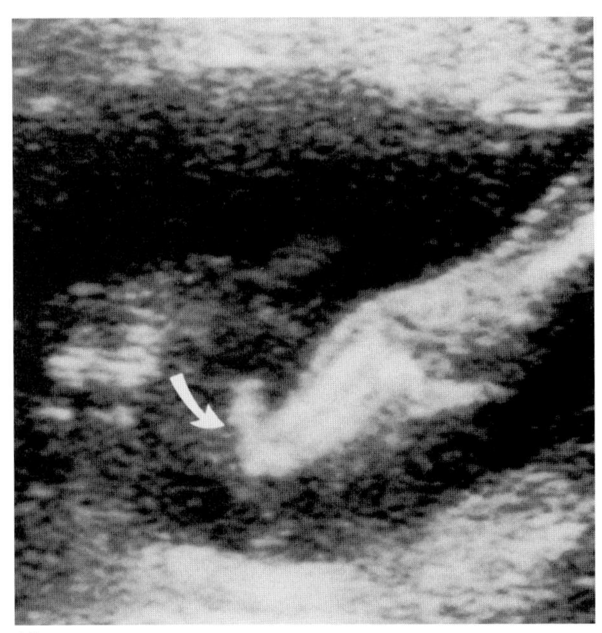

(B)

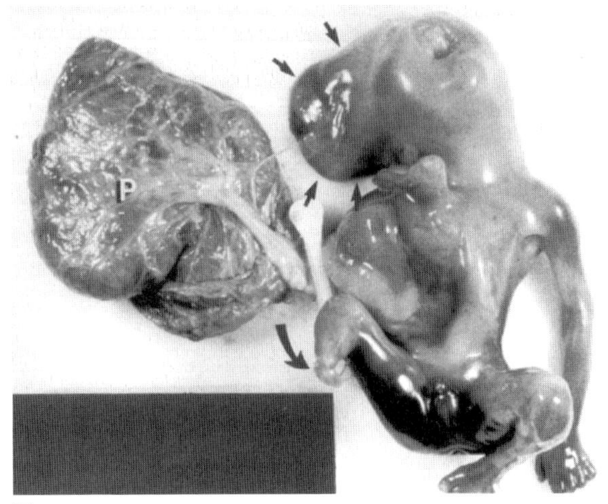

(C)

FIGURE 34 A–C ■ Limb–body wall complex—severe amniotic bands. (**A**) Coronal ultrasound scan demonstrates orbits (O), an open fetal mouth (M), and disorganized brain tissue (*arrows*). (**B**) Amputational defect of the foot of the lower extremity (*curved arrow*) is visible. (**C**) Corresponding anatomic specimen demonstrates an asymmetric cranial defect (*straight arrows*) and amputation of the upper and lower limbs (*curved arrow*), with an abdominal defect contiguous with the placenta (P) and umbilical cord.

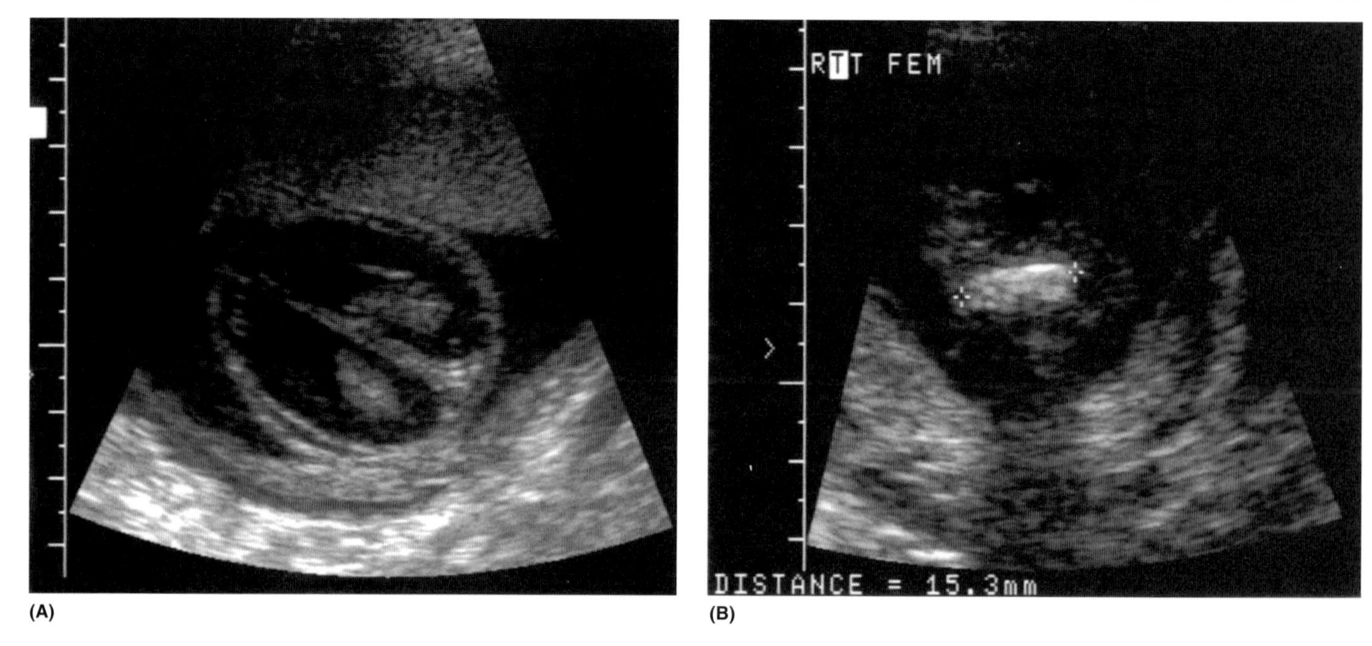

(A) (B)

FIGURE 35 A AND B ■ Osteogenesis Imperfecta. (A) "Super visualization" of the intracranial structures on this axial scan of the head resulting from poor cranial ossification in the fetus with osteogenesis imperfecta. (B) Ultrasound of the femur demonstrates a short and thickened femur.

TABLE 12 ■ Mean and SD of Head Perimeter as a Function of Gestational Age

	SD above mean			SD below mean				
Week no.	+2	+1	Mean	−1	−2	−3	−4	−5
20	204	189	175	160	145	131	116	101
21	216	201	187	172	157	143	128	113
22	228	213	198	184	169	154	140	125
23	239	224	210	195	180	166	151	136
24	250	235	221	206	191	177	162	147
25	261	246	232	217	202	188	173	158
26	271	257	242	227	213	198	183	169
27	282	267	252	238	223	208	194	179
28	291	277	262	247	233	218	203	189
29	301	286	271	257	242	227	213	198
30	310	295	281	266	251	236	222	207
31	318	304	289	274	260	245	230	216
32	327	312	297	283	268	253	239	224
33	334	320	305	290	276	261	246	232
34	341	327	312	297	283	268	253	239
35	348	333	319	304	289	275	260	245
36	354	339	325	310	295	281	266	251
37	360	345	330	316	301	286	272	257
38	364	350	335	320	306	291	276	262
39	369	354	339	325	310	295	281	266
40	372	358	343	328	314	299	284	270
41	375	360	346	331	316	302	287	272
42	377	363	348	333	319	304	289	275

Abbreviation: SD, standard deviation.
Source: From Ref. 65.

TABLE 13 ■ Mean and SD of Head Perimeter to Abdominal Perimeter as a Function of Gestational Age

	SD above mean			SD below mean				
Week no.	+2	+1	Mean	−1	−2	−3	−4	−5
20	1.43	1.34	1.25	1.16	1.07	0.98	0.89	0.8
21	1.42	1.33	1.24	1.15	1.06	0.97	0.88	0.79
22	1.41	1.32	1.23	1.14	1.05	0.96	0.87	0.78
23	1.4	1.31	1.22	1.13	1.04	0.95	0.86	0.78
24	1.39	1.3	1.21	1.12	1.03	0.94	0.86	0.77
25	1.38	1.29	1.2	1.11	1.02	0.94	0.85	0.76
26	1.37	1.28	1.19	1.1	1.02	0.93	0.84	0.75
27	1.36	1.27	1.18	1.1	1.01	0.92	0.83	0.74
28	1.35	1.26	1.17	1.09	1	0.91	0.82	0.73
29	1.34	1.25	1.17	1.08	0.99	0.9	0.81	0.72
30	1.33	1.25	1.16	1.07	0.98	0.89	0.8	0.71
31	1.33	1.24	1.15	1.06	0.97	0.88	0.79	0.7
32	1.32	1.23	1.14	1.05	0.96	0.87	0.78	0.69
33	1.31	1.22	1.13	1.04	0.95	0.86	0.77	0.68
34	1.3	1.21	1.12	1.03	0.94	0.85	0.76	0.68
35	1.29	1.2	1.11	1.02	0.93	0.84	0.76	0.67
36	1.28	1.19	1.1	1.01	0.92	0.84	0.75	0.66
37	1.27	1.18	1.09	1.00	0.92	0.83	0.74	0.65
38	1.26	1.17	1.08	1.00	0.91	0.82	0.73	0.64
39	1.25	1.16	1.08	0.99	0.90	0.81	0.72	0.63
40	1.24	1.16	1.07	0.98	0.89	0.80	0.71	0.62
41	1.24	1.15	1.06	0.97	0.88	0.79	0.70	0.61
42	1.23	1.14	1.05	0.96	0.87	0.78	0.69	0.60

Abbreviation: SD, standard deviation.
Source: From Ref. 65.

TABLE 14 ■ Mean and SD of Femur Length to Head Perimeter Ratio as a Function of Gestation Age

	SD below mean						SD above mean				
Week no.	−5	−4	−3	−2	−1	Mean	+1	+2	+3	+4	+5
20	0.107	0.122	0.137	0.152	0.167	0.180	0.197	0.212	0.227	0.242	0.257
21	0.111	0.126	0.141	0.156	0.171	0.190	0.201	0.216	0.231	0.246	0.261
22	0.115	0.130	0.145	0.160	0.175	0.190	0.205	0.220	0.235	0.250	0.265
23	0.118	0.133	0.148	0.163	0.178	0.190	0.208	0.223	0.238	0.253	0.268
24	0.121	0.136	0.151	0.166	0.181	0.200	0.211	0.226	0.241	0.256	0.271
25	0.123	0.138	0.153	0.168	0.183	0.200	0.213	0.228	0.243	0.258	0.273
26	0.125	0.140	0.155	0.170	0.185	0.200	0.215	0.230	0.245	0.260	0.275
27	0.127	0.142	0.157	0.172	0.187	0.200	0.217	0.232	0.247	0.262	0.277
28	0.129	0.144	0.159	0.174	0.189	0.200	0.219	0.234	0.249	0.264	0.279
29	0.130	0.145	0.160	0.175	0.190	0.200	0.220	0.235	0.250	0.265	0.280
30	0.131	0.146	0.161	0.176	0.191	0.210	0.221	0.236	0.251	0.266	0.281
31	0.132	0.147	0.162	0.177	0.192	0.210	0.222	0.237	0.252	0.267	0.282

Abbreviation: SD, standard deviation.
Source: From Ref. 65.

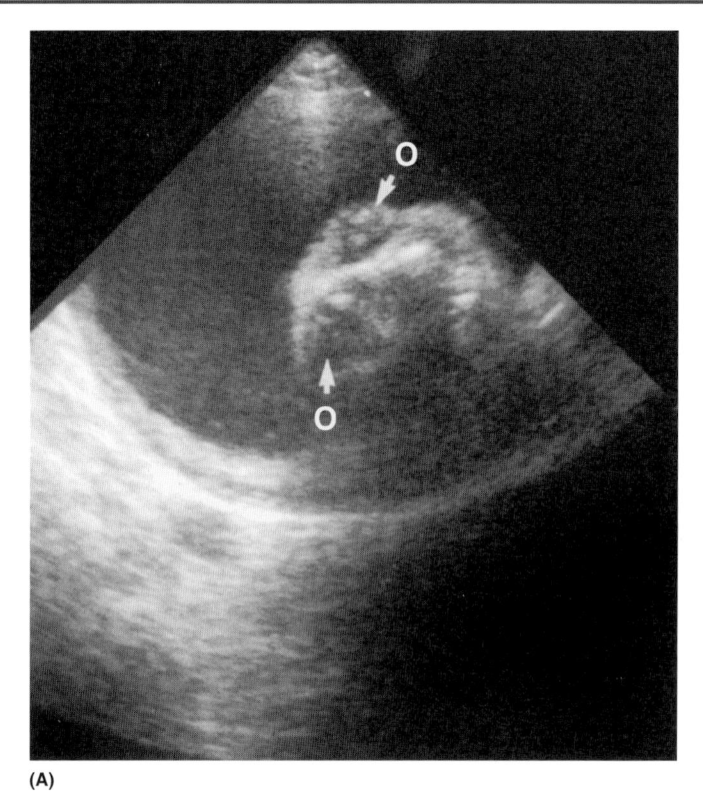

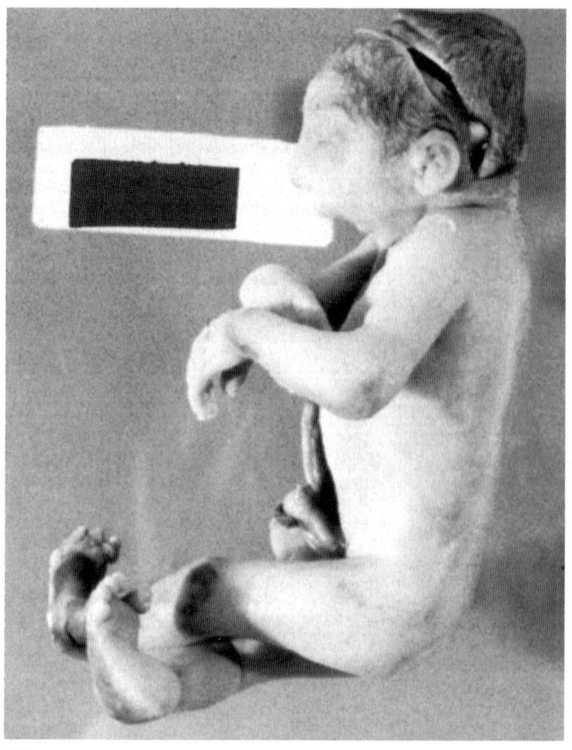

(A) (B)

FIGURE 36 A AND B ■ Microcephaly. (**A**) Coronal ultrasound through the fetal face demonstrates question. Lack of bony calvarium above orbits (O). This is similar in appearance to anencephaly. (**B**) Specimen shows intact calvarium with microcephaly.

Causes of microcephaly are numerous and include in utero infections, anoxia, and chromosomal abnormalities (Fig. 36).

Fetal Demise

When fetal demise has occurred, one may note poor visualization of the intracranial structures and overlapping of the cranial sutures, or the Spalding sign (Fig. 37) (66). Other more definitive signs of fetal demise, including absence of fetal movement and lack cardiac motion, confirm the diagnosis.

"Lemon" Head

Mild overlapping of the cranial sutures occurs most commonly as a variant of normal pregnancy; however, it also may occur as previously described with open neural tube defects. Careful scanning of the posterior fossa and the spine is indicated in these cases. Even with an open neural tube defect in the third trimester of pregnancy, the cranial shape reverts to a more normal oval shape, rather than that of a "lemon" head. As previously described, the lemon-head appearance may occur in association with fetal cephaloceles (Tables 10 and 11).

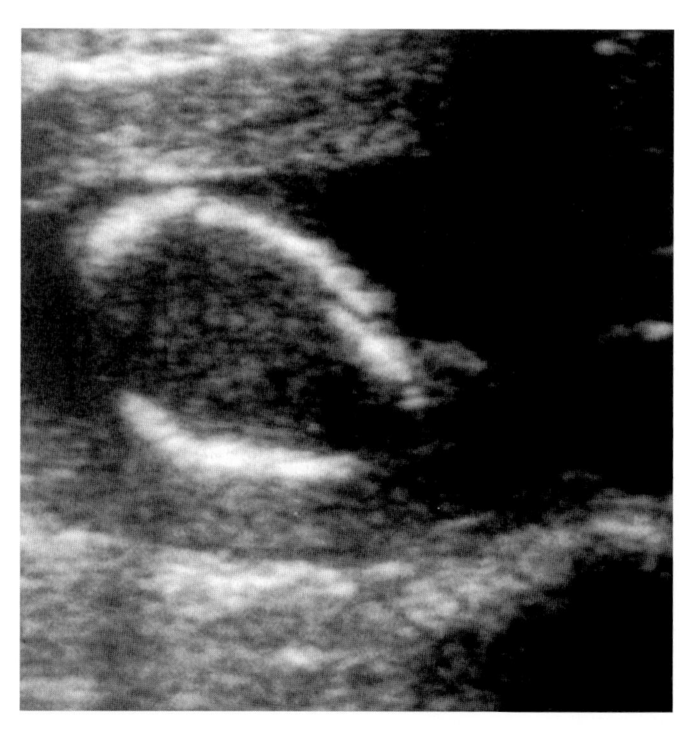

FIGURE 37 ■ Fetal demise. Ultrasound image of the fetal cranium demonstrates poor visualization of the fetal cranial structure and overlapping of the cranial sutures (Spalding's sign), as occurs with in utero fetal demise.

Cloverleaf Skull

The term cloverleaf refers to a skull characterized by frontal and bitemporal bulging of the calvarium. This

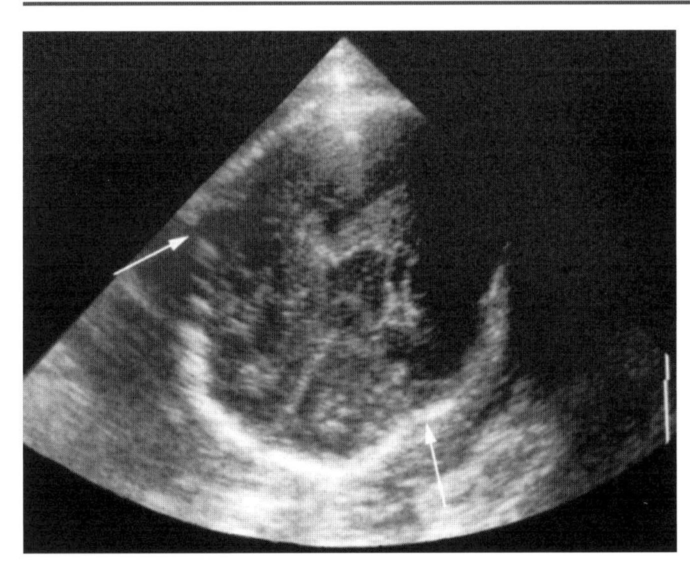

FIGURE 38 ■ Thanatophoric dwarfism. Mild "cloverleaf" deformity of the fetal cranium with prominent bitemporal (*arrows*) and frontal regions. This defect may be seen in other malformations. Cloverleaf deformity of the skull identified with limb shortening and small thorax is diagnostic of thanatophoric dwarfism.

pronounced deformity of the fetal cranium (cloverleaf) may occur in thanatophoric dwarfism (Fig. 38) or homozygous achondroplasia (67). A cloverleaf appearance is not pathognomonic of these dwarfisms, however, and it may occur with Apert, Carpenter, Crouzon, or Pfeiffer syndromes. Cloverleaf skull associated with limb shortening and abnormally small thoracic circumference is nearly pathognomonic of thanatophoric dwarfism. This condition is discussed in more detail in the chapter on the Fetal Skeleton.

Intracranial Calcifications

In utero cerebral calcifications are rare, subtle, and difficult to recognize on prenatal ultrasound. Most commonly, these calcifications are a result of in utero infection (68), but they may be secondary to fetal brain tumors (54,55). In utero infections are most commonly viral and are secondary to cytomegalic inclusion disease. In utero infections may manifest as a faint subependymal calcification with accompanying ventriculomegaly and possible microcephaly (Fig. 39). Herpes simplex type II can be acquired in utero and may have a similar appearance (68). Alternatively, calcifications from toxoplasmosis in the neonate are scattered, and calcifications from rubella are rare. Intracranial calcifications as a result of intracranial tumor are associated with a mass effect and are localized to one portion of the brain (Fig. 39) (54,55).

INTRACRANIAL DOPPLER IMAGING

Doppler ultrasound has been used to investigate the fetal circulation to assess various different fetal conditions. Blood flow analysis of the umbilical arteries, descending thoracic aorta, middle cerebral artery (MCA), tricuspid and mitral inflow, ductus venosus, and inferior vena cava have all been assessed (69). Within the fetal brain, Doppler imaging has been used to investigate not only the MCA but also the internal carotid artery and the anterior cerebral artery in various fetal and obstetric populations (70,71). Some authors have been site specific in studying differences of the origin (M1) and the distal (M2) portion of the MCA.

MCA for Fetal Distress

An area of interest in Doppler analysis of intracranial vascularity is to detect fetal distress. The theory behind performing Doppler imaging of the intracranial vessels is the concept of a brain-sparing phenomenon associated with any form of fetal distress. In situations such as intrauterine growth retardation, resistance to the umbilical arteries flowing into the placenta may be increased.

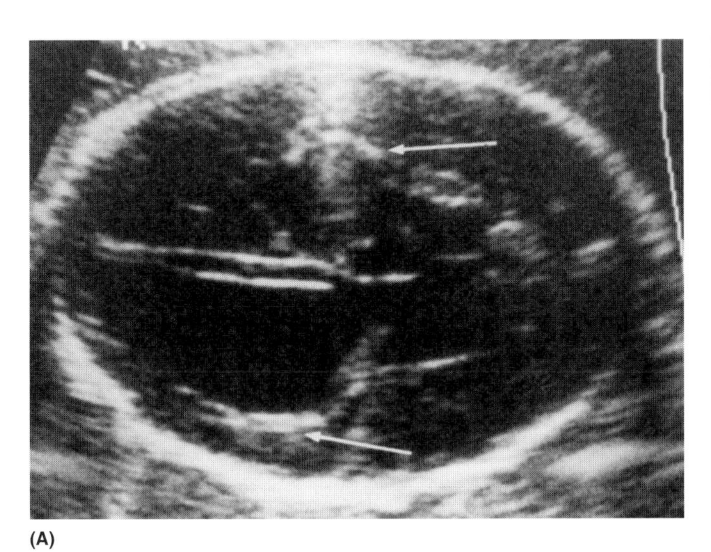

(A)

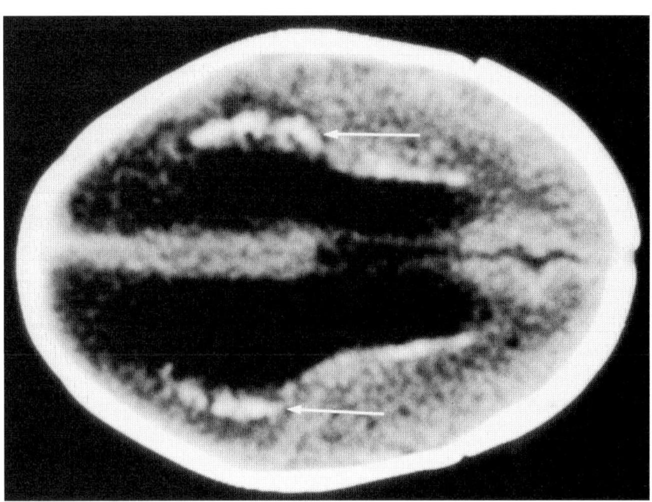

(B)

FIGURE 39 A AND B ■ Intracranial infection. (**A**) Ultrasound image of the fetal head shows dilated lateral ventricles with bilateral periventricular echogenicities (*arrows*) that correspond to calcifications. (**B**) Postnatal nonenhanced CT scanning demonstrates periventricular calcifications (*arrows*).

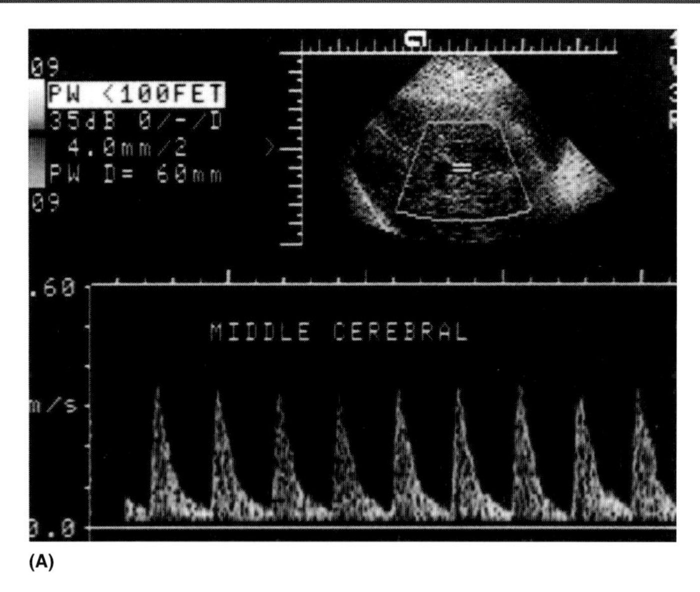

(A)

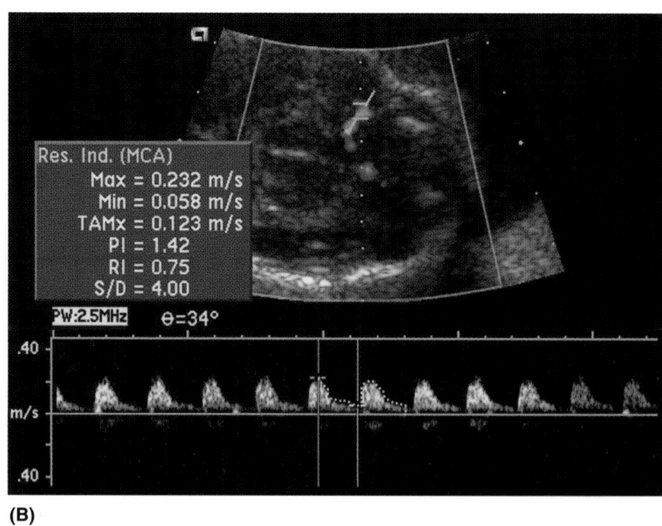

(B)

FIGURE 40 A AND B ■ Middle cerebral artery Doppler. (**A**) Color flow ultrasound demonstrating circle of Willis with middle cerebral artery (*arrow*). (**B**) Angled corrected middle cerebral artery demonstrating peak systolic velocity, the PI, RI and *S/D* ratio. PI, pulsatility index; RI, resistive index; *S*, peak systolic velocity; *D*, end-diastolic velocity.

This phenomenon is observed as a decrease or even a reversal of the end-diastolic velocity of the umbilical artery. With decreased circulation to the fetus, however, comes a resultant decreased vascular resistance to the vessels feeding the brain to preserve cerebral blood flow. This change is observed as increased end-diastolic velocity. Thus, several different parameters have been used to evaluate the systolic-to-diastolic ratios of the vessels of the brain.

Within the literature, few investigators agree on which Doppler parameters of the MCA should be used. Some have advocated use of peak systolic velocity (PSV) (*S*), whereas others use end-diastolic velocity (*D*) or time-average velocities. Other investigators have used the pulsatility index, the resistive index, or the *S/D* ratio. These indices are listed in Figures 40 and 41. Some workers have evaluated these indices in the MCA and have compared them with indices in other vessels, most notably the umbilical artery (73–77).

Sources of error inherent to measurement of blood flow velocity are often due to the angle of insonation. Thus, absolute velocity measurements are usually not used, but instead, parameters that are independent of angle of insonation of the vessels have been advocated. These ratios include the S/D ratio, the Pourcelot (resistive) index, and the pulsatility index (Fig. 40). The *S/D* ratio is the PSV (*S*) within the MCA divided by the end-diastolic velocity (*D*) in the MCA. The second ratio used is the Pourcelot index, also termed the resistive index. This resistive index is $[(S - D)/S)]$ where *S* is equal to the PSV and *D* corresponds to the end-diastolic velocity. A third measurement used is the pulsatility index, which is the PSV minus the end-diastolic velocity, and this number is divided by the mean velocity. The pulsatility index is the least convenient index to use because it requires digitization of the entire waveform to obtain the mean velocity.

The lower references of pulsatility and resistance indices are of importance in assessment of fetal intrauterine growth retardation. In reviewing these different ratios, the resistive indices are only represented from a scale from 0 to 1. However, the pulsatility index provides a waveform analysis of a wide range of waveform patterns. Measurements of resistive index have been reported in a number of large studies. Recently, a study by Kurmanavicius et al. (73) evaluated the umbilical and middle cerebral arteries and computed the ratios between the two vessels, the umbilical artery resistive index to MCA index (Table 15).

A number of studies have evaluated the pulsatility index with published ranges from 20 weeks until term. In

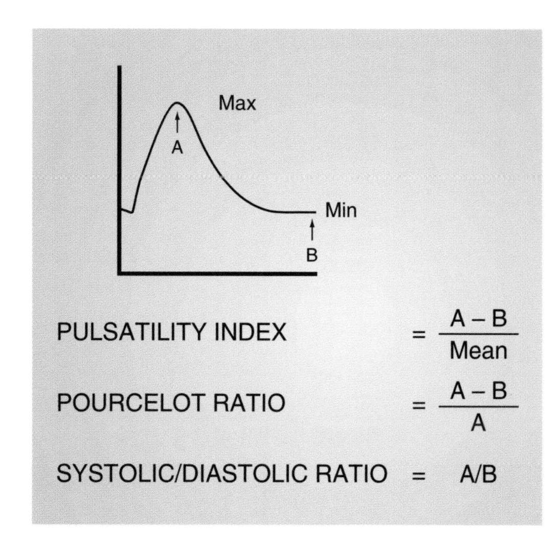

FIGURE 41 ■ Methods of quantification of arterial waveform. The resistive index is synonymous with the Pourcelot ratio. A, systolic velocity; B, diastolic velocity; Mean, mean velocity. *Source*: From Ref. 72.

TABLE 15 ■ Fitted RI Centiles in the Umbilical Artery, Middle Cerebral Artery and Placentocerebral Ratio

	Umbilical artery			Middle cerebral artery			Placentocerebral ratio		
Gestation week	5th	50th	95th	5th	50th	95th	5th	50th	95th
24	0.61	0.72	0.83	0.78	0.87	–	0.70	0.81	0.97
25	0.60	0.71	0.82	0.79	0.88	–	0.68	0.79	0.96
26	0.59	0.70	0.81	0.80	0.89	–	0.66	0.78	0.95
27	0.58	0.69	0.80	0.80	0.90	–	0.64	0.76	0.94
28	0.57	0.68	0.79	0.80	0.90	–	0.63	0.75	0.93
29	0.56	0.67	0.79	0.79	0.90	–	0.62	0.74	0.93
30	0.55	0.66	0.78	0.79	0.90	–	0.60	0.73	0.93
31	0.54	0.65	0.77	0.78	0.89	–	0.60	0.73	0.93
32	0.53	0.64	0.76	0.76	0.88	–	0.59	0.72	0.93
33	0.52	0.63	0.75	0.75	0.87	–	0.58	0.72	0.94
34	0.51	0.62	0.74	0.73	0.86	–	0.58	0.72	0.94
35	0.50	0.61	0.73	0.72	0.85	–	0.58	0.72	0.95
36	0.49	0.60	0.73	0.70	0.83	–	0.57	0.72	0.96
37	0.47	0.59	0.72	0.68	0.81	–	0.57	0.72	0.97
38	0.46	0.58	0.71	0.66	0.80	–	0.57	0.72	0.98
39	0.45	0.57	0.70	0.63	0.78	–	0.58	0.73	0.99
40	0.44	0.56	0.69	0.61	0.76	–	0.58	0.73	1.00
41	0.43	0.55	0.68	0.58	0.73	–	0.58	0.74	1.02
42	0.42	0.54	0.67	0.56	0.71	–	0.59	0.75	1.03

Abbreviation: RI, resistance index.
Source: From Ref. 73.

TABLE 16 ■ Pulsatility Index in the Umbilical and Middle Cerebral Arteries and CPR in 306 Normal Fetuses

		Umbilical artery		Middle cerebral artery		CPR	
Gestation week	N	Mean	SD	Mean	SD	Mean	SD
20	258	1.31	0.26	1.76	0.24	1.37	0.40
21	15	1.27	0.18	1.79	0.20	1.44	0.25
22	9	1.28	0.17	1.87	0.33	1.48	0.29
23	11	1.12	0.12	1.65	0.16	1.49	0.23
24	21	1.21	0.14	1.85	0.21	4.53	0.22
25	13	1.13	0.16	2.03	0.41	1.83	0.48
26	14	1.11	0.13	2.09	0.43	1.92	0.55
27	17	1.07	0.17	2.18	0.68	2.12	0.61
28	17	1.05	0.13	2.21	0.41	2.13	0.52
29	17	1.11	0.19	2.02	0.31	1.86	0.43
30	12	1.04	0.23	2.34	0.33	2.34	0.55
31	19	0.99	0.13	2.21	0.31	2.29	0.34
32	10	0.93	0.19	1.81	0.19	2.036	0.48
33	17	0.92	0.17	1.90	0.38	2.10	0.40
34	21	0.89	0.13	1.79	0.27	2.10	0.45
35	13	0.91	0.11	1.81	0.31	2.01	0.34
36	19	0.93	0.18	1.80	0.27	2.01	0.46
37	6	0.95	0.24	2.06	0.68	2.25	1.66
38	11	0.89	0.16	1.66	0.30	1.90	0.41
39	8	1.01	0.17	1.64	0.26	1.64	0.29
40	11	0.75	0.16	1.29	0.21	1.80	0.44

Abbreviations: CPR, cerebroplacental Doppler ratio; SD, standard deviation.
Source: From Ref. 78.

a study by Baschat and Gembruch (78) the pulsatility index was used to evaluate the mean and standard deviation of the MCA from 20 weeks until term. The lower limits of normal are the most important values used. The cerebroplacental Doppler ratio may be used to evaluate the umbilical artery pulsatility index in the MCA (Table 16).

MCA Doppler for Anemia

Fetal anemia may be secondary to a number of ideologies including maternal sensitization to the D antigen of the Rhesus (Rh) blood group (79). Cordocentesis is considered the gold standard for diagnosis of fetal anemia. However, there is considerable risk associated with cord punctures. Therefore, other less-invasive tests have been advocated. Amniocentesis using amniotic fluid delta od_{450} has been used to select patients for cordocentesis. Amniocentesis has its drawbacks and it too is an invasive procedure. Authors showed that the MCA-PSV was better than any other Doppler assessments for fetal anemia. Later, a meta-analysis demonstrated that this optimal MCA-PSV cutoff for identifying moderate anemia was 1.5 multiples of the median (MoM) (sensitivity of 100% and a false-positive rate of 12%) and that the cutoff for predicting severe anemia was 1.55 MoMs (sensitivity of 100% and a 15% false-positive rate) (Table 17).

Proper techniques for sampling the MCA-PSV include: (*i*) Examination of the fetal head with the long axis parallel to the face of the transducer so that angle of insonation of the MCA is very close to 0°. (*ii*) The fetus needs to be in a period of rest with no breathing movements for a least two minutes. (*iii*) The circle of Willis is imaged with color Doppler. (*iv*) The sonographer zooms the area of the MCA so that it occupies more than 50% of the screen. The MCA should be visualized for its entire length. (*v*) The sample volume (1 mm) is placed closed to the origin of the MCA from the internal carotid artery in such a way the angle between the direction of blood flow and the ultrasound beam is as close as possible to 0°. (*vi*) The waveforms (between 15 and 360) should be similar to each other. The highest PSV is measured. (*vii*) The above steps are repeated at least three times.

In summary, when evaluating intracranial vascularity, various parameters have been used. The most common measurement for intracranial vessel is the MCA. The systolic velocity within this artery can be used to help detect fetal anemia. Also, resistive index and pulsatility index as well as the ratio between umbilical artery and MCA indices may be used to help detect interuteral restricted fetuses.

TABLE 17 ■ Middle Cerebral Artery Peak Systolic Velocity (Mean) at Different Gestational Age; 1.5 and 1.55 MoM Represent the Cutoff Points for Moderate and Severe Anemia, Respectively

Weeks	Mean	1.5 MoM	1.55 MoM
18	23.2	34.8	36.0
19	23.3	36.5	37.7
20	25.5	38.2	39.5
21	26.7	40.0	41.3
22	27.9	41.9	43.3
23	29.3	53.9	45.4
24	30.7	46.0	47.5
25	32.1	48.2	49.8
26	33.6	50.4	52.1
27	35.2	52.8	54.6
28	36.9	55.4	57.2
29	38.7	58.0	59.9
30	40.5	60.7	62.8
31	42.4	63.6	65.7
32	44.4	66.6	68.9
33	46.5	69.8	72.4
34	48.7	73.1	75.6
35	51.1	76.6	79.1
36	53.5	80.2	82.9
37	56.0	81.0	86.8
38	58.7	88.0	91.0
39	61.5	92.2	95.3
40	64.4	96.6	99.8

Abbreviation: MoM, multiples of the median.
Source: From Ref. 79.

REFERENCES

1. Cyr DR, Mack LA, Nyberg DA, et al. Fetal rhombencephalon: normal US findings. Radiology 1988; 166:691.
2. Babcook CJ, Chong BW, Salamat MS, et al. Sonographic anatomy of the developing cerebellum: normal embryology can resemble pathology. AJR Am J Roentgenol 1996; 166:427.
3. Bromley B, Nadel AS, Pauker S, et al. Closure of the cerebellar vermis: evaluation with second trimester US. Radiology 1994; 193:761.
4. Guidelines for performance of the antepartum obstetrical ultrasound examination. Laural, MD, American Institute of Ultrasound in Medicine.
5. Filly RA, Cardoza JD, Goldstein RB, et al. Detection of fetal central nervous system anomalies: a practical level of effort for a routine sonogram. Radiology 1989; 172:403.
6. Cardoza JD, Goldstein RB, Filly RA. Exclusion of fetal ventriculomegaly with a single measurement: the width of the lateral ventricular atrium. Radiology 1988; 169:711.
7. Alagappan R, Browning PD, Laorr A, et al. Distal lateral ventricular atrium: reevaluation of normal range. Radiology 1994; 193:405.
8. Cronan MS, McGahan JP. A new ultrasound technique to visualize the proximal fetal cerebral ventricle. J Diagn Med Sonogr 1991; 6:33.
9. Browning BP, Laorr A, McGahan JP, et al. Proximal fetal cerebral ventricle: description of US technique and initial results. Radiology 1994; 192:337.
10. Mahony BS, Callen PW, Filly RA, et al. The fetal cisterna magna. Radiology 1984; 153:773.
11. Knutzon R, McGahan JP, Salamat MS, et al. Fetal cisterna magna septa: a normal anatomic finding. Radiology 1991; 180:799.
12. Nicolaides KH, Campbell S, Gabbe SG, et al. Ultrasound screening for spina bifida: cranial and cerebellar signs. Lancet 1986; 2:72.
13. Filly RA. Fetal neural axis: practical approach to identify anomalous development. In: Rifkin MD, Charboneau JW, Laing FC, eds. Syllabus: special course, ultrasound, 1991, presented at 77th Scientific Assembly and Annual Meeting of the Radiological Society of North America, Chicago, Illinois 1991:103.

14. Schoenecker S, Pretorius D, Manco-Johnson M. Artifacts seen commonly on ultrasonography of the fetal cranium. J Reprod Med 1985; 30:541.
15. Laing F, Stamler C, Jeffrey B. Ultrasonography of the fetal subarachnoid space. J Ultrasound Med 1983; 2:29.
16. Cardoza JD, Filly RA, Podrasky AE. The dangling choroid plexus: a sonographic observation of value in excluding ventriculomegaly. AJR Am J Roentgenol 1988; 151:767.
17. Nelson NL, Callen PW, Filly RA. The choroid plexus pseudocyst: sonographic identification and characterization. J Ultrasound Med 1992; 11:597.
18. Heiserman J, Filly RA, Goldstein RB. The effect of measurement errors on the sonographic evaluation of ventriculomegaly. J Ultrasound Med 1991; 10:121.
19. Russ P, Pretorius D, Johnson M. Dandy-Walker syndrome: a review of 15 cases evaluated by prenatal sonography. Am J Obstet Gynecol 1989; 161:401.
20. Laing FC, Frates MC, Brown DL, et al. Sonography of the fetal posterior fossa: false appearance of mega-cisterna magna and Dandy-Walker variant. Radiology 1994; 192:247.
21. Levine D, Barnes PD, Madsen JR, et al. Central nervous system abnormalities assessed with prenatal magnetic resonance imaging. Obstet Gynecol 1999; 92:1011–1019.
22. Levine D, Barnes P, Korf B, Edelman R. Tuberous sclerosis: second trimester diagnosis of subependymal tubers with fast MRI. AJR Am J Roentgenol 2000; 175:1067–1069.
23. Hertzberg BS, Lile R, Foosaner DE, et al. Choroid plexus-ventricular wall separation in fetuses with normal-sized cerebral ventricles at sonography: postnatal outcome. AJR Am J Roentgenol 1994; 163:405.
24. Bromley B, Frigoletto FD Jr, Benacerraf BR. Mild fetal lateral cerebral ventriculomegaly: clinical course and outcome. Am J Obstet Gynecol 1991; 164:863.
25. Nadel AS, Benacceraf BR. Lateral ventricular atrium: larger in male then female fetuses. Int J Gynecol Obstet 1995; 51:1234.
26. Patel MD, Goldstein RB, Tung S, Filly RA. Fetal cerebral ventricular atrium: difference in size according to sex. Radiology 1995; 194:713.
27. Pilu G, Falco P, Gabrielli S, et al. The clinical significance of fetal isolated cerebral borderline ventriculomegaly: report of 31 cases and review of the literature. Ultrasound Obstet Gynecol 1999; 14:320.
28. Nyberg et al. Cerebral malformations. In: Nyberg DA, Mahony BS, Pretorius DH, eds. Diagnostic Ultrasound of Fetal Anomalies: Text and Atlas. Chicago: Year Book, 1990:93.
29. Goldstein RB, LaPidus AS, Filly RA. Fetal cephaloceles: diagnosis with ultrasound. Radiology 1991; 180:803.
30. Estroff JA, Scott MR, Benacerraf BR. Dandy-Walker variant: prenatal sonographic features and clinical outcome. Radiology 1992; 185:755.
31. Bertino RE, Nyberg DA, Cyr DR, et al. Prenatal diagnosis of agenesis of the corpus callosum. J Ultrasound Med 1988; 7:251.
32. Barkovich AJ, Gressen P, Evrard P. Formation, maturation and disorders of brain neocortex. AJNR Am J Neuroradiol 1992; 13:423.
33. McGahan JP, Grix A, Gerscovich EO. Prenatal diagnosis of lissencephaly: Miller-Dieker syndrome. J Clin Ultrasound 1994; 22:560.
34. Friedman JM, Santos-Ramos R. Natural history of X-linked aqueductal stenosis in the second and third trimesters of pregnancy. Am J Obstet Gynecol 1984; 150:104.
35. McGahan JP, Ellis W, Lindfors KK, et al. Congenital cerebrospinal fluid-containing intracranial abnormalities: a sonographic classification. J Clin Ultrasound 1988; 16:531.
36. Jones KL. Trisomy 13 syndrome. In: Smith's Recognizable Patterns of Human Malformation. 5th ed. Philadelphia: WB Saunders, 1997:18–23.
37. Dublin AB, French BN. Diagnostic image evaluation of hydranencephaly and pictorially similar entities, with emphasis on computed tomography. Radiology 1980; 137:81.
38. Chang MC, Russell SA, Callen PW, et al. Sonographic detection of inferior vermian agenesis in Dandy-Walker malformations: prognostic implications. Radiology 1994; 193:765.
39. Dan U, Shalev E, Greif M, et al. Prenatal diagnosis of fetal brain arteriovenous malformation: the use of color Doppler imaging. J Clin Ultrasound 1992; 20:149.
40. Sauerbrei EE, Cooperberg PL. Cystic tumors of the fetal and neonatal cerebrum: ultrasound and computed tomographic evaluation. Radiology 1983; 147:689.
41. Benacerraf BR, Harlow B, Figoletto FD. Are choroid plexus cysts an indication for second trimester amniocentesis? Am J Obstet Gynecol 1990; 162:1001.
42. Porto M, Murata Y, Warneke L, et al. Fetal choroid plexus cysts: an independent risk factor for chromosomal anomalies. J Clin Ultrasound 1993; 21:103.
43. Jones KL. Trisomy 18 syndrome. In: Smith's Recognizable Patterns of Human Malformation. 4th ed. Philadelphia: WB Saunders, 1988:16.
44. Nyberg DA, Kramer D, Resta RG, et al. Prenatal sonographic findings of trisomy 18: review of 47 cases. J Ultrasound Med 1993; 12:103.
45. Sahinoglu Z, Uludogan M, Sayar C, Turkover B, Toksoy G. Second trimester choroid plexus cysts and trisomy 18. Int J Gynaecol Obstet 2004; 85(1):24–29.
46. Bronsteen R, Lee W, Vettraino IM, Huang R, Comstock CH. Second trimester sonography and trisomy 18. J Ultrasound Med 2003; 22(11):1219–1227.
47. Yeo L, Guzman ER, Day-Salvatore D, Walters C, Chavez D, Vintzileos AM. Prenatal detection of fetal trisomy 18 through abnormal sonographic features. J Ultrasound Med 2003; 22(6): 581–590; quiz 591–592.
48. Bronsteen R, Lee W, Vettraino IM, Huang R, Comstock CH. Second trimester sonography and trisomy 18: the significance of isolated choroid plexus cysts after an examination that includes the fetal hands. J Ultrasound Med 2004; (2):241–245.
49. Walkinshaw S, Pilling D, Spriggs A. Isolated choroid plexus cysts: the need for routine offer of karyotyping. Prenat Diagn 1994; 14:663.
50. Benacerraf BR, Nadel A, Bromley B. Identification of second-trimester fetuses with autosomal trisomy by use of a sonographic scoring index. Radiology 1994; 193:135.
51. Barkovich AJ. Metabolic and destructive brain disorders. In: Barkovich AJ, ed. Contemporary Neuroimaging: Pediatric Neuroimaging. Vol. 1. New York: Raven Press, 1990:35.
52. Barkovich AJ. Metabolic and destructive brain disorders. In: Barkovich AJ, ed. Contemporary neuroimaging: Pediatric Neuroimaging. Vol. 1. New York, Raven Press, 1990:60, 99.
53. Yakovlev PI, Wadsworth RC. Schizencephalies. II. Clefts with hydrocephalus and lips separated. J Neuropathol Exp Neurol 1946; 5:169.
54. Crade M. Ultrasonic demonstration in utero of an intracranial teratoma. JAMA 1982; 247:1173.
55. Osborn RA, McGahan JP, Dublin AB. Sonographic appearance of congenital malignant astrocytoma. AJNR Am J Neuroradiol 1984; 5:814.
56. McGahan JP, Hasslein HC, Meyers M, et al. Sonographic recognition of in utero intraventricular hemorrhage. AJR Am J Roentgenol 1984; 142:171.
57. Hiller L IV, McGahan JP, Bijan B, Melendres G, Towner D. Sonographic detection of in utero isolated cerebellar hemorrhage. J Ultrasound Med 2003; 22(6):649–652.
58. Goldstein RB, Filly RA. Prenatal diagnosis of anencephaly: spectrum of sonographic appearances and distinction from the amniotic band syndrome. AJR Am J Roentgenol 1988; 151:547.
59. Cox GG, Rosenthal SJ, Holsapple JW. Exencephaly: sonographic findings and radiologic-pathologic correlation. Radiology 1985; 155:755.
60. Naidich TP, Altman NR, Braffman BH, et al. Cephaloceles and related malformations. AJNR Am J Neuroradiol 1992; 13:655–690.
61. Mahony BS, Filly RA, Callen PW, et al. The amniotic band syndrome: antenatal diagnosis and potential pitfalls. Am J Obstet Gynecol 1985; 152:63.
62. Gorczyca DP, Lindfors KK, McGahan JP, et al. Limb-body-wall complex: another cause for elevated maternal serum alpha fetoprotein. J Clin Ultrasound 1990; 18:198.
63. Andrews M, Amparo EG. In utero clue to congenital lethal osteogenesis imperfecta. AJR Am J Roentgenol 1983; 160:212.
64. Kurtz AB, Wapner RJ, Rubin CS, et al. Ultrasound criteria for in utero diagnosis of microcephaly. J Clin Ultrasound 1980; 8:11.

65. Chervenak FA, Jeanty P, Cantraine F, et al. The diagnosis of fetal microcephaly. Am J Obstet Gynecol 1984; 149:512.
66. Platt LD, Manning FA, Murata Y, et al. Diagnosis of fetal death in utero by real-time ultrasound. Obstet Gynecol 1980; 55:191.
67. Issacson G, Blakemore KJ, Chervenak FA, Thanatophoric dysplasia with cloverleaf skull. Am J Dis Child 1983; 137:896.
68. Dublin AB, Merten DF. Computed tomography in the evaluation of herpes simplex encephalitis. Radiology 1977; 125:133.
69. Hecher K, Campbell S, Doyle P, et al. Assessment of fetal compromise by Doppler ultrasound investigation of the fetal circulation: arterial intracardiac, and venous blood flow velocity studies. Circulation 1995; 91:129.
70. Figueras F, Lanna M, Placacio M, et al. Middle cerebral artery Doppler indices at different sites: prediction of umbilical cord gases in prolonged pregnancies. Ultrasound Obstet Gynecol 2004; 24:529–533.
71. Devine PA, Bracero LA, Lysikiewicz A, Evans R, Womak S, Byrne DW. Middle cerebral to umbilical artery Doppler ratio in post-date pregnancies. Obstet Gynecol 1994; 84:856–860.
72. Trudinger B. Doppler ultrasonography and fetal well-being. In: Reece EA, Hobbins JC, Mahoney MJ, et al., eds. Medicine of the Fetus and Mother. Philadelphia: JB Lippincott, 1992:704.
73. Kurmanavicius J, Florio I, Wisser J, et al. Reference resistance indices of the umbilical, fetal middle cerebral and uterine arteries at 24–42 weeks of gestation. Ultrasound Obstet Gynecol 1997; 10:112–120.
74. Mimica M, Pejkovic L, Furlan I, et al. Middle cerebral artery velocity waveforms in fetuses with absent umbilical artery end-diastolic flow. Biol Neonate 1995; 67:21.
75. Noordam MJ, Heydanus R, Hop WC, et al. Doppler colour flow imaging of fetal intracerebral arteries and umbilical artery in the small for gestational age fetus. Br J Obstet Gynaecol 1994; 101:504.
76. Palacio M, Figueras F, Zamora L, et al. Reference ranges for umbilical and middle cerebral artery pulsatility index and cerebroplacental ratio in prolonged pregnancies. Ultrasound Obstet Gynecol 2004; 24:647–653.
77. Yoshimura S, Masuzaki H, Gotoh H, et al. The relationship between blood flow redistribution in umbilical artery and middle cerebral artery and fetal growth in intrauterine growth retardation. Nippon Sanka Fujinka Gakkai Zasshi 1995; 47:1352.
78. Baschat AA, Gembruch U. The cerebroplacental Doppler ratio revisited. Ultrasound Obstet Gynecol 2003; 21:124–127.
79. Detti L, Mari G. Noninvasive diagnosis of fetal anemia. Clin Obstet Gynecol 2003; 46(4):923–930.

44

Fetal Face

Terry L. Coates and John P. McGahan

The American Institute of Ultrasound in Medicine has published guidelines for performance of the antepartum obstetrical ultrasound examination, with specific recommendations for evaluation of second- and third-trimester fetuses (1). Although these official guidelines do not specifically recommend an examination of the fetal face, they do state that suspected abnormalities may require a specialized evaluation to permit diagnosis of numerous anomalies that could significantly alter perinatal management.

Fetal facial anomalies may be isolated, but most facial abnormalities are associated with more complex fetal malformation syndromes or abnormal karyotypes. Therefore, identification of a facial malformation may suggest a specific diagnosis when found in association with other anomalies (2,3). Likewise, when a fetal anomaly is detected, a careful ultrasonographic evaluation of the fetal face may also contribute to a specific diagnosis. In high-risk pregnancies, evaluation of the fetal face has become an essential part of the ultrasonographic examination. Continuing improvements in the resolution of real-time ultrasound equipment and operator technique allow for a better survey of the fetal face and should improve the detection of subtle structural defects and allow a more accurate diagnosis of anomalous pregnancies (2–4).

EMBRYOLOGY

Facial development begins at about four to five menstrual weeks and is almost complete by the end of the embryonic period, at about 10 menstrual weeks (5,6). Therefore, the embryo has acquired all its basic morphologic characteristics before the face has attained adequate size to permit ultrasonographic examination. Although a detailed knowledge of the complex embryogenesis of the facial region is not necessary to diagnose most facial abnormalities, a brief review of normal facial embryology is appropriate in predicting patterns of malformation.

Most malformations of the face originate from anomalous development of the branchial apparatus, optic vesicles, pharyngeal pouches, and facial prominences, as listed by days after fertilization (Fig. 1) (5–8). By 14 days after fertilization, the early embryo has formed a circular bilaminar disk containing Hensen's node. Hensen's node eventually gives rise to many of the primordia of the craniofacial complex (6).

The branchial arches appear in the fourth to fifth weeks of development as neural crest cells migrate into the future head and neck region. Each branchial arch consists of mesenchyme derived from the intraembryonic mesoderm and is covered externally by ectoderm and internally by endoderm (5). The neural crest cells migrate into the branchial arches surrounding the central core of mesenchymal cells, and their proliferation produces the discrete swellings that demarcate each branchial arch.

The first branchial arch gives rise to the paired mandibular prominences, which form the lower jaw, or mandible, and to the paired smaller maxillary prominences, which give rise to the upper jaw, or maxilla. The primitive mouth, or stomodeum, is a slight depression on the surface of

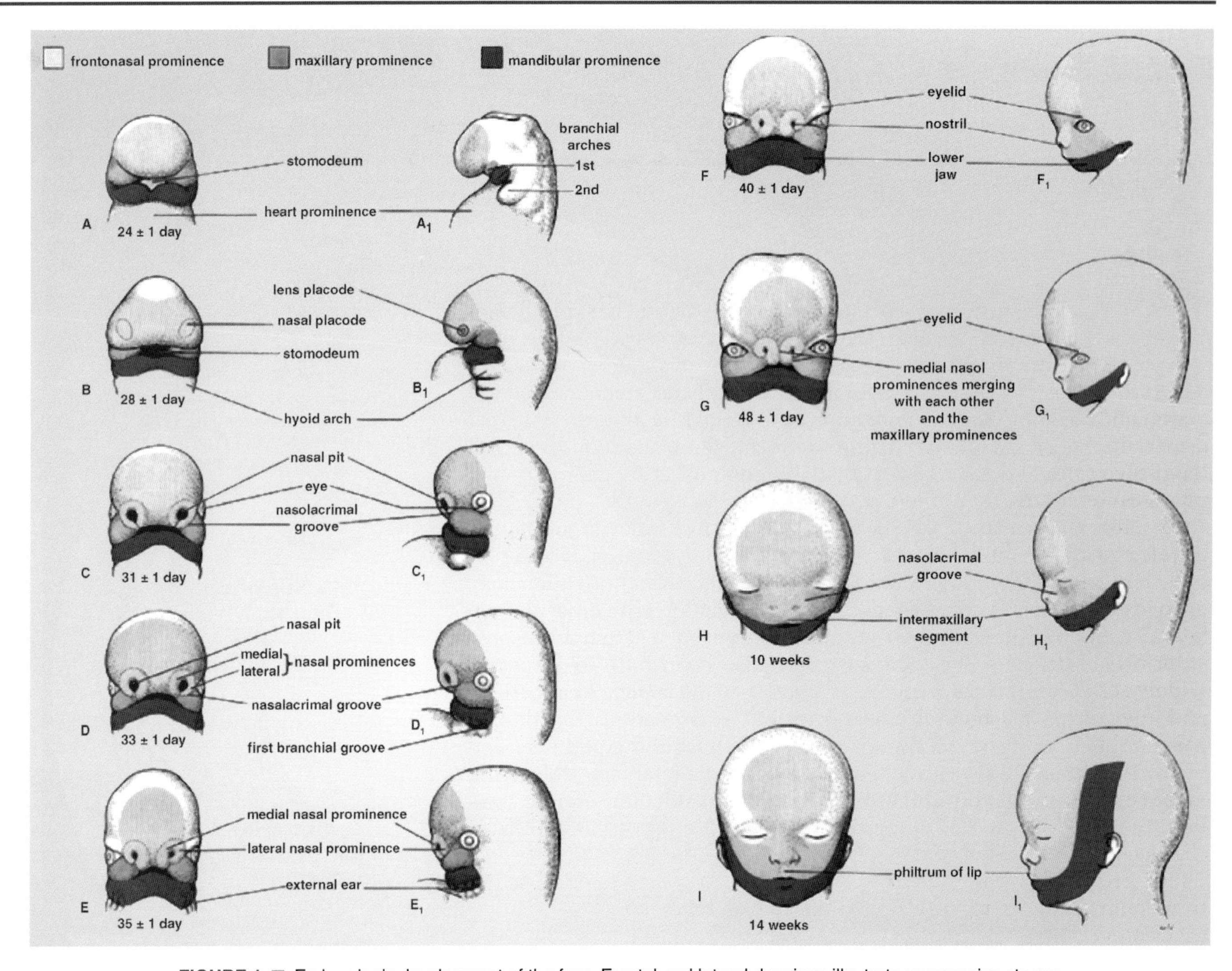

FIGURE 1 ■ Embryologic development of the face. Frontal and lateral drawings illustrate progressive stages in the development of the human face from 40 days to 14 weeks. Hatched areas indicate frontonasal prominence; light shaded areas, maxillary prominence; dark shaded areas, mandibular prominence. *Source*: From Ref. 7.

the ectoderm. Early in the fourth week after fertilization, five facial primordia appear around the stomodeum. The frontonasal prominence constitutes the cranial border of the stomodeum; the paired maxillary prominences of the first branchial arch form the lateral boundaries of the stomodeum; and the paired mandibular prominences of the first branchial arch form the caudal boundary. Bilateral oval thickenings of the surface ectoderm, called nasal placodes, then develop on each side of the inferior part of the frontonasal prominence. Mesenchyme proliferates into elevations at the margins of these placodes, the sides of which are called the medial and lateral nasal prominences. These placodes invaginate into depressions called the nasal pits. The maxillary prominences grow medially toward each other and the medial nasal prominences, which results in each lateral nasal prominence being separated from the maxillary prominence by a cleft called the nasolacrimal groove. By the end of the fifth week after fertilization, the oracles of the external ears have begun to develop. During the sixth and seventh weeks, the medial nasal prominences merge with each other and the maxillary prominences, forming an intermaxillary segment. This gives rise to the middle portion, or philtrum of the lip, the premaxillary part of the maxilla and its associated gums, and the primary palate. The lateral parts of the upper lip, most of the maxilla, and the secondary palate form from the maxillary prominences. The frontonasal prominence forms the forehead and the dorsum and apex of the nose. The lateral nasal prominences result in the sides of the nose, and the nasal septum is formed by the medial nasal prominences. Maxillary prominences form the upper cheek and upper lip. The mandibular prominences give rise to the lower lip, chin, and lower cheeks.

With the differential growth of these mesodermal masses, the ectodermal grooves soon become obliterated; however, it is along these grooves that facial clefts commonly develop (Table 1) (6). The maxillary, mandibular, and frontonasal prominences of the facial processes are of great importance because they determine through

TABLE 1 ■ Facial Prominences Contributing to Facial Formation

Prominence	Derivative
Frontonasal	Forehead, nasal bridge, and apex of the nose
Lateral nasal	Sides of the nose
Medial nasal	Philtrum of the upper lip, nasal septum, premaxillary part of the maxilla and gum, and primary palate
Maxillary	Upper cheek regions and most of the upper lip
Mandibular	Lower cheek regions, lower lip, and chin

fusion and differential growth the size of the mandible, upper lip, palate, and nose (8). Development of the tongue, face, lips, jaws, palate, pharynx, and neck largely involves transformation of the branchial apparatus into adult structures. During the late embryonic period, development of the face is characterized mainly by changes in proportion and relative position of individual structures (6). The eyes, initially directed laterally, gradually move medially to become directed anteriorly. The nose begins above the orbits as two widely separated nasal placodes, which move medially and inferiorly as the nasal septum thins and the medial nasal folds fuse. The external ears initially develop below the level of the mandible but gradually assume a more cephalic orientation. About 60 days after fertilization, the embryo has acquired all its basic morphologic characteristics and then enters the fetal period during which growth and maturation of these primordia continue along with reorganization of the spatial relations between various structures (6).

ULTRASONOGRAPHIC APPROACH TO THE FETAL FACE AND NORMAL ANATOMY

Major features of the fetal face, such as the orbit, forehead, nose, and mouth, can be identified as early as 11 to 12 menstrual weeks; but it is usually not until 14 to 16 menstrual weeks that the ultrasonographer can observe more detailed facial anatomy (3,9). In pregnancies complicated by polyhydramnios, concomitant extrafacial structural anomalies, a maternal history of teratogen exposure, or a family history of previous craniofacial malformation (e.g., facial cleft), a targeted ultrasonographic examination of the face may be particularly useful (10,11). Detailed facial anatomy, including specific muscles, nerves, and arteries, can be observed (10,11). Visibility may be limited in the presence of oligohydramnios, maternal obesity, or fetal position, especially in the occiput anterior position (2,11). For detailed facial anatomy, the following views should be obtained:

- Axial views of the orbits, nose, lips, anterior palate, tongue, and oropharynx
- Coronal views of the orbits, nose, lips, maxilla, and anterior portion of the mandible
- Profile view of soft tissues and facial bones, including the nasal bones and mandible
- Views of the ears (3)

If available, three-dimensional (3D) imaging of the face may be useful to better define facial abnormalities. Ultrasonographic evaluation of the fetal face not only identifies facial anomalies that may signify anomalies and chromosomal abnormalities but also may serve as an indicator of fetal well-being and behavioral changes (3,12,13). For example, fetal vomiting or regurgitation can be identified by watching the fetal mouth, using color-flow imaging to indicate regurgitation in fetuses with upper gastrointestinal obstruction (14).

The following sections discuss in more detail the ultrasonographic approach to the various anatomic regions of the fetal face. Later sections present the more common facial anomalies that may be encountered.

Orbit and Periorbital Regions

Orbital imaging is not performed routinely during obstetric sonography, but the discovery of abnormal orbital diameters provides evidence of fetal dysgenesis (15). In most cases, the size and location of the orbits are evaluated subjectively during the real-time sonogram. An axial view of the orbits allows visualization of the different components of the eyes, including the globe, vitreous body, lens, anterior chamber, extraocular muscles, optic nerve, and hyaloid artery (16). Documentation of the presence and size of two eyes and of the distance between them probably provides the most useful diagnostic information (15,17,18). Jeanty (18,19), Mayden (17), Trout (15), and their coauthors tabulated normal ultrasound values for ocular biometry throughout pregnancy. These charts (biometric tables) for normal ocular, intraorbital, and interorbital diameters help to differentiate between normal ocular size, hypertelorism, and hypotelorism (Table 2). The intraorbital distance is approximately equal to the two interorbital diameters. The biocular distance is measured from the lateral wall of each orbit.

The axial plane of the orbits is a section parallel but caudal to the section used for the biparietal diameter (Fig. 2). The following criteria are used for this plane:

- Symmetric image
- Both eyes imaged and of equal diameter
- Largest possible diameter of the eye (3,18).

The tabulated measurements for the orbits include the following:

- Ocular diameter (OD), measured from the medial to the lateral wall of the bony orbit
- Interocular distance (ID), measured from the medial wall of one orbit to the medial wall of the other orbit
- Binocular distance (BD), determined by measurement between the lateral orbital rims

These measurements may be useful in predicting such abnormalities as ocular hypertelorism, ocular hypotelorism, and other orbital abnormalities; or less frequently,

TABLE 2 ■ Growth of the Ocular Parameters

	Binocular distance (mm)			Interocular distance (mm)			Ocular diameter (mm)		
Age (wk)	5th	50th	95th	5th	50th	95th	5th	50th	95th
11	5	13	20	–	–	–	–	–	–
12	8	15	23	4	9	13	1	3	6
13	10	18	25	5	9	14	2	4	7
14	13	20	28	5	10	14	3	5	8
15	15	22	30	6	10	14	4	6	9
16	17	25	32	6	10	15	5	7	9
17	19	27	34	6	11	15	5	8	10
18	22	29	37	7	11	16	6	9	11
19	24	31	39	7	12	16	7	9	12
20	26	33	41	8	12	18	8	10	13
21	28	35	43	8	13	17	8	11	13
22	30	37	44	9	13	18	9	12	14
23	31	39	46	9	14	18	10	12	15
24	33	41	48	10	14	19	10	13	15
25	35	42	50	10	15	19	11	13	16
26	36	44	51	11	15	20	12	14	16
27	38	45	53	11	16	20	12	14	17
28	39	47	54	12	16	21	13	15	17
29	41	48	56	12	17	21	13	15	18
30	42	50	57	13	17	22	14	16	18
31	43	51	58	13	18	22	14	16	19
32	45	52	60	14	18	23	14	17	19
33	46	53	61	14	19	23	15	17	19
34	47	54	62	15	19	24	15	17	20
35	48	55	63	15	20	24	15	18	20
36	49	56	64	16	20	25	16	18	20
37	50	57	65	16	21	25	16	18	21
38	50	58	65	17	21	26	16	18	21
39	51	59	66	17	22	26	16	19	21
40	52	59	67	18	22	26	16	19	21

Source: From Ref. 20.

they may assist in obstetric dating when there is severe distortion of other biometric parameters (15,17–20).

The hyaloid artery, a branch of the main ophthalmic artery, can be visualized ultrasonographically in the second trimester of pregnancy as an echogenic line coursing through the vitreous of the globe to the posterior surface of the lens. It normally regresses before the middle of the third trimester. Birnholz and Farrel (21) suggested that temporarily delayed regression of the hyaloid artery may occur with trisomy 21 syndrome and other forms of retarded brain development (21). The lens of the eye is identifiable as a small echogenic circle within the bony orbit.

The following questions may prove useful in the detection of anomalies involving the orbital and periorbital region:

- Is hypotelorism, hypertelorism, microphthalmia, or anophthalmia present?
- Is a proboscis or cebocephaly present?
- Are any periorbital masses or intraocular abnormalities present? (3)

Lips, Mouth, and Tongue

The scanning plane that has been proposed for most optimal evaluations of the fetal lips and mouth is a coronal scan that approximates the frontal planes through the fetal nose, upper and lower lips, and chin (Fig. 3) (2,3,22). The lower face of the fetus is imaged coronally through the lips, particularly the upper lip and maxilla; in a longitudinal plane through the fetal profile; and in a transverse plane through the mandible and maxilla.

The normal hypoechoic orbicularis oris muscles of the lip can be visualized in continuity with the lips and nose after about 16 weeks' gestation. Subtle changes in transducer angulation from the coronal plane permit sequential scans of the soft tissues of the lips, chin, alveolar ridge, and nose. The following questions may

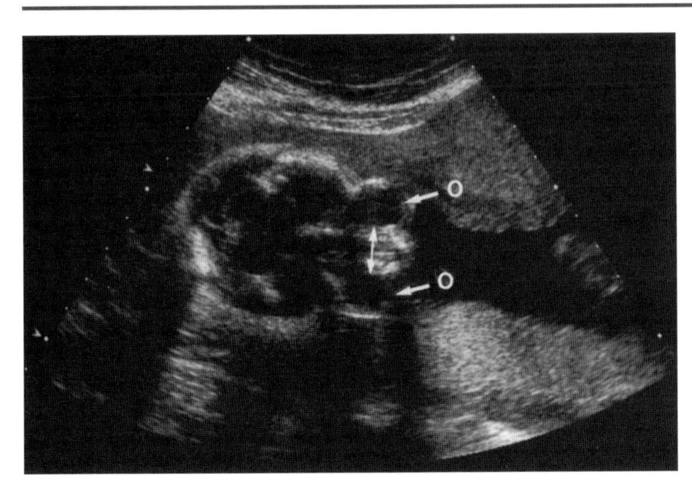

FIGURE 2 ■ Axial scan through the level of the normal orbits at 20 menstrual weeks demonstrates the nasal bridge, the medial wall of each orbit (O), and the lateral wall of each orbit. The ocular diameter is the distance from the medial to the lateral wall of the orbit and is about equal to the interocular distance, which is measured from one medial wall to the other (*arrows*). The binocular distance is the sum of the two ocular diameters and the interocular distance or the measurement from one lateral wall to the other lateral wall of the orbit.

prove useful in the detection of abnormalities involving the upper lip and mouth:

- Is the upper lip cleft?
- Is the tongue protuberant?
- Is the mouth widely open or fixed in position?

Oligohydramnios, maternal body habitus, and fetal positioning can interfere with obtaining the optimal images of these landmarks.

Babcook et al. (23) reviewed the ultrasonographic appearances of normal fetal midface anatomy in three planes in 12 in vitro fetal specimens ranging in gestational age from 16 to 22 weeks. The authors correlated these appearances with examination of 100 normal in utero fetuses in the coronal, axial, and sagittal planes. Visualization was characterized as excellent in (*i*) the coronal plane, when the entire upper lip was visualized in one scan with optimal resolution; (*ii*) the axial plane, when the soft tissues of the upper lip overlying the maxilla were optimally resolved and the anterior six-tooth sockets, the continuous, smooth, echogenic C-shaped curve of the tooth-bearing alveolar ridge, and the maxilla could be identified; and (*iii*) the sagittal plane, when a true sagittal view demonstrated the soft-tissue profile of the fetal face and a continuous straight echogenic line extended posteriorly at the level of the hard palate. Anatomy was visualized in the coronal plane (upper lip) in 95% of the in utero fetuses; in the axial plane (alveolar ridge, tooth sockets, and point of fusion of primary and secondary palates) in 97%; and in the sagittal plane (fusion line of the secondary palate) in 26%.

A variety of intermittent fetal movements often associated with swallowing can be seen normally in utero and should be considered a part of normal fetal activity (14). Detection of fetal regurgitation or vomiting in a setting of otherwise unexplained polyhydramnios may be associated with fetal abnormalities that are most likely high gastrointestinal obstruction, for example, tracheoesophageal fistula or duodenal atresia (14). In the transverse plane behind the tongue, the oropharynx and its connection to the larynx can be identified.

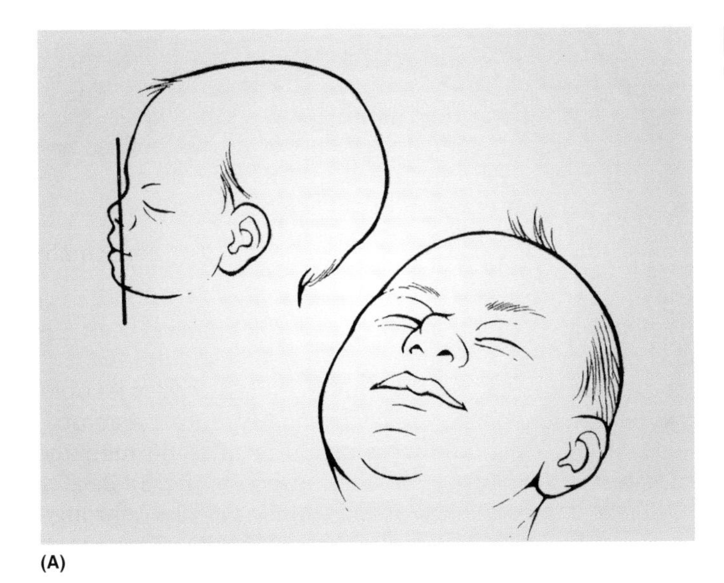

(A)

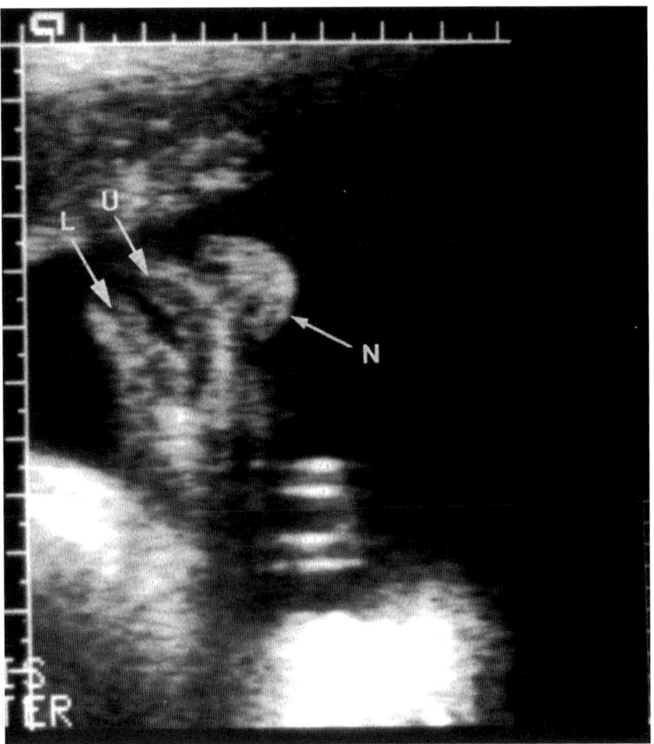

(B)

FIGURE 3 A AND B ■ (**A**) The approximate frontal plane of section through the upper lip, nose, and chin. (**B**) Coronal ultrasound of the lips demonstrates the nose, upper lip, lower lip, and hypoechoic linear region in the midline between the upper lip and the nose corresponding to the normal philtrum. N, nose; U, upper lip; L, lower lip.

Otto and Platt (24) published an ultrasonographic method for objective measurement of the fetal jaw. Watson and Katz (25) generated a biometric table of transverse inner-to-inner mandible and anteroposterior measurements using the fetal hypopharynx as a reference point in relation to gestational age. These standards for fetal jaw measurement may be useful in the ultrasonographic identification of micrognathia; however, it is uncertain whether measurement of the fetal mandible early in pregnancy can identify fetal micrognathia (26). Chitty et al. (27) constructed percentile charts for fetal mandibular size from 12 until 28 weeks' gestation, which may aid the prenatal diagnosis of micrognathia. Measurement of the fetal mandible in the third trimester, however, is difficult because of changes in fetal position and bony shadowing from adjacent bony structures. Otto and Platt (24) derived a curve of mandibular size from 15 to 40 weeks' gestation.

Views of the lips, mouth, and tongue are especially useful in the setting of polyhydramnios, concomitant anomalies, or a positive family history, when detection of facial clefts, macroglossia, or an unusually shaped mouth suggests the presence of a specific syndrome (3).

Pitfalls

The soft tissues of the fetal upper lip can be visualized in the coronal and axial planes in most fetuses on prenatal ultrasonography. Visualization of anatomy improves with advancing gestational age secondary to the increasing soft-tissue bulk of the fetal upper lip. Poor visualization of the fetal midface anatomy, particularly in the coronal and axial planes, is usually secondary to large maternal body habitus or fetal positioning. In the sagittal plane, fetal positioning is the most common cause of suboptimal visualization of the true midline view. Babcook et al. (23) found that the ability to visualize the hard palate decreased with advancing gestational age owing to increasing acoustic shadowing from overlying fetal extremities and from the upside alveolar ridge obscuring the downside structures. Ultrasonographic evaluation is ideally made before 24 weeks' gestation.

A false-positive diagnosis of a cleft lip can be made when visualization of the normal philtrum (midline groove of the upper lip) or frenulum labii superioris is confused for a cleft (3).

Facial Profile View

The fetal profile is optimally demonstrated with the mid-sagittal view of the face (Fig. 4). This view clearly demonstrates the nasal bones, nasal bridge, soft tissues, and mandible. When ultrasonographically evaluating the facial profile, the following questions may prove useful in the detection of facial anomalies:

- Is the nose flattened or the nasal bridge absent?
- Is the chin abnormally small?
- Is an abnormal mass present? (3).

The mid-sagittal view may reveal large facial or neck masses and facial clefting as well as more subtle abnormalities, including abnormal or absent nose, micrognathia, enlarged tongue, absent nasal bridge, or perhaps a clue to the presence of a chromosomal abnormality such as trisomy 13, 18, or 21 (3,25). The profile view aids in demonstrating most large masses, including epignathus, anterior cephalocele, cervical teratomas, and hemangiomas. Parasagittal scans may be necessary to detect small tumors that do not extend to the midline, as may be the case with some facial hemangiomas or sarcomas (3). The sagittal view also assists in the diagnosis of the holoprosencephaly complex when a proboscis is demonstrated or in the diagnosis of bilateral cleft lip and cleft palate with demonstration of a small premaxillary protrusion. Obtaining routine mid-sagittal images of the fetal face may be especially helpful in a pregnancy complicated by a family history of facial abnormalities, the presence of other anomalies, or polyhydramnios and maternal teratogen exposure (3). Abnormalities seen on a facial profile view may assist in the accurate diagnosis of one of the multiple malformation syndromes, even in the absence of genetic or teratogen risk.

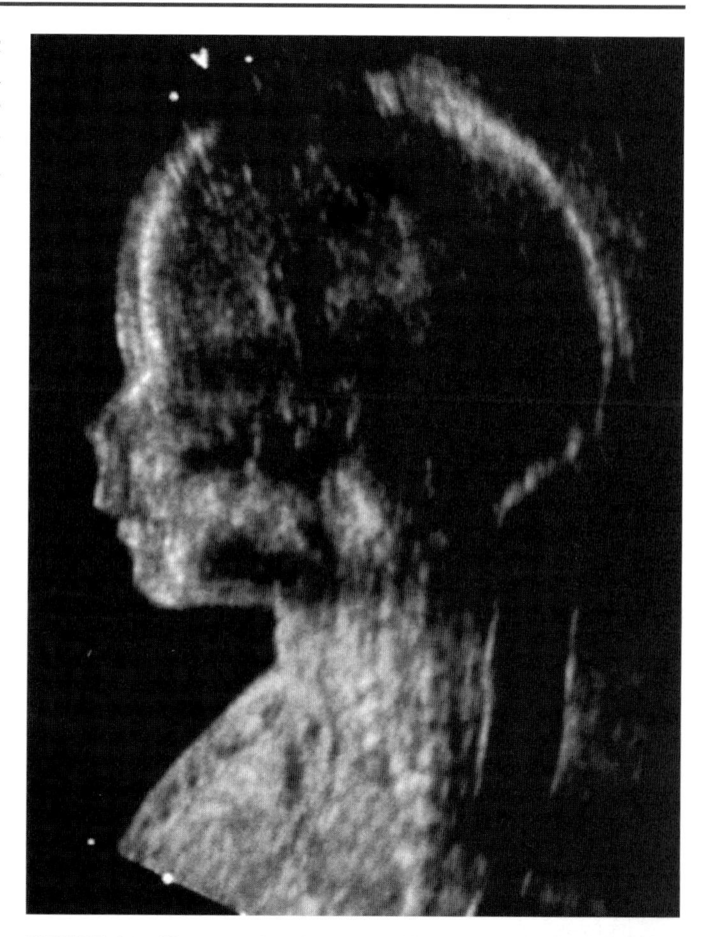

FIGURE 4 ■ The normal mid-sagittal facial profile view. Note the echogenic nasal bridge, the normal maxilla, and the normally formed mandible.

Pitfalls

When scanning, the ultrasonographer must be careful that the true sagittal plane is obtained because hypoplasia of the nasal bridge and mandibular hypoplasia may be simulated when scanning in the parasagittal axis (Fig. 5) (3).

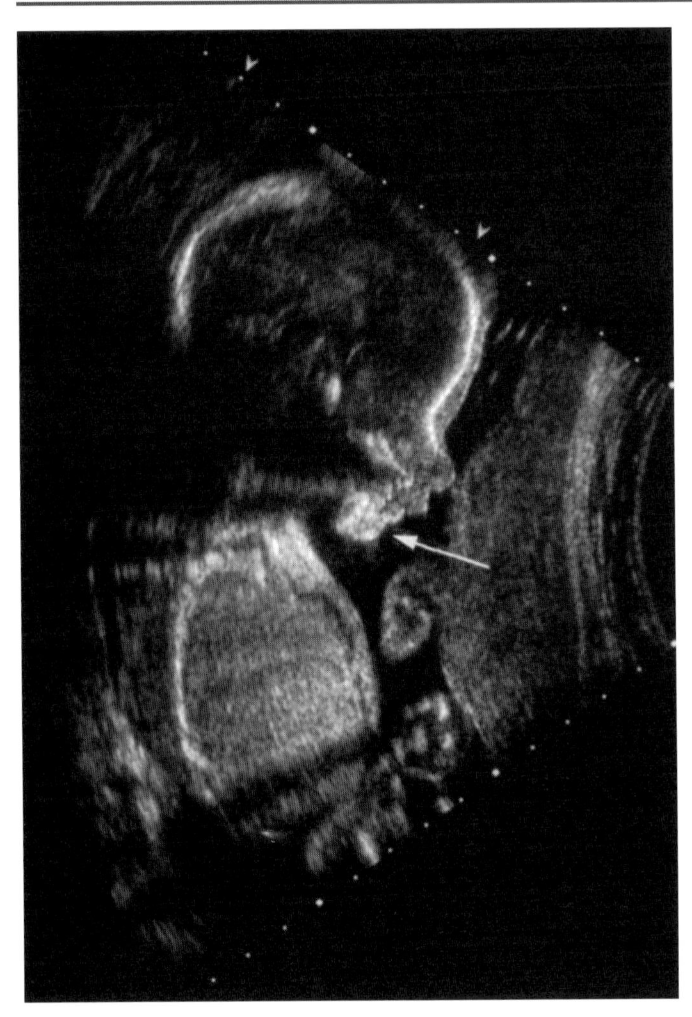

FIGURE 5 ■ Off-axis (parasagittal) profile view of pseudomandibular hypoplasia. Parasagittal view demonstrates apparent mandibular hypoplasia (*arrow*) when scanning off to one side of the face.

The Fetal Mandible

Mandibular anomalies are commonly encountered with fetal facial defects. It has been estimated that there are more than 100 genetic syndromes that have mandibular abnormalities (26). Authors emphasize the importance of differentiating anomalies of position, retrognathia, from micrognathia, retrognathia implies recession of the chin from micrognathia, which implies insufficient size of the mandible. However, in most published reports, the term micrognathia is used for mandibular position or size abnormalities. The different syndromes that may be associated with micrognathia are too many to be reviewed within this text. Micrognathia may be associated with a variety of syndromes such as Pierre Robin syndrome, Treacher Collins syndrome, and various chromosomal abnormalities including trisomy 13 and trisomy 18. There have been a number of other syndromes that have been associated with fetal facial abnormalities. Its recognition may be important, as fetuses with mandibular anomalies are at risk of acute neonatal respiratory distress syndrome. This is a neonatal emergency as the tongue may obstruct the upper airway, leading to neonatal suffocation.

Mandibular abnormalities may be recognized by a number of different methods. The simplest method is to obtain a true sagittal image of the fetal face in which there is a prominent upper lip with an impression of a small mandible, resulting in a receding chin. Off-axis imaging may give the false-positive impression of micrognathia. Others have utilized more objective measurements, which may be obtained from the reference list and are too lengthy for a discussion within this chapter. Both Chitty, and Otto and Platt measured the one ramus of the mandible with the ultrasound beam at right angles to the jaw (24,27). Watson and Katz measured the anterior-posterior and the transverse-mandibular measurements using the hypopharynx as the reference point (28). Others, such as Rotten, have proposed using an inferior facial angle as measured on the sagittal view of the face (29). Rotten demonstrated the use of 3D sonography may be useful in detection of micrognathia as it is easy to obtain perfectly symmetrical views that are computer generated of the face and mandible. A frontal 3D image is presented in Figure 6. Other 3D images are presented elsewhere in this chapter.

Fetal Nasal Bone

Flattening of the facial profile is a feature of Down syndrome. Cicero et al. (30) have reported the absence of the

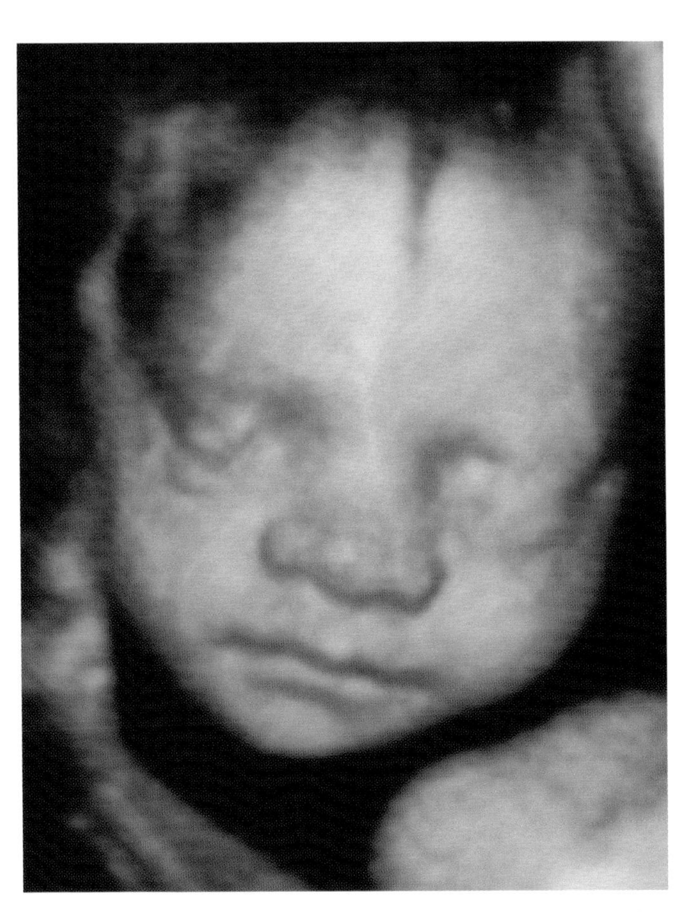

FIGURE 6 ■ 3D face. 3D image showing normal facial features and no evidence of cleft lip or palate. Diagnosis of mandibular hypoversia would require 3D profile view.

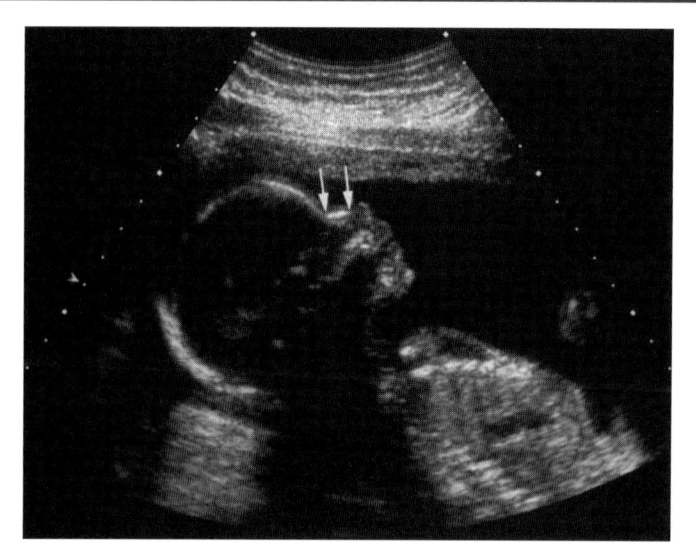

FIGURE 7 ■ Normal nasal bone. Nineteen week facial profile demonstrating nasal bone measured from echogenic base to echogenic tip (*arrows*).

nasal bone occurred in 43 of 59 fetuses with trisomy 21 given a sensitivity of 73%. Also, in that series, only 3 of 603 normal fetuses had absence of the nasal bone between 11 and 14 weeks given a false-positive read of 0.5%. This presence or absence of the nasal bone gave a likelihood ratio of 146 and combined with the maternal age and nuchal translucency gave an overall sensitivity of 85% (30).

A midsagittal facial profile must be obtained in order to measure the nasal bone. If there is off axis from this midsagittal plane, this will produce an erroneous measurement. The nasal bone is measured from its base to its tip (Fig. 7). Other authors such as Bunduki et al. (31) have produced tabular data of measurements of the fetal nasal bone from 16 to 24 weeks. They show that screening for trisomy 21 using the fifth percentile as a cutoff value resulted in a sensitivity of 59.1% and a 5.1 false-positive rate. Their likelihood ratio was 11.6. Others such as Sonek et al. (32) have produced normal range for the nasal bone length in millimeters from 11 to 40 weeks. This is produced in Table 3.

TABLE 3 ■ Normal Ranges for Nasal Bone Lengths (in mm) (*n* = 3537)

	Percentile				
Gestational age (wk)	**2.5%**	**5%**	**50%**	**95%**	**97.5%**
11	1.3	1.4	2.3	3.3	3.4
12	1.7	1.8	2.8	4.2	4.3
13	2.2	2.3	3.1	4.6	4.8
14	2.2	2.5	3.8	5.3	5.7
15	2.8	3.0	4.3	5.7	6.0
16	3.2	3.4	4.7	6.2	6.2
17	3.7	4.0	5.3	6.6	6.9
18	4.0	4.3	5.7	7.0	7.3
19	4.6	5.0	6.3	7.9	8.2
20	5.0	5.2	6.7	8.3	8.6
21	5.1	5.6	7.1	9.0	9.3
22	5.6	5.8	7.5	9.3	10.2
23	6.0	6.4	7.9	9.6	9.9
24	6.6	6.8	8.3	10.0	10.3
25	6.3	6.5	8.5	10.7	10.8
26	6.8	7.4	8.9	10.9	11.3
27	7.0	7.5	9.2	11.3	11.6
28	7.2	7.6	9.8	12.1	13.4
29	7.2	7.7	9.8	11.8	12.3
30	7.3	7.9	10.0	12.6	13.2
31	7.9	8.2	10.4	12.6	13.2
32	8.1	8.6	10.5	13.6	13.7
33	8.6	8.7	10.8	12.8	13.0
34	9.0	9.1	10.9	12.8	13.5
35	7.5	8.5	11.0	14.1	15.0
36	7.3	7.8	10.8	12.8	13.6
37	8.4	8.7	11.4	14.5	15.0
38	9.2	9.3	11.7	15.7	16.6
39	9.1	9.2	10.9	14.0	14.8
40	10.3	10.4	12.1	14.5	14.7

Source: From Ref. 32.

They demonstrated that there was not any significant increased incidence of nasal bone hypoplasia in Afro-Caribbean population as compared to Caucasians. Chen et al. demonstrated that there was a shorter nasal bone length in Chinese fetuses as compared to other populations (33).

Ears

Routine ultrasonographic examination of the fetal external ears rarely provides useful information in the absence of other or obvious abnormalities (3,34). At prenatal ultrasonography, anatomic detail of at least one external ear is usually visible when amniotic fluid outlines the ear structure.

Coronal and parasagittal views of the fetal external ear may prove useful in the detection of anomalies, allowing the physician to answer the following questions:

- Are the ears obviously deformed?
- Are the ears low set or of abnormal size? (3).

The fetal ear may have an abnormal shape, location, or size in a number of fetal abnormalities. For instance, low-set ears may be identified in fetal syndromes including Noonan syndrome, Nager syndrome, Carpenter syndrome and Klippel–Feil syndrome. Various chromosomal abnormalities including trisomy 18 may be associated with low-set ears. Certain teratogens, including aminopterin, phenylephrine, and valproic acid may be associated with low-set ears. Some fetal chromosomal abnormalities have a shortened ear length.

The ear is usually measured as the maximum distance from the helix to the end of the lobe in a coronal or a parasagittal view (Fig. 8) (35,36). The helix is the most cephalic portion of the fetal ear. Sacchini has demonstrated that in fetuses from 11 to 14 weeks of gestation, the ear length in trisomy 21 fetuses is significantly reduced, but the degree of deviation from the normal is too small for this measurement to be useful in screening for trisomy 21 (37).

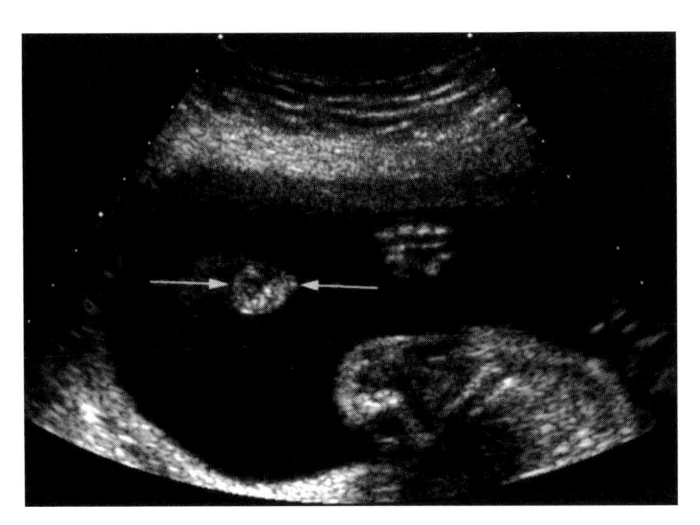

FIGURE 8 ■ Normal ear. Parasagittal scan of the normal ear that is measured from the upper helix of the ear to the end of the lobe (*arrows*).

A number of authors have demonstrated that fetal ear length may be shortened in Turner syndrome, trisomy 13, trisomy 18, and trisomy 21 (35,36,38,39). In an article by Yeo (39) using the 10th percentile, nearly all fetuses with trisomy 13 and trisomy 18 had an ear length percentile less than the 10th percent based upon gestational age (39). For trisomy 21 using the 10th percentile only 41% of fetuses had an ear length less than that percentile. The table reproduced by Yeo (39) is included as Table 4. Utilizing the 10th percentile for a short fetal ear length measurement, either alone or in combination with other sonographically detected structural markers, may be a useful parameter in predicting fetal aneuploidy.

Maternal obesity, fetal positioning, and oligohydramnios frequently limit visibility of one or both ears. A potential pitfall was described by Fink et al. (40) in which a prominent external ear could be mistaken for a rare parietal cephalocele. By documenting the characteristic external ear anatomy in the absence of a calvarial defect, this error can be avoided.

TABLE 4 ■ Smoothed Fetal Ear Length (mm) Percentiles by GA

Gestational age (wk)	Percentile		
	10th	50th	90th
14	6	7	9
15	7	9	10
16	9	10	11
17	10	11	13
18	11	13	14
19	13	14	15
20	14	15	17
21	15	17	18
22	16	18	20
23	17	19	21
24	19	20	22
25	20	22	24
26	21	23	25
27	22	24	26
28	23	25	27
29	24	26	29
30	25	27	30
31	26	28	31
32	27	29	32
33	27	30	33
34	28	31	34
35	29	32	35
36	30	33	36
37	30	33	36
38	31	34	37
39	31	34	38
40	32	35	38
41	32	35	39

Source: From Ref. 39.

OVERVIEW OF FACIAL ANOMALIES

The sensitivity of ultrasound screening for facial defects in unselected populations has not yet been determined. Although a large number of malformations and syndromes involving the face have been documented, prevalence figures for many specific anomalies are either unknown or low (41–44). The literature concerning the fetal face predominantly describes small series or case reports. Two prenatal ultrasonographic series have discussed the accuracy of detecting facial abnormalities and the frequency of detection of these anomalies with prenatal ultrasound. The data for both these series were obtained during the period from 1980 to 1985, and therefore, it is likely that technologic innovations have improved the accuracy of detecting these abnormalities (3).

Hegge et al. (10) detected 17 facial abnormalities in 11 (0.15%) fetuses from 7100 low- and high-risk obstetric ultrasonographic examinations, examined from 20 to 39 weeks' gestation. All affected fetuses had coexistent structural abnormalities, polyhydramnios, or a history of maternal teratogen exposure. Pilu et al. (11) examined the accuracy of prenatal ultrasound in detecting craniofacial malformations among 223 fetuses between 18 and 40 weeks' gestation, who were at risk for craniofacial malformations. The incidence of facial defects was 8%, with prenatal ultrasound examinations successfully detecting craniofacial malformations in 14 of 18 (78%) fetuses. Nicolaides et al. (44) observed facial defects in 146 (7%) of 2086 fetuses in a high-risk population who were karyotyped because of ultrasonographically detected fetal malformations and/or growth retardation. This study found other malformations and chromosomal abnormalities in a substantial proportion of fetuses with prenatally diagnosed facial abnormalities, which was not surprising in view of the high-risk study population. In the total series of 2086 fetuses with malformations or growth retardation, there were 31 with trisomy 13, 83 with trisomy 18, and 69 with trisomy 21; facial defects were found in 71%, 36%, and 14% of these fetuses, respectively. These studies demonstrate that prenatal ultrasound can be an accurate and reliable tool for the prenatal diagnosis of facial anomalies.

Under optimal conditions, the complex curvature of the face makes it difficult to obtain adequate images with two-dimensional ultrasonography, and many cross-sectional images may be required to obtain a complete impression. Preliminary work by Pretorius and Nelson (45) has demonstrated that three-dimensional ultrasonography (3DUS) has the potential to offer clearer visualization and understanding of fetal facial anatomy. 3DUS offers the potential advantage of improving visualization of anatomic spatial relations, which is particularly important in cases involving the complex anatomy encountered with fetal anomalies (e.g., cleft palate, limb–body wall complex, and amniotic band syndrome), in which the standard landmarks may be distorted or missing. An additional advantage is that the diagnostician can review the volume data interactively, employing orientations beyond the range of the imaging system for optimal visualization of the relevant anatomy.

ORBITAL AND PERIORBITAL ANOMALIES

Hypotelorism

Clinical Features

The prenatal ultrasonographic diagnosis of a decreased interorbital distance (hypotelorism) is most commonly associated with holoprosencephaly. Most cases of holoprosencephaly are sporadic, but genetic, chromosomal, and teratogenic causes have been described (20). Less common causes of hypotelorism are listed in Table 5 (3,42,46).

Holoprosencephaly is a complex spectrum of intracranial abnormalities resulting from absent or incomplete median cleavage of the forebrain (prosencephalon) during early embryonic development (47,48). Depending on the degree of forebrain cleavage, holoprosencephaly is categorized as alobar, semilobar, or lobar (47–50). Alobar holoprosencephaly, the most severe form, is associated with severe facial deformities and is characterized by complete absence of cleavage of the forebrain, resulting in a monoventricular cavity and absent midline structures. Holoprosencephaly is associated with a number of characteristic intracranial and facial abnormalities, including cyclopia, ethmocephaly, cebocephaly, median cleft lip and palate, lateral cleft lip and palate, bilateral cleft lip, and mild hypotelorism (Fig. 9) (49).

Cyclopia. Median orbit with various degrees of ocular fusion, absent nose, proboscis from the lower forehead, absent philtrum of the upper lip, absent facial bones, and low-set ears (Fig. 10).

TABLE 5 ■ Syndromes Associated with Hypotelorism

Common
- Holoprosencephaly

Uncommon
- Trigonocephaly
- Oculodentodigital dysplasia
- Microcephaly
- Meckel–Gruber syndrome
- William's syndrome
- Baller–Gerold syndrome
- Maternal phenylketonuria
- Myotic dystrophy
- Hemifacial microsomia (Goldenhar's syndrome)
- Chromosomal aberrations:
  - Trisomy 13
  - Trisomy 21
  - 18p–
  - 5p–
  - 14q+

Source: From Ref. 46.

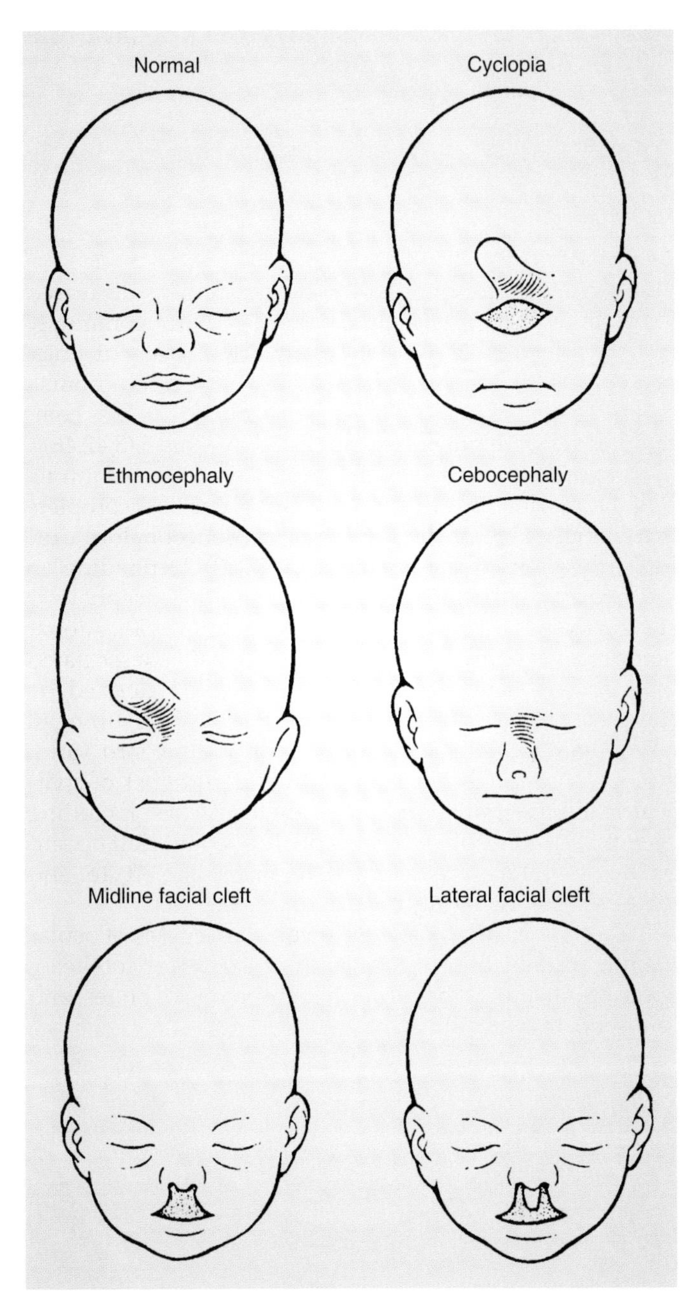

FIGURE 9 ■ The variable facial features of holoprosencephaly in contrast to the normal facial features.

Ethmocephaly. Marked hypotelorism, absent nose with proboscis present at the level of the orbits, and low-set, malformed ears.

Cebocephaly. Flat and rudimentary nose, hypotelorism, single flat nostril, and absent philtrum of the upper lip.

Median cleft lip. Mild and subtle hypotelorism, flat nose, and median cleft lip with or without cleft palate (Figs. 11 and 12).

Lateral cleft lip. Unilateral or bilateral lateral cleft lip and palate.

Intracranial features of holoprosencephaly, usually of the alobar type, are present in all cases of cyclopia, ethmocephaly, cebocephaly, and hypotelorism with median cleft lip (47,48,51). The converse, however, is not

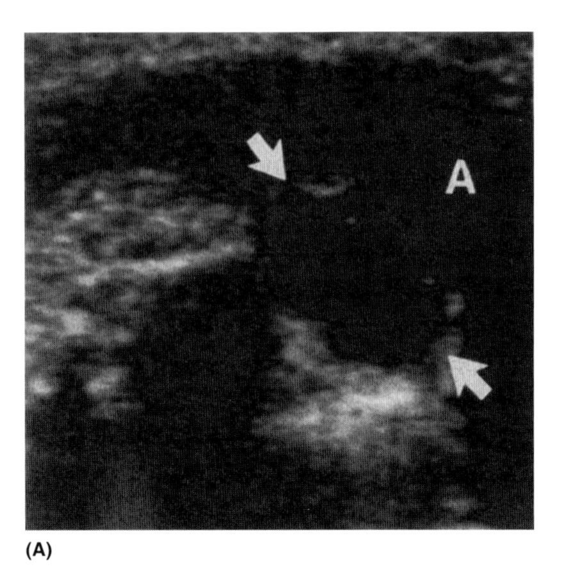

(A)

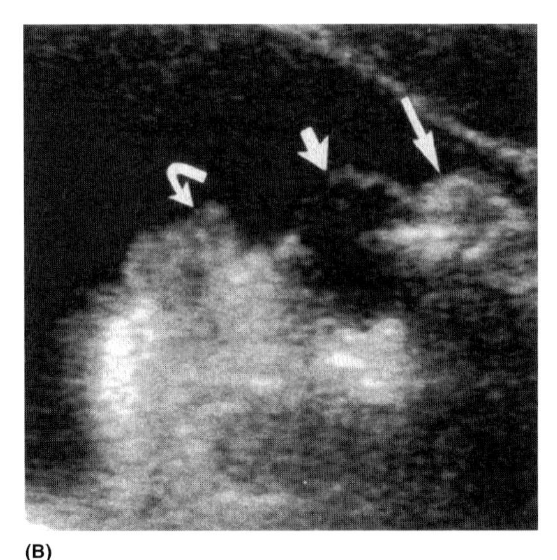

(B)

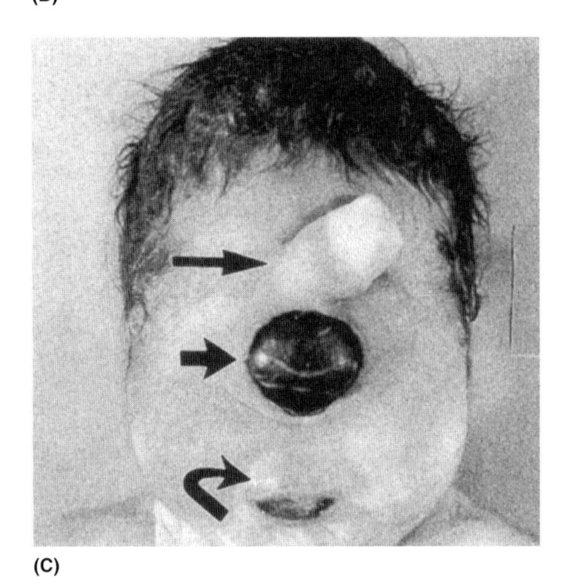

(C)

FIGURE 10 A–C ■ Cyclopia. (**A**) In utero axial sonogram of face shows fused but widely separated bony orbit (*arrows*). Longitudinal scan in utero (**B**) and gross specimen (**C**) show cyclopia with fused orbits (*short straight arrows*), proboscis cephalad to fused orbit (*long straight arrows*), and fetal lips (*curved arrows*). Normal medial fold of upper lip, or philtrum, is absent. A, anterior. *Source*: From Ref. 47.

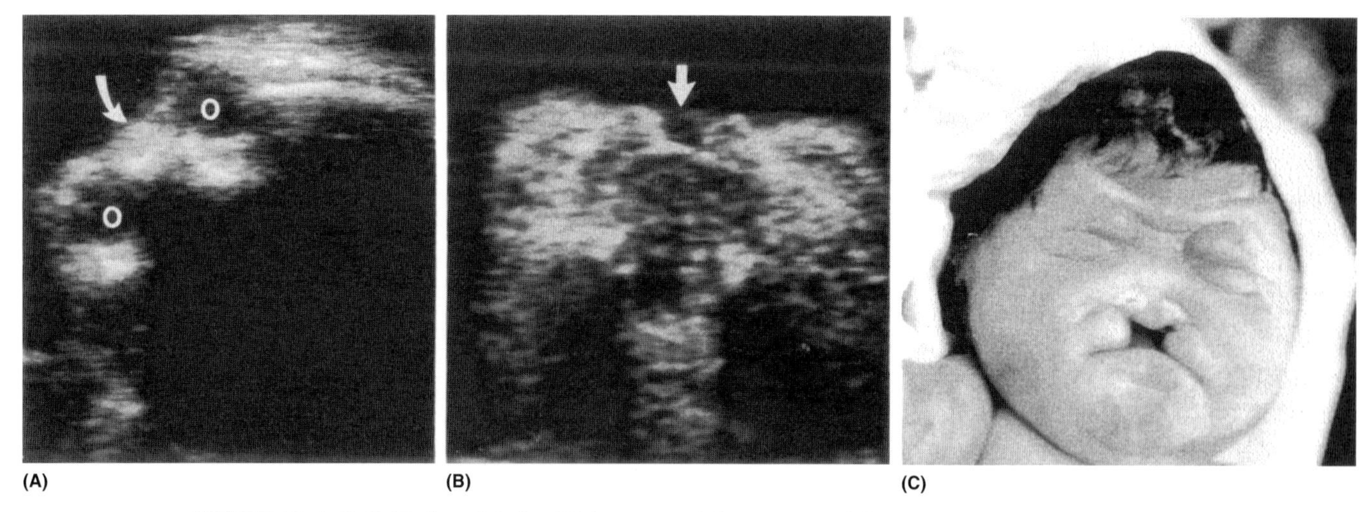

FIGURE 11 A–C ■ Median cleft lip. (**A**) In utero axial sonogram of face shows closely spaced orbits (hypotelorism) and absence of nasal bridge (*arrows*). Axial sonogram (**B**) and corresponding gross specimen (**C**) of midface show median cleft lip (*arrow*). O, orbits. *Source*: From Ref. 47.

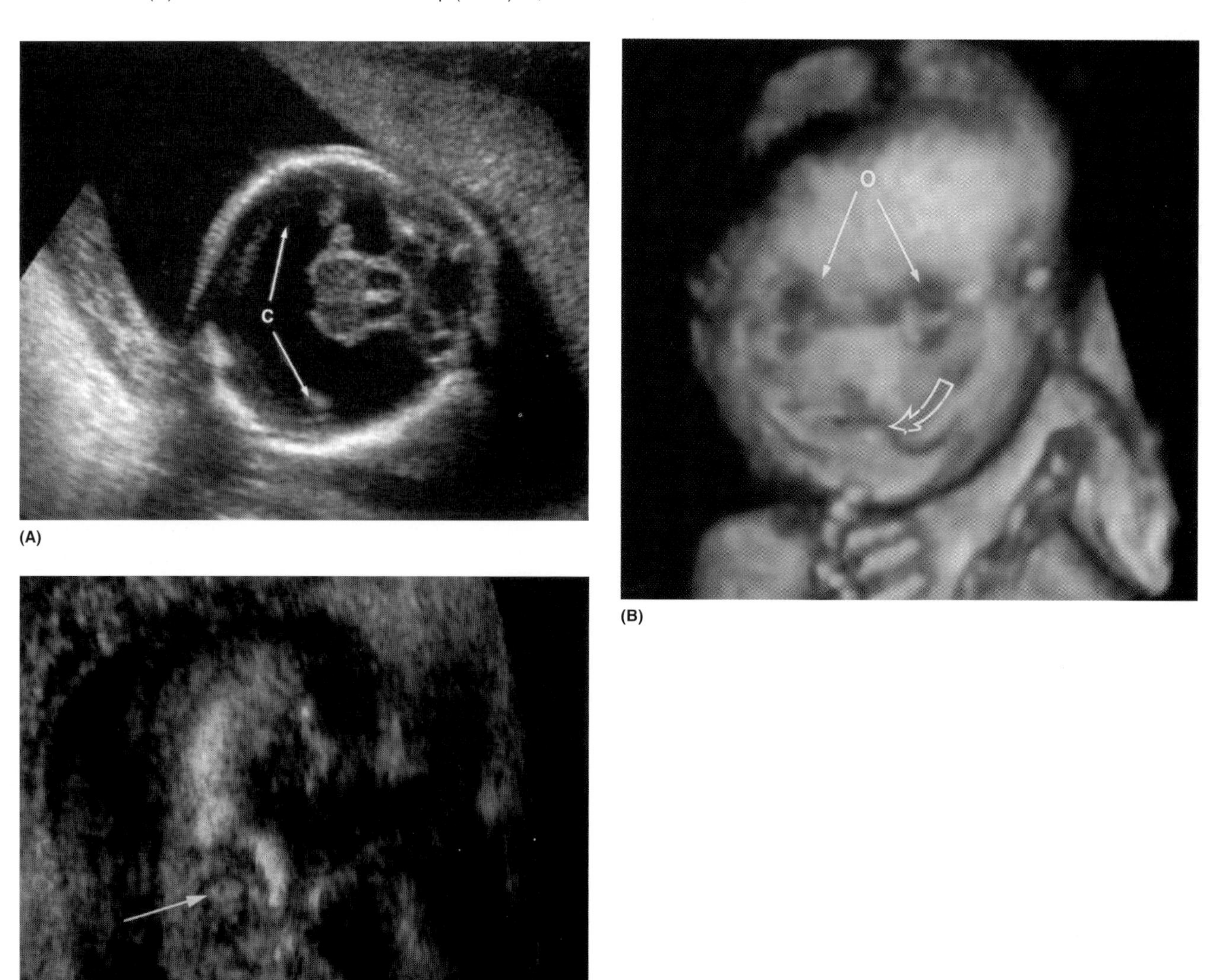

FIGURE 12 A–C ■ Holoprosencephaly—(**A**) Scan through the fetal head demonstrating a monoventricular cavity (C) and fused thalami without other midline structures consistent with alobar holoprosencephaly. (**B**) This 3D image demonstrates midline cleft lip and palate (*curved arrow*) with hypotelorism and small or near absent nares in this fetus with alobar holoprosencephaly. O, orbits. (**C**) Coronal image of the fetal face and orbits demonstrating echogenic lens of the one orbit, consistent with a cataract (*arrow*).

true. DeMyer (50) reported that 17% of patients with alobar holoprosencephaly had a normal face. The semilobar and lobar types are less severe forms of holoprosencephaly and are often associated with lesser degrees of facial abnormality or no facial deformities (47).

Ultrasonographic Findings

In fetuses with hypotelorism, generally both the ID and the BD fall well below 2 standard deviations of the mean (15). In an ultrasonographic review of facial features of 27 fetuses with holoprosencephaly (22 of the alobar type and five of the semilobar type), facial abnormalities were present in 24 (89%), and 14 (58%) of these were detected on prenatal ultrasound (47). The prenatal abnormalities detected by ultrasonography included cyclopia (four of five), ethmocephaly (two of three), cebocephaly (one of three), midline cleft lip (four of eight), lateral cleft lip (two of two), and mild hypotelorism (one of three). An abnormal karyotype was revealed in 13 of the 26 fetuses in whom it was performed. The most common abnormality was trisomy 13, but other chromosomal abnormalities included triploidy (XXY), partial chromosomal deletions, and chromosome translocation. The menstrual age at the time of diagnosis of holoprosencephaly was less than 24 weeks in 56% of cases.

Associated Anomalies and Prognosis

Numerous extrafacial anomalies are associated with holoprosencephaly and are listed in Table 6 (47,48,52). Extracranial malformations have been identified in about half the fetuses with alobar or semilobar holoprosencephaly. Associated abnormalities in the fetus may be overlooked on prenatal ultrasound, particularly when there has been a confident prenatal diagnosis of holoprosencephaly and therapeutic options have been chosen, so that the associated findings may be deemed less important (3).

The most severely affected neonates die at birth or in the first six months of life. In mild cases, survival with variable mental retardation occurs. When there is demonstration of facial and extra central nervous system anomalies, additional prognostic information regarding the fetus is provided. Cyclopia and ethmocephaly are uniformly associated with a fatal outcome, while fetuses with cebocephaly or premaxillary agenesis rarely survive beyond one year (20,47–49,52–55).

TABLE 6 ■ Extracranial Malformations Associated with Holoprosencephaly

Abdomen
■ Renal dysplasia
■ Omphalocele
■ Esophageal atresia
■ Intestinal abnormalities
■ Bladder exstrophy
Thorax
■ Pulmonary hypoplasia
■ Cardiac defects
Ventricular septal defect
Atrial septal defect
Two-chamber heart
Transposition of great vessels
Amniotic fluid
■ Polyhydramnios
Other
■ Meningomyelocele
■ Polydactyly
■ Clubfoot
■ Intrauterine growth retardation

Source: From Ref. 46.

Hypertelorism

Clinical Features

Hypertelorism, or an increased interorbital distance, can be an isolated primary defect, or it can occur secondarily as part of multiple syndromes, many of which are associated with chromosomal abnormalities or occur as a result of maternal teratogen exposure (e.g., phenytoin) (Table 7) (20,43,56,57). The following are some of the syndromes and malformations most commonly associated with hypertelorism:

- Frontal, ethmoidal, or sphenoidal cephalocele
- Median cleft face syndrome
- Craniosynostosis (Apert's syndrome, Crouzon's syndrome, and Carpenter's syndrome)

Frontonasal cephaloceles are rare and appear to arise from intrinsic calvarial defects, projecting through a defect in the ethmoid, sphenoid, or frontal bone (57). Frontonasal cephaloceles are generally smaller than cephaloceles in other locations and may also be associated with other midline facial defects such as cleft lip and palate (3,57). Frontal cephaloceles result in hypertelorism by displacing the orbital globe downward and outward (8).

The median cleft face syndrome is characterized by hypertelorism; median clefting of the nose; various degrees of clefting of the lip, maxilla, and palate; a V-shaped anterior hairline; and cranium bifidum occulta (58).

Ultrasonographic Findings

In a small series of fetuses with hypertelorism reviewed by Trout et al. (15), all cases of hypertelorism were characterized by an inner orbital diameter (IOD) above 2 standard deviations from the mean, whereas the outer orbital measurement fell just at the 95th percentile. Therefore, the IOD may be more reliable than the outer orbital diameter as a measure of hypertelorism. The ultrasonographic diagnosis of a cephalocele requires the demonstration of a calvarial defect (57). A frontal cephalocele protruding through a frontal calvarial defect may produce hypertelorism and a mass visible on the sagittal profile view or on an axial view through the periorbital region (34). Identification of the calvarial

TABLE 7 ■ Malformations and Syndromes Associated with Hypertelorism

Median plane facial malformations
■ Median cleft face syndrome
■ Frontal, ethmoid, or sphenoid meningoencephalocele
■ Frontal, ethmoid, or sphenoid (dermolipoma) teratoma
■ Nasal glioma
■ Nasofrontal mucocele
Miscellaneous facial defects
■ Proboscis lateralis
■ Facial clefts other than median
■ Facial hemangioma
■ Extra nares
Prominent skull dysplasias
■ Craniosynostosis
Apert's syndrome
Crouzon's syndrome
Carpenter's syndrome
Pfeiffer's syndrome
Kleeblattschädel
■ Thickened skull
Albers–Schönberg disease
■ Metopism
Teeth defects
■ Rieger's syndrome
Prominent neurologic and brain defects
■ Hydrocephalus (any type)
■ Megalencephaly with skull enlargement
■ Familial neurovisceral lipidosis syndrome
■ Lissencephaly
■ Agenesis of the septum pellucidum
■ Agenesis of the corpus callosum
Chromosomal anomalies
■ Wolf's syndrome of chromosome 4p–
■ Chromosome 5p– cri du chat
■ Various translocations
■ Chromosome 18p, 18q–syndromes
■ Turner's XO syndrome
■ 48,XXXX
■ 49,XXXXY
■ Various trisomies (10, partial 7,9,14)
Others
■ Ocular defects
■ Cleft lip with or without palate
■ Prominent skin manifestations
■ Prominent skeletal manifestations
■ Sexual organ malformations
Miscellaneous
■ Hypercalcemia with supravalvular aortic stenosis
■ Potter's syndrome
■ Inguinal hernia
■ Lymphedema and yellow nails
■ Acquired immunodeficiency syndrome dysmorphism
■ Noonan's syndrome

Source: From Ref. 20.

defect assists in distinguishing a cephalocele from the following other frontonasal masses (3,59):

- Lacrimal duct cysts
- Orbital duplication
- Facial hemangioma
- Dermoid cyst
- Retinoblastoma

The detection of hypertelorism in the fetus at risk for a specific syndrome or in association with other malformations may provide convincing evidence for its diagnosis (3). Because the list of syndromes associated with hypertelorism is extensive, the associated anomalies depend on the underlying malformation syndrome, which characteristically determines the prognosis and reoccurrence risk as well as the prenatal ultrasonographic findings (15,56,57). Frontonasal cephaloceles have a better prognosis than cephaloceles in other locations, perhaps because they usually are smaller. However, when the cephalocele is associated with major multiple malformation syndromes that themselves carry a poor prognosis (e.g., atelosteogenesis—a generalized chondrodysplasia), a poor outcome can be expected (57).

Microphthalmia and Anophthalmia

Clinical Features

Microphthalmia (decreased orbital size) and anophthalmia (absence of the eye or vestigial eye) are rare disorders that are associated with at least 25 syndromes, chromosomal abnormalities, and intrauterine infections (3,42,43,60,61), including some of the following:

- Chromosomal abnormalities (trisomy 13, trisomy 18q–)
- Fetal infections (rubella, varicella, toxoplasmosis)
- Fraser's syndrome
- Goldenhar's syndrome (hemifacial microsomia)
- Meckel–Gruber syndrome
- Neulaxova's syndrome
- Oculodentodigital syndrome
- Warburg's syndrome

Routine measurement of the intraocular diameter would not be expected to provide more than a low diagnostic yield, although it may detect recurrence of these syndromes. Assessment for the presence or absence of microphthalmia may become important when other intracranial abnormalities are present or in the fetus with a familial risk of orbital abnormalities (61). When an orbital abnormality is detected, a detailed survey for other anatomic anomalies and karyotype analysis usually is warranted.

Anophthalmia occurs more rarely than microphthalmia and results from failure of the optic vesical to form (61). Taybi and Lachman (42) associated anophthalmia with Goldenhar–Gorlin syndrome (hemifacial microsomia), trisomy 13, Lenz's syndrome (X-linked microphthalmia), and microphthalmia with digital anomalies.

Ultrasonographic Findings

The use of ocular measurements and the potential usefulness of nomograms to confirm and recognize an anomaly are emphasized in the literature (15,60). The diagnosis of microphthalmia during the second trimester may be considerably more difficult than in the third trimester, unless the orbits are unequivocally small. The OD increases progressively but not linearly during the second and third trimesters; however, the normal globe size becomes progressively more variable in the third trimester (62).

Prenatal detection of anophthalmia was described in a case with the Goldenhar–Gorlin syndrome (61) (hemifacial microsomia; Fig. 13). This is best demonstrated in axial and frontal coronal sections through the expected level of the orbits and is manifested by absence of the globe and often the orbit, with associated ipsilateral ear and facial abnormalities.

The frequent association between anophthalmia or microphthalmia and other chromosomal, craniofacial, vertebral, and extremity abnormalities should prompt a careful search of the fetus for other subtle malformations and consideration of chromosomal analysis whenever the absence of the globe or severe microphthalmia is detected.

Intraocular Abnormalities

Because the eye is an embryologic derivative of the forebrain, fetal and neonatal ocular findings, including fetal eye movement patterns and regression of the hyaloid artery of the eye, may provide useful information about the general pattern of cerebral development and function (21,63). For example, Birnholz (63) reported that normal slow eye movements typically occur by 16 weeks' gestation. Rapid eye movements begin at 23 weeks and are more frequent between 24 and 35 weeks, after which eye inactivity becomes more common. Birnholz (63) detected abnormal fetal eye movement patterns in fetuses between 16 and 39 weeks' gestation, all of whom had obvious abnormalities of brain structure.

Hyaloid Artery

The hyaloid artery, which originates from the main ophthalmic artery, is seen in fetuses of 20 weeks' gestational age or younger and regresses spontaneously at the start of the third trimester (21). It appears as an echogenic line coursing centrally through the eyes and demonstrates a beaded appearance as it begins to regress. Delayed hyaloid artery regression may indicate retardation in early third-trimester cerebral development. The hyaloid artery remained visible beyond 30 weeks' gestation in 9 of 25 abnormal pregnancies (trisomy 21, microcephaly without chromosomal abnormality, trisomy 13, trisomy 18, fetal alcohol syndrome, and fetal hydantoin syndrome) (21).

Walker–Warburg Syndrome

Clinical Features ■ Walker–Warburg syndrome is a rare autosomal-recessive condition characterized by type II lissencephaly with agyria, hydrocephalus, other brain

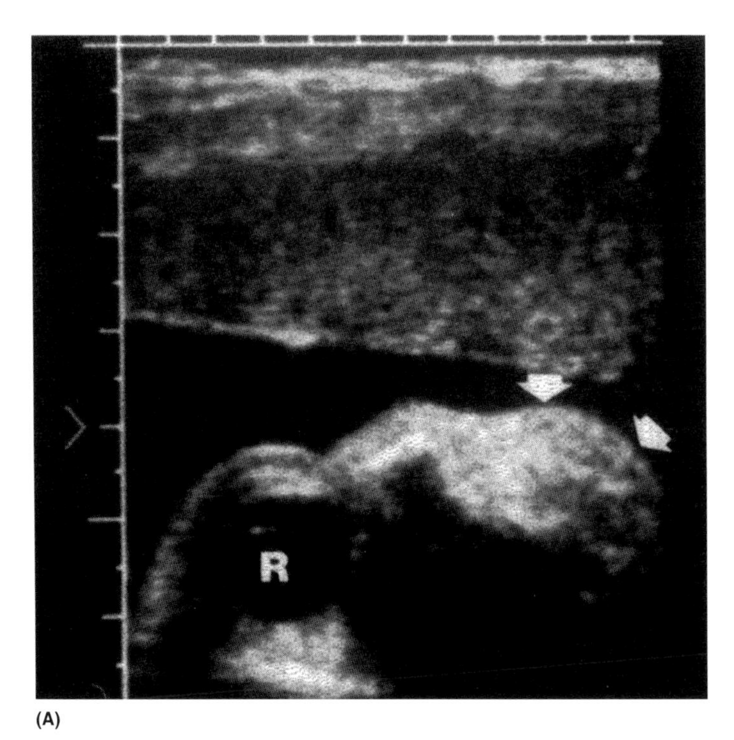

(A)

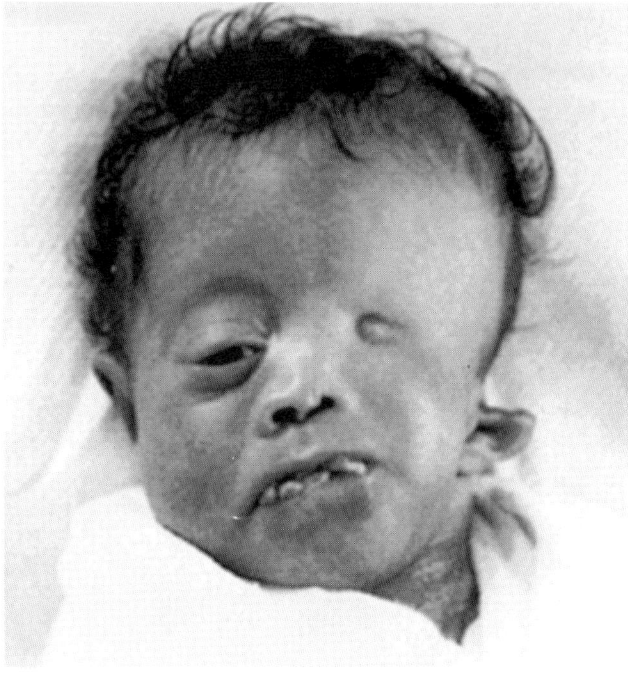

(B)

FIGURE 13 A AND B ■ Goldenhar–Gorlin syndrome (hemifacial microsomia) at 30 weeks' gestation. (**A**) Axial sonogram at the mid-orbital level demonstrates absence of the left orbit (*arrows*). (**B**) Postnatal photograph confirms the prenatal ultrasonographic features, including facial asymmetry with absence of the left orbit and low-set, malformed left ear. The left maxilla and mandible are hypoplastic. R, right eye. *Source*: From Ref. 61.

malformations, and eye abnormalities. Half of cases have associated occipital encephalocele (64). The eye abnormalities comprise a wide spectrum of malformations, including severe microphthalmia; anterior and posterior chamber defects, including persistent hypoplastic primary vitreous and retinal dysplasia and detachment; corneal clouding; cataracts; and colobomatous malformations (64,65).

Ultrasonographic Findings ■ Of the many ocular findings seen postnatally in patients with Walker–Warburg syndrome, only–retinal dysplasia has been diagnosed prenatally. Farrell et al. (64) diagnosed Walker–Warburg syndrome on the basis of the detection of retinal detachment in the setting of a small occipital cephalocele, absent cerebellar vermis, and hydrocephalus. The retinal detachment was not apparent until 32 weeks' gestation, when an abnormal concentric echogenic ring was demonstrated, which initially was thought to represent an unusually small globe but which on further followup at 35 weeks' gestation assumed a conical echogenic appearance attributed to retinal detachment.

Retinoblastoma

Retinoblastomas are embryonic neoplasms that can be inherited (chromosome 13q14) or can occur as a spontaneous mutation (59). A single case report is described within the literature of a large retinoblastoma detected in a prenatal sonogram of a fetus at 21 weeks' gestation. The tumor consisted of an irregular echogenic mass arising from the orbit surrounded by a transonic area and covered by a thin membrane. The tumor deformed the normal anatomy of the facial bones (48). Prenatal ultrasonographic examination of the fetal eyes can be offered in those cases of retinoblastomas with a significant recurrence risk, particularly when prenatal DNA diagnosis is not possible. The use of high-frequency transvaginal transducers has allowed the visualization of even small abnormalities of the fetal eyes early in pregnancy (59,66).

Periorbital and Facial Masses

Masses involving and distorting the fetal facial architecture are rare. They can be cystic or solid and can involve the orbits, nose, oropharynx, upper neck, and brain (59,67–77). These masses range from relatively minor congenital abnormalities, such as dacryocystoceles, to severe craniofacial teratomas (59,67–77). The differential diagnosis of fetal facial masses includes the following:

- Dacryocystocele
- Hemangioma
- Teratoma
- Anterior cephalocele
- Dermoid cyst
- Lymphangioma
- Isolated cystic hygroma
- Soft-tissue sarcoma
- Granular-cell tumor
- Malignant melanoma
- Retinoblastoma
- Neurofibroma
- Mucocele

Unfortunately, the ultrasonographic appearance is often not helpful in differentiating among these entities because most are solid or complex. The location of the mass may be helpful in narrowing the differential diagnosis.

Dacryocystoceles

Clinical Features ■ Dacryocystoceles (lacrimal duct cysts) result from congenital impatency of the distal portion of the nasolacrimal duct, leading to cystic dilation of the lacrimal duct (68,69). The nasolacrimal duct usually becomes patent by the seventh or eighth intrauterine month. Dacryocystoceles are located inferomedial to the orbit and are caused by a thin obstructing mucous membrane near Hasner's valve. About 30% of all newborns have an impatent nasolacrimal duct, but only 2% develop symptomatology. The obstructing membrane spontaneously ruptures in 78% of cases by three months of age and in 91% by six months (68).

The differential diagnosis of a hypoechoic lesion inferomedial to the globe includes anterior encephalocele, hemangioma, and periorbital dermoid cyst.

Ultrasonographic Findings ■ Prenatal ultrasonographic findings of dacryocystoceles include the demonstration of a hypoechoic mass in the typical location inferomedial to the fetal globe, without displacement of the globe but with synchronous eye movements (Fig. 14) (68). In four reported cases in the literature, these cysts measured less than 1.5 cm in diameter and were not evident until after 30 weeks' gestation (67–69).

The ultrasonographic features of periorbital and facial masses are listed in Table 8. The typical location of a dacryocystocele should help to suggest its diagnosis because entities such as cephaloceles rarely occur medial to the orbit, often displace the globe downward and outward, and are associated with an underlying calvarial defect and often with hydrocephalus. Periorbital dermoid cysts typically occur superolateral to the globe, as opposed to inferomedial. An excellent prognosis can be expected when a dacryocystocele is discovered on prenatal ultrasound.

Hemangiomas

Clinical Features ■ Hemangiomas are one of the most common benign neoplasms seen in the neonatal age. Any organ can be affected, but most are cutaneous in origin and remain small and clinically insignificant (3). Although histologically benign, large, rapidly growing hemangiomas can result in grotesque deformity in the head and neck area and can progressively distort normal structures, resulting in airway or esophageal obstruction, pressure and necrosis of surrounding structures, and obstruction of the auditory canal (70,71).

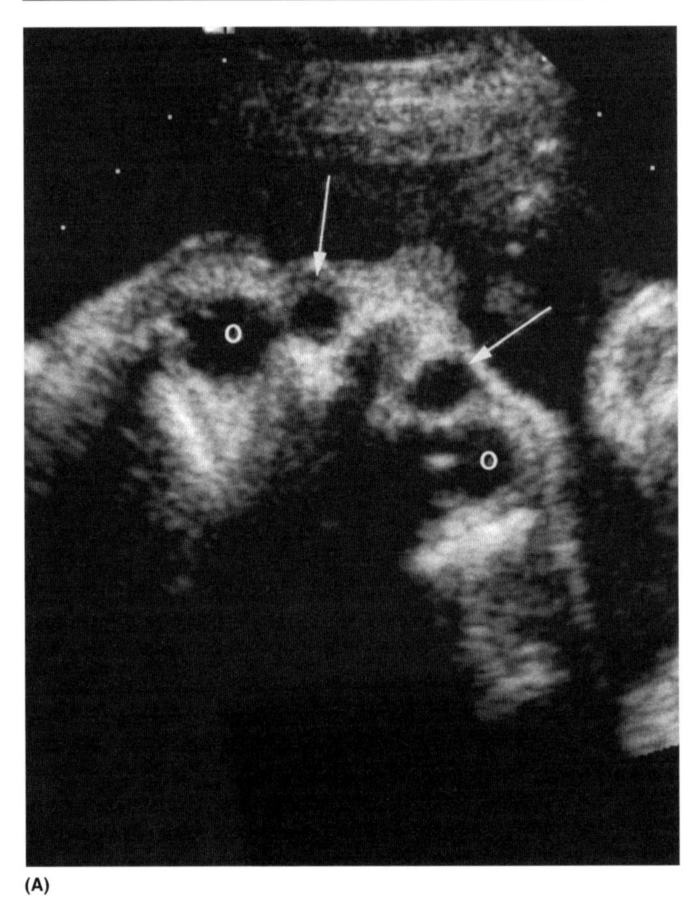

(A)

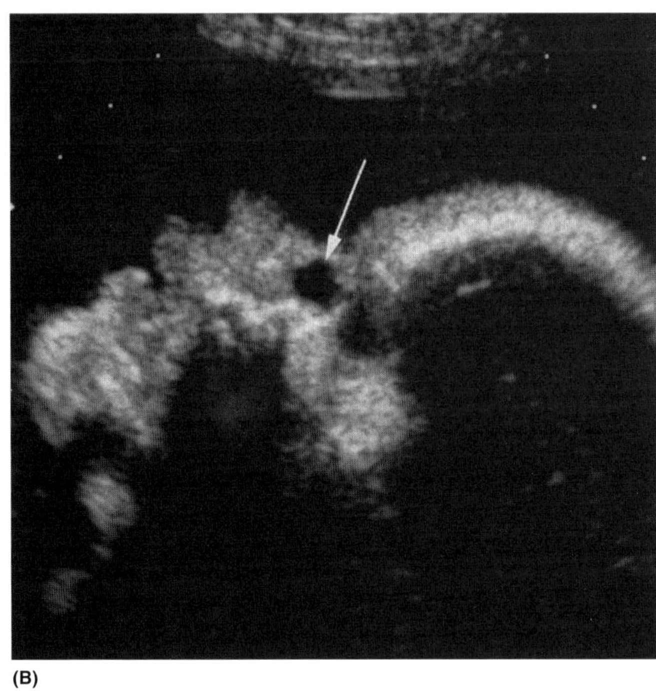

(B)

FIGURE 14 A AND B ■ Dacryocystocele. (**A**) Dacro 42. Axial scan through the orbits (O) demonstrating dacryocystoceles (*arrows*) medial to both orbits. (**B**) Parasagittal scan demonstrating cystic mass (*arrow*) close to the midline corresponding to dacryocystocele.

TABLE 8 ■ Differential Diagnosis of Periorbital and Facial Masses

Lesion	Location	Ultrasonographic features
Dacryocystocele	Inferomedial to orbital globe	Hypoechoic or anechoic mass
Hemangioma	Cutaneous head and neck	Cystic or solid
		Similar or greater echogenicity to placenta
		Obtuse angle with skull
		Doppler pulsations or mixed arterial and venous waveform
		Occasionally scattered calcifications
Teratoma	All parts of the head and neck	Cystic, solid, or both
	Palate	Highly echogenic areas
	Nasopharynx	Irregular shape
	If periorbital, superolateral to orbital globe	Possibly coarse calcifications
Anterior cephalocele	Usually midline	Small
	Projects through ethmoid, sphenoid, or frontal calvarial defect	Displaces orbit inferiorly and laterally
		Calvarial defect
		Acute angle with skull
		Smooth, rounded shape
		Possibly hydrocephalus or other intracranial abnormality
Myoblastoma	Oral cavity	Echogenic with small anechoic areas
		Multilobular shape
Retinoblastoma	Orbital globe	Irregular echogenic mass surrounded by anechoic area
		Covering membrane
Cystic hygroma	Posterolateral neck	Cystic and may contain septa
		May include hydrops

Ultrasonographic Findings ■ Periorbital and facial hemangiomas can present as either cystic or solid masses but frequently demonstrate a homogeneously echogenic pattern similar to that of the placenta or are slightly more echogenic (Fig. 15) (70–73). The differential diagnosis for hemangiomas is similar to that for dacryocystoceles (Table 8). Hemangiomas are larger and usually demonstrate well-defined vascular spaces. Doppler flow patterns vary depending on the size of the vessels and the degree of arteriovenous shunting (70–73). Several cases of hemangiomas examined prenatally with Doppler ultrasound have been reported, with variable ultrasonographic and Doppler characteristics (70–73). The various reports described on pulsed (duplex) Doppler included a low-resistance wave pattern, arterial and venous pulsations, and central pulsating vessels within confluent cystic areas. Bulas et al. (73), with color-flow Doppler imaging, identified vessels at the periphery of the mass but noted an avascular center.

Whenever a solid facial tumor is detected, the ultrasonographer should look for areas of calcification. Gross calcifications suggest the possibility of a teratoma, whereas small, widely scattered calcifications giving rise to homogeneous echogenicities may suggest a hemangioma. (Fig. 16) (3). Other solid head or neck masses can include a thyroid mass or enlargement (Fig. 17). Color Doppler ultrasound may be helpful in evaluating the vascularity of these lesions to distinguish them from less vascular lesions (67).

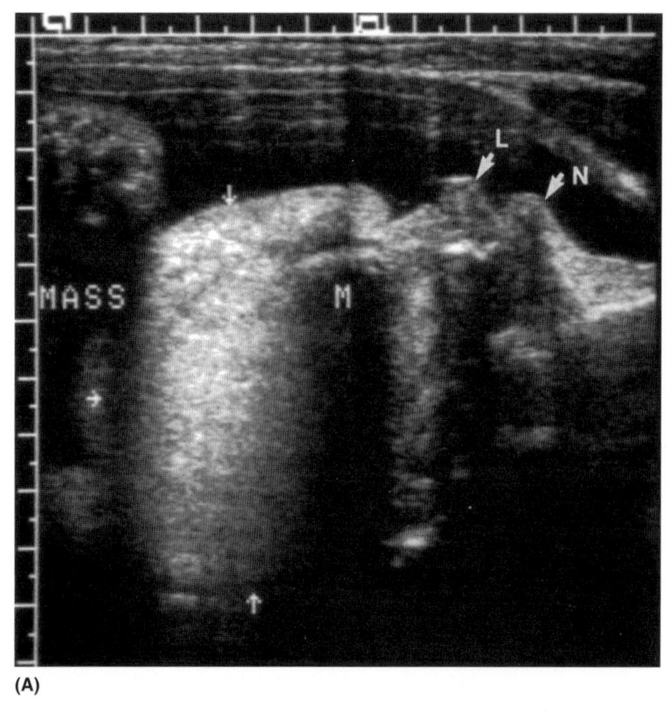

(A)

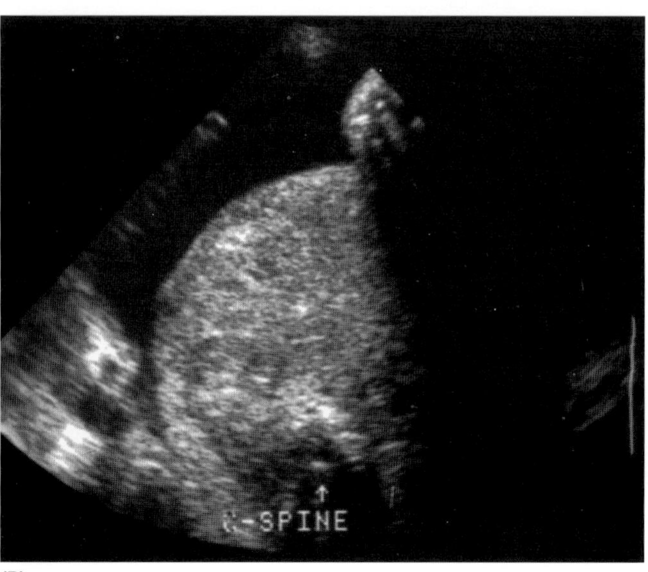

(B)

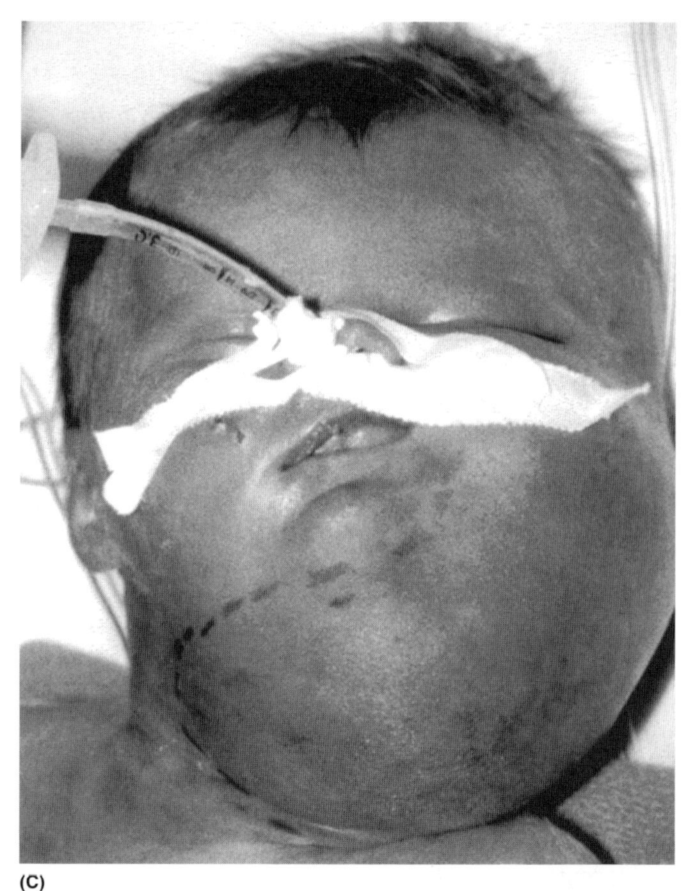

(C)

FIGURE 15 A–C ■ Facial hemangioma. (**A**) Profile view through the fetal face demonstrates the nose (N), upper lip (L), mandible (M), and large echogenic mass (*arrows*). (**B**) Axial scan through the mass (*arrows*) demonstrates echogenic appearance of the mass. (C-spine, cervical spine posteriorly). (**C**) Preoperative photograph of the newborn demonstrates marked deformity of the face and neck due to the large hemangioma. *Source*: (**C**) Courtesy of Marshal Schwartz, MD, Children's National Hospital, Washington, DC.

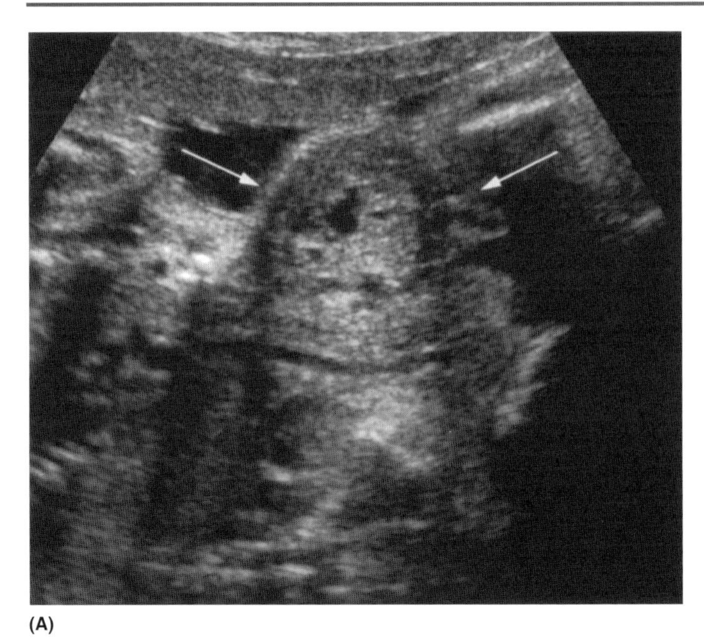

(A)

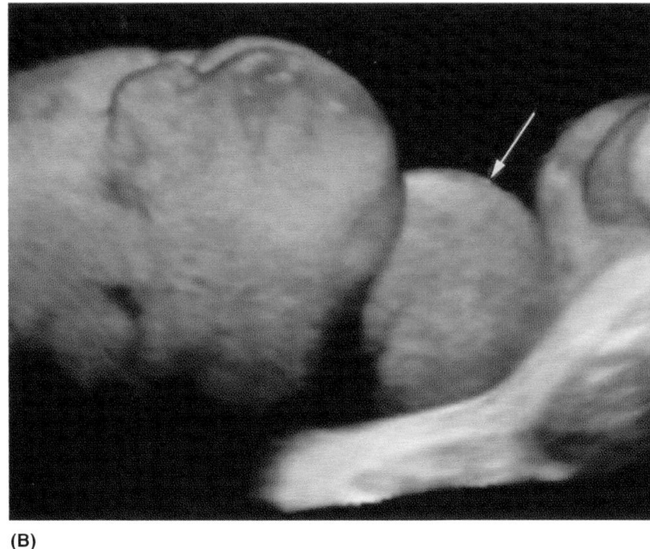

(B)

FIGURE 16 A AND B ■ Thyroid teratoma. (**A**) Parasagittal scan of the neck demonstrating mainly solid mass with a few cystic components in the anterior portion of the neck (*arrows*). (**B**) 3D image demonstrating fetal neck with large soft-tissue mass noted in the anterior portion of the neck corresponding to thyroid teratoma (*arrow*).

Larger hemangiomas may be more readily diagnosed prenatally. As the detection rate increases, it is important to be able to differentiate these benign lesions from other, more serious fetal craniofacial and nuchal masses to avoid the possibility of suggesting a therapeutic termination procedure (3). Hemangiomas are more likely to arise from the scalp and skin rather than from the brain, and although the ultrasonographic texture of the mass cannot be used to differentiate among the various entities, the location is important in narrowing the differential diagnosis (67). Early detection is also important with larger lesions because they may cause hydrops fetalis, dystocia, or fetal injury during labor. Larger hemangiomas may cause platelet sequestration in the neonatal period, and it is valuable to recognize these in advance to observe the child for any hematologic consequences (71). Prenatal ultrasonographic identification of an extracalvarial mass with features atypical of a cephalocele should suggest benign processes such as scalp hematoma, edema, or hemangioma. Prenatal diagnosis of benign processes may change management of patients in the third trimester (allowing normal vaginal delivery instead of cesarean section) and result in changes in management of less common second-trimester cases (obviating pregnancy termination) (78). Surgical excision in neonates and young children is frequently successful (67–73).

FIGURE 17 ■ Goiter. Coronal image of the neck demonstrates fairly symmetrical, solid-appearing mass in the anterior neck corresponding to a fetal goiter (*arrows*).

Other cystic masses may include cystic hygromas. These may be associated with generalized fetal edema if they present as a nuchal mass and are associated with Turner's syndrome. They may also present as nuchal translucency or thickening associated with trisomy 21 (Fig. 18). Occasionally, they may present as a focal cystic mass in the neck (Fig. 19).

LIPS, MOUTH, AND TONGUE

Coronal and sagittal views of the fetal mouth, tongue, and lips, especially the upper lip, probably provide the most useful information regarding cleft lip or macroglossia. They are especially useful in the setting of polyhydramnios, concomitant anomalies, or a positive family history when detection of facial clefts, macroglossia, or an

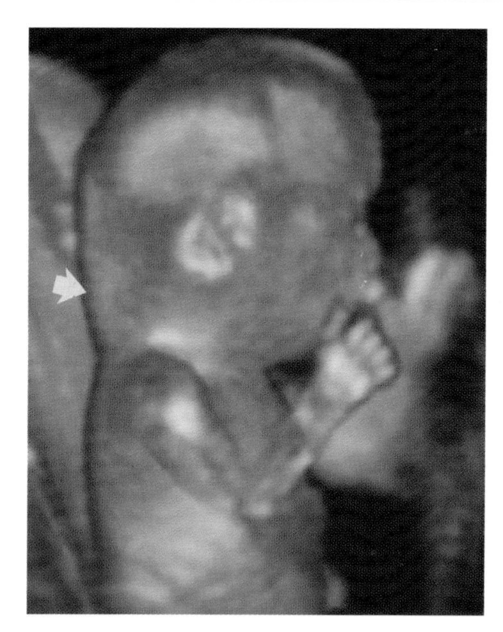

FIGURE 18 ■ Trisomy 21. This 3D image of a fetus with Down syndrome demonstrates thickened nuchal fold (*arrow*). Also, note a relatively flat face and small ears.

unusually shaped mouth suggests the presence of a specific syndrome (3).

Cleft Lip and Cleft Palate

Clinical Features

Cleft lip with or without cleft palate is the most common congenital malformation involving the face. Although often associated, cleft lip and cleft palate are embryologically and etiologically distinct malformations, originate at different times during development, and involve different developmental processes (5). Cleft lip and cleft palate may be incomplete or complete, unilateral or bilateral, lateral or midline, and symmetric or asymmetric (79).

The incidence of cleft lip and palate is about 1 per 1000 live births in the white population (80). Marked racial and geographic variability has been observed, with a higher prevalence among Asians (1.5–2.0 per 1000) and Native Americans (3.6 per 1000) and a lower frequency among African-Americans (0.5 per 1000) (80). About 80% of infants with cleft lip also have cleft palate (5,8).

The two major groups of cleft lip and palate are (*i*) clefts involving the upper lip and the anterior part of the maxilla, (primary palate), with or without involvement of parts of the remaining hard and soft regions of the palate (secondary palate) and (*ii*) clefts involving the hard and soft regions of the palate [secondary palate (Fig. 20) (5)]. Cleft lip with or without cleft palate results from failure of fusion of the maxillary prominence with the medial nasal prominence on one or both sides at about the seventh week of development. The palate forms from the fusion of the primary and secondary palates between the fifth and the 12th weeks. The primary palate becomes the premaxillary part of the maxilla, the portion of the hard palate that contains the four incisor teeth. The premaxillary segment forms from paired premaxillary bones joined in the midline by the interpremaxillary suture, which represents the anterior third of the normal mid-palatal suture (5,8,81,82). The secondary palate is the primordium of the remainder of the hard and soft palates and contains the remaining tooth sockets. It consists of two lateral palatine processes that extend medially and that progressively fuse in the midline from anterior to posterior. When development is complete, the line of fusion between the primary and the secondary palates runs in an arch-like configuration. Perpendicular to this in the midline, the fusion line of the secondary palate extends from the incisive foramen anteriorly to the uvula posteriorly. Cleft palate results from a

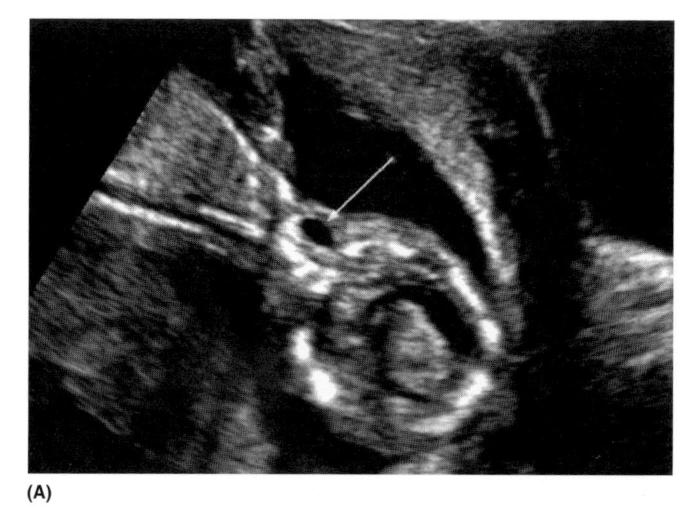

(A)

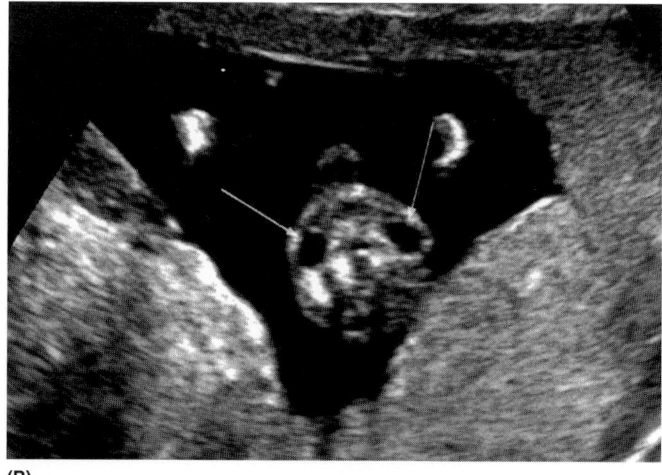

(B)

FIGURE 19 A AND B ■ (**A**) Longitudinal ultrasound in the region of the fetal neck showing small cystic structure within the neck, corresponding to a small cystic hygroma (*arrow*). (**B**) This axial scan through the neck demonstrates that these cystic structures are symmetrical and bilateral and correspond to small cystic hygromas (*arrows*).

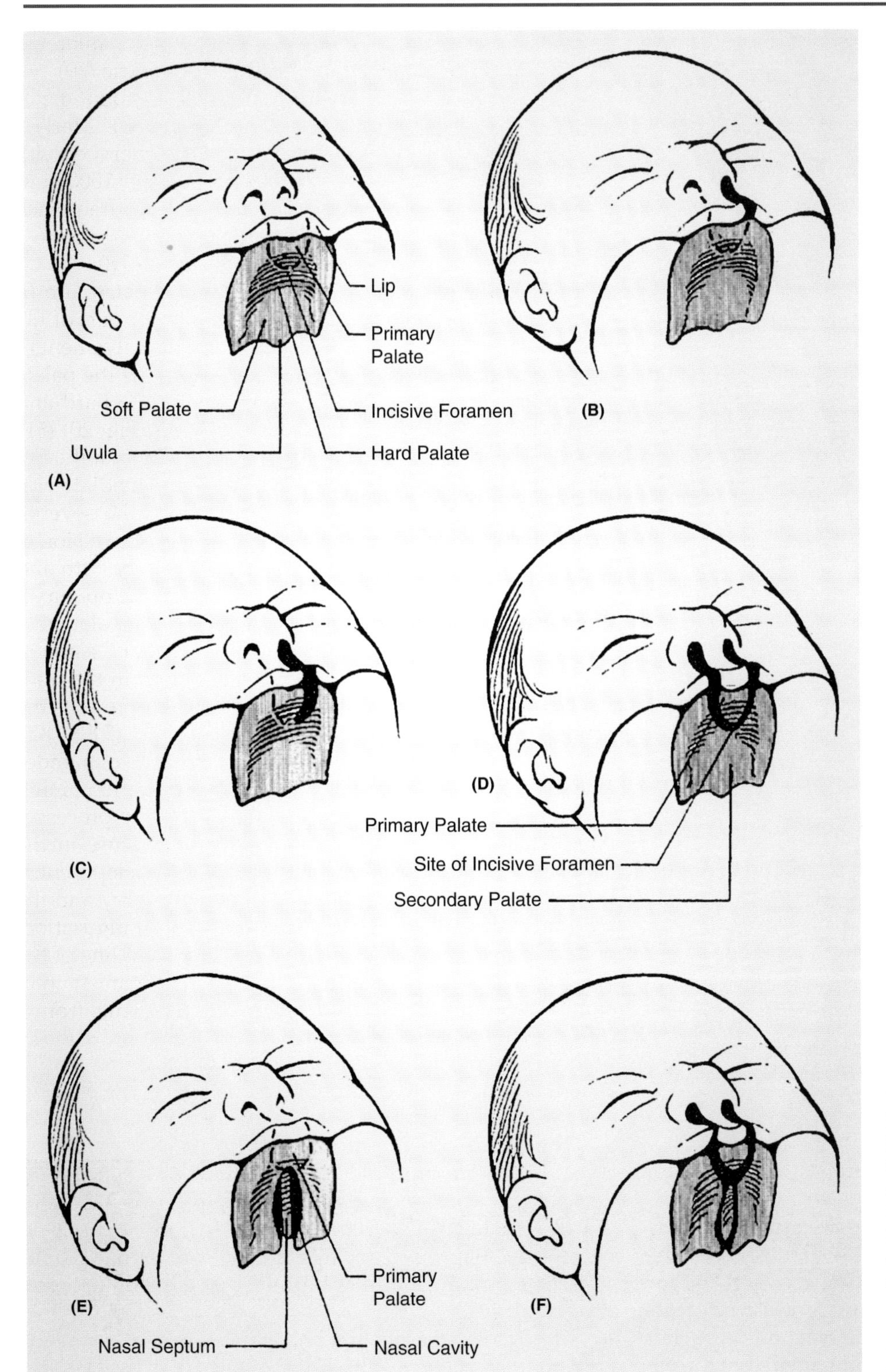

FIGURE 20 A–F ■ Various types of cleft lip and cleft palate. (**A**) Normal lip and palate. (**B**) Unilateral cleft lip extending into the nose. (**C**) Complete unilateral cleft of the lip and alveolar process with unilateral cleft of the anterior or primary palate. (**D**) Bilateral cleft of the lip and alveolar process with bilateral cleft of the anterior palate. (**E**) Isolated cleft palate. (**F**) Bilateral cleft of the lip and alveolar process with bilateral cleft of the anterior and posterior palate. It can also occur with unilateral anterior cleft.

failure of the mesenchymal masses and the lateral palatine processes to meet and fuse with each other, with the nasal septum, or with the posterior margin of the median palatine process (Fig. 21) (5,8,23,82). The clefting abnormality of the anterior or primary palate disrupts the continuity of the arch-like suture between the primary and the secondary palates on one side (unilateral cleft) or both sides (bilateral cleft) and disturbs the normally smooth, C-shaped contour of the alveolar ridge of the maxilla. A posterior cleft abnormality disrupts the continuity of the midline suture between the lateral palatine processes, which may be limited posteriorly to the soft palate or uvula or extend anteriorly to the incisive foramen (5). The median cleft lip is caused by incomplete merging of the two medial

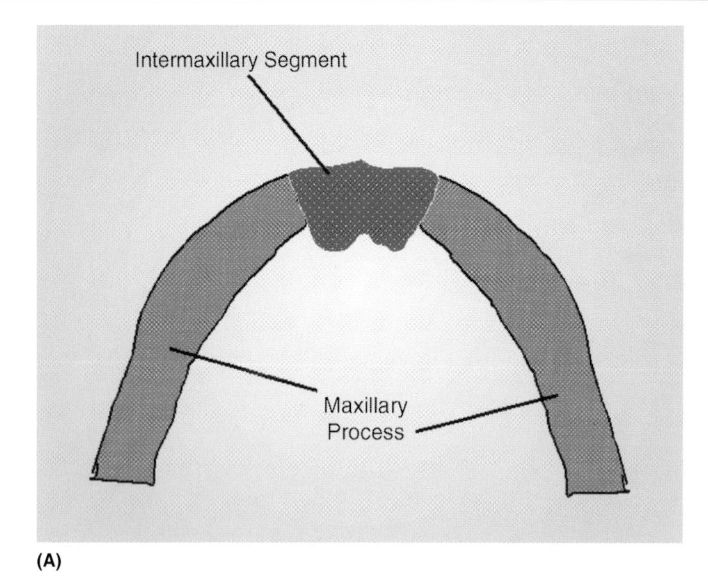

(A)

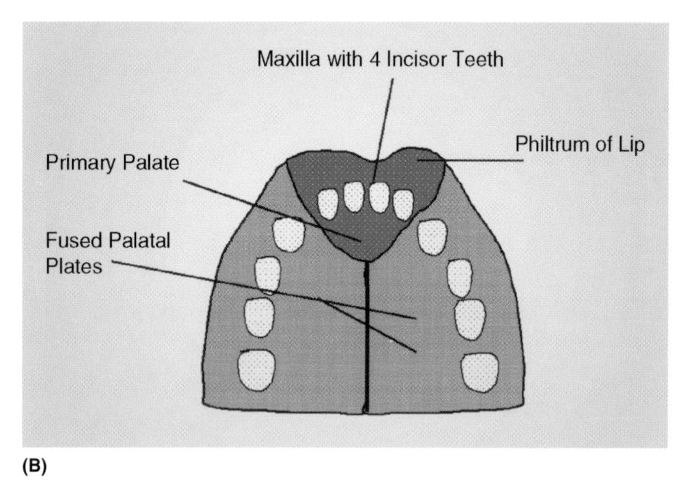

(B)

FIGURE 21 A AND B ■ Normal development of maxilla. (**A**) The intermaxillary segment fuses with the two maxillary processes. (**B**) The intermaxillary segment will form the soft tissues of the lip, the four maxillary incisors, the philtrum, and the primary palate. The maxillary process will form the palatal plates and the rest of the teeth.

nasal prominences in the midline and is a much rarer abnormality.

At least 100 syndromes have been described in association with cleft lip and palate (20,42,43). In most cases, the cause is idiopathic, and syndromes probably account for less than 10% of all cases (Table 9) (28). Maternal teratogens that have been reported in association with cleft lip and palate include alcohol, maternal phenylketonuria, hyperthermia, hydantoin, trimethadione, aminopterin, and methotrexate. In those cases not associated with a syndrome, the cause is most likely a multifactorial combination of environmental and genetic factors with a threshold effect. The incidence is increased in patients with an affected close relative. Most isolated cleft lips or cleft palates are of multifactorial inheritance, but dominant, recessive, and X-linked syndromes have been described. The recurrence risk is 4% with one affected sibling, 4% with one affected parent, 9% with two affected siblings, and 17% with one affected sibling and one affected parent (20). There is a high incidence of chromosomal abnormalities among fetuses with facial clefts, in particular trisomy 13 and 18 (79). Siblings of patients with cleft lip with or without cleft palate have an increased frequency of cleft lip with or without cleft palate, but not of cleft palate alone; whereas siblings of patients with cleft palate alone have an increased frequency of cleft palate but not of cleft lip with or without cleft palate (80). Cleft lip with or without cleft palate affects male fetuses twice as frequently as females; however, cleft palate alone affects males 25% less than female fetuses.

TABLE 9 ■ Some Common Associations, Malformations, and Syndromes Associated with Cleft Lip and Palate

Familial

Chromosomal abnormalities

- Trisomy 13
- Trisomy 18
- Trisomy 21
- Trisomy 22
- XXXXY syndrome
- Various translocations
- Triploidy

Autosomal-dominant, autosomal-recessive, and x-linked cleft syndromes

- Multiple syndromes

Nongenetic cleft syndromes

- Amniotic band syndrome
- Anencephaly
- Congenital heart disease
- Holoprosencephaly
- Encephaloceles
- Medial cleft face syndrome
- Congenital oral teratoma

Source: From Ref. 20.

The likelihood of having cleft lip or cleft palate in association with multiple malformations or a syndrome depends on the associated malformations or syndrome (3). In a series by Nicolaides et al. (44), of 64 cases of prenatally diagnosed facial cleft, 88% of fetuses had additional malformations and 48% had chromosomal abnormalities.

Most isolated cleft lips and palates can be repaired with good cosmetic results, but the severity of the cleft lip and palate influences the repair. Associated anomalies may be difficult to detect ultrasonographically, and frequently, the other anomalies tend to be more conspicuous than the facial cleft. Chromosomal analysis may be offered once a cleft is detected because of the potential for chromosomal abnormalities and associated anomalies. Even if the prenatal detection of a facial cleft is an isolated malformation, prepartum knowledge of the cleft may help to prepare the parents psychologically for the visually disturbing deformity at birth, particularly if a treatment plan has been discussed (3).

Ultrasonographic Findings

Few studies have reported the prenatal detection of cleft lip with or without cleft palate despite the frequency among newborns (81). This may reflect the difficulty in diagnosis and the fact that evaluation of the face is not routine in low-risk obstetric ultrasonography. Most clefts detected prenatally with ultrasound are not subtle and tend to be large. Whenever a labial cleft is identified, an attempt is made to determine laterality and extent of the clefting and to detect other associated malformations. Nyberg et al. (81) developed a prenatal ultrasonographic classification system of cleft lip with or without cleft palate and correlated the classification with fetal outcome. They emphasized the importance of correctly classifying fetal cleft lip and cleft palate because the fetal outcome and the presence of additional fetal structural and chromosomal abnormalities are correlated with the type of cleft. The type of cleft impacts on the degree of cosmetic deformity and surgical result even when it occurs as an isolated fetal abnormality. Nyberg et al. (81) classified clefts into one of five categories.

- *Type I:* cleft lip alone
- *Type II:* unilateral cleft lip and palate
- *Type III:* bilateral cleft lip and palate
- *Type IV:* midline cleft lip and palate
- *Type V:* facial defects associated with amniotic bands or limb–body wall complex

The results are summarized in Table 10.

Unilateral Cleft Lip/Unilateral Cleft Lip and Palate

Most clefts detected prenatally with ultrasound are not subtle and tend to be large. Unilateral cleft lip and palate or incomplete cleft lip, however, may be more subtle

TABLE 10 ■ Prenatal Ultrasonographic Classification of CL with or without Cleft Palate

	Type I	Type II	Type III	Type IV	Type V
Definition	CL alone	Unilateral CL-P	Bilateral CL-P Premaxillary protrusion Hypoplastic midface	Midline CL-P	Facial cleft associated with amniotic bands or LBWC
Embryology	Incomplete fusion of the lip Normally formed palate	Incomplete fusion of the lip Variable degree of incomplete fusion of the primary and secondary palates Left affected more than right Deeper and longer cleft than type I	Failure of fusion of primary and secondary palates Growth and displacement of maxilla anteriorly Absence of primary palate with preservation of midline lip	Absence of primary palate and overlying lip	Bizarre slash defects secondary to fibrous bands
Most optimal ultrasonographic views	Coronal, transverse, or both	Coronal, transverse, or both	Midline sagittal best Coronal and transverse	Midline sagittal best Coronal and transverse	Coronal Transverse Sagittal
Concurrent anomalies	Unusual	Frequent	Frequent	Universally associated	Universally associated Nonembryologic distribution
Chromosome abnormalities	Unusual	20%	30%	52% Trisomy 13 Trisomy 18	0%
Ultrasonography	Oblique gap in the lip with variable extension to the nose	Same as type I but with gap in the maxilla and palate	Intermaxillary echogenic area or soft-tissue mass protruding from upper lip Bilateral clefts with hypoplastic midface May be confused with type IV	Gaping midline CL-P Hypoplastic midface	Facial cleft not in embryologic paramedian or median distribution
Outcome and prognosis	Favorable	Variable outcome dependent on associated anomalies	Variable outcome dependent on associated anomalies	Frequent associated anomalies and fatal	Frequent associated anomalies and fatal

Abbreviations: CL, cleft lip; CL-P, cleft lip and palate; LBWC, limb–body wall complex.
Source: From Ref. 81.

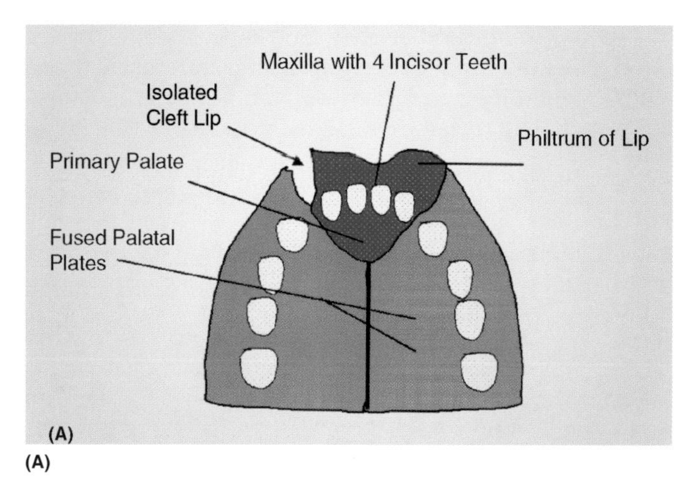

(A)

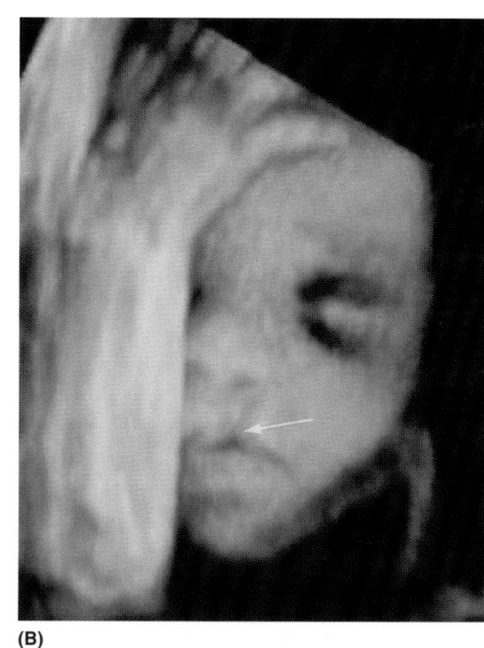

(B)

FIGURE 22 A AND B ■ Lateral cleft lip. (**A**) Diagram showing nonfusion of intermaxillary segment and maxillary process. (**B**) Three-dimensional image through the nose demonstrates a small defect lateral to the midline in the upper lip, which extends to the nares and corresponds to a lateral cleft lip. N, nose; C, cleft lip.

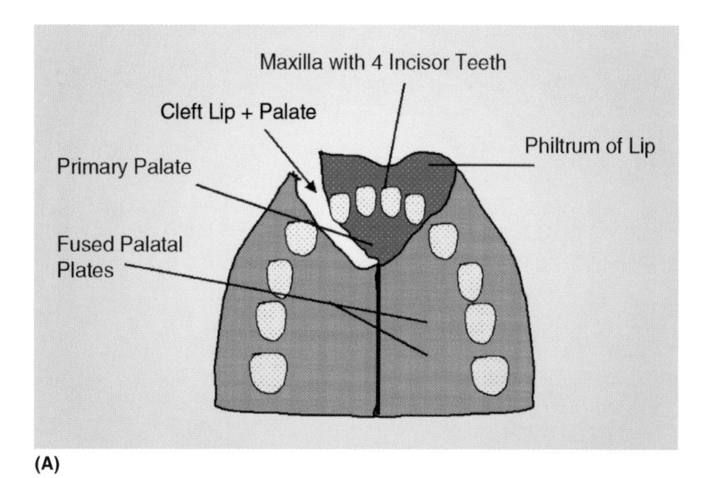

(A)

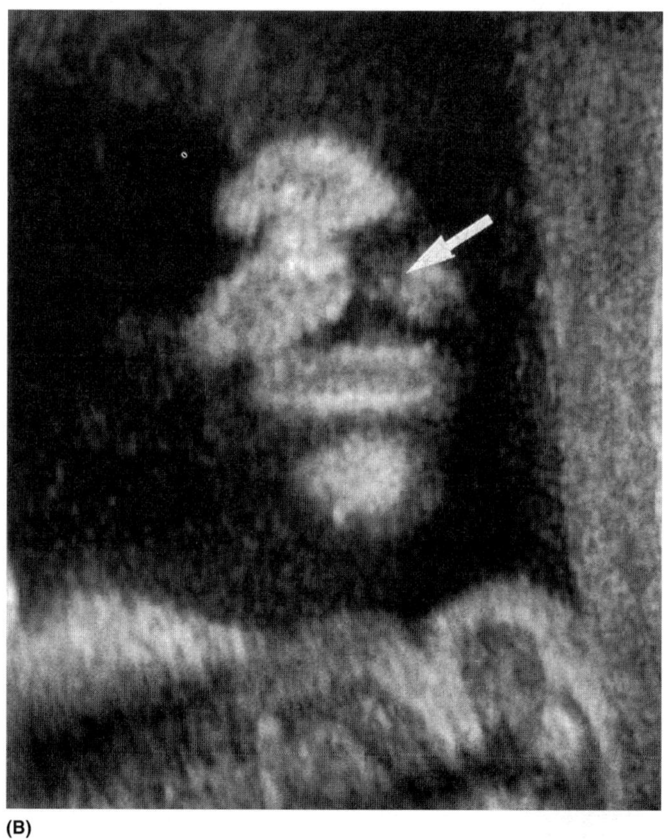

(B)

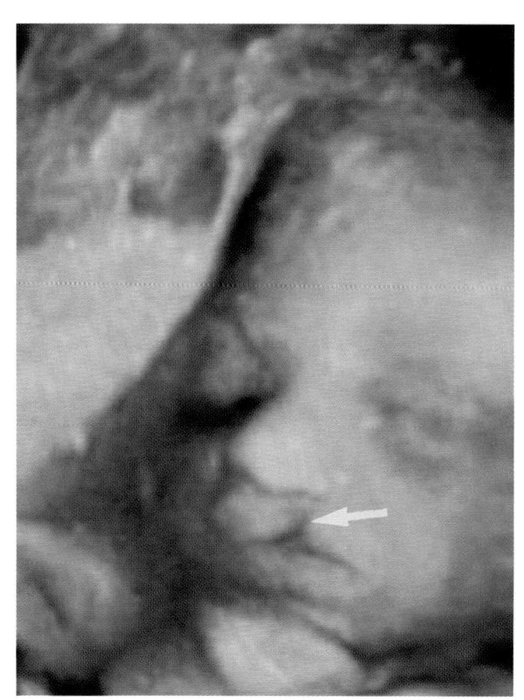

(C)

FIGURE 23 A–C ■ Lateral cleft lip and palate. (**A**, **B**, and **C**) This defect is similar to isolated cleft hip but extends posterior into the palate as shown on this diagram in **A**, and the 2D and 3D images in **B** and **C**. 3D, three dimensional; 2D, two dimensional.

and is best identified in the coronal view (Fig. 22). Typical ultrasound features include an anechoic region in the upper lip just lateral to the midline, which extends to the nares with a unilateral cleft lip and palate (Fig. 23). There is often a flattened appearance to the affected side, with widening of the nostril and communication between the nostril and the mouth (3,83). The sagittal profile view frequently demonstrates a nose with a hooked appearance. A gap in the maxilla and palate is sometimes present. Although complete unilateral and bilateral cleft lip and palate can be identified as early as 16 weeks' gestation, the incomplete cleft lip may be considerably more subtle and not as easily visualized until late in the second or third trimester of pregnancy (22,83). Therefore, families at risk for cleft lip and palate should be informed that an incomplete cleft lip may not be detected as early as a complete cleft lip and palate. The incomplete cleft is visualized best in the coronal view of the lip and may not be recognized in other planes because the integrity of the nose is usually not sufficiently disturbed. The sensitivity with which smaller defects are detected is not known. Incomplete cleft lip is less likely to be associated with chromosomal abnormalities, and surgical repair yields excellent results (22,81). With larger lateral abnormalities, the defect often appears to be midline (3,81). Ultrasonographic assessment of the fetal face should always include both coronal and sagittal scanning so that subtle facial clefting is not missed.

In about 25% of cases, only the lip is cleft (Fig. 22); in 50%, there is combined clefting of the lip and palate (Fig. 23); and in less than 25%, there is clefting of the palate only (84).

Bilateral Cleft Lip and Palate

Bilateral cleft lip and palate may be identified ultrasonographically before 16 weeks' gestation (22). Complete bilateral cleft lip and palate may be easily recognizable ultrasonographically because the appearance of the face is sufficiently disrupted, even in the early second trimester (83). Parasagittal and sagittal profile views may assist in identifying premaxillary protrusion, which may be the first or primary clue to the presence of a facial abnormality, permitting early or prenatal identification of bilateral cleft lip and cleft palate (Fig. 24) (85–87). Premaxillary protrusion occurs only with bilateral complete cleft lip and palate. The paranasal echogenic mass represents protruding bone and alveolar structures within the premaxillary protrusion. The mass typically is inferior to the nose, irregular in shape, and of similar echogenicity to bone and alveolar structures. A soft-tissue mass protruding outward and upward from the upper lip on sagittal views of the face may also be seen in some fetuses, with premaxillary protrusion corresponding to an everted upper lip (86,87). Nyberg et al. (81) categorized bilateral cleft lip and palate into those with premaxillary protrusion and those with a hypoplastic midface. Premaxillary protrusion is most apparent on midline sagittal views but can also be seen on coronal or transverse views (81). The paranasal echogenic mass appears most prominent during the

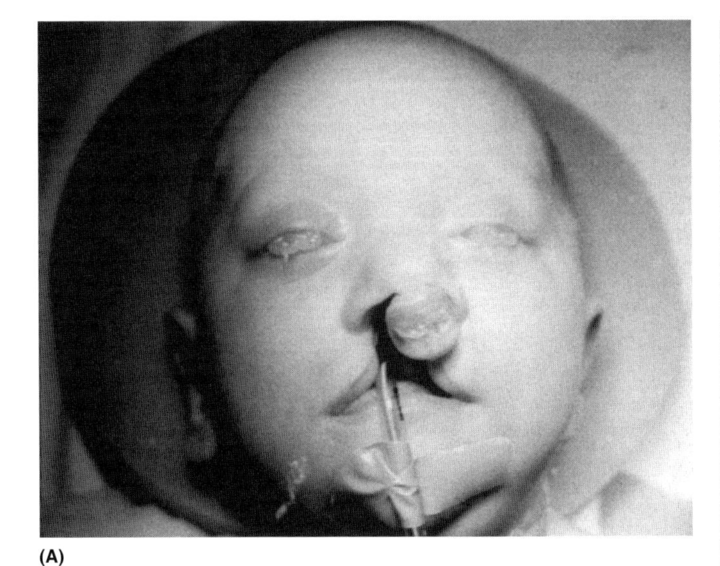

(A)

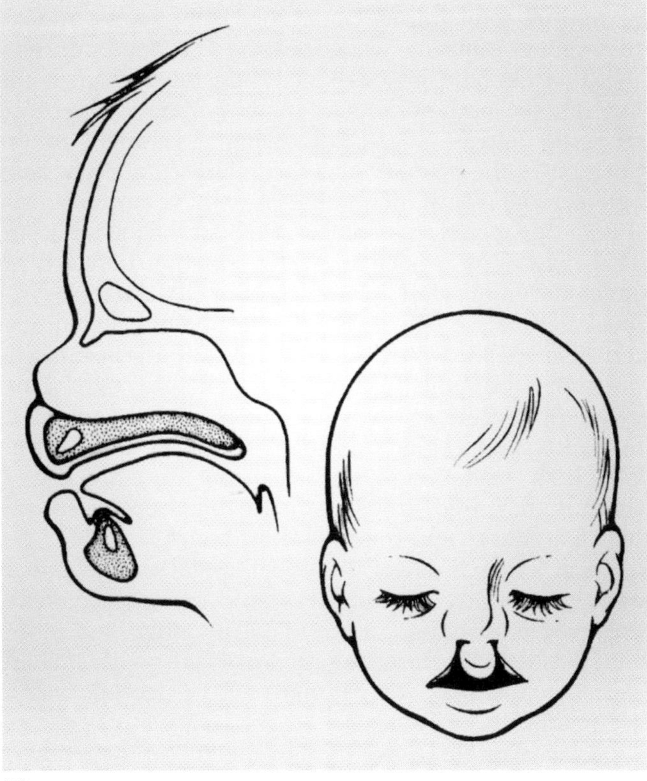

(B)

FIGURE 24 A AND B ■ Bilateral cleft lip and palate—premaxillary protrusion. (**A**) Photo of fetus with bilateral cleft lips and palate, leaving nonfused premaxillary protrusion. (**B**) Corresponding drawing.

second trimester, at a time when the cleft is small and often difficult to detect. With advancing gestational age, the mass may become less prominent (87). In cases of bilateral cleft lip and palate associated with a hypoplastic midface, premaxillary protrusion is conspicuously absent. The ultrasonographic appearance is consistent with absence of the primary bony palate but preservation of the midline lip (81). At times, these may be difficult to distinguish from median cleft lip and palate, and this appearance has a high rate of associated chromosomal abnormalities similar to median cleft lip and palate.

The differential diagnosis for premaxillary protrusion includes the following:

- Anterior cephalocele
- Hemangioma
- Teratoma
- Enlarged protruding tongue
- Proboscis (associated with holoprosencephaly) (3)

Premaxillary protrusion should be differentiated from these masses by its position, size, and ultrasonographic appearance because only premaxillary protrusion contains an echogenic bony component within the mass (86).

Cleft Palate

Isolated cleft palate is much more difficult to observe and diagnose ultrasonographically and is frequently missed on prenatal sonograms because of shadowing from facial bones and the identification of an intact upper lip (3). Frequently, only the soft palate is affected, and the cleft cannot be recognized because only hard palate clefts are detectable prenatally (79). In an in vivo and in vitro study of normal fetal midface anatomy, Babcook et al. (23) ultrasonographically visualized the C-shaped curve of the anterior fetal maxilla, including the region of the anterior six-tooth sockets, in more than 85% of second-trimester fetuses, thereby allowing assessment of the junction of the primary and secondary palate that runs between the lateral incisor and the canine tooth sockets. It is in this region that anterior clefts of the palate occur. This may allow systematic differentiation between an isolated cleft lip, cleft lip with anterior cleft palate, or isolated cleft palate. The prenatal detection of isolated posterior cleft palate may remain difficult. These investigators noted that the ability to visualize the anatomy of the hard palate decreased with advancing gestational age; therefore, ultrasonographic assessment should be made before 24 weeks' gestation. With advancing gestational age, there is increased acoustic shadowing from overlying fetal extremities and from the facial bones.

If a cleft lip is detected, coronal frontal scans obtained more posteriorly and axial views may demonstrate incomplete formation of the maxillary ridge, indicating cleft palate (3). Features suggestive of cleft palate include cleft lip, maxillary interruption, and increased tongue excursion (79).

Color Doppler ultrasound evaluation of the study of nonvascular slow amniotic fluid flow into the nasal and buccal cavities may be considered a useful diagnostic aid in identification of fetal palate defects (88). It is possible to follow the flow of amniotic fluid into the buccal cavity and sometimes into the pharynx, trachea, and esophagus by observing the facial profile in an anterior sagittal plane during swallowing. Color Doppler ultrasound demonstrates the swallowed fluid as an outgoing flow always below the plane of the palate. Parasagittal scans show amniotic fluid flowing in and out of the nasal fossae. During visualization, the flow of amniotic fluid is normally seen solely in the oral cavity or in the nasal fossae but not in both at the same time. A clear image of the color in the oral cavity, with no trace of color in the nasal cavity, allows a palate defect to be excluded. Demonstration by color Doppler ultrasound of flowing amniotic fluid in the mouth and the nasal fossae simultaneously ascertains the involvement of the palate in the defect (89). Ancillary signs of the diagnosis of cleft lip and palate also are important. A small, absent, or transiently visualized fetal stomach and polyhydramnios may be ultrasonographic clues that the fetus has cleft palate (3). These findings are related to impaired fetal swallowing, but the diagnosis is not made unless the cleft is observed because these ancillary signs may alternatively result from any proximal gastrointestinal tract obstruction, either from an anatomic or from a functional cause.

Median Cleft Lip

Midline clefts should be distinguished from lateral clefts because the two forms of clefting are distinct pathologic entities (3). The median cleft lip and palate, or premaxillary agenesis, is the rare form and occurs primarily in association with two syndromes: the median cleft face syndrome (frontonasal dysplasia) and the holoprosencephaly complex (Fig. 11) (42,49). The median cleft lip is caused by the incomplete merging of the two medial nasal prominences in the midline. The ultrasonographic findings can be explained by the absence of the primary palate and overlying lip. The coronal, transverse, or sagittal view of the face demonstrates a gaping midline cleft lip and palate and a hypoplastic midface characterized by a flattened nose and maxilla (81). These fetuses have a poor prognosis; in one series, these clefts were universally associated with concurrent anomalies and a fatal outcome (81). More than half of affected fetuses in this series had either trisomy 13 or trisomy 18.

Associated Findings and Prognosis

The causes of cleft lip and cleft palate are multifactorial and include syndromes, chromosomal abnormalities (autosomal trisomies), and teratogens. Most cases of cleft lip and cleft palate diagnosed prenatally have been detected as a manifestation of a syndrome (90–92). Nyberg et al. (81) emphasized the importance of identifying the specific type of facial cleft, owing to the strong correlation between the type of facial cleft and both mortality rate and frequency of additional fetal anatomic and chromosomal anomalies.

Kraus et al. (93) found a 60% incidence of associated anomalies, most commonly clubfoot and polydactyly. The prenatal detection of a facial cleft is associated with a high frequency of aneuploidy, especially when other abnormalities are demonstrated (81). In a series of 64 cases of antenatally diagnosed facial cleft (44), 56 (88%) of the fetuses had additional malformations, and 31 (48%) had chromosomal abnormalities. In a series of 32 fetuses (22), associated anomalies were noted in 17 (56%) fetuses, and five (16%) had reported trisomic conditions. Nyberg et al. (81) noted chromosomal abnormalities in 20 of 65 (31%) fetuses. The frequency and type of chromosomal abnormality varied with the type of cleft. The highest rate of chromosomal abnormalities was found in median clefts, which were dominated by trisomy 13. Trisomy 13 is associated with cleft lip and cleft palate or isolated cleft palate in 60% of cases. Other common features of trisomy 13 include the holoprosencephaly complex with distinct facial features (44,91). Forty percent of cases with trisomy 18 have cleft lip with or without cleft palate. Both these trisomies are associated with poor prognosis. Cleft lip with or without cleft palate also occurs in about 0.5% of cases with trisomy 21 and in 30% of triploidy cases. Compared with findings in postnatal studies, prenatal detection of cleft lip and palate shows a higher frequency of associated anomalies, chromosome abnormalities, and fetal mortality. When no other anomalies are identified, a favorable prognosis can be offered. When cleft lip with or without cleft palate is detected prenatally, a thorough search for other anomalies should be undertaken, and chromosome analysis should be considered especially when the clefts are known to be associated with a high risk for chromosome abnormalities (types II to IV). In a series of 33 fetuses with trisomy 13, Lehman et al. (94) reported that cleft lip and palate was the most common facial anomaly identified, occurring in 15 fetuses (45%). Central nervous system and facial anomalies frequently coexisted.

Marked progress has been made in the structural and functional repair of isolated cleft lip and palate.

Amniotic Band Syndrome

Clinical Features

The amniotic band syndrome, also known as the amniotic band disruption complex or limb–body wall complex, is a common, usually sporadic cause of various fetal malformations involving the craniofacial region, limbs, and trunk. It is estimated to occur in one in 1200 live births (95). The degree of developmental disruption by the bands determines the clinical outcome, with the malformations ranging from mild deformities to severe anomalies that are incompatible with postnatal life. Bizarre facial clefts, fissures, or slash defects in nonembryologic distributions are common. The pathogenesis is thought to be related to rupture of the amnion, with subsequent entrapment and entanglement of the fetus by fibrous mesodermic bands that emanate from the chorionic side of the amnion. The fetal head, trunk, and extremities may be involved individually or in combination, with entrapment of fetal parts by the bands causing lymphedema, amputation, or slash defects in nonembryologic distributions (95,96). The facial defects probably result from fetal swallowing of the fibrous bands (96). The swallowed portion of the band most likely tethers the unswallowed portion, which then cuts across the face in a random fashion, producing bizarre slash defects of the face or mandible. The prognosis for the amniotic band syndrome varies depending on the level and degree of fetal entrapment, ranging from death secondary to disruption of the umbilical cord or discordant anencephaly to minor facial clefts, lymphedema, or extremity amputations (95,97).

Ultrasonographic Findings

Detection of bizarre facial clefts in unusual locations warrants a careful search for other manifestations of the amniotic band syndrome (Fig. 25) (98). These include asymmetric encephalocele, gastropleuroschisis, asymmetric amputations, and focal constrictions with distal lymphedema (95,98). The facial problems can include cleft lip and occasionally cleft palate, asymmetric microphthalmia, and severe nasal deformity. A search for the amniotic bands also should be made, although they may be subtle and difficult to visualize. Visualization of the fibrous bands attached to the fetus, with characteristic deformities and restriction of motion, is diagnostic of the amniotic band syndrome. Sequelae of the amniotic band syndrome may be observed as early as 12 to 13 weeks' gestation on endovaginal examination (99). Identification of a band is not necessary to make the diagnosis, and a diagnosis of amniotic band syndrome should never be made on observation of these bands in the absence of fetal deformities because several types of membranes may be seen in normal pregnancies (95).

Macroglossia

Clinical Features

Prenatal ultrasonography frequently demonstrates normal fetal tongue movement. Macroglossia is rare but can be seen in at least 20 syndromes, described by Taybi and Lachman (42) and Jones (43). The most common cause for an enlarged tongue prenatally is the Beckwith–Wiedemann syndrome. The incidence of this syndrome has been estimated to be one in 13,700. It is more common in females and usually occurs sporadically but may occur as an autosomal-dominant trait with incomplete penetrance and variable expressivity (43,91,100). Beckwith–Wiedemann syndrome is most commonly associated with somatic macrosomia, macroglossia, gigantism, and omphalocele. Other commonly present abnormalities include variable visceromegaly such as hepatomegaly, renal hyperplasia and dysplasia, distinctive horizontal earlobe creases, prolonged neonatal hypoglycemia, and increased incidence of intra-abdominal malignancies such as Wilms' tumor and hepatoblastoma (100–103). Macroglossia is the most frequent anomaly present and appears in 97.5% of cases, whereas 60% have other anomalies (103).

Macroglossia has also been reported in association with chromosomal abnormalities, particularly trisomy 21 (83), and in the rare cases of congenital ichthyosis (harlequin fetus), a lethal autosomal-recessive disorder (104–106).

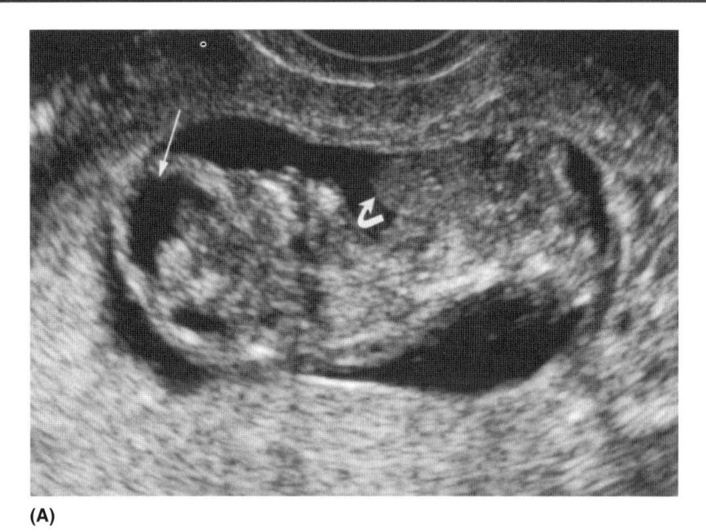

(A)

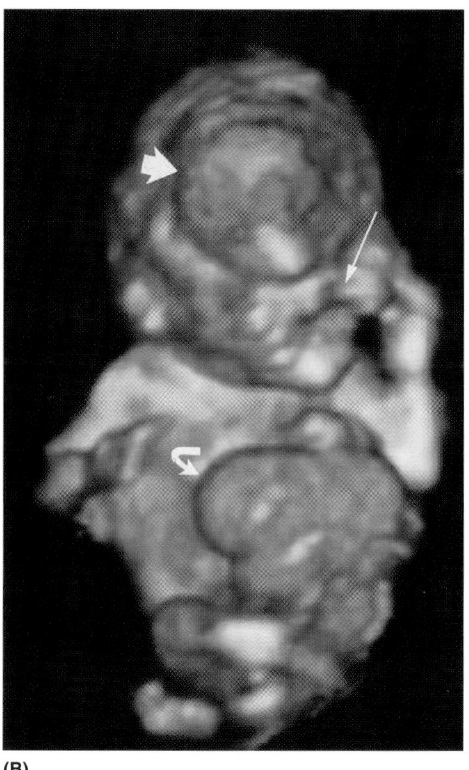

(B)

FIGURE 25 A AND B ■ Amniotic bands 3D. (**A**) This coronal image of the fetus demonstrates an asymmetrical cranial defect with evidence of a hydrocephalus (*arrow*) and an anterior abdominal wall defect (*curved arrow*). (**B**) This 3D image demonstrates asymmetrical anterior calvarial defect (*wide arrow*), a severe midfacial cleft (*long arrow*), and an anterior wall defect (*curved arrow*) in this fetus with amniotic bands syndrome. 3D, three dimensional.

Harlequin fetus is characterized by massive overgrowth of the keratin layer of fetal skin. Mild and moderately severe forms of ichthyosis are relatively common disorders, but the harlequin fetus represents a rare, severe, and dramatic form of this disorder. It is associated with high perinatal mortality, thick fissured skin, marked ectropion (eyelid eversion), eclabium (eversion of the lips), and flexion deformities.

Ultrasonographic Findings

Prenatal ultrasonography can detect macroglossia using both the mid-sagittal and the coronal views. The mid-sagittal view allows the examiner to delineate the tongue from its origin to its tip. The ultrasonographic detection of an enlarged and protuberant tongue, especially when associated with large-for-gestational-age measurements, polyhydramnios, visceromegaly, or omphalocele, has permitted the prenatal diagnosis of Beckwith–Wiedemann syndrome (Fig. 26) (103). A detailed evaluation of the fetal face, especially the fetal facial profile, is important when there is a positive family history.

Although the incidence of macroglossia in Beckwith–Wiedemann syndrome approaches 100%, the prenatal diagnosis based on the detection of macroglossia has been reported in only one case (103,107). Cobellis et al. (103) were the first to report the prenatal diagnosis of macroglossia at 30 menstrual weeks in a large-for-gestational-age fetus and in a family at risk for Beckwith–Wiedemann syndrome. Shah and Metlay (101) did not demonstrate evidence of macroglossia in a fetus who had multiple serial ultrasound examinations between 16 and 39 menstrual weeks, but examination of the neonate did demonstrate a large and protuberant tongue. The diagnosis of Beckwith–Wiedemann syndrome was made at 26 menstrual weeks and was based on the presence of omphalocele, fetal macrosomia, and polyhydramnios. At 26 menstrual weeks, the fetal facial profile appeared normal. Therefore, macroglossia may be a late ultrasonographic feature (103).

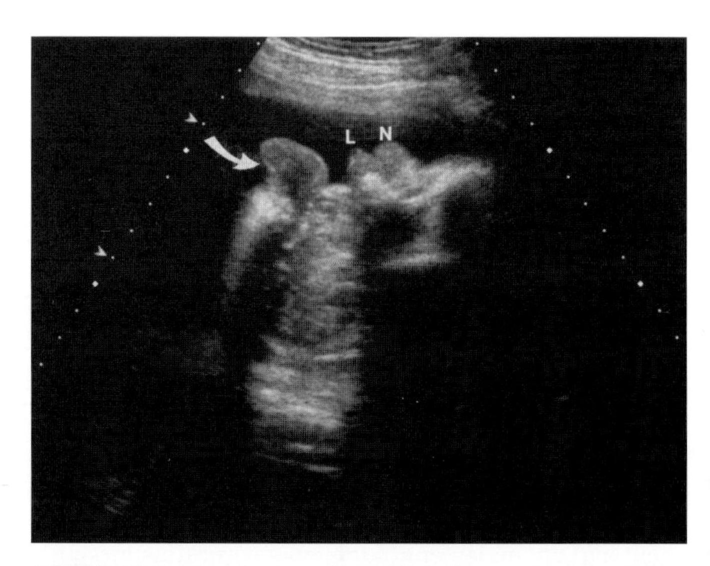

FIGURE 26 ■ Lingual mass. Parasagittal ultrasound exam, demonstrating a soft-tissue mass protruding from the mouth, consistent with a teratoma (*curved arrow*). This fetus was delivered using the EXIT procedure (ex utero intrapartum treatment) to control the airway at the time of delivery. The mass was removed after birth. L, upper lip; N, nares.

Placental enlargement is another manifestation of the Beckwith–Wiedemann syndrome (101).

Macroglossia can also occur with the following disorders (108):

- Congenital ichthyosis
- Chromosomal abnormalities, particularly trisomy 21
- Mucopolysaccharide syndromes
- Athyrotic hypothyroidism

The differential diagnosis for macroglossia includes:

- anterior cephalocele,
- tumors of the tongue, and
- epignathus.

Occasionally, an anterior cephalocele may extend through the ethmoid sinuses into the mouth, resembling an enlarged tongue. Epignathus is a pharyngeal teratoma that extends through the mouth and is explained in more detail in the next section (Fig. 26).

The prenatal ultrasonographic diagnosis of marked eclabium (eversion of the lips) and macroglossia has been reported in cases of the rare disorder congenital ichthyosis (104–106) (harlequin fetus). The ultrasonographic signs leading to prenatal diagnosis of harlequin fetus may include widely open mouth, marked eclabium and ectropion with large, bulging eyes, hypoplastic nose with evident apertures corresponding to the nostrils, macroglossia, lack of fetal breathing movements, edematous-like limbs, and intrauterine membranes (104–106). One significant problem with early prenatal diagnosis is that in the normal fetus, keratinization does not begin until 22 to 24 weeks' gestation; therefore, prenatal diagnosis of congenital ichthyosis by ultrasonography probably is impossible before the late second trimester of pregnancy (105). Thus, ultrasonography can be used for the prenatal diagnosis of even rare conditions. Based on the knowledge of the neonatal characteristics of a particular condition, the ultrasonographer can make a detailed search for alterations in fetal structure or surface anatomy, giving particular attention to these known neonatal features (105).

Epignathus

Clinical Features

Epignathus is a rare pharyngeal teratoma arising from the oral cavity, usually the palate in the region of Rathke's pouch. Teratomas are the largest group of congenital neoplasms, accounting for about 36% of the tumors seen in the first month of life. The incidence of congenital teratomas of all sites is one in 4000 live births, and less than 2% of these are oropharyngeal (108). The tumor generally extends through the mouth and creates an anterior mass that varies greatly in size and texture. Most arise from the hard palate, but other sites include the posterior nasopharynx, upper lip and nose, soft palate, tonsil, or tongue. There is no known genetic risk or predisposing factor, but there is a female predilection. Histologically, most are benign and composed of all three germ-cell layers (ectoderm, endoderm, and mesoderm). When an epignathus is identified antenatally, fetal surveillance with ultrasonography and biophysical monitoring is indicated. If there is no evidence of hydropic change, perinatal management involves planned cesarean section to avoid dystocia and fetal trauma, with immediate establishment of an airway, followed by surgical excision. Survival rates are only about 30% to 40% (Fig. 26) (75,76,109–111).

TABLE 11 ■ Congenital Lesions Presenting with Oropharyngeal Mass

- Epignathus
- Myoblastoma
- Hemangioma
- Gingival granular-cell tumor
- Gastrointestinal cyst
- Dermoid cyst
- Encephalocele
- Lingual thyroid
- Hamartoma of the nasopharynx

Ultrasonographic Findings

Demonstration of a variable-sized complex cystic and solid mass adjacent to the anterior portion of the face should suggest the possible diagnosis of epignathus (Table 11). Of the reported prenatally detected cases of epignathus, most are diagnosed during the early third trimester, but prenatal diagnosis has been reported as early as 19 weeks' gestation [in association with an elevated maternal serum α-fetoprotein level (AFP)] (Fig. 26) (75,112). These masses are generally large, ranging from 3 to 25 cm in diameter (75,109–117). Calcification has been reported (76), as has variable neck posturing, which was demonstrated in one case with noticeable hyperextension of the head (116). Other cystic masses of the tongue and mouth have been reported (Fig. 27).

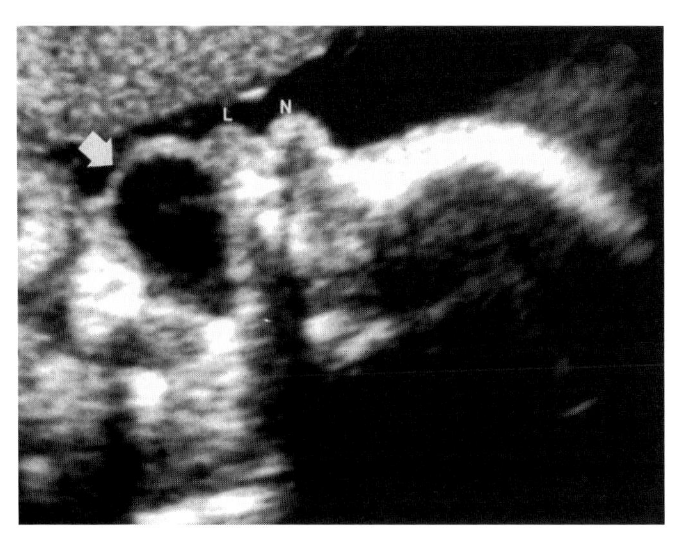

FIGURE 27 ■ Tongue cyst. Parasagittal ultrasound demonstrating cyst originating from tongue corresponding to a salivary gland cyst (*arrow*). It was aspirated at birth and contained large amounts of amylase. L, upper lip; N, nares.

Because of the nonspecific ultrasonographic appearance of these masses, the location of the mass may be helpful in narrowing the differential diagnosis. Masses originating in the brain and extending through the mouth are likely to be aggressive teratomas, whereas isolated masses inside the mouth are more likely to be teratomas, gingival granular-cell tumors, or simple cysts (67,77).

Associated Anomalies and Prognosis

Less than 10% of oropharyngeal teratomas have associated anomalies; those reported include facial clefts, congenital heart defects, umbilical hernias, and facial hemangiomas (117). Concomitant polyhydramnios has been detected in most reported cases, presumably because of obstruction of the fetal oropharynx causing a failure of fetal swallowing. Polyhydramnios may also occur secondary to high-output cardiac failure and hydrops fetalis in the presence of a large and vascular epignathus (75,111). Survival may be predicted by the extent of the mass and the presence or absence of polyhydramnios. The presence of polyhydramnios implies an obstruction to fetal swallowing and may predict neonatal respiratory compromise (77). The outcome is variable, but polyhydramnios is associated with a poor outcome. In the literature, cases in which there was no attempt at excision or partial excision resulted in death. In cases in which complete excision was achieved, and in all cases in which it was attempted even if multiple procedures were required, there was long-term survival with no evidence of disease recurrence (67,75, 109–111). Functional problems with speech, deglutition, and cosmetic appearance can be present. Management may involve cesarean section to avoid dystocia and immediate resuscitation of the neonate to establish an open airway. Prenatal ultrasonography is critical in predicting the extent of involvement of the hypopharynx and oropharynx, and it optimizes postnatal management.

Micrognathia and the Fetal Mandible

Mandibular malformations are associated with the first branchial arch syndrome. This syndrome consists of several malformations resulting from the disappearance or abnormal development of various components of the first pharyngeal arch (5,6,8). The first arch syndrome is caused by deficiency or insufficient migration of the neural crest cells. Because neural crest cells contribute to the formation of aortic and pulmonary arteries, the first arch syndrome is often associated with cardiac anomalies such as transposition of the great vessels. Either genetic or environmental factors may be responsible for these changes. The main syndromes encountered include Treacher Collins syndrome (mandibulofacial dysostosis), Nager's syndrome, and Pierre Robin syndrome.

Micrognathia is seen in more than 50 conditions or syndromes, including bone dysplasias, fetal chromosomal abnormalities, fetal drug exposure (retinoic acid, aminopterin, and trimethadione), and maternal phenylketonuria (Table 12) (25,43,118). These conditions or syndromes may be associated with other ultrasonographically recognizable fetal pathologic conditions, or micrognathia may be the only

TABLE 12 ■ Conditions Associated with Micrognathia

- Achondrogenesis
- Acrofacial dysostoses—various types
- Arthrogryposis
- Atelosteogenesis
- Beckwith–Wiedemann syndrome
- Camptomelic dysplasia
- CHARGE association
- Chromosome
    - Trisomy 13
    - Trisomy 9
    - Trisomy 18
- Diastrophic dysplasia
- DiGeorge syndrome (includes velocardiofacial syndrome)
- Fetal alcohol syndrome
- Fetal drug exposure
    - Valproic acid
    - Retinoic acid (vitamin A syndrome)
    - Trimethadione
- Focal femoral deficiency—unusual facies
- Freeman–Sheldon syndrome (whistling facies)
- Fryns' syndrome
- Goldenhar–Gorlin syndrome (hemifacial microsomia)
- Harlequin syndrome
- Hypoglossia—hypodactyly
- Maternal phenylketonuria
- Multiple pterygium syndrome
- Nager's acrodysostosis
- Otocephaly
- Pena–Shokeir syndrome

Pierre Robin sequence

- Seckel's syndrome
- Short-rib, polydactyly syndromes—various types
- Smith–Lemli–Opitz syndrome

Stickler's syndrome
Treacher Collins syndrome (mandibulofacial dysostosis)

Note: Boldface indicates most common condition.
Abbreviation: CHARGE, coloboma, heart disease, atresia choanae, retarded growth and development, hypergonadism, ear anomalies and/or deafness.
Source: From Ref.118.

fetal abnormality that is ultrasonographically detectable. In a review of 20 fetuses with ultrasonographically identified micrognathia (119), 5 of 20 fetuses (25%) had abnormal karyotypes, including three with trisomy 18 and one each with trisomy 13 and trisomy 9. Nicolaides et al. (44) detected chromosomal abnormalities in 37 of 56 (66%) fetuses with micrognathia. Other investigators detected micrognathia in 9 of 24 fetuses with facial anomalies; 38% of these had an abnormal karyotype, predominantly trisomy 18. These authors also noted a high incidence of skeletal dysplasias when micrognathia was discovered.

When micrognathia is detected on prenatal ultrasound, the prognosis is poor because of the frequently

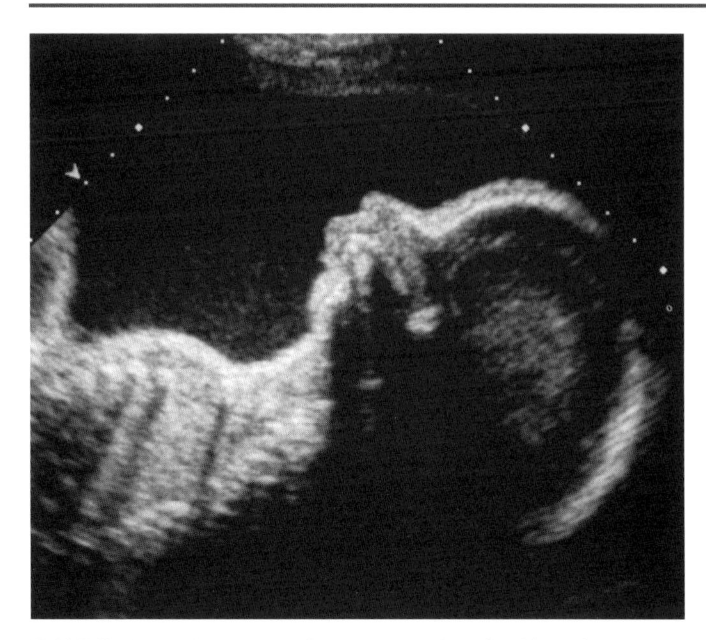

FIGURE 28 ■ Micrognathia. Sagittal scan of the fetal face demonstrating a severe micrognathia in this fetus. Micrognathia can be identified in a number of syndromes, skeletal malformations, and chromosomal abnormalities, including trisomy 18.

associated karyotypic and lethal anomalies (27,44,119). Bromley and Benacerraf (119) noted polyhydramnios in 70% of cases. The etiology of polyhydramnios is uncertain but may be related to difficulty in swallowing associated with the micrognathia or to other structural abnormalities.

When micrognathia is detected prenatally, a diverse set of syndromes must be considered, and the following

TABLE 13 ■ Conditions Associated with Frontal Bossing

- Achondroplasia
- Dyssegmental dysplasia
- Thanatophoric dysplasia
- Short-rib, polydactyly syndrome types I through III

steps should be considered: (*i*) karyotyping; (*ii*) search for other anomalies, including cardiac defects and skeletal dysplasias; (*iii*) evaluation of the fetal ears to assist in diagnosing Treacher Collins or Goldenhar–Gorlin syndrome; (*iv*) maternal history of drug exposure or familial syndromes; and (*v*) parental counseling regarding the poor prognosis, particularly when other ultrasonographic anomalies are identified.

At birth, micrognathia may result in respiratory distress and difficult intubation (119).

Ultrasonographic Findings

In the mid-sagittal view of the fetal face (facial profile), micrognathia presents as an unusually small mandible with a receding chin. Although percentile charts exist for the measurement of the mandible, the diagnosis of micrognathia is usually made subjectively when there is a visibly abnormal profile with a markedly receding chin (Fig. 28) (25,26). 3D imaging may be helpful to better diagnose mandibular hypoplasia (Fig. 29). One must be careful when scanning in the parasagittal axis because hypoplasia of the nasal bridge and mandibular hypoplasia may be simulated. Unfortunately, hypoplasia of the mandible may not be obvious during the second trimester. Pilu et al. (120) reported a case with Pierre Robin syndrome in which a mid-sagittal facial view at 23 weeks'

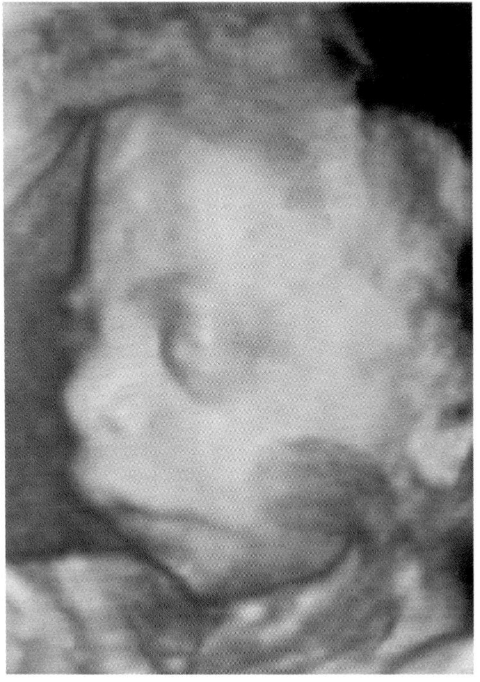

(A)

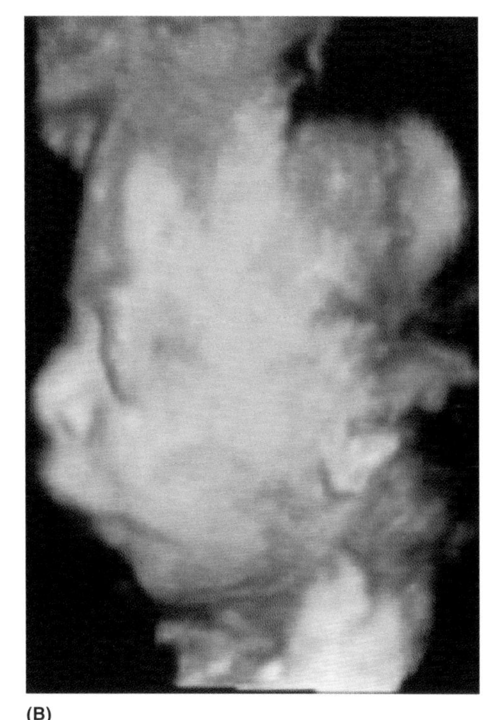

(B)

FIGURE 29 A AND B ■ 25-week micrognathia. (**A**) In this 3D image of the face at 25 weeks, there is hypoplasia of the mandible consistent with micrognathia on this oblique view. (**B**) On this lateral 3D image, the micrognathia is more obvious than on the previous image.

TABLE 14 ■ Conditions Associated with Nasal Bridge Defects

- Fetal warfarin syndrome
- Conradi's syndrome (chondrodysplasia punctata)
- Thanatophoric dwarfism
- Trisomy 21
- Saethre–Chotzen syndrome

gestation was interpreted as normal; however, a similar view at 35 weeks demonstrated marked micrognathia. Unequivocal facial deformity of a fetus at risk for a specific syndrome allows confirmation of a diagnosis and permits accurate diagnosis. Normal ultrasonographic findings, however, do not absolutely exclude an affected fetus because the reliability of ultrasonography in detecting a syndrome depends on severity of expression. For example, there is great variability in the degree of micrognathia and facial clefting in the autosomal-dominant mandibulofacial dysostosis (Treacher Collins syndrome) (121). Abnormalities seen on a facial profile view may assist in accurate diagnosis in multiple malformation syndromes, even in the absence of genetic or teratogen risk. Of the 56 cases with sufficiently severe micrognathia to be diagnosed prenatally by ultrasonography by Nicolaides et al. (44), additional malformations or growth retardation were demonstrated. These were associated with a poor perinatal outcome. Although micrognathia is present in more than 80% of fetuses with trisomy 18 or triploidy in pathologic series, Nicolaides et al. (44) detected micrognathia in only 21 of 83 fetuses (25% with trisomy 18) and in nine of 83 fetuses (21% with triploidy). Therefore, the sensitivity of detection of micrognathia on ultrasound screening remains to be determined. The underestimation of cases may be related to unfavorable fetal positioning or oligohydramnios (27,119).

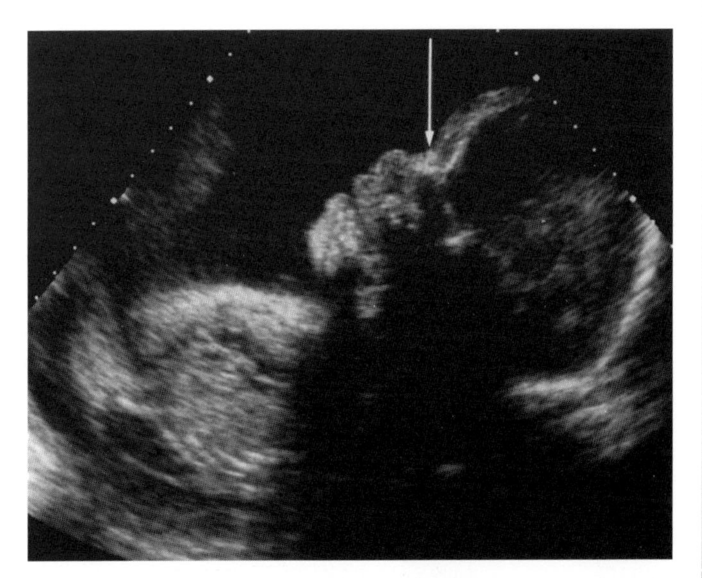

FIGURE 30 ■ Apert syndrome. Sagittal view of the face in a fetus with Apert syndrome shows a depressed nasal bridge (*arrow*). Also, note the flat appearance to the face, which occurs in this syndrome.

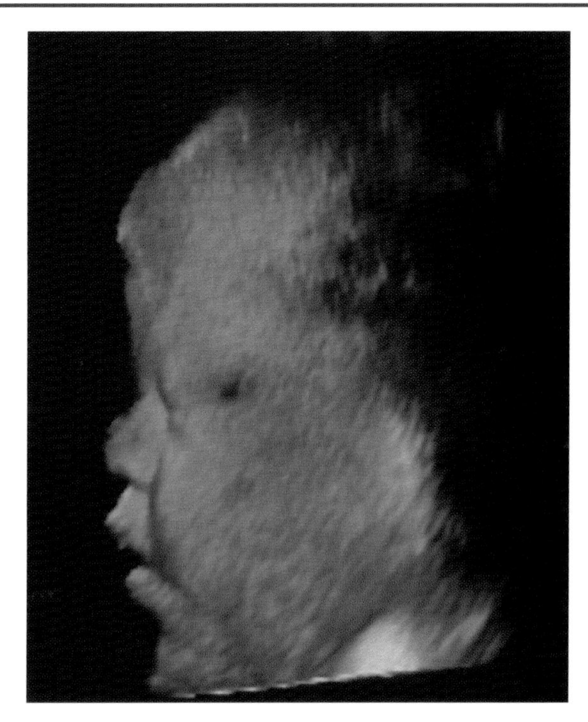

FIGURE 31 ■ Down syndrome. Sagittal view of the face of a fetus with Down syndrome demonstrates the relatively flattened nose. Also note the depressed nasal bridge and flattened appearance to the face on these views.

The sagittal profile view may also be useful in the antenatal detection of fetal frontal bossing or a depressed nasal bridge (Tables 13 and 14). These anomalies may be detected antenatally in skeletal dysplasias (27). A flattened facial profile, in which the nose is small or recessed, has

TABLE 15 ■ Conditions Associated with Flattened Facial Profile (Short Nose)

Aarskog's syndrome
Achondroplasia
Apert's and Crouzon's syndrome
Chromosome—trisomy 18
Chromosome—trisomy 21 (Down syndrome)
Cleft lip and cleft palate
Cornelia de Lange's syndrome
Fetal alcohol syndrome
Fetal anticonvulsant syndrome
Fetal warfarin syndrome
Holoprosencephaly
Miller–Dieker syndrome
Osteogenesis imperfecta type II
Pfeiffer's syndrome
Robinow's syndrome
Smith–Lemli–Opitz syndrome
Stickler's syndrome

Note: Boldface indicates most common condition.
Source: From Ref. 118.

been seen in association with chromosomal abnormalities, particularly trisomy 18 and 21, but it has also been reported in many more syndromes (Figs. 30 and 31) (Table 15) (99).

REFERENCES

1. American Institute of Ultrasound in Medicine. Guidelines for performance of the antepartum ultrasound examination. J Ultrasound Med 1991; 10:576.
2. Benacerraf BF, Frigoletto FD Jr, Bieber FR. The fetal face: ultrasound examination. Radiology 1984; 153:495.
3. Coates TL, McGahan JP. The fetal face. In: McGahan JP, Porto M, eds. Diagnostic Obstetrical Ultrasound. Philadelphia: JB Lippincott, 1994:227.
4. Mahony BS, Nyberg DA. The fetal face. In: Chervenak FA, Isaacson GC, Campbell S, eds. Ultrasound in Obstetrics and Gynecology. Chicago: Little, Brown, 1993:859.
5. Moore KL. The Developing Human: Clinically Oriented Embryology. 4th ed. Toronto: WB Saunders, 1988:170.
6. Stewart RE. Craniofacial malformations: clinical and genetic considerations. Pediatr Clin North Am 1978; 25:485.
7. Moore KL. The Developing Human: Clinically Oriented Embryology. 5th ed. Philadelphia: WB Saunders, 1993.
8. Sadler TW. Head and neck. In: Langman's Medical Embryology. 6th ed. Baltimore: Williams & Wilkins, 1990:297.
9. Escobar LF, Bixler D, Padilla LM, et al. Fetal craniofacial morphometries: in utero evaluation at 16 weeks gestation. Obstet Gynecol 1988; 72:674.
10. Hegge FN, Prescott GH, Watson PT. Fetal facial abnormalities identified during obstetric sonography. J Ultrasound Med 1986; 5:679.
11. Pilu G, Reece EA, Romero R, et al. Prenatal diagnosis of craniofacial malformations with ultrasonography. Am J Obstet Gynecol 1986; 155:45.
12. Birnholz JC. The development of human fetal eye movement patterns. Science 1981; 72:674.
13. Birnholz JC, Benacerraf BR. The development of human fetal hearing. Science 1983; 222:516.
14. Bowie JD, Clair MR. Fetal swallowing and regurgitation: observation of normal and abnormal activity. Radiology 1982; 144:877.
15. Trout T, Budorick NE, Pretorius DH, et al. Significance of orbital measurements in the fetus. J Ultrasound Med 1994; 13:937.
16. Jeanty P, Romero R, Staudach A, et al. Facial anatomy of the fetus. J Ultrasound Med 1986; 5:607.
17. Mayden KL, Tortora M, Berkowitz RL, et al. Orbital diameters: a new parameter for prenatal diagnosis and dating. Am J Obstet Gynecol 1992; 144:289.
18. Jeanty P, Cantraine F, Cousaert MS, et al. The binocular distance: a new way to estimate fetal age. J Ultrasound Med 1984; 3:241.
19. Jeanty P, Dramaix-Wilmet M, Van Gansbeke D, et al. Fetal ocular biometry by ultrasound. Radiology 1982; 143:513.
20. Romero R, Pilu G, Jeanty P, et al. Prenatal diagnosis of congenital anomalies. Norwalk, CT: Appleton & Lange, 1988.
21. Birnholz JC, Farrel EE. Fetal hyaloid artery: timing of regression with US. Radiology 1988; 166:781.
22. Benacerraf BR, Mulliken JB. Fetal cleft lip and palate: sonographic diagnosis and postnatal outcome. Plast Reconstr Surg 1993; 92:1045.
23. Babcook CJ, McGahan JP, Chong BW, et al. Evaluation of fetal midface anatomy related to facial clefts: use of ultrasound. Radiology 1996; 201:113.
24. Otto C, Platt LD. The fetal mandible measurement: an objective determination of fetal jaw size. Ultrasound Obstet Gynecol 1991; 1(1):12–17.
25. Turner GM, Twining P. The facial profile in the diagnosis of fetal abnormalities. Clin Radiol 1993; 47:389.
26. Benacerraf BR, Nyberg DA. The face and neck. In: Nyberg DA, McGahan JP, Pretorius DH, Pilu G, eds. Diagnostic Imaging of Fetal Anomalies. Philadelphia: Lippincott Williams & Wilkins, 2003:335–379.
27. Chitty LS, Campbell S, Altman DG. Measurement of the fetal mandible: feasibility and construction of a centile chart. Prenat Diagn 1993; 13(8):749–756.
28. Watson WJ, Katz VL. Sonographic measurement of the fetal mandible: standards for normal pregnancy. Am J Perinatol 1993; 10(3):226–228.
29. Rotten D, Levaillant JM, Martinez H, Ducou le Pointe H, Vicaut E. The fetal mandible: a 2D and 3D sonographic approach to the diagnosis of retrognathia and micrognathia. Ultrasound Obstet Gynecol 2002; 19(2):122–130.
30. Cicero S, Curcio P, Papageorghiou A, Sonek J, Nicolaides K. Absence of nasal bone in fetuses with trisomy 21 at 11–14 weeks of gestation: an observational study. Lancet 200117; 358(9294): 1665–1667.
31. Bunduki V, Ruano R, Miguelez J, Yoshizaki CT, Kahhale S, Zugaib M. Fetal nasal bone length: reference range and clinical application in ultrasound screening for trisomy 21. Ultrasound Obstet Gynecol 2003; 21(2):156–160.
32. Sonek JD, McKenna D, Webb D, Croom C, Nicolaides K. Nasal bone length throughout gestation: normal ranges based on 3537 fetal ultrasound measurements. Ultrasound Obstet Gynecol 2003; 21(2):152–155.
33. Chen M, Lee CP, Leung KY, Hui PW, Tang MH. Pilot study on the midsecond trimester examination of fetal nasal bone in the Chinese population. Prenat Diagn 2004; 24(2):87–91.
34. Birnholz JC. The fetal external ear. Radiology 1983; 147:819.
35. Chitkara U, Lee L, Oehlert JW, Bloch DA, Holbrook RH Jr, El-Sayed YY, Druzin ML. Fetal ear length measurement: a useful predictor of aneuploidy? Ultrasound Obstet Gynecol 2002; 19(2):131–135.
36. Dudarewicz L, Kaluzewski B. Prenatal screening for fetal chromosomal abnormalities using ear length and shape as an ultrasound marker. Med Sci Monit 2000; 6(4):801–806.
37. Sacchini C, El-Sheikhah A, Cicero S, Rembouskos G, Nicolaides KH. Ear length in trisomy 21 fetuses at 11–14 weeks of gestation. Ultrasound Obstet Gynecol 2003; 22(5):460–463.
38. Wang RY, Earl DL, Ruder RO, Graham JM Jr. Syndromic ear anomalies and renal ultrasounds. Pediatrics 2001; 108(2):E32.
39. Yeo L, Guzman ER, Ananth CV, Walters C, Day-Salvatore D, Vintzileos AM. Prenatal detection of fetal aneuploidy by sonographic ear length. J Ultrasound Med 2003; 22(6):565–576; quiz 578–579.
40. Fink IJ, Chinn DH, Callen PW. A potential pitfall in the ultrasonographic diagnosis of fetal encephalocele. J Ultrasound Med 1983; 2:313.
41. Mahony BS, Hegge FN. The face and neck. In: Mahony BS, Nyberg DA, Pretorius DH, eds. Diagnostic Ultrasound of Fetal Anomalies: Text and Atlas. Chicago: Year Book, 1990:203.
42. Taybi H, Lachman RS. Radiology of Syndromes, Metabolic Disorders, and Skeletal Dysplasias. 3rd ed. Chicago: Year Book, 1990.
43. Jones KL. Smith's Recognizable Patterns of Human Malformation. 4th ed. Philadelphia: WB Saunders, 1988.
44. Nicolaides KH, Salvesen DR, Snijders RJ, et al. Fetal facial defects: associated malformations and chromosomal abnormalities. Fetal Diagn Ther 1993; 8:1.
45. Pretorius DH, Nelson TR. Fetal face visualization using three-dimensional ultrasonography. J Ultrasound Med 1995; 14:349.
46. Coates TL, McGahan JP. The fetal face. In: McGahan JP, Porto M, eds. Diagnostic Obstetrical Ultrasound. 1st ed. Philadelphia: JB Lippincott, 1994:207.
47. McGahan JP, Nyberg DA, Mack LA. Sonography of facial features of alobar and semilobar holoprosencephaly. AJR Am J Roentgenol 1990; 154:143.
48. Nyberg DA, Mack LA, Bronstein A, et al. Holoprosencephaly: prenatal sonographic diagnosis. AJR Am J Roentgenol 1987; 149:1051.
49. DeMyer W, Zeman W, Palmer CG. The face predicts the brain: diagnostic significance of median facial anomalies for holoprosencephaly (arhinencephaly). Pediatrics 1964; 34:256.
50. DeMyer W. Classification of cerebral malformations. Birth Defects 1971; 7:78.

51. Filly RA, Chinn DH, Callen PW. Alobar holoprosencephaly: ultrasonographic prenatal diagnosis. Radiology 1984; 151:455.
52. Berry SM, Gosden C, Snijders RJ, et al. Fetal holoprosencephaly: associated malformations and chromosomal defects. Fetal Diagn Ther 1990; 5:92.
53. Greene MF, Benacerraf BR, Frigoletto FD Jr. Reliable criteria for the prenatal sonographic diagnosis of alobar holoprosencephaly. Am J Obstet Gynecol 1987; 156:687.
54. Schinzel A, Savoldelli G, Briner J, et al. Prenatal ultrasonographic diagnosis of holoprosencephaly: two cases of cebocephaly and two of cyclopia. Arch Gynecol 1984; 236:47.
55. Pilu G, Romero R, Rizzo N, et al. Criteria for the prenatal diagnosis of holoprosencephaly. Am J Perinatol 1987; 4:41.
56. Chervenak FA, Tortora M, Mayden K, et al. Antenatal diagnosis of median cleft face syndrome: sonographic demonstration or cleft lip and hypertelorism. Am J Obstet Gynecol 1984; 149:94.
57. Chervenak FA, Isaacson G, Rosenberg JC, et al. Antenatal diagnosis of frontal cephalocele in a fetus with atelosteogenesis. J Ultrasound Med 1986; 5:111.
58. DeMeyer W. The median cleft face syndrome. Neurology 1967; 17:961.
59. Maat-Kievit JA, Oepkes D, Hartwig NG, et al. A large retinoblastoma detected in a fetus at 21 weeks of gestation. Prenat Diagn 1993; 13:377.
60. Feldman E, Shalev E, Weiner E, et al. Microphthalmia—prenatal ultrasonic diagnosis: a case report. Prenat Diagn 1985; 5:205.
61. Tamas DE, Mahony BS, Bowie JB, et al. Prenatal sonographic diagnosis of hemifacial microsomia (Goldenhar-Gorlin syndrome). J Ultrasound Med 1986; 5:461.
62. Birnholz JC. Ultrasonic fetal ophthalmology. Early Hum Dev 1985; 12:199.
63. Birnholz JD. Fetal eye movement pattern. Science 1981; 213:679.
64. Farrell SA, Toi A, Leadman ML, Davidson RG, Caco C. Prenatal diagnosis of retinal detachment in Walker-Warburg syndrome. Am J Med Genet 1987; 28:619.
65. Maynor CH, Hertzberg BS, Ellington KS. Antenatal sonographic features of Walker-Warburg syndrome: value of endovaginal sonography. J Ultrasound Med 1992; 11:301.
66. Bronshtein M, Zimmer E, Gershoni-Baruch R, et al. First- and second-trimester diagnosis of fetal ocular defect and associated anomalies: report of eight cases. Obstet Gynecol 1991; 77:443.
67. Shipp TD, Bromley B, Benacerraf B. The ultrasonographic appearance and outcome for fetuses with masses distorting the fetal face. J Ultrasound Med 1995; 14:673.
68. Davis WK, Mahony BS, Carroll BA, et al. Antenatal sonographic detection of benign dacrocystoceles (lacrimal duct cysts). J Ultrasound Med 1987; 6:461.
69. Walsh G, Dubbins PA. Antenatal sonographic diagnosis of a dacryocystocele. J Clin Ultrasound 1994; 22:457.
70. Meizner I, Bar-Ziv J, Holcberg G, et al. In utero prenatal diagnosis of fetal facial tumor-hemangioma. J Clin Ultrasound 1985; 13:435.
71. Pennell RG, Baltarowich OH. Prenatal sonographic diagnosis of a fetal facial hemangioma. J Ultrasound Med 1986; 5:525.
72. Lasser D, Preis O, Dor N, Tancer ML. Antenatal diagnosis of giant cystic cavernous hemangioma by Doppler velocimetry. Obstet Gynecol 1988; 72:476.
73. Bulas DI, Johnson D, Allen JF, Kapur S. Fetal hemangioma: sonographic and color flow Doppler findings. J Ultrasound Med 1992; 11:499.
74. Smart PJ, Schwarz C, Kelsey A. Ultrasonographic and biochemical abnormalities associated with the prenatal diagnosis of epignathus. Prenat Diagn 1990; 10:327.
75. Conran RM, Kent SG, Wargotz ES. Oropharyngeal teratomas: a clinicopathologic study of four cases. Am J Perinatol 1993; 10:71.
76. Chervenak FA, Tortora M, Moya FR, et al. Antenatal sonographic diagnosis of epignathus. J Ultrasound Med 1984; 3:235.
77. Hulett RL, Bowerman RA, Marks T, et al. Prenatal ultrasound detection of congenital gingival granular cell tumor. J Ultrasound Med 1991; 10:185.
78. Winter TC, Mack LA, Cyr DR. Prenatal sonographic diagnosis of scalp edema/cephalohematoma mimicking an encephalocele. AJR Am J Roentgenol 1993; 161:1247.
79. Bundy AL, Saltzman DH, Emerson D, et al. Sonographic features associated with cleft palate. J Clin Ultrasound 1986; 14:486.
80. Melnick M. Cleft lip and cleft palate: etiology and pathogenesis. In: Kernahan DA, Rosenstein SW, Dado DV, eds. Cleft Lip and Palate: a System of Management. Baltimore: Williams & Wilkins, 1990:3.
81. Nyberg DA, Sickler GK, Hegge FN, et al. Fetal cleft lip with and without cleft palate: US classification and correlation with outcome. Radiology 1995; 195:677.
82. Latham RA. Anatomy of the facial skeleton in cleft lip and palate. In: McCarthy JG, ed. Plastic Surgery: Cleft Lip and Palate and Craniofacial Anomalies. Vol. 4. Philadelphia: WB Saunders, 1990:2581.
83. Benacerraf BR. Ultrasound evaluation of the fetal face. In: Callen PW, ed. Ultrasonography in Obstetrics and Gynecology. 3rd ed. Philadelphia: WB Saunders, 1994:235.
84. Gorlin RJ, Cervenka J, Pruzansky S. Facial clefting and its syndromes. Birth Defects 1971; 7:3.
85. Seeds JW, Cefalo RC. Technique of early sonographic diagnosis of bilateral cleft lip and palate. Obstet Gynecol 1983; 62:2S.
86. Nyberg DA, Mahony BS, Kramer D. Paranasal echogenic mass: sonographic sign of bilateral complete cleft lip and palate before 20 menstrual weeks. Radiology 1992; 184:757.
87. Nyberg DA, Hegge FN, Kramer D, et al. Premaxillary protrusion: a sonographic clue to bilateral cleft lip and palate. J Ultrasound Med 1993; 12:331.
88. Aubry MC, Aubry JP. Prenatal diagnosis of cleft palate: contribution of color Doppler ultrasound. Ultrasound Obstet Gynecol 1992; 2:221.
89. Monni G, Ibba RM, Olla G, et al. Color Doppler ultrasound and prenatal diagnosis of cleft palate. J Clin Ultrasound 1995; 23:189.
90. Benacerraf BR, Miller WA, Frigoletto FD Jr. Sonographic detection of fetuses with trisomies 13 and 18: accuracy and limitations. Am J Obstet Gynecol 1988; 158:404.
91. Saltzman DH, Benacerraf BR, Frigoletto FD. Diagnosis and management of fetal facial clefts. Am J Obstet Gynecol 1986; 155:377.
92. Benacerraf BR, Frigoletto FD, Greene MF. Abnormal facial features and extremities in human trisomy syndromes: prenatal US appearance. Radiology 1986; 159:243.
93. Kraus BS, Kitamura H, Ooe T. Malformations associated with cleft lip and palate in human embryos and fetuses. Am J Obstet Gynecol 1963; 86:321.
94. Lehman CD, Nyberg DA, Winter TC, et al. Trisomy 13 syndrome: prenatal US findings in a review of 33 cases. Radiology 1995; 194:217.
95. Burton DJ, Filly RA. Sonographic diagnosis of the amniotic band syndrome. AJR Am J Roentgenol 1991; 156:555.
96. Fiske CE, Filly RA, Golbus MS. Prenatal ultrasound diagnosis of amniotic band syndrome. J Ultrasound Med 1982; 1:45.
97. Kalousek DK, Bamforth S. Amnion rupture sequence in previable fetuses. Am J Med Genet 1988; 31:63.
98. Mahony BS, Filly RA, Callen PW, et al. The amniotic band syndrome: antenatal sonographic diagnosis and potential pitfalls. Am J Obstet Gynecol 1985; 152:63.
99. Amniotic band syndrome. In: Sanders RC, ed. Structural Fetal Abnormalities: the Total Picture. St Louis: Mosby Year Book, 1996:169.
100. Whisson CC, Whyte A, Ziesing P. Beckwith-Wiedemann syndrome: antenatal diagnosis. Australas Radiol 1994; 38:130.
101. Shah YG, Metlay L. Prenatal ultrasound diagnosis of Beckwith-Wiedemann syndrome. J Clin Ultrasound 1990; 18:597.
102. Koontz WL, Shaw LA, Lavery JP. Antenatal sonographic appearance of Beckwith-Wiedemann syndrome. J Clin Ultrasound 1986; 14:57.
103. Cobellis G, Iannoto P, Stabile M, et al. Prenatal ultrasound diagnosis of macroglossia in the Wiedemann-Beckwith syndrome. (Short communication) Prenat Diagn 1988; 8:79.
104. Meizner I. Prenatal ultrasonic features in a rare case of congenital ichthyosis (harlequin fetus). J Clin Ultrasound 1992; 20:132.
105. Watson WJ, Mabee LM. Prenatal diagnosis of severe congenital ichthyosis (harlequin fetus) by ultrasonography. J Ultrasound Med 1995; 14:241.

106. Mihalko M, Lindfors KK, Grix AW, et al. Prenatal sonographic diagnosis of harlequin ichthyosis. AJR Am J Roentgenol 1989; 153:827.
107. Wieacker P, Wilhelm C, Greiner P, et al. Prenatal diagnosis of Wiedemann-Beckwith syndrome. J Perinat Med 1989; 17:351.
108. Zanetti B, Signori E, Consolaro G, et al. Congenital fibrosarcoma of the tongue. Z Kinderchir 1982; 35:7.
109. Smith NM, Chambers SE, Billson VR, et al. Oral teratoma (epignathus) with intracranial extension: a report of two cases. Prenat Diagn 1993; 13:945.
110. Holmgren G, Rydnert J. Male fetus with epignathus originating from the ethmoidal sinus. Eur J Obstet Gynecol Reprod Biol 1987; 24:69.
111. Alter AD, Cove JK. Congenital nasopharyngeal teratoma: report of a case and review of the literature. J Pediatr Surg 1987; 22:179.
112. Chervenak FA, Isaacson G, Touloukian R, et al. Diagnosis and management of fetal teratomas. Obstet Gynecol 1985; 66:666.
113. Kang KW, Hissong SL, Langer A. Prenatal ultrasound diagnosis of epignathus. J Clin Ultrasound 1978; 6:330.
114. Kaplan C, Perimutter S, Molinoff S. Epignathus with placental hydrops. Arch Pathol Lab Med 1980; 104:374.
115. Teal LN, Angtuaco TL, Jimenez JF, et al. Fetal teratomas: antenatal diagnosis and clinical management. J Clin Ultrasound 1988; 16:329.
116. Gaucherand P, Rudigoz RC, Chappuis JP. Epignathus: clinical and sonographic observations of two cases. Ultrasound Obstet Gynecol 1994; 4:241.
117. Carney JA, Thompson DP, Johnson CL, et al. Teratomas in children: clinical and pathologic aspects. J Pediatr Surg 1972; 7:271.
118. Sanders C, Blackman L, Hogge W, et al. Structural Fetal Abnormalities: the Total Picture. St Louis: Mosby Year Book, 1996.
119. Bromley B, Benacerraf BR. Fetal micrognathia: associated anomalies and outcome. J Ultrasound Med 1994; 13:529.
120. Pilu G, Romero R, Reece EA, et al. The prenatal diagnosis of Robin anomalad. Am J Obstet Gynecol 1986; 154:630.
121. Meizner I, Carmi R, Katz M. Prenatal ultrasonic diagnosis of mandibulofacial dysostosis (Treacher Collins syndrome). J Clin Ultrasound 1991; 19:124.

Fetal Neck and Spine

Mladen Predanic, John P. McGahan, and Frank A. Chervenak

45

NECK

The fetal neck represents anatomically a small part of the human body. Due to unique location, it conveys vital structures that communicate between the fetal head and the rest of the human body. It contains conduits of respiration, deglutition, and blood to and from the brain as well as important endocrine and neural structures. As such, any anomalies of the neck that are fairly infrequent often are associated with poor fetal outcome. Regardless of its importance, the ultrasound evaluation of the fetal neck is limited. During the fetal anatomic survey in the second trimester of pregnancy, neck is evaluated as a part of the head and neck and as a part of the spine examination. American Institute of Ultrasound in Medicine (AIUM) recommends a more detailed fetal anatomic examination only if an abnormality or suspected abnormality is found on the standard examination (1).

Embryology and Normal Anatomy

Development of the head and neck begins early in embryonic life and continues until the cessation of postnatal growth in the late teens. Anterior compartment of the neck develops as the part of fundamental organization of pharyngeal region. The lower portion of the face with the anterior neck structures is developed from the four pharyngeal arches, where the third and the fourth arches are responsible for development of the neck musculature and skeletal derivatives, such as laryngeal cartilages. Midline structures arising from the pharynx include thyroid gland and thymus. An assessment of the anterior and anterolateral neck compartments is important to detect abnormalities of this region, such as goiter, hemangioma, or teratoma. In contrast to the anterior and lateral aspects of the neck (Figs. 1 and 2), the posterior compartment contains the cervical spine that is closely associated with the base of the skull. As such, it is an important part of the neck and cervical spine ultrasound examination that may detect abnormalities such as cystic hygroma, occipital cephalocele, or cervical meningomyelocele (2). This may be achieved with both transverse and longitudinal scanning. In addition, a specific transverse view through the posterior fossa and the base of the occiput has been used to assess the thickness of the skin and soft tissues over the neck (Fig. 3).

Normal Anatomic Variations

Certain normal anatomic variations may be encountered when examining the fetal neck in the first, second, or third trimester of pregnancy.

Nonfused Amnion vs. Cystic Neck Mass

When examining the fetal neck region for nuchal translucency or occiput and spine evaluation in the first trimester of pregnancy, nonfused amnion may overlap the fetal neck and give the artificial appearance of a cystic mass in or around the occiput. This may be remedied by checking elsewhere within the uterus for incomplete fusion of the amnion and the chorion or by checking the fetus when it has moved into a different anatomic position in which the neck is separate from the amnion.

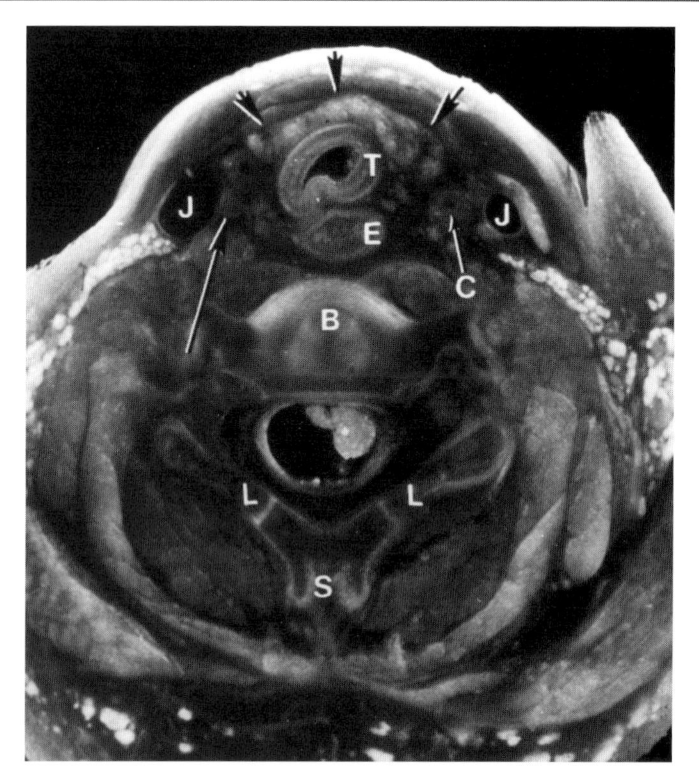

FIGURE 1 ■ Transverse section of lower neck in fetal specimen at 20 weeks' gestation. *Arrowheads*, thyroid gland; *long arrow*, points to vagus nerve. T, trachea; E, esophagus; J, jugular vein; C, carotid artery; B, body of cervical vertebra; L, laminar ossification centers; S, spinous process. *Source*: From Ref. 2.

Pseudonuchal Thickening

When examining the fetal neck for nuchal thickening in the second trimester of pregnancy, a specific anatomic plane must be obtained. If angulation is too steep and a more coronal plane is obtained, the artificial appearance of increased thickness of the skin over the posterior occiput results. In addition, the overlying cord around the fetal neck may be mistaken for increased skin thickness of the neck (Figs. 4 and 5). This pitfall may be overcome by using either pulsed Doppler or color Doppler over the presumed skin thickening to ascertain whether this represents the umbilical cord.

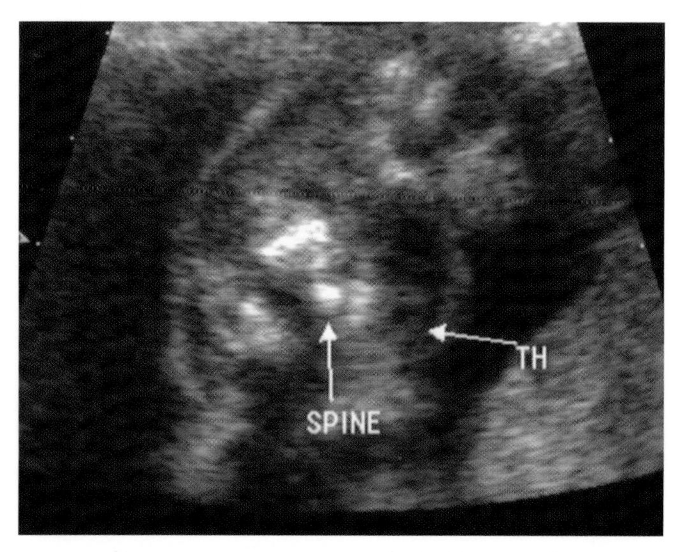

FIGURE 2 ■ Transverse sonogram of the neck with smooth anterior contours. Note three ossification centers of the cervical spine: two lateral and one midline for vertebral body. TH, trachea.

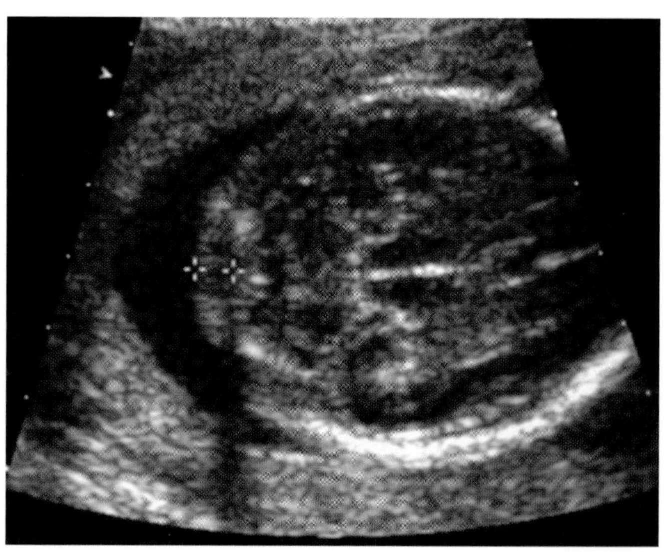

FIGURE 3 ■ Nuchal fold/skin thickness. Correct plane and technique for measuring nuchal skin fold thickness encompasses critical landmarks: cavum septi pellucidi, cerebral peduncles, and cerebellar hemispheres. Calipers are placed from the outer skull table to the outer skin surface.

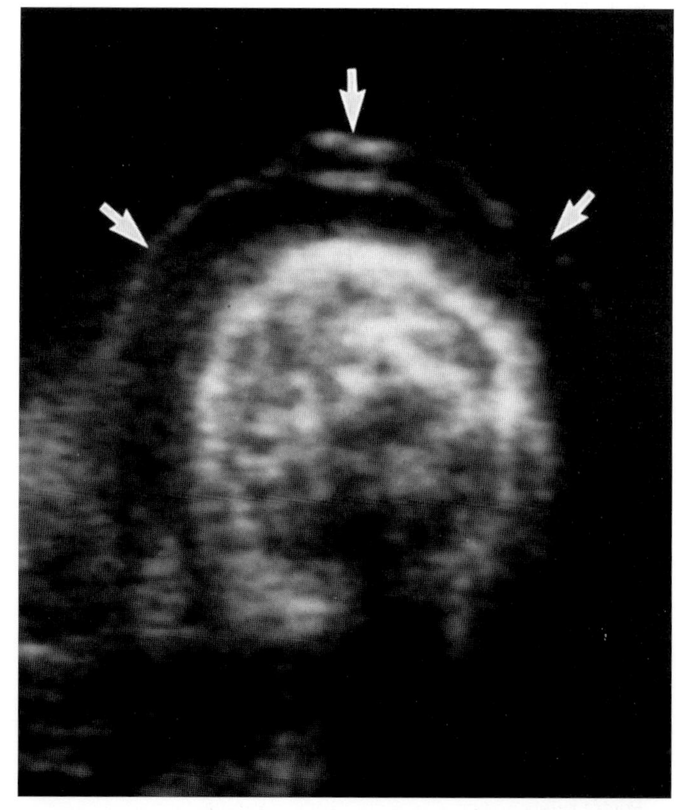

FIGURE 4 ■ Pseudo neck mass visible as a hypoechoic area (*arrows*) posterior to the neck should not be mistaken for increased skin thickness. This mass represents nuchal cord and can be easily recognized by color Doppler ultrasonography (Fig. 5).

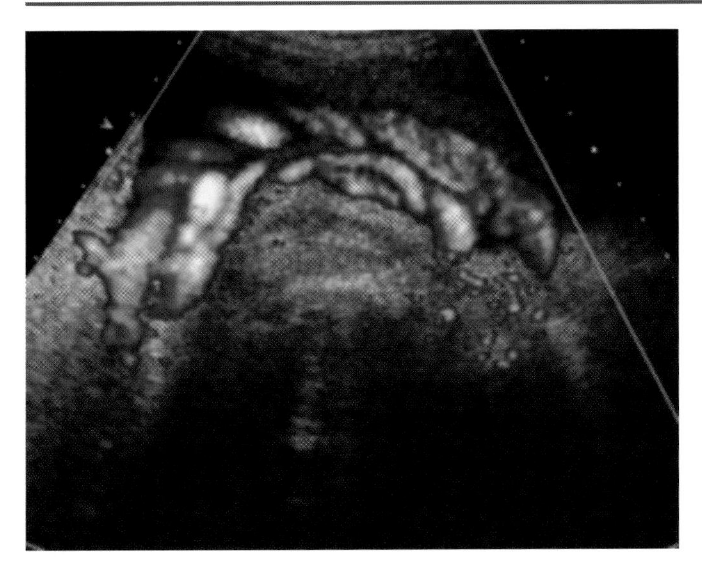

FIGURE 5 ■ Nuchal cord demonstrated by power Doppler ultrasonography.

Nuchal Cord

In the third trimester of pregnancy, the fetal umbilical cord may overlie or wrap around the fetal neck causing so called nuchal cord. The exact significance of nuchal cord and its in utero frequency are unknown. No difference in perinatal mortality was found when fetuses with the nuchal cord and control group were compared (3). However, there was a significantly higher incidence of several complications in the nuchal-cord group. These included increased incidence of lower Apgar score at one and five minutes, meconium-stained amniotic fluid, emergency cesarean section, neonatal resuscitation, and admittance to the newborn intensive care unit in the nuchal-cord group (3–5). Therefore, ultrasonographic identification of a nuchal cord could be considered as an important observation during the third trimester, particularly when there are multiple loops around the neck and then when there is associated decreased fetal movement.

Abnormalities

Neck abnormalities may be classified by a number of criteria, including location, ultrasound characteristics, and associated findings. The most likely anterior neck masses are goiter and teratoma (Table 1). In contrast, the most likely posterior-midline neck lesions are cervical meningomyelocele or occipital cephaloceles (6). Occasionally, however, cephaloceles occur in the frontal-ethmoid, parietal, or other regions of the skull.

TABLE 1 ■ Characteristic Location of Fetal Neck Masses

Location	Neck lesion
Anterior (bilateral)	Goiter
Anterolateral (unilateral)	Cervical teratoma
Posterior (midline)	Cervical meningomyelocele
	Occipital cephalocele
Posterolateral (bilateral)	Cystic hygroma
Variable	Hemangioma

Source: From Ref. 6.

TABLE 2 ■ Frequency of the Fetal Neck Masses

Common	Cystic hygroma
Less common	Cervical meningomyelocele
	Occipital cephaloceles
	Goiter
	Teratoma, hemangioma
Rare	Neuroblastoma
	Hemangioendothelioma
	Lipoma, fibroma
	Branchial cleft cyst
	Thyroglossal duct cyst
	Metastases

Source: From Ref. 6.

In regard to neck lesion frequency, the most common visualized fetal neck mass is cystic hygroma (Table 2). A less common neck mass of major concern is occipital cephalocele. Regardless of its location, majority of the neck abnormalities represent growth of the lesion that produces mass effect due to limited space of the neck compartments.

Ultrasonic features of the neck masses range from cystic, multiseptated cystic to solid lesions. The same lesion may have different ultrasonic appearance depending on the contents of the sac. Whereas cystic hygromas typically are multiseptated cystic structures, cephaloceles can be cystic, complex, or solid, depending on the presence of the brain tissue in the herniated sac. Nevertheless, majority of the neck masses have typical features that may assist in differentiating the neck lesion origins (Table 3).

Cystic Hygroma

The most common neck mass identified in the fetus is cystic hygroma. The incidence varies with gestational age, and ranges from one in 184 to one in 200 if diagnosed during first trimester of pregnancy, compared to one per 1000 live births (7,8). Cystic hygroma, or "moist tumor," is considered a congenital malformation of the lymphatic system, likely due to defect in the formation of lymphatic vessels. The fetal lymphatic vessels drain into two large sacs lateral to the jugular veins. If the lymphatic and venous structures fail to connect, the jugular lymph sacs enlarge, resulting in cystic hygromas of the posterior triangles of the neck (Fig. 6) (10). A dense midline septum

TABLE 3 ■ Ultrasound Features of Fetal Neck Masses

Ultrasound appearance	Neck lesion
Multicystic with midline septation	Cystic hygroma
Cystic/complex/solid (depending on contents or histology)	Cervical meningomyelocele
	Occipital cephalocele
	Hemangioma
Solid (generalized hypoechoic)	Goiter
Solid with cystic components	Cervical teratoma

Source: From Ref. 6.

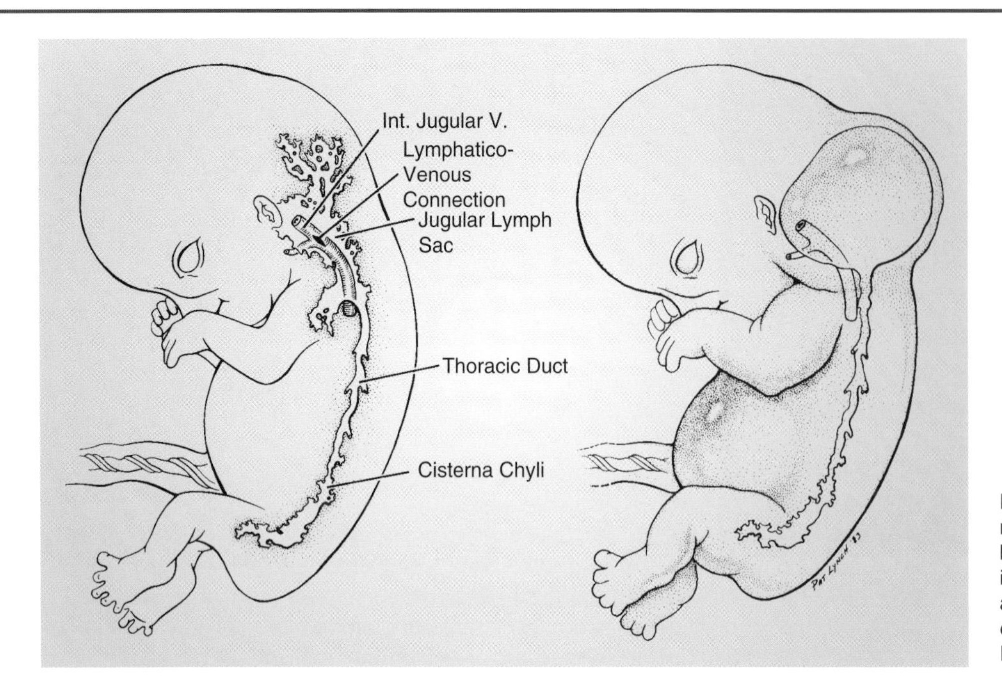

FIGURE 6 ■ Lymphatic system in a normal fetus with a patent connection between the jugular lymph sac and the internal jugular vein (*left*). Cystic hygroma and hydrops from a failed lymphatovenous connection (*right*). *Source*: From Ref. 9.

extending from the fetal neck across the full width of the hygroma is characteristically observed (Figs. 7 and 8). This septum represents the nuchal ligament (9).

Within the cystic structure, thinner septae are seen and thought to derive from either fibrous structure of the neck or deposits of fibrin (Figs. 9 and 10). Resolution of the cystic hygroma may result in redundant skin and clinical appearance of webbed neck (pterygium colli), which is a common feature of Turner syndrome (10). Cystic hygroma can be variable in size. Large and septated hygromas are usually easily recognized, though the smaller ones can be misdiagnosed for occipital cephalocele or cervical meningomyelocele (Table 4) while presence of midline septum is fairly pathognomic for cystic hygroma (11).

Once a cystic hygroma is detected, a careful search is made for associated skin edema, ascites, and pleural or pericardial effusions, because there is often concomitant fetal hydrops (Figs. 11 and 12). The outcomes of fetuses with cystic hygromas are variable and are associated with the presence of fetal aneuploidy or hydrops. The frequency of aneuploidy and associated anomalies varies with gestational age, the appearance, and probably the size of the cystic hygroma. Large and cystic hygromas, diagnosed in the second trimester of pregnancy, are usually associated with chromosome abnormality in 65% of cases (12).

In fetuses diagnosed during the first trimester (Fig. 13), the incidence of abnormal karyotype is about 50% (13). The

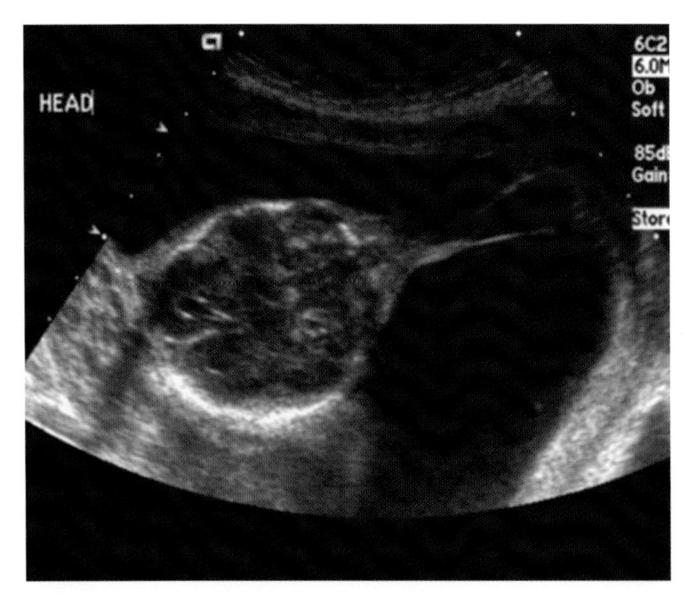

FIGURE 7 ■ A large posterior neck cystic mass characterized with midline septum represents cystic hygroma. Midline septation is nuchal ligament.

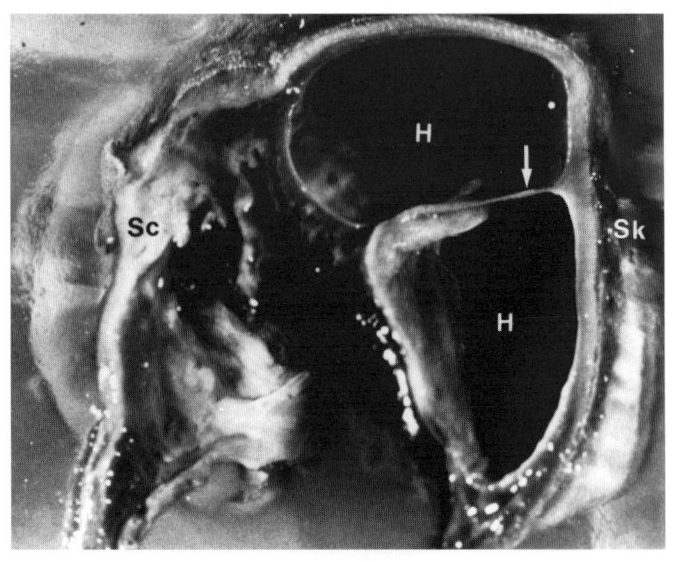

FIGURE 8 ■ Section of hygroma embedded in gelatin with arrow pointing to midline septum. H, hygroma; Sc, scalp; Sk, skin covering hygroma. *Source*: From Ref. 11.

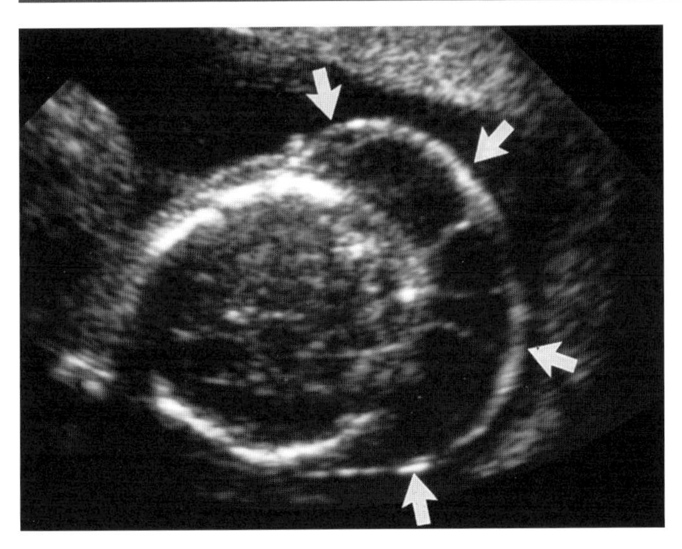

FIGURE 9 ■ Cystic hygroma with multiple septations.

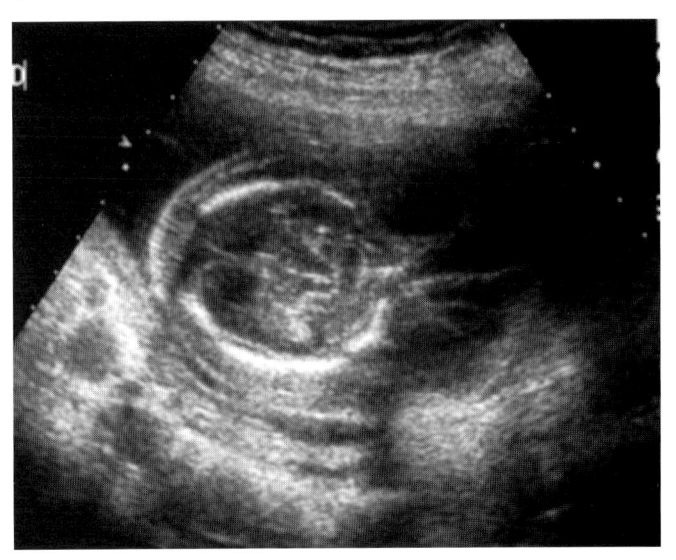

FIGURE 10 ■ Another example of large cystic hygroma with multiple septations and associated skin edema visible as a thick skin over fetal scalp.

most common chromosomal abnormality is Turner's syndrome (45,X), though trisomy is often encountered (14).

Presence of cystic hygroma and aneuploidy can result in in utero demise or partial regression, leaving a webbed neck (Fig. 14). When first-trimester cystic hygroma occurs in the presence of normal karyotype, an entirely normal outcome is possible (13–15). However, when hydrops is present, the outlook is grave. In reports of cases in which the hydropic fetuses were alive at initial ultrasound examination and the pregnancies were not electively terminated, all fetuses died within the next few weeks (9,16,17).

Some cystic hygromas may regress in utero, leaving only a webbed neck or possibly the nuchal thickening of Down syndrome (Fig. 14). Isolated cystic hygromas (i.e., those not occurring as part of a jugular lymphatic obstruction sequence) can be surgically corrected, and these fetuses have a good prognosis (Fig. 15).

Cephaloceles

A cephalocele is a defect in the skull associated with a protrusion of the meninges and/or brain substance. Due to location of the occipital cephaloceles at the base of the skull and the upper posterior neck, it is usually considered as a part of the spectrum of neural tube defects (NTDs) in the neck region. A number of different types of cephaloceles have been described depending on the sac contents: cranial gliocele is protrusion of a glial-lined cyst; atretic cephalocele is residue of a cephalocele; cranial meningocele is protrusion of cerebrospinal fluid (CSF) and meninges; cranial meningoencephaloceles is protrusion of CSF, leptomeninges, and brain; and meningoencephalocystocele is protrusion of CSF, leptomeninges, brain, and ventricles. Most commonly, cephaloceles are either meningoceles or meningoencephaloceles.

The incidence is about one in 2000 live births, with more than two-thirds occurring in the occiput (18). Occipital cephaloceles are most common in the European population, whereas frontal cephaloceles are frequently seen in patients of Southeast Asian descent (19). Cephaloceles also may occur in the parietal, frontal, ethmoid (sincipital), or nasopharyngeal region. Cephaloceles may be completely skin covered and present with normal maternal serum and amniotic fluid α-fetoprotein (AFP) levels.

TABLE 4 ■ Differential Ultrasound Features Between Cystic Hygroma, Occipital Cephalocele, and Cervical Meningomyelocele

Abnormality	Fetal cystic hygroma (common)	Occipital cephalocele (uncommon)	Cervical meningomyelocele (rare)
Bony defect	No	Yes (skull)	Yes (spine)
Intracranial abnormalities	No	Common	Common
Symmetric abnormalities	No	Yes	Yes
Septations	Yes	No	No
Acute angle with skin	No	Yes	Yes
Hydrops	Common	No	No

Source: From Ref. 6.

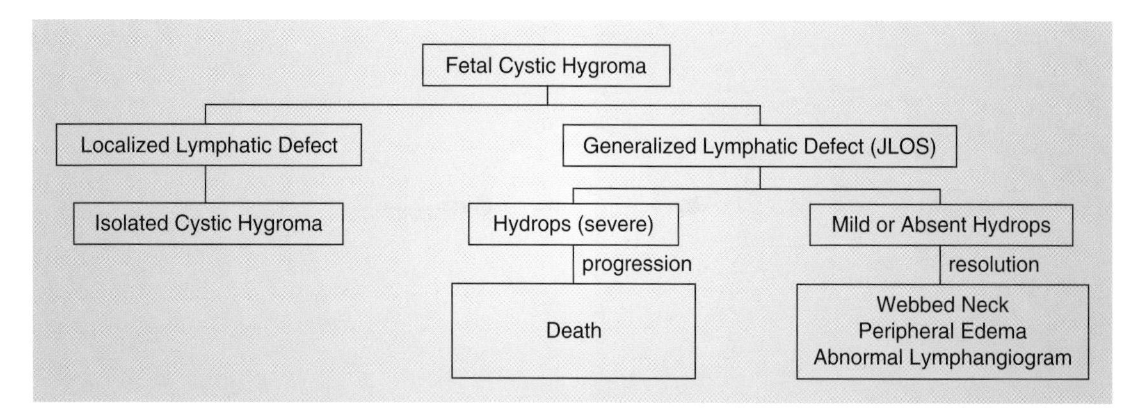

FIGURE 11 ■ Natural history of cystic hygroma. Generalized hydrops results from JLOS, which may result in either isolated hygroma, generalized edema and death, or partial regression of hygroma. JLOS, jugular lymphatic obstruction sequence. *Source*: From Ref. 11.

Ultrasonographically, a cephalocele appears as a saclike protrusion around the head that is not covered by bone. The diagnosis can be made with certainty only if a bony defect in the skull is detected (Fig. 16).

Cephaloceles are highly variable in size and ultrasonic appearance. The position of the defect may be determined using the bony structures of the face, the spine, and if possible, the midline echo of the brain for orientation. If brain has herniated, the contents of the sac have usually a heterogeneous appearance, though the pathology can be diagnosed only with certainty if magnetic resonance imaging (MRI) is utilized. A diligent search for skull defects, using serial examinations if necessary, may help to improve diagnostic accuracy, because a transition from solid to fluid patterns and transient disappearance of the cephaloceles have been described (20,21). If a defect in the skull is not found, the differential diagnosis includes cystic hygroma, teratoma, hemangioma, and branchial cleft cyst. The common features that help to differentiate cephalocele from cystic hygroma include a bony defect in the occipital vault (indicating cephaloceles); septation (indicating hygroma); and the echogenic appearance of the malformation (indicating cephalocele) (22).

Cephaloceles are often associated with ventriculomegaly; microcephaly; tectal beaking, and flattened basiocciput (20). In addition, in 27% and 19% of pregnancies, poly- and oligohydramnios was observed, respectively (20). In this study, 80% (four out five) pregnancies associated with

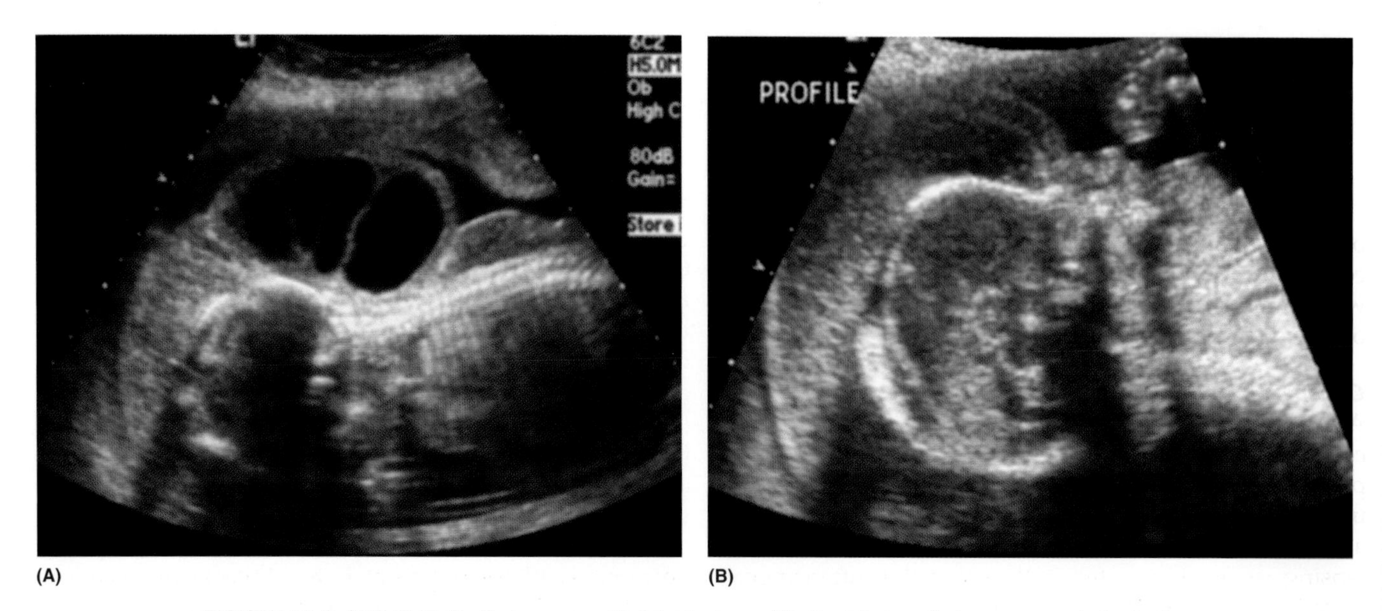

FIGURE 12 A AND B ■ Cystic hygroma with fetal hydrops. (**A**) Septated cystic hygroma posterior to the occiput with associated skin thickening demonstrated in the coronal fetal view. (**B**) Profile of the fetus with visible enormously thick and edematous skin.

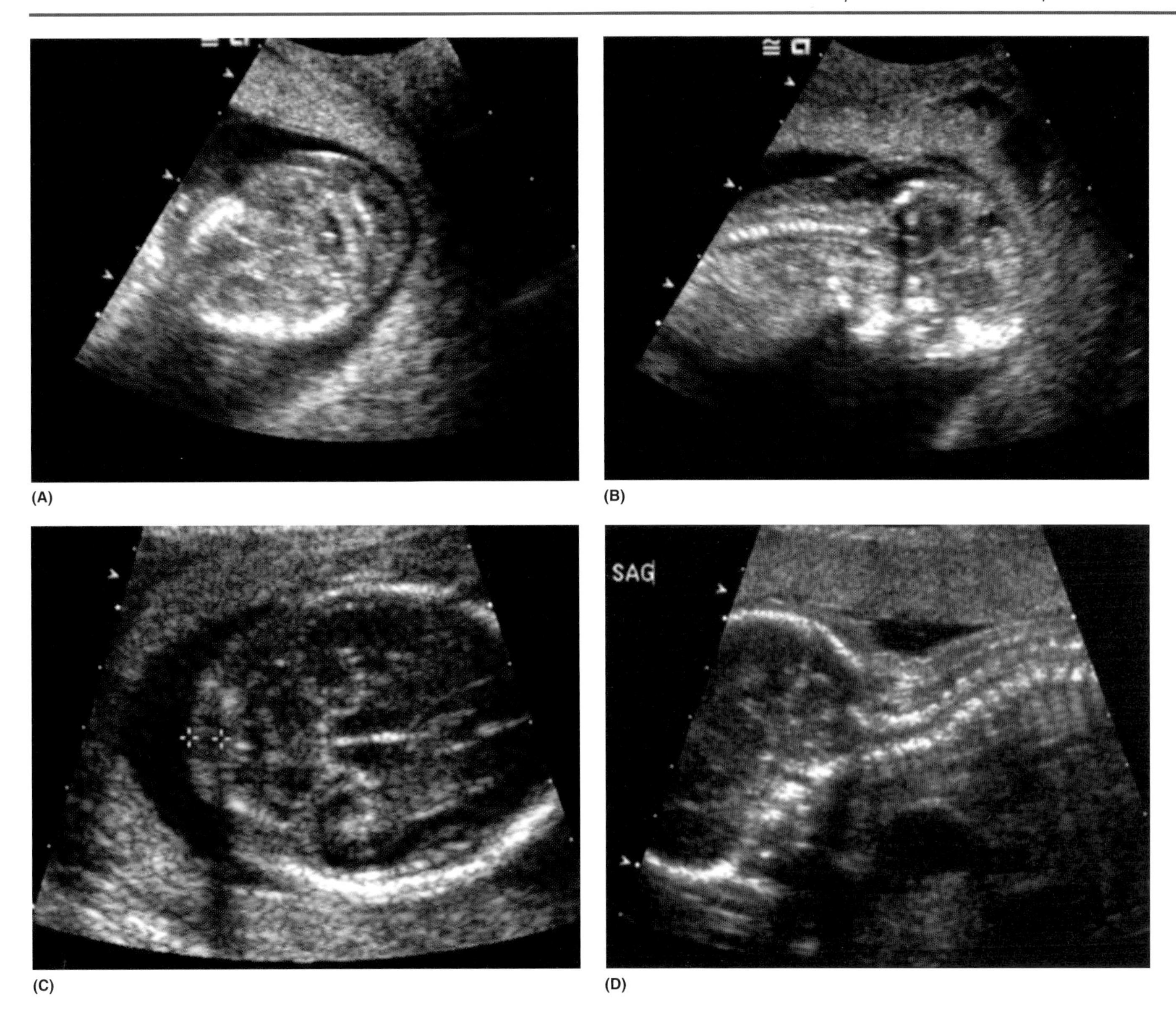

FIGURE 13 A–D ■ First trimester and increased nuchal translucency representing cystic hygroma (**A**) transverse; (**B**). **C** and **D** represent same fetus at 21 weeks of gestational age with slightly increased nuchal fold thickness, though no associated anomalies are observed (**C**) transverse; (**D**) sagittal scan.

oligohydramnios had other structural abnormalities, most commonly Meckel–Gruber syndrome and three patients with amniotic band syndrome. Table 5 demonstrates some of the conditions associated with cephaloceles.

Cervical Meningomyelocele

Although far less common than cystic hygroma and cephalocele, cervical meningomyeloceles may present as posterior neck masses. Cervical meningomyeloceles are usually located in the midline of the dorsum of the neck and most often are cystic in nature; but because of prolapsing ossification centers of the spine, they often appear complex. Splaying of the posterior ossification centers of the spine distinguishes this lesion from other neck masses (Fig. 17).

Goiter

Goiter, a massively enlarged thyroid gland, can result from either hypo- or hyperthyroidism. It is associated with variable maternal thyroid states but most often occurs in pregnancies in which maternal ingestion of iodides or other thyroid-blocking agents have been used (23). Goiter has also occurred secondary to maternal Graves' disease as a result of transplacental passage of a thyroid-stimulating substance such as a long-acting thyroid stimulant. Congenital hypothyroidism occurs in about one in 3700 births (23). Because the fetal thyroid hormone synthesis begins at 11 to 12 weeks of gestation and is critical for normal development, fetal production and hypothyroid state can be easily produced iatrogenically during the maternal treatment with antithyroid

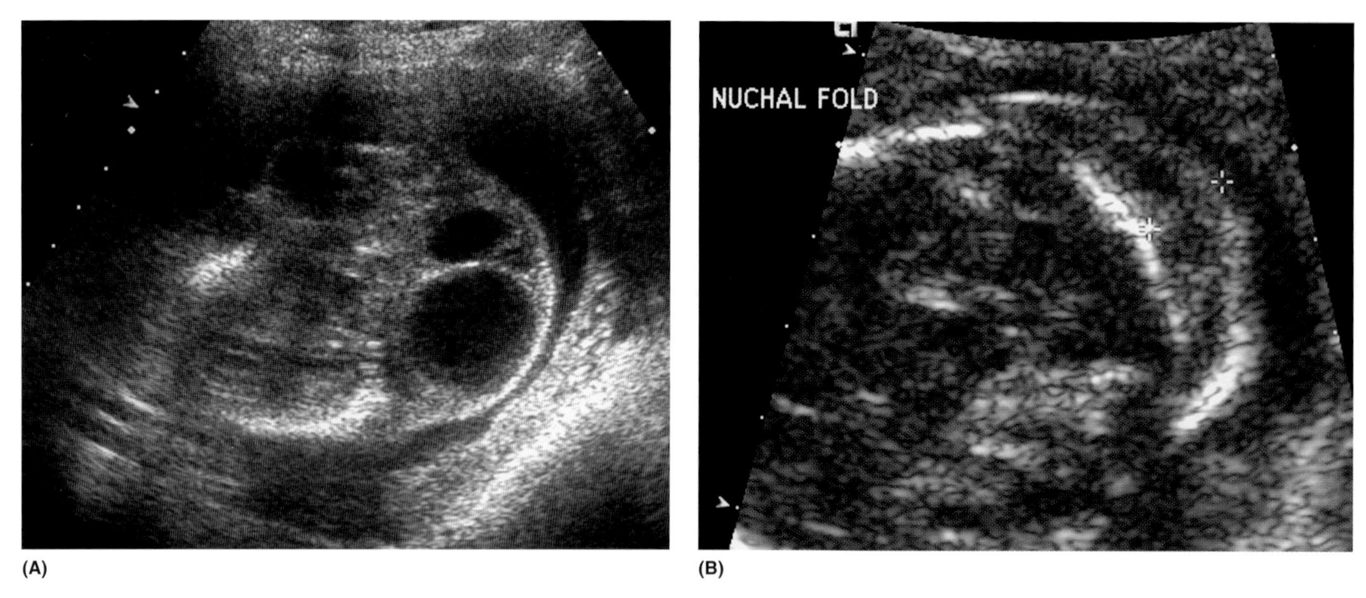

(A) (B)

FIGURE 14 A AND B ■ Regression of fetal cystic hygroma. (**A**) Ultrasound of the head demonstrating septated fetal cystic hygroma at 20 weeks' gestation. (**B**) Follow-up ultrasound obtained at 31 weeks of gestation demonstrates partial regression of the cystic hygroma.

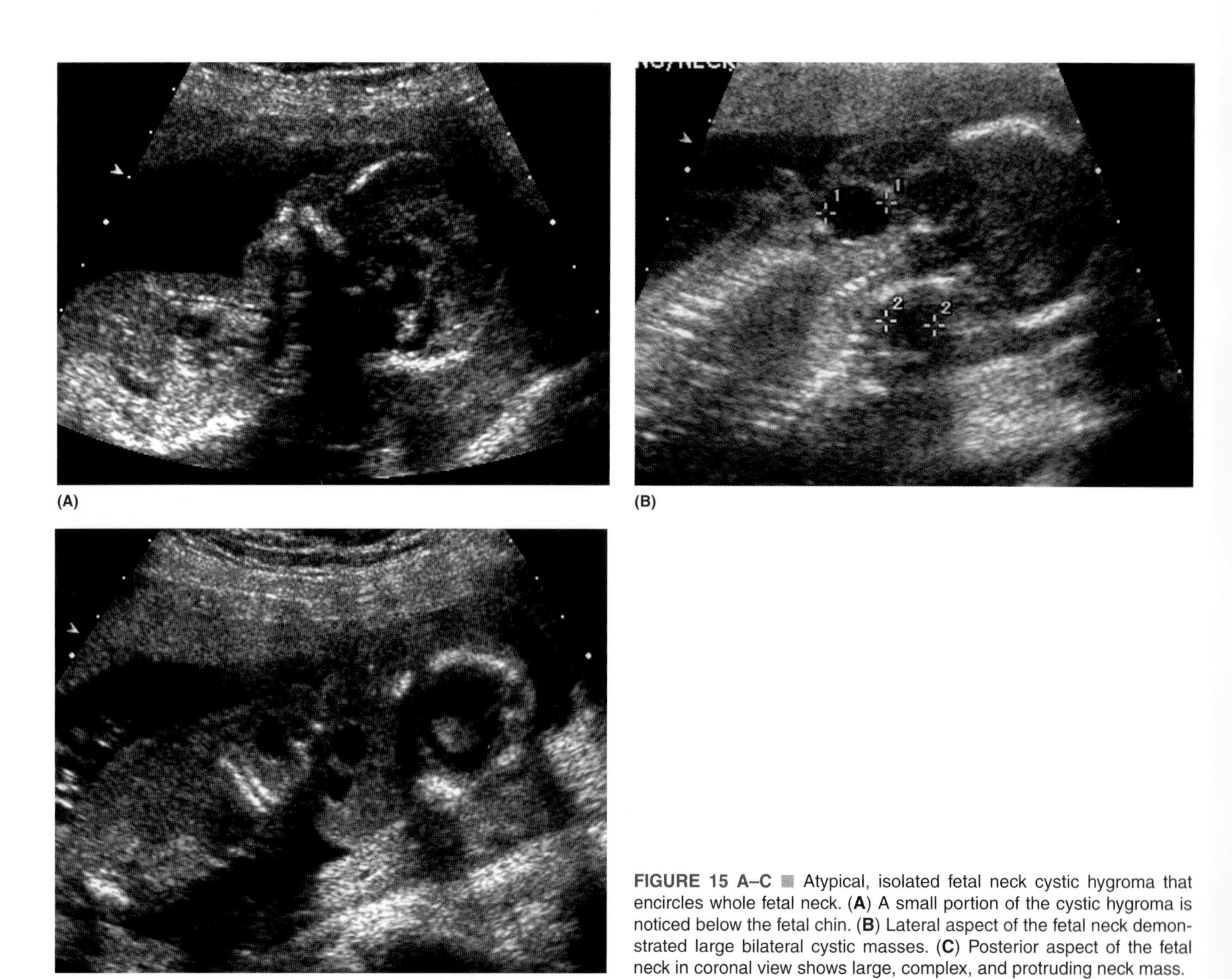

(A) (B) (C)

FIGURE 15 A–C ■ Atypical, isolated fetal neck cystic hygroma that encircles whole fetal neck. (**A**) A small portion of the cystic hygroma is noticed below the fetal chin. (**B**) Lateral aspect of the fetal neck demonstrated large bilateral cystic masses. (**C**) Posterior aspect of the fetal neck in coronal view shows large, complex, and protruding neck mass.

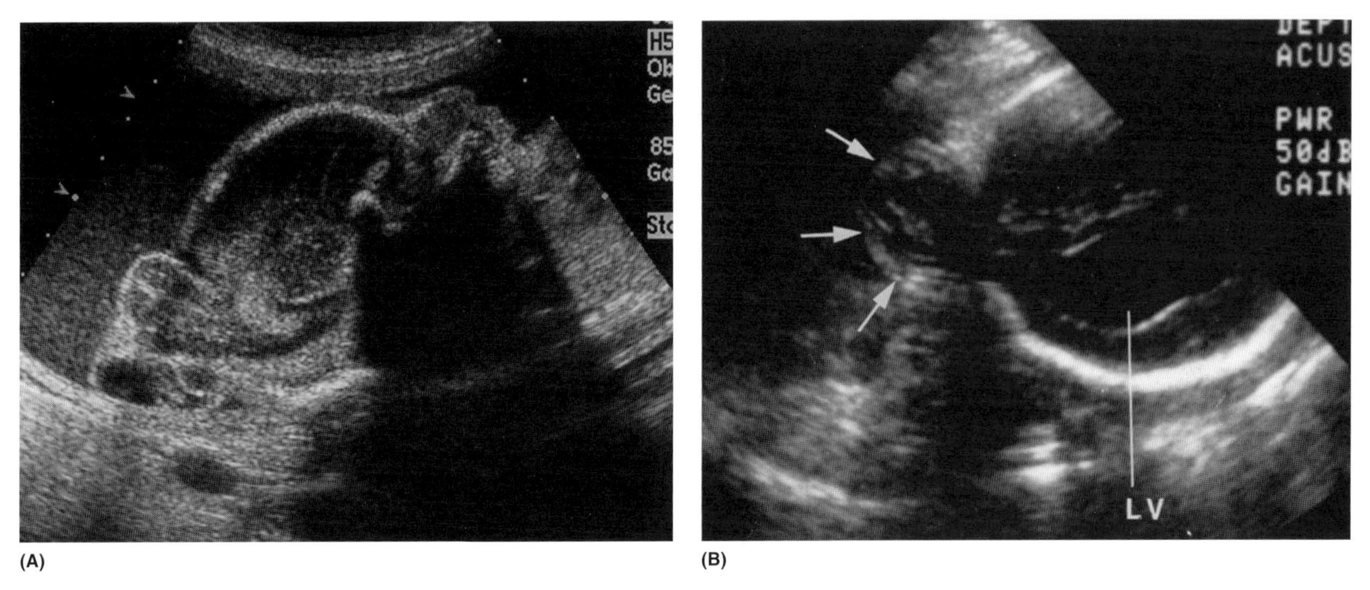

(A) (B)

FIGURE 16 A AND B ■ Various appearances of fetal cephaloceles. (**A**) Complex mass representing cephalocele with herniation of meninges and large portion of the brain. Note acute angulation of cephalocele with bony calvarium. (**B**) Another example of smaller occipital cephalocele (outlined by *arrows*) in association with dilated lateral ventricles. LV, lateral ventricle.

drugs (commonly propylthiouracil or carbimazole for Graves' disease).

Fetal goiter is typically characterized ultrasonographically as a solid, bilobed, anterior neck mass (Fig. 18). The mass is usually homogeneous and, although solid, may appear hypoechoic (24). Displacement of the common carotids posteriorly or hyperextension of the fetal head also may be noted. Associated findings include hydramnios, presumably due to impaired fetal swallowing. Nevertheless, the most important differential diagnoses are cervical teratoma and hemangioma (see below).

The accurate fetal thyroid status can be obtained by fetal blood sampling, whereas amniotic fluid analysis is considered inadequate for measurement of the fetal thyroid hormone or thyroid-stimulating hormone (TSH) (25,26).

Cervical Teratoma

Teratomas are neoplasms derived from pluripotent cells. They are composed of a diversity of tissues foreign to the anatomic site in which they arise (27). These tissues are derived from precursors in two or three different embryonic germ layers. Teratomas are rare neoplasms and occur in one in 20,000 to one in 40,000 live births, with about 5% of tumors that appear in the neck (28). Cervical teratomas are frequently detected prenatally. These tumors are typically unilateral with an anterolateral neck location and are solid or mixed solid-cystic in echostructure, though may change

TABLE 5 ■ Some Conditions Associated with Cephaloceles

Condition	Striking features	Cause
Meckel's syndrome	Polydactyly, multicystic kidneys	Autosomal recessive
Knobloch's syndrome	Myopia, vitreoretinal degeneration	Autosomal recessive
Chemke's syndrome	Hydrocephaly, agyria	Autosomal recessive
Dyssegmental dwarfism	Lethal dwarfism	Autosomal recessive
Pseudo Meckel's syndrome	Arhinencephalia, absent corpus callosum	Autosomal recessive t (p+)
Frontonasal dysplasia	Ocular hypertelorism	? Sporadic
Amniotic band syndrome	Amputational defects	? Sporadic
Iniencephaly	? Typical appearance of iniencephaly	? Sporadic

Source: From Ref. 6.

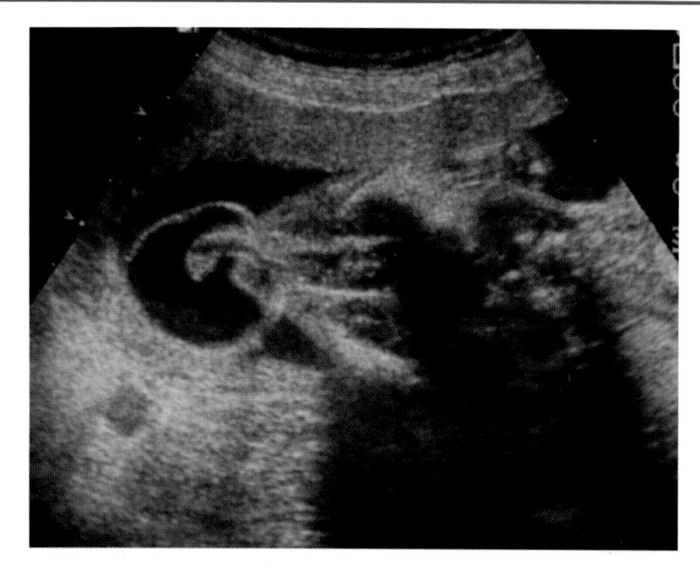

FIGURE 17 ■ Transverse ultrasound scan of the large, predominantly cystic meningomyelocele. Note small portion of the spinal tissue protruding into the sac.

appearance with advancing gestation. They are generally cystic masses early in gestation, becoming larger and more complex as the gestation progresses. Calcification is present in about 45% of cases. Cervical teratomas are associated with polyhydramnios in 30% of cases (29).

Hemangioma

Hemangiomas are localized proliferations of vascular endothelium that can occur anywhere in the body; they rarely present as neck masses prenatally. Ultrasonographic appearance of the hemangiomas varies largely due to number of different types (e.g., capillary, arteriovenous, and venous angiomas). Most hemangiomas appear solid and uniformly echogenic owing to innumerable small vascular channels that act as multiple interfaces (30). Occasionally, small internal hypoechoic spaces may be observed (Fig. 19). Due to nature of present blood vessels within the tumor, pulsed Doppler ultrasound may reveal significant blood flow and helps to distinguish hemangioma from other neck masses (31). Occasionally, the hemangioma that contains large vascular spaces may be encountered. In these cases, the mass appears hypoechoic owing to the dilated vascular channels. In addition, although there are no associated findings with neck hemangiomas, tumor itself may produce mass effect in terms of high-output congestive heart failure that can lead to subsequent fetal hydrops and death in utero. Therefore, close ultrasound followup of these pregnancies is suggested. Mode of delivery depends on the size of the hemangioma; if it is large, prompt airway management and debulking surgery are necessary.

Nuchal Thickening

Redundant skin at the posterior aspect of the fetal neck is related to fetal aneuploidy, mainly trisomy 21. This posterior neck thickening can be noted in the first, and/or second trimester of pregnancy (Fig. 13). It was originally described as a marker for detecting second-trimester fetuses with Down syndrome (32). The reason for nuchal thickening is not known, but it may be secondary to resolution of cystic hygroma. Later, it was noted that nuchal thickening can be also noted in first trimester and was described as a nuchal translucency (33,34) that was associated with fetal aneuploidy. Nuchal skin fold thickness can be measured on axial view of the skull in which the following internal landmarks are identified: cavum septi pellucidi, cerebral peduncles, cerebellar hemispheres, and cisterna magna (35). A soft-tissue thickening of 6 mm or greater between 16 and 20 weeks' gestation is considered abnormal. The role of nuchal translucency and nuchal fold thickness in screening for fetal aneuploidy will be described elsewhere in this text.

Iniencephaly

Iniencephaly is a rare, deadly, and heterogeneous malformation consisting of (*i*) a defect of the occipital bone with accompanying enlarged foramen magnum, (*ii*) partial or total absence of thoracic and cervical vertebrae, (*iii*) significant shortening of the spine as a result of an extreme lordosis and hyperextension (retroflexion) or the malformed spinal column at the neck and thorax (Fig. 20), and (*iv*) an upturned face (36). It is usually combined with a cervical NTD and sometimes with fusion to the cranium. The resultant extreme cervical retroflexion, which may be fixed, gives affected infants their characteristic appearance. Diagnosis is based on ultrasonographic inability to locate the entire fetal spine on longitudinal scan in conjunction with visualization of the head when scanning the thorax transversely (37,38).

SPINE

The major skeletal support to the body is provided by the spine. It is composed of a series of bones and houses a major portion of the central nervous system composed of spinal cord and nerve rootlets. To begin to approach anomalies of the spine in a systematic way requires an appreciation of the embryologic events leading to the formation of these structures.

Embryology and Normal Anatomy

The neural tube is the basic embryologic structure, giving rise to the central nervous system. It arises from an infolding of the neural plate, a midline thickening of ectoderm, during the third week of intrauterine life. This infolding, or neurulation, starts from the region of the fourth somite (the center of the plate) and proceeds both rostrally and caudally. The rostral end of the tube (anterior neuropore) closes on or about day 23 of embryonic life. The caudal end of the tube (posterior neuropore) closes on or about day 28. If this process is interrupted, defects

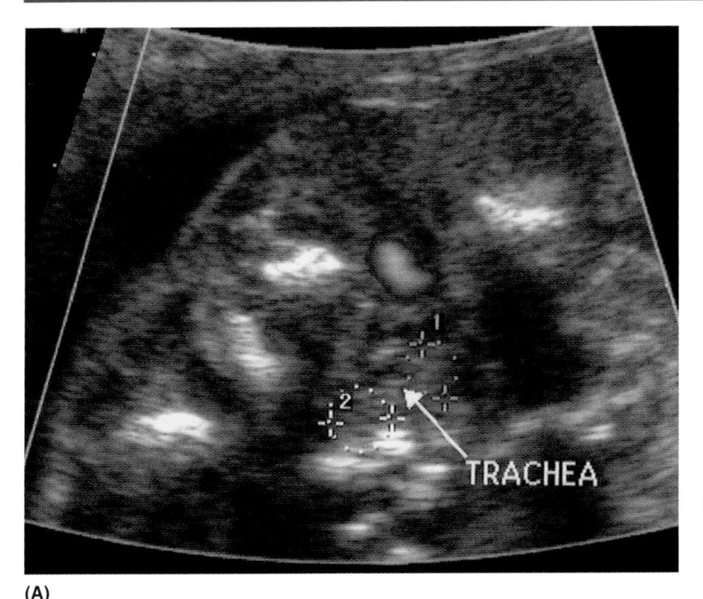

(A)

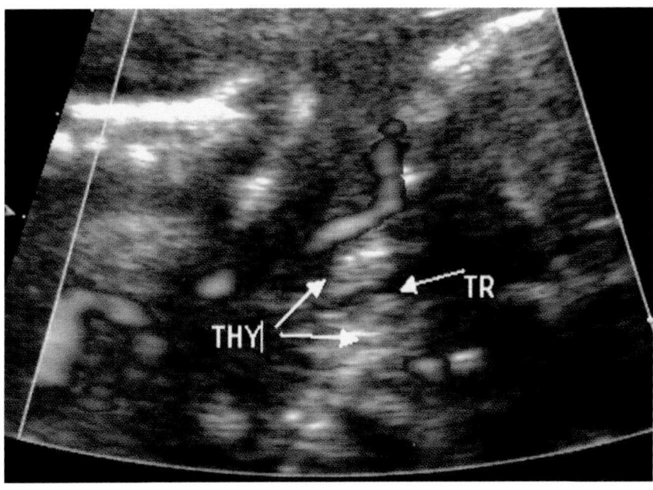

(B)

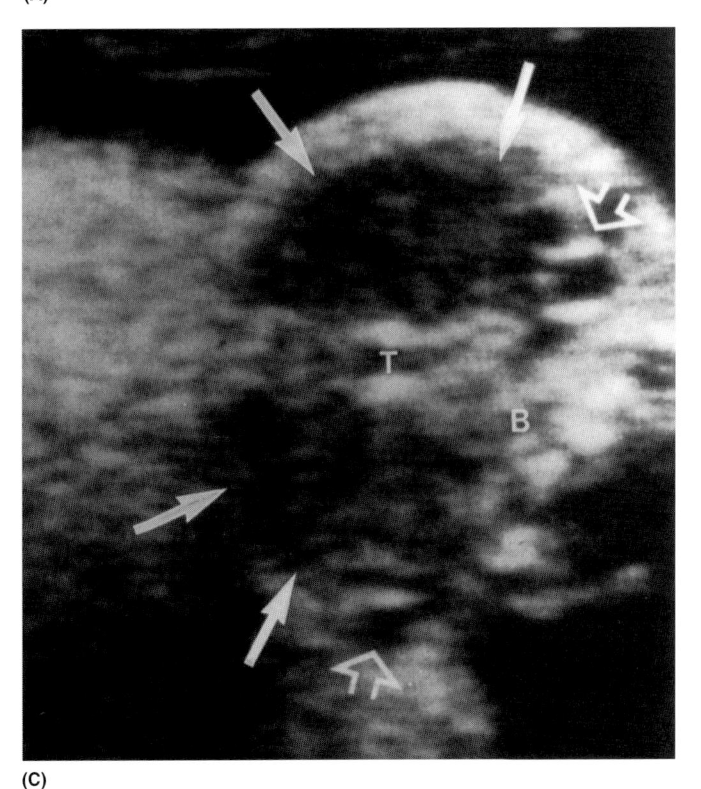

(C)

FIGURE 18 A–C ■ (**A**) Fetal thyroid gland in transverse ultrasound view: midline cystic structure with echogenic ring represents fetal trachea, while laterally, two mildly hyperechogenic masses represent fetal thyroid gland (both lobes; encircled with dotted lines). Posteriorly to the fetal trachea, two lateral ossification centers of the cervical spine are noted. (**B**) Fetal thyroid gland in coronal view. Midline tubular structure represents trachea (TH), with two hyperechogenic masses on both tracheal sides that represent fetal thyroid gland (*arrows*). Note blood flow through both carotid arteries (located at the neck lateral aspects) depicted by power Doppler. (**C**) Fetal goiter. Axial scan of the neck (*closed arrows*) demonstrating a goiter. The neck vessels are indicated by open arrows. The trachea can be seen between the lobes of the thyroid. *Abbreviations*: (**B**) THY, thyroid gland; TR. (**C**) T, trachea; B, vertebral body of spine. *Source*: From Ref. 24.

can occur in both the lumbosacral and the cranial ends of the neural tube, thus explaining the association of cephalocele and anencephaly with spina bifida.

Mineralization of the spine vertebrae starts about the eighth menstrual week and continues with development of the three ossification centers: vertebral body (central ossification center) and the lateral masses (two lateral ossification centers) around 13 menstrual weeks (39,40). The variability in the shapes of the vertebral bodies is present and recognized as the change in ultrasonographic appearance of the spine during gestation. By 16 weeks' gestation, individual vertebrae may be identified ultrasonographically by the observation of their echogenic ossification centers in the transverse plane. Two of these are posterior to the spinal canal in the laminae, and one is anterior, representing the vertebral body. During the third trimester, the vertebral body, pedicles, transverse processes, laminae, and spinous processes may all be identified as echogenic structures on transverse scans. In contrast to the vertebral bodies, the spinal canal and intervertebral foramina may be seen as nonechogenic areas.

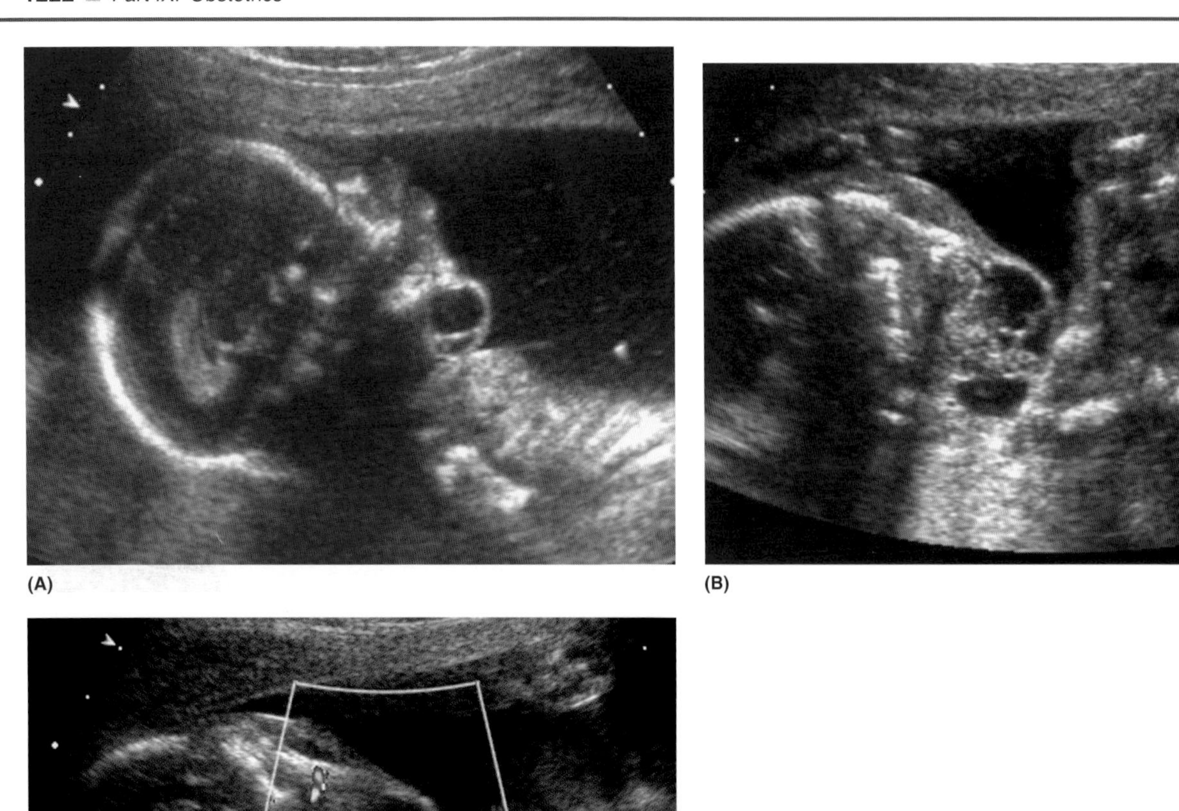

(A) (B) (C)

FIGURE 19 A–C ■ Fetal neck hemangioma. (**A**) Prenatal ultrasound demonstrates a well-circumscribed cystic mass located anterior to the neck below the fetal chin. (**B**) Transverse image of the neck hemangioma with marked cystic avascular areas. (**C**) Transverse view of the same mass with corresponding color Doppler depiction of the hemangioma's blood vessels.

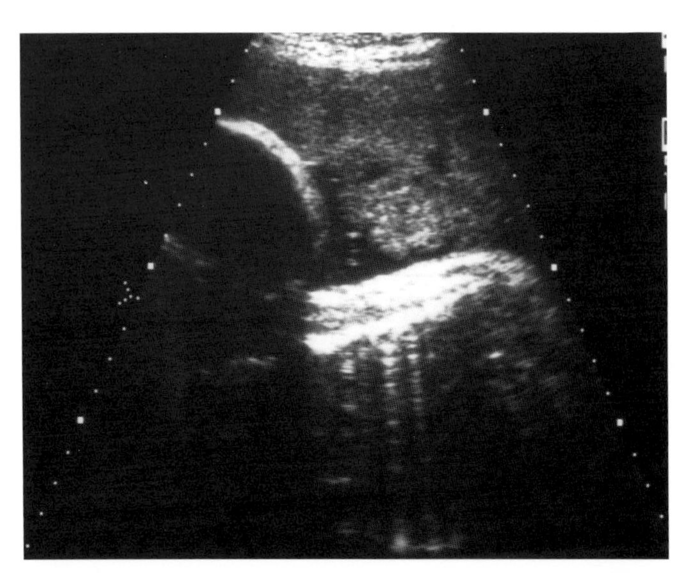

FIGURE 20 ■ Iniencephaly observed by ultrasound at 22 weeks of gestational age. Note sharp angulation between the fetal head and the cervical spine.

The fetal spine can be assessed by three basic scanning planes: transverse or axial, sagittal, off-axis sagittal and rarely coronal view (Fig. 21). Transverse plane of the spine and individual vertebral body reveals all three ossification centers. In the sagittal plane, the posterior aspects of the vertebral bodies become more complex (Fig. 22). The spinal cord appears hypoechoic, and the central canal of the spinal canal can be identified with careful scanning. Scanned sagittally (longitudinally) and off-axis, a parallel line of lateral vertebral bodies may be identified anterior to the nonechogenic spinal canal (Fig. 21). In the second trimester of pregnancy, the posterior elements are not identified if scanned in a true sagittal plane because the spinous process is incompletely ossified. The posterior elements are identified only if there is angulation of the sagittal image through the two posterior ossification centers. Note that a small spinal defect may be missed when scanning in an off-axis rather than a true sagittal plane.

Due to complexity of the spine, a systemic approach to ultrasonographic examination is necessary to reveal

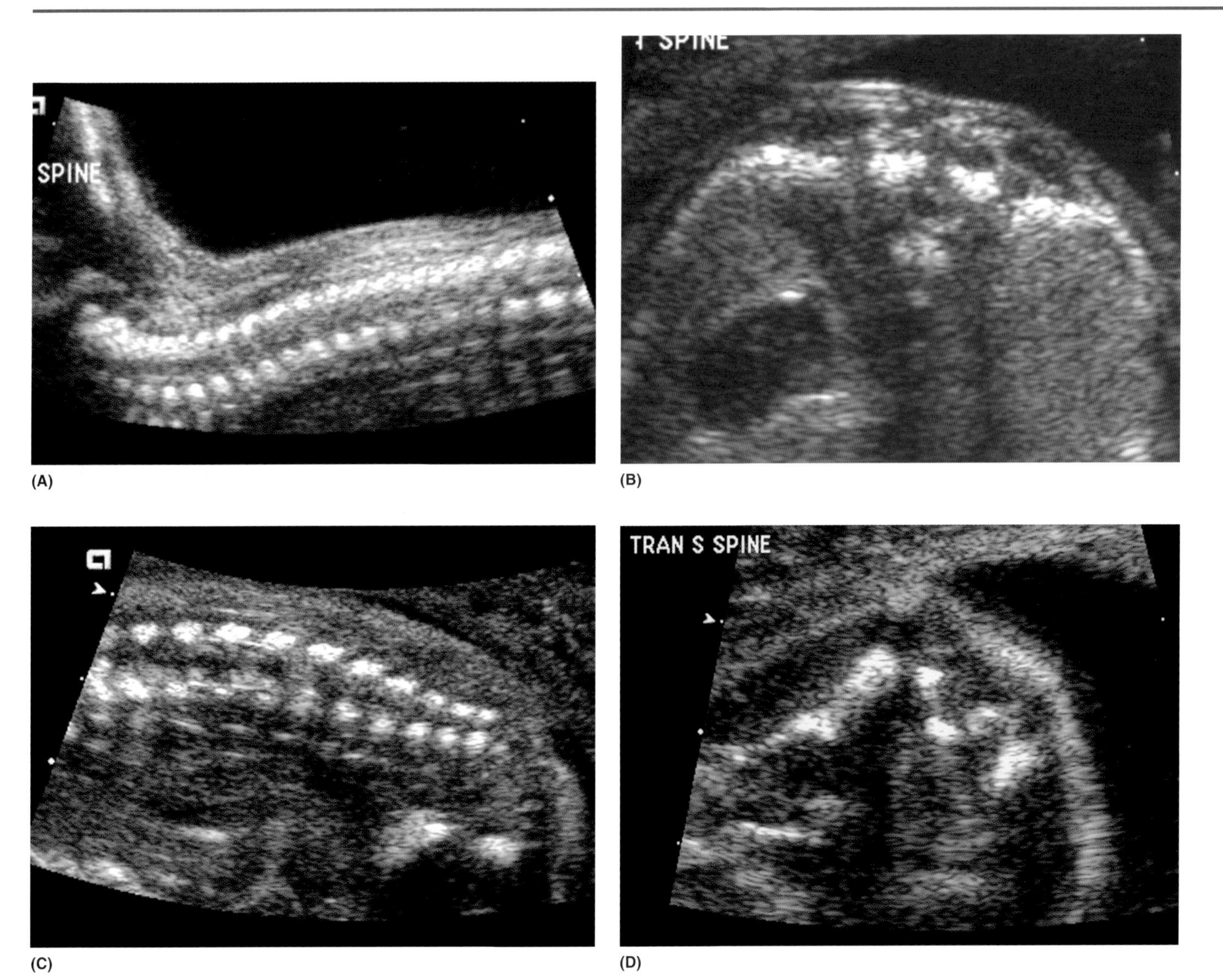

(A) (B) (C) (D)

FIGURE 21 A–D ■ Normal fetal spine. (**A**) Sagittal, off-axis view of the thoracic and cervical spine. Note parallel lateral ossification centers of the vertebral bodies. (**B**) Corresponding transverse scan of the thoracic spine. Two lateral and one central (vertebral body) ossification centers can be observed. Intact skin covers neural tube. (**C**) Sacral spine demonstrated in sagittal, off-axis view. Note converging lateral ossification centers of the sacral body toward the end of the coccygeal bone. (**D**) Corresponding transverse view of the sacrum covered with intact skin and laterally long bone that represents iliac bone.

the numerous possible abnormalities in this region. To avoid distortions in the spinal anatomy, a continuous scanning of the entire spine in both transverse and sagittal planes is essential.

Normal Anatomic Variations

Pseudosplaying of the Cervical Spine

In the coronal plane, the two rows of echogenic posterior ossification centers are seen to diverge progressively in both the cervical and the lumbar regions. This is especially pronounced close to the base of the skull and should not be misinterpreted as spinal dysraphism (Fig. 23).

Pseudodysraphism of the Lumbosacral Spine

Possible pitfalls encountered when scanning the fetal spine include incorrect assessment of fetal position and errors in maintaining the proper scanning angle (41). Oblique transverse angulation of the ultrasound transducer creates a plane in which the anterior vertebral elements are included but posterior elements are missed, thus simulating spina bifida (41). This re-emphasizes the import of maintaining a true transverse orientation along the normal curvature of the spine.

Incomplete Ossification

Early in pregnancy, especially in the early second trimester, ossification of the lateral masses occurs first in

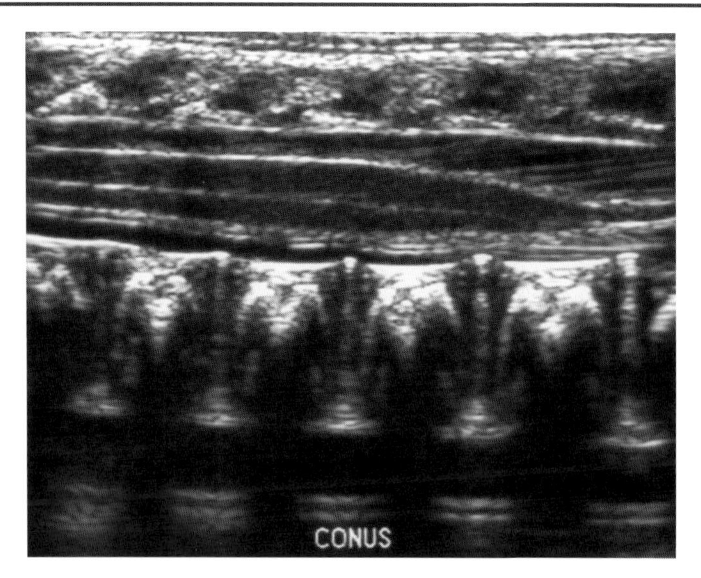

FIGURE 22 ■ Sagittal view of the spine, note the hypoechoic spinal canal, above clearly separated individual vertebral bodies, that contains spinal cord.

the region of the laminae. Later, ossification progresses both anteriorly toward the pedicle and posteriorly toward the spinous process. Early scanning reveals incomplete ossification of the lateral masses, which may appear parallel rather than converging posteriorly; this may be erroneously misinterpreted as spinal dysraphism.

Abnormalities

Open NTDs

The evaluation of the fetal spine is complex in nature and requires optimal fetal position as well as gestational age to appreciate all spinal structures. Therefore, small, flat, or cystic open NTDs may be easily missed when scanning in the early second trimester. Meticulous transverse and sagittal imaging is needed with optimally focused transducers to determine whether the posterior ossification centers are diverging rather than converging toward the midline. The presence of the bony defect is often associated with the presence of intracranial abnormalities such as Arnold–Chiari type II malformation, including the lemon sign or banana sign (42). Recognition of such ultrasonic signs appears to be very helpful in recognition of NTDs (43). Regardless that most spinal abnormalities can be detected with ultrasound, not all chromosomal or structural abnormalities may be identified. Therefore, it was proposed that amniocentesis with analysis of amniotic fluid should be offered to patients (44). On the other hand, it was shown that quality ultrasonographic visualization of the spine and cranial signs of spina bifida produces excellent detection of spina bifida (45). When a pregnant patient presents with an elevated maternal serum AFP level, she should be given the option of a thorough ultrasound examination.

Nevertheless, in addition to bone defect, the very shape and curvature of the spine also contribute in detection of open NTD (46). The majority of the abnormalities that cause malformation of spinal curvature are open NTD, followed by myelomeningocele, limb–body wall complex, and multiple congenital anomalies, together with a whole array of other spinal and head malformations (Table 6). Anomalies such as isolated hemivertebra may be missed in utero; therefore, their true incidence may be higher than is reflected in this table.

Between the NTD and spinal deformities, the most common malformation is spina bifida aperta (Table 7).

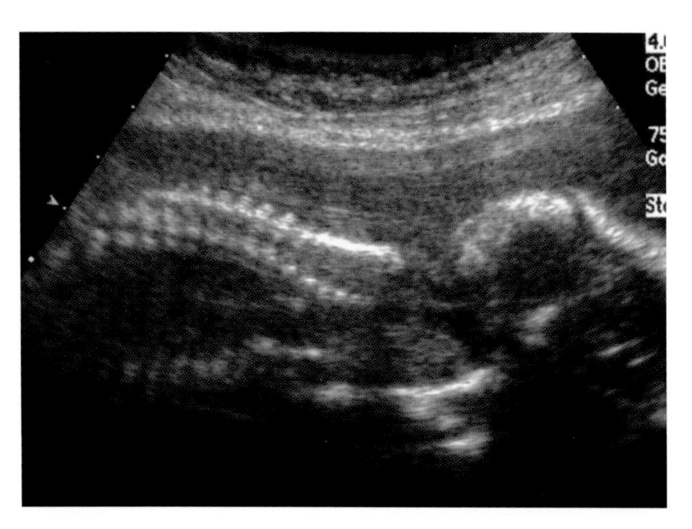

FIGURE 23 ■ Coronal view of the cervical and thoracic fetal spine. Lateral ossification centers of the cervical spine diverge toward the base of the skull. This is normal finding in contrast to the diverging lateral ossification centers of the sacrum that usually represents spina bifida.

TABLE 6 ■ Diagnoses in Fetuses with Abnormal Spinal Curvature

Diagnosis	Number (*N* = 20)
Neural tube defect	12
Myelomeningocele	6
Frontal encephalocele with ectrodactyly	1
Exencephaly	1
Anencephaly	1
Anencephaly with trisomy 18	1
Iniencephaly	1
Craniorachischisis	1
LBWC	2
ABS	1
Multiple congenital anomalies, likely LBWC or ABS	2
Caudal regression syndrome	1
Thoracic dysplasia with multiple anomalies	1
Hemivertebra, no other anomalies	1

Abbreviations: LBWC, limb–body wall complex; ABS, amniotic band syndrome.

Source: From Ref. 46.

TABLE 7 ■ Classification of Cranial and Spinal Dysraphism

- Spina bifida aperta
- Myeloschisis
- Myelomeningocele
- Hemimyelomeningocele
- Syringomyelomeningocele
- Spinal meningocele
- Arnold–Chiari malformation
- Dandy–Walker malformation
- Cranium bifidum
- Cranial meningocele
- Encephalomeningocele
- Occult cranial dysraphism
- Cranial dermal sinus
- Occult spinal dysraphism
- Spinal dermal sinus
- Tethered cord syndrome
- Lumbosacral lipoma
- Diastematomyelia
- Neurenteric cyst
- Combined anterior and posterior spina bifida
- Anterior sacral meningocele
- Occult intrasacral meningocele
- Nondysraphic malformations
- Perineurial (Tarlov's) cyst
- Spinal extradural cyst
- Nondysraphic spinal meningocele
- Caudal regression syndrome
- Sacrococcygeal teratoma

Source: From Ref. 47.

This table presents a broad classification of cranial and spinal dysraphism that can be encountered during the ultrasound evaluation of the fetal neck and spine.

Because spina bifida may appear as a closed cystic mass, the differential diagnosis for lumbosacral mass is given in Table 8. Most commonly, these masses are open NTDs, but they may include a symmetric defect such as amniotic band syndrome or limb–body wall complex. Early tumors, such as sacrococcygeal teratomas or other tumors, may occur in this region.

Spina Bifida

Spina bifida is a general term used to describe the open forms of spinal dysraphism that result from failure of the posterior neuropore to close. It is a defect of the vertebrae resulting in exposure of the contents of the neural canal. In majority of cases, the defect is localized to the posterior arch of the vertebrae, although, in rare instances, splitting of the vertebral body causes the defect. Dorsal defects of the spina bifida are categorized as occulta and aperta. Spina bifida occulta represents a small spinal defect that is completely covered by skin, and it can be observed in approximately 15% of cases (Figs. 24 and 25). Most of the time, this is a small abnormality that is usually incidentally discovered at radiographic examination of the spine later in life. In contrast to skin-covered spina bifida defect, aperta, derived from the Latin meaning open, describes the communication between the spinal canal and the amniotic space/fluid (Fig. 26). It is observed in 85% of cases, and it is associated with infection and inflammation of the spinal neural contents that produces adverse neurologic symptoms (47).

Spina bifida is the most common malformation of cranial and spinal dysraphism and was considered as one

TABLE 8 ■ Typical Lumbosacral Masses and Associated Ultrasound Features

	Type of mass	Ultrasound features
Common	Open neural tube defects	Symmetric cystic, complex, or solid mass in the lumbosacral region
		Associated with cranial defect and lemon or banana sign
Uncommon	Amniotic band syndrome	Asymmetric defect with scoliosis
		Cystic to complex mass
		Associated amputation defects
	Limb–body wall complex	Limb–body wall complex
		Similar to amniotic band syndrome
		Plus fetus in continuity with placental surface and umbilical cord
	Sacrococcygeal teratoma	Usually solid mass from buttocks
		Often intrapelvic components
		Rarely a cystic mass
Rare	Tumors	Rare tumors such as lipomas that are echogenic, or other neural or cutaneous tumors
		Ultrasound appearance depending on tumor type

Source: From Ref. 6.

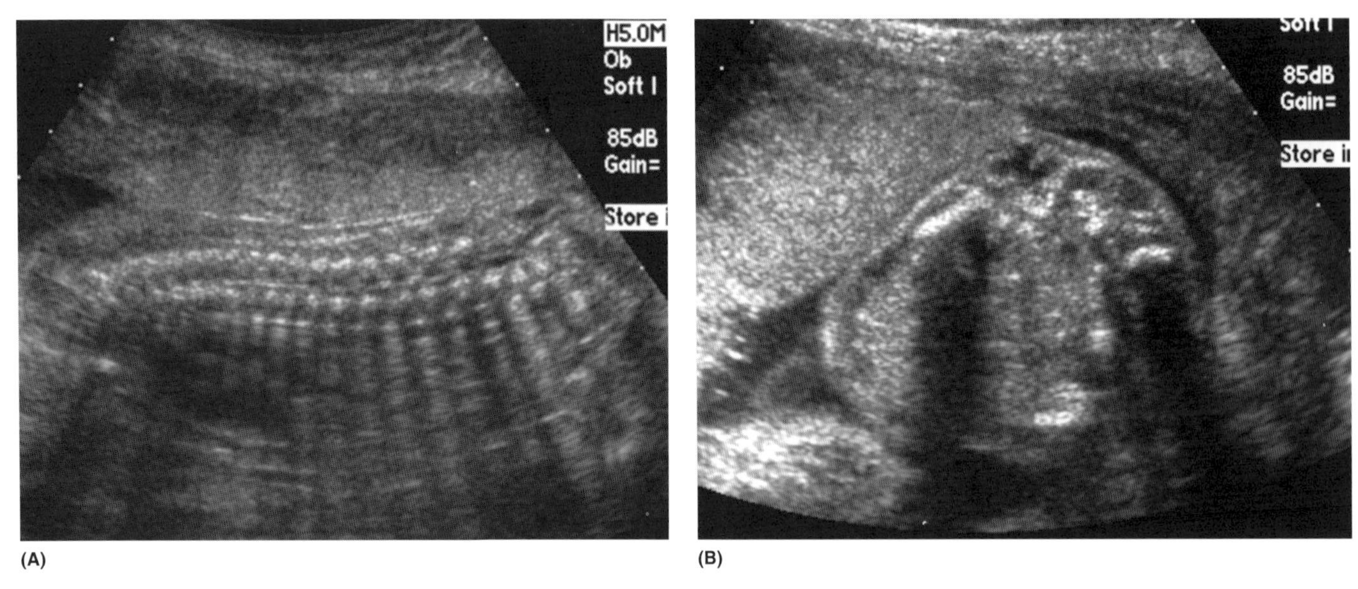

FIGURE 24 A AND B ■ (**A**) Sagittal, off-axis view of the lumbosacral spine that reveals the bony defect toward the sacral region (diverging lateral vertebral ossification centers). (**B**) Corresponding transverse view of the lumbosacral spine region with poorly seen lateral ossification centers, though cystic structure covered with intact skin can be clearly visualized.

of the most common congenital malformations in the Western world. Its incidence is approximately one per 1000, though it varies greatly, ranging from 0.3 per 1000 births in Japan to more than four to eight per 1000 births in parts of Great Britain. The average incidence of spina bifida in the United States is about 0.5 to 2 per 1000 births (48–50). Incidence also varies depending on specific risk factors (Table 9).

Although spina bifida aperta is the most frequent lesion, sometimes the defect is not completely open or completely covered by skin. If the lesion contains purely meninges, and ultrasonically appears as a cystic lesion, it is categorized as a meningocele.

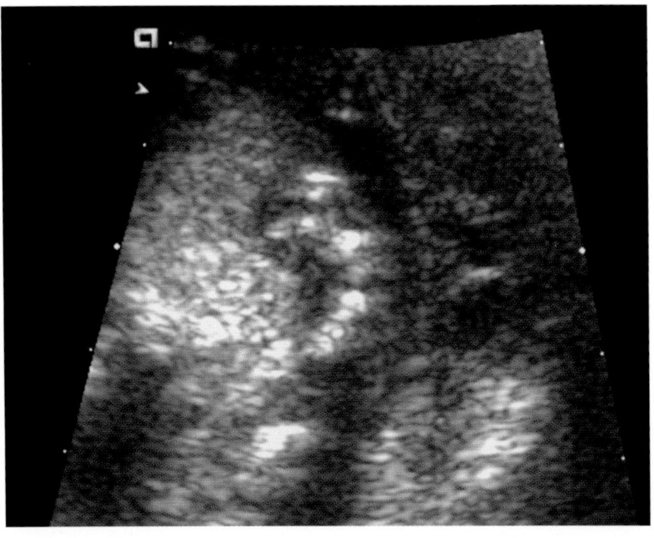

FIGURE 25 ■ Another example of the spina bifida occulta. In this transverse view of the sacrum, bony defect with diverging lateral ossification centers covered with intact skin can be noted.

Meningoceles are cystic lesions that result from protrusion of the dura and the arachnoid through the spinal defect (Fig. 27). They usually occur at the upper and lower ends of the neural axis (i.e., occipital, cervical, upper thoracic, and low sacral positions). In contrast, meningomyeloceles (or myelomeningoceles) contain abnormal central nervous system tissue (i.e., a malformed, maldeveloped spinal cord) in the defect. These lesions are usually large and covered by a neural plaque composed of a membrane of meningeal origin and malformed spinal cord substance. If the neural tissue is part of the mass, the lesion is referred as a meningomyelocele that typically occupies the lower thoracolumbar, lumbar, and lumbosacral areas. These lesions usually are single and only rarely are multiple.

Ultrasonographically, spina bifida is seen as a splaying of the posterior ossification centers of the spine, giving the vertebra a U-shaped appearance instead of the normal triangular appearance. The posterior ossification centers in a defective vertebra should be more widely spaced than those in vertebrae above and below the defect. Although spina bifida may be visualized on longitudinal scanning, meticulous transverse examination of the entire vertebral ossification centers is necessary to detect smaller defects (Fig. 28).

Associated anomalies include the Arnold–Chiari malformation that includes hindbrain anomalies, with displacement of the inferior cerebellar vermis into the upper cervical canal and caudal dislocation of the medulla and the fourth ventricle. This probably is caused by tension due to the spinal dysraphism, resulting in the spinal cord pulling the cerebellum and the medulla caudally. Two characteristic ultrasonic abnormalities

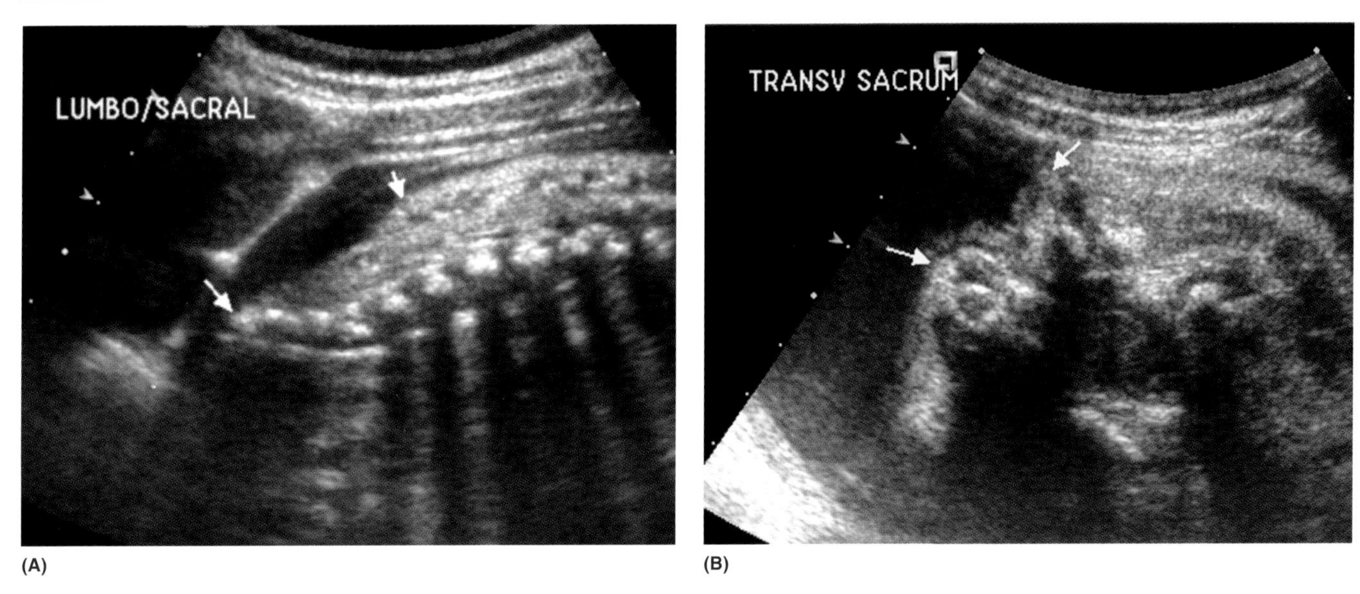

FIGURE 26 A AND B ■ Spina bifida aperta. (**A**) Sagittal, off-axis view of the lumbosacral spine region that shows bony defect (between the arrows). (**B**) Transverse corresponding view of the spina bifida aperta (area between the arrows) that clearly demonstrates absence of the skin above the lesion.

were described in this relation (42). A scalloping of the frontal bones gives a lemon-like configuration to the skull of an affected fetus in axial section during the second trimester (the lemon sign) (Fig. 29). The caudal displacement of the cranial contents within a pliable skull is thought to produce this scalloping effect. Similarly, as the cerebellar hemispheres are displaced into the cervical canal, they are flattened rostrocaudally, and the cisternal magna is obliterated. This produces a flattened, centrally curved, banana-like ultrasonographic appearance (the banana sign) (Fig. 30). The third finding that can be noted in approximately 90% of cases of open spina bifida at birth is borderline or mild ventriculomegaly. The sensitivity of 99% and higher while using ultrasonic cranial findings in identifying spina bifida was reported (51). False-positive cranial ultrasonic findings are almost not observed, though the lemon sign can be found in 1% of normal fetuses.

TABLE 9 ■ Estimated Incidence of Neural Tube Defects Based on Specific Risk Factors in the United States

Population	Incidence per 1000 live births
Mother as reference	
General incidence	1.4–1.6
Women undergoing amniocentesis for advanced maternal age	1.5–3.0
Women with diabetes mellitus	20
Women on valproic acid in first trimester	10–20
Fetus as reference	
One sibling with NTD	15–30
Two siblings with NTD[a]	57
Parent with NTD	11
Half sibling with NTD	8
First cousin (mother's sister's child)	10
Other first cousins	3
Sibling with severe scoliosis secondary to multiple vertebral defects	15–30
Sibling with occult spinal dysraphism	15–30
Sibling with sacrococcygeal teratoma or hamartoma	≤15–30

[a]Risk is higher in British studies. Risk increases further for three or more siblings or combinations of other close relatives.

Abbreviation: NTD, neural tube defect.

Source: From Ref. 50.

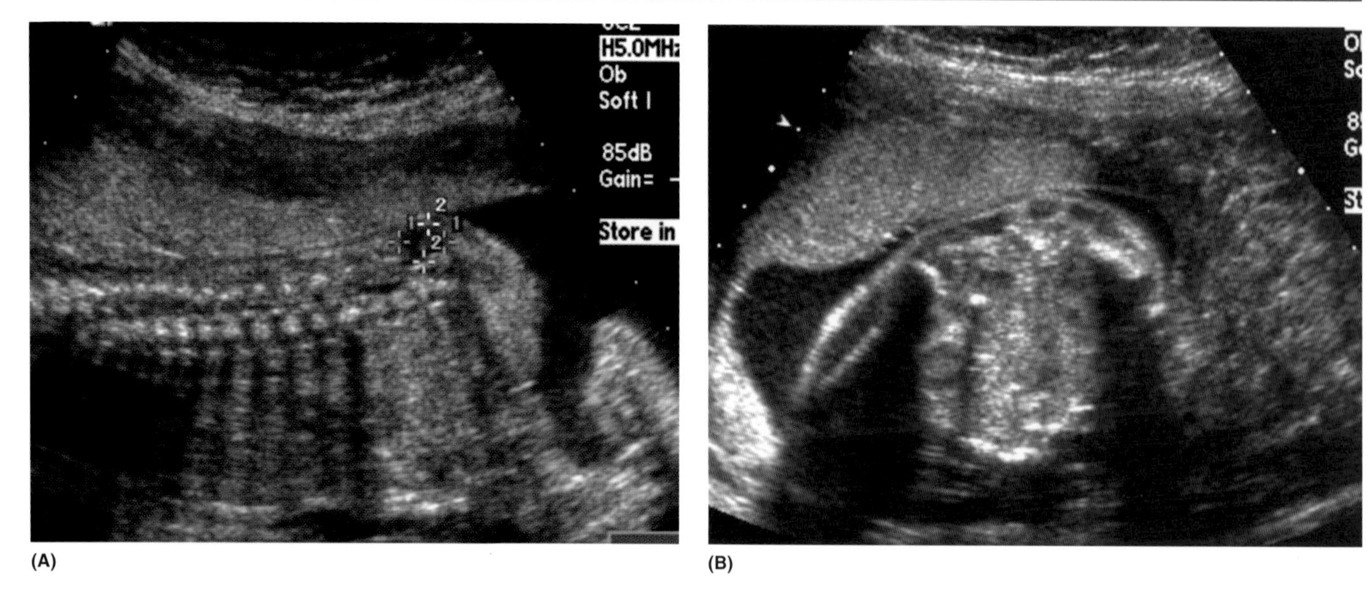

FIGURE 27 A AND B ■ An ultrasonic example of a meningocele located in the lumbosacral region of the fetal spine. (**A**) Sagittal, off-axis view of the lumbosacral fetal spine demonstrates small cystic lesion (0.8 × 0.7 cm large). Corresponding transverse scan of the same lesion (**B**) shows cystic lesions localized just below the intact skin and above the vertebral body.

The ultrasound accuracy in detection of spina bifida is highly dependant on the operator experience as well as the quality of the equipment. In the most quoted routine antenatal diagnostic imaging with ultrasound (RADIUS) study, a sensitivity of 80%, though in conjunction with maternal AFP screening, was reported (52). In other centers, a large range of sensitivity from 33% to 100% was observed (53).

Diastematomyelia

Diastematomyelia represents a split spinal cord. A longitudinal clefting of spinal cord into halves, usually by a bony spicule or fibrous band in an area of spina bifida, produces two hemicords. This defect may or may not be covered by intact skin (54). The defect can involve a single vertebra or extend to several vertebrae. The pathophysiology of the diastematomyelia is unknown, though neurenteric canal "fistula" formation, splitting of the neural plate near the midline, and excessive dilatation of the neural tube have been implicated.

As a rare spinal deformity, few reports demonstrated ultrasound features of diastematomyelia, mainly broadening of the lateral masses of the spine, an identification of the extra-echogenic posterior focus in the spinal canal,

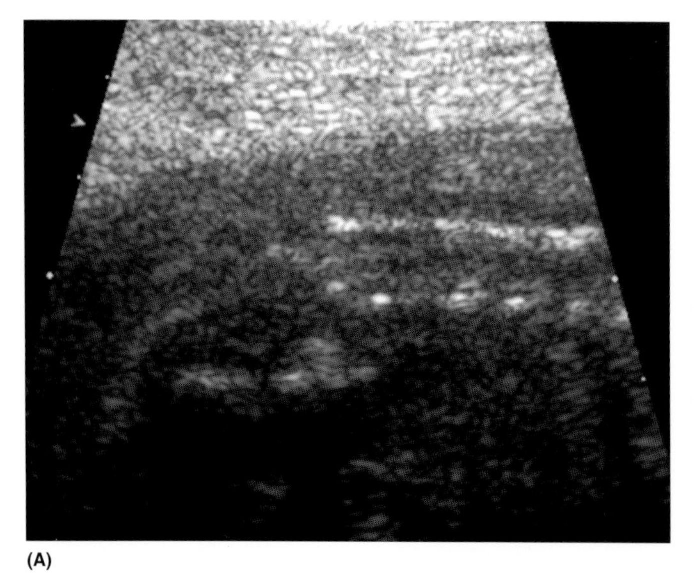

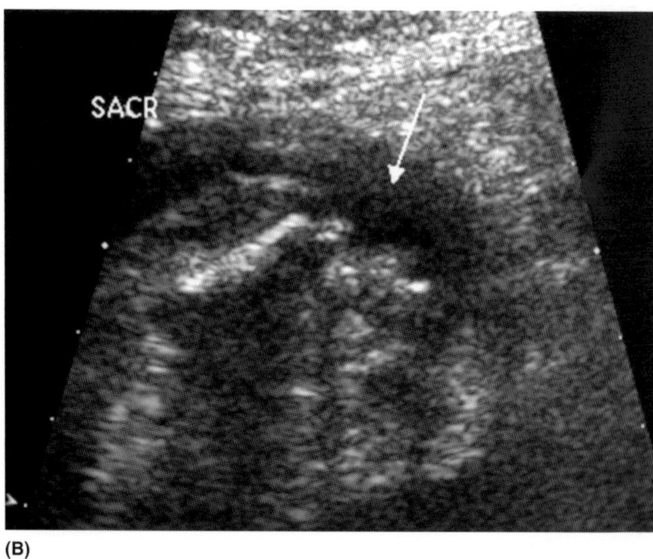

FIGURE 28 A AND B ■ (**A**) A sagittal, off-axis view of the fetal sacrum with diverging lateral ossification centers denoting possible spina bifida. (**B**) A transverse scan of the same lesion clearly demonstrates small, open neural tube defect (*arrow*)—spina bifida aperta.

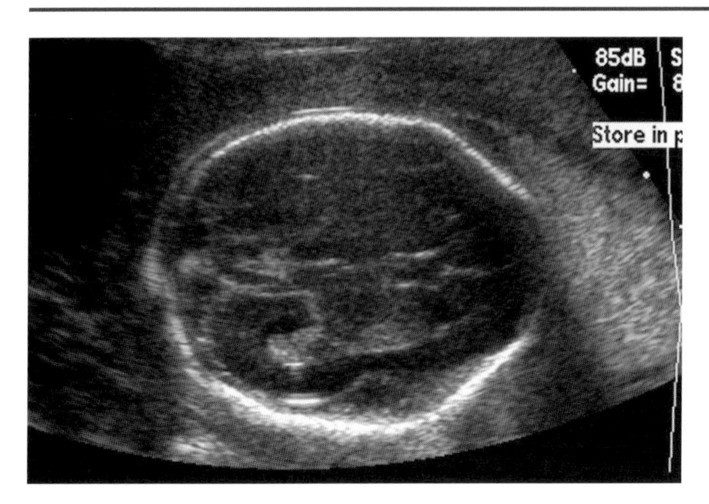

FIGURE 29 ■ A sonographic presentation of the lemon sign. Note the scalloping of the frontal head bones into the shape of the lemon (transverse view of the fetal head).

with overlying soft tissues and skin that appear intact (55,56). The point of greatest diastasis may be identified in the images of the double vertebral column, and this confirms the diagnosis.

When diastematomyelia presents as a part of the NTD with open spina bifida, prognosis is related to the spina bifida symptoms and not diastematomyelia. However, if this condition is presented as a closed NTD, the prognosis for neurologic function may be enhanced by early surgical removal of the septum dividing the spinal cord (54).

Lipomyelomeningocele

Lipomyelomeningocele (intraspinal lipoma) is a histologically benign fatty tumor associated with occult spina bifida and represents a variation of spinal dysraphism. The fatty tissue often originates in the spinal cord or cauda equina and is covered with skin. It is nearly always asymmetric (57) and ultrasonically appears as a discrete echogenic mass without structural organization in an area of spina bifida occulta (58). Sometimes, this midline mass has the appearance of the fetal tail. Because it may be associated with other spinal anomalies, central nervous system defects, careful scanning of the rest of the fetus should be performed.

Early diagnosis and early resection, with unfettering of the cord, remain the principles of management. Because the defect is typically covered, vaginal delivery has been suggested, but because the ability to discriminate open and closed NTDs by ultrasonography has not been demonstrated, we advise a more conservative approach, with elective cesarean delivery at term.

Sacrococcygeal Teratoma

Sacrococcygeal teratoma is a germ-cell tumor that arises from pluripotential embryonic cells of the coccyx, which may mimic a spinal defect. It is believed that these pluri- or totipotent embryonic cells are nested in primitive knot or Hensen's node that migrates caudally inside the tail of the embryo during the first week post conception, eventually resting anterior to the coccyx. Therefore, the occurrence of the teratomas in the sacral area rather than in other parts of the body is seen most frequently (59).

Sacrococcygeal teratoma is considered to be the most commonly occurring tumor among neonates, reported in one in 35,000 births, and about 75% occur in girls (60). Four types of sacrococcygeal teratomas are recognized: Type I, predominantly external with minimal presacral component; Type II, predominantly external with significant intrapelvic component; Types 3 and 4, predominantly

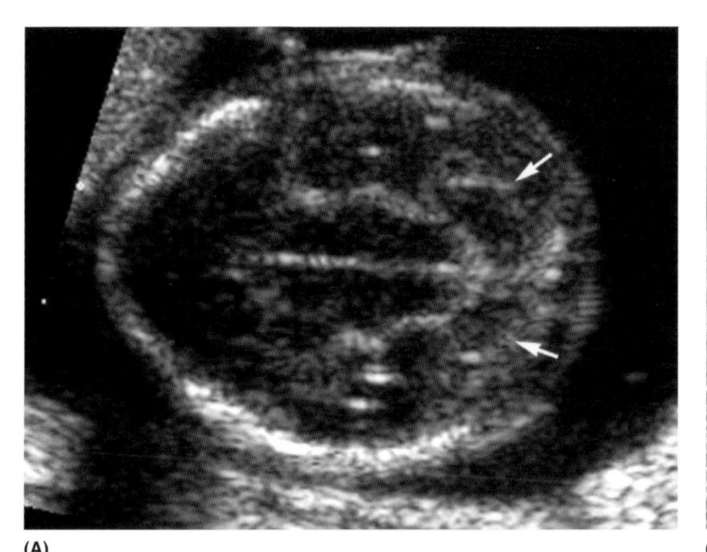

(A)

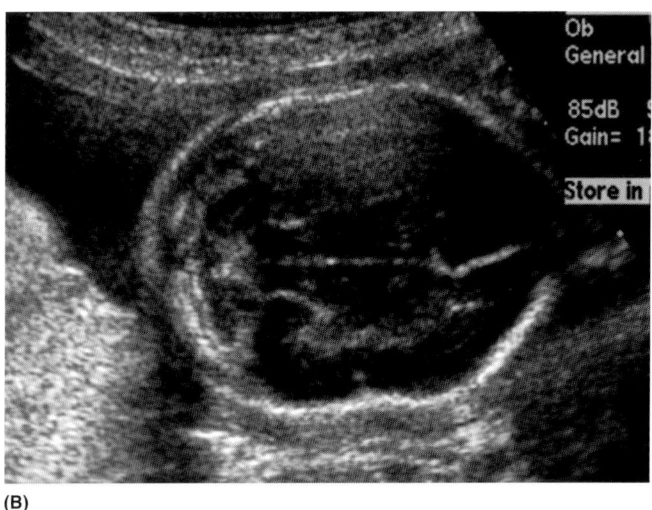

(B)

FIGURE 30 A AND B ■ An oblique-transverse view of the fetal head at the level of the cerebellum. (**A**) Fetal cerebellum is distorted in shape and appears flattened and centrally curved into the banana-like configuration (banana sign, marked by *arrows*). Another example of the banana sign and distorted cerebellum (*arrows*) that occludes cisterna magna space (**B**).

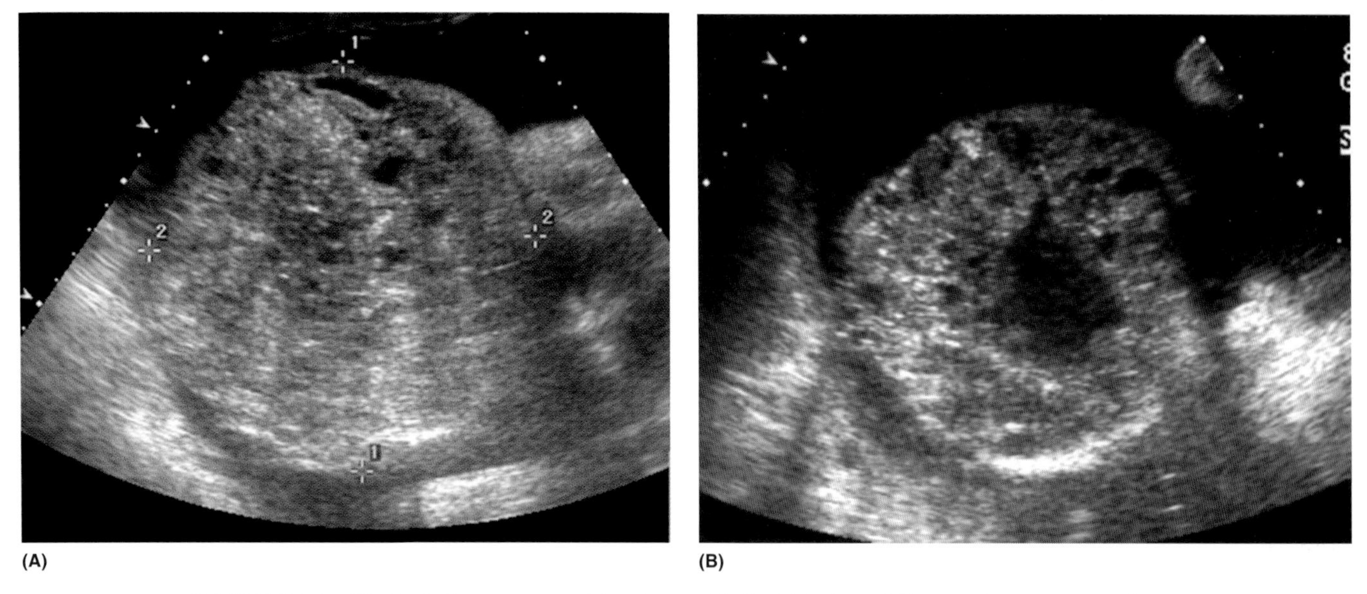

(A) (B)

FIGURE 31 A AND B ■ Sacrococcygeal teratoma. (**A**) Sagittal view of the 9 × 8 cm large teratoma, predominantly solid in echotexture, though transverse scan of the same tumor reveals centrally located cystic component (**B**).

internal with abdominal extension, and entirely internal, respectively. Types 1 and 2 include 80% of cases, though Types 3 and 4 have the highest incidence of malignancy. Most tumors are benign at birth, and the incidence of malignancy increases with increasing age (61).

Ultrasonographically, most sacrococcygeal teratomas are solid or mixed, and only 15% are entirely cystic (Fig. 31). The solid areas are composed of tissues of different densities (e.g., cartilage and liver), and they may also include calcified areas of tooth and bone. Large and solid teratomas have significant vascular component that is easily recognized by dolor Doppler ultrasound modality (Fig. 32). Therefore, these tumors may grow rapidly. In addition, tumors with predominantly cystic components have mainly irregular and angular borders formed by cavities lined with neural, respiratory, gastrointestinal, and squamous epithelium.

Anomalies, especially of the musculoskeletal, renal, and nervous systems, have been reported in association with 18% of sacrococcygeal teratomas in newborns.

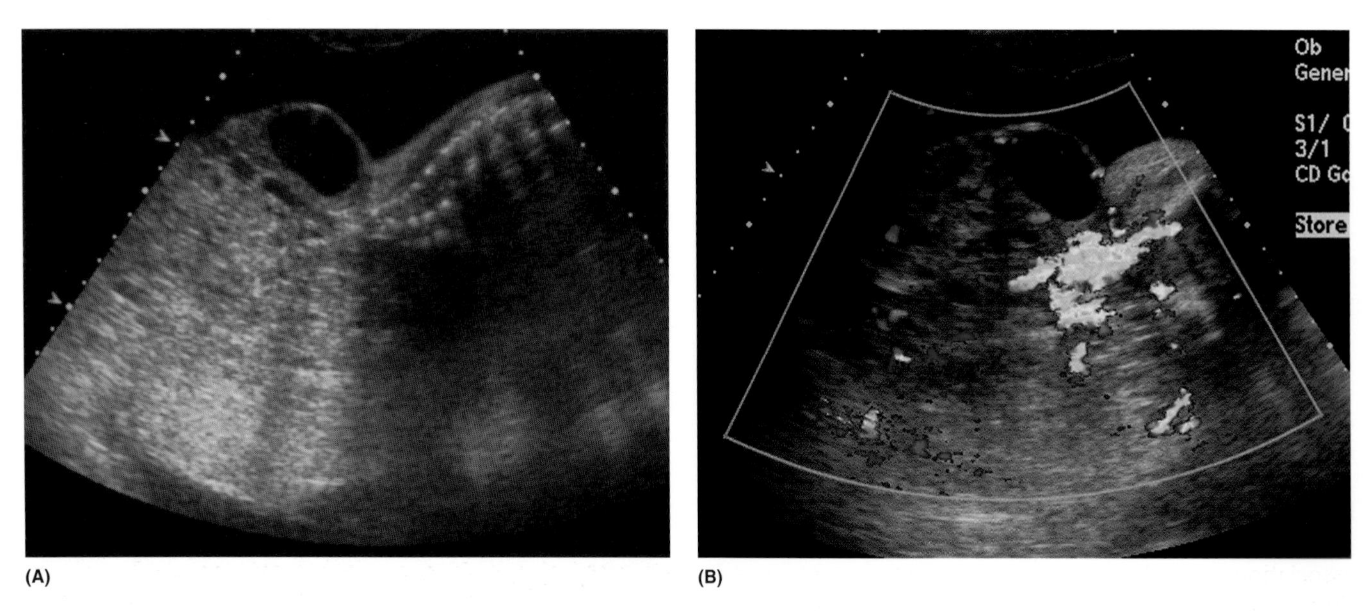

(A) (B)

FIGURE 32 A AND B ■ (**A**) Sagittal view of the sacrococcygeal, predominantly solid, teratoma that can be observed in continuation of the fetal sacral spine (sagittal, off-axis view). (**B**) Color Doppler evaluation of the sacrococcygeal teratoma reveals a significant vascular component, mainly limited to a small area of the lesion's attachment to the fetal body.

Hydrops, polyhydramnios, and placentomegaly are also commonly associated findings, especially in large tumors with significant vascular component that may cause fetal anemia due to intratumoral hemorrhage and consequently high-output cardiac failure.

The prognosis depends on associated anomalies, presence of hydrops, and histology of the lesion. Perinatal mortality rate is significant, at the range of 35%, which is more associated with solid vascular lesions and prominent internal component (52).

Scoliosis

Scoliosis is an abnormal lateral curvature of the spinal column of any cause, whereas an abnormal anterior angulation of the spine that may occur in conjunction with scoliosis is categorized as kyphosis. In the newborn, scoliosis is almost always related to congenital hemivertebra. It is commonly associated with NTD, mainly meningomyelocele. Other causes of scoliosis may include fused or splinted vertebrae, skeletal dysplasias, and the *v*ertebral, *a*nal, *c*ardiac, *t*racheal, *e*sophageal, *r*enal, and *l*imb (VACTERL) association (62). The fetal skeleton is flexible and subject to strong deformation forces within the uterus. Therefore, the pathologic diagnosis of scoliosis should be made with caution. When gross skeletal defects, including absent ribs or hemivertebra, are detected, associated lateral spinal deflection probably represents true scoliosis (63). In the absence of such defects, bends in the vertebral column should reach the extreme of 90° and should be demonstrated in serial scans as unchanging before the diagnosis of scoliosis is made.

The prognosis for fetuses with abnormal spinal curvature generally is determined from the prognosis of the underlying malformation and is usually poor. Hypoplastic thorax with chest restriction may lead to pulmonary hypoplasia and is a leading cause of death (46). There is a high frequency of stillborn fetuses and neonatal deaths.

Hemivertebra

Hemivertebrae represents a congenital anomaly of the spine in which one-half of the vertebral body mass is developed. It results from the failure (aplasia or hypoplasia) of one of the two chondrification centers to develop. During the first trimester, the vertebral body forms from parallel right and left chondrification centers. These centers coalesce during the second fetal month. The defective vertebra causes spinal axis wedge and leads to excessive lateral curvature of the spine or scoliosis (Figs. 33–35). Therefore, an ultrasound examination reveals hemivertebra as lateral displacement of the anterior ossification center from the straight-line arrangement of the other anterior ossification centers with or without apparent scoliosis (64,65).

Hemivertebra may occur as an isolated entity but may be associated with NTDs as well as with VATER (*v*ertebral, *a*norectal, *t*racheo*e*sophageal, *r*adial/*r*enal) or VACTERL associations (66). A careful study of posterior elements in three perpendicular planes is required before this relatively benign disorder is diagnosed. Amniocentesis for AFP and acetylcholinesterase determination should be considered in the search for small NTDs.

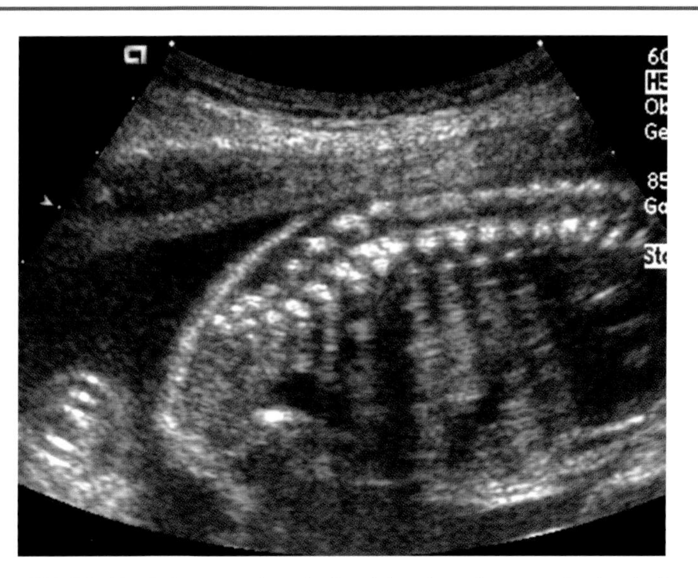

FIGURE 33 ■ A case of an isolated hemivertebrae at the level of lumbosacral spine. It appears that two lateral ossification centers on one side are fused, while one of the lateral ossification centers on the opposite side is missing. This anomaly creates scoliosis and lateral displacement of the sacral spine in relation to the rest of lumbar spine.

Although isolated congenital hemivertebra is not a life-threatening malformation, the prenatal diagnosis of this abnormality should raise the index of suspicion for other congenital defects and for orthopedic complications of vertebral body malformations in childhood. Severe scoliosis or kyphoscoliosis can develop later in childhood; therefore, the pediatrician should be alerted to watch for early signs of problems.

Caudal Regression Syndrome

Caudal regression syndrome or sirenomelic sequence represents a fusion of the early lower limb buds, with absent or incompletely developed intervening caudal structures. This is an embryologic defect that occurs during the third week of development, usually caused

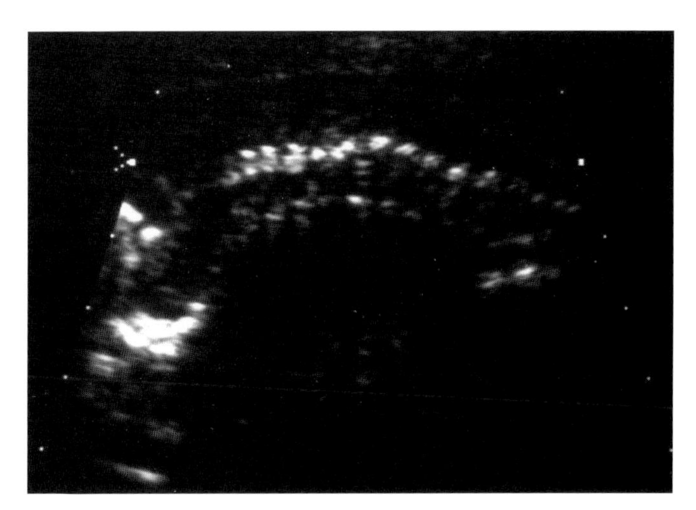

FIGURE 34 ■ Another example of hemivertebrae where several of lateral ossification centers in thoracic portion of the fetal spine are absent.

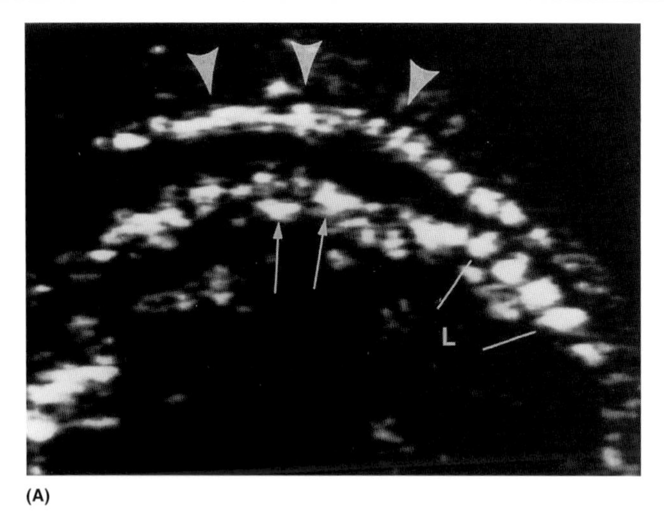

(A)

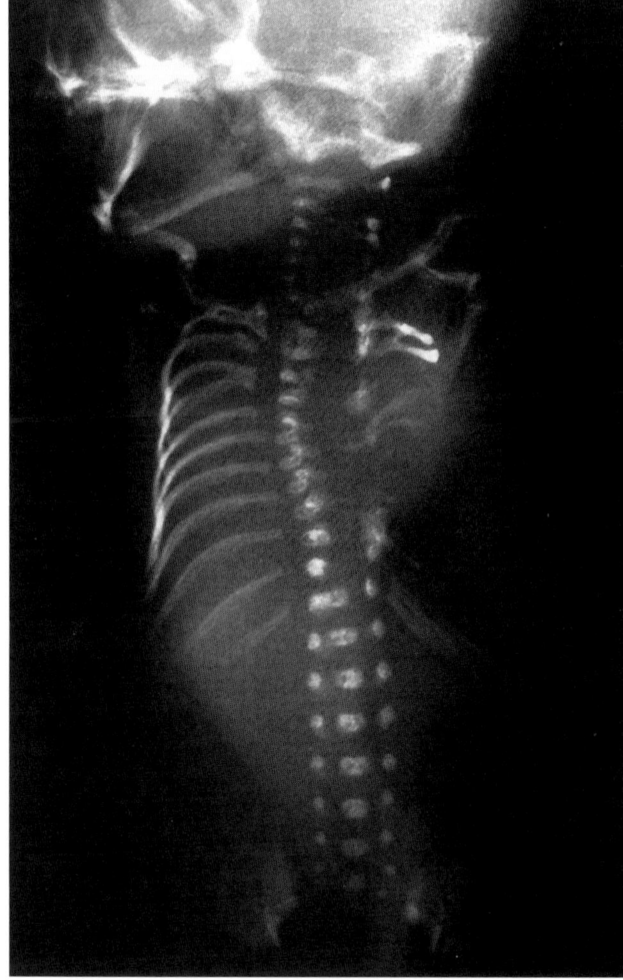

(B)

FIGURE 35 A AND B ■ Spinal scoliosis vertebral anomalies. (**A**) Longitudinal, coronal scan of fetal spine at 18 weeks confirms irregularity of the vertebral bodies and mild scoliosis. Note splaying of the spine at this level (*arrowheads*) and only a few posterior ossification centers on one side of the scoliosis (*long arrows*). (**B**) Postmortem radiograph shows marked thoracic cage deformity with mild thoracic scoliosis. L, lumbar spine. *Source*: From Ref. 46.

by wedge-shaped defect in the posterior-axis caudal blastema, which can be observed with ultrasound later in pregnancy (Fig. 36). This sirenomelic sequence is considered by some to be the most severe form of caudal regression syndrome, a defect occurring in about one in 60,000 newborns. It may result in a single umbilical artery arising directly from the aorta, lower limb fusion, imperforate anus, urologic deficits, and lower vertebral and pelvic abnormalities (including agenesis). There is a wide spectrum of severity in this disorder; imperforate anus alone represents its mildest form. Abnormalities of the abdominal wall and the genitourinary, cardiac, and pulmonary systems also have been identified ultrasonographically (67,68). Caudal regression syndrome is strongly associated with poorly controlled insulin-dependent diabetes mellitus and monozygotic twin gestation. The relative risk for this rare lesion is increased 200-fold in infants of diabetic women but still it has been observed in two in 1000 pregnancies complicated by diabetes mellitus (69). The potential pitfall of caudal regression is various forms

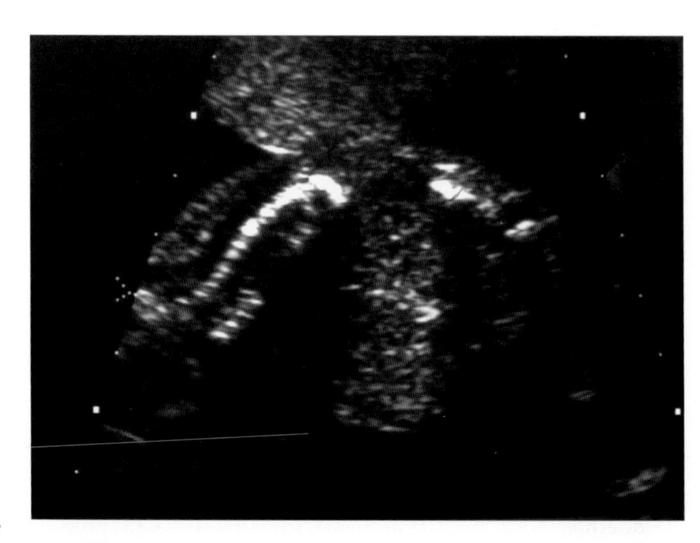

FIGURE 36 ■ Sagittal, off-axis, ultrasound view of the lumbosacral spine. Note abrupt ending of the thoracolumbar spine ("U" shaped), and absence of lumbosacral segment of the fetal spine (empty space between the spine and iliac bone).

of conjoined twins, who may be separate at the skull base but fused from the lumbosacral spine caudally.

Depending on the presence of associated anomalies, the prognosis of fetuses with caudal regression is poor. The majority of the perinatal deaths are related to the associated anomalies of the renal and cardiorespiratory systems (68). Because survivors sustain major degrees of neurologic impairment and bladder dysfunction, appropriate counseling and intervention should be performed at the time of diagnosis.

REFERENCES

1. AIUM practice guideline for the performance of an antepartum obstetric ultrasound examination. J Ultrasound Med 2003; 22:1116–1125.
2. Chervenak FA, Isaacson G, Campbell S. Ultrasound in Obstetrics and Gynecology. Boston: Little, Brown, 1992.
3. Jauniaux E, Ramsay B, Peellaerts C, et al. Perinatal features of pregnancies complicated by nuchal cord. Am J Perinatol 1995; 12:255.
4. Assimakopoulos E, Zafrakas M, Garmiris P, et al. Nuchal cord detected by ultrasound at term is associated with mode of delivery and perinatal outcome. Eur J Obstet Gynecol Reprod Biol 2005. Epub ahead of print.
5. Peregrine E, O'Brien P, Jauniaux E. Ultrasound detection of nuchal cord prior to labor induction and the risk of Cesarean section. Ultrasound Obstet Gynecol 2005; 25(2):160–164.
6. LeScale KB, Eddleman KA, Chervenak FA. The fetal neck and spine. In: McGahan JP, Porto M, eds. Diagnostic Obstetrical Ultrasound. 1st ed. Philadelphia: JB Lippincott, 1994:172.
7. Byrne J, Blank WA, Warburton D, et al. The significance of cystic hygroma in fetuses. Hum Pathol 1984; 15:61–67.
8. Trauffer PM, Anderson CE, Johnson A, et al. The natural history of euploid pregnancies with first-trimester cystic hygromas. Am J Obstet Gynecol 1994; 170:1279–1284.
9. Chervenak FA, Isaacson G, Blakemore KJ, et al. Fetal cystic hygroma: cause and natural history. N Engl J Med 1983; 309:822.
10. Van der Putte SCJ. The development of the lymphatic system in man. Adv Anat Embryol Cell Biol 1975; 51:3.
11. Chervenak FA, Isaacson G, Lorber J. Anomalies of the Fetal Head, Neck and Spine: Ultrasound Diagnosis and Management. Philadelphia: WB Saunders, 1988.
12. Bernstein HS, Filly RA, Goldberg JD, et al. Prognosis of fetuses with a cystic hygroma. Prenatal Diagn 1991; 11:349.
13. Cullen MT, Gabrielli S, Green JJ, et al. Diagnosis and significance of hygroma in the first trimester. Prenat Diagn 1990; 10:643.
14. Shulman LP, Emerson DS, Felker RE, et al. High frequency of cytogenetic abnormalities in fetuses with cystic hygroma diagnosed in the first trimester. Obstet Gynecol 1992; 80:80.
15. Johnson MP, Johnson A, Holzgreve W, et al. First trimester simple hygroma: cause and outcome. Am J Obstet Gynecol 1993; 168:156.
16. Shaub M, Wilson R, Collea J. Fetal cystic lymphangioma (cystic hygroma) ultrasound findings. Radiology 1976; 121:449.
17. Lee CY, Madrazo BL, Van Dyke DL, et al. Prenatal diagnosis of fetal cystic hygromas associated with generalized lymphangiectasis. Henry Ford Hosp Med J 1981; 29:93.
18. Ingraham FD, Swah H. Spina bifida and cranium bifidum. I. A survey of five hundred and forty six cases. N Engl J Med 1943; 228:559.
19. David D, Proudman T. Cephaloceles: classification, pathology and management. World J Surg 1989; 13:349.
20. Budorick NE, Pretorius DH, McGahan MC, et al. Cephalocele detection in utero: sonographic and clinical features. Ultrasound Obstet Gynecol 1995; 55:77.
21. Bronshtein M, Zimmer EZ. Transvaginal sonographic follow-up on the formation of fetal cephaloceles at 13–19 weeks' gestation. Obstet Gynecol 1991; 78:528.
22. Van Zalen-Sprock MM, van Vugt JMG, van der Harten HJ, et al. Cephalocele and cystic hygroma: diagnosis and differentiation in the first trimester of pregnancy with transvaginal sonography. Report of two cases. Ultrasound Obstet Gynecol 1992; 2:289.
23. Mehta PS, Mehta SJ, Vorherr H. Congenital iodide goiter and hypothyroidism: a review. Obstet Gynecol Surv 1983; 38:237.
24. Bromley B, Frigoletto FD Jr, Cramer D, et al. The fetal thyroid: normal and abnormal sonographic measurements. J Ultrasound Med 1992; 11:25.
25. Wenstrom KD, Weiner CP, Williamson RA, et al. Prenatal diagnosis of fetal hyperthyroidism using funipuncture. Obstet Gynecol 1990; 76:809–814.
26. Polak M, Leger J, Luton D, et al. Fetal cord blood sampling in the diagnosis and the treatment of fetal hyperthyroidism in the offsprings of an euthyroid mother, producing thyroid stimulating immunoglobulins. Ann Endocrinol 1997; 58:338–342.
27. Gonzalez-Crussi F. Extragonadal teratomas. In: Atlas of Tumor Pathology, Series 2. Bethesda: Armed Forces Institute of Pathology, 1982:1.
28. Teal LN, Angtuaco TL, Himenez JF, et al. Fetal teratomas: antenatal diagnosis and clinical management. J Clin Ultrasound 1988; 16:29.
29. Rosenfeld CR, Coln CD, Duenhoelter JH. Fetal cervical teratoma as a cause of polyhydramnios. Pediatrics 1979; 64:176.
30. Shipp TD, Bromley B, Benacerraf B. The ultrasonographic appearance and outcome for fetuses with masses distorting the fetal face. J Ultrasound Med 1995; 14:673.
31. McGahan JP, Schneider JM. Fetal neck hemangioendothelioma with secondary hydrops fetalis: sonographic diagnosis. J Clin Ultrasound 1986; 14:384.
32. Benacerraf BR, Frigoletto FD Jr, Laboda LA. Sonographic diagnosis of Down's syndrome in the second trimester. Am J Obstet Gynecol 1985; 153:49.
33. Nicolaides KH, Azar G, Snijders RJM, et al. Fetal nuchal oedema: associated malformations and chromosomal defects. Fetal Diagn Ther 1992; 7:123.
34. Ville Y, Lalondrelle C, Doumerc S, et al. First-trimester diagnosis of nuchal anomalies: significance and fetal outcome. Ultrasound Obstet Gynecol 1992; 2:314.
35. Pajkrt E, Bilardo CM, Van Lith JM, et al. Nuchal translucency measurement in normal fetuses. Obstet Gynecol 1995; 86:994.
36. Hrgovic Z, Panitz HG, Kurjak A, et al. Contribution to the recognition of iniencephaly on the basis of a new case. J Perinat Med 1989; 17:375.
37. Foderaro AE, Abu-Yousef MM, Benda JA, et al. Antenatal ultrasound diagnosis of iniencephaly. J Clin Ultrasound 1987; 15:550.
38. Meizner I, Bar-Ziv J. Prenatal ultrasonic diagnosis of a rare case of iniencephaly apertus. J Clin Ultrasound 1987; 15:200.
39. O'Rahilly R, Muller F, Meyre DB, The human vertebral column at the end of the embryonic period proper: the column as a while. J Anat 1980; 131:565–575.
40. Cochlin DL. Ultrasound of the fetal spine. Clin Radiol 1982; 33:641–650.
41. Dennis MA, Drose JA, Pretorius DH, et al. Normal fetal sacrum simulating spina bifida: "pseudodysraphism." Radiology 1985; 155:751.
42. Nicolaides KH, Campbell S, Gabbe SG, et al. Ultrasound screening for spina bifida: cranial and cerebellar signs. Lancet 1986; 2:72.
43. Budorick NE, Pretorius DH, Nelson TR. Sonography of the fetal spine: technique, imaging findings, and clinical implications. AJR Am J Roentgenol 1995; 164:421.
44. Miller CE. Elevated maternal serum AFP and normal ultrasound: what next? Semin Ultrasound CT MR 1993; 14:31.
45. Nadel AS, Green JK, Holmes LB, et al. Absence of need for amniocentesis in patients with elevated levels of maternal serum alpha fetoprotein and normal ultrasonographic examinations. N Engl J Med 1990; 323:559.
46. Harrison LA, Pretorius DH, Budorick NE. Abnormal spinal curvature in the fetus. J Ultrasound Med 1992; 11:473.
47. Youmans JR. Neurological Surgery. Philadelphia: WB Saunders, 1982:1082.
48. Lorber J, Ward AM. Spina bifida: a vanishing nightmare? Arch Dis Child 1985; 60:1086.

49. Stein SC, Feldman JG, Friedlander M, et al. Is myelomeningocele a disappearing disease? Pediatrics 1982; 69:511.
50. Main DM, Menmuti MT. Neural tube defects: issues in prenatal diagnosis and counseling. Obstet Gynecol 1986; 67:1.
51. Watson WJ, Cheschier NC, Katz VL, et al. The role of ultrasound in the evaluation of patients with elevated maternal serum alpha-fetoprotein: a review. Obstet Gynecol 1991; 76:123–128.
52. Crane JP, LeFevre ML, Winborn RC, et al. A randomized trial of prenatal ultrasonographic screening: impact on the detection, management, and outcome of anomalous fetuses. The RADIUS Study Group. AM J Obstet Gynecol 1994; 171:392–399.
53. Boyd PA, Wellesley DG, De Walle HE, et al. Evaluation of the prenatal diagnosis of neural tube defects by fetal ultrasonographic examination in different centers across Europe. J Med Screen 2000; 7:169–174.
54. Silverman FN. Caffey's pediatric x-ray diagnosis: an integrated teaching approach. Chicago: Year Book Medical Publication, 1985, 295:298.
55. Winter RK, Mc Knight L, Byrne RA, et al. Diastematomyelia: prenatal ultrasonic appearances. Clin Radiol 1989; 40:291–294.
56. Anderson NG, Jordan S, McFarlane MR, et al. Diastematomyelia: diagnosis by prenatal sonography. AJR Am J Roentgenol 1994; 163:911–914.
57. Dubowitz W, Lorber J, Zachary RBZ. Lipoma of the cauda equina. Arch Dis Child 1965; 40:207.
58. Seeds JW, Jones FD. Lipomyelomeningocele: prenatal diagnosis and management. Obstet Gynecol 1986; 67:34S.
59. Gross RE, Clatworthy HW Jr, Meeker IA. Sacrococcygeal teratomas in infants and children. A report of 40 cases. Surg Gynecol Obstet 1951; 92:341.
60. Flake AW, Harrison MR, Adzick NS, et al. Fetal sacrococcygeal teratoma. J Pediatr Surg 1986; 21:563.
61. Donnellan WA, Sewnson O. Benign and malignant sacrococcygeal teratomas. Surgery 1968; 64:834.
62. Nyberg DA, Mack LA. The spine defects. In: Nyberg DA, Mahoney BS, Pretorius DH, eds. Diagnostic Ultrasound of Fetal Anomalies. Chicago: Year Book Medical Publication, 1990:192.
63. Patten RM, Van Allen M, Mack LA, et al. Limb-body wall complex: in utero sonographic diagnosis of a complicated fetal malformation. AJR Am J Roentgenol 1986; 146:1019.
64. Benacerraf BR, Greene MF, Barss VA. Prenatal sonographic diagnosis of congenital hemivertebrae. J Ultrasound Med 1986; 5:257.
65. Abrams SL, Filly RA. Congenital malformations: prenatal diagnosis using ultrasonography. Radiology 1985; 155:762.
66. McGahan JP, Leeba JM, Lindfors KK. Prenatal sonographic diagnosis of VATER association. J Clin Ultrasound 1988; 16:588.
67. Elejalde MM, Elejalde BF. Visualization of the fetal spine: a proposal of a standard to increase reliability. Am J Med Genet 1985; 21:445.
68. Loewy JA, Richards DG, Toi A. In-utero diagnosis of the caudal regression syndrome: report of three cases. J Clin Ultrasound 1987; 15:469.
69. Sonek JD, Gabbe SG, Landon MB, et al. Antenatal diagnosis of sacral agenesis syndrome in a pregnancy complicated by diabetes mellitus. Am J Obstet Gynecol 1990; 162:806.

Fetal Thorax

Ashley J. Robinson and Ruth B. Goldstein

46

INTRODUCTION

Some important pathologic conditions of the fetal thorax are detectable with antenatal sonography. Published guidelines for the second- and third-trimester obstetric sonogram recommend, at a minimum, a limited examination of the fetal chest that includes a four-chamber view of the heart (1). Besides the detection of cardiac malformations, an important benefit of including this view is in enhancing the detection of noncardiac thoracic malformations. This chapter focuses on antenatal detection of fetal chest abnormalities using ultrasound, with an emphasis on differential diagnosis, characteristic and distinguishing features, and associated malformations.

EMBRYOLOGY

Many sonographically detectable pulmonary malformations can be linked to the interruption of normal developmental sequences. During the first five weeks of gestation, the lung buds grow out of the ventral aspect of the primitive foregut. If these buds do not form, lung agenesis results. The trachea and esophagus become separated by the fifth gestational week. Buds from the early trachea form and penetrate the mesenchymal masses destined to become the lungs. An abnormal budding of a segment of the tracheobronchial tree may result in formation of a bronchogenic cyst. Between 5 and 16 weeks (the pseudoglandular period), the bronchial tree is formed. Through a series of divisions and budding, bronchi give rise to bronchioles, and each terminal bronchiole later gives rise to alveolar ducts and alveoli. By 16 weeks, the formation of the bronchial tree is essentially complete (2,3). Insults to the lung before then result in fewer-than-expected bronchi. Bronchial airway generations are reduced in such conditions as congenital diaphragmatic hernia (CDH), renal agenesis, absent phrenic nerve, rhesus isoimmunization, and idiopathic lung hypoplasia (3).

Between 16 and 24 weeks (the canalicular period), one sees a dramatic increase in the number and complexity of air spaces, large blood vessels, and capillaries. Insults to the lungs during this phase result in smaller airways and a reduction in the number and size of acini.

After 24 weeks, terminal sacs and alveoli continue to develop and mature (the alveolar period), and the number and complexity of the airspaces are further increased. The last 16 weeks of gestation are referred to as the terminal sac period (4). Type II pneumocytes, the surfactant-producing cells, mature between 32 and 36 weeks. The number of alveoli continues to grow during childhood through alveolar multiplication, after which alveolar expansion becomes the major means of lung growth until adolescence (5). Investigators have speculated that if an abnormality were removed or interrupted before this stage (the alveolar period), the normal process of air space maturation could potentially be renewed. Fetal surgical resections of lung masses and repair of CDHs have been performed for this purpose.

NORMAL ANATOMY OF THE FETAL THORAX

Bones and Soft Tissues

The chest wall is lean and is made up of skin, muscles, and minimal fat. The ribs, smoothly marginated and regularly spaced, form the lateral boundaries and extend anteriorly more than halfway around the thorax from their dorsal attachments (Fig. 1). The ribs also extend inferiorly to the upper abdomen. Therefore, observation of the ribs and abdominal structures in the same transverse plane should not be misinterpreted as pathologic.

Diaphragm

The fetal thoracic cavity is bell shaped and is bordered by the clavicles at the apex and by the smooth, hypoechoic diaphragm inferiorly (Fig. 2). The hypoechoic muscular diaphragm separates the lungs from the abdominal contents. It is most easily seen on the coronal and sagittal views (Fig. 3).

Intrathoracic Contents

On a four-chamber view, the heart occupies approximately one-third of the thoracic volume, and most of the cardiac volume is located in the left anterior quadrant of the chest on an axial plane at the level of the four-chamber view (Fig. 4) (6). On this image, no abdominal viscera should be visualized. Oblique images through the chest and abdomen, however, may include abdominal viscera and heart on the same image and, therefore, may be misleading and may cause the examiner to misinterpret normal findings as pathologic, such as false diagnosis of a diaphragmatic hernia. A coronal or sagittal image should be obtained in these cases to confirm the normal abdominal location of the fetal stomach. With real-time imaging, mediastinal and pulmonary vessels can be easily

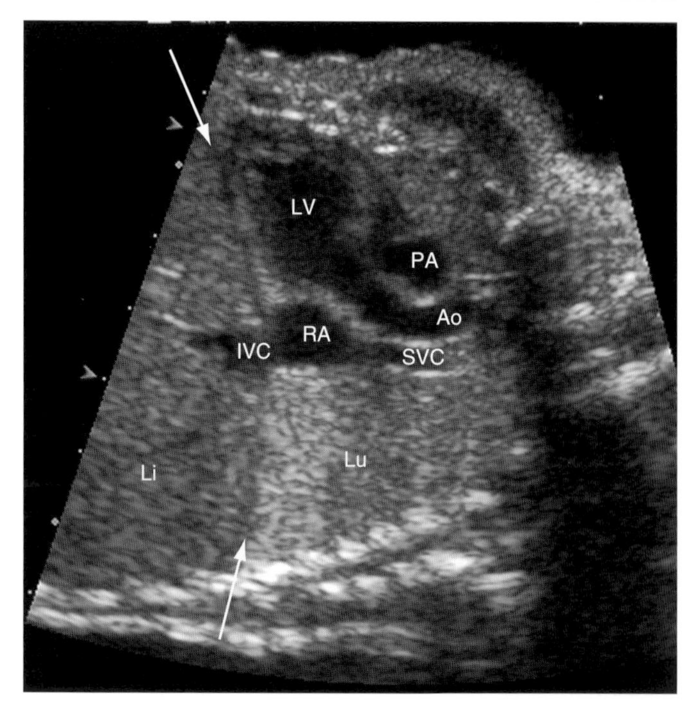

FIGURE 2 ■ Normal fetal chest: coronal oblique. The muscular diaphragm is hypoechoic (*arrows*) and is seen to divide the thorax from the abdomen, with a gap at the hiatus for the inferior vena cava. LV, left ventricle; Ao, aorta; SVC, superior vena cava; RA, right atrium; PA, pulmonary artery; Lu, lung; Li, liver; IVC, inferior vena cava.

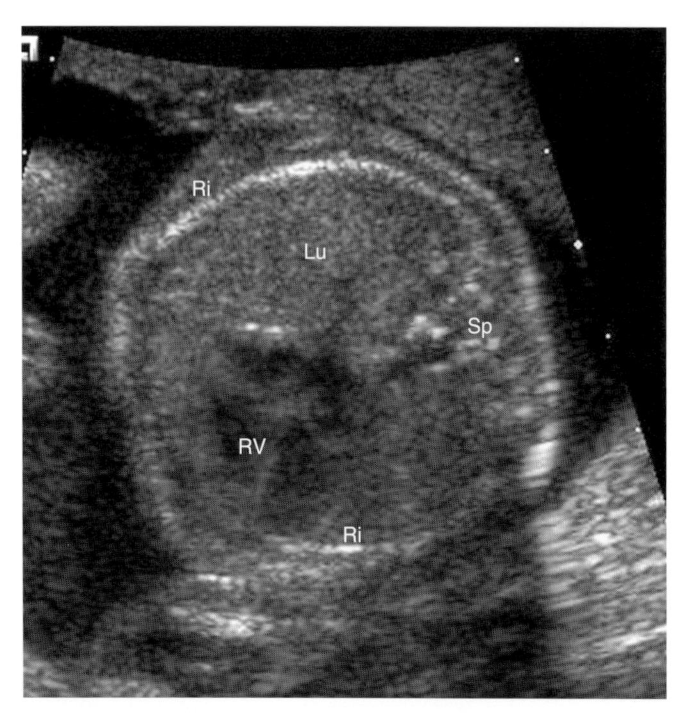

FIGURE 1 ■ Normal fetal chest: axial. The smoothly marginated ribs extend anteriorly more than halfway around the thorax. Ri, ribs; RV, Right ventricle; Sp, spine; Lu, lungs.

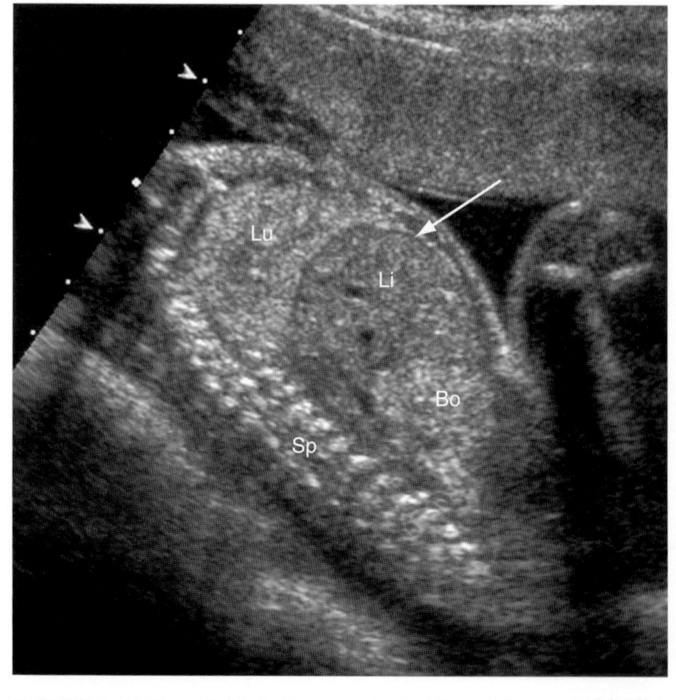

FIGURE 3 ■ Normal fetal chest: sagittal oblique. The muscular diaphragm (*arrow*) is visible completely separating the liver from the lung. Bo, bowel; Sp, spine; Li, liver; Lu, lung.

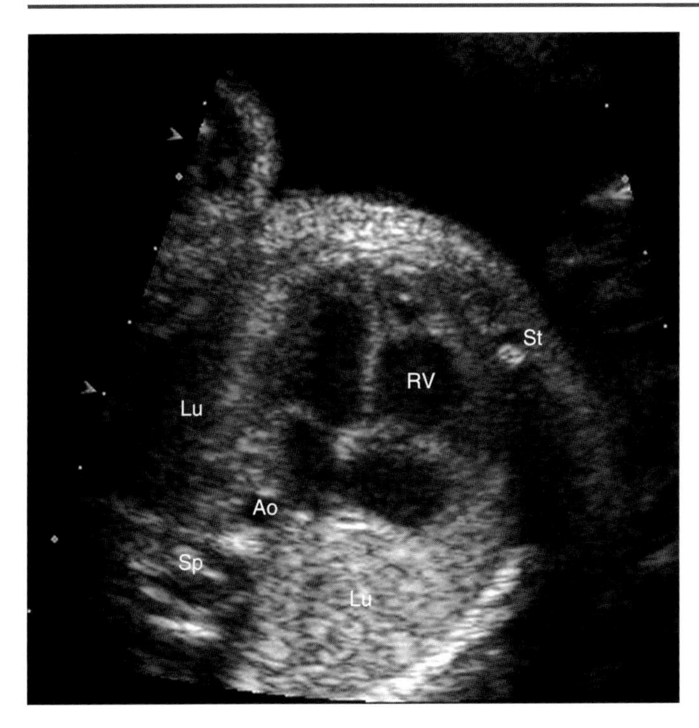

FIGURE 4 ■ Normal fetal chest: axial. The heart occupies approximately one-third of the total area and most of the cardiac volume lies within the left anterior quadrant. Ossification centers in the sternum are visible anteriorly, and those in the spine posteriorly. St, sternum; Sp, spine posteriorly; RV, Right ventricle; Ao, aorta; Lu, lungs.

appreciated; pulmonary veins can be followed to the left atrium to confirm normal venous connections to the heart. The fetal thymus is usually not distinguishable unless large pleural effusions are present.

Normal fetal lungs are homogeneous in echotexture, and echogenicity may be greater than, less than, or equal to the echogenicity of the liver. As a general observation, the echogenicity of fetal lungs tends to increase progressively to an echogenicity greater than that of the liver during gestation (7).

ABNORMAL ANATOMY OF THE FETAL THORAX

A few clinically important thoracic malformations have been observed prenatally. Both extrinsic (i.e., abnormalities of the bony thorax) and intrathoracic abnormalities can interfere with normal lung growth and development.

Important intrathoracic abnormalities include those originating in the mediastinum, pleural effusions, intrinsic masses of the lung, and diaphragmatic hernias. Among intrathoracic abnormalities, CDH is the most common, followed by cystic adenomatoid malformation (CAM), bronchopulmonary sequestration (BPS), bronchogenic cyst, and bronchial atresia.

Although specific antenatal diagnosis is not always possible, diagnostic consideration can be prioritized based on the location, sonographic characteristics of the lesion, and associated malformations. The natural history, associated malformations, and characteristic sonographic appearances of these intrathoracic abnormalities are discussed in the following paragraphs.

Thoracic Wall and Mediastinal Abnormalities

Bony Abnormalities

Bony thoracic abnormalities usually present as part of a multisystemic or generalized fetal abnormality. The fetal thorax may be abnormally small or misshapen, with fractured or shortened ribs (8–14). The latter should raise the suspicion of a primary skeletal dysplasia. A small but otherwise normal-appearing thorax usually has another cause such as severe and prolonged oligohydramnios or severe fetal neurologic disorder.

Soft-Tissue Abnormalities

The most common cause of diffuse thickening of the subcutaneous tissues is integumentary edema associated with hydrops or lymphangiectasia. Skin thickening is also seen in fetuses with thanatophoric dysplasia, harlequin ichthyosis (15,16) and neurologic disorders such as myotonic dystrophy and Pena Shokeir. Examiners should interpret the thickening of the subcutaneous tissues of the middle to upper chest cautiously, especially thickening of tissues that include the scapula, because normal fetal tissues in this region may appear thicker than those of the abdomen or scalp.

Masses of the subcutaneous tissue of the chest are rare and include hemangiomas, lymphangiomas (cystic hygromas), teratomas, hamartomas, breast tissue, and thoracic myelomeningoceles (17–19). With the exception of the last, these masses may be difficult to distinguish antenatally. Hamartoma of the chest wall arises from the ribs and has been detected prenatally as an intrathoracic, highly echogenic, heterogeneous mass with calcifications, associated with pleural effusion (Table 1) (20–24).

Masses in the mediastinum are rare, but fetal goiter, cystic hygroma (Fig. 5), pericardial rhabdomyoma and teratoma (Figs. 6 and 7) and hemangioma, thoracic neuroblastoma, and esophageal duplication cyst have been observed prenatally (25–29).

Finally, several abnormalities cause primary or acquired defects in the thoracic cavity, for example, limb–body wall complex and Pentalogy of Cantrell (Fig. 8).

Pleural Cavity Abnormalities

The origin of an intrathoracic mass is nearly always pulmonary, gastrointestinal (CDH or foregut malformation),

TABLE 1 ■ Abnormal Subcutaneous Tissues of the Fetal Chest

- Integumentary edema (hydrops)
- Cystic hygroma (lymphangioma)
- Hemangioma
- Hamartoma
- Fetal breasts
- Thoracic myelomeningocele

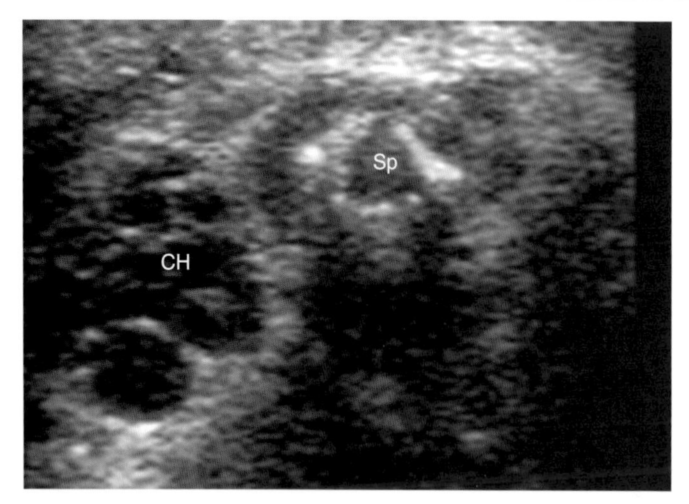

FIGURE 5 ■ Thorax wall mass: cystic hygroma. A focal loculated fluid-filled echogenic mass is seen in the axilla. CH, cystic hygroma; Sp, spine.

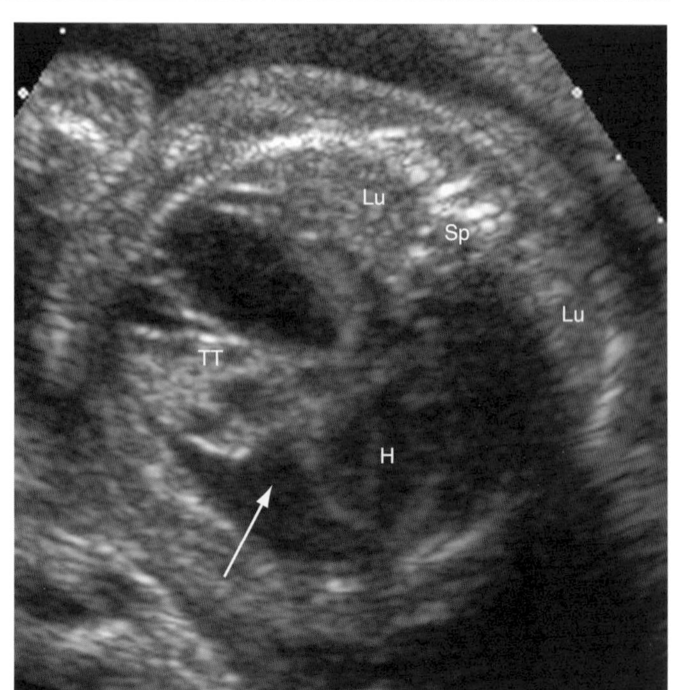

FIGURE 7 ■ Cardiac mass: teratoma. Large exophytic cystic mass arising from the heart, with pericardial effusion (*arrow*), and compression of the lungs posteriorly. TT, cystic mass; H, heart; Lu, lungs; Sp, spine.

or mediastinal. These masses frequently require surgical removal postnatally, and, as a result of mass effect on the developing lung and cardiomediastinal structures, they often result in pulmonary hypoplasia and fetal hydrops. Nevertheless, roughly two-thirds of all fetuses with antenatally detected chest masses survive the prenatal period (80%, if terminated pregnancies are excluded) (28). In general, poor prognostic indicators in fetuses with chest masses include a large lesion, hydrops fetalis, polyhydramnios, and concomitant malformations occurring in approximately 10% (Table 2) (28,30).

Marked mediastinal shift probably contributes to the development of hydrops by impeding cardiac venous return and elevating central venous pressure (31). Rice et al. produced an animal model of hydrops by inflating a

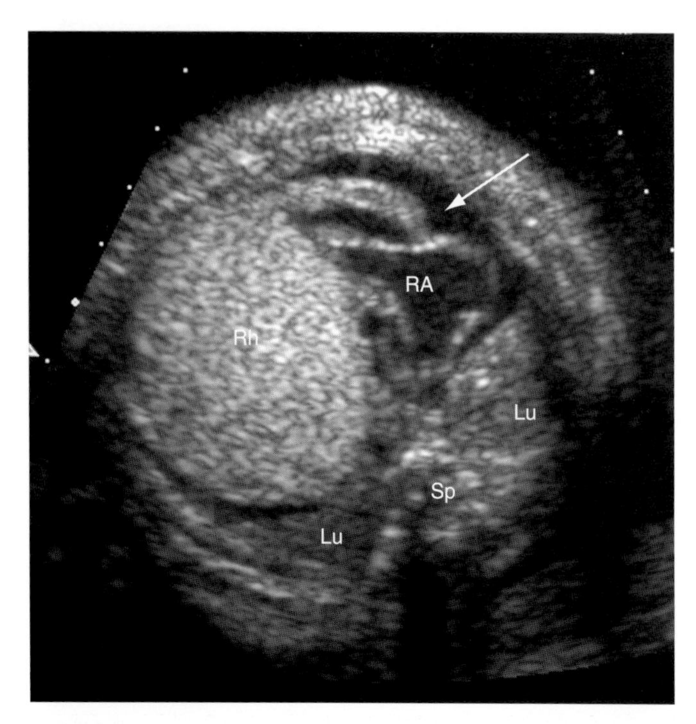

FIGURE 6 ■ Cardiac mass: rhabdomyoma. Large homogeneous echogenic mass clearly contiguous with the myocardium. There is a pericardial effusion (*arrow*) and the lungs are compressed posteriorly. Lu, lungs; Rh, rhabdomyoma; RA, right atrium; Sp, spine.

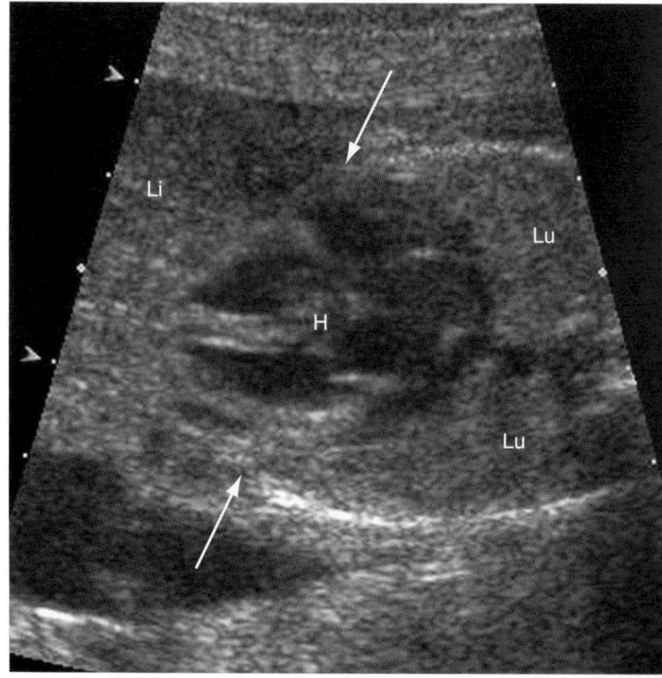

FIGURE 8 ■ Thorax wall defect: Pentalogy of Cantrell. Defect of sternum, pericardium, diaphragm, and abdominal wall allowing heart and liver to herniate through the ventral defect (*arrows*). H, heart; Li, liver; Lu, lungs.

TABLE 2 ■ Fetal Chest Masses: Prognostic Features

- Large lesion (>67% hemithorax)
- Severe cardiomediastinal shift
- Hydrops
- Associated congenital malformations
- Chromosomal abnormalities

tissue expander in the chest of fetal lambs (31). Hydrops developed as the tissue expander was inflated within the chest of the fetal lambs when the central venous pressure exceeded 14 mmHg. All evidence of hydrops disappeared when the tissue expander was deflated, in combination with the return of the central venous pressure to baseline (4–7 mmHg).

Congenital Diaphragmatic Hernia

CDH describes a defect in the diaphragm thought to result from incomplete fusion of the pleuroperitoneal membrane at 6 to 10 weeks. The developing lungs are severely compressed by the herniated viscera, resulting in nearly universal pulmonary hypoplasia, which is often lethal. The incidence in live births is estimated to be one in 3000 to 5000, but, owing to unrecognized CDH in fetal and neonatal deaths, the overall incidence is thought to be as high as one in 2200 births (32,33). CDH occurs more commonly in females (3:2) (34). The most common location of CDH is posterolateral and on the left (85–90%) (35). The size of the defect in the diaphragm varies, and in 1% to 2%, the diaphragm is completely absent (36). Associated malformations are common (25–57%), most notably cardiac defects (9–23%) (37,38) neural tube defects (28%), spinal defects, trisomies (trisomy 21 and 18, in 10% diagnosed prenatally), intestinal atresias, extralobar sequestrations (ELS), hydronephrosis and renal agenesis, anencephaly, spina bifida, and certain well-defined syndromes (i.e., Fryn's, DiGeorge, Cornelia de Lange, Apert, Goldenhar, and Beckwith–Wiedemann) (33,39–41). Most cases occur sporadically, with anecdotal reports of familial CDH (42). Sonographically, mediastinal shift is one of the first abnormalities observed. If a fluid-filled stomach cannot be detected below the diaphragm, CDH should be strongly considered in the differential diagnosis of a cystic chest mass. An intrathoracic stomach and peristalsis of the herniated small bowel are helpful in confirming that the chest mass is a hernia, although bowel loops are usually collapsed and are difficult to discern individually (Fig. 9). The herniated viscera most commonly include the stomach and bowel, and in large defects, the spleen, the left kidney, and the left lobe of the liver (Figs. 10–13).

Prognosis of fetuses with CDH is guarded. Overall mortality in prenatally diagnosed cases is variable, but it is often reported to be more than 70% (33,36,43,44) or approximately 60% if the CDH is isolated and extracorporeal membrane oxygenation (ECMO) is available (45). Poor prognostic indicators include preterm birth, additional malformations, and large hernias. Stringer et al. reported outcomes of a small group of fetuses diagnosed with CDH after birth, presumably without visceral herniations, or at least without large hernias. The

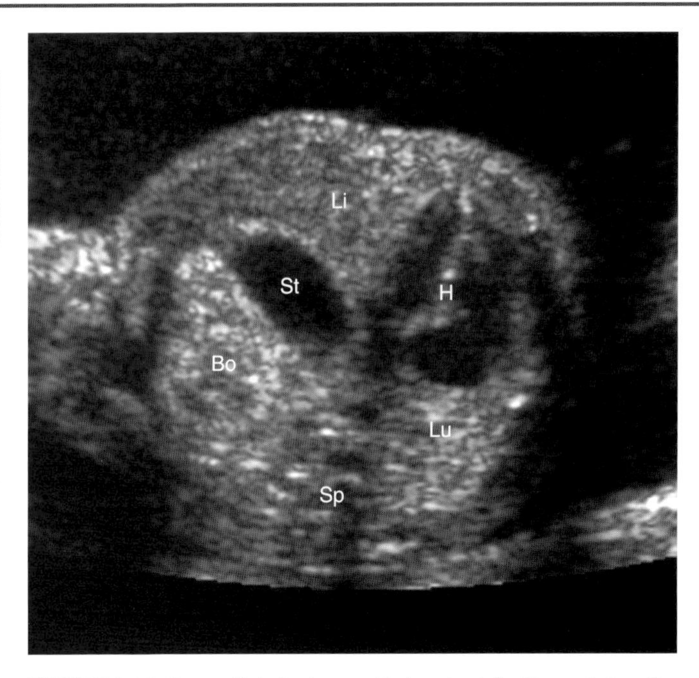

FIGURE 9 ■ Congenital diaphragmatic hernia: left. Stomach is adjacent to the heart, which is displaced to the right. Bowel is seen posteriorly to the stomach. The liver is also up in the chest. The lung ipsilateral to the defect is usually not visible. The contralateral lung is small. H, heart; Li, liver; Lu, lung; Sp, spine; St, stomach; Bo, bowel.

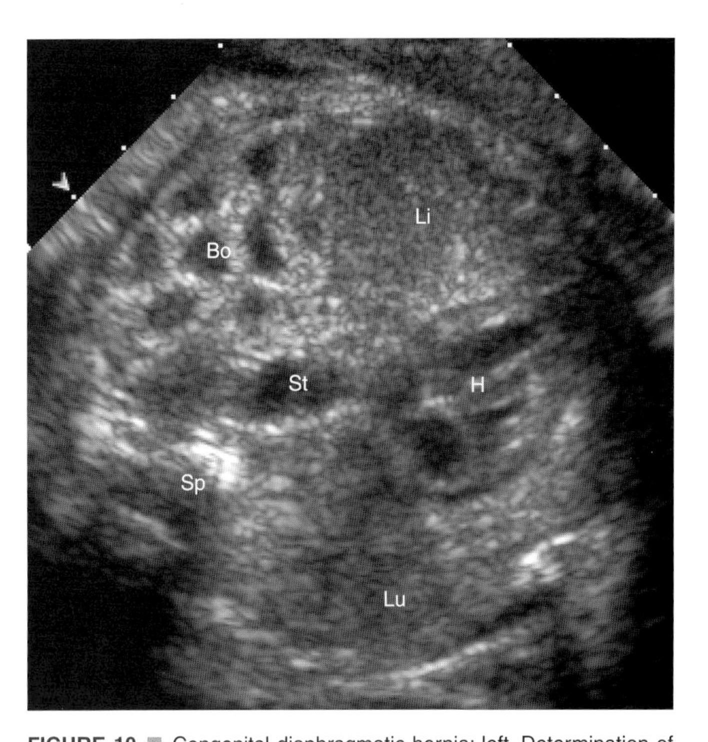

FIGURE 10 ■ Congenital diaphragmatic hernia: left. Determination of whether or not liver is herniated can be difficult. The echogenic submucosa of bowel appears as echogenic rings, even when it is empty, whereas liver is more homogeneous. Stomach displaced behind the heart usually indicates a more severe defect. H, heart; Li, liver; Lu, lung; Sp, spine; St, stomach; Bo, bowel.

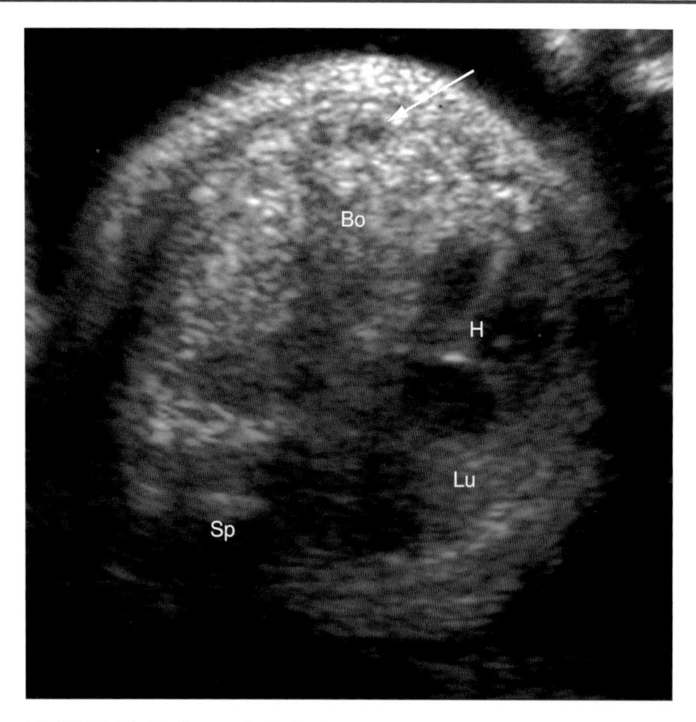

FIGURE 11 ■ Congenital diaphragmatic hernia: left. Bowel only is seen above the diaphragm. Again, the echogenic submucosa appears as echogenic rings (*arrow*). H, heart; Lu, right lung; Sp, spine; Bo, bowel.

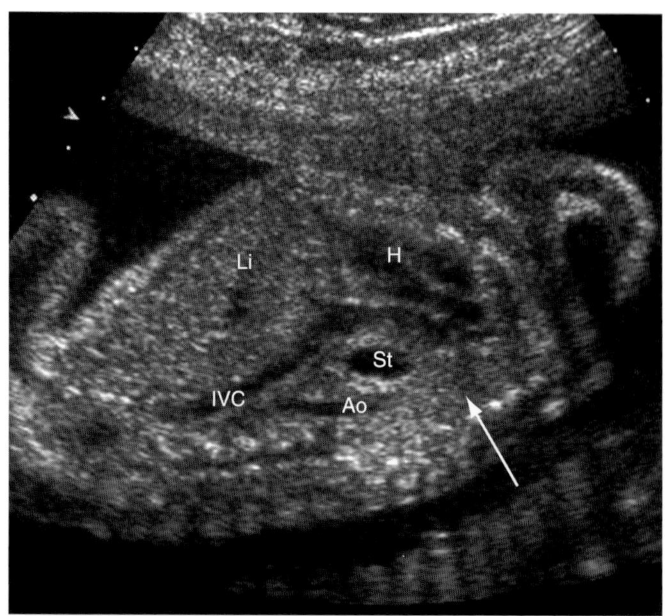

FIGURE 13 ■ Congenital diaphragmatic hernia: left. Sagittal view demonstrating liver is down in the abdomen. Stomach and bowel (*arrow*) only are herniated. Li, liver; St, stomach; Ao, aorta; H, heart; IVC, inferior vena cava.

prenatal sonograms, even retrospectively reviewed, failed to show any evidence of CDH. Outcomes in these fetuses were much better than in controls with visible visceral herniation. The survival rate among fetuses without obvious visceral herniation was 100%, and only 11% required ECMO or prosthetic diaphragmatic repair (46). Gestational age of less than 25 weeks at diagnosis also appears to be associated with poor outcome (47). Fetuses in whom the initial diagnosis is made after birth have much better outcomes than those in whom the CDH is detected antenatally. Cannon et al. reported a 78% survival rate among infants diagnosed in the neonatal period, compared with only 35% survival among those diagnosed prenatally (33).

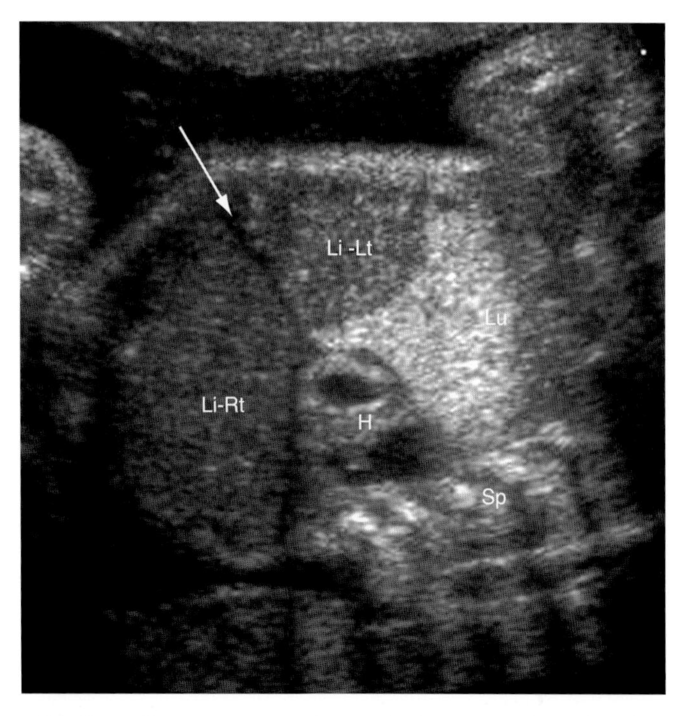

FIGURE 12 ■ Congenital diaphragmatic hernia: left. A sagittal view demonstrating the left lobe of liver above the diaphragm (*arrow*), and right lobe below. This demonstrates how the liver becomes "locked" in position by the ledge of diaphragm anteriorly. Li-Lt, left lobe of liver; Li-Rt, right lobe; H, heart; Lu, lung; Sp, spine.

Prognosis of fetuses with CDH has also been linked to the size of the contralateral lung observed on the sonogram. The right lung area has been estimated on a transaxial image of the chest that includes the four-chamber heart (Fig. 14). Survival rates in one study were higher when the contralateral (right) lung area was at least half the area of the right hemithorax (86% survival), compared with those in whom the lung area was less than half the right thoracic area (25% survival) (48). Similarly, Metkus et al. determined a ratio of right lung area to head circumference (HC) (to account for gestational age-dependent lung size) in 55 patients with prenatally diagnosed CDH. All fetuses with a lung-to-head ratio (LHR) less than 0.6 died, and all fetuses with LHRs greater than 1.35 survived (47). Other researchers have found that an LHR below 1.0 was universally fatal (49); however, there is significant intra- and interobserver variability for this measurement. It is now possible to stratify fetuses into high-risk and lower-risk groups, thus triaging those who may benefit from fetal intervention (Table 3) (47,50). Fetuses that do not have liver herniation have an excellent chance of survival

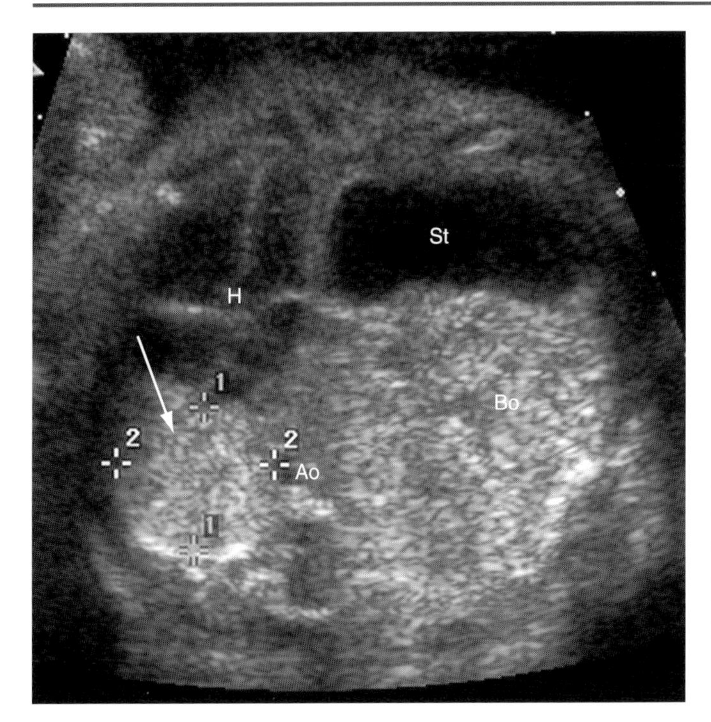

FIGURE 14 ■ Congenital diaphragmatic hernia: left. Axial view showing absence of liver from the thorax—only stomach and bowel are herniated. This is still a severe defect with heart displaced to the right. The contralateral lung (*arrow*) area is being measured to determine the lung: head ratio. Note that the coronal dimension should be measured from the side of the aorta to the chest wall, and the sagittal dimension from the posterior chest wall to the posterior aspect of the cardiac atria. Bo, bowel; St, stomach; Li, liver; H, heart.

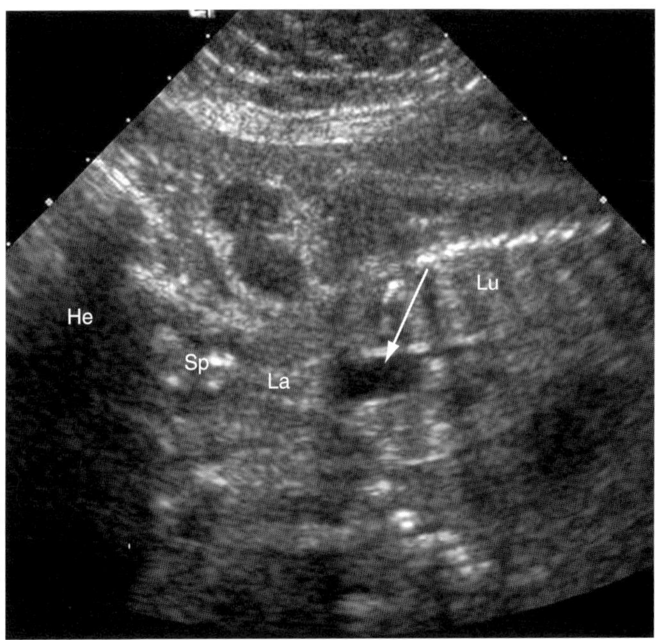

FIGURE 15 ■ Congenital diaphragmatic hernia. Post occlusion of the trachea by endoscopically placed balloon (*arrow*). La, larynx; Sp, cervical spine; He, head; Lu, lung.

with postnatal therapy (51) and are not considered for antenatal therapy. If the liver is demonstrated to be within the abdomen, the LHR is not predictive (230). Fetuses with liver herniation and an unfavorable LHR have a grim prognosis. However, fetal endoscopic (FETENDO) temporary occlusion (FETO) using a detachable tracheal balloon is a promising new technique for improving lung growth and development (52,53), leading to a possible decrease in mortality in fetuses with an LHR below 1.0 (Figs. 15 and 16) (51).

TABLE 3 ■ Congenital Diaphragmatic Hernia: Poor Prognostic Indicators

- Large defect
- Dilated intrathoracic stomach
- Prenatal diagnosis, especially <24 weeks
- Intrathoracic liver (liver up)
- Hydrops
- Other congenital malformations
- Small contralateral lung (LHR < 1.0)
- Bilateral defects
- Preterm delivery

Abbreviation: LHR, lung-to-head ratio.

Above an LHR of 1.0, the improvement in survival may plateau (54). Placement of a balloon has been shown to lead to an increase in LHR from a median of 0.7 to 1.8 within two weeks following surgery, with a possible

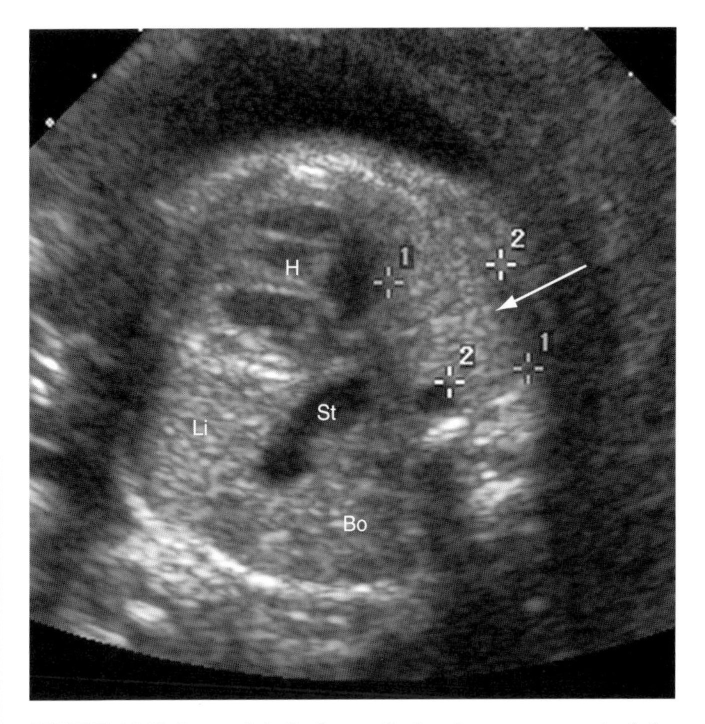

FIGURE 16 ■ Congenital diaphragmatic hernia: measurement of the contralateral lung area (*arrow*) as part of the lung: head ratio, post balloon occlusion in a severe hernia with liver up in the thorax. The ratio significantly increased post occlusion. Bo, bowel; St, stomach; Li, liver; H, heart.

improvement in postnatal survival, although numbers are currently too small to be certain (55).

Some degree of pulmonary hypoplasia is usually present in fetuses with CDHs, and it is the leading cause of perinatal mortality. As with CAMs and BPSs, most investigators believe that compression of the developing lung by the herniated viscera interrupts normal pulmonary development. Morphometric analyses of neonates with CDH confirm an arrest in pulmonary development (reduced number of bronchial branches and alveoli), and infants dying of a large CDH usually have lung weights 15% to 40% of those expected for gestational age. The pulmonary vascular bed is also abnormal, with a reduced number of vessels and increased medial muscular hyperplasia in small peripheral arteries. The latter is thought to contribute to persistent fetal circulation, which is often seen in newborns with CDH. The greatest effect is on the lung ipsilateral to the hernia, but significant and similar abnormalities are also noted on the contralateral lung.

Despite improved postnatal care and ECMO, pulmonary hypoplasia and persistent fetal circulation remain important causes of neonatal mortality among newborns with CDH. Surgical repair of CDH by hyperinflating the developing lungs have been attempted in utero to promote lung development. Thus far, however, in utero surgical repairs of the diaphragmatic defect have not improved survival rates, and this method has been abandoned in favor of tracheal occlusion. When the liver was herniated into the chest, a prohibitively high intraoperative mortality was reported in fetal diaphragmatic repairs, because of kinking of the sinus venosus during intraoperative manipulation of the liver (56). As a result, intrathoracic liver is now believed to be a contraindication to in utero repair of the diaphragmatic defect. Sonographically, liver usually herniates anterior to the stomach. Herniated liver can be definitively identified within the chest most accurately by the presence and path of portal and hepatic veins within the thorax in association with a posteriorly positioned stomach within the thorax (Fig. 17) (57,58). The presence of intrathoracic liver is worth noting during the prenatal sonogram even if prenatal surgical correction is not contemplated, because intrathoracic liver is a poor prognostic indicator and has also been reported to adversely influence prognosis and predict the necessity for ECMO postnatally (50).

Right-sided posterolateral hernias are much less common than those on the left. The right lobe of the liver most commonly herniates (59), and the small bowel and kidney are thus affected less commonly (Fig. 18). Similar survival rates are seen between fetuses with right-sided CDHs and those with left-sided CDHs (48,59). Right-sided hernias are less conspicuous, owing to similar echotexture of the liver and lung, and therefore right-sided CDHs are more difficult to detect with sonography than those on the left. The presence of a "solid" right-sided chest mass displacing the heart to the left is usually the first sign of a right-sided CDH. Ipsilateral "pleural effusions," which represent a herniated peritoneal sac and ascites, can be associated (60). A helpful observation is a small, single

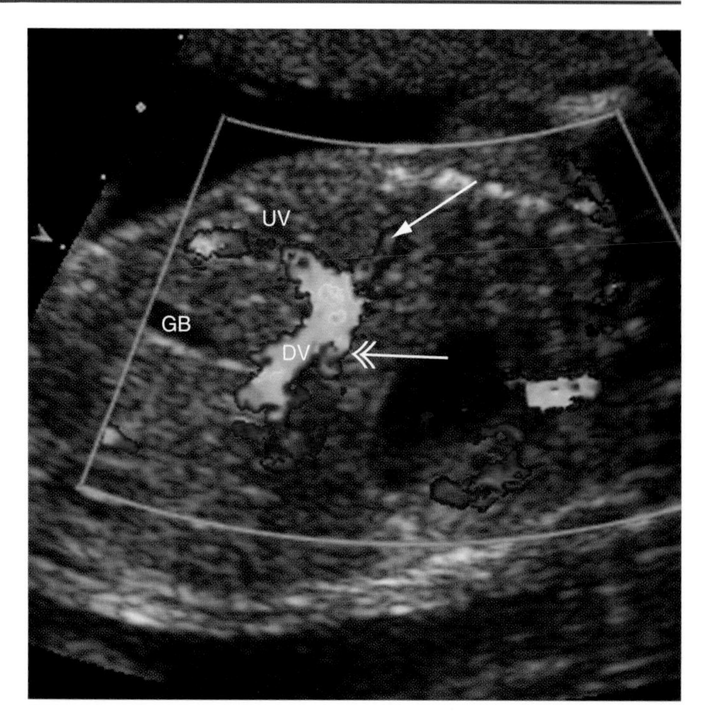

FIGURE 17 ■ Congenital diaphragmatic hernia: left. Sagittal color Doppler image demonstrating superior bowing of the umbilical vein and ductus venosus, an indirect indicator that the liver is herniated up in the thorax. The umbilical vein is fixed at the abdominal wall and the ductus venosus is fixed at the hepatic venous confluence; therefore, only the vessel in between these points is free to move. The portal vein branch to hepatic segment 3 is pointing superiorly (*single arrow*), and the branch to segment 2 is also seen (*double arrow*). UV; umbilical vein; DV, ductus venosus.

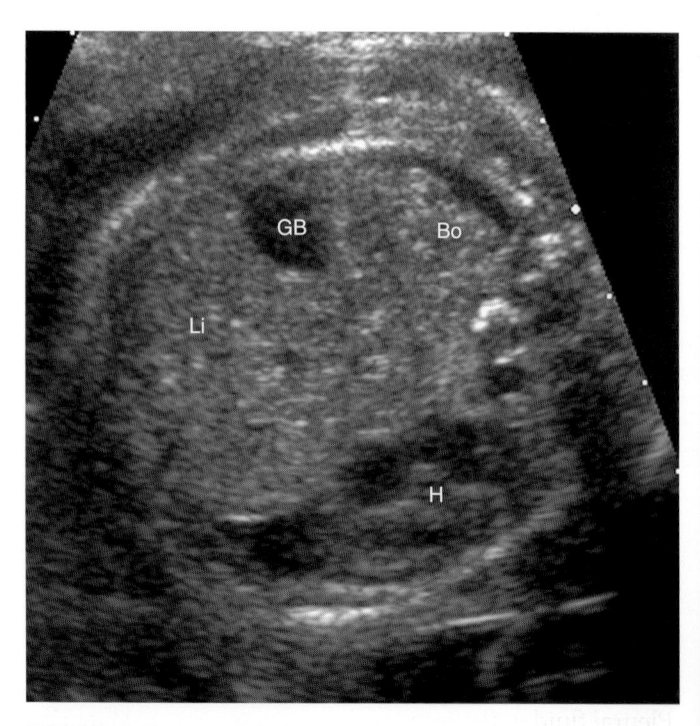

FIGURE 18 ■ Congenital diaphragmatic hernia: right sided defect. The heart is displaced to the left, and the gallbladder is seen in the thorax, with bowel seen posteriorly. Neither lung is visible. H, heart; GB, gallbladder; Bo, bowel; Li, liver.

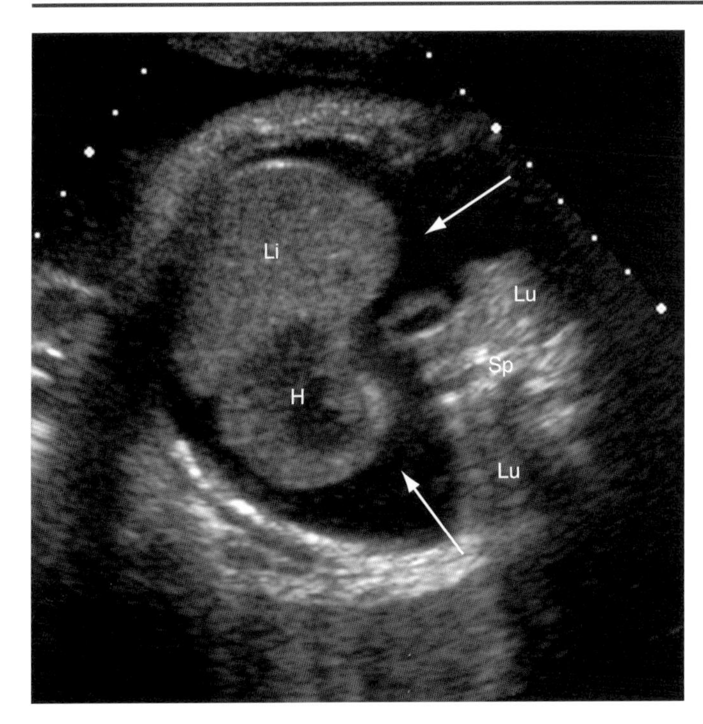

FIGURE 19 ■ Morgagni hernia. The diaphragmatic defect is typically anterior and to the right. In this case, liver is adjacent to the right heart border, and there is a pericardial effusion (*arrows*), with the lungs compressed posteriorly. Li, liver; H, heart; Lu, lungs; Sp, spine.

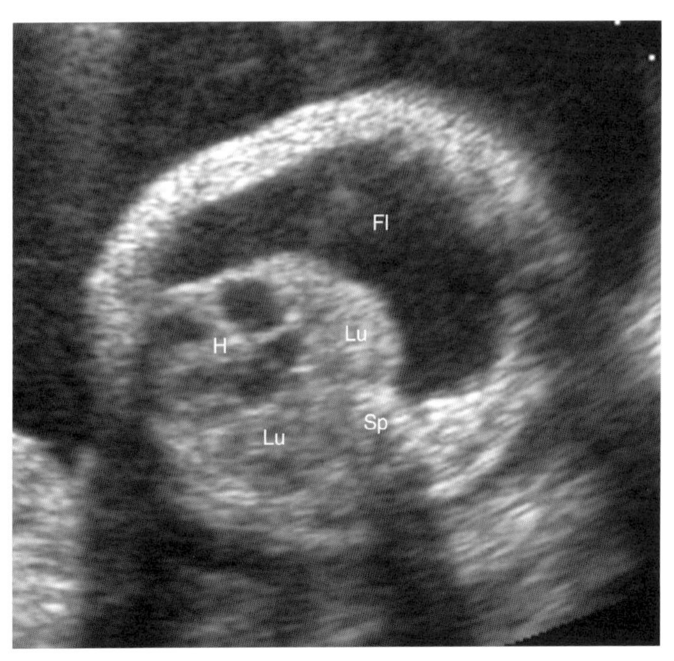

FIGURE 20 ■ Unilateral hydrothorax: axial view. Heart and lungs are displaced to the left by the large fluid collection in the right hemithorax. H, heart; Lu, lungs; Sp, spine; Fl, pleural fluid.

cystic structure within the right-sided mass that is the gallbladder. Unlike in left-sided hernias, the fetal stomach is usually subdiaphragmatic.

Anteromedial hernias are the rarest form of hernia (1 in 100,000); they occur in the retrosternal area (foramen of Morgagni), and they are thought to result from malformation of the septum transversum (Fig. 19). An associated defect in the pericardium may allow the viscera to herniate into the pericardium or the heart into the peritoneal cavity. Associated malformations are common, and chromosomal abnormalities, mental retardation, and heart defects are found in more than 50%.

Eventrations of the diaphragm are much rarer diaphragmatic defects (5%), characterized by thinning of the diaphragm that results in upward migration of viscera and shift of the heart and mediastinum. A fetal diaphragmatic eventration may be sonographically indistinguishable from a true hernia, and magnetic resonance imaging (MRI) can be helpful in their differentiation (61). Unilateral eventration may be associated with Beckwith–Wiedemann syndrome, and it is frequently associated with chromosomal abnormalities, especially trisomies 13 to 15 and 18. Bilateral eventrations have been associated with toxoplasmosis, cytomegalovirus infection, arthrogryposis, and other syndromes (62,63).

Pleural Effusions

Pleural fluid in the fetus is always abnormal. Fetal hydrothorax (Figs. 20 and 21) may be an isolated primary abnormality (fetal chylothorax) or more commonly a secondary manifestation of one of many possible disorders, including immune and nonimmune fetal anemia, chromosomal (especially trisomy 21 and 45 XO), cardiac, metabolic abnormalities, pulmonary masses, and abnormalities of the placenta and umbilical cord (Table 4).

In many cases, detailed anatomic survey of the fetus and karyotype testing suggest the cause of the

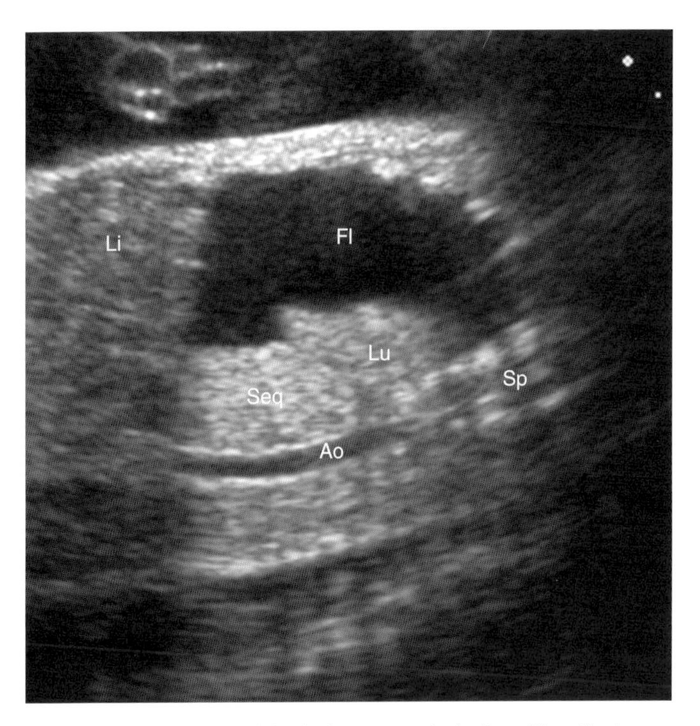

FIGURE 21 ■ Unilateral hydrothorax: sagittal view. The diaphragm as delineated by the superior surface of the liver is flattened due to the mass effect of the fluid. Li, liver; Lu, lung; Sp, spine; Ao aorta; Fl, pleural fluid; Seq, sequestration.

TABLE 4 ■ Causes of Pleural Effusion

- Idiopathic primary chylothorax
- Hydrops fetalis (multiple causes)
- Abnormal karyotype, mostly 45 XO, trisomy 21
- Lung mass: sequestration, congenital diaphragmatic hernia

hydrothorax, but in many cases, the precise origin cannot be determined. Secondary fetal hydrothorax may occur as often as in one in 1400 live births (64), whereas primary fetal hydrothorax is less common (one in 12,000 live births) (65). Prognosis is related to underlying cause, associated morphologic (38%) and chromosomal (8%) abnormalities, hydrops fetalis (70% perinatal mortality) (66), and gestational age (Table 5) (67). Bilateral effusions (Figs. 22 and 23) are associated with worse outcomes than unilateral effusions. The mortality rate among fetuses with hydrothoraces is considerably higher (50%) than that in newborns with chylothorax (15%) (66,68). Fetal pleural effusions may resolve spontaneously in utero in as many as 10% (65,66,68,69). This resolution is more likely if the effusion is unilateral, small, and unassociated with hydrops or other malformations. In the latter situation, a period of observation is warranted before therapeutic intervention.

Most isolated or primary hydrothoraces are chylous in origin (Fig. 24). The cause is unknown (70–74). Males are affected twice as often as females, and unilateral hydrothoraces are more common on the left (71,75). In the newborn, chylothorax is the most frequent cause of isolated pleural effusion, leading to respiratory distress (70,71). In the feeding infant or adult, the diagnosis of a chylous effusion is based on the milky appearance of aspirated chest fluid, which contains chylomicrons. In contradistinction, chest aspirates in the fetus are generally clear and straw-colored, and chylomicrons are not found (because the fetus is "fasting"). A pleural fluid cell count of more than 80% lymphocytes in the fetus is considered suggestive of primary chylothorax (65).

Sonographically, pleural effusions appear as anechoic fluid collections in the fetal chest that conform to the normal chest and diaphragmatic contour (76) and should be differentiated from a pericardial effusion (Fig. 25). When large, a hydrothorax may be associated with bulging of the chest and flattening or inversion of the diaphragm. Although the distinction between primary and secondary fetal pleural effusion is important, it is nearly impossible based solely on the appearance of the aspirated fluid (70–73,77). Sonographic features that suggest that the pleural effusion is the primary abnormality include the following:

TABLE 5 ■ Pleural Effusion: Good Prognostic Features

- Unilateral (vs. bilateral)
- Isolated (vs. associated with infection or chromosomal abnormality)
- No hydrops
- Spontaneous resolution (vs. progression)

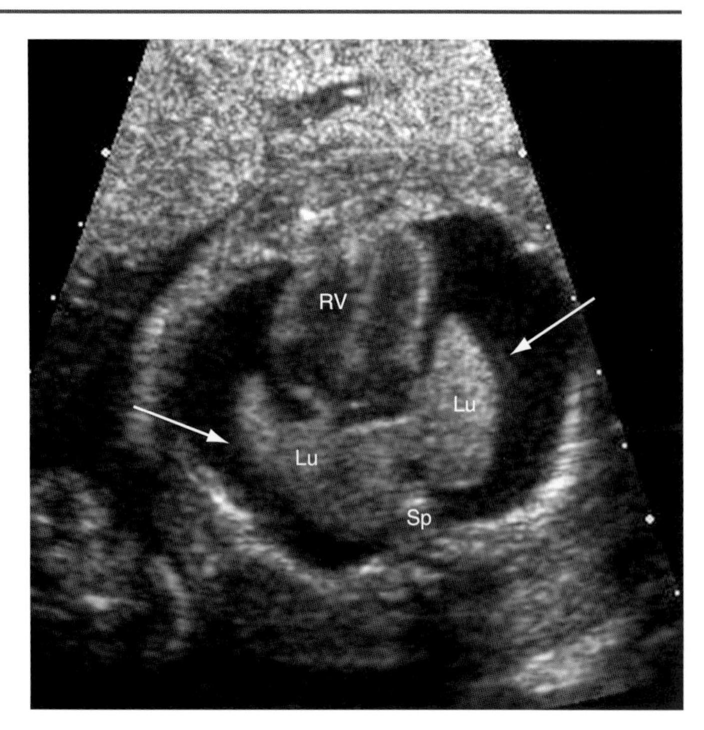

FIGURE 22 ■ Bilateral pleural effusions in a case of nonimmune hydrops: axial view. The lungs are floating like little "angel wings" surrounded by fluid (*arrows*). The heart is not displaced. RV, right ventricle; Sp, spine; Lu, lungs.

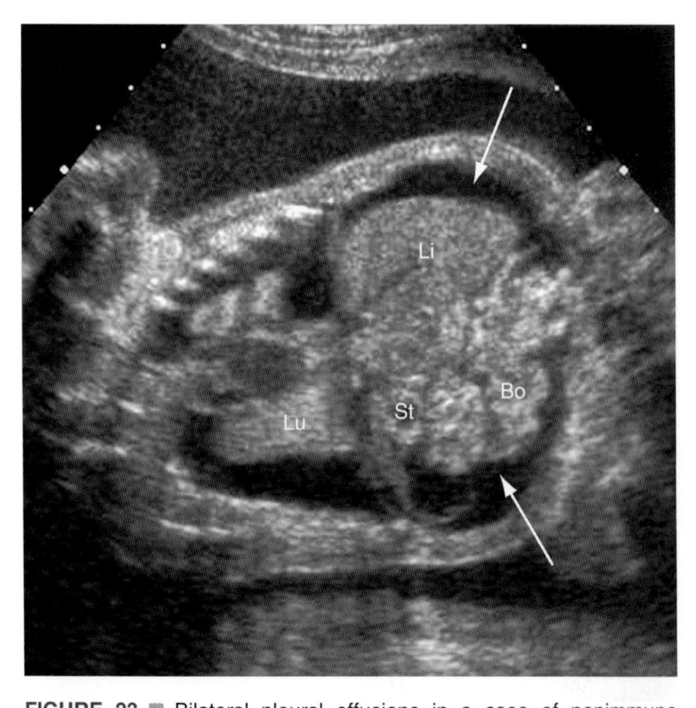

FIGURE 23 ■ Bilateral pleural effusions in a case of nonimmune hydrops: coronal view. Lungs seen as before. There is also ascites (*arrows*) with stomach and bowel floating in the fluid. The liver is enlarged probably due to its accelerated erythropoietic activity. St, stomach; Bo, bowel; Li, liver; Lu, lungs.

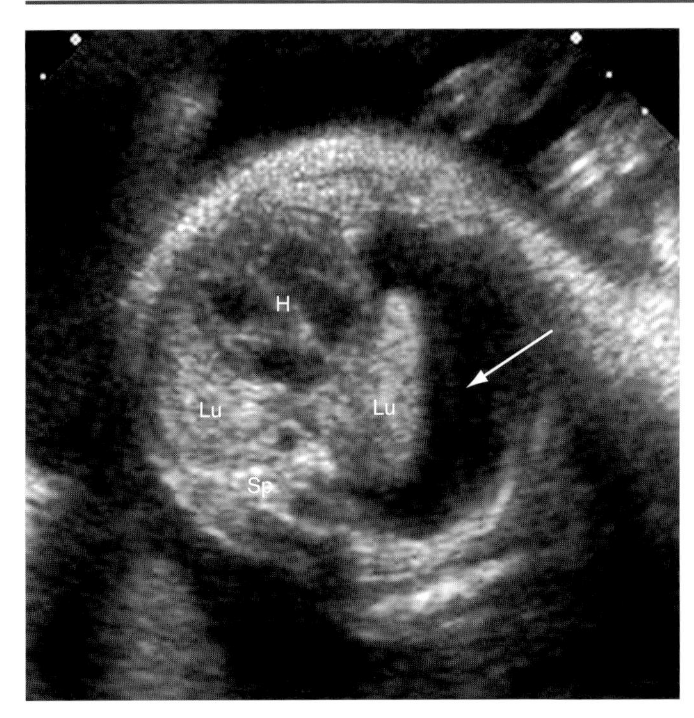

FIGURE 24 ■ Unilateral primary chylothorax. Chyle (*arrow*) is anechoic. There is mass effect as in the previous case but less severe. The lungs are compressed. Lu, lungs; H, heart; Sp, spine.

The pleural effusion is the only abnormal finding (infectious screen negative and chromosomes are normal).

The effusion occurs first as an isolated finding with evidence of mass effect, and it is later followed by the development of hydrops fetalis.

The size of the effusion is disproportionately large compared with the contralateral effusion, and it exerts a clear mass effect on the mediastinal structures.

A large, or enlarging, unilateral hydrothorax can function just like a chest mass in the fetus. By distorting the mediastinal structures, the effusion may have sufficient mass effect to produce hydrops. Because a previously unilateral chylothorax may eventually cause a contralateral hydrothorax, which is usually smaller than the primary chylothorax, the examiner should be sensitive to disproportionate size of the effusions when they are bilateral. Even when effusions are unilateral, signs of mass effect such as mediastinal shift or flattening of the diaphragm are noteworthy, and they warrant followup at close intervals.

Further evaluation of a fetus with a pleural effusion includes a detailed anatomic morphologic survey, fetal echocardiogram, and karyotype testing. Amniotic fluid is also tested for potential infection. If anemia is considered, fetal blood sampling may be performed. Because fetal hydrothorax is associated with a poor outcome and a significantly increased risk of pulmonary hypoplasia, aspiration of the effusion can be performed percutaneously (Fig. 26). Additional pleurodesis with OK-432 has been successful in preventing reaccumulation (78–81). If no other causes can be found and the effusion is small and unilateral, conservative management is appropriate, as long as the hydrothorax is stable in size and signs of hydrops are absent. As mentioned above, some effusions resolve spontaneously. If a contralateral effusion develops, or if other evidence signals

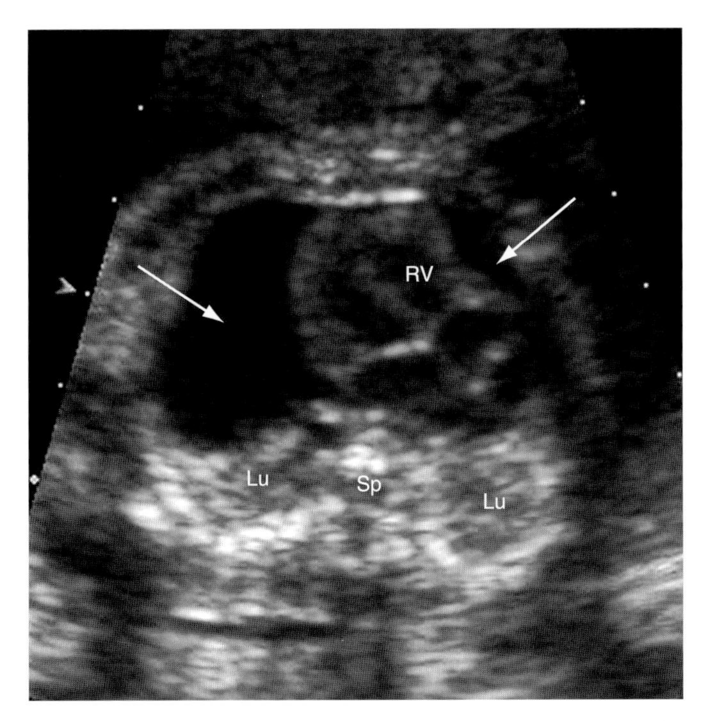

FIGURE 25 ■ Large pericardial effusion (*arrows*). Contrast this with the pleural effusion seen previously; in this case, the heart is surrounded by fluid, not the lungs, and the lungs are compressed posteriorly. Lu, lungs; Sp, spine; RV, right ventricle.

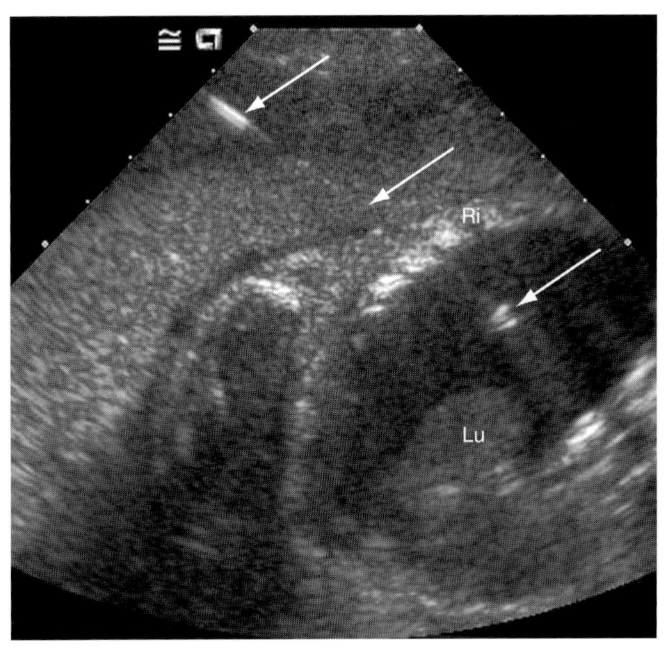

FIGURE 26 ■ Thoracocentesis. The needle (*arrows*) can be seen traversing the maternal body wall, uterus, placenta, and fetal thoracic wall, with its tip in the fluid. Lung can be seen pushed to one side. Lu, lung; Ri, ribs.

the development of hydrops, however, expeditious intervention should be considered. If the effusion is large, reaccumulates, or appears to be under tension, thoracoamniotic shunting can be performed. This procedure has been accomplished antenatally in many cases, with moderate success (72,82–86). Best results have been obtained in fetuses with unilateral effusions and no evidence of hydrops at the time of catheter placement (83,84). If hydrops is present, survival rates of 57% and 75% have been achieved (87,88), with preterm delivery the main source of morbidity. Some investigators have suggested performing in utero thoracentesis just before delivery to improve ventilation in the immediate postpartum period, although this approach is controversial (89,90). Drainage during EXIT (ex utero intrapartum treatment) procedure has also been described, thus avoiding neonatal hypoxia (91). Postnatally, these effusions are treated with chest tube drainage for days to weeks, and in a series of cases diagnosed at birth, the condition eventually resolved with normal outcome in almost all (70,74).

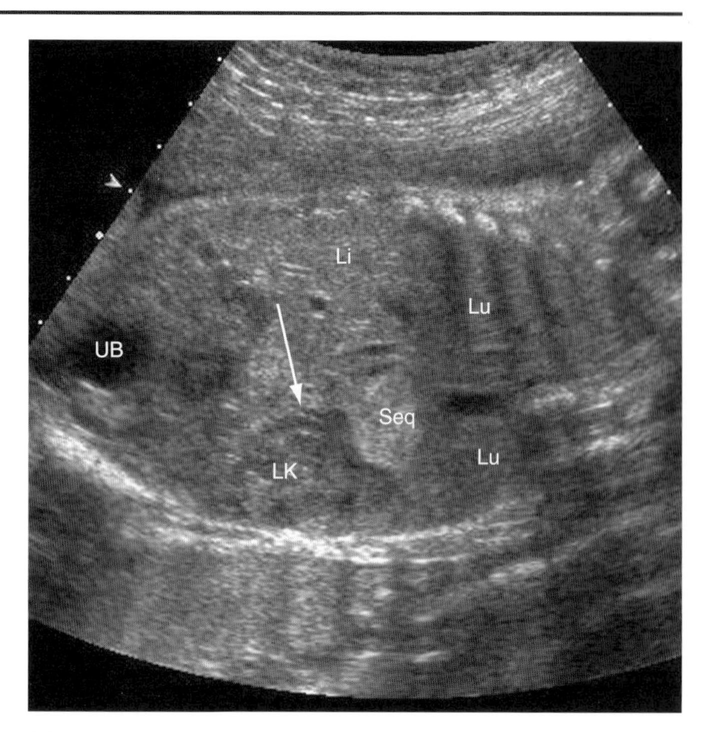

FIGURE 27 ■ Bronchopulmonary sequestration. A small left sided infradiaphragmatic mass is seen, with normal lungs in the thorax. Normal liver and left kidney are demonstrated. The mass can be differentiated from a neuroblastoma because a normal adrenal gland is demonstrated (*arrow*). Seq, sequestration; Lu, lungs; Li, liver; UB, urinary bladder; LK, left kidney.

Bronchopulmonary–Foregut Malformations

Because the bronchial tree arises from the primitive foregut, a close association exists between malformations of these two systems. BPS, bronchogenic cysts, and neurenteric cysts are considered within the spectrum of bronchopulmonary foregut malformations.

Bronchopulmonary Sequestration

A BPS is a mass of pulmonary tissue separated from its normal bronchial and vascular connections and supplied by systemic arteries. Many believe that BPSs are formed when a supernumerary lung bud arises caudal to the normal lung bud, and, as development progresses, the BPS is carried caudally with the esophagus. Alternatively, a BPS may result from failure of obliteration of a systemic arterial connection to the base of the lung (92).

Two forms of sequestrations have been described, intralobar and extralobar. Both are supplied by a systemic artery from the thoracic or abdominal aorta. Intralobar sequestrations (ILS) share a common pleural investment with the rest of the lung but are separated from the bronchial tree. ELS are invested by a separate pleural envelope. Venous drainage from an ILS is usually to the pulmonary veins. Venous drainage from an ELS is to systemic veins (azygous system, cava) or portal veins. This systemic drainage produces a left-to-right shunt in an ELS. Postnatally, cysts may be found in the bronchiectatic form of ILS.

Overall, few (approximately 25%) BPSs diagnosed after birth are ELS, but, for reasons that are unclear, most BPSs diagnosed before birth are ELS (93). ELS occur more commonly in males (4:1), and 90% occur in the left posterior basal thorax (92). Five percent occur below the diaphragm and can be confused with an adrenal tumor (Fig. 27). Pathologically, lymphatic dilatation within an ELS is a common finding. Focal areas containing microscopic features of type II CAM can be found within 15% to 25% of ELS (94,95). The incidence of concomitant anomalies is increased in association with ELS (not in ILS); such anomalies are reported to occur in 15% to 60%, including diaphragmatic hernia (96–98), cardiac abnormalities (99,100), esophageal, gastric, and colonic duplication cysts (101–106), cervical vertebral anomalies (107), bronchial atresia of the right upper lobe with anomalous pulmonary venous drainage (Scimitar syndrome) (108,109), and foregut communications (110–113).

Sonographically, a BPS appears as a homogeneous, echogenic lung mass (Figs. 28–30). The hyperechogenicity is thought to result from multiple reflecting interfaces of tiny dilated bronchioles and air spaces. BPSs are usually small to moderate in size, but they can be large. Color Doppler imaging can be helpful in confirming that the echogenic lung mass is a BPS by demonstrating a feeding artery from the thoracic or abdominal aorta (Figs. 31 and 32). Because the arterial supply to a CAM is usually from the pulmonary artery, Doppler findings can distinguish a CAM from a BPS in the fetus (114–116). On occasion, a large hydrothorax occurs in association with a BPS in the fetus (Figs. 32 and 33) (114,116–118). Investigators have speculated that tension hydrothorax results from torsion of the ELS on its pedicle (the attachment to the lung, mediastinum, or diaphragm) (119). In this way, even a small or moderate-sized ELS, associated with a large ipsilateral tension hydrothorax, may cause severe cardiomediastinal shift and hydrops fetalis.

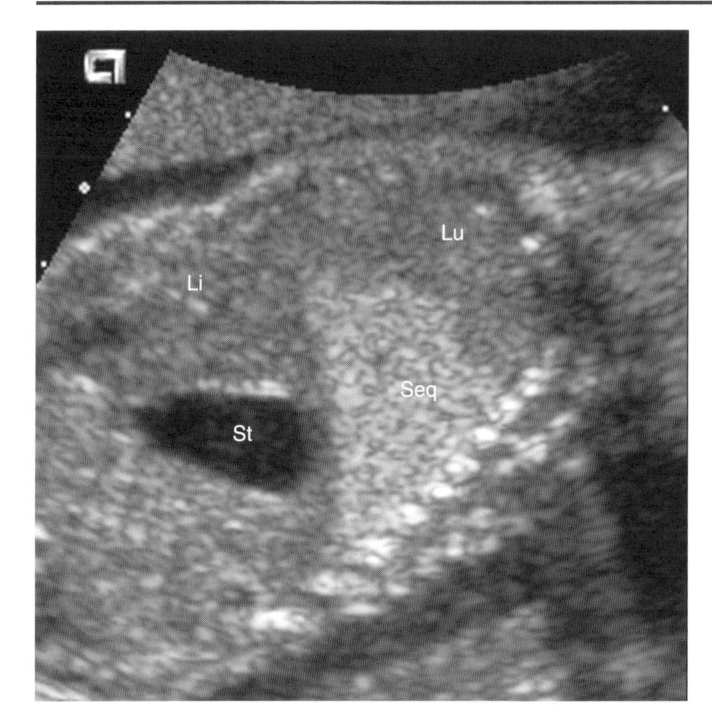

FIGURE 28 ■ Bronchopulmonary sequestration. The sequestration conforms to a segment of lung and is relatively echogenic. Seq, sequestration; Lu, lung; Li, liver; St, stomach.

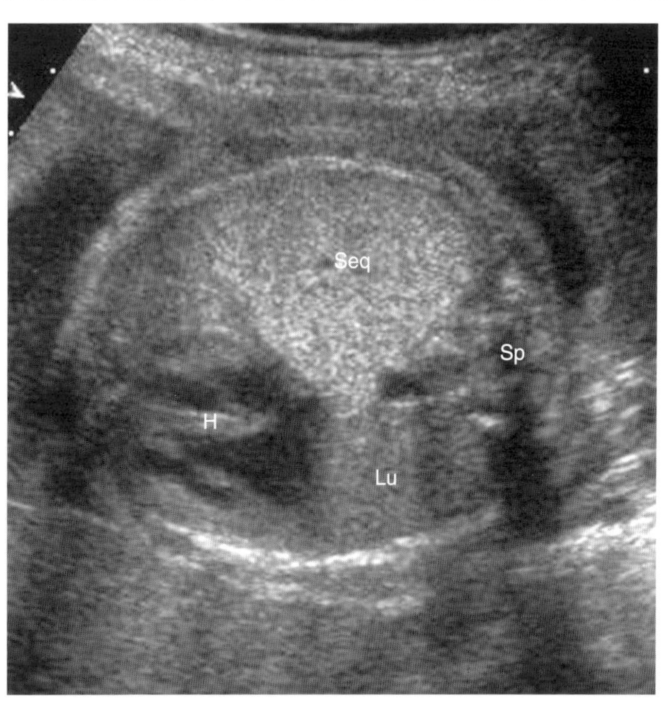

FIGURE 30 ■ Bronchopulmonary sequestration. In this axial view, the echogenic mass is again causing mass effect with displacement of the heart and compression of the contralateral lung. Small hypoechoic areas may be confused with the cysts of CCAM (see next). Seq, sequestration; Lu, lung; H, heart; Sp, spine; CCAM, congenital cystic adenomatoid malformation.

Sometimes, a sequestered lobe may communicate, through a fistula, with the esophagus or stomach (120–122), and at least one fetus with a bronchoesophageal communication within a sequestration has been described prenatally (123). In the latter case, the abnormality detected was a large, echogenic mass secondary to the esophagobronchial communication (obstructed lung), associated with mediastinal shift.

Prognostic features of BPS are similar to those of CAM. Large lesions (with or without tension hydrothoraces)

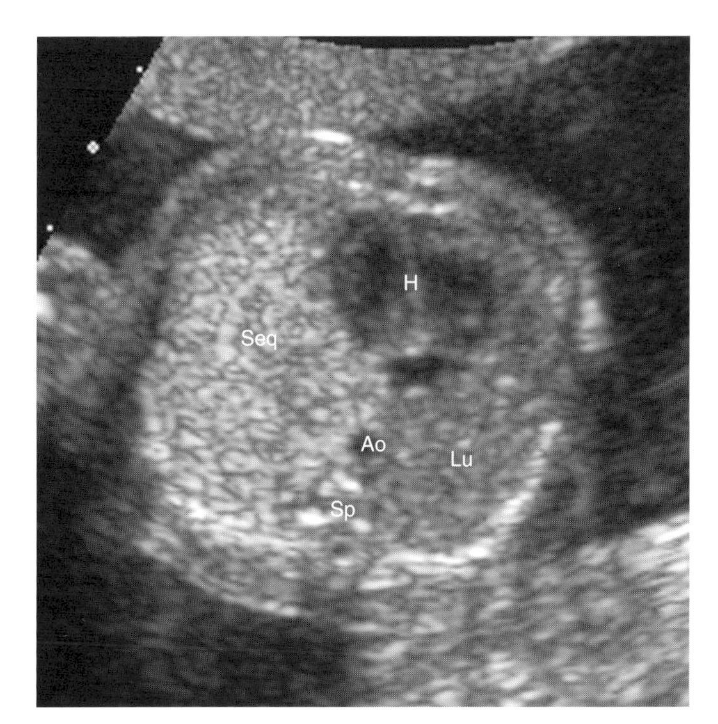

FIGURE 29 ■ Bronchopulmonary sequestration. Axial view of the same case shows mass effect is causing contralateral shift of the heart and aorta, and compression of the contralateral lung. H, heart; Ao, aorta; Lu, lung; Sp, spine; Seq, sequestration.

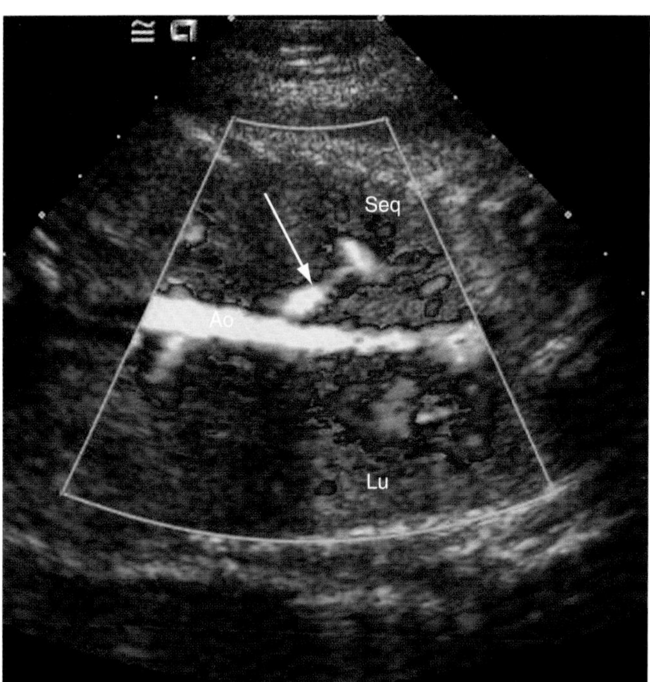

FIGURE 31 ■ Bronchopulmonary sequestration. Coronal view of the same case shows an infradiaphragmatic vessel from the aorta feeding into the mass. Normal lung. Seq, sequestration; Lu, lung.

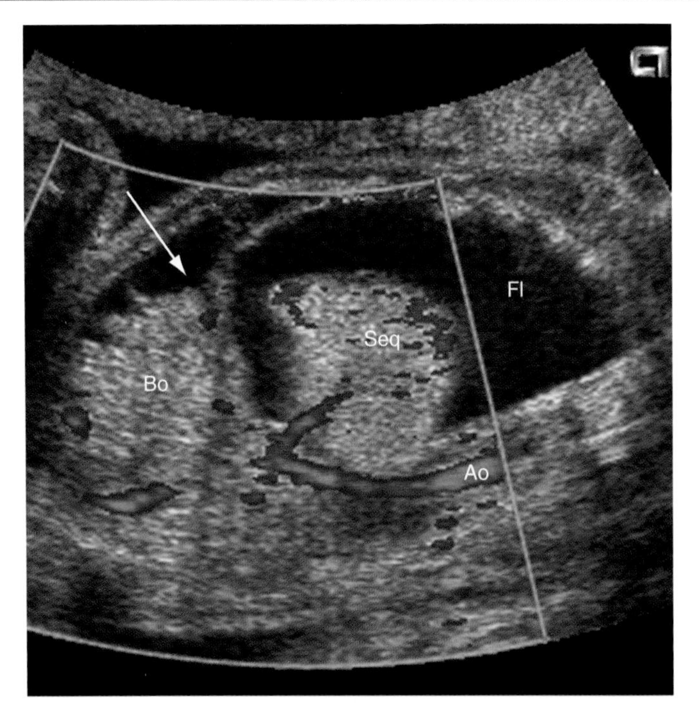

FIGURE 32 ■ Bronchopulmonary sequestration. A semicoronal view of the same mass again shows an infradiaphragmatic feeder artery from the aorta. Ascites is also demonstrated with bowel floating in it. Fl, pleural fluid; Ao, aorta; Seq, sequestration; Bo, bowel.

and associated malformations are poor prognostic indicators. The degree of pulmonary hypoplasia is the primary determinant of outcome. BPSs are probably more common than published childhood studies indicate, because sequestrations appear as incidental findings in 1% to 2% of all patients undergoing pulmonary resection (124).

Bronchogenic Cyst

Bronchogenic cysts are bronchopulmonary foregut malformations thought to occur as a result of abnormal budding of the ventral diverticulum of the foregut. Therefore, most such cysts probably develop between the 26th and the 40th day of fetal life. The location depends on the time at which the abnormal bud separates from the bronchi. If separation occurs earlier in gestation, these cysts are usually found in or near the mediastinum, with most located between the trachea and the esophagus. If separation occurs later, the cyst may develop peripherally, within the lung substance, usually within the lower lobes. In either case, these cysts likely occur before bronchial development is complete (i.e., before 16 weeks) (4).

Prenatal detection with ultrasound is rare, but it has been accomplished by observing (*i*) a single unilocular pulmonary cyst (125,126) or (*ii*) an echogenic, distended lung obstructed by a small bronchogenic cyst (127). Rarely, these cysts are found in extrathoracic locations (92). A fetus with an ectopic bronchogenic cyst in the neck has also been described (128). Bronchogenic cysts are not usually associated with other congenital anomalies (129,130).

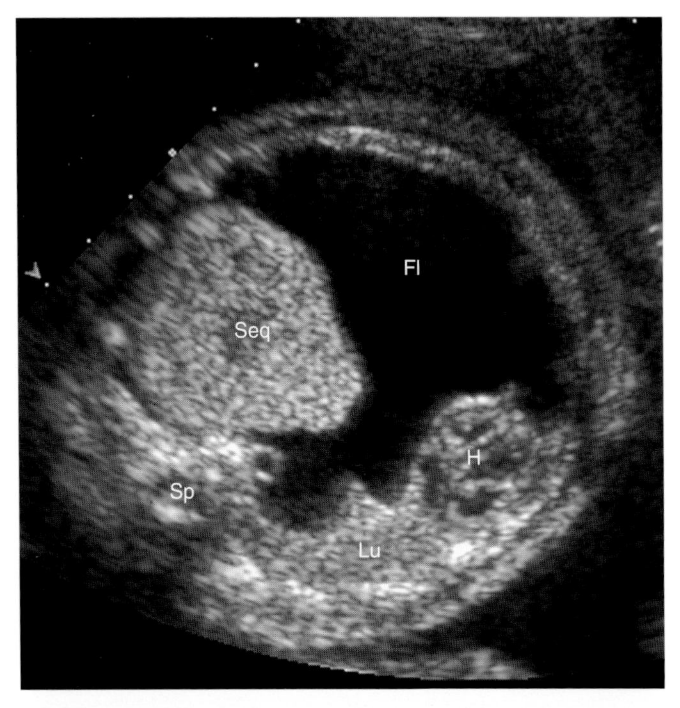

FIGURE 33 ■ Bronchopulmonary sequestration. This echogenic lung mass associated with a large pleural effusion could again be confused with a CCAM (see next). Heart and lungs are displaced and compressed by the mass effect from the fluid. Sp, spine; Fl, pleural effusion; H, heart; Lu, lungs; CCAM, congenital cystic adenomatoid malformation; Seq, sequestration.

Enteric and Neurenteric Cyst

Alimentary tract duplication can occur anywhere from mouth to anus, and often go undetected unless they become symptomatic (131). Simple duplication of the esophagus is a rare form of foregut duplication and probably originates in the fourth gestational week. These lesions are usually centrally located near the mediastinum, but an unusual esophageal duplication cyst was confused with a chest mass when it appeared as a cyst within the posterior left lung (28). If the cyst communicates dorsally with the spinal canal, the malformation is referred to as a neurenteric cyst (Figs. 34 and 35). Neurenteric cysts, the rarest of bronchopulmonary foregut malformations, are thought to originate from incomplete separation of the foregut and notochord. They are invariably associated with spinal dysraphisms including hemivertebra, allowing communication of the cyst with the spinal canal, thoracic myelomeningocele, and absent vertebra. These lesions may also communicate with bowel below the diaphragm through a diaphragmatic defect. Although a cyst within the thorax is a nonspecific observation, when it is associated with a thoracic spinal dysraphism, the diagnosis of neurenteric cyst should be strongly considered (132,133).

Pulmonary Abnormalities

Agenesis

Pulmonary agenesis results from a unilateral lack of primitive mesenchyme, with complete (or almost complete) failure of development of the lung. It is associated

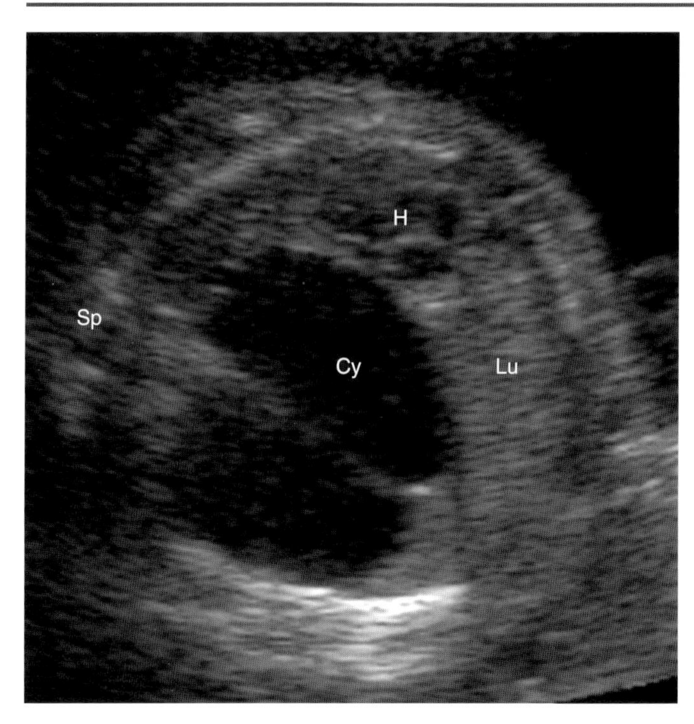

FIGURE 34 ■ Large septated cystic mass in the thorax causing compression of the heart and lung. The wide differential can be narrowed by additional features (see next). Sp, spine; Cy, cystic mass; H, heart; Lu, lung.

with other anomalies in approximately 60% including persistent ductus arteriosus (PDA), anomalies of great vessels, tetralogy of Fallot, bronchogenic cyst, CDH, and bone anomalies (134–136). Pulmonary agenesis is also seen in congenital pulmonary venolobar syndrome (Scimitar syndrome), most commonly with associated partial anomalous pulmonary venous return (Scimitar vein) and/or other congenital thoracic abnormalities. Bilateral pulmonary agenesis should be considered where appearances suggest severe CDH (137,138).

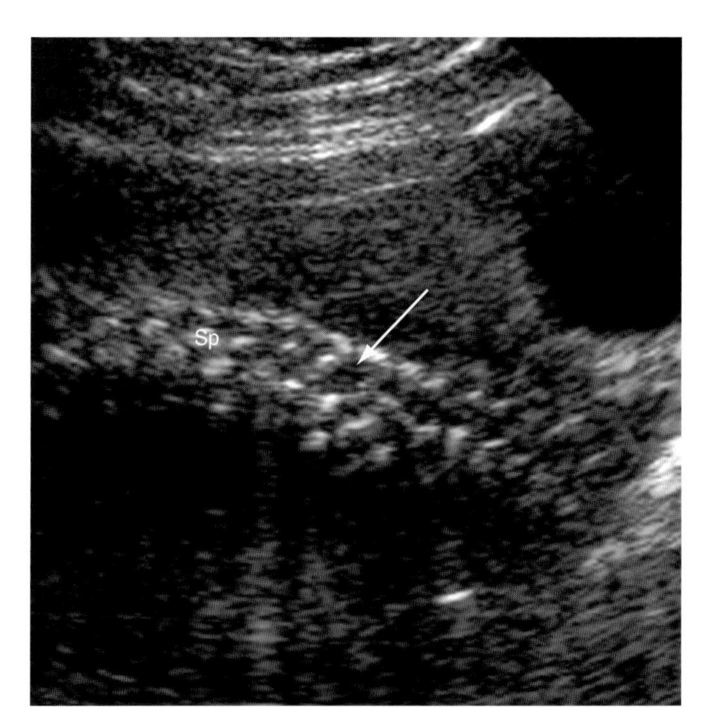

FIGURE 35 ■ View of the spine in the same fetus show cervicothoracic dysraphism (*arrow*) at a higher level than the thoracic cyst. These findings are all explained by a neurenteric cyst. Sp, spine.

Hypoplasia

Pulmonary hypoplasia, pathologically defined as a low ratio of lung to body weight and a low radial alveolar count, remains an important source of postnatal morbidity and mortality. Four factors are most important for normal lung development: (*i*) adequate gestational duration for lung maturation; (*ii*) adequate amniotic fluid volume; (*iii*) intrathoracic space; and (*iv*) fetal breathing movements (139–141). The pathogenesis of pulmonary hypoplasia is often obscure, but experiments confirm that both distension of the lung with the fluid produced by the lungs and fetal respiratory movements are necessary for normal lung growth. Fetal conditions that adversely influence any of these factors, and therefore compromise lung growth and development, can be grouped into several categories (Table 6).

Lung development can be compromised by a skeletal dysplasia (small thorax) or an intrathoracic mass that compresses the developing lung and functionally reduces thoracic space available for lung expansion and growth. Depressed fetal breathing movements resulting from a severe neurologic defect or experimental sectioning of the phrenic nerve in animals also have been reported to cause pulmonary hypoplasia. Although breathing movements are important for normal lung development, they are not sufficient to prevent pulmonary hypoplasia. Finally, severe prolonged oligohydramnios, whether from chest compression, low amniotic fluid pressure (142), or reduction of fluid within the fetal lung (143), is almost always associated with pulmonary hypoplasia (Fig. 36). The critical importance of the presence of amniotic fluid for lung development is emphasized in monoamniotic twins with discordant anomalies of the genitourinary tract. In this case, although renal agenesis is virtually always associated with lethal pulmonary hypoplasia, if the co-twin is normal and the pregnancy is monoamniotic, the shared amniotic fluid allows improved lung development of the twin with renal agenesis (144).

The most common clinical situation where pulmonary hypoplasia is of utmost concern is in preterm premature rupture of membranes, where perinatal survival rates depend on the gestational age at onset, being dismal prior to 23 weeks gestation (145,146), although successful

TABLE 6 ■ Causes of Pulmonary Hypoplasia

- Prolonged oligohydramnios (ruptured membranes, renal anomalies)
- Skeletal malformations (small thorax)
- Neurologic (decreased breathing movements)
- Space-occupying lesions in the chest (lung compression)
- Chromosomal abnormalities (trisomies 12, 18, and 21)
- Decreased pulmonary arterial perfusion (cardiac-pulmonary artery)

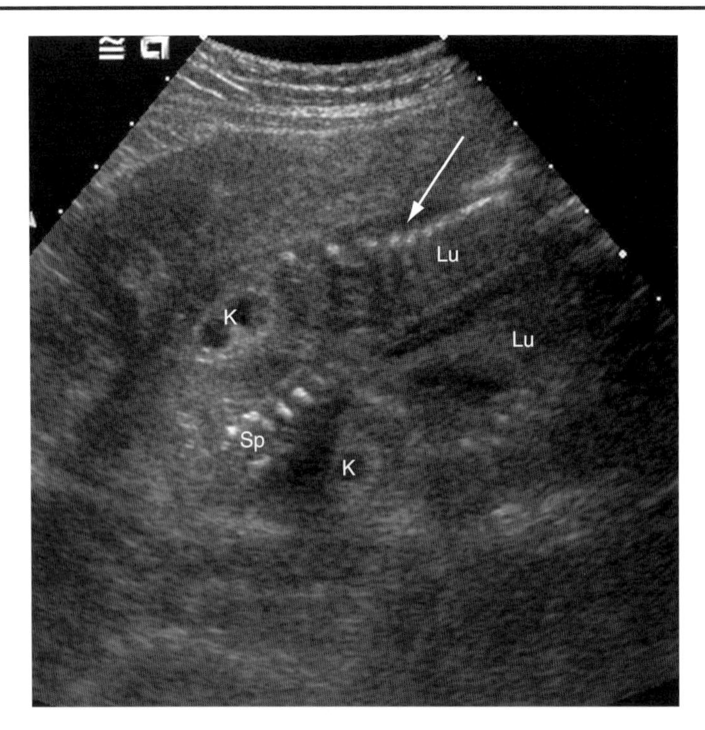

FIGURE 36 ■ Severe pulmonary hypoplasia. The chest wall is bell-shaped (*arrow*) and the lungs are small. There is oligohydramnios. The kidneys are cystic with echogenic parenchyma, in keeping with dysplasia. Sp, spine; Lu, lung; K, kidneys.

treatment of oligohydramnios by amnioinfusion is associated with significantly better outcome (147–149).

Although it would be ideal for obstetric management and parental counseling to be able to predict the severity of pulmonary underdevelopment based on the observed sonographic abnormality, this is not always possible. The time of onset of the lung insult may not be known, and the severity of the abnormality that interferes with normal pulmonary development (e.g., oligohydramnios or chest mass) may vary during gestation. Further, precisely how much amniotic fluid volume or chest space is necessary for normal lung development is not known. Nevertheless, concern about potentially significant pulmonary hypoplasia is warranted in any pregnancy affected by prolonged oligohydramnios (especially if it occurs before 26 weeks and lasts more than 5 weeks) or by a large space-occupying fetal chest lesion.

Although the sonographic appearance of the fetal lung changes with gestational age, attempts to predict lung maturity based on observations such as pulmonary echogenicity, coarsening of the lung echotexture, and progressive increase in sound transmission have largely been unsuccessful (7,150). Semiquantitative methods of sonographic estimation of lung growth are more reliable, especially in the setting of oligohydramnios.

The ultrasonographer routinely makes a subjective assessment of cardiac and thoracic size in each fetus examined. If the examiner has a subjective impression of cardiothoracic disproportion, it is helpful to measure cardiac circumference and thoracic circumference (TC). References to published nomograms of normal fetal heart and chest size according to gestational age are helpful in determining whether the subjective impression is due to a small chest or to an enlarged fetal heart (151). TC is measured on an axial plane of the chest that includes the four-chamber view of the heart. The subcutaneous tissues are generally not included in the measurement, or only minimally so, to avoid overestimating chest size in the fetus with integumentary edema or a skeletal dysplasia associated with redundant skin folds. Small TC has been correlated with pulmonary hypoplasia (152,153) with accurate sensitivities, specificities, and normal or abnormal predictive values (i.e., at least 80%, 80–90%, and 85–90%, respectively). The rate of growth of the fetal TC is linear between 16 and 40 weeks (Table 7), similar to other biometric parameters such as abdominal circumference (AC), HC, and femur length (FL). Therefore, the ratio of TC to AC, HC, or FL is constant, and these ratios may function as useful gestational age-independent parameters (Table 8). TC:AC seems to be the ratio with the smallest variability in normal fetuses (154), and it is more than 0.80 [mean, 0.89 (154), 0.85 (155)] in nearly all normal pregnancies beyond 20 weeks. D'Alton et al. (156) reported 75% sensitivity and 100% specificity using the TC:AC ratio for the prediction of neonatal death from pulmonary disease after premature rupture of membranes (and oligohydramnios) in otherwise normal fetuses. Unfortunately, predictors of lung maturity based on the size of the thorax are far from perfect. For example, when the chest is expanded by a lung mass or by hydrothorax, the size of the bony thorax does not reflect lung growth.

Other potential but less extensively investigative parameters for prediction of pulmonary hypoplasia are Doppler velocimetry of the pulmonary arteries (157–160), branch pulmonary artery size (161), and lung lengths (162). Recently, highly accurate nomograms have been produced for calculating expected lung volume in normal fetuses over a wide range of gestational ages by MR imaging (163–165) and can be used to quantify fetal pulmonary hypoplasia. Lung signal intensity by MR is proving a variably useful predictor of pulmonary hypoplasia (166–169). Similar work has been published using three-dimensional (3D) ultrasonographic volumetry (170).

Despite evaluating other parameters, amniotic fluid lecithin:sphingomyelin ratio and phosphatidyl-glycerol determination remains the most reliable measurement for the determination of fetal lung maturity when premature delivery is contemplated. Whereas these measurements are traditionally performed on samples obtained at amniocentesis, magnetic resonance spectroscopy may provide a future noninvasive tool for their determination in vivo (171).

Bronchial Atresia

Atresia of the segmental, lobar, or main stem bronchus is thought to occur secondary to a vascular accident during development. The bronchi distal to the atresia are usually normal in number. The insult is thought to occur after the 16th week of gestation (when bronchial branching is complete) because an earlier insult would be more likely to result in lung agenesis. This abnormality has been detected in the fetus sonographically as an echogenic fetal lung mass (presumably the fluid-filled lung distal to the bronchial obstruction) (172,173). A dilated fluid-filled "bronchus" may also be observed. The abnormality may not be

TABLE 7 ■ Fetal Thoracic Circumference Measurements (cm)

		Predictive percentiles								
Gestational age (wk)	**Number**	**2.5**	**5**	**10**	**25**	**50**	**75**	**90**	**95**	**97.5**
16	6	5.9	6.4	7.0	8.0	9.1	10.3	11.3	11.9	12.4
17	22	6.8	7.3	7.9	8.9	10.0	11.2	12.2	12.8	13.3
18	31	7.7	8.2	8.8	9.8	11.0	12.1	13.1	13.7	14.2
19	21	8.6	9.1	9.7	10.7	11.9	13.0	14.0	14.6	15.1
20	20	9.5	10.0	10.6	11.7	12.8	13.9	15.0	15.5	16.0
21	30	10.4	11.0	11.6	12.6	13.7	14.8	15.8	16.4	16.9
22	18	11.3	11.9	12.5	13.5	14.6	15.7	16.7	17.3	17.8
23	21	12.2	12.8	13.4	14.4	15.5	16.6	17.6	18.2	18.8
24	27	13.2	13.7	14.3	15.3	16.4	17.5	18.5	19.1	19.7
25	20	14.1	14.6	15.2	16.2	17.3	18.4	19.4	20.0	20.6
26	25	15.0	15.5	16.1	17.1	18.2	19.3	20.3	21.0	21.5
27	24	15.9	16.4	17.0	18.0	19.1	20.2	21.3	21.9	22.4
28	24	16.8	17.3	17.9	18.9	20.0	21.2	22.2	22.8	23.3
29	24	17.7	18.2	18.8	19.8	21.0	22.1	23.1	23.7	24.2
30	27	18.6	19.1	19.7	20.7	21.9	23.0	24.0	24.6	25.1
31	24	19.5	20.0	20.6	21.6	22.8	23.9	24.9	25.5	26.0
32	28	20.4	20.9	21.5	22.6	23.7	24.8	25.8	26.4	26.9
33	27	21.3	21.8	22.5	23.5	24.6	25.7	26.7	27.3	27.8
34	25	22.2	22.8	23.4	24.4	25.5	26.6	27.6	28.2	28.7
35	20	23.1	23.7	24.3	25.3	26.4	27.5	28.5	29.1	29.6
36	23	24.0	24.6	25.2	26.2	27.3	28.4	29.4	30.0	30.6
37	22	24.9	25.5	26.1	27.1	28.2	29.3	30.3	30.9	31.5
38	21	25.9	26.4	27.0	28.0	29.1	30.2	31.2	31.9	32.4
39	7	26.8	27.3	27.9	28.9	30.0	31.1	32.2	32.8	33.3
40	6	27.7	28.2	28.8	29.8	30.9	32.1	33.1	33.7	34.2

Source: From Ref. 154.

visible before 24 weeks (174). That other fetal and neonatal pulmonary lesions such as lobar emphysema (175) and CAM (176) have also been associated with bronchial atresia raises speculation that lobar emphysema and CAM may be linked in some way to the insult causing the bronchial atresia or to the bronchial obstruction itself. Mainstem bronchial atresia is associated with a poor prognosis (173).

TABLE 8 ■ Mean Values with Standard Deviations for Various Thoracic Ratios

Ratio	**Mean predicted value**	**Standard deviation (*N* = 543)**
TC:AC[a]	0.89	0.06
TC:HC[a]	0.80	0.12
TC:HL	4.31	0.36
TC:FL	4.03	0.33

[a]Ratios did not vary significantly with gestational age.

Abbreviations: TC, thoracic circumference; AC, abdominal circumference; HC, head circumference; HL, humerus length; FL, femur length.

Source: From Ref. 154.

Congenital High Airway Obstruction

The antenatal sonographic findings in fetuses with upper respiratory tract severe stenosis or atresia (laryngeal or tracheal) are bilateral enlarged echogenic lungs usually associated with ascites, with or without hydrops (Figs. 37 and 38) (177–179). Localization of the obstruction/stenosis can be determined by scanning the upper airway in the coronal plane, and lack of flow is seen on color Doppler during fetal respiratory movements (180). The heart appears small in the chest relative to the enlarged lungs. Symmetrically positioned tubular anechoic structures reflecting the mucus-filled bronchi are also seen. The lungs are enlarged because of retention of lung fluid, and their increased echogenicity is likely due to reflections from the multiple tiny fluid-filled air spaces and bronchioles within them. In contradistinction to the hypoplastic lungs often present in fetuses with other "chest masses," lungs associated with laryngotracheal atresia are histologically normal or hyperplastic (181). Ascites and hydrops are thought to occur secondary to cardiomediastinal compression by the enlarged lungs. Although the defect occurs early, sonographic detection may not be possible before 20 weeks (179).

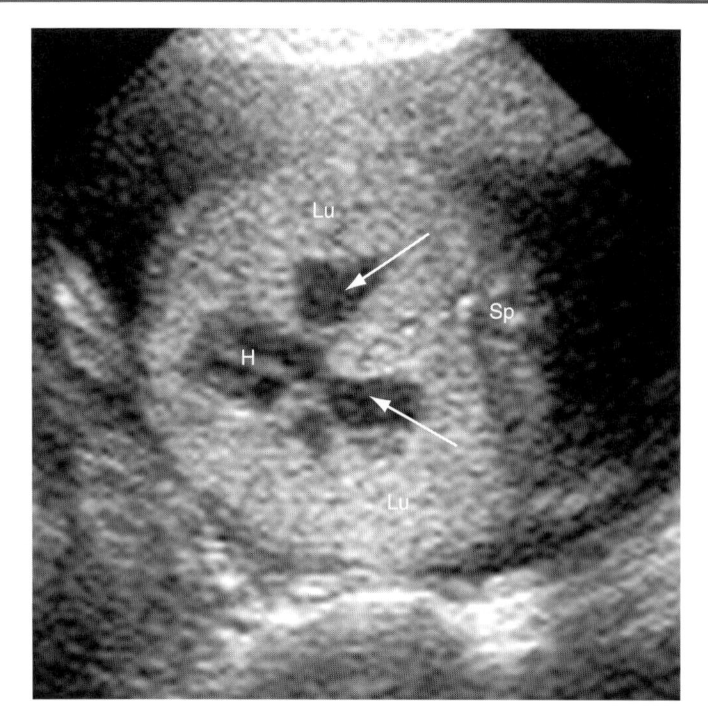

FIGURE 37 ■ Congenital high airway obstruction. The lungs are very echogenic and large, with cystic spaces seen centrally (*arrows*). The heart is compressed by both lungs. Lu, lungs; H, heart; Sp, spine.

Laryngeal atresia is more common, and tracheal agenesis is associated in 40%. The cause of this defect is not known, but investigators have speculated that the insult occurs at five to seven weeks' gestation. Neonatal series have demonstrated associated major congenital anomalies in up to 50% (182), including esophageal and tracheal atresias and fistulas, and multiple other malformations. Mortality is close to 100% without intervention, depending on whether the obstruction is complete (183); however, the EXIT procedure allows potential salvage of an affected fetus (184–189), and in-utero fetal tracheostomy has been successful where there is impending fetal demise secondary to hydrops (190).

Congenital CAM

CAM is described pathologically as a hamartoma or focal dysplasia of the lung, and it accounts for approximately 25% of congenital lung lesions (94,191–193). Investigators have speculated that this malformation results from an early embryonic insult (earlier than eight weeks), because the bronchial system within the lesion is poorly developed. CAM is thought to occur from a failure of the endodermal bronchiolar epithelium to induce surrounding mesenchyme to form normal bronchopulmonary segments. Stocker et al. described three pathologic categories of CAM: type I, macrocystic with cysts 2 to 10 cm (Figs. 39 and 40); type II, medium-sized cysts (Figs. 41 and 42); and type III, microcystic with cysts 0.3 to 0.5 cm (Figs. 43–45) (191). Reports of associated renal and chromosomal anomalies (3–8%), especially with type II CAM, were reported by Stocker et al. (28,94,191,194–196), but such reports have not been confirmed by all investigators (197,198). A careful fetal anatomic survey in search of potential associated malformations is recommended after detection of a suspected CAM. Fetal chromosomal analysis should also be considered.

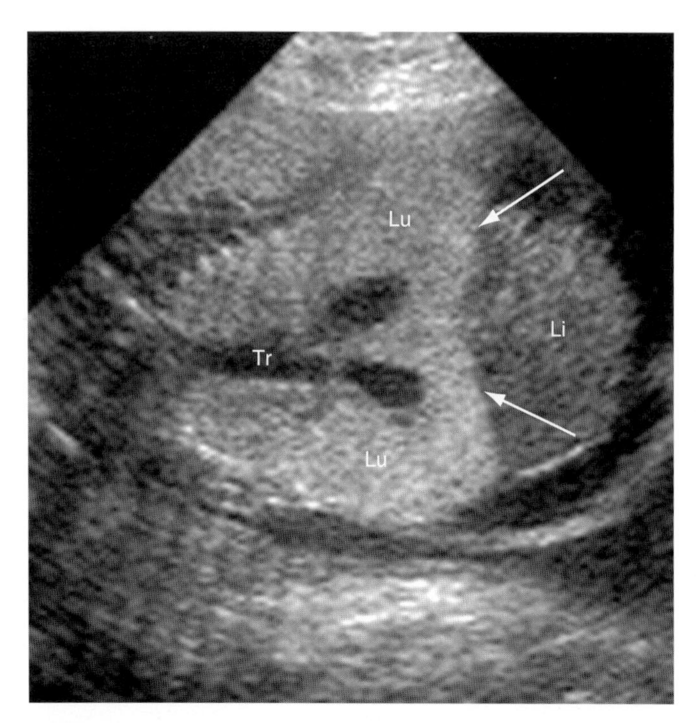

FIGURE 38 ■ Congenital high airway obstruction. Coronal view in the same fetus shows the cystic spaces centrally within the lungs to be the dilated tracheobronchial tree. The mass effect has caused the diaphragms to evert (*arrows*). Li, liver; Lu, lungs; Tr, tracheobronchial tree.

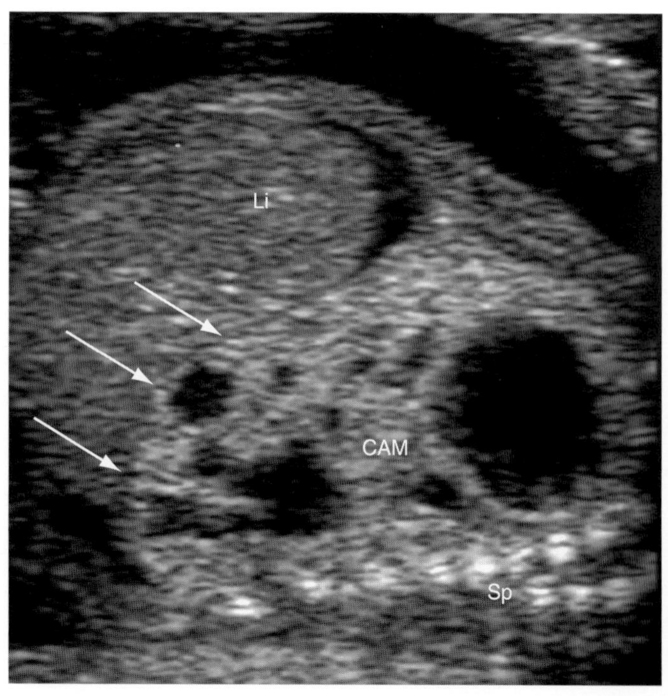

FIGURE 39 ■ Congenital CAM: type I. The mass (CAM) is echogenic with multiple large cysts. There is mass effect with eversion of the diaphragm (*arrows*) and inferior displacement of the liver. CAM, cystic adenomatoid malformation; Li, liver; Sp, spine.

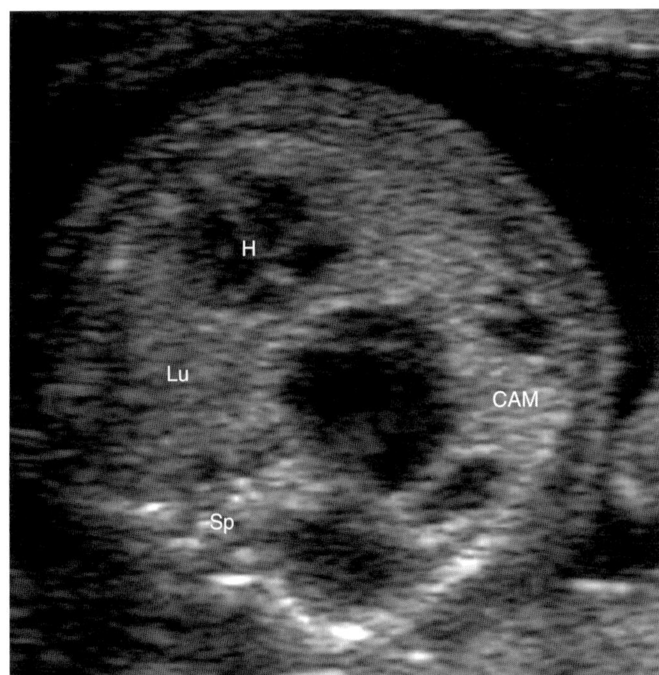

FIGURE 40 ■ Congenital cystic adenomatoid malformation: type I. Axial view of the previous case shows mass effect with displacement of the heart and compression of the contralateral lung. H, heart; Lu, lung; Sp, spine.

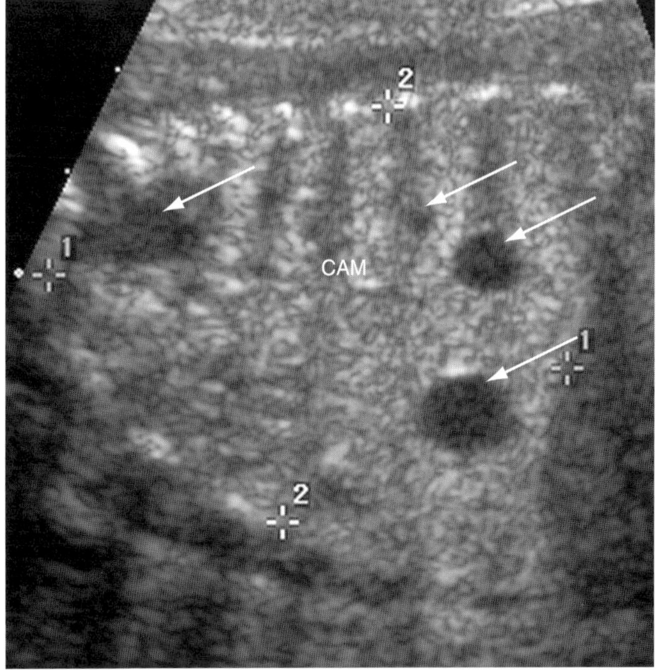

FIGURE 42 ■ Congenital CAM: type II. Coronal view of the previous case demonstrating multiple smaller cysts (*arrows*) within the mass (CAM). CAM, cystic adenomatoid malformation.

The sonographic appearance of a CAM ranges from a multicystic mass to a solid pulmonary lesion. Most CAMs contain at least one sonographically detectable cyst. Type III CAMs, however, appear solid on the sonogram because of reflections from the walls of many tiny cysts (similar to infantile polycystic kidney). The differential diagnosis of a cystic or solid fetal chest mass includes CAM, pulmonary sequestration, bronchogenic cyst, and bronchial atresia. Neurenteric cyst, esophageal duplication, and neuroblastoma should also be considered if the mass appears centrally located in the chest or if thoracic spinal abnormalities are present. Calcifications are not observed in CAM. If the

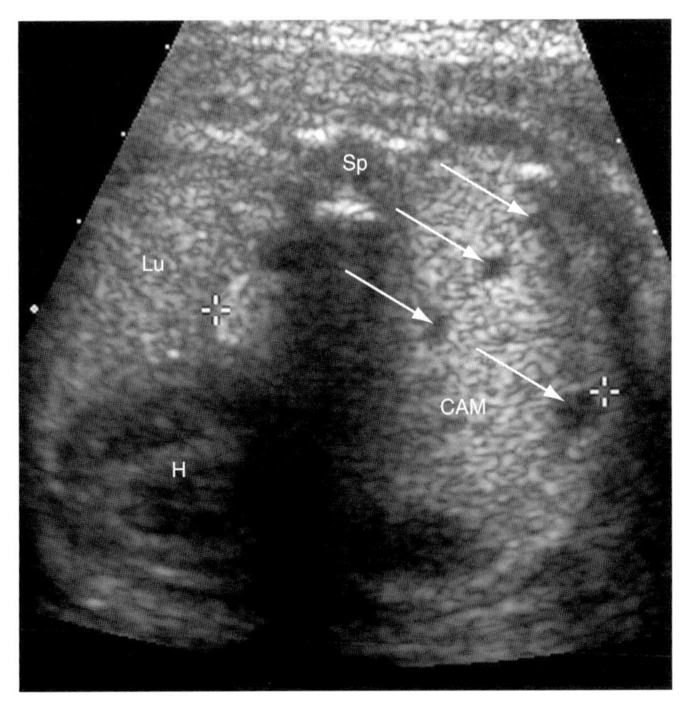

FIGURE 41 ■ Congenital CAM: type II. The mass (CAM) contains a few smaller cysts (*arrows*) within echogenic parenchyma. There is mass effect on the heart and contralateral lung. CAM, cystic adenomatoid malformation; Sp, spine; H, heart; Lu, lung.

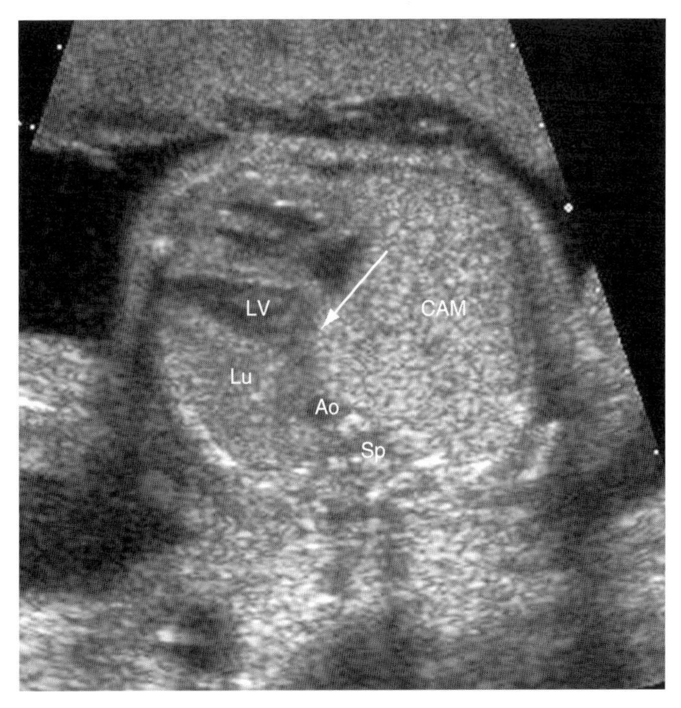

FIGURE 43 ■ Congenital CAM: type III. The mass (CAM) is homogeneously echogenic and is causing mass effect. CAM, cystic adenomatoid malformation; LV, left ventricle; Lu, lung; Ao, aorta; Sp, spine.

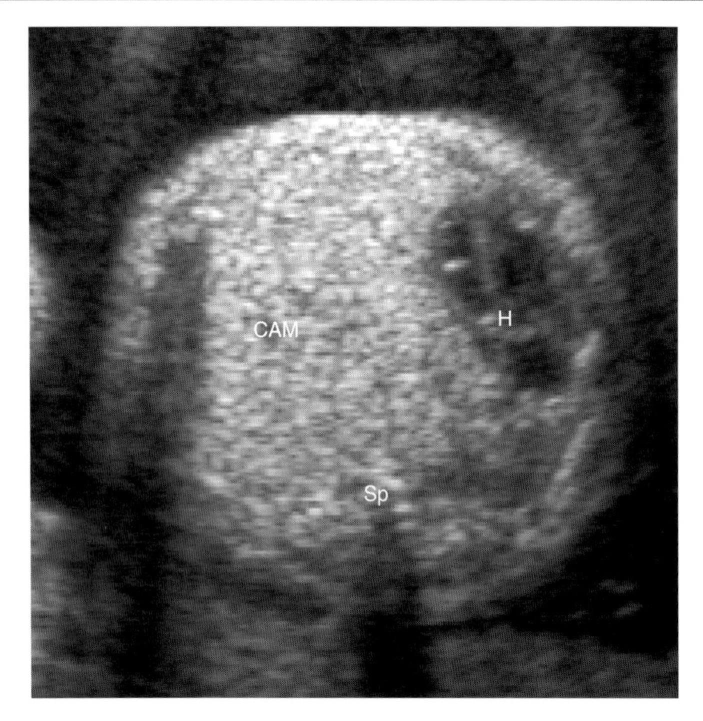

FIGURE 44 ■ Congenital CAM: type III. This case demonstrates severe mass effect (CAM) with compression of the heart. CAM, cystic adenomatoid malformation; H, heart; Sp, spine.

mass contains multiple large cysts, and if CDH has been excluded, the most likely diagnosis is CAM.

Most CAMs are unilateral, without preference for side, and they usually involve one lobe or segment. All lobes are affected equally, and CAMs occur slightly more commonly in males. Rarely, CAM is bilateral, or a single whole lung may be involved (30). In contrast to sequestrations, CAMs are typically supplied by the normal pulmonary circulation (pulmonary arteries and veins), although rare cases with a systemic arterial supply have been reported (199). The lesion has no known risk of recurrence. CAMs lack well-formed bronchi, but most communicate with the tracheobronchial tree. Postnatally (or possibly prenatally), CAMs may become obstructed (air trapping postnatally) because of the absence of cartilaginous bronchi within the lesion (200).

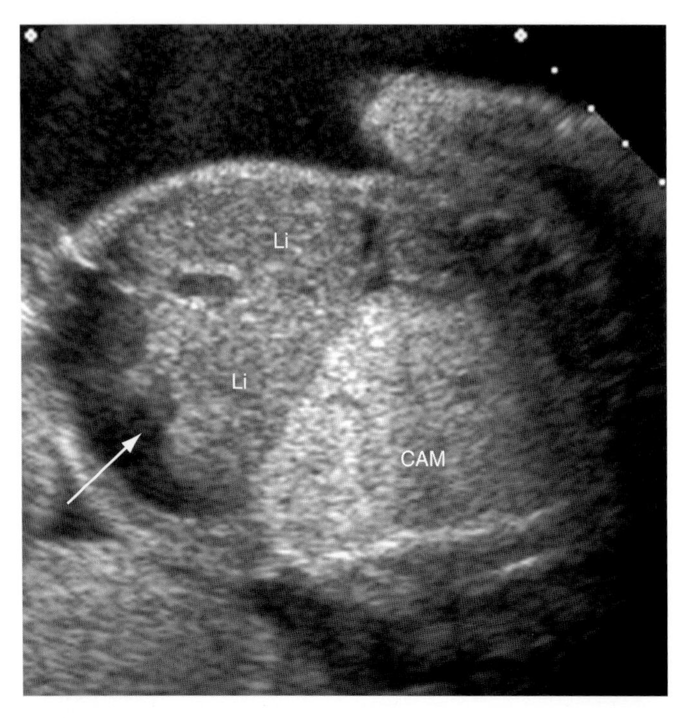

FIGURE 45 ■ Congenital cystic adenomatoid malformation: type III. A coronal view in the same fetus demonstrates ascites (*arrow*) around the liver most likely secondary to impaired venous return to the heart. Li, liver.

The larger the lesion, the more likely are pulmonary hypoplasia and hydrops. Small lesions rarely produce life-threatening pulmonary hypoplasia or hydrops. Although hydrops fetalis is usually associated with a poor prognosis (at least 90% mortality) spontaneous resolution of fetal hydrops associated with a CAM has been observed (200–206). This resolution probably occurs secondary to spontaneous diminution in size of the tumor. CAMs may also cause polyhydramnios, speculated to result from compression of the fetal esophagus, impairing fetal swallowing, or possibly from excess lung fluid produced by the lesion. Prognosis is worse if polyhydramnios is present, as it is if the lesion is large and associated with severe shift of the cardiomediastinal structures (28,30,207).

If the CAM is small, many neonates have marginal or no respiratory symptoms at birth. Even if the lesion is small, however, the child may become symptomatic in the first year of life because of infection of the mass, which communicates with the bronchial tree (30,193). Fewer than 10% of children diagnosed with CAMs present after the first year of life (64). Surviving neonates with larger lesions usually require earlier resection, owing to neonatal respiratory distress. If the child is asymptomatic, some pediatric surgeons opt for conservative management, following the child clinically [including serial chest radiographs or computed tomography (CT) scans] in lieu of routine surgical resection (118,208).

When a CAM is large and associated with hydrops, successful fetal surgical resection of the lesion in utero has also been accomplished (209,210). Successful management and improved survival has been achieved with fetal thoracentesis or thoracoamniotic shunting (211–213) and cyst aspiration (117,214). Aspiration alone may not be curative because fluid often reaccumulates within the cyst (215,216). The natural history of a CAM can be variable. Generally, CAMs remain stable and as many as 50% regress spontaneously during gestation (28,217). Fetal intervention is therefore not usually recommended unless fetal compromise (e.g., hydrops) is evident (218,219), however even in those cases, complete resolution can occur without intervention (220).

Congenital Lobar Emphysema

Congenital lobar emphysema is characterized by a progressive overdistension of one or multiple lobes, most often the left upper lobe, followed by the right middle and right upper lobes. Causes are usually a primary abnormality of the bronchial cartilage, or secondary to endobronchial obstruction or extrinsic bronchial compression. It is most often seen in males (134).

Few cases are reported in the literature (221–225), and the sonographic findings include increased lung

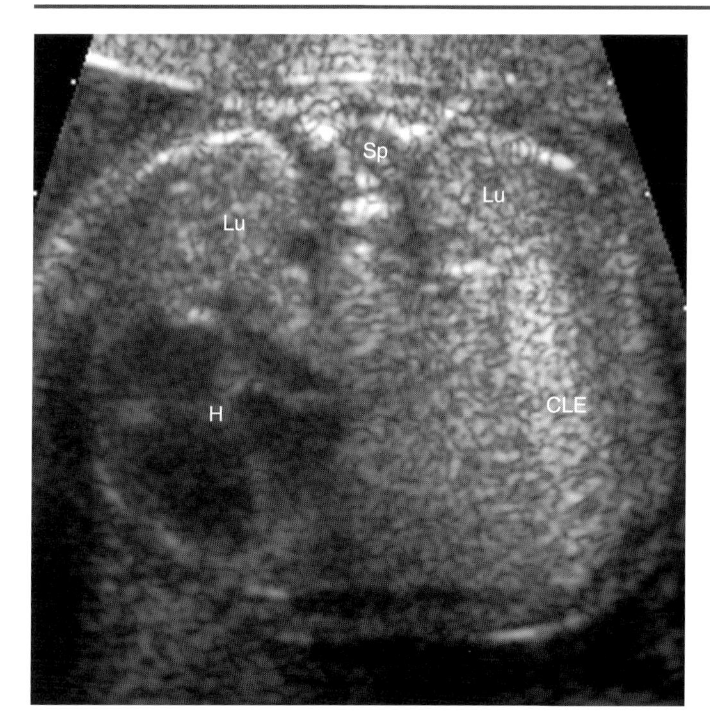

FIGURE 46 ■ CLE. This mass (CLE) can be difficult if not impossible to differentiate from other echogenic lung masses. Normal lungs are compressed posteriorly and the heart is displaced due to mass effect. Lu, lungs; H, heart; CLE, congenital lobar emphysema; Sp, spine.

echogenicity and cystic lesions (Figs. 46 and 47). It can be almost impossible to differentiate prenatally from other echogenic or cystic chest masses (222), and final diagnosis is often made through postnatal followup, usually postlobectomy. Respiratory distress can occur

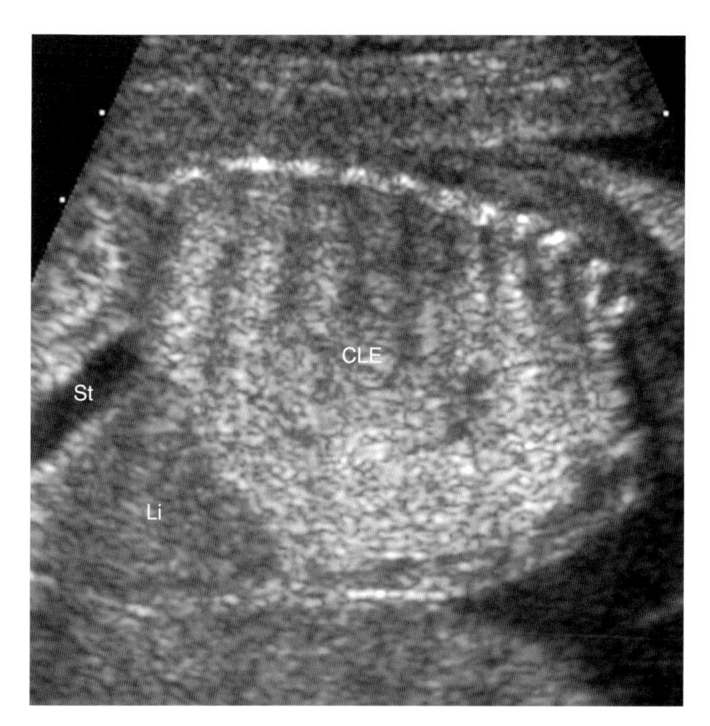

FIGURE 47 ■ CLE. Semisagittal view in the same case showing mass (CLE) with inversion of the diaphragm. CLE, congenital lobar emphysema; Li, liver; St, stomach.

even though the lesion may appear to resolve during gestation (223,225,226).

APPROACH TO A FETUS WITH A CHEST MASS

Although specific antenatal diagnosis cannot always be made, sonographic characteristics of the mass can often help to narrow the diagnostic considerations. A predominantly cystic mass is statistically most likely to be a CAM, but CDH, bronchogenic cyst and enteric cyst should also be considered. Rarely, single or multiple cysts are observed in an ELS. Further, elements of both CAM and ELS are found in the same masses on occasion (227). Normal infradiaphragmatic anatomy usually excludes the diagnosis of left-sided CDH. If the thoracic spine is dysraphic, neurenteric cyst is most likely. A completely solid lesion may be a CAM, sequestration, or obstructed lung or segment. Because sequestration occurs most commonly in the basilar portion of the left lower lobe, a solid lesion in the right upper lobe is more likely to be a CAM or, less likely, the result of bronchial atresia or stenosis. These latter three may be indistinguishable on the sonogram (228). If an anomalous artery directly from the aorta feeding the solid mass is observed, regardless of side, pulmonary sequestration is the most likely cause. Mixed cystic and solid lesions are usually CAMs, but teratoma or CDH should also be considered (Table 9).

Unilateral Fetal Chest Masses

Intrathoracic masses are detected on the sonogram as a result of altered lung echogenicity or mediastinal shift. These masses may appear cystic, solid, or mixed, and most are unilateral. Three lesions constitute most unilateral fetal chest masses: CDH, CAM, and BPS. Bronchogenic cyst, unilateral bronchial atresia, and bronchial stenosis are rarer. Bronchial atresias or stenoses are detected because an obstructed bronchus prevents normal efflux of fetal lung fluid and causes the lung distal to the bronchial atresia or stenosis to become enlarged and echogenic. The "obstructed" lung itself may simulate a chest mass in these cases. Sonographic features such as the presence of large cysts, the location of the lesion in the upper chest (upper

TABLE 9 ■ Fetal Chest Masses: Narrowing the Differential Diagnosis

- Solid and left lower lobe: usually pulmonary sequestration
- Discernible cysts: suggest CAM; also consider CDH
- Right lung: most likely CAM; if solid also consider sequestration
- Bilateral solid: upper respiratory tract atresia
- Vascular supply: CAM fed by pulmonary artery, ELS fed by vessel from aorta
- Calcifications: hamartoma, neuroblastoma, or CDH with meconium

Abbreviations: CAM, cystic adenomatoid malformation; CDH, congenital diaphragmatic hernia; ELS, extralobar sequestration.

TABLE 10 ■ Unilateral Fetal Chest Masses

Cystic:
- CAM
- Bronchogenic cyst
- CDH
- Enteric cyst
- Mediastinal meningocele
- Esophageal duplication cyst

Cystic and solid:
- CAM
- Enteric cyst
- Teratoma (pericardial)
- CDH
- Mixed CAM and sequestration
- Bronchopulmonary sequestration (predominantly solid)
- Thoracic neuroblastoma

Solid:
- CAM (microcystic, type III)
- Bronchopulmonary sequestration
- Bronchial atresia
- Bronchogenic cyst with obstructed bronchus (solid mass is obstructed lung or lobe)
- CDH (bowel or liver)

Abbreviations: CAM, cystic adenomatoid malformation; CDH, congenital diaphragmatic hernia.

vs. lower lung vs. mediastinal), vascular supply to the lesion, and associated spinal abnormalities may be helpful in prioritizing the differential diagnosis (Table 10).

Bilateral Chest Masses

Bilateral fetal chest masses may be CAMs or BPSs, although these lesions are rarely bilateral. The differential diagnosis in these fetuses is limited. CAM is rarely bilateral, but it would be unlikely to be so symmetric. A centrally placed mass such as a bronchogenic cyst or teratoma could conceivably obstruct the trachea and produce this appearance, although this has not been described. Finally, bilateral diaphragmatic hernias can result in bilateral chest masses, but they would not be expected to be homogeneous (Table 11).

Regressing Fetal Chest Masses

Our understanding of the natural history of fetal chest masses is still evolving, but it is crucial for informed parental counseling and obstetric management. The size of a fetal chest mass may dramatically diminish over the course of gestation (Table 12), and is estimated to occur in as many as 30% to 50% of fetuses serially studied (28,94). In contrast, few lung lesions increase in size over time (28,200,229). The likelihood that a chest mass will spontaneously regress in utero has great impact on how pregnancies are managed and how parents are counseled. All three types of CAM and pulmonary sequestrations have been observed to regress in utero. Unfortunately, clinicians are not able to predict which and management decisions are further complicated by the finding that lesions may not regress spontaneously until the third trimester. Postnatally, most affected neonates survive and many have minimal respiratory symptoms, especially if the mass is small.

TABLE 11 ■ Bilateral Chest Masses

- Laryngeal or tracheal atresia
- Bilateral cystic adenomatoid malformation or bronchopulmonary sequestration (rare)
- Bilateral congenital diaphragmatic hernia

TABLE 12 ■ Characteristics of Fetal Chest Masses that Regress During Gestation[a]

- Possible regression of cystic adenomatoid malformations
- First sign of improvement possible at 22–37 wk (wide range)
  - Decrease in absolute or relative size
  - Decrease in echogenicity
  - Decrease in mediastinal shift

[a]May not begin to regress until the third trimester.

REFERENCES

1. American Institute of Ultrasound in Medicine. AIUM Practice Guideline for the Performance of an Antepartum Obstetric Ultrasound Examination. Laurel, MD: American Institute of Ultrasound in Medicine, 2003.
2. Murray JF. The Normal Lung. 2nd ed. Philadelphia: WB Saunders, 1986:1.
3. Reid LM. Lung growth in health and disease. Br J Dis Chest 1984; 78:113.
4. Kravitz RM. Congenital malformations of the lung. Pediatr Clin North Am 1994; 41:453.
5. Hislop AA, Wigglesworth JS, Desai R. Alveolar development in the human fetus and infant. Early Hum Develop 1986; 13:1.
6. Comstock CH. Normal fetal heart axis and position. Obstet Gynecol 1987; 70:255.
7. Fried AM, Loh FK, Umer MA, et al. Echogenicity of fetal lung: relation to fetal age and maturity. AJR Am J Roentgenol 1985; 145:591.
8. Parilla BV, Leeth EA, Kambich MP, Chilis P, MacGregor SN. Antenatal detection of skeletal dysplasias. J Ultrasound Med 2003; 22(3):255–258; quiz 259–261.
9. Das BB, Nagaraj A, Fayemi A, Rajegowda BK, Giampietro PF. Fetal thoracic measurements in prenatal diagnosis of Jeune syndrome. Indian J Pediatr 2002 Jan; 69(1):101–103.
10. Chen CP, Chern SR, Shih JC, et al. Prenatal diagnosis and genetic analysis of type I and type II thanatophoric dysplasia. Prenat Diagn 2001; 21(2):89–95.
11. Chen CP, Lin SP, Liu FF, Jan SW, Lin SY, Lan CC. Prenatal diagnosis of asphyxiating thoracic dysplasia (Jeune syndrome). Am J Perinatol 1996; 13(8):495–498.
12. Merz E, Miric-Tesanic D, Bahlmann F, Weber G, Hallermann C. Prenatal sonographic chest and lung measurements for predicting severe pulmonary hypoplasia. Prenat Diagn 1999; 19(7):614–619.
13. Hill LM, Leary J. Transvaginal sonographic diagnosis of short-rib polydactyly dysplasia at 13 weeks' gestation. Prenat Diagn 1998; 18(11):1198–1201.
14. Hersh JH, Angle B, Pietrantoni M, et al. Predictive value of fetal ultrasonography in the diagnosis of a lethal skeletal dysplasia. South Med J 1998; 91(12):1137–1142.

15. Akiyama M, Dale BA, Smith LT, Shimizu H, Holbrook KA. Regional difference in expression of characteristic abnormality of harlequin ichthyosis in affected fetuses. Prenat Diagn 1998; 18(5):425–436.
16. Akiyama M, Kim DK, Main DM, Otto CE, Holbrook KA. Characteristic morphologic abnormality of harlequin ichthyosis detected in amniotic fluid cells. J Invest Dermatol 1994; 102(2):210–213.
17. Sharara FI, Khoury AN. Prenatal diagnosis of a giant cavernous hemangioma in association with nonimmune hydrops. A case report. J Reprod Med 1994; 39(7):547–549.
18. Tseng JJ, Chou MM, Ho ES. Fetal axillary hemangiolymphangioma with secondary intralesional bleeding: serial ultrasound findings. Ultrasound Obstet Gynecol 2002; 19(4):403–406.
19. Bezzi M, Mitchell DG, Kurtz AB, et al. Prominent fetal breasts: a normal variant. J Ultrasound Med 1987; 6:655.
20. Sbragia L, Paek BW, Feldstein VA, et al. Outcome of prenatally diagnosed solid fetal tumors. J Pediatr Surg 2001; 36(8):1244–1247.
21. Brar MK, Cubberley DA, Baty BJ, et al. Chest wall hamartoma in a fetus. J Ultrasound Med 1988; 7:217.
22. Smith LG Jr, Carpenter RJ Jr, Gonsoulin W, et al. Prenatal diagnosis of a chest wall mass with ultrasonography and Doppler velocimetry. A case report. Am J Obstet Gynecol 1990; 163:567.
23. D'Ercole C, Boubli L, Potier A, Borrione CL, Leclaire M, Blanc B. Fetal chest wall hamartoma: a case report. Fetal Diagn Ther 1994; 9(4):261–263.
24. Jain SK, Afzal M, Mathew M, Ramani SK. Malignant mesenchymoma of the chest wall in an adult. RAX 1993; 48(4):407–408.
25. Matsumoto T, Miyakoshi K, Kasai K, et al. Fetal goitrous hypothyroidism followed by neonatal transient hyperthyroidism. A case report. Fetal Diagn Ther 2003; 18(6):459–462.
26. Serreau R, Polack M, Leger J, et al. Fetal thyroid goiter after massive iodine exposure. Prenat Diagn 2004; 24(9):751–753.
27. de Filippi G, Canestri G, Bosio U, et al. Thoracic neuroblastoma: antenatal demonstration in a case with unusual postnatal radiographic findings. Br J Radiol 1986; 59:704.
28. Bromley B, Parad R, Estroff JA, et al. Fetal lung masses: prenatal course and outcome. J Ultrasound Med 1995; 14:927.
29. Puligandla PS, Kay S, Morin L, et al. Pericardial hemangioma presenting as thoracic mass in utero. Fetal Diagn Ther 2004; 19(2):178–181.
30. Thorpe-Beeston JG, Nicolaides KH. Cystic adenomatoid malformation of the lung: prenatal diagnosis and outcome. Prenat Diagn 1994; 14:677.
31. Rice HE, Estes JM, Hedrick MH, et al. Congenital cystic adenomatoid malformation: a sheep model of fetal hydrops. J Pediatr Surg 1994; 29:692.
32. Harrison MR, Bjordal RI, Langmark F, et al. Congenital diaphragmatic hernia: hidden mortality. J Pediatr Surg 1978; 13:227.
33. Cannon C, Dildy GA, Ward R, et al. A population-based study of congenital diaphragmatic hernia in Utah: 1988–1994. Obstet Gynecol 1996; 87:959.
34. Butler N, Claireax AE. Congenital diaphragmatic hernia as a cause of perinatal mortality. Lancet 1962; 1:659.
35. Schumacher RE, Farrell PM. Congenital diaphragmatic hernia: a major remaining challenge in neonatal respiratory care. Perinatol Neonatol 1985; 9:29.
36. Wenstrom KD, Weiner CP, Hanson JW. A five-year statewide experience with congenital diaphragmatic hernia. Am J Obstet Gynecol 1991; 165:838.
37. Greenwood RD, Rosenthal A, Nadas AS. Cardiovascular abnormalities associated with congenital diaphragmatic hernia. Pediatrics 1976; 57:92.
38. David TJ, Illingsworth CA. Diaphragmatic hernia in the southwest of England. J Med Genet 1976; 13:253.
39. Nakayama DK, Harrison MR, Chinn DH, et al. Prenatal diagnosis and natural history of the fetus with a congenital diaphragmatic hernia: initial clinical experience. J Pediatr Surg 1985; 20:118.
40. Puri P, Gorman F. Lethal nonpulmonary anomalies associated with congenital surgery. J Pediatr Surg 1984; 19:29.
41. Sharland GK, Lockhart SM, Heward AJ, et al. Prognosis in fetal diaphragmatic hernia. Am J Obstet Gynecol 1992; 166:9.
42. Crane JP. Familial congenital diaphragmatic hernia: prenatal diagnostic approach and analysis of twelve families. Clin Genet 1979; 16:244.
43. Torfs CP, Curry CJ, Bateson TF, et al. A population-based study of congenital diaphragmatic hernia. Teratology 1992; 45:555.
44. Au-Yeung JY, Chan KL. Prenatal surgery for congenital diaphragmatic hernia. Asian J Surg 2003; 26(4):240–243.
45. Harrison MR, Adzick NS, Estes JM, et al. A prospective study of the outcome for fetuses with diaphragmatic hernia. JAMA 1994; 271:382.
46. Stringer MD, Goldstein RB, Filly RA, et al. Fetal diaphragmatic hernia without visceral herniation. J Pediatr Surg 1995; 30:1264.
47. Metkus AP, Filly RA, Stringer MD, et al. Sonographic predictors of survival in fetal diaphragmatic hernia. J Pediatr Surg 1996; 31:148.
48. Guibaud L, Filiatrault D, Garel L, et al. Fetal congenital diaphragmatic hernia: accuracy of sonography in the diagnosis and prediction of the outcome after birth. AJR Am J Roentgenol 1996; 166:1195.
49. Laudy JA, Van Gucht M, Van Dooren MF, Wladimiroff JW, Tibboel D. Congenital diaphragmatic hernia: an evaluation of the prognostic value of the lung-to-head ratio and other prenatal parameters. Prenat Diagn 2003; 23(8):634–639.
50. Pfleghaar KM, Wapner RJ, Kuhlman KA, et al. Congenital diaphragmatic hernia: prognosis and prenatal detection. Fetal Diagn Ther 1995; 10:393.
51. Sydorak RM, Harrison MR. Congenital diaphragmatic hernia: advances in prenatal therapy. Clin Perinatol 2003; 30(3):465–479.
52. Harrison MR, Albanese CT, Hawgood SB, et al. Fetoscopic temporary tracheal occlusion by means of detachable balloon for congenital diaphragmatic hernia. Am J Obstet Gynecol 2001; 185(3):730–733.
53. Chiba T, Albanese CT, Farmer DL, et al. Balloon tracheal occlusion for congenital diaphragmatic hernia: experimental studies. J Pediatr Surg 2000; 35(11):1566–1570.
54. Keller RL, Glidden DV, Paek BW, et al. The lung-to-head ratio and fetoscopic temporary tracheal occlusion: prediction of survival in severe left congenital diaphragmatic hernia. Ultrasound Obstet Gynecol 2003; 21(3):244–249.
55. Deprest J, Gratacos E, Nicolaides KH; FETO Task Group. Fetoscopic tracheal occlusion (FETO) for severe congenital diaphragmatic hernia: evolution of a technique and preliminary results. Ultrasound Obstet Gynecol 2004; 24(2):121–126.
56. Harrison MR, Adzick NS, Flake AW, et al. Correction of congenital diaphragmatic hernia in utero. VI. Hard-earn lessons. J Pediatr Surg 1993; 28:1411.
57. Bootstaylor BS, Filly RA, Harrison MR, et al. Prenatal sonographic predicators of liver herniation in congenital diaphragmatic hernia. J Ultrasound Med 1995; 14:515.
58. Beaudoin S, Bargy F, Mahieu D, Barbet P. Anatomic study of the umbilical vein and ductus venosus in human fetuses: ultrasound application in prenatal examination of left congenital diaphragmatic hernia. Surg Radiol Anat 1998; 20(2):99–103.
59. Hedrick HL, Crombleholme TM, Flake AW, et al. Right congenital diaphragmatic hernia: prenatal assessment and outcome. J Pediatr Surg 2004; 39(3):319–323; discussion 319–323.
60. Gilsanz V, Emons D, Hansmann M, et al. Hydrothorax, ascites and right diaphragmatic hernia. Radiology 1986; 158:243.
61. Tsukahara Y, Ohno Y, Itakura A, Mizutani S. Prenatal diagnosis of congenital diaphragmatic eventration by magnetic resonance imaging. Am J Perinatol 2001; 18(5):241–244.
62. Spock A, Schneider S, Baylin GJ. Mediastinal gastric cysts: a case report and review of English literature. Am Rev Respir Dis 1966; 94:97.
63. Knochel JQ, Lee TG, Melendez MG, et al. Fetal anomalies involving the thorax and abdomen. Radiol Clin North Am 1982; 20:297.
64. Morin L, Crombleholme TM, D'Alton ME. Prenatal diagnosis and management of fetal thoracic lesions. Semin Perinatol 1994; 18:228.
65. Longaker MT, Laberge JM, Dansereau J, et al. Primary fetal hydrothorax: natural history and management. J Pediatr Surg 1989; 24:573.
66. Weber AM, Philipson EH. Fetal pleural effusion: a review and meta-analysis for prognostic indicators. Obstet Gynecol 1992; 79:281.
67. Hashimoto K, Shimizu T, Fukuda M, et al. Pregnancy outcome of embryonic/fetal pleural effusion in the first trimester. J Ultrasound Med 2003; 22(5):501–505.

68. Pijpers L, Reuss A, Stewart PA, Wladimiroff JW. Noninvasive management of isolated bilateral fetal hydrothorax. Am J Obstet Gynecol 1989; 161(2):330–332.
69. Estroff JA, Parad RB, Frigoletto FD Jr, et al. The natural history of isolated fetal hydrothorax. Ultrasound Obstet Gynecol 1992; 2:162.
70. Vain NE, Swarner OW, Cha CC. Neonatal chylothorax: a report and discussion of nine consecutive cases. J Pediatr Surg 1980; 15:261.
71. Chernick V, Reed MH. Pneumothorax and chylothorax in the neonatal period. J Pediatr 1970; 76:624.
72. Petres RE, Redwine FO, Cruikshank DP. Congenital bilateral chylothorax: antepartum diagnosis and successful intrauterine surgical management. JAMA 1982; 248:1360.
73. Lange IR, Manning FA. Antenatal diagnosis of congenital pleural effusion. Am J Obstet Gynecol 1981; 140:839.
74. Broadman RF. Congenital chylothorax. NY State J Med 1975; 75:553.
75. Nicolaides KH, Azar GB. Thoraco-amniotic shunting. Fetal Diagn Ther 1990; 5:153.
76. Mahony BS, Filly RA, Callen PW, et al. Severe nonimmune hydrops fetalis: sonographic evaluation. Radiology 1984; 151:757.
77. Benacerraf BR, Frigoletto FD. Mid-trimester fetal thoracentesis. J Clin Ultrasound 1985; 13:202.
78. Tsukihara A, Tanemura M, Suzuki Y, Sato T, Tanaka T, Suzumori K. Reduction of pleural effusion by OK-432 in a fetus complicated with congenital hydrothorax. Fetal Diagn Ther 2004; 19(4):327–331.
79. Jorgensen C, Brocks V, Bang J, Jorgensen FS, Ronsbro L. Treatment of severe fetal chylothorax associated with pronounced hydrops with intrapleural injection of OK-432. Ultrasound Obstet Gynecol 2003; 21(1):66–69.
80. Okawa T, Takano Y, Fujimori K, Yanagida K, Sato A. A new fetal therapy for chylothorax: pleurodesis with OK-432. Ultrasound Obstet Gynecol 2001; 18(4):376–377.
81. Tanemura M, Nishikawa N, Kojima K, Suzuki Y, Suzumori K. A case of successful fetal therapy for congenital chylothorax by intrapleural injection of OK-432. Ultrasound Obstet Gynecol 2001; 18(4):371–375.
82. Nicolaides KH, Rodeck CH, Lange I, et al. Fetoscopy in the assessment of unexplained fetal hydrops. Br J Obstet Gynaecol 1985; 92:671.
83. Blott M, Nicolaides KH, Greenough A. Pleuroamniotic shunting for decompression of fetal pleural effusions. Obstet Gynecol 1988; 71:798.
84. Rodeck CH, Fisk NM, Fraser DI, et al. Long-term in utero drainage of fetal hydrothorax. N Engl J Med 1988; 319:1135.
85. Chao AS, Chung CL, Cheng PJ, Lien R, Soong YK. Thoracoamniotic shunting for treatment of fetal bilateral hydrothorax with hydrops. J Formos Med Assoc 1998; 97(9):646–648.
86. Wittman BK, Martin KA, Wilson RD, Peacock D. Complications of long-term drainage of fetal pleural effusion: case report and review of the literature. Am J Perinatol 1997; 14(8):443–447.
87. Picone O, Benachi A, Mandelbrot L, Ruano R, Dumez Y, Dommergues M. Thoracoamniotic shunting for fetal pleural effusions with hydrops. Am J Obstet Gynecol 2004; 191(6):2047–2050.
88. Wilson RD, Johnson MP. Prenatal ultrasound guided percutaneous shunts for obstructive uropathy and thoracic disease. Semin Pediatr Surg 2003; 12(3):182–189.
89. Seeds JW, Bowes WA. Results of treatment of severe fetal hydrothorax with bilateral pleuroamniotic catheters. Obstet Gynecol 1986; 68:577.
90. Gonen R, Degani S, Kugelman A, Abend M, Bader D. Intrapartum drainage of fetal pleural effusion. Prenat Diagn 1999; 19(12):1124–1126.
91. Prontera W, Jaeggi ET, Pfizenmaier M, Tassaux D, Pfister RE. Ex utero intrapartum treatment (EXIT) of severe fetal hydrothorax. Dis Child Fetal Neonatal Ed 2002; 86(1):F58–F60.
92. Haddon MJ, Bowen A. Bronchopulmonary and neurenteric forms of foregut anomalies. Radiol Clin North Am 1991; 29:241.
93. Maulik D, Robinson L, Daily D, et al. Prenatal sonographic depiction of intralobar pulmonary sequestration. J Ultrasound Med 1987; 6:703.
94. Revillon Y, Jan D, Plattner V, et al. Congenital cystic adenomatoid malformation of the lung: prenatal management and prognosis. J Pediatr Surg 1993; 28:1009.
95. Rosado-de-Christenson ML, Stocker JT. Congenital cystic adenomatoid malformation. Radiographics 1991; 11:865.
96. Harris K. Extralobar sequestration with congenital diaphragmatic hernia: a complicated case study. Neonatal Netw 2004; 23(6):7–24.
97. Hamrick SE, Brook MM, Farmer DL. Fetal surgery for congenital diaphragmatic hernia and pulmonary sequestration complicated by postnatal diagnosis of transposition of the great arteries. Fetal Diagn Ther 2004; 19(1):40–42.
98. Luet'ic T, Crombleholme TM, Semple JP, D'Alton M. Early prenatal diagnosis of bronchopulmonary sequestration with associated diaphragmatic hernia. J Ultrasound Med 1995; 14(7):533–535.
99. Devine WA, Webber SA, Anderson RH. Congenitally malformed hearts from a population of children undergoing cardiac transplantation: comments on sequential segmental analysis and dissection. Pediatr Dev Pathol 2000; 3(2):140–154.
100. Kimbrell B, Degner T, Glatleider P, Applebaum H. Pulmonary sequestration presenting as mitral valve insufficiency. J Pediatr Surg 1998; 33(11):1648–1650.
101. Menon P, Rao KL, Saxena AK. Duplication cyst of the stomach presenting as hemoptysis. Eur J Pediatr Surg 2004; 14(6):429–431.
102. Saggese A, Carbonara A, Russo R, Ciancia G, Ardimento G. Intra-abdominal extra lobar pulmonary sequestration communicating with gastric duplication—a case report. Eur J Pediatr Surg 2002; 12(6):426–428.
103. Hadley GP, Egner J. Gastric duplication with extralobar pulmonary sequestration: an uncommon cause of "colic". Clin Pediatr (Phila) 2001; 40(6):364.
104. Fenton LZ, Williams JL. Bronchopulmonary foregut malformation mimicking neuroblastoma. Pediatr Radiol 1996; 26(10): 729–730.
105. Yasufuku M, Hatakeyama T, Maeda K, Yamamoto T, Iwai Y. Bronchopulmonary foregut malformation: a large bronchogenic cyst communicating with an esophageal duplication cyst. J Pediatr Surg 2003; 38(2):e2.
106. Kim KW, Kim WS, Cheon JE, et al. Complex bronchopulmonary foregut malformation: extralobar pulmonary sequestration associated with a duplication cyst of mixed bronchogenic and oesophageal type. Pediatr Radiol 2001; 31(4):265–268.
107. Derbent M, Orun UA, Varan B, et al. A new syndrome within the oculo-auriculo-vertebral spectrum: microtia, atresia of the external auditory canal, vertebral anomaly, and complex cardiac defects. Clin Dysmorphol 2005; 14(1):27–30.
108. Brown JW, Ruzmetov M, Minnich DJ, et al. Surgical management of scimitar syndrome: an alternative approach. J Thorac Cardiovasc Surg 2003; 125(2):238–245.
109. Zylak CJ, Eyler WR, Spizarny DL, Stone CH. Developmental lung anomalies in the adult: radiologic-pathologic correlation. Radiographics 2002; 22 Spec No:S25–S43.
110. Ryckman PC, Rosenkrantz JG. Thoracic surgical problems in infancy and childhood. Surg Clin North Am 1985; 65:1423.
111. Savic B, Birtel FJ, Tholen W, et al. Lung sequestration: report of seven cases and review of 540 published cases. Thorax 1979; 34:96.
112. Landing BH, Dixon LG. Congenital malformations and genetic disorder of the respiratory tract (larynx, trachea, bronchi, and lung). Am Rev Respir Dis 1979; 120:151.
113. Bratu I, Flageole H, Chen MF, Di Lorenzo M, Yazbeck S, Laberge JM. The multiple facets of pulmonary sequestration. J Pediatr Surg 2001; 36(5):784–790.
114. Becmeur F, Horta-Geraud P, Donato L, Sauvage P. Pulmonary sequestrations: prenatal ultrasound diagnosis, treatment, and outcome. J Pediatr Surg 1998; 33(3):492–496.
115. Walford N, Htun K, Chen J, Liu YY, Teo H, Yeo GS. Intralobar sequestration of the lung is a congenital anomaly: anatomopathological analysis of four cases diagnosed in fetal life. Pediatr Dev Pathol 2003; 6(4):314–321.
116. Lopoo JB, Goldstein RB, Lipshutz GS, Goldberg JD, Harrison MR, Albanese CT. Fetal pulmonary sequestration: a favorable congenital lung lesion. Obstet Gynecol 1999; 94(4):567–571.

117. Pumberger W, Hormann M, Deutinger J, Bernaschek G, Bistricky E, Horcher E. Longitudinal observation of antenatally detected congenital lung malformations (CLM): natural history, clinical outcome and long-term follow-up. Eur J Cardiothorac Surg 2003; 24(5):703–711.
118. Adzick NS, Harrison MR, Crombleholme TM, Flake AW, Howell LJ. Fetal lung lesions: management and outcome. Am J Obstet Gynecol 1998; 179(4):884–889.
119. Hernanz-Schulman M, Stein SM, Neblett WW, et al. Pulmonary sequestration: diagnosis with color Doppler sonography and a new theory of associated hydrothorax. Radiology 1991; 180:818.
120. Kato H, Yoshikawa M, Saito T, Fukuchi M, Kato R, Kuwano H. Congenital bronchoesophageal fistula with Crohn's disease in an adult: report of a case. Surg Today 2001; 31(5):446–449.
121. Weitzman JJ, Brennan LP. Bronchogastric fistula, pulmonary sequestration, malrotation of the intestine, and Meckel's diverticulum—a new association. J Pediatr Surg 1998; 33(11): 1655–1677.
122. Gomendoza M, Padmanabhan K, Nicolas A, Sheka KP. CT appearance of esophagobronchial fistula: a clue to lung sequestration. AJR Am J Roentgenol 1997; 169(2):601–602.
123. Siffring PA, Forrest TS, Hill WC, et al. Prenatal sonographic diagnosis of bronchopulmonary foregut malformations. J Ultrasound Med 1989; 8:277.
124. Carter R. Pulmonary sequestration. Ann Thorac Surg 1969; 7:68.
125. Albright EB, Crane JP, Schackelford GD. Prenatal diagnosis of a bronchogenic cyst. J Ultrasound Med 1988; 7:91.
126. Rahmani MR, Filler RM, Shuckett B. Bronchogenic cyst occurring in the antenatal period. J Ultrasound Med 1995; 14:971.
127. Young G, L'Heureux PR, Krueckenberg ST, et al. Mediastinal bronchogenic cyst: prenatal sonographic diagnosis. AJR Am J Roentgenol 1989; 152:125.
128. De Catte L, De Backer T, Delhove O, et al. Ectopic bronchogenic cyst: sonographic findings and differential diagnosis. J Ultrasound Med 1995; 14:321.
129. Fraser RG, Pare JAP. Pulmonary abnormalities of developmental origin. In: Diagnosis of Diseases of the Chest. 2nd ed. Philadelphia: WB Saunders, 1977:602.
130. Dumontier C, Graviss ER, Silberstein MJ, et al. Bronchogenic cysts in children. Clin Radiol 1985; 36:431.
131. McCullagh M, Bhuller AS, Pierro A, Spitz L. Antenatal identification of a cervical oesophageal duplication. Pediatr Surg Int 2000; 16(3):204–205.
132. Uludag S, Madazli R, Erdogan E, Dervisoglu S, Celik E, Ocak V. A case of prenatally diagnosed fetal neurenteric cyst. Ultrasound Obstet Gynecol 2001; 18(3):277–279.
133. Wilkinson CC, Albanese CT, Jennings RW, et al. Fetal neurenteric cyst causing hydrops: case report and review of the literature. Prenat Diagn 1999; 19(2):118–121.
134. Masumoto K, Arima T, Nakatsuji T, Kukita J, Toyoshima S. Duodenal atresia with a deletion of midgut associated with left lung, kidney, and upper limb absences and right upper limb malformation. J Pediatr Surg 2003; 38(11):E1–E4.
135. Berkenstadt M, Lev D, Achiron R, Rosner M, Barkai G. Pulmonary agenesis, microphthalmia, and diaphragmatic defect (PMD): new syndrome or association? Am J Med Genet 1999; 86(1):6–8.
136. Dahnert WF. Radiology Review Manual. 4th ed. Philadelphia: Lippincott Williams & Wilkins, 1998.
137. Vettraino IM, Tawil A, Comstock CH. Bilateral pulmonary agenesis: prenatal sonographic appearance simulates diaphragmatic hernia. J Ultrasound Med 2003; 22(7):723–726.
138. Song MS, Yoo SJ, Smallhorn JF, Mullen JB, Ryan G, Hornberger LK. Bilateral congenital diaphragmatic hernia: diagnostic clues at fetal sonography. Ultrasound Obstet Gynecol 2001; 17(3):255–258.
139. Harrison MR, Jester JA, Ross NA. Correction of congenital diaphragmatic hernia in utero. I. The model: intrathoracic balloon produces fatal pulmonary hypoplasia. Surgery 1980; 88:174.
140. Harrison MR, Bressack MA, Chung AM, et al. Correction of congenital diaphragmatic hernia in utero. II. Simulated correction permits fetal lung growth with survival at birth. Surgery 1980; 88:260.
141. Keslar P, Newman B. Radiographic manifestations of anomalies of the lung. Radiol Clin North Am 1991; 29:255.
142. Kizilcan F, Tanyel FC, Cakar N, et al. The effect of low amniotic pressure without oligohydramnios on fetal lung development in a rabbit model. Am J Obstet Gynecol 1995; 173:36.
143. Harding R, Hooper SB, Dickson KA. A mechanism leading to reduced lung expansion and lung hypoplasia in fetal sheep during oligohydramnios. Am J Obstet Gynecol 1990; 163:1904.
144. McNamara MF, McCurdy CM, Reed KL, et al. The relation between pulmonary hypoplasia and amniotic fluid volume: lessons learned from discordant urinary tract anomalies in monoamniotic twins. Obstet Gynecol 1995; 85:867.
145. Yang LC, Taylor DR, Kaufman HH, Hume R, Calhoun B. Maternal and fetal outcomes of spontaneous preterm premature rupture of membranes. J Am Osteopath Assoc 2004; 104(12):537–542.
146. Falk SJ, Campbell LJ, Lee-Parritz A, et al. Expectant management in spontaneous preterm premature rupture of membranes between 14 and 24 weeks' gestation. J Perinatol 2004; 24(10): 611–616.
147. Vergani P, Locatelli A, Verderio M, Assi F. Premature rupture of the membranes at <26 weeks' gestation: role of amnioinfusion in the management of oligohydramnios. Acta Biomed Ateneo Parmense 2004; 75(suppl 1):62–66.
148. De Carolis MP, Romagnoli C, De Santis M, Piersigilli F, Vento G, Caruso A. Is there significant improvement in neonatal outcome after treating pPROM mothers with amnio-infusion? Biol Neonate 2004; 86(4):222–229. Epub 2004.
149. De Santis M, Scavo M, Noia G, et al. Transabdominal amnioinfusion treatment of severe oligohydramnios in preterm premature rupture of membranes at less than 26 gestational weeks. Fetal Diagn Ther 2003; 18(6):412–417.
150. Gayea PD, Grant DC, Doubilet PM, et al. Prediction of fetal lung maturity: inaccuracy of study using conventional ultrasound instruments. Radiology 1985; 155:473.
151. Jordan HVF. Cardiac size during prenatal development. Obstet Gynecol 1987; 69:854.
152. Nimrod C, Davies D, Iwanicki S, et al. Ultrasound prediction of pulmonary hypoplasia. Obstet Gynecol 1986; 68:495.
153. DeVore GR, Horenstein J, Platt LD. Fetal echocardiography: assessment of cardiothoracic disproportion—a new technique for the diagnosis of thoracic hypoplasia. Am J Obstet Gynecol 1986; 155:1066.
154. Chitkara U, Rosenberg J, Chervenak FA, et al. Prenatal sonographic assessment of the fetal thorax: normal values. Am J Obstet Gynecol 1987; 156:1069.
155. Vintzileos AM, Campbell WA, Rodis JF, et al. Comparison of six different ultrasonographic methods for predicting lethal fetal pulmonary hypoplasia. Am J Obstet Gynecol 1989; 161:606.
156. D'Alton M, Mercer B, Riddick E, et al. Serial thoracic versus abdominal circumference ratios for the prediction of pulmonary hypoplasia in premature rupture of the membranes remote from term. Am J Obstet Gynecol 1992; 166:658.
157. Fuke S, Kanzaki T, Mu J, et al. Antenatal prediction of pulmonary hypoplasia by acceleration time/ejection time ratio of fetal pulmonary arteries by Doppler blood flow velocimetry. Am J Obstet Gynecol 2003; 188(1):228–233.
158. Laudy JA, Tibboel D, Robben SG, de Krijger RR, de Ridder MA, Wladimiroff JW. Prenatal prediction of pulmonary hypoplasia: clinical, biometric, and Doppler velocity correlates. Pediatrics 2002; 109(2):250–258.
159. Yoshimura S, Masuzaki H, Miura K, Muta K, Gotoh H, Ishimaru T. Diagnosis of fetal pulmonary hypoplasia by measurement of blood flow velocity waveforms of pulmonary arteries with Doppler ultrasonography. Am J Obstet Gynecol 1999; 180(2 Pt 1):441–446.
160. Mitchell JM, Roberts AB, Lee A. Doppler waveforms from the pulmonary arterial system in normal fetuses and those with pulmonary hypoplasia. Ultrasound Obstet Gynecol 1998; 11(3):167–172.
161. Sokol J, Bohn D, Lacro RV, et al. Fetal pulmonary artery diameters and their association with lung hypoplasia and postnatal outcome in congenital diaphragmatic hernia. Am J Obstet Gynecol 2002; 186(5):1085–1090.

162. Roberts AB, Mitchell JM. Direct ultrasonographic measurement of fetal lung length in normal pregnancies and pregnancies complicated by prolonged rupture of membranes. Am J Obstet Gynecol 1990; 163:1560.
163. Williams G, Coakley FV, Qayyum A, Farmer DL, Joe BN, Filly RA. Fetal relative lung volume: quantification by using prenatal MR imaging lung volumetry. Radiology 2004; 233(2):457–462. Epub 2004.
164. Tanigaki S, Miyakoshi K, Tanaka M, et al. Pulmonary hypoplasia: prediction with use of ratio of MR imaging-measured fetal lung volume to US-estimated fetal body weight. Radiology 2004; 232(3):767–772.
165. Coakley FV, Lopoo JB, Lu Y, Hricak H, Albanese CT, Harrison MR, Filly RA. Normal and hypoplastic fetal lungs: volumetric assessment with prenatal single-shot rapid acquisition with relaxation enhancement MR imaging. Radiology 2000; 216(1):107–111.
166. Osada H, Kaku K, Masuda K, Iitsuka Y, Seki K, Sekiya S. Quantitative and qualitative evaluations of fetal lung with MR imaging. Radiology 2004; 231(3):887–892. Epub 2004.
167. Bhargava R, Brewerton L, Chari R, Liang Y. Lung-to-liver signal intensities: development of a normal scale and possible role in predicting pulmonary hypoplasia in utero. In: Radiological Society of North America scientific assembly and annual meeting program. Oak Brook, Ill: Radiological Society of North America, 2004.
168. Keller TM, Rake A, Michel SC, et al. MR assessment of fetal lung development using lung volumes and signal intensities. Eur Radiol 2004; 14(6):984–989. Epub 2004.
169. Kuwashima S, Nishimura G, Iimura F, et al. Low-intensity fetal lungs on MRI may suggest the diagnosis of pulmonary hypoplasia. Pediatr Radiol 2001; 31(9):669–672.
170. Ruano R, Benachi A, Joubin L, et al. Three-dimensional ultrasonographic assessment of fetal lung volume as prognostic factor in isolated congenital diaphragmatic hernia. BJOG 2004; 111(5): 423–429.
171. Fenton BW, Lin CS, Seydel F, Macedonia C. Lecithin can be detected by volume-selected proton MR spectroscopy using a 1.5T whole body scanner: a potentially non-invasive method for the prenatal assessment of fetal lung maturity. Prenat Diagn 1998; 18(12):1263–1266.
172. Kamata S, Sawai T, Usui N, et al. Case of congenital bronchial atresia detected by fetal ultrasound. Pediatr Pulmonol 2003; 35(3):227–229.
173. Keswani SG, Crombleholme TM, Pawel BR, et al. Prenatal diagnosis and management of mainstem bronchial atresia. Fetal Diagn Ther 2005; 20(1):74–78.
174. McAlister WH, Wright JR, Crane JP. Main-stem bronchial atresia: intrauterine sonographic diagnosis. AJR Am J Roentgenol 1987; 148:364.
175. Mendoza A, Wolf P, Edwards DK, et al. Prenatal ultrasonic diagnosis of cystic adenomatoid malformation of the lung. Arch Pathol Med 1986; 110:402.
176. Cachia R, Sobornya RE. Congenital cystic adenomatoid malformation of the lung with bronchial atresia. Hum Pathol 1981; 12:947.
177. Watson WJ, Thorp JM, Miller RC, et al. Prenatal diagnosis of laryngeal atresia. Am J Obstet Gynecol 1990; 163:1456.
178. Dolkart LA, Reimers FT, Wertheimer IS, et al. Prenatal diagnosis of laryngeal atresia. J Ultrasound Med 1992; 11:496.
179. Weston MJ, Porter HJ, Berry PJ, et al. Ultrasonographic prenatal diagnosis of upper respiratory tract atresia. J Ultrasound Med 1992; 11:673.
180. Kalache KD, Chaoui R, Tennstedt C, Bollmann R. Prenatal diagnosis of laryngeal atresia in two cases of congenital high airway obstruction syndrome (CHAOS). Prenat Diagn 1997; 17(6):577–581.
181. Silver MM, Thurston WA, Patrick JE. Perinatal pulmonary hyperplasia due to laryngeal atresia. Hum Pathol 1988; 19:110.
182. Fox H, Locher J. Laryngeal atresia. Arch Dis Child 1964; 7:515.
183. Richards DS, Yancey MK, Duff P, et al. The perinatal management of severe laryngeal stenosis. Obstet Gynecol 1992; 80:537.
184. Lim FY, Crombleholme TM, Hedrick HL, et al. Congenital high airway obstruction syndrome: natural history and management. J Pediatr Surg 2003; 38(6):940–945.
185. Hirose S, Farmer DL, Lee H, Nobuhara KK, Harrison MR. The ex utero intrapartum treatment procedure: looking back at the EXIT. J Pediatr Surg 2004; 39(3):375–380; discussion 375–380.
186. Oepkes D, Teunissen AK, Van De Velde M, Devlieger H, Delaere P, Deprest J. Congenital high airway obstruction syndrome successfully managed with ex-utero intrapartum treatment. Ultrasound Obstet Gynecol 2003; 22(4):437–439.
187. Bouchard S, Johnson MP, Flake AW, et al. The EXIT procedure: experience and outcome in 31 cases. J Pediatr Surg 2002; 37(3):418–426.
188. Bui TH, Grunewald C, Frenckner B, et al. Successful EXIT (ex utero intrapartum treatment) procedure in a fetus diagnosed prenatally with congenital high-airway obstruction syndrome due to laryngeal atresia. Eur J Pediatr Surg 2000; 10(5):328–333.
189. Crombleholme TM, Sylvester K, Flake AW, Adzick NS. Salvage of a fetus with congenital high airway obstruction syndrome by ex utero intrapartum treatment (EXIT) procedure. Fetal Diagn Ther 2000; 15(5):280–282.
190. Paek BW, Callen PW, Kitterman J, et al. Successful fetal intervention for congenital high airway obstruction syndrome. Fetal Diagn Ther 2002; 17(5):272–276.
191. Stocker JT, Madewell JE, Drake RM. Congenital cystic adenomatoid malformation of the lung. Hum Pathol 1977; 8:155.
192. Van Dijk C, Wagenvoort CA. The various types of congenital adenomatoid malformations of the lung. J Pathol 1973; 110:131.
193. Wolf SA, Hertzler JH, Philippart AI. Cystic adenomatoid dysplasia of the lung. J Pediatr Surg 1980; 15:925.
194. Heling KS, Tennstedt C, Chaoui R. Unusual case of a fetus with congenital cystic adenomatoid malformation of the lung associated with trisomy 13. Prenat Diagn 2003; 23(4):315–318.
195. Pham TT, Benirschke K, Masliah E, Stocker JT, Yi ES. Congenital pulmonary airway malformation (congenital cystic adenomatoid malformation) with multiple extrapulmonary anomalies: autopsy report of a fetus at 19 weeks of gestation. Pediatr Dev Pathol 2004; 7(6):661–666. Epub 2004.
196. Roberts D, Sweeney E, Walkinshaw S. Congenital cystic adenomatoid malformation of the lung coexisting with recombinant chromosome 18. A case report. Fetal Diagn Ther 2001; 16(2):65–67.
197. Bale PM. Congenital cystic malformation of the lung. Am J Clin Pathol 1979; 71:411.
198. Oster AG, Fortune DW. Congenital cystic adenomatoid malformation of the lung. Am J Clin Pathol 1978; 70:595.
199. Miller RK, Sieber WK, Yunis EJ. Congenital adenomatoid malformations of the lung: a report of 17 cases, and review of the literature. Pathol Annu 1980; 15:387.
200. MacGillivray TE, Harrison MR, Goldstein RB, et al. Disappearing fetal lung lesions. J Pediatr Surg 1993; 28:1321.
201. Fine C, Adzick NS, Doubilet PM. Decreasing size of a congenital cystic adenomatoid malformation in utero. J Ultrasound Med 1988; 7:405.
202. Glaves J, Baker JL. Spontaneous resolution of maternal polyhydramnios in congenital cystic adenomatoid malformation of the lung. Br J Obstet Gynaecol 1983; 90:1065.
203. Diamond IR, Wales PW, Smith SD, Fecteau A. Survival after CCAM associated with ascites: a report of a case and review of the literature. J Pediatr Surg 2003; 38(9):E1–E3.
204. De Santis M, Masini L, Noia G, Cavaliere AF, Oliva N, Caruso A. Congenital cystic adenomatoid malformation of the lung: antenatal ultrasound findings and fetal-neonatal outcome. Fifteen years of experience. Fetal Diagn Ther 2000; 15(4):246–250.
205. Bunduki V, Ruano R, da Silva MM, et al. Prognostic factors associated with congenital cystic adenomatoid malformation of the lung. Prenat Diagn 2000; 20(6):459–464.
206. Higby K, Melendez BA, Heiman HS. Spontaneous resolution of nonimmune hydrops in a fetus with a cystic adenomatoid malformation. J Perinatol 1998; 18(4):308–310.
207. Laberge JM, Flageole H, Pugash D, et al. Outcome of the prenatally diagnosed congenital cystic adenomatoid lung malformation: a Canadian experience. Fetal Diagn Ther 2001; 16(3):178–186.
208. van Leeuwen K, Teitelbaum DH, Hirschl RB, et al. Prenatal diagnosis of congenital cystic adenomatoid malformation and its postnatal presentation, surgical indications, and natural history. J Pediatr Surg 1999; 34(5):794–798; discussion 798–799.

209. Harrison MR, Adzick NS, Jennings R, et al. Antenatal intervention for congenital cystic adenomatoid malformation. Lancet 1990; 336:965.
210. Adzick NS, Flake AW, Crombleholme TM. Management of congenital lung lesions. Semin Pediatr Surg 2003; 12(1):10–16.
211. Brown MF, Lewis D, Brouillette RM, Hilman B, Brown EG. Successful prenatal management of hydrops, caused by congenital cystic adenomatoid malformation, using serial aspirations. J Pediatr Surg 1995; 30(7):1098–1099.
212. Meagher SE, Simon DR, Hodges S, Glasson MJ. Successful outcome with serial amniocenteses for polyhydramnios complicating cystic adenomatoid malformation of the lung. Aust N Z J Obstet Gynaecol 1995; 35(3):326–328.
213. Wilson RD, Baxter JK, Johnson MP, et al. Thoracoamniotic shunts: fetal treatment of pleural effusions and congenital cystic adenomatoid malformations. Fetal Diagn Ther 2004; 19(5):413–420.
214. Morikawa M, Yamada H, Okuyama K, et al. Prenatal diagnosis and fetal therapy of congenital cystic adenomatoid malformation type I of the lung: a report of five cases. Congenit Anom (Kyoto) 2003; 43(1):72–78.
215. Adzick NS, Harrison MR, Glick PL. Fetal cystic adenomatoid malformation: prenatal diagnosis and natural history. J Pediatr Surg 1985; 20:483.
216. Clark SL, Vitale DJ, Minton SD, et al. Successful fetal therapy for cystic adenomatoid malformation associated with second trimester hydrops. Am J Obstet Gynecol 1987; 157:294.
217. Kuller AJ, Yankowitz J, Goldberg JD, et al. Outcome of antenatally diagnosed cystic adenomatoid malformation. Am J Obstet Gynecol 1992; 167:1038.
218. Dommergues M, Louis-Sylvestre C, Mandelbrot L, et al. Congenital adenomatoid malformation of the lung: when is active fetal therapy indicated? Am J Obstet Gynecol 1997; 177(4):953–958.
219. Miller JA, Corteville JE, Langer JC. Congenital cystic adenomatoid malformation in the fetus: natural history and predictors of outcome. J Pediatr Surg 1996; 31(6):805–808.
220. Entezami M, Halis G, Waldschmidt J, Opri F, Runkel S. Congenital cystic adenomatoid malformation of the lung and fetal hydrops—a case with favourable outcome. Eur J Obstet Gynecol Reprod Biol 1998; 79(1):99–101.
221. Kasales CJ, Coulson CC, Meilstrup JW, Ambrose A, Botti JJ, Holley GP. Diagnosis and differentiation of congenital diaphragmatic hernia from other noncardiac thoracic fetal masses. Am J Perinatol 1998; 15(11):623–628.
222. Lacy DE, Shaw NJ, Pilling DW, Walkinshaw S. Outcome of congenital lung abnormalities detected antenatally. Acta Paediatr 1999; 88(4):454–458.
223. Wansaicheong GK, Ong CL. Congenital lobar emphysema: antenatal diagnosis and follow up. Australas Radiol 1999; 43(2):243–245.
224. Babu R, Kyle P, Spicer RD. Prenatal sonographic features of congenital lobar emphysema. Fetal Diagn Ther 2001; 16(4): 200–202.
225. Quinton AE, Smoleniec JS. Congenital lobar emphysema—the disappearing chest mass: antenatal ultrasound appearance. Ultrasound Obstet Gynecol 2001; 17(2):169–171.
226. Olutoye OO, Coleman BG, Hubbard AM, Adzick NS. Prenatal diagnosis and management of congenital lobar emphysema. J Pediatr Surg 2000; 35(5):792–795.
227. MacKenzie TC, Guttenberg ME, Nisenbaum HL, Johnson MP, Adzick NS. A fetal lung lesion consisting of bronchogenic cyst, bronchopulmonary sequestration, and congenital cystic adenomatoid malformation: the missing link? Fetal Diagn Ther 2001; 16(4):193–195.
228. McCullagh M, MacConnachie I, D Garvie, et al. Accuracy of prenatal diagnosis of congenital cystic adenomatoid malformation. Arch Dis Child 1994; 71:F111.
229. Budorick NE, Pretorius DH, Leopold GR, et al. Spontaneous improvement of intrathoracic masses diagnosed in utero. J Ultrasound Med 1992; 11:653.
230. Albanese CT, Lopoo J, Goldstein RB, et al. Fetal liver position and perinatal outcome for congenital diaphragmatic hernia. Prenat Diag 1998; 18(11):1138–1142.

Fetal Heart

John P. McGahan and Beryl R. Benacerraf

47

INTRODUCTION

Congenital heart disease is a common congenital anomaly, estimated to be present in 8 of 1000 of all live births (1,2). Prenatal diagnosis of congenital heart disease is especially important when considering that fetal cardiac malformations often have a high associated rate of morbidity and mortality in both the fetus and the neonate. Often cardiac defects are associated with chromosomal abnormalities or other noncardiac fetal abnormalities. Complete examination of the fetal heart includes use of high-resolution real-time ultrasound examination of the heart in several anatomic planes, M-mode sonography, pulsed Doppler ultrasound, color flow mapping and, more recently, use of three-dimensional or four-dimensional cardiac imaging.

INDICATIONS

Certain fetuses are at high risk of congenital heart disease. Risk factors may be divided into two separate groups: fetal risk factors and maternal or familial risk factors (Table 1) (3,4). Furthermore, if there has been a cardiac defect in a sibling or a parent, this is associated with an increased risk of cardiac malformations in subsequent pregnancies. Although the risk of congenital heart disease in a neonate is 8 in 1000, this risk increases up to 12% if a parent has congenital heart disease (5). When any fetal anomaly is detected, the fetal heart should be carefully scrutinized and examined because of the high risk of associated cardiac malformations in these cases (Table 2) (6,7).

TECHNIQUE AND NORMAL ANATOMY

When evaluating the fetal heart, the examiner should first document the fetal position. Second, the examiner must determine if the left side of the heart is up or down. Third, the stomach side and the relation of the suprahepatic portion of the inferior vena cava to the right atrium should be noted. Fetal cardiac malposition is extremely complex. For more details, the reader should refer to an excellent review of malposition of the heart by Van Praagh et al. (8).

The ten cardiac segments may be greatly simplified into three main cardiac segments that are diagnostically important: (*i*) the visceroatrial situs, which is important in localization of the atrium; (*ii*) the ventricular loop, which is important in diagnosing the relation of the ventricles to the atrium; and (*iii*) the truncus arteriosus, which is important for diagnostic understanding of the relation between the great arteries and the ventricle.

Visceroatrial situs abnormalities may be divided into three separate types:

Situs solitus is the normal, noninverted type in which the visceral situs is normal, with the stomach to the left. The morphologic right atrium is right sided, and the morphologic left atrium is left sided.

TABLE 1 ■ Risk Factors for Congenital Heart Disease

Fetal risk factors	Maternal or familial risk factors
Extracardiac anomalies	Congenital heart disease
Chromosomal abnormality	Sibling
Fetal cardiac arrhythmia	Maternal
Heart block	Paternal
Premature atrial or ventricular contractions	Teratogenic exposures
Tachycardia (>200 beats/min)	Alcohol
Bradycardia	Lithium carbonate
Nonimmune hydrops fetalis	Progestins
Question of cardiac anomaly on prior ultrasound	Amphetamines
Intrauterine growth retardation	Others
Polyhydramnios	Maternal disorders
	Diabetes mellitus
	Collagen vascular disease
	Phenylketonuria
	Maternal infections
	Rubella
	Toxoplasmosis
	Coxsackievirus infection
	Cytomegalovirus infection
	Mumps
	Familial syndromes

Source: Adapted from Refs. 3, 4.

Situs inversus is the exact mirror image of situs solitus. The stomach is to the left, but the morphologic right atrium is left sided, and the morphologic left atrium is right sided in an inverted anatomic pattern.

TABLE 2 ■ Systemic Malformations Associated with Congenital Heart Disease

Central nervous system	Hydrocephalus
	Dandy–Walker syndrome
	Agenesis of corpus callosum
	Meckel–Gruber syndrome
	Microcephaly
	Holoprosencephaly
Mediastinum	Tracheoesophageal fistula
	Esophageal atresia
Gastrointestinal tract	Duodenal atresia
	Jejunal atresia
	Anorectal anomalies
	Imperforate anus
Ventral wall	Omphalocele
	Ectopia cordis
Diaphragmatic hernia	
Genitourinary tract	Renal agenesis
	Horseshoe kidney
	Renal dysplasia

Source: Adapted from Ref. 7.

Situs ambiguus is an anatomically indeterminate type of visceral situs in which the liver is usually located in the midline, with stomach located either to the left or to the right. This type of situs is part of the heterotaxy syndrome or cardiosplenic syndrome. This has been typically divided into what can be thought of as the asplenia syndrome and the polysplenia syndrome. Asplenia syndrome is often thought morphologically to be two right atria. Although this is not completely true, this abnormality is important to recognize because most of these fetuses

TABLE 3 ■ Classification of Visceroatrial Situs and Associated CHD

Situs solitus
- Thoracic and abdominal viscera normal position
- Levocardia: less than 1% incidence of CHD
- Dextrocardia: 95% incidence of CHD

Situs inversus
- Thoracic and abdominal viscera completely inverted
- Levocardia: extremely rare, with greater than 95% incidence of CHD
- Dextrocardia: approximately 5% incidence of CHD

Situs ambiguous
- Liver midline
- Asplenia: 95% incidence of CHD
- Polysplenia: approximately 75% incidence of CHD

Abbreviation: CHD, congenital heart disease.

develop severe cyanosis as newborns from congenital heart disease. Alternatively, polysplenia, often thought of as bilateral left sidedness, is also associated with cardiac problems (Table 3). In these syndromes, the liver may be midline. There may be a common atrium, a common atrioventricular valve, and a single ventricle (Fig. 1).

The relation of the atria to the ventricles may be important to recognize prenatally. Anatomically, the right atrium may open into the right ventricle in a normal or concordant relation, or it may be discordant, with the right atrium opening into the left ventricle. Furthermore, other abnormalities of the relation of the atria to the ventricles

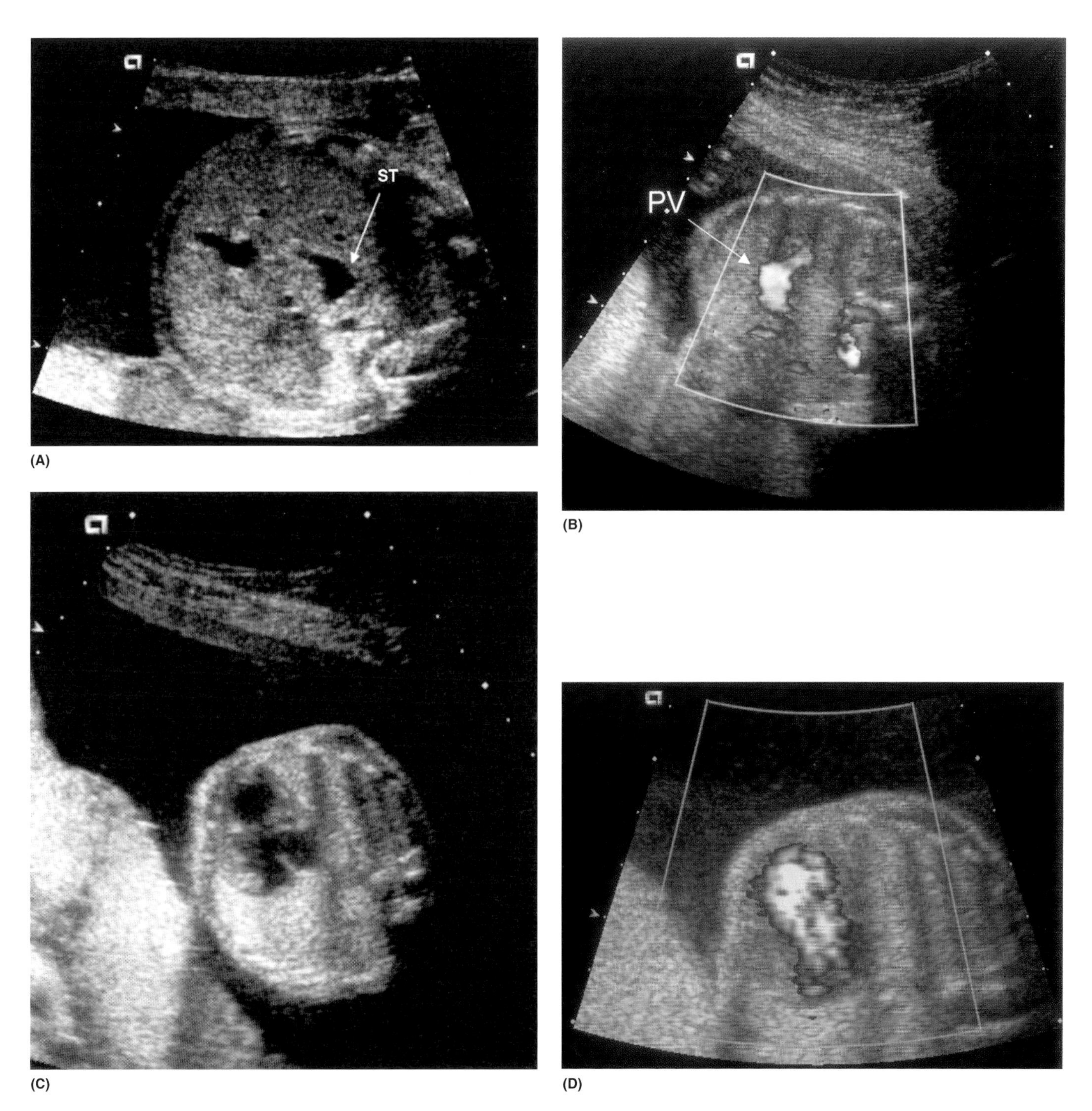

FIGURE 1 A–D ■ Cardiac heterotaxy cardiosplenic syndrome. (**A**) A transverse view of the upper abdomen; the aorta demonstrates the aorta anterior to the spine and the stomach (ST) is noted midline. (**B**) A transverse view of the upper abdomen demonstrates the liver as noted midline and is documented by this color Doppler image of the portal vein (PV). (**C**) This four-chamber view of the heart demonstrates the cardiac apex pointed to the right and associated large atrioventricular septal defect. (**D**) Power Doppler image of the heart better demonstrates the large atrioventricular septal defect.

may be present, including abnormal positions of the tricuspid valve within the right ventricle, such as in Ebstein's anomaly.

Finally, an abnormal relation of the outflow tracts may be diagnosed prenatally. The most common abnormality of this type is transposition of the great vessels in which the right ventricle gives rise to the aorta and the left ventricle supplies the pulmonary artery.

Four-Chamber View of the Heart

After determining that the heart is located within the left side of the chest, the examiner identifies the fetal thoracic spine, and a scan is obtained transverse to the thorax to obtain a four-chamber view of the heart (Fig. 2). Anatomically, the right ventricle is behind the sternum, and the left ventricle is inferior and to the left of the right ventricle. Differential

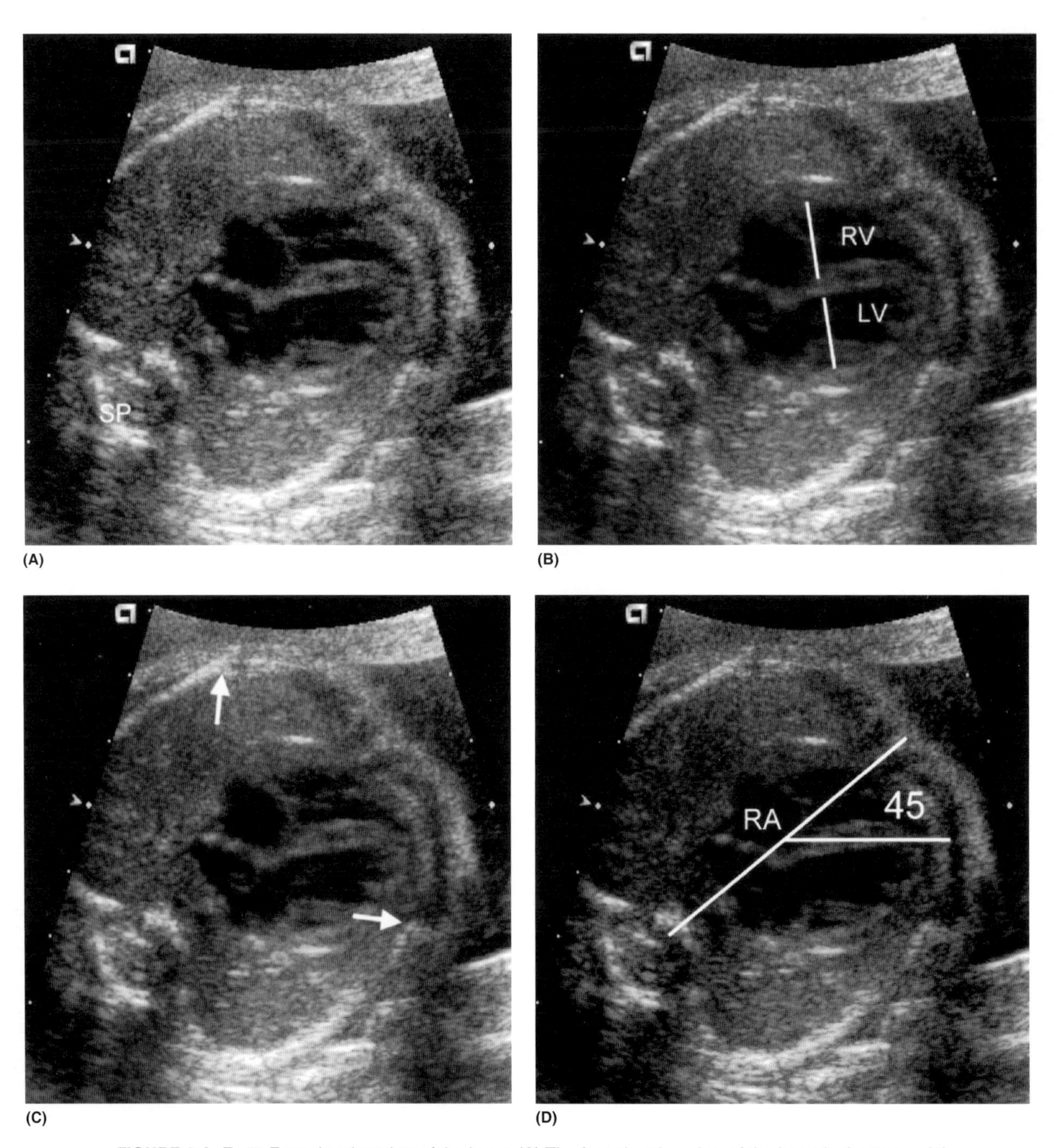

FIGURE 2 A–E ■ Four-chamber view of the heart. (**A**) The four-chamber view of the heart in the transaxial plane shows the spine (SP) noted posteriorly. (**B**) Visual inspection of the diameter of the right ventricle (RV) and the left ventricle (LV) is performed to ensure that they are in a ratio of approximately 1 to 1. (**C**) The anterior margin of both ribs is located (*arrows*). (**D**) A line is drawn from the spine to the anterior sternum. The interventricular septum intersects that line at approximately 45°. Note that the right atrium (RA) lies to the right side of the spinal sternal line. (*Continued*)

(E)

FIGURE 2 A–E ■ (*Continued*) (**E**) The heart can be noted to occupy approximately one-third of the fetal thorax.

points in identification of the right and left ventricles are given in Table 4 (9). Generally, the diameters of the right and left ventricles maintain about a 1:1 ratio, as identified on sonograms during diastole at the atrioventricular valve region (Table 5). This ratio may be assessed on real-time ultrasound, and if precise measurements are to be made, documentation may be done with M-mode ultrasound. M-mode biventricular measurements obtained during ventricular diastole are compared with the thoracic circumference and may be helpful to predict cardiomegaly or pulmonary hypoplasia (10). Regardless of the fetal position, the interventricular septum traverses a plane of about 40° to 45°, with a line drawn between the spine and the sternum (Fig. 2). The heart should occupy about one-third of the fetal thorax. Features of the four-chamber view of the heart are detailed in Table 5.

Specific heart chambers can be identified as follows: The right atrium may be identified when scanning in different anatomic planes and noting the hepatic veins, inferior vena cava, and superior vena cava draining into that structure. The foramen ovale is noted opening from the right atrium into the left atrium.

The left atrium is posterior in location in comparison with the right atrium, with the foramen ovale opening into this chamber. The position of the spine is noted, with the left atrium lying close to the vertebral column.

The right atrium drains into the right ventricle. The right ventricle is retrosternal in location, with the tricuspid valve lower in position within the right ventricle than is the mitral valve within the left ventricle. One also sees a large echogenic structure lying within the right ventricle, the muscular moderator band (Fig. 2).

The left atrium drains into the left ventricle. Papillary muscles are identified within the left ventricle. Echogenic bright spots seen in the left ventricle are thought to be attachments of the papillary muscles and are not in themselves abnormal. They will be discussed later in this chapter. The mitral valve is farther from the apex within the left ventricle than is the tricuspid within the right ventricle (Fig. 3). The apex of the heart and the ventricular septum are just cephalad to the fetal stomach.

TABLE 5 ■ Features Found in the Normal Four-Chamber View of the Fetal Heart

- Left ventricle to right ventricle ratio approximately 1:1
- Left atrium to right atrium ratio approximately 1:1
- Cardiac apex approximately 45°
- Cardiac area approximately one-third of thoracic area
- Right ventricle retrosternal
- Left ventricle-left heart border
- Descending aorta to the left/anterior to spine
- Left atrium anterior to descending aorta
- Foramen ovale protrudes into left atrium
- Muscle of moderator band in right ventricle thicker than muscles in left ventricle
- Tricuspid valve insertion more toward apex vs. mitral valve

TABLE 4 ■ Identification of Right and Left Ventricles from the Four-Chamber View

View	Right ventricle	Left ventricle
Position within thorax	Right ventricle retrosternal	Left border, same side as stomach
Flap of foramen ovale	–	Present within left atrium
Insertion of atrioventricular valve leaflets on interventricular septum	Tricuspid valve inserted lower than mitral valve	Mitral valve inserted higher than tricuspid valve
Muscle	Thicker moderator band	–

Source: Modified from Ref. 9.

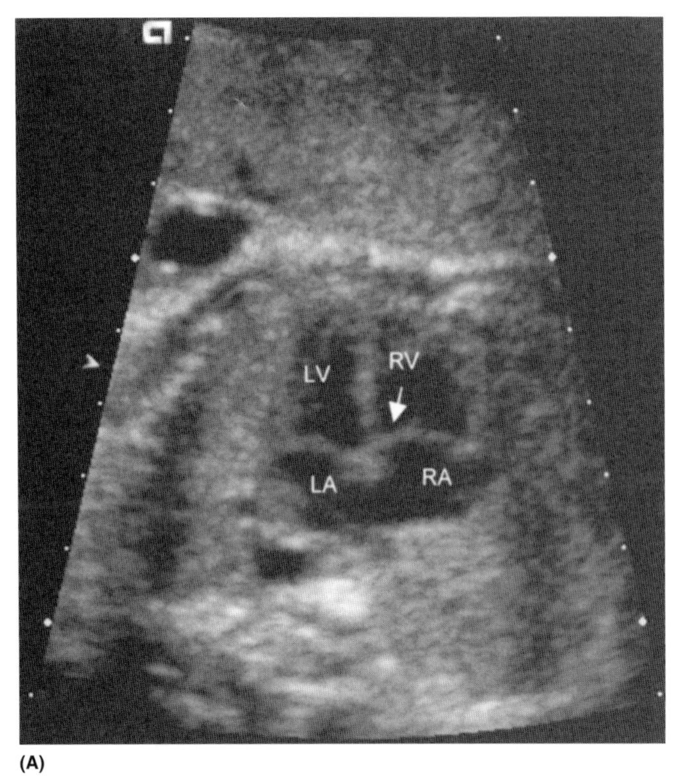

(A)

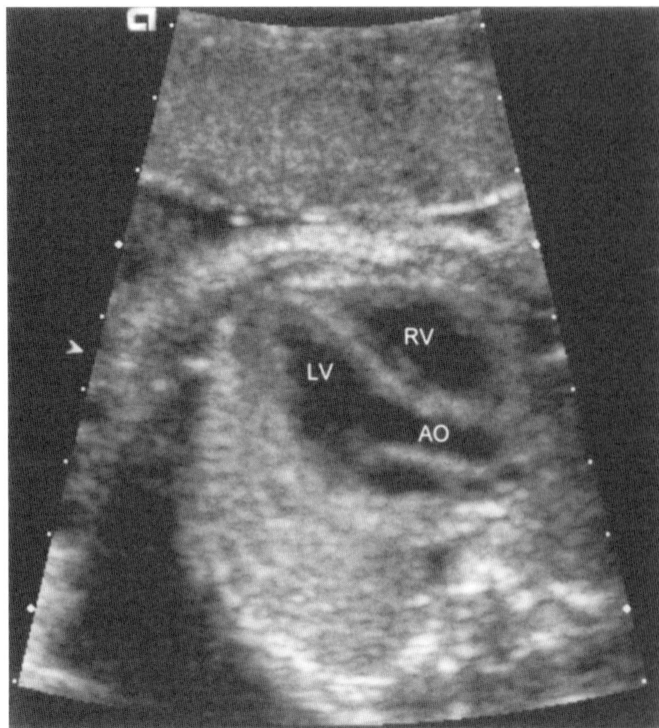

(B)

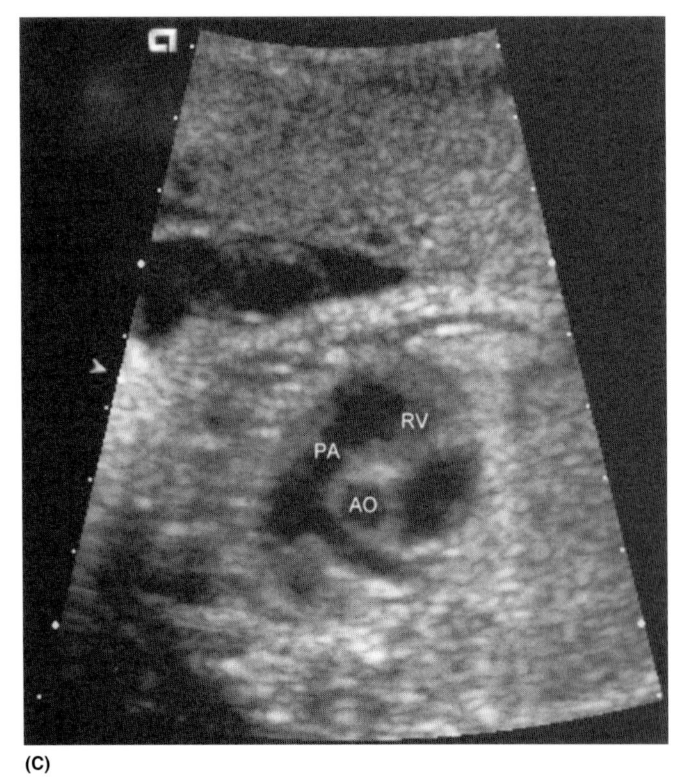

(C)

FIGURE 3 A–C ■ Outflow tracts—apex up. (**A**) The four-chamber view of the heart with the apex up. Note that tricuspid valve is closer to the apex (*arrow*) compared to the mitral valve. (**B**) After performing a four-chamber view of the heart, the transducer is placed in an angle between the left upper quadrant of the abdomen and the right shoulder. This demonstrates the retrosternal location of the right ventricle (RV) as well as the aorta (AO) originating from left ventricle (LV). This view is also helpful to detect membranous ventricular septal defects. (**C**) With the apex of the heart again pointed toward the transducer, the transducer is angled at almost 90° from the long axis plane. In this view, the circular AO is noted centrally. The RV gives rise to the main pulmonary artery (PA).

Long-Axis View of the Heart

In most fetuses, the interventricular septum lies perpendicular to the transducer beam. The relation of the aorta to the left ventricle is best evaluated by the left ventricular long-axis view of the fetal heart. This view is obtained by rotating the transducer from the four-chamber view into a plane-angled one from the fetal stomach toward the right shoulder of the fetus (Figs. 3 and 4). This view is helpful for evaluating the relation of the proximal aortic arch exiting the left ventricle.

Pulmonary Outflow Tract

Once the aortic outflow tract is identified as described previously, the transducer is "rocked" into nearly a straight

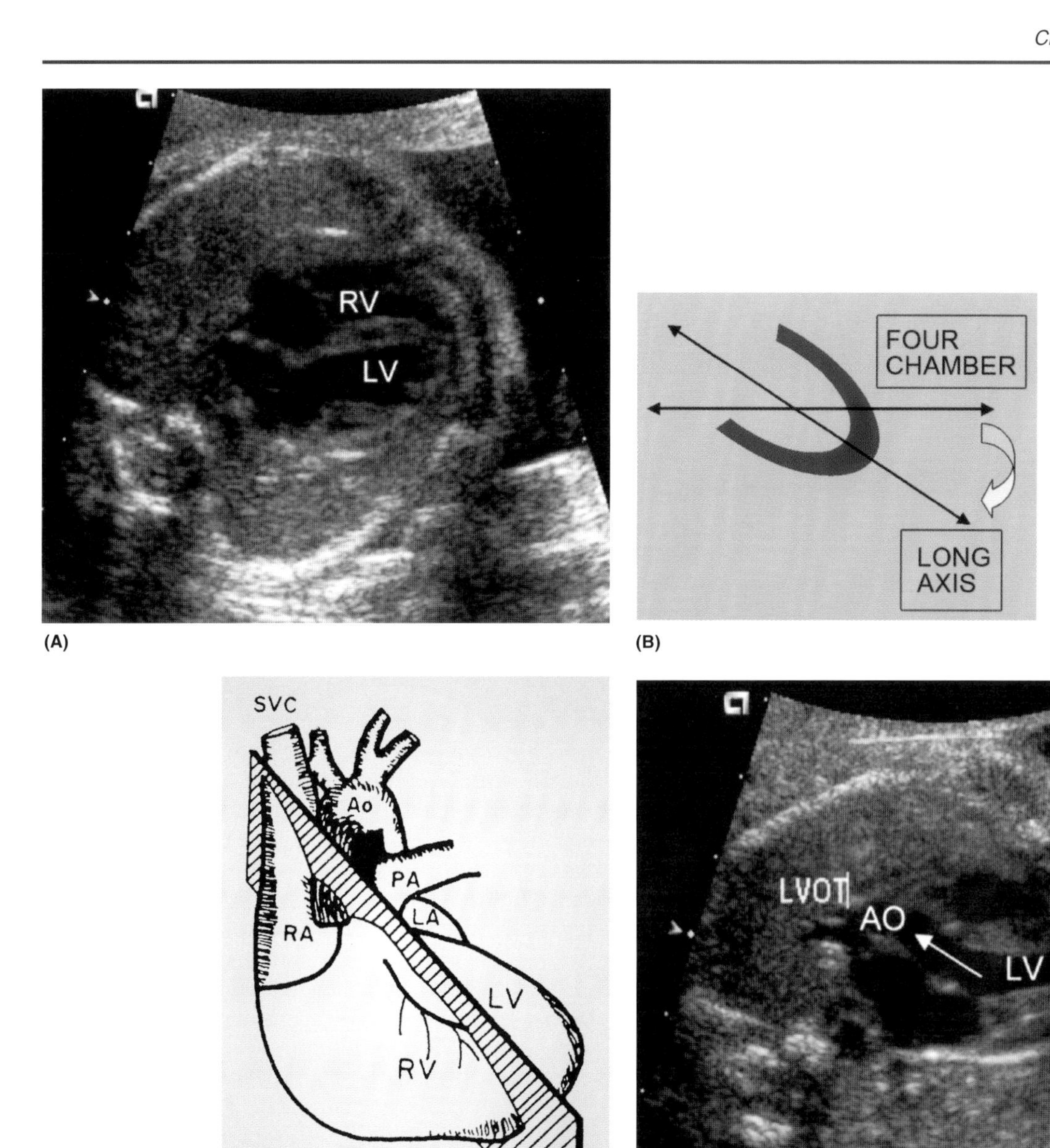

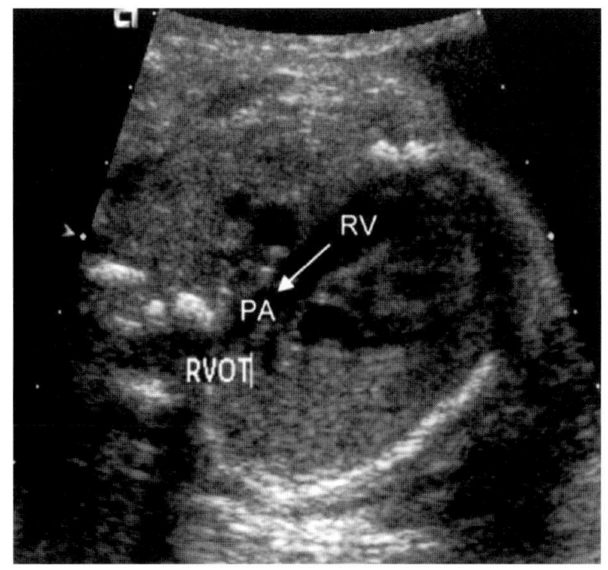

FIGURE 4 A–E ■ Outflow tracts. (**A**) Normal four chambers of the heart. (**B**) By changing from the four-chamber view of the heart to a more oblique scan plane angling from the fetal left upper quadrant and abdomen to the right shoulder, one is in the correct anatomic plane for the long-axis view of the heart. (**C**) Anatomic drawing showing the plane used to obtain the four-chamber view of the heart through the left ventricle (LV) and the aortic (AO) outflow tract. (**D**) Ultrasound shows the LVOT with the AO exiting the LV (*arrow*). (**E**) By rocking the transducer, the RVOT is seen with the PA exiting the RV. Note the RVOT is at right angle to the LVOT obtained in (**D**). RV, right ventricle; LV, left ventricle; AO, aorta; PA, pulmonary artery; RVOT, right ventricular outflow tract; LVOT, left ventricular outflow tract.

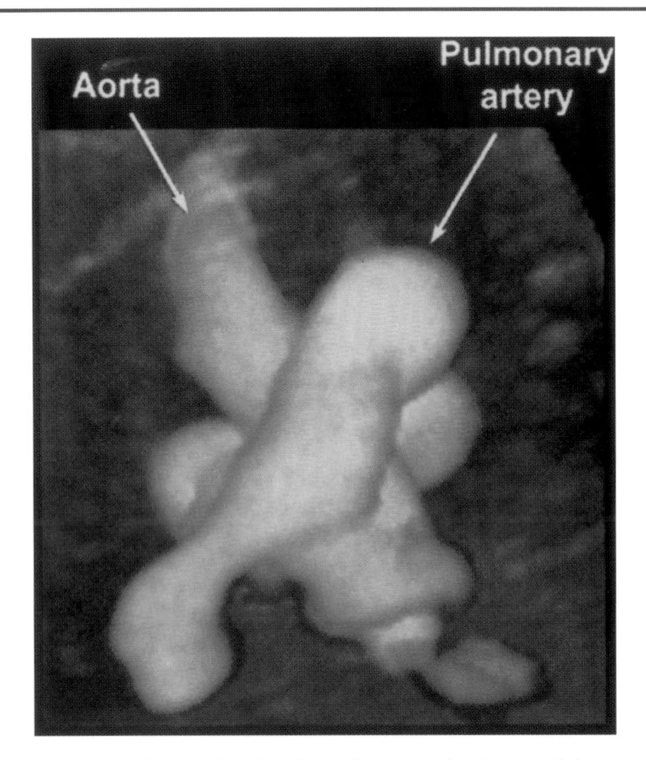

FIGURE 5 ■ Power Doppler four dimensional ultrasound imaging demonstrates normal crisscross relationship of the pulmonary artery and aorta. *Source*: From Ref. 11.

sagittal plane. This view identifies the main pulmonary artery exiting the right ventricle perpendicular to the ascending aorta. This crisscross relation of the pulmonary artery and the aorta is the result of normal rotation of the great vessels early in embryogenesis (Figs. 4 and 5) (11). If the pulmonary artery and aorta are parallel rather than intersecting, a rotational abnormality of the great vessel is present, most commonly transposition of the great vessels or double outlet of the right ventricle.

Short-Axis View of the Heart

When the fetal cardiac apex is directed toward the transducer, the examiner can best identify the pulmonary outflow tracts by placing the transducer parallel to the spine and moving to the left side of the fetus (Fig. 3). In this view, a circular structure is identified (aorta) with a similar-size vessel draping over it (pulmonary artery). Lateral to the aorta, the tricuspid valve leaflets can be observed opening and closing. Within the aorta and the pulmonary artery, movement of the valves can be observed. The pulmonary artery bifurcates near the spine (9).

Five Views of the Heart

Yagel et al. proposed an examination of a comprehensive examination of the fetal heart. These views including the following (Fig. 6) (12). These five views include a view through the stomach to establish situs. The second view is the fourth chamber view of the heart. In the five-chamber view, the aorta centrally is the third view, followed by a slightly more cephalad view showing the pulmonary artery bifurcation. The fifth and most cephalad view is of the pulmonary artery, aorta, and superior vena cava and other anatomy as listed in Figure 6.

Aortic Arch

The longitudinal view of the aortic arch can be obtained when the transducer is positioned so that a parasagittal scan through a right anterior approach or a left posterior approach can identify the aortic arch and the descending aorta. In this longitudinal view of the aortic arch, the great vessels are identified originating from the aortic arch (Fig. 7) (9). A view of the ductus arteriosus entering the descending aorta can be obtained by moving the transducer slightly from this anatomic plane (Fig. 7). The arch of the aorta makes a smooth curve with a small radius described as a "candy cane." The pulmonary artery makes a curve with a wider radius, and the right ventricle lies more anteriorly than the left ventricles. The pulmonary artery forces the ductus and the descending aorta, and this has been described as a "hockey stick" (Fig. 7).

ULTRASOUND PITFALLS

Pitfalls normally encountered during an examination of the fetal heart include the entities described in the following paragraphs.

FIGURE 6 ■ (*Facing page*) Five short-axis views for optimal fetal heart screening. The color image shows the trachea, heart and great vessels, liver, and stomach with the five planes of insonation superimposed. Polygons show the angle of the transducer and are assigned to the relevant gray-scale images. (**I**) The most caudal plane, showing the fetal stomach, cross-section of the abdominal aorta, spine, and liver. (**II**) Four-chamber view of the fetal heart, showing the RV and LV and atria (RA, LA), foramen ovale, and pulmonary veins to the right and left of the aorta. (**III**) "Five-chamber" view, showing the aortic root, LVs, RVs and atria (LA, RA), and a cross-section of the descending aorta (AO with *arrow*). (**IV**) The slightly more cephalad view showing the MPA and the bifurcation of LPA and RPA and cross-sections of the ascending and descending aorta (AO and AO with *arrow*, respectively). (**V**) The three-vessel tracheal plane of insonation showing the pulmonary trunk proximal aorta, DA, distal aorta, SVC, and the trachea. AO, aorta; DA, ductus arteriosus; PA, pulmonary artery; SVC, superior vena cava; (**I**) AO, abdominal aorta; SP, spine; LI, liver; LT, left; RT, right; (**II**) RV, right ventricle; LV, left ventricle; FO, foramen ovale; PV, pulmonary veins; AO, aorta; (**III**) AO, aortic root; RV, right ventricle; LV, left ventricle; MPA, main pulmonary artery; LPA, left pulmonary artery; RPA, right pulmonary artery; (**V**) P, pulmonary trunk; (P)Ao, proximal aorta; DA, ductus arteriosus; (D)Ao, distal aorta; SVC, superior vena cava; T, trachea. *Source*: From Ref. 12.

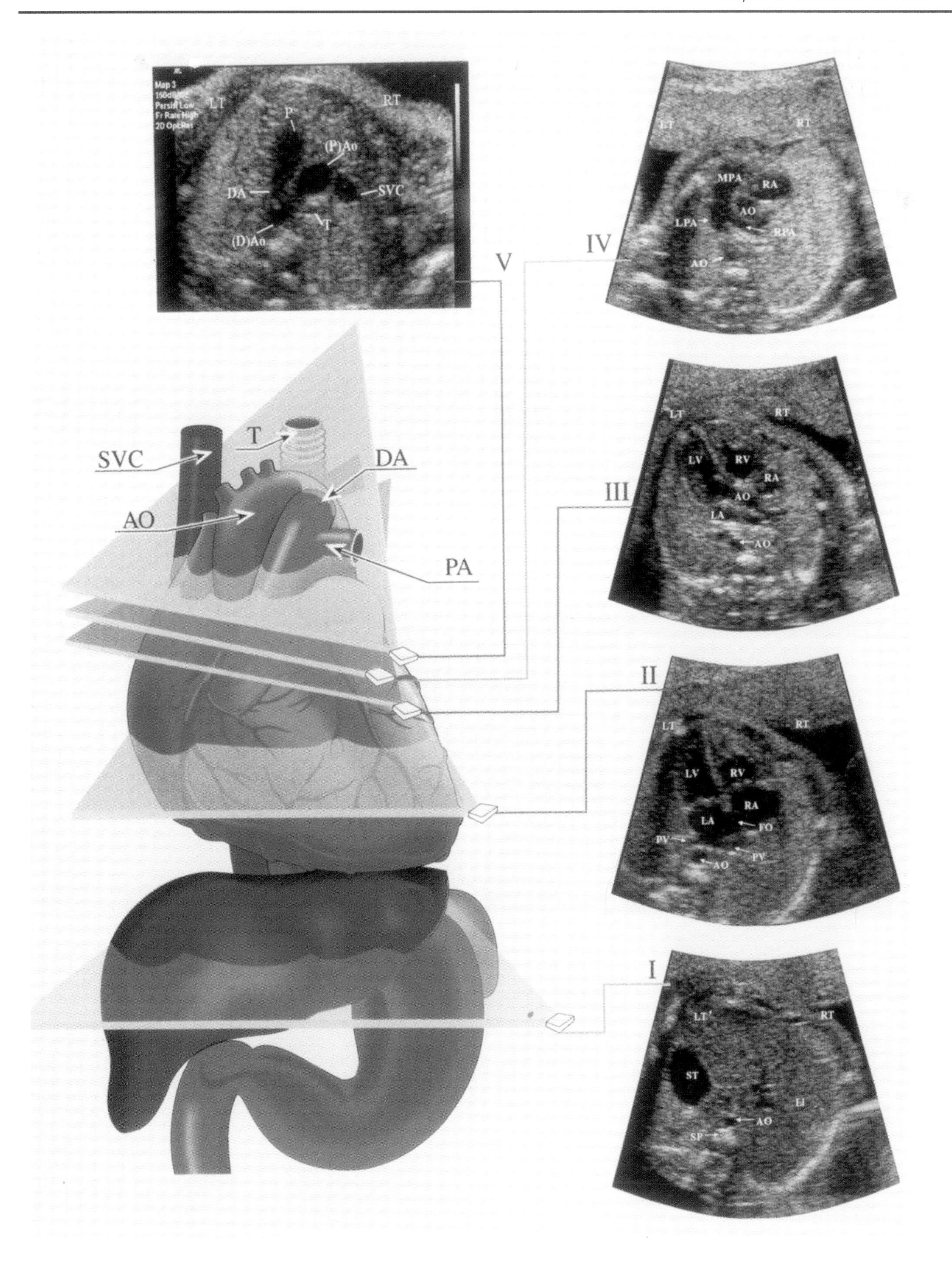

LT
RT
P
(P)Ao
DA
SVC
T
(D)Ao
V
IV
MPA
RA
AO
LPA
RPA
SVC
T
DA
AO
PA
III
LV
RV
LA
II
FO
PV
I
ST
SP

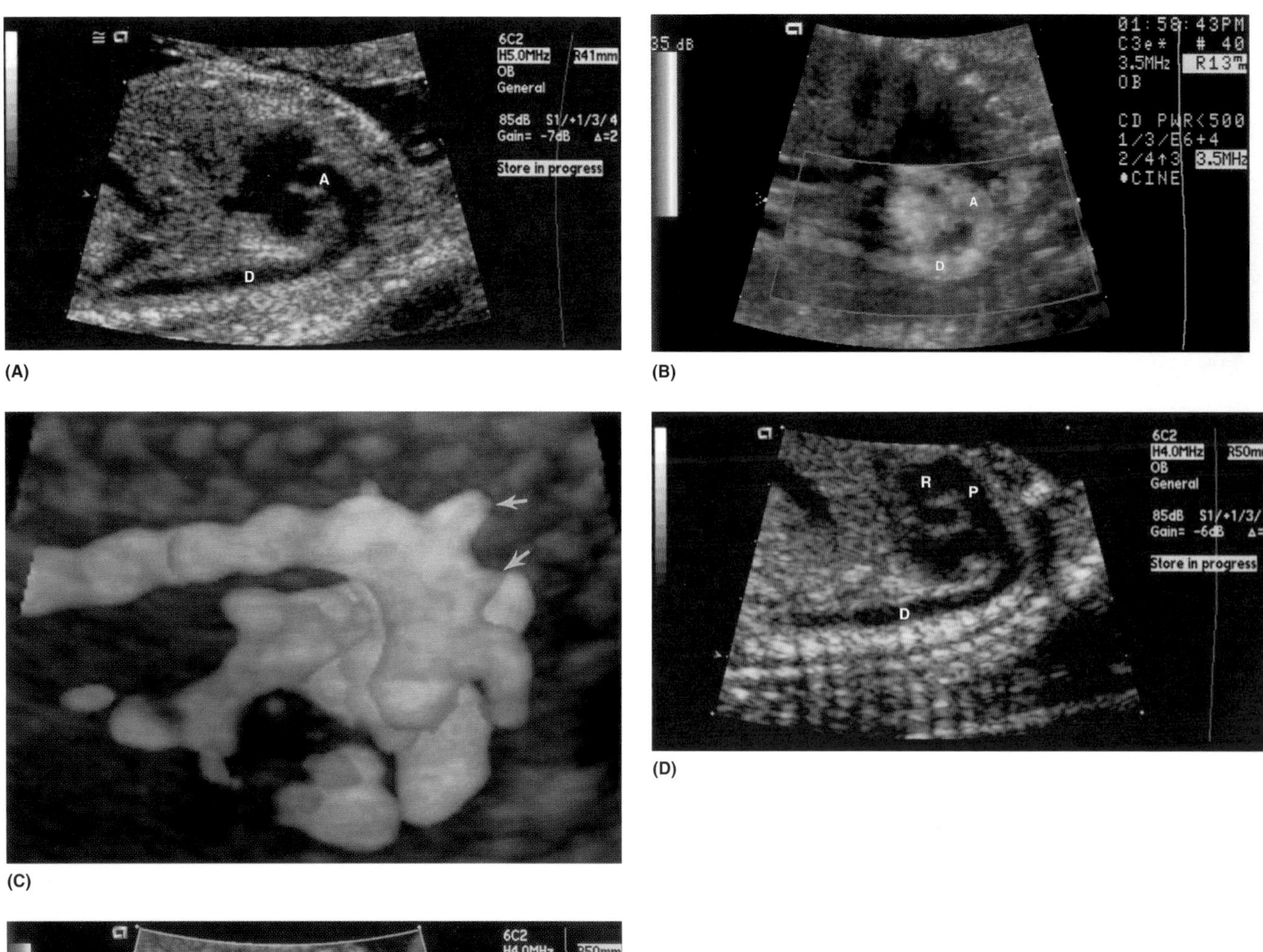

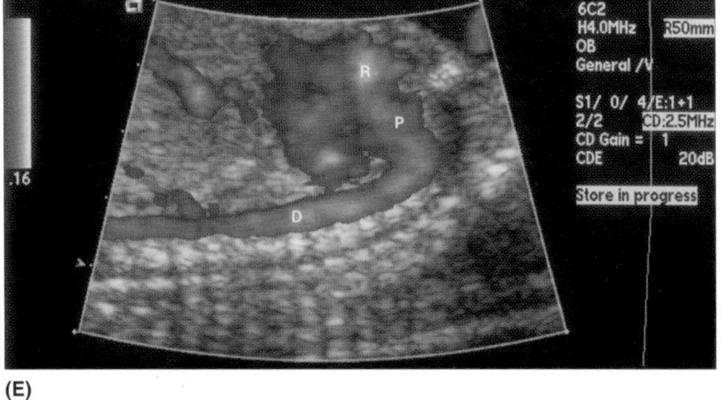

FIGURE 7 A–E ■ Aortic and pulmonic artery. (**A**,**B**) Real-time and power Doppler longitudinal-view images of the aortic arch, which gives a "candy cane" appearance. (**C**) Three-dimensional power Doppler image of aortic arch demonstrating takeoff of great vessel from the arch (*curved arrows*). (**D**,**E**) Real-time and power Doppler image of the pulmonary artery with ductal connection to the descending aorta giving a "hockey stick" configuration. R, right ventricle; MPA, main pulmonary artery; A, ascending aorta; D, descending aorta; p, pulmonary artery.

Septation within the Right Atrium

Normally, a valve is present where the inferior vena cava and the coronary sinus empty into the right atrium. This may appear as a septation within the right atrium. Another potential explanation of septation within the right atrium is the small remnants of the valve of the sinus venosus, called Chiari's network. In any case, these structures are normal and should not be mistaken for abnormal folds within the right atrium (13).

Bright Spot Apex of Ventricle

Occasionally, a "bright spot" may be observed at the apex of the left ventricle within the ventricular wall. This phenomenon is probably caused by echogenic focus of the overlying rib or sternum (Fig. 8) (13). Moreover, an intraventricular "bright spot" or echogenic focus may be observed in either ventricle, most commonly the left, as discussed later in this chapter (Fig. 9).

Pseudoventricular Septal Defect

The membranous portion of the ventricular septum is located just below the atrioventricular valves and is normally thin. It may be so thin that it appears to be a small ventricular septal defect (VSD). Color-flow imaging may be helpful to further ascertain whether a defect is actually present in the ventricular septum.

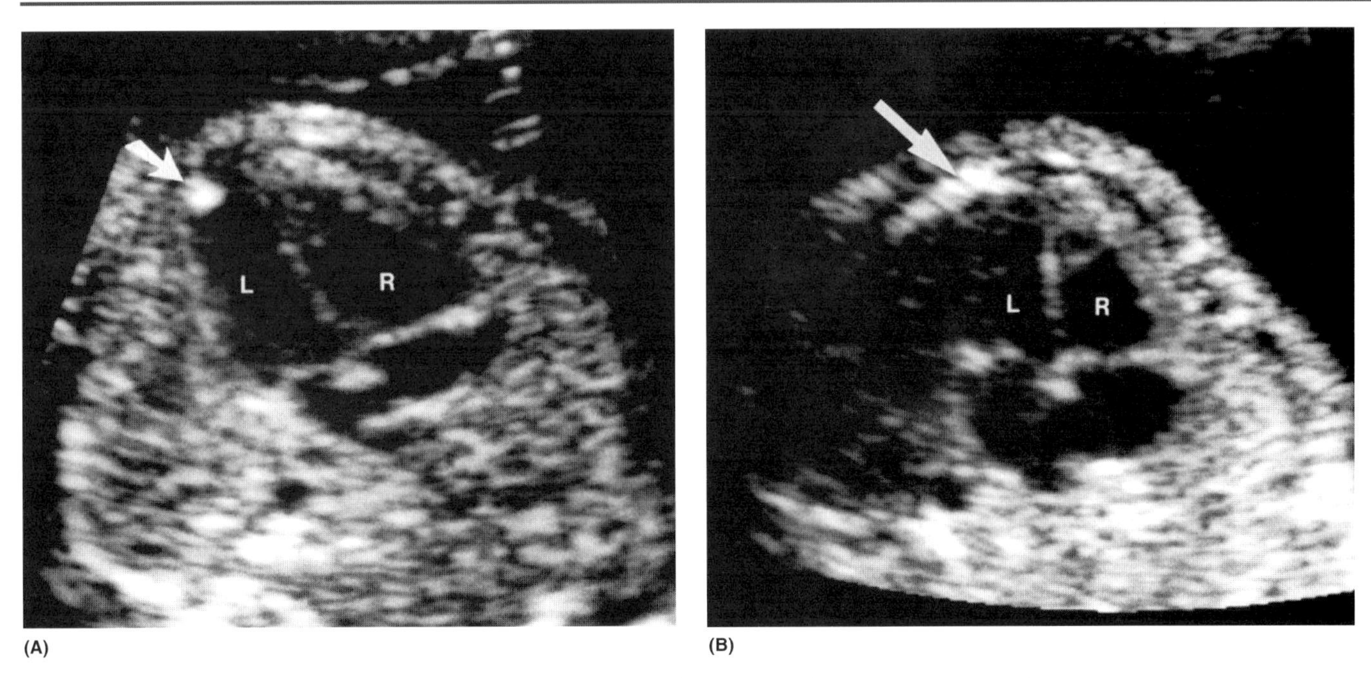

FIGURE 8 A AND B ■ Sonogram of an echogenic focus at the edge of the myocardium. (**A**) Four-chamber view shows a small echogenic focus (*arrow*) at the apex of the left ventricle (L). (**B**) If the transducer is rotated into a plane that shows a longer section of the rib (*arrow*), then a pitfall can be recognized. R, right ventricle. *Source*: From Ref. 13.

Pseudoatrial Septal Defect

Because of the rapid movement of both the septum primum and the septum secundum, these structures may not be visualized when parallel to the ultrasound beam and thus may appear to be an atrial septal defect (ASD). Review of the real-time images in a number of planes may better demonstrate normal motion of the foramen ovale.

Pseudopericardial Effusion

The normal hypoechoic myocardium should not be mistaken for a pericardial effusion. The myocardium of the ventricles during systole may appear hypoechoic and may be mistaken for a pericardial effusion (14). Normally, a small amount of pericardial fluid may be observed around the heart. This fluid is observed more prominently during ventricular systole (13,15).

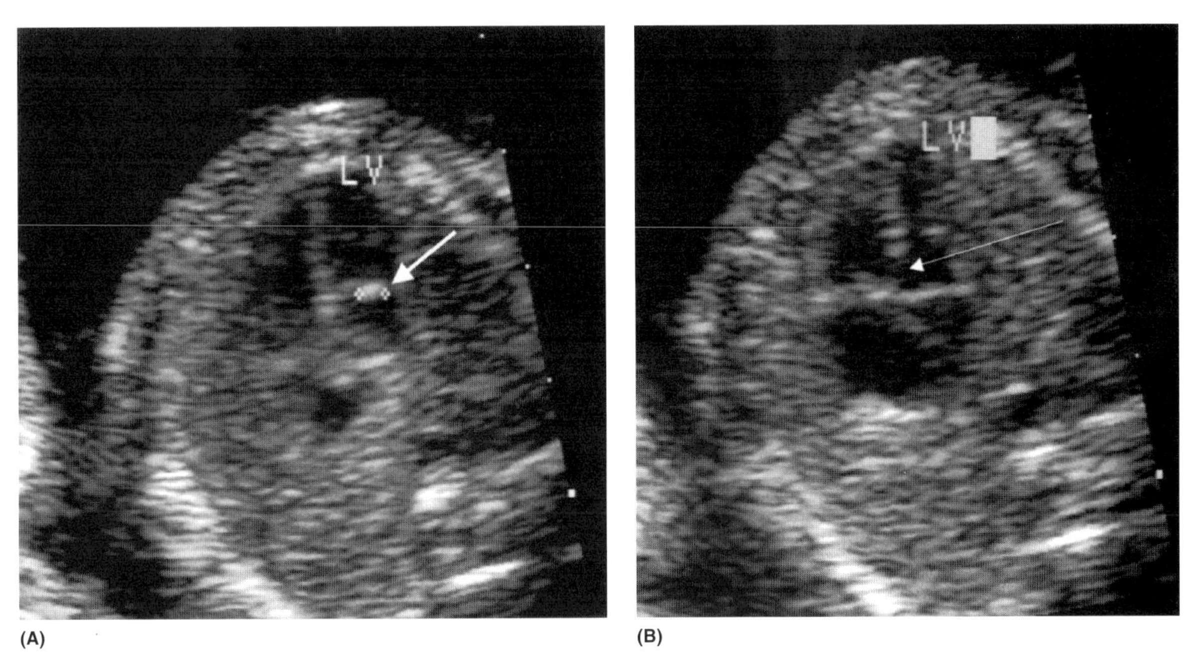

FIGURE 9 A AND B ■ Echogenic cardiac focus—Trisomy 21. (**A**) Echogenic cardiac focus is noted within the left ventricle (LV) of the heart (*arrow*). (**B**) Also noted was a small ventricular septal defect (*arrow*). There was nuchal thickening greater than 8 mm on this fetus that had Trisomy 21.

TABLE 6 ■ Causes of Fetal Cardiac Death

Cardiac abnormalities	Incidence (%)[a]
Hypoplastic left ventricle	29
Coarctation of the aorta	10
Transposition of the great vessels	8.5
Pulmonic stenosis or pulmonic atresia	7.5
Tetralogy of Fallot	6
Truncus arteriosus	3
Ventricular septal defect	2
Atrial ventricular canal	1
Tricuspid atresia	0.5

[a]This table details the percentage of different cardiac abnormalities identified in newborns that died within the first month of life; data collected from 1967 through 1978 at the University of Minnesota.

Source: From Ref. 2.

TABLE 8 ■ Potential Causes of Fetal Cardiomegaly

Hydrops
- Immune
- Nonimmune

Increased flow
- Arteriovenous malformations
- Vein of Galen aneurysm
- Hemangioma of the liver

Intrinsic cardiac anomaly
- Ebstein's (tricuspid valve) abnormality
- Other isolated valve stenosis with intact septum (rare)

Cardiomyopathy

Cardiac tumor
- Rhabdomyoma

DETECTION OF CARDIAC ANOMALIES: FOUR-CHAMBER VIEW

Controversy exists about the accuracy of the four-chamber view of the heart in detecting cardiac malformations, with the percentage varying from 0% in a nontertiary care center to as high as 96% (3,16). Other studies have demonstrated the accuracy of the four-chamber view in detecting cardiac malformations varying between 48% and 63% (17,18). A review of an autopsy series of newborns from the University of Minnesota indicates that common cardiac anomalies include hypoplastic left ventricle, pulmonary atresia, large VSDs, atrioventricular canal, and tricuspid atresia. Investigators may reasonably believe that most of these abnormalities may be detected on a four-chamber view of the heart (Table 6) (18). When performing a four-chamber view of the heart, two questions are often helpful in detecting intrathoracic extracardiac disease (19).

Intrathoracic Extracardiac Disease

Is the Heart in Normal Position?

The heart may be used as a normal anatomic marker for extracardiac intrathoracic abnormalities. For instance, a large pulmonary cystic adenomatoid malformation or a diaphragmatic hernia displaces the heart into an abnormal location. In this case, the four-chamber view may be used as an anatomic marker for extracardiac intrathoracic abnormalities. The heart also may lie in an ectopic location, a condition called "thoracic ectopia cordis" (20). Ectopia cordis may be an isolated abnormality, or it may be associated with a more complex abnormality, such as pentalogy of Cantrell (Table 7). Moreover, some studies have evaluated the normal cardiac axis in large populations. For instance, Shipp et al. (21) found a normal cardiac axis of 43° with a standard deviation of 7°. Two standard deviations above the mean set the upper limits of normal at 57°. In this study, 44% of the intrinsic cardiac defects had a cardiac axis greater than 57°. This phenomenon occurred most commonly in Ebstein's anomaly, tetralogy of Fallot, pulmonary stenosis, truncus arteriosus, and coarctation of the aorta. Some of these anomalies and their abnormal cardiac axis are detailed later in this chapter.

TABLE 7 ■ Classification of Ectopia Cordis

- Isolated finding
- Pentalogy of Cantrell
  - Omphalocele
  - Ectopia cordis
  - Diaphragmatic defect
  - Pericardial defect
  - Intrinsic cardiac abnormality

Is the Heart Normal in Size in Comparison with the Fetal Thorax?

The four-chamber view of the heart may be used to assess fetal cardiothoracic disproportion. Such measurements may be obtained from real-time images, but they are more accurately obtained from M-mode ultrasound. Standard measurements of the ratio of the cardiac biventricular outer dimensions to the chest circumference obtained by real-time M-mode measurements have been tabulated throughout pregnancy. These measurements may be helpful to predict cardiomegaly, or they may be used as an indirect measure of fetal pulmonary hypoplasia (22). I, more commonly, use rules of threes. This is simply of a four-chamber view; approximately three fetal hearts can fit into the fetal thorax (Fig. 2). When there is a small heart to thorax, it may be due to an extrinsic mass compressing the heart. If there is an increased size if the heart, this may be due to cardiomegaly (Table 8). When evaluating the fetal heart, asking four specific questions is helpful and these are listed in Table 9 (19).

TABLE 9 ■ Congenital Anomalies Detected in Utero Because of an Abnormal Four-Chamber View of the Heart

Are the ventricular chambers approximately equal in size?
Hypoplastic left ventricle
Hypoplastic right ventricle
Hypoplastic aortic arch
Aortic stenosis
Coarctation of the aorta (simple)
Coarctation of the aorta (complex)
Subaortic stenosis
Ostium primum defect
Single ventricle
Double outlet of right ventricle
Ebstein's anomaly
Is a septal defect present?
Atrial septal defect
Endocardial cushion defect
Ventricular septal defect
Simple
Tetralogy of Fallot
Truncus arteriosus
Double outlet of right ventricle
Is the relationship of the atrioventricular valve position abnormal?
Ebstein's anomaly
Is there an abnormality of the endocardium or the myocardium?
Focal increased thickness
Focal valvular atresia
Diffuse increased thickness: cardiomyopathy
Focal masses: rhabdomyoma

Source: Adapted from Refs. 9, 19.

Intrinsic Cardiac Disease

When evaluating the fetal heart, asking four specific questions is useful to detect intrinsic cardiac defects and these are listed in Table 9 (19).

Are the Ventricular Chambers About Equal in Size?

In general, the right and left ventricles should be about equal in size when studying the four-chamber view of the heart (19). This view is most accurately seen when an M-mode sonogram is obtained during diastole of the atrioventricular valve region. Real-time examination can be also used as a rough estimate of ventricular chamber size, however.

Most common anomalies can be classified as either hypoplasia of the left ventricle or hypoplasia of the right ventricle (Tables 10 and 11). Hypoplastic left ventricle is usually called the "hypoplastic left heart syndrome" (23). This syndrome may comprise a spectrum of findings, including underdevelopment of the aorta, the aortic valve, the left ventricle, or the mitral valve. Usually, the ventricular septum is intact (Fig. 10).

TABLE 10 ■ Possible Causes of Smaller Left Than Right ventricle

- Hypoplastic left ventricle
- Coarctation of the aorta
- Hypoplastic aortic arch
- Mitral stenosis

TABLE 11 ■ Possible Causes of Smaller Right Than Left Ventricle

- Pulmonary atresia, with or without ventricular septal defect
- Tricuspid atresia, with or without ventricular septal defect
- Aortic stenosis or insufficiency (left ventricle enlarged)

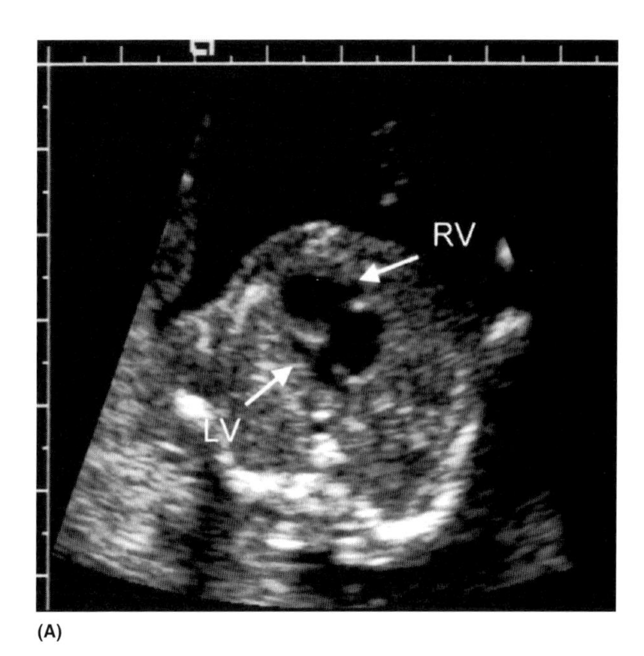

(A)

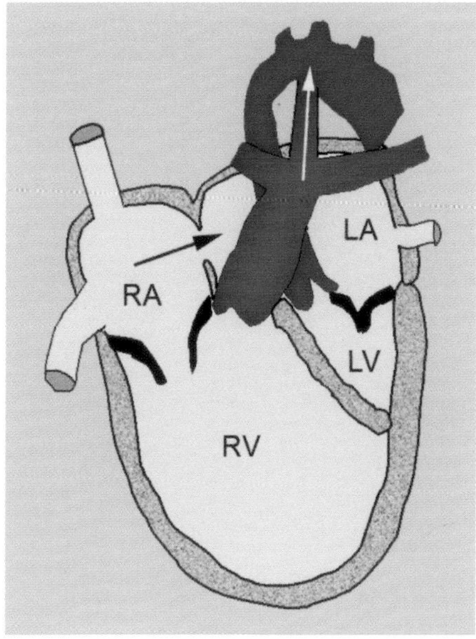

(B)

FIGURE 10 A AND B ■ Hypoplastic left heart. (**A**) There is hypoplasia of the left ventricle (LV) as compared to right ventricle (RV). The interventricular septum was intact. (**B**) Corresponding drawing.

Similarly, the only clue to an abnormality of the aortic arch may be the discrepancy in size between left and right ventricles. For instance, in coarctation of the aorta, blood flowing from the left ventricle is decreased and is shunted from the right ventricle into the ductus and the descending aorta. Because blood flow in the left ventricle is decreased, there is resultant underdevelopment of the left ventricle, and, thus, the left ventricle is smaller than the right ventricle. A rare cause of ventricular asymmetry is valve stenosis (Fig. 11).

Considerably more confusion concerns the diagnosis of hypoplasia of the right ventricle. This may be considered either one of two anomalies: pulmonary atresia or tricuspid atresia with or without an intact ventricular septum (Figs. 12 and 13). Usually, severe hypoplasia of the right ventricle is associated with pulmonary atresia and intact ventricular septum, whereas the amount of hypoplasia of the right ventricle varies with tricuspid atresia when a VSD is present. The degree of hypoplasia of the right ventricle depends on the size of the tricuspid or pulmonary atresia and is inversely dependent on the size of the VSD (Figs. 12 and 13) (19,23). Typical causes for the right ventricle to be smaller than the left are listed in Table 11. In some situations, two atria may enter into a single ventricle. Alternatively, both the atrial and the ventricular septa may be absent, resulting in a common atria and a common ventricle, the so-called two-chamber heart. Finally, the heart should occupy approximately one-third of the fetal thorax. When the cardiothoracic ratio is increased, this may be secondary

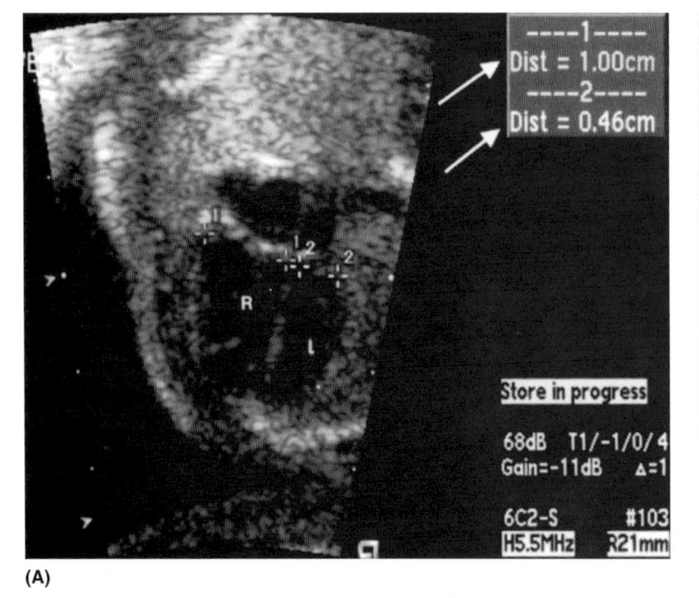

(A)

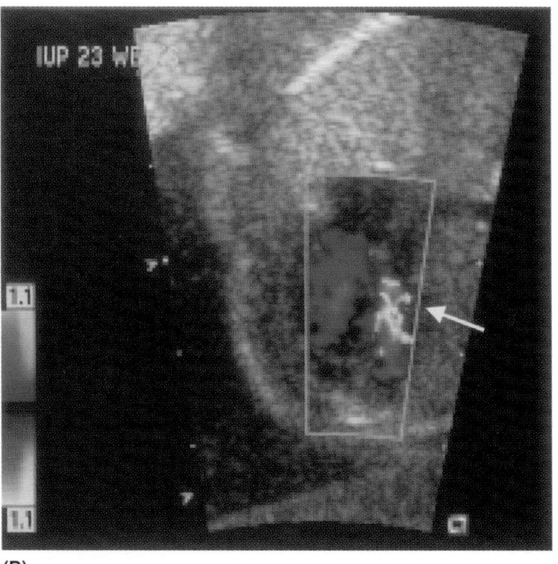

(B)

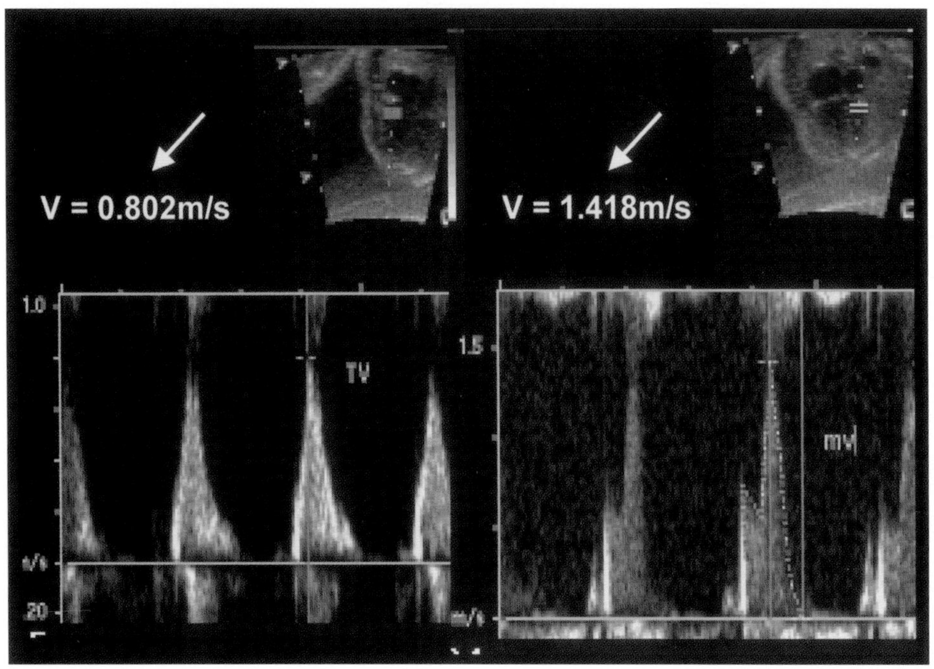

(C)

FIGURE 11 A–C ■ Mitral valve stenosis. (**A**) On this four-chamber view of the heart, there is discrepancy in size between the smaller left ventricle (L) and the larger right ventricle (R). Also, the mitral valve annulus marked by calipers is approximately one-half the diameter of the tricuspid valve annulus. (**B**) The color flow ultrasound demonstrates high-velocity jet through the mitral valve (*arrow*). (**C**) Pulsed Doppler demonstrates velocity through the tricuspid valve of approximately 80 cm per second as compared to the velocity through mitral valve of 141 cm per second.

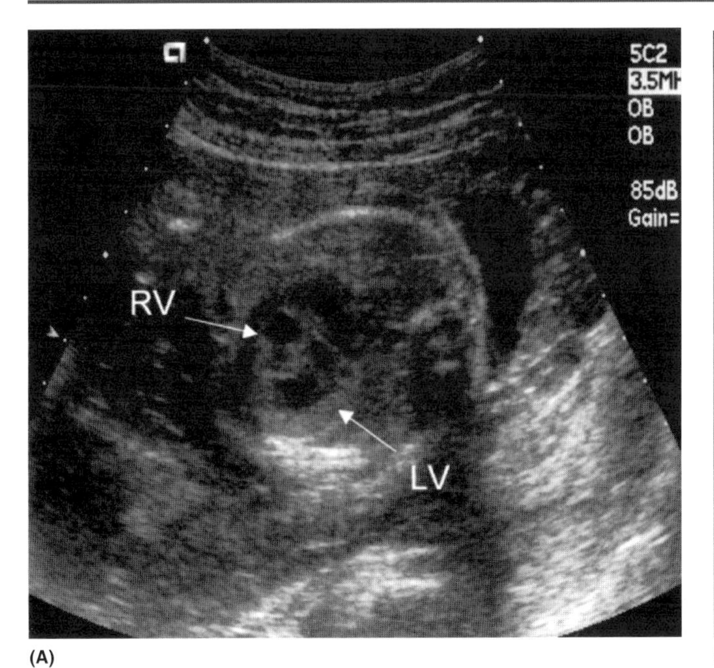

(A)

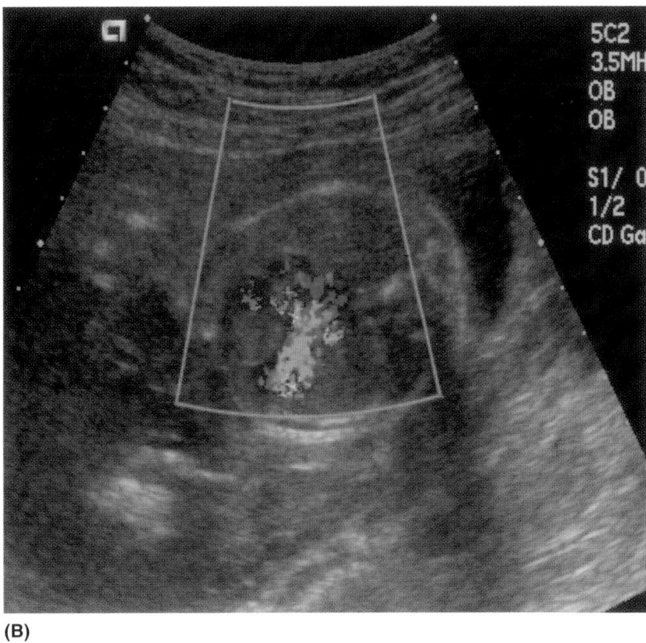

(B)

FIGURE 12 A AND B ■ Pulmonary atresia with intact ventricular septum. (**A**) Four-chamber view of the heart demonstrates small right ventricle (RV) as compared to the left ventricle (LV). Also note intact ventricular septum. (**B**) Color Doppler demonstrates flow through the LV with no significant flow through the right ventricle.

to fetal hydrops or secondary to certain valvular abnormalities such as Ebstein's anomaly (Fig. 14). In Ebstein's anomaly, there is displacement of the tricuspid leaflets into the right ventricle, creating an atrialized portion of the right atrium. There may also be tricuspid valve dysplasia where little tricuspid tissue is present. There is usually tricuspid valve regurgitation and an enlarged right atrium (Fig. 14).

Is There a Septal Defect?

Septal defects may be divided into three basic types: ASD, atrioventricular canal, and VSD. ASD can be further

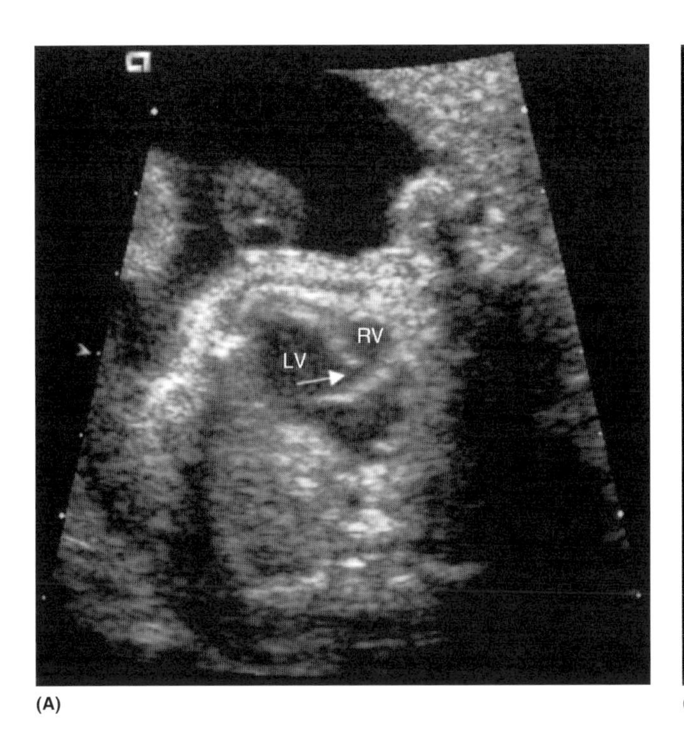

(A)

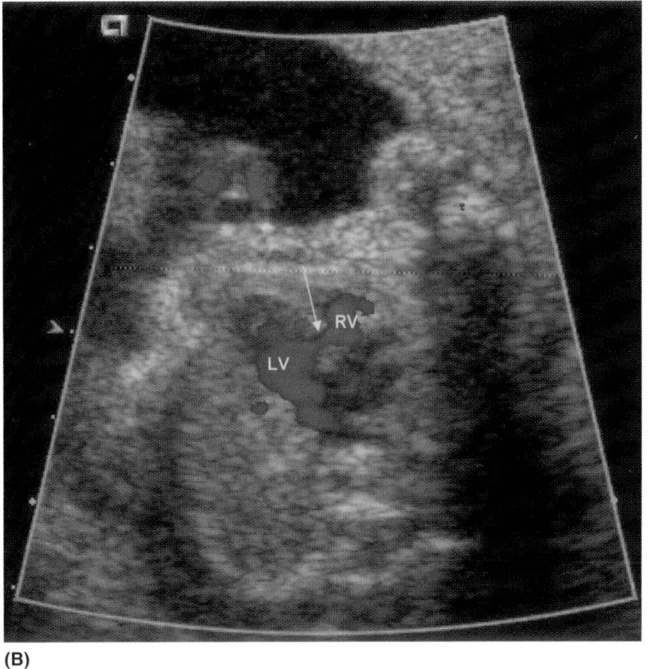

(B)

FIGURE 13 A–C ■ Tricuspid atresia with ventricular septal defect (VSD). (**A**)This four-chamber view of the heart demonstrates a small right ventricle (RV) as compared to the larger left ventricle (LV). Also note small VSD (*arrow*). (**B**) Color Doppler ultrasound again demonstrates small RV as compared to LV. The color Doppler demonstrates the VSD (*arrow*). (*Continued*)

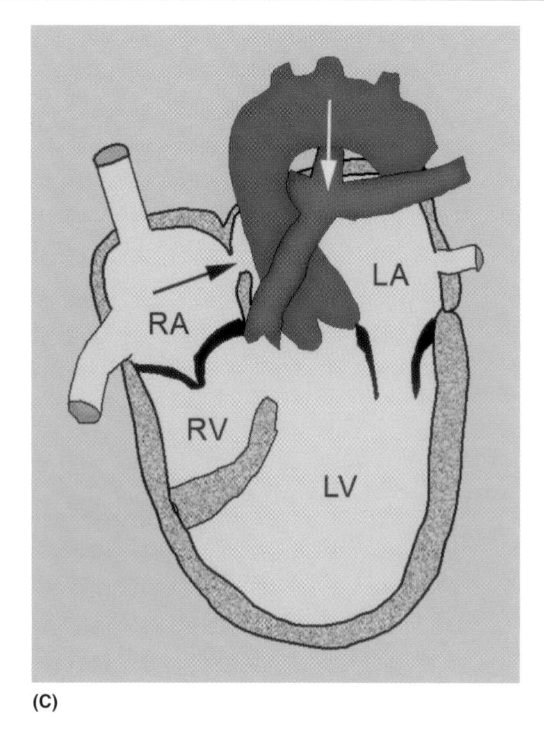

(C)

FIGURE 13 A–C ■ (*Continued*) (**C**) Corresponding line drawing. RA, right atrium; LA, left atrium; RV, right ventricle; LV, left ventricle.

divided into ostium primum and ostium secundum defects. Ostium secundum defects are located high in the atrial septum; ostium primum defects are low in the septum and often are associated with an abnormality of the atrioventricular valve. Small defects may be difficult to detect in utero, whereas large defects may be more easily identified (Figs. 13, 15–19).

A VSD may be small or large. Smaller defects occur in the membranous portion of the ventricular septum. Larger defects involve larger portions of the ventricular septum and are usually more easily detectable but may be challenging (Figs. 1, 17–19). Large VSDs are often associated with other intrinsic cardiac abnormalities. A small VSD may be overlooked on in utero scanning; so careful scanning of the entire ventricular septum is needed to detect these defects (Fig. 15) Color flow imaging allows easier recognition of ventricular defects (Fig. 16) Perimembranous VSDs are often better visualized on a long axis view of the heart with the defect just below the aortic valve. An ASD or VSD noted on a four-chamber view should prompt a more complete cardiac examination.

An atrioventricular canal defect is more easily recognized in utero as complete absence of the endocardial cushion (Figs. 1 and 18). Color flow imaging may be used to identify this significant abnormality (Figs. 1 and 19). There may be a partial or a complete defect. The more common complete defect includes an ASD, a VSD, and abnormal atrioventricluar valves. The defect can show symmetry in the size of the ventricles or asymmetry between the two ventricles as with an unbalanced defect. Outflow tract anomalies are often associated with atrioventricular canal defects.

Are the Atrioventricular Valves in Normal Position?

Usually the tricuspid valve is located closer to the cardiac apex than the mitral valve (Fig. 3). A right-sided abnormality of the atrioventricular valve position is usually associated with the malformation of the tricuspid valve in which the septal leaflet is displaced into the cavity of the right ventricle. This defect is called "Ebstein's anomaly." Sonographically, this anomaly is identified

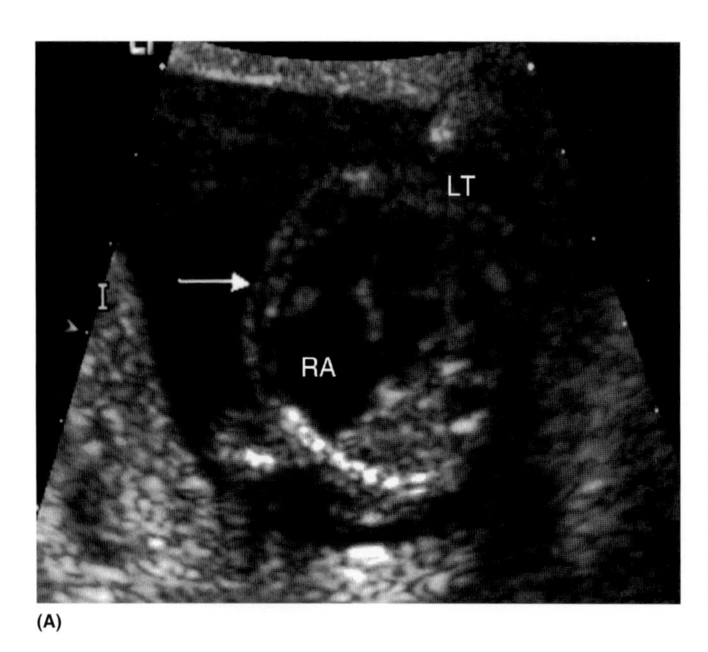

(A)

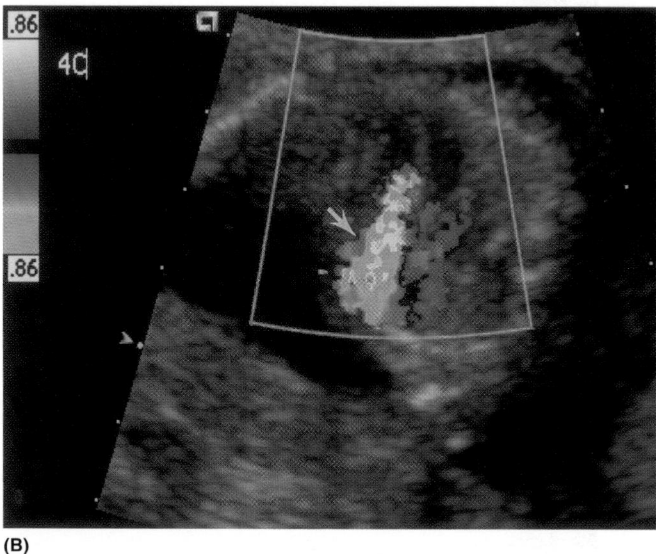

(B)

FIGURE 14 A AND B ■ Ebstein's anomaly. (**A**) There is a dysplastic tricuspid valve with thickened leaflets, which is not displaced into the right ventricle. However, there is an enlargement of the right atrium (RA). (**B**) Note tricuspid valve regurgitation (*arrow*) identified by color Doppler in blue. Flow within the left side of the heart is in the normal direction, detected in red. LT, left.

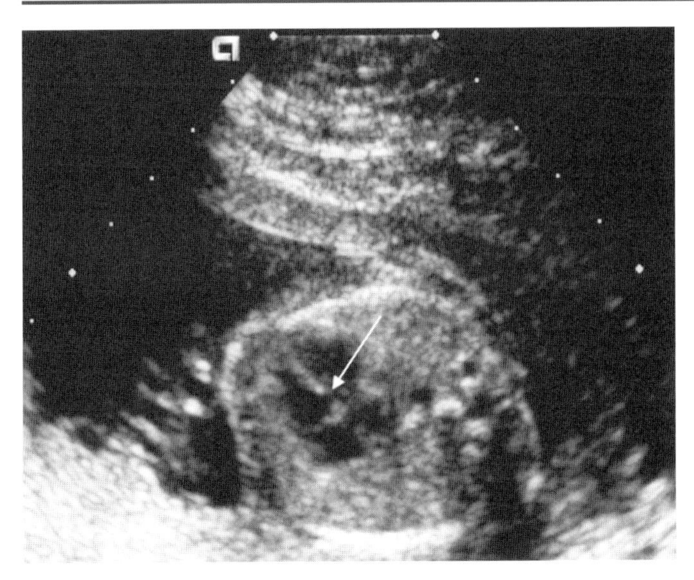

FIGURE 15 ■ VSD. Four-chamber view of the heart demonstrating a small ventricular septal defect (*arrow*) in the membranous portion of the septum.

as a large right atrium with a characteristic appearance (Fig. 14). Similar enlargement of the right atrium can result from tricuspid dysplasia, with tricuspid regurgitation (19). Also, with an endocardial cushion defect, the residual tricuspid and mitral valves appear to lie at an equal distance from the cardiac apex (Fig. 19).

Is There an Abnormality of the Endocardium or Myocardium?

Focal increased thickening of the endocardium or myocardium is usually associated with isolated valvular atresia or severe stenosis. This usually occurs with valvular atresia with an intact ventricular septum and hypertrophy of the corresponding ventricle.

Most commonly, diffuse thickening of the heart muscle is associated with cardiomyopathy. Causes of increased endocardial or myocardial thickness are listed in Table 12. A rare cause is multiple cardiac rhabdomyomas that cause outflow obstruction and, therefore, diffuse hypertrophy of the cardiac myometrium.

Increased echogenicity within the ventricles is a normal finding in the right ventricle. This phenomenon is due to the thickened muscle in the right ventricle called the moderator band. Focal areas of increased echogenicity are usually most commonly caused by echogenicity of the chordae tendineae or the papillary muscles. Rarely, focal increased echogenicity may be due to a fetal cardiac tumor (Figs. 20 and 21).

An echogenic focus has previously been identified to occur most commonly in the left ventricle but also within the right ventricle (Fig. 9). In 1988, Levy described an echogenic focus in the left ventricle of the fetal heart a normal finding (24). However, others described this echogenic focus to occur with greater frequency in fetuses with Trisomy 13 and Trisomy 21. For instance, Roberts reported this (25) echogenic focus to be present as a discrete calcification of the papillary muscle in 16% of

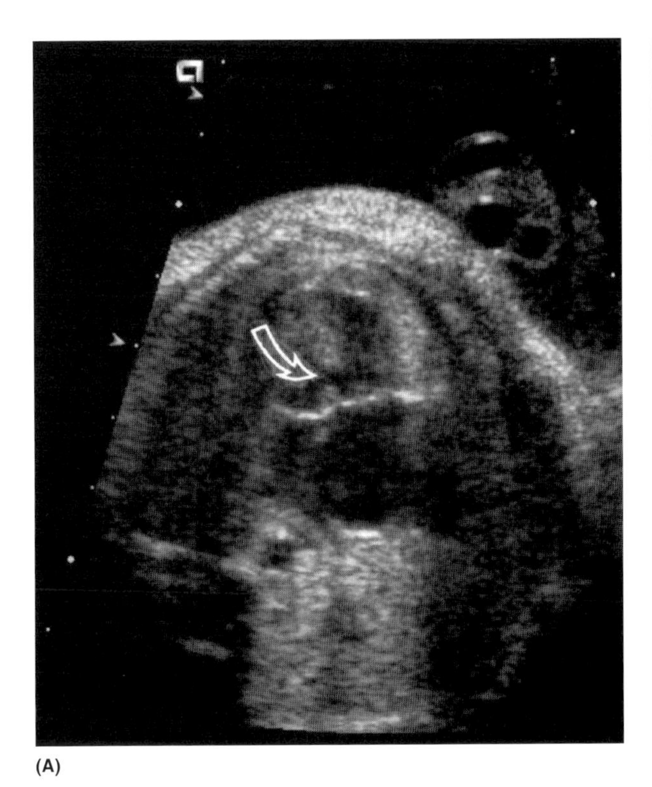

(A)

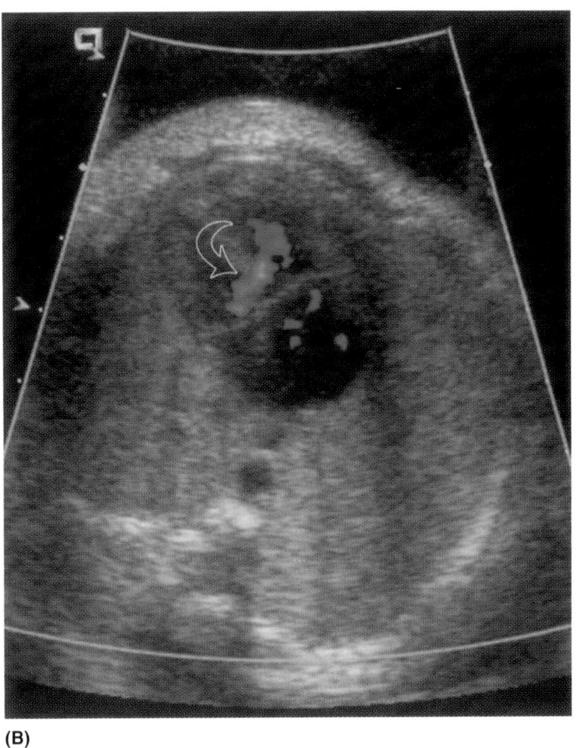

(B)

FIGURE 16 A AND B ■ Small ventricular septal defect (VSD). (**A**) Real-time image showing small VSD (*curved arrow*). (**B**) This is better identified using color Doppler (*curved arrow*).

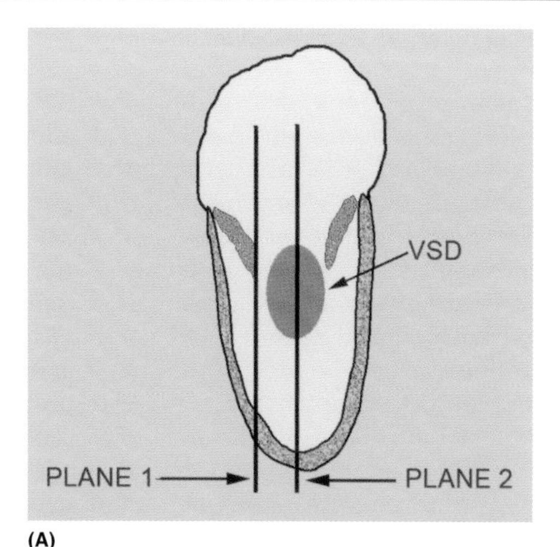

(A)

FIGURE 17 A–C ■ Ventricular septal defect (VSD). (**A**) Complete scanning of ventricular septum is needed to exclude a VSD, In this diagram, scanning through the ventricular septum in plane 1 appears normal. Scanning in plane 2 would optimally visualize the defect. (**B**) In this four-chamber view of the heart, the ventricular septum appears intact (*arrow*) while scanning through plane 1. (**C**) By obtaining a scan in plane 2 through the interventricular septum, a large VSD (*open arrow*) is identified between the ventricles. O, omphalocele; S, spine; V, ventricle. *Source*: From Ref. 19.

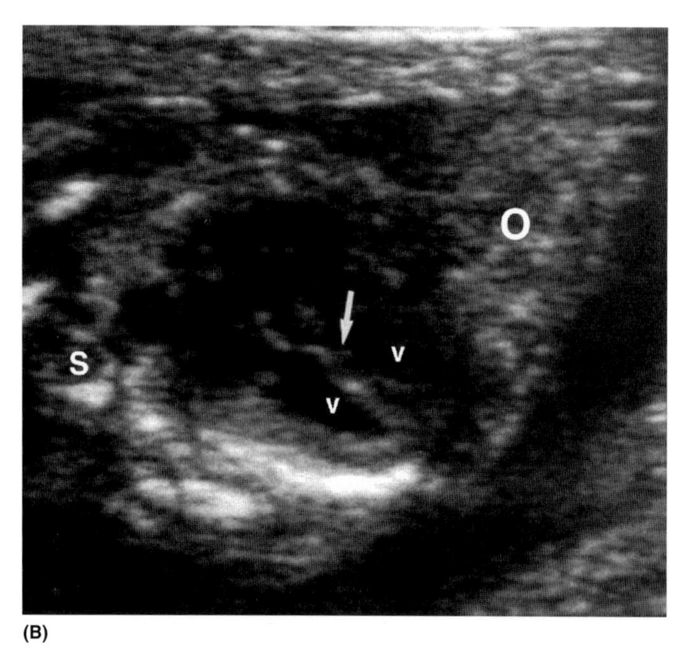

(B)

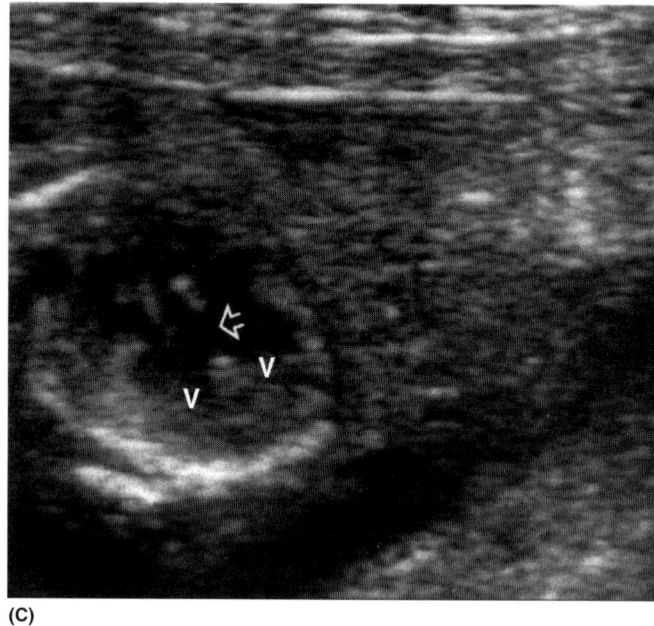

(C)

fetuses with Trisomy 21, 39% of fetuses with Trisomy 13, and 2% of euploidy fetuses (25). Other studies have questioned the validity or use of soft markers, such as this echogenic focus, as risk factors for chromosomal abnormalities (26). A meta-analysis of 11 studies within the literature has shown a sensitivity of 26% and specificity of 95.8% of the echogenic focus for detecting Trisomy 21 (27). This overall study group presents a likelihood ratio of 6.2. With a presumed 0.8% risk of fetal loss with amniocentesis, one fetus is lost per Trisomy 21 case detected when the background risk is 1:770. If the echogenic focus is isolated and not associated with any other fetal anomalies, the likelihood ratio is less at 5.4. The authors of this study also point out for women in their mid-30s, the risk for Down syndrome after a typical negative triple marker biochemical screen is less. Perhaps with likelihood ratios in the range of 2 to 3. Other studies (28) have shown maternal and technical factors may account for a high rate of occurrence of echogenic focus in some normal pregnancies. These factors include thinner patients, patients of African-American or Asian descent, and the use of the apical four-chamber view of the heart (29).

In summary, an echogenic cardiac focus occurs in a higher frequency in Trisomy 21 and Trisomy 13 fetuses, compared to the normal population. However, maternal and technical factors may account for a higher rate of occurrences in certain normal pregnancies. The occurrences of an echogenic cardiac focus should be correlated with a comprehensive obstetrical ultrasound exam, patient age, other risk factors, and biochemical screening.

The most common in utero cardiac tumor is a cardiac rhabdomyoma, which is often secondary to tuberous sclerosis (30). This tumor can be distinguished sonographically from the normal echogenic focus usually observed within the left ventricle. Tumors are much larger, usually show mildly increased echogenicity, and are generally close to the ventricular wall (Figs. 20 and 21). Possible cardiac tumors are listed in Table 13.

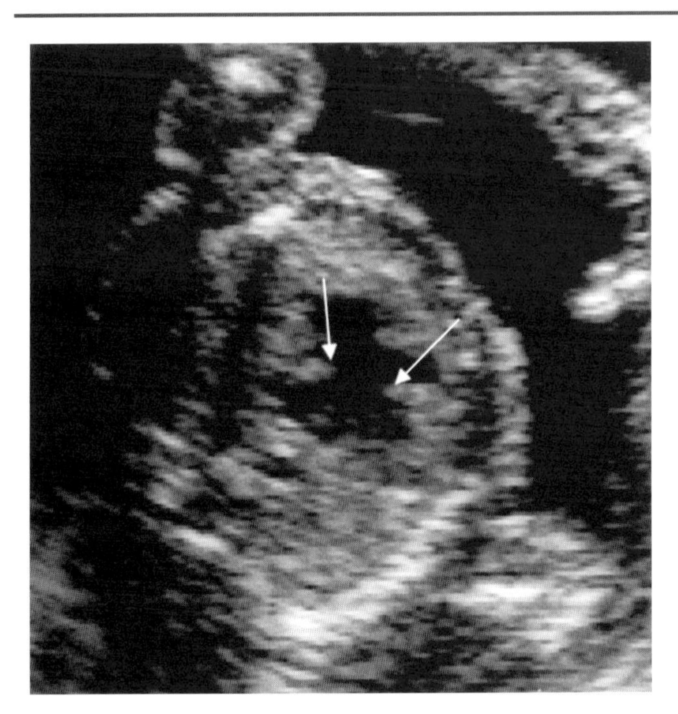

FIGURE 18 ■ Atrioventricular septal defect. Four-chamber view of the heart demonstrates large atrioventricular canal defect (*arrows*).

The four-chamber view of the heart may be used to detect other abnormalities. For instance, recognition of a pericardial effusion may be a clue to a systemic disorder such as fetal hydrops. Causes of possible fluid around the heart are included in Table 14.

TABLE 12 ■ Causes of Increased Thickness of the Endocardium or Myocardium

Focal causes
Critical valve disease
Cardiac tumor
Bright spot of papillary muscle
Diffuse causes
Cardiomyopathy
Endocardial fibroelastosis
Glycogen storage disease
Hypertrophic cardiomyopathy
Cardiac tumor

DETECTION OF AORTIC AND PULMONARY OUTFLOW TRACTS

Other significant fetal cardiac malformations can be detected with an examination of the fetal cardiac outflow tracts in addition to four-chamber view of the heart (17,18). These defects include transposition of the great arteries, tetralogy of Fallot, and truncus arteriosus. In fact, about

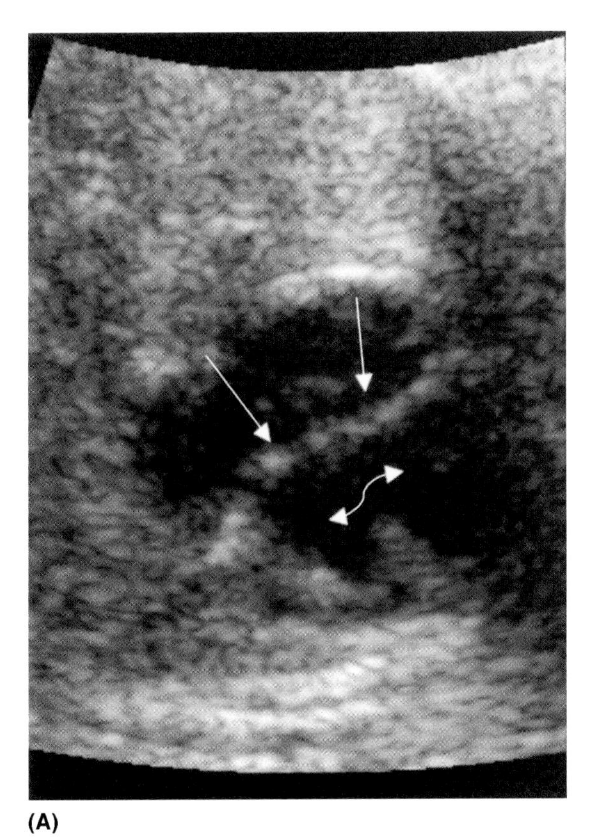

(A)

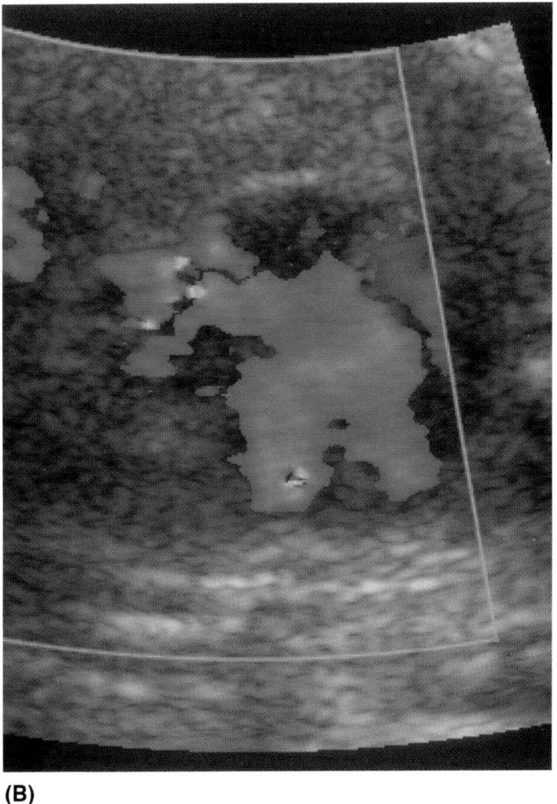

(B)

FIGURE 19 A AND B ■ Atrioventricular septal defect. (**A**) In this four-chamber view of the heart, a ventricular septum defect is noted (*curved arrow*). The mitral/tricuspid valve leaflets are parallel (*arrows*). (**B**) By obtaining a scan with color flow, a large atrioventricular defect is identified.

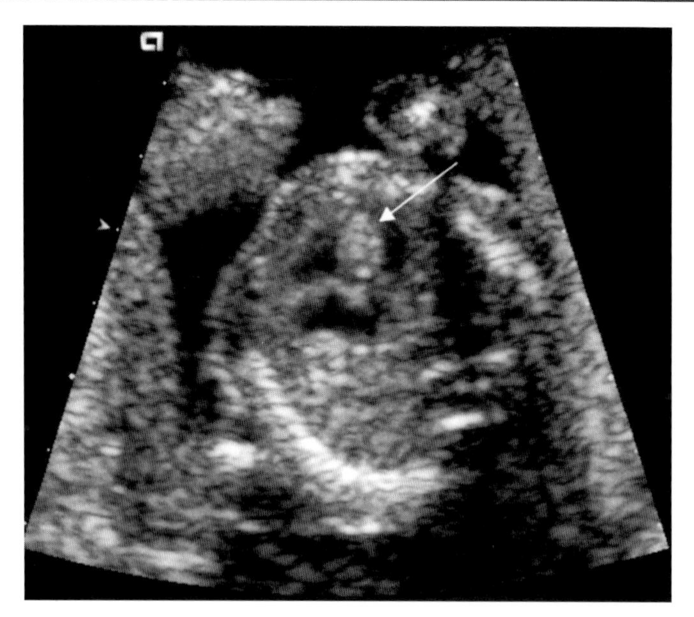

FIGURE 20 ■ Rhabdomyoma of the heart. Four-chamber view of the heart demonstrates a well circumscribed mass arising from interventricular septum and protruding into the left ventricle (*arrow*) corresponding to cardiac rhabdomyoma in this fetus with tuberous sclerosis.

85% of all cardiac abnormalities may be diagnosed routinely by obtaining a four-chamber view of the heart and examining outflow tracts (18). This detection rate (86%) has been validated using an extended examination in a

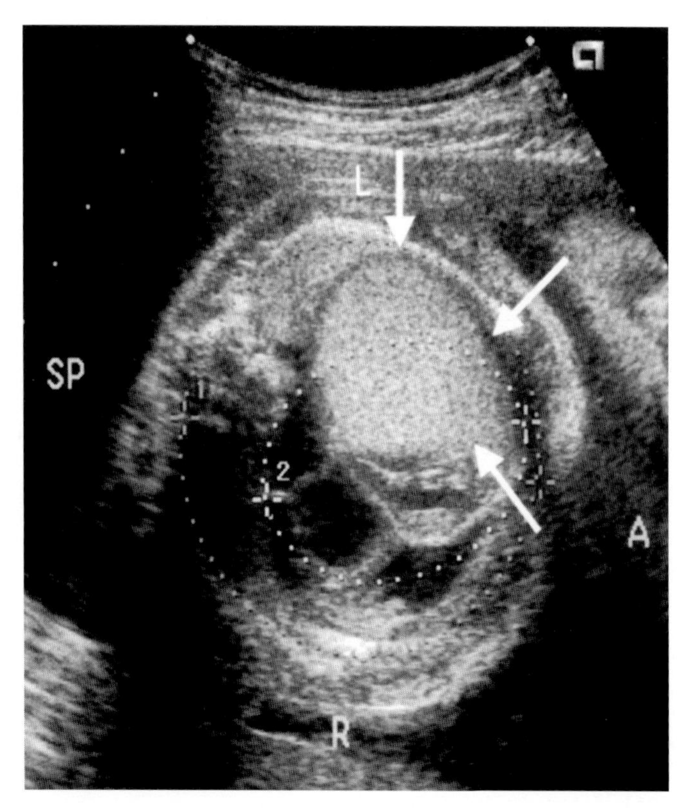

FIGURE 21 ■ Fetal cardiac rhabdomyoma. Large echogenic mass (*arrow*) within the left ventricle of the heart, which distorted the entire cardiac anatomy and occupies a large portion of the chest. SP, spine; R, right; A, anterior.

TABLE 13 ■ Possible Causes of Fetal and Newborn Cardiac Tumors

Uncommon
Rhabdomyoma (may be associated with tuberous sclerosis)
Extremely rare
Teratoma
Fibroma
Myxoma
Hemangioma

low-risk population (17). Similarly, Giancotti et al. (31) detected 89% of cardiac defects prenatally in a high-risk population using an extended fetal echocardiogram. Other investigators, however, have shown problems in detecting certain complex fetal cardiac anomalies, even with a comprehensive fetal echocardiogram. For instance, Lee et al. (34) detected only 12 of 17 abnormalities (70% of tetralogy of Fallot). Commonly missed in this series was the discrepancy between aortic and pulmonary artery size.

Potential pitfalls still exist in this schema, including missed diagnosis of ASD, small VSD, isolated valvular stenosis, coarctation of the aorta, and total anomalous pulmonary venous return. When reviewing the data from the University of Minnesota series, however, abnormalities that most often present within the first days to the first month of life should be detected with the combination of the four-chamber view of the heart and outflow tract views (Table 6) (17,18,31–35). A check-off list, when evaluating the outflow tract, is presented in Table 15.

Three questions should be asked when evaluating the relation of the great arteries to the right and left ventricles:

Does a Normal Crisscross Relation of the Pulmonary Artery and Aorta Exist?

If this relation does not exist, and the aorta and pulmonary artery lie parallel rather than crisscross, this may be a clue to the diagnosis of double outlet of the right ventricle transposition of the great vessels or, in rare cases, corrected transposition (Figs. 3–5, 22–24).

Is There a Discrepancy in the Size of the Aorta and Main Pulmonary Artery?

The assessment of the relation of the sizes of these two vessels is helpful in diagnosing such abnormalities as tetralogy of Fallot, in which the aortic root is larger than the atretic pulmonary artery (19,32). Lee et al. (34)

TABLE 14 ■ Possible Causes of Fluid Around the Heart

- Pericardial effusion
- Pleural effusion
- Variant: hypoechoic myocardium
- Normal: pericardial fluid identified during diastole

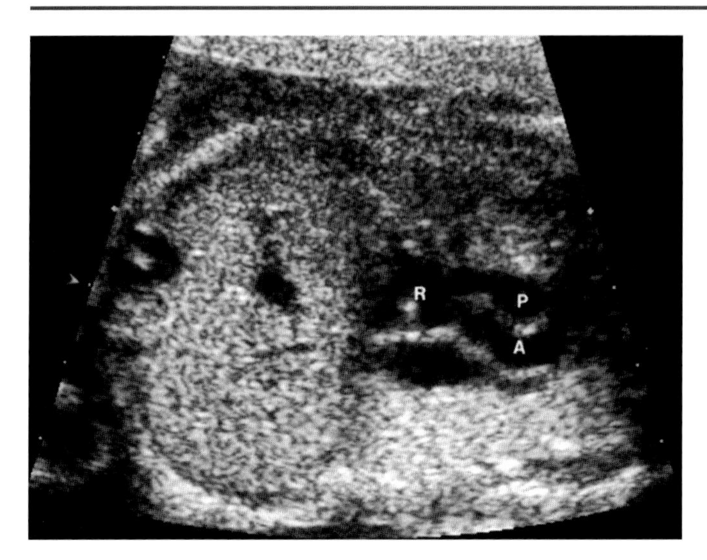

FIGURE 22 ■ Double outlet right ventricle—the aorta (A). Aorta and the pulmonary artery (P) are parallel and originated from the right ventricle (R).

(Figs. 25 and 26), however, have shown that this discrepancy is difficult to identify prenatally. Similarly, in truncus arteriosus, one large or common trunk is identified, rather than a separate aorta and pulmonary artery (Fig. 27). Causes of increased and decreased aortic diameter are listed in Table 16. Aortic measurements and femoral length measurements have been tabulated throughout pregnancy, as has the size ratio of the aorta to the main pulmonary artery (Fig. 28).

TABLE 15 ■ List for Outflow Tract in the Normal Fetal Heart

Two semilunar valves are present
Two arterial roots are present
Left ventricle is connected to the aorta posteriorly
Right ventricle is connected to the pulmonary artery anteriorly
Proximal great arteries cross each other's courses
Pulmonary artery diameter exceeds slightly that of the aorta
Pulmonary artery branches proximally, gives off branches laterally

Is the Aorta or Pulmonary Artery in Normal Relation to the Ventricular Septum?

The aorta overrides the ventricular septum in tetralogy of Fallot; therefore, this clue is important for diagnosis (Figs. 25 and 26). Also in other anomalies such as truncus arteriosus, the aorta may override the ventricular septum. Possible anomalies associated with an overriding aorta are presented in Table 17.

A checklist for the outflow tracts is presented in Table 15.

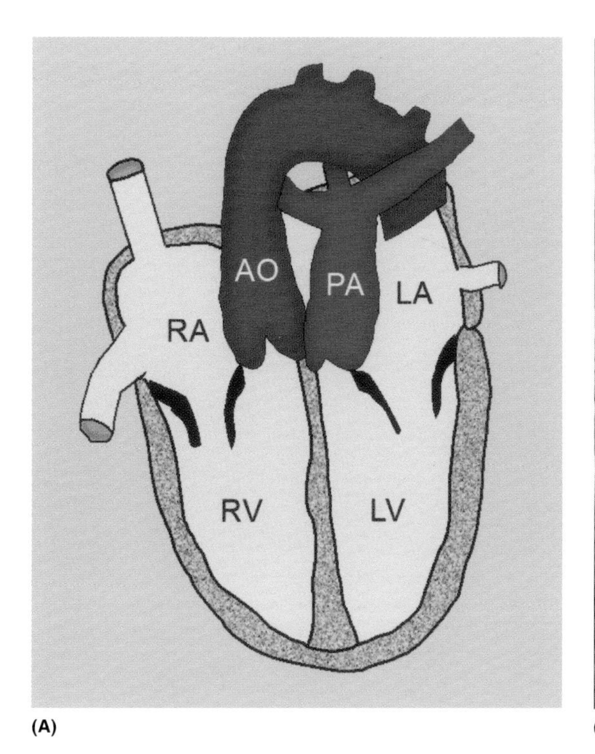

(A)

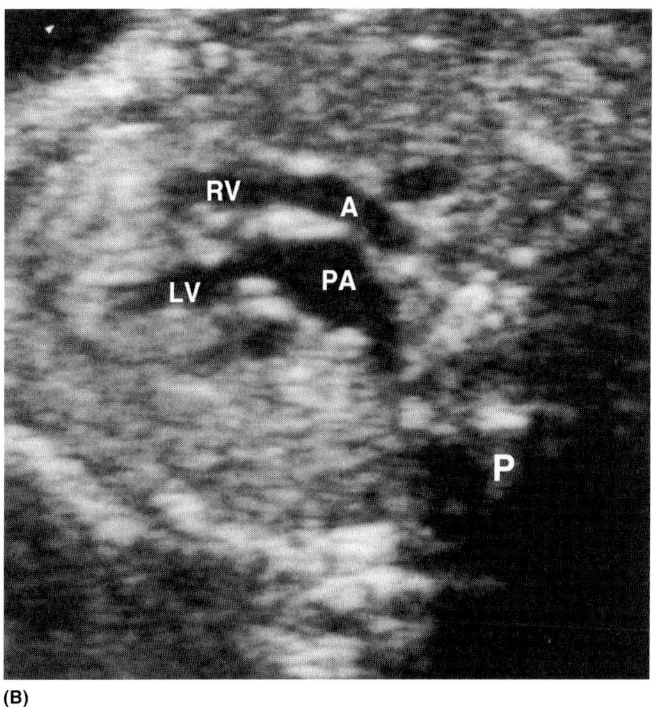

(B)

FIGURE 23 A AND B ■ Transposition of the great arteries. (**A**) Diagram shows transposition of the great vessels, with the aorta originating from the right ventricle and the pulmonary artery originating from the left ventricle. An accompanying septal defect may be present. (**B**) An outflow view shows the parallel course of the aorta and the pulmonary artery, rather than the normal perpendicular course of these vessels. A, aorta; PA, pulmonary artery; LV, left ventricle; RA, right atrium; LA, left atrium; RV, right ventricle. *Source*: From Ref. 33.

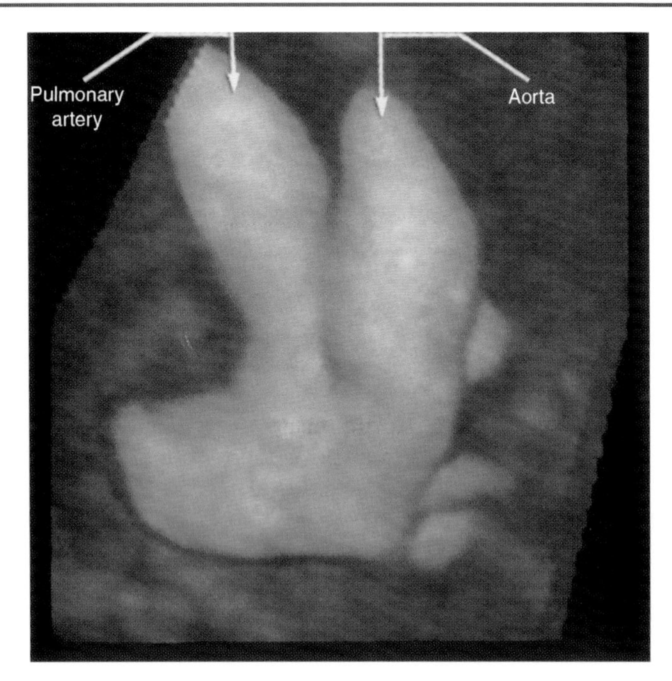

FIGURE 24 ■ Transposition of the great vessels—four-dimensional ultrasound. This 4D ultrasound demonstrates parallel position of the pulmonary artery and the aorta in this case of transposition of the great vessel. *Source*: From Ref. 11.

COLOR DOPPLER SONOGRAPHY

Color Doppler sonography, widely used in the evaluation of the fetal heart, is superimposed as Doppler information colored-encoded format on the standard real-time ultrasound image. Color-encoded information is recorded in any of various color-encoded formats. Typically, for peripheral arterial and venous imaging, flow toward the transducer is encoded in red, whereas flow away from the transducer is encoded in blue. For peripheral arterial and venous imaging, red is arbitrarily assigned to arteries and blue is assigned to veins. For fetal cardiac imaging, the color provides information about direction of flow and some information concerning flow velocity by noting the color scale. For instance, when blood is moving too fast, the velocity exceeds the ability to record the Doppler and thus results in aliasing of the color Doppler signal. In this situation, the color encoded changes from red to blue, even though the flow may be toward the transducer. This occurs when the Nyquist limit (i.e., the maximum velocity setting) is too low for high-velocity flow (36). This situation commonly occurs in fetal cardiac imaging because of the high velocities encountered in the fetal heart.

Color Doppler imaging may be helpful in examination of the fetal heart. Copel et al. (37) found color flow useful in several circumstances, including

- Direction of flow
- Vessel identification
- Left-to-right shunts
- Hypoplastic heart

For instance, in hypoplastic left heart, one sees atresia of the left ventricle and aorta. Therefore, color flow may demonstrate the absence of flow within the left ventricle (Fig. 10). Moreover, color imaging may demonstrate that

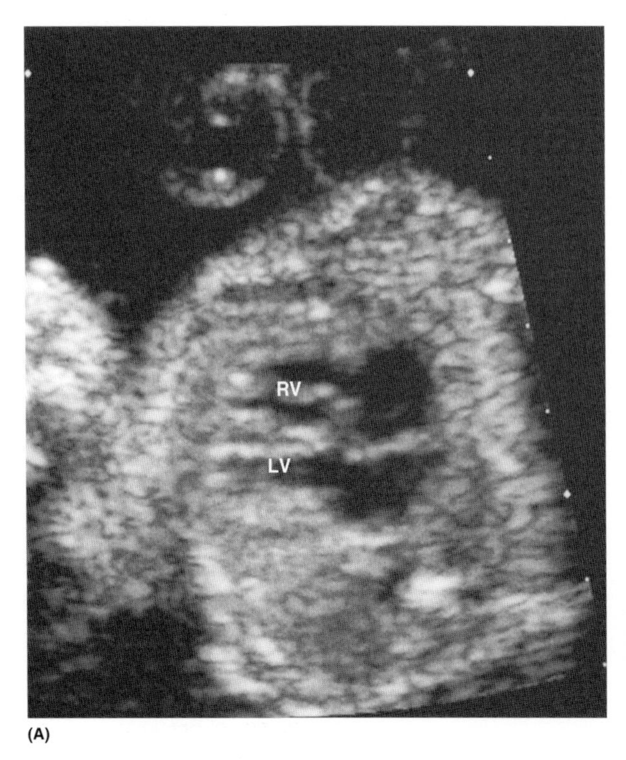

(A)

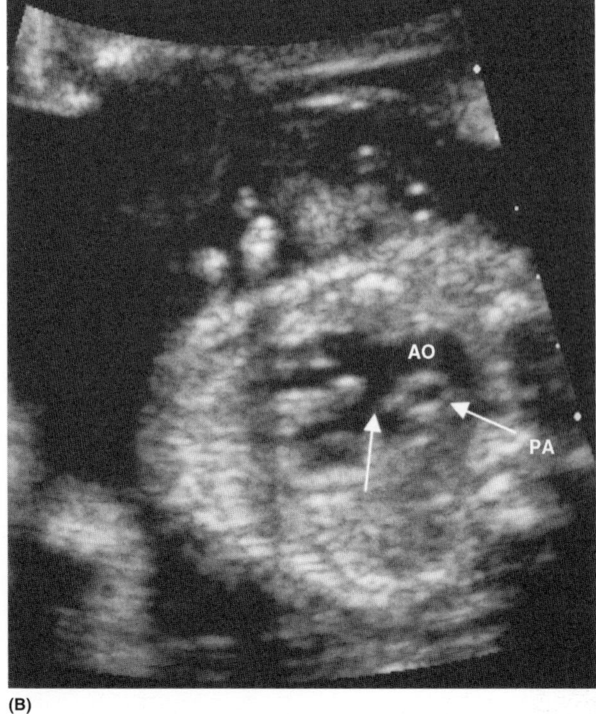

(B)

FIGURE 25 A AND B ■ Tetralogy of fallot. The four-chamber view of the heart is unremarkable except for deviation of the cardiac axis. Long-axis view of the heart demonstrates a large aorta (AO) overriding the ventricular septal defect (*arrow*). There is a small pulmonary artery (PA). RV, right ventricle; LV, left ventricle.

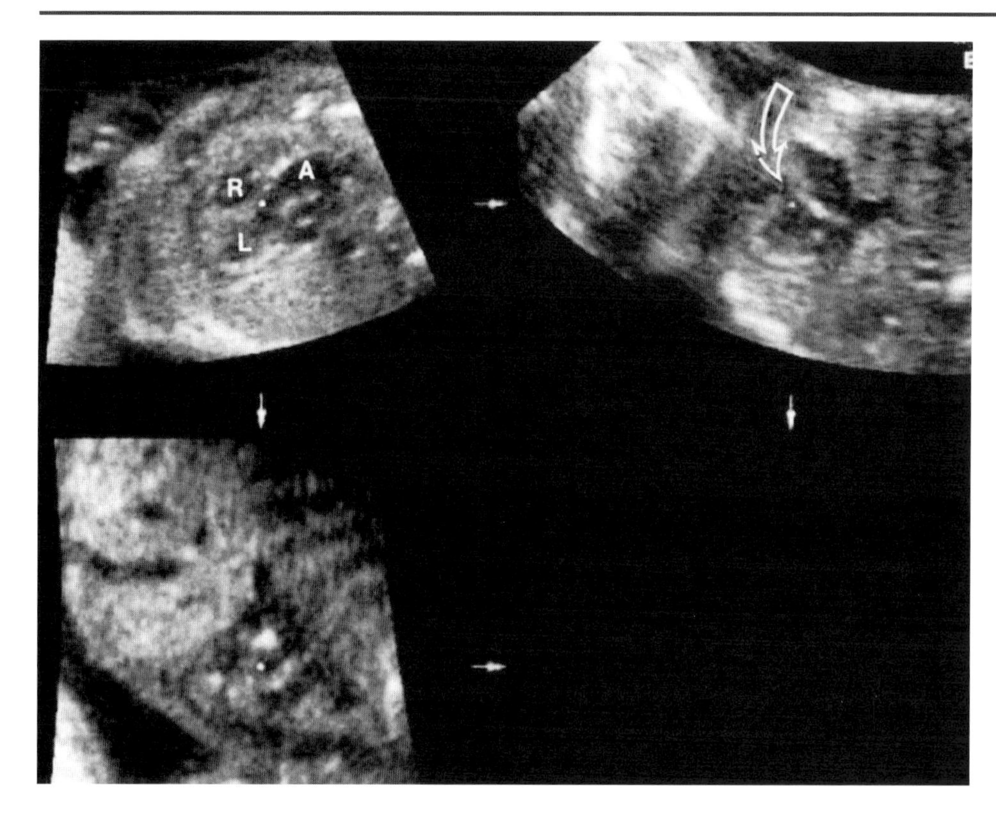

FIGURE 26 ■ Tetralogy of fallot. Using STIC, three orthogonal planes of the heart are rendered simultaneously rotating about the central dot. These demonstrate overriding aorta (A) in first image with ventricular septal defect (VSD) seen in second image (*arrow*). STIC, spatiotemporal image correlation; R, right ventricle; L, left ventricle.

flow through the arch is reversed from the pulmonary artery through the ductus arteriosus into the aorta. Color imaging may be helpful to identify this retrograde flow. In some situations such as coarctation of the aorta, color flow may allow examiners to be confident in the diagnosis. For instance, coarctation of the aorta may be recognized with a real-time examination, but color imaging may confirm the diagnosis.

FIGURE 27 ■ Truncus arteriosus. A single large vessel (T) is identified arising form the base of the heart. It partially overrides the ventricular septal defect (*arrow*).

In addition to the foregoing areas, we have found color imaging useful in evaluation of the anatomically difficult patient. For instance, on occasion, it may be difficult in a large patient to obtain even a normal four-chamber view of the heart. Color flow may be helpful in such a situation. In such circumstances, if the maximum velocity is too low, aliasing may occur, giving the appearance of multiple colors within the ventricle. Alternatively, by altering the baseline of the color scale and increasing the maximum velocity setting, the entire ventricle may appear in one color format. In addition to identifying normal chambers within the heart, outflow tracks may occasionally be difficult to visualize. Color may be useful to identify the aorta and the pulmonary artery more competently.

Color Doppler imaging may be helpful in other situations (37,38). For instance, real-time ultrasound may be used to diagnose VSDs (Figs. 13, 16, and 17). On occasion, the membranous portion of the ventricular septum may

TABLE 16 ■ Causes of Increased and Decreased Aortic Diameter

Increased aortic diameter
Common: tetralogy of Fallot
Less common: truncus arteriosus (common trunk)
Very uncommon: hypoplastic left ventricle with transposition
Decreased aortic diameter
Common: hypoplastic left ventricle
Uncommon: coarctation of the aorta

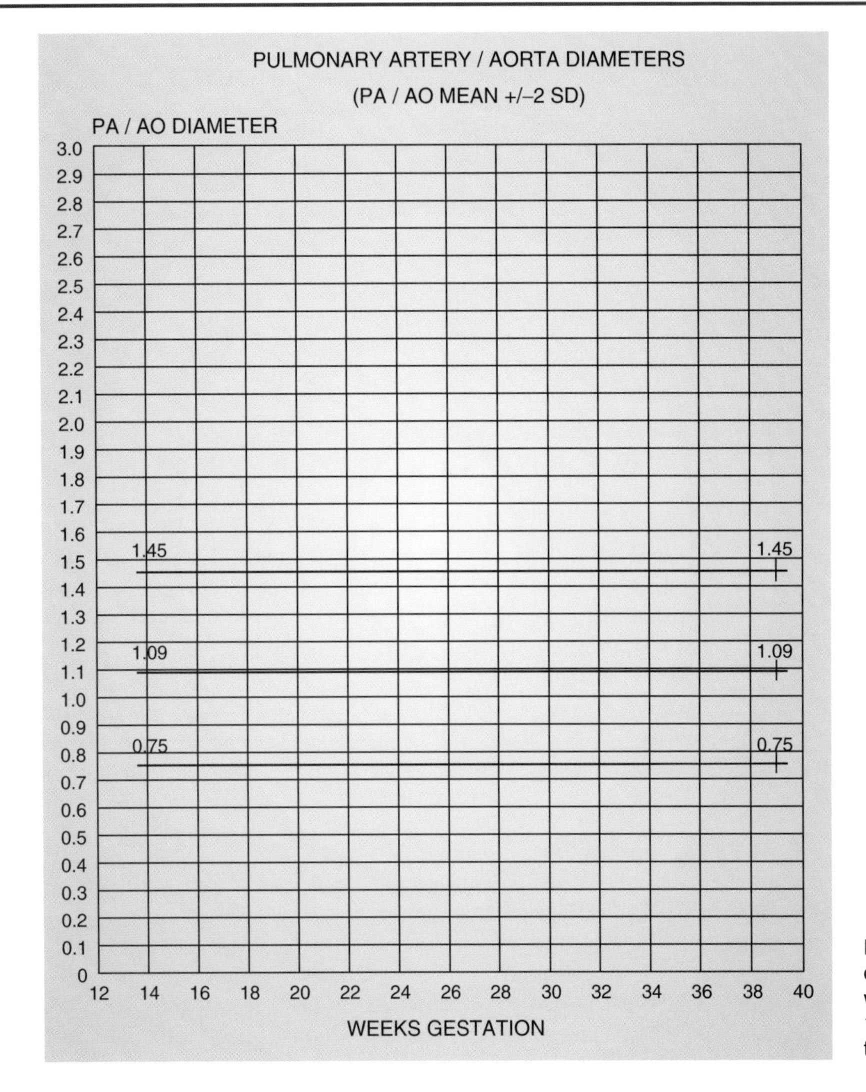

FIGURE 28 ■ Pulmonary artery-to-aortic (PA/AO) dimensions. The root of the pulmonary artery compared with the aorta contains a constant relationship of nearly 1.09 to 1, with a standard deviation of 0.75 to 1.45 throughout pregnancy. *Source*: From Ref. 35.

be difficult to identify. Trying to position the septum parallel and then perpendicular to the ultrasound beam may be useful to identify VSD. Questions may still arise, however, and color Doppler imaging may be helpful. Using multiple angles, color may be useful to identify flow from one ventricle to the other during both systole and diastole.

Similarly, large defects such as endocardial cushion defects (atrioventricular canal) may be difficult to interpret on real-time examinations. Activation of color imaging

TABLE 17 ■ Causes of Aorta Override of the Ventricular Septum

Common
Tetralogy of Fallot
Uncommon
Pulmonary atresia with ventricular septal defect
Rare
Truncus arteriosus
Double outlet of right ventricle

TABLE 18 ■ Doppler Measurements

Measurement[a]	Method	Units
Peak or maximal velocity	Zero line to peak	cm/sec
Mean velocity	Time velocity integral/time of cardiac cycle	cm/sec
Volume flow	Mean velocity × area[b] × 60	mL/min
Acceleration time	Time from onset to peak	msec
Deceleration time	Time from peak to zero line along slope of descent	msec
A/E ratio	Peak velocity with (A)/peak velocity during early diastole (E)	–

[a]Velocities should be measured within 30° of estimated direction of flow or should be angle corrected.

[b]Area obtained from diameters measured with two-dimensional ultrasound (39).

Abbreviation: A/E, atrial contraction/early diastole.

TABLE 19 ■ Normal Doppler Echocardiographic Features in the Fetus

Valve	Tricuspid	Mitral	Pulmonary	Aortic
Maximal velocity (cm/sec)	51 ± 4	47 ± 4	60 ± 4	70 ± 3
Mean velocity (cm/sec)	12 ± 1	11 ± 1	16 ± 2	18 ± 2
Valve diameter (mm)[a]	8 ± 0.5	6.6 ± 0.4	7.6 ± 0.3	6.7 ± 0.2
Cardiac output (mL/kg/min)[a]	307 ± 30	232 ± 25	312 ± 11	250 ± 9
A/E ratio[a]	1.29 ± 0.04	1.35 ± 0.01	–	–
Deceleration time (msec)[a]	97 ± 29	110 ± 31	–	–
Acceleration time (msec)[a]	–	–	50.6 ± 12.0	46.7 ± 9.1

[a]Varies with gestational age.
Abbreviation: A/E, atrial contraction/early diastole.
Source: From Ref. 39.

clearly identifies the communication between the ventricles (Figs. 1 and 19).

A final area in which color imaging is useful is examining regurgitant or stenotic flow within the tricuspid or mitral valve (Figs. 11 and 14). Tricuspid regurgitation occurs with right-sided heart failure and is the most common type of valvular regurgitation identified. Tricuspid regurgitation has various causes, including construction of the ductus arteriosus resulting from indomethacin use. Usually, the atrioventricular axis should be parallel to the ultrasound beam and the velocity setting and the color scale increased to filter out normal atrial flow. In these situations, regurgitant flow may be observed during ventricular systole, into the right atrium through the tricuspid valve.

Power color doppler is different from conventional color Doppler in that it uses amplitude changes rather than the Doppler frequency shift. Therefore, power Doppler does not display the direction of flow but only the presence or absence of flow. The advantage of power Doppler is that it is angle dependent and has a better edge discrimination than conventional color Doppler. It may be used in a similar fashion as conventional Doppler in better identifying normal or abnormal fetal cardiac anatomy (Figs. 1, 5, and 7).

PULSED DOPPLER SONOGRAPHY

Valvular Doppler Imaging

Duplex ultrasound uses both range-gated Doppler and real-time imaging. The real-time image usually demonstrates the source of the Doppler signal, and a separate image depicts the Doppler signal. It is best to have the Doppler cursor placed parallel to the direction of blood flow to avoid the angle correction necessary when the Doppler cursor is not placed parallel to flow direction (Fig. 29). Although regurgitant flow may be diagnosed with Doppler imaging, this technique is usually used to measure areas of stenosis. To assess tricuspid or mitral valve stenosis, the Doppler cursor (sample volume) may be placed just distal to those valves. Doppler parameters for evaluating the heart are listed in Table 18. Aortic or pulmonary blood velocities are measured in a similar fashion by placing the Doppler sample volume just beyond the valve leaflets. Valvular stenosis is associated with increased velocity (40). Normal velocities for the mitral valve, tricuspid valve, aortic valve, and pulmonary valve are listed in Table 19 (39,40). The Doppler tracings for the aorta and pulmonary artery show a single peak, whereas the Doppler tracings for the tricuspid and mitral valves show twin peaks. The first Doppler peak of the semilunar valve results from flow caused by the early diastole, and the second peak is due to flow caused by the atrial contraction.

Aortic velocities are usually obtained from the long-axis view, and pulmonary artery velocities may be obtained from the short-axis view. Tricuspid valve and mitral valve velocities are best obtained when these valves are perpendicular to the ultrasound beam so the valve leaflets are parallel to the plane of the transducer. In addition, in situations such as hypoplastic ventricle,

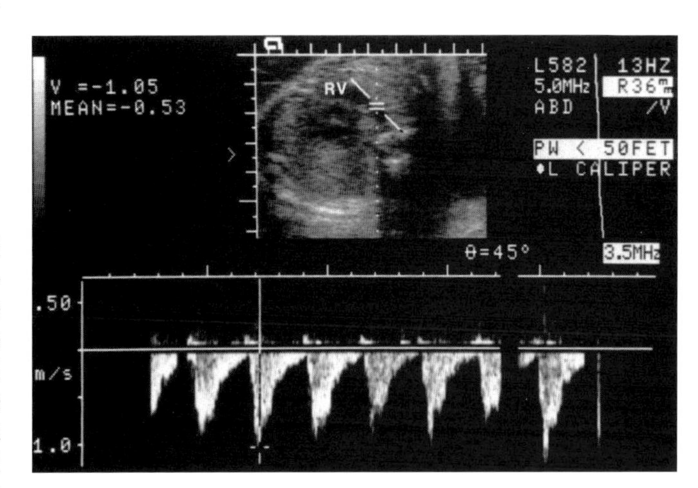

FIGURE 29 ■ Doppler imaging of ductus arteriosus.Doppler cursor angle corrected through the ductus demonstrates a maximum velocity (V) of 105 cm per second.

Doppler imaging may show increased velocities in one outflow tract versus another. For instance, in hypoplastic left heart syndrome, the velocities in the aorta may not be detectable, whereas the velocities in the pulmonary artery are increased.

Valve stenosis is associated with increased velocities through the affected valves. Grading stenosis is usually not performed, but velocities are usually compared with normal values (Fig. 11) (Tables 18 and 19) (40,41).

Doppler Imaging of the Ductus Arteriosus

The longitudinal view of the aortic arch can be obtained from a parasagittal scan. The aortic arch, pulmonary outflow, and ductus arteriosus may be visualized when the fetus is imaged in a sagittal plane. The pulmonary artery lies more anteriorly in the chest than the aorta. The arch of the aorta makes a smooth curve like a candy cane, whereas the pulmonary artery to the ductus has a wider path and has been likened to a hockey stick (Fig. 7). Ductus arteriosus velocities may be obtained by placing the Doppler cursor parallel to the long axis of the ductus arteriosus. Normal systolic ductal velocities vary from 91 to 135 cm per second, and diastolic velocities range from 14 to 25 cm per second (Fig. 26) (39,40). Fetuses may experience constriction in the ductus arteriosus under certain conditions. These include the use of certain medications, most commonly indomethacin (42,43). Increased ductal constriction increases the flow velocity. Abnormal systolic velocities of 141 to 235 cm per second and elevated diastolic velocities of 37 to 168 cm per second have been reported with ductal constriction. More recent data have been used to establish the upper limit of normal of peak diastolic velocity of 35 cm per second for the cutoff level for constriction of the ductus arteriosus (42,43).

CARDIAC ARRHYTHMIAS

Doppler Ultrasound

Cardiac Arrhythmias can be diagnosed by M-Mode and Doppler ultrasound. In its simplest use, a Doppler cursor placed across the aorta or aortic outflow can detect a premature ventricular contraction by noting the early Doppler pulsation (44).

Pulsed Doppler may obtain a better analysis of an irregular rhythm by placing the Doppler cursor between inflow and outflow tracts of the left ventricle. This allows detection of simultaneous arterial and ventricular activity. By positioning of the sample volume in the left ventricle, the outflow signals from the aorta are displayed simultaneously with the outflow inflow signals from the mitral valves.

M-Mode Ultrasound

Although M-mode ultrasound has been used to document ventricular wall thickness, chamber size, atrioventricular valve size, and the measurements of the aortic and pulmonary outflow tracts, it is most commonly used to document fetal cardiac rate and rhythm. M-mode line placement is not particularly crucial when documenting normal fetal cardiac activity. However, to better check a rhythm disturbance of the fetus, the m-mode cursor is placed perpendicular to the atrial septum. This will record the wall motion of the right and left atrial walls and can document any atrial rhythm disturbance. To check to see if the atrial contraction is conducted to the ventricles, the m-mode of the atrium and the ventricle must be displayed simultaneously. The m-mode is placed through both atrial and ventricular walls (45).

REFERENCES

1. Allan LD, Crawford DC, Chita SK, et al. Prenatal screening for congenital heart disease. Br Med J 1986; 92:1717.
2. Moller JH, Neal WA. Incidence of cardiac malformation. Heart Disease in Infancy. New York: Appleton-Century-Crofts, 1981:1.
3. Copel JA, Pilu G, Green J, et al. Fetal echocardiographic screening for congenital heart disease: the importance of the four-chamber view. Am J Obstet Gynecol 1987; 157:648.
4. Schmidt KG, Silverman NH. Evaluation of the fetal heart by ultrasound. In: Callen PW, ed. Ultrasonography in Obstetrics and Gynecology. 2nd ed. Philadelphia: WB Saunders, 1988:165.
5. Benacerraf BR, Sanders SP. Fetal echocardiography. Radiol Clin North Am 1990; 28:131.
6. McGahan JP, Nyberg DA, Mack LA. Sonography of facial features of alobar and semilobar holoprosencephaly. AJR Am J Roentgenol 1990; 154:143.
7. Copel JA, Pilu G, Kleinman CS. Congenital heart disease and extracardiac anomalies: associations and indications for fetal echocardiography. Am J Obstet Gynecol 1986; 154:1121.
8. Van Praagh R, Weinberg PM, Smith SD, et al. Malpositions of the heart. In: Adams FH, Emmanouilides GC, Riemenschneider TA, eds. Moss' Heart Disease in Infants, Children, and Adolescents. 4th ed. Baltimore: Williams & Wilkins, 1980:530.
9. DeVore GR. The prenatal diagnosis of congenital heart disease: a practical approach for the fetal sonographer. J Clin Ultrasound 1985; 13:229.
10. DeVore GR, Siassi B, Platt LD. Fetal echocardiography. IV. M-mode assessment of ventricular size and contractility during the second and third trimesters of pregnancy in the normal fetus. Am J Obstet Gynecol 1984; 150:981.
11. Goncalves LF et al. Four-dimensional ultrasonography of the fetal heart using color Doppler spatiotemporal image correlation. J Ultrasound Med 2004; 23:473.
12. Yagel S, Cohen SM, Achiron R. Examination of the fetal heart by five short-axis views: a proposed screening method for comprehensive cardiac evaluation. Ultrasound Obstet Gynecol 2001; 17:367–369.
13. Brown DL, DiSalvo DN, Frates MC, et al. Sonography of the fetal heart: normal variants and pitfalls. AJR Am J Roentgenol 1993; 160:1251.
14. Brown DL, Cartier MS, Emerson DS, et al. The peripheral hypoechoic rim of the fetal heart. J Ultrasound Med 1989; 149:529.
15. Jeanty P, Romero R, Hobbins JC. Fetal pericardial fluid: a normal finding of the second half of gestation. Am J Obstet Gynecol 1984; 149:529.
16. Ewigman BG, Crane JP, Frigoletto FD, et al. Effect of prenatal ultrasound screening on perinatal outcome: RADIUS Study Group. N Engl J Med 1993; 329:821.
17. Achiron R, Glaser J, Gelernter I, et al. Extended fetal echocardiographic examination for detecting cardiac malformations in low risk pregnancies. Br Med J 1992; 304:671.
18. Bromley B, Estroff JA, Sanders SP, et al. Fetal echocardiographic accuracy: accuracy and limitations in a population at high and low risk for heart defects. Am J Obstet Gynecol 1992; 166:1473.
19. McGahan JP. Sonography of the fetal heart: findings on the four-chamber view. AJR Am J Roentgenol 1991; 156:547.

20. Wicks JD, Levine MD, Mettler FA. Intrauterine sonography of thoracic ectopia cordis. AJR Am J Roentgenol 1981; 137:619.
21. Shipp TB, Bromley B, Hornberger LK, et al. Levorotation of the fetal cardiac axis: a clue for the presence of congenital heart disease. Obstet Gynecol 1995; 85:97.
22. DeVore GR, Horenstein J, Platt LD. Fetal echocardiography. VI. Assessment of cardiothoracic disproportion: a new technique for the diagnosis of thoracic hypoplasia. Am J Obstet Gynecol 1986; 155:1066.
23. McGahan JP, Choy M, Parrish MD, et al. Sonographic spectrum of fetal cardiac hypoplasia. J Ultrasound Med 1991; 10:539.
24. Levy DW, Mintz MC. The left ventricular echogenic focus: a normal finding. AJR Am J Roentgenol 1988; 150:85.
25. Roberts DJ, Genest D. Cardiac histological pathology characteristic of trisomies 13 and 21. Hum Path 1992; 23:1130–1140.
26. Smith-Bindman R, Hosmer W, Feldstein VA, Deeks JJ, Goldberg JD. Second-trimester ultrasound to detect fetuses with Down syndrome. A meta-analysis. JAMA 2001; 285:1044–1055.
27. Sotiriadis A, Makrydimas G, Ioannidis JP. Diagnostic performance of intracardiac echogenic foci for Down syndrome: meta-analysis. Obstet Gynecol 2003; 101 (5 Pt 1):1009–1016.
28. Cuckle H. Biochemical screening for Down syndrome. Eur J Obstet Gynecol Reprod Biol 2000; 92:97–101.
29. Ehrenberg HM, Fischer RL, Hediger ML, Hansen C, Stine D. Are maternal and sonographic factors associated with the detection of a fetal echogenic cardiac focus? J Ultrasound Med 2001; 20 (10):1047–1052.
30. Coates T, Mcgahan HO. Fetal cardiac rhabdomyomas presenting as diffuse myocardial thickening. J Ultrasound Med 1994; 13:813.
31. Giancotti R, Torcia F, Giampa F, et al. Prenatal evaluation of congenital heart disease in high-risk pregnancies. Clin Exp Obstet Gynecol 1995; 22:225.
32. DeVore GR. The aortic and pulmonary outflow tract screening examination in the human fetus. J Ultrasound Med 1992; 11:345.
33. Nyberg DA, Emerson SD. Cardiac malformations. In: Nyberg DA, Mahoney BS, Pretorius DH, eds. Diagnostic Ultrasound of Fetal Anomalies: Text and Atlas. Chicago: Year Book, 1990:300.
34. Lee W, Smith RS, Comstock CH, et al. Teratology of Fallot: prenatal diagnosis and postnatal survival. Obstet Gynecol 1995; 86:583.
35. Comstock CH, Riggs T, Lee W, et al. Pulmonary-to-aorta diameter ratio in the normal and abnormal fetal heart. Am J Obstet Gynecol 1991; 165:1038.
36. Merritt CRB. Doppler color flow imaging. J Clin Ultrasound 1987; 15:591.
37. Copel JR, Morotti R, Hobbins JC, et al. The antenatal diagnosis of congenital heart disease using fetal echocardiography: is color flow mapping necessary? Obstet Gynecol 1991; 78:1.
38. DeVore GR. The role of color Doppler in the screening examination of the fetal heart. In: McGahan JP, Porto M, eds. Diagnostic Obstetrical Ultrasound. Philadelphia: JB Lippincott, 1994:282.
39. Reed KL. Fetal Doppler echocardiography. Clin Obstet Gynecol 1989; 32:728.
40. Reed KL, Duerbeck NB. Doppler ultrasound in obstetrics. In: McGahan JP, Porto M, eds. Diagnostic Obstetrical Ultrasound. Philadelphia: JB Lippincott, 1994:297.
41. Shenker L, Reed KL, Marx GR, et al. Fetal cardiac Doppler flow studies in prenatal diagnosis of heart disease. Am J Obstet 1988; 158:1267.
42. Moise KJ, Huhta JG, Sharif DS. Indomethacin in the treatment of premature labor. N Engl J Med 1988; 319:327.
43. Van den Veyver IB, Moise KJ, Ou CN, et al. The effect of gestational age and fetal indomethacin levels on the incidence of constriction of the fetal ductus arteriosus. Obstet Gynecol 1993; 82:500.
44. Reed KL, Sahn DJ, Marx GR. Cardiac Doppler flow during fetal arrhythmias: physiologic consequences. Obstet Gynecol 1987; 70:1.
45. Humes RA. Diagnosis of fetal cardiac arrhythmias. In: McGahan JP, Porto M, eds. Diagnostic Obstetrical Ultrasound. Philadelphia: JB Lippincott, 1994:315.

Fetal Abdomen and Pelvis

48

John P. McGahan, M. Porto, R. M. Steiger, and Wilbert Fortson

INTRODUCTION

This chapter focuses on a practical approach to the diagnosis of the major abdominal anomalies, using a differential diagnosis format based primarily on their ultrasonographic appearance. It is divided into four major sections—the first addresses the ventral abdominal wall and cord insertion and the group of anomalies that result from its defective closure, the second primarily discusses the gastrointestinal tract and its anomalies, the third details the genitourinary system and its anomalies, and the fourth briefly discusses the use of ultrasound to determine fetal gender.

ANTERIOR ABDOMINAL WALL

Anomalies of the anterior abdominal wall include omphalocele, gastroschisis, and rare defects such as body stalk anomaly and exstrophy of the bladder and cloaca. Patients with cloacal exstrophy usually have an omphalocele. The incidence of these disorders (Table 1) (1–6) increases significantly when the mother participates in a selective screening program, such as α-fetoprotein (AFP) screening. Ventral wall defects (particularly gastroschisis) are the second most common lesions, after neural tube defects, identified by these screening methods (7–9). Chi et al. (8) reported the combined incidence of ventral wall defects to be 5.4 in 10,000 births in Glasgow, U.K., since obstetric ultrasound and serum screening became routine—considerably higher than that reported in the literature published before the mid-1980s. Most recent data from the State of California Birth Defects monitoring program suggest that the frequency of gastroschisis is about 0.35 per 1000 births, and the frequency of omphaloceles is about 0.09 per 1000 births (9). In addition, the frequency is significantly higher, 0.64 per 1000, among mothers less than 20 years old (10). Some cystic lesions of the proximal umbilical cord and anterior abdominal wall that may be confused with omphalocele also are discussed; these include persistent urachus, allantoic cysts, and omphalomesenteric cysts.

The fetal abdomen is bound superiorly by the diaphragm, posteriorly by the back and retroperitoneum, inferiorly by the pelvic bones and pelvic diaphragm, and anteriorly by the ventral abdominal wall (11,12). Open defects of these boundaries are unique to the diaphragm and ventral wall; for example, a diaphragmatic hernia may coexist with an omphalocele. Abnormalities of the pelvic floor are extremely uncommon, although it is possible to see the separation of the pubic bones with absence of the pubic symphysis in bladder and cloacal exstrophy (12).

It is difficult to approach the subject of abdominal wall defects without a brief discussion of the abnormal development involved. *Body stalk anomaly* can arise from failed closure of the lateral body folds in embryogenesis (13). An alternate hypothesis for the development of body stalk anomaly is early rupture of the amnion at four to five weeks' gestation, possibly leading to mechanical compression of the body against the placenta before closure of the lateral folds (14). A third theory is that body stalk anomaly is a severe form of amniotic band syndrome (13). Although

TABLE 1 ■ Abdominal Wall Defects and Associated Rate of Occurrence

Defect	Frequency	Investigators
Omphalocele	1/5000	Carpenter et al., 1984 (1); Lindham, 1981 (2)
Gastroschisis	1/4500–10,000	Carpenter et al., 1984 (1); Lindham, 1981 (2); Baird and MacDonald, 1981 (3); Torfs et al., 1996 (9)
Body stalk anomaly	1/15,000	Mann et al., 1984 (4)
Exstrophy of bladder	1/30,000	Engel, 1974 (5)
Exstrophy of cloaca	1/200,000	Soper and Kilger, 1964 (6)

Source: From Ref. 178.

it is likely that some amniotic band disruptions can mimic the findings of body stalk anomaly, well-documented cases give greater support to the other theories.

Incomplete closure of the lateral body folds may also be partially responsible for omphalocele; more typically, failure of the bowel to return to the abdomen after the normal herniation into the cord between the 8th and 12th weeks of gestation is held to be responsible (Figs. 1–3) (15). Gastroschisis is thought to arise from the occlusion of the right omphalomesenteric artery early in embryonic life, leading to infarction of the right anterior abdominal wall (15). Exstrophy of the bladder and cloaca is thought to result from persistence of the cloacal membrane, inhibiting normal development of the lower anterior abdominal wall, symphysis pubis, and perineum (Fig. 4). If the urorectal septum fails to develop completely as well, the bladder and hindgut continue to communicate with each other in a common chamber (the cloaca), and the anus remains imperforate (Fig. 5). Cloacal and bladder exstrophy or eversions result when the cloacal membrane finally breaks down (16).

Normal Anatomy

Standards

The latest guidelines, issued in 2003 by the American Institute of Ultrasound in Medicine for second- and

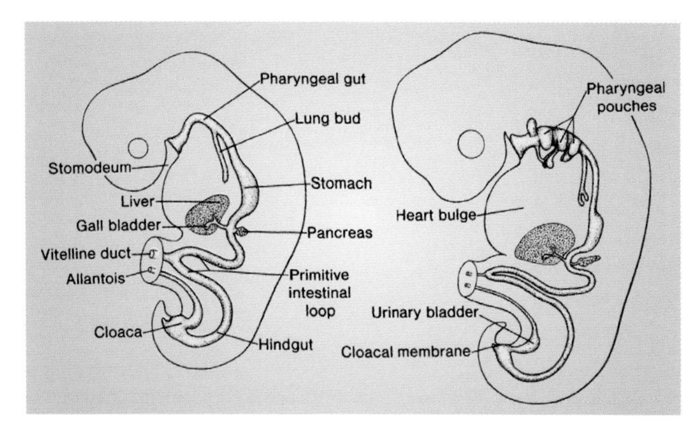

FIGURE 1 ■ Gut migration/cloaca. Early in development, the primitive intestine communicates with the Vitelline duct as it herniates into the umbilical cord. The urinary bladder is derived from the cloaca and connects with the allantois in the umbilical cord. *Source*: From Ref. 179.

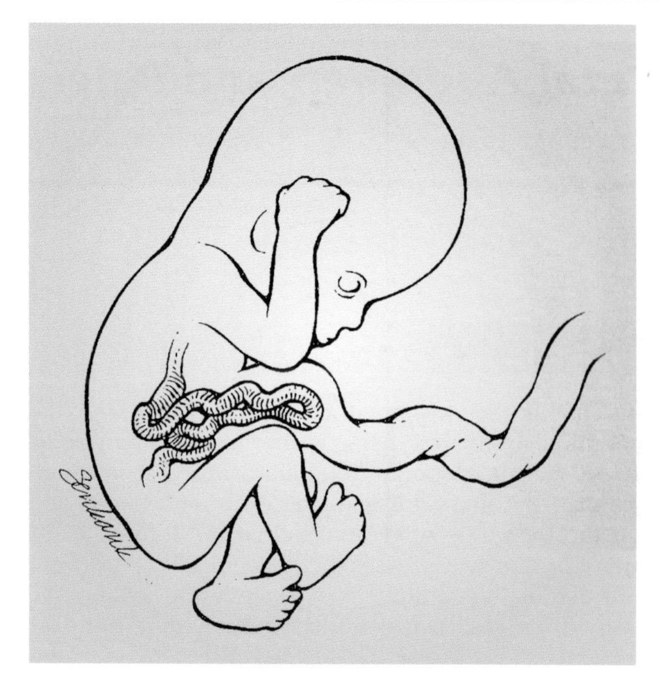

FIGURE 2 ■ Normal gut migration. The bowel normally herniates into the base of the umbilical cord between 8 and 12 menstrual weeks and returns to the abdominal cavity by 12 weeks. This early herniation should not be confused with an omphalocele. *Source*: Modified from Ref. 180.

third-trimester ultrasonography, states the following: "The study should include, but not necessarily be limited to, assessment of the following fetal anatomy: ... stomach, kidneys, urinary bladder, fetal umbilical cord insertion site and intactness of the anterior abdominal wall" (17). This statement, as well as similar guidelines put forth by the American Colleges of Radiology and Obstetrics and Gynecology, delineates the minimum standard views of the abdomen that are necessary to visualize the defects discussed in this chapter (18). The stomach, kidneys, bladder, and umbilical cord insertion must be seen. The contour of the abdomen should also be studied. The normal approach is as follows.

The fetal abdomen is generally viewed in serial transverse planes. The smooth contour of the abdominal wall should be noted while scanning from one end of the abdomen to the other. On the inner aspect of the abdominal wall is a thin, 1- to 3-mm hypoechoic zone, which is the muscle tissue of the abdominal wall. This should not be confused with ascites, because it is smooth and not crescent shaped, as are ascites that surround the liver or bowel (Figs. 6 and 7). This hypoechoic zone can also be noted to be continuous with the ribs. After identifying the fetal stomach and bladder, the umbilicus should be identified. This can be difficult in situations in which the extremities are close to the abdominal wall. The level of difficulty is increased when the fetus is prone and in the third trimester, when there is relatively less amniotic fluid. The location of the umbilicus can be inferred in these situations by identifying the umbilical vein in the upper abdomen and following it caudally

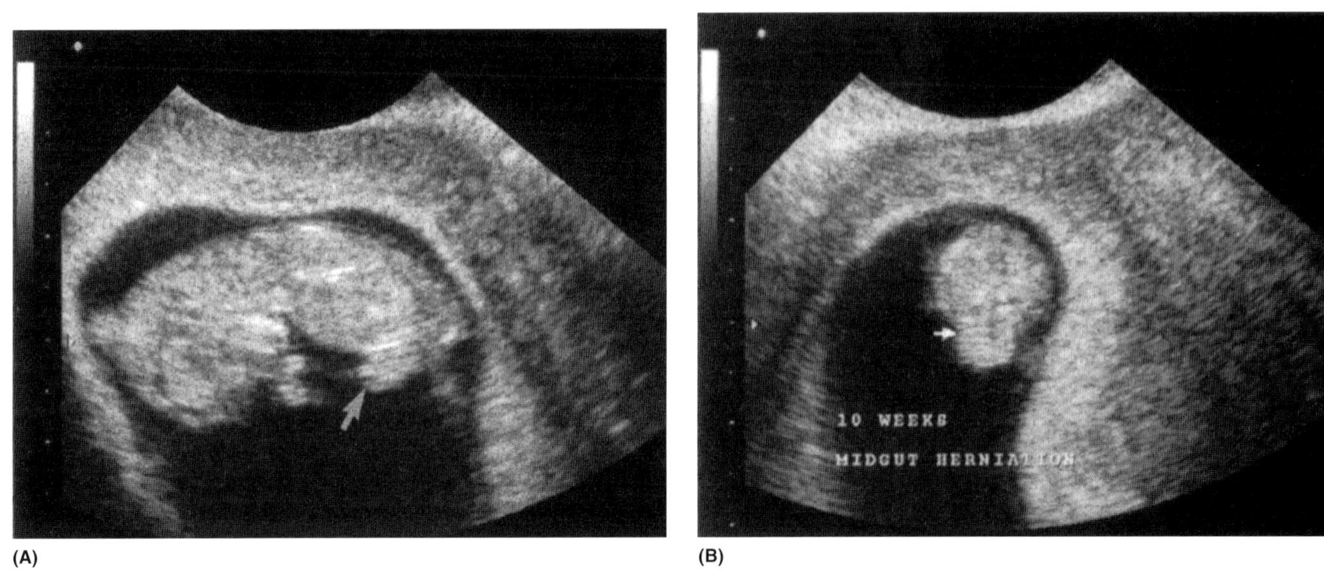

FIGURE 3 A AND B ■ Normal midgut herniation (*arrows*) in a 10-week fetus.

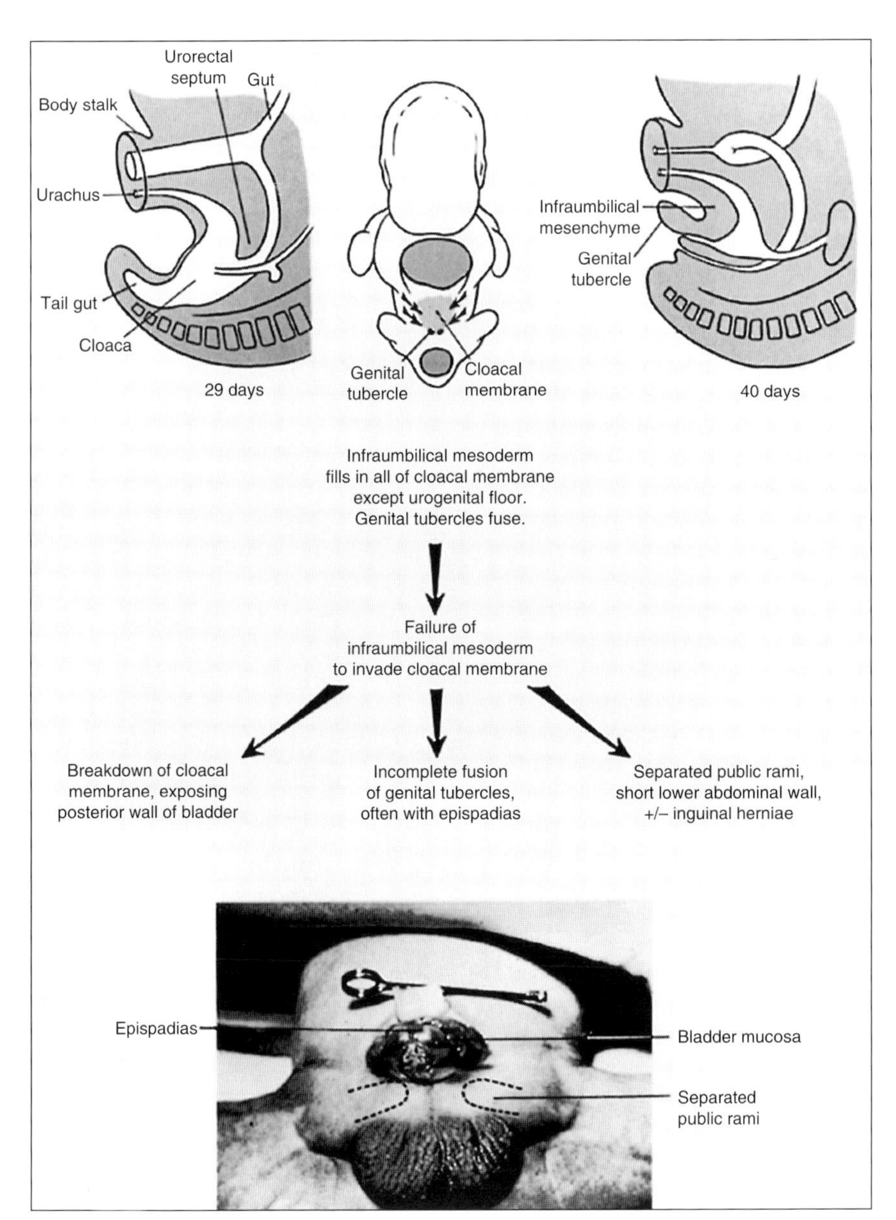

FIGURE 4 ■ Exstrophy of the bladder sequence. Normally, the infraumbilical mesenchyme migrates to give rise to the abdominal wall, genital tubercle, and pubic rami. Failure of migration leads to exposed bladder wall, incomplete fusion of the genital tubercle, and separation of the pubic rami. *Source*: From Ref. 182.

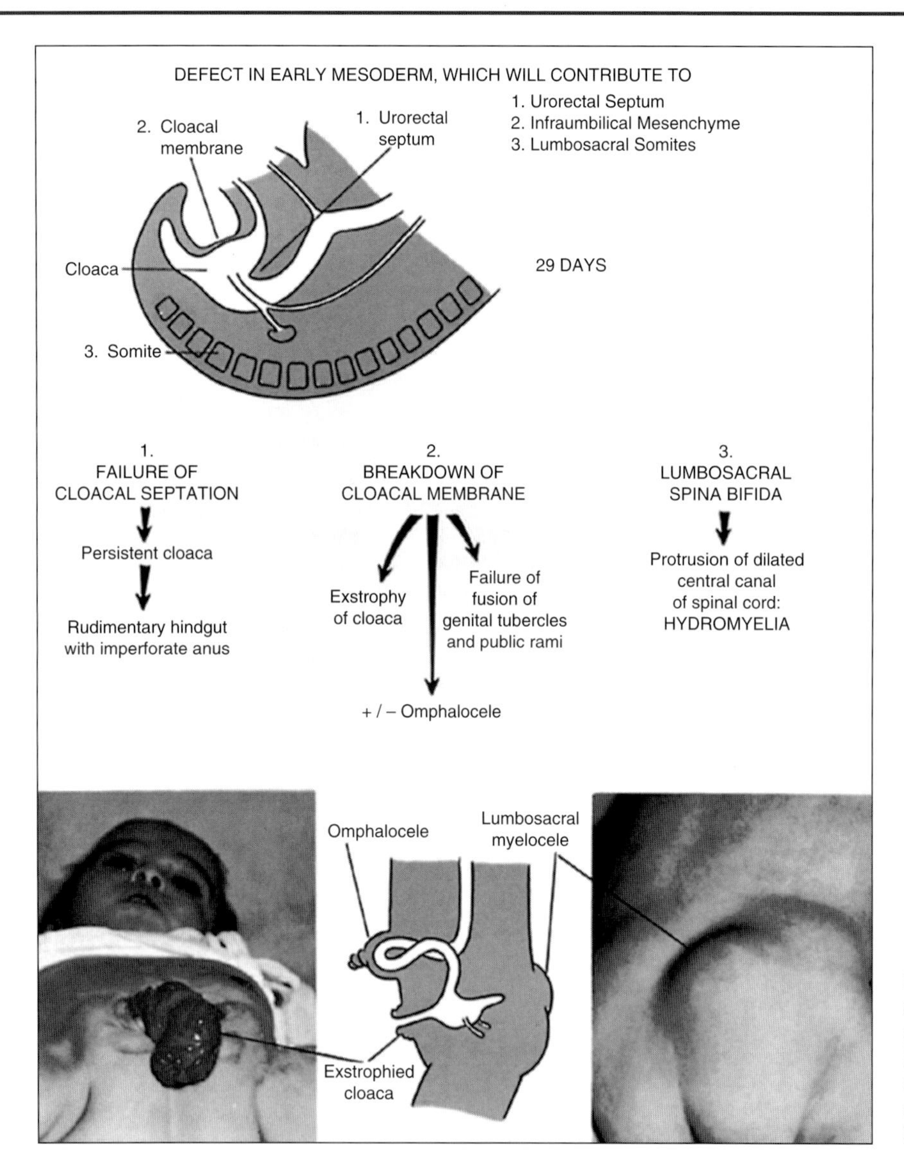

FIGURE 5 ■ Developmental sequence of cloacal exstrophy. Normally, the mesoderm contributes to the (*i*) infraumbilical mesenchyme, (*ii*) cloacal septum, and (*iii*) caudal vertebra. Defective development of the early mesoderm leads to exstrophy of the cloacae with omphalocele and lumbosacral myelocele. *Source*: From Ref. 182 .

until it approaches the abdominal wall and finally disappears from the abdomen. This point of disappearance should be at or just below the umbilicus. Sagittal views can also be helpful in following the umbilical vein and studying the contour of the abdominal wall. When possible, visualization of the proximal umbilical cord with enumeration of the vessels is ideal to rule out cystic lesions of the umbilical cord, such as a patent urachus or allantoic cysts. The umbilicus can be a good place to count the vessels, with the two umbilical arteries coming in from the lower abdomen and the vein entering from the upper abdomen. Cloacal exstrophy is usually associated with a single umbilical artery (15). As a word of caution, smaller omphaloceles (<3 cm in diameter) are far more likely to be associated with chromosomal anomalies than larger lesions (Fig. 8) (19).

Artifacts

Common artifacts that can be mistaken for abdominal wall defects are listed in Table 2.

1. The bowel normally herniates into the proximal umbilical cord in the eighth menstrual week but returns to the abdomen during the 12th week (Figs. 2 and 3) (14). For this reason, it has often been stated that the diagnosis of omphalocele or gastroschisis cannot be made with certainty before 13 weeks of gestation (20,21). Aided by high-resolution transvaginal transducers at 12 to 16 weeks, however, Blazer et al. (22) reported on a series of 38 cases of early pregnancy diagnosis of omphalocele. Sixteen of these cases were diagnosed with the omphalocele being an isolated finding, including a normal karyotype. In 8 of the 16 cases, however, no omphaloceles were identified. These data stress that it is important to use caution when considering the early diagnosis of omphalocele.
2. Oblique views of the fetus can give the false impression of an omphalocele. This is especially true when the fetus is in a flexed position with decreased amniotic fluid (Fig. 9). Viewing the abdomen in the proper

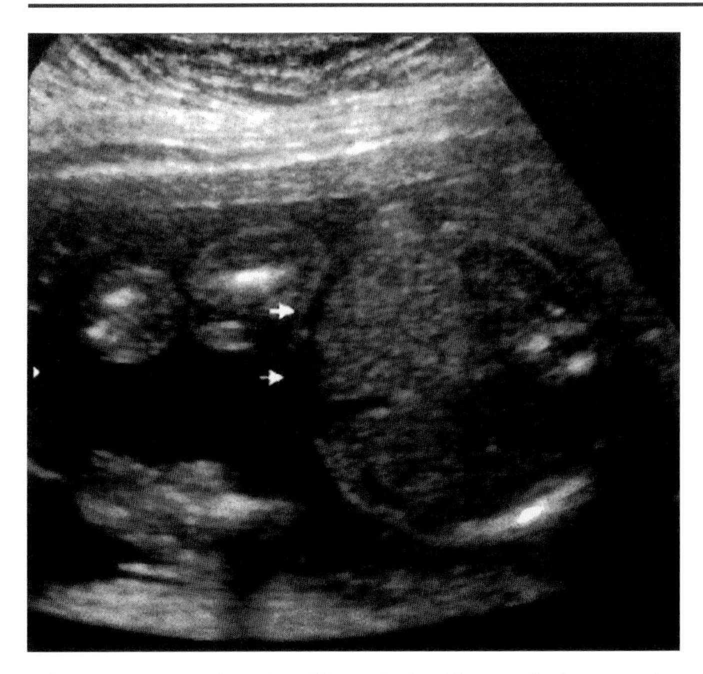

FIGURE 6 ■ Pseudoascites. Hypoechoic, thin muscle layer on the inner aspect of the anterior abdominal wall mimics ascites.

plane, when the fetus is at rest, with the spine relatively straight, while noting the subcutaneous fat covering the anterior abdominal wall, usually resolves this problem.

3. Normal umbilical cord can be confused with gastroschisis in the fetus when the umbilical cord is bunched and compressed against the abdomen. This confusion may be worsened if the cord is large or has a large umbilical vein (23). The use of color Doppler or color power angiography quickly resolves this if flow cannot be appreciated.
4. Localized deposition of Wharton's jelly at the umbilicus can lead to the impression of a small omphalocele (24); but the omphalocele and the vessels run through it, not around it. Omphaloceles with predominately liver contents are also hyperechoic, but these defects are larger.
5. Focal masses of the umbilical cord, such as omphalomesenteric cysts or allantoic cysts, may be confused with abdominal wall defects. These cysts can occur in conjunction with abdominal wall defects.

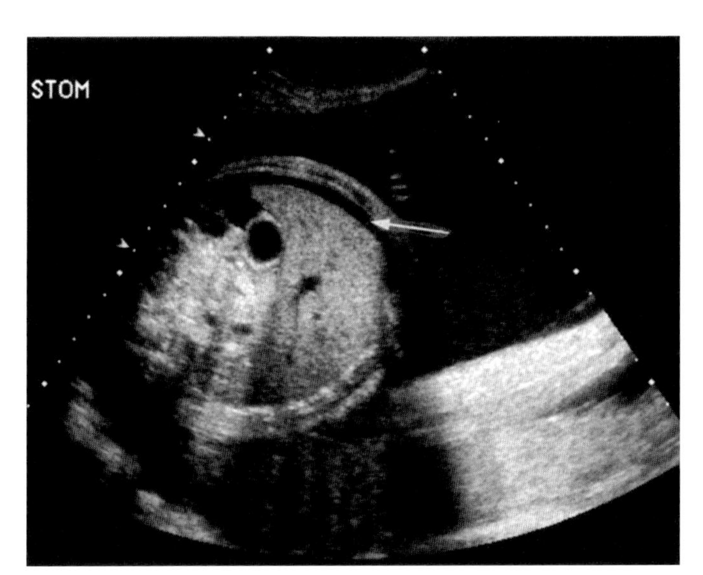

FIGURE 7 ■ True ascites. Arrow points to ascites surrounding the liver.

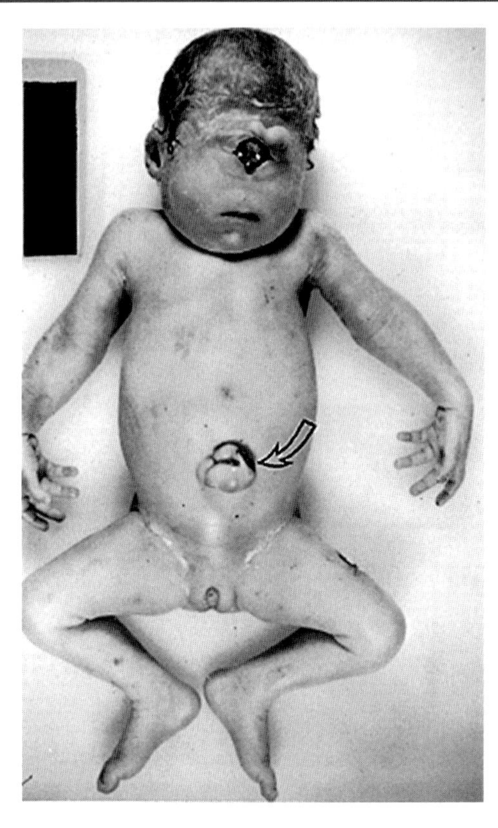

FIGURE 8 ■ Typical facial features of trisomy 13 with cyclopia, proboscis, and very small bowel containing omphalocele (*arrow*).

Abnormal Ventral Wall

The two levels of approach to the diagnosis of ventral wall defects are detection and diagnosis. The first level is the

TABLE 2 ■ Artifacts Commonly Confused with Ventral Wall Defects

Artifact	Comment
Bowel herniation into umbilical cord	Normally present at 8–12 menstrual weeks
Oblique fetal abdomen (pseudo-omphalocele)	Note elongated rather than oval shape of abdomen
Normal umbilical cord	Doppler or color to check for flow
Localized Wharton's jelly	Hyperechoic appearance with vessels traversing
Cord cysts or masses	Focal lesions of cord

Source: From Ref. 11.

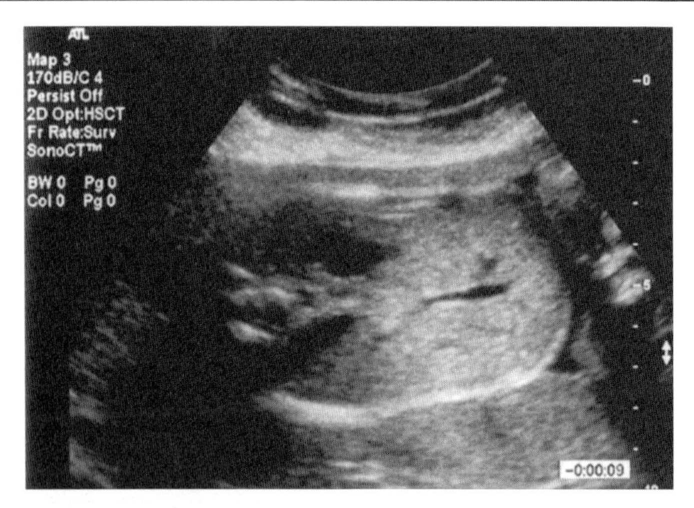

FIGURE 9 ■ Pseudo-omphalocele. Oblique scan through the compressed abdomen demonstrates an apparent omphalocele in the anterior abdominal wall, which was not in fact an omphalocele but compression of the anterior abdominal wall.

TABLE 3 ■ Site of Ventral Wall Defects in Relation to Umbilicus

- *Above umbilicus*: consider pentalogy of Cantrell
- *At umbilicus*: consider gastroschisis or omphalocele
- *Below umbilicus*: consider exstrophy of bladder or cloaca
- *Difficult to tell because of size*: consider body stalk anomaly

Source: From Ref. 10.

outcome of a careful and thorough survey of fetal anatomy as outlined previously. Detection can be greatly aided by an organized AFP screening program; this is discussed separately. Once a lesion is detected, the examination is targeted to determine the exact nature and extent of the lesion as well as to determine the presence of associated anomalies. This is critical in predicting the prognosis of the fetus and the relative need for karyotype determination.

An outline to the approach of ventral wall defects is given in Table 3 and is discussed in more detail here. First, it is important to localize the site of the defect in relation to the umbilicus. If a break in the contour of the abdominal wall or a herniation of abdominal contents is detected, then the approximate site of the defect must be determined. When it is at the level of the umbilicus, it is either a gastroschisis or omphalocele. Lesions that extend below the umbilicus suggest exstrophy of the bladder or cloaca (Fig. 10). Pentalogy of Cantrell is an omphalocele with a defect extending upward toward the thorax, including a herniated heart (ectopia cordis) and diaphragmatic and sternal defects (25). This defect is discussed elsewhere in this text. When the ventral wall defect is so large that it is hard to tell where it begins or ends, then a body stalk anomaly should be suspected (Fig. 11).

When the herniated abdominal contents include the bowel, the lesion is either a gastroschisis, an omphalocele, or a body stalk anomaly (Table 4). Omphaloceles and gastroschisis are the most common abnormalities and have distinguishing features and important differential points (Table 5). Omphaloceles occasionally may contain only liver but more typically also include portions of bowel (Figs. 12–14). Omphaloceles may contain liver that appears solid with hepatic vessels, small bowel that appears echogenic, or other intra-abdominal contents, such as the stomach or gallbladder (Fig. 12). Omphaloceles that are small (<3 cm in diameter) and that contain only bowel and no liver are more often associated with karyotypic abnormalities (Fig. 8) (19).

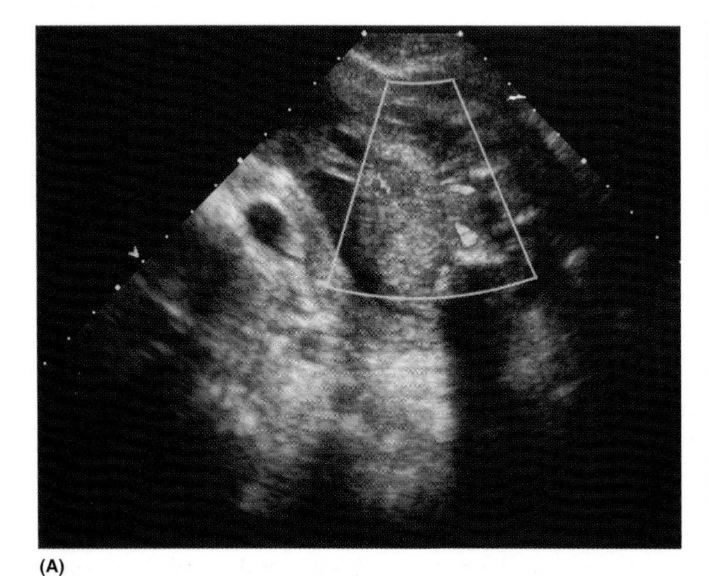
(A)

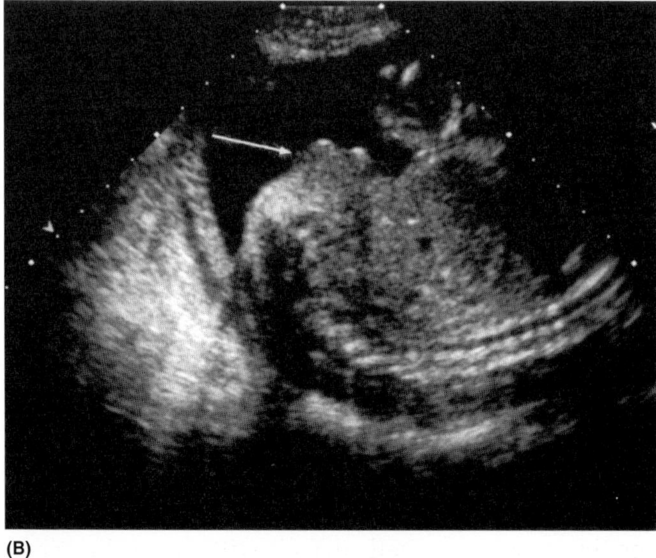
(B)

FIGURE 10 A AND B ■ Bladder exstrophy. (**A**) Scan showing umbilical arteries, but no bladder. (**B**) Longitudinal scan of pelvis showing soft tissue mass in anterior pelvis, corresponding to exstrophy of bladder.

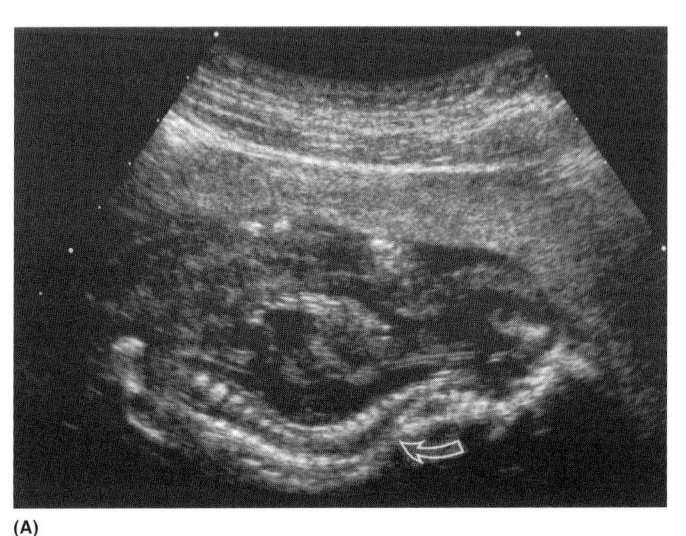

(A)

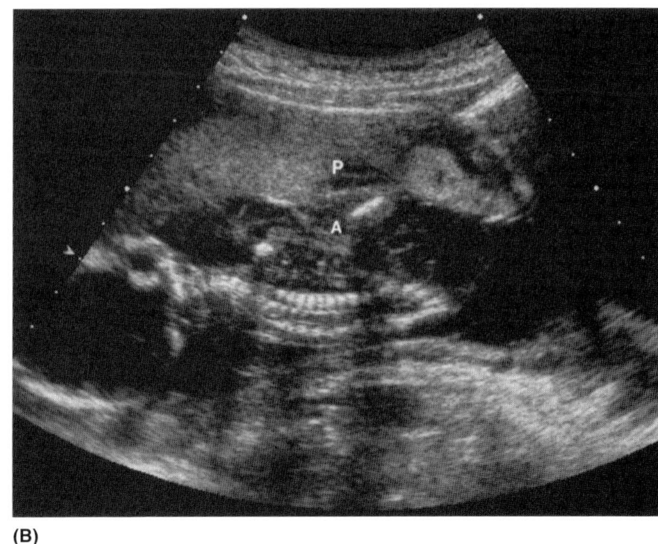

(B)

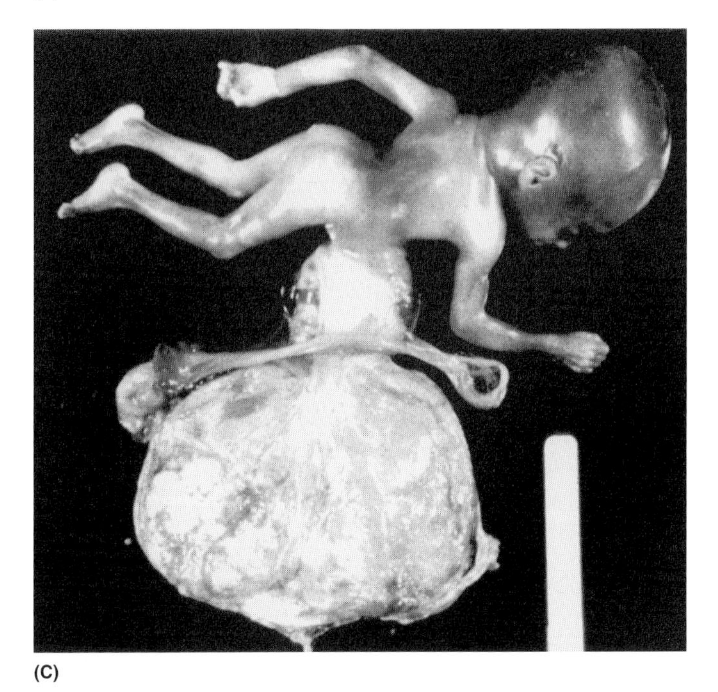

(C)

FIGURE 11 A–C ■ Body stalk anomaly. (**A**) Coronal view through the fetal spine demonstrates typical scoliosis (*arrow*). (B) Longitudinal scan showing anterior abdomen (**A**) to be contiguous with the placenta. (**C**) Postmortem specimen in another fetus demonstrates defect in the anterior abdominal wall, which is contiguous with the placenta. P, placenta.

Ascites may be seen with omphaloceles (Fig. 13). An omphalocele herniates through the umbilicus into the cord, creating a sac that is bounded by a thin membrane consisting of peritoneum and amnion. The presence of this sac is the easiest way to differentiate an omphalocele from gastroschisis (Fig. 13). Unfortunately, this sac ruptures about 15% of the time (26). Localization of the umbilicus and its vessels in relation to the defect is then important. The ventral wall defect is on the right side of the umbilical vessels in gastroschisis (Figs. 15–19). The umbilical vessels come off the top of an omphalocele. A herniated liver in a gastroschisis has only been reported infrequently (27). The presence of a herniated liver in a ventral wall defect without a limiting membrane is more suggestive of a ruptured omphalocele. Extra-abdominal and intra-abdominal bowel obstruction is more common in gastroschisis, so dilated bowel loops suggest gastroschisis. Controversy exists about whether a critical bowel diameter (1.7–2 cm) is a sufficiently sensitive and specific marker of bowel compromise to suggest the need for intervention (delivery). Investigative reports found no reliable correlation between maximal bowel diameter and outcome in gastroschisis (28). However, Aina-Mumuney et al. reported that stomach dilation (mean transverse diameter of greater than 2 cm) was associated with abnormal antepartum testing, longer time to full feeds, and longer hospital stay (29). Distinguishing features that separate gastroschisis from omphalocele are listed in Table 5.

The diagnostic features of body stalk anomaly are a large ventral wall defect, the absence of the umbilical cord (the body stalk), and the presence of fetal scoliosis from immobilization of the midtrunk (Fig. 11). The absence of the umbilical cord can be difficult to confirm because the

TABLE 4 ■ Initial Approach to Ventral Wall Defect

Clinical features	Suspected anomaly
I. Bowel involved	
A. Contained in sac	Omphalocele, unless:
1. Large defect, no separate cord, fetal scoliosis	Body stalk anomaly
2. Bladder persistently not seen, neural tube defect present	Cloacal exstrophy
B. Not contained in sac—cord to left side	Gastroschisis, unless:
1. Liver involved	Possible ruptured omphalocele
2. Malformations other than bowel (Dilated bowel loops are consistent with obstruction and imply gastroschisis)	Probable ruptured omphalocele
II. Solid mass	
A. Smooth border	
1. Large	Probably omphalocele with liver
2. Relatively small with umbilical vessels, no bowel	Umbilical cord with localized Wharton's jelly
B. Irregular border, below umbilicus	Bladder exstrophy
III. Large septate cystic mass in lower abdomen or pelvis; neural tube defect present	Cloacal exstrophy

Source: From Ref. 178.

umbilical vessels usually travel together along the edge of the membranes. The vessels, however, are amazingly straight because torsion of the cord is not possible (30). First-trimester diagnosis was reported with transvaginal ultrasonography (31). The high incidence of amputation defects of extremities from amniotic bands in cases considered to be body stalk anomalies has led to the synonym *limb-body wall complex* (Fig. 11).

A defect below the umbilicus suggests exstrophy of the bladder. The presence of this defect and a persistently absent fetal bladder are diagnostic of this condition (Fig. 4).

TABLE 5 ■ Important Distinguishing Factors Between Gastroschisis and Omphalocele

Gastroschisis	Omphalocele
Right-sided, paraumbilical	Umbilical
Free-floating bowel	Defect covered by peritoneum and amnion
Liver in defect—extremely rare	Liver in defect—common
Bowel obstruction (intra-abdominal or extra-abdominal)—common	Bowel obstruction—rare
Associated anomalies—extremely rare	Associated anomalies—common, especially cardiac
Chromosomes—aneuploidy not increased	Chromosomes—frequently abnormal
Associated syndromes—rare	Associated syndromes—common (Tables 1–6)

Source: From Ref. 178.

The diagnosis should also be actively pursued when a normal bladder is persistently absent and the kidneys appear normal. Although bladder exstrophy is several times more common than cloacal exstrophy, it has been diagnosed antenatally only rarely (32–37). There is separation of the pubic bones with absence of the symphysis

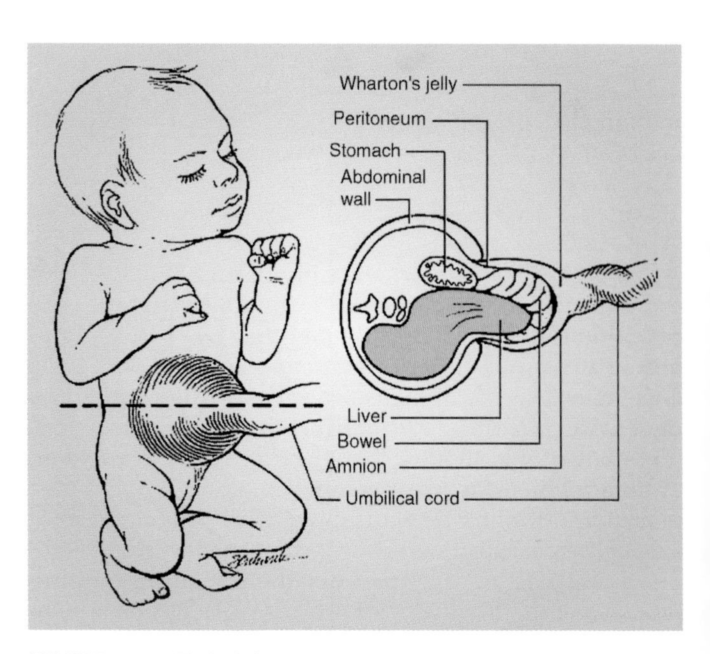

FIGURE 12 ■ Typical features of omphalocele containing liver and bowel on external examination and cross-sectional view. This defect, unlike gastroschisis, is covered by both the amnion and the peritoneum and can contain a number of structures, including the liver, bowel, and fetal stomach. *Source*: Modified from Ref. 182.

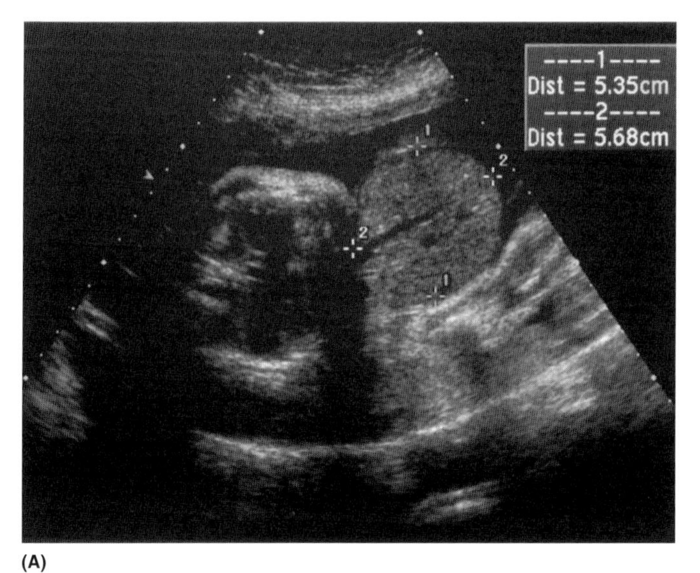

(A)

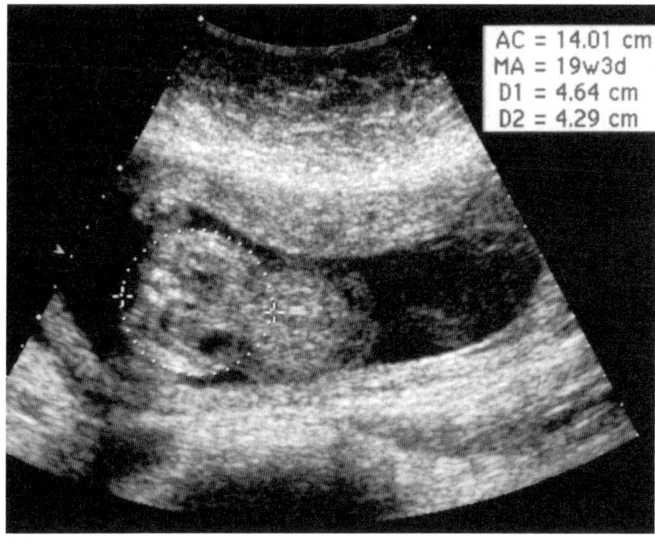

(B)

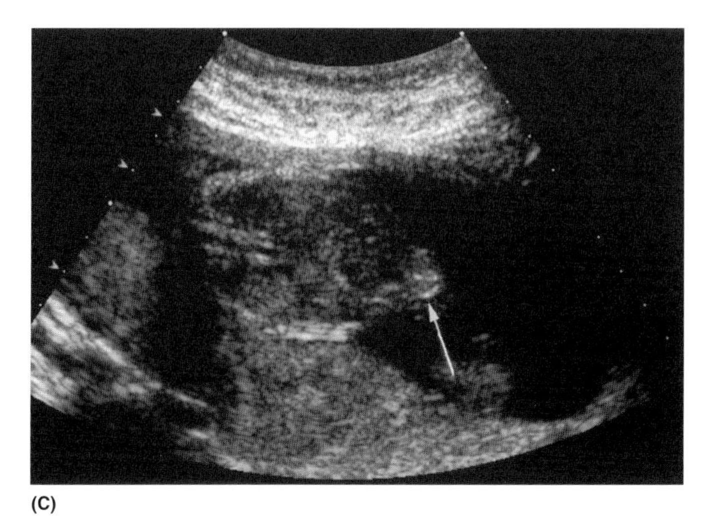

(C)

FIGURE 13 A–C ■ Variable appearance of omphalocele. (**A** and **B**) Typical appearance of liver containing omphalocele with a figure-eight appearance of the omphalocele, as noted on transverse scan of the abdomen. Note the acute angle of the omphalocele with the anterior abdominal wall in both **A** and **B** (*arrow*). (**C**) Small-bowel-only omphalocele (*arrow*).

in both cloacal and bladder exstrophy, but this has only occasionally been suggested antenatally.

This presentation of cloacal exstrophy is much more complex. Cloacal exstrophy, in addition to being the rarest of ventral wall defects, is also the most complex and varied in its ultrasonographic appearance. Most of the reported cases had a large cystic mass in the pelvis or abdomen, frequently noted to be septate and not usually protruding from the anterior abdominal wall. This cyst may be accompanied by hydronephrosis and oligohydramnios in most cases and is frequently misinterpreted as an obstructed bladder outflow tract. These cystic structures technically are more likely to represent a persistent cloaca than cloacal exstrophy, but the diagnosis is based on the findings at the time of birth. Ascites and neural tube defects are frequent findings as well. Neural tube defects are thought to be uniformly present in cloacal exstrophy. The presence of a lumbosacral neural tube defect in association with an omphalocele or in the absence of a bladder should lead to suspicion of cloacal exstrophy. In at least one case, ascites were noted to predate the development of the abdominal cyst (38).

A sequence of events may be used to attempt to explain the varied ultrasonographic presentations of cloacal exstrophy. In cases of cloacal exstrophy, the cloacal membrane initially remains intact. The cloaca may leak as urine fills it and cause ascites, possibly through the fallopian tubes attached to the cloaca, as proposed by Petrikovsky et al. (38). If there is no leak, or after the leak seals, the cloaca becomes distended, and in turn, the kidneys show changes consistent with hydronephrosis. The incomplete urorectal septum causes the cyst to appear septate. The cloacal membrane may then rupture, leading to the classic findings of cloacal exstrophy. When the membrane does not rupture, it is preferable to refer to the abdominal cyst as a persistent cloaca or cloacal cyst. There is documentation by serial ultrasound of rupture of the persistent cloacal membrane (11). Omphaloceles have not been seen in the presence of the cloacal cysts. A normal fetal bladder was seen only once, and it was subsequently replaced by a large septate cystic structure (38). Beyond this isolated case, a normal bladder has never been seen—it is either absent or thought to be very large. The large cloacal cyst is not

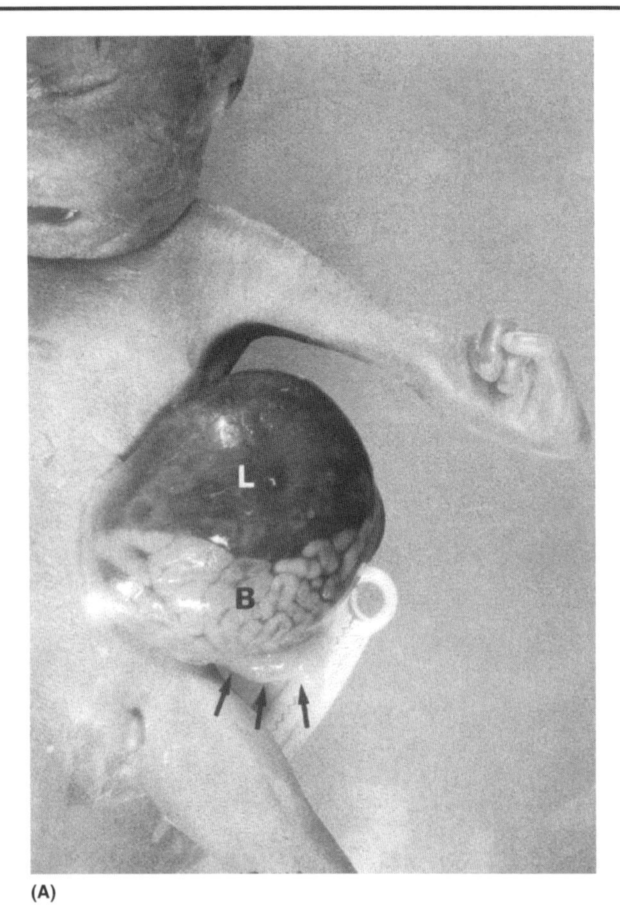

(A)

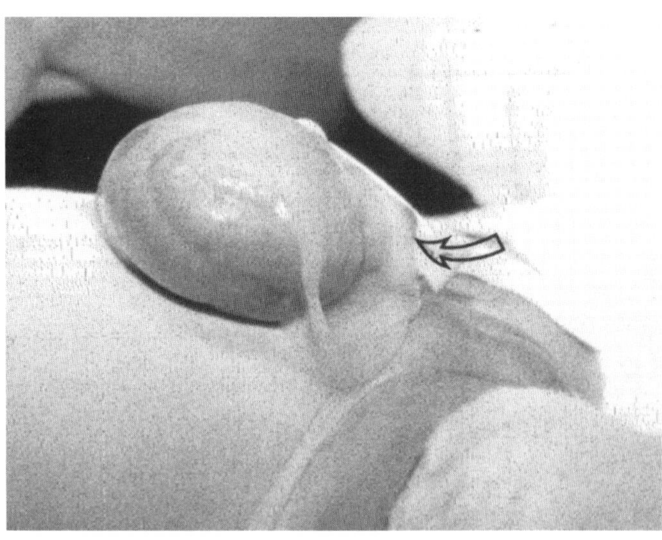

(B)

FIGURE 14 A AND B ■ (**A**) Omphalocele containing both liver and bowel. Umbilical cord is marked with arrows. (**B**) Small bowel containing omphalocele. Umbilical cord is marked with curved arrow. L, liver; B, bowel. *Source*: Courtesy of Marshall Z. Schwartz, MD, Washington, DC.

infrequently mistaken for megalocystis secondary to posterior urethral valve (PUV). The spinal lesions seen with cloacal exstrophy are discussed in the sections that follow.

FIGURE 15 ■ Typical features of gastroschisis demonstrated on external examination and cross-sectional view. The ventral wall defect is right paraumbilical, with bowel loops floating free in the amniotic fluid. *Source*: Modified from Ref. 182.

Maternal Serum AFP and Ventral Wall Defects

Maternal serum AFP levels are higher in gastroschisis than in omphalocele, probably because the former defect is not covered by a membrane (39). Another possibility is the strong association between trisomies and omphalocele. Trisomic fetuses, especially trisomy 18, are known to result in lower maternal serum AFP levels (40). The

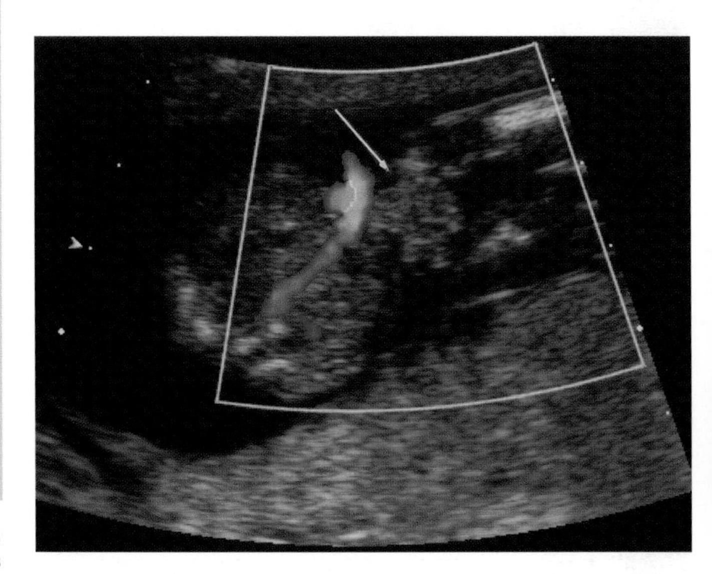

FIGURE 16 ■ Gastroschisis. Free loops of bowel herniate to the right (*arrows*) of the umbilical cord insertion (*red*).

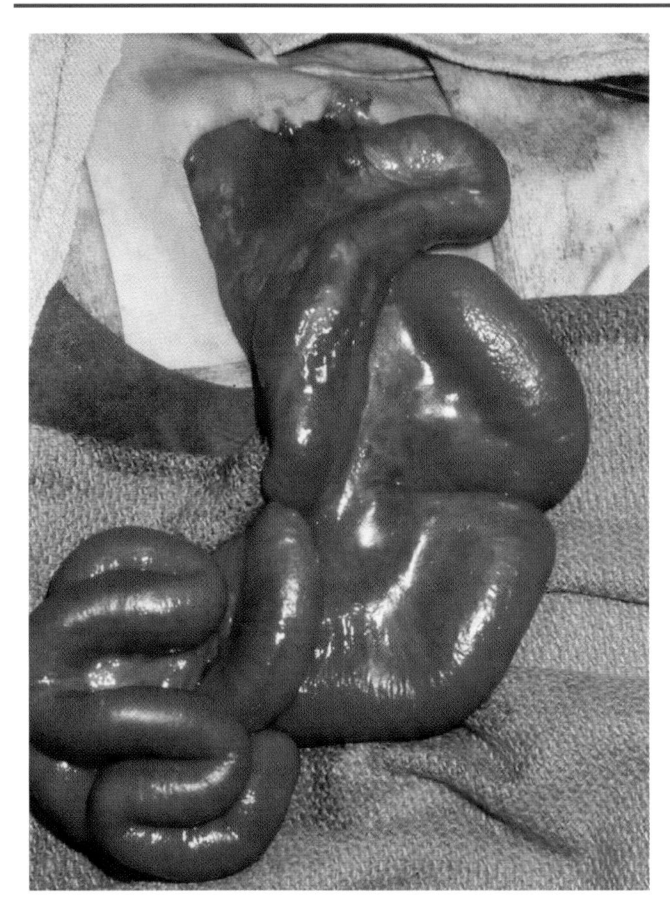

FIGURE 17 ■ Gross examination of gastroschisis at the time of delivery demonstrates right paraumbilical defect with loops of bowel. *Source*: Courtesy of Marshall Z. Schwartz, MD, Washington, DC.

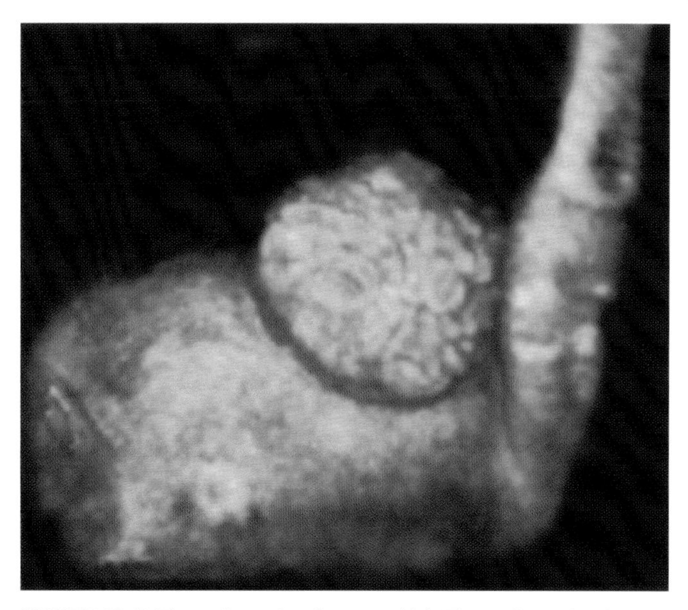

FIGURE 19 ■ Three-dimensional gastroschisis. Three-dimensional image showing free-floating loops of bowel to the right side of the umbilical cord. *Source*: Courtesy of George Bega, MD, Philadelphia, PA.

combined effect of these two trends may result in normalization of the maternal serum AFP values. This has been specifically addressed in trisomy 18 fetuses (41). The authors of this study compared multiples of the median (MoM) in trisomy 18 fetuses without ventral wall defects and neural tube defects to those in fetuses with omphalocele and to a small group of three fetuses with both defects.

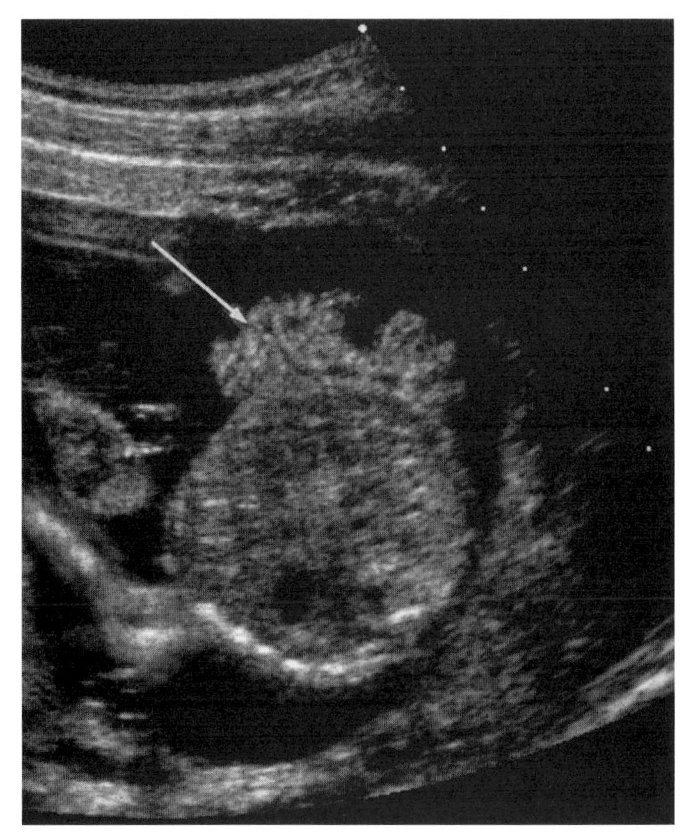

(A)

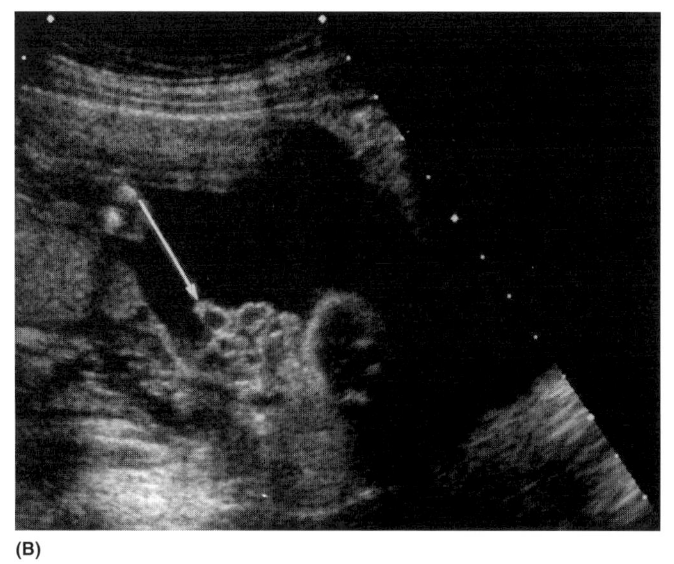

(B)

FIGURE 18 A AND B ■ Gastroschisis. Variable appearance of free-floating bowel, which can appear echogenic and collapse as in (**A**) (*arrow*) or with bowl dilatation as in (**B**) (*arrow*).

The median MoM with uncomplicated trisomy 18 was 0.6; with an omphalocele, it was 1.1 MoM; and the trisomy 18 fetuses with both omphalocele and neural tube defects had a median MoM of 4.5. The fetus with trisomy 18 and omphalocele may not be picked up with a screening maternal serum AFP, and as previously stated, the omphalocele in the aneuploid fetus may be small.

Body stalk anomalies are associated with very high maternal serum AFP levels, as would be expected from the size of the defect. In the one reported case in which an AFP value was available, the value was 10.1 MoM (42). The last three body stalk anomalies from our facility yielded an average AFP level of 9.9 MoM, with a range of 6.8 to 11.6 MoM. None of our cases had an open neural tube defect.

Cloacal exstrophy is generally thought to be associated with an elevated maternal serum AFP level, based on a series of six patients reported to have exstrophy of the cloaca (43). The description of five of the six cases is more consistent with body stalk anomaly. This leaves only one reported case of cloacal exstrophy with an elevated AFP level from Gosden and Brock's series. There are few reported cases in which only an amniotic fluid AFP level was obtained with cloacal exstrophy, and all were reported to be normal (38,44). An elevated AFP level would be expected in this disorder, which is associated with omphalocele and neural tube defects. The neural tube defect, however, may be a hydromyelia, which is not an open defect and is skin covered (16). The cloacal and bladder exstrophy themselves do not cause contact between the peritoneal cavity and the amniotic fluid. The omphalocele may be small, as it was in the case from our facility. To our knowledge, there are no publications that report maternal serum AFP levels in bladder exstrophy.

TABLE 6 ■ Syndromes Associated with Omphalocele

Syndrome	Ultrasonographically detectable features
Beckwith–Wiedemann	Macrosomia (increased kidneys, liver), enlarged protuberant tongue (macroglossia)
Cloacal exstrophy	Absent bladder, lumbar sacral neutral tube defect, single umbilical artery
Pentalogy of Cantrell	Omphalocele, ectopia cordis, diaphragmatic hernia, pericardial defect, intrinsic cardiac abnormality
Meckel–Gruber	Encephalocele, enlarged echogenic kidneys, polydactyly, cleft lip, Dandy–Walker malformation
Trisomy 13	Multiple cardiac defects, midline facial defects, microcephaly, holoprosencephaly, polydactyly, rocker-bottom feet, single umbilical artery
Trisomy 18	Multiple cardiac defects, intrauterine growth restriction, cleft lip, single umbilical artery, cystic hygroma, hydrocephaly, overlapping fingers, rocker-bottom feet
Trisomy 21	Multiple cardiac defects, duodenal atresia, cystic hygroma
Polyploidy	Severe early onset, asymmetric intrauterine placenta, growth restriction, molar degeneration of the multiple defects of the heart and central nervous system, oligohydramnios

Source: From Ref. 178.

Omphalocele

In addition to being one of the most frequent ventral wall defects, omphalocele is the most complex in terms of its causes. Some of the syndromes in which one of the manifestations is omphalocele are listed in Table 6, along with other ultrasonographically detectable features seen in the particular syndrome. Most of these syndromes are covered in greater detail in other chapters.

The large number of karyotypic abnormalities linked to this disorder is of special importance. The incidence of abnormal karyotypes reported in fetuses with omphalocele ranges from 18% to 54% (19,45–47). This risk has been shown to increase from 53% to as high as 100% if the omphalocele contains only bowel as opposed to liver (Figs. 8, 13, and 14) (19,46,47). The incidence of chromosomal abnormalities, although much lower when the liver is involved in the omphalocele, is not zero and should not preclude the need for an amniocentesis. The presence of other malformations also increases the risk of chromosomal abnormalities (19,46).

Omphalocele is the only type of ventral wall defect that has been proved to be associated with increased risk of chromosomal anomalies (Table 5). Unfortunately, the number of reported cases of bladder and cloacal exstrophy is too small for us to be comfortable with the lack of an association. An additional concern with cloacal exstrophy is the fact that omphalocele is a major feature of the disorder. Gastroschisis, therefore, is the only lesion with which we can be comfortable that there is a lack of associated risk of chromosomal abnormalities. Occasionally, however, there remains some concern about the possibility of a ruptured omphalocele.

One of the most important prognostic factors in omphalocele is the presence of congenital heart defects. As many as half of the neonatal deaths in babies born with omphaloceles may be secondary to heart defects (48). Although most cases of omphalocele with congenital heart defects are linked to aneuploidy, the incidence of congenital heart defects is more than expected even after the exclusion of fetuses with chromosomal abnormalities (45,49).

In addition to an increased risk of congenital heart disease, fetuses with omphaloceles have also been reported to have an increased risk of pernatal cardiac abnormalities including persistent pulmonary hypertension (48).

The prognosis for omphaloceles may be further divided based on the location: i.e., epigastric, central, or hypogastric type. The central type is the most common, followed by the epigastric type. In one series, 69% of the central type had an abnormal karyotype compared to only 12.5% of the epigastric type (50). However, even if there were a normal karyotype, there is a high incidence of associated abnormalities, with 89% of the central type and 71% of the epigastric type with other anomalies. In fact, of 90 in utero fetuses in which an omphalocele was detected only 8 of 90 (9%) were alive and healthy. Others had been terminted, died or had associated abnormalties.

A variety of other abnormalities have been linked to omphalocele, but it is difficult to determine whether these occur from the common causes listed in Table 2. Ascites are frequently seen in association with omphalocele, but this does not appear to worsen the prognosis (51).

Large omphaloceles can be further divided based on the size of the waist of the omphalocele. A giant omphalocele usually has a wide waist. A giant omphalocele with a small waist is rarer, but may pose additional problems. These omphaloceles may cause vascular compromise of the herniated bowel or liver. This can be evaluated with serial sonography (52).

Gastroschisis

The most common conditions associated with gastroschisis are the various types of bowel atresia, most commonly ileal atresia (51). This is consistent with the theory that gastroschisis arises from early loss of the omphalomesenteric artery, because this also develops into the superior mesenteric artery (15). The atresia can lead to bowel obstruction, gangrene, and necrosis. This has led many authors to recommend early delivery if marked dilation of the bowel is present (variously defined as: more than 1.0 cm, more than 1.7 cm, or more than 2.0 cm in largest diameter) (53). Although these data suggest poorer neonatal outcome and higher complication rates with small bowel dilation above these thresholds, other authors (28) have found no justification that this approach improves fetal outcome in retrospective series, and no prospective comparative trials have been reported. As stated earlier, a dilated stomach has been associated with abnormal antepartum testing, longer time to full feeds, and a longer hospital stay in one series (54). Also, gastric dilation was associated with a higher incidence of catastrophic bowel complications such as midgut volvulus and subsequent neonatal morbidity and death. The route of delivery of fetuses diagnosed with gastroschisis remains controversial (55–57).

However, a more recent series followed 30 consecutive fetuses with gastroschisis. Fetuses were followed by sonography for bowel dilatation, thickness, motility, amniotic fluid volume, and fetal development. Babies were delivered by cesarian section between 36 and 38 weeks of gestation. Gastroschisis repair was scheduled 90 minutes after birth. Primary repair without additional incision was achieved in 83% (58).

Body Stalk Anomaly

Body stalk anomaly is frequently associated with several schisis-type defects of the neural tube, such as exencephaly and encephalocele (59,60). A variety of defects of the limbs and heart are present. The ventral wall defect may involve the chest wall as well as the abdomen. The diaphragm is frequently also missing (74%) (60). The heart may be extracorporeal but is contained in the defect (Fig. 11). The additional defects do not add substantially to the diagnosis because the defect is easily recognizable and the mortality rate is 100% regardless of associated defects. Prenatal diagnosis of body stalk anomaly has been reported in as early as the first trimester (59). Maternal serum AFP is abnormal in 100% of the cases (60). Chromosomal abnormalities were not noted in our review of the literature on body stalk anomaly.

Exstrophy of the Bladder

A variety of genital defects may be seen in both bladder and cloacal exstrophy, but these lesions are of minimal significance to the ultrasonographer because they cannot be reliably seen with ultrasound. Separation of the symphysis is common to both types of exstrophy (Figs. 4 and 10).

Exstrophy of the Cloaca

Disagreement can be found in the literature about whether omphalocele is a constant feature of cloacal exstrophy. *Smith's Recognizable Patterns of Human Malformations* states that there is "often omphalocele" (16). There is also confusion about whether the lumbosacral spinal defect always seen in association with cloacal exstrophy is a meningomyelocele. Smith's textbook also points out that a hydromyelia with incomplete development of the lumbosacral vertebra is part of cloacal exstrophy. Hydromyelia, however, is a closed spinal cord abnormality with a large dilated central canal. Whether this lesion should correctly be referred to as a meningomyelocele is unclear. The lesion is covered with skin and can be large. During antenatal sonograms, it may be difficult to differentiate hydromyelia from meningomyelocele. To simplify matters in regard to what is included under the term *cloacal exstrophy*, one might use the term *OEIS syndrome* (omphalocele, exstrophy, imperforate anus, and spinal defects) instead (61).

Defects of the anterior abdominal are often complex. A number of malformations can be identified to occur together as the OEIS complex or cloacal exstrophy. Thus the OEIS complex or cloacal exstrophy should be considered in fetuses with an absent bladder combined with an anterior abdominal wall mass or other defect (62).

Cystic Lesions of the Anterior Abdominal Wall and Proximal Cord

Urachal cysts, or a patent urachus, are the most common cystic lesions of the anterior abdominal wall and proximal cord and should be considered in the differential diagnosis of common ventral wall defects. These are anechoic pockets that form in a line along the inner aspect of the anterior abdominal wall from the umbilical cord to the bladder. They are the remnants of the allantois, which is an embryonic precursor of the bladder and urachus. The cyst may be directly contiguous with or separate from the bladder. There can be more than one urachal cyst. The intervening urachus may be ligamentous or patent. Most urachal cysts resolve by the time of delivery, but some newborns develop a patent urachus despite apparent resolution on the antenatal ultrasound (63). The pediatrician should be made aware of this possibility.

Omphalomesenteric cysts and the more common allantoic cysts have also been noted as anechoic cysts on antenatal ultrasound examinations (64–66). They are embryologically different from urachal cysts. Although they generally are thought to be of no clinical significance, a near-term stillbirth resulting from intrauterine rupture of an omphalomesenteric cyst has been reported. Most recently, the transvaginal ultrasonographic diagnosis of intra-abdominal omphalomesenteric cysts was documented at eight weeks of gestation (62). Interestingly, two weeks later, the cysts had apparently migrated into the proximal umbilical cord, and a first-trimester demise was confirmed.

Cystic structures of the umbilical cord can generally be differentiated from omphalocele in that there is usually only one anechoic area as opposed to the multiple anechoic areas seen with the bowel loops associated with an omphalocele. Cord lesions are discussed in more detail in Chapter 38.

Three-Dimensional Ultrasound

Three-dimensional (3D) sonography may be used to identify the normal and abnormal appearance of the anterior abdominal wall. In a report by Bonilla-Musoles et al., the use of 3D sonography provided additional information when evaluating abdominal wall defects such as gastroschisis or omphalocele (Fig. 19) (67). This information may provide for more efficient counseling and postnatal therapeutic planning. Other abnormalities such as limb–body wall complex have been evaluated with 3D sonography (68).

GASTROINTESTINAL SYSTEM

The abdomen houses a large number of organs from several different organ systems; as a result, the ultrasonographic diagnosis and identification of normal structures and anomalies is potentially difficult. This section focuses on a practical approach to the diagnosis of the major gastrointestinal tract anomalies, using a differential diagnosis format based primarily on their ultrasonographic appearance. The following section details the genitourinary tract and its anomalies. Because the differential diagnosis of abdominal anomalies so often encompasses both systems, there will be considerable overlap in the discussion of individual anomalies and artifacts.

Embryology and Normal Ultrasonographic Appearance

The embryonic gut has three major parts: the foregut (esophagus, stomach, proximal duodenum, liver, and pancreas), the midgut (distal duodenum, jejunum, ileum, and proximal colon), and the hindgut (distal colon, rectum, anus, and parts of the vagina and bladder). Dilation of the foregut during the sixth menstrual week leads to the formation of the stomach, which then descends into the abdominal cavity by the eighth to ninth week. The midgut typically returns to the abdominal cavity by the 11th week. Although intestinal peristalsis begins by week 11, it can rarely be visualized ultrasonographically before the second trimester.

Fetal swallowing is evident before the end of the first trimester, and beyond the 14th week, the stomach should be visualized in nearly all normal fetuses (69). With a transvaginal probe, the stomach can be visualized as early as the ninth week of gestation. In contrast, the normal fetal esophagus is not generally visualized ultrasonographically.

The small bowel and large bowel can be imaged and distinguished from each other, particularly in the third trimester (Fig. 20). The more centrally located small bowel is often difficult to discern in normal fetuses, appearing as a relatively homogeneous, slightly echogenic mass. The small bowel remains more echogenic than the large bowel until near term, displaying active peristalsis during the third trimester. Late in the third trimester, small bowel loops are commonly identified, generally measuring less than 15 mm in length and rarely more than 5 mm in diameter. Improvements in ultrasound imaging technology have permitted the imaging of small bowel loops at as early as 14 weeks' gestation.

The colon frames the fetal abdomen, appearing as a continuous tube with a hypoechoic lumen (Fig. 20). (Meconium creates this hypoechoic appearance, in contrast to the echogenic bowel wall.) In fact, normal colon with liquid meconium is often mistaken for pathology (e.g., dilated bowel, cysts). The colon exhibits far less peristalsis than the small intestine. By the middle of the third trimester, colonic haustra can be identified in nearly all fetuses, despite the fact that clinically, haustra are generally poorly developed in newborns. The diameter of the large bowel increases in linear fashion from 3 to 5 mm at 20 weeks' gestation to up to 20 mm at term. Although colonic grading has been attempted based on comparative echogenicity, the clinical utility of this approach has not been demonstrated.

The liver is the dominant organ in the upper fetal abdomen, the area for standard abdominal circumference

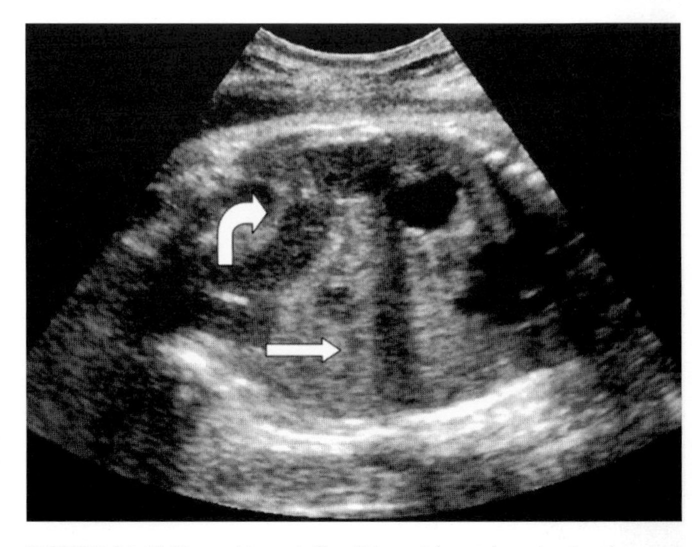

FIGURE 20 ■ Normal bowel. Small bowel (*arrow*) and colon (*curved arrow*) can be identified in this 36-week fetus.

measurements (Fig. 21). The gallbladder is often overlooked or misrepresented ultrasonographically (under the assumption that it represents an intrahepatic vein). Although it is a sonolucent structure, its ovoid or conical shape and lack of flow on color Doppler or power angiography, as well as its location inferior and to the right of the intrahepatic segment of the umbilical vein, should help distinguish it (Fig. 21). The spleen is another upper abdominal organ that is often overlooked or mistaken for an abnormal solid left-sided mass. Although not always easily visualized, it can usually be seen throughout the second half of gestation, posterior to the stomach and lateral and superior to the left kidney. It is homogeneous in appearance and slightly less echogenic than the liver, similar to the kidney. The pancreas is sometimes visualized in the third trimester as well, especially with the fetus in a supine position.

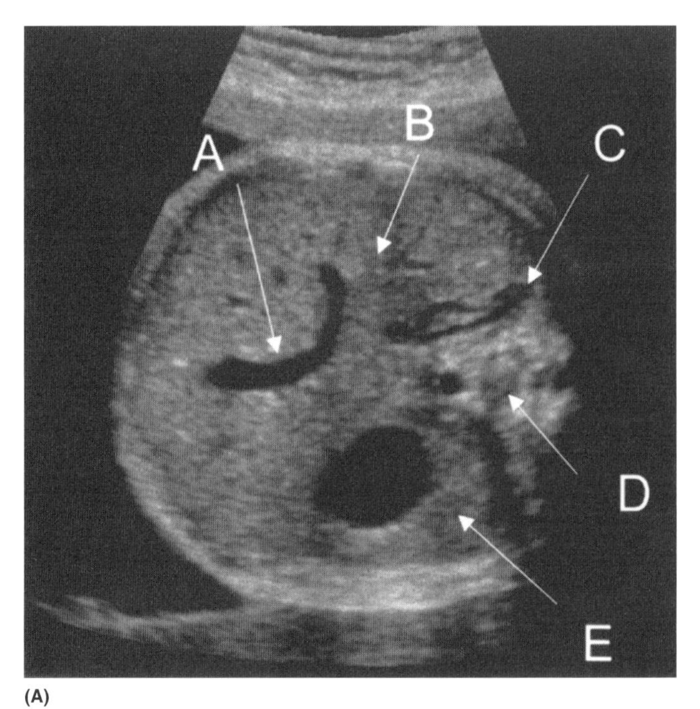

(A)

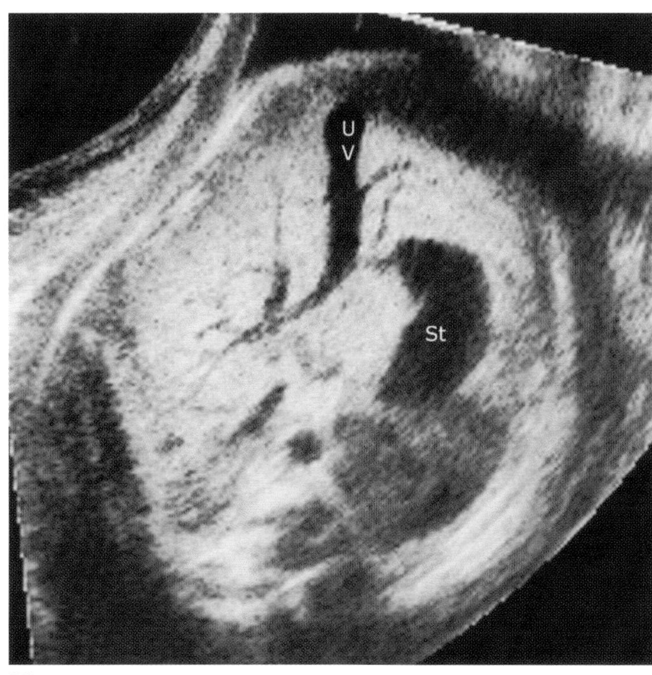

(B)

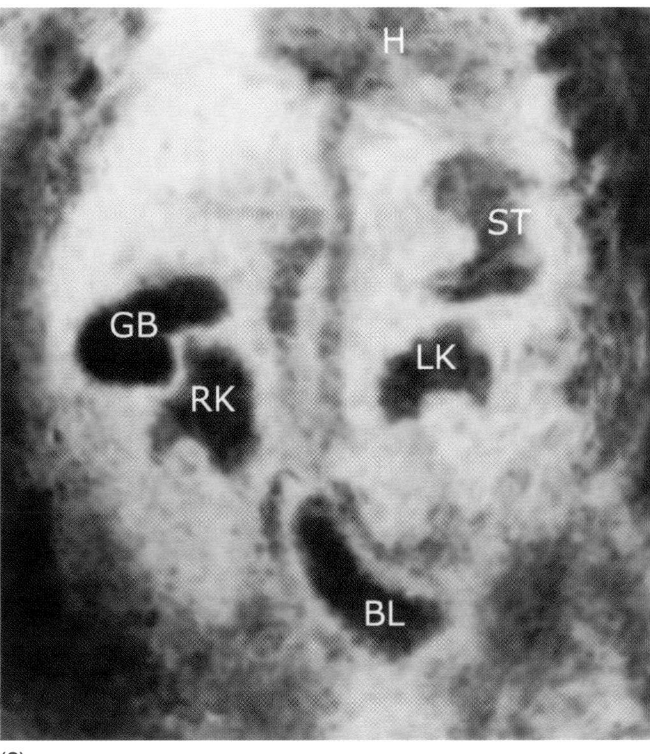

(C)

FIGURE 21 A–C ■ Fetal abdomen. (**A**) Transverse scans of the abdomen of the fetus demonstrate umbilical vein (**A**), fetal liver (**B**), fetal adrenal (**C**), spine (**D**), and spleen (**E**), posterior to stomach. (**B** and **C**) 3-D reconstruction in transverse plane (**B**) and coronal plane (**C**) show dilatation of right and left renal collecting system (RK, LK). *Abbreviations*: H, heart; GB, gallbladder; ST, stomach; UV, umbilical vein; BL, bladder. *Source*: (**B**) and (**C**) courtesy of George Baga, MD, Philadelphia, PA.

Pitfalls and Artifacts

Absent Stomach

Inability to visualize the stomach in the second and third trimesters of pregnancy requires further examination. Repeated examination over time (several hours or days later) confirms a normal fetal stomach in about half of the cases. Esophageal atresia (Figs. 22 and 23) should always be considered when polyhydramnios is accompanied by failure to visualize the stomach bubble. The differential diagnosis for this problem is summarized in Table 7.

Gastric Pseudomass

A well-defined echogenic focus can be identified within the fluid-filled fetal stomach in about 1% of ultrasonographic evaluations (Fig. 24). These echogenic pseudomasses are thought to represent swallowed cellular debris (although this has not been proved), and they generally resolve spontaneously. No association with clinical problems has been confirmed (70).

Pseudoascites

A common pitfall in obstetrical ultrasound is the over-diagnosis of fetal ascites. This can occur even with a relatively experienced ultrasonographer, especially when evaluating a patient at risk for hydrops (e.g., Rh isoimmunized). Typically, a hypoechoic band that parallels the anterior abdominal wall is interpreted to represent free fluid (Fig. 6). This band is actually the abdominal wall musculature (71). True ascites usually displays an irregular, hypoechoic area within the fetal abdomen (Fig. 7) and, even in its earliest stages, allows visualization of both sides of the bowel wall (72). Moreover, pseudoascites rarely exhibits some of the other signs of early hydrops fetalis (Table 8). Nevertheless, it is important to avoid the pitfall of diagnosing early ascites when visualizing normal bowel in the late third trimester.

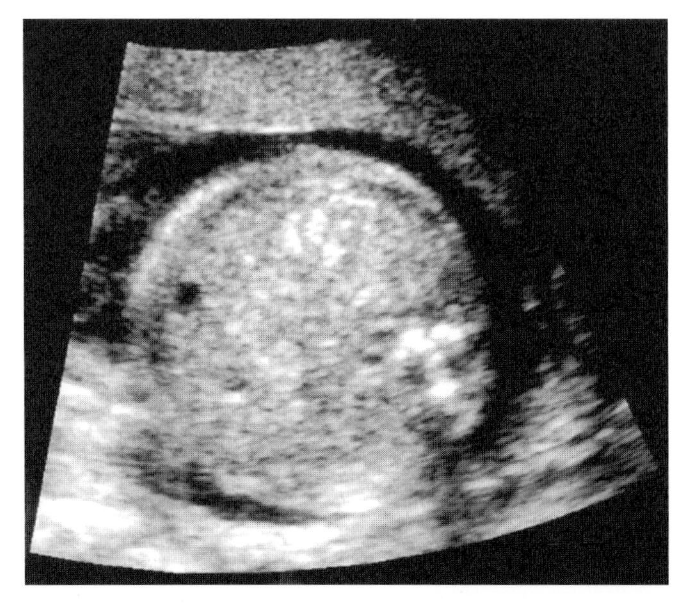

FIGURE 22 ■ Nonvisualization of the fetal stomach. No stomach was visualized in this case of esophageal atresia. *Source*: Courtesy of Beryl Benacerraf, MD, Boston, MA.

Anomalies

Situs Inversus and Heterotaxy Syndrome

The diagnosis of situs inversus totalis requires a solid sense of orientation to fetal position because the two key landmarks (stomach and heart) are on the same side of the body. By routine orientation, however, the fact that the stomach, heart axis, and aortic arch are on the right should be apparent. In addition, the liver and spleen are transposed, with the gallbladder on the left. This relatively rare disorder generally carries a good prognosis, with few associated anomalies, including Kartagener's syndrome. Rare instances occur of situs inversus (thoracic and abdominal visceral situs are inverted), however, with the stomach on the right and the cardiac apex pointed to the left. This carries a high incidence of associated cardiac anomalies, which are explained in more detail in Chapter 14.

By way of contrast, situs ambiguous is associated with severe concurrent anomalies and a mortality rate of 50% to 90%. This disorder has been classically divided into two types, each with a distinct set of problems: asplenia, or bilateral right-sidedness (Fig. 25); and polysplenia, or bilateral left-sidedness. Reference to the presence or absence of the spleen is misleading because, clinically, the organ is often difficult to identify. Most recently, the addition of fetal MR has been used to aid in clarifying the diagnosis (73). In addition, the spleens in polysplenia are often small and located in multiple locations throughout the abdomen. The key ultrasonographic feature of both subtypes is the presence of the stomach on the right side of the abdomen, discordant from a relatively normally positioned heart. Complex cardiac anomalies are common with both but generally are more severe with the asplenia syndrome. Other ultrasonographic clues to this anomaly include (*i*) an abnormal liver position (midline and horizontal), (*ii*) midline gallbladder, (*iii*) the aorta and inferior vena cava on the same side, and (*iv*) as the name suggests, absence of the spleen.

In the polysplenia syndrome, the liver is on the left, the gallbladder is absent, and the vena cava is interrupted at the liver. Sheley et al. (74) described the double-vessel sign for polysplenia syndrome, consisting of two similar-sized vessels in a paraspinous location posterior to the heart. They represent the aorta, which is normally the only vessel seen at this level, and the dilated continuation of the azygous vein. The azygous vein appears more posterior than the usual vena cava and ascends to the right of the descending aorta (75). The more common cardiac lesions seen with these cardiosplenic syndromes include endocardial cushion (atrioventricular canal) defect, outflow tract, and great vessel anomalies (see Chapter 43 for further details). Biliary atresia is seen in asplenia and polyspenia (76). Neural tube defects and genitourinary anomalies have also been reported with these conditions.

Esophageal Atresia

Esophageal atresia should always be considered when polyhydramnios is accompanied by failure to visualize

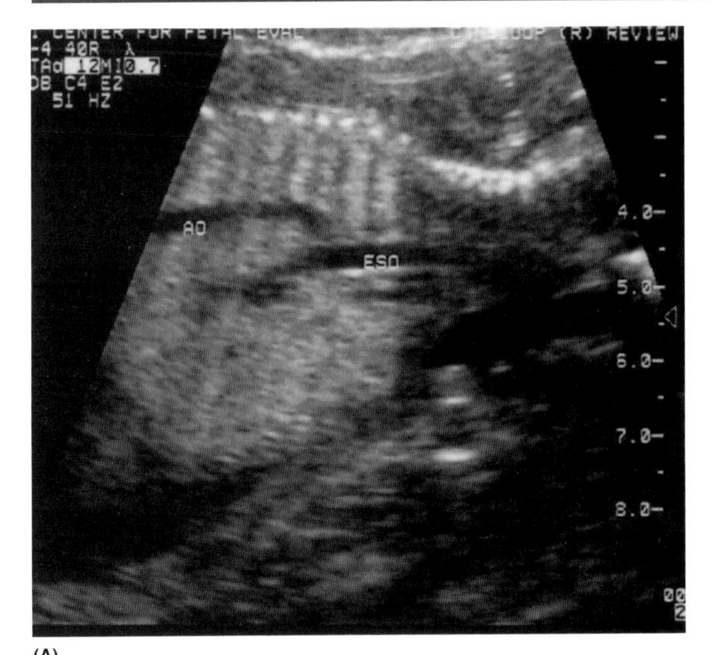

(A)

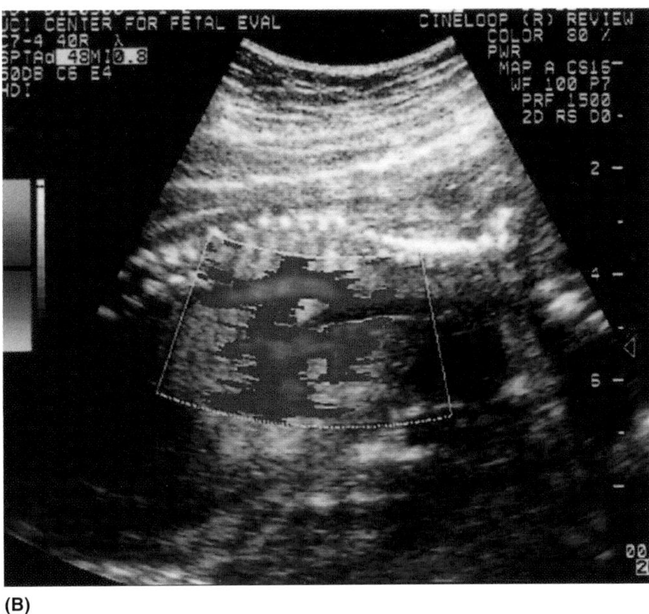

(B)

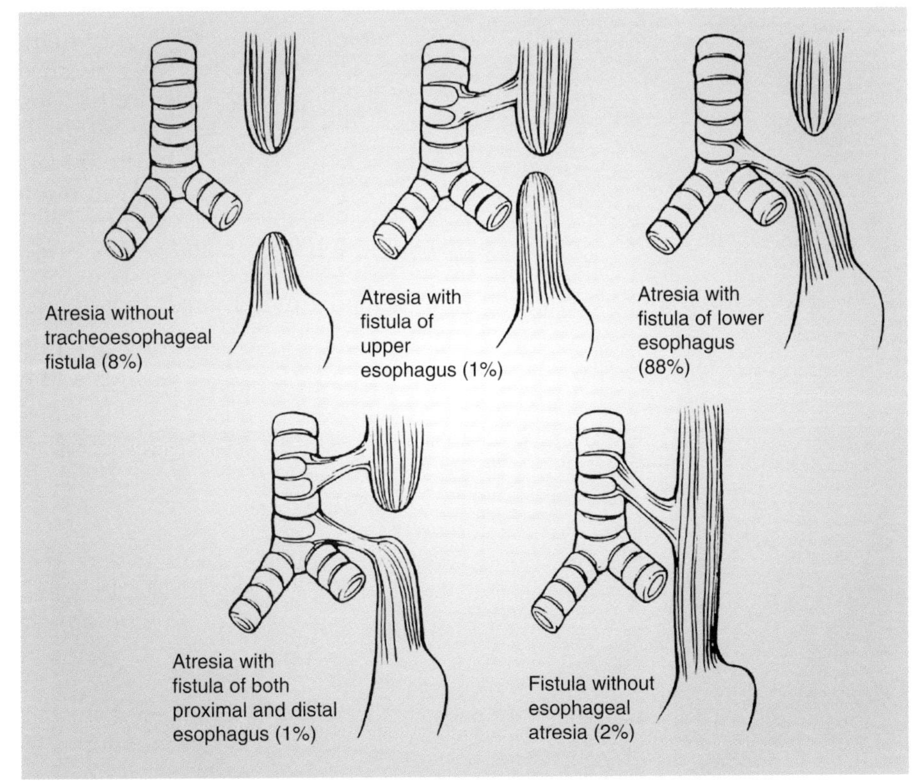

(C)

FIGURE 23 A–C ■ (**A**) Blind esophageal pouch in fetal thorax. (**B**) Color power angiography of the same fetus confirming nonvascular esophageal. (**C**) Diagram representing different types of tracheoesophageal fistulas occurring with esophageal atresia. Most commonly, there is esophageal atresia with a tracheoesophageal fistula to the stomach. AO, aorta; ESO, esophagus. *Source*: Modified from Refs. 183,184.

TABLE 7 ■ Differential Diagnosis of Absent Stomach Bubble

Empty stomach (time)
Esophageal atresia with or without tracheoesophageal fistula
Diaphragmatic hernia
Situs inversus
Oligohydramnios
Facial clefts
Central nervous system disorders

Source: From Ref. 186.

the stomach bubble (Figs. 22 and 23). However, 90% of cases are associated with a communicating tracheoesophageal fistula, usually with a visible stomach. The gastric mucosa alone may produce enough fluid to allow visualization of the stomach in some cases of atresia without fistula. Pretorius et al. (77) reviewed 22 cases of tracheoesophageal fistula, noting polyhydramnios in two-thirds and nonvisualization of the stomach in one-third. However, Borsellino et al. reported three false-positive prenatal diagnoses in a series of 11 cases referred for polyhydramnios and microgastria. When the stomach bubble

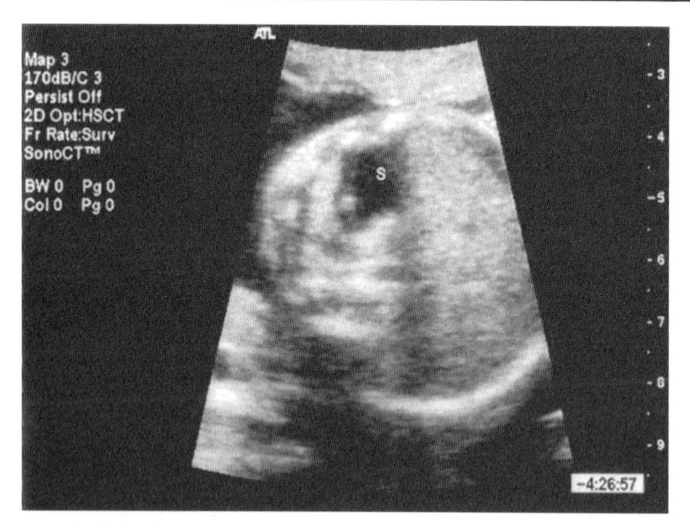

FIGURE 24 ■ Gastric pseudomass. Echogenic foci along the wall of the stomach are thought to represent swallowed echogenic debris. These are of no significance. S, stomach.

TABLE 8 ■ Differentiating Ascites from Pseudoascites

	Pseudoascites	Ascites
Polyhydramnios	No	Often
Other hydropic signs (anasarca, effusions)	No	Often
Fluid between bowel loops	No	Yes
Fluid surrounding liver or omentum or in the pelvis	No	Often
Symmetric hypoechoic rim of anterior abdominal wall	Yes	No
Outline around umbilical vein and falciform ligament	No	Often

Source: From Ref. 186.

was persistently absent, they reported no false-positive diagnoses (78). The blind pouch of the proximal esophageal segment has been seen infrequently, as has regurgitation after fetal swallowing during real-time observations. Associated anomalies are seen in nearly 60% of cases, with other gastrointestinal (28%), cardiac (24%), and chromosomal (19%) anomalies being the most common (79). As with most intestinal atresias, esophageal atresia is rarely diagnosed before the third trimester. Although the survival rate is particularly poor overall (<50%), the prognosis is much improved with isolated esophageal atresia (85% survival rate). (Sites of bowel obstruction and ultrasound findings are presented in Fig. 26.)

Duodenal Atresia

Duodenal atresia, with an incidence of about 1 in 5000 births, is the most common intestinal obstruction encountered in the perinatal period. The lesion is most commonly caused by one or more membranes interrupting the duodenal lumen. The classic double-bubble appearance (Fig. 27), almost invariably associated with polyhydramnios, makes the ultrasonographic diagnosis relatively straightforward. In a recent series of 29 cases diagnosed prenatally, 83% exhibited polyhydramnios, though the

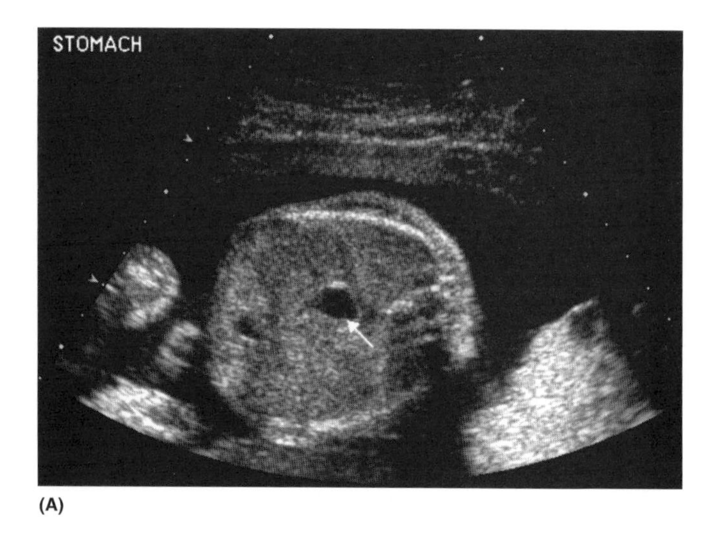

(A)

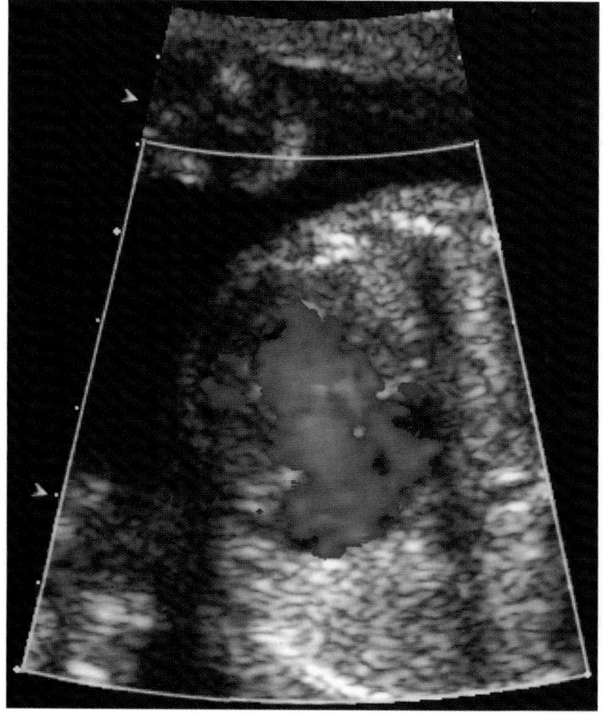

(B)

FIGURE 25 A AND B ■ Fetus with heterotaxia syndrome. (**A**) The stomach (*arrow*) is midline. (**B**) Color-flow image of the fetal heart demonstrating complex cardiac defect including endocardial cushion defect.

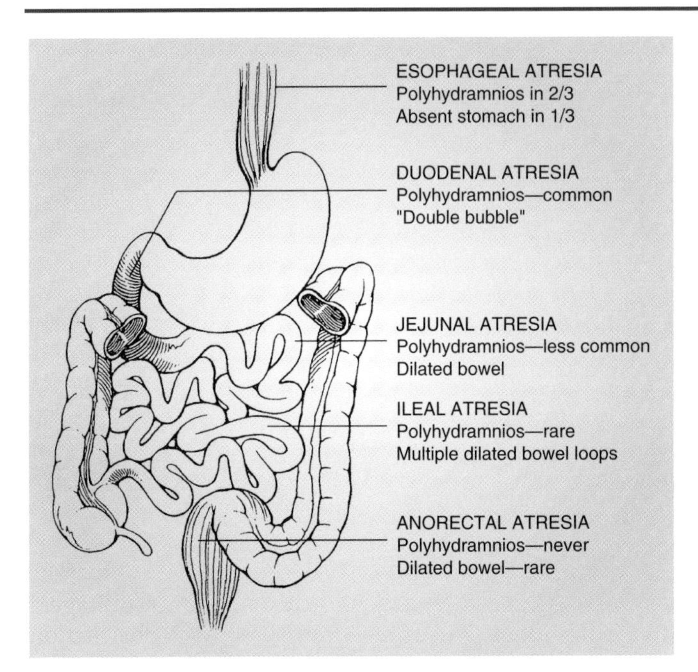

FIGURE 26 ■ Sites of gastrointestinal obstruction and typical ultrasound findings. *Source*: From Ref. 186.

mean gestational age at diagnosis was 29 weeks (80). It is extremely important, however, to attempt to demonstrate the communication between the stomach and the first part of the duodenum.

Associated anomalies are the rule rather than the exception with this disorder, occurring in more than half of the cases. In particular, 30% of fetuses with duodenal atresia have Down syndrome; hence, prenatal diagnosis should be offered. By way of contrast, however, only 10% of Down syndrome fetuses have duodenal atresia. Other anomalies commonly associated with Down syndrome include skeletal, cardiac, and other gastrointestinal malformations. Renal anomalies are seen in about 8% of cases, and growth restriction is common. While in the past, isolated duodenal atresia (without trisomy 21) has generally had an excellent postnatal prognosis, the report of four cases of sudden uterine demise at 31 to 35 weeks of gestation despite reassuring fetal heart rate testing is of concern (81).

Although it is extremely uncommon to make the diagnosis of duodenal atresia before 24 weeks' gestation, it is important to reevaluate patients with a prominent stomach bubble during an early second-trimester sonogram. Several authors have reported the diagnosis of unusual cases of duodenal atresia at 16 weeks' gestation (82), and even in the late first trimester (83). As with our experience, such cases are generally associated with other intestinal atresias (e.g., esophageal, tracheoesophageal fistula) leading to early development of polyhydramnios and the double-bubble sign.

A prominent stomach bubble may on occasion be mistaken for duodenal atresia if imaged in a coronal or oblique rather than a transverse plane. Another key to avoiding this pitfall is to note that the smaller duodenal bubble is generally located to the right of the midline. Pitfalls for duodenal atresia include mistaking it for a choledochal cyst, a hepatic cyst, or the gallbladder (Table 9). The lack of a communication between the two cystic structures, as well as the absence of polyhydramnios, should help clarify the diagnosis.

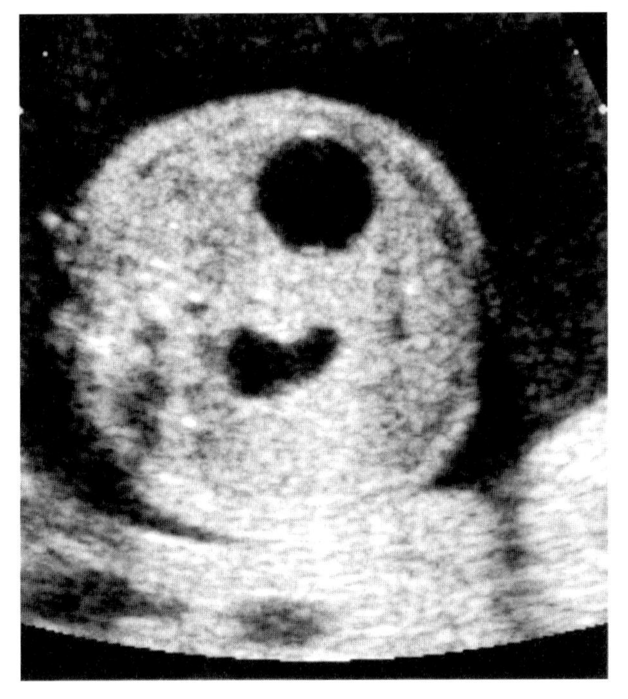

FIGURE 27 ■ Duodenal atresia. Classic double-bubble sign of dilated stomach and duodenum associated with duodenal atresia. *Source*: Courtesy of Beryl Benacerraf, MD, Boston, MA.

Small Bowel Obstruction

The incidence of small bowel obstruction is about 1 in 5000 births, but the prevalence of in utero or ultrasonographic detection may be somewhat higher. It is often difficult, if not impossible, to differentiate jejunal from ileal atresia, although the extent of bowel dilation is often a clue (more loops are present with ileal abnormalities; Fig. 28). A midgut volvulus is difficult, if not impossible, to differentiate from jejunoileal atresia;

TABLE 9 ■ Common and Less Common Causes of Double-Bubble Sign Identified on Prenatal Ultrasound

Common
Duodenal atresia (associated with trisomy 21)
Less common
Duodenal stenosis
Duodenal web
Annular pancreas
Proximal jejunal atresia
Bowel malrotation

Source: From Ref. 186.

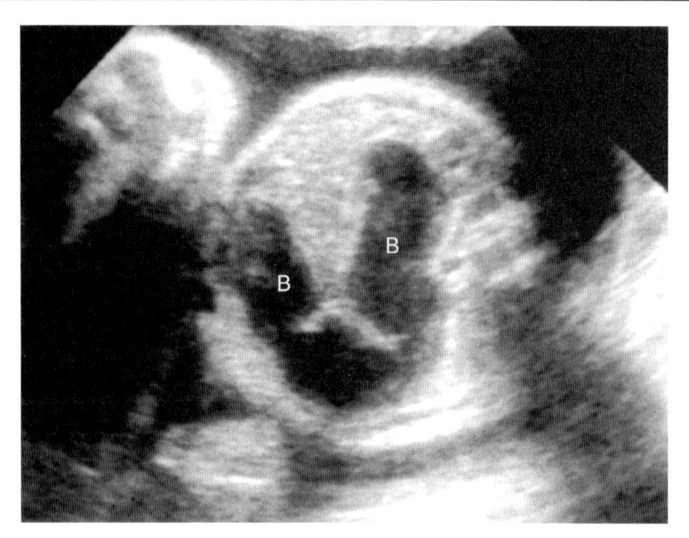

FIGURE 28 ■ Jejunal atresia. Dilated small bowel loops (B) in a fetus with proximal jejunal atresia. *Source*: Courtesy of Beryl Benacerraf, MD, Boston, MA.

TABLE 10 ■ Malformations Associated with VACTERL Syndrome

Vertebral abnormalities
Hemivertebra
Sacral agenesis
Syringomyelia
Anorectal atresia
Cardiovascular malformation
Central nervous system anomalies
Chromosomal abnormalities
Tracheoesophageal fistula
Esophageal atresia
Renal abnormalities
Renal agenesis
Renal dysplasia
Horseshoe kidney
Limb abnormality
Radial ray malformations

Source: From Ref. 186.

therefore, it is most accurate to refer to these disorders as *small bowel obstructions* rather than by the specific location (Fig. 26).

Polyhydramnios is often the initial clue to a small bowel obstruction and is present in most cases. In fact, in the absence of polyhydramnios, one recent report had a 41% false-positive diagnosis of small bowel obstruction when dilated and/or echogenic bowel with or without ascites was used to make the diagnosis (84). Unfortunately, as with duodenal atresia, jejunal and ileal obstructive problems are invariably diagnosed late in pregnancy, specifically in the third trimester. Unlike duodenal atresia, however, nongastrointestinal anomalies are rare in these cases. Other intestinal tract anomalies are common, with up to 45% exhibiting malrotations or enteric duplications, for example. Distal ileal obstructions appear to have the most favorable prognosis. In the absence of other malformations, dilated bowel loops and polyhydramnios together are highly predictive of small bowel atresia (85).

Multiple dilated loops of small bowel (more than 7 mm in diameter) are the typical features in these cases, often with increased peristalsis (Fig. 28). Differentiating small bowel from large bowel obstruction is usually possible from the location of the loops, the absence of haustra in small bowel, and the presence of polyhydramnios. Colonic obstructions, however, are uncommonly diagnosed in utero, and Hirschsprung's disease (congenital aganglionic megacolon) is likely to present with dilated small bowel and polyhydramnios, making it difficult to differentiate it from a jejunoileal lesion (86).

Anorectal Atresia

The incidence of anorectal malformations is about 1 in 5000 live births. The prenatal diagnosis of imperforate anus and other anorectal anomalies has been reported about 15 times. Generally, the diagnosis has been as part of a syndrome of multiple anomalies (e.g., caudal regression syndrome, VACTERL or VACTER syndromes, McKusick–Kaufman syndrome) (87,88). VACTERL syndrome may have a number of concurrent abnormalities, including vertebral abnormalities, anorectal atresia, cardiac anomalies, tracheoesophageal fistulas, esophageal atresia, renal anomalies, and limb abnormalities (Table 10). Harris et al. (89) reported several cases with the salient features of dilated colon in the pelvis or lower abdominal cavity with normal amniotic fluid. They stressed, however, that concomitant renal or upper gastrointestinal anomalies may lead to oligohydramnios or polyhydramnios, respectively.

Meconium Ileus

Meconium ileus results from a distal ileal obstruction with abnormally thick, viscous meconium. When this is present, the fetus is almost certain to have cystic fibrosis (CF), although only 10% to 15% of infants with CF have the condition (90). The typical appearance of dilated ileum, with a normal jejunum, collapsed colon, and polyhydramnios, may be indistinguishable from other small bowel obstructions (91). In some cases, however, the bowel presents as an echogenic mass. Initially thought to be a highly specific finding for meconium ileus and CF, echogenic bowel has been associated with a number of other diagnoses, as discussed next (Table 11).

Meconium plug syndrome (92), or an obstructed colon from intraluminal meconium, represents the large bowel equivalent of meconium ileus. Although it is not as closely linked to CF (less than 25% of cases), genetic counseling should be strongly advised for patients suspected of having a pregnancy with these findings.

TABLE 11 ■ Differential Diagnosis of Echogenic Bowel

Normal second-trimester variant
High gain or low dynamic range settings
Bloody amniotic fluid (intraluminal red blood cells)
Meconium ileus (cystic fibrosis)
Down syndrome
Cytomegalovirus infection
Source: From Ref. 186.

Meconium Peritonitis

Meconium peritonitis is a condition resulting from a small bowel perforation. Although it is a relatively rare condition in neonates, this entity has been well described for more than a century. The perforation generally leads to a chemical peritonitis, which results in intra-abdominal calcifications that are the characteristic ultrasonographic feature in this condition. The calcifications can occur anywhere within the peritoneal cavity but must be distinguished from intraluminal or intrahepatic calcifications. Typically, the calcifications are linear in nature, sometimes rimming a hypoechoic mass or pseudocyst, and are usually associated with ascites and polyhydramnios. Meconium pseudocyst (Fig. 29) is usually the result of a contained bowel perforation, and nearly half of all cases of meconium peritonitis are the result of an underlying obstructive process (93). Ascites is an extremely common feature of the condition; it is usually echogenic in appearance (intraperitoneal meconium) (94). The ascites in conjunction with an inflamed abdominal wall, however, can mimic the appearance of nonimmune hydrops (95). The lack of generalized anasarca and the subsequent development of peritoneal calcifications are helpful clues in avoiding this pitfall. In male fetuses, scrotal calcifications have been reported on occasion (96).

Meconium peritonitis is sometimes confused with meconium ileus, a problem that is fostered by the fact that up to 40% of infants with the condition may also have CF (meconium ileus) (97). In utero series, however, suggest that this association is far less common (98). Therefore, despite a 10% to 15% risk of CF, most cases of meconium peritonitis have a good prognosis, sometimes without postnatal detection of the primary perforation. As with many of the anomalies discussed in this section, the diagnosis of meconium peritonitis is typically made in the third trimester.

Echogenic Bowel

A wide range of associated fetal conditions may be suggested when the bowel displays a hyperechoic appearance (Fig. 30). These range from artifacts to devastating abnormalities (Table 11). How to approach the differential diagnosis of this not infrequent finding has remained a controversy for the better part of the past decade. One of the biggest problems is the somewhat subjective nature of

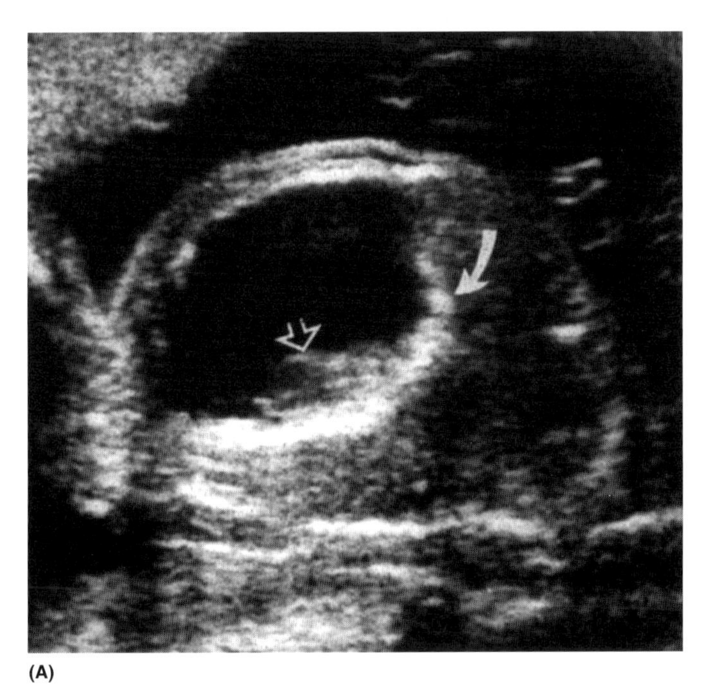

(A)

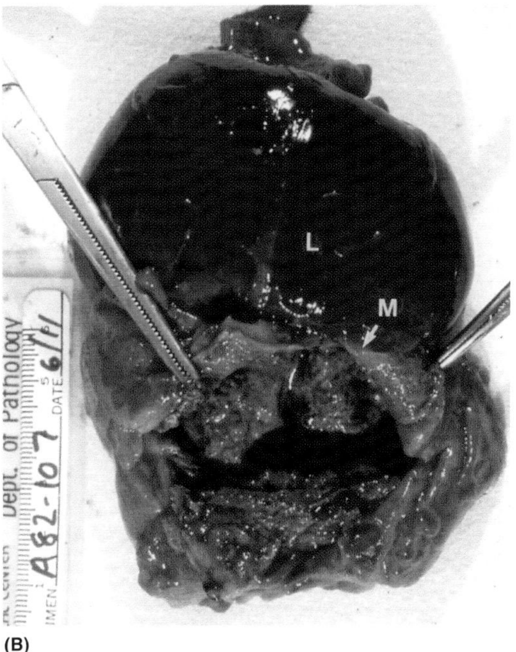

(B)

FIGURE 29 A AND B ■ Meconium pseudocyst. (**A**) Oblique scan of the fetal abdomen demonstrates a large, mid-abdominal anechoic mass with thick echogenic walls (*curved arrows*). Open arrows depict echogenic material layered in the dependent portion of the cyst. (**B**) Gross specimen demonstrates an opened, thick-walled meconium pseudocyst adherent to the surrounding bowel, bladder, and liver. L, liver; M, meconium pseudocyst. *Source*: From Ref. 93.

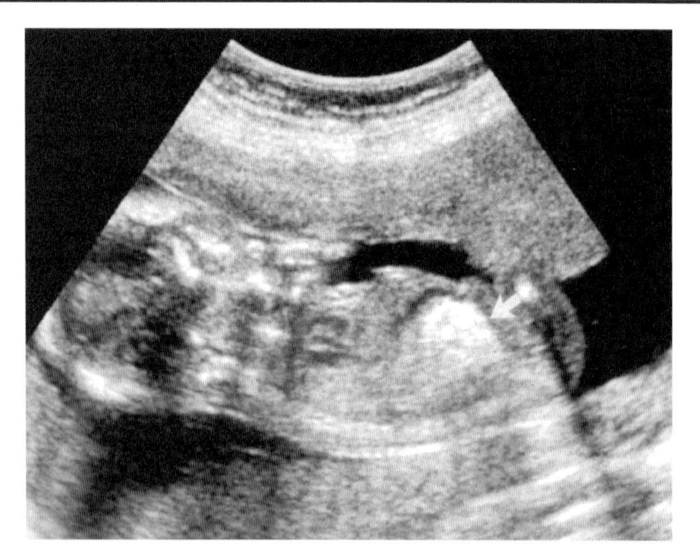

FIGURE 30 ■ Echogenic bowel. Transverse scan of the fetal abdomen in a 20-week fetus with echogenic bowel, as echogenic as fetal bone. *Source*: Courtesy of Beryl Benacerraf, MD, Boston, MA.

the finding. The bowel commonly displays some relative echogenicity; this raises the question of how echogenic the bowel must be for classification as abnormal. At times, certain ultrasound units routinely present the bowel with a somewhat echogenic appearance. Some settings with high gain or low dynamic range can simulate a hyperechoic appearance as well. Nyberg et al. (99) established criteria and a grading definition to describe the bowel. In grade one, the bowel has increased echogenicity that is less than that of surrounding bone; in grade 2, it is equivalent to that of surrounding bone; and in grade 3, the bowel is more echogenic than that of the surrounding bone. Despite this, several authors have described echogenic bowel as a normal variant in the second trimester (100). As mentioned earlier, meconium ileus (CF) and meconium peritonitis can be described as displaying an echogenic bowel appearance. Although this finding typically occurs in the third trimester (101), a few second-trimester cases of CF have been diagnosed prenatally with hyperechoic bowel as the only ultrasonographic finding (102).

TABLE 12 ■ Differential Diagnosis of Abdominal Cyst

Anomaly	Location	Appearance	Other anomalies	Gender
Choledochal	Right upper	Single cyst near gallbladder, dilated adjacent hepatic ducts	Rare	Usually female
Hepatic	In liver	Usually single	None	Usually female
Multicystic renal disease	Dorsal, left, or right	Cysts of variable size, which are noncommunicating	Various syndromes	Either
Hydronephrosis	Dorsal, left, or right	Often multiple communicating cysts within renal fossa	Other genitourinary anomalies	Usually male
Hydroureter	Lateral	Often tubular, communicates with kidney or bladder	Other genitourinary anomalies; associated obstructed, enlarged bladder	Usually male
Ureterocele	In bladder	Single cyst or septated bladder	Hydronephrosis, hydroureter, multicystic dysplastic kidney disease	Usually male
Megacystic–microcolon–intestinal hypoperistalsis syndrome	In mid-abdomen	Increased bladder, hydronephrosis, hydroureter	±Dilated bowel	Either
Meconium pseudocyst	In mid-abdomen	Thick or calcified wall; echogenic debris	Associated bowel obstruction	Either
Bowel atresia	Dependent on site of obstruction	Double-bubble sign to dilated bowel	Common; trisomy 21	Either
Mesenteric or omental	Mobile, middle	Variable: small to large, unilocular to septated	None	Either
Ovarian	Pelvis, lower	Unilocular, round, occasional septae; bilateral uncommon	None	Female
Umbilical vein varix	Cord insertion	Single cyst; Doppler venous flow	High α-fetoprotein level, stillbirth	Either
Urachal	Ventral	Smooth cyst, communicates with bladder	None	Either
Sacrococcygeal teratoma	Off coccyx with both internal and external extension	Usually solid but may be cystic	None	Either
Anterior meningocele	Sacral	Cystic to complex	CNS malformations	Either
Hydrometrocolpos	Pelvis, retrovesical anomalies	Cystic or solid	Frequent genitourinary anomalies	Female

Source: From Ref. 186.

Scioscia et al. (103) reported on a series of 22 fetuses with echogenic bowel, suggesting a strong association with Down syndrome. Since that report, a number of investigators have confirmed some association between echogenic bowel and trisomy 21, although the frequency of this association remains in question. When the bowel exhibits the same echogenicity as the pelvic bones, a detailed search for other anomalies associated with aneuploidy should be undertaken and prenatal diagnosis considered.

Several cases of congenital cytomegalovirus have been diagnosed as a result of the ultrasonographic identification of hyperechoic bowel (104,105), one as early as 14 weeks' gestation.

In our experience, red blood cells in the intestinal tract may be a more common cause of echogenic bowel. This has been reported after intra-amniotic bleeding (106) and association with elevated maternal serum AFP levels (107). Achiron et al. (107) found echogenic bowel in 6 of 396 (1.5%) fetuses with AFP levels higher than 2.5 MoM. Although none of these fetuses had a chromosomal anomaly, CF, or cytomegalovirus, the perinatal outcome was dismal, with four intrauterine deaths and one neonatal death. All six had evidence of growth restriction.

Abdominal Cyst: Differential Diagnosis

When faced with the dilemma of an anechoic, presumably cystic mass in the fetal abdomen, a systematic approach is critical to making an appropriate diagnosis (Tables 12 and 13). Preliminary assessments to make include:

- Location and orientation of the mass
- Relation to other abdominal organs
- Size and shape of the lesion
- Fetal gender
- Wall and contents of the cyst

In addition, the change in appearance over time may be extremely helpful in difficult cases. (Some of these cystic-appearing abnormalities were discussed in the preceding sections on bowel obstruction and meconium pseudocyst.)

Ovarian cysts are among the more common anechoic abdominal anomalies seen in a female fetus in the third trimester. They are usually, although not exclusively, located within the lower fetal abdomen. They may change in position in the abdomen or pelvis and may undergo torsion because they are often on a pedicle (Fig. 31). Although most fetal ovarian cysts are simple in nature, it is important to look for septa in these typically benign lesions. It is thought that these cysts arise as a result of hormonal stimulation from the mother and placenta. A series of 32 cases of fetal and neonatal ovarian cysts revealed that size may be an important prognostic factor in determining outcome. In cysts less than 4 cm in diameter, there was a tendency toward regression. In cysts greater than 4 cm, cystectomy was always necessary due to torsion in 37.5% or due to intracystic hemorrhage in 62.5% (108).

More commonly, cystic lesions in the abdomen are related to bowel or urinary tract pathology, making it critical to examine the apparent cyst in multiple planes, addressing its relation to adjoining structures—particularly

TABLE 13 ■ Cystic Abdominal Masses by Location

Cystic Abdominal Masses by Location
Right upper quadrant
■ Hepatic cyst
■ Choledochal cyst
Left upper quadrant
■ Splenic cyst
Posterior (renal)
■ Renal cyst (Tables 1–21 for types)
■ Hydronephrosis
■ Urinoma or urine ascites (secondary to rupture of calyx in utero with hydronephrosis)
Anterior or mid-abdomen
■ Mesenteric cyst
■ Omental cyst
■ Meconium pseudocyst
■ Umbilical vein varix
Lower abdomen
■ Ovarian cyst (but may migrate to mid-abdomen)
■ Ureterocele
■ Urachal cyst
■ Hydrometrocolpos
■ Sacrococcygeal teratoma (most commonly solid but may be cystic)
■ Anterior meningocele

Note: Many cystic abdominal masses may be in variable locations other than those listed.
Source: From Ref. 186.

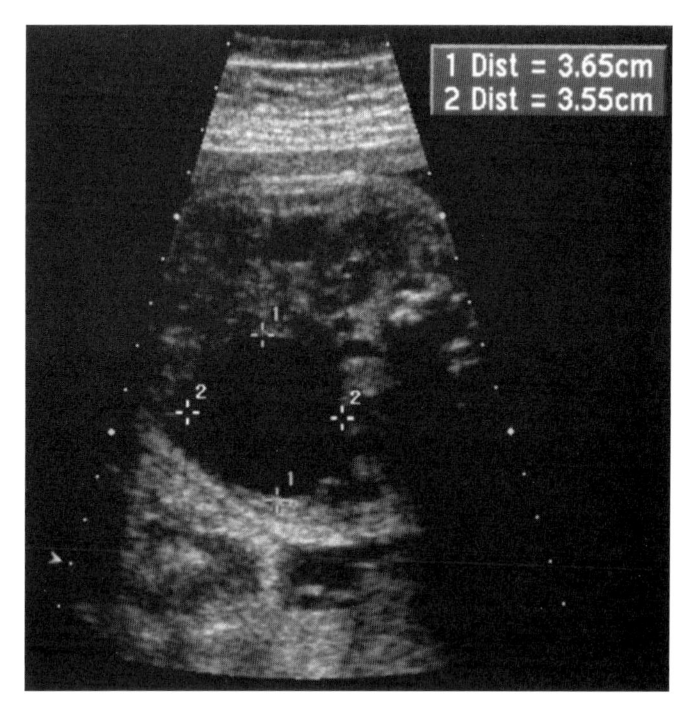

FIGURE 31 ■ Ovarian cyst. Transverse scan of the upper abdomen shows a cyst (*calipers*). This corresponds to a pedunculated ovarian cyst. The cyst was on a pedicle and changed in position throughout the abdomen and pelvis in utero.

the kidneys, bladder, and stomach. For example, a unilateral or bilateral cystic structure in the dorsal, paraspinal region of the mid-abdomen most likely represents some form of obstructive uropathy (e.g., hydronephrosis, hydroureter). In many cases, involvement with the renal fossa can be demonstrated and the diagnosis clarified. Other cystic masses localized to the kidney, such as multicystic dysplastic kidney disease (MCDK), are discussed later.

The urachus represents the communication from the embryonic allantois (umbilical cord) to the urogenital sinus (urinary bladder), a structure that usually fibroses to form the umbilical ligament. Although often confused with an abdominal wall defect, a urachal cyst or diverticulum can be located within the anterior abdomen (Fig. 32). The key to the diagnosis is the cyst's communication with the bladder and often with the umbilicus as well. Obviously, it can occur in males or females and is unilocular and singular in nature.

A urachal cyst should not be confused with a persistent right umbilical vein. A urachal cyst is located caudal to the umbilical cord insertion, whereas a persistent right umbilical vein is located cephalad to the umbilical cord insertion. The umbilical veins are paired at between two and four weeks' gestation, usually with the right umbilical vein regressing and the left umbilical vein persisting. The left umbilical vein then drains into the portal venous system. If the reverse occurs, however, the persistent right umbilical vein will not drain directly into the portal vein but will instead drain into variable locations, including the right atrium, the inferior vena cava, or the iliac vein (Fig. 33). The importance of this finding lies in the fact that persistent right umbilical vein is associated with a number of different anomalies (109).

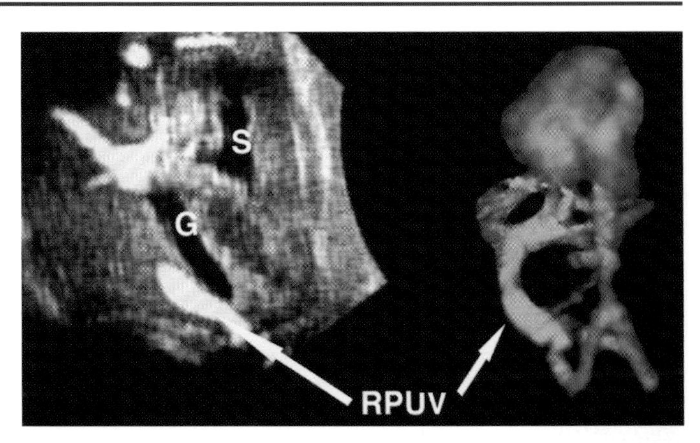

FIGURE 33 ■ Right persistent umbilical vein. Reconstructed 2-D and 3-D surface rendering showing right persistent umbilical vein draining to the right of the gallbladder. RPUV, right persistent umbilical vein; G, gallbladder; S, stomach. *Source*: Courtesy of George Bega, MD, Philadelphia, PA.

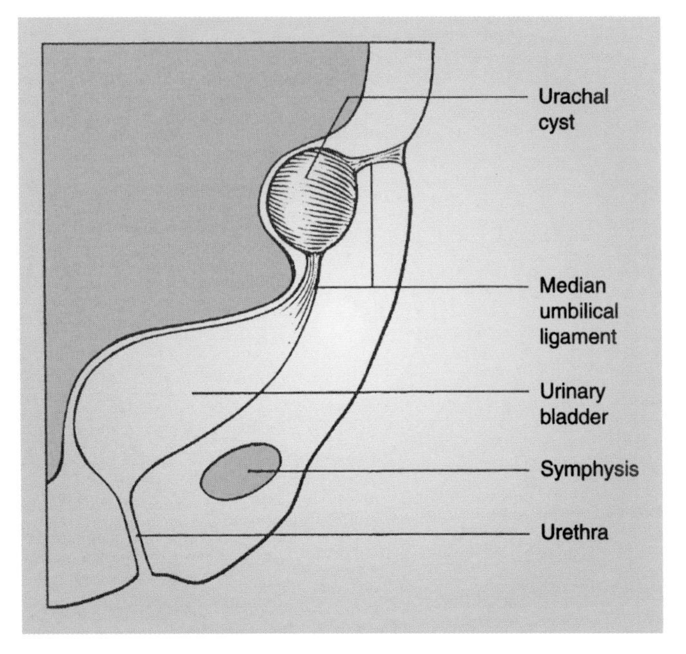

FIGURE 32 ■ Sagittal diagram of a urachal cyst lying cephalad to the bladder in the median umbilical ligament, which is somewhere along its course between the umbilicus and the most cephalad portion of the bladder. *Source*: From Ref.186.

Mesenteric or omental cysts are usually located in the mid-abdomen and are frequently mobile. Enteric duplication cysts are generally tubular in appearance, do not communicate with the bowel, and are more common in male fetuses. These anomalies are extremely difficult, if not impossible, to differentiate from an ovarian cyst or from each other. Hydrometrocolpos is a rare anomaly characterized by a retrovesical cystic mass in a female fetus. This represents hormonal secretions filling the vagina as a result of a hymenal obstruction or a vagina anomaly (atresia or septum). Other genitourinary anomalies, such as renal atresia and intestinal atresias, may be seen in association with hydrometrocolpos. McKusick–Kaufman syndrome is an autosomal recessive triad of hydrometrocolpos, congenital heart disease, and polydactyly.

Another relatively uncommon lesion is the umbilical vein varix (Fig. 34). Although not truly a cyst, it may be confused with one located in the anterior mid-abdomen at the cord insertion site. The key to the diagnosis is the communication with the umbilical and portal vasculature; in difficult cases, color Doppler or power angiography should clarify the vascular nature of this cystic-appearing lesion.

A more recent series reviewed a total of 91 cases of umbilical vein varix collected from their institution or from the literature. This revealed that 31.9% had additional sonographic abnormalities. The most common abnormality was cardiovascular. There was a 9.9% rate of chromosomal abnormalities and a 13% perinatal loss rate. In cases with isolated umbilical vein varix, there was an 8.1% incidence of unexplained fetal death. Thus, umbilical vein varix seems to be associated with increased risk to the fetus (110).

Outside Japan, prenatal detection of choledochal cysts is rare (Fig. 35). These cysts of the biliary tree are

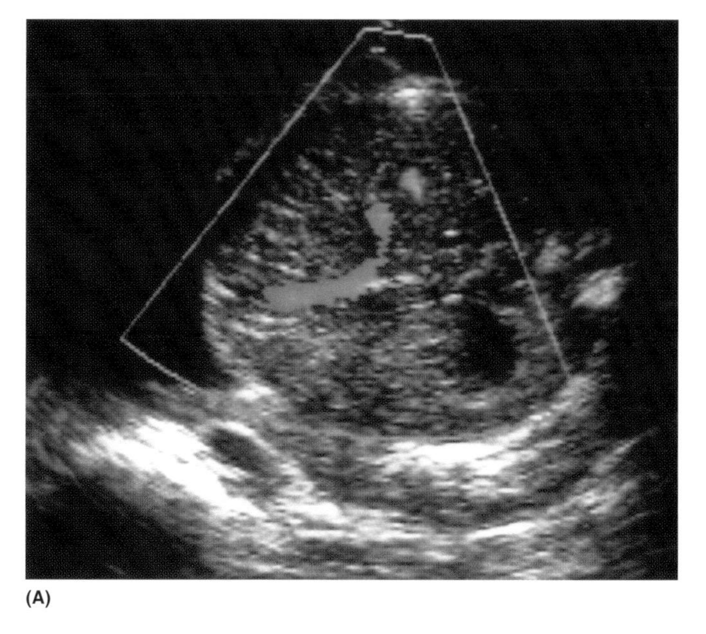

(A)

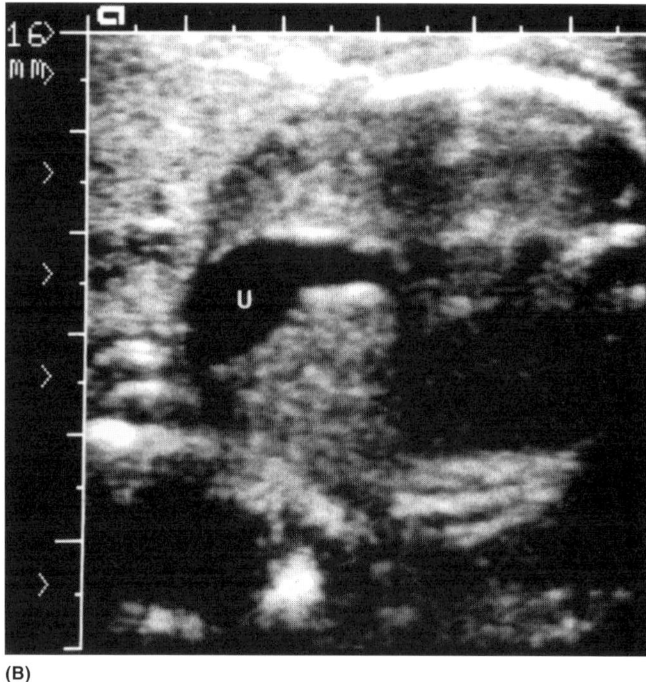

(B)

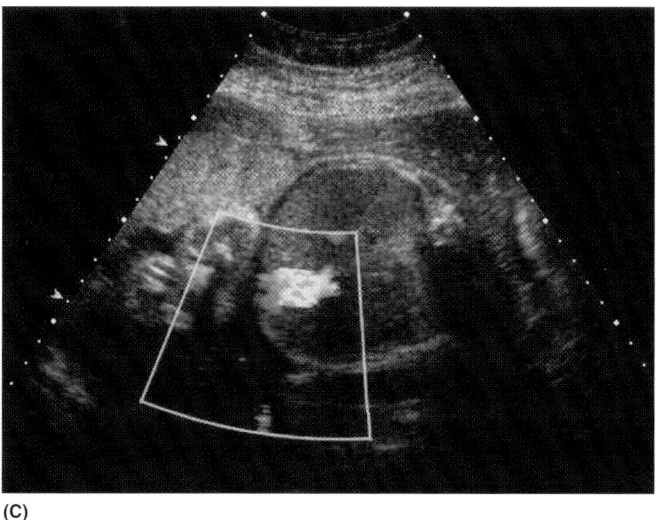

(C)

FIGURE 34 A–C ■ Umbilical vein varix. (**A**) Normal color flow of umbilical vein. (**B**) Umbilical vein varix with 2-D ultrasound. (**C**) In another fetus color-flow ultrasound of umbilical vein varix. *Abbreviations*: U, umbilical vein.

usually single and, as expected, are found in the right upper quadrant near the gallbladder. The keys to differentiating this cyst from a gastrointestinal anomaly are documenting the hepatic communication, the absence of peristalsis, the lack of polyhydramnios, or any communication with the stomach or small bowel. A hepatic cyst is another rare lesion of the right upper quadrant. Like choledochal cysts, these lesions are far more common in female fetuses.

Sacrococcygeal teratomas are discussed in more detail elsewhere in this text. Usually, they are solid masses, but they may have cystic components. They originally arise from the deep pelvis coccyx and may have both intrapelvic and extrapelvic components (Fig. 36).

One pitfall in the differential diagnosis is mistaking the gallbladder for an abdominal cyst. This structure is usually visible throughout the second half of pregnancy, and the ultrasonographer should become comfortable with its location and appearance.

Ascites

The presence of fluid within the fetal abdomen, or ascites, is always abnormal. As discussed earlier, however, the overdiagnosis (pseudoascites) is common (Figs. 7 and 37). Although often seen as an isolated finding, it frequently represents an early stage of hydrops fetalis, immune or nonimmune; as such, an accurate, early diagnosis is of critical importance in the management of patients at risk.

Truly isolated ascites may be caused by a myriad of conditions, including infections and a large number of idiopathic causes. It is important to perform a detailed search for anomalies not only to discover those associated with hydrops but also to rule out bowel obstructions or urinary ascites associated with obstructive uropathy.

Awareness of the more common sites for the detection of ascites can aid in early diagnosis: surrounding the liver, in the flanks, surrounding the bowel, and in the pelvis. As stated earlier, the earliest sign of ascites may be the appearance of prominent fetal bowel loops, particularly

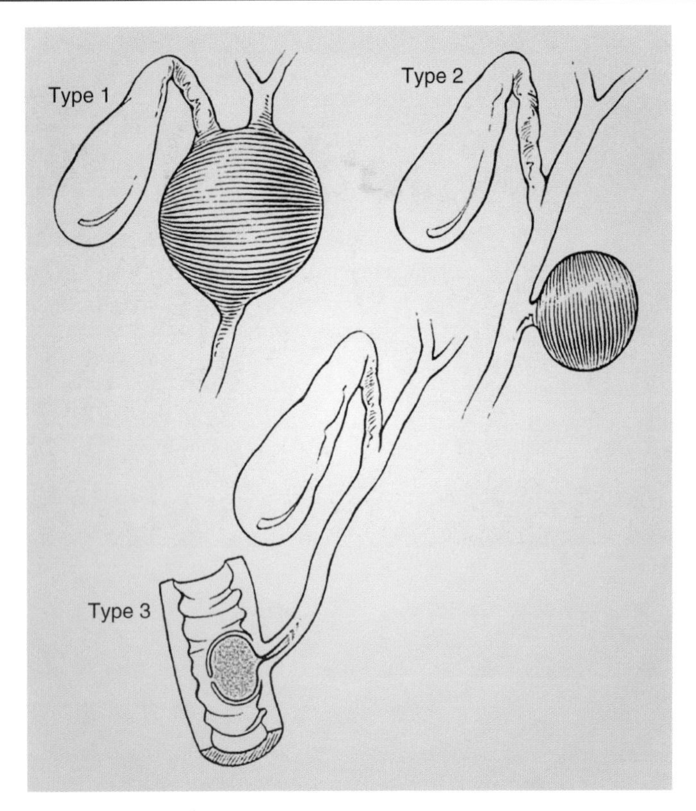

FIGURE 35 ■ Different types of choledochal cyst. Type 1 is the most common and represents dilation of the common duct, which may extend into the common hepatic duct and the proximal cystic duct. Type 2 is a focal diverticulum of the common bile duct. Type 3 is a smaller diverticulum of the common bile duct. As it enters the duodenum types 4 and 5 can occur with multiple intra- or extrahepatic cysts. *Source*: Modified from Ref. 185.

in the second and early third trimesters (75). This finding may be of critical importance in patients at risk for treatable fetal anemias (e.g., Rh isoimmunization, parvovirus infection). In its more advanced stages, the liver and spleen are completely outlined by fluid, and the falciform ligament, umbilical vein, mesentery, and omentum can also be seen.

Abdominal Calcifications

Abdominal calcifications are relatively common ultrasonographic findings in pregnancy; it is also common to misrepresent a bright echogenic area as an abnormality. To avoid this pitfall, it is important to document the acoustic shadow cast by the reflecting mass (Fig. 38). Meconium peritonitis is the most common cause of calcific lesions in the subhepatic peritoneal cavity (Table 14).

Intrahepatic calcifications are seen with some frequency. Although they may occur as a result of a viral or other congenital infection, such as cytomegalovirus, rubella, toxoplasmosis, or varicella, many of these TORCH (toxoplasmosis, other agents, rubella, cytomegalovirus, herpes simplex) cases also manifest with periventricular calcifications in the brain and intrauterine growth restriction. Vascular lesions such as hemangiomas, hepatoblastomas, and ischemic hepatic infarcts may be seen as well.

Intra-abdominal tumors must be considered in the differential diagnosis (Table 14), particularly when the calcification is associated with a mass; neuroblastomas and teratomas should be high on the differential list because they can occur in a variety of sites within the abdomen.

Intraluminal calcifications in the right upper quadrant may represent fetal gallstones. Although uncommon, these lesions usually resolve postnatally without any symptoms or treatment. Adrenal hemorrhage can present as suprarenal calcifications as well (Fig. 47).

Although it is important to rule out the potentially significant diagnoses mentioned earlier, when liver calcifications are an isolated finding in an otherwise normal fetus, the outcome is almost always favorable.

Echogenic intra-abdominal masses should be further classified as to their exact location, the lesion

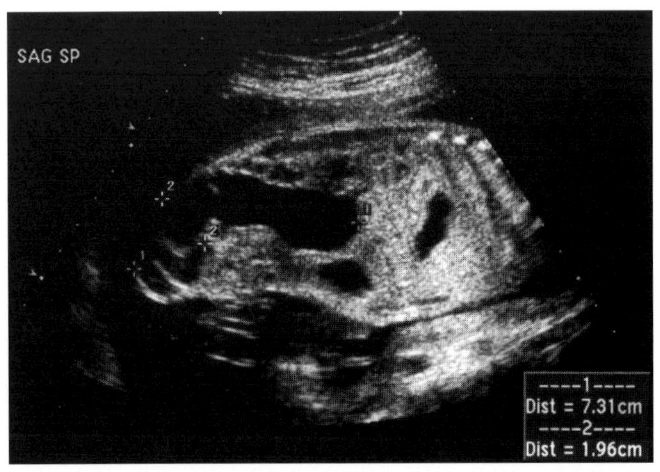

(A)

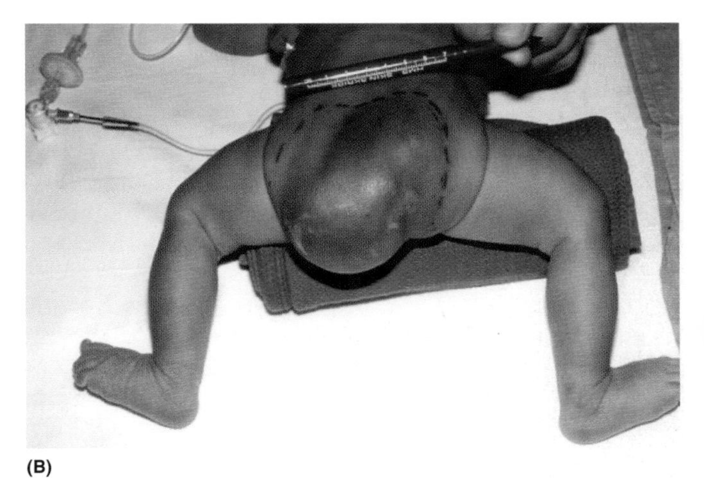

(B)

FIGURE 36 A AND B ■ Sacrococcygeal teratoma. (**A**) Longitudinal ultrasound shows large cystic pelvic mass, which extended from upper abdomen and through coccyx. (**B**) In another patient, sacrococcygeal teratoma is marked before surgical resection *Source*: Courtesy of Marshal Swartz, MD, Washington, DC.

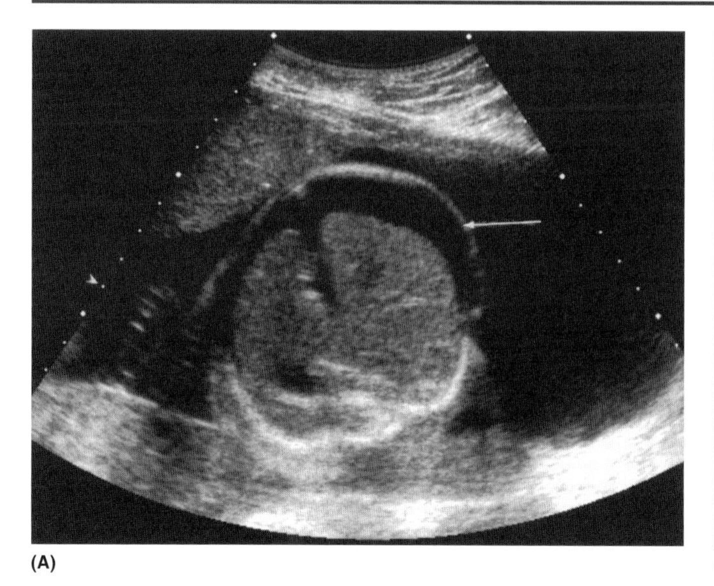

(A)

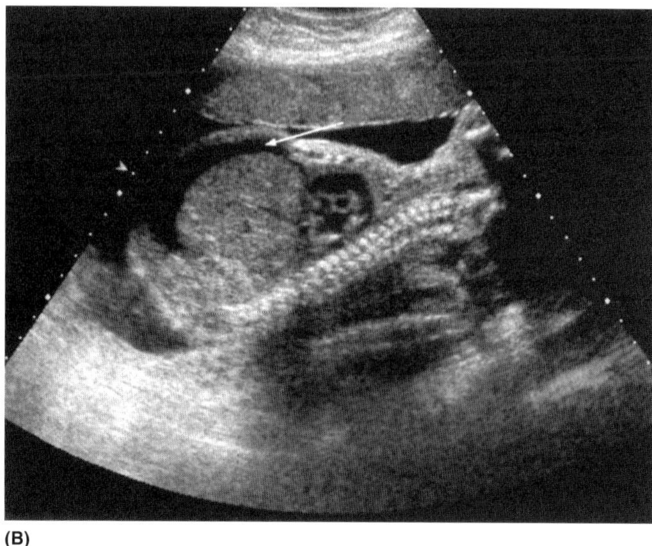

(B)

FIGURE 37 A AND B ■ (**A**) Ascites (*arrow*) surrounding the fetal liver. (**B**) Sagittal view of the fetal torso with ascites (*arrow*). This was secondary to parvovirus. L, liver.

echotexture, associated calcifications, additional findings and evolution over time. In most cases, conservative management is sufficient, but in others, a more detailed investigation is necessary (111). For instance, in calcifications not associated with a mass isolated to the left upper quadrant, the location of the echogenic foci was along the stomach, spleen, or left lobe of the liver. Many of these foci disappeared with in utero scanning. Only one of 26 fetuses that were tested had elevated immunoglobulin G levels when evaluated with TORCH titers (112).

Hepatosplenomegaly

Enlargement of the liver or spleen may occur secondary to an isolated finding such as a liver tumor (hemangioendothelioma) or systemic abnormalities such as fetal hydrops (Table 15).

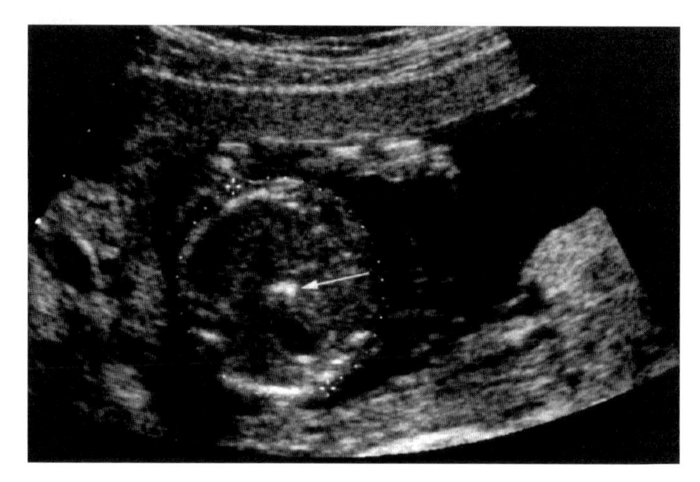

FIGURE 38 ■ Intra-abdominal calcifications. The arrow depicts a densely echogenic calcification in the upper abdomen of a 20-week fetus. Other calcifications were noted in the abdomen shadow cast by the lesion. A workup for infection proved negative.

3D Ultrasound

3D ultrasound has been shown to be helpful in evaluating the fetal abdomen. For instance, reliable fetal volumes can be obtained early in pregnancy. Interestingly, the crown rump length doubles from a mean of 48 mm at 11 weeks to 79 mm at 14 weeks. This is associated with a five- to sixfold increase in fetal volume (113). 3D has also been used to measure liver volume. For instance, in

TABLE 14 ■ Differential Diagnosis of Intra-abdominal Calcifications

Meconium (extraluminal)
■ Focal or diffuse (meconium peritonitis)
■ Associated cyst (meconium pseudocyst)
Meconium (intraluminal)
■ Anorectal atresia
■ Small bowel atresia
■ Meconium ileus
Idiopathic (focal)
■ No etiology determined
Infection
■ Cytomegalovirus, toxoplasmosis
Tumors
■ Neuroblastoma
■ Teratoma
■ Hemangioma
■ Hepatoblastoma
Parenchymal
■ Liver
■ Spleen
Cholelithiasis (gallbladder)
■ Usually regresses postnatally

Source: From Ref. 186.

TABLE 15 ■ Fetal Hepatosplenomegaly

Common
- Hydrops
- Immune
- Nonimmune
- Infection

TORCH
- Hepatitis (liver)
- Congestive heart failure
- Congenital anemias

Uncommon
- Errors of metabolism
- Tumors
- Infantile hemangioendothelioma (liver)
- Hepatoblastoma (liver)
- Beckwith–Wiedemann syndrome
- Systemic malignancy
- Metastatic neuroblastoma
- Leukemia

Abbreviation: TORCH, toxoplasmosis, other agents, rubella, cytomegalovirus, herpes simplex.
Source: From Ref. 186.

fetuses with intrauterine growth retardation (IUGR), the liver length may be normal, but liver volume is small (114). Likewise, 3D power Doppler has been helpful in predicting a case of splenomegaly secondary to a cytomegalovirus infection (115).

GENITOURINARY SYSTEM

Embryology and Normal Ultrasonographic Appearance

The pronephros and mesonephros are the two primitive urinary systems of the human embryo. The pronephros

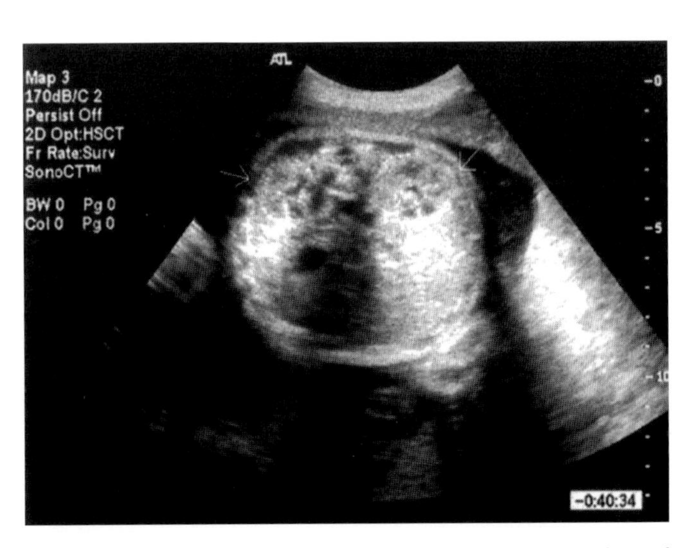

FIGURE 39 ■ Normal kidneys in a 20-week fetus. Transverse views of both kidneys (*arrows*).

TABLE 16 ■ Normal Length of Fetal Kidneys at Different Gestational Ages

Menstrual age (wk)	−2 SD[a] (mm)	Predicted value[b] (mm)	+2 SD[a] (mm)
24	22.0	24.5	27.0
25	22.6	25.1	27.7
26	23.3	25.8	28.3
27	24.0	26.5	29.0
28	24.7	27.2	29.8
29	25.5	28.0	30.5
30	26.3	28.8	31.3
31	27.1	29.6	32.1
32	27.9	30.4	32.9
33	28.8	31.3	33.8
34	29.6	32.2	34.7
35	30.5	33.1	35.6
36	31.5	34.0	36.5
37	32.4	35.0	37.5
38	33.4	36.0	38.5
39	34.5	37.0	39.5
40	35.5	38.0	40.5
41	36.6	39.1	41.6
42	37.7	40.2	42.7

[a]Standard deviation (SD) = 1.259 mm.
[b]Length = 16.8933 + 0.0132 (menstrual age).
Source: From Ref. 187.

TABLE 17 ■ Normal Anteroposterior Diameter of Kidneys at Different Gestational Ages

Menstrual age (wk)	−2 SD[a] (mm)	Predicted value[b] (mm)	+2 SD[a] (mm)
22	8.9	11.3	13.7
23	9.3	11.7	14.1
24	9.7	12.1	14.5
25	10.2	12.6	15.0
26	10.7	13.1	15.5
27	11.3	13.7	16.1
28	11.9	14.3	16.7
29	12.5	15.0	17.4
30	13.2	15.6	18.0
31	14.0	16.4	18.8
32	14.8	17.2	19.6
33	15.6	18.0	20.4
34	16.5	18.9	21.3
35	17.5	19.9	22.3
36	18.5	20.9	23.3
37	19.5	21.9	24.4
38	20.7	23.1	25.5
39	21.8	24.3	26.7
40	23.1	25.5	27.9

[a]Standard deviation (SD) = 1.209.
[b]Predicted value = 8.457278951 + 0.00026630314 (menstrual age).
Source: From Ref. 187.

regresses by week 6; at that time, the mesonephros gives rise to the ureteral bud, preceding and inducing the development of the metanephros by the 8th to 10th week of gestation. The metanephros gives rise to the permanent kidney (nephron), whereas the ureteral bud is responsible for the formation of the renal collecting system (i.e., pelvis, calyces).

Although the kidneys and bladder can be imaged ultrasonographically as early as 10 to 12 weeks' gestation, especially with newer equipment and transducers, the internal architecture of the kidneys cannot reliably be assessed before 14 to 16 weeks (Fig. 39). The kidneys appear as oval, hypoechoic, paraspinal structures in the dorsal area of the mid-abdomen. In the latter half of pregnancy, the renal capsule becomes more echogenic, and the kidneys are best viewed as circular, paraspinal structures in the transverse view of the middle to lower abdomen. The kidneys grow steadily throughout gestation, ranging from 2 cm in length at 20 weeks to nearly 4 cm at term (Table 16). Anterior-to-posterior dimension is about 1.0 to 1.1 cm at 20 weeks, and increases to 2.5 cm at 40 weeks (Table 17). A good rule of thumb is that in the second and third trimesters, the kidneys increase in length by about 1 mm per week and increase in diameter by about 0.5 mm per week. Grannum et al. (116) reported the ratio of kidney circumference to abdominal circumference in normal gestation, a useful ratio that remains relatively constant throughout the latter half of pregnancy, at about 28% to 30% (Table 18). The normal ureter is not seen ultrasonographically. The normal adrenal gland can often be visualized at the superior border of the kidney (Fig. 40).

Although the fetal kidneys are functionally immature throughout fetal life, urine production in the fetus begins by 10 weeks' gestation. Fetal urine, however, does not become the primary source of amniotic fluid volume until 16 to 18 weeks' gestation. This point is critical in the evaluation of possible urinary tract anomalies because amniotic fluid volume, particularly in the latter half of pregnancy, is a reliable indicator of fetal urine output.

TABLE 18 ■ Fetal Kidney and Abdominal Parameters by Gestational Age Group in Normal Fetuses

	Gestational age (wk)					
Variable[a]	<16 (*n* = 9)	17–20 (*n* = 18)	21–25 (*n* = 7)	26–30 (*n* = 11)	31–35 (*n* = 19)	>36 (*n* = 25)
Fetal kidney						
Anteroposterior (cm)						
Mean	0.84	1.16	1.49	1.93	2.20	2.32
SD	0.24	0.24	0.37	0.19	0.32	0.32
Transverse (cm)						
Mean	0.86	1.13	1.64	2.00	2.34	2.63
SD	0.14	0.25	0.40	0.28	0.42	0.50
Circumference (cm)						
Mean	2.79	3.80	5.40	6.58	7.86	8.42
SD	0.64	0.72	0.68	0.67	0.86	1.39
Fetal abdomen						
Anteroposterior (cm)						
Mean	2.92	3.73	5.12	6.74	8.50	8.88
SD	0.62	0.72	0.80	0.58	0.82	0.94
Transverse (cm)						
Mean	2.93	3.68	5.12	7.09	8.76	9.68
SD	0.59	0.56	0.49	0.61	0.89	1.44
Circumference (cm)						
Mean	9.66	12.37	17.36	22.03	28.11	30.45
SD	1.88	2.28	1.77	1.77	2.31	3.45
KC/AC ratio						
Mean	0.28	0.30	0.30	0.29	0.28	0.27
SD	0.02	0.03	0.02	0.02	0.03	0.04

[a]Calculations of ratio used variables measured to eight decimal places.

Abbreviations: SD, standard deviation; KC, kidney circumference; AC, abdominal circumference.

Source: From Ref. 188.

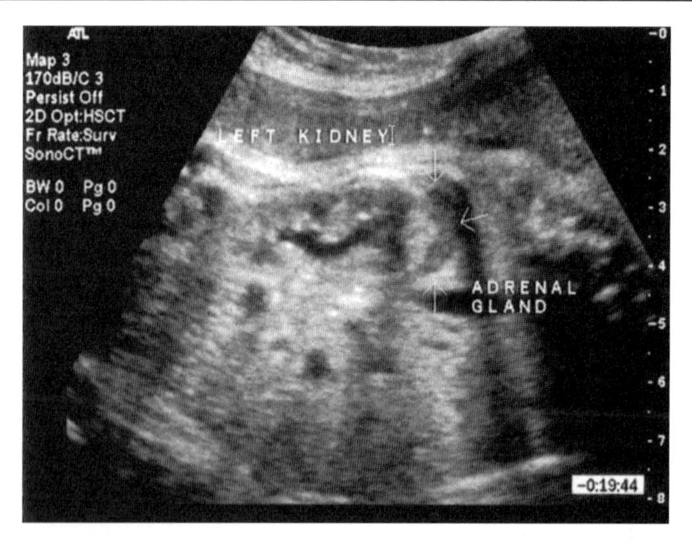

FIGURE 40 ■ Normal kidney—adrenal gland. Longitudinal scan of the kidney demonstrates cortical medullary distinction. At the most cephalad border is the crescent-shaped adrenal gland (*arrows*).

TABLE 19 ■ Nonvisualization of the Fetal Bladder

Renal abnormality
Bilateral renal agenesis
Bilateral ureteral pelvic junction obstruction
Bilateral multicystic dysplastic kidney
Bilateral combinations of any of the above
Infantile polycystic kidney disease
Bladder abnormality
Cloacal (or bladder) exstrophy
Persistent cloaca
Systemic
Severe intrauterine growth restriction

Source: From Ref. 186.

Oligohydramnios

A detailed discussion of amniotic fluid volume is beyond the scope of this chapter; however, some understanding of oligohydramnios is critical to the evaluation of urinary tract malformations. Although the subjective impression of oligohydramnios by experienced ultrasonographers is generally accurate, a semiquantitative estimate of amniotic fluid can be accomplished using a four-quadrant measurement technique, the *amniotic fluid index* (117). With this technique, the sums of the largest vertical pockets in each quadrant are added together to form the index. A value below 5cm is consistent with severe oligohydramnios throughout the second and third trimesters, before 32 to 34 weeks' gestation; a value below 8cm raises concern of decreased amniotic fluid volume as well. The index is a useful tool for serial examination and management of patients by multiple examiners of varying experience using different equipment. The test is reproducible as well, with interobserver and intraobserver error, of less than 7% (119).

Anomalies

Urinary tract anomalies represent some of the most common anomalies seen in fetal and neonatal life. In our practical scheme, the basic questions to ask include the following:

1. Are the bladder and both kidneys documented in their normal locations (Table 19)?
2. Are the renal pelvises, ureters, or bladder dilated?
3. Are renal cysts present? If so, do they communicate?
4. Are the kidneys abnormal in size or echogenicity?
5. Is the abnormality bilateral or unilateral?
6. Is the amniotic fluid volume normal?

Are the Bladder and Both Kidneys Documented in Their Normal Locations?

Renal Agenesis: Bilateral ■ The incidence of complete bilateral renal agenesis (Potter's syndrome) is about 1 in 4000 births (120). The malformation is more common in males (as are most urinary tract anomalies) and is invariably accompanied by severe oligohydramnios beyond 16 weeks' gestation. In the early part of gestation, amniotic fluid volume is not dependent on urine production and may be normal despite absent renal function. The diagnosis of bilateral renal agenesis should be suspected when severe oligohydramnios is seen and the kidneys and bladder cannot be visualized. The differential diagnosis includes symmetric intrauterine growth restriction with oligohydramnios, small dysplastic kidneys, and early, preterm premature rupture of membranes.

Potential pitfalls in the diagnosis include:

- assuming that amniotic fluid volume is normal despite nonvisualization of kidneys and bladder
- inability to image the kidneys and bladder owing to the poor image quality associated with severe oligohydramnios (especially with an anterior placenta or a thick maternal abdominal wall in the early second trimester)
- adrenal hypertrophy associated with renal agenesis (Fig. 41) (121)

The adrenal glands are almost always present in renal agenesis because the kidneys and adrenals are parts of different organ systems embryologically. The adrenals often appear larger and assume a discoid shape, have a distinct cortex and medulla, and thus may be mistaken for renal tissue in these cases.

In more difficult cases, a transabdominal infusion of saline by amniocentesis has been employed with increasing enthusiasm and great success to enhance the image quality of the anatomic survey, and the kidneys in particular (122). In addition, this technique facilitates the instillation of indigo carmine dye to confirm or rule out premature rupture of the membranes in appropriate cases.

When the adrenal glands are enlarged with Potter's sequence, color-flow Doppler and color power angiography (color Doppler energy) are sensitive enough to depict the renal vasculature with some reliability (Fig. 41) (124). Inability to visualize renal arterial blood flow supports the diagnosis of renal agenesis; but perhaps more

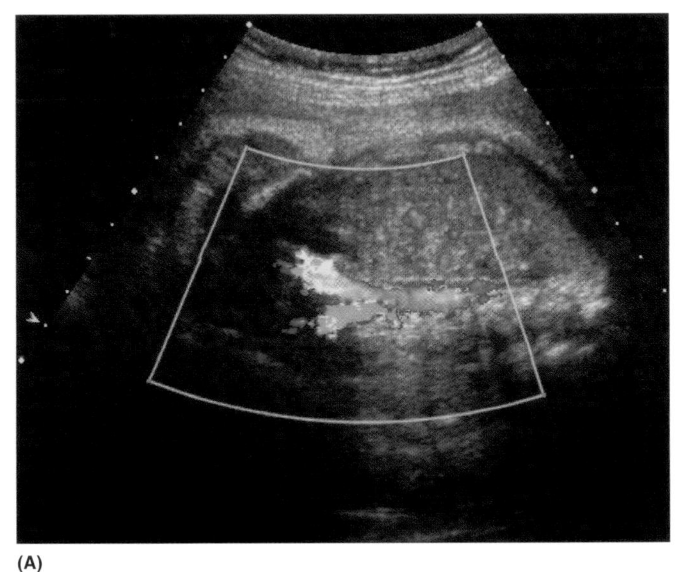
(A)

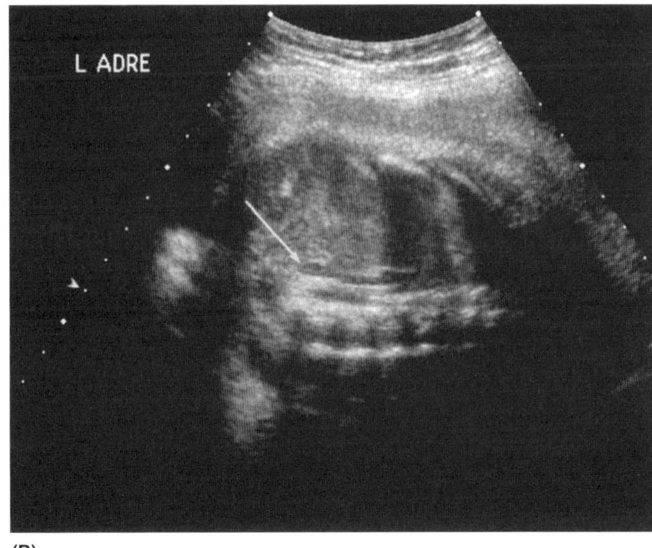

(B)

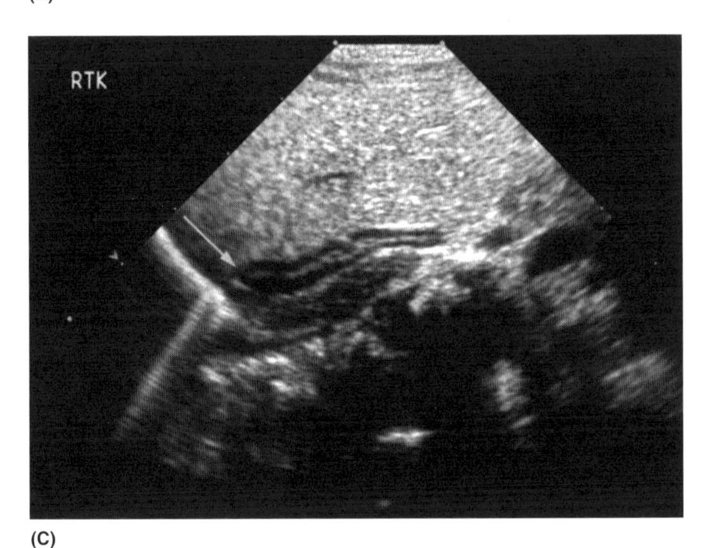

(C)

FIGURE 41 A–C ■ Adrenal gland with renal agenesis. (**A**) Longitudinal ultrasound of the fetal abdomen demonstrating the spine (S) and the pancake-shaped adrenal gland (*arrow*). The adrenal gland is nearly 2 cm in length, with distinct hypoechoic adrenal cortex and echogenic adrenal medulla. (**B**) Coronal color-flow image showing no color to the renal fossa. (**C**) After delivery, ultrasound of the right upper quadrant demonstrates "pancake" adrenal (*arrow*).

importantly, color-flow mapping can often help delineate the kidneys in difficult cases of oligohydramnios. Although this raises the concern of confusing the adrenal arteries for renal vessels, this has not posed a problem in our experience because the adrenal vessels are rarely imaged (124,125). Devore (124) notes that they originate on the ventral side of the aorta and are only seen with low velocity (less than 0.12 m/sec).

Associated anomalies with bilateral renal agenesis are common, although many of these represent deformations rather than malformations, the consequences of oligohydramnios (Table 19). The abnormal facies so typical of Potter's sequence, the limb deformities and pulmonary hypoplasia invariably seen, are a direct result of the prolonged lack of amniotic fluid volume (120). The small chest circumference and abnormal chest/abdominal circumference ratio (126) support the diagnosis of pulmonary hypoplasia. No reliable ultrasonographic marker has been developed, however, that can diagnose pulmonary hypoplasia definitively in utero. In addition, musculoskeletal anomalies are reported in 40% of fetuses with Potter's syndrome; cardiac anomalies are seen in nearly 15%; and central nervous system and gastrointestinal anomalies have been reported as well.

If the diagnosis is certain, pregnancy termination should be offered, given the uniform lethality of the condition. Although severe oligohydramnios alone carries a dismal prognosis, it is critical, particularly in the later part of gestation, to be certain of the diagnosis of bilateral renal agenesis before offering nonintervention for the fetus in these cases. With this in mind, it is important to avoid the pitfalls noted earlier and certainly appropriate to use vaginal transducers, color-flow Doppler, and amnioinfusion if necessary to confirm the diagnosis.

Unilateral renal agenesis is a much more common disorder, although the diagnosis may be far more difficult (Table 20). This may be explained in part by the fact that in some cases, severely dysplastic kidneys involute prenatally, creating the postnatal appearance of agenesis. The presence of a normal-appearing bladder and amniotic fluid volume, coupled with the hyperplastic adrenal gland occupying the renal fossa on the affected side,

TABLE 20 ■ Differential Diagnosis of Absent Fetal Kidney (Unilateral)

■ Shadowing from the fetal spine
■ Pelvic kidney
■ Unilateral renal agenesis
■ Crossed renal ectopia

Source: From Ref. 186.

makes it easy to see why this diagnosis is often overlooked. In many cases, other anomalies are present (127). The normal contralateral kidney exhibits compensatory growth (130) and generally is significantly longer than expected for gestational age.

A common pitfall in this diagnosis results from failure to image one kidney in a transverse plane owing to acoustic shadowing from the spine (Table 20). Rotating the transducer to image both renal fossa is critical (Fig. 42).

Ectopic Kidney ■ Pelvic kidney is a relatively frequent occurrence (about 1 in 1000 births) and a common pitfall in the diagnosis of unilateral renal agenesis (129). When both kidneys are not clearly identified in their usual paraspinous locations, a high index of suspicion is required to make the diagnosis. Further support for this is found in the fact that the clinical diagnosis of renal ectopia is only made in less than 10% of cases. Two cases of bilateral fetal pelvic kidneys diagnosed in the third trimester have been reported. Ectopic kidneys can also be found in the thorax and other areas within the abdomen.

Less commonly, crossed renal ectopia may be confused with unilateral renal agenesis (130). In this condition, the kidney is located on the opposite side of its ureter. Typically, the kidney is bilobed, enlarged, and S- or L-shaped, often with findings of obstructive uropathy.

Horseshoe Kidney ■ Horseshoe kidney is a relatively common renal anomaly, with an incidence of about 1 in 400 births. The kidneys are usually joined at the lower pole by an isthmus adjacent to the third or fourth lumbar vertebra. Associated anomalies, including cardiac, central nervous system, and chromosomal anomalies, such as Turner's syndrome (45, XO) and trisomy 18, are common with this disorder. In the absence of other anomalies, horseshoe kidney is a relatively benign disorder that often goes undetected.

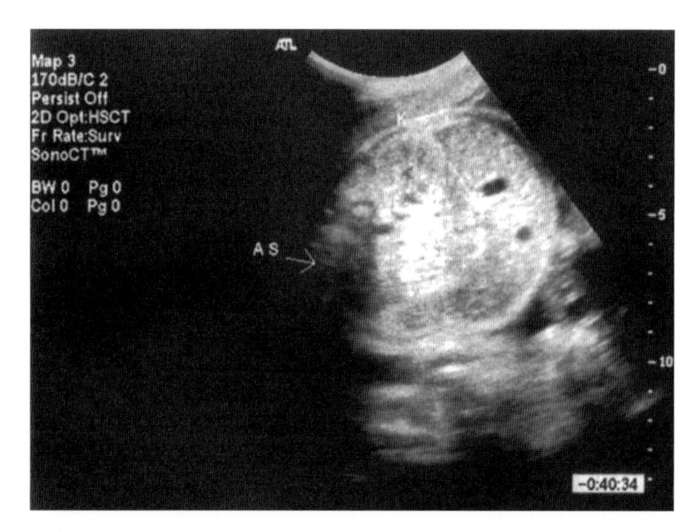

FIGURE 42 ■ Kidney visualization—pitfall. Acoustic shadowing from the spine obliterates the view of one kidney in this transverse view of a 32-week fetus. AS, acoustic shadowing; K, kidney.

Are the Renal Pelvises, Ureters, or Bladder Dilated?

Obstructive lesions of the genitourinary system are among the most common ultrasonographic findings in obstetrics. In a study of 142 neonates with obstructive uropathy, more than 75% were initially diagnosed prenatally (131). Fifty percent of the cases involved a ureteropelvic junction (UPJ) obstruction, 25% exhibited ureterovesical junction (UVJ) obstruction, 15% had a duplicate collecting system, and 5% had a bladder obstruction (PUV). It seems appropriate to discuss these anomalies in their order of frequency.

UPJ Obstruction ■ UPJ obstruction is by far the most common form of obstructive uropathy, representing fully two-thirds of fetal hydronephrosis cases. It is also important because congenital hydronephrosis often has a subtle clinical presentation for neonates; in some series, less than 25% of cases are diagnosed by one year of age, and only 55% of cases are diagnosed by age of five years. Given the fact that diagnostic delay may increase the risk of permanent renal impairment, it is crucial to make an early and accurate diagnosis whenever possible.

Although an early diagnosis of fetal UPJ obstruction is clearly important for optimal outcome, minimal or mild renal collecting system dilation, or pyelectasis, generally resolves without clinical sequelae (132). Frequent rescanning of these patients increases health care costs and patient anxiety, all without any benefit to the fetus. There is no consensus on management of patients with minimal pyelectasis.

A study by Odibo et al. challenged this dictum (133). Based upon their ROC curve, they found that an anterior posterior (AP) renal pelvic diameter of less than 7.0 mm after 32 weeks was predictive of normal postnatal renal function (likelihood ratio of 5.8). In addition, Odibo et al. (133) evaluated the initial renal pelvic diameters at 18–30 menstrual weeks to determine the thresholds for normal postnatal renal function. They found that the optimal renal AP diameter associated with normal postnatal outcome was less then 6 mm, with sensitivity, specificity, and positive and negative predictive values of 71%, 80%, 89%, and 53%, respectively (133). The effect of bladder filling on the size of the renal pelvis may be a contributing factor to the diagnostic dilemma. In one report (134), more than half of the fetuses with pyelectasis of 5 mm or more when the fetal bladder was full reverted to normal after the bladder emptied.

More recent publications have emphasized the importance of follow-up third-trimester sonography for cases of mild second-trimester pyelectasis. Mild second-trimester pyelectasis is usually defined as an AP renal pelvis

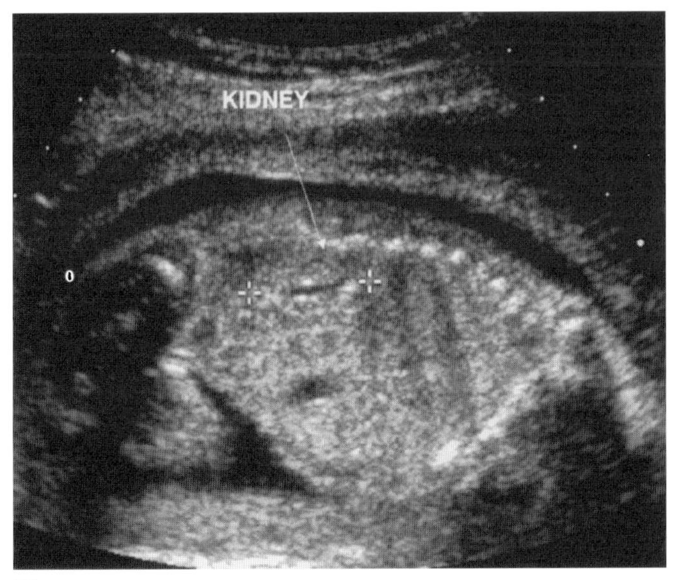

(A)

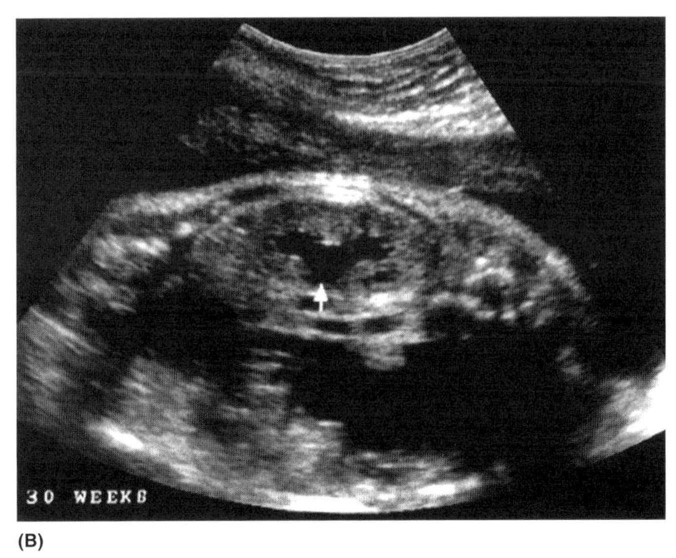

(B)

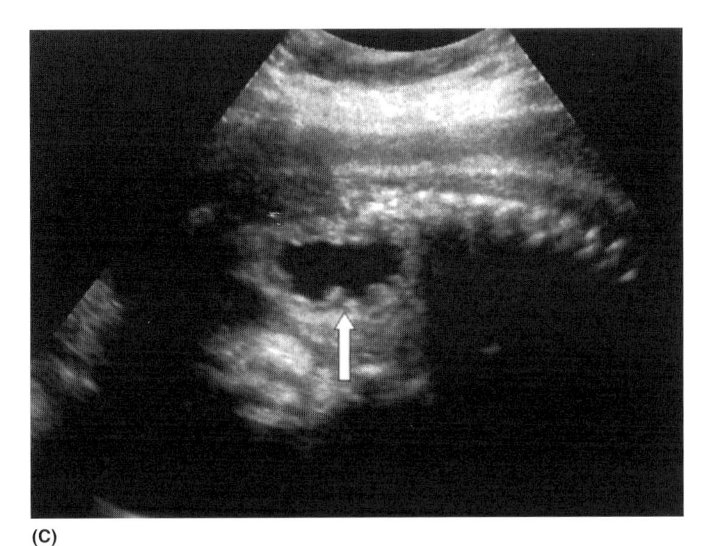

(C)

FIGURE 43 A–C ■ Hydronephrosis. (**A**) Normal fetal kidney. (**B**) In another fetus, there is renal pelvis (*arrow*) and calyceal dilatation to a moderate degree. (**C**) In another fetus, the renal pelvis and coccygeal dilatation (*arrow*) is more severe.

diameter of 4 mm or more (Fig. 43). In one study, they concluded that significant neonatal nephrouropathy is best predicted with an 8 mm threshold in the third trimester. Using an AP diameter of 8 mm at that time in pregnancy had an odds ratio of 128.33 and it almost always predicts neonatal nephrouropathy requiring surgery (135). Likewise, another study used a level of 8 mm in the third trimester scan as the cut-off value to predict most cases of renal uropathy. The sensitivity of the 8 mm diameter was 80%, while the specificity was 79% (136,137). On simply stratifying the fetal renal pelvis diameters on third-trimester sonography as to mild (7–9.9 mm), moderate (10–14.9 mm), and severe (≥15 mm), they found that none of the mild-dilatation children had follow-up surgery, while 23% of the moderate and 64% of the severe-dilatation children either had a neonatal urinary tract infection or required surgery. Therefore, third-trimester renal sonography is established as important in follow-up of mild second-trimester pyelectasis (Fig. 44). Cut-off values will be determined so as to balance sensitivities and specificities.

Second-trimester renal pyelectasis has been associated with chromosomal abnormalities, particularly Down syndrome (138). Benacerraf et al. (139), using similar criteria for pyelectasis as shown earlier, noted retrospectively that 25% of 44 Down syndrome fetuses exhibited mild pyelectasis. Snijders et al. (140) reported a 7.3% prevalence of aneuploidy among 1177 fetuses with mild hydronephrosis in the second trimester. Among the 805 fetuses with *isolated* hydronephrosis, however, there were only five (0.62%) with trisomy 21; this was not significantly different from the expected frequency in that population (0.4%) based on maternal and gestational age. Conflicting data have been reported by Wickstrom et al. (141). These authors advocate prenatal diagnosis for isolated pyelectasis, citing a 3.3-fold increase in risk for all chromosomal abnormalities in their series of 121 cases. However, based on series by Nyberg et al. (142) and Bromley et al. (143),

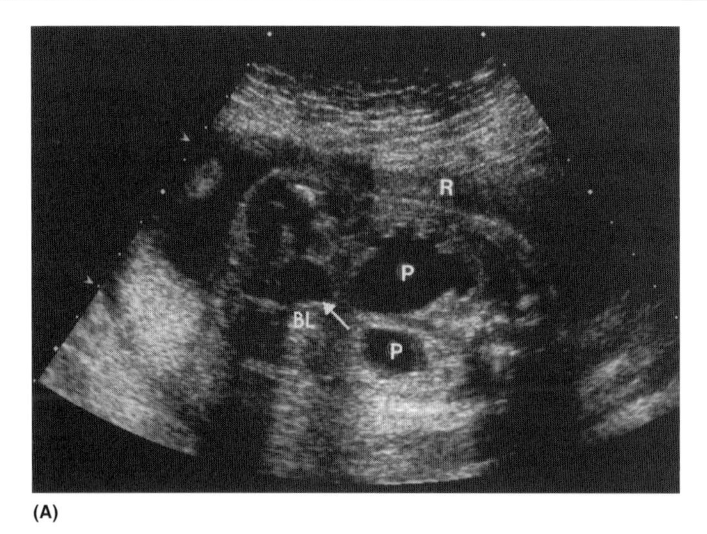

(A)

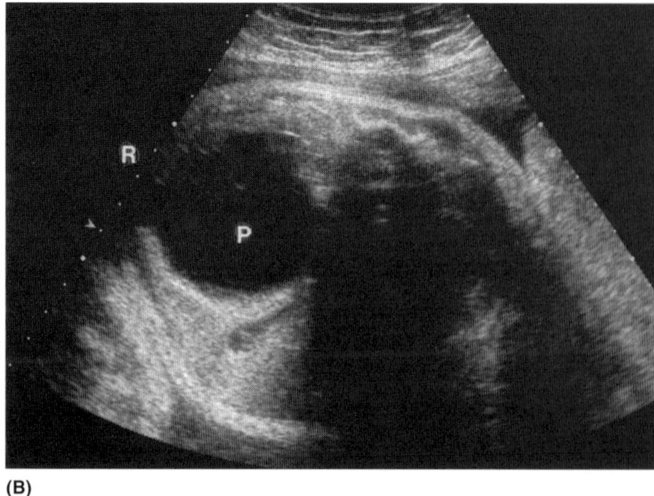

(B)

FIGURE 44 A AND B ■ Bilateral hydroureter. (**A**) Transverse scan at 20 weeks revealing right and left pyelectasis. (**B**) At 36 weeks, the right-sided dilatation (P) increased, while the left-sided dilatation stabilized. R, right; L, left; BL, bladder.

Nicolaides (144) calculated the likelihood of trisomy 21 with the finding of isolated fetal hydronephrosis. The likelihood ratio of trisomy 21 given isolated fetal hydronephrosis is 1.0 based upon Nicolaides' calculations (144). Given the frequency with which pyelectasis is diagnosed ultrasonographically in the second trimester (2–3%), it would appear prudent to offer prenatal diagnosis when renal collecting system dilation is accompanied by other ultrasonographic signs of aneuploidy.

Coco and Jeanty (145) studied the association of fetal pyelectasis (as diameter ≥ mm) and chromosomal abnormalities. They concluded that a careful scan of the fetus is warranted in these cases, but in the absence of other findings, isolated pyelectasis is not a justification for performance of an amniocentesis.

Most commonly, the obstruction is unilateral, maintaining normal amniotic fluid volume or even compensatory polyhydramnios, and requires no prenatal intervention. When unilateral hydronephrosis is accompanied by oligohydramnios, a search for contralateral pathology must be made (e.g., agenesis, dysplasia). With bilateral hydronephrosis, serial ultrasound evaluations are necessary in the third trimester to address amniotic fluid volume and progression of renal pelvic dilation.

Severe oligohydramnios in the second trimester carries an extremely poor prognosis, and termination of pregnancy is a definite consideration. In the third trimester, decisions regarding delivery versus conservative management must be made based on the overall clinical circumstances and the patient's wishes. When bilateral UPJ obstruction is associated with normal or increased amniotic fluid, a targeted search for a concomitant gastrointestinal anomaly is warranted. Of course, the obstetrical ultrasound axiom cannot be overstated: *Whenever one anomaly is encountered, another should be expected*. The presence of ascites in conjunction with severe UPJ obstruction strongly suggests urinary ascites and generally carries a poor prognosis for the kidney, with a presumed ruptured collecting system (Fig. 45). Open surgical intervention for severe bilateral hydronephrosis, using a hysterotomy approach to perform open cutaneous ureterostomies, has been attempted only a few times (146). This procedure clearly remains experimental.

Hydroureter and UVJ Obstruction ■ Ureteral dilation is typically the result of urinary obstruction or reflux, often beginning at the UVJ. This disorder is several times more common in males and is more frequently encountered on the left side. As mentioned previously, the normal fetal

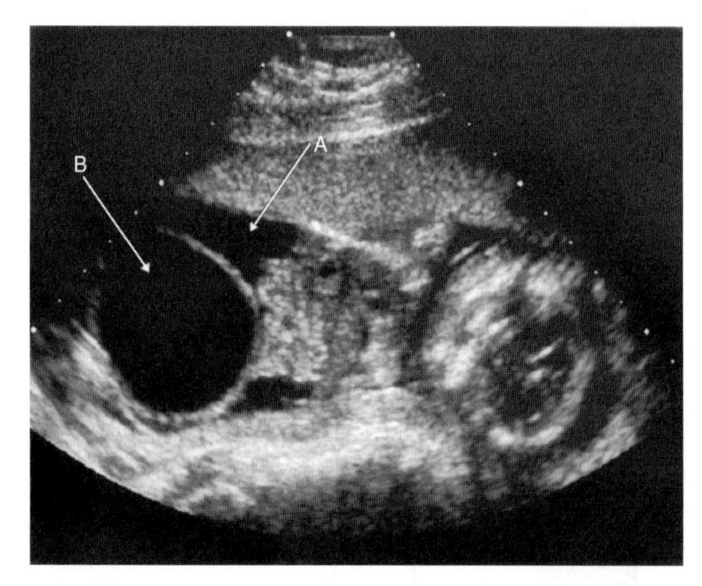

FIGURE 45 ■ Sagittal ultrasound of the fetal abdomen demonstrating massive dilation of the bladder (B) with urine ascites (A) in a fetus with bladder outlet obstruction.

ureter is generally 1 to 2 mm in diameter and is not readily visualized ultrasonographically. *Primary megaureter* refers to a defect in the ureter, as in ureteral valve stenosis and ureterocele. *Secondary megaureter* refers to an anomaly elsewhere resulting in megaureter, such as in the classic PUV syndrome. Hydronephrosis and hydroureter are concomitant findings in typical UVJ obstruction cases, although the ureter tends to be more dilated than the renal pelvis (Fig. 46). The diagnosis of UVJ obstruction is secure when one encounters dilated ureters in the absence of megalocystis.

The ultrasonographic appearance of a hydroureter may be difficult to distinguish from that of bowel. The presence of active peristalsis of a sonolucent tubular structure on real-time ultrasonography does not confirm bowel; it is often seen with hydroureter as well. The snake-like sonolucent segments must be meticulously tracked to demonstrate their communication with the kidney or bladder (147). In addition, the ureter generally comes into close contact with the spine, but the small bowel does not. These structures can be several centimeters wide, so that size alone does not preclude megaureter as a possible diagnosis (Table 21).

Ureterocele ■ Ureterocele is a dilated ureter that has prolapsed into the bladder either at the normal ureteral

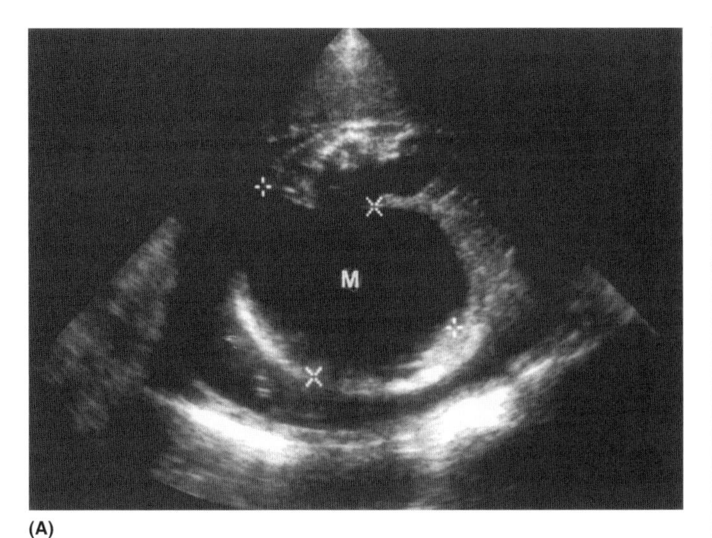

(A)

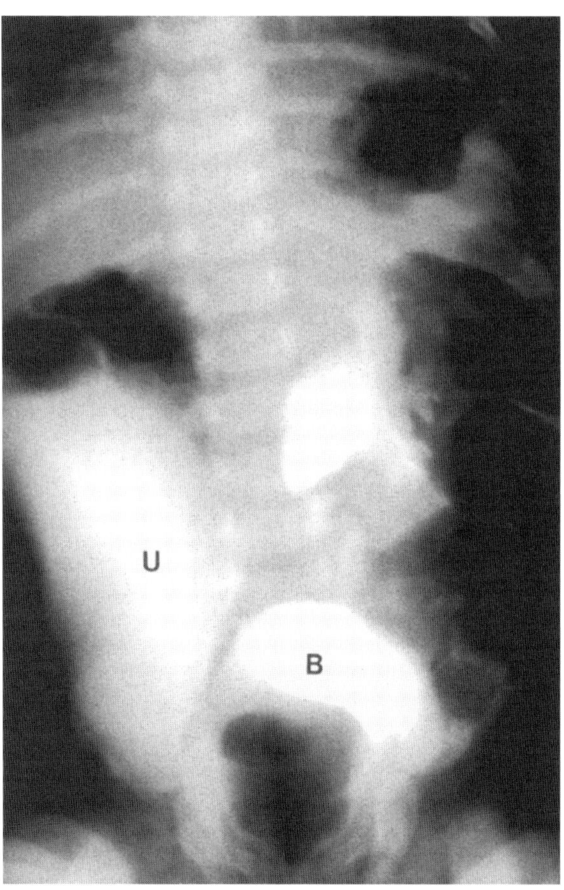

(B)

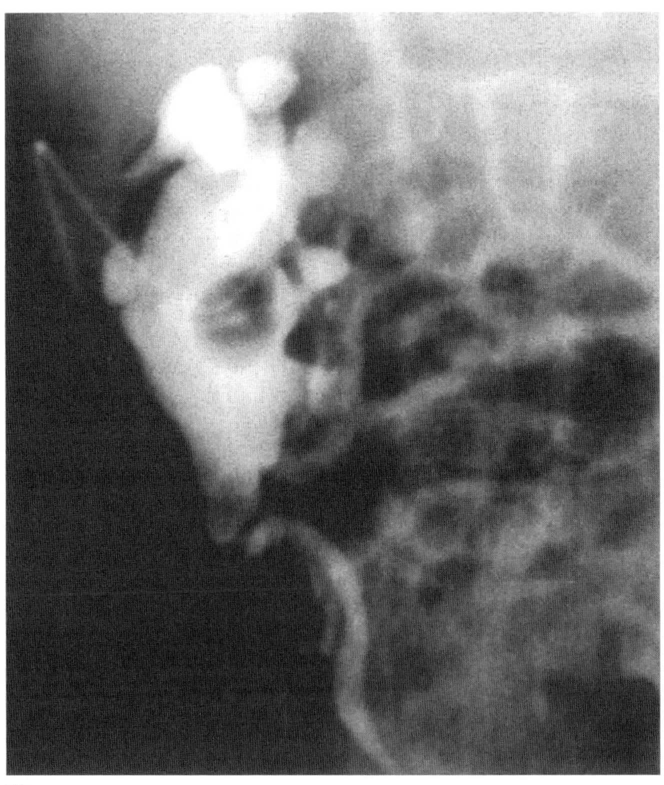

(C)

FIGURE 46 A–C ■ Cystic mass. (**A**) Ultrasound showing large cystic abdominal mass. (**B**) Postnatal urogram showing massively dilated renal pelvis and ureter. (**C**) Postnephrostomy showing decompression of the renal collecting system. M, mass; U, ureter.

TABLE 21 ■ Differentiating Hydroureter from Bowel

	Hydroureter	Bowel
Appearance	Anechoic, tubular	Anechoic, tubular
Kidneys and bladder	Often dilated	Usually normal
Location	Posterior	Anterior
Communication	Renal pelvis	Other loops
Peristalsis	Yes	Yes
Intraluminal particles	No	Yes
Gender	Usually male	Either

Source: From Ref. 186.

TABLE 22 ■ Causes of Megacystis

- Posterior urethral valves
- Urethral diaphragmatic membrane
- Urethral atresia (agenesis)
- Urethral stricture
- Persistent cloaca
- Megacystis–microcolon–intestinal hypoperistalsis syndrome

Source: From Ref. 186.

orifice (simple) or at an atypical orifice location at the bladder neck or urethra (ectopic; Fig. 47). Unlike most urinary tract disorders, ureterocele is nearly four times more common in females and is generally left sided. A duplicate renal collecting system is a commonly associated anomaly, typically with the upper pole obstructed in an ectopic ureterocele (148). Therefore, if focal hydronephrosis is detected, a careful search of the urinary bladder is necessary to exclude a ureterocele. The kidney on the side of the ureterocele tends to be significantly affected, often with dysplastic or multicystic changes.

The ultrasonographic diagnosis is fairly straightforward when the bladder is at least partially full. Under these circumstances, a cyst-like structure can be readily detected within the bladder. If the bladder is empty or only minimally distended, the diagnosis can often be overlooked. Therefore, it is important to evaluate the bladder over a prolonged period of time when upper urinary tract pathology is detected, to avoid missing a possible ureterocele.

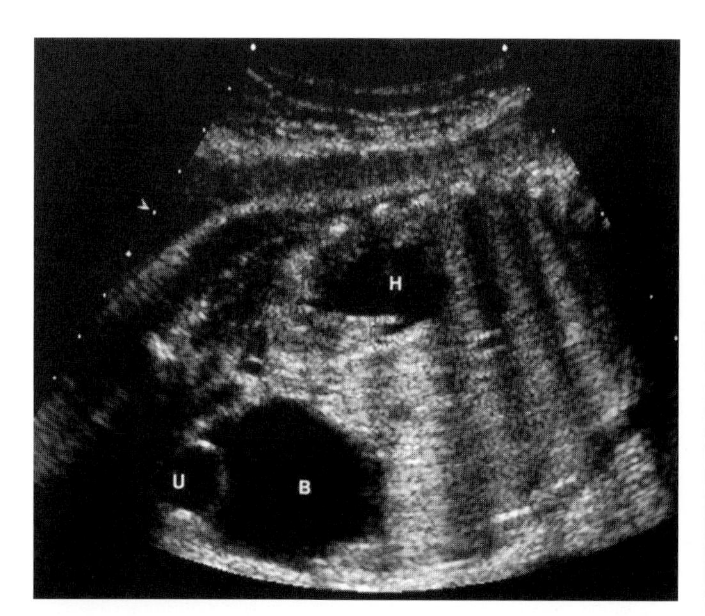

FIGURE 47 ■ Ureterocele. Parasagittal scan of the fetal abdomen demonstrating a prominent bladder with a ureterocele appearing within the bladder. There is hydronephrosis of the upper pole of the kidney. B, bladder; U, ureterocele; H, hydronephrosis.

Megacystis ■ The most likely cause of a persistently distended bladder is a urethral obstruction. Among the various causes, PUV is the most common, followed by a urethral membrane, urethral atresia, stricture, and persistent cloaca (Table 22). A nonobstructive cause for megalocystis, the megacystis–microcolon–intestinal hypoperistalsis syndrome (MMIH), is discussed separately later. Complete urethral obstruction, with its bilateral destructive effects and pulmonary hypoplasia secondary to the oligohydramnios, is potentially one of the most devastating fetal anomalies.

Classically, the diagnosis of PUV consists of a distended fetal bladder with a dilated proximal urethra (keyhole appearance) in a male fetus, often with oligohydramnios (Fig. 48) (149). If the bladder does not completely obscure more detailed abdominal examination, the kidneys and ureters display evidence of bilateral obstruction as well, in the form of hydroureter and hydronephrosis. If the obstruction is longstanding, the kidneys may actually be small and echogenic, consistent with dysplasia, an extremely poor prognostic sign. Once decompressed, either by spontaneous rupture or by percutaneous aspiration (vesicocentesis), the bladder usually displays a thickened muscular wall (more than 2 mm).

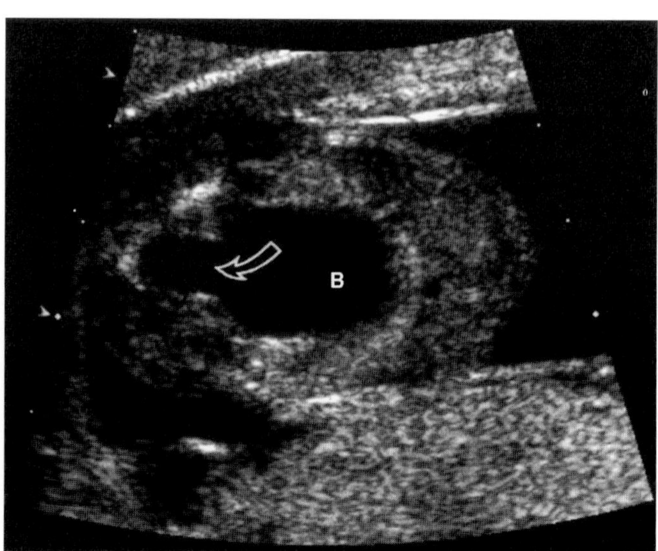

FIGURE 48 ■ Keyhole bladder. Typical keyhole (*arrow*) appearance to the (B) bladder secondary to posterior urethral valves.

Even though PUV and urethral membrane often cause an intermittent obstruction, urethral atresia with its complete obstruction has been detected in the first trimester (150). Bulic et al. (151) noted a 2-cm sonolucent mass in the fetal pelvis at as early as 11 weeks' gestation, providing strong support for early fetal urine production. More disturbing, however, was the fact that both cases showed serious renal lesions at pathologic examination despite delivery at 14 and 15 weeks. Although these data raise questions regarding the timing and feasibility of fetal therapy in this disorder, a similar case from our institution, with first-trimester diagnosis and in utero shunt placement at 14 weeks, had a successful outcome.

Fetal therapy for urethral obstruction has been a controversial subject of intense interest for more than a decade. Although much has been learned over that time, major questions remain about patient selection criteria, timing of intervention, and surgical technique. On the surface, it would appear straightforward to conclude that if urethral obstruction leads to renal dysplasia and pulmonary hypoplasia over time, then diverting the urine to the amniotic cavity through a timely vesicoamniotic shunt should prevent further renal damage and pulmonary hypoplasia. Data from the International Fetal Surgery Registry noted that nearly all neonatal deaths after vesicoamniotic shunt were secondary to pulmonary hypoplasia (151). As with most urinary tract anomalies, the presence or absence and duration of oligohydramnios are critical factors in the fetal prognosis. Mahony et al. (152) evaluated 40 bladder outlet obstructions to determine the important prognostic prenatal ultrasound features. More than half of survivors had a normal amniotic fluid volume, whereas only 7% survived with oligohydramnios. In contrast, 80% of those suffering a subsequent demise had decreased amniotic fluid, whereas only 12% had a normal volume. The only other predictor in their series appeared to be the degree of caliectasis. More than 70% of survivors had moderate to marked collecting system dilation, whereas 80% of demises had little or no caliectasis. Other authors have found the presence of renal cortical cysts to have a 100% specificity for renal dysplasia but less than 60% sensitivity (153). The presence of increased renal echogenicity has had a similar predictability.

Fetal urine analysis obtained by percutaneous aspiration has been used extensively to help predict renal function in utero (154,155). Normally, fetal urine is very hypotonic, but damage to the renal tubules from prolonged obstruction impairs renal resorption of certain electrolytes, leading to hypertonic urine. For this reason, measuring urine electrolytes has produced some useful cutoff values for prognosis. The thresholds suggestive of good renal function include: sodium less than or equal to 100 mEq/L, chloride less than or equal to 90 mEq/L, osmolality less than or equal to 210 mOsm/L, calcium less than 1.6 mmol/L, and urine output greater than 2 mL/hr. Although urine parameters above these values suggest a poor prognosis, other investigators (156,157) have challenged the predictive value of these thresholds. Measurement of a low-molecular-weight plasma protein, urinary β_2-microglobulin, has been touted as a reliable predictor of fetal renal function and candidacy for in utero intervention (155). In their series, Lipitz et al. (155) found that when β_2-microglobulin levels exceeded 13 mg/L, the outcome was uniformly dismal. The best prognosis was found with levels less than or equal to 1.5 mg/L, but survivors with various degrees of renal impairment were found with values between these two thresholds. Further data are required to determine the ultimate role of the various fetal urinary assessments.

The fresh fetal urine analysis, obtained by vesicocentesis, is combined with the ultrasound appearance of the kidneys and amniotic fluid volume to counsel the patient regarding the prognosis and options (Table 23). The need for a fresh urine sample from the fetal bladder cannot be emphasized enough. Johnson et al. (158) performed sequential vesicocenteses in 29 fetuses with complete urinary obstruction. Fetuses with good renal function had both patterns of decreasing hypertonicity and last specimens with urine analytes below the thresholds for favorable prognosis. Their data provide convincing evidence that analysis of urine that has resided in the fetal bladder for many days may not accurately depict the current status of renal function.

Once the patient is determined to be a good candidate for intervention, one fetal surgical choice is the percutaneous placement (under ultrasound guidance) of a pig-tailed catheter into the bladder, with the distal end on the anterior abdominal wall draining into the amniotic cavity. Among 98 vesicoamniotic shunt cases reported to the International Registry for Fetal Surgery, a 70% survival rate has been noted in patients with proven PUV syndrome (159). For candidates with normal urinary electrolytes and preoperative ultrasound findings, the survival rate may be even higher. The fact that 7 of 98 patients were subsequently found to have major chromosomal anomalies stresses the importance of case selection and restricting management to centers with ongoing experience in fetal therapy.

Problems with catheter placement include the potential for complications such as preterm labor, ruptured membranes, fetal trauma, infection, and, above all, shunt

TABLE 23 ■ Preoperative Prognostic Parameters for Fetuses with Posterior Urethral Valve Syndrome

Parameter	Favorable	Unfavorable
Urinary sodium	≤100 mEq/L	>100 mEq/L or 95% percentile for gestational age
Urinary chloride	≤90 mEq/L	>90 mEq/L or 95% percentile for gestational age
Urinary calcium	≤1.5 mmol/L	>1.5 mmol/L
Urine osmolality	≤210 mOsm/L	>210 mOsm/L
β_2-Microglobulin	<1.5 mg/L	>13 mg/L
Renal appearance	Normal, mild hydronephrosis	Echogenic, cystic
Amniotic fluid	Normal volume	Oligohydramnios

migration or occlusion. As a result of some of these concerns, Harrison et al. (146) performed a series of open vesicostomies through a hysterotomy incision in the late second trimester. Although this procedure avoids the technical problems associated with a foreign body, it carries the risks of major surgery and anesthesia as well as those associated with an open uterine cavity (preterm delivery, infection, fetal hypoxia, hypothermia, and premature rupture of membranes). Much more work is needed to evaluate the relative efficacy of these approaches.

Megacystic–Microcolon–Intestinal Hypoperistalsis Syndrome ■ MMIH is another cause of an enlarged bladder. Although MMIH is far less common than PUV, differentiating between these two distinct lesions is critical, especially if fetal therapy is considered. MMIH carries a dismal prognosis, with most neonates dying soon after birth. The syndrome involves not only an unobstructed distended bladder but also a dilated small bowel and distal microcolon. The key features differentiating megalocystis from urethral obstruction are the following:

Amniotic fluid is normal or increased in MMIH.

In MMIH, the fetus is usually female; in PUV, the fetus is almost always male.

Dilated small bowel is present with MMIH.

Although fetal surgery is of no benefit to the fetus with MMIH, the overdistended bladder may cause a soft tissue dystocia, which vesicocentesis before delivery may prevent.

Are Renal Cysts Present?

Among neonatal renal anomalies, cystic lesions of the kidney are second only to hydronephrosis and hydroureter in frequency of diagnosis (160). The various conditions are often confused, given the similarity of the names involved: polycystic, multicystic dysplastic, cystic dysplasia, and Potter's syndrome types I, II, III, and IV. This section attempts to clarify these anomalies (Table 24).

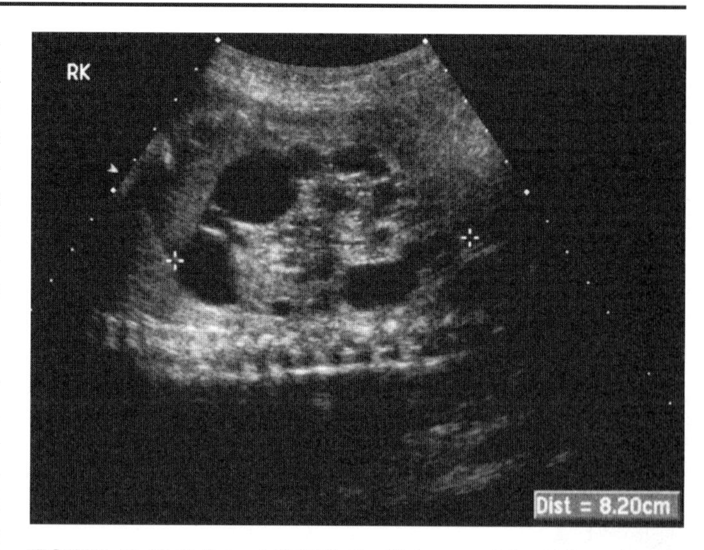

FIGURE 49 ■ Unilateral MCDK. Sagittal scan of the kidney demonstrating multiple cysts of variable size noted within this MCDK or Potter's syndrome type II. Note that the cysts are haphazardly arranged in this kidney and are of variable size, which helps to distinguish this from hydronephrosis. MCDK, multicystic dysplastic kidney.

MCDK (Potter's Syndrome Type II) ■ MCDK, or Potter's syndrome type II, usually represents the end result of an early, severe urinary obstruction, such as a ureterocele. This embryonic insult leaves a functionless kidney replaced by multiple cysts, both grossly and ultrasonographically (Figs. 49 and 50). The lesion is unilateral in 70% to 80% of patients, although 20% to 30% have a renal anomaly of the contralateral kidney (163). Bilateral processes such as bilateral MCDK or unilateral MCDK with

TABLE 24 ■ Classification of Major Renal Cystic Diseases (Potter's Syndromes)

	Type I	Type II	Type III	Type IV
Nomenclature	Infantile polycystic kidney disease	Multicystic dysplastic kidney disease (MCDK)	Adult polycystic kidney disease	Renal cystic disease associated with hydronephrosis
Location	Bilateral	Unilateral; 30% with contralateral renal disease including hydronephrosis, renal agenesis, or MCDK	Bilateral	Dependent on etiology of hydronephrosis
Ultrasound appearance	Enlarged echogenic	Cysts of variable size; noncommunicating	Rare in utero, enlarged echogenic	Hydronephrosis with cortical cysts, may change in appearance during pregnancy
Amniotic fluid	Often oligohydramnios	Normal to oligohydramnios (if bilateral)	Usually normal	Normal to oligohydramnios depending on underlying cause
Prognosis	Poor	Good, dependent on contralateral kidney and associated malformations	Adult onset	Dependent on underlying cause
Risk in subsequent pregnancies	25%	Probably <5%	50%	Dependent on underlying cause

Abbreviation: MCDK, multicystic dysplastic kidney disease.
Source: From Ref. 186.

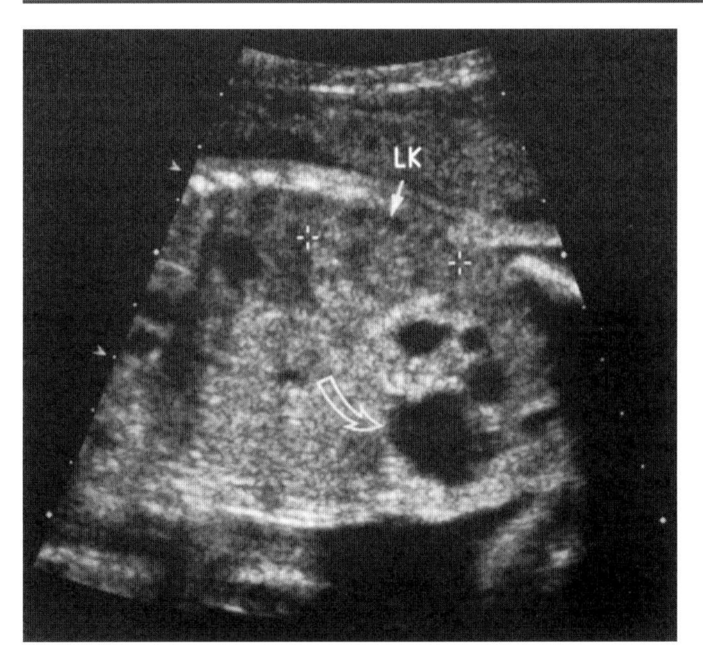

FIGURE 50 ■ Ectopic multicystic dysplastic kidney. Semicoronal scan showing normal left kidney and right multicystic dysplastic right kidney (*arrow*), located in fetal pelvis. LK, left kidney.

TABLE 25 ■ Differential Diagnosis of Severe UPJO and MCDK

	UPJO	MCDK
Renal parenchyma visible	Usually	No
Cyst characteristics	Oval, irregular; communicating with each other and the renal pelvis peripheral	Round; noncommunicating with each other or pelvis central
Ureteral dilation	Often	No
Contralateral kidney	10–40% with UPJO	40% abnormal; 20% bilateral; 10% agenesis; 10% with UPJO

Abbreviations: UPJO, ureteropelvic junction obstruction; MCDK, multicystic dysplastic kidney disease.
Source: From Ref. 186.

contralateral renal agenesis are invariably fatal to the fetus. Unilateral MCDK, however, most commonly occurs with some form of contralateral hydronephrosis and therefore has a much better prognosis. Even when the contralateral kidney appears normal, significant reflux can be found postnatally in about 20% of unilateral cases of MCDK (162). Given the lack of function of MCDK, the presence of oligohydramnios strongly suggests a serious contralateral renal anomaly. Unilateral renal agenesis occurs in 10% of cases, and a similar percentage of patients develop hydronephrosis. The corollary to this statement is also true, in that bilateral cystic disease of the kidneys in the presence of a normal amniotic fluid volume essentially precludes the diagnosis of complete bilateral MCDK. These patients may have a duplex collecting system with a functioning upper or lower pole system (163). The importance of obtaining postnatal renal scans and voiding cystourethrograms in such cases cannot be stressed enough.

Differentiating severe hydronephrosis from MCDK is sometimes less than straightforward (164). The key to the differential diagnosis (Table 25) is the cysts themselves (165); if they communicate and are of similar size, they most likely represent calyceal dilation of hydronephrosis. In MCDK, the cysts tend to be round and of varying size, occasionally giving the kidney the appearance of a cluster of grapes (Fig. 49). Although the appearance and size of MCDK may change markedly over time (166), the kidney generally loses its shape and is larger than normal, with no organizational pattern to the cysts or recognizable renal parenchyma.

Cystic Renal Dysplasia (Potter's Syndrome Type IV) ■ Cystic renal dysplasia, or Potter's syndrome type IV, is the consequence of an obstructive uropathy occurring in the second or third trimester of pregnancy. This contrasts with MCDK, which results from an early first-trimester insult. A PUV obstruction frequently leads to this lesion, and given the lack of renal function of such dysplastic kidneys, it is critical to recognize them before considering in utero intervention. Less commonly, UPJ and UVJ obstructive uropathies can also lead to cystic renal dysplasia.

The ultrasonographic appearance is that of parenchymal echogenicity and subcapsular cysts. These cysts may change in size and appearance during pregnancy. The diagnosis is not always so straightforward, however, because the kidneys can be dysplastic without visible cysts and with normal-appearing parenchyma (Table 24).

Renal Cysts ■ Renal cysts can be seen on occasion in the absence of other obvious genitourinary anomalies. Often, they are the result of an anomalous duplication of the collecting system with a nonfunctional pole. More often, however, the correct diagnosis relates to an obstructive uropathy or cystic disease, as discussed earlier.

Are the Kidneys Abnormal in Size or Echogenicity?

Infantile Polycystic Kidney Disease (Potter's Syndrome Type I) ■ Infantile polycystic kidney disease, or Potter's syndrome type I, is an autosomal recessive disorder with an incidence of 1 in 40,000 births. From the ultrasonographer's point of view, the name is a misnomer in that the kidneys generally do not appear cystic when imaged (167). The disease is characterized by microscopic cysts, but ultrasonographically, the kidneys appear as bilaterally enlarged echogenic masses in association with oligohydramnios and nonvisualization of the bladder (Fig. 51; Table 24). In fetuses at risk on the basis of family history, it is important to initiate ultrasonographic evaluation by 16 weeks of gestation because the enlarged kidneys may be diagnosed early in some cases, particularly in those known to be at risk (168).

The disease carries a dismal prognosis owing to the poor renal function and pulmonary hypoplasia that can occur. Several different subtypes of the disorder are now recognized, however, each with its own distinct presentation and prognosis. The foregoing discussion relates to the perinatal variety, which carries the worst prognosis and is unfortunately the most common. In the neonatal subtype, an in utero diagnosis can sometimes be suggested based on enlarged, echogenic kidneys; there often is hepatic fibrosis (probably not ultrasonographically detectable), usually without pulmonary hypoplasia, and a relatively normal amniotic fluid volume. Renal failure may be delayed for several weeks after birth, and death typically occurs within the first year. The later-appearing subtypes have progressively more benign clinical courses from the renal standpoint, with increasingly severe hepatic involvement. Now that the gene for infantile polycystic kidney disease has been isolated (chromosomal region 6p21-cen) (169), the importance of genetic consultation for suspected cases by history or ultrasonographic appearance must be stressed because earlier and more precise prenatal diagnosis is often possible.

Patients with infantile polycystic kidney disease must have a detailed anatomic survey, with special scrutiny of the cranium for an encephalocele, of the hands and feet for polydactyly, and for the triad of Meckel–Gruber syndrome, another autosomal recessive condition.

Adult Polycystic Kidney Disease (Potter's Syndrome Type III) ■ Adult polycystic kidney disease, or Potter's syndrome type III, is an autosomal dominant condition that is only rarely reported prenatally (170). Most cases are diagnosed in early adult life. In families at risk, a fetus with symmetrically large, echogenic kidneys, similar in appearance to those of infantile polycystic kidney disease, should raise a suspicion of Potter's syndrome type III. On occasion, the kidneys may appear cystic (Table 24).

Renal Fossa Mass ■ Fetal renal tumors usually present ultrasonographically as a solid mass in the paraspinous area or renal fossa. The most common tumors are mesoblastic nephroma (or leiomyomatous hamartoma; Fig. 52) and, less frequently, Wilms' tumor (or nephroblastoma). The two are indistinguishable by current ultrasonographic

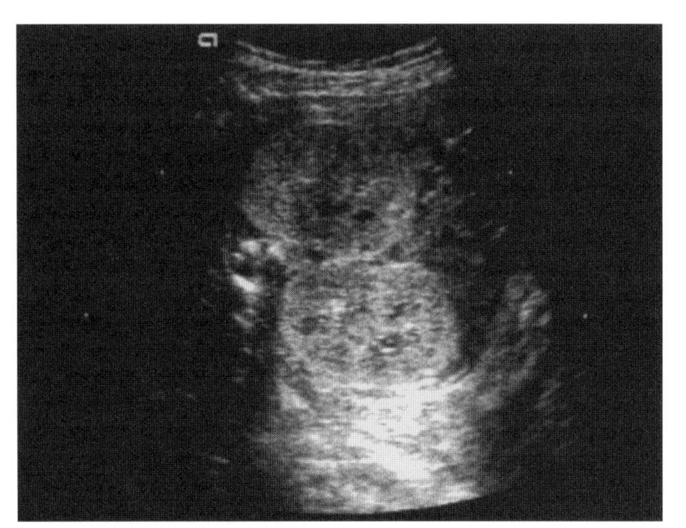
(A)

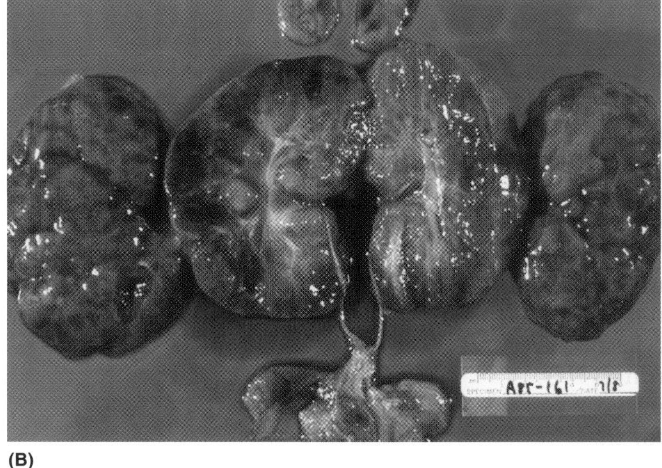
(B)

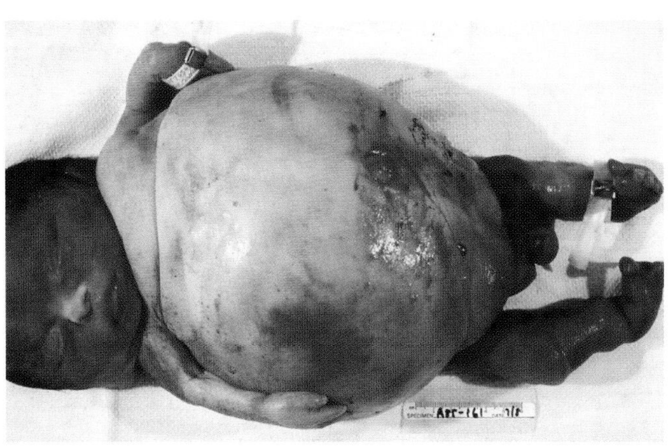
(C)

FIGURE 51 A–C ■ Infantile polycystic kidney disease. (**A**) Transverse ultrasound demonstrating echogenic, bilaterally enlarged kidneys. (**B**) Postmortem exam showed bilaterally enlarged kidneys. (**C**) In another fetus, there is massive enlargement of the fetal abdomen owing to the enlarged kidneys.

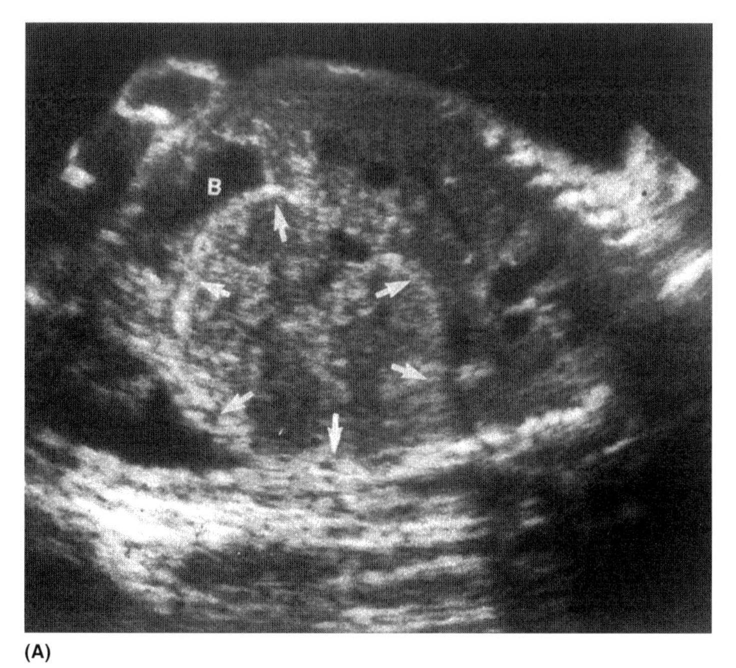

(A)

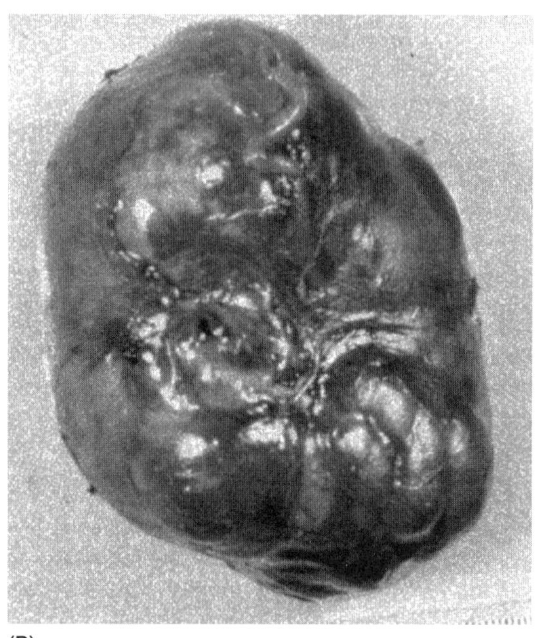

(B)

FIGURE 52 A AND B ■ Mesoblastic nephroma. (**A**) Sagittal ultrasound of the fetus showing echogenic mass (*arrows*) arising from the left renal fossa. (**B**) After delivery, a well-encapsulated mass, which corresponded to a mesoblastic nephroma, was removed from the left perinephric space. B, bladder. *Source*: From Ref. 170.

techniques (171). Although the ultrasonographic distinction between the two lesions cannot be made, their prognoses are dramatically different. Mesoblastic nephroma is a benign entity, in contrast to Wilms' tumor, which is a malignant lesion. The prognosis for Wilms' tumor varies with the grade and extent of the lesion.

The differential diagnosis for a unilateral solid mass in the renal fossa includes the aforementioned renal lesions as well as an adrenal neuroblastoma and adrenal hemorrhage (Table 26). The mesoblastic nephromas reported have consistently been associated with polyhydramnios (171–173), although the mechanism for this finding is not readily apparent.

Adrenal neuroblastoma is the most common adrenal lesion in prenatal and neonatal life. The usual ultrasonographic appearance is of a mixed cystic and solid lesion, occasionally with calcifications as well. Although adrenal neuroblastomas are most commonly found in the superior renal fossa area, they can also be metastatic lesions in a variety of locations. Besides the usual location, the differential diagnosis can be aided by the endocrine nature of these lesions. Because they usually produce catecholamines, maternal symptoms may support the ultrasonographic diagnosis. We found amniotic fluid catecholamines to be markedly elevated in a recent case; thus, amniocentesis may also aid in the prenatal diagnosis.

Adrenal hemorrhage can have a variable ultrasonographic appearance similar to that of an adrenal or renal neoplasm (174,175). It can present as prominent sonolucent or mixed suprarenal masses. Later in their evolution, these masses may present with calcifications. The key to the diagnosis is the evolution of the lesion over time. Postnatal confirmation may be difficult because they are typically benign lesions that resolve spontaneously, but urinary catecholamines should be normal. Fetal risks, as with neonatal cases, are related to the underlying cause: hypoxia, sepsis, renal vein thrombosis, or blood dyscrasias (174). Rare masses in the renal region is an extra lobar pulmonary sequestration (Fig. 53).

TABLE 26 ■ Differential Diagnosis of Renal Fossa Mass

- Mesoblastic nephroma
- Adrenal neuroblastoma
- Adrenal hemorrhage
- Wilms' tumor
- Retroperitoneal teratoma

Source: From Ref. 186.

FETAL GENDER

Ultrasound can be used to assess fetal gender. In a study in the early 1980s, about one third of all fetuses could not be sexed because of fetal position (176). One third were correctly identified as male and one third correctly identified as female (Fig. 54). Although enhanced image resolution has improved the diagnostic

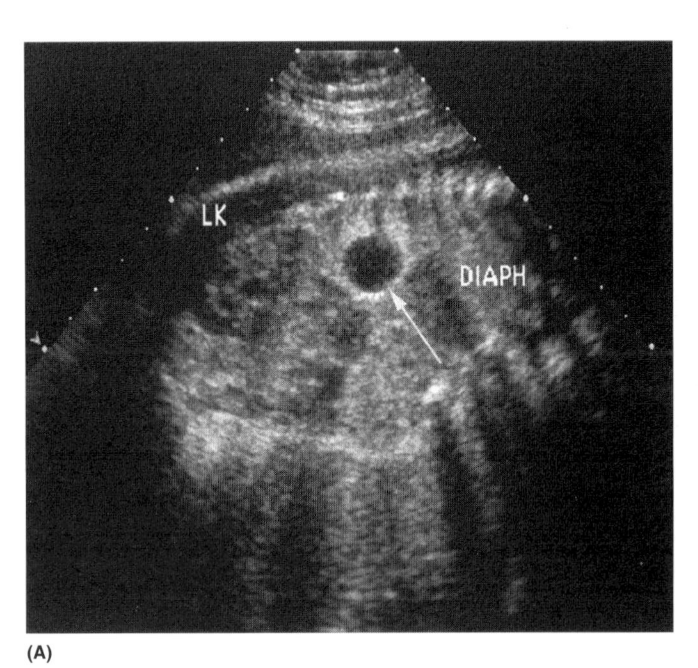

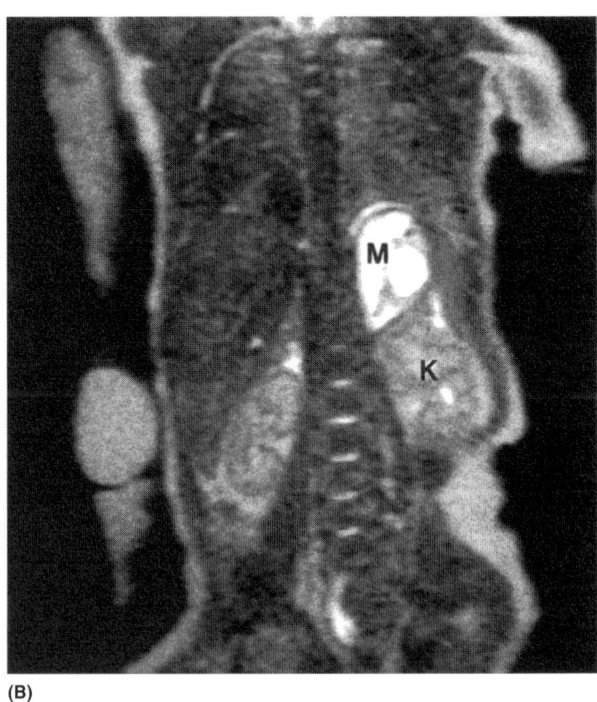

(A) (B)

FIGURE 53 A AND B ■ Subdiaphragmatic pulmonary sequestration. (**A**) Ultrasound showing solid mass with central cyst above left kidney and below diaphragm. (**B**) Postnatal T-2 weighted MRI showing this mass which corresponded to subdiaphragmatic pulmonary sequestration. LK, left kidney; M, mass.

accuracy of gender identification in the late first and early second trimesters, incorrect gender assignment is still a common pitfall. This inaccuracy often is due to the fact that the labia and clitoris are fairly prominent at this stage of pregnancy and may be misinterpreted as male genitalia.

Hydrocele is commonly observed during third-trimester ultrasonographic examinations. Most of these have no clinical significance; some may occur secondary to intra-abdominal ascites that has tracked into the scrotum (Fig. 55). In either case, there is usually little significance to the finding of a fetal hydrocele in a male fetus.

A number of reports have been made of successful prenatal diagnoses of hypospadias and ambiguous genitalia (177,178). These cases invariably involve third-trimester diagnoses, usually in cases at particular risk because of associated anomalies or risks. Given the frequency of hypospadias (1 in 250 live births) and the rarity

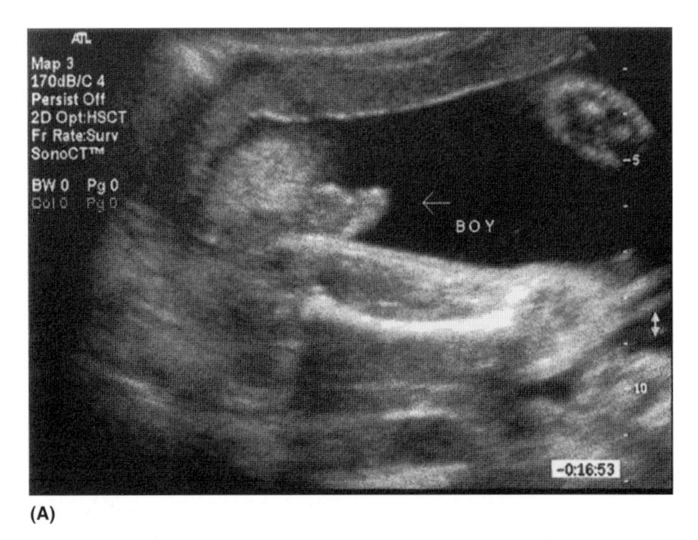

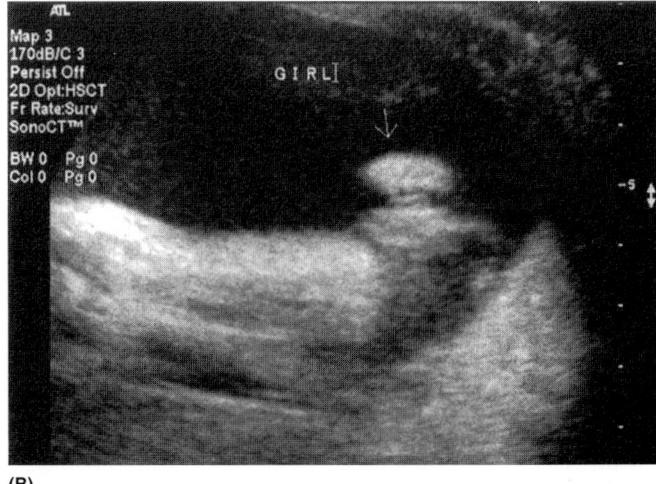

(A) (B)

FIGURE 54 A AND B ■ (**A**) Normal male genitalia. (**B**) Normal female genitalia (*arrows* point to labia) at 23 weeks' gestation.

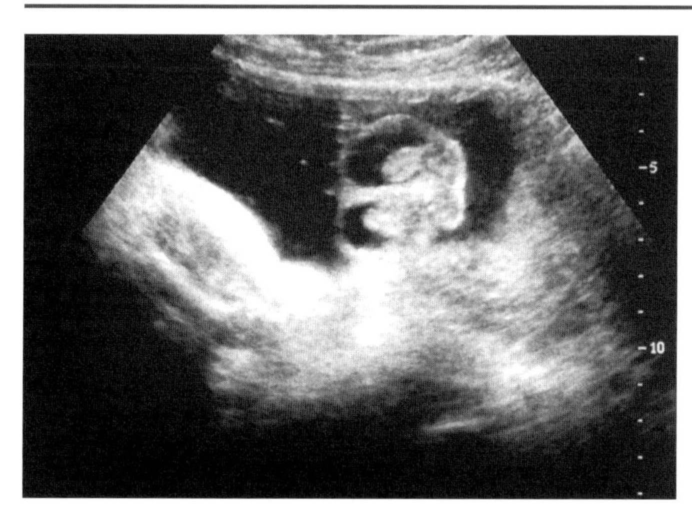

FIGURE 55 ■ Scan of the scrotal sac demonstrating bilateral hydroceles.

with which its prenatal diagnosis has been reported, it is clear that most cases will continue to go undiagnosed prenatally.

REFERENCES

1. Carpenter MW, Curci MR, Dibbins AW, Haddow JE. Perinatal management of ventral wall defects. Obstet Gynecol 1984; 64:646–651.
2. Lindham S. Omphalocele and gastroschisis in Sweden: 1965–1976. Acta Pediatr Scand 1981; 70:55–60.
3. Baird PA, MacDonald EC. An epidemiologic study of congenital malformations in the anterior abdominal wall in more than a half million consecutive live births. Am J Hum Genet 1981; 33: 470–78.
4. Mann L, Ferguson-Smith MA, Desai M, Gibson AA, Raine PA. Prenatal assessment of anterior wall defects and their prognosis. Prenat Diagn 1984; 4:427–435.
5. Engel RM. Exstrophy of the bladder and associated anomalies. Birth Defects 1974; 10:146–149.
6. Soper RT, Kilger K. Vesico-intestinal fissure. J Urol 1964; 92: 490–501.
7. Brock JH, Barron L, Duncan P. Significance of elevated mid-trimester maternal plasma-alpha-fetoprotein values. Lancet 1979; 1:1281–1282.
8. Chi LH, Stone DH, Gilmour WH. Impact of prenatal screening and diagnosis on the epidemiology of structural congenital anomalies. J Med Screen 1995; 2:67–70.
9. Torfs CP, Katz EA, Bateson TF, Lam PK, Carry CJ. Maternal medications. Tetralogy 1996; Aug 54(2):84–92.
10. California Birth Defects Monitoring Program, 2006.
11. Langer JC, Brennan B, Lappalainen RE, et al. Cloacal exstrophy: prenatal diagnosis before rupture of cloacal membrane. J Pediatr Surg 1992; 27:1352–1355.
12. Potter C, ed. Diaphragmatic and abdominal hernias. In: Pathology of the Fetus and the Infant. 3rd ed. Chicago: Year Book Medical Publishers, 1975:384.
13. Higginbottom MC, Jones KL, Hall BD, Smith DW. The amniotic band disruption complex: timing of amniotic rupture and variable spectra of consequent defects. J Pediatr 1979; 95:544–549.
14. Moore KL. The digestive system. In: The Developing Human. 4th ed. Philadelphia: WB Saunders, 1988.
15. Hoyme HE, Higginbottom MC, Jones KL. The vascular pathogenesis of gastroschisis: intrauterine interruption of the omphalomesenteric artery. J Pediatr 1981; 98:228–231.
16. Jones KL. Exstrophy of cloaca sequence. In: Smith's Recognizable Patterns of Human Malformation. 4th ed. Philadelphia: WB Saunders, 1988:567.
17. AIUM Practice Guideline for the performance of an antepartum obstetric ultrasound examination. J Ultrasound Med 2003; 22:1116–1125.
18. Ultrasound in Pregnancy: ACOG Practice Bulletin Number 58 (59), 2004, ACOG.
19. DeVeciana M, Major CA, Porto M. Prediction of an abnormal karyotype in fetuses with omphalocele. Prenat Diagn 1994; 14; 487–492.
20. Kushnir O, Izquierdo L, Vigil D, Curet LB. Early transvaginal sonographic diagnosis of gastroschisis. J Clin Ultrasound 1990; 18:194–197.
21. Gray DL, Martin CM, Crane JP. Differential diagnosis of first trimester ventral wall defect. J Ultrasound Med 1989; 8:255–258.
22. Blazer S, Zimmer EZ, Gover A, Bronshtein M. Fetal omphalocele detected early in pregnancy: associated anomalies and outcomes. Radiology 2004; 232:191–195.
23. Casola G, Scheible W, Leopold GR. Large umbilical cord: a normal finding in some fetuses. Radiology 1985; 156:181–182.
24. Ramanathan K, Epstein S, Yaghoobian J. Localized deposition of Wharton's jelly: sonographic findings. J Ultrasound Med 1986; 5:339–340.
25. Cantrell JR, Haller JA, Ravitch MM. A syndrome of congenital defects involving the abdominal wall, sternum, diaphragm, pericardium, and heart. Surg Gynecol Obstet 1958; 107:602–614.
26. Schwaitzberg SD, Pokorny WJ, McGill CW, Harberg FJ. Gastroschisis and omphalocele. Am J Surg 1982; 144:650–654.
27. Argyle JC. Pulmonary hypoplasia in infants with giant abdominal wall defects. Pediatr Pathol 1989; 9:43–55.
28. Lenke RR, Persutte WH, Nemes J. Ultrasonographic assessment of intestinal damage in fetuses with gastroschisis: is it of clinical value? Am J Obstet Gynecol 1990; 163:995–998.
29. Aina-Mumuney AJ, Fischer AC, Blakemore KJ, et al. A dilated fetal stomach predicts a complicated postnatal course in cases of prenatally diagnosed gastroschisis. Am J Obstet Gynecol 2004; 190: 1326–1330.
30. Lockwood CL, Scioscia AL, Hobbins JC. Congenital absence of the umbilical cord resulting from maldevelopment of embryonic body folding. Am J Obstet Gynecol 1986; 155:1049–1051.
31. Shalev E, Eliyahu S, Battino S, Weiner E. First trimester transvaginal sonographic diagnosis of body stalk anomaly [correction of anatomy] J Ultrasound Med 1995; 14:641–642.
32. Mirk P, Calisti A, Fileni A. Prenatal sonographic diagnosis of bladder exstrophy. J Ultrasound Med 1986; 5:291–293.
33. Goldstein I, Shalev E, Nisman D. The dilemma of prenatal diagnosis of bladder exstrophy: a case report and a review of the literature. Ultrasound Obstet Gynecol 2001; 17:357–359.
34. Bronshtein M, Bar-Hava I, Blumenfeld Z. Differential diagnosis of the nonvisualized fetal urinary bladder by transvaginal sonography in the early second trimester. Obstet Gynecol 1993; 82:490–493.
35. Messelink EJ, Aronson DC, Knuist M, Heij HA, Vos A. Four cases of bladder exstrophy in two families. J Med Genet 1994; 31: 490–492.
36. Gearhart JP, Ben-Chaim J, Jeffs RD, Sanders RC. Criteria for the prenatal diagnosis of classic bladder exstrophy. Obstet Gynecol 1995; 85:961–964.
37. Froster UG, Heinritz W, Bennek J, Horn LC, Faber R. Another case of autosomal dominant exstrophy of the bladder. Prenat Diagn 2004; 24:375–377.
38. Petrikovsky BM, Walzak MP, D'Addario PF. Fetal cloacal anomalies: prenatal sonographic findings and differential diagnosis. Obstet Gynecol 1988; 72:464–469.
39. Palomaki GE, Hill LE, Knight GJ, Haddow JE, Carpenter M. Second-trimester maternal serum alpha-fetoprotein levels in pregnancies associated with gastroschisis and omphalocele. Obstet Gynecol 1988; 71:906–909.
40. Merkatz IR, Nitowshy HM, Macri JN, Johnson WE. An association between low maternal serum alpha-fetoprotein and fetal chromosomal abnormalities. Am J Obstet Gynecol 1984; 148:886–894.
41. Lindenbaum RH, Ryynanen M, Holmes-Siedle M, Puhakainen E, Jonasson J, Keenan J. Trisomy 18 and maternal serum and amniotic fluid alpha-fetoprotein. Prenat Diagn 1987; 7:511–519.

42. Gorczyca DP, Lindfors KK, McGahan JP, Hanson FW. Limb-body-wall complex: another cause for elevated maternal serum alpha-fetoprotein. J Clin Ultrasound 1990; 18:198–201.
43. Gosden C, Brock DJ. Prenatal diagnosis of exstrophy of the cloaca. Am J Med Genet 1981; 8:95–109.
44. Hesser JW, Murata Y, Swalwell CI. Exstrophy of cloaca with omphalocele: two cases. Am J Obstet Gynecol 1984; 150:1004–1006.
45. Gilbert WM, Nicolaides KH. Fetal omphalocele: associated malformations and chromosomal defects. Obstet Gynecol 1987; 70:633–635.
46. Nyberg DA, Fitzsimmons J, Mack LA, et al. Chromosomal abnormalities in fetuses with omphalocele: significance of omphalocele contents. J Ultrasound Med 1989; 8:299–308.
47. Benacerraf BR, Saltzman DH, Estroff JA, Frigoletto FD. Abnormal karyotype of fetuses with omphalocele: prediction based on omphalocele contents. Obstet Gynecol 1990; 75:317–319.
48. Gibbin C, Touch S, Broth RE, Berghella V. Abdominal wall defects and congenital heart disease. Ultrasound Obstet Gynecol 2003; 21:334–337.
49. Hughes MD, Nyberg DA, Mack LA, Pretorius DH. Fetal omphalocele: prenatal detection of concurrent anomalies and other predictors of outcome. Radiology 1989; 173:371–376.
51. Moore TC. Gastroschisis and omphalocele: clinical differences. Surgery 1977; 82:561–568.
52. Pelizzo G, Maso G, Dell'Oste C, et al. Giant omphaloceles with a small abdominal defect: prenatal diagnosis and neonatal management. Ultrasound Obstet Gynecol 2005; 26:786–788.
53. Pryde PG, Bardicef M, Treadwell MC, Klein M, Isada NB, Evans MI. Gastroschisis: can antenatal ultrasound predict infant outcomes? Obstet Gynecol 1994; 84:505–510.
54. Strauss RA, Balu R, Kuller JA, McMahon MJ. Gastroschisis: the effect of labor and ruptured membranes on neonatal outcome. Am J Obstet Gynecol 2003; 189:1672–1678.
55. Sakala EP, Erhard LN, White JJ. Elective cesarean section improves outcomes of neonates with gastroschisis. Am J Obstet Gynecol 1993; 169(4):1050–1053.
56. Quirk JG Jr, Fortney J, Collins HB II, West J, Hassad SJ, Wagner C. Outcomes of newborns with gastroschisis: the effects of mode of delivery, site of delivery, and interval from birth to surgery. Am J Obstet Gynecol 1996; 174:1134–1138.
57. Segel SY, Marder SJ, Parry S, Macones GA. Fetal abdominal wall defects and mode of delivery: a systematic review. Obstet Gynecol 2001; 98:867–873.
58. Vegunta RK, Wallace LJ, Leonardi MR, et al. Perinatal management of gastroschisis: analysis of a newly established clinical pathway. J Pediatr Surg 2005; 40:528–534.
59. Ginsberg NE, Cadkin A, Strom C. Prenatal diagnosis of body stalk anomaly in the first trimester of pregnancy. Ultrasound Obstet Gynecol 1997; 10:419–421.
60. Morrow RJ, Whittle MJ, McNay MB, Raine PA, Gibson AA, Crossley J. Prenatal diagnosis and management of anterior abdominal wall defects in the west of Scotland. Prenat Diagn 1993; 13:111–115.
61. Carey JC, Greenbaum B, Hall BD. The OEIS complex (omphalocele, exstrophy, imperforate anus, spinal defects). Birth Defects Orig Artic Ser 1978; 14:253–263.
62. Wu JL, Fang KH, Yeh GP, Chou PH, Hsieh CT. Using color Doppler sonography to identify the perivesical umbilical arteries: a useful method in the prenatal diagnosis of omphalocele-exstrophy-imperforate anus-spinal defects complex. J Ultrasound Med 2004; 23:1211–1215.
63. Persutte WH, Lenke RR, Kropp K, Ghareeb C. Antenatal diagnosis of fetal patent urachus. J Ultrasound Med 1988; 7:399–403.
64. Rosenberg JC, Chervenak FA, Walker BA, Chitkara U, Berkowitz RL. Antenatal sonographic appearance of omphalomesenteric duct cyst. J Ultrasound Med 1986; 5:719–720.
65. Skibo LK, Lyons EA, Levi CS. First trimester umbilical cord cysts. Radiology 1992; 182:719–722.
66. McCalla CO, Lajinian S, DeSouza D, Rottem S. Natural history of antenatal omphalomesenteric duct cyst. J Ultrasound Med 1995; 14:639–640.
67. Bonilla-Musoles F, Machado LE, Bailao LA, Osborne NG, Raga F. Abdominal wall defects: two- versus three-dimensional ultrasonographic diagnosis. J Ultrasound Med 2001; 20:379–389.
68. Solerte L. Three-dimensional multiplanar ultrasound in a limb-body wall complex fetus: clinical evidence for counseling. J Matern Fetal Neonatal Med 2006; 19:109–112.
69. Souka AP, Pilalis A, Kavalakis Y, Kosmas Y, Antsaklis P, Antsaklis A. Assessment of fetal anatomy at the 11–14-week ultrasound examination. Ultrasound Obstet Gynecol 2004; 24:730–734.
70. Fakhry J, Shapiro LR, Schecter A, Weingarten M, Glennon A. Fetal gastric pseudomasses. J Ultrasound Med 1987; 6:177–180.
71. Hashimoto BE, Filly RA, Callen PW. Fetal pseudoascites: further anatomic observations. J Ultrasound Med 1986; 5:151–152.
72. Benacerraf BR, Frigoletto FD, Jr. Sonographic sign for the detection of early fetal ascites in the management of severe isoimmune disease without intrauterine transfusion. Am J Obstet Gynecol 1985; 153:635 and 1985; 152:1039–1041.
73. Salomon LJ, Baumann C, Delezoide AL, Oury JF, Pariente D, Sebag G, Garel C. Abnormal abdominal situs: what and how should we look for? Prenat Diagn 2006; 26:282–285.
74. Sheley RC, Nyberg DA, Kapur R. Azygous continuation of the interrupted inferior vena cava: a clue to prenatal diagnosis of the cardiosplenic syndromes. J Ultrasound Med 1995; 14:381–387.
75. Hinds R, Davenport M, Mieli-Vergani G, Hadzic N. Antenatal presentation of biliary atresia. J Pediatr 2004; 144:43–46.
76. Salomon LJ, Baumann C, Delezoide AL, et al. Abnormal abdominal situs: what and how should we look for? Prenat Diagn 2006; 26:282–285.
77. Pretorius DH, Drose JA, Dennis MA, Manchester DK, Manco-Johnson ML. Tracheoesophageal fistula in utero: 22 cases. J Ultrasound Med 1987; 6:509–513.
78. Borsellino A, Zaccara A, Nahom A, et al. False-positive rate in prenatal diagnosis of surgical anomalies. J Pediatr Surg 2006; 41:826–829.
79. Holder TM, Cloud DT, Lewis JE Jr, Pilling GP. Esophageal atresia and tracheoesophageal fistula: a survey of its members by the surgical section of the American Academy of Pediatrics. Pediatrics 1964; 34:542–549.
80. Brantberg A, Blaas HG, Salvesen KA, Haugen SE, Mollerlokken G, Eik-Nes SH. Fetal duodenal obstructions: increased risk of prenatal sudden death. Ultrasound Obstet Gynecol 2002; 20:439–446.
81. Iacobelli BD, Zaccara A, Spirydakis I, et al. Prenatal counselling of small bowel atresia: watch the fluid! Prenat Diagn 2006; 26:214–217.
82. Estroff JA, Parad RB, Share JC, Benacerraf BR. Second trimester prenatal findings in duodenal and esophageal atresia without tracheoesophageal fistula. J Ultrasound Med 1994; 13:375–379.
83. Petrikovsky BM. First-trimester diagnosis of duodenal atresia. Am J Obstet Gynecol 1994; 171:569–570.
84. Lawrence MJ, Ford WD, Furness ME, Hayward T, Wilson T. Congenital duodenal obstruction: early antenatal ultrasound diagnosis. Pediatr Surg Int 2000; 16:342–345.
85. Haeusler MC, Berghold A, Stoll C, Barisic I, Clementi M; EUROSCAN Study Group. Prenatal ultrasonographic detection of gastrointestinal obstruction: results from 18 European congenital anomaly registries. Prenat Diagn 2002; 22:616–623.
86. Vermesh M, Mayden KL, Confino E, Giglia RV, Gleicher N. Prenatal sonographic diagnosis of Hirschprung's disease. J Ultrasound Med 1986; 5:37–39.
87. Robinow M, Shaw A. The McKusick-Kaufman syndrome: recessively inherited vaginal atresia, hydrometrocolpos, utero-vaginal duplications, anorectal anomalies, postaxial polydactyly and congenital heart disease. J Pediatr 1979; 94:776–778.
88. Guzman ER, Ranzini A, Day-Salvatore D, Weinberger B, Spigland N, Vintzileos A. The prenatal ultrasonographic visualization of imperforate anus in monoamniotic twins. J Ultrasound Med 1995; 14:547–551.
89. Harris RD, Nyberg DA, Mack LA, Weinberger A. Anorectal atresia: prenatal sonographic diagnosis. AJR Am J Roentgenol 1987; 149:395–400.
90. Caspi N, Elchalal U, Lancet U, Chemke J. Prenatal diagnosis of cystic fibrosis: ultrasonographic appearance of meconium ileus in the fetus. Prenat Diagn 1988; 8:379–382.
91. Goldstein RB, Filly RA, Callen PW. Sonographic diagnosis of meconium ileus in utero. J Ultrasound Med 1987; 6:663–666.

92. Nyberg DA, Hastrup W, Watts H, Mack LA. Dilated fetal bowel: a sonographic sign of cystic fibrosis. J Ultrasound Med 1987; 6:257–260.
93. McGahan JP, Hanson F. Meconium peritonitis with accompanying pseudocyst: prenatal sonographic diagnosis. Radiology 1983; 148:125–126.
94. Chan KL, Tang MH, Tse HY, Tang RY, Tam PK. Meconium peritonitis: prenatal diagnosis, postnatal management and outcome. Prenat Diagn 2005; 25:676–682.
95. Dillard JP, Edwards DU, Leopold GR. Meconium peritonitis masquerading as fetal hydrops. J Ultrasound Med 1987; 6:49–51.
96. Heydenrych JJ, Marcus PB. Meconium granuloma of the tunica vaginalis. J Urol 1976; 115:596–598.
97. Finkel LI, Slovis TL. Meconium peritonitis, intraperitoneal calcifications, anmd cystic fibrosis. Pediatr Radiol 1982; 12:92–93.
98. Foster MA, Nyberg DA, Mahony BS, Mack LA, Marks WM, Raabe RD. Meconium peritonitis: prenatal sonographic findings and clinical significance. Radiology 1987; 165:661–665.
99. Nyberg DA, Dubinsky T, Resta RG, Mahony BS, Hickok DE, Luthy DA. Echogenic fetal bowel during the second trimester: clinical importance. Radiology 1993; 188:527–531.
100. Fakhry J, Reiser M, Shapiro LR, Schechter A, Pait LP, Glennon A. Increased echogenicity in the lower fetal abdomen: a common normal variant in the second trimester. J Ultrasound Med 1986; 5:489–492.
101. Benacerraf B, Chaudhury AK. Echogenic fetal bowel in the third trimester associated with meconium ileus secondary to cystic fibrosis. J Reprod Med 1989; 34:299–300.
102. Dicke JM, Crane JP. Sonographically detected hyperechoic fetal bowel: significance and implication for pregnancy management. Obstet Gynecol 1992; 80:778–782.
103. Scioscia AL, Pretorius DH, Budorick NE, Cahill TC, Axelrod FT, Leopold GR. Second-trimester echogenic bowel and chromosomal abnormalities. Am J Obstet Gynecol 1992; 167:889–894.
104. Bromley B, Doubilet P, Frigoletto FD Jr., Krauss C, Estroff JA, Benacerraf BR. Is fetal hyperechoic bowel on second-trimester sonogram an indication for amniocentesis? Obstet Gynecol 1994; 83:647–651.
105. Weiner Z. Congenital cytomegalovirus infection with oligohydramnios and echogenic bowel at 14 weeks' gestation. J Ultrasound Med 1995; 14:617–618.
106. Petrikovsky B, Smith-Levitin M, Holsten N. Intra-amniotic bleeding and fetal echogenic bowel. Obstet Gynecol 1999; 93:684–686.
107. Achiron R, Seidman DS, Horowitz A, Mashiach S, Goldman B, Lipitz S. Hyperechoic fetal bowel and elevated serum alpha-fetoprotein: a poor fetal prognosis. Obstet Gynecol 1996; 88:369–371.
108. Comparetto C, Giudici S, Coccia ME, Scarselli G, Borruto F. Fetal and neonatal ovarian cysts: what's their real meaning? Clin Exp Obstet Gynecol 2005; 32:123–125.
109. Mahony BS, McGahan JP, Nyberg DA, Reisner DP. Varix of the fetal intra-abdominal umbilical vein: comparison with normal. J Ultrasound Med 1992; 11:73–76.
110. Fung TY, Leung TN, Leung TY, Lau TK. Fetal intra-abdominal umbilical vein varix: what is the clinical significance? Ultrasound Obstet Gynecol 2005; 25:149–154.
111. McNamara A, Levine D. Intraabdominal fetal echogenic masses: a practical guide to diagnosis and management. Radiographics 2005; 25:633–645.
112. Ji EK, Lee EK, Kwon TH. Isolated echogenic foci in the left upper quadrant of the fetal abdomen: are they significant? J Ultrasound Med 2004; 23:483–488.
113. Falcon O, Peralta CF, Cavoretto P, Faiola S, Nicolaides KH. Fetal trunk and head volume measured by three-dimensional ultrasound at 11 + 0 to 13 + 6 weeks of gestation in chromosomally normal pregnancies. Ultrasound Obstet Gynecol 2005; 26:263–266.
114. Kuno A, Hayashi Y, Akiyama M, et al. Three-dimensional sonographic measurement of liver volume in the small-for-gestational-age fetus. J Ultrasound Med 2002; 21:361–366.
115. Chaoui R, Zodan-Marin T, Wisser J. Marked splenomegaly in fetal cytomegalovirus infection: detection supported by three-dimensional power Doppler ultrasound. Ultrasound Obstet Gynecol 2002; 20:299–302.
116. Grannum P, Bracken M, Silverman, Hobbins JC. Assessment of kidney size in normal gestation by comparison of kidney circumference to abdominal circumference. Am J Obstet Gynecol 1980; 136:249–254.
117. Rutherford SE, Smith CV, Phelan JP, Kawakami K, Ahn MO. Four-quadrant assessment of amniotic fluid volume: interobserver and intraobserver variation. J Reprod Med 1987; 32:587–589.
118. Moore TR, Cayle JE. The amniotic fluid index in normal pregnancy. Am J Obstet Gynecol 1990; 162:1168–1173.
119. Potter EL. Bilateral absence of ureters and kidneys: a report of 50 cases. Obstet Gynecol 1965; 25:3–12.
120. McGahan JP, Myracle MR. Adrenal hypertrophy: potential pitfall in the sonographic diagnosis of renal agenesis. J Ultrasound Med 1986; 5:265–268.
121. Nicolaides K, Rodeck C, Gosden C. Rapid karyotyping in non-lethal fetal malformations. Lancet 1986; 1:283–287.
122. Mackenzie FM, Kingston GO, Oppenheimer L. The early prenatal diagnosis of bilateral renal agenesis using transvaginal sonography and color Doppler ultrasonography. J Ultrasound Med 1994; 13:49–51.
123. DeVore GR. The value of color Doppler sonography in the diagnosis of renal agenesis. J Ultrasound Med 1995; 14:443–449.
124. Fortunato SJ. The use of power Doppler and color power angiography in fetal imaging. Am J Obstet Gynecol 1996; 174:1828–1831.
125. Johnson A, Callan NA, Bhutani VK, Colmorgen GH, Weiner S, Bolognese RJ. Ultrasonic ratio of fetal thoracic to abdominal circumference: an association with fetal pulmonary hypoplasia. Am J Obstet Gynecol 1987; 157:764–769.
126. Cascio S, Paran S, Puri P. Associated urological anomalies in children with unilateral renal agenesis. J Urol 1999; 162:1081–1083.
127. Glazebrook KN, McGrath FP, Steele BT. Prenatal compensatory renal growth: documentation with US. Radiology 1993; 189: 733–735.
128. Hill LM, Peterson CM. Antenatal diagnosis of fetal pelvic kidneys. J Ultrasound Med 1987; 6:393–396.
129. Greenblatt AM, Beretsky I, Lankin DH, Phelan L. In uterodiagnosis of crossed renal ectopia using high resolution real-time ultrasound. J Ultrasound Med 1985; 4:105–107.
130. Brown T, Mandell J, Lebowitz RL. Neonatal hydronephrosis in the era of sonography. AJR Am J Roentgenol 1987; 148:959–963.
131. Odibo AO, Marchiano D, Quinones JN, Riesch D, Egan JF, Macones GA. Mild pyelectasis: evaluating the relationship between gestational age and renal pelvic anterior-posterior diameter. Prenat Diagn 2003; 23:824–827.
132. Odibo AO, Raab E, Elovitz M, Merrill JD, Macones GA. Prenatal mild pyelectasis: evaluating the thresholds of renal pelvic diameter associated with normal postnatal renal function. J Ultrasound Med 2004; 23:513–517.
133. Petrikovsky BM, Cuomo MI, Schneider EP, Wyse LJ, Cohen HL, Lesser M. Isolated fetal hydronephrosis: beware the effect of bladder filling. Prenat Diagn 1995; 15:827–829.
134. Gramellini D, Fieni S, Caforio E, Benassi G, Bedocchi L, Beseghi U, Benassi L. Diagnostic accuracy of fetal renal pelvis anteroposterior diameter as a predictor of significant postnatal nephrouropathy: second versus third trimester of pregnancy. Am J Obstet Gynecol 2006; 194:167–173.
135. Cohen-Overbeek TE, Wijngaard-Boom P, Ursem NT, Hop WC, Wladimiroff JW, Wolffenbuttel KP. Mild renal pyelectasis in the second trimester: determination of cut-off levels for postnatal referral. Ultrasound Obstet Gynecol 2005; 25:378–383.
136. Wollenberg A, Neuhaus TJ, Willi UV, Wisser J. Outcome of fetal renal pelvic dilatation diagnosed during the third trimester. Ultrasound Obstet Gynecol 2005; 25:483–488.
137. Bornstein E, Barnhard Y, Donnenfeld A, Ferber A, Divon MY. Fetal pyelectasis: does fetal gender modify the risk of major trisomies? Obstet Gynecol 2006; 107:877–879.
138. Benacerraf BR, Mandell J, Estroff JA, Harlow BL, Frigoletto FD Jr. Fetal pyelectasis: a possible association with Down syndrome. Obstet Gynecol 1990; 76:58–60.
139. Snijders RJ, Sebire NJ, Faria M, Patel F, Nicolaides KH. Fetal mild hydronephrosis and chromosomal defects: relation to maternal age and gestation. Fetal Diagn Ther 1995; 10:349–355.

140. Wickstrom EA, Thangavelu M, Parilla BV, Tamura RK, Sabbagha RE. A prospective study of the association between isolated fetal pyelectasis and chromosomal abnormality. Obstet Gynecol 1996; 88:379–382.
141. Nyberg DA, Souter VL, El-Bastawissi A, Young S, Luthhardt F, Luthy DA. Isolated sonographic markers for detection of fetal Down syndrome in the second trimester of pregnancy. J Ultrasound Med 2001; 20:1053–1063.
142. Bromley B, Lieberman E, Shipp TD, Benacerraf BR. The genetic sonogram: a method of risk assessment for Down syndrome in the second trimester. J Ultrasound Med 2002; 21:1087–1096.
143. Nicolaides KH. Screening for chromosomal defects. Ultrasound Obstet Gynecol 2003; 21:313–321.
144. Coco C, Jeanty P. Isolated fetal pyelectasis and chromosomal abnormalities. Am J Obstet Gynecol 2005; 193:732–738.
145. Harrison MR, Globus MS, Filly RA. The Unborn Patient. 2, 282 1991. Serial (Book, Monograph)
146. Montana MA, Cyr DR, Lenke RR, Shuman WP, Mack LA. Sonographic detection of fetal ureteral obstruction. AJR Am J Roentgenol 1985; 145:595–596.
147. Nussbaum AR, Dorst JP, Jeffs RD, Gearhart JP, Sanders RC. Ectopic ureter and ureterocele: their varied sonographic manifestations. Radiology 1986; 159:227–235.
148. Eckoldt F, Heling KS, Woderich R, Wolke S. Posterior urethral valves: prenatal diagnostic signs and outcome. Urol Int 2004; 73:296–301.
149. Bulic M, Podobnik M, Korenie B, Bistricki J. First trimester diagnosis of low obstructive uropathy: an indicator of initial renal function in the fetus. J Clin Ultrasound 1987; 15:537–541.
150. Manning FA, Harison MR, Rodeck C. Catheter shunts for fetal hydronephrosis and hydrocephalus. Report of the International Fetal Surgery Registry. N Engl J Med 1986; 315:336–340.
151. Mahony BS, Callen PW, Filly RA. Fetal urethral obstruction: US evaluation. Radiology 1985; 157:221–224.
152. Sanders RC, Nussbaum AR, Solez K. Renal dysplasia: sonographic findings. Radiology 1988; 167:623–626.
153. Nicolaides KH, Chen HH, Snijders RJM, Moniz CF. Fetal urine biochemistry in the assessment of obstructive uropathy. Am J Obstet Gynecol 1992; 166:932–937.
154. Lipitz S, Ryan G, Samuell C, et al. Fetal urine analysis for the assessment of renal function in obstructive uropathy. Am J Obstet Gynecol 1993; 168:174–179.
155. Wilkins IA, Chitkara U, Lynch L, Goldberg JD, Mehalek KE, Berkowitz RL. The nonpredictive value of fetal urinary electrolytes: preliminary report of outcomes and correlation with pathologic diagnosis. Am J Obstet Gynecol 1987; 157:694–698.
156. Elder JS, O'Grady JP, Ashmead G, Duckett JW, Philipson E. Evaluation of fetal renal function: unreliability of fetal electrolytes. J Urol 1990; 144:574–578.
157. Johnson MP, Corsi P, Bradfield W, et al. Sequential urinalysis improves evaluation of fetal renal function in obstructive uropathy. Am J Obstet Gynecol 1995; 173:59–65.
158. Manning FA. Ultrasound guided fetal invasive therapy: current status. In: Fleischer AC, et al., eds. Sonography in Obstetrics and Gynecology: Principles and Practice. 5th ed. Norwalk, CT: Appleton & Lange, 1996:699.
159. Helin I, Persson PH. Prenatal diagnosis of urinary tract abnormalities by ultrasound. Pediatrics 1986; 78:879–883.
160. Kleiner B, Filly RA, Mack L, Callen PW. Multicystic dysplastic kidney: observations of contralateral disease in the fetal population. Radiology 1986; 161:27–29.
161. al Khaldi N, Watson AR, Zuccollo J, Twining P, Rose DH. Outcome of antenatally detected cystic dysplastic kidney disease. Arch Dis Child 1994; 70:520–522.
162. Corrales JG, Elder JS. Segmental multicystic kidney and ipsilateral duplication anomalies. J Urol 1996; 155:1398–401.
163. Rizzo N, Gabrielli S, Pilu G, et al. Prenatal diagnosis and obstetrical management of multicystic dysplastic kidney disease. Prenat Diagn 1987; 7:109–118.
164. Beretsky I, Labkin DH, Rusoff JH, Phelan L. Sonographic differentiation between the multicystic dysplastic kidney and ureteropelvic junction obstruction in utero using high resolution real-time scanners employing digital detection. J Clin Ultrasound 1984; 12:429–433.
165. Hashimoto BE, Filly RA, Callen PW. Multicystic dysplastic kidney: changing appearance on ultrasound. Radiology 1986; 159:107–109.
166. Luthy DAM, Hirsch JH. Infantile polycystic kidney disease: observations from attempts at prenatal diagnosis. Am J Med Genet 1985; 20:505–517.
167. Wisser J, Hebisch G, Froster U, et al. Prenatal sonographic diagnosis of autosomal recessive polycystic kidney disease (ARPKD) during the early second trimester. Prenat Diagn 1995; 15:868–871.
168. Zerres K, Mucher G, Bachner L, et al. Mapping of the gene for autosomal recessive polycystic kidney disease (ARPKD) to chromosome 6p21-cen. Nat Genet 1994; 7:429–432.
169. Pretorius DH, Lee ME, Manco-Johnson ML, Weingast GR, Sedman AB, Gabow PA. Diagnosis of autosomal dominant polycystic kidney disease in utero and in the young infant. J Ultrasound Med 1987; 6:249–255.
170. Walter JP, McGahan JP. Mesoblastic nephroma: prenatal sonographic detection. J Clin Ultrasound 1985; 13:686–689.
171. Romano WL. Neonatal renal tumor with polyhydramnios. J Ultrasound Med 1984; 3:475–476.
172. Apuzzio JJ, Unwin W, Adhate A, Nichols R. Prenatal diagnosis of fetal renal mesoblastic nephroma. Am J Obstet Gynecol 1986; 154:636–637.
173. Lee W, Comstock CH, Jurcak-Zaleski S. Prenatal diagnosis of adrenal hemorrhage by ultrasonography. J Ultrasound Med 1992; 11:369–371.
174. Strouse PJ, Bowerman RA, Schlesinger AE. Antenatal sonographic findings of fetal adrenal hemorrhage. J Clin Ultrasound 1995; 23:442–446.
175. Elejalde BR, de Elejalde MM, Heitman T. Visualization of the fetal genitalia by ultrasonography: a review of the literature and analysis of its accuracy and ethical implications. J Ultrasound Med 1985; 4:633–639.
176. Bronshtein M, Riechler A, Zimmer EZ. Prenatal sonographic signs of possible fetal genital anomalies. Prenat Diagn 1995; 15:215–219.
177. Sides D, Goldstein RB, Baskin L, Kleiner BC. Prenatal diagnosis of hypospadias. J Ultrasound Med 1996; 15:741–746.
178. Steiger RM, Porto M. The anterior abdominal wall. In: McGahan JP, Porto M, eds. Diagnostic obstetrical ultrasound. Philadelphia, JB Lippincott, 1994.
179. Sadler TW; Langman's Medical Embryology. 6th ed. Baltimore: Williams & Wilkins:76.
180. Cyr DR, Bach LA, Schoeneker SA, et al. Bowel migration in the normal fetus: US detection. Radiology 1986; 1961:119.
181. Jones KL, ed. Smith's Recognizable Pattern of Human Malformations. 4th ed. Philadelphia: WB Saunders, 1988.
182. Callen PW, ed. Ultrasonography in Obstetrics and Gynecology. 2nd ed. Philadelphia: WB Saunders, 1988.
183. Gwinn JL. Tracheoesophageal fistula with and without esophageal atresia: special aspects. In: Kaufman HJ, ed. Progress in Pediatric Radiology. Chicago: Year Book, 1969.
184. Singleton EB. X-Ray Diagnosis of the Alimentary Tract in Infants and Children. Chicago: Year Book, 1959.
185. Gray SW, Skandalakis JF. Embryology for Surgeons. Philadelphia: WB Saunders, 1972.
186. Porto M, McGahan JP. The fetal abdomen and pelvis. In: McGahan JP, Porto M, eds. Diagnostic Obstetrical Ultrasound. Philadelphia: JB Lippincott, 1994.
187. Bertagnoli L, Lalatta F, Gallicchio R, et al. Quantitative characterization of growth of the fetal kidney. J Clin Ultrasound 1983; 11:349–356.
188. Grannum P, Bracken M, Silverman R, et al. Assessment of fetal kidney size in normal gestation by comparison of ratio of kidney circumference to abdominal circumference. Am J Obstet Gynecol 1980; 136:253.

Fetal Skeleton

Lyndon M. Hill

49

INTRODUCTION

Accurate examination of the fetal skeleton became possible with the development of high-resolution real-time ultrasound equipment. The severity of the effect on the long bones, as well as the ease with which the bones can be imaged, makes possible the second-trimester diagnosis of lethal skeletal dysplasias. Other prenatal tests may then be used to confirm, or exclude, the presence of a specific skeletal dysplasia. For example, DNA testing of amniotic fluid or chorionic villi can be diagnostic when achondroplasia is suspected (1). The prevalence rate of the skeletal dysplasias has been estimated at between 2.4 and 4.7 per 10,000 births (2). Seventy percent of the skeletal dysplasias are one of the four following disorders: achondroplasia, thanatophoric dysplasia, achondrogenesis, or osteogenesis imperfecta. A detailed review of the sonographic findings in these four conditions will, therefore, be provided.

An understanding of normal bone development is necessary before evaluating specific pathologic conditions. The two types of bone formation are membranous and endochondral ossification. The cranial vault undergoes membranous ossification (i.e., mesenchymal cells are directly transformed into osteoblasts without a cartilage stage). A preexisting cartilaginous base is used by the long bones for endochondral ossification. The upper limb buds appear at approximately 5.5 weeks' menstrual age, and the lower limb buds appear a few days later. Cartilage first appears at seven weeks' menstrual age. By 12 weeks, primary centers of ossification are present in the diaphysis or body of the long bones. Osteoblasts deposit bone matrix on cartilage. The length of the long bones is determined by growth at the diaphyseal–epiphyseal junction. At birth, the diaphyses are almost completely ossified, whereas the epiphyses are still cartilaginous. Once the epiphyseal plates ossify, the bone can no longer grow in length (3). Growth in bone diameter is due to the deposition of calcium at the periosteum.

The osteochondrodysplasias are abnormalities of cartilage and bone formation, resulting in faulty development of the long bones and vertebrae. The Paris classification of the osteochondrodysplasias was initially proposed in 1969 and was revised in 1977 and 1983 (Table 1). A correct diagnosis of a lethal skeletal dysplasia is important because of its implications for subsequent genetic counseling concerning further pregnancies. The modes of genetic transmission for the more common skeletal dysplasias are provided in Table 1.

This chapter is not a compendium of all known skeletal dysplasias. Rather, it is a stepwise approach to the assessment of the fetus with a suspected skeletal dysplasia. The primary distinction that must be made by the sonologist or sonographer is whether a particular skeletal malformation is lethal. Definitive diagnosis or syndrome identification in most cases must await evaluation by a geneticist or pathologist after delivery. As previously mentioned, in some instances, prenatal diagnosis is possible (1).

The classification of the skeletal dysplasias is based on clinical history, specific radiographic findings, and the genetic features of the fetus or neonate. Alizarin stain of the fetal skeleton (5) and xeroradiography (Fig. 1) (6) have been used in attempts to establish a definitive diagnosis. A descriptive classification divides the skeletal dysplasias into rhizoid

TABLE 1 ■ Constitutional Disease of Bone Identifiable at Birth and that May Require Pathologic Examination

Osteochondrodysplasias

Chondrodysplasia (defects of growth of tubular bones and/or spine)

- Usually lethal before or shortly after birth
  - Achondrogenesis type I (Parenti–Fraccaro)[a] (AR)[b]
  - Achondrogenesis type II (Langer–Saldino)[a] (AR)
  - Hypochondrogenesis[a]
  - Fibrochondrogenesis[a] (AR)
  - Thanatophoric dysplasia with cloverleaf skull[a]
  - Atelosteogenesis[a]
  - Short-rib syndrome (with or without polydactyly)
    - Type I (Saldino–Noonan)[a] (AR)
    - Type II (Majewski)[a] (AR)
    - Type III (Verma–Naumoff)[a] (AR)
    - Type IV (Beemer–Langer)[a] (AR)
- Usually nonlethal dysplasia
  - Chondrodysplasia punctata
    - Rhizomelic form, autosomal recessive[a] (AR)
    - Dominant X-linked form[a] (XLD, lethal in male)
    - Common mild form (Sheffield)
    - Excluded: symptomatic stippling (warfarin, chromosomal aberration)
  - Camptomelic dysplasia[a] (AR)
  - Kyphomelic dysplasia (AR)
  - Achondroplasia (homozygous and heterozygous)[a] (AD)
  - Diastrophic dysplasia[a] (AR)
  - Metatrophic dysplasia (several forms)[a] (AR, AD)
  - Chondroectodermal dysplasia (Ellis–van Creveld)[a] (AR)
  - Asphyxiating thoracic dysplasia (Jeune)[a] (AR)
  - Spondyloepiphyseal dysplasia congenita[a]
    - Autosomal-dominant form (AD)
    - Autosomal-recessive form (AR)
  - Kniest dysplasia[a] (AR)
  - Dyssegmental dysplasia[a] (AR)
  - Larsen syndrome[a] (AR, AD)

Abnormalities of density of cortical diaphyseal structure or metaphyseal modeling

- Osteogenesis imperfecta (several forms)[a] (AR)
- Osteopetrosis (several forms)[a] (AR)
- Pylenodysostosis (AR)
- Dominant osteosclerosis, type Stanescu (AD)
- Infantile cortical hyperostosis[a] (Caffey disease) (AD)

Primary metabolic abnormalities

- Hypophosphatasis (several forms)[a] (AR)
- Mucolipidosis II (I-cell disease)[a] (AR)
- Other

[a]Osteochondral pathology reported in the literature.

[b]The mode of genetic transmission in parentheses.

Abbreviations: AR, autosomal recessive; AD, autosomal dominant; XLD, X-linked dominant.

Source: From Ref. 4.

(root or proximatel), mesial (middle), and micral (generalized) shortening of the extremities (Fig. 2). Acral (distal) shortening involves only the hands and feet.

SONOGRAPHIC EVALUATION OF THE NORMAL SKELETON

Transvaginal sonography can clearly image the normally developing skeleton in the first trimester. At eight weeks' menstrual age, the upper and lower limbs are first visualized (Fig. 3). It is possible to examine the digits of the hands and feet at 12 weeks' menstrual age (Fig. 4) (8).

Measurement of the femur is an established part of the standard second and third trimester ultrasound examination. With transvaginal sonography, first trimester bone lengths can also be obtained (Fig. 5). Depending on the severity of a skeletal dysplasia and the gestational age at which it becomes manifest, measurement of the femur length (FL) alone enables one to detect most severe skeletal dysplasias (9).

DIAGNOSTIC APPROACH TO SKELETAL DYSPLASIAS

As with any congenital malformation, a detailed fetal anatomic survey is mandatory whenever a skeletal dysplasia is suspected. However, specific key measurements, ratios, and anatomic landmarks should be assessed in every case. The following features of the anatomic survey assist the sonologist or sonographer in narrowing the differential diagnosis of a fetus with a skeletal dysplasia.

INITIAL QUESTIONS

When a skeletal dysplasia is suspected, the initial evaluation should be of the femur. Table 2 outlines the three most important questions that should be answered in the sonographic assessment of a skeletal dysplasia.

What Is the FL?

Numerous studies have correlated the length of specific long bones and menstrual age (Table 3) (11–13). Even with excellent dating criteria, single and multiple long-bone lengths below 2 standard deviations (SD) of the mean for gestational age are not necessarily diagnostic for a skeletal dysplasia. However, the number of millimeters the FL is below 2 SD differentiates a constitutionally small fetus from one with a skeletal dysplasia. In one study (14), all 12 fetuses with an FL at least 5 mm below 2 SD of the mean were affected by a skeletal dysplasia. Table 4 provides other possible diagnoses for a short FL.

What Has Been the Interval Growth of the FL?

The growth rate of the long bones decreases before the absolute length falls below the criteria outlined earlier

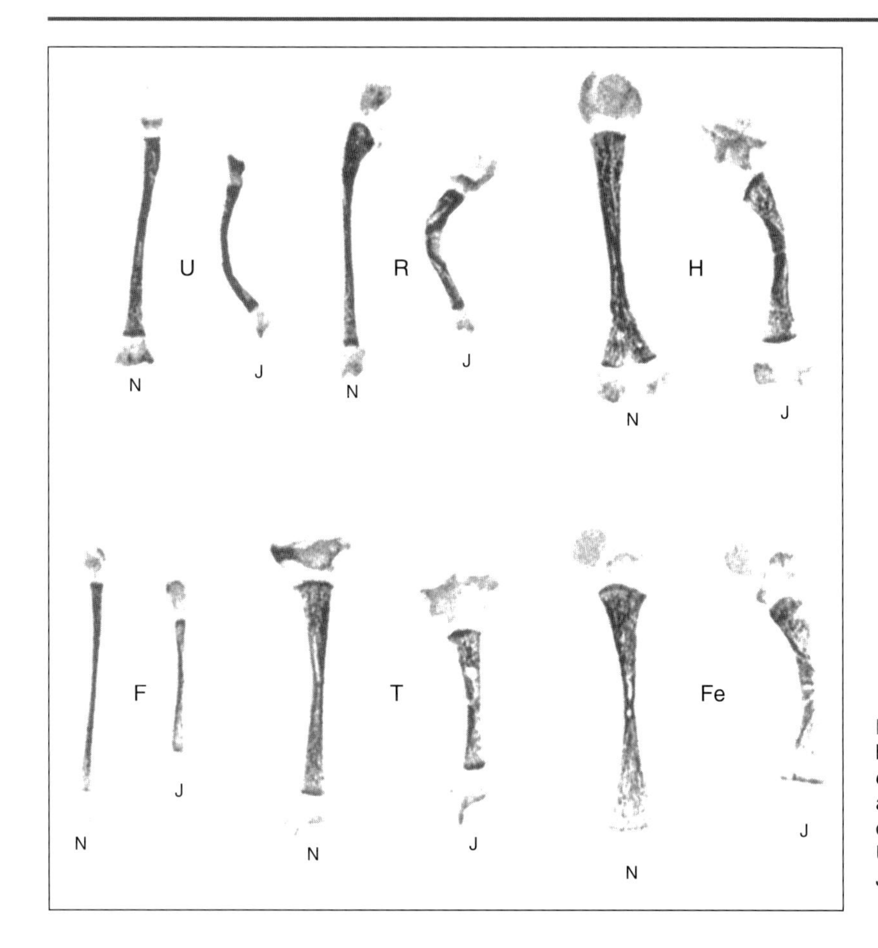

FIGURE 1 ■ Bones at 20 weeks' gestation, one affected by Jeune syndrome and the other normal. Note the differences in shape (bones in Jeune syndrome are shortened and bowed) and structure of the endochondral and perichondral bone. Note the epiphyses and growth plate. U, ulna; R, radius; H, humerus; Fe, femur; T, tibia; F, fibula; J, Jeune syndrome; N, normal. *Source*: From Ref. 6.

to diagnose a skeletal dysplasia. From 16 to 22 weeks' gestation, the mean length of all the long bones increases between 2.5 and 2.7mm per week. Limb growth must be assessed critically, with the history of specific skeletal dysplasias taken into account. For example, a fetus with osteogenesis imperfecta type II may have an abnormal FL at 15 weeks' gestation, whereas a fetus with heterozygous achondroplasia may not have an abnormally short femur until 21 to 27 weeks' menstrual age (15).

Is the Ratio Between the FL and Other Body Measurements Appropriate?

The proportions among specific body parts help to confirm the diagnosis of a skeletal dysplasia and also provide

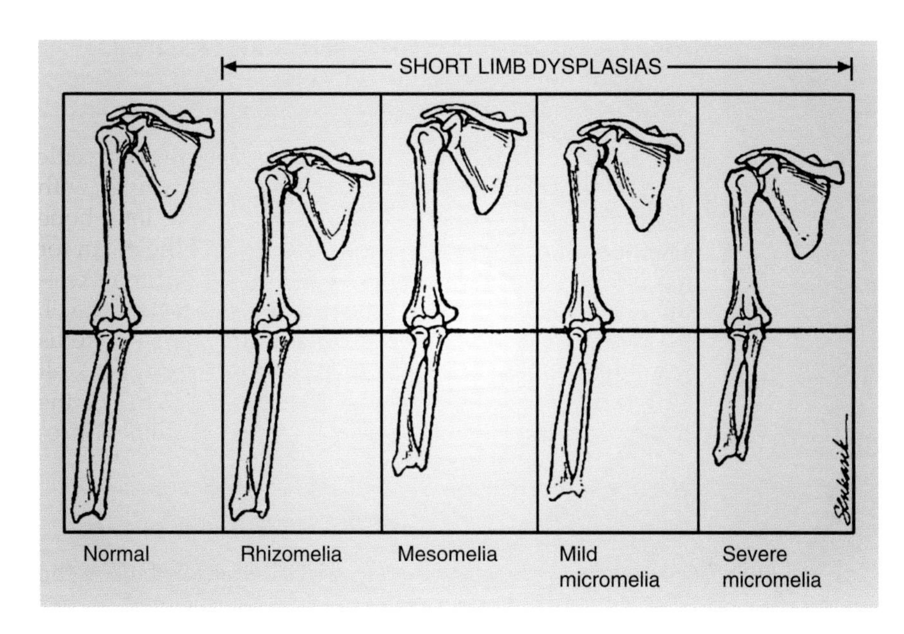

FIGURE 2 ■ Classification on long-bone shortening as it affects rhizoid (proximal), mesial (middle), or micral (generalized) shortening of the extremities. *Source*: From Ref. 7.

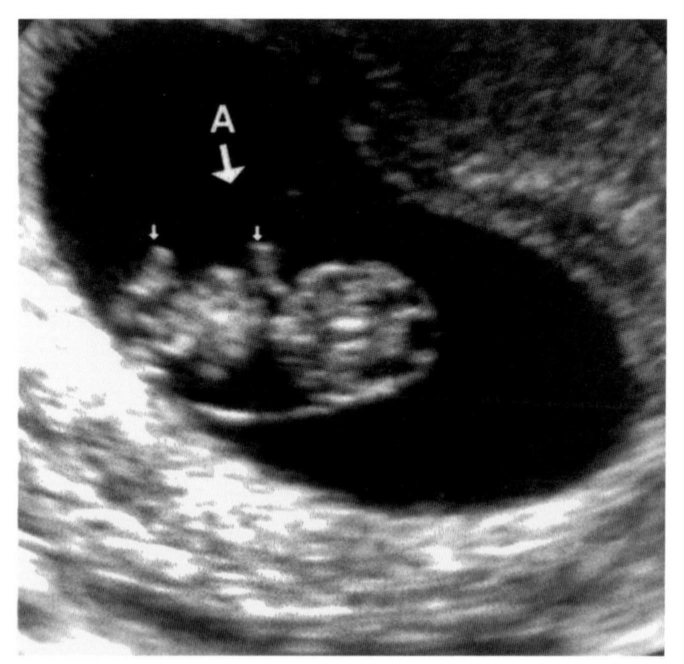

FIGURE 3 ■ Fetal extremities (*arrows*) at eight weeks' menstrual age. The diameter of the amniotic cavity (A) is equivalent to the crown–rump length.

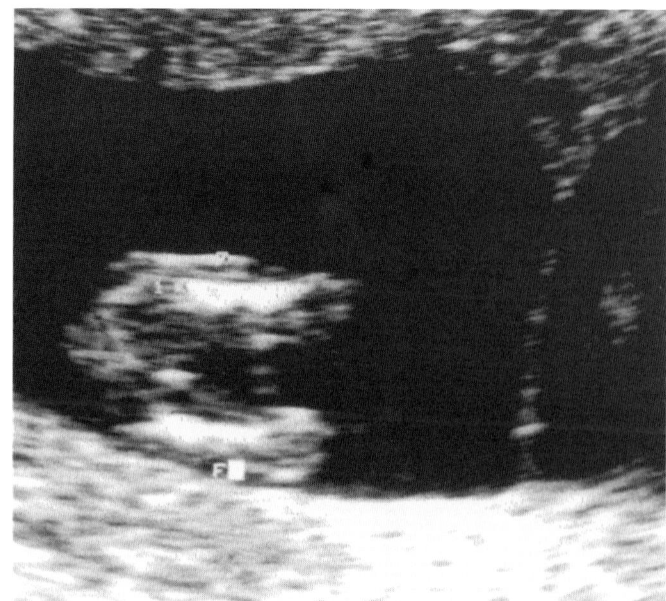

FIGURE 5 ■ Bilateral femurs (F) at 12 weeks' gestation.

important diagnostic clues. The ratio of FL to head circumference of more than 3 SD below the mean suggests a skeletal dysplasia (Fig. 6) (16). The FL/abdominal circumference (AC) ratio is normally between 0.20 and 0.24 (17). In two studies, 25 of 26 fetuses with a ratio below 0.16 had a lethal skeletal dysplasia (18,19). When there is severe bowing of the femur, its length must be traced. The shorter value obtained from a linear measurement may result in a spuriously low FL/AC ratio. It must be emphasized that these results were reported in a population suspected of having a skeletal dysplasia. The utilization of an FL/AC ratio of below 0.16 in the general population would result in an increase in the false positive rate.

The FL and foot are of comparable length in the normal fetus. Because this ratio (femur to foot) is not affected by gestational age, it may be used even when the patient's dates are in doubt (20). In one study, a lower limit of 0.87 for the femur-to-foot ratio enabled clinicians to discriminate satisfactorily between fetuses with a skeletal dysplasia and those with intrauterine growth retardation (Fig. 7) (16).

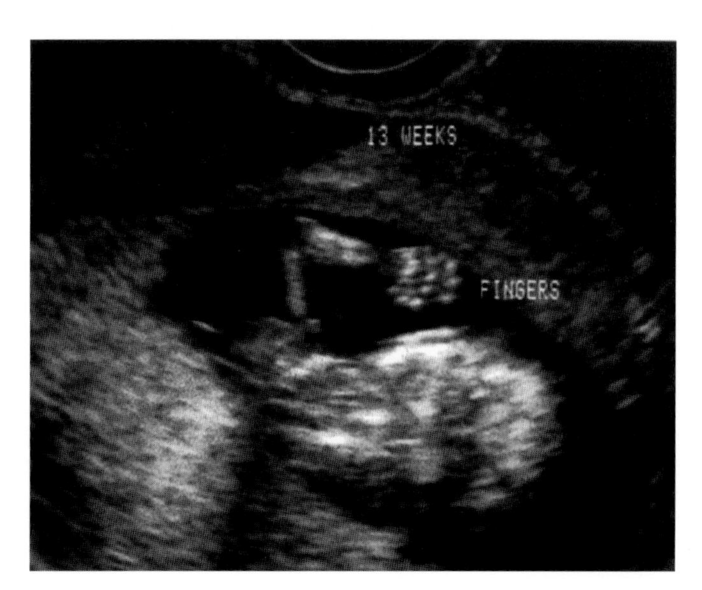

FIGURE 4 ■ Fetal hand radius, ulna, and humerus at 13 weeks' menstrual age.

OTHER QUESTIONS

After an abnormality of the long bones is identified, other questions must be asked to categorize the type of skeletal dysplasia (Table 5). These questions include those found in the following paragraphs.

What Is the Appearance of the Fetal Skull or Face?

Micrognathia (Fig. 8), frontal bossing, or a depressed nasal bridge may be detected on a facial profile. Although micrognathia is generally diagnosed subjectively, nomograms for mandible length by gestational age are available (22). Frontal bossing is characteristic of fetuses with

TABLE 2 ■ Initial Questions in Evaluating a Skeletal Dysplasia

- What is the femur length?
- What has been the interval growth of the femur length?
- Is the ratio between the femur length and the other body measurements appropriate?

TABLE 3 ■ Nomogram of Normal Values of Fetal Long Bones vs. Menstrual Age with 5th and 95th Percentiles (mm)

	Tibia			Fibula			Femur			Humerus			Ulna			Radius		
	Percentile			Percentile			Percentile			Percentile			Percentile			Percentile		
Week	5th	50th	95th	5th	50th	95th	5th	50th	95th	5th	50th	95th	5th	50th	95th	5th	50th	95th
12	–	7	–	–	6	–	4	8	13	–	9	–	–	7	–	–	7	–
13	–	10	–	–	9	–	6	11	16	6	11	16	5	10	15	6	10	14
14	7	12	17	6	12	19	9	14	18	9	14	19	8	13	18	8	13	17
15	9	15	20	9	15	21	17	21	12	17	22	11	11	16	21	11	15	20
16	12	17	22	13	18	23	15	20	24	15	20	25	13	18	23	13	18	22
17	15	20	25	13	21	28	18	23	27	18	22	27	16	21	26	14	20	26
18	17	22	27	15	23	31	21	25	30	20	25	30	19	24	29	15	22	29
19	20	25	30	19	26	33	24	28	33	23	28	33	21	26	31	20	24	29
20	22	27	33	21	28	36	26	31	36	25	30	35	24	29	34	22	27	32
21	25	30	35	24	31	37	29	34	38	28	33	38	26	31	36	24	29	33
22	27	32	38	27	33	39	32	36	41	30	35	40	28	33	38	27	31	34
23	30	35	40	28	35	42	35	39	44	33	38	42	31	36	41	26	32	39
24	32	37	42	29	37	45	37	42	46	35	40	45	33	38	43	26	34	42
25	34	40	45	34	40	45	40	44	49	37	42	47	35	40	45	31	36	41
26	37	42	47	36	42	47	42	47	51	39	44	49	37	42	47	32	37	43
27	39	44	49	37	44	50	45	49	54	41	46	51	39	44	49	33	39	45
28	41	46	51	38	45	53	47	52	56	43	48	53	41	46	51	33	40	48
29	43	48	53	41	47	54	50	54	59	45	50	55	43	48	53	36	42	47
30	45	50	55	43	49	56	52	56	61	47	51	56	44	49	54	36	42	47
31	47	52	57	42	51	59	54	59	63	48	53	58	46	51	56	38	44	50
32	48	54	59	42	52	63	56	61	65	50	55	60	48	53	58	37	45	53
33	50	55	60	46	54	62	58	63	67	51	56	61	49	54	59	41	46	51
34	52	57	62	46	55	65	60	65	69	53	58	63	51	56	61	40	47	53
35	53	58	64	51	57	62	62	67	71	54	59	64	52	57	62	41	48	54
36	55	60	65	54	58	63	64	68	73	56	61	65	53	58	63	39	48	57
37	56	61	67	54	59	65	65	70	74	57	62	67	55	60	65	45	49	53
38	58	63	68	56	61	65	67	71	76	59	63	68	56	61	66	45	49	54
39	59	64	69	56	62	67	68	73	77	60	65	70	57	62	67	45	50	54
40	61	66	71	59	63	67	70	74	79	61	66	71	58	63	68	46	50	55

Source: From Ref. 10.

TABLE 4 ■ Possible Causes of Long-Bone Length Less Than Two Standard Deviations Below the Mean for Gestational Age Below the Mean

- Normal physiologic variation
- Intrauterine growth restriction
- Abnormal karyotypes
- Focal skeletal abnormality
- Skeletal dysplasia

thanatophoric dysplasia and osteogenesis imperfecta. Fourteen percent of fetuses with thanatophoric dysplasia, as well as fetuses with homozygous achondroplasia, may have a cloverleaf skull (Table 6). This deformity is due either to premature closure of the coronal and lambdoidal sutures or to abnormal membranous calcification (26). Hydrocephalus may be associated with a cloverleaf cranial deformity. In its milder form, a cloverleaf skull can resemble an encephalocele (27).

What Is the Appearance of the Thorax?

Tables of thoracic circumference percentiles at specific gestational ages have been developed from cross-sectional data (Table 7) (28). In one study, a chest circumference below the fifth percentile for gestational age has a positive predictive value of 94% for detecting pulmonary hypoplasia (Fig. 9) (29).

Sonographically, the chest circumference is measured perpendicular to the spine at the level of the four-chamber view (Fig. 10). In some severe skeletal dysplasias, the chest circumference is significantly smaller than expected for gestational age. Hence, the ratio between thoracic circumference and AC or between thoracic circumference and head circumference is abnormal. The normal ratios between thoracic and AC and between thoracic and head circumference (± 2SD) are 0.89 (± 0.06) and 0.80 (± 0.12), respectively. An additional advantage of these ratios is that they do not vary significantly with gestational age (21). DeVore et al. (30) developed confidence limit graphs for chest circumference regressed against the biparietal diameter, head circumference, AC, and FL (Fig. 11).

The shape of the chest, as well as the configuration of the ribs, should also be assessed (Fig. 12). In lethal skeletal dysplasias, a narrow chest is characteristic (Fig. 13). Consequently, the heart fills the chest cavity and, in the longitudinal plane, the abdomen is protuberant.

A short trunk with a decreased number of fused ribs, a protuberant abdomen, and a shortened spine suggest a diagnosis of spondylothoracic dysplasia (Jarcho–Levin syndrome) (Fig. 14). In this syndrome,

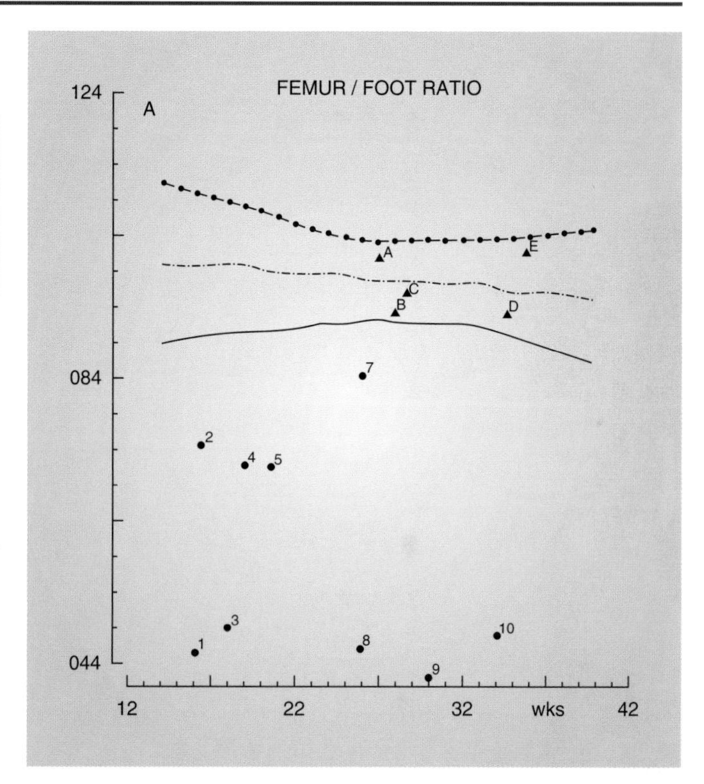

FIGURE 7 ■ Femur-to-foot ratios of 10 fetuses affected by skeletal dysplasia (circles, 1–10) and five growth-retarded fetuses (A–E). *Source*: From Ref. 21.

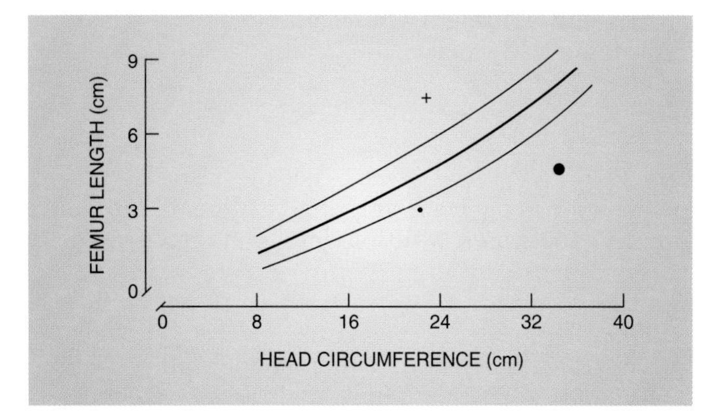

FIGURE 6 ■ Plot of femur lengths against head circumference in 361 healthy patients. The centerline is the 50th percentile, and the upper and lower lines indicate the 99th confidence intervals. The black square represents a fetus at 25 weeks, who had an unclassified form of short-limb dwarfism at delivery. The circle represents a fetus with short lower extremities secondary to osteogenesis imperfecta at 34 weeks. The plus mark indicates a fetus with severe microcephaly at 32 weeks. *Source*: From Ref. 16.

TABLE 5 ■ Additional Questions in Evaluating a Skeletal Dysplasia

- What is the appearance of the fetal head or face?
- What is the appearance of the fetal thorax?
- What is the appearance of the fetal spine?
- What is the appearance of the other long bones?
  - Are other long bones affected? (rhizomelic, mesomelic, micromelic)
  - Is there associated bowing or fracture?
  - Is there hypomineralization?
- Are there abnormalities of the hands or feet?
- Is there polyhydramnios?
- Is there fetal hydrops?

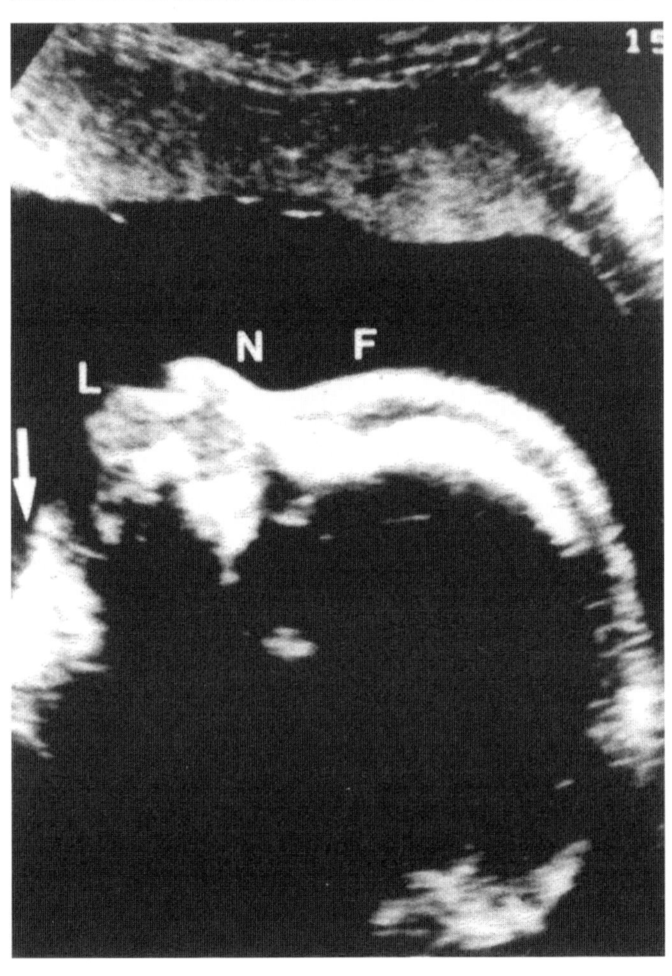

FIGURE 8 ■ Profile of a fetal face with micrognathia (*arrow*) at 32 weeks' gestation. F, forehead; N, nose; L, upper lip.

the limbs are of normal length, but multiple defects are present in the vertebral column and thorax. Type I spondylothoracic dysplasia has an autosomal-recessive pattern of inheritance and is usually fatal by 15 months of age because of respiratory failure. Type II spondylothoracic dysplasia is autosomal dominant, has a milder involvement of the thorax and vertebral column, and is compatible with a normal life span. Associated malformations include microcephaly, cleft palate, abdominal wall defects, syndactyly, and camptodactyly (31–33).

TABLE 6 ■ Skeletal Dysplasias with an Abnormal Skull Contour

Frontal bossing
- Thanatophoric dysplasia
- Osteogenesis

Cloverleaf skull
- Homozygous achondroplasia
- Thanatophoric dysplasia

Craniosynostosis
- Cloverleaf skull
    - Homozygous achondroplasia
    - Thanatophoric dysplasia
- Apert syndrome (23)
- Crouzan syndrome (24)
- Pfeiffer syndrome (25)

What Is the Appearance of the Fetal Spine?

The spine may be poorly mineralized or abnormally angulated (kyphosis or scoliosis). Hypomineralization of the spine is characteristic of achondrogenesis type II. Platyspondyly or flattening of the spine with wide intervening spaces occurs with either thanatophoric dysplasia or achondroplasia (34). The ratio between vertebral body height and vertebral interface (Figs. 15 and 16) is constant between 20 weeks' menstrual age and term.

$$\frac{\text{Vertebral body}}{\text{Vertebral interface}} = 0.5 \text{ to } 0.75$$

Because widening of the vertebral interface with thanatophoric dysplasia is significant, the ratio of vertebral body to vertebral interface is well below normal. It is also usually below normal in achondroplasia. In osteogenesis imperfecta type II, this ratio varies widely and may be normal.

What Is the Appearance of the Long Bones?

Are Other Bones Affected?

This may be an important clue in classifying the type of skeletal dysplasia. For example, proximal long-bone shortening (rhizomelia) is present in achondroplasia. A generalized shortening of all the long bones (micromelia) occurs in achondrogenesis and osteogenesis imperfecta type II (Table 8).

Long Bone Shape: Bowing and Fractures?

The normal diaphysis of the femur has a curved medial border and a straight lateral border (Fig. 17) (35). In camptomelic dysplasia, the limbs are bowed anteriorly, especially the lower extremities (Table 9). The bowing may be due to primary shortness of the calf muscles, or it may result from a defect in the embryonic cartilage (36,37). Fetuses with thanatophoric dysplasia or osteogenesis imperfecta type II also have bowed extremities. Multiple long bone fractures are a pathognomonic finding in osteogenesis imperfecta type II. Isolated femoral fractures at the point of maximal bowing are characteristic of the Antley–Bixler syndrome (Fig. 18A) (38). Other sonographic findings associated with this autosomal-recessive condition include the following: craniosynostosis, resulting in trigonocephaly (Fig. 18B) or a cloverleaf skull; frontal bossing; a depressed nasal bridge (Fig. 18C); and joint contractures. As illustrated in Figure 18A, three-dimensional imaging has been helpful in detecting facial dysmorphology associated with specific skeletal dysplasias (39).

TABLE 7 ■ Fetal Thoracic Circumference Measurements (cm)

Gestational age (wk)	No.	Predictive percentiles								
		2.5	5	10	25	50	75	90	95	97.5
16	6	5.9	6.4	7.0	8.0	9.1	10.3	11.3	11.9	12.4
17	22	6.8	7.3	7.9	8.9	10.0	11.2	12.2	12.8	13.3
18	31	7.7	8.2	8.8	9.8	11.0	12.1	13.1	13.7	14.2
19	21	8.6	9.1	9.7	10.7	11.9	13.0	14.0	14.6	15.1
20	20	9.5	10.0	10.6	11.7	12.8	13.9	15.0	15.5	16.0
21	30	10.4	11.0	11.6	12.6	13.7	14.8	15.8	16.4	16.9
22	18	11.3	11.9	12.5	13.5	14.6	15.7	16.7	17.3	17.8
23	21	12.2	12.8	13.4	14.4	15.5	16.6	17.6	18.2	18.8
24	27	13.2	13.7	14.3	15.3	16.4	17.5	18.5	19.1	19.7
25	20	14.1	14.6	15.2	16.2	17.3	18.4	19.4	20.0	20.6
26	29	15.0	15.5	16.1	17.1	18.2	19.3	20.3	21.0	21.5
27	24	15.9	16.4	17.0	18.0	19.1	20.2	21.3	21.9	22.4
28	24	16.8	17.3	17.9	18.9	20.0	21.2	22.2	22.8	23.3
29	24	17.7	18.2	18.8	19.8	21.0	22.1	23.1	23.7	24.2
30	27	18.6	19.1	19.7	20.7	21.9	23.0	24.0	24.6	25.1
31	24	19.5	20.0	20.6	21.6	22.8	23.9	24.9	25.5	26.0
32	28	20.4	20.9	21.5	22.6	23.7	24.8	25.8	26.4	26.9
33	27	21.3	21.8	22.5	23.5	24.6	25.7	26.7	27.3	27.8
34	25	22.2	22.8	23.4	24.4	25.5	26.6	27.6	28.2	28.7
35	20	23.1	23.7	24.3	25.3	26.4	27.5	28.5	29.1	29.6
36	23	24.0	24.6	25.2	26.2	27.3	28.4	29.4	30.0	30.6
37	22	24.9	25.5	26.1	27.1	28.2	29.3	30.03	30.9	31.5
38	21	25.9	26.4	27.0	28.0	29.1	30.2	31.2	31.9	32.4
39	7	26.8	27.3	27.9	28.9	30.0	31.1	32.2	32.8	33.3
40	6	27.7	28.2	28.8	29.8	30.9	32.1	33.1	33.7	34.2

Source: From Ref. 28.

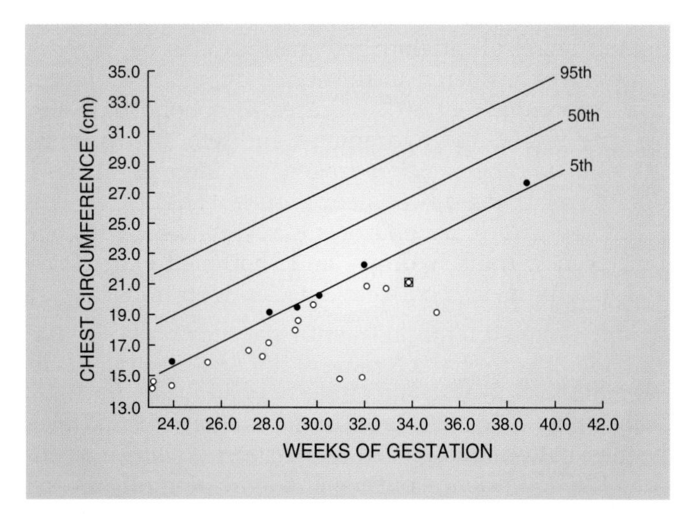

FIGURE 9 ■ Chest circumference nomogram showing gestational age distribution of fetuses at (*closed circles*) and below (*open circles*) the fifth percentile. Of 16 fetuses with a chest circumference below the fifth percentile, 15 had pulmonary hypoplasia. *Source*: From Ref. 29.

What Is the Degree of Mineralization?

The degree of skeletal ossification should be assessed (Table 10). Acoustic shadowing is characteristically present from the calvarium, ribs, and long bones of normal fetuses (Fig. 19). A poorly ossified skull is slightly echoic, with little or no acoustic shadow. In addition, intracranial anatomy is visible in both near and far fields. Flattening of the fetal calvarium with the real-time transducer is an additional sign of hypomineralization. The detection of rib or long bone fractures, in association with severe micromelia, suggests a diagnosis of osteogenesis imperfecta type II.

Are There Any Abnormalities of the Hands or Feet?

The anatomy of the hands and feet should be assessed whenever a skeletal dysplasia is suspected. If an extra digit is present on the side of the thumb or great toe, it is referred to as preaxial; postaxial polydactyly is on the side of the fifth finger or toe. The extra digit may be

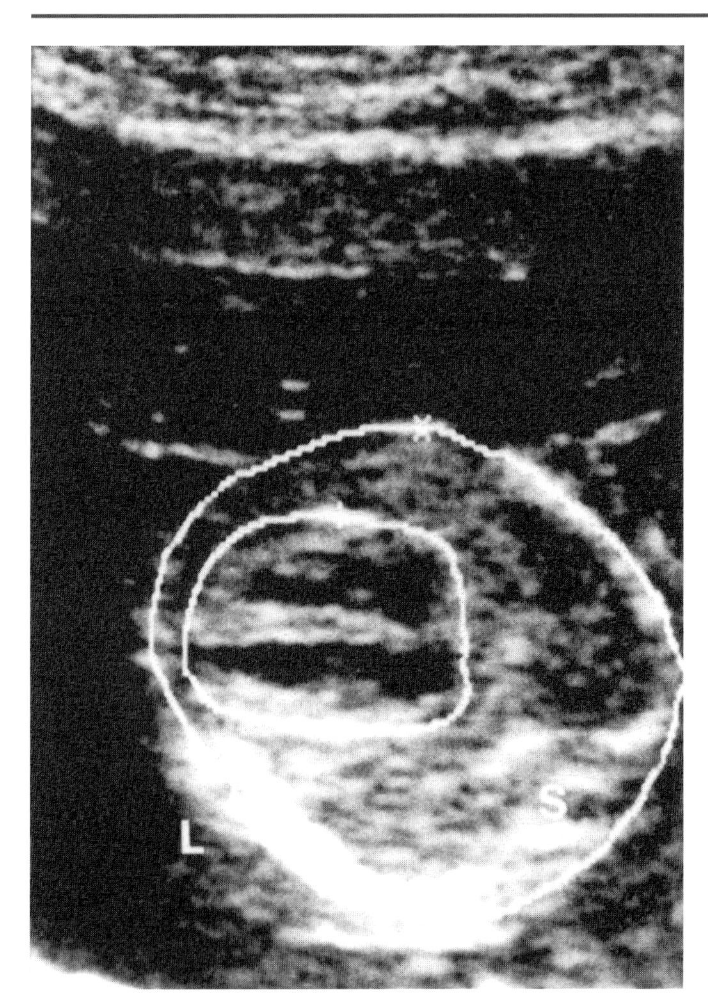

FIGURE 10 ■ Chest circumference and heart circumference at the level of the four-chamber view in a normal 17-week fetus. S, spine; L, left.

no more than a skin tag and therefore difficult to image (Fig. 20) (Table 11). Syndactyly is observed in certain rare skeletal dysplasias including Apert's syndrome, Jarcho–Levin syndrome, and Robert's syndrome.

Clubfeet (talipes equinovarus) are frequently found in camptomelic dysplasia, diastrophic dysplasia, and osteogenesis imperfecta.

Is There Polyhydramnios?

Polyhydramnios is frequently associated with severe skeletal dysplasias. In the late second or third trimester, 58% of fetuses with thanatophoric dysplasia and 27% of fetuses with achondroplasia have polyhydramnios (40). Other skeletal dysplasias associated with polyhydramnios are listed in Table 12. The severity of chest restriction and the ability of the maternal fetal unit to compensate for the excess amniotic fluid by using other pathways for fluid removal determine which fetuses will have polyhydramnios and which fetuses will have normal amniotic fluid volume.

Is There Nonimmune Hydrops?

Fetal ascites or nonimmune hydrops has been reported in chondrodysplasia punctata (43), short-rib-polydactyly syndrome (SRPS) (42), and achondrogenesis (44).

First-Trimester Nuchal Translucency

An increased nuchal translucency (Fig. 21) in the first trimester has primarily been associated with chromosomal abnormalities (49). It has also been reported with a number of skeletal dysplasias (Table 13). Achondroplasia, osteogenesis imperfecta, and thanatophoric dysplasia can now be diagnosed by first trimester chorionic villus sampling (50).

SONOGRAPHIC IDENTIFICATION OF SPECIFIC SKELETAL DYSPLASIAS

The more common lethal skeletal dysplasias and their characteristic sonographic findings are outlined in Table 14.

Lethal Skeletal Dysplasias

Achondrogenesis

Achondrogenesis is due to a generalized disorganization of cartilage, resulting in a failure of ossification. The two classic groups of achondrogenesis are based on the severity of micromelia and the degree of vertebral body and calvarial ossification (52). Type I has been further divided into type IA (Houston–Harris) and type IB (Parenti–Fraccaro). Both IA and IB have an autosomal-recessive pattern of inheritance. Type IA and IB have severely short limbs, a short neck, and a short trunk. Type IA is also characterized by deficient cranial calcification and rib fractures. In type IB, there is cranial ossification and no rib fractures (53). Type II achondrogenesis (Langer–Saldino) is characterized by more variable limb shortening, an absence of rib fractures, and relatively normal skull ossification (52). Three subtypes of type II achondrogenesis are distinguished by the severity of extremity shortening and the degree of vertebral body ossifications (54). Type II is currently considered an autosomal-dominant mutation (55). Approximately 80% of cases of achondrogenesis are type II (56). Polyhydramnios and fetal hydrops are common in type II achondrogenesis; there is marked redundancy of the subcutaneous tissue (Fig. 22).

In one case of achondrogenesis, a thickened nuchal translucency, frank hydrops, and shortened extremities held in a fixed position were detected with transvaginal sonography as early as 11 to 12 weeks' gestation (58).

Thanatophoric Dysplasia

Thanatophoric dysplasia is the most common lethal skeletal dysplasia; it was recognized as a distinct skeletal dysplasia in 1967 (59). Hence, molecular diagnostic methods can now be used to diagnose some cases of thanatophoric dysplasia by chorionic villus sampling or amniocentesis. Histologically, thanatophoric dysplasia is characterized by severe disruption of endochondral ossification (60).

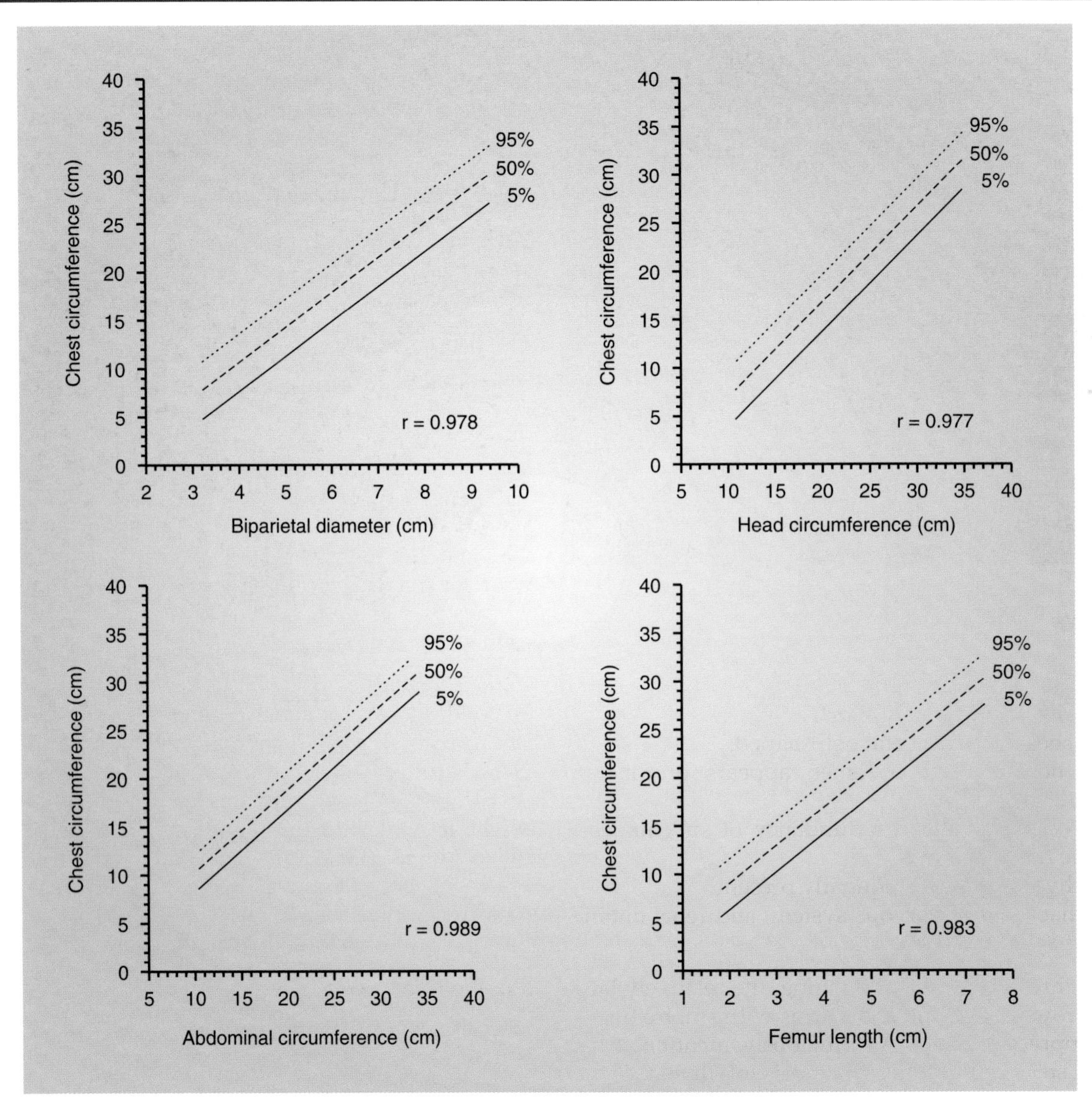

FIGURE 11 ■ Confidence limit graphs for the chest circumference. The mean, 95%, and 5% confidence limits for individual predictions of the chest circumference regressed against the biparietal diameter, head circumference, abdominal circumference, and femur length. *Source*: From Ref. 30.

The limbs are markedly shortened, the thorax is narrowed, and the abdomen is protuberant. There is usually mild frontal bossing. Although documented cases of family occurrences suggest an autosomal-recessive inheritance, the majority of cases are considered autosomal dominant. Type I thanatophoric dysplasia is characterized by severely shortened long bones, "telephone-receiver" femora, and a significantly narrowed chest. In type II, the degree of long-bone shortening is less. A severe cloverleaf skull is generally present with type II thanatophoric dysplasia, whereas individuals with type I thanatophoric dysplasia may or may not have a cloverleaf skull. Mutations in fibroblast growth factor receptor 3 cause most or all cases of type II thanatophoric dysplasia and 50% of cases of type I. The presence or absence of a cloverleaf skull is not related to the type of mutation but rather is due to variable expressivity. Since fibroblast growth factor receptor 3 is also present in the central nervous system, individuals with thanatophoric dysplasia, who survive to the neonatal period will have profound metal handicaps (59). Sonographic findings associated with thanatophoric dysplasia include the following:

1. There are markedly shortened and bowed limbs (Fig. 23A). The femurs have a characteristic "telephone receiver" appearance (Fig. 23B and C).
2. There is frontal bossing and a depressed nasal bridge (Fig. 23D and E).
3. A cloverleaf skull (Fig. 23F and G) with secondary hydrocephalus may be present.
4. Wide disc spaces are present between the flattened vertebrae (platyspondyly).

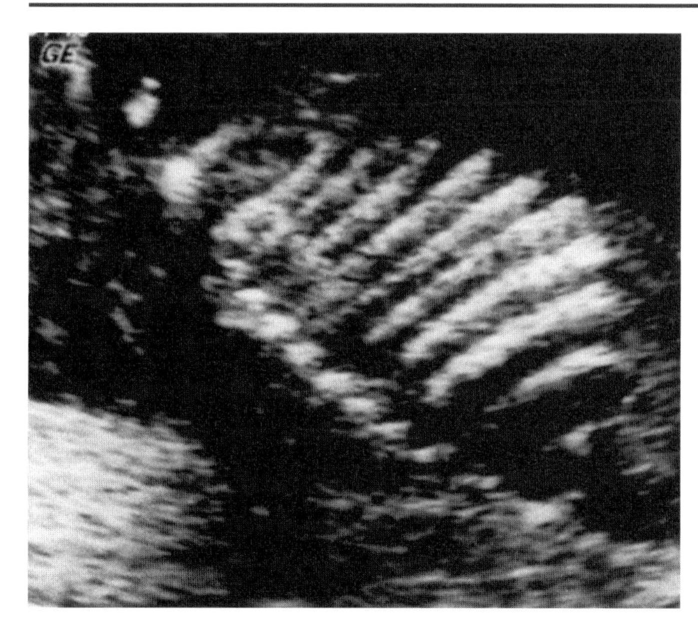

FIGURE 12 ■ Normal shape, contour, and spacing of ribs at 25 weeks' menstrual age.

5. The ribs are extremely short.
6. The chest is narrow and bell shaped.
7. The normal-sized abdomen appears protuberant (Fig. 23H and I).
8. There is generalized redundancy of subcutaneous tissue (Fig. 23J and K).
9. Polyhydramnios is frequently present.
10. Cardiac, central nervous system, and renal anomalies may occur.

With three-dimensional imaging, the relatively large head, narrow thorax, and short arms with stubby fingers can be appreciated (61). As previously mentioned, the first-trimester sonographic diagnosis of thanatophoric dysplasia can frequently be confirmed with chorionic villus sampling (62).

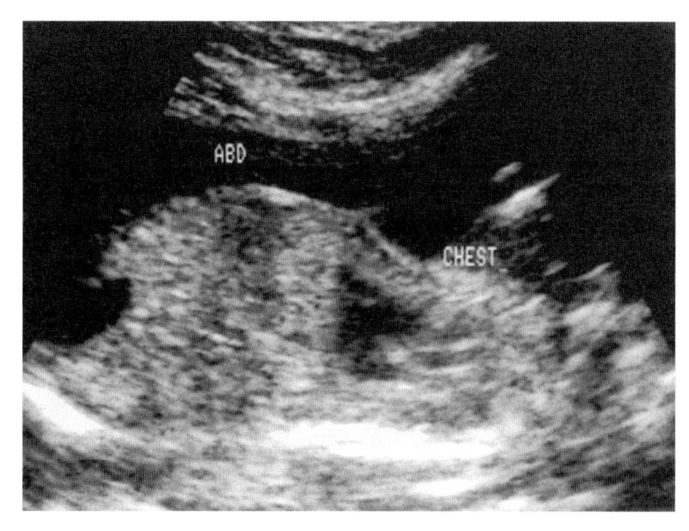

FIGURE 13 ■ Jeune syndrome at 20 weeks' gestation. The chest diameter is narrowed compared with the abdominal diameter.

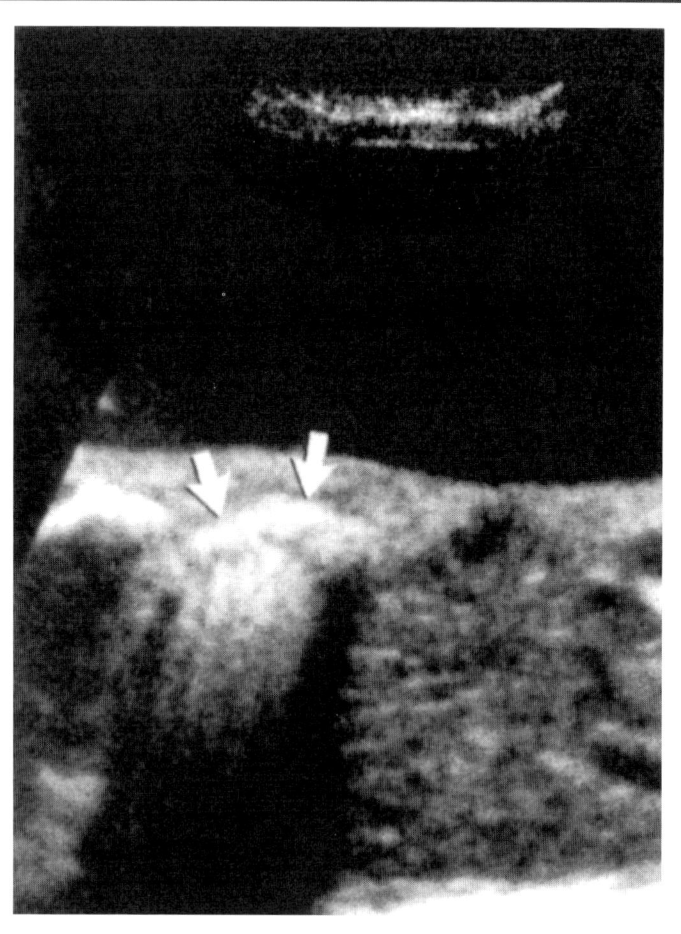

FIGURE 14 ■ Jarcho–Levin syndrome. Fusion and crowding of the ribs (*arrows*) at 23 weeks' gestation. *Source*: From Ref. 31.

Osteogenesis Imperfecta

Osteogenesis imperfecta is a group of autosomal-dominant and autosomal-recessive disorders of collagen formation. Molecular genetic studies have shown that the defect involves one of the genes that encode the chains for type I procollagen. Consequently, the collagen produced is unstable and rapidly degrades (63). Reliable prenatal

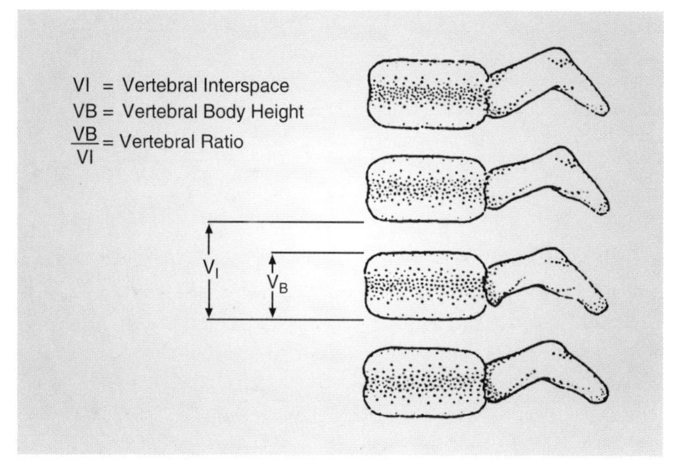

FIGURE 15 ■ Anatomic drawing of lumbar spine segments (*lateral view*) with definitions of vertebral ratio and vertebral interface. *Source*: From Ref. 34.

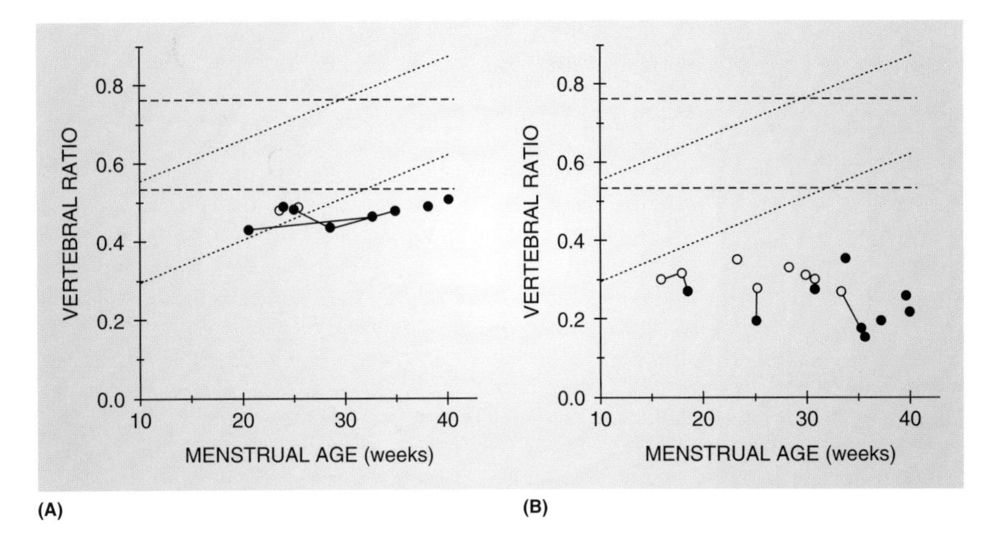

FIGURE 16 A AND B ■ Sonographic (*open circles*) and radiographic (*closed circles*) vertebral ratios observed in the second and third trimesters in (**A**) achondroplasia and (**B**) thanatophoric dwarfism. Two or more examinations of the same fetus are connected by lines. Also shown are 95% confidence bands obtained from normal sonographic data (*long dashes*) and normal radiographic data (*short dashes*). *Source*: From Ref. 34.

diagnosis by means of chorionic villus sampling is now possible (64,65).

Sillence et al. (66) classified osteogenesis imperfecta into four types based on genetic, phenotypic, and radiographic criteria. Type II has been further subdivided into three subgroups: A, B, and C (Table 15). By far, the most common subtype is A. Fetuses in this subgroup have crumpled, irregular femurs and short, beaded ribs secondary to fractures. In subtype B, the femurs have a similar appearance to subtype A, but there is minimal or no rib fractures. Subtype C fetuses have narrowed, fractured femurs and thin, beaded ribs.

The age of onset and severity of skeletal malformations differ among the various types of osteogenesis imperfecta. Maroteaux et al. (68) therefore added prognosis to the Sillence classification and identified three forms: lethal (type II); severe (type III); and regressive (types I and IV). Skeletal hypoechogenicity and limb bowing are frequently not detected until after 24 weeks' gestation in type I and type IV osteogenesis imperfecta (69). A diagnosis of nonlethal osteogenesis imperfecta should be considered when normally mineralized bones with congenital bowing are identified prenatally (70). In some milder cases, the bowed legs detected in the neonatal period straighten spontaneously. On the other hand, type II osteogenesis imperfecta can be diagnosed even in the first trimester with transvaginal sonography (70–72). In type III osteogenesis imperfecta, long-bone shortening and deformity may not become apparent until 19 to 22 weeks' gestation (Fig. 24) (73).

The sonographic characteristics of osteogenesis type IIA include the following:

1. Micromelia is early in onset and severe (Fig. 25A).
2. Long bones are deformed because of multiple intrauterine fractures (Fig. 25B).
3. Limb movement is limited.
4. Multiple fractures result in ribs that are short, beaded, and concave (Fig. 25C–E).
5. The abdomen is protuberant.
6. Hypomineralization of the skull results in excellent visualization of fetal intracranial anatomy in the near field (Fig. 25F).
7. The skull is compressible by the ultrasound transducer (Fig. 25G).

TABLE 8 ■ Descriptive Classification of the Skeletal Dysplasias

Type	Dysplasia
Rhizomelic (proximal)	Asphyxiating thoracic dystrophy
	Chondrodysplasia punctata
	Heterozygous achondroplasia
Mesomelic (middle)	Ellis–van Creveld
Micromelic (generalized)	Achondrogenesis
	Diastrophic dysplasia
	Osteogenesis imperfecta, type II
	Short-rib-polydactyly syndrome
	Thanatophoric dysplasia

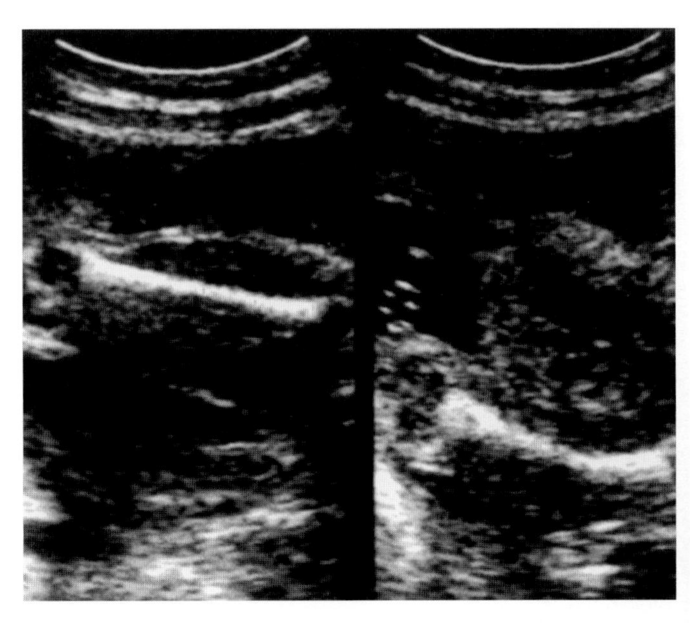

FIGURE 17 ■ The straight lateral border (*left*) and curved medial border (*right*) of a normal femur.

TABLE 9 ■ Causes of Significantly Bowed Long Bones

Prevalence	Disorder
Common	Thanatophoric dysplasia
	Camptomelic dysplasia
Less common	Achondrogenesis
	Hypophosphatasia

During the first trimester, three-dimensional ultrasound provides a global view of the fetus. As a result, the images obtained are easier to interpret and review with the patient (75). A three-dimensional profile in the second trimester helps to differentiate osteogenesis imperfecta type IIA (normal facies) from achondrogenesis (micrognathia and a flat face) (76) and short-rib polydactyly syndrome (SPRS) (hypertelorism and a broad nose) (77).

Rare Lethal Skeletal Dysplasias

Hypophosphatasia

This primary metabolic abnormality is due to defective regulation of the alkaline phosphatase gene on chromosome 1 (78). As a result, serum and tissue alkaline phosphatase activity is low. Hypophosphatasia is a heterogeneous disorder with recessive inheritance and variable presentation (79). The three types of hypophosphatasia—congenital, childhood, and adult—are defined by age of onset and the severity of the disease (80). In the congenital (lethal) type, the calvarium is hypomineralized; the long bones are shortened with midshaft angulation, metaphyseal cupping, and fractures; and the chest is small because of multiple rib fractures (Fig. 26). Polyhydramnios is an additional sonographic finding. The differential diagnosis includes achondrogenesis type I and osteogenesis imperfecta type IIC. The prevalence of hypophosphatasia is about half that of osteogenesis

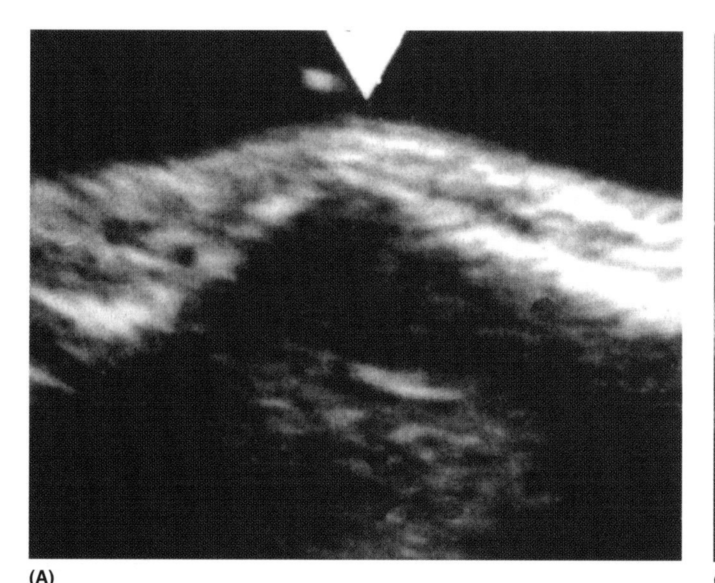

(A)

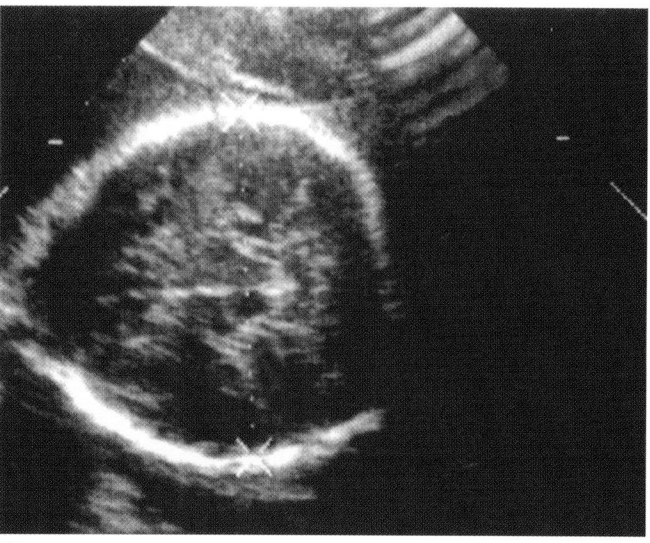

(B)

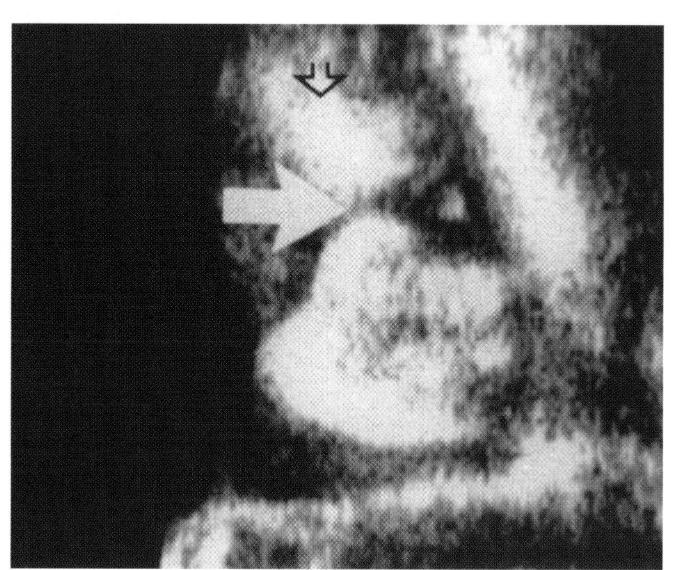

(C)

FIGURE 18 A–C ■ Antley–Bixler syndrome. (**A**) Femoral bowing (*arrow*). (**B**) Trigonocephaly. (**C**) Depressed nasal bridge (*arrow*), frontal bossing (*open arrow*), and hypoplastic midfacies. *Source*: From Ref. 38.

TABLE 10 ■ Skeletal Dysplasias with Hypomineralization

- Achondrogenesis
- Hypophosphatasia
- Osteogenesis imperfecta

imperfecta type IIC. First trimester diagnosis by means of chorionic villous sampling is possible through the measurement of cellular alkaline phosphatase (82).

Short-Rib-Polydactyly Syndrome

The four types of SRPS are autosomal-recessive chondrodysplasias that are invariably fatal. They are characterized by severely narrow chests, significant micromelia, and preaxial or postaxial polydactyly (Table 16) (84,85). Type II can be distinguished from types I and III by a severely shortened tibia and a cleft lip; the fetus may be hydropic. The absence of cystic renal disease in type III distinguishes it from types I and II. Type IV SRPS is characterized by a median cleft lip/palate, anophthalmia, and hydrops (83). Polyhydramnios and congenital heart defects, specifically atrial septal defects, are common with all the SRPSs (42,86).

SRPS has been diagnosed in the first trimester (87).

Since SRPS is an autosomal-recessive condition, a late first-trimester ultrasound examination should, therefore, be performed in any subsequent pregnancy.

Three-dimensional ultrasound in the second trimester can be used to specifically evaluate the fetal face and limbs. Characteristic facial features that can readily be appreciated with three-dimensional imaging include hypertelorism, a broad nose, and low-set ears; polydactyly is also more easily detected (88).

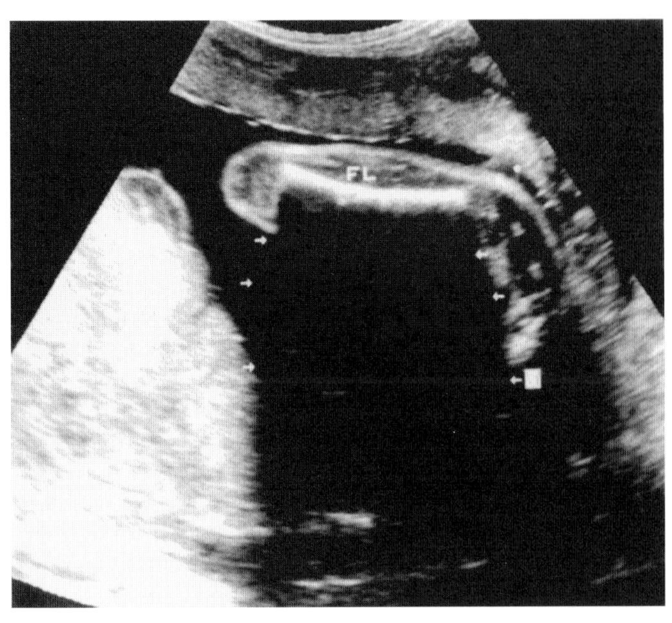

FIGURE 19 ■ Acoustic shadowing (*arrows*) from a femur length of a 24-week fetus. FL, femur length.

Camptomelic Dysplasia

The altered collagen synthesis that results in camptomelic dysplasia is due to a mutation with autosomal-recessive inheritance that has been localized to chromosome 17.

Camptomelia (bent-limb) dysplasia is characterized by prominent bowing of the long bones, particularly the lower extremities. The bowing of the femurs occurs in the upper half of the bone; the tibias are bowed in their lower half.

Although camptomelic dysplasia had been considered uniformly lethal, there are reports of survivors up to

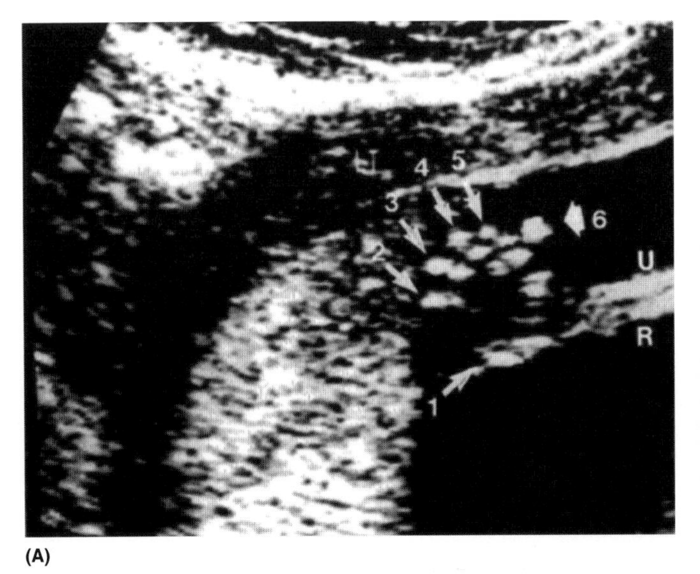

(A)

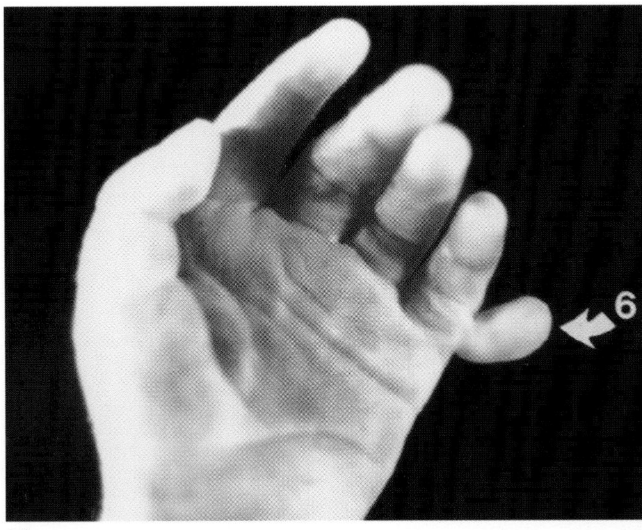

(B)

FIGURE 20 A AND B ■ Postaxial polydactyly of the left hand. (**A**) Ultrasound image demonstrating thumb (1) and digits two through five with postaxial polydactyly. (**B**) Corresponding fetal specimen. R, radius; U, ulna.

TABLE 11 ■ Skeletal Dysplasias Associated with Polydactyly

- Asphyxiating thoracic dystrophy (Jeune syndrome)
- Short-rib-polydactyly syndrome
- Chondroectodermal dysplasia

the age of 17 years (37). Three types of camptomelic dysplasia can be distinguished radiographically. Patients with short-limbed camptomelia are either normocephalic or have a cloverleaf skull. In both types, the extremities are short and bent. The third type of camptomelia (long-limbed) is characterized by a normal calvarium and bowing of a slightly shortened femur and tibia. The upper extremities are minimally affected. Polyhydramnios has been noted in cases of short-limbed camptomelia diagnosed prenatally (45). Absent or hypoplastic scapulae and fibulae, a cleft palate, micrognathia, clubbed feet, and hydronephrosis may also be present (89,90).

Chondrodysplasia Punctata: Rhizomelic Type

This skeletal dysplasia is an autosomal-recessive condition with severe rhizomelic (proximal) shortening of the long bones. Multiple punctate calcifications at the epiphyses of the long bones can be identified sonographically. Laryngeal stippling due to calcifications may also be detected (91). Joint contractures are common. Additional sonographic findings are provided in Table 17. Most neonates with the rhizomelic form usually die within the first few weeks of life, and those who survive beyond one year are severely retarded (92).

There are four other types of chondrodysplasia punctata: x-linked dominant (Conradi–Hunermann); x-linked recessive; tibia-metacarpal (chondrodysplasia punctata, tibial-metacarpal, CP-MT); and those associated with other syndromes. The x-linked dominant form is only seen in females; it is lethal in affected males. The degree of limb shortening with Conradi–Hunermann is less severe than with the rhizomelic type and in some cases can be

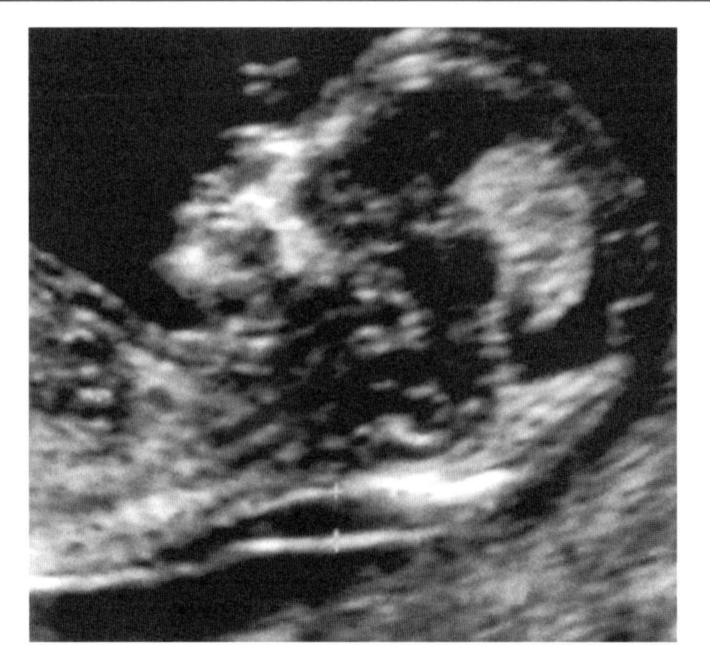

FIGURE 21 ■ First trimester thickened nuchal translucency (between the markers).

quite mild; normal intelligence is expected. The x-linked recessive type would only affect males. In addition to shortened extremities and microcephaly, these infants are generally mentally handicapped. The characteristic stippled calcification is generally absent after 18 months of age (93). In a case report of CP-MT, the sonographic findings at 16 to 18 weeks' menstrual age revealed shortened long bones, curved radii and ulnae, micrognathia, and a sacral tail (94).

Nonlethal Skeletal Dysplasias

Only a few of the nonlethal skeletal dysplasias are reviewed (Table 18). The interested reader is referred to a more definitive text (95) for additional information.

Diastrophic Dysplasia

This condition is an autosomal-recessive disorder of cartilage that is due to an abnormality in cellular sulfate

TABLE 12 ■ Skeletal Dysplasias Associated With Polyhydramnios

Prevalence	Dysplasia	Investigators
Common	Thanatophoric dysplasia	Thomas et al. (40)
	Achondroplasia	Thomas et al. (40)
Less common	Hypophosphatasia	Wladimiroff et al. (41)
	Short-rib-polydactyly syndrome	Meizner and Bar-Ziv (42)
	Chondrodysplasia punctata	Straub et al. (43)
	Achondrogenesis	Benacerraf et al. (44)
	Camptomelic dysplasia	Winter et al. (45)
	Schneckenbecken dysplasia	Borochowitz et al. (46)
	Metatrophic dysplasia	Manouvrier-Hanu et al. (47)
	Spondyloepiphyseal dysplasia congenita	Kirk and Comstock (48)

TABLE 13 ■ Skeletal dysplasias associated with a thickened nuchal translucency (50,51)

- Short-rib polydactyly
- Arthrogryposis
- Asphyxiating thoracic dystrophy
- Achondrogenesis
- Camptomelia
- Jarcho–Levin syndrome
- Osteogenesis imperfecta
- Roberts syndrome
- Thanatophoric dysplasia
- Ellis–van Creveld syndrome

TABLE 14 ■ Sonographic Features of the More Common Lethal Skeletal Dysplasias

Skeletal dysplasia	Sonographic findings
Achondrogenesis	Deficient ossification of vertebral bodies and skull
Thanatophoric dysplasia	Cloverleaf skull
	Narrowed thorax
	"Telephone receiver" femur
Osteogenesis imperfecta	Multiple factures
	Hypomineralized calvarium
Hypophosphatasia	Hypomineralized calvarium
	Demineralized long bones with fractures
Short-rib-polydactyly syndrome	Narrow chest
	Polydactyly
Camptomelic dysplasia	Bowing of the upper half of the femur and lower half of the tibia
Chondrodysplasia punctata	Punctate epiphyseal calcifications of the long bones

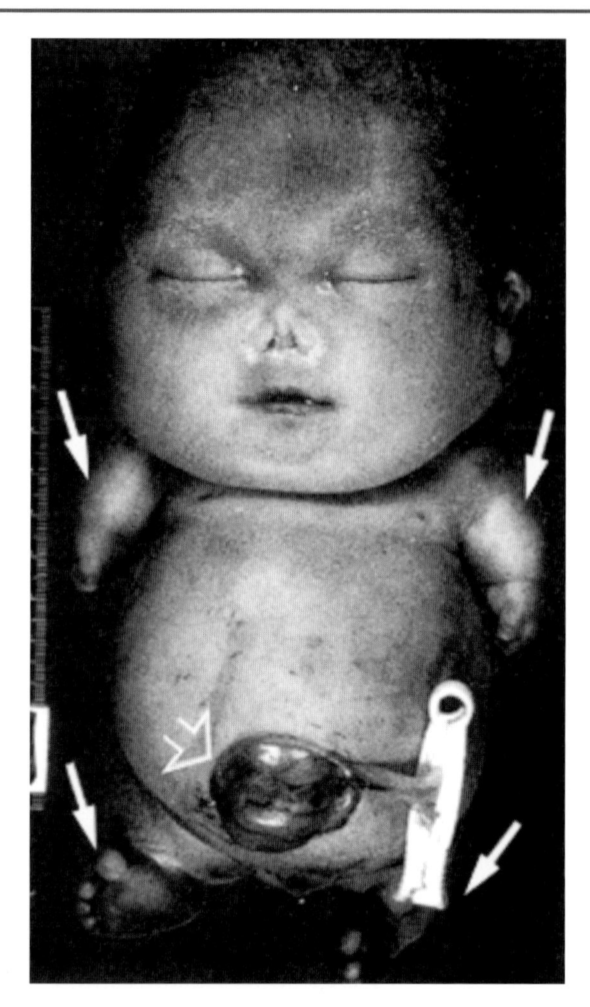

FIGURE 22 ■ Achondrogenesis: stillborn, 36-wk-old infant with extremely short limbs (*arrows*), narrow thorax, omphalocele (*open arrows*), relative macrocephaly, and a depressed nasal bridge. *Source*: From Ref. 57.

transport. In addition to severe micromelia, patients have micrognathia, cleft palate, bilateral clubbed feet, kyphosis, and polyhydramnios. A characteristic sonographic finding is a subluxated thumb ("hitchhiker" thumb) or big toe. Because the thumb is not in the sample plane as the fingers, a "hitchhiker" thumb must be specifically sought sonographically (Fig. 27). Polyhydramnios may occur (97,98). The mortality rate in the neonatal period is approximately 25% and is due to respiratory insufficiency. Progressive kyphoscoliosis and contractures result in severe physical disability (96).

Three-dimensional ultrasound can delineate not only the characteristic hitchhiker thumbs but also the facial dysmorphism associated with this skeletal dysplasia (Fig. 28A and B) (99).

Heterozygous Achondroplasia

Heterozygous achondroplasia is one of the more common skeletal dysplasias. Eighty percent of cases occur as spontaneous mutations; the remaining 20% are autosomal dominant (100). The cause in over 90% of cases is a mutation in the fibroblast growth factor receptor 3 gene (1). The affected infant has a large head with frontal bossing and a depressed nasal bridge; the chest is normal. Rhizomelic (proximal) shortening of the limbs is generally not significant until after 24 weeks' gestation (Fig. 29). The long bones in the lower extremities may be affected before those in the upper extremities (101). The fingers are of equal length, and the third and fourth fingers do not approximate (102). This three-pronged (trident) appearance of the hands is a characteristic finding in heterozygous achondroplasia (Fig. 30) (101,103). If one parent is achondroplastic, or if both are, or if a couple has had a previously affected child, prenatal diagnosis through chorionic villus sampling or amniocentesis is available (1).

Growth hormone has been shown to have significant dose-dependent effects on skeletal growth in neonates with achondroplasia (104).

Homozygous achondroplasia is usually incompatible with life. The thorax is significantly smaller than in heterozygous achondroplasia, resulting in respiratory insufficiency. Pathologically, the cartilage in heterozygous achondroplasia shows reduced but organized growth. However, the growth zone in fetuses with homozygous achondroplasia is markedly disorganized. As a result, an abnormally short femur is detected approximately one month sooner in homozygous than in heterozygous achondroplasia (105).

Three-dimensional ultrasound has been utilized to visualize the characteristic facial features (mid-face hypoplasia and a small flat nasal bridge) as well as the pointed aspect of the upper end of the femur in fetuses with achondroplasia (106).

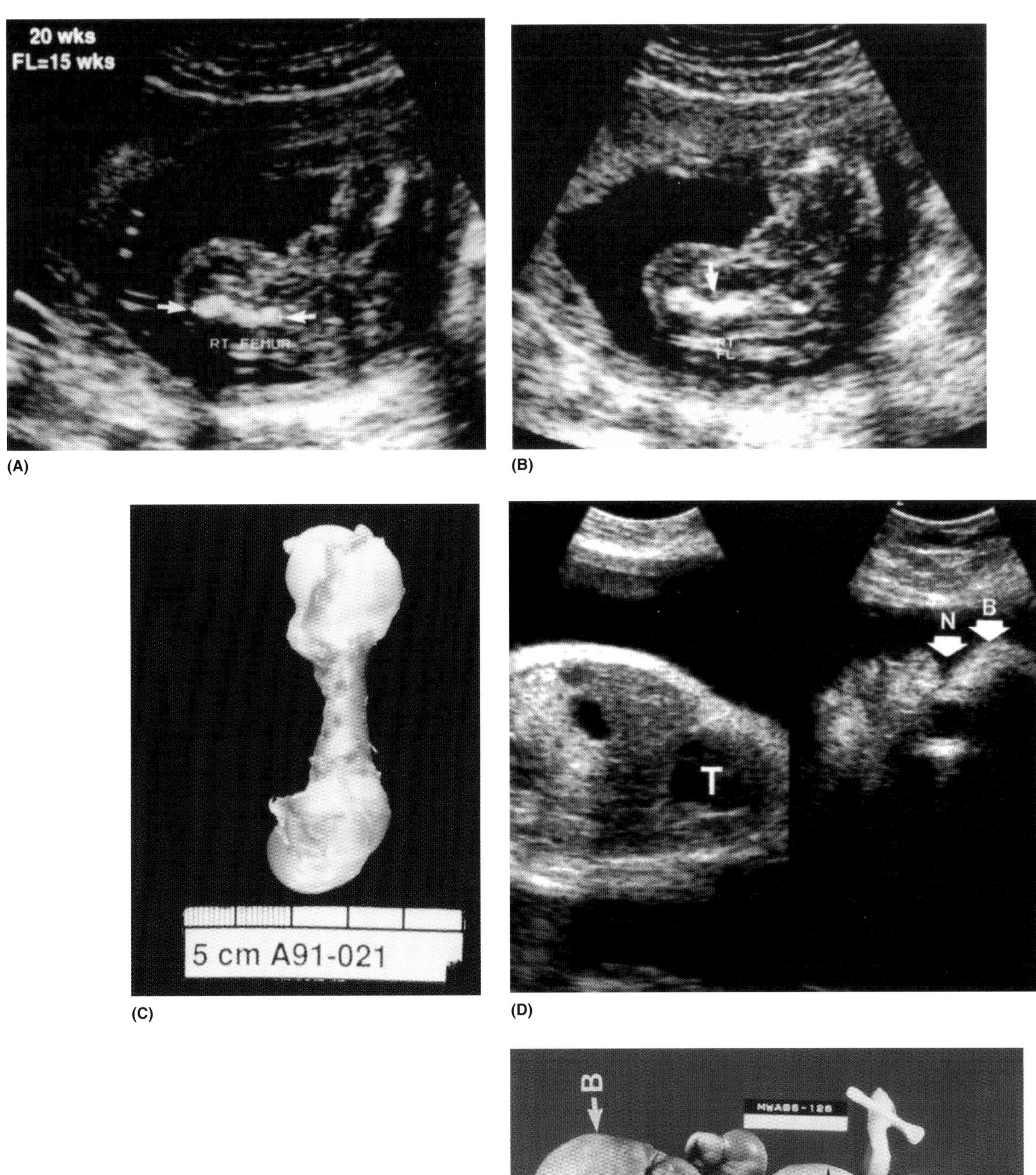

FIGURE 23 A–K ■ Thanatophoric dysplasia. (**A**) Femur length (*arrow*) equivalent to 15 weeks' gestation at 20 weeks' menstrual age. (**B**) Thickened distal femurs or so-called "telephone receiver" femur (*arrow*) at 24 weeks' gestation. (**C**) Telephone receiver femur; pathologic specimen. (**D**) Frontal bossing (B) and depressed nasal bridge (N). (**E**) Frontal bossing (B) and depressed nasal bridge on autopsy. Also note narrow thorax (T) and protuberant abdomen (A). (*Continued*)

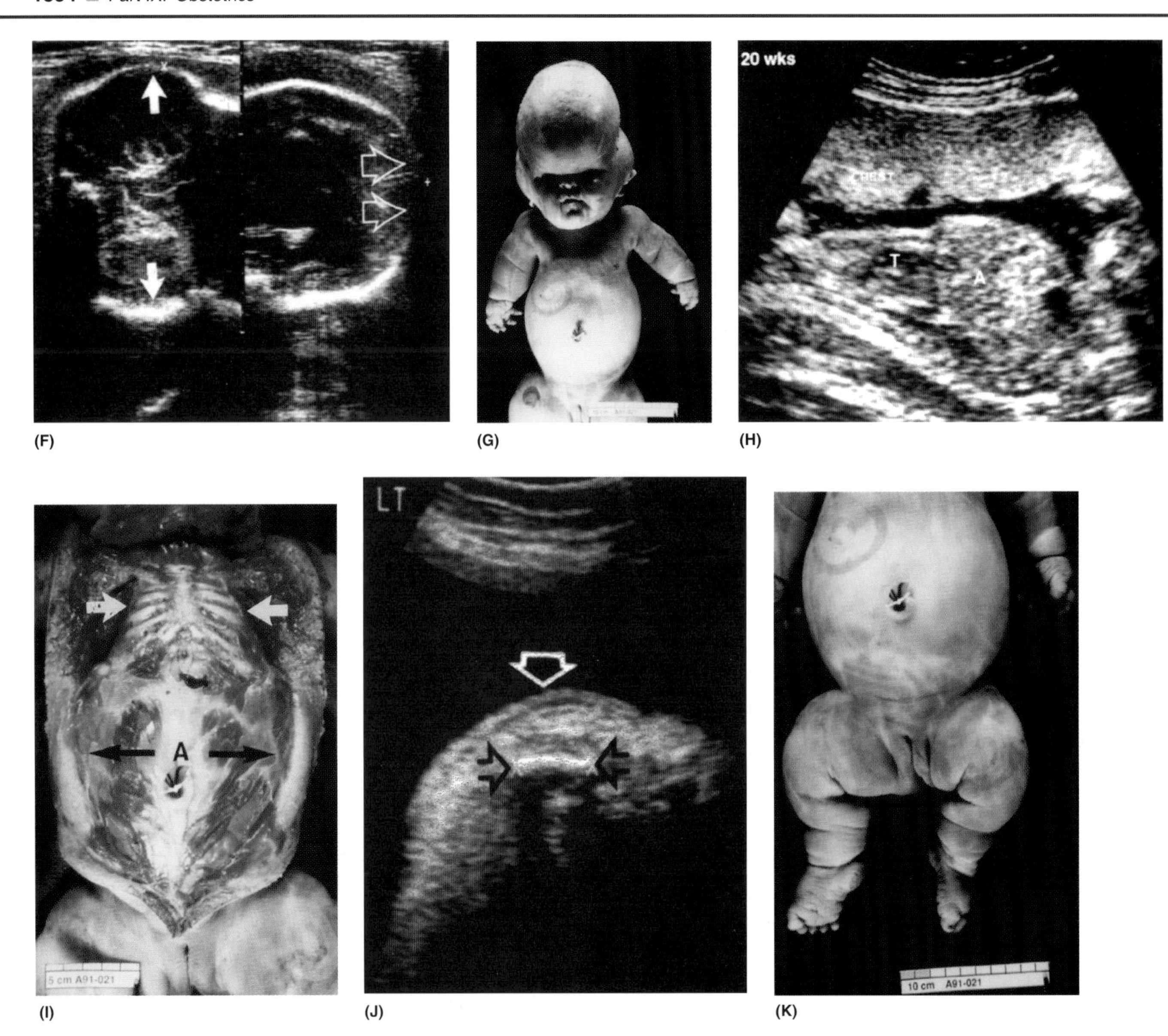

(F) (G) (H) (I) (J) (K)

FIGURE 23 A–K ■ (*Continued*) (**F**) Coronal ultrasound demonstrating outward bulging in the temporal region (*solid arrows*) and frontal bossing (*open arrows*). (G) Corresponding autopsy specimen of cloverleaf skull. (**H**) Narrowed chest (**T**) and protuberant abdomen at 20 weeks' gestation. (**I**) Narrowed chest (*arrow*) and protuberant abdomen; autopsy after dissection. (**J**) Increased subcutaneous tissue (*white open arrow*) about a severely shortened ulna (*open arrows*). (**K**) Redundancy of skin about lower extremities on autopsy.

Hypochondroplasia is an autosomal-dominant disorder. The skull is not affected, and shortening of the long bones may not be apparent until the neonatal period. Prenatal sonographic detection has been reported (107).

Asphyxiating Thoracic Dystrophy (Jeune Syndrome)

This skeletal disorder has an autosomal-recessive pattern of inheritance and is characterized by an extremely narrow chest with secondary pulmonary hypoplasia. The limbs are mildly shortened or of normal length. Although Jeune syndrome has been diagnosed as early as 17 weeks' gestation, long-bone shortening may not occur until the late second or early third trimester. Cleft lip or palate and polydactyly may occur. Unlike the other SRPSs, congenital heart disease is not an associated finding with asphyxiating thoracic dystrophy; renal dysplasia may also occur. Approximately 60% of neonates with asphyxiating thoracic dystrophy die in infancy from pulmonary hypoplasia (108–111).

Chondroectodermal Dysplasia (Ellis–Van Creveld Syndrome)

Short limbs with polydactyly and congenital heart disease (50% of cases) are the characteristic findings in this skeletal dysplasia. It is present in inbred Amish communities in Pennsylvania. Patients have acromesomelic shortening; the hands, feet, radius and ulna, and

TABLE 15 ■ Classification of Osteogenesis Imperfecta

Type[a]	Qualifying suffixes	Course	Inheritance
I	Congenita tarda with blue sclerae	Usual onset in childhood or puberty; 96% of patients able to walk independently; deafness (35%) experienced as greatest handicap	AD (most frequent)
II	Perinatally lethal	Death before or shortly after birth	
A			AD (mostly new mutations, rarely AR)
B			AR, but possible new dominant mutations
C			AR
III	Progressively deforming	Progressive deformity of limbs and spine leading to severe handicap and often early death	AR, but possible new dominant mutations
IV	Congenita tarda with normal sclerae	Like type I but no deafness; dentinogenesis imperfecta	AD (most frequent)

[a]Types defined from Ref. 66.
Abbreviations: AD, autosomal dominant; AR, autosomal recessive.
Source: From Ref. 67.

tibia and fibular are shorter than the humerus and femur. The fetal chest is long and narrow, and the ribs are shortened. Postaxial polydactyly of the hands is present in almost every case; 10% of fetuses also have postaxial polydactyly of the feet. An atrial septal defect is the most common type of congenital heart defect (112). Table 16 outlines the findings in the six types of SRPSs.

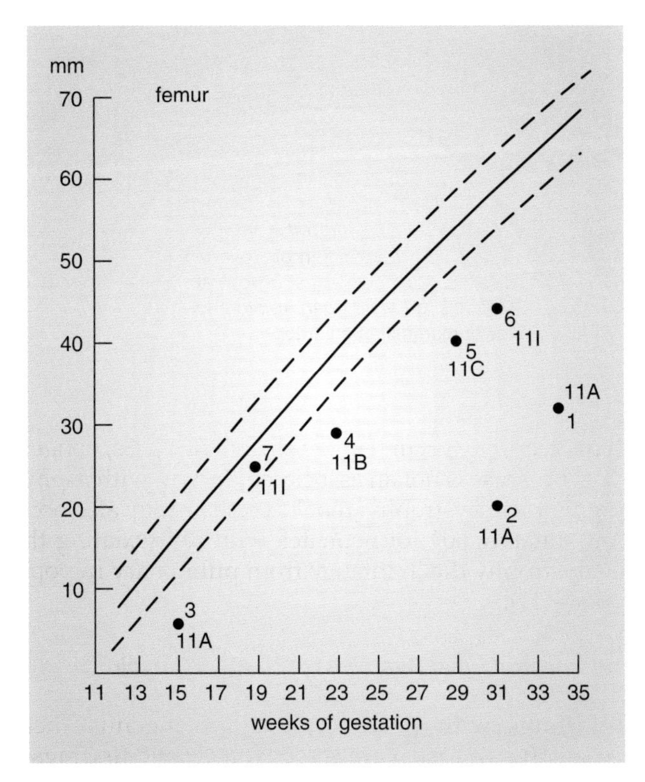

FIGURE 24 ■ Femur length in seven fetuses with osteogenesis imperfecta types IIA, IIB, and IIC, and III plotted against normal curves. *Source*: From Ref. 67.

Metatropic Dysplasia

The term metatropic means changeable. Although these neonates have a slightly elongated trunk at birth, severe kyphoscoliosis results in a shortened trunk later in life. In type I metatropic dysplasia, the face is normal, whereas the clinical features of type II metatropic dysplasia (Kniest syndrome) include a flat forehead and nose. Sonographically, the long bones are short in both types I and II metatropic dysplasia. However, the lag in FL growth may not be apparent until the third trimester (113). An autosomal-recessive lethal type of metatropic dysplasia is also recognized in which the long bones are significantly shortened in the second trimester. Whereas the femurs and humeri classically have broad, rounded ends (dumbbell appearance) (Fig. 31), this is not always the case (114). Although the bones of the hands and feet are also short, widening of the joint spaces makes the hands and feet longer than expected for gestational age. Polyhydramnios has also been described in patients with metatropic dysplasia.

Spondyloepiphyseal Dysplasia Congenita

Patients with this autosomal-dominant condition have delayed ossification of the spine, skull, and proximal epiphyses. The femurs are shortened and mildly bowed. Because of the characteristically shortened trunk, intrathoracic volume can be significantly decreased, even in the presence of an apparently normal chest circumference. Hypertelorism, a cleft palate, and polyhydramnios are associated findings in some cases (48).

MALFORMATIONS OF INDIVIDUAL BONES, SINGLY OR IN COMBINATION

Limb-Reduction Abnormalities

Limb-reduction deformities involve either the absence or reduction of one or more limbs or parts of a limb. The overall incidence of congenital limb reduction is 0.49 per 10,000 births (115). Approximately three-fourths of these malformations affect the upper extremities and one-fourth the lower limbs (116). The radius is most frequently absent.

Radial aplasia or hypoplasia (Fig. 32) is often associated with absence or hypoplasia of the navicular bone, first metacarpal, and thumb. This combination of defects is referred to as the radial-ray reduction malformations (117). Possible causes of this constellation of malformations include chromosomal abnormalities, genetic or nongenetic syndromes, and teratogens. For example, trisomy 13 and trisomy 18, as well as alcohol or valproic acid ingestion, have been associated with radial-ray reduction malformations. Other possible diagnoses when a radial-ray defect is detected sonographically are outlined in Table 19.

Congenital absence of the ulna also results in a clubbed hand. Although congenital absence of the ulna may be an isolated event, it also occurs in various different syndromes.

A congenitally short femur (proximal focal femoral deficiency) is a rare congenital anomaly, occurring once in every 50,000 deliveries (124). Proximal femoral focal deficiency has been classified by Aitkin (125).

1. A small area of deficient femur below the femoral head
2. Absence of a large segment of the femur
3. Absence of part of the femur as well as the femoral head and deformation of the acetabulum
4. Absence of both the acetabulum and femoral head

A reduction in the length or complete absence of the fibula is the most common long-bone deficiency (Fig. 33). It is usually associated with anteromedial bowing and shortening of the tibia (126). If there is a significant leg length discrepancy, amputation is usually recommended. The ipsilateral femur may also be shortened. Of infants with proximal femoral focal deficiency, 50% will have fibular hemimelia (127).

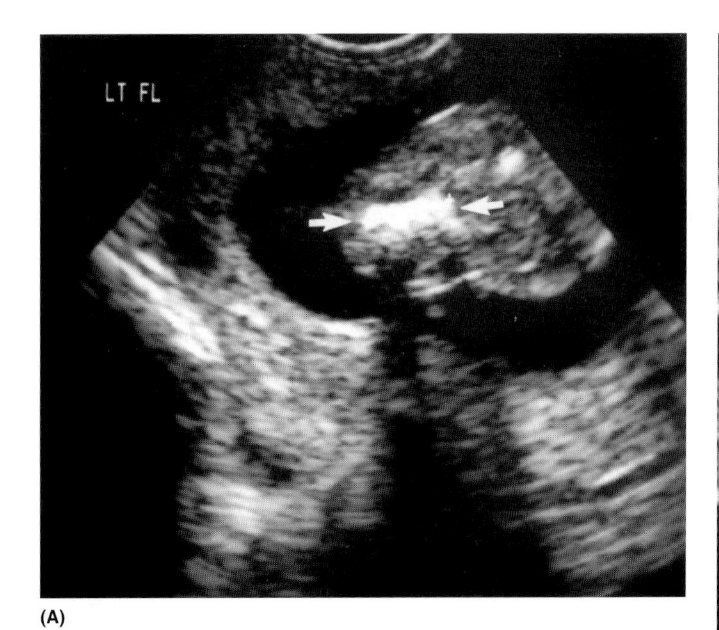

(A)

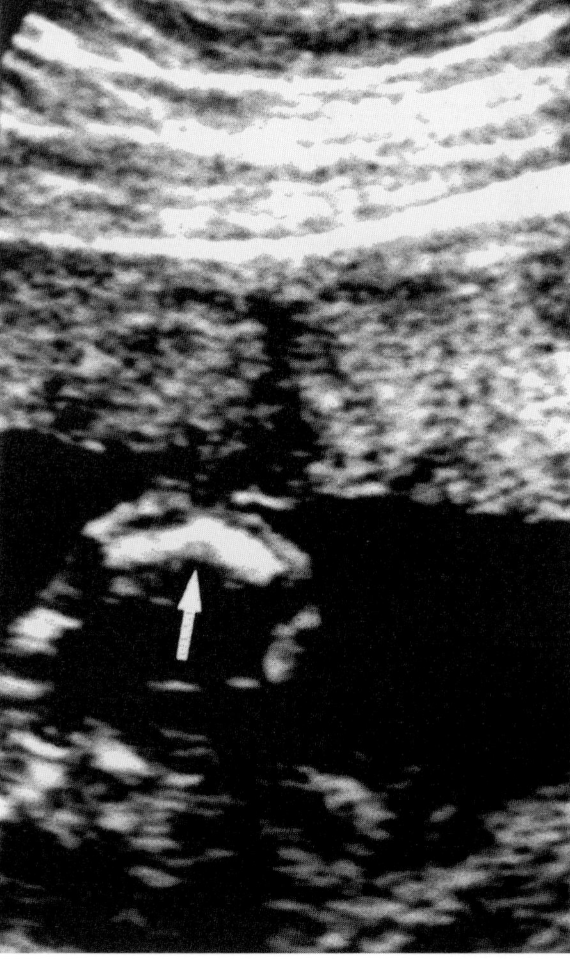
(B)

FIGURE 25 A–G ■ Osteogenesis imperfecta type IIA. (**A**) Thickened, mottled, severely shortened femur (*arrows*). (**B**) Fractured ulna (*arrow*) at 20 weeks' gestation. (*Continued*)

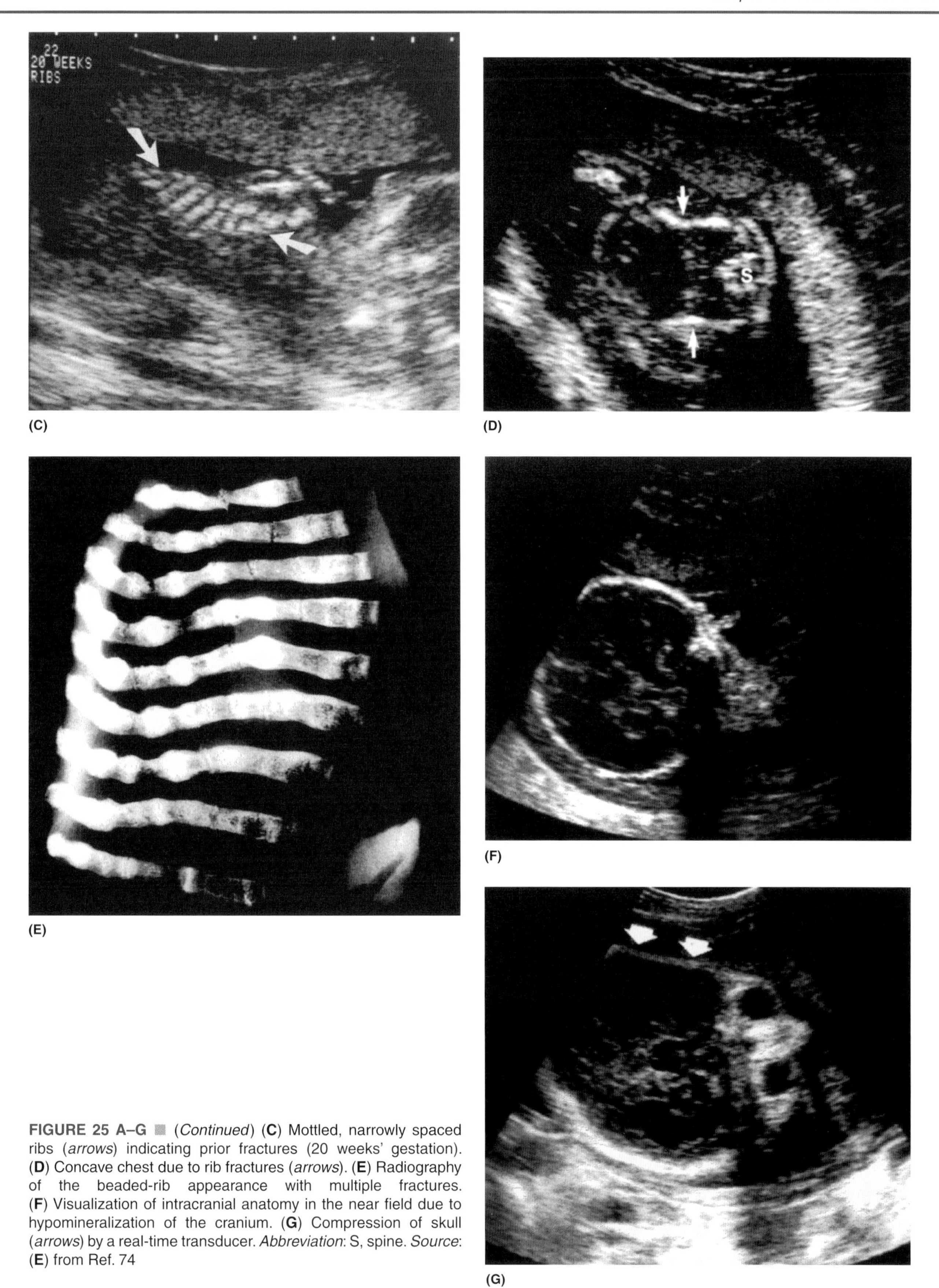

FIGURE 25 A–G ■ (*Continued*) (**C**) Mottled, narrowly spaced ribs (*arrows*) indicating prior fractures (20 weeks' gestation). (**D**) Concave chest due to rib fractures (*arrows*). (**E**) Radiography of the beaded-rib appearance with multiple fractures. (**F**) Visualization of intracranial anatomy in the near field due to hypomineralization of the cranium. (**G**) Compression of skull (*arrows*) by a real-time transducer. *Abbreviation*: S, spine. *Source*: (**E**) from Ref. 74

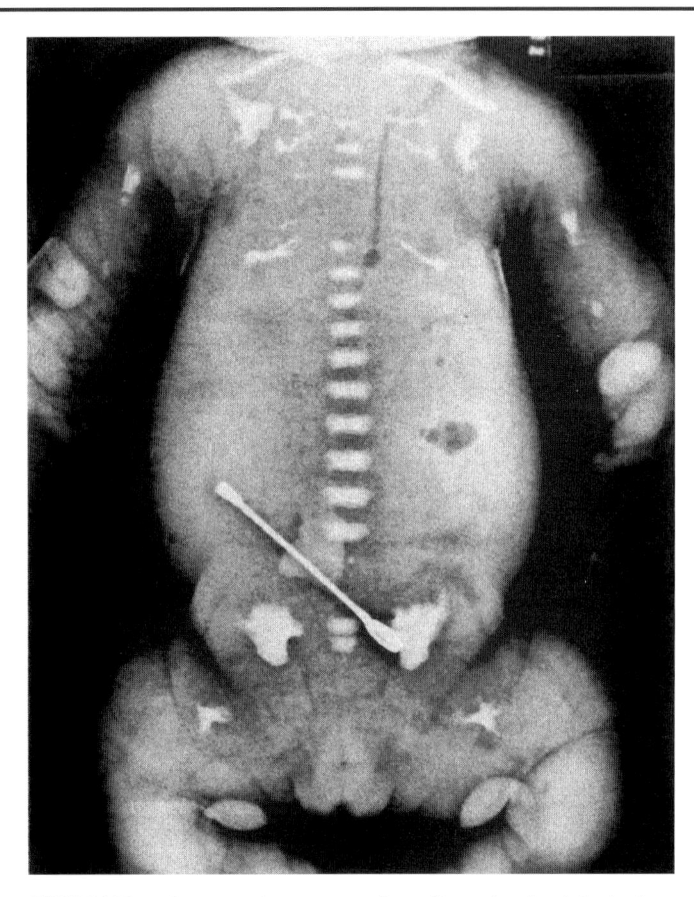

FIGURE 26 ■ A postnatal anteroposterior radiography of an infant's chest and abdomen demonstrating severe hypophosphatasia. There is incomplete mineralization of the ribs and long bones. *Source*: From Ref. 81.

The femur-fibula-ulna syndrome is characterized by a hypoplastic to absent femur in association with the absence of the fibular (128). Because the tibia supports the body, its absence is more serious, and treatment is generally unsatisfactory.

TABLE 17 ■ Associated Sonographic Findings with Chondrodysplasia Punctata, Rhizomelic Type

- Microcephaly
- Micrognathia
- Cleft palate
- Cataracts
- Congenital heart disease
- Kyphoscoliosis
- Ascites
- Polyhydramnios

Amniotic Band Syndrome

The prevalence of amniotic band syndrome is approximately 7.8 per 10,000 live births. It is much more frequent in spontaneous miscarriages (178 per 10,000) (129). Torpin (130) theorized that the amniotic band syndrome is due to amnion rupture, with oligohydramnios and subsequent contact between the fetus and the chorion. However, many anomalies associated with the amniotic band syndrome remain unexplained by this theory (Table 20). Lockwood et al. (131) proposed that the amniotic band syndrome is multifactorial in origin, with a final common pathway of vascular compromise. The disruption of tissue from hemorrhage then leads to various malformations. The amniotic bands are, therefore, secondary to vascular damage and tissue disruption and are not primarily responsible for the fetal malformations. These

TABLE 16 ■ Ultrasonic Approach to the Diagnosis of the SRPS

–	SRPS I (Saldino–Noonan)	SRPS II (Majewski)	SRPS III (Naumoff)	SRPS IV (Beemer–Langer)	Chondroectodermal dysplasia	Asphyxiating thoracic dystrophy
Short limbs	++	+	++	+	(+)	(+)
Narrow thorax	++	++	++	+	+	+
Polydactyly	++	++	++	–	++	+
Cleft lip and palate	–	++	–	–	–	–
Cardiovascular abnormalities	+	+	+	–	+	–
Polycystic kidneys	+	+	–	+	–	+
Genital anomalies	++	+	++	+	–	–
Metaphyseal dysplasia:						
Pointed metaphyses	++	–	–	–	–	–
Widened metaphyses with marginal spurs	–	–	++	–	–	–
Disproportional short tibia	–	++	–	–	–	–

Key: ++, present; +, inconsistently present; –, absent; (+) present to a mild degree.
Abbreviation: SRPS, Short-rib-polydactyly syndrome.
Source: From Refs. 42, 83.

TABLE 18 ■ Sonographic Features of the More Common Nonlethal Skeletal Dysplasias

Skeletal dysplasia	Sonographic findings
Diastrophic dysplasia	Hitchhiker thumbs
Heterozygous achondroplasia	"Trident" hand
Asphyxiating thoracic dystrophy (Jeune syndrome)	Extremely narrow chest
Chondroectodermal dysplasia (Ellis–van Creveld syndrome)	Postaxial polydactyly: atrial septal defect
Metatropic dwarfism	"Dumbbell" femur
Spondyloepiphyseal dysplasia congenita	Bowed femurs; delayed spinal and skull ossification

investigators proposed that the amniotic band syndrome be renamed the "fetal disruption complex."

Isacsohn et al. (132) classified annular constrictions (Fig. 34) of the extremities into five stages of severity: (*i*) the presence of a shallow groove; (*ii*) encroachment on underlying subcutaneous tissue and muscle by the constriction band; (*iii*) extension of the constriction band down to the bone; (*iv*) pseudoarthrosis of underlying bone; and (*v*) intrauterine amputation. The amniotic bands may be imaged sonographically restricting fetal movement. However, a diagnosis of amniotic band syndrome cannot be excluded sonographically if amniotic bands are not visualized.

Sirenomelia

Sirenomelia (mermaid syndrome) denotes fusion of the lower extremities that ranges from a membranous attachment to total fusion with a single midline femur and fibula (Fig. 35A and B). The fetal kidneys are usually absent or dysplastic, resulting in oligohydramnios, chest compression, and pulmonary hypoplasia (134).

Sonographic diagnosis of the lower limb abnormalities are limited by the anhydramnios associated with bilateral renal agenesis. Color and power Doppler have been used to evaluate the peculiar vascular pattern associated with this malformation. The aorta ends abruptly and does not divide into two iliac arteries. The vitelline artery originates from the aorta, transverses the abdomen and exits at the umbilicus (135).

Malformations of the Hands and Feet

Polydactyly

Hexadactyly (six fingers or toes) is a frequent anomaly (Fig. 36A and B). Polydactyly with more than six digits is

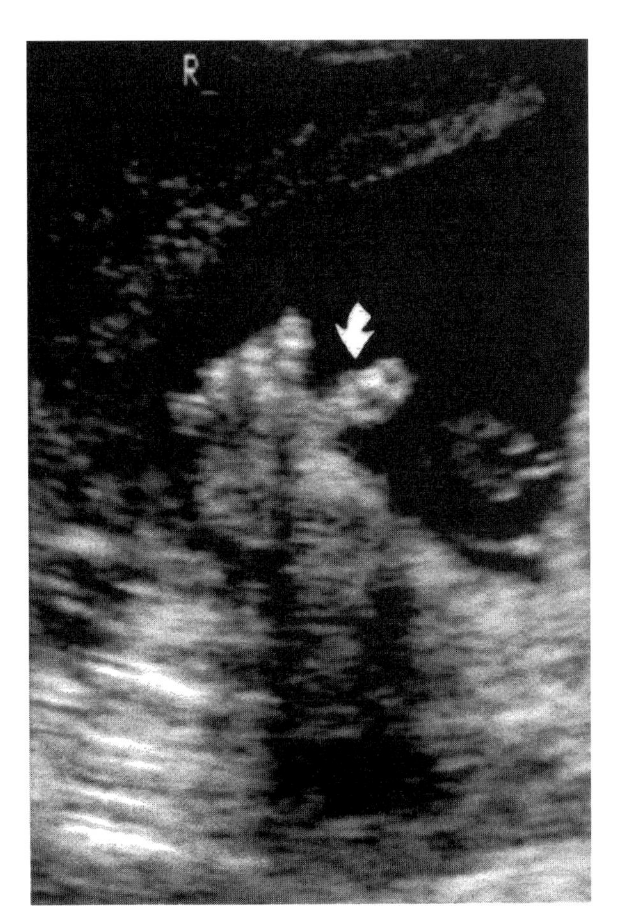

(A)

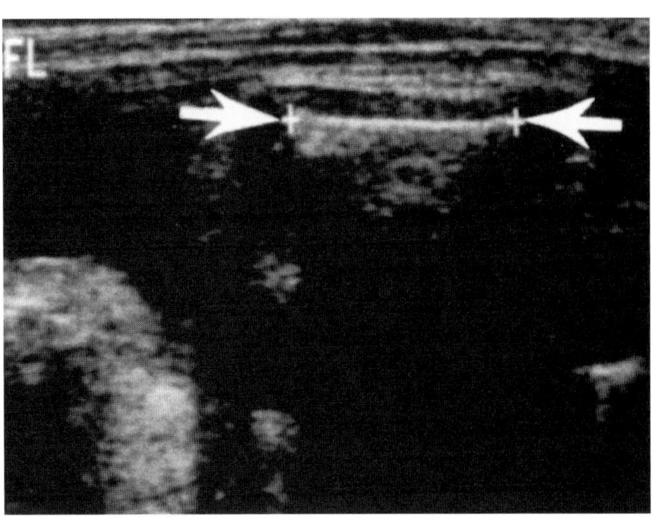

(B)

FIGURE 27 A–F ■ Diastrophic dysplasia at 33 weeks' gestation. (**A**) Subluxation of the big toe (*arrow*). (**B**) The femur length of 4.1 cm is consistent with a 23.5-week gestation. (*Continued*)

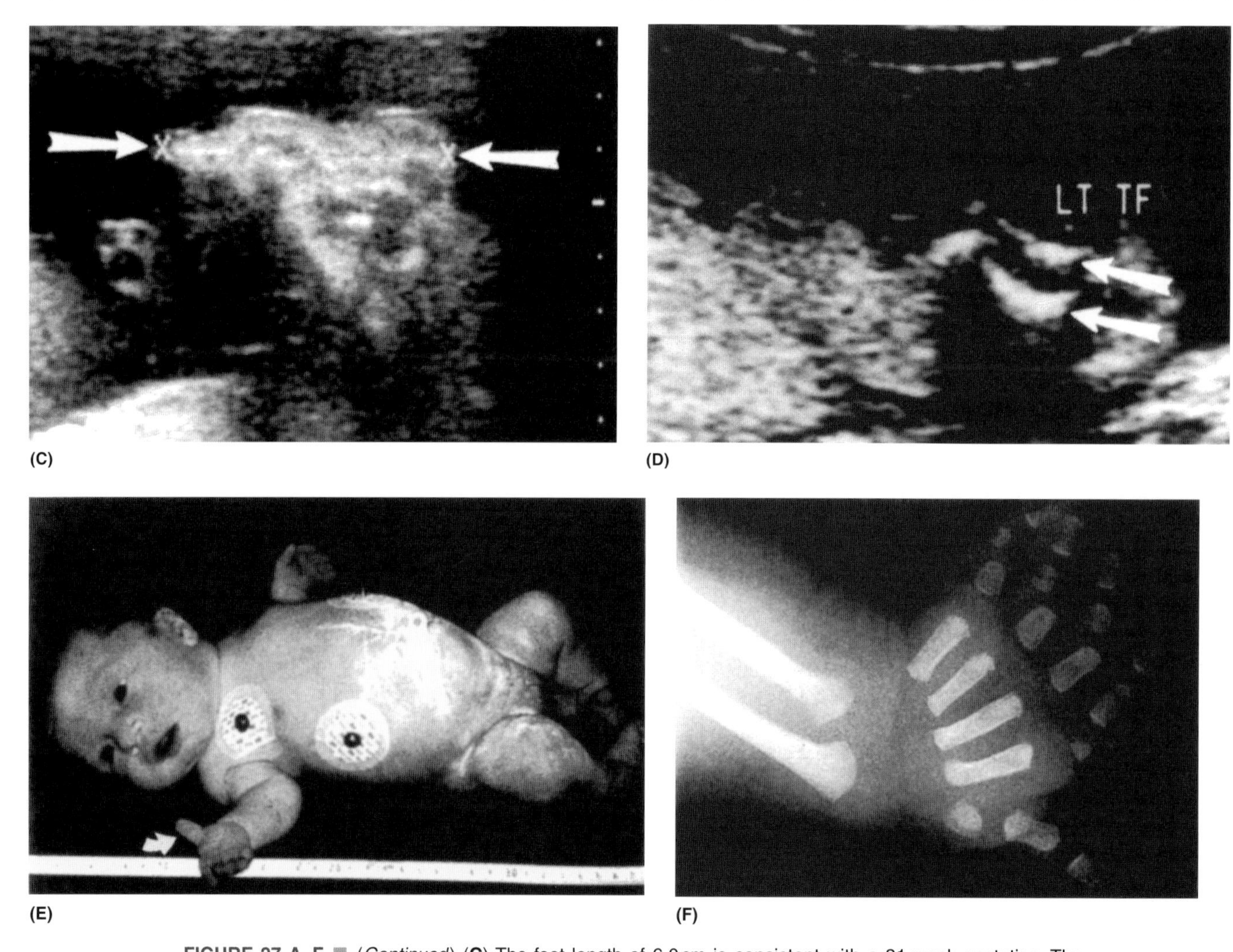

FIGURE 27 A–F ■ (*Continued*) (**C**) The foot length of 6.0 cm is consistent with a 31-week gestation. The femur-to-foot ratio of 0.68 indicates a skeletal dysplasia. (**D**) Short, thick, and clubbed tibia and fibula (*arrows*). (**E**) Newborn with short limbs, clubfeet, and hitchhiker thumb (*arrow*). (**F**) Ulnar deviation of the hand, irregular length, and form of metacarpals and phalanges, and short first metacarpal with proximally inserted and abducted thumb. *Source*: (**E**) and (**F**) from Ref. 96.

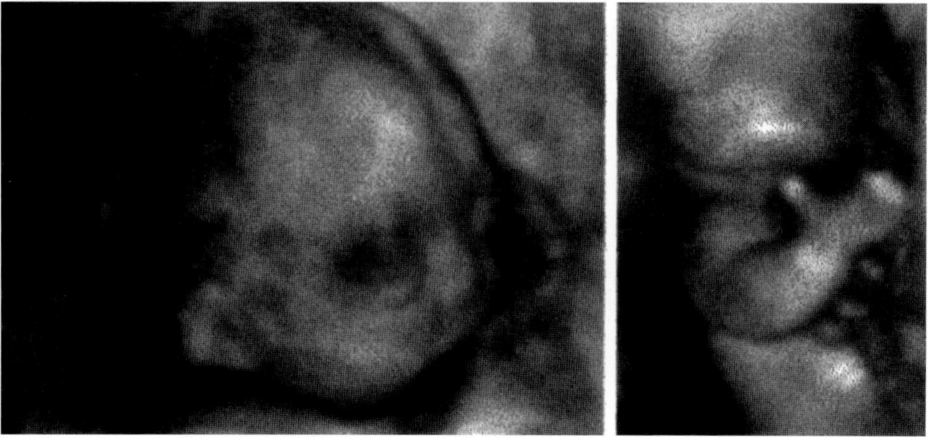

FIGURE 28 A AND B ■ Three-dimensional sonographic features of diastrophic dysplasia at 23 weeks' gestation. (**A**) Facial dysmorphism with severe micrognathia and small nose. (**B**) Shortening of the upper limb with hitchhiker thumb. *Source*: From Ref. 99.

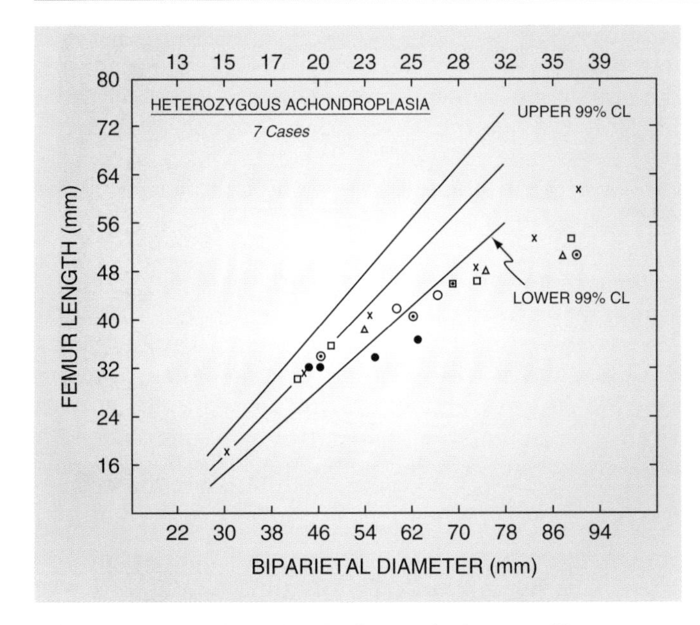

FIGURE 29 ■ The femur length of seven fetal cases of heterozygous achondroplasia are plotted, showing abnormal growth of the femur when compared with the biparietal diameter: open square, case 1; dotted square, case 2; open circle, case 3; dotted circle, case 4; triangle, case 5; solid circle, case 6; and x, case 7. *Source*: From Ref. 15.

less common. Polydactyly of the hands is more frequent than that of the feet. In New York City, the incidence of polydactyly is one in every 713 live births (136).

Polydactyly is categorized as preaxial or duplication of the thumb; central with duplication of one of the three middle fingers; and postaxial with duplication of the fifth digit. Postaxial polydactyly may have a well-developed digit (type A) or a rudimentary structure without a bony component (type B). Preaxial polydactyly occurs more commonly in Caucasians and postaxial polydactyly occurs significantly more frequently in African-Americans. Overall, postaxial polydactyly is more common.

In one report, 85.4% of polydactyly cases were isolated (137). When polydactyly occurs with another limb defect, it is usually syndactyly (53.4% of cases). Polydactyly is rarely associated with other congenital anomalies, except in specific syndromes. Additional fetal anomalies occur more frequently when polydactyly has an osseous component (138). Trisomy 21 is associated with preaxial polydactyly. There is a sevenfold lower frequency of trisomy 21 with postaxial polydactyly (137).

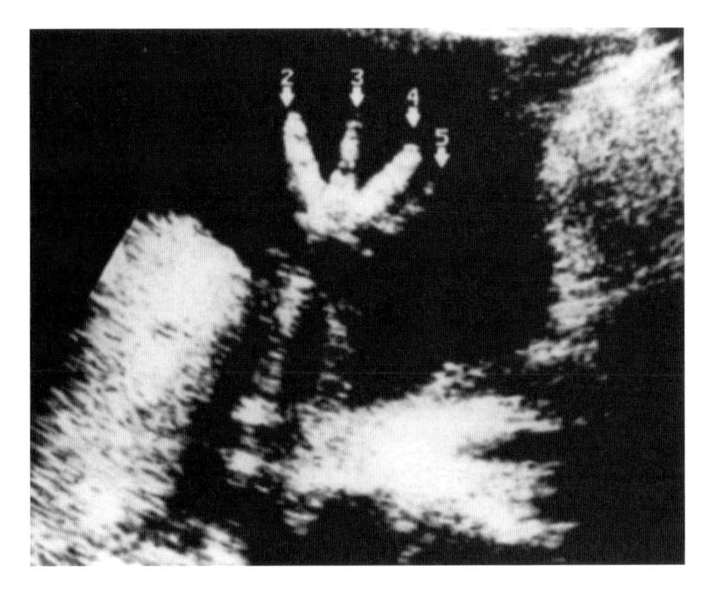

FIGURE 30 ■ Three-pronged (trident) appearance of second, third, and fourth digits at 31 weeks' gestation. *Source*: From Ref. 103.

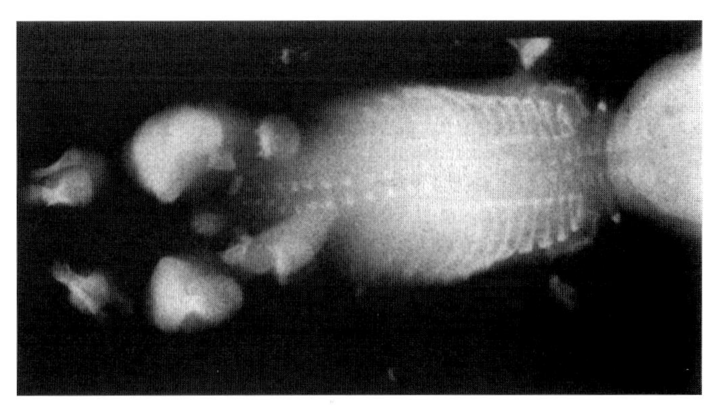

FIGURE 31 ■ Fetal X-ray study of a metatropic dwarf. Note the characteristic dumbbell appearance of the long bones, the narrow thorax, and the flattened vertebral bodies. *Source*: From Ref. 47.

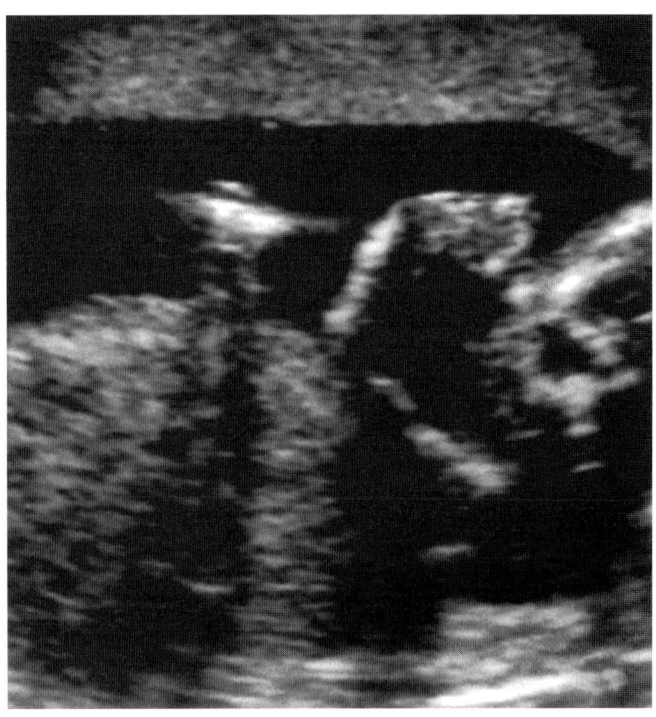

FIGURE 32 ■ Radial aplasia at 16 weeks' gestation.

TABLE 19 ■ Syndrome Associated with Radial-Ray Defects

Syndrome	Other sonographic findings
Thrombocytopenia with absent radii	None
Holt–Oram syndrome	Polydactyly
	Atrial septal defect
	Ventricular septal defect
Nager acrofacial dysostosis	Micrognathia
	Abnormal external ear
Goldenhar syndrome	Micrognathia
	Ventricular septal defect
	Occipital encephalocele
	Intrauterine growth restriction
	Scoliosis
Seckel's syndrome	Microcephaly
	Intrauterine growth restriction
Acrorenal syndrome	Renal anomalies
Robert's syndrome (pseudothalidomide)	Micrognathia
	Cleft lip or palate
	Cardiac defects
	Polyhydramnios
VACTERL association	*Vertebral*
	Hemivertebra
	Kyphosis
	Cardiac
	Ventricular septal defect
	Single umbilical artery
	Renal
	Agenesis
	Cystic disease
	Limb
	Polydactyly
	Other
	Cleft lip and palate
	Omphalocele
	Neural tube defect

Abbreviation: VACTERL., vertebral anomalies, anal atresia, congenital cardiac disease, tracheoesophageal fistula, renal anomalies, radial dysplasia, and other limb defects.

Source: From Refs. 118–123.

Syndactyly

A deficient separation of the mesenchymal digital primordium results in syndactyly. Fusion of the digits is either cutaneous or osseous. Syndactyly of the fingers may be suspected when the digits remain together (Fig. 37); fusion of the toes is much more difficult to discern sonographically. Syndactyly is occasionally an isolated anomaly. It is also an associated finding in various syndromes (Table 21).

Clinodactyly

Clinodactyly refers to a permanent bend in one or more digits. It is present in 60% of neonates with Down syndrome (143). However, autosomal-dominant clinodactyly is far more common than Down syndrome.

Ectrodactyly

Ectrodactyly refers to the absence of one or more digits. The split hand deformity is due to the absence of the third digit. Since there is syndactyly of the remaining digits, the hand resembles a claw. This anomaly is associated with a number of genetic and nongenetic syndromes (144).

Clubfoot (Talipes)

A clubfoot is one of the most frequent congenital anomalies, occurring in one in 250 to one in 1000 live births (Fig. 38A and B) (145). About 15% of patients have a family history of clubfoot. With anatomically normal parents, the siblings of an affected boy or girl have a 2% or 5% chance, respectively, of having a clubfoot. If a parent also has a clubfoot, the risk approaches 25% (146). In most cases, the foot is turned medially so the lateral aspect of the foot points to the ground (talipes equinovarus). The majority of cases of talipes with additional anomalies are bilateral (147). The false positive rate of antenatal sonography is significantly higher with unilateral, in contrast to bilateral, clubbed feet (148). Approximately 19% of fetuses initially diagnosed with an isolated clubbed foot will subsequently be found to have additional anomalies (147). Hence, if a clubbed foot is diagnosed at the 18- to 20-week scan, a follow-up detailed fetal anatomic survey should be scheduled for approximately 28 weeks' gestation to look for additional malformations.

Many causes of clubfoot have been identified (Table 22). For example, there is a 32% incidence of clubfeet with prolonged oligohydramnios (149). Other causes include central nervous system malformations, muscular abnormalities, and abnormal bone formation. A clubfoot is associated with various syndromes that affect the musculoskeletal system such as arthrogryposis (multiple contractures), Pena–Shokeir syndrome (joint contractures, intrauterine growth restriction, and pulmonary hypoplasia), and Larsen's syndrome (multiple joint dislocations). Other syndromes also have a clubfoot as an associated finding. For example, the Smith–Lemli–Opitz syndrome is a lethal condition characterized by microcephaly, congenital heart disease, and renal dysplasia. A clubfoot is associated with this condition because of the renal disease that results in oligohydramnios. A clubfoot occurs in approximately 30% of fetuses with trisomy 18; it is less often associated with trisomy 13 (150). Between 14.3% (151) and 22.2% (152) of prenatally diagnosed fetuses with a clubbed foot will have a karyotypic abnormality. In one study (148), all of the fetuses with a clubbed foot and aneuploidy had additional anomalies detected on antenatal sonography. Even when chromosomal abnormalities and syndromes are excluded, 10.3% of fetuses with a clubfoot have other malformations (e.g., cleft lip or palate and congenital heart disease) (153). The type and severity of these additional anomalies determine whether fetal karyotyping is warranted.

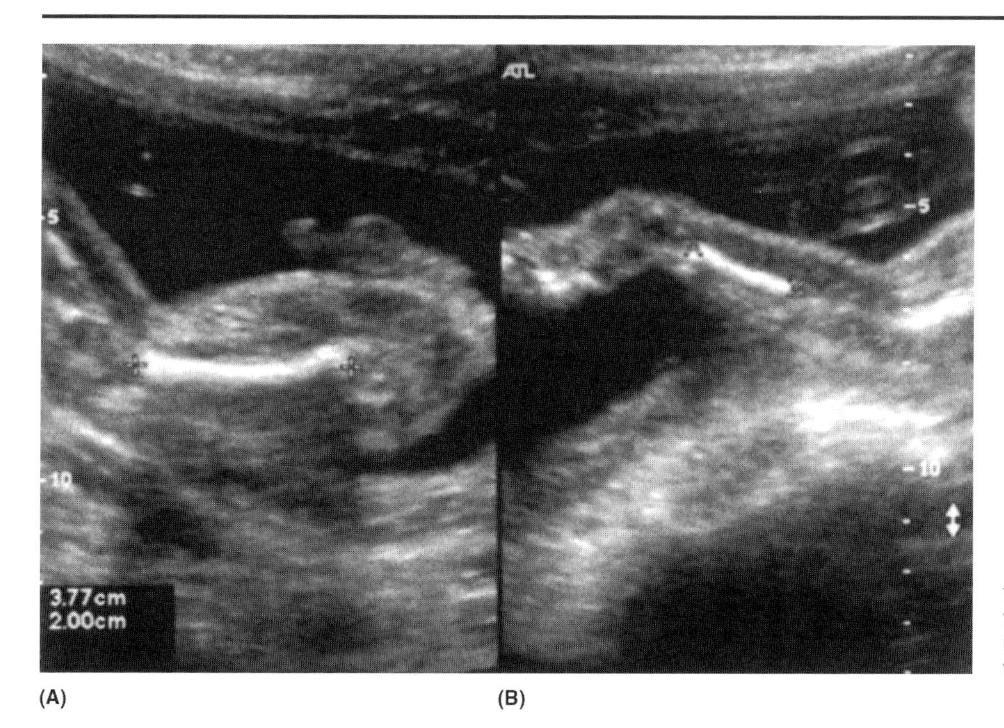

(A) (B)

FIGURE 33 A AND B ■ (**A**) Normal femur length. Fibular hypoplasia at 22 weeks' gestation. (**B**) The fibula is hypoplastic and shortened (consistent with 17 weeks).

Clubbed feet have been diagnosed with transvaginal ultrasound at 12.3 weeks' menstrual age (144).

Of the cases of talipes, 75% are structural and will require surgical correction. The remaining 25% are positional and can be corrected without surgery (154). When talipes equinovarus is isolated, pregnancy outcome is usually excellent (147).

TABLE 20 ■ Anomalies Associated with Amniotic Band Syndrome

Explainable by exogenous theory	Lacking explanation by exogenous theory
Scalp adhesions	Holoprosencephaly
Skull defects	Cerebellar dysplasia
Asymmetric facial clefts	Absent olfactory bulbs
Eye disruptions	Anophthalmia
Abdominal wall disruptions	Hypertelorism
Syndactyly	Migrational defects
Amputation	Heterotopic brain
Constriction rings	Cardiac anomalies
Hip dislocation	Tracheoesophageal fistula
Sacral rotation	Renal agenesis
	Accessory spleens
	Gallbladder agenesis
	Malrotation of the gut
	Single umbilical artery
	Internal genital malformation
	Anal atresia
	Simian crease

Source: From Ref. 129.

THREE-DIMENSIONAL IMAGING

Three-dimensional ultrasound has begun to play an important role in the assessment of the fetus suspected of having a skeletal dysplasia. The spatial reconstruction of three-dimensional ultrasound permits simultaneous visualization of the eyes, nose, lips, and ears. Hence, midface hypoplasia, cleft lip, micrognathia, hypo or hypertelorism, and auricular abnormalities of size, shape, and location can be appreciated (77). Specific facial phenotypic characteristics can, therefore, be used to help differentiate between specific skeletal dysplasias. Three-dimensional sonography is also quite useful in the evaluation of the hands and feet. The number of digits and the relative position of the hands and feet are better appreciated with three-dimensional sonography. Three-dimensional surface mode quickly demonstrates clubbed feet and splayed digits. With respect to the limbs, three-dimensional sonography shows the relative disproportion of limb segments that cannot be appreciated with two-dimensional imaging (91). Three-dimensional ultrasound is particularly suited for detecting the scapular abnormalities associated with camptomelic dysplasia (155). After data acquisition has occurred, the clinician has no limitations concerning the angle, scale, or rotation used to evaluate three-dimensional images. The technical limitations of this modality include movement artifacts and the limited quality of surface reconstruction in the presence of oligohydramnios.

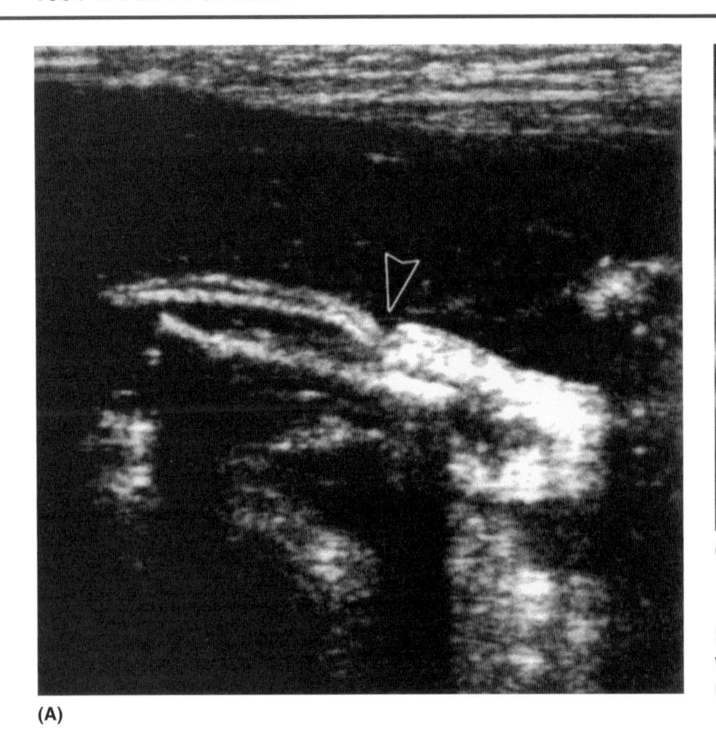

(A)

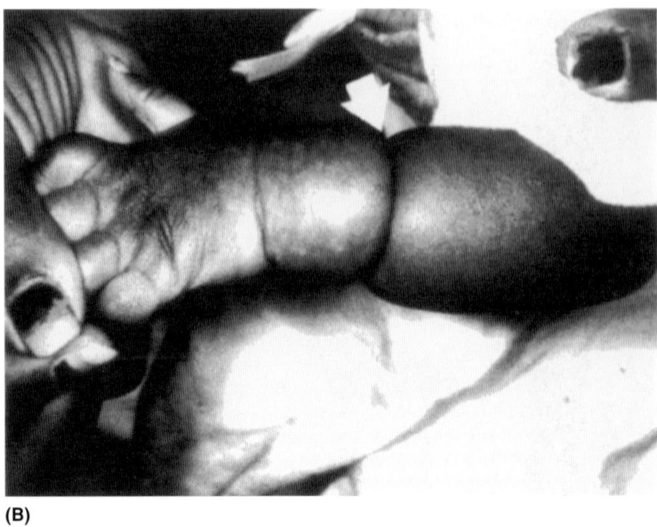

(B)

FIGURE 34 A AND B ■ (**A**) Annular constriction in the forearm at 38.6 weeks' gestation. (**B**) Left arm at six weeks of age illustrating the prominent constriction band. *Source*: From Ref. 133.

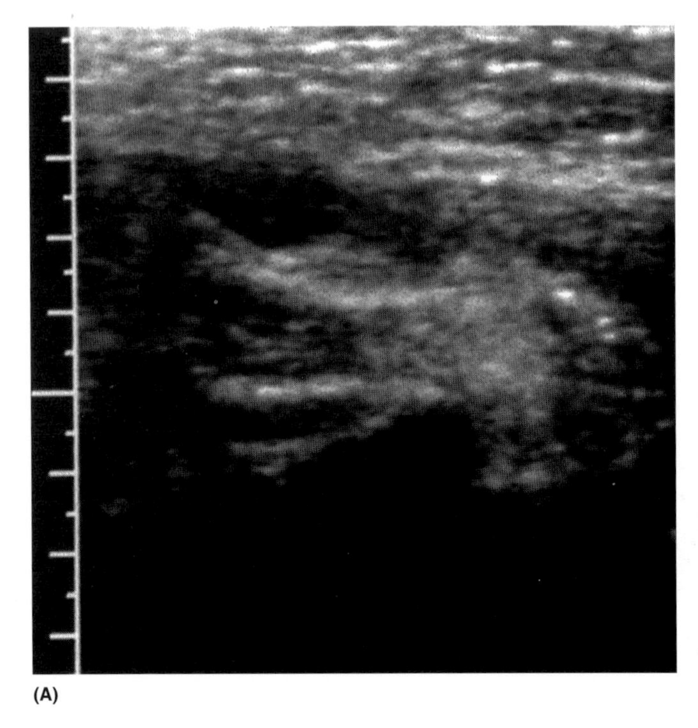

(A)

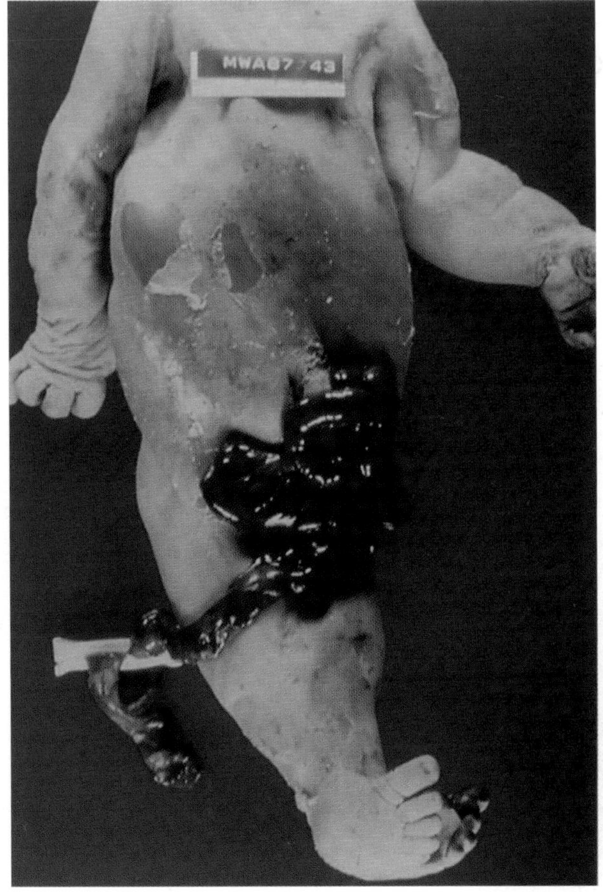

(B)

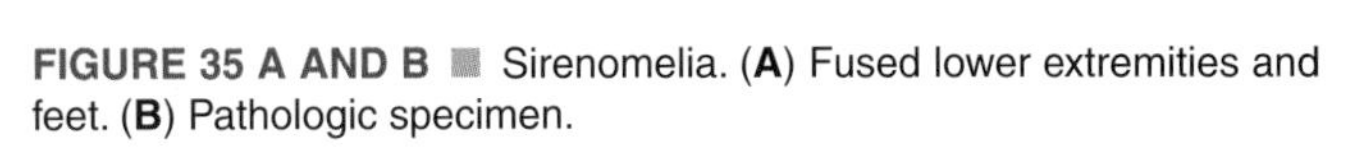

FIGURE 35 A AND B ■ Sirenomelia. (**A**) Fused lower extremities and feet. (**B**) Pathologic specimen.

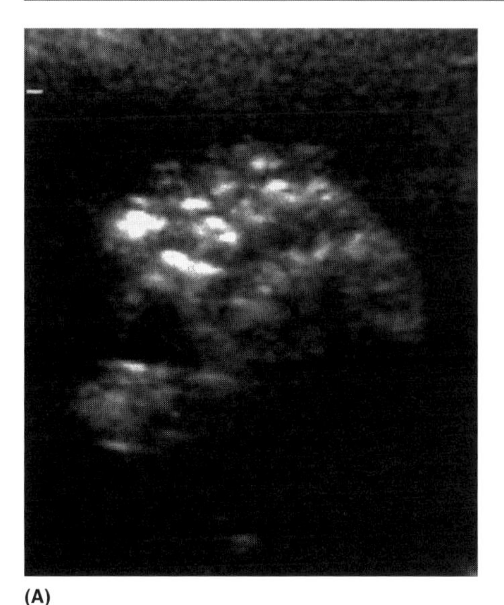
(A)

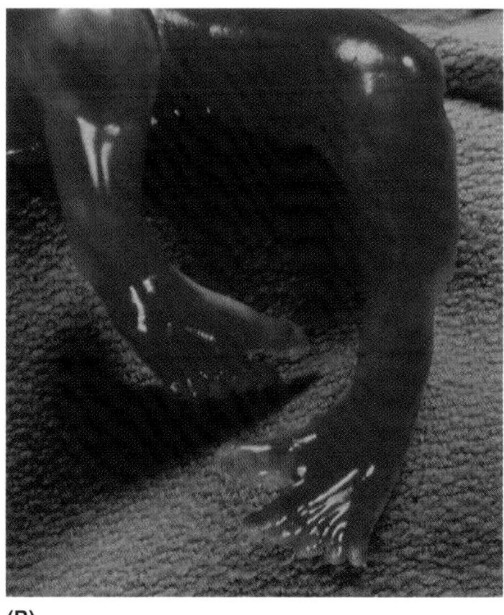
(B)

FIGURE 36 A AND B ■ Hexadactyly of the feet. (A) Ultrasound. (B) Pathology of another case.

The reduced fetal movement associated with some of the severe skeletal dysplasias and the relative excess of amniotic fluid make these particular malformations particularly suitable for three-dimensional imaging. Real-time three-dimensional imaging may also be employed.

CONCLUSIONS

An evaluation of a suspected fetal skeletal dysplasia can be divided into three components: identification, diagnosis, and prognosis. The accuracy of prenatal diagnosis varies with the type of skeletal dysplasia. From 48% (156) to 65% (19) of skeletal dysplasias are accurately diagnosed before birth. However, the prognostic accuracy of prenatal ultrasound in predicting lethality is between 92% (156) and 100% (19,157).

The sonographic evaluation of a fetus with a suspected skeletal abnormality requires a thorough assessment not only of the extremities but also of the cranium, spine, thorax, and abdomen. Some characteristic features

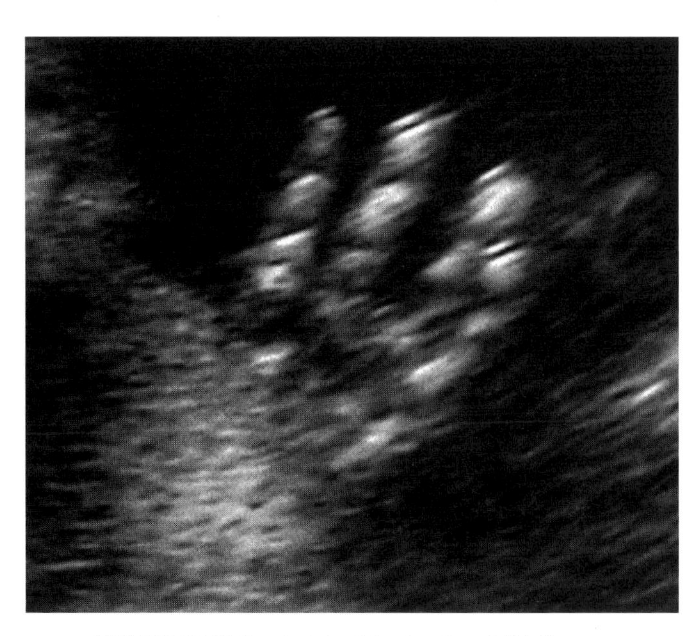
FIGURE 37 ■ Syndactyly of the fourth and fifth fingers.

TABLE 21 ■ Syndrome Associated with Syndactyly

Syndrome	Investigators	Characteristics
Carpenter	Robinson et al. (139)	Acrocephaly due to craniosynostosis
		Preaxial polydactyly
Apert	Hill et al. (23)	Coronal craniosynostosis
		Hypertelorism
		Cleft palate
Fraser	Burn and Marwood (140)	Renal agenesis
		Laryngeal atresia
Pfeiffer	Hill and Grzybek (25)	Craniosynthosis
		Medially displaced big toe
		Large thumb
Smith–Lemli–Opitz	Cherstovy et al. (141)	Microcephaly
		Congenital heart disease
		Renal dysplasia
Orofaciodigital	Gorlin and Psaume (142)	Midline cleft lip
		Cleft tongue
		Cleft palate
		Porencephalic cysts

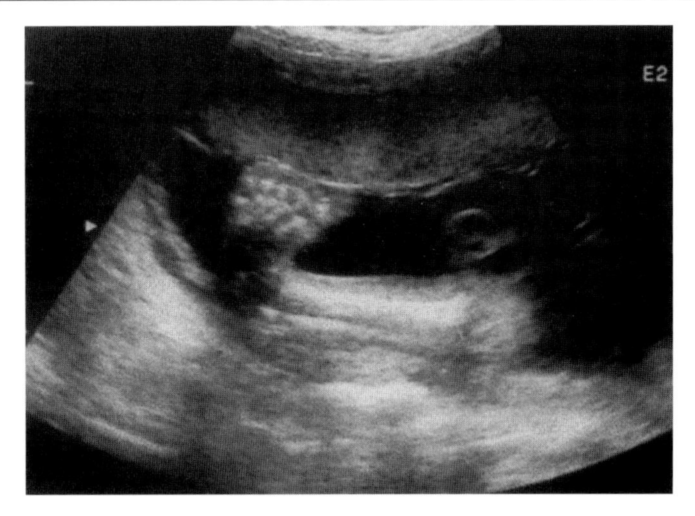

FIGURE 38 ■ Left clubbed foot.

of the more common skeletal dysplasias have been presented. When a sonographer or sonologist is confronted with a fetus with a skeletal malformation, the principles and sonographic techniques outlined in this chapter should help to narrow the differential diagnosis. Although certain sonographic signs may be characteristic, they are not pathognomonic for any one skeletal dysplasia. A team approach of the sonologist or sonographer, clinician, geneticist, and perinatal pathologist is essential in the management of dysmorphic fetuses. A postmortem examination that includes karyotyping, photographs, radiographs, and, frequently, bone histology is often necessary before a definitive diagnosis can be made.

TABLE 22 ■ Causes of Equinovarus (Clubfoot) Anomaly

Causes
Chromosomal abnormalities
Trisomy 13
Trisomy 18
Triploidy
Isolated clubfoot
Musculoskeletal
Arthrogryposis
Larsen's syndrome
Pena–Shokeir syndrome
Neural tube defects
Oligohydramnios sequence
Skeletal dysplasias
Camptomelic dysplasia
Diastrophic dysplasia
Osteogenesis imperfecta
Other syndromes
Thrombocytopenia with absent radius
Robert's syndrome
Smith–Lemli–Opitz syndrome

REFERENCES

1. Shiang R, Thompson LM, Zhu Y-Z, et al. Mutations in the transmembrane domain of FGFR3 cause the most common genetic form of dwarfism, achondroplasia. Cell 1994; 78(2):335–342.
2. Orioli IM, Castilla EE, Barbosa-Neto JG. The birth prevalence ratios for the skeletal dysplasias. J Med Genet 1986; 23(4):328–332.
3. Moore KL. The Developing Human: Clinically Oriented Embryology. 4th ed. Philadelphia: W. B. Saunders, 1988:336.
4. Wigglesworth JS, Singer DB. Textbook of Fetal and Perinatal Pathology. Oxford: Blackwell Scientific, 1991:1176.
5. Staples RE, Schnell VL. Refinements in rapid clearing technique in the Koh-Alizarin Red-S method for fetal bone. Stain Technol 1964; 39:61–63.
6. Elejalde BR, de Elejalde MM, Gilman M. Analysis of the human fetal skeleton and organs with xeroradiography. Am J Obstet Gynecol 1985; 151(5):666–670.
7. Levi CS, Reed MH, Harman CR. The fetal musculoskeletal system. In: Rumack CM, Wilson SR, Sharboneau JW, eds. Diagnostic Ultrasound. St. Louis: Mosby-Year Book, 1991:849.
8. Timor-Tritsch IE, Farine D, Rosen MG. A close look at early embryonic development with the high-frequency transvaginal transducer. Am J Obstet Gynecol 1988; 159(3):676–681.
9. Hegge FN, Prescott GH, Watson PT. Utility of a screening examination of the fetal extremities during obstetrical sonography. J Ultrasound Med 1986; 5(11):639–645.
10. Romero R, Althanassiadis AP, Jeanty P: fetal skeletal anomalies. Radiol Clin North Am 1989; 28:75.
11. Hadlock FP, Harrist RB, Deter Rl, et al. Fetal femur length as a predictor of menstrual age: sonographcially measured. AJR Am J Roentgenol 1982; 138(5):875–878.
12. Jeanty P, Rodesch F, Delbeke D, et al. Estimation of gestational age from measurements of fetal long bones. J Ultrasound Med 1984; 3(2):75–79.
13. Hill LM, Guzick D, Thomas ML, et al. Fetal radius length: a critical evaluation of race as a factor in gestational age assessment. Am J Obstet Gynecol 1989; 161(1):193–199.
14. Kurtz AB, Needleman L, Wapner RJ, et al. Usefulness of a short femur in the in utero detection of skeletal dysplasia. Radiology 1990; 177(1):197–200.
15. Kurtz AB, Filly RA, Wapner RJ, et al. In utero analysis of heterozygous achondroplasia: variable time of onset as detected by femur length measurements. J Ultrasound Med 1986; 5(3): 137–140.
16. Hadlock FP, Harrist RB, Shah Y, et al. The femur length/head circumference relation in obstetric sonography. J Ultrasound Med 1984; 3(1):439–442.
17. Hadlock FP, Harrist RB, Fearneyhough TC, et al. Use of femur length/abdominal circumference ratio in detecting the macrosomic fetus. Radiology 1985; 154(2):503–505.
18. Ramus RM, Martin LB, Twickler DM. Ultrasonographic prediction of fetal outcome in suspected skeletal dysplasias with use of the femur length-to-abdominal circumference ratio. Am J Obstet Gynecol 1998; 179(5):1348–1352.
19. Parilla BV, Leeth EA, Kambich MP, et al. Antenatal detection of skeletal dysplasias. J Ultrasound Med 2003; 22(3):255–258.
20. Campbell J, Henderson RGN, Campbell S. The fetal/foot length ratio: a new parameter to assess dysplastic limb reduction. Obstet Gynecol 1988; 72(2):181–184.
21. Brons JTJ, van der Harten JJ, Van Geijn HP, et al. Ratios between growth parameters for the prenatal ultrasonographic diagnosis of skeletal dysplasias. Eur J Obstet Gynecol Reprod Biol 1990; 34(1–2):37–46.
22. Otto C, Platt LD. The fetal mandible measurement: an objective determination of fetal jaw size. Ultrasound Obstet Gynecol 1991; 1(1):12–17.
23. Hill LM, Thomas ML, Peterson CS. The ultrasonic detection of Apert syndrome. J Ultrasound Med 1987; 6(10):601–604.
24. Cohen MM Jr, Kreiborg S. Birth prevalence studies of the Crouzon syndrome: comparison of direct and indirect methods. Clin Genet 1992; 41(1):12–15.

25. Hill LM, Grzybek PC. Sonographic findings with Pfeiffer syndrome. Prenat Diagn 1994; 14(1):47–49.
26. Mahony BS, Filly RA, Callen PW, et al. Thanatophoric dwarfism with the cloverleaf skull: a specific antenatal sonographic diagnosis. J Ultrasound Med 1985; 4(3):151–154.
27. Stamm ER, Pretorius DH, Rumack CM, et al. Kleeblattschadel anomaly: in utero sonographic appearance. J Ultrasound Med 1987; 6(16):319–324.
28. Chitkara U, Rosenberg J, Chervenak FA, et al. Prenatal sonographic assessment of the fetal thorax: normal values. Am J Obstet Gynecol 1987; 156(5):1069–1074.
29. Nimrod C, Nicholson S, Davies D, et al. Pulmonary hypoplasia testing in clinical obstetrics. Am J Obstet Gynecol 1988; 158(2):277–280.
30. DeVore GR, Horenstein J, Platt LD. Fetal echocardiography. VI. Assessment of cardiothoracic disproportion: a new technique for the diagnosis of thoracic hypoplasia. Am J Obstet Gynecol 1986; 155(5):1066–1071.
31. Marks F, Hernanz-SchulmanM, Horii S, et al. Spondylothoracic dysplasia: clinical and sonographic diagnosis. J Ultrasound Med 1989; 8(1):1–5.
32. Romero R, Chidini A, Eswara MS, et al. Prenatal findings in a case of spondylocostal dysplasia type I (Jarcho-Levin syndrome). Obstet Gynecol 1988; 71(6 Pt 2):988–991.
33. Aymé S, Preus M. Spondylocostal/spondylothoracic dysostosis: the clinical basis for prognosticating and genetic counseling. Am J Med Genet 1986; 24(4):599–606.
34. Rouse GA, Filly RA, Toomey F, et al. Short-limb skeletal dysplasias: evaluation of the fetal spine with sonography and radiography. Radiology 1990; 174(1):177–180.
35. Abrams SL, Filly RA. Curvature of the fetal femur: a normal sonographic finding. Radiology 1985; 156(2):490.
36. Bain AD, Barrett HS. Congenital bowing of the long bones: report of a case. Arch Dis Child 1959; 34:516–524.
37. Houston CS, Opitz JM, Spranger JW, et al. The campomelic syndrome: review, report of 17 cases and follow-up on the currently 17 year old boy first reported by Maroteaux et al. in 1971. Am J Med Genet 1983; 15(1):3–28.
38. Jacobson RL, Dignan P, Miodovnik M, et al. Antley-Bixler syndrome. J Ultrasound Med 1992; 11(4):161–164.
39. Machado LE, Osborne NG, Bonilla-Musoles F. Antley-Bixler syndrome. Report of a case. J Ultrasound Med 2001; 20(1):73–77.
40. Thomas RL, Hess LW, Johnson TRB. Prepartum diagnosis of limb-shortening defects with associated hydramnios. Am J Perinatol 1987; 4(4):293–299.
41. Wladimiroff JW, Niermeijer MF, Van der Harten JJ. Early prenatal diagnosis of congenital hypophosphatasia: case report. Prenat Diagn 1985; 5(1):47–52.
42. Meizner I, Bar-Ziv J. Prenatal ultrasonic diagnosis of short-rib polydactyly syndrome (SRPS) type III: a case report and a proposed approach to the diagnosis of SRPS and related conditions. J Clin Ultrasound 1985; 13(4):284–287.
43. Straub W, Zarabi M, Mazer J. Fetal ascites associated with Conradi's disease (chondrodysplasia punctata): report of a case. J Clin Ultrasound 1983; 11(4):234–236.
44. Benacerraf B, Ostharondth R, Rieber FR. Achondrogenesis type I: ultrasound diagnosis in utero. J Clin Ultrasound 1984; 12(6): 357–359.
45. Winter R, Rosencranz W, Hofmann H, et al. Prenatal diagnosis of camptomelic dysplasia by ultrasonography. Prenat Diagn 1985; 5(1):1–8.
46. Borochowitz Z, Jones KL, Silberg R, et al. A distinct lethal neonatal chondrodysplasia with a snail-like pelvis: schneckenbecken dysplasia. Am J Med Genet 1986; 25(1):47–59.
47. Manouvrier-Hanu S, Devisme L, Zelasko MC, et al. Prenatal diagnosis of metatropic dwarfism. Prenat Diagn 1995; 15(8):753–756.
48. Kirk JS, Comstock CH. Antenatal sonographic appearance of spondyloepiphyseal dysplasia. J Ultrasound Med 1990; 9(3):173–175.
49. Snijders RJM, Noble P, Sebire N, et al. UK multicenter project: an assessment of risk of trisomy 21 by maternal age and fetal nuchal translucency thickness at 10–14 weeks of gestation. Lancet 1998; 352(9125):343–346.
50. Souka AP, Krampl E, Bakalis S, et al. Outcome of pregnancy in chromosomally normal fetuses with increased nuchal translucency in the first trimester. Ultrasound Obstet Gynecol 2001; 18(1):9–17.
51. Makrydimas G, Souka A, Skentou H, et al. Osteogenesis imperfecta and other skeletal dysplasias presenting with increased nuchal translucency in the first trimester. Am J Med Genet 2001; 98(2):117–120.
52. Soothill PW, Vuthiwong C, Rees AH. Achondrogenesis type 2 diagnosed by transvaginal ultrasound at 12 weeks' gestation. Prenat Diagn 1993; 13(6):523–528.
53. Borochowitz Z, Lachman R, Adomian GE, et al. Achondrogenesis type I delineation of further heterogeneity and identification of two distinct subgroups. J Pediatr 1988; 112:23–32.
54. Mahony BS, Filly RA, Cooperberg PL. Antenatal sonographic diagnosis of achondrogenesis. J Ultrasound Med 1984; 3(7):333–335.
55. Beighton P, Giedion A, Corlin R, et al. International classification of osteochondrodysplasias. Am J Med Genet 1992; 44(2):223–229.
56. Whitley CB, Gorlin RJ. Achondrogenesis: new nosology with evidence of genetic heterogeneity. Radiology 1983; 148(3):693–698.
57. Eyre DR, Upton MP, Shapiro FD, et al. Nonexpression of cartilage type II collagen in a case of Langer-Saldino achondrogenesis. Am J Hum Genet 1986; 39:52.
58. Fisk NM, Vaughan J, Smidt M, et al. Transvaginal ultrasound recognition of nuchal edema in the first trimester diagnosis of achondrogenesis. J Clin Ultrasound 1991; 19(9):586–590.
59. Tavormina PL, Shiang R, Thompson LM, et al. Thanatophoric dysplasia (types I and II) caused by distinct mutations in fibroblast growth factor receptor 3. Nat Genet 1995; 9(3):321–328.
60. Elejalde BR, de Elejalde MM. Thanatophoric dysplasia: fetal manifestations and prenatal diagnosis. Am J Med Genet 1984; 22(4):669–683.
61. Machado LE, Bonilla-Musoles F, Osborne NG. Thanatrophoric dysplasia. (Picture of the month). Ultrasound Obstet Gynecol 2001; 18(1):85–86.
62. Benacerraf BR, Lister JE, DuPonte BL. First-trimester diagnosis of fetal abnormalities. A report of three cases. J Reprod Med 1988; 33(9):777–780.
63. Prockop DJ, Kivirikko KI. Heritable diseases of collagen. N Engl J Med 1984; 311(6):376–386.
64. Byers PH, Wallis GA, Willing MC. Osteogenesis imperfecta: translation of mutation to phenotype. J Med Genet 1991; 28(7):433–442.
65. Byers PH. Osteogenesis imperfecta: an update. Growth Genet Horm 1988; 4:1–5.
66. Sillence DO, Senn A, Danks DM. Genetic heterogeneity in osteogenesis imperfecta. J Med Genet 1979; 16(2):101–116.
67. Brons JTJ, et al. Prenatal ultrasonographic diagnosis of osteogenesis imperfecta. Am J Obstet Gynecol 1988; 159:176.
68. Maroteaux P, Stanescu V, Stanescu R. The lethal chondrodysplasias. Clin Orthop 1976; 114:31–45.
69. Chervenak FA, Romero R, Berkowitz RL, et al. Antenatal sonographic findings of osteogenesis imperfecta. Am J Obstet Gynecol 1982; 143(2):228–230.
70. Bulas DI, Stern HJ, Rosenbaum KN, et al. Variable prenatal appearance of osteogenesis imperfecta. J Ultrasound Med 1994; 13(6):419–427.
71. Berge LN, Marton V, Tranebjaerg L, et al. Prenatal diagnosis of osteogenesis imperfecta. Acta Obstet Gynecol Scand 1995; 74(4):321–323.
72. DiMaio MS, Barth R, Koprivnikar KE, et al. First trimester prenatal diagnosis of osteogenesis imperfecta type II by DNA analysis and sonography. Prenat Diagn 1993; 13(7):589–596.
73. Robinson LP, Worten NJ, Lachman RS, et al. Prenatal diagnosis of osteogenesis imperfecta type III. Prenat Diagn 1987; 7(1):7–15.
74. Winter RM, Knowles SAD, Bieber FR, et al. The Malformed Fetus and Stillbirth: a Diagnostic Approach. New York: John Wiley & Sons, 1988:34.
75. Ruano R, Picone O, Benachi A, et al. First-trimester diagnosis of osteogenesis imperfecta associated with encephalocele by conventional and three-dimensional ultrasound. Prenat Diagn 2003; 23(7):539–542.
76. Chang L-W, Chang C-H, Yu C-H. Three-dimensional ultrasonography of osteogenesis imperfecta at early pregnancy (letter to the editor). Prenat Diagn 2002; 22(1):77–78.

77. Azumendi G, Kurjak A. Three-dimensional and four-dimensional sonography in the study of the fetal face. Ultrasound Rev Obstet Gynecol 2003; 3:160–169.
78. Weiss MJ, Ray K, Fallon MD, et al. Analysis of liver/bone/kidney alkaline phosphatase mRNA, DNA and enzymatic activity in cultured skin fibroblasts in 14 unrelated patients with severe hypophosphatasia. Am J Hum Genet 1989; 44(5):686–694.
79. Fallon MD, Teitelbaum SL, Weinstein RS, et al. Hypophosphatasia: clinico-pathologic comparison of the infantile, childhood and adult forms. Medicine 1984; 63(1):12–24.
80. Fraser D. Hypophosphatasia. Am J Med 1957; 22(5):730–746.
81. DeLange M, Rouse GA. Prenatal diagnosis of hypophosphatasia. J Ultrasound Med 1990; 9:115.
82. Brock DJ, Barron L. First trimester prenatal diagnosis of hypophosphatasia: experience with 16 cases. Prenat Diagn 1991; 11(6):387–391.
83. Golombeck K, Jacobs VR, von Karisenberg C, et al. Short rib-polydactyly syndrome type III: comparison of ultrasound, radiology, and pathology findings. Fetal Diagn Ther 2001; 16(3):133–138.
84. Richardson MM, Beaudet AL, Wagner ML, et al. Prenatal diagnosis of recurrence of Saldino-Noonan dwarfism. J Pediatr 1977; 91(3):467–471.
85. Gembruch U, Hansmann M, Fodisch HJ. Early prenatal diagnosis of short rib-polydactyly (SRP) syndrome type I (Majewski) by ultrasound in a case at risk. Prenat Diagn 1985; 5(5):357–362.
86. Benacerraf BR. Prenatal sonographic diagnosis of short rib-polydactyly syndrome: type II. Majewski type. J Ultrasound Med 1993; 12(9):552–555.
87. Hill LM, Leary J. Transvaginal sonographic diagnosis of short-rib polydactyly dysplasia at 13 weeks' gestation. Prenat Diagn 1998; 18(11):1198–1201.
88. Viora E, Sciarrone A, Bastonero S, et al. Three-dimensional ultrasound evaluation of short-rib polydactyly syndrome type II in the second trimester: a case report. Ultrasound Obstet Gynecol 2002; 19(1):88–91.
89. Cordone M, Littiana M, Zampatti C. In utero ultrasonographic features of camptomelic dysplasia. Prenat Diagn 1989; 9(11):745–750.
90. Sanders RC, Greyson-Fleg RT, Hogge WA, et al. Osteogenesis imperfecta and camptomelic dysplasia: difficulties in prenatal diagnosis. J Ultrasound Med 1994; 13(9):691–700.
91. Krakow D, Williams J, Poehl M, et al. Use of three-dimensional ultrasound imaging in the diagnosis of prenatal-onset skeletal dysplasias. Ultrasound Obstet Gynecol 2003; 21(5):467–472.
92. Duff P, Harlass FE, Milligan DA. Prenatal diagnosis of chondrodysplasia punctata by sonography. Obstet Gynecol 1990; 76(3 Pt 2):497–500.
93. Curry CJ, Magenis RE, Brown M, et al. Inherited chondrodysplasia punctata due to a deletion of the terminal short arm of an X chromosome. N Engl J Med 1984; 311(16):1010–1015.
94. Jansen V, Sarafoglou K, Rebarber A, et al. Chondrodysplasia punctata, tibial-metacarpal type in 16 week fetus. J Ultrasound Med 2000; 19(10):719–722.
95. Meizner I, Bar-Ziv J. In utero diagnosis of skeletal disorders. In: An Atlas of Prenatal Sonographic and Post-Natal Radiologic Correlation. Boca Raton, FL: CRC Press, 1993.
96. Gembruch U, Niesen M, Kehrberg H, et al. Diastrophic dysplasia: a specific prenatal diagnosis by ultrasound. Prenat Diagn 1988; 8(7):539–545.
97. Kaitila I, Ammala P, Karjalainen O, et al. Early prenatal detection of diastrophic dysplasia. Prenat Diagn 1984; 3(3):237–244.
98. Mantagos S, Weiss RR, Mahoney M, et al. Prenatal diagnosis of diastrophic dwarfism. Am J Obstet Gynecol 1981; 139(1):111–113.
99. Sepulveda W, Sepulveda-Swatson E, Sanchez J. Diastrophic dysplasia: prenatal three-dimensional ultrasound findings. Ultrasound Obstet Gynecol 2004; 23(3):312–314.
100. Tyson JE, Barnes AC, McKusick VA, et al. Obstetric and gynecologic considerations of dwarfism. Am J Obstet Gynecol 1970; 108(5):688–704.
101. Cordone M, Lituania M, Bocchino G, et al. Ultrasonographic features in a case of heterozygous achondroplasia at 25 weeks' gestation. Prenat Diagn 1993; 13(5):395–401.
102. Filly RA, Golbus MS, Cary JC, et al. Short limbed dwarfism: ultrasonographic diagnosis by mensuration of fetal femoral length. Radiology 1981; 138(3):653–656.
103. Guzman ER, Day-Salvatore D, Westover T, et al. Prenatal ultrasonographic demonstration of the trident hand in heterozygous achondroplasia. J Ultrasound Med 1994; 13(1):63–66.
104. Seino Y, Marïwake T, Tanaka H, et al. Molecular defects in achondroplasia and the effects of growth hormone treatment. Acta Paediatr Suppl 1999; 88(428):118–120.
105. Donnenfeld AE, Mennuti MT. Second trimester diagnosis of fetal skeletal dysplasias. Obstet Gynecol Surv 1987; 42(4):199–217.
106. Hutchon DJR. Three-dimensional sonographic aspects in the antenatal diagnosis of achondroplasia (letter to the editor). Ultrasound Obstet Gynecol 2001; 18:81.
107. Stoll C, Manini P, Block J, et al. Prenatal diagnosis of hypochondroplasia. Prenat Diagn 1985; 5(6):423–426.
108. Elejalde BR, de Elejalde MM, Pansch D. Prenatal diagnosis of Jeune syndrome. Am J Med Genet 1985; 21(3):433–438.
109. Muller LM, Cremin BJ. Ultrasonic demonstration of fetal skeletal dysplasia. S Afr Med J 1985; 67(6):222–226.
110. Schinzel A, Savoldelli G, Briner J, et al. Prenatal sonographic diagnosis of Jeune syndrome. Radiology 1988; 154(3):777–778.
111. Yang SS, Heidelberger KP, Brough A, et al. Lethal short-limbed chondrodysplasia in early infancy. Perspect Pediatr Pathol 1976; 3:1–40.
112. Mahoney MJ, Hobbins JC. Prenatal diagnosis of chondroectodermal dysplasia (Ellis-van Crevald syndrome) with fetoscopy and ultrasound. N Engl J Med 1977; 297(5):258–260.
113. Bromley B, Miller W, Foster SC, et al. The prenatal sonographic features of Kniest syndrome. J Ultrasound Med 1991; 10(12):705–707.
114. Beck M, Roubicek M, Rogers JG, et al. Heterogeneity of metatropic dysplasia. Eur J Pediatr 1983; 140(31):231–237.
115. Bod M, Criezel A, Lenz W. Incidence at birth of different types of limb reduction abnormalities in Hungary, 1975–1977. Hum Genet 1983; 65(1):27–33.
116. Froster-Iskenius UG, Baird PA. Limb reduction defects in over one million consecutive live births. Teratology 1989; 39(2):127–135.
117. Brons JT, van der Harten HJ, Van Gejin HP, et al. Prenatal ultrasonographic diagnosis of radial-ray reduction malformations. Prenat Diagn 1990; 10(5):279–288.
118. Weinblatt M, Petrikovsky B, Bialer M, et al. Prenatal evaluation and in utero platelet transfusion for thrombocytopenia absent radii syndrome. Prenat Diagn 1994; 14(9):892–896.
119. Brons JTJ, van Geijn HP, Wladimiroff JW, et al. Prenatal ultrasound diagnosis of the Holt-Oram syndrome. Prenat Diagn 1988; 8(3):175–181.
120. Benson CB, Pober BR, Hirsch MP, et al. Sonography of Nager acrofacial dysostosis syndrome in utero. J Ultrasound Med 1988; 7(3):163–167.
121. Benacerraf BR, Frigoletto FD. Prenatal ultrasonographic recognition of Goldenhar's syndrome. Am J Obstet Gynecol 1988; 159(4):950–952.
122. Grundy HO, Burlbaw J, Watson S. Roberts syndrome: antenatal ultrasound. A case report. J Perinat Med 1988; 16(1):71–75.
123. Khoury MJ, Cordero JF, Greenberg F, et al. A population study of the VACTERL Association: evidence for its etiologic heterogeneity. Pediatrics 1983; 71(5):815–820.
124. Graham M. Congenital short femur: prenatal sonographic diagnosis. J Ultrasound Med 1985; 4(7):361–363.
125. Jeanty P, Kleinman G. Proximal femoral focal deficiency. J Ultrasound Med 1989; 8(11):639–642.
126. Sepulveda W, Weinder E, Bridger JE, et al. Prenatal diagnosis of congenital absence of the fibula. J Ultrasound Med 1994; 13(8):655–657.
127. Uffelman J, Woo R, Richards DS. Prenatal diagnosis of bilateral fibular hemimelia. J Ultrasound Med 2000; 19(5):341–344.
128. Hirose K, Koyanagi T, Hara K, et al. Antenatal ultrasound diagnosis of the femur- fibular-ulna syndrome. J Clin Ultrasound 1988; 16(3):199–203.
129. Concalves LV, Jeanty P. Ultrasound evaluation of fetal abdominal wall defects. In: Callen PW, ed. Ultrasonography in Obstetrics and Gynecology. Philadelphia: W. B. Saunders, 1994.
130. Torpin R. Amniochorionic mesoblastic fibrous strings and amniotic bands. Am J Obstet Gynecol 1965; 91:65–75.

131. Lockwood C, Ghidini A, Romero R, et al. Amniotic band syndrome: re-evaluation of its pathogenesis. Am J Obstet Gynecol 1989; 160(5 Pt 1):1030–1033.
132. Isacsohn M, Aboulafia Y, Horwitz B, et al. Congenital annular constriction due to amniotic bands. Acta Obstet Gynecol Scand 1976; 55(2):179–182.
133. Hill LM, Kislak S. Prenatal ultrasound diagnosis of a forearm constriction band. J Ultrasound Med 1988; 7:293.
134. Sirtori M, Ghidini A, Romero R, et al. Prenatal diagnosis of sirenomelia. J Ultrasound Med 1989; 8(2):83–88.
135. Patel S, Suchet I. The role of color and power Doppler ultrasound in the prenatal diagnosis of sirenomelia. Ultrasound Obstet Gynecol 2004; 24(6):684–691.
136. Sesgin MZ, Stark BB. The incidence of congenital defects. Plast Reconstr Surg 1961; 27:26–27.
137. Castilla EE, Lugarinho R, Dutra M, et al. Associated anomalies in individuals with polydactyly. Am J Med Genet 1998; 80(5):459–465.
138. Zimmer EZ, Bronshtein M. Fetal polydactyly diagnosis during early pregnancy: clinical applications. Am J Obstet Gynecol 2000; 183(3):755–758.
139. Robinson LK, James JE, Mubarak SJ, et al. Carpenter syndrome: natural history and clinical spectrum. Am J Med Genet 1985; 20(3):461–469.
140. Burn J, Marwood RP. Fraser syndrome presenting as bilateral renal agenesis in three sibs. J Med Genet 1982; 19(5):360–361.
141. Cherstovy ED, Lazjuk GI, Lurie IW. The pathological anatomy of the Smith-Lemli- Opitz syndrome. Clin Genet 1975; 7(5):382–387.
142. Gorlin RJ, Psaume J. Orodigitofacial dysostosis: a new syndrome. J Pediat 1962; 61:520–530.
143. Benacerraf BR, Osathanondh R, Frigoletto FD. Sonographic demonstration of hypoplasia of the middle phalanx of the fifth digit: a finding associated with Down syndrome. Am J Obstet Gynecol 1988; 159(1):181–183.
144. Leung KY, MacLaclan NA, Sepulveda W. Prenatal diagnosis of ectrodactyly: the "lobster claw" anomaly. Ultrasound Obstet Gynecol 1995; 6(6):443–446.
145. Chervenak FA, Tortora M, Hobbins JC. Antenatal sonographic diagnosis of clubfoot. Ultrasound Med 1985; 4(1):49–50.
146. Wynne-Davies R. Genetic and environmental factors in the etiology of talipes equinovarus. Clin Orthop 1972; 84:9–13.
147. Bakalis S, Sairam S, Homfray T, et al. Outcome of antenatally diagnosed talipes equinovarus in an unselected obstetric population. Ultrasound Obstet Gynecol 2002; 20(3):226–229.
148. Mammen L, Benson CB. Outcome of fetuses with clubfeet diagnosed by prenatal sonography. J Ultrasound Med 2004; 23(4):497–500.
149. Torpin R. Fetal malformations caused by amnion rupture during gestation. Springfield, IL: Charles C. Thomas, 1968.
150. Benacerraf BR, Miller WA, Frigoletto FD. Sonographic detection of fetuses with trisomies 13 and 18: accuracy and limitations. Am J Obstet Gynecol 1988; 158(2):404–409.
151. Rijhsinghani A, Yankowitz J, Kanis AB, et al. Antenatal sonographic diagnosis of club foot with particular attention to the implications and outcomes of isolated club foot. Ultrasound Obstet Gynecol 1998; 12(2):103–106.
152. Benacerraf BR. Antenatal sonographic diagnosis of congenital club foot: a possible indication for amniocentesis. J Clin Ultrasound 1986; 14(9):703–706.
153. Yamamoto H. A clinical genetic and epidemiologic study of congenital clubfoot. Jpn J Hum Genet 1979; 24(1):37–44.
154. Carroll SGM, Lockyer H, Andrews H, et al. Outcomes of fetal talipes following in utero sonographic diagnosis. Ultrasound Obstet Gynecol 2001; 18(5):437–440.
155. Garjian KV, Pretorius DH, Budorick NE, et al. Fetal skeletal dysplasia: three-dimensional US—initial experience. Radiology 2000; 214(3):717–723.
156. Hersh JH, Angle B, Pietrantoni M, et al. Predictive value of fetal ultrasonography in the diagnosis of a lethal skeletal dysplasia. South Med J 1998; 91(12):1137–1142.
157. Doray B, Favre R, Viville B, et al. Prenatal sonographic diagnosis of skeletal dysplasias. A report of 47 cases. Annales de Génét 2000; 43(3–4):163–169.

50

Twin Pregnancy

Carol B. Benson and Peter M. Doubilet

INTRODUCTION

Sonographic evaluation of a twin pregnancy involves more than the examination of two single fetuses. It includes an assessment of features specific to twins—amnionicity and chorionicity—and a search for disorders unique to twins, such as twin–twin transfusion syndrome. Ultrasound is an important tool throughout a twin pregnancy, helpful for detecting and characterizing twins early in gestation and for monitoring for complications later in pregnancy.

FREQUENCY AND TYPES OF TWINS

The incidence of twins at birth of naturally conceived pregnancies is approximately one in 80 to 90 deliveries in the United States. Seventy percent are dizygotic ("fraternal"), arising from fertilization of two separate ova, and 30% are monozygotic ("identical"), resulting from division of a single fertilized ovum. The frequency of monozygotic twins is the same in all populations, occurring at a rate of 3 to 5 per 1000 pregnancies delivered. In contrast, several factors influence the frequency of dizygotic twinning, including race, parity, and the use of ovulation-induction therapy for infertility. Among these last patients, dizygotic twin gestations occur in up to 50% of resulting pregnancies (Table 1) (1–4).

SONOGRAPHIC ASSESSMENT OF PREGNANCY NUMBER

From six weeks of gestation onward, sonographic assessment of pregnancy number is straightforward: count the number of embryos or fetuses with heartbeats. Prior to six weeks, however, assessment of pregnancy number is less straightforward and more prone to miscounting, since the embryonic heartbeat is not consistently visible at this very early stage of pregnancy. In women with multiple gestations, who are scanned at 5.0 to 5.9 weeks, the early scan will undercount pregnancy number in 14% of cases. The rate of undercounting is especially high in 5.0- to 5.4-week sonograms (5).

Sonograms prior to six weeks' gestation can also diagnose a higher pregnancy number than is present on subsequent scans. In some cases, this is due to the "vanishing twin" phenomenon, in which a pregnancy begins as a twin or higher-order multiple gestation, but one or more embryos fail to develop beyond an early stage and their gestational sacs involute (6). In other cases, it is due to erroneously diagnosing a fluid collection (e.g., hematoma) as a gestational sac.

PLACENTATION

Types

The placentation type of twin pregnancy refers to its chorionicity, the number of placentas, and its amnionicity, the number of amniotic sacs. Three placentation types occur: dichorionic–diamniotic, monochorionic–diamniotic,

TABLE 1 ■ Factors Influencing Frequency of Dizygotic Twins

Factor	Effect on frequency
Race	Blacks > Whites > Asians
Maternal family history	Increased if family history of dizygotic twins
Prior dizygotic twins	Increased in mothers with prior dizygotic twins
Maternal age	Increased with increasing age
Parity	Multiparous > nulliparous
Therapy for infertility	Increased with ovulation-induction medication

Source: From Ref. 4.

and monochorionic–monoamniotic. The placentation type of twin pregnancy depends on the time of twinning (Table 2). The further from fertilization that twinning occurs, the lesser the separation of chorion and amnion. All dizygotic twins, which are separate from the time of ovulation and fertilization, are dichorionic–diamniotic. Monozygotic twins divide at variable times after fertilization and, therefore, may have one of several placentation types (1,3). Monozygotic twin pregnancies are monochorionic–diamniotic in approximately two-thirds of cases, dichorionic–diamniotic in about one-third of cases, and monochorionic–monoamniotic less than 1% of the time (Fig. 1).

Sonographic Evaluation of Placentation

Several sonographic features can be used to determine placentation type (Table 3) (3,7). The definitive diagnosis of a diamniotic gestation can be made when a membrane is identified. Because the dividing membrane cannot be visualized by ultrasound in some cases, inability to demonstrate a membrane does not prove monoamnionicity. A monoamniotic gestation can be diagnosed with certainty only in the presence of both nonvisualization of a membrane and intermingling of the two umbilical cords. In the early first trimester, the number of yolk sacs is generally concordant with amnionicity, so that twins with a single yolk sac are likely to be monoamniotic (8).

TABLE 2 ■ Zygosity and Time of Twinning vs. Placentation Type

Zygosity	Time of twinning	Placentation type
Dizygotic	Fertilization	DC–DA
Monozygotic	After fertilization:	
	0–4 day (~33% of MZ twins)	DC–DA
	4–8 day (~67% of MZ twins)	MC–DA
	8–13 day (~1% of MZ twins)	MC–MA
	>13 day (rare)	MC–MA, conjoined twins

Abbreviations: MZ, monozygotic; DC, dichorionic; DA, diamniotic; MC, monochorionic; MA, monoamniotic.

The definitive diagnosis of a dichorionic gestation can be made if two placental masses or different fetal sexes are noted. In the absence of these findings, membrane thickness can be used to predict chorionicity. A thick membrane suggests a dichorionic gestation, whereas a thin, wispy membrane suggests monochorionicity (Fig. 2). Membrane thickness is highly accurate for predicting chorionicity in the first trimester (Fig. 3) and has a predictive accuracy of 83% thereafter (7).

Prognosis

The prognosis of a twin pregnancy depends on placentation type, because the nature and frequency of complications vary with chorionicity and amnionicity. The more separate the membranes, the fewer the complications, so dichorionic–diamniotic twins have the best prognosis, monochorionic–diamniotic twins have a poorer prognosis, and monochorionic–monoamniotic twins have the worst prognosis (Table 4). At the time of a first trimester sonogram demonstrating two heartbeats, the likelihood of the pregnancy's resulting in zero, one, or two liveborn infants can be predicted based on the placentation type, the gestational age at the time of the sonogram, and the presence or absence of sonographic abnormalities (1,9).

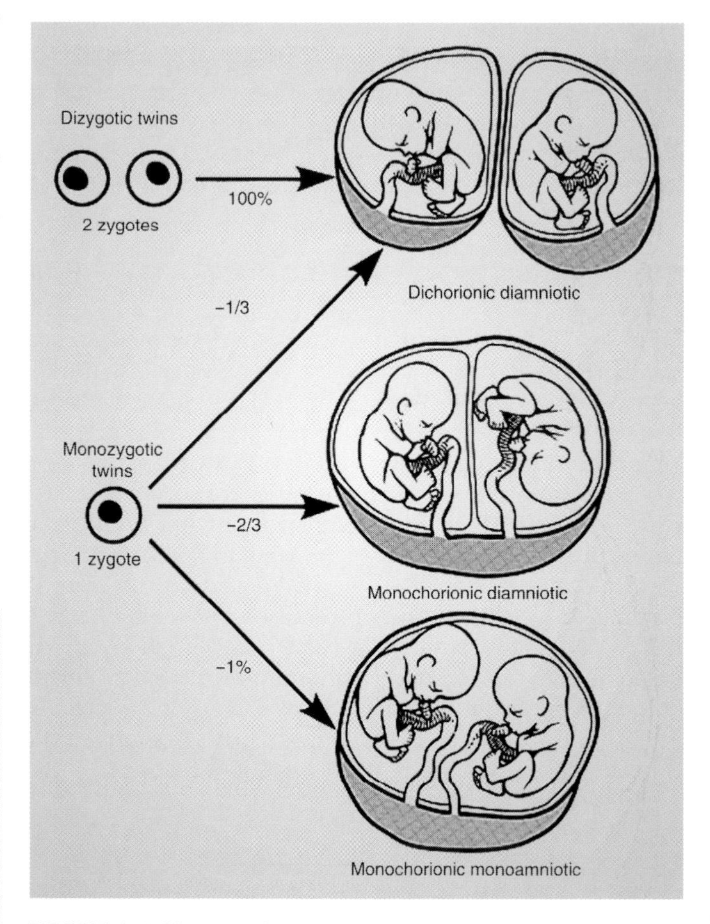

FIGURE 1 ■ Monozygotic versus dizygotic placentation. All dizygotic twins have dichorionic–diamniotic placentation. Monozygotic twins may be of any placentation type; about two-thirds are monochorionic–diamniotic. *Source*: From Ref. 4.

TABLE 3 ■ Placentation: Distinguishing Sonographic Features

Feature	Placentation type		
	DC–DA	MC–DA	MC–MA
Number of placentas	2 (may be contiguous)	1	1
Fetal sex	Same or different	Same	Same
Membrane thickness	Thick	Thin, wispy	None
Umbilical cords	Separate	Separate	May be intermingled
Number of yolk sacs	2	2	1

Abbreviations: DC, dichorionic; DA, diamniotic; MC, monochorionic; MA, monoamniotic.
Source: From Ref. 4.

COMPLICATIONS OF TWIN PREGNANCIES

Twin gestations have a higher rate of complications than do singletons, both because some obstetric complications that occur in singletons occur more frequently with twins and because certain complications are unique to twins (Table 5). Multiple gestations are more likely to result in premature delivery than are singletons and are thus at elevated risk for the neonatal morbidity and mortality associated with prematurity. Other general obstetric complications also occur with increased frequency in twins (Table 5). For example, monozygotic twins have an

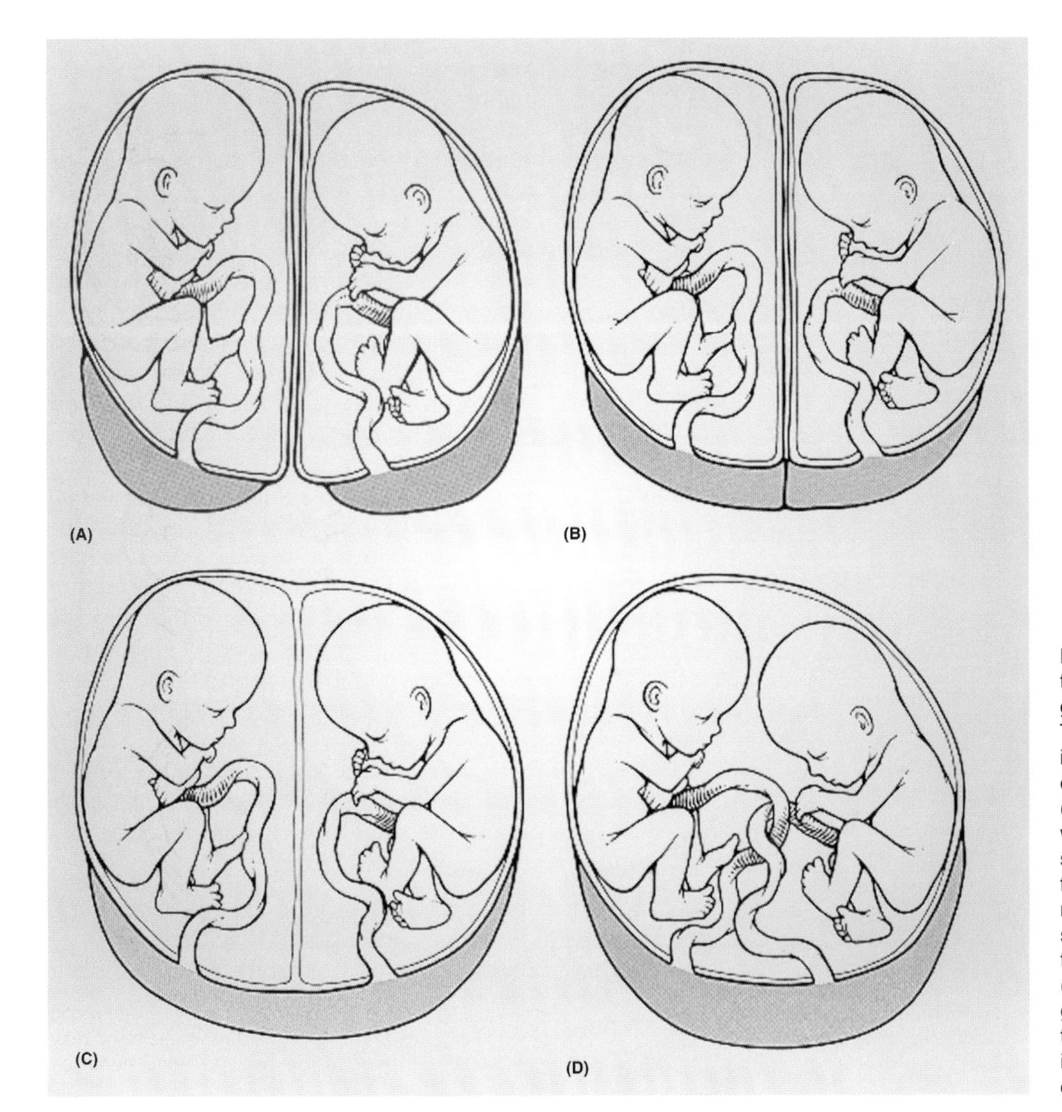

FIGURE 2 A–F ■ Placentation types. (**A**) Dichorionic–diamniotic gestation with separate placentas. The membrane separating the twins is thick, with four layers: two chorionic and two amniotic membranes. (**B**) Dichorionic–diamniotic gestation with fused placentas. The membrane separating the twins is thick, with four layers. (**C**) Monochorionic–diamniotic gestation. The membrane separating the twins is thin, with two layers of amniotic membrane. (**D**) Monochorionic–monoamniotic gestation. No membrane separates the twins. The two umbilical cords intermingle in the single amniotic cavity. (*Continued*)

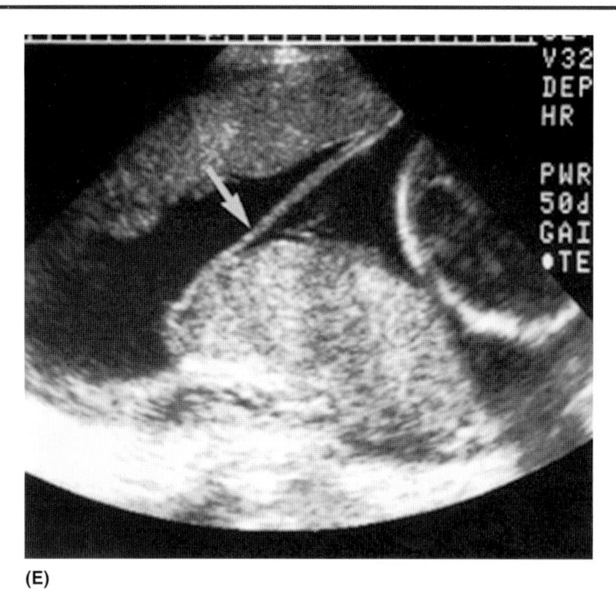

(E)

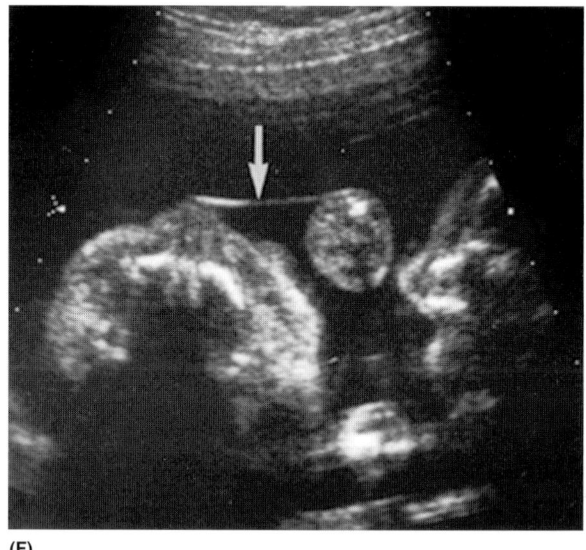

(F)

FIGURE 2 A–F ■ (*Continued*) (**E**) Thick membrane. Sonogram demonstrates thick membrane (*arrow*) of a dichorionic–diamniotic gestation. The two placentas are clearly separate. (**F**) Thin membrane. Sonogram demonstrates thin, wispy membrane (*arrow*) of a monochorionic–diamniotic twin gestation. *Source*: From Ref. 4.

increased incidence of fetal anomalies, including anencephaly, hydrocephalus, holoprosencephaly, sirenomelia, and sacrococcygeal teratoma. These anomalies are often discordant, affecting only one twin, even though both twins arise from a single fertilized egg (Fig. 4).

Complications specific to twins depend on placentation type (3,10). Monochorionic twins may develop complications resulting from vascular anastomoses through the common placenta. Monoamniotic twins are at high risk for umbilical cord entanglement and serious cord accidents within their common amniotic cavity.

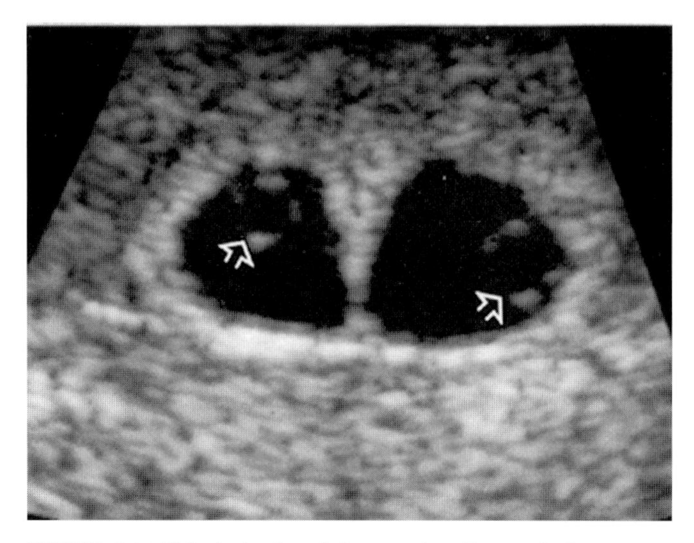

FIGURE 3 ■ Dichorionic–diamniotic gestation. Transvaginal sonogram demonstrates two gestational sacs with thick separating membranes. A yolk sac is seen in each gestational sac (*open arrows*). *Source*: From Ref. 4.

Twin–Twin Transfusion Syndrome

Twin–twin transfusion syndrome is a complication of monochorionic twins that results from unbalanced exchange of blood from one fetus (donor twin) to the other (recipient twin) across anastomoses in their common placenta. The donor twin becomes anemic, the recipient twin polycythemic. The sonographic features of twin–twin transfusion syndrome include discrepant amniotic fluid volumes and discordant fetal sizes (Fig. 5) (11–14). Discordance is best diagnosed when the difference in the estimated fetal weights between the twins is greater than 25% of the estimated weight of the larger twin (15,16). The anemic donor twin is small and has oligohydramnios. The oligohydramnios may be so severe that the donor twin is compressed against the uterine wall by the diamniotic membrane. When this occurs, the affected fetus is termed a "stuck" twin. The polycythemic recipient twin usually is appropriate in size, has polyhydramnios, and may be hydropic. Both the donor and the recipient twin carry a poor prognosis, with a mortality rate (intrauterine and perinatal) of approximately 70%. The prognosis is especially poor when the donor twin is a stuck twin.

A system of grading the sonographic and Doppler findings of twin–twin transfusion syndrome has been used to help predict the prognosis of this complication. The grading is as follows:

- Grade I: Polyhydramnios around the recipient, bladder seen in the donor, and normal Doppler
- Grade II: Polyhydramnios around the recipient, bladder not seen in the donor, and normal Doppler
- Grade III: Abnormal umbilical artery, umbilical vein, or ductus venosus Doppler, e.g., absent or reversed

TABLE 4 ■ Outcome of Twin Gestations

	Twins with heartbeats by first-trimester ultrasound			
Placentation	Two (%) liveborn	One (%) liveborn	No liveborn (%)	Survival rate per fetus (%)
DC–DA	83	12	5	89
MC–DA	56	11	33	61

	Twins Diagnosed Later in Pregnancy
Placentation	Survival rate per fetus (%)
DC–DA	91
MC–DA	75
MC–MA	50

Abbreviations: DC, dichorionic; DA, diamniotic; MC, monochorionic; MA, monoamniotic.
Source: From Ref. 4.

diastolic flow in the umbilical artery, pulsatile umbilical venous flow, or reversed flow in the ductus venosus

- Grade IV: Hydrops of one twin
- Grade V: Death of one or both twins

The higher the sonographic grade, the worse the prognosis (17). Furthermore, twin pregnancies complicated by twin–twin transfusion syndrome can be followed to determine if the condition is worsening, as indicated by an increase in grade, or improving, when the grade decreases.

TABLE 5 ■ Complications of Twin Gestations

Complications more frequent in twin than singleton gestations

- Preterm delivery
- Intrauterine growth retardation
- Preeclampsia
- Premature rupture of membranes
- Placenta previa
- Abruptio placentae
- Postpartum hemorrhage
- Umbilical cord accidents
- Congenital anomalies (increased frequency in monozygotic twins)

Complications unique to twins

- Twin–twin transfusion syndrome (monochorionic twins)
- Acardiac anomaly (monochorionic twins)
- Co-twin demise with twin embolization syndrome (monochorionic twins)
- Umbilical cord entanglement (monoamniotic twins)
- Conjoined twins

Source: From Ref. 4.

Acardiac Anomaly

The acardiac anomaly is a rare complication of monochorionic twins that results from large artery-to-artery and vein-to-vein anastomoses in their common placenta. In such cases, one twin becomes the "pump" twin, with its heart supplying blood both to itself and to the acardiac co-twin. The direction of blood flow in the co-twin is reversed, with blood entering this twin through its umbilical artery and exiting through its umbilical vein. This circulatory pattern leads to severe anomalies in the acardiac twin, including failure of development of a heart and poor or absent development of the upper extremities and head. The acardiac twin often has a two-vessel umbilical cord. The sonographic findings in acardiac twins are pathognomonic (Fig. 6, Table 6). The reversed direction of blood flow in the acardiac twin's umbilical vessels can be documented with Doppler (18–20).

Cord Entanglement

Monoamniotic twins and their umbilical cords reside in a common amniotic sac, leading to potential intertwining of the two cords that may cut off blood flow to one or both fetuses. This complication occurs frequently in monoamniotic twins and contributes to the high mortality rate in such twins (1).

Conjoined Twins

Conjoined twins are rare, occurring in approximately 1 in 50,000 births. They are classified according to the location of conjoining, with the most common sites being the anterior thoracic and abdominal walls. Conjoined twins are easily identified by ultrasound. Sonography is particularly useful for determining the location of conjoining and the extent of organ sharing (Figs. 7 and 8). This technique allows assessment of prognosis, surgical planning, and parental counseling before delivery (21,22).

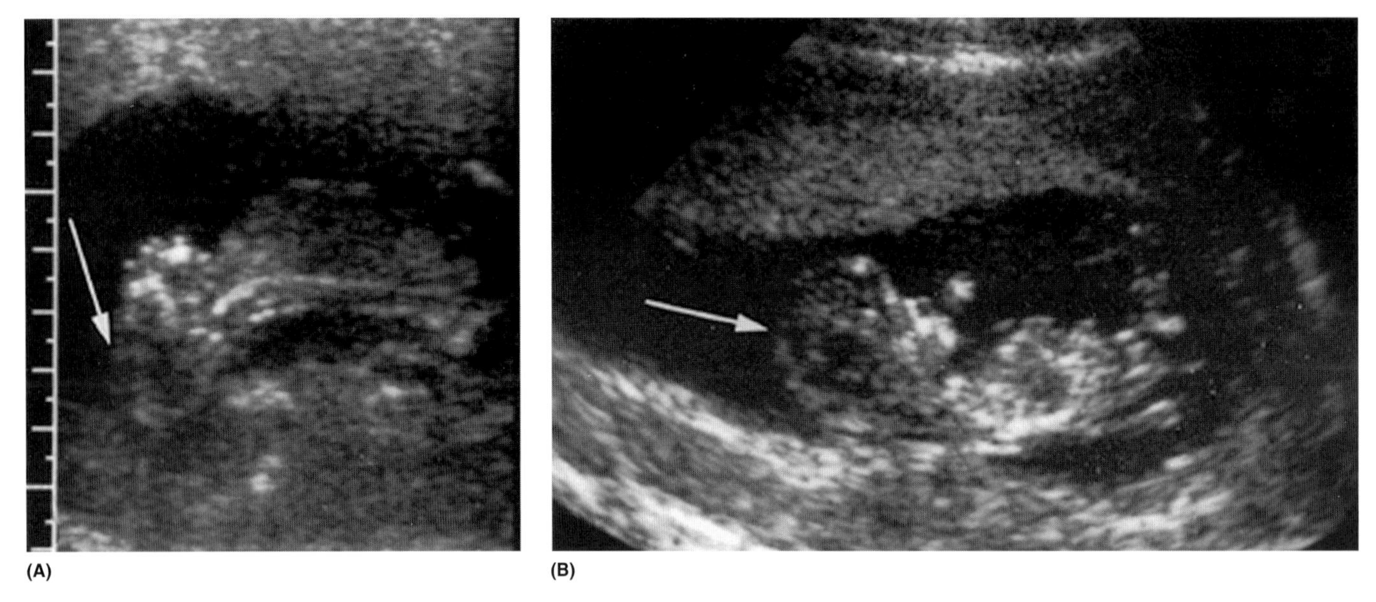

(A) (B)

FIGURE 4 A AND B ■ Discordant anomaly in monozygotic twins. (**A**) Absent cranium in anencephalic twin A (*arrow*). (**B**) Twin B with normal cranium (*arrow*). *Source*: From Ref. 4.

Effect on Survivor Following Demise of One Twin

When one twin dies in utero, the sonographic findings and risk to the surviving co-twin depend on the chorionicity and the gestational age at the time of the demise. When demise of one of a dichorionic pair occurs in the first trimester, the ultrasound appearance becomes that of a normal singleton pregnancy after several weeks, and the surviving co-twin is typically normal. This situation is sometimes referred to as the "vanishing twin" phenomenon (6). With monochorionic twins, when one

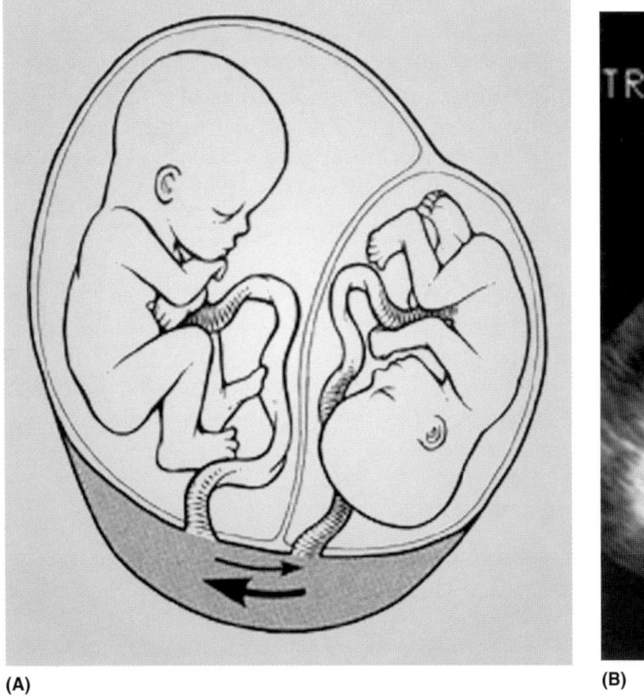

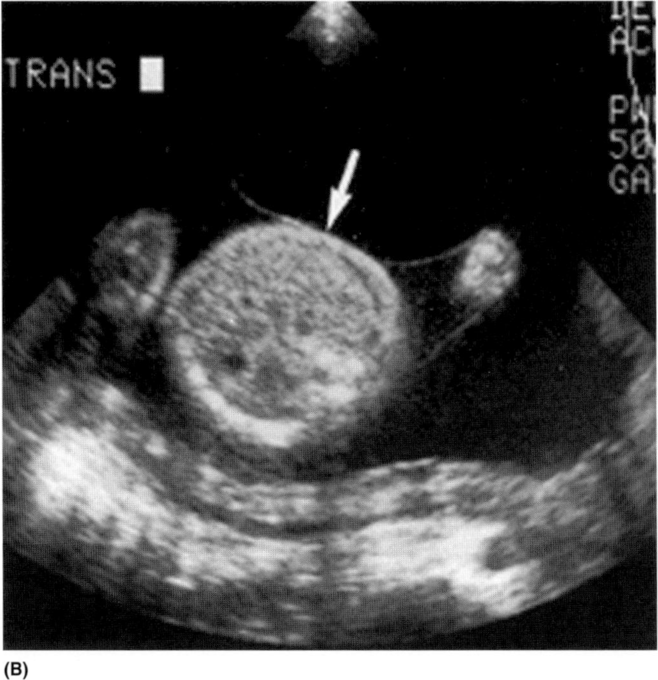

(A) (B)

FIGURE 5 A–D ■ Twin–twin transfusion syndrome. (**A**) Arterial-to-venous anastomoses across the single placenta permit unbalanced exchange of blood between the twins, such that one becomes anemic and small and the other plethoric. The smaller "donor" twin is surrounded by oligohydramnios, the larger "recipient" twin by polyhydramnios. (**B**) Discrepant amniotic fluid volumes. Oligohydramnios surrounds the donor twin (*arrow*), whereas polyhydramnios is found in the recipient twin sac. (*Continued*)

(C)

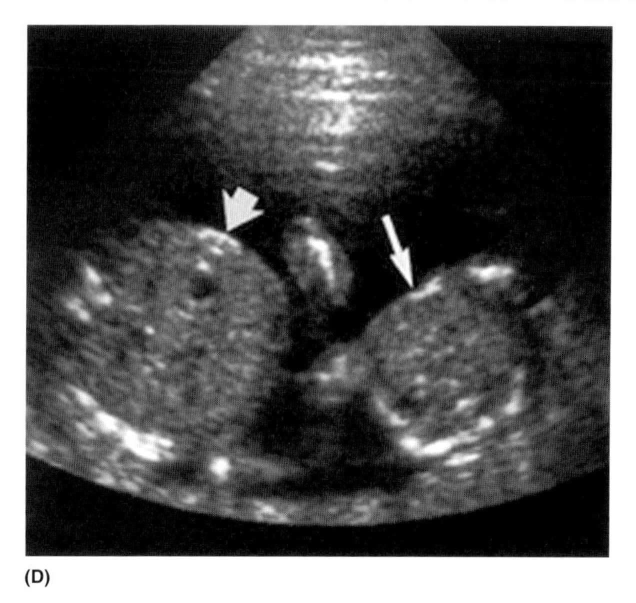

(D)

FIGURE 5 A–D ■ (*Continued*) (**C**) Stuck twin. Sonogram demonstrating smaller twin (*arrows*) held against uterine wall by tightly applied membrane resulting from severe oligohydramnios. Severe polyhydramnios surrounds the co-twin. (**D**) Discordant fetal abdomens. Sonogram of twin abdomens shows that the donor twin's abdomen (*long arrow*) is much smaller than recipient twin's (*short arrow*). *Source*: From Refs. 4, 14.

twin dies during the first trimester, its co-twin usually dies, as well (9).

After demise of a twin in the second or third trimesters, the dead twin remains visible as a fetus papyraceus (paper-thin fetus) (Fig. 9) or a macerated fetus for the remainder of pregnancy. If the pregnancy is dichorionic, the surviving co-twin will generally suffer no sequelae, but if it is monochorionic, the surviving co-twin may suffer ischemic damage to various organs or, in severe cases, may even die. This damage to the surviving monochorionic co-twin is termed the "twin embolization syndrome," and occurs more often

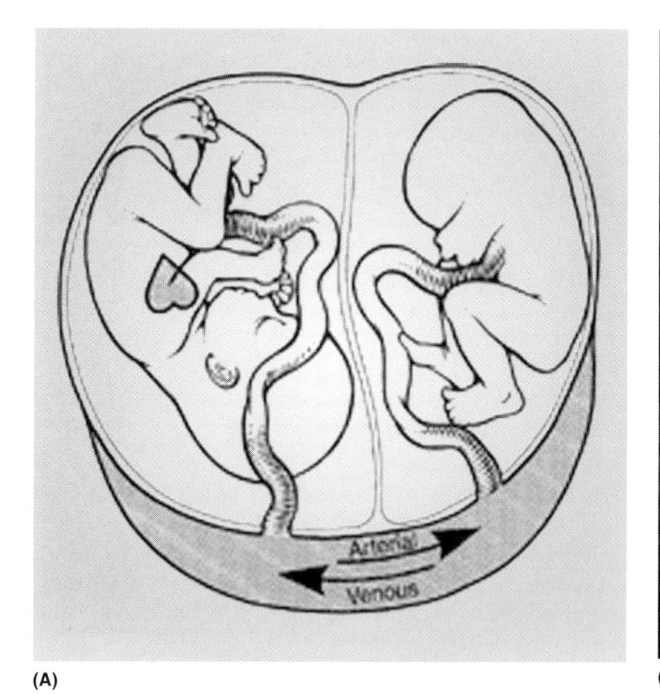

(A)

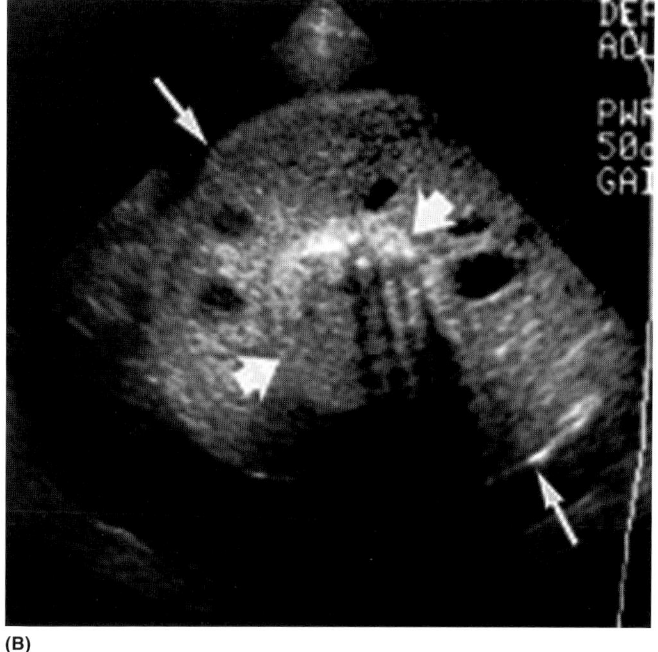

(B)

FIGURE 6 A–D ■ Acardiac twinning. (**A**) Large artery-to-artery and vein-to-vein anastomoses across the common placenta result in reversed flow in the umbilical vessels, failure of development of the heart, and structural abnormalities in the acardiac twin. (**B**) Massive edema (*large arrow*) surrounds the acardiac twin abdomen (*short arrows*). (*Continued*)

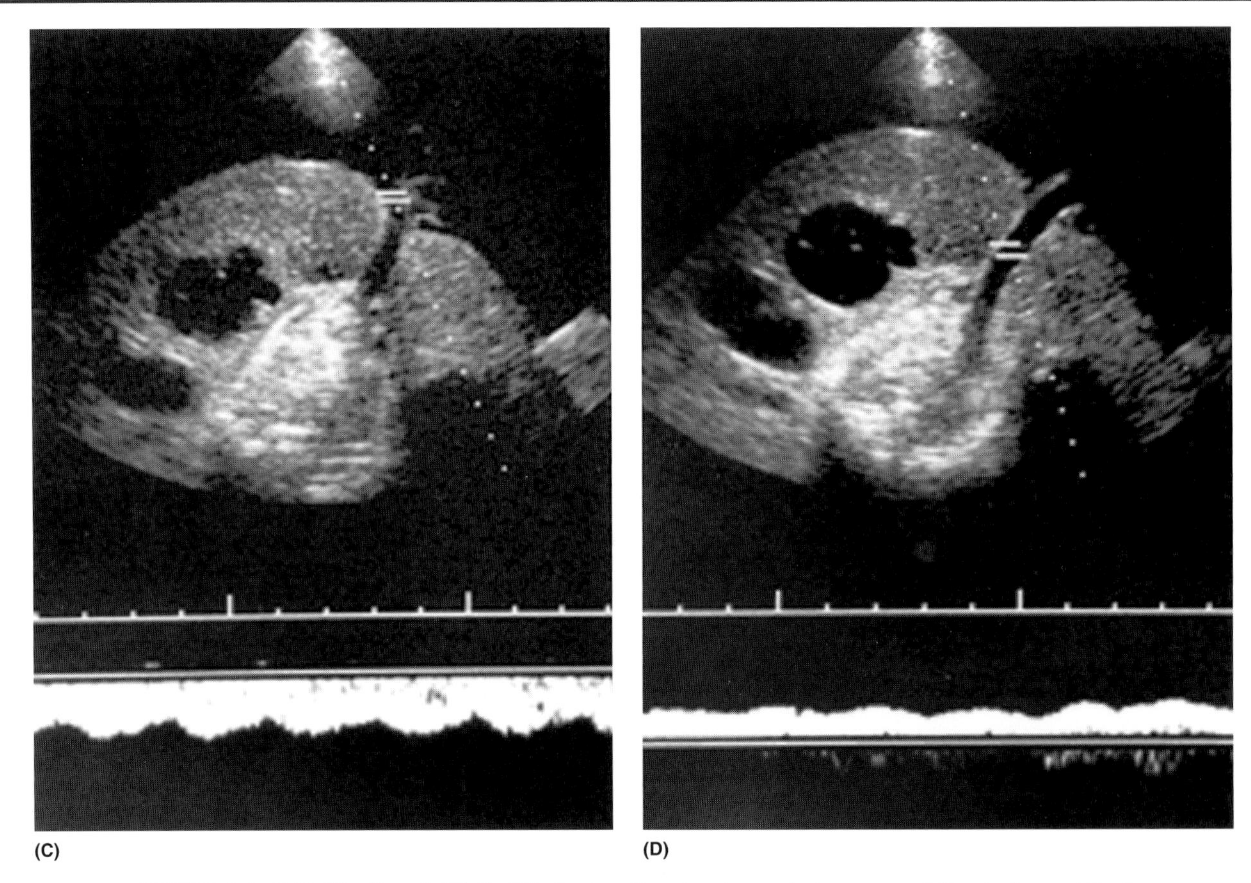

(C) (D)

FIGURE 6 A–D ■ (*Continued*) (**C**) Umbilical artery Doppler demonstrates reversed flow in the umbilical artery, toward the acardiac twin. (**D**) Umbilical vein Doppler demonstrates reversed flow in the umbilical vein, away from the acardiac twin, toward the placenta. *Source*: From Refs. 4, 18.

following second trimester death than after third trimester death. Twin embolization syndrome occurs either because emboli pass from the dead to the living twin through placental anastomoses or because the survivor suffers an acute hypotensive episode at the time of death of the co-twin. (3,23–25). Sonographically visible abnormalities include porencephaly, microcephaly, intestinal atresias, and renal cortical defects.

TABLE 6 ■ Sonographic Features of Acardiac Twinning

Acardiac twin

- Absent or rudimentary heart
- Absent or malformed head
- Absent or malformed upper extremities
- Subcutaneous edema (often massive)
- Two-vessel umbilical cord (common)
- Reversed flow in umbilical artery (toward fetus) and vein (away from fetus)

Pump twin

- Normal or increased amniotic fluid
- Possible hydrops

Source: From Ref. 4.

FETAL MEASUREMENTS

Fetal measurements serve the same purposes in a twin pregnancy as in a singleton pregnancy: to assign gestational age at the time of the first sonogram during the pregnancy and to assess normality of the size and growth of each fetus. In addition, measurements of the two fetuses should be compared with one another to evaluate for discordance.

In the first and second trimesters, the size of fetal body parts is similar in twins and singletons of the same gestational age. In the third trimester, this remains true for the femur length, but data are conflicting concerning the biparietal diameter. Some studies have found no difference between twin and singleton biparietal diameter measurements, whereas others have found that the biparietal diameter in twins becomes progressively smaller than that of singletons as the third trimester proceeds. When dating twin gestations, therefore, singleton tables or formulas for femur length are accurate for dating throughout the second and third trimesters. Those for

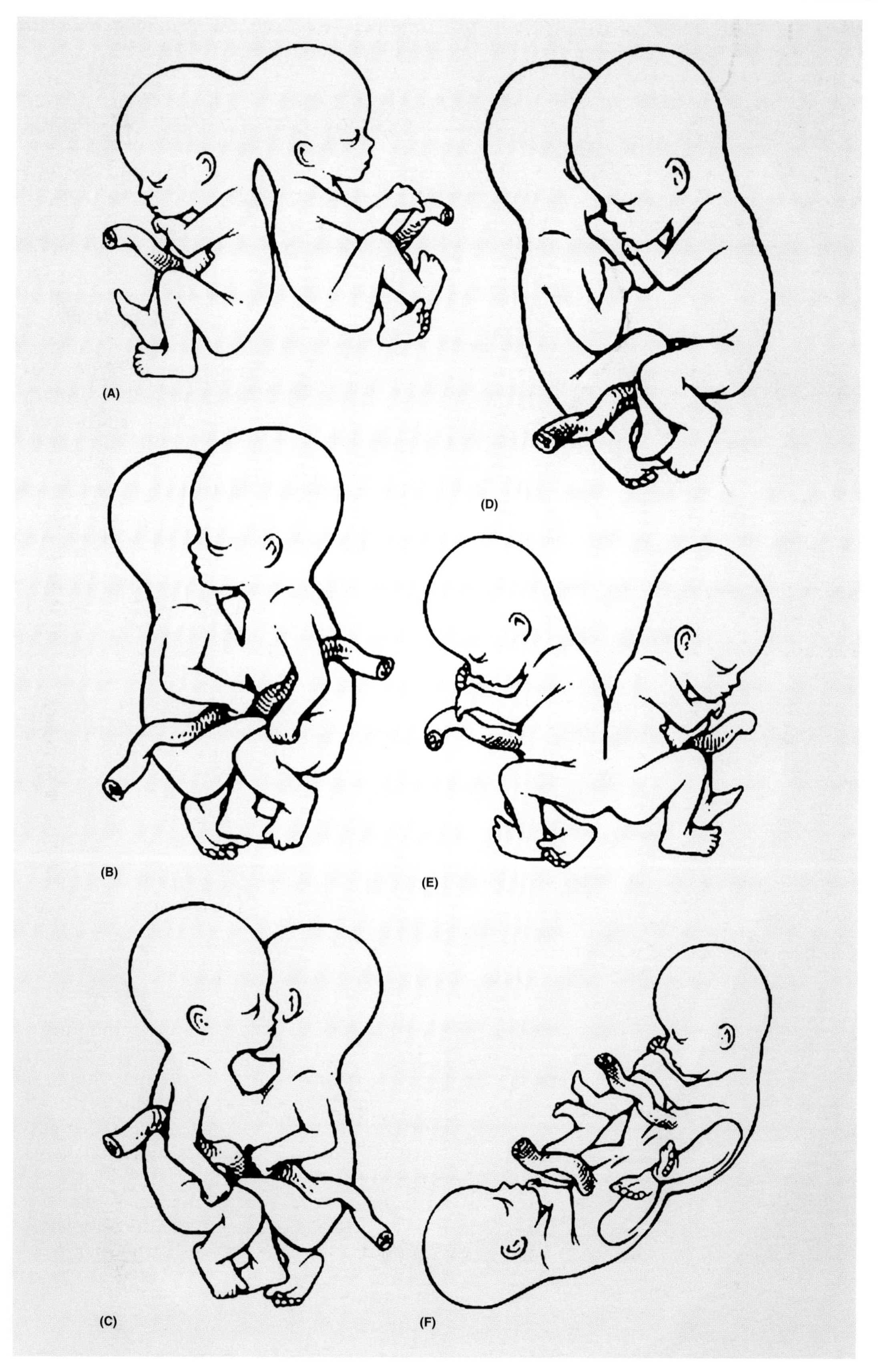

FIGURE 7 A–F ■ Most common forms of conjoined twins. Incomplete division of the fertilized ovum leads to conjoined twins with shared organs. (**A**) Craniopagus twins are joined at the head. (**B**) Xiphopagus twins are joined at the xiphoid. (**C**) Thoracopagus twins are joined at the thorax. (**D**) Omphalopagus twins are joined across the anterior abdomen. (**E**) Pygopagus twins are joined at the pelvis, facing away. (**F**) Ischiopagus twins are joined at the pelvis, opposed. *Source*: From Ref. 4.

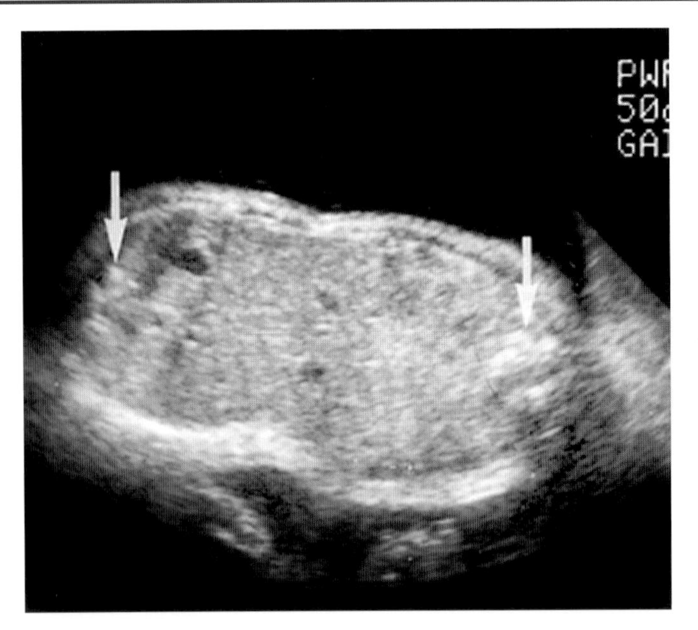

FIGURE 8 ■ Omphalopagus twins. Transverse sonogram of conjoined twin abdomens demonstrates separate spines (*arrows*) with joining across the anterior abdomen. *Source*: From Ref. 14.

biparietal diameter are accurate at least through the end of the second trimester (3,26–29).

ULTRASOUND-GUIDED INVASIVE PROCEDURES

Ultrasound can be used to guide certain procedures in multiple gestations. Twin amniocenteses are usually accomplished by two separate needle insertions, with injection of dye after the first insertion to confirm that the second needle has entered the second sac. In multiple gestations with three or more fetuses, selective reduction by intracardiac potassium chloride injection may be used to reduce the number of fetuses to enhance the survival chances of the remaining ones (14,30). This procedure is most often used in pregnancies that result from ovulation-induction therapy, because such pregnancies are frequently multiple and each fetus usually has its own placenta. Separate placentas reduce the risks to the surviving fetuses. The technique of selective termination by intracardiac injection can also be used in dichorionic twin gestations with one abnormal fetus.

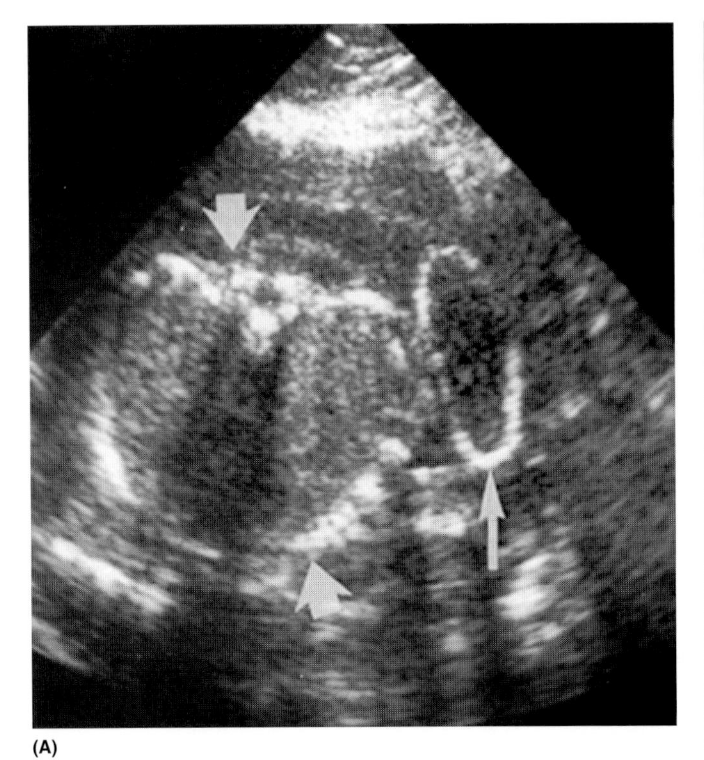

(A)

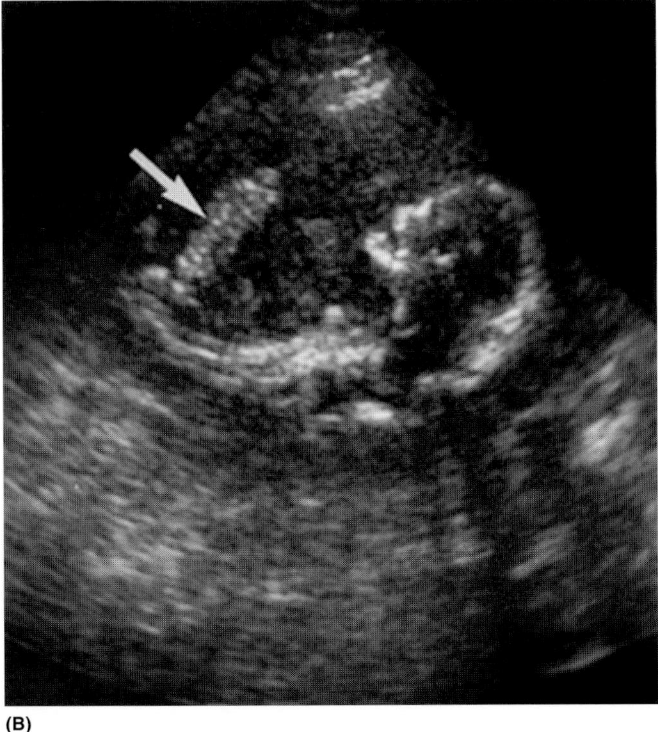

(B)

FIGURE 9 AND B ■ Fetus papyraceus. (**A**) Head of paper-thin fetus (*long arrow*) is compressed against the uterine wall by the surviving co-twin (*short arrows*). (**B**) Longitudinal view of fetus papyraceus demonstrates the head and spine still visible with the echogenic (thrombosed) umbilical cord (*arrow*) extending anteriorly. *Source*: From Ref. 14.

REFERENCES

1. Benirschke K, Kim CK. Multiple pregnancy. I. N Engl J Med 1973; 288:1276.
2. Benirschke K, Kim CK. Multiple pregnancy. II. N Engl J Med 1973; 288:1329.
3. Benson CB, Doubilet PM. Ultrasound of multiple gestations. Semin Roentgenol 1991; 26:50.
4. Benson C, Doubilet P. Twin pregnancy. In: McGahan J, Porto M, eds. Diagnostic Obstetrical Ultrasound. Philadelphia: JB Lippincott, 1994:434.
5. Doubilet PM, Benson CB. Appearing twin: undercounting of multiple gestations on early first trimester sonograms. J Ultrasound Med 1998; 17:199.
6. Landy HJ, Weiner S, Corson SL, et al. The "vanishing twin": ultrasonographic assessment of fetal disappearance in the first trimester. Am J Obstet Gynecol 1986; 155:14.
7. Townsend RR, Simpson GF, Filly RA. Membrane thickness in ultrasound prediction of chorionicity of twin gestations. J Ultrasound Med 1988; 7:327.
8. Levi CS, Lyons EA, Dashefsky SM, Lindsay DJ, Holt SC. Yolk sac number, size and morphologic features in monochorionic monoamniotic twin pregnancy. Can Assoc Radiol J 1996; 47:98.
9. Benson CB, Doubilet PM. Outcome of twin gestations following sonographic demonstration of two heartbeats in the first trimester. J Ultrasound Obstet Gynecol 1993; 3:343.
10. Mariona FG. Anomalies specific to multiple gestations. In: Chervenak FA, Isaacson VS, Campbell S, eds. Ultrasound in Obstetrics And Gynecology. Boston: Little, Brown, 1993:1051.
11. Brown DL, Benson CB, Driscoll SG, et al. Twin–twin transfusion syndrome: sonographic findings. Radiology 1989; 170:61.
12. Arts VFT, Lohman AH. The vascular anatomy of monochorionic diamniotic twin placentas and the transfusion syndrome. Eur J Obstet Gynecol 1971; 3:85.
13. Pretorius DH, Manchester D, Barkin S, et al. Doppler ultrasound of twin transfusion syndrome. J Ultrasound Med 1988; 7:117.
14. Benson CB, Doubilet PM. Sonography of multiple gestations. Radiol Clin North Am 1990; 28:149.
15. Storlazzi E, Vintzileos AM, Campbell WA, et al. Ultrasonic diagnosis of discordant fetal growth in twin gestations. Obstet Gynecol 1987; 69:363.
16. Divon MY, Girz BA, Sklar A, et al. Discordant twins: a prospective study of the diagnostic value of real-time ultrasonography combined with umbilical artery velocimetry. Am J Obstet Gynecol 1989; 161:757.
17. Quintero RA, Morales, WJ, Allen MH, et al. Staging of Twin-Twin Transfusion Syndrome. J Perinatol 1999; 9:550.
18. Benson CB, Bieber FR, Genest DR, et al. Doppler demonstration of reversed umbilical blood flow in an acardiac twin. J Clin Ultrasound 1989; 17:291.
19. Benirschke K, Harper VDR. The acardiac anomaly. Teratology 1977; 15:311.
20. Pretorius DH, Leopold GR, Moore TR, et al. Acardiac twin report of Doppler sonography. J Ultrasound Med 1988; 7:413.
21. Fitzgerald EJ, Toi A, Cochlin DL. Conjoined twins: antenatal ultrasound diagnosis and a review of the literature. Br J Radiol 1985; 58:1053.
22. Sakala EP. Obstetric management of conjoined twins. Obstet Gynecol 1986; 67:21S.
23. Landy HJ, Weingold AB. Management of a multiple gestation complicated by an antepartum fetal demise. Obstet Gynecol Surv 1989; 44:171.
24. Jauniaux E, Elkazen N, Leroy F, et al. Clinical and morphologic aspects of the vanishing twin phenomenon. Obstet Gynecol 1988; 72:577.
25. Patten RM, Mack LA, Nyberg DA, et al. Twin embolization syndrome: prenatal sonographic detection and significance. Radiology 1989; 173:685.
26. Grumback K, Coleman BG, Arger PH, et al. Twin and singleton growth patterns compared using ultrasound. Radiology 1986; 158:237.
27. Socol ML, Tamura RK, Sabbagha RE, et al. Diminished biparietal diameter and abdominal circumference growth in twins. Obstet Gynecol 1984; 64:235.
28. Reece EA, Yarkoni S, Abdalla M, et al. A prospective longitudinal study of growth in twin gestations compared with growth in singleton pregnancies. I. The fetal head. J Ultrasound Med 1991; 10:439.
29. Reece EA, Yarkoni S, Abdalla M, et al. A prospective longitudinal study of growth in twin gestations compared with growth in singleton pregnancies. II. The fetal limbs. J Ultrasound Med 1991; 10:445.
30. Golbus MS. Selective termination. In: Harrison MR, Golbus MS, Filly RA, eds. The Unborn Patient: Prenatal Diagnosis and Treatment. Philadelphia: WB Saunders, 1991:166.

Growth Disturbances: Large-for-Date and Small-for-Date Fetuses

Lawrence D. Platt, Greggory R. Devore, and Dru E. Carlson

51

LARGE-FOR-DATE FETUSES

A maternal fundal height that is 3 cm or greater than the estimated gestational age warrants a detailed ultrasonographic evaluation of fetal size, anatomy, and amniotic fluid volume (1). The body habitus of the obese pregnancy woman can confound the practitioner's ability to measure the fund height accurately. A list of the most common causes of large-for-date pregnancies is found in Table 1 (2).

INACCURATE DATING

In about 20% to 40% of all pregnancies, the true menstrual age is uncertain (3). Thus, the most common explanation for the large-for-date pregnancy is inaccurate dating. Patients with unclear normal last menstrual periods, who present for care in the first trimester, should have a dating ultrasound performed to measure crown–rump length. This is accurate within three days of true gestation (4). As pregnancy progresses, the accuracy of estimating age diminishes owing to normal human variation of fetal size. Performance of an early ultrasound often alleviates confusion about size and date later in gestation.

MULTIPLE GESTATIONS

The era of delivery room surprise twins (or more) should be of historic interest. Unfortunately, that is not the case. One important reason to obtain an ultrasound evaluation of the large-for-date pregnancy is to rule out multiple gestations (5). The added concern with multiple gestations is the increased risk of congenital anomalies secondary to the twinning process. Also, the finding of polyhydramnios in one sac and oligohydramnios or a stuck twin in the other brings forth a host of management decisions that are discussed elsewhere in this text, particularly in relationship to twin-to-twin transfusion syndrome.

MATERNAL OBESITY

The morbidly obese woman is defined as having a body weight 135% greater than her prepregnant ideal body weight. The pregnancy concerns of these women are myriad. The risk of diabetes is increased seven times above normal, and perinatal loss is increased 30% owing to diabetes, hypertension, and macrosomic fetuses with traumatic deliveries (6).

Many physicians believe that if the blood glucose level is well controlled in an insulin-dependent diabetic mother, there is no risk for a macrosomic

TABLE 1 ■ Causes of Large-for-Dates Pregnancies

- Inaccurate dating
- Multiple gestation
- Hydatidiform mole
- Diabetes mellitus
- Macrosomia
- Genetic syndromes featuring macrosomia
- Polyhydramnios
- Pelvic or abdominal mass

infant. Unfortunately, this is not always true. In one study (7), 43% of insulin-dependent diabetic mothers gave birth to macrosomic infants even when their blood glucose levels were considered well controlled. The unifying pattern in these patients was a large amount of weight gain in the third trimester, which correlated with a higher hemoglobin A1C level. Thus, the physician should suspect macrosomia in all diabetic pregnancies and perform frequent ultrasound assessments to monitor growth.

Most fetal weight gain occurs in the third trimester, and early fetal growth does not appear to be influenced by diabetic control. Patients with poorly controlled diabetes, who lack any vasculopathy, however, have been in longitudinal studies to be the ideal candidates for having macrosomic infants (8).

Besides macrosomia, the examiner should also be aware of the abnormalities associated with fetuses of diabetic mothers. A list of common anomalies is given in Table 2. A disturbing study from Italy showed that strict maternal control does not exclude accelerated fetal cardiac growth and abnormal development of cardiac function even though cardiac structure may be normal (9). Therefore, careful attention to the size of the fetal heart may be warranted throughout gestation.

In the ultrasound evaluation of these difficult-to-image patients, several scanning tips are helpful in trying to assess fetal size and anatomy.

1. Use the maternal umbilicus as a window to the fetus. Put a large amount of imaging gel into the umbilicus and then insert a vaginal probe or small transducer. This eliminates several centimeters of adipose, and manipulation of the angle of the probe often replaces the previously blurred image with one of relative clarity.
2. To obtain a good biparietal diameter measurement in the vertex-positioned fetus, place the sheathed vaginal probe transducer just slightly into the vaginal introitus. The head often is well outlined and more easily measured.
3. Have an assistant elevate the abdominal pannus to allow an improved image through the lower abdominal crease.

TABLE 2 ■ Prenatally Diagnosed Syndromes Associated with Macrosomia

Syndrome	Bone maturation	Macrocephaly
Beckwith–Wiedemann	No acceleration	Yes, plus omphalocele
Marshall–Smith	Marked acceleration	Yes, plus occasional absent corpus callosum
Weaver's	Marked acceleration	No, plus camptodactly
Sotos'	No acceleration	Marked plus large hands and feet
Ruvalcaba–Myhre	No acceleration	Marked

The greatest concern is that the physician will underestimate the weight of the fetus and not be prepared for a macrosomic neonate with the risk of shoulder dystocia. Obtaining the clearest image is the best help in estimating weight. Regardless, it is important to inform the patient that the quality of ultrasound imaging and exact estimation of weight are highly dependent on the distance from the probe to the fetus.

Using both pre- and postprocessing can also help optimize the image. Using a transducer with a range of frequencies will also allow you to improve the image in an obese patient. Often, removing harmonic imaging may improve the image in some patients.

GENETIC SYNDROMES AND FETAL MACROSOMIA

Various methods have been developed to achieve accurate weight estimates of the average-sized fetus. Unfortunately, there is no excellent way of precisely predicting macrosomia. The problem lies in the fact that the best ultrasound estimation of weight can be as much as 10% discrepant of the actual weight. This is an acceptable variance in the average-weight fetus, but in the macrosomic fetus, this may amount to several hundred grams of potentially damaging error. Watson and Seed (10) calculated that to achieve a 90% confidence that a newborn will actually weigh more than 4000 g, one must estimate the fetal weight by ultrasound at 4750 g. Thus, it is not surprising that a study showed that intrapartum ultrasound evaluation of possible macrosomia (estimated fetal weight of more than 4000 g) was judged no better than clinical impression (11). Recently, Lee has suggested that using three-dimensional (3D) volume may aid in the diagnosis of macrosomia.

The following genetic syndromes are associated with large-for-gestational-age newborns; these are summarized in Table 2.

Beckwith–Wiedemann syndrome usually is a sporadic abnormality (12). It is seen with macroglossia and, occasionally, umbilical hernia, which may appear on ultrasound as a small omphalocele. The unusual fissures of the ears are not easily appreciated on a prenatal examination except by 3D image. The diagnosis is confirmed at birth by adrenocortical cytomegaly, often with pancreatic hyperplasia manifested by severe neonatal hypoglycemia.

Marshall–Smith syndrome is a sporadic disease in which markedly accelerated skeletal maturation results in a long, relatively thin newborn (13). An overestimation of weight is made because the head and the femur are so much larger than anticipated. Occasionally, these neonates demonstrate an absent corpus callosum or an omphalocele. They are often mentally retarded and thrive poorly.

Weaver's syndrome also shows accelerated bone maturation but has the added problems of camptodactyly of the hands and foot deformities, such as talipes equinovarus, which can be seen on ultrasound (13,14).

Sotos' syndrome is associated with profound macrocephaly and mild dilation of the cerebral ventricles, but there is no consistent pattern of brain malformation (15). The hands and feet are also very large for gestational age (LGA). Mental retardation is present in up to 80% of these children.

Ruvalcaba–Myhre syndrome also presents with profound macrocephaly but the cerebral ventricles are always of normal size (16). The newborn has macrosomia and poor tone, but normal adult stature is seen. Unusual pigmentation of the skin is seen in early childhood. Polyhydramnios is seen on ultrasound, probably owing to the poor fetal tone.

The physician should be aware of these syndromes and search for the associated anomalies. Above all, the physician should always look at the parents when assessing a fetus with macrocephaly or isolated large hands or feet. Often, large size is simply an expression of familial propensity.

The suspicion of macrosomia starts with the third trimester examination of an enlarged fundal height and should always include a careful interview of the patient to elicit a history of other large babies, diabetes in this or other pregnancies, and current weight gain as well as a review of any specific genetic diseases in the family. The ultrasound examination should always be performed with attempts to visualize the fetus through various maternal windows. Finally, the physician should accept that our ability to estimate fetal weight of the LGA newborn is not as accurate as we would hope. The limitation should be understood by the patient and her family.

POLYHYDRAMNIOS

The ultrasound evaluation of a large-for-date fetus should always include accurate measurement of the amniotic fluid volume. During the past decade, several attempts have been made to clarify the most accurate ultrasound method of defining clinically significant polyhydramnios. These methods include the following:

1. A single largest vertical pocket of more than 8 cm (17,18)
2. The qualitative impression of polyhydramnios (19)
3. The four-quadrant technique of measuring fluid, called the amniotic fluid index (AFI) (20)

The AFI involves holding the transducer perpendicular to the floor and dividing the maternal uterus into four quadrants. The linea nigra and umbilicus serve as longitudinal and vertical markers of the measuring pockets. These are then added together to give the AFI (Fig. 1), which is normally 13 ± 5 cm (21). Because 2 standard deviations above the average 13 cm are 23 cm, we used 24 cm as the lower limit of clinically significant polyhydramnios. One hundred and twelve nondiabetic women were referred with the descriptive diagnosis of polyhydramnios made by experienced ultrasonographers. There was poor correlation between these descriptions and fetal outcome. Of the 112, 49 met the AFI of 24 cm-or-more definition of polyhydramnios. This allowed the inclusion of all fetuses with structural or chromosomal abnormalities. Interestingly, seven patients had AFIs of less than 24 cm but had the classic definition of polyhydramnios of one fluid pocket of 8 cm or greater. None of these patients had poor fetal outcome or polyhydramnios diagnosed at delivery. Several authors have endorsed this method of amniotic fluid measurements (22,23).

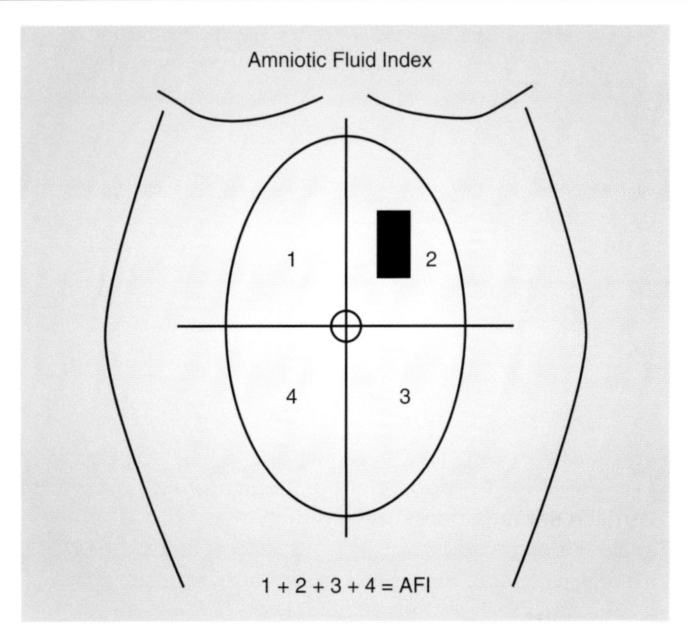

FIGURE 1 ■ The AFI is calculated by adding the largest vertical pocket of amniotic fluid in the four quadrants of the pregnant uterus. AFI, amniotic fluid index.

When the examiner determines that the AFI is 24 cm or greater, a careful, detailed examination of the fetal anatomy is warranted. More than 100 fetal anomalies are associated with polyhydramnios; however, this task can be daunting. A systematic, calculated physical examination of the fetus with polyhydramnios allows the examiner to identify the structural abnormalities commonly associated with this condition. In one study, 66 fetuses between 26 weeks' gestation and term were examined using this seven-point system. All abnormalities were diagnosed, with an average examination time of 17 minutes. The examiner begins at the fetal head and works caudally down the fetal body to answer these seven questions (Table 3).

1. *Head*. Is there a structural brain malformation that is causing poor swallowing and thus increased fluid?

TABLE 3 ■ Known Clauses of Intrauterine Growth Retardation

Intrinsic
- Aneuploidy
  - Trisomy
  - Triploidy
- Fetal structural anomalies
  - Disruption
  - Skeletal dysplasias
  - Syndromes
- Constitutional

Extrinsic
- Infection
  - Syphilis
  - Toxoplasmosis
  - Rubella
  - Cytomegalovirus
  - Herpes
  - Malaria
  - Chagas' disease
  - *Listeria monocytogenes* infection
- Maternal disease
  - Chronic hypertension
  - Collagen vascular disease
  - Hemoglobinopathies
  - Thyrotoxicosis
  - Cyanotic cardiopulmonary disease
  - Vascularly involved diabetes
  - Renal failure
  - Antiphospholipid antibodies
  - Active malabsorption diseases
- Drugs
  - Alcohol
  - Aminoproterin
  - Amphetamines
  - Cigarettes
  - Cocaine
  - Warfarin sodium (Coumadin)
  - Cytotoxic drugs
  - Heroin, methadone
  - Isotretinoin
  - Phenytoin

This can be seen with fetal hydrocephaly, holoprosencephaly, anencephaly, intracranial teratomas, and almost all other brain malformations.

2. *Throat and mouth.* Is swallowed amniotic fluid obstructed before arriving in the esophagus? This can be seen with choanal atresia (Figs. 2 and 3), cleft palate, median facial cleft, goiter, and teratoma.
3. *Heart.* Is there heart failure? Pathologically increased output by the fetal heart can be caused by a vascular teratoma in another part of the body or by a condition resulting in severe anemia, such as isoimmunization or fetal viral infection. Decreased output can also

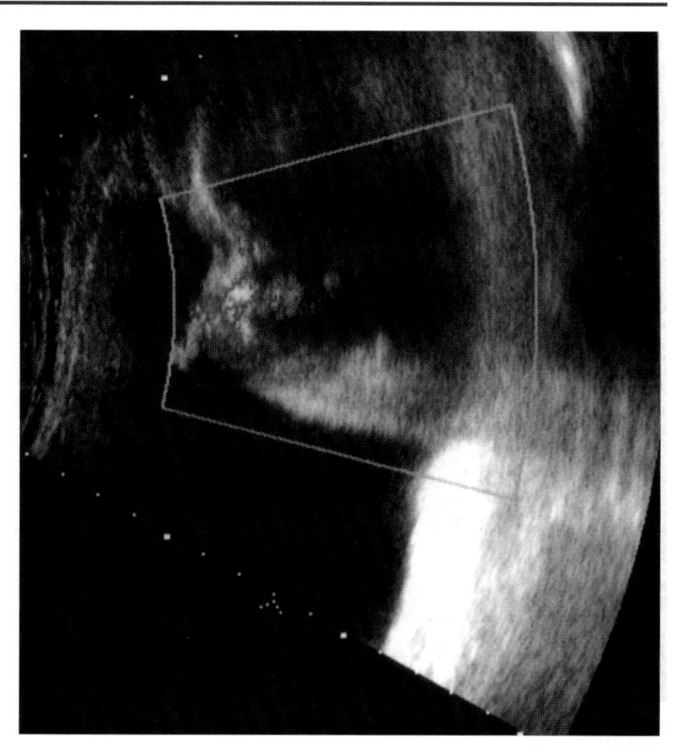

FIGURE 2 ■ Note the fluid flowing through the nares in a fetus with polyhydramnios and normal nares and pharynx.

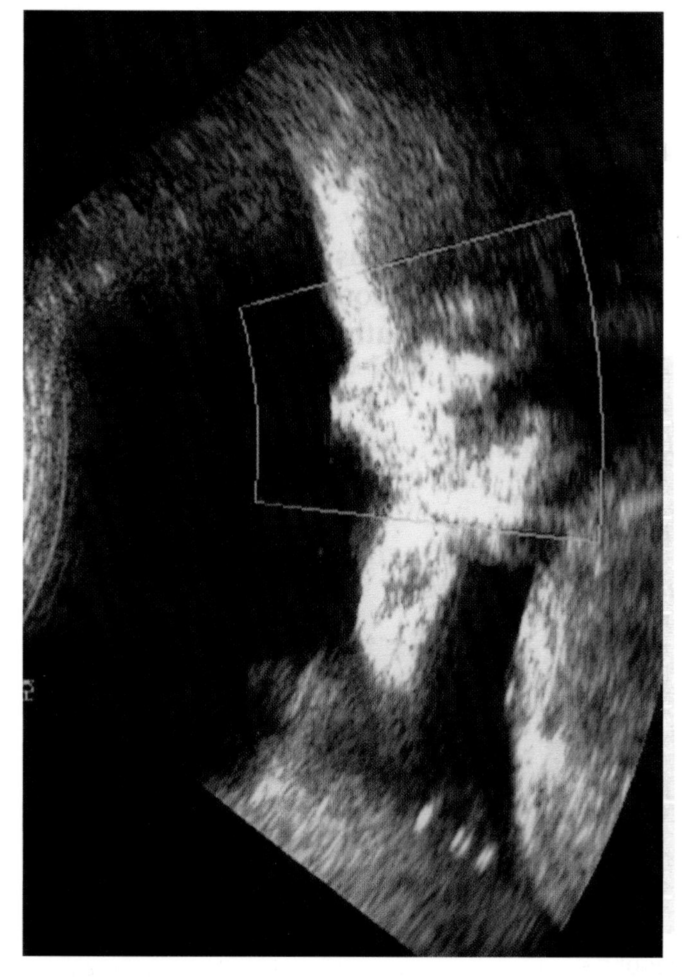

FIGURE 3 ■ Note the lack of fluid through the nares in a fetus with choanal atresia.

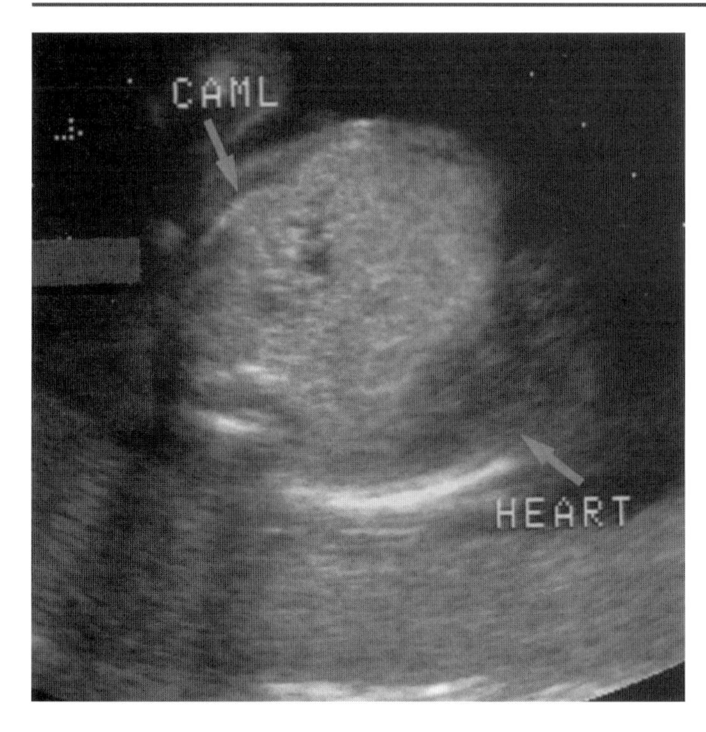

FIGURE 4 ■ Cross-section of fetal chest with cystic adenomatoid malformation of the lung (CAML) pushing the heart to the side.

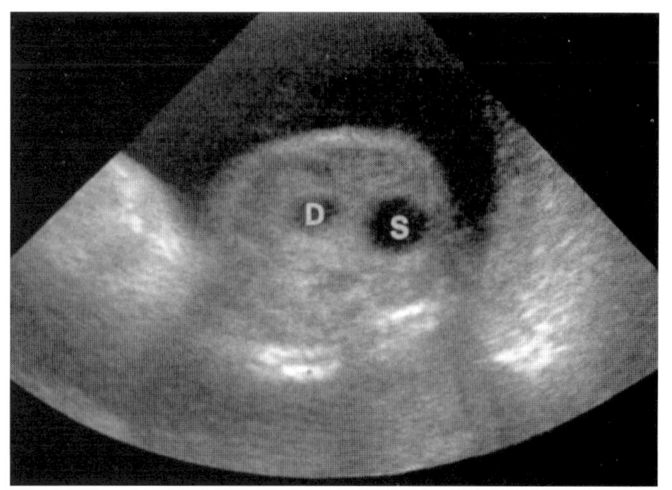

FIGURE 6 ■ Duodenal atresia with double-bubble sign. D, duodenum; S, stomach.

result in cardiac failure originating from fetal arrhythmias (heart block or supraventricular tachycardia) or from complex cardiac malformations.

4. *Chest*. Is a lesion in the fetal chest causing compression or diversion of blood? This can be seen with cystic adenomatoid malformations of the lung (Fig. 4), pulmonary sequestration, and diaphragmatic hernia (Fig. 5).
5. *Esophagus*. Is there a stomach bubble? If not, esophageal atresia or tracheoesophageal fistula should be considered.
6. *Upper gastrointestinal tract*. Is an obstruction distal to the stomach causing fetal regurgitation? This is seen with duodenal atresia (double-bubble sign; Fig. 6) and small intestine obstruction, resulting in multiple, circular, fluid-filled loops of bowel (Fig. 7).
7. *Overall tone*. Is an abnormality in the neurological function of the fetus (but not an appreciated intracranial lesion) causing poor movement and swallowing? This is associated with arthrogryposis multiforme, Neu-Laxova, and sometimes aneuploidy. This important observation centers on the evaluation of the overall

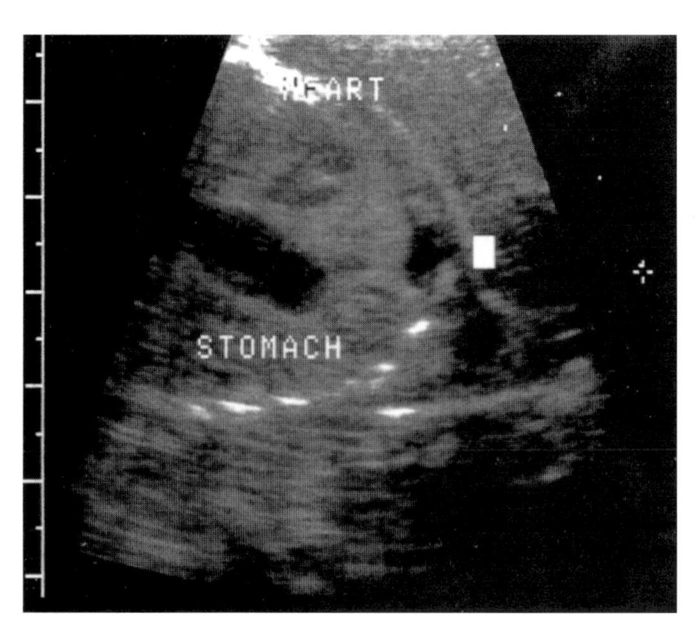

FIGURE 5 ■ Longitudinal section of fetal chest showing diaphragmatic hernia with stomach in thorax next to heart. Fetal abdomen is on the left side of the image.

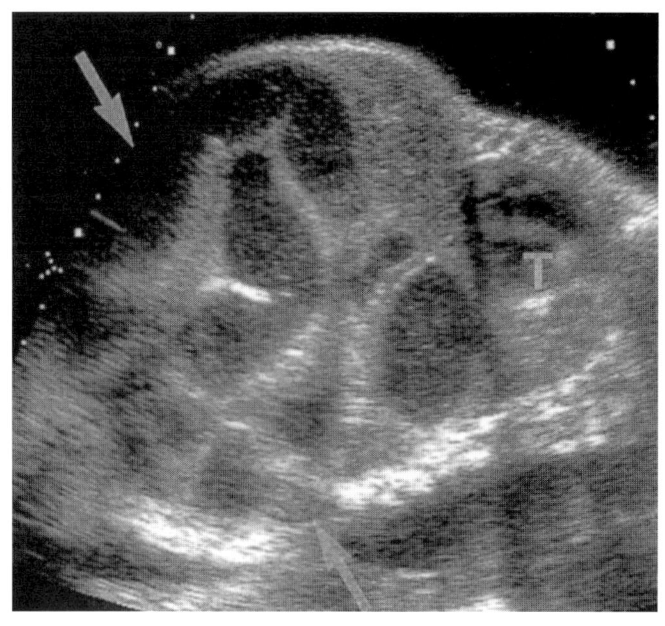

FIGURE 7 ■ Meconium ileus with multiple loops of dilated bowel (*arrows*) in this sagittal scan of the fetus. T, thorax.

tone of the fetus. Is the fetus moving normally? The examiner should watch the hands open and shut clearly. Are the feet strangely postured and fixed or do they turn and flex easily? Is there normal movement of the fetal mouth, with easily observed sucking and opening and shutting?

One easy way to determine the difference between fetal regurgitation secondary to obstruction and a neurological movement disorder is to place a color-flow Doppler window over the mouth of the fetus, with the color velocity set to visualize the umbilical cord. A vomiting fetus with a bowel obstruction demonstrates repeated bouts of explosive fluid eruptions from its open mouth and protruding tongue. A neurologically abnormal fetus either never opens or never closes its mouth in a normal manner. While looking at the movement of the fetal mouth, the examiner should also measure the mandible (24). Micrognathia is often observed with neurologically abnormal fetuses that have never moved their mouths normally (Fig. 8).

If a fetus displays the triad of polyhydramnios, a structural abnormality, and abnormal hand posturing, the parents should be offered rapid fetal karyotyping to rule out aneuploidy. In a study of 49 patients with polyhydramnios, six fetuses demonstrated this triad, and all had trisomic karyotypes (25). Three had trisomy 18, one had trisomy 13, and two had trisomy 21. The fetuses with trisomy 18 or 13 demonstrated classic clenched hands. Fetuses with trisomy 18 or 13 often have overlapping bones in their hands that never open and shut normally, but their arms still move and thus there is the appearance of "bright lights" on the ultrasound screen (Figs. 9 and 10). (It is important to stress that this is different from fetuses with movement disorders, such as arthrogryposis multiforme, who display an inability to move the arms and legs as well as the hands and have polyhydramnios.) In addition, fetuses with trisomy 13 can have polydactyly. The anomalies seen in the fetuses in this study included ventricular septal defect, hydrocephaly (25) with cleft lip and palate, and diaphragmatic hernia. The fetuses with trisomy 21 showed gastrointestinal atresias that accounted for the polyhydramnios. In addition, these fetuses with trisomy 21 and polyhydramnios had wide-open hands with sluggish closure of the fingers. Thus, fetuses with this triad should undergo rapid karyotyping such as umbilical blood sampling or placental biopsy.

When fetal hand movement is normal but an anatomic lesion can be seen, the families should still be

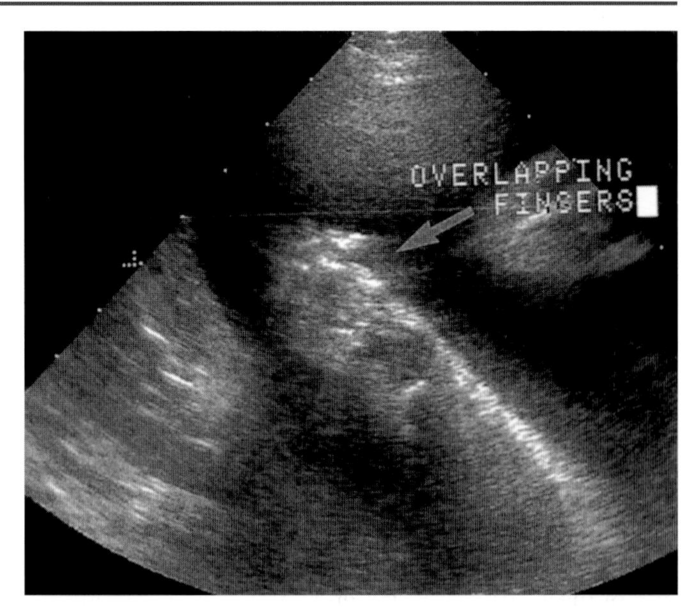

FIGURE 9 ■ Clenched, overlapping fingers of fetus with trisomy 18, seen as a bright light on ultrasound secondary to overlapping bone.

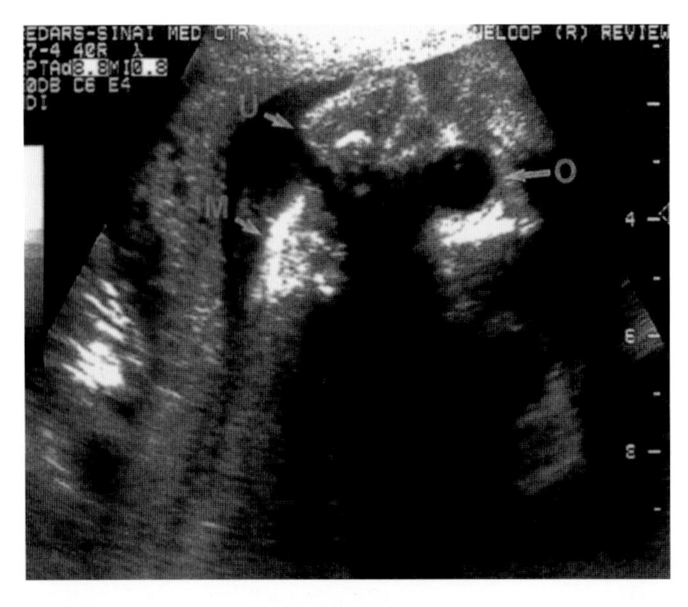

FIGURE 8 ■ Micrognathia in profile of fetus with Neu-Laxova (movement disorder). Note the open mouth. This syndrome is associated with poor movement, clenched hands, and polyhydramnios. M, micrognathia; U, upper lip; O, orbit.

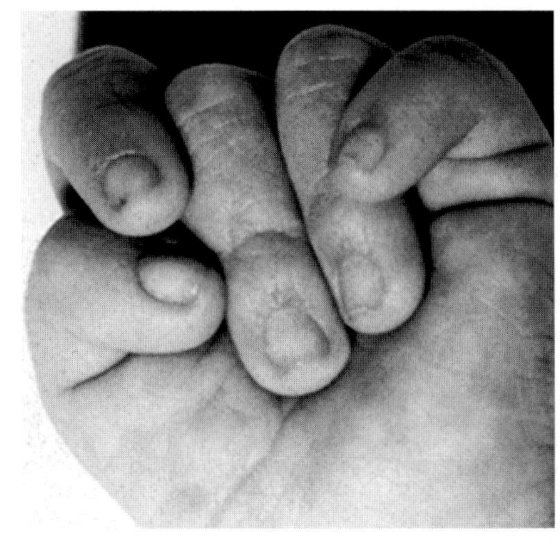

FIGURE 10 ■ Newborn with the same clenched hands.

offered karyotyping, although the urgency of the result can be tempered by the gestational age of the fetus. All fetuses with structural abnormalities should undergo chromosomal evaluation, regardless of the amniotic fluid status (26).

Whether every fetus that has polyhydramnios but no obvious anatomic abnormalities should be offered genetic karyotyping is controversial. The confidence and experience of the ultrasonographer is paramount in this situation. If there is no anatomic lesion, the hands are moving normally, and the fetus is growing normally, there is apparently little quantifiable increased risk for trisomy (27). The examination should include the seven questions listed earlier, however, and should stress echocardiography of the heart, excellent evaluation of the fetal brain, and documentation of normal movement of the hands and a normal stomach bubble and abdomen. If the examiner is uncertain about the adequate evaluation of any of these structures, genetic amniocentesis should be offered. Fetuses with Down syndrome typically do not have polyhydramnios unless there is a physical abnormality that predisposes them to that condition, such as bowel obstruction or a brain abnormality. Fetuses with lethal trisomies usually have obvious hand clenching and several anatomic abnormalities. Answering the seven questions listed earlier should ensure the opportunity to diagnose a chromosomally abnormal fetus with polyhydramnios.

SMALL-FOR-DATE FETUSES

The practitioner should be concerned about a small fetus or decreased amniotic fluid volume when the prenatal examination reveals the fundal height to be more than 2 cm less than the expected gestational age. The amount of adipose in the mother, the position of the fetus, and accurate information about last menstrual period should be considered when evaluating this discrepancy. The provider should be aware that ultrasound examination of the fetus is indicated to ensure fetal well-being.

CAUSES OF INTRAUTERINE GROWTH RESTRICTION

Intrinsic and extrinsic factors are involved in the cause of poor fetal growth (Table 3). The physician should begin with a careful maternal history of medication, illicit drug use (particular cocaine and amphetamines), alcohol, and cigarette smoking or other exposures. The medical history of the patient should include concerns with any disease that may cause vascular constriction, including insulin-dependent diabetes mellitus, chronic hypertension, collagen vascular disease, hypercoagulability, or multiple miscarriages that may indicate underlying antiphospholipid antibody syndrome, severe hypoxemic asthma, and renal insufficiency. Other physical abnormalities in the mother should be apparent (e.g., congenital heart disease, active Crohn's disease, ulcerative colitis, and gastrointestinal bypass surgery).

A search for maternal infection is the next step. Viruses, parasites, and bacteria, including toxoplasmosis, rubella, cytomegalovirus, syphilis, *Listeria monocytogenes* infection, herpes, Chagas' disease, and malaria, have all been implicated in intrauterine growth retardation (IUGR).

When there are no maternal diseases or ingestions that may cause a small fetus, the ultrasonographer should pay careful attention to the structure of the fetus.

Intrinsic Causes

Aneuploidy should always be considered in the small fetus with any structural abnormality. From 11% to 35% of fetuses with abnormalities have aneuploidy (26). Overlapping fingers or strangely postured hands in a small fetus should immediately elicit a concern for trisomy (28). The poor growth of fetuses with trisomy is partially or totally related to the fact that the placenta and fetus initially arise from the same cell mass; thus, trisomy in the fetus means trisomy in the placenta, and the resulting poor function of the placenta begets the poorly grown fetus.

A partial molar pregnancy has fetal tissue present. Triploidy (69,XXX or 69,XXY) is seen 70% of the time. If the karyotype is derived from an extra set of maternal chromosomes, the phenotypic picture is that of severe symmetric IUGR. Paternal extra sets results in macrocephaly and the classic cystic placenta (29). Other anomalies seen with both types are included in Table 4.

Associated Structural and Genetic Abnormalities

In the absence of a chromosomal abnormality, a limited number of isolated fetal anomalies can cause IUGR. Anything that disrupts the fetal abdomen leads to apparent IUGR because the abdominal circumference is altered. This may include gastroschisis (Fig. 11) and omphalocele (Figs. 2 and 3). The limb–body wall complex and amniotic band sequence can also lead to a small, deformed fetus.

Certain skeletal dysplasias can lead to an overall small growth. This should be readily apparent on measuring the long bones of the fetus (Fig. 22). Although achondroplasia with rhizomelic (proximal) shortening of the limbs is the most common nonlethal skeletal dysplasia, this shortening is not always apparent by ultrasound

TABLE 4 ■ Ultrasound Features of Fetal Triploidy

Site	Ultrasound abnormality
Hands and feet	Toes at an extreme angle to the foot
	Syndactyly of fingers and toes
	Ulna and radius malformations
Brain	Dandy–Walker malformation
Face	Cleft lip and cleft palate
Gastrointestinal	Omphalocele
Genitourinary	Renal dysgenesis (resulting in extreme oligohydramnios)

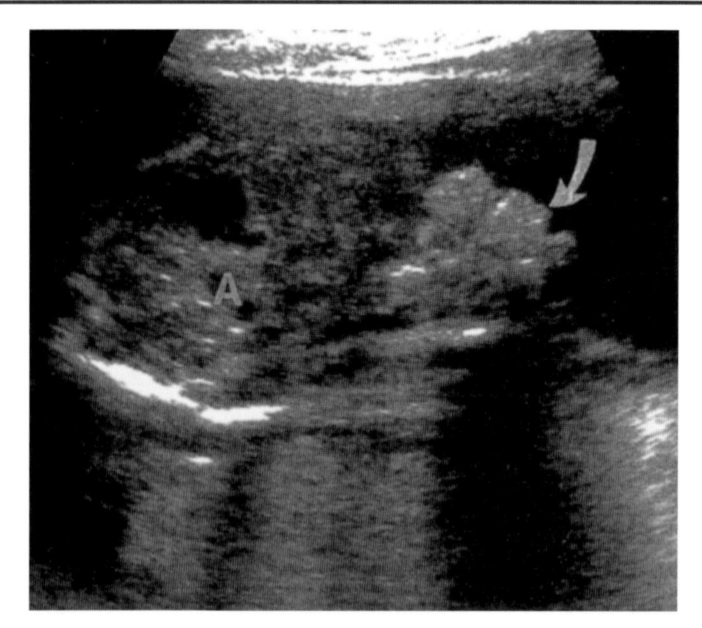

FIGURE 11 ■ Gastroschisis with small bowel protruding through the abdominal wall (*arrow*) and floating in the amniotic fluid. A, fetal abdomen.

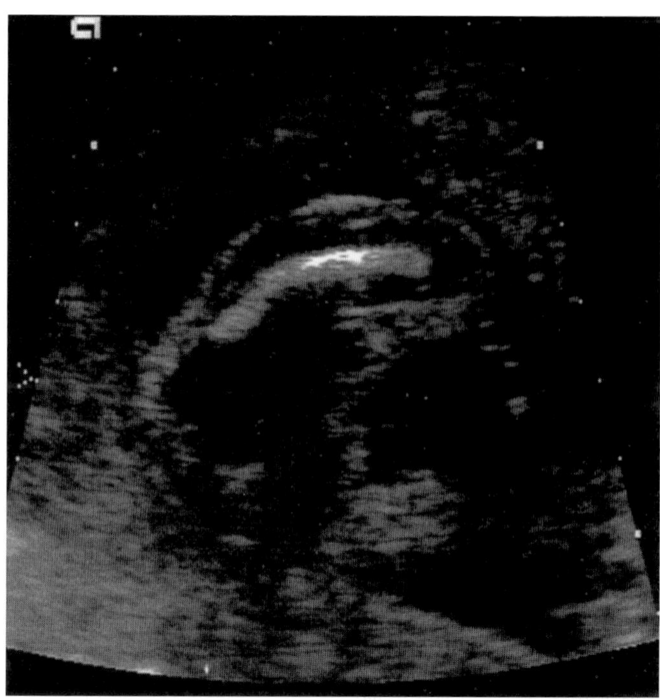

FIGURE 12 ■ Bent femur of fetus with thanatophoric dysplasia.

until 26 weeks' gestation. The gene for this disease is on chromosome 4 and can be diagnosed by obtaining fetal DNA by amniocentesis, percutaneous umbilical sampling, or placental biopsies (30). Most cases of achondroplasia are new dominant mutations and thus are not diagnosed until after 26 weeks or until well after birth. In parents with achondroplasia, chorionic villus sampling can be performed to identify the presence or absence of the gene. In parents who are both affected, this information can be helpful in identifying the fetus that has received a double copy of the dominant achondroplasia gene. This condition, known as double-dominant achondroplasia, is uniformly fatal (Figs. 12 and 13).

A variety of genetic syndromes results in poor growth of the fetus (Table 5). The greatest clue to diagnosing these syndromes is simply to consider the possibility initially. Some of the more commonly found syndromes are the following (31):

Russell–Silver syndrome is an overall undergrowth of the skeleton that results in an intellectually normal but small adult. (Several horse jockeys are suspected of having this condition.) The other features are asymmetry of the arms or legs, a triangular face, and a high-pitched voice. This syndrome can be autosomal dominant or recessive.

Seckel's syndrome is highlighted by severe IUGR. These newborns have a bird-like facial appearance with mandibular hypoplasia, small eyes, and a large nose. They are mentally impaired, probably secondarily to their macrocephaly, and have a shortened life span. This is an autosomal-recessive syndrome.

Smith–Lemli–Opitz (SLO) syndrome is a severe autosomal-recessive disease that can be diagnosed by evaluating the level of 7-dehydrocholesterol in amniotic fluid (32). The prenatal onset of IUGR is present in about 90% of patients. The anomalies associated with this are gonadal hypoplasia (apparent in males), syndactyly of toes two and three, polydactyly, and heart abnormalities. Occasionally, the only finding is severe IUGR. Although maternal serum screening can help identify fetuses for SLO, it must be recognized that it is only a screen, not diagnostic.

De Lange's syndrome is associated with micromelia of the feet and often the hands (31). In addition, there is macrocephaly, synophrys (a single eyebrow), and a characteristic face with a small mouth. These patients are mentally retarded, and the inheritance pattern is unknown.

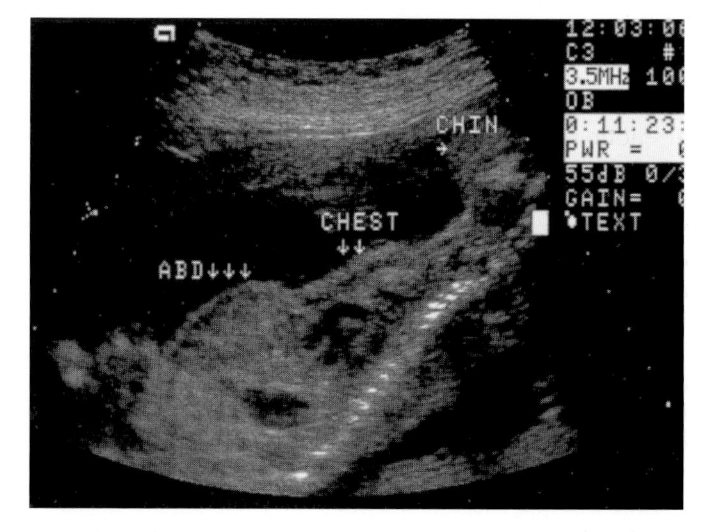

FIGURE 13 ■ Extremely small thoracic cavity of fetus with double-dominant achondroplasia.

TABLE 5 ■ Syndromes Associated with Prenatal Intrauterine Growth Restriction

Syndrome	Common ultrasound features	Inheritance
De Lange's	Micromelia	Unknown
	Micrognathia	
	Microcephaly	
Russell–Silver	Mild limb asymmetry	AD, AR
Mulibrey nanism	(none)	AR
	Postnatal pericardial constriction	
Dubowitz's	Mild microcephaly	AR
Bloom's	(none)	AR
	Postnatal cancers	
Johanson–Blizzard	Hypoplastic nares	AR
Seckel's	Severe microcephaly	AR
	Micrognathia	
Smith–Lemli–Opitz	Microcephaly	AR
	Micrognathia	
	Male: genital absence	
Williams'	Congenital heart disease: aortic stenosis, pulmonic stenosis	Unknown (gene known)
Lethal multiple pterygia	Flexion contractures Often polyhydramnios	AR
Neu–Laxova	Microcephaly	AR
	Lissencephaly	
	Evolving akinesia	
	Lethal postnatal	
Pallister–Hall	Hypothalamic hamartoma	Unknown
	Postaxial polydactyly	
	Lethal postnatal	
Ruvalcaba	Hypoplastic nares	Unknown
	Microcephaly	

Abbreviations: AD, autosomal dominant; AR, autosomal recessive.

OLIGOHYDRAMNIOS IN A NORMAL-GROWTH FETUS

Occasionally, the normal-growth fetus measures small by fundal height because of abnormally low amounts of amniotic fluid. A variety of ultrasound methods can be used to quantify decreased amniotic fluid volume. The four-quadrant technique of measurement has gained wide acceptance. An AFI of less than 5 cm has been associated with perinatal mortality in the post-date fetus (33), and this level of oligohydramnios has been applied to imply concern at any gestational age.

The first consideration should be rupture of the amniotic sac. A sterile pelvic examination for pooling, ferning, and alkaline pH should be done. If there is inconclusive evidence but concern still remains, amniocentesis with instillation of indigo carmine can be attempted. A blue-stained menstrual pad is sufficient evidence for overt leakage, and appropriate obstetric care should ensue. This invasive procedure is rarely needed and rarely

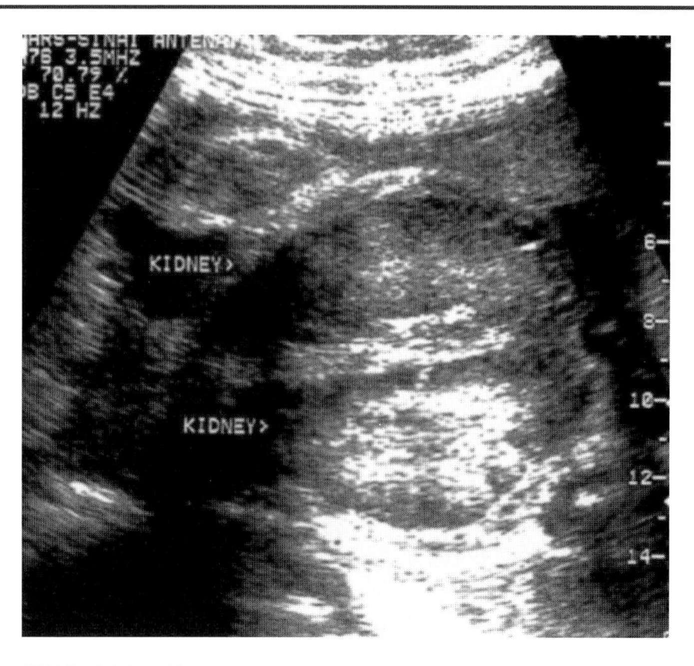

FIGURE 14 ■ Fetus with bilateral infantile polycystic kidney disease and oligohydramnios. The kidneys are enlarged and echogenic.

done if a careful history and physical examination are performed adequately.

The fetal genitourinary system should be evaluated by ultrasound for structural abnormalities, including bilateral renal agenesis, cystic renal disease (Fig. 14), severe bilateral hydronephrosis, and urethral obstruction. These abnormalities are not subtle and are discussed in greater detail elsewhere in this book.

SUMMARY OF IDENTIFIABLE CAUSES

Ultrasound is the most important tool for initial identification of the poor-growth fetus and then for identifying intrinsic causes of IUGR. Unusual genetic disease states should always be considered and a careful family pedigree and maternal medical history should be obtained. Oligohydramnios without evidence of growth retardation should cause concern for renal anomalies and dysfunction. After all of these abnormalities have been evaluated, however, only 40% of cases of poor fetal growth are clearly explained (34), and the remaining 60% are considered idiopathic IUGR.

BACKGROUND: WHAT IS THE CONCERN WITH IDIOPATHIC INTRAUTERINE GROWTH RETARDATION?

The most common cause of IUGR is simply poor placental perfusion. Why do we search for and worry about the poorly grown fetus? It is clear that severe IUGR is associated with poor perinatal outcome. Severely growth-restricted fetuses who were followed up for two years were found to have 3.5 times increased risk of cerebral palsy, 2.4 times increased risk of neonatal death, and 1.6

times increased risk of minor neurodevelopmental impairment. Bacterial sepsis in the newborn period was 10 times as likely (35).

The Problem with Definitions of the Poor-Growth Newborn

The traditional newborn growth outcome classification system separates infants into small-for-gestational-age (SGA), appropriate-for-gestational-age, and LGA categories (fetuses with birth weights below the 10th percentile, between the 10th and the 90th percentiles, or above the 90th percentile, respectively) (36). The controversy arises when a fetus followed up for IUGR is reclassified after birth as appropriately grown, or a normal-sized fetus behaves as a newborn who is poorly grown (37). The proper measure of IUGR is function, not size. Size is a physical dimension than can be measured at any given moment in time. In contrast, growth is a dynamic process of change in size over time and therefore can only be assessed by serial observations. Growth is an important aspect of function in the developing fetus. IUGR is a persistent suppression of genetic growth potential. Thus, IUGR is not synonymous with SGA (38), in that not all fetuses with evidence of poor in utero growth have birth weights below the 10th percentiles (39). Thus, another type of measurement of the newborn has been used, called the Ponderal Index (40). The Ponderal Index measures soft tissue and muscle mass with the formula: weight (in grams) divided by one-third of the length (in centimeters), times 100. The idea is that a small, normal fetus would theoretically manifest normal soft-tissue mass for length, whereas the malnourished fetus would experience soft-tissue wasting with relative sparing of skeletal growth and would be asymmetric. Thus, a low Ponderal Index indicates an asymmetrically growth-retarded newborn; a normal Ponderal Index indicates a symmetrically small infant; and an increased Ponderal Index represents a macrosomic infant. Ott (38) has shown a poor correlation between Ponderal Index and predicted birth weight for gestational age and projected ideal weight. The problem with the Ponderal Index involves strictly dividing IUGR fetuses into asymmetric and symmetric IUGR. Growth retardation is more of a continuum of loss; thus, early asymmetric IUGR fetuses can evolve into symmetrically small fetuses with loss of skeletal mass as well as subcutaneous fat.

Mathematical Modeling

Out of this confusion has arisen the Rossavik mathematical model (41) that emphasizes individual growth curve standards for various fetal anatomic parameters using data from second-trimester ultrasound measurements. This mathematic model relies on many coefficient variables, which have limited its use. In these small studies using the Rossavik growth model, it appears that normally growing, healthy fetuses can have their birth weights accurately predicted by two second-trimester scans separated by an interval of at least five weeks (42–44). Much larger studies need to be done to fully assess these mathematical predictors of abnormal growth. Ariyuki et al. (45) used this model to predict the weights and, more important, the outcomes of newborns that equaled, exceeded, or fell below their predicted ideal growth potential compared with traditional growth parameters of SGA and Ponderal Index. They concluded that the Ponderal Index is insensitive in identifying the poorly grown fetus that has a poor perinatal outcome. The individualized growth assessment appeared to perform as well or slightly better than the traditional identification of the SGA baby using less than 10% in utero estimated fetal weight.

Traditional Ultrasound Assessment

Barring use of these elaborate mathematical models, an accurate diagnosis of IUGR depends on two things: (*i*) a well-dated pregnancy and (*ii*) the ability to estimate fetal weight accurately. Divon et al. (46) estimated fetal weight using Hadlock's formulas based on biparietal diameter, abdominal circumference, and femur length and had a sensitivity and specificity of 87%, positive predictive value of 78%, and a negative predictive value of 92% for estimating fetal weight less than the 10th percentile for gestational age. An ultrasound-estimated fetal weight of less than 10% for a specific gestational age is the definition of IUGR that is used in the United States. A one-time assessment of any fetus is an inadequate predictor of poor growth; thus, it is important to perform serial ultrasound assessment of growth at least two and preferably three weeks apart. It is imperative that the examiner use growth curve parameters that are specific for the practitioner's area and patient population. For example, the Denver growth curve charts reflect the smaller fetuses produced at that high altitude and are not appropriate for sea-level cities.

Other Measures and Modalities

Placental grading, transcerebellar diameter, thigh circumference, magnetic resonance imaging, and other modalities have been suggested for the diagnosis of IUGR. None of these have proved to be clearly reproducible or beneficial (47).

When there is a suspicion of IUGR at an initial scan in a viable fetus, or when there is poor interval growth, other modalities can be used to assess fetal well-being.

Antepartum Testing and Outcome

Several surveillance modalities have been suggested to follow IUGR. These include the biophysical profile, amniotic fluid assessment alone, nonstress test alone, Doppler velocimetry of a variety of fetal and maternal vessels, or a combination of these. The idea is to separate the poorly grown fetus in distress from the poorly grown fetus who is either constitutionally small or is simply not stressed.

Biophysical profile consists of five parts that receive two points each if present: fetal breathing, fetal tone, fetal movement, AFI, and reactive nonstress test (48,49). Fluid assessment by AFI is normal if it is greater than 5cm and less than 24cm, with an average of 13 ± 5cm after 24 weeks (21).

Doppler Velocimetry

This ultrasound modality has been used to examine primarily three vessels in fetuses at risk for intrauterine growth restriction. This section will review the utilization of evaluation of the uterine arteries to identify fetuses at risk for IUGR, followed by Doppler examination of the umbilical arteries and the middle cerebral artery to manage fetuses with growth restriction.

UTERINE ARTERY

The right and left uterine arteries originate from their respective internal iliac arteries, course upward over the external iliac artery, and enter the uterus bilaterally (Fig. 15). The vascular bed that the uterine arteries provide blood flow to consists of the following structures (Fig. 16):

- *Arcuate artery*. This is a vascular plexus that circumscribes the uterus and is located just below the serosa in the myometrium.
- *Radial arteries*. There are approximately 100 radial arteries that originate from the arcuate plexus and enter the myometrium perpendicular to the uterine surface. They give branches to the basal artery and the spiral arteries.
- *Basal arteries*. These remain in the myometrium.

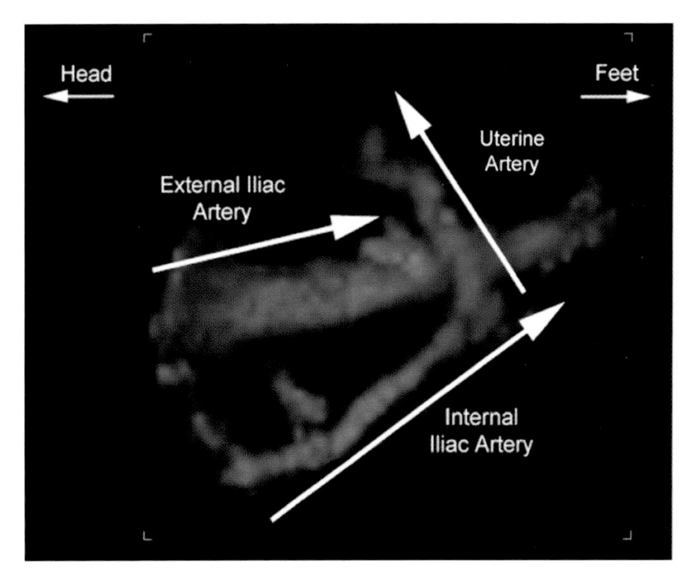

FIGURE 15 ■ This is a B-flow image of the internal and external iliac arteries demonstrating the uterine artery coursing over the external iliac artery.

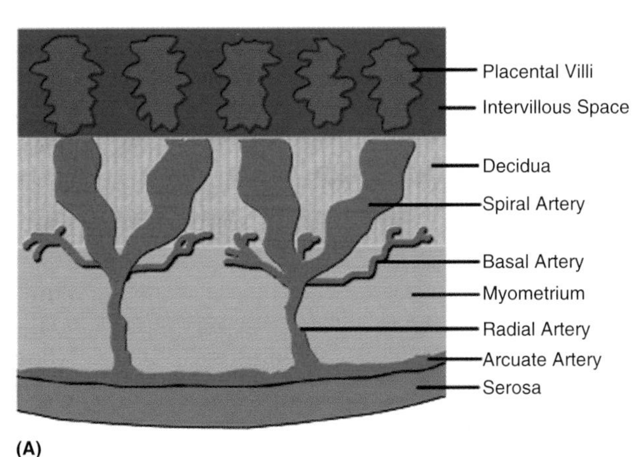

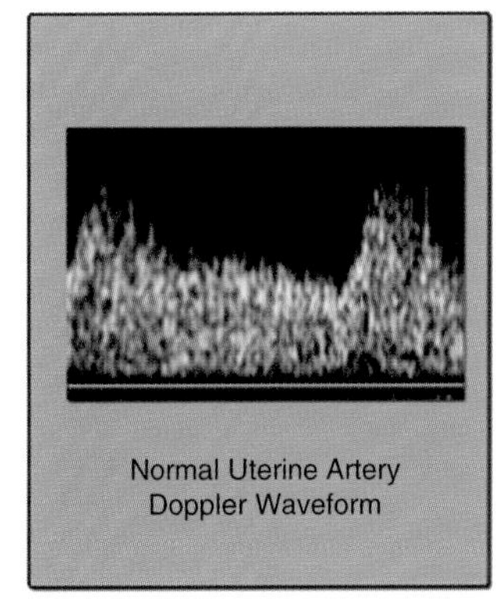

(A)

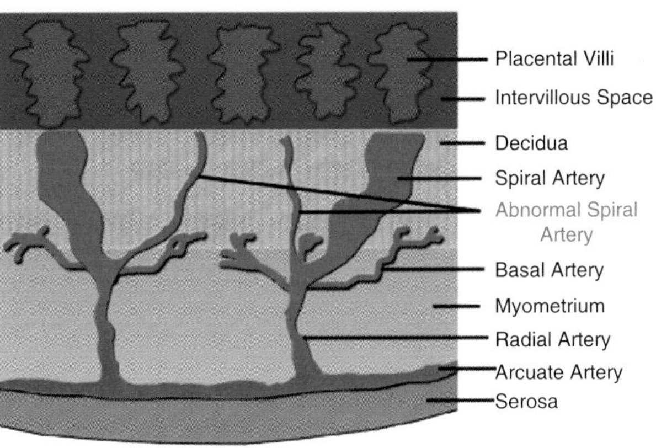

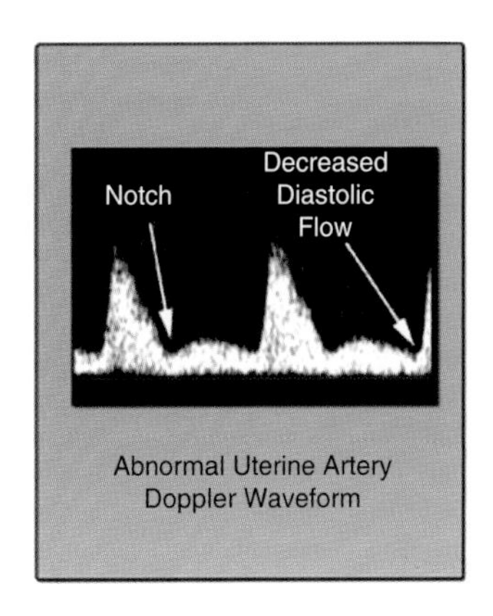

(B)

FIGURE 16 A AND B ■ Schematic of normal and abnormal development of the spiral arteries. Panel A illustrates normal development of the spiral arteries. Panel B demonstrates abnormal development of the spiral arteries, resulting in an abnormal uterine artery Doppler waveform.

- *Spiral arteries*. The spiral arteries enter the deciduas and then branch to the intervillous space.
- *Cotyledons*. These are supplied by the spiral arteries.

Pathological Studies

The key development of uterine blood flow during pregnancy requires changes within the spiral arteries in which they are transformed from maternal vessels to uteroplacental vessels. This occurs during two stages in pregnancy. In the first trimester, trophoblastic changes occur when these cells enter the decidual layer of the spiral arteries (50). This results in an initial decrease in uterine artery pressure and is manifest in the uterine artery waveform. In the second trimester, this process continues with trophoblastic cells entering the myometrial portion of the spiral arteries, resulting in dilation of these vessels (50–53). The diameters of the spiral arteries undergo a four- to sixfold increase in size. This results in a 10-fold increase in uterine artery blood flow (Fig. 16) (53). When there is failure of the spiral arteries to undergo trophoblastic invasion, an increase in resistance to blood flow occurs, resulting in a decrease in blood flow to the placenta (Fig. 16). This has been associated with adverse pregnancy outcome (54–56).

Doppler Studies

In 1983, Campbell et al. were the first to report the association between an abnormal second and a third trimester uterine artery waveform and an adverse pregnancy outcome (57). Since the first report by Campbell et al., a number of studies have identified an abnormal uterine artery Doppler waveform with adverse pregnancy outcome (58–81).

Qualitative Evaluation of the Doppler Waveform

In the early first trimester, notching of the uterine artery may occur (Fig. 17). However, as the first trimester progresses, the notch often disappears (82). As the pregnancy continues, the uterine artery Doppler waveform demonstrates an increase in diastolic flow, without evidence of postsystolic notching (Fig. 18). If, however, notching of the uterine arteries persists, this has been associated with adverse pregnancy outcome, especially if it persists after 24 to 26 weeks of gestation (Fig. 19).

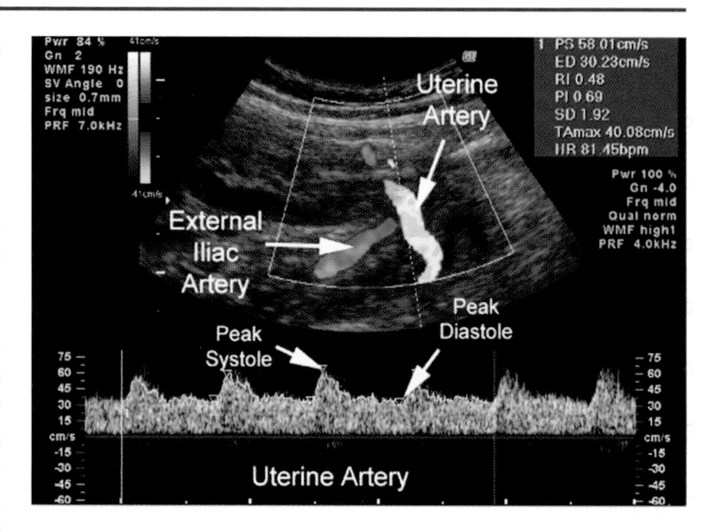

FIGURE 18 ■ Normal uterine artery at 22 weeks' gestation.

Quantitative Evaluation of the Doppler Waveform

Another option for evaluation of the uterine artery Doppler waveform is to measure the resistance index or the pulsatility index. Reference curves have been published for ranges that include the 90th percentile and the 95th percentiles (6,83). Other investigators have suggested a resistance index greater than 0.58 to 0.70 as pathological (6,84–86). Recently, Merz published a reference curve for the resistance index (Fig. 20) and the pulsatility index (Fig. 21) (87).

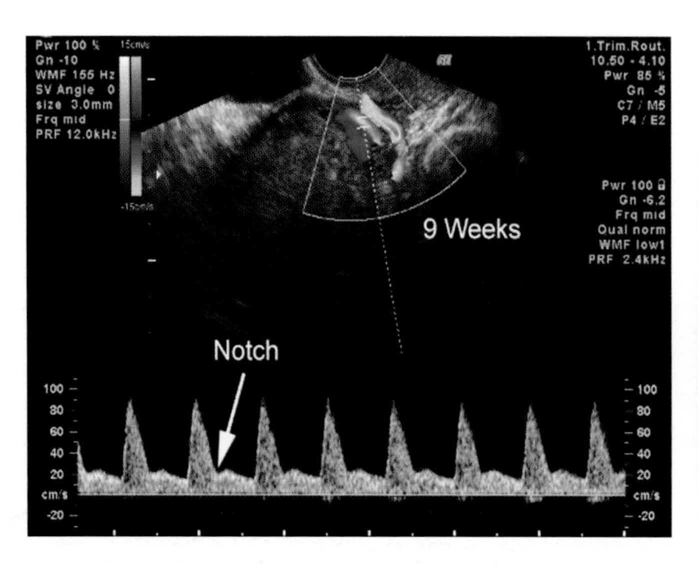

FIGURE 17 ■ Uterine artery Doppler waveform from a nine-week pregnancy. A notch is present.

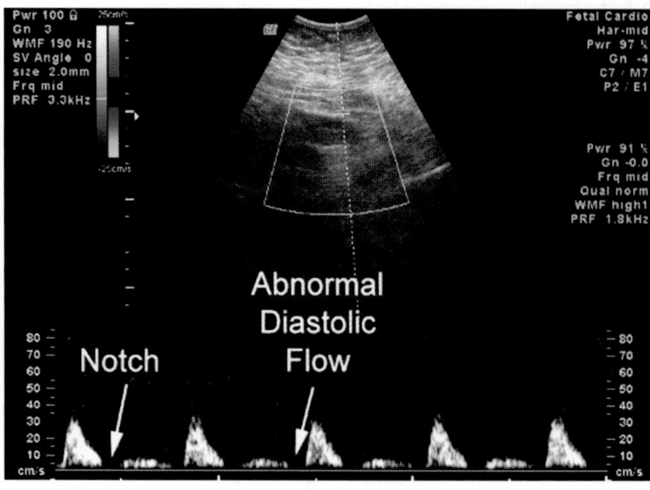

FIGURE 19 ■ Notching of the uterine artery in a fetus with intrauterine growth restriction at 26 weeks' gestation.

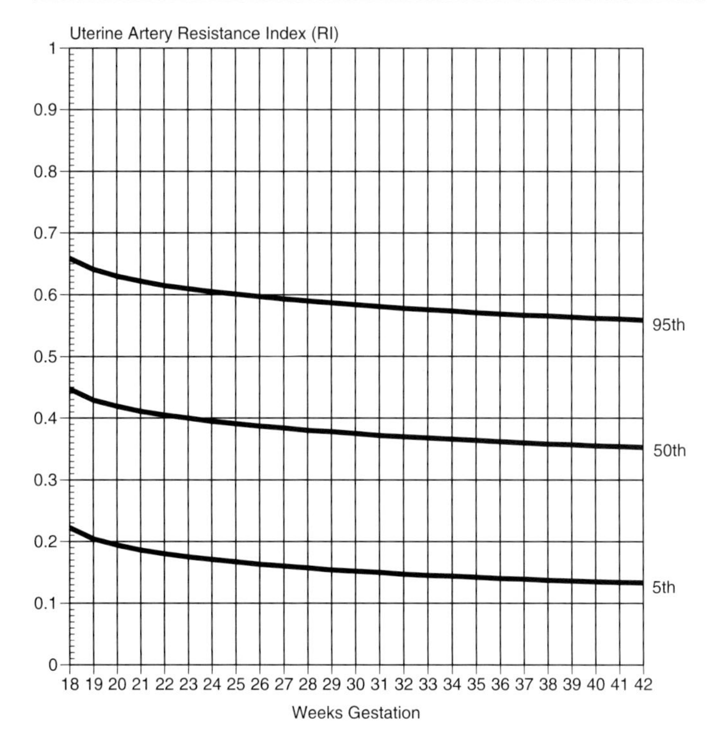

FIGURE 20 ■ The RI for the uterine artery [(systolic maximal velocity–diastolic maximal velocity)/systolic maximal velocity]. RI, resistance index.

UMBILICAL ARTERY

Each of two umbilical arteries originate from their respective internal iliac arteries, course around the bladder, and exit the fetus, forming two of the three components of the umbilical cord. The umbilical cord floats freely in the amniotic cavity before entering the placenta. Most examiners have elected to obtain a pulsed Doppler waveform from the umbilical artery floating within the amniotic cavity.

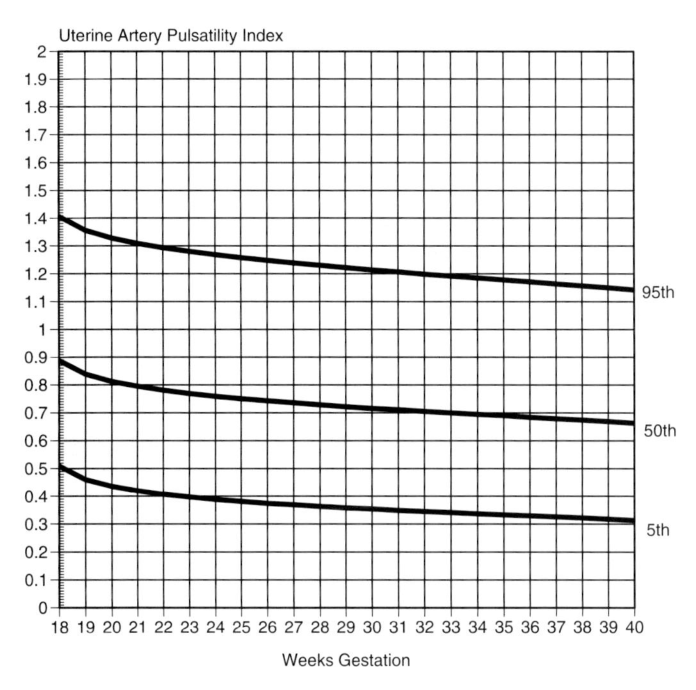

FIGURE 21 ■ Pulsatility index for the uterine artery [(Systolic maximal velocity–diastolic maximal velocity)/Time average maximal velocity].

Identification and Recording of the Umbilical Artery

Originally, the umbilical artery waveform was recorded with continuous wave Doppler. However, with the introduction of duplex pulsed wave imaging, the vessel could be identified with B-mode ultrasound and the Doppler waveform recorded (88,89). The umbilical artery can be identified in one of several orientations: cross-section, longitudinal, or oblique using B-mode, or in combination with color or power Doppler ultrasound (Fig. 22). The benefit of using a combination of modalities is that the examiner can easily identify the umbilical arteries and place the Doppler sample volume parallel to blood flow and the waveform recorded (Fig. 23). Depending upon the size of the sample volume, either the arterial waveform is recorded in isolation or the umbilical venous waveform is recorded in conjunction with the arterial waveform (Fig. 24).

Normal Umbilical Artery Waveform

During pregnancy, the resistance to flow from the fetus to the placenta decreases, resulting in increased diastolic flow. Therefore, the Doppler waveform in the first trimester of pregnancy may demonstrate absent flow (first trimester) or minimal flow during diastole, but the presence of reverse diastolic flow may be associated with an increased risk for adverse outcome (83,90–94). However, after the 14th week of gestation, diastolic flow should be present in all fetuses (93,95).

Qualitative Evaluation of the Umbilical Artery

Investigators have used a qualitative approach in which velocity indices are not measured, but the waveform is inspected visually (96,97). An abnormal waveform demonstrates either absent or reverse flow during ventricular diastole (Fig. 25). As the waveform becomes progressively worse, perinatal and neonatal complications become more frequent (98).

Quantitative Evaluation of the Umbilical Artery

Investigators have quantified umbilical artery resistance by measuring the peak systolic velocity/peak diastolic velocity (S/D) ratio, resistance index, or the pulsatility index (Fig. 26). In 1996, Coppens et al. measured the pulsatility index in fetuses between 8 and 14 weeks' gestation (Fig. 27) and found that it significantly decreased with gestational age. DeVore et al. reported that the S/D ratio and the resistance index did not change a function of whether the fetus was examined at sea level or at an altitude between 4200 and 4500 feet above sea level (99). Clinically, most physicians measure the resistance index (Fig. 28) or the pulsatility index (Fig. 29) (100). Studies have shown that first-trimester abnormal umbilical waveforms, coupled with abnormal uterine artery waveforms, may predict adverse outcome (83,92).

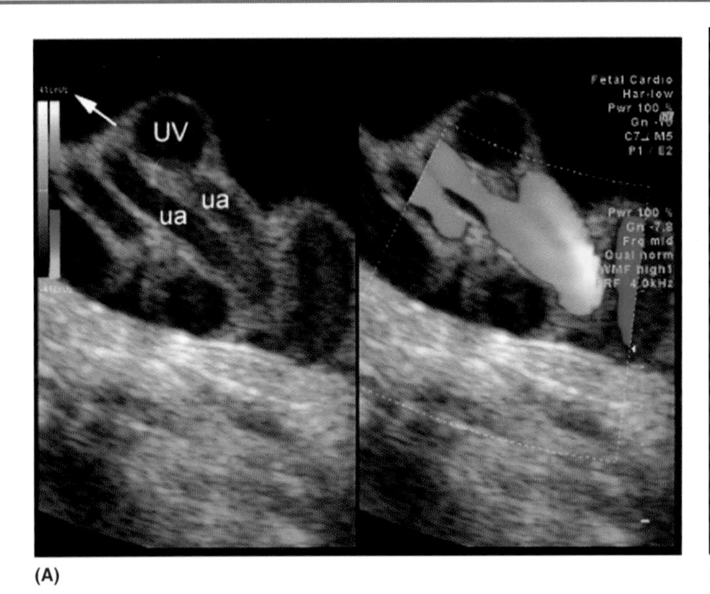

(A)

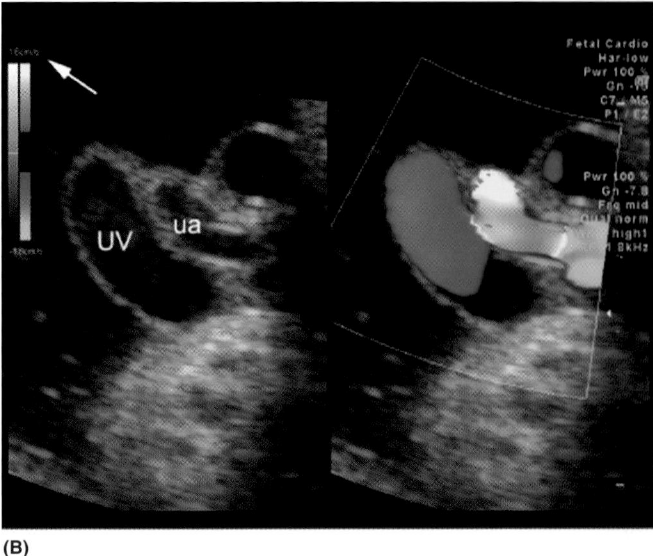

(B)

FIGURE 22 A AND B ■ B-mode and color Doppler imaging of the umbilical cord. Panel A illustrates identification of the umbilical artery tangential to the ultrasound beam with B-mode and color Doppler. As the result of the velocity setting of 41 cm/sec, only the umbilical artery is identified with color Doppler. Panel B illustrates the umbilical artery tangential and perpendicular to the ultrasound beam. As the result of the velocity setting of 18 cm/sec, both the umbilical vein and umbilical artery are identified with color Doppler. ua, umbilical artery; UV, umbilical vein.

Clinical Implications of an Abnormal Umbilical Artery Doppler Waveform

The umbilical artery Doppler waveform has undergone more scrutiny than any other form of noninvasive test for fetal well-being. Meta-analysis has demonstrated that evaluation of the umbilical artery Doppler waveform, when compared to the nonstress test, results in better outcome and often precedes abnormalities of the nonstress test or the biophysical profile (101–104). However, once an abnormal umbilical artery waveform is identified, other Doppler parameters should be examined (i.e., middle cerebral artery, inferior vena cava, and ductus venosus). A survey of Maternal–Fetal Medicine specialists in 1994 by DeVore indicated that as the umbilical artery Doppler waveform became more severe, specialists increased the frequency of antepartum testing, increased the total number of hours of maternal bed rest, and increased the frequency of ultrasound surveillance (105). Recent studies have identified the relationship between abnormal umbilical artery waveforms and adverse outcome (106–116).

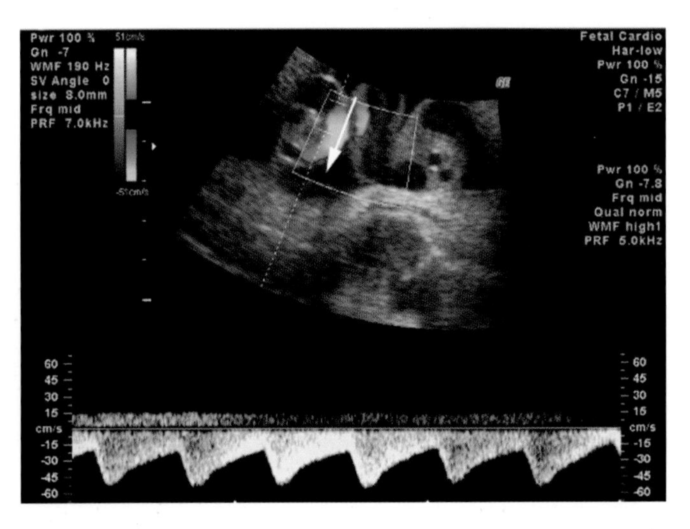

FIGURE 23 ■ Imaging of the umbilical artery by placing the sample volume parallel to blood flow (*white arrow*).

MIDDLE CEREBRAL ARTERY

Historical Review

Pulsed Doppler evaluation of the middle cerebral artery was first reported in 1987 to demonstrate a decrease in the S/D ratio in compromised fetuses (117). In 1989, Arstrom et al. were the first to report the use of the ratio between the middle cerebral artery and the umbilical artery (118). This same year, Satoh et al. reported the association between abnormal middle cerebral artery Doppler waveform indices and abnormal fetal heart rate patterns (119). In 1990, investigators reported that when the middle cerebral artery was examined in fetuses with anemia, there was no change in the pulsatility index, but there was an increase in the systolic peak velocity when anemia was present (120,121). In 1991, Rizzo and Arduini were the first to study the middle cerebral artery in conjunction with cardiac Doppler evaluation of normal and growth-restricted fetuses (122). In 1992, Arduini and Rizzo reported that although the middle cerebral artery pulsatility index predicted adverse outcome in fetuses with

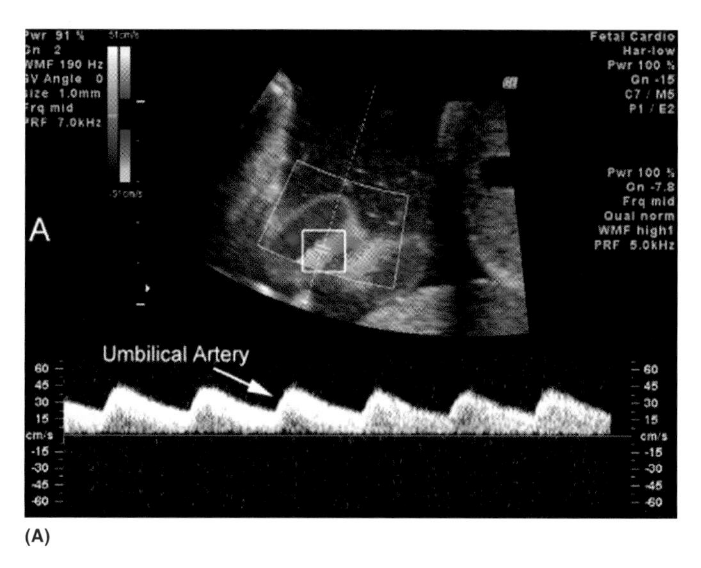

(A)

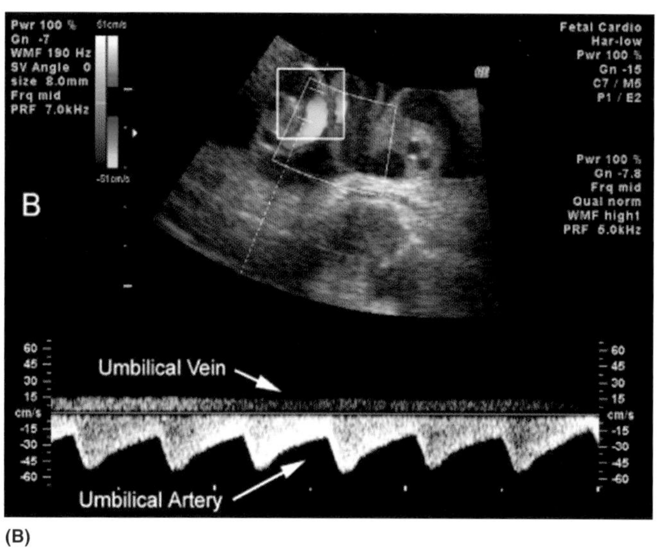

(B)

FIGURE 24 A AND B ■ (**A**) The sample volume is small; therefore, only the umbilical artery waveform is recorded. (**B**) The sample volume is increased in size, resulting in the recording of the Doppler waveform of the umbilical artery and vein.

growth restriction, the ratio of the umbilical artery/middle cerebral artery was a better predictor (123). In 1993, Veille et al. performed a longitudinal study of middle cerebral artery hemodynamics and found the time velocity integral, peak flow velocity, diameter, and blood flow of the middle cerebral artery increased significantly with gestational age, whereas the percent of the total cardiac output to the middle cerebral artery did not significantly change with gestational age (124). In 1994, Weiner et al. studied growth-restricted fetuses with absent end diastolic flow of the umbilical artery and found that as the computerized heart rate tracing became abnormal, the pulsatility index of the middle cerebral artery increased, losing its initial compensatory mechanism (125). In 1994, several investigators reported that Doppler evaluation of the middle cerebral artery was a better predictor of adverse outcome during labor in post-term fetuses when compared to the nonstress test, biophysical profile, and AFI (126,127). In 1995, Hecher et al. were the first to report the relationship between the middle cerebral artery and the venous system in the evaluation of fetuses with intrauterine growth restriction (128). In 1995, Ghezzi et al. reported the association between a decreased pulsatility index and a premature delivery as the result of preterm labor (129). In 1995, Harrington et al. reported the association between abnormal Doppler indices of the

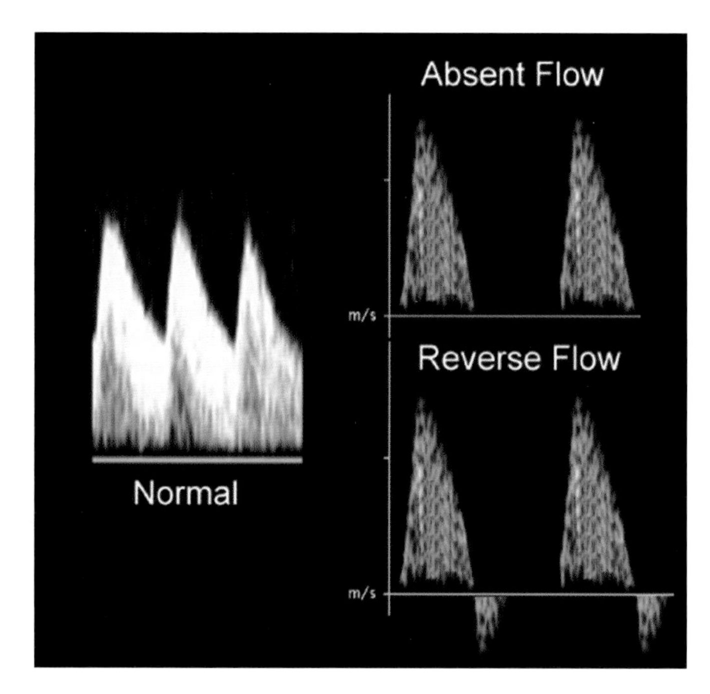

FIGURE 25 ■ Normal and abnormal umbilical artery Doppler waveforms illustrating absent and reverse diastolic flow.

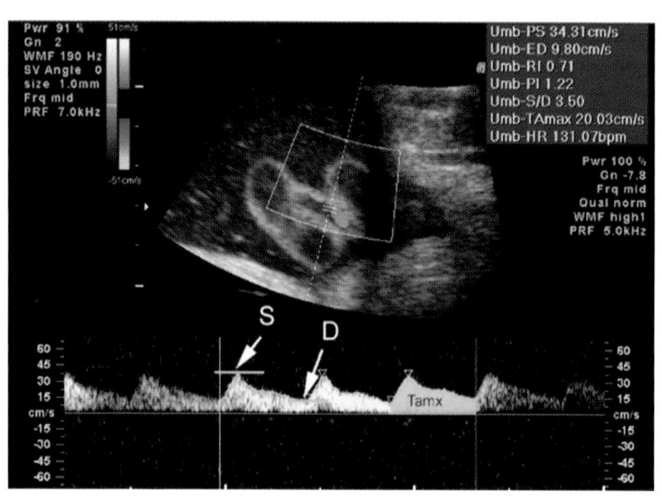

FIGURE 26 ■ Quantification of the umbilical artery Doppler waveform. S/D ratio = peak systolic velocity/peak diastolic velocity, resistance index = (S–D)/S, pulsatility index = (S–D)/Tamx. S, peak systolic velocity; D, peak diastolic velocity; Tamx, total mean velocity.

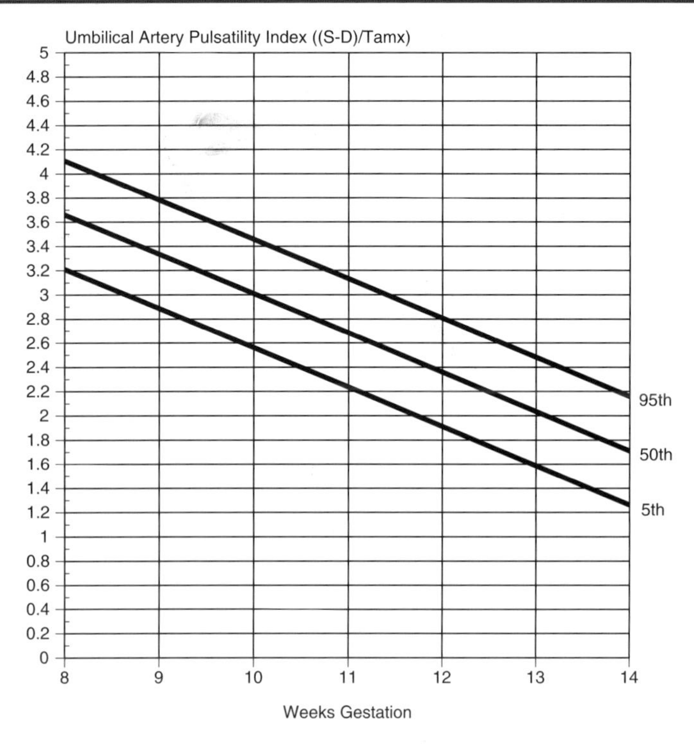

FIGURE 27 ■ Pulsatility index between 8 and 14 weeks' gestation. S, peak systolic velocity; D, peak diastolic velocity; Tamx, total mean velocity.

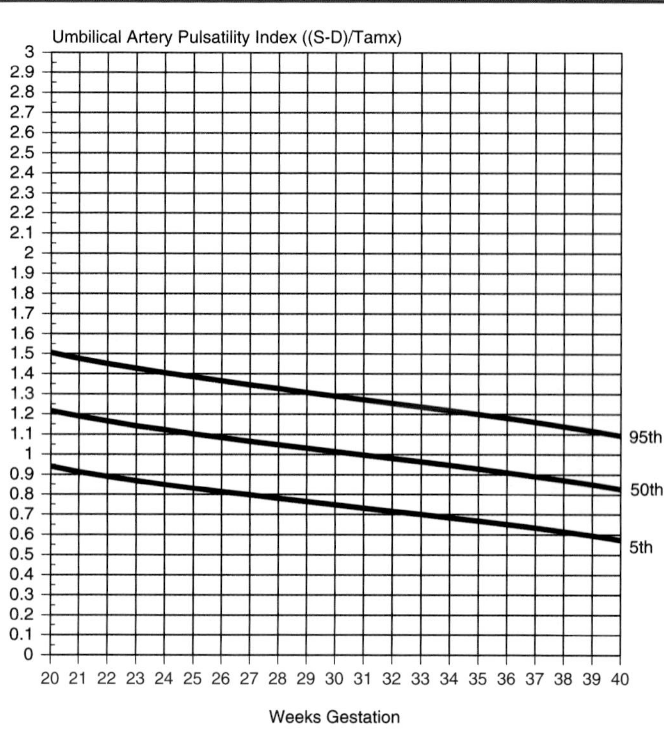

FIGURE 29 ■ Pulsatility index of the umbilical artery. S, peak systolic velocity; D, peak diastolic velocity; Tamx, total mean velocity.

middle cerebral artery, preeclampsia, and intrauterine growth restriction (130). In 1995, Hecher et al. and Rizzo et al. reported the association between abnormal Doppler indices of the middle cerebral artery and abnormal blood gases obtained from cordocentesis in fetuses with intrauterine growth restriction (131,132). In 1996, Almstrom and Sonesson reported the association between maternal hyperoxygenation and an increase in the pulsatility index of the middle cerebral artery (133). In 2000, Hata et al. reported that fetuses SGA who only had an abnormal middle cerebral artery pulsatility index also had an increase in adverse outcome when compared to similar fetuses with normal middle cerebral artery and umbilical artery pulsatility indices (134).

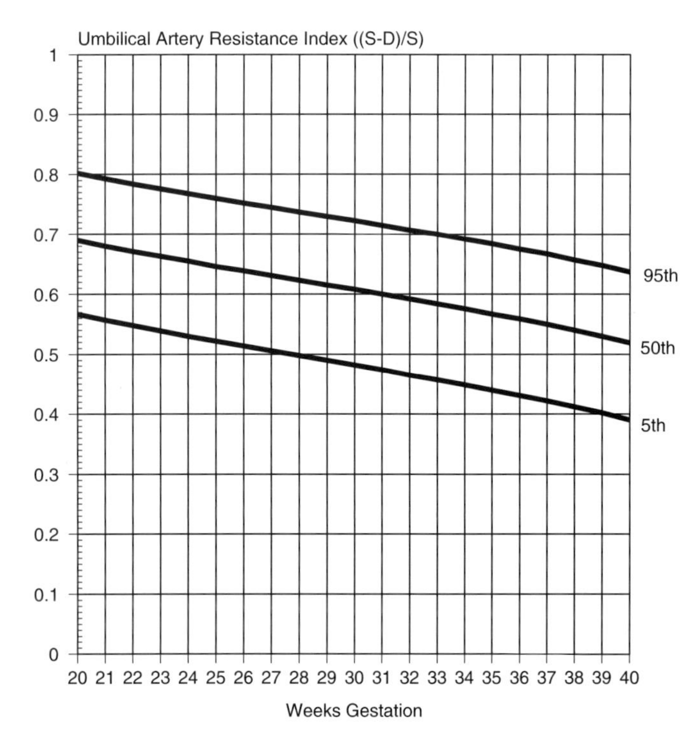

FIGURE 28 ■ Resistance index of the umbilical artery. S, peak systolic velocity; D, peak diastolic velocity.

Imaging the Middle Cerebral Artery

The middle cerebral artery courses anteriolaterally from the circle of Willis in a circular direction (Fig. 30). Because of its location, it is easily imaged when the fetal head is in the position from which biometry (biparietal diameter, head circumference) is obtained.

Identification and Recording of the Middle Cerebral Artery

Although the examiner can identify the middle cerebral artery using B-mode ultrasound, it is easier if color Doppler is used to identify this vessel (Fig. 30). To optimally image the middle cerebral artery with color Doppler, the examiner does the following:

1. Image the fetal head in the transverse plane, at the level in which biometry is measured (Fig. 30).
2. After identifying the midline, direct the ultrasound beam toward the fetal neck.
3. Identify the pons (Fig. 30). The circle of Willis is anterior to this structure.

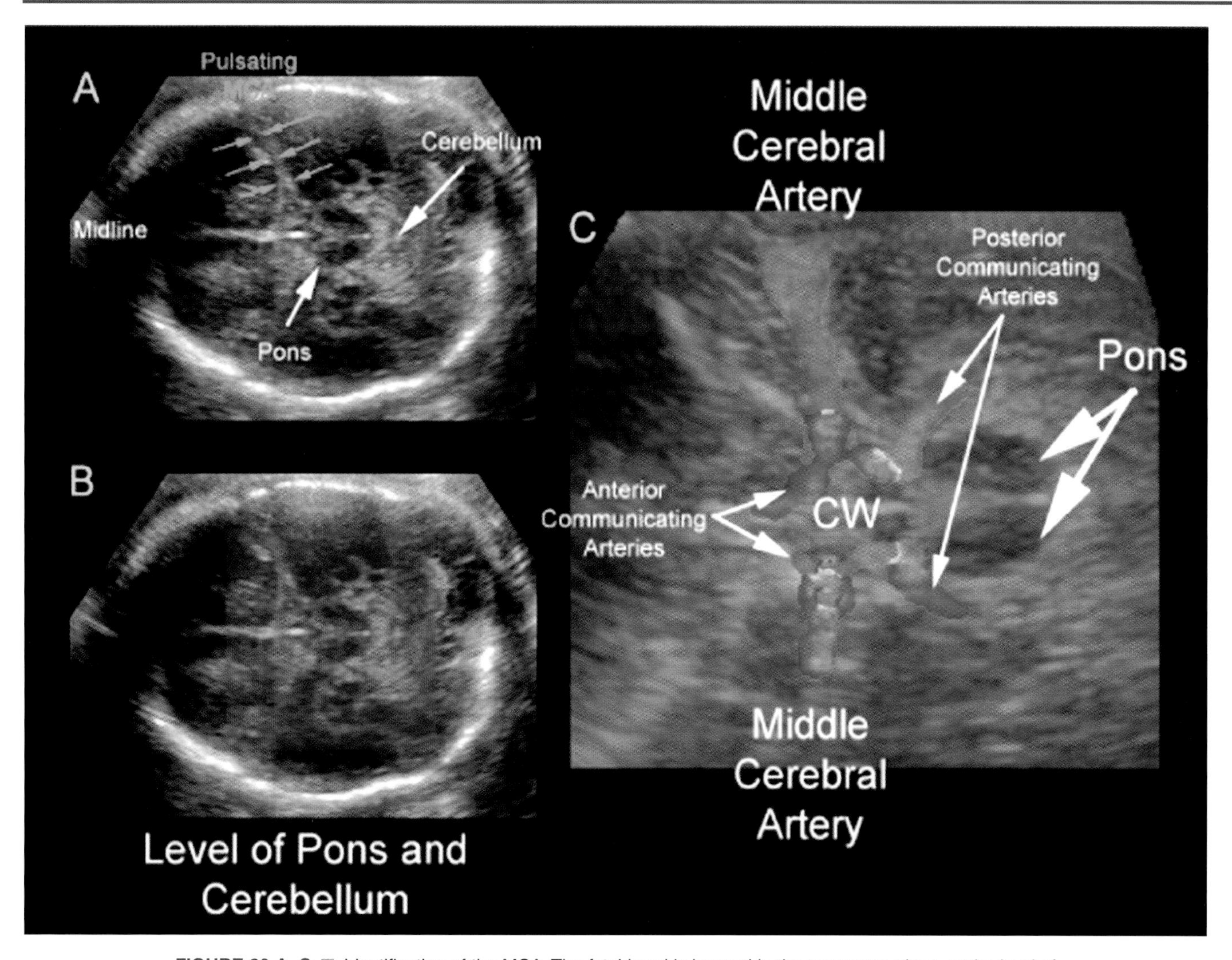

FIGURE 30 A–C ■ Identification of the MCA. The fetal head is imaged in the transverse plane, at the level of the pons **(A)**. The pulsating MCA can be observed in the B-mode image as it lies in a groove of the sphenoid bone (*green arrows*) **(A)**. **(B)** illustrates the area that is displayed in **(C)**. The middle cerebral artery runs in a plane anterior-lateral to the midline, originating from the circle of Willis **(C)**. MCA, middle cerebral artery; CW, circle of Willis.

4. Adjust the color Doppler velocity setting to a lower setting than what may be used to image the umbilical artery.
5. Rock the transducer in a cephalad–caudal direction to visualize the middle cerebral artery (Fig. 30).
6. Place the Doppler sample volume in the proximal mid portion of the middle cerebral artery to record the pulsed Doppler waveform (Fig. 31). If the sample volume is placed in the distal portion of this vessel, there may be differences in the peak velocity as well as quantitative indices (vida infra) (Fig. 31).

Normal Middle Cerebral Artery Waveform

Throughout most of pregnancy, there is a high resistance to blood flow in the cerebral vascular system. This is manifest by low or absent diastolic flow early in gestation. After 32 weeks of gestation, the resistance decreases, resulting in an increase in diastolic flow. The change in the middle cerebral blood flow is almost a mirror image of what occurs in the umbilical artery in which there is low resistance to blood flow associated with an increase in diastolic flow (Fig. 32).

Qualitative Evaluation of the Middle Cerebral Artery

Qualitative evaluation of the middle cerebral artery can be performed if the examiner identifies minimal or absent diastolic flow (Fig. 32). The reason for this is because only when diastolic flow is increased is this an indicator of increased blood flow to the brain. However, there have been several reports of reverse flow during diastole being associated with adverse outcome (135,136).

Quantitative Evaluation of the Middle Cerebral Artery

Intrauterine Growth Restriction

Several indices have been measured from the middle cerebral artery waveform in the evaluation of the fetus for intrauterine growth restriction because it may reflect left ventricular after-load (109,110,116,130–132,137–161). The most widely used clinically, however, are the resistance index and the pulsatility index. A number of

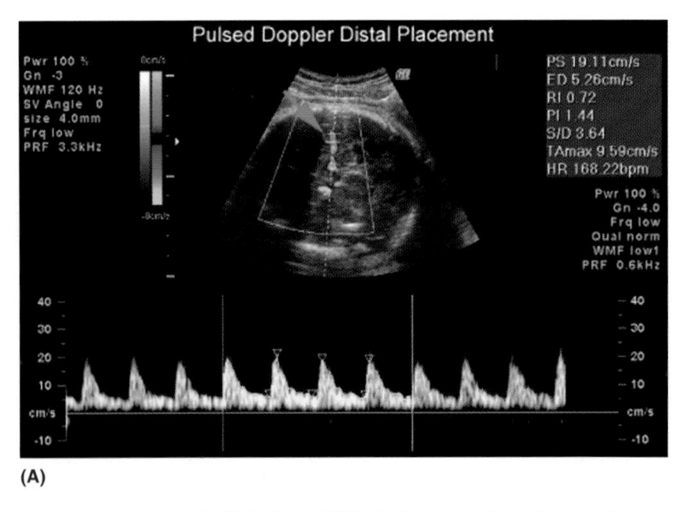

(A)

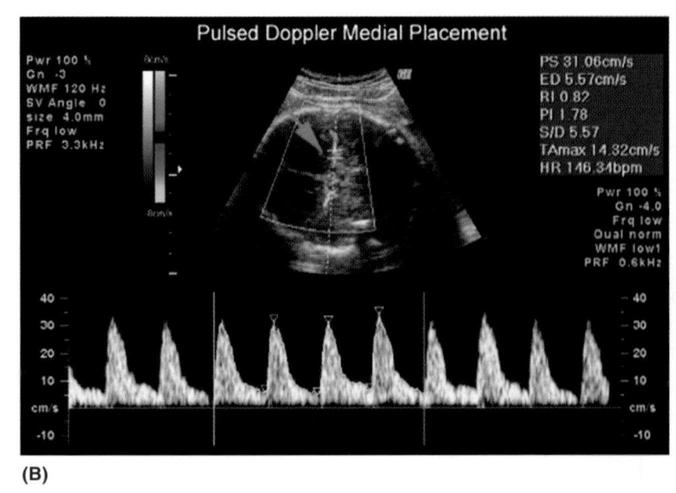

(B)

FIGURE 31 A AND B ■ Pulsed Doppler waveforms obtained from the medial and distal portions of the MCA. (*Panel A*) The pulsed Doppler sample volume is placed in the distal portion of the MCA. (*Panel B*). The pulsed Doppler sample volume is placed in the medial portion of the MCA. When comparing the peak velocity, RI, the PI, the S/D ratio, and the Tamx, notice that there are differences between the two sample sites. MCA, middle cerebral artery; RI, resistance index; PI, pulsatility index; S, peak systolic velocity; D, peak diastolic velocity; Tamx, total mean velocity.

studies have provided Doppler reference curves for the middle cerebral artery (118,124,130,159,161–166). However, not all studies provide a reference range covering the second and third trimesters of pregnancy (118,124,130,159,161–166). Mari and Deter compared a cross-sectional with a longitudinal study and found no significant differences between the two groups (159). Hsieh et al. reported a difference between the proximal third and the middle and distal third of the middle cerebral artery when the pulsatility index was measured (167). However, there were no differences between the sites of sampling when the resistance index was measured (167).

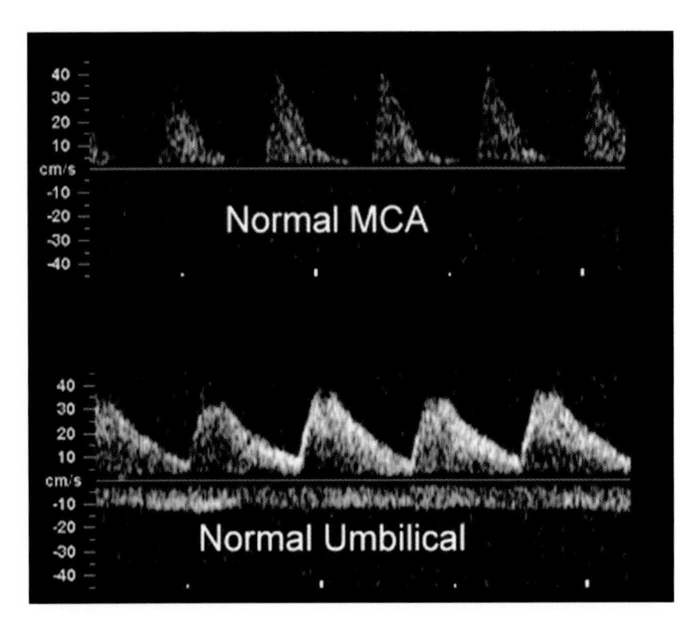

FIGURE 32 ■ Comparison of diastolic flow of the umbilical artery and MCAs. Because of low resistance, the diastolic flow in the umbilical artery is increased. As the result of increased resistance, the diastolic flow in the MCA is decreased. When pathology occurs, the waveforms change in opposite directions. The umbilical artery demonstrates decreased diastolic flow and the MCA demonstrates increased diastolic flow. MCA, middle cerebral artery.

Resistance Index ■ Measurement of the resistance index has been reported in two large studies (165,168). The study by Kurmanavicius et al. evaluated umbilical and middle cerebral arteries and computed the ratio between the two vessels (umbilical artery resistance index/middle cerebral artery resistance index) (Fig. 33) (168). Bahlman et al. only computed the resistance index but found their fifth percentile to be much lower than that of Kurmanavicius et al. (Figs. 34 and 35).

Pulsatility Index ■ A number of studies have evaluated the pulsatility index, but only a few have published reference ranges from 20 weeks until term (120,146,159,162,165,169). Figure 36 compares the pulsatility index between three studies, Mari and Deter, Bahlman et al., and Baschat and Gembruch (159,165,169). Although the mean and upper range of normal varies between the studies, the lower range of normal is very similar. Since the majority of fetuses with an abnormal middle cerebral blood will demonstrate pulsatility index values below the lower limits of normal (2.5 or 5 percentile), any of these three graphs can be used for analysis (Fig. 37). Baschat and Gembruch published the only study that has evaluated the relationship between the umbilical artery pulsatility index and the middle cerebral artery pulsatility index (Fig. 38).

Growth in Multiple-Gestation Fetuses

Lynch et al. (170) examined their retrospective experiences with estimating fetal weight in singletons versus twins or triplets one week before delivery. At birth weights

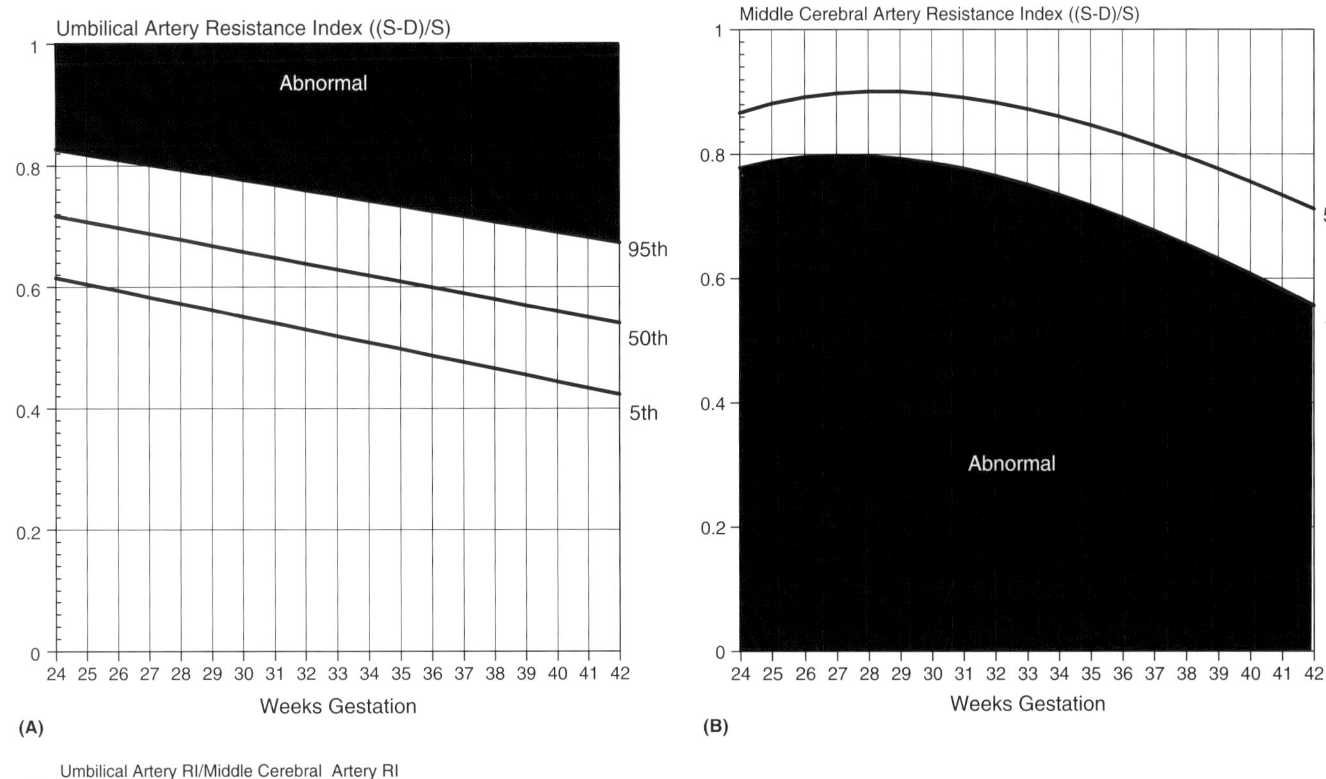

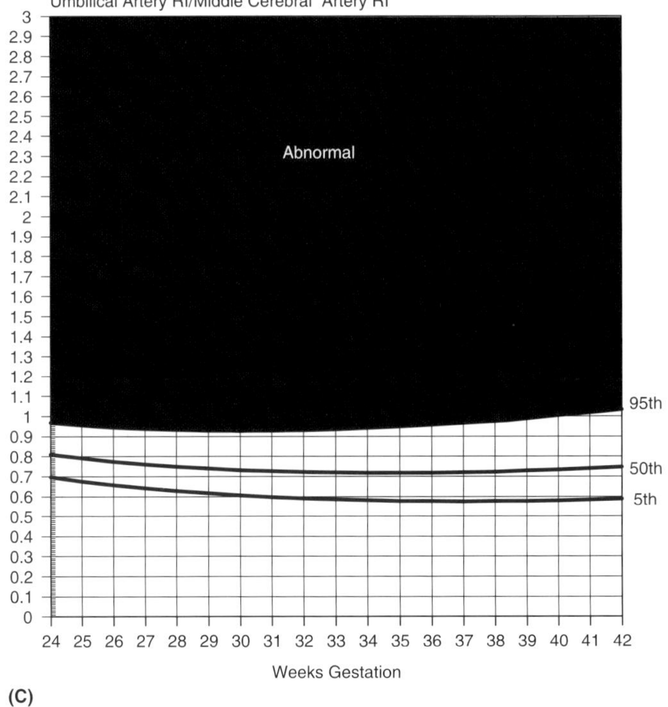

FIGURE 33 A–C ■ RI for the UA, MCA, and the UA/MCA ratio as reported by Kurmanavicius et al. MCA, middle cerebral artery; RI, resistance index; S, peak systolic velocity; D, peak diastolic velocity; UA, umbilical artery.

below 2500 g, there was significant overestimation of fetal weight in twins versus singletons, but the accuracy of the estimates was the same. At birth weights of more than 2500 g, no difference was detected in accuracy between singletons and twins. There was not enough statistical power in the study to reach conclusions concerning triplet weight estimates.

New Causes of Intrauterine Growth Retardation

Confined placental mosaicism, uniparental disomy, and other genetic problems have begun to be implicated in the pathogenesis of IUGR (171,172). Confined placental mosaicism implies that there is trisomy in some parts of the placenta and not in the fetus. Thus, the placenta does not

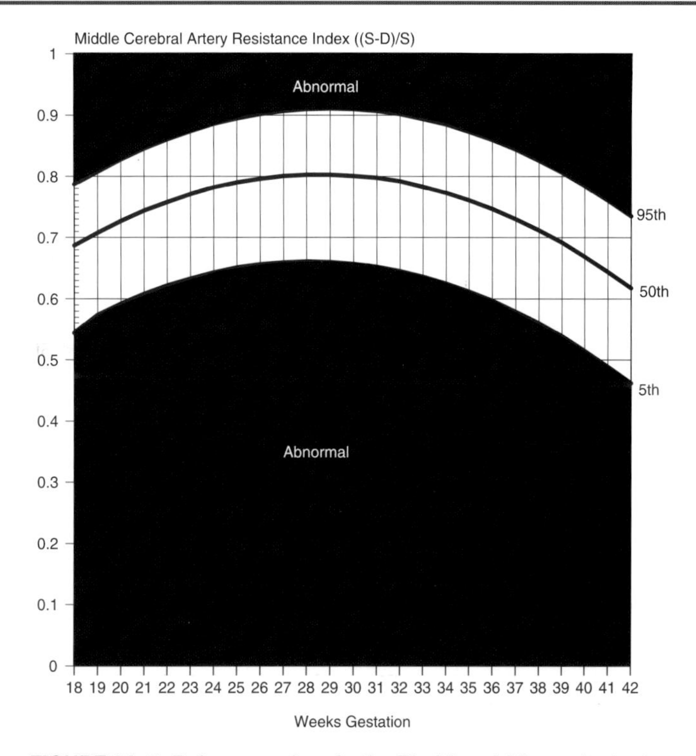

FIGURE 34 ■ Reference values for the RI of the middle cerebral artery from Bahlman et al. (165). If there is brain sparing, then the RI will be below the fifth percentile. If the RI is above the 95th percentile, this could represent abnormal flow from vasoconstriction. RI, resistance index; S, peak systolic velocity; D, peak diastolic velocity.

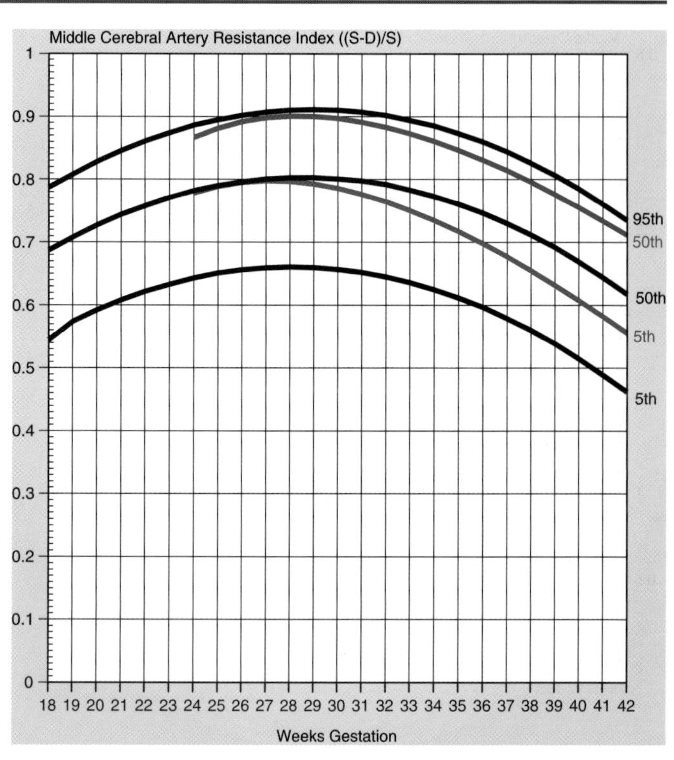

FIGURE 35 ■ Reference values for the resistance index of the middle cerebral artery from Bahlman et al. (*black line*) and Kurmanavicius et al. (*blue lines*). The fifth percentile reference line is lower for Bahlman et al. and the 50th percentile for Kurmanavicius et al. approaches the 95th reference line for Bahlman et al. S, peak systolic velocity; D, peak diastolic velocity. *Source*: From Refs. 165 and 168.

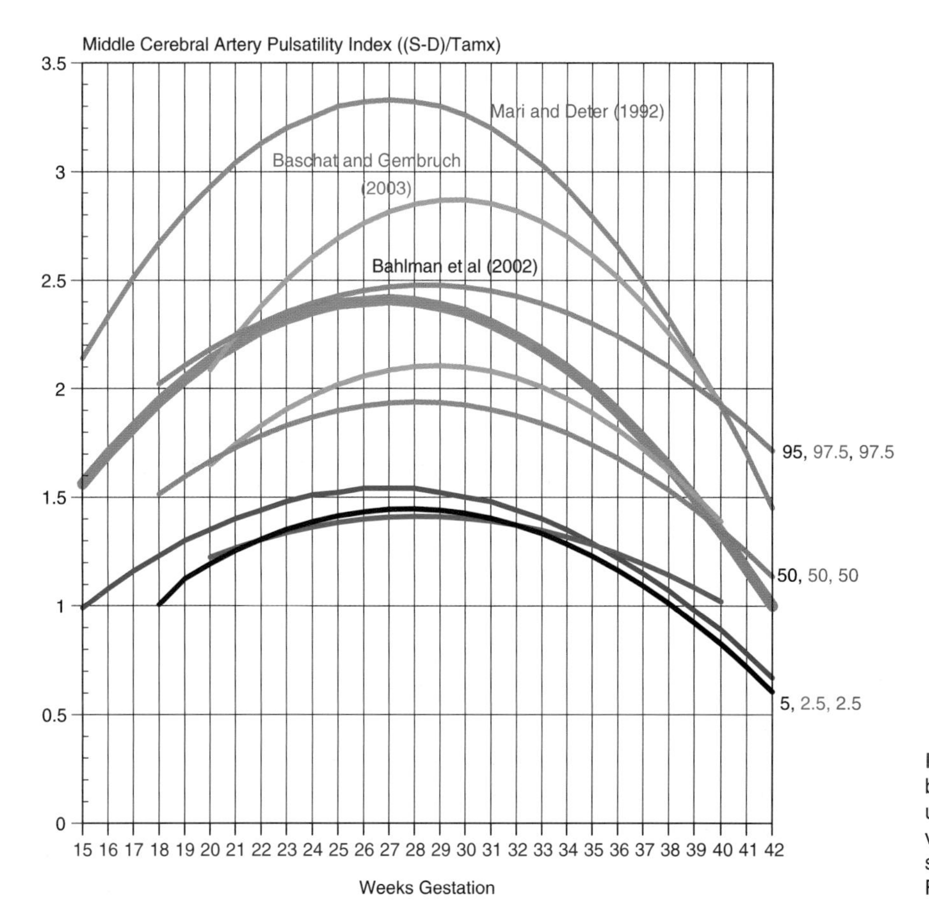

FIGURE 36 ■ Comparison of pulsatility index between three studies. Although the mean and upper limits differ, the lower range of normal is very similar. S, peak systolic velocity; D, peak diastolic velocity; Tamx, total mean velocity. *Source*: From Refs. 159, 165, 169.

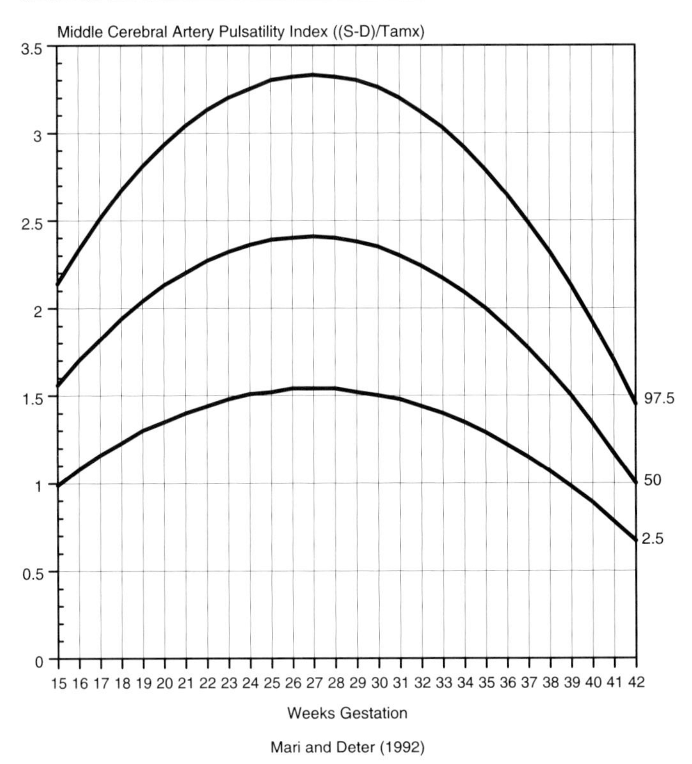

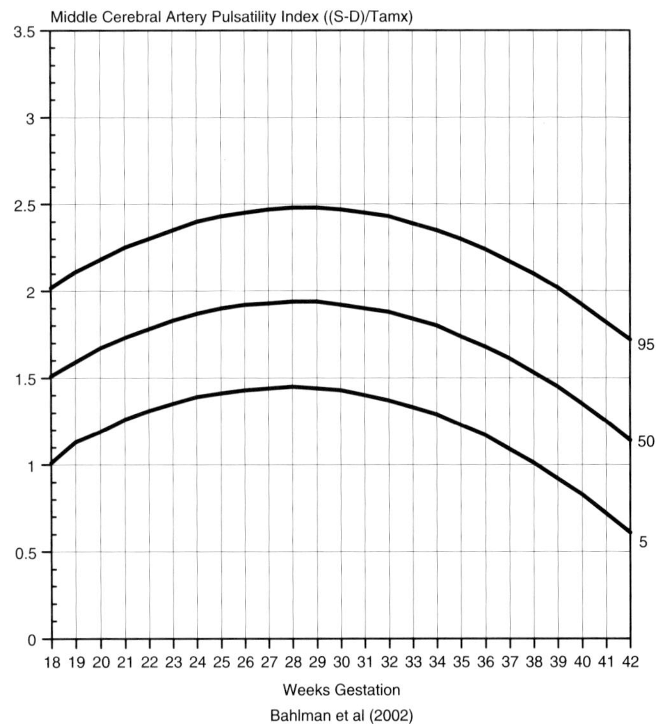

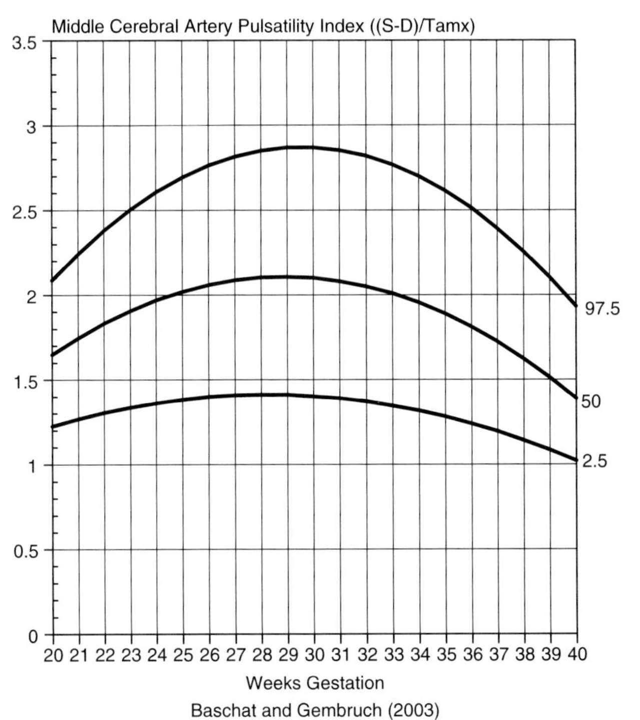

FIGURE 37 ■ Pulsatility index of the middle cerebral artery. S, peak systolic velocity; D, peak diastolic velocity; Tamx, total mean velocity. *Source*: From Refs. 159, 165, 168.

function efficiently in these trisomic areas, and poor fetal perfusion may result. Uniparental disomy usually means that there was trisomy at some point and that there was a loss of one of the extra chromosomes during gestation, resulting in a double copy of maternal or paternal chromosome. This has also been implicated in poor fetal growth (172). The understanding of these processes and their overall impact on the etiology and demographics of IUGR is just beginning. In the future, these findings may be helpful in counseling patients about the risk of recurrence.

CONCLUSION

Poor fetal growth is a potentially serious complication of pregnancy than can result in poor neonatal outcome. The initial ultrasound examination should look for intrinsic and extrinsic explanations of poor fetal growth. When no structural or chromosomal abnormalities are seen, idiopathic poor growth should be the diagnosis of exclusion. The largest errors in diagnosing IUGR involve wrong dates, poor technique, and not understanding that it is

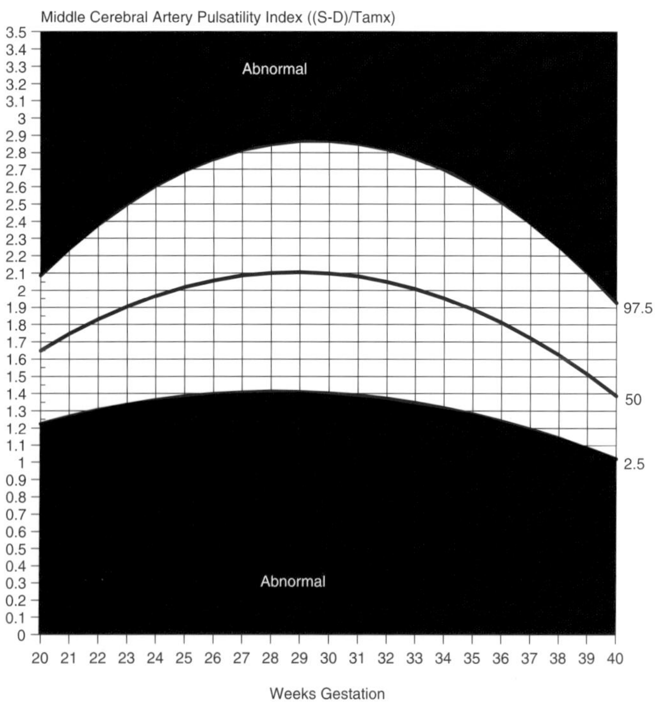

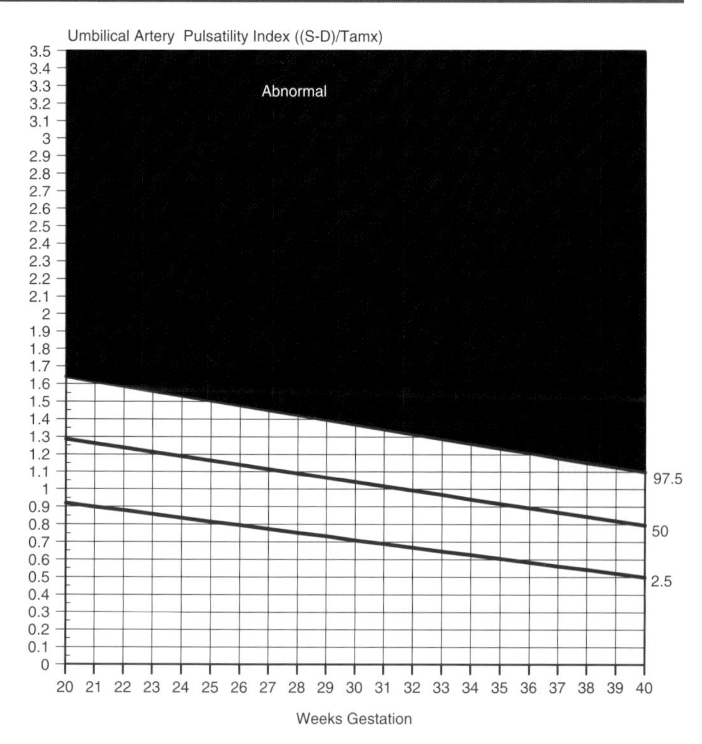

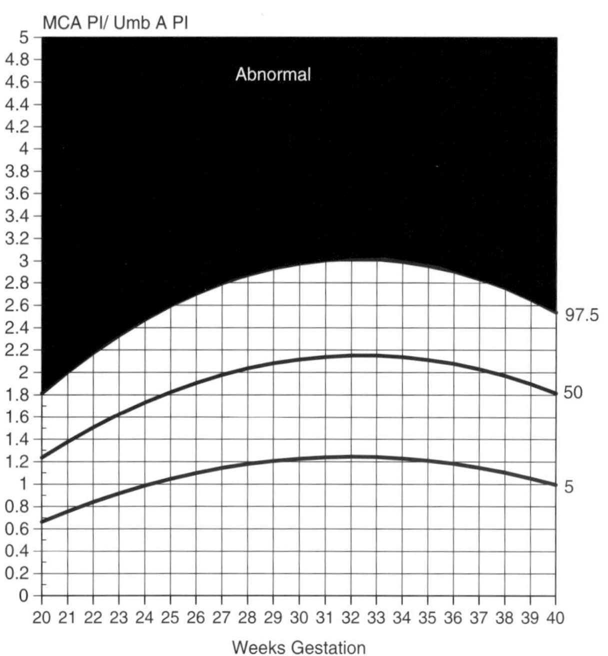

FIGURE 38 ■ UA, MCA, and the MCA/UA ratio as reported by Baschat and Gembruch. S, peak systolic velocity; D, peak diastolic velocity; Tamx, total mean velocity; Umb A, umbilical artery; MCA, middle cerebral artery; PI, pulsatility index.

the pattern of growth that is important to fetal well-being, not merely weight. In fetuses at increased risk, ultrasound parameters of estimating fetal weight should be serially performed, looking for a lag in interval growth. The use of antenatal surveillance, including Doppler velocimetry, is a necessary adjunct in assessing well-being. The neonatologist or pediatrician caring for the newborn should be educated about the importance of poor growth versus SGA or abnormal PI. New thoughts on the etiology of unexplained IUGR are being investigated, and data about their importance in recurrence risks and long-term outcome are awaiting large clinical trials.

REFERENCES

1. Jimenez JM, Tyson JE, Reisch JS. Clinical measures of gestational age in normal pregnancies. Obstet Gynecol 1983; 61:348.
2. McGahan JP, Osborn A. Sonographic evaluation of the large-for-dates pregnancy. Perinat Neonat 1985; 9:45.
3. Bowie JD, Andreotti RF. Estimating gestational age in utero. Radiol Clin North Am 1980; 20:325.
4. Robinson HP, Fleming JEE. A critical evaluation of sonar crown-rump length measurements. Br J Obstet Gynaecol 1975; 82:702.
5. Levi S. Ultrasonic assessment of high rate of human multiple pregnancy in the first trimester. J Clin Ultrasound 1976; 14:3.
6. Abrams B, Parker J. Overweight and pregnancy complications. Int J Obesity 1988; 12:293.

7. Berk MA, Mimouni F, Miodovnik M, et al. Macrosomia in infants of insulin-dependent diabetic mothers. Pediatrics 1989; 3:1029.
8. Reece EA, Smikle C, O'Connor TZ, et al. A longitudinal study comparing growth in diabetic pregnancies with growth in normal gestations. I. The fetal weight. Obstet Gynecol Surv 1990; 35:160.
9. Desmedt EJ, Henry OA, Beischer N. Polyhydramnios and associated maternal and fetal complications in singleton pregnancies. Br J Obstet Gynaecol 1990; 97:1115.
10. Watson WJ, Seed JW. Sonographic diagnosis of macrosomia. In: Divon MY, ed. Abnormal Fetal Growth. New York, NY: Elsevier, 1991:233.
11. Chauhan SP, Cowan BD, Magann EF, et al. Intrapartum detection of a macrosomic fetus: clinical versus 8 sonographic models. Aust N Z J Obstet Gynecol 1995; 35:266.
12. Wieacker P, Wilhelm C, Greiner P, et al. Prenatal diagnosis of Wiedemann-Beckwith syndrome. J Perinatol Med 1989; 17:351.
13. Charon A, Gillerot Y, van Maldergem LO, et al. The Marshall-Smith syndrome. Eur J Pediatric 1990; 150:54.
14. Ardinger HH, Hanson JW, Harrow MJ, et al. Further delineation of Weaver syndrome. J Pediatr 1986; 108:228.
15. Cole TR, Hughes HE. Soto's syndrome. J Med Genet 1990; 27:571.
16. DiLiberti JH. Correlation of skeletal muscle biopsy with phenotype in the familial macrocephaly syndromes. J Med Genet 1992; 29:46.
17. Schifrin BS, Guntres V, Gergely RC, et al. The role of real-time scanning in antenatal fetal surveillance. Am J Obstet Gynecol 1981: 140–525.
18. Ellis JW. Disorders of the umbilical cord, placenta, membranes, and amniotic fluid. Curr Prob Obstet Gynecol 1981; 4:18.
19. Landy HJ, Isada NB, Larsen JW. Genetic implications of idiopathic hydramnios. Am J Obstet Gynecol 1987; 157:114.
20. Carlson DE, Platt LD, Medearis AL, et al. Quantifiable polyhydramnios: diagnosis and management. Obstet Gynecol 1990; 75:989.
21. Phelan JP, Smith CV, Broussard P, et al. Amniotic fluid volume assessment with the four quadrant technique at 36–42 weeks gestation. J Reprod Med 1987; 32:540.
22. Hoskins IA, McGovern PGT, Ordorica SA, et al. Amniotic fluid index: correlation with amniotic fluid volume. Am J Perinatol 1992; 9:315.
23. Magann EF, Nolan TE, Hess LW, et al. Measurement of amniotic fluid volume: accuracy of ultrasonography techniques. Am J Obstet Gynecol 1992; 167:1533.
24. Otto CE, Platt LD. The fetal mandible measurement: an objective determination of fetal jaw size. Ultrasound Obstet Gynecol 1991; 1:12.
25. Carlson DE, Platt LD, Medearis AL. The ultrasound triad of fetal hydramnios, abnormal hand posturing and any other anomaly predict autosomal trisomy. Obstet Gynecol 1992; 79:731.
26. Platt LD, DeVore GR, Lopez E, et al. Role of amniocentesis in ultrasound-detected fetal malformations. Obstet Gynecol 1986; 68:153.
27. Barnhard Y, Bar-Hava I, Divon MY. Is polyhydramnios in an ultrasonographically normal fetus an indication for genetic evaluation? Am J Obstet Gynecol 1995; 173:1523.
28. Seoud MA, Alley DC, Smith Dl, et al. Prenatal sonographic findings in trisomy 13, 18, 21, and 22: a review of 46 cases. J Reprod Med 1994; 39:781.
29. McFadden De, Lkalousek DK. Two different phenotypes with fetuses with chromosomal triploidy: correlation with parental origin of the extra haploid set. Am J Med Genet 1991; 38:535.
30. Bellus GA, Escallon CS, Ortiz de Luna R, et al. First-trimester prenatal diagnosis I couple at risk for homozygous achondroplasia. (Letter) Lancet 1994; 344:1511.
31. Jones KL, ed. Smith's recognizable patterns of human malformation. Philadelphia, PA: W. B. Saunders, 1988.
32. Dallaire L, Mitchell G, Gigure R, et al. Prenatal diagnosis of Smith-Lemli-Opitz syndrome is possible by measurement of 7-dehydrocholesterol in amniotic fluid. Prenat Diagn 1995; 15:855.
33. Rutherford SE, Phelan JP, Smith CU, et al. The four quadrant assessment of amniotic fluid volume: an adjunct to antepartum fetal heart rate testing. Obstet Gynecol 1987; 70:353.
34. Hoffman H, Bakketeig L. Heterogeneity of intrauterine growth retardation and recurrence risks. Semin Perinatol 1984; 8:15.
35. Spinillo A, Capusso E, Egbe TO, et al. Pregnancies complicated by idiopathic intrauterine growth retardation. J Reprod Med 1995; 40:209.
36. Battaglia FA, Lubchenco LO. A practical classification of newborn infants by weight and gestational age. J Pediatric 1967; 71:159.
37. Mahadevan N. The proper measure of intrauterine growth retardation is function, not size. (Editorial) Br J Obstet Gynaecol 1994; 101:1032.
38. Ott WJ. Defining altered fetal growth by second-trimester sonography. Obstet Gynecol 1990; 75:1053.
39. Deter RL, Hadlock FP, Harrist RB. Evaluation of normal fetal growth and the detection of intrauterine growth retardation. In: Callen PW, ed. Ultrasonography in Obstetrics and Gynecology. Philadelphia, PA: W. B. Saunders, 1983:113.
40. Patterson RM, Pouliot MR. Neonatal morphometrics and perinatal outcome: who is growth retarded? Am J Obstet Gynecol 1987; 157:691.
41. Deter RL, Rossavik IK, Harrist RB, et al. Mathematic modeling of fetal growth: development of individual growth curve standards. Obstet Gynecol 1986; 68:156.
42. Deter RL, Harrist RB, Hill RM. Neonatal growth assessment score; a new approach to the detection of intrauterine growth retardation in the newborn. Am J Obstet Gynecol 1990; 162:1030.
43. Ariyuki Y, Hata T, Kitao M. Growth assessment at birth using individual growth curve standards. Jpn J Med Ultrasonics 1993; 20(suppl 1):591.
44. Simon NV, Deter RL, Kofinas AD, et al. Small-for-menstrual-age infants: different subgroups detected using individualized fetal growth assessment. J Clin Ultrasound 1994; 22:3.
45. Ariyuki Y, Hata T, Kitao M. Evaluation of perinatal outcome using individualized growth assessment: comparison with conventional methods. Pediatrics 1995; 96:36.
46. Divon MY, Guidetti DA, Braverman JJ, et al. Intrauterine growth retardation: a prospective study of the diagnostic value of real time sonography combined with umbilical artery flow velocimetry. Obstet Gynecol 1988; 72:611.
47. Snijders RJM, De Courcy-Wheeler RHB, Nicolaides KH. Intrauterine growth retardation and fetal transverse cerebellar diameter. Prenat Diagn 1994; 14:1101.
48. Manning FA, Platt LD, Sipos L. Antepartum fetal evaluation: development of a fetal biophysical profile. Am J Obstet Gynecol 1980; 136:787.
49. Platt LD, Walla CA, Paul RH, et al. A prospective trial of the fetal biophysical profile versus the nonstress test in the management of high risk pregnancies. Am J Obstet Gynecol 1985; 153:624.
50. Pijnenborg R, Dixon G, Robertson WB, Brosens I. Trophoblastic invasion of human decidua from 8 to 18 weeks of pregnancy. Placenta 1980; 1:3–19.
51. Brosens I, Robertson WB, Dixon HG. The physiological response of the vessels of the placental bed to normal pregnancy. J Pathol Bacteriol 1967; 93:569–579.
52. Brosens I, Robertson WB, Dixon HG. Morphologic changes in uteroplacental arteries during pregnancy. Am J Obstet Gynecol 1978; 132:233.
53. Freese UE. The uteroplacental vascular relationship in the human. Am J Obstet Gynecol 1968; 101:8–16.
54. Brosens I, Dixon HG, Robertson WB. Fetal growth retardation and the arteries of the placental bed. Br J Obstet Gynaecol 1977; 84:656–663.
55. Brosens IA, Robertson WB, Dixon HG. The role of the spiral arteries in the pathogenesis of pre-eclampsia. J Pathol 1970; 101:vi.
56. Robertson WB, Brosens I, Dixon G. Uteroplacental vascular pathology. Eur J Obstet Gynecol Reprod Biol 1975; 5:47–65.
57. Robertson WB, Brosens I, Dixon G. Maternal uterine vascular lesions in the hypertensive complications of pregnancy. Perspect Nephrol Hypertens 1976; 5:115–127.
58. Campbell S, Diaz-Recasens J, Griffin DR, et al. New Doppler technique for assessing uteroplacental blood flow. Lancet 1983; 1:675–677.
59. Madazli R, Somunkiran A, Calay Z, Ilvan S, Aksu MF. Histomorphology of the placenta and the placental bed of growth

restricted foetuses and correlation with the Doppler velocimetries of the uterine and umbilical arteries. Placenta 2003; 24:510–516.
60. Martin AM, Bindra R, Curcio P, Cicero S, Nicolaides KH. Screening for pre-eclampsia and fetal growth restriction by uterine artery Doppler at 11–14 weeks of gestation. Ultrasound Obstet Gynecol 2001; 18:583–586.
61. McCowan LM, North RA, Harding JE. Abnormal uterine artery Doppler in small-for-gestational-age pregnancies is associated with later hypertension. Aust N Z J Obstet Gynaecol 2001; 41:56–60.
62. Missfelder-Lobos H, Teran E, Lees C, Albaiges G, Nicolaides KH. Platelet changes and subsequent development of pre-eclampsia and fetal growth restriction in women with abnormal uterine artery Doppler screening. Ultrasound Obstet Gynecol 2002; 19:443–448.
63. Miyakoshi K, Tanaka M, Gabionza D, Ishimoto H, Miyazaki T, Yoshimura Y. Prediction of smallness for gestational age by maternal serum human chorionic gonadotropin levels and by uterine artery Doppler study. Fetal Diagn Ther 2001; 16:42–46.
64. North RA, Ferrier C, Long D, Townsend K, Kincaid-Smith P. Uterine artery Doppler flow velocity waveforms in the second trimester for the prediction of preeclampsia and fetal growth retardation. Obstet Gynecol 1994; 83:378–386.
65. Olofsson P, Laurini RN, Marsal K. A high uterine artery pulsatility index reflects a defective development of placental bed spiral arteries in pregnancies complicated by hypertension and fetal growth retardation. Eur J Obstet Gynecol Reprod Biol 1993; 49:161–168.
66. Papageorghiou AT, Yu CK, Bindra R, Pandis G, Nicolaides KH. Multicenter screening for pre-eclampsia and fetal growth restriction by transvaginal uterine artery Doppler at 23 weeks of gestation. Ultrasound Obstet Gynecol 2001; 18:441–449.
67. Papageorghiou AT, Yu CK, Cicero S, Bower S, Nicolaides KH. Second-trimester uterine artery Doppler screening in unselected populations: a review. J Matern Fetal Neonatal Med 2002; 12:78–88.
68. Papageorghiou AT, Yu CK, Nicolaides KH. The role of uterine artery Doppler in predicting adverse pregnancy outcome. Best Pract Res Clin Obstet Gynaecol 2004; 18:383–396.
69. Park YW, Cho JS, Choi HM, et al. Clinical significance of early diastolic notch depth: uterine artery Doppler velocimetry in the third trimester. Am J Obstet Gynecol 2000; 182:1204–1209.
70. Parretti E, Mealli F, Magrini A, et al. Cross-sectional and longitudinal evaluation of uterine artery Doppler velocimetry for the prediction of pre-eclampsia in normotensive women with specific risk factors. Ultrasound Obstet Gynecol 2003; 22:160–165.
71. Phupong V, Dejthevaporn T, Tanawattanacharoen S, Manotaya S, Tannirandorn Y, Charoenvidhya D. Predicting the risk of preeclampsia and small for gestational age infants by uterine artery Doppler in low-risk women. Arch Gynecol Obstet 2003; 268:158–161.
72. Prefumo F, Sebire NJ, Thilaganathan B. Decreased endovascular trophoblast invasion in first trimester pregnancies with high-resistance uterine artery Doppler indices. Hum Reprod 2004; 19:206–209.
73. Sagol S, Ozkinay E, Oztekin K, Ozdemir N. The comparison of uterine artery Doppler velocimetry with the histopathology of the placental bed. Aust N Z J Obstet Gynaecol 1999; 39:324–329.
74. Schuchter K, Metzenbauer M, Hafner E, Philipp K. Uterine artery Doppler and placental volume in the first trimester in the prediction of pregnancy complications. Ultrasound Obstet Gynecol 2001; 18:590–592.
75. Sebire NJ. Routine uterine artery Doppler screening in twin pregnancies? Ultrasound Obstet Gynecol 2002; 20:532–534.
76. Simanaviciute D. Significance of early diastolic notch in the prognosis of unfavorable pregnancy outcome. Medicina (Kaunas) 2003; 39:659–668.
77. Simmons LA, Hennessy A, Gillin AG, Jeremy RW. Uteroplacental blood flow and placental vascular endothelial growth factor in normotensive and pre-eclamptic pregnancy. BJOG 2000; 107:678–685.
78. Strigini FA, Lencioni G, De LG, Lombardo M, Bianchi F, Genazzani AR. Uterine artery velocimetry and spontaneous preterm delivery. Obstet Gynecol 1995; 85:374–377.
79. Valensise H, Vasapollo B, Novelli GP, et al. Maternal diastolic function in asymptomatic pregnant women with bilateral notching of the uterine artery waveform at 24 weeks' gestation: a pilot study. Ultrasound Obstet Gynecol 2001; 18:450–455.
80. van den Elzen HJ, Cohen-Overbeek TE, Grobbee DE, Quartero RW, Wladimiroff JW. Early uterine artery Doppler velocimetry and the outcome of pregnancy in women aged 35 years and older. Ultrasound Obstet Gynecol 1995; 5:328–333.
81. Venkat-Raman N, Backos M, Teoh TG, Lo WT, Regan L. Uterine artery Doppler in predicting pregnancy outcome in women with antiphospholipid syndrome. Obstet Gynecol 2001; 98:235–242.
82. Vergani P, Roncaglia N, Andreotti C, et al. Prognostic value of uterine artery Doppler velocimetry in growth-restricted fetuses delivered near term. Am J Obstet Gynecol 2002; 187:932–936.
83. Harrington K, Goldfrad C, Carpenter RG, Campbell S. Transvaginal uterine and umbilical artery Doppler examination of 12–16 weeks and the subsequent development of pre-eclampsia and intrauterine growth retardation. Ultrasound Obstet Gynecol 1997; 9:94–100.
84. Bower S, Bewley S, Campbell S. Improved prediction of preeclampsia by two-stage screening of uterine arteries using the early diastolic notch and color Doppler imaging. Obstet Gynecol 1993; 82:78–83.
85. Steel SA, Pearce JM, McParland P, Chamberlain GV. Early Doppler ultrasound screening in prediction of hypertensive disorders of pregnancy. Lancet 1990; 335:1548–1551.
86. Valensise H, Bezzeccheri V, Rizzo G, Tranquilli AL, Garzetti GG, Romanini C. Doppler velocimetry of the uterine artery as a screening test for gestational hypertension. Ultrasound Obstet Gynecol 1993; 3:18–22.
87. Mertz E. Uteroplacental circulation. Volume: 1. Ultrasonography in Obstetrics and Gynecology. Stuttgart, New York: Thieme, 2005:469–480, 613.
88. Brar HS, Medearis AL, DeVore GR, Platt LD. A comparative study of fetal umbilical velocimetry with continuous- and pulsed-wave Doppler ultrasonography in high-risk pregnancies: relationship to outcome. Am J Obstet Gynecol 1989; 160:375–378.
89. Brar HS, Medearis AL, DeVore GR, Platt LD. Fetal umbilical velocimetry using continuous-wave and pulsed-wave Doppler ultrasound in high-risk pregnancies: a comparison of systolic to diastolic ratios. Obstet Gynecol 1988; 72:607–610.
90. Borrell A, Martinez JM, Farre MT, Azulay M, Cararach V, Fortuny A. Reversed end-diastolic flow in first-trimester umbilical artery: an ominous new sign for fetal outcome. Am J Obstet Gynecol 2001; 185:204–207.
91. Borrell A, Costa D, Martinez JM, et al. Reversed end-diastolic umbilical flow in a first-trimester fetus with congenital heart disease. Prenat Diagn 1998; 18:1001–1005.
92. Harrington K, Carpenter RG, Goldfrad C, Campbell S. Transvaginal Doppler ultrasound of the uteroplacental circulation in the early prediction of pre-eclampsia and intrauterine growth retardation. Br J Obstet Gynaecol 1997; 104:674–681.
93. Coppens M, Loquet P, Kollen M, De Neubourg F, Buytaert P. Longitudinal evaluation of uteroplacental and umbilical blood flow changes in normal early pregnancy. Ultrasound Obstet Gynecol 1996; 7:114–121.
94. den Ouden M, Cohen-Overbeek TE, Wladimiroff JW. Uterine and fetal umbilical artery flow velocity waveforms in normal first trimester pregnancies. Br J Obstet Gynaecol 1990; 97:716–719.
95. Zalen-Sprock MM, van Vugt JM, Colenbrander GJ, van Geijn HP. First-trimester uteroplacental and fetal blood flow velocity waveforms in normally developing fetuses: a longitudinal study. Ultrasound Obstet Gynecol 1994; 4:284–288.
96. Malcus P, Andersson J, Marsal K, Olofsson PA. Waveform pattern recognition—a new semiquantitative method for analysis of fetal aortic and umbilical artery blood flow velocity recorded by Doppler ultrasound. Ultrasound Med Biol 1991; 17:453–460.
97. Laurin J, Lingman G, Marsal K, Persson PH. Fetal blood flow in pregnancies complicated by intrauterine growth retardation. Obstet Gynecol 1987; 69:895–902.
98. Soregaroli M, Bonera R, Danti L, et al. Prognostic role of umbilical artery Doppler velocimetry in growth-restricted fetuses. J Matern Fetal Neonatal Med 2002; 11:199–203.

99. DeVore GR, Medearis AL, Platt LD. The effect of altitude on the umbilical artery Doppler resistance. J Ultrasound Med 1992; 11:317–320.
100. Uteroplacental Circulation. In: Merz E, ed. Volume: 1 Ultrasonography in Obstetrics and Gynecology. Stuttgart, New York: Thieme, 2005:469–480, 614.
101. Divon MY, Ferber A. Umbilical artery Doppler velocimetry—an update. Semin Perinatol 2001; 25:44–47.
102. Divon MY. Umbilical artery Doppler velocimetry: clinical utility in high-risk pregnancies. Am J Obstet Gynecol 1996; 174:10–14.
103. Alfirevic Z, Neilson JP. Doppler ultrasonography in high-risk pregnancies: systematic review with meta-analysis. Am J Obstet Gynecol 1995; 172:1379–1387.
104. Marsal K. Rational use of Doppler ultrasound in perinatal medicine. J Perinat Med 1994; 22:463–474.
105. DeVore GR. The effect of an abnormal umbilical artery Doppler on the management of fetal growth restriction: a survey of maternal-fetal medicine specialists who perform fetal ultrasound. Ultrasound Obstet Gynecol 1994; 4:294–303.
106. Valcamonico A, Accorsi P, Battaglia S, Soregaroli M, Beretta D, Frusca T. Absent or reverse end-diastolic flow in the umbilical artery: intellectual development at school age. Eur J Obstet Gynecol Reprod Biol 2004; 114:23–28.
107. Baschat AA. Doppler application in the delivery timing of the preterm growth-restricted fetus: another step in the right direction. Ultrasound Obstet Gynecol 2004; 23:111–118.
108. Martinez JM, Bermudez C, Becerra C, Lopez J, Morales WJ, Quintero RA. The role of Doppler studies in predicting individual intrauterine fetal demise after laser therapy for twin-twin transfusion syndrome. Ultrasound Obstet Gynecol 2003; 22:246–251.
109. Gagnon R, Van den HM. The use of fetal Doppler in obstetrics. J Obstet Gynaecol Can 2003; 25:601–614.
110. Muller T, Nanan R, Rehn M, Kristen P, Dietl J. Arterial and ductus venosus Doppler in fetuses with absent or reverse end-diastolic flow in the umbilical artery: longitudinal analysis. Fetal Diagn Ther 2003; 18:163–169.
111. Kutschera J, Tomaselli J, Urlesberger B, et al. Absent or reversed end-diastolic blood flow in the umbilical artery and abnormal Doppler cerebroplacental ratio—cognitive, neurological and somatic development at 3 to 6 years. Early Hum Dev 2002; 69:47–56.
112. Bhatt AB, Tank PD, Barmade KB, Damania KR. Abnormal Doppler flow velocimetry in the growth restricted foetus as a predictor for necrotising enterocolitis. J Postgrad Med 2002; 48:182–185.
113. Seyam YS, Al Mahmeid MS, Al Tamimi HK. Umbilical artery Doppler flow velocimetry in intrauterine growth restriction and its relation to perinatal outcome. Int J Gynaecol Obstet 2002; 77:131–137.
114. Axt-Fliedner R, Hendrik HJ, Schmidt W. Nucleated red blood cell counts in growth-restricted neonates with absent or reversed-end-diastolic umbilical artery velocity. Clin Exp Obstet Gynecol 2002; 29:242–246.
115. Bush KD, O'Brien JM, Barton JR. The utility of umbilical artery Doppler investigation in women with the HELLP (hemolysis, elevated liver enzymes, and low platelets) syndrome. Am J Obstet Gynecol 2001; 184:1087–1089.
116. Gramellini D, Piantelli G, Verrotti C, Fieni S, Chiaie LD, Kaihura C. Doppler velocimetry and non stress test in severe fetal growth restriction. Clin Exp Obstet Gynecol 2001; 28:33–39.
117. Woo JS, Liang ST, Lo RL, Chan FY. Middle cerebral artery Doppler flow velocity waveforms. Obstet Gynecol 1987; 70:613–616.
118. Arstrom K, Eliasson A, Hareide JH, Marsal K. Fetal blood velocity waveforms in normal pregnancies. A longitudinal study. Acta Obstet Gynecol Scand 1989; 68:171–178.
119. Satoh S, Koyanagi T, Fukuhara M, Hara K, Nakano H. Changes in vascular resistance in the umbilical and middle cerebral arteries in the human intrauterine growth-retarded fetus, measured with pulsed Doppler ultrasound. Early Hum Dev 1989; 20:213–220.
120. Vyas S, Nicolaides KH, Campbell S. Doppler examination of the middle cerebral artery in anemic fetuses. Am J Obstet Gynecol 1990; 162:1066–1068.
121. Mari G, Moise KJ Jr, Deter RL, Kirshon B, Stefos T, Carpenter RJ Jr. Flow velocity waveforms of the vascular system in the anemic fetus before and after intravascular transfusion for severe red blood cell alloimmunization. Am J Obstet Gynecol 1990; 162:1060–1064.
122. Rizzo G, Arduini D. Fetal cardiac function in intrauterine growth retardation. Am J Obstet Gynecol 1991; 165:876–882.
123. Arduini D, Rizzo G. Prediction of fetal outcome in small for gestational age fetuses: comparison of Doppler measurements obtained from different fetal vessels. J Perinat Med 1992; 20:29–38.
124. Veille JC, Hanson R, Tatum K. Longitudinal quantitation of middle cerebral artery blood flow in normal human fetuses. Am J Obstet Gynecol 1993; 169:1393–1398.
125. Weiner Z, Farmakides G, Schulman H, Penny B. Central and peripheral hemodynamic changes in fetuses with absent end-diastolic velocity in umbilical artery: correlation with computerized fetal heart rate pattern. Am J Obstet Gynecol 1994; 170:509–515.
126. Devine PA, Bracero LA, Lysikiewicz A, Evans R, Womack S, Byrne DW. Middle cerebral to umbilical artery Doppler ratio in post-date pregnancies. Obstet Gynecol 1994; 84:856–860.
127. Anteby EY, Tadmor O, Revel A, Yagel S. Post-term pregnancies with normal cardiotocographs and amniotic fluid columns: the role of Doppler evaluation in predicting perinatal outcome. Eur J Obstet Gynecol Reprod Biol 1994; 54:93–98.
128. Hecher K, Campbell S, Doyle P, Harrington K, Nicolaides K. Assessment of fetal compromise by Doppler ultrasound investigation of the fetal circulation. Arterial, intracardiac, and venous blood flow velocity studies. Circulation 1995; 91:129–138.
129. Ghezzi F, Ghidini A, Romero R, et al. Doppler velocimetry of the fetal middle cerebral artery in patients with preterm labor and intact membranes. J Ultrasound Med 1995; 14:361–366.
130. Harrington K, Carpenter RG, Nguyen M, Campbell S. Changes observed in Doppler studies of the fetal circulation in pregnancies complicated by pre-eclampsia or the delivery of a small-for-gestational-age baby. I. Cross-sectional analysis. Ultrasound Obstet Gynecol 1995; 6:19–28.
131. Hecher K, Snijders R, Campbell S, Nicolaides K. Fetal venous, intracardiac, and arterial blood flow measurements in intrauterine growth retardation: relationship with fetal blood gases. Am J Obstet Gynecol 1995; 173:10–15.
132. Rizzo G, Capponi A, Arduini D, Romanini C. The value of fetal arterial, cardiac and venous flows in predicting pH and blood gases measured in umbilical blood at cordocentesis in growth retarded fetuses. Br J Obstet Gynaecol 1995; 102:963–969.
133. Almstrom H, Sonesson SE. Doppler echocardiographic assessment of fetal blood flow redistribution during maternal hyperoxygenation. Ultrasound Obstet Gynecol 1996; 8:256–261.
134. Hata T, Aoki S, Manabe A, et al. Subclassification of small-for-gestational-age fetus using fetal Doppler velocimetry. Gynecol Obstet Invest 2000; 49:236–239.
135. Respondek M, Woch A, Kaczmarek P, Borowski D. Reversal of diastolic flow in the middle cerebral artery of the fetus during the second half of pregnancy. Ultrasound Obstet Gynecol 1997; 9:324–329.
136. Sepulveda W, Shennan AH, Peek MJ. Reverse end-diastolic flow in the middle cerebral artery: an agonal pattern in the human fetus. Am J Obstet Gynecol 1996; 174:1645–1647.
137. Harman CR, Baschat AA. Comprehensive assessment of fetal wellbeing: which Doppler tests should be performed? Curr Opin Obstet Gynecol 2003; 15:147–157.
138. Detti L, Akiyama M, Mari G. Doppler blood flow in obstetrics. Curr Opin Obstet Gynecol 2002; 14:587–593.
139. Severi FM, Bocchi C, Visentin A, et al. Uterine and fetal cerebral Doppler predict the outcome of third-trimester small-for-gestational age fetuses with normal umbilical artery Doppler. Ultrasound Obstet Gynecol 2002; 19:225–228.
140. Baschat AA, Gembruch U, Harman CR. The sequence of changes in Doppler and biophysical parameters as severe fetal growth restriction worsens. Ultrasound Obstet Gynecol 2001; 18:571–577.

141. Konje JC, Bell SC, Taylor DJ. Abnormal Doppler velocimetry and blood flow volume in the middle cerebral artery in very severe intrauterine growth restriction: is the occurrence of reversal of compensatory flow too late? BJOG 2001; 108:973–979.
142. Takahashi Y, Kawabata I, Tamaya T. Characterization of growth-restricted fetuses with breakdown of the brain-sparing effect diagnosed by spectral Doppler. J Matern Fetal Med 2001; 10:122–126.
143. Machlitt A, Wauer RR, Chaoui R. Longitudinal observation of deterioration of Doppler parameters, computerized cardiotocogram and clinical course in a fetus with growth restriction. J Perinat Med 2001; 29:71–76.
144. Baschat AA, Gembruch U, Reiss I, Gortner L, Weiner CP, Harman CR. Relationship between arterial and venous Doppler and perinatal outcome in fetal growth restriction. Ultrasound Obstet Gynecol 2000; 16:407–413.
145. Brantberg A, Sonesson SE. Central arterial hemodynamics in small-for-gestational-age fetuses before and during maternal hyperoxygenation: a Doppler velocimetric study with particular attention to the aortic isthmus. Ultrasound Obstet Gynecol 1999; 14:237–243.
146. Harrington K, Thompson MO, Carpenter RG, Nguyen M, Campbell S. Doppler fetal circulation in pregnancies complicated by pre-eclampsia or delivery of a small for gestational age baby: 2. Longitudinal analysis. Br J Obstet Gynaecol 1999; 106:453–466.
147. Ozeren M, Dinc H, Ekmen U, Senekayli C, Aydemir V. Umbilical and middle cerebral artery Doppler indices in patients with preeclampsia. Eur J Obstet Gynecol Reprod Biol 1999; 82:11–16.
148. Ozcan T, Sbracia M, d'Ancona RL, Copel JA, Mari G. Arterial and venous Doppler velocimetry in the severely growth-restricted fetus and associations with adverse perinatal outcome. Ultrasound Obstet Gynecol 1998; 12:39–44.
149. Areias JC, Matias A, Montenegro N. Venous return and right ventricular diastolic function in ARED flow fetuses. J Perinat Med 1998; 26:157–167.
150. Arbeille P. Fetal arterial Doppler-IUGR and hypoxia. Eur J Obstet Gynecol Reprod Biol 1997; 75:51–53.
151. van Splunder P, Stijnen T, Wladimiroff JW. Fetal atrioventricular, venous, and arterial flow velocity waveforms in the small for gestational age fetus. Pediatr Res 1997; 42:765–775.
152. Hecher K, Hackeloer BJ. Cardiotocogram compared to Doppler investigation of the fetal circulation in the premature growth-retarded fetus: longitudinal observations. Ultrasound Obstet Gynecol 1997; 9:152–161.
153. Yoshimura S, Masuzaki H, Gotoh H, Ishimaru T. Fetal redistribution of blood flow and amniotic fluid volume in growth-retarded fetuses. Early Hum Dev 1997; 47:297–304.
154. Forouzan I, Tian ZY. Fetal middle cerebral artery blood flow velocities in pregnancies complicated by intrauterine growth restriction and extreme abnormality in umbilical artery Doppler velocity. Am J Perinatol 1996; 13:139–142.
155. Manabe A, Hata T, Kitao M. Longitudinal Doppler ultrasonographic assessment of alterations in regional vascular resistance of arteries in normal and growth-retarded fetuses. Gynecol Obstet Invest 1995; 39:171–179.
156. Hata T, Manabe A, Hata K, Kitao M. Fetal circulatory system in growth-retarded fetus with late decelerations and oligohydramnios. Gynecol Obstet Invest 1994; 37:96–98.
157. Okagaki A, Sagawa N, Ihara Y, et al. Clinical application of pulsatility index of flow volume to detect the hemodynamic changes in IUGR fetus. J Perinat Med 1994; 22:243–251.
158. Mori A, Iwashita M, Takeda Y. Haemodynamic changes in IUGR fetus with chronic hypoxia evaluated by fetal heart-rate monitoring and Doppler measurement of blood flow velocity. Med Biol Eng Comput 1993; 31(suppl):S49–S58.
159. Mari G, Deter RL. Middle cerebral artery flow velocity waveforms in normal and small-for-gestational-age fetuses. Am J Obstet Gynecol 1992; 166:1262–1270.
160. Veille JC, Cohen I. Middle cerebral artery blood flow in normal and growth-retarded fetuses. Am J Obstet Gynecol 1990; 162:391–396.
161. van den Wijngaard JA, Groenenberg IA, Wladimiroff JW, Hop WC. Cerebral Doppler ultrasound of the human fetus. Br J Obstet Gynaecol 1989; 96:845–849.
162. Arduini D, Rizzo G. Normal values of Pulsatility Index from fetal vessels: a cross-sectional study on 1556 healthy fetuses. J Perinat Med 1990; 18:165–172.
163. Ferrazzi E, Gementi P, Bellotti M, et al. Doppler velocimetry: critical analysis of umbilical, cerebral and aortic reference values. Eur J Obstet Gynecol Reprod Biol 1991; 38:189–196.
164. Kurmanavicius J, Streicher A, Wright EM, et al. Reference values of fetal peak systolic blood flow velocity in the middle cerebral artery at 19–40 weeks of gestation. Ultrasound Obstet Gynecol 2001; 17:50–53.
165. Bahlman F, Reinhard I, Krummenauer F, Neubert S, Macchiella D, Wellek S. Blood flow velocity waveforms of the fetal middle cerebral artery in a normal population: reference values from 18 weeks to 42 weeks of gestation. J Perinat Med 2002; 30:490–501.
166. Selam B, Koksal R, Ozcan T. Fetal arterial and venous Doppler parameters in the interpretation of oligohydramnios in postterm pregnancies. Ultrasound Obstet Gynecol 2000; 15:403–406.
167. Hsieh YY, Chang CC, Tsai HD, Tsai CH. Longitudinal survey of blood flow at three different locations in the middle cerebral artery in normal fetuses. Ultrasound Obstet Gynecol 2001; 17:125–128.
168. Kurmanavicius J, Florio I, Wisser J, et al. Reference resistance indices of the umbilical, fetal middle cerebral and uterine arteries at 24–42 weeks of gestation. Ultrasound Obstet Gynecol 1997; 10:112–120.
169. Baschat AA, Gembruch U. The cerebroplacental Doppler ratio revisited. Ultrasound Obstet Gynecol 2003; 21:124–127.
170. Lynch L, Lapinski R, Alvarez M, et al. Accuracy of ultrasound estimation of fetal weight in multiple pregnancies. Ultrasound Obstet Gynecol 1995; 6:349.
171. Wilkins-Haug L, Roberts DJ, Morton CC. Confined placental mosaicism and intrauterine growth retardation: a case-control analysis of placentas at delivery. Am J Obstet Gynecol 1995; 172:44.
172. Kalousek DK, Harrison K. Uniparental disomy and unexplained intrauterine growth retardation. Contemp Obstet Gynecol 1995; September:41.

Index

An f following an entry indicates a page containing a figure; a t indicates a page containing a table.
This is a cumulative index for both volumes (Volume 1, pages 1–770; Volume 2, pages 771–1408).

An f following an entry indicates a page containing a figure; a t indicates a page containing a table.
This is a cumulative index for both volumes (Volume 1, pages 1–770; Volume 2, pages 771–1408).

An f following an entry indicates a page containing a figure; a t indicates a page containing a table.
This is a cumulative index for both volumes (Volume 1, pages 1–770; Volume 2, pages 771–1408).

An f following an entry indicates a page containing a figure; a t indicates a page containing a table.
This is a cumulative index for both volumes (Volume 1, pages 1–770; Volume 2, pages 771–1408).

An f following an entry indicates a page containing a figure; a t indicates a page containing a table.
This is a cumulative index for both volumes (Volume 1, pages 1–770; Volume 2, pages 771–1408).

An f following an entry indicates a page containing a figure; a t indicates a page containing a table.
This is a cumulative index for both volumes (Volume 1, pages 1–770; Volume 2, pages 771–1408).

An f following an entry indicates a page containing a figure; a t indicates a page containing a table.
This is a cumulative index for both volumes (Volume 1, pages 1–770; Volume 2, pages 771–1408).

An f following an entry indicates a page containing a figure; a t indicates a page containing a table.
This is a cumulative index for both volumes (Volume 1, pages 1–770; Volume 2, pages 771–1408).

An f following an entry indicates a page containing a figure; a t indicates a page containing a table.
This is a cumulative index for both volumes (Volume 1, pages 1–770; Volume 2, pages 771–1408).

An f following an entry indicates a page containing a figure; a t indicates a page containing a table.
This is a cumulative index for both volumes (Volume 1, pages 1–770; Volume 2, pages 771–1408).

An f following an entry indicates a page containing a figure; a t indicates a page containing a table.
This is a cumulative index for both volumes (Volume 1, pages 1–770; Volume 2, pages 771–1408).

An f following an entry indicates a page containing a figure; a t indicates a page containing a table.
This is a cumulative index for both volumes (Volume 1, pages 1–770; Volume 2, pages 771–1408).

An f following an entry indicates a page containing a figure; a t indicates a page containing a table.
This is a cumulative index for both volumes (Volume 1, pages 1–770; Volume 2, pages 771–1408).

An f following an entry indicates a page containing a figure; a t indicates a page containing a table.
This is a cumulative index for both volumes (Volume 1, pages 1–770; Volume 2, pages 771–1408).

An f following an entry indicates a page containing a figure; a t indicates a page containing a table.
This is a cumulative index for both volumes (Volume 1, pages 1–770; Volume 2, pages 771–1408).

An f following an entry indicates a page containing a figure; a t indicates a page containing a table.
This is a cumulative index for both volumes (Volume 1, pages 1–770; Volume 2, pages 771–1408).

An f following an entry indicates a page containing a figure; a t indicates a page containing a table.
This is a cumulative index for both volumes (Volume 1, pages 1–770; Volume 2, pages 771–1408).

An f following an entry indicates a page containing a figure; a t indicates a page containing a table.
This is a cumulative index for both volumes (Volume 1, pages 1–770; Volume 2, pages 771–1408).

An f following an entry indicates a page containing a figure; a t indicates a page containing a table.
This is a cumulative index for both volumes (Volume 1, pages 1–770; Volume 2, pages 771–1408).

An f following an entry indicates a page containing a figure; a t indicates a page containing a table.
This is a cumulative index for both volumes (Volume 1, pages 1–770; Volume 2, pages 771–1408).

An f following an entry indicates a page containing a figure; a t indicates a page containing a table.
This is a cumulative index for both volumes (Volume 1, pages 1–770; Volume 2, pages 771–1408).

An f following an entry indicates a page containing a figure; a t indicates a page containing a table.
This is a cumulative index for both volumes (Volume 1, pages 1–770; Volume 2, pages 771–1408).

An f following an entry indicates a page containing a figure; a t indicates a page containing a table.
This is a cumulative index for both volumes (Volume 1, pages 1–770; Volume 2, pages 771–1408).

An f following an entry indicates a page containing a figure; a t indicates a page containing a table.
This is a cumulative index for both volumes (Volume 1, pages 1–770; Volume 2, pages 771–1408).

An f following an entry indicates a page containing a figure; a t indicates a page containing a table.
This is a cumulative index for both volumes (Volume 1, pages 1–770; Volume 2, pages 771–1408).

An f following an entry indicates a page containing a figure; a t indicates a page containing a table.
This is a cumulative index for both volumes (Volume 1, pages 1–770; Volume 2, pages 771–1408).